Living in the Environment 19e

ABOUT THE COVER PHOTO

A major new theme for this edition of *Living in the Environment* is *biomimicry* or *learning from nature*. In recent years, scientists have been studying nature in an effort to learn how a variety of life has existed on the earth for 3.8 billion years despite several catastrophic changes in the planet's environmental conditions. They include strikes by huge meteorites, long warming periods and ice ages, and five mass extinctions—each wiping out 60% to 95% of the world's species.

Examples of how life on Earth has sustained itself for 3.8 billion years are being used to help us develop technologies and solutions to the environmental problems we face and to learn how to live more sustainably. For example, the front cover of this book shows a kingfisher diving toward a body of water. The bird's long pointed beak allows it to dive into water at a high speed without making a splash to catch fish. In Japan, designers used this lesson from nature to increase the speed of and reduce the noise from high-speed bullet trains by modeling the train's front end after the kingfisher's beak.

Throughout this book, we provide a number of other examples of biomimicry, or learning from the earth.

Charlie Hamilton James/Nature Picture Library

Living in the Environment 19e

G. Tyler Miller

Scott E. Spoolman

NATIONAL GEOGRAPHIC LEARNING | CENGAGE

Australia • Brazil • Canada • Mexico • Singapore • United Kingdom • United States

NATIONAL GEOGRAPHIC LEARNING | CENGAGE

***Living in the Environment*, Nineteenth Edition**
G. Tyler Miller, Scott E. Spoolman

Product Director: Dawn M. Giovanniello
Product Manager: April Cognato
Senior Content Developer: Jake Warde
Marketing Manager: Ana Albinson
Content Project Manager: Harold P. Humphrey
Senior Designer: Michael C. Cook
Manufacturing Planner: Karen Hunt
Production Service: Matt Rosenquist, Graphic World Inc.
Photo Researcher: Venkat Narayanan, Lumina Datamatics
Text Researcher: Manjula Subramanian, Lumina Datamatics
Copy Editor: Graphic World Inc.
Illustrator: Patrick Lane, ScEYEnce Studios
Text Designer: Jeanne Calabrese
Cover Designer: Michael C. Cook
Cover Image: Charlie Hamilton James/Nature Picture Library
Compositor: Graphic World Inc.

Library of Congress Control Number: 2016952661

Student Edition:

ISBN: 978-1-337-09415-3

Loose-leaf Edition:

ISBN: 978-1-337-10010-6

Cengage
200 Pier 4 Boulevard
Boston, MA 02210
USA

Printed in Mexico
Print Number: 07 Print Year: 2020

Brief Contents

Contents

Supplements

Preface

For Instructors

We wrote this book to help you achieve three important goals: *first,* to explain to your students the basics of environmental science; *second,* to help your students in using this scientific foundation to understand the environmental problems that we face and to evaluate possible solutions to them; and *third,* to inspire your students to make a difference in how we treat the earth on which our lives and economies depend, and thus in how we treat ourselves and our descendants.

We view environmental problems and possible solutions to them through the lens of *sustainability*—the integrating theme of this book. We believe that most people can live comfortable and fulfilling lives, and that societies will be more stable and prosperous when sustainability becomes one of the chief measures by which personal choices and public policies are made.

In this new edition, we are happy to be continuing our partnership with *National Geographic Learning.* One result has been the addition of many stunning and informative photographs, numerous maps, and several stories of National Geographic Explorers—people who are making a positive difference in the world. With these tools, we continue to tell of the good news from various fields of environmental science, hoping to inspire young people to commit themselves to making our world a more sustainable place to live for their own and future generations.

What's New in This Edition?

- *An emphasis on learning from nature*: We establish this in the Core Case Study for Chapter 1, *Learning from the Earth,* which introduces the principles of biomimicry. We further explore the principles and applications of biomimicry in a Science Focus box and a feature article on biomimicry pioneer Janine Benyus later in the chapter. In our research, we have found that biomimicry presents a growing number of opportunities for using nature's genius, as Benyus puts it, to make our own economies and lifestyles more sustainable.
- A new feature called *Learning from Nature*—a set of brief summaries of specific applications of biomimicry in various industries and fields of research, appearing in most chapters.
- An *attractive and efficient new design* with visual elements inspired by National Geographic Learning to capture and hold the student's attention.
- *New Core Case Studies* for 11 of the book's 25 chapters bring important real-world stories to the forefront for use in applying those chapters' concepts and principles.
- A *heavier emphasis on data analysis,* with new questions added to the captions of all figures that involve data graphs, designed to get students to analyze the data represented in the figure. These complement the exercises we provide at ends of chapters and in Supplement 5.
- A new feature called *Eco-numbers,* which highlight key statistics that will be helpful for students to remember.
- *New treatment of the history* of environmental conservation and protection in the United States.

Sustainability Is the Integrating Theme of This Book

Sustainability, a watchword of the 21st century for those concerned about the environment, is the overarching theme of this textbook. You can see the sustainability emphasis by looking at the Brief Contents (p. v).

Six **principles of sustainability** play a major role in carrying out this book's sustainability theme. These principles are introduced in Chapter 1. They are depicted in Figure 1.2 (p. 6), in Figure 1.6 (p. 9), and on the inside back cover of the book and are used throughout the book, with each reference marked in the margin by (see pp. 5 and 8).

We use the following five major subthemes to integrate material throughout this book.

- ***Natural capital.*** Sustainability depends on the natural resources and ecosystem services that support all life and economies. See Figures 1.3, p. 7, and 8.4, p. 170.
- ***Natural capital degradation.*** We describe how human activities can degrade natural capital. See Figures 6.4, p. 123, and 12.14, p. 294.
- ***Solutions.*** We present existing and proposed solutions to environmental problems in a balanced manner and challenge students to use critical thinking to evaluate them. See Figures 10.14, p. 232, and 11.17, p. 272.
- ***Trade-offs.*** The search for solutions involves trade-offs, because any solution requires weighing advantages against disadvantages. Our Trade-offs diagrams located in several chapters present the benefits and drawbacks of various environmental technologies and solutions to environmental problems. See Figures 12.22, p. 301; 12.26, p. 304; and 16.10, p. 419.

- ***Individuals Matter.*** Throughout the book, Individuals Matter boxes and some of the Case Studies describe what various scientists and concerned citizens (including several National Geographic Explorers) have done to help us work toward sustainability (see pp. 10, 186, and 517). Also, a number of What Can You Do? diagrams describe how readers can deal with the environmental problems we face (see Figures 9.13, p. 204; 12.28, p. 306; and 13.25, p. 348). Especially important ways in which individuals can live more lightly on the earth are summarized in Figure 25.13 (p. 693).

Other Successful Features of This Textbook

- ***Up-to-Date Coverage.*** Our textbooks have been widely praised for keeping users up-to-date in the rapidly changing field of environmental science. Since the last edition, we have updated the information and concepts in this book using thousands of articles and reports published between 2013 and 2016. Major new or updated topics include biomimicry, fracking, ocean acidification, and developments in battery technology. Other such topics include: edge effects and ecotones; the phylogenetic tree of life; synthetic biology; threats to the Monarch butterfly; survivorship curves; importance of girl's education, globally; age structure diagrams for China and India; Chinese, Indian, and U.S. population trends; African Savanna; elephants as keystone species; ecosystem services provided by oceans; mangrove degradation; wild boar invasions; climate change and species extinction; wildfires in the western United States; insect threats to forest ecosystems; Amazon and Indonesian forest loss; Green Belt Movement; jellyfish populations explosion; marine protected areas and marine reserves; ways to keep Asian carp from threatening Great Lakes fisheries; Bt crops; effects of over-fertilization; nutritional value of feedlot vs. grass-fed beef; aquaculture effects on mangroves; effects of neonicotinoids and glyphosate; organic no-till farming; insects as a protein source; CAFO production; embedded water; desalination research and development; deep-sea mining; potential hazards of nanotech; rare earth metals reserves and production; U.S. oil sands deposits; costs of producing heavy oil from tar sands; increased natural gas production in the United States; methane leaks from natural gas production; coal burning and air pollution in China; future of all-electric vehicles; microgrids in less developed countries; shared (community) solar power; China's progress in renewable energy; e-waste: new types and growth; new recyclable thermoset plastic; C. diff superbug; ebola virus; endocrine disrupters PFOA in cookware and PBDE flame retardant; MRSA death rates; TB rates; smoking and e-cig use among youth; deaths from air pollution in China and India; new regulations on coal plants in U.S. (Clean Power Plan); melting of Asian glaciers; effects of warming on El Nino and La Nina; changing range of lodge pole pines; case study on climate change in Alaska; rocketing rate of dengue fever; good news on overall drop in coal use; U.S.-China carbon emissions agreement; embedded carbon in consumer goods; water pollution by mining wastes in the U.S.; Northern Pacific Garbage Patch; lead in drinking water (in Flint, Michigan and elsewhere).
- ***Concept-Centered Approach.*** To help students focus on the main ideas, we built each major chapter section around a key question and one to three key concepts, which state the section's most important take-away messages. In each chapter, all key questions are listed at the front of the chapter, and each chapter section begins with its key question and concepts (see pp. 5 and 10). Also, the concept applications are highlighted and referenced throughout each chapter.
- ***Science-Based Approach.*** Chapters 2 through 8 cover scientific principles important to the course and discuss how scientists work (see Brief Contents, p. v). Important environmental science topics are explored in depth in Science Focus boxes distributed among the chapters throughout the book (see pp. 33 and 92) and integrated throughout the book in various Case Studies (see pp. 93 and 177) and in numerous figures.
- ***Global Coverage.*** This book also provides a global perspective, first on the ecological level, revealing how all the world's life is connected and sustained within the biosphere, and second, through the use of information and images from around the world. This includes more than 80 maps in the basic text and in Supplement 4. At the end of each chapter is a Global Environment Watch exercise that applies this global perspective.
- ***Core Case Studies.*** Each chapter opens with a Core Case Study (see pp. 4, 192, and 284), which is applied

throughout the chapter. These applications are indicated by the notation (**Core Case Study**) wherever they occur (see pp. 83 and 91). Each chapter ends with a Tying it All Together box (see pp. 95 and 187), which connects the Core Case Study and other material in the chapter to some or all of the principles of sustainability.

- ***Case Studies.*** In addition to the 25 Core Case Studies, more than 70 additional Case Studies (see pp. 177, 202, and 522) appear throughout the book (and are listed in the Detailed Contents, pp. vi–xviii). Each of these provides an in-depth look at specific environmental problems and their possible solutions.
- ***Critical Thinking.*** The Learning Skills section (p. xxvii) describes critical thinking skills, and specific critical thinking exercises are used throughout the book in several ways:
    - As more than 100 *Thinking About* exercises that ask students to analyze material immediately after it is presented (see pp. 85 and 184).
    - In all *Science Focus* boxes.
    - In dozens of *Connections* boxes that stimulate critical thinking by exploring often surprising connections related to environmental problems (see pp. 207 and 256).
    - In the captions of many of the book's figures (see Figures 9.7, p. 199 and 11.4, p. 258).
    - In end-of-chapter questions (see pp. 96 and 136).
- ***Data Analysis.*** We include a data or ecological footprint analysis exercise at the end of each chapter, and we have added data analysis questions to the captions of many figures that include data graphs (see Figures 1.11, p. 16 and 5.17, p. 111). We also include such questions on each of our Data Analysis exercises in Supplement 5.
- ***Visual Learning.*** With a new design heavily influenced by material from National Geographic Learning and more than 400 photographs, this is the most visually interesting environmental science textbook available (see Figure 1.5, p. 8; Chapter 8 opening photo, pp. 166-67; and Figure 9.4, p. 196). Add in the more than 130 diagrams, each designed to present complex ideas in understandable ways relating to the real world (see Figures 5.12, p. 108, and 9.7, p. 199), and you also have one of the most visually informative textbooks available.
- ***Flexibility.*** To meet these diverse needs of hundreds of widely varying environmental science courses, we have designed a highly flexible book that allows instructors to vary the order of chapters and sections within chapters without exposing students to terms and concepts that could confuse them. We recommend that instructors start with Chapter 1, which defines basic terms and gives an overview of sustainability, population, pollution, resources, and economic development issues that are discussed throughout the book. This provides a springboard for instructors to use other chapters in almost any order. One often-used strategy is to follow Chapter 1 with Chapters 2 through 8, which introduce basic science and ecological concepts. Instructors can then use the remaining chapters in any order desired. Some instructors follow Chapter 1 with any or all of Chapters 23, 24, and 25 on environmental economics, politics, and worldviews, respectively, before proceeding to the chapters on basic science and ecological concepts. We provide a second level of flexibility in five Supplements (see p. xviii in the Detailed Contents and p. S1), which instructors can assign as desired to meet the needs of their specific courses. Examples include basic chemistry (Supplement 3), maps and map analysis (Supplement 4), and environmental data and data analysis (Supplement 5).
- ***In-Text Study Aids.*** Each chapter begins with a list of *Key Questions* showing how the chapter is organized (see p. 3). Wherever a new *key term* is introduced and defined, it appears in boldface type and all such terms are summarized in the glossary at the end of the book. More than 100 *Thinking About* exercises reinforce learning by asking students to think critically about the implications of various environmental issues and solutions immediately after they are discussed in the text (see pp. 272 and 528). The captions of many figures contain similar questions that get students to think about the figure content (see pp. 121 and 199). In their reading, students also encounter *Connections* boxes, which briefly describe connections between human activities and environmental consequences, environmental and social issues, and environmental issues and solutions (see pp. 62 and 300). New to this edition is a set of *Learning from Nature* boxes that give quick summaries of biomimicry applications (see pp. 52 and 173). Also new in this edition are *Eco-numbers*, which highlight important statistics (see pp. 14, 122, and 200). The text of each chapter concludes with three *Big Ideas* (see pp. 44 and 318), which summarize and reinforce three of the major take-away messages from each chapter. Finally, a *Tying It All Together*

section relates the Core Case Study and other chapter content to the principles of sustainability (see pp. 45 and 318). These concluding features reinforce the main messages of the chapter along with the themes of sustainability to give students a stronger understanding of how they all tie together.

Each chapter ends with a *Chapter Review* section containing a detailed set of review questions that include all the chapter's key terms in bold type; *Critical Thinking* questions that encourage students to think about and apply what they have learned to their lives; *Doing Environmental Science*—a research exercise that will help students to get a feel for what it is like to work as environmental scientists; a *Global Environment Watch* exercise taking students to Cengage's GREENR site where they can use this tool for interesting research related to chapter content; and a *Data Analysis* or *Ecological Footprint Analysis* problem built around ecological footprint data or some other environmental data set (see pp. 354 and 439).

Supplements for Instructors

- ***MindTap*** MindTap is a new approach to highly personalized online learning. Beyond an eBook, homework solution, digital supplement, or premium website, MindTap is a digital learning platform that works alongside your campus Learning Management System (LMS) to deliver course curriculum across the range of electronic devices in your life. MindTap is built on an "app" model allowing enhanced digital collaboration and delivery of engaging content across a spectrum of Cengage and non-Cengage resources. Visit the Instructor's Companion Site for tips on maximizing your MindTap course.
- ***Instructor's Companion Site.*** Everything you need for your course in one place! This collection of book-specific lecture and class tools is available online via www.cengage.com/login. Access and download PowerPoint presentations, images, instructor's manual, videos, and more.
- ***Cognero Test Bank.*** Available to adopters. Cengage Learning Testing Powered by Cognero is a flexible, online system that allows you to:
  - author, edit, and manage test bank content from multiple Cengage Learning solutions;
  - create multiple test versions in an instant; and
  - deliver tests from your LMS, your classroom, or wherever you want.

Help Us Improve This Book or Its Supplements

Let us know how you think this book can be improved. If you find any errors, bias, or confusing explanations, please e-mail us about them at:

- mtg89@hotmail.com
- sspoolman@gmail.com

Most errors can be corrected in subsequent printings of this edition, as well as in future editions.

Acknowledgments

We wish to thank the many students and teachers who have responded so favorably to the 18 previous editions of *Living in the Environment*, the 15 editions of *Environmental Science*, the 11 editions of *Sustaining the Earth*, and the 7 editions of *Essentials of Ecology*, and who have corrected errors and offered many helpful suggestions for improvement. We are also deeply indebted to the more than 300 reviewers, who pointed out errors and suggested many important improvements in the various editions of these four books.

It takes a village to produce a textbook, and the members of the talented production team, listed on the copyright page, have made vital contributions. Our special thanks go to development editor Jake Warde; production editors Hal Humphrey, Chris Waller, and Matt Rosenquist; compositor Graphic World Inc.; photo researcher Venkat Narayanan of Lumina Datamatics; artist Patrick Lane; content development manager Alexandria Brady; product assistant Marina Starkey; and Cengage Learning's hard-working sales staff. Finally, we have been fortunate to have the guidance, inspiration, and unfailing support of Morgan Carney, Dawn Giovanniello, and April Cognato and their dedicated team of highly talented people who have made this and our other Cengage Learning book projects such a pleasure to work on.

G. Tyler Miller

Scott E. Spoolman

Pedagogy Contributors

Dr. Dean Goodwin and his colleagues, Berry Cobb, Deborah Stevens, Jeannette Adkins, Jim Lehner, Judy Treharne, Lonnie Miller, and Tom Mowbray provided excellent contributions to the Data Analysis and Ecological Footprint Analysis exercises. Mary Jo Burchart of Oakland Community College wrote the in-text Global Environment Watch exercises.

Cumulative List of Reviewers

Barbara J. Abraham, Hampton College; Donald D. Adams, State University of New York at Plattsburgh; Larry G. Allen, California State University, Northridge; Susan Allen-Gil, Ithaca College; James R. Anderson, U.S. Geological Survey; Mark W. Anderson, University of Maine; Kenneth B. Armitage, University of Kansas; Samuel Arthur, Bowling Green State University; Gary J. Atchison, Iowa State University; Thomas W. H. Backman, Lewis-Clark State College; Marvin W. Baker, Jr., University of Oklahoma; Virgil R. Baker, Arizona State University; Stephen W. Banks, Louisiana State University in Shreveport; Ian G. Barbour, Carleton College; Albert J. Beck, California State University, Chico; Marilynn Bartels, Black Hawk College; Eugene C. Beckham, Northwood University; Diane B. Beechinor, Northeast Lakeview College; W. Behan, Northern Arizona University; David Belt, Johnson County Community College; Keith L. Bildstein, Winthrop College; Andrea Bixler, Clarke College; Jeff Bland, University of Puget Sound; Roger G. Bland, Central Michigan University; Grady Blount II, Texas A&M University, Corpus Christi; Barbara I. Bonder, Flagler College; Lisa K. Bonneau, University of Missouri–Kansas City; Georg Borgstrom, Michigan State University; Arthur C. Borror, University of New Hampshire; John H. Bounds, Sam Houston State University; Leon F. Bouvier, Population Reference Bureau; Daniel J. Bovin, Université Laval; Jan Boyle, University of Great Falls; James A. Brenneman, University of Evansville; Michael F. Brewer, Resources for the Future, Inc.; Mark M. Brinson, East Carolina University; Dale Brown, University of Hartford; Patrick E. Brunelle, Contra Costa College; Terrence J. Burgess, Saddleback College North; David Byman, Pennsylvania State University Worthington Scranton; Michael L. Cain, Bowdoin College; Lynton K. Caldwell, Indiana University; Faith Thompson Campbell, Natural Resources Defense Council, Inc.; John S. Campbell, Northwest College; Ray Canterbery, Florida State University; Deborah L. Carr, Texas Tech University; Ted J. Case, University of San Diego; Ann Causey, Auburn University; Richard A. Cellarius, Evergreen State University; William U. Chandler, Worldwatch Institute; F. Christman, University of North Carolina, Chapel Hill; Peter Chen, College of DuPage; Lu Anne Clark, Lansing Community College; Preston Cloud, University of California, Santa Barbara; Bernard C. Cohen, University of Pittsburgh; Richard A. Cooley, University of California, Santa Cruz; Dennis J. Corrigan; George Cox, San Diego State University; John D. Cunningham, Keene State College; Herman E. Daly, University of Maryland; Raymond F. Dasmann, University of California, Santa Cruz; Kingsley Davis, Hoover Institution; Edward E. DeMartini, University of California, Santa Barbara; James Demastes, University of Northern Iowa; Robert L. Dennison, Heartland Community College; Charles E. DePoe, Northeast Louisiana University; Thomas R. Detwyler, University of Wisconsin; Bruce DeVantier, Southern Illinois University at Carbondale; Peter H. Diage, University of California, Riverside; Stephanie Dockstader, Monroe Community College; Lon D. Drake, University of Iowa; Michael Draney, University of Wisconsin–Green Bay; David DuBose, Shasta College; Dietrich Earnhart, University of Kansas; Robert East, Washington & Jefferson College; T. Edmonson, University of Washington; Thomas Eisner, Cornell University; Michael Esler, Southern Illinois University; David E. Fairbrothers, Rutgers University; Paul P. Feeny, Cornell University; Richard S. Feldman, Marist College; Vicki Fella-Pleier, La Salle University; Nancy Field, Bellevue Community College; Allan Fitzsimmons, University of Kentucky; Andrew J. Friedland, Dartmouth College; Kenneth O. Fulgham, Humboldt State University; Lowell L. Getz, University of Illinois at Urbana-Champaign; Frederick F. Gilbert, Washington State University; Jay Glassman, Los Angeles Valley College; Harold Goetz, North Dakota State University; Srikanth Gogineni, Axia College of University of Phoenix; Jeffery J. Gordon, Bowling Green State University; Eville Gorham, University of Minnesota; Michael Gough, Resources for the Future; Ernest M. Gould, Jr., Harvard University; Peter Green, Golden West College; Katharine B. Gregg, West Virginia Wesleyan College; Stelian Grigoras, Northwood University; Paul K. Grogger, University of Colorado at Colorado Springs; L. Guernsey, Indiana State University; Ralph Guzman, University of California, Santa Cruz; Raymond Hames, University of Nebraska, Lincoln; Robert Hamilton IV, Kent State University, Stark Campus; Raymond E. Hampton, Central Michigan University; Ted L. Hanes, California State University, Fullerton; William S. Hardenbergh, Southern Illinois University at Carbondale; John P. Harley, Eastern Kentucky University; Cindy Harmon, State Fair Community College; Neil A. Harriman, University of Wisconsin, Oshkosh; Grant A. Harris, Washington State University; Harry S. Hass, San Jose City College; Arthur N. Haupt, Population Reference Bureau; Denis A. Hayes, environmental consultant; Stephen Heard, University of Iowa; Gene Heinze-Fry, Department of Utilities, Commonwealth of Massachusetts; Jane Heinze-Fry, environmental educator; Keith R. Hench, Kirkwood Community College; John G. Hewston, Humboldt State University; David L. Hicks, Whitworth College; Kenneth M. Hinkel, University of Cincinnati; Eric Hirst, Oak Ridge National Laboratory; Doug Hix, University of Hartford; Kelley Hodges, Gulf Coast State College; S. Holling, University of British Columbia; Sue Holt, Cabrillo College; Donald Holtgrieve, California State University, Hayward; Michelle Homan, Gannon University; Michael H. Horn, California State University, Fullerton; Mark A. Hornberger, Bloomsberg University; Marilyn Houck, Pennsylvania State University; Richard D. Houk, Winthrop College; Robert J. Huggett, College of William and Mary; Donald Huisingh, North Carolina State University; Catherine Hurlbut, Florida Community College at Jacksonville; Marlene K. Hutt, IBM; David R. Inglis, University of Massachusetts; Robert Janiskee, University of South Carolina; Hugo H. John, University of Connecticut; Brian A. Johnson, University of Pennsylvania, Bloomsburg; David I. Johnson, Michigan State University; Mark Jonasson, Crafton Hills College; Zoghlul Kabir, Rutgers, New Brunswick; Agnes Kadar, Nassau Community College; Thomas L. Keefe, Eastern Kentucky University; David Kelley, University of St. Thomas; William E. Kelso, Louisiana State University; Nathan Keyfitz, Harvard University; David Kidd, University of New Mexico; Pamela S. Kimbrough; Jesse Klingebiel, Kent School; Edward J. Kormondy, University of Hawaii–Hilo/West Oahu College; John V. Krutilla, Resources for the Future, Inc.; Judith Kunofsky, Sierra Club; E. Kurtz; Theodore Kury, State University of New York at Buffalo; Troy A. Ladine, East Texas Baptist University; Steve Ladochy, University of Winnipeg; Anna J. Lang, Weber State University; Mark B. Lapping, Kansas State University; Michael L. Larsen, Campbell University; Linda Lee, University of Connecticut; Tom Leege, Idaho Department of Fish and Game; Maureen Leupold, Genesee Community College; William S. Lindsay, Monterey Peninsula College; E. S. Lindstrom, Pennsylvania State University; M. Lippiman, New York University Medical Center; Valerie A. Liston, University of Minnesota; Dennis Livingston, Rensselaer Polytechnic Institute; James P. Lodge, air pollution consultant; Raymond C. Loehr, University of Texas at Austin; Ruth Logan, Santa Monica City College; Robert D. Loring, DePauw University; Paul F. Love, Angelo State University; Thomas Lovering, University of California, Santa Barbara; Amory B. Lovins, Rocky Mountain Institute; Hunter Lovins, Rocky Mountain Institute; Gene A. Lucas, Drake University; Claudia Luke, University of California, Berkeley; David Lynn; Timothy F. Lyon, Ball State University; Stephen Malcolm, Western Michigan University; Melvin G. Marcus, Arizona State University; Gordon E. Matzke, Oregon State University; Parker Mauldin, Rockefeller Foundation; Marie McClune, The Agnes Irwin School (Rosemont, Pennsylvania); Theodore R. McDowell, California State University; Vincent E. McKelvey, U.S. Geological Survey; Robert T. McMaster, Smith College; John G. Merriam, Bowling Green State University; A. Steven Messenger, Northern Illinois University; John Meyers, Middlesex Community College; Raymond W. Miller, Utah State University; Arthur B. Millman, University of Massachusetts, Boston; Sheila Miracle, Southeast Kentucky Community & Technical College; Fred Montague, University of Utah; Rolf Monteen, California Polytechnic State University; Debbie Moore, Troy University Dothan Campus; Michael K. Moore, Mercer University; Ralph Morris, Brock University, St. Catherine's, Ontario, Canada; Angela Morrow, Auburn University; William W. Murdoch, University of California, Santa Barbara; Norman Myers, environmental consultant; Brian C. Myres, Cypress College; A. Neale, Illinois State University; Duane Nellis, Kansas State University; Jan Newhouse, University of Hawaii, Manoa; Jim Norwine, Texas A&M University, Kingsville; John E. Oliver, Indiana State University; Mark Olsen, University of Notre Dame; Bruce Olszewski, San Jose State University; Carol Page, copy editor; Bill Paletski, Penn State University; Eric Pallant, Allegheny College; Charles F. Park, Stanford University; Richard J. Pedersen, U.S. Department of Agriculture, Forest Service; David Pelliam, Bureau of Land Management, U.S. Department of the Interior; Barry Perlmutter, College of Southern Nevada; Murray Paton Pendarvis, Southeastern Louisiana University; Dave Perault, Lynchburg College; Carolyn J. Peters, Spoon River College; Rodney Peterson, Colorado State University; Julie Phillips, De Anza College; John Pichtel, Ball State University; William S. Pierce, Case Western Reserve University; David Pimentel, Cornell University; Peter Pizor, Northwest Community College; Mark D. Plunkett, Bellevue Community College; Grace L. Powell, University of Akron; James H. Price, Oklahoma College; Alan D. Redmond, East Tennessee State University; Marian E. Reeve, Merritt College; Carl H. Reidel, University of Vermont; Charles C. Reith, Tulane University; Erin C. Rempala, San Diego City College; Roger Revelle, California State University, San Diego; L. Reynolds, University of Central Arkansas; Ronald R. Rhein, Kutztown University of Pennsylvania; Charles Rhyne, Jackson State University; Robert A. Richardson, University of Wisconsin; Benjamin F.

Richason III, St. Cloud State University; Jennifer Rivers, Northeastern University; Ronald Robberecht, University of Idaho; William Van B. Robertson, School of Medicine, Stanford University; C. Lee Rockett, Bowling Green State University; Terry D. Roelofs, Humboldt State University; Daniel Ropek, Columbia George Community College; Christopher Rose, California Polytechnic State University; Richard G. Rose, West Valley College; Stephen T. Ross, University of Southern Mississippi; Robert E. Roth, Ohio State University; Dorna Sakurai, Santa Monica College; Arthur N. Samel, Bowling Green State University; Shamili Sandiford, College of DuPage; Floyd Sanford, Coe College; David Satterthwaite, I.E.E.D., London; Stephen W. Sawyer, University of Maryland; Arnold Schecter, State University of New York; Frank Schiavo, San Jose State University; William H. Schlesinger, Ecological Society of America; Stephen H. Schneider, National Center for Atmospheric Research; Clarence A. Schoenfeld, University of Wisconsin, Madison; Madeline Schreiber, Virginia Polytechnic Institute; Henry A. Schroeder, Dartmouth Medical School; Lauren A. Schroeder, Youngstown State University; Norman B. Schwartz, University of Delaware; George Sessions, Sierra College; David J. Severn, Clement Associates; Don Sheets, Gardner-Webb University; Paul Shepard, Pitzer College and Claremont Graduate School; Michael P. Shields, Southern Illinois University at Carbondale; Kenneth Shiovitz; F. Siewert, Ball State University; E. K. Silbergold, Environmental Defense Fund; Joseph L. Simon, University of South Florida; William E. Sloey, University of Wisconsin, Oshkosh; Michelle Smith, Windward Community College; Robert L. Smith, West Virginia University; Val Smith, University of Kansas; Howard M. Smolkin, U.S. Environmental Protection Agency; Patricia M. Sparks, Glassboro State College; John E. Stanley, University of Virginia; Mel Stanley, California State Polytechnic University, Pomona; Richard Stevens, Monroe Community College; Norman R. Stewart, University of Wisconsin, Milwaukee; Frank E. Studnicka, University of Wisconsin, Platteville; Chris Tarp, Contra Costa College; Roger E. Thibault, Bowling Green State University; Nathan E. Thomas, University of South Dakota; William L. Thomas, California State University, Hayward; Jamey Thompson, Hudson Valley Community College; Kip R. Thompson, Ozarks Technical Community College; Shari Turney, copy editor; John D. Usis, Youngstown State University; Tinco E. A. van Hylckama, Texas Tech University; Robert R. Van Kirk, Humboldt State University; Donald E. Van Meter, Ball State University; Rick Van Schoik, San Diego State University; Gary Varner, Texas A&M University; John D. Vitek, Oklahoma State University; Harry A. Wagner, Victoria College; Lee B. Waian, Saddleback College; Warren C. Walker, Stephen F. Austin State University; Thomas D. Warner, South Dakota State University; Kenneth E. F. Watt, University of California, Davis; Alvin M. Weinberg, Institute of Energy Analysis, Oak Ridge Associated Universities; John F. Weishampel, University of Central Florida; Brian Weiss; Margery Weitkamp, James Monroe High School (Granada Hills, California); Anthony Weston, State University of New York at Stony Brook; Raymond White, San Francisco City College; Douglas Wickum, University of Wisconsin, Stout; Charles G. Wilber, Colorado State University; Nancy Lee Wilkinson, San Francisco State University; John C. Williams, College of San Mateo; Ray Williams, Rio Hondo College; Roberta Williams, University of Nevada, Las Vegas; Samuel J. Williamson, New York University; Dwina Willis, Freed-Hardeman University; Ted L. Willrich, Oregon State University; James Winsor, Pennsylvania State University; Fred Witzig, University of Minnesota at Duluth; Martha Wolfe, Elizabethtown Community and Technical College; George M. Woodwell, Woods Hole Research Center; Peggy J. Wright, Columbia College; Todd Yetter, University of the Cumberlands; Robert Yoerg, Belmont Hills Hospital; Hideo Yonenaka, San Francisco State University; Brenda Young, Daemen College; Anita Závodská, Barry University; Malcolm J. Zwolinski, University of Arizona.

Learning Skills

Study nature, love nature, stay close to nature. It will never fail you.

Frank Lloyd Wright

Why Is It Important to Study Environmental Science?

Welcome to **environmental science**—an *interdisciplinary* study of how the earth works, how we interact with the earth, and how we can deal with the environmental problems we face. Because environmental issues affect every part of your life, the concepts, information, and issues discussed in this book and the course you are taking will be useful to you now and throughout your life.

Understandably, we are biased, but *we strongly believe that environmental science is the single most important course that you could take*. What could be more important than learning about the earth's life-support system, how our choices and activities affect it, and how we can reduce our growing environmental impact? Evidence indicates strongly that we will have to learn to live more sustainably by reducing our degradation of the planet's life-support system. We hope this book will inspire you to become involved in this change in the way we view and treat the earth, which sustains us, our economies, and all other living things.

You Can Maximize Your Study and Learning Skills

Making the most of your ability to learn might involve improving your study and learning skills. Here are some suggestions for doing so:

Make daily to-do lists. Put items in order of importance, focus on the most important tasks, and assign a time to work on these items. Shift your schedule as needed to accomplish the most important items.

Set up a study routine in a distraction-free environment. Study in a quiet, well-lit space. Take breaks every hour or so. During each break, take several deep breaths and move around; this will help you to stay more alert and focused.

Avoid procrastination. Do not fall behind on your reading and other assignments. Set aside a particular time for studying each day and make it a part of your daily routine.

Make hills out of mountains. It is psychologically difficult to read an entire book, read a chapter in a book, write a paper, or cram to study for a test. Instead, break these large tasks (mountains) down into a series of small tasks (hills). Each day, read a few pages of a book or chapter, write a few paragraphs of a paper, and review what you have studied and learned.

Ask and answer questions as you read. For example, "What is the main point of a particular subsection or paragraph?" "How does it relate to the key question and key concepts addressed in each major chapter section?"

Focus on key terms. Use the glossary in your textbook to look up the meaning of terms or words you do not understand. This book shows all key terms in **bold** type and lesser, but still important, terms in *italicized* type. The *Chapter Review* questions at the end of each chapter also include the chapter's key terms in bold. Flash cards for testing your mastery of key terms for each chapter are available on the website for this book, or you can make your own.

Interact with what you read. You could highlight key sentences and paragraphs and make notes in the margins. You might also mark important pages that you want to return to.

Review to reinforce learning. Before each class session, review the material you learned in the previous session and read the assigned material.

Become a good note taker. Learn to write down the main points and key information from any lecture using your own shorthand system. Review, fill in, and organize your notes as soon as possible after each class.

Check what you have learned. At the end of each chapter, you will find review questions that cover all of the key material in each chapter section. We suggest that you try to answer each of these questions after studying each chapter section.

Write out answers to questions to focus and reinforce learning. Write down your answers to the critical thinking questions found in the *Thinking About* boxes throughout the chapters, in many figure captions, and at the end of each chapter. These questions are designed to inspire you to think critically about key ideas and connect them to other ideas and to your own life. Also, write down your answers to all chapter-ending review questions. The website for each chapter has an additional detailed list of review questions for that chapter. Save your answers for review and test preparation.

Use the buddy system. Study with a friend or become a member of a study group to compare notes, review material, and prepare for tests. Explaining something to someone else is a great way to focus your thoughts and reinforce your learning. Attend any review sessions offered by instructors or teaching assistants.

Learn your instructor's test style. Does your instructor emphasize multiple-choice, fill-in-the-blank, true-or-false, factual, or essay questions? How much of the test will come from the textbook and how much from lecture material? Adapt your learning and studying methods to this style.

Become a good test taker. Avoid cramming. Eat well and get plenty of sleep before a test. Arrive on time or early. Calm yourself and increase your oxygen intake by taking several deep breaths. (Do this also about every 10–15 minutes while taking the test.) Look over the test and answer the questions you know well first. Then work on the harder ones. Use the process of elimination to narrow down the choices for multiple-choice questions. For essay questions, organize your thoughts before you start writing. If you have no idea what a question means, make an educated guess. You might earn some partial credit and avoid getting a zero. Another strategy for getting some credit is to show your knowledge and reasoning by writing something like this: "If this question means so and so, then my answer is ________."

Take time to enjoy life. Every day, take time to laugh and enjoy nature, beauty, and friendship.

You Can Improve Your Critical Thinking Skills

Critical thinking involves developing skills to analyze information and ideas, judge their validity, and make decisions. Critical thinking helps you to distinguish between facts and opinions, evaluate evidence and arguments, and take and defend informed positions on issues. It also helps you to integrate information and see relationships and to apply your knowledge to dealing with new and different problems, as well as to your own lifestyle choices. Here are some basic skills for learning how to think more critically.

Question everything and everybody. Be skeptical, as any good scientist is. Do not believe everything you hear and read, including the content of this textbook, without evaluating the information you receive. Seek other sources and opinions.

Identify and evaluate your personal biases and beliefs. Each of us has biases and beliefs taught to us by our parents, teachers, friends, role models, and our own experience. What are your basic beliefs, values, and biases? Where did they come from? What assumptions are they based on? How sure are you that your beliefs, values, and assumptions are right and why? According to the American psychologist and philosopher William James, "A great many people think they are thinking when they are merely rearranging their prejudices."

Be open-minded and flexible. Be open to considering different points of view. Suspend judgment until you gather more evidence, and be willing to change your mind. Recognize that there may be a number of useful and acceptable solutions to a problem, and that very few issues are either black or white. Understand that there are trade-offs involved in dealing with any environmental issue, as you will learn in reading this book.

Be humble about what you know. Some people are so confident in what they know that they stop thinking and questioning. To paraphrase American writer Mark Twain, "It's what we know is true, but just ain't so, that hurts us."

Find out how the information related to an issue was obtained. Are the statements you heard or read based on firsthand knowledge and research or on hearsay? Are unnamed sources used? Is the information based on reproducible and widely accepted scientific studies or on preliminary scientific results that may be valid but need further testing? Is the information based on a few isolated stories or experiences or on carefully controlled studies that have been reviewed by experts in the field involved? Is it based on unsubstantiated and dubious scientific information or beliefs?

Question the evidence and conclusions presented. What are the conclusions or claims based on the information you're considering? What evidence is presented to support them? Does the evidence support them? Is there a need to gather more evidence to test the conclusions? Are there other, more reasonable conclusions?

Try to uncover differences in basic beliefs and assumptions. On the surface, most arguments or disagreements involve differences of opinion about the validity or meaning of certain facts or conclusions. Scratch a little deeper and you will find that many disagreements are based on different (and often hidden) basic assumptions concerning how we look at and interpret the world around us. Uncovering these basic differences can allow the parties involved to understand each other's viewpoints and to agree to disagree about their basic assumptions, beliefs, or principles.

Try to identify and assess any motives on the part of those presenting evidence and drawing conclusions. What is their expertise in this area? Do they have any unstated assumptions, beliefs, biases, or values? Do they have a personal agenda? Can they benefit financially or politically from acceptance of their evidence and conclusions? Would investigators with different basic assumptions or beliefs take the same data and come to different conclusions?

Expect and tolerate uncertainty. Recognize that scientists cannot establish absolute proof or certainty about anything. However, the reliable results of science have a high degree of certainty.

Check the arguments you hear and read for logical fallacies and debating tricks. Here are six of many examples of such debating tricks: *First,* attack the presenter of an argument rather than the argument itself. *Second,* appeal to emotion rather than facts and logic. *Third,* claim that if one piece of evidence or one conclusion is false, then all other related pieces of evidence and conclusions are false. *Fourth,* say that a conclusion is false because it has not been scientifically proven. (Scientists never prove anything absolutely, but they can establish high degrees of certainty.) *Fifth,* inject irrelevant or misleading information to divert attention from important points. *Sixth,* present only either/or alternatives when there may be a number of options.

Do not believe everything you read on the Internet. The Internet is a wonderful and easily accessible source of information that includes alternative explanations and opinions on almost any subject or issue—much of it not available in the mainstream media and scholarly articles. Blogs of all sorts have become a major source of information, even more important than standard news media for some people. However, because the Internet is so open, anyone can post anything they want to some blogs and other websites with no editorial control or review by experts. As a result, evaluating information on the Internet is one of the best ways to put into practice the principles of critical thinking discussed here. Use and enjoy the Internet, but think critically and proceed with caution.

Develop principles or rules for evaluating evidence. Develop a written list of principles to serve as guidelines for evaluating evidence and claims. Continually evaluate and modify this list on the basis of your experience.

Become a seeker of wisdom, not a vessel of information. Many people believe that the main goal of their education is to learn as much as they can by gathering more and more information. We believe that the primary goal is to learn how to sift through mountains of facts and ideas to find the few nuggets of wisdom that are especially useful for understanding the world and for making decisions. This book is full of facts and numbers, but they are useful only to the extent that they lead to an understanding of key ideas, scientific laws, theories, concepts, and connections. The major goals of the study of environmental science are to find out how nature works and sustains itself (*environmental wisdom*) and to use *principles of environmental wisdom* to help make human societies and economies more sustainable, more just, and more beneficial and enjoyable for all. As writer Sandra Carey observed, "Never mistake knowledge for wisdom. One helps you make a living; the other helps you make a life."

To help you practice critical thinking, we have supplied questions throughout this book, found within each chapter in brief boxes labeled *Thinking About,* in the captions of many figures, at the end of each Science Focus box, and at the end of each chapter. There are no right or wrong answers to many of these questions. A good way to improve your critical thinking skills is to compare your answers with those of your classmates and to discuss how you arrived at your answers.

Use the Learning Tools We Offer in This Book

We have included a number of tools throughout this textbook that are intended to help you improve your learning skills and apply them. First, consider the *Key Questions* list at the beginning of each chapter section. You can use these to preview a chapter and to review the material after you have read it.

Next, note that we use three different special notations throughout the text. Each chapter opens with a **Core Case Study**, and each time we tie material within the chapter back to this core case, we note it in bold, colored type as we did in this sentence. You will also see two icons appearing regularly in the text margins. When you see the *sustainability* icon, you will know that you have just read something that relates directly to the overarching theme of this text, summarized by our six **principles of sustainability**, which are introduced in Figures 1.2, p. 6, and 1.6, p. 9, and which appear on the inside back cover of this book. The *Good News* icon appears near each of many examples of successes that people have had in dealing with the environmental challenges we face.

We also include several brief *Connections* boxes to show you some of the often surprising connections between environmental problems or processes and some of the products and services we use every day or some of the activities we partake in. Also, look for *Learning from Nature* boxes, which are brief summaries of how we can apply lessons from nature through biomimicry (see p. 173). These, along with the *Thinking About* boxes scattered throughout the text (all designated by the *Consider This ...* heading), are intended to get you to think carefully about activities and choices we take for granted and how they might be affecting the environment.

At the end of each chapter, we list what we consider to be the *three big ideas* that you should take away from the chapter. Following that list in each chapter is a *Tying It All Together* box. This feature quickly reviews the Core Case Study and how chapter material relates to it, and it explains how the principles of sustainability can be applied to deal with challenges discussed in the core case study and throughout the chapter.

Finally, we have included a *Chapter Review* section at the end of each chapter, with questions listed for each chapter section. These questions cover all of the key material and key terms in each chapter. A variety of other exercises and projects follow this review section at the end of each chapter.

Know Your Own Learning Style

People have different ways of learning and it can be helpful to know your own learning style. *Visual learners* learn best by reading and viewing illustrations and diagrams. *Auditory learners* learn best by listening and discussing. They might benefit from reading aloud while studying and using a tape recorder in lectures for study and review. *Logical learners* learn best by using concepts and logic to uncover and understand a subject rather than relying mostly on memory.

This book and its supporting website materials contain plenty of tools for all types of learners. Visual learners can benefit from using flash cards (available on the website) to memorize key terms and ideas. This is a highly visual book with many carefully selected photographs and diagrams designed to illustrate important ideas, concepts, and processes. Auditory learners can make use of our ReaderSpeak app in MindTap, which can read the chapter aloud at different speeds and in different voices. For logical learners, the book is organized by key concepts that are revisited throughout any chapter and related carefully to other concepts, major principles, and case studies and other examples. We urge you to become aware of your own learning style and make the most of these various tools.

This Book Presents a Positive, Realistic Environmental Vision of the Future

Our goal is to present a positive vision of our environmental future based on realistic optimism. To do so, we strive not only to present the facts about environmental issues, but also to give a balanced presentation of different viewpoints. We consider the advantages and disadvantages of various technologies and proposed solutions to environmental problems. We argue that environmental solutions usually require *trade-offs* among opposing parties, and that the best solutions are *win-win* solutions. Such solutions are achieved when people with different viewpoints work together to come up with a solution that both sides can live with. And we present the good news as well as the bad news about efforts to deal with environmental problems.

One cannot study a subject as important and complex as environmental science without forming conclusions, opinions, and beliefs. However, we argue that any such results should be based on use of critical thinking to evaluate conflicting positions and to understand the trade-offs involved in most environmental solutions. To that end, we emphasize critical thinking throughout this textbook, and we encourage you to develop a practice of applying critical thinking to everything you read and hear, both in school and throughout your life.

Help Us Improve This Book

Researching and writing a book that covers and connects the numerous major concepts from the wide variety of environmental science disciplines is a challenging and exciting task. Almost every day, we learn about some new connection in nature. However, in a book this complex, there are bound to be some errors—some typographical mistakes that slip through and some statements that you might question, based on your knowledge and research. We invite you to contact us to correct any errors you find, point out any bias you see, and suggest ways to improve this book. Please e-mail your suggestions to Tyler Miller at mtg89@hotmail.com or Scott Spoolman at sspoolman@gmail.com.

Now start your journey into this fascinating and important study of how the earth's life-support system works and how we can leave our planet in a condition at least as good as what we now enjoy. Have fun.

Supplements for Students

You have a large variety of electronic and other supplemental materials available to you to help you take your learning experience beyond this textbook:

- ***Environmental Science MindTap.*** MindTap provides you with the tools you need to better manage your limited time—you can complete assignments whenever and wherever you are ready to learn with course material specially customized for you by your instructor and streamlined in one proven, easy-to-use interface. MindTap includes an online homework solution that helps you learn and understand key concepts through focused assignments, exceptional text-art integration, and immediate feedback. With these resources and an array of tools and apps—from note taking to flashcards—you'll get a true understanding of course concepts, helping you to achieve better grades and setting the groundwork for your future courses.
- ***Global Environment Watch.*** Integrated within MindTap and updated several times a day, the Global Environment Watch is a focused portal into GREENR—the Global Reference on the Environment, Energy, and Natural Resources—an ideal one-stop site for classroom discussion and research projects. This resource center keeps courses up-to-date with the most current news on the environment. Users get access to information from trusted academic journals, news outlets, and magazines, as well as statistics, an interactive world map, videos, primary sources, case studies, podcasts, and much more.

  Other student learning tools include:
- ***Essential Study Skills for Science Students* by Daniel D. Chiras.** This book includes chapters on developing good study habits, sharpening memory, getting the most out of lectures, labs, and reading assignments, improving test-taking abilities, and becoming a critical thinker. Available for students on instructor's request.

About the Authors

G. TYLER MILLER

G. Tyler Miller has written 62 textbooks for introductory courses in environmental science, basic ecology, energy, and environmental chemistry. Since 1975, Miller's books have been the most widely used textbooks for environmental science in the United States and throughout the world. They have been used by almost 3 million students and have been translated into eight languages.

Miller has a professional background in chemistry, physics, and ecology. He has a PhD from the University of Virginia and has received two honorary doctoral degrees for his contributions to environmental education. He taught college for 20 years, developed one of the nation's first environmental studies programs, and developed an innovative interdisciplinary undergraduate science program before deciding to write environmental science textbooks full time in 1975. Currently, he is the president of Earth Education and Research, devoted to improving environmental education.

He describes his hopes for the future as follows:

If I had to pick a time to be alive, it would be the next 75 years. Why? First, there is overwhelming scientific evidence that we are in the process of seriously degrading our own life-support system. In other words, we are living unsustainably. Second, within your lifetime we have the opportunity to learn how to live more sustainably by working with the rest of nature, as described in this book.

I am fortunate to have three smart, talented, and wonderful sons—Greg, David, and Bill. I am especially privileged to have Kathleen as my wife, best friend, and research associate. It is inspiring to have a brilliant, beautiful (inside and out), and strong woman who cares deeply about nature as a lifemate. She is my hero. I dedicate this book to her and to the earth.

SCOTT E. SPOOLMAN

Scott Spoolman is a writer with more than 30 years of experience in educational publishing. He has worked with Tyler Miller since 2003 as coauthor of *Living in the Environment, Environmental Science,* and *Sustaining the Earth.* With Norman Myers, he coauthored *Environmental Issues and Solutions: A Modular Approach.*

Spoolman holds a master's degree in science journalism from the University of Minnesota. He has authored numerous articles in the fields of science, environmental engineering, politics, and business. He worked as an acquisitions editor on a series of college forestry textbooks. He has also worked as a consulting editor in the development of over 70 college and high school textbooks in fields of the natural and social sciences.

In his free time, he enjoys exploring the forests and waters of his native Wisconsin along with his family—his wife, environmental educator Gail Martinelli, and his son Will and daughter Katie.

Spoolman has the following to say about his collaboration with Tyler Miller.

I am honored to be working with Tyler Miller as a coauthor to continue the Miller tradition of thorough, clear, and engaging writing about the vast and complex field of environmental science. I share Tyler Miller's passion for ensuring that these textbooks and their multimedia supplements will be valuable tools for students and instructors. To that end, we strive to introduce this interdisciplinary field in ways that will be informative and sobering, but also tantalizing and motivational.

If the flip side of any problem is indeed an opportunity, then this truly is one of the most exciting times in history for students to start an environmental career. Environmental problems are numerous, serious, and daunting, but their possible solutions generate exciting new career opportunities. We place high priorities on inspiring students with these possibilities, challenging them to maintain a scientific focus, pointing them toward rewarding and fulfilling careers, and in doing so, working to help sustain life on the earth.

From the Authors

My Environmental Journey—*G. Tyler Miller*

My environmental journey began in 1966 when I heard a lecture on population and pollution problems by Dean Cowie, a biophysicist with the U.S. Geological Survey. It changed my life. I told him that if even half of what he said was valid, I would feel ethically obligated to spend the rest of my career teaching and writing to help students learn about the basics of environmental science. After spending 6 months studying the environmental literature, I concluded that he had greatly underestimated the seriousness of these problems.

I developed an undergraduate environmental studies program and in 1971 published my first introductory environmental science book, an interdisciplinary study of the connections between energy laws (thermodynamics), chemistry, and ecology. In 1975 I published the first edition of *Living in the Environment*. Since then, I have completed multiple editions of this textbook, and of three others derived from it, along with other books.

Beginning in 1985, I spent 10 years in the deep woods living in an adapted school bus that I used as an environmental science laboratory and writing environmental science textbooks. I evaluated the use of passive solar energy design to heat the structure; buried earth tubes to bring in air cooled by the earth (geothermal cooling) at a cost of about $1 per summer; set up active and passive systems to provide hot water; installed an energy-efficient instant hot water heater powered by liquefied petroleum gas (LPG); installed energy-efficient windows and appliances and a composting (waterless) toilet; employed biological pest control; composted food wastes; used natural planting (no grass or lawnmowers); gardened organically; and experimented with a host of other potential solutions to major environmental problems that we face.

I also used this time to learn and think about how nature works by studying the plants and animals around me. My experience from living in nature is reflected in much of the material in this book. It also helped me to develop the six simple **principles of sustainability** that serve as the integrating theme for this textbook and to apply these principles to living my life more sustainably.

I came out of the woods in 1995 to learn about how to live more sustainably in an urban setting where most people live. Since then, I have lived in two urban villages, one in a small town and one within a large metropolitan area.

Since 1970, my goal has been to use a car as little as possible. Since I work at home, I have a "low-pollute commute" from my bedroom to a chair and a laptop computer. I usually take one airplane trip a year to visit my sister and my publisher.

As you will learn in this book, life involves a series of environmental trade-offs. Like most people, I still have a large environmental impact, but I continue to struggle to reduce it. I hope you will join me in striving to live more sustainably and sharing what you learn with others. It is not always easy, but it sure is fun.

Cengage Learning's Commitment to Sustainable Practices

We the authors of this textbook and Cengage Learning, the publisher, are committed to making the publishing process as sustainable as possible. This involves four basic strategies:

- *Using sustainably produced paper.* The book publishing industry is committed to increasing the use of recycled fibers, and Cengage Learning is always looking for ways to increase this content. Cengage Learning works with paper suppliers to maximize the use of paper that contains only wood fibers that are certified as sustainably produced, from the growing and cutting of trees all the way through paper production.
- *Reducing resources used per book.* The publisher has an ongoing program to reduce the amount of wood pulp, virgin fibers, and other materials that go into each sheet of paper used. New, specially designed printing presses also reduce the amount of scrap paper produced per book.
- *Recycling.* Printers recycle the scrap paper that is produced as part of the printing process. Cengage Learning also recycles waste cardboard from shipping cartons, along with other materials used in the publishing process.
- *Process improvements.* In years past, publishing has involved using a great deal of paper and ink for the writing and editing of manuscripts, copyediting, reviewing page proofs, and creating illustrations. Almost all of these materials are now saved through the use of electronic files. Very little paper and ink were used in the preparation of this textbook.

Living in the Environment 19e

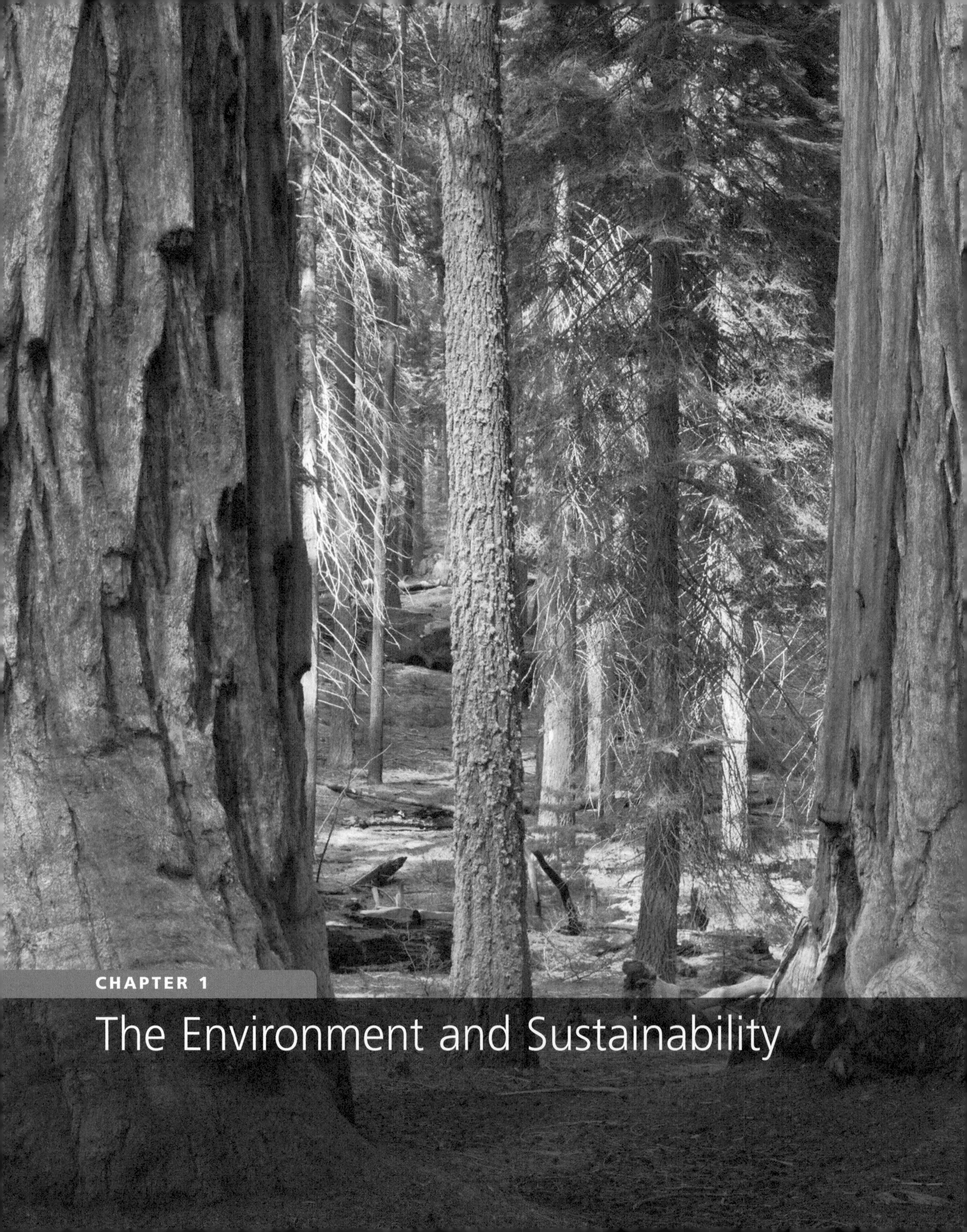

CHAPTER 1

The Environment and Sustainability

No civilization has survived the ongoing destruction of its natural support system. Nor will ours.

LESTER R. BROWN

Key Questions

1.1 What are some key principles of sustainability?

1.2 How are our ecological footprints affecting the earth?

1.3 What causes environmental problems and why do they persist?

1.4 What is an environmentally sustainable society?

Forests such as this one in California's Sequoia National Park help to sustain all life and economies.

Robert Harding World Imagery/Alamy Stock Photo

Learning from the Earth

Sustainability is the capacity of the earth's natural systems that support life and human social systems to survive or adapt to changing environmental conditions indefinitely. Sustainability is the big idea and the integrating theme of this book.

The earth is a remarkable example of a sustainable system. Life on the earth has existed for around 3.8 billion years. During this time, the planet has experienced several catastrophic environmental changes. They include gigantic meteorite impacts, ice ages lasting millions of years, long warming periods that melted land-based ice and raised sea levels by hundreds of feet, and five mass extinctions—each wiping out 60–95% of the world's species. Despite these dramatic environmental changes, an astonishing variety of life has survived.

How has life survived such challenges? Long before humans arrived, organisms had developed abilities to use sunlight to make their food and to recycle all of the nutrients they needed for survival. Organisms have developed amazing abilities to find food and survive. Spiders create webs strong enough to capture fast-moving flying insects. Bats have a radar system for finding prey and avoiding collisions. These and many other abilities and materials were developed without the use of the high-temperature or high-pressure processes or the harmful chemicals that we employ in manufacturing.

This explains why many scientists call for us to focus on learning from the earth about how to live more sustainably. In recent years, there have been efforts to make people more aware of such earth wisdom. Biologist Janine Benyus (Individuals Matter 1.1, p. 10) is a pioneer in this area. In 1997 she coined the term **biomimicry** to describe the rapidly growing scientific effort to understand, mimic, and catalog the ingenious ways in which nature has sustained life on the earth for 3.8 billion years. She views the earth's life-support system as the world's longest and most successful research and development laboratory.

How do geckos (Figure 1.1, left) cling to and walk on windows, walls, and ceilings? Scientists have learned that these little lizards have many thousands of tiny hairs growing in ridges on the toes of their feet and that each hair is divided into a number of segments that geckos use to grasp the tiniest ridges and cracks on a surface (Figure 1.1, right). They release their iron grip by tipping their foot until the hairs let go.

This discovery led to the development of a sticky, toxin-free "gecko tape" that could replace toxin-containing glues and tapes. It is an excellent example of biomimicry and you will see many more of such examples throughout this book.

Nature can teach us how to live more sustainably on the amazing planet that is our only home. As Benyus puts it, after billions of years of trial-and-error research and development: "Nature knows what works, what is appropriate, and what lasts here on Earth."

FIGURE 1.1 The gecko (left) has an amazing ability to cling to surfaces because of projections from many thousands of tiny hairs on its toes (right).

1.1 WHAT ARE SOME KEY PRINCIPLES OF SUSTAINABILITY?

CONCEPT 1.1A Life on the earth has been sustained for billions of years by solar energy, biodiversity, and chemical cycling.

CONCEPT 1.1B Our lives and economies depend on energy from the sun and on natural resources and ecosystem services (*natural capital*) provided by the earth.

CONCEPT 1.1C We could live more sustainably by following six principles of sustainability.

Environmental Science Is a Study of Connections in Nature

The **environment** is everything around you. It includes all the living things (such as plants and animals) and the nonliving things (such as air, water, and sunlight) with which you interact. You are part of nature and live in the environment, as reflected in the title of this textbook. Despite humankind's many scientific and technological advances, our lives depend on sunlight and the earth for clean air and water, food, shelter, energy, fertile soil, a livable climate, and other components of the planet's *life-support system*.

Environmental science is a study of connections nature. It is an interdisciplinary study of **(1)** how the earth (nature) works and has survived and thrived, **(2)** how humans interact with the environment, and **(3)** how we can live more sustainably. It strives to answer several questions: What environmental problems do we face? How serious are they? How do they interact? What are their causes? How has nature solved such problems? How can we solve such problems? To answer such questions, environmental science integrates information and ideas from fields such as biology, chemistry, geology, geography, economics, political science, and ethics.

A key component of environmental science is **ecology**, the branch of biology that focuses on how living organisms interact with the living and nonliving parts of their environment. Each of the earth's organisms, or living things, belongs to a **species**, or a group of organisms having a unique set of characteristics that set it apart from other groups.

A major focus of ecology is the study of ecosystems. An **ecosystem** is a set of organisms within a defined area of land or volume of water that interact with one another and with their environment of nonliving matter and energy. For example, a forest ecosystem consists of plants (especially trees; see chapter-opening photo), animals, and other organisms that decompose organic materials. These organisms interact with one another, with solar energy, and with the chemicals in the forest's air, water, and soil.

Environmental science and ecology should not be confused with **environmentalism** or **environmental activism**, which is a social movement dedicated to protecting the earth's life and its resources. Environmentalism is practiced more in the realms of politics and ethics than in science. However, the findings of environmental scientists can provide evidence to back the claims and activities of environmentalists.

Learning from the Earth: Three Scientific Principles of Sustainability

The latest version of our species has been around for about 200,000 years—less than the blink of an eye, relative to the 3.8 billion years life has existed on the earth (see the Geologic and Biological Time Scale in Supplement 6, p. S46). During our short time on the earth, and especially since 1900, we have expanded into and dominated almost all of the earth's ecosystems.

We have cleared forests and plowed grasslands to grow food on 40% of the earth's land and built cities that are home for more than half of the world's population. We use many of the world's natural resources and add pollution and wastes to the environment. We control 75% of the world's freshwater and most of ocean that covers 71% of the earth's surface. This large and growing human impact threatens the existence of many species and biological centers of life such as tropical rainforests and coral reefs. It also adds pollutants to the earth's air, water, and soil.

Three scientific natural factors play key roles in the long-term sustainability of the planet's life, as summarized below and in Figure 1.2 (**Concept 1.1A**). Understanding these three **scientific principles of sustainability**, or major *lessons from nature*, can help us move toward a more sustainable future.

- **Dependence on solar energy**: The sun's energy warms the planet and provides energy that plants use to produce **nutrients**, the chemicals that plants and animals need to survive.
- **Biodiversity**: The variety of genes, species, ecosystems, and ecosystem processes are referred to as **biodiversity** (short for *biological diversity*). Interactions among species provide vital ecosystem services and keep any population from growing too large. Biodiversity also provides ways for species to adapt to changing environmental conditions and replace species wiped out by catastrophic environmental changes with new species.
- **Chemical cycling**: The circulation of chemicals or nutrients needed to sustain life from the environment (mostly from soil and water) through various organisms and back to the environment is called **chemical cycling**, or **nutrient cycling**. The earth receives a continuous supply of energy from the sun but it receives no new supplies of life-supporting chemicals. Through billions of years of interactions with their living and nonliving environment, organisms have developed ways to continually recycle the

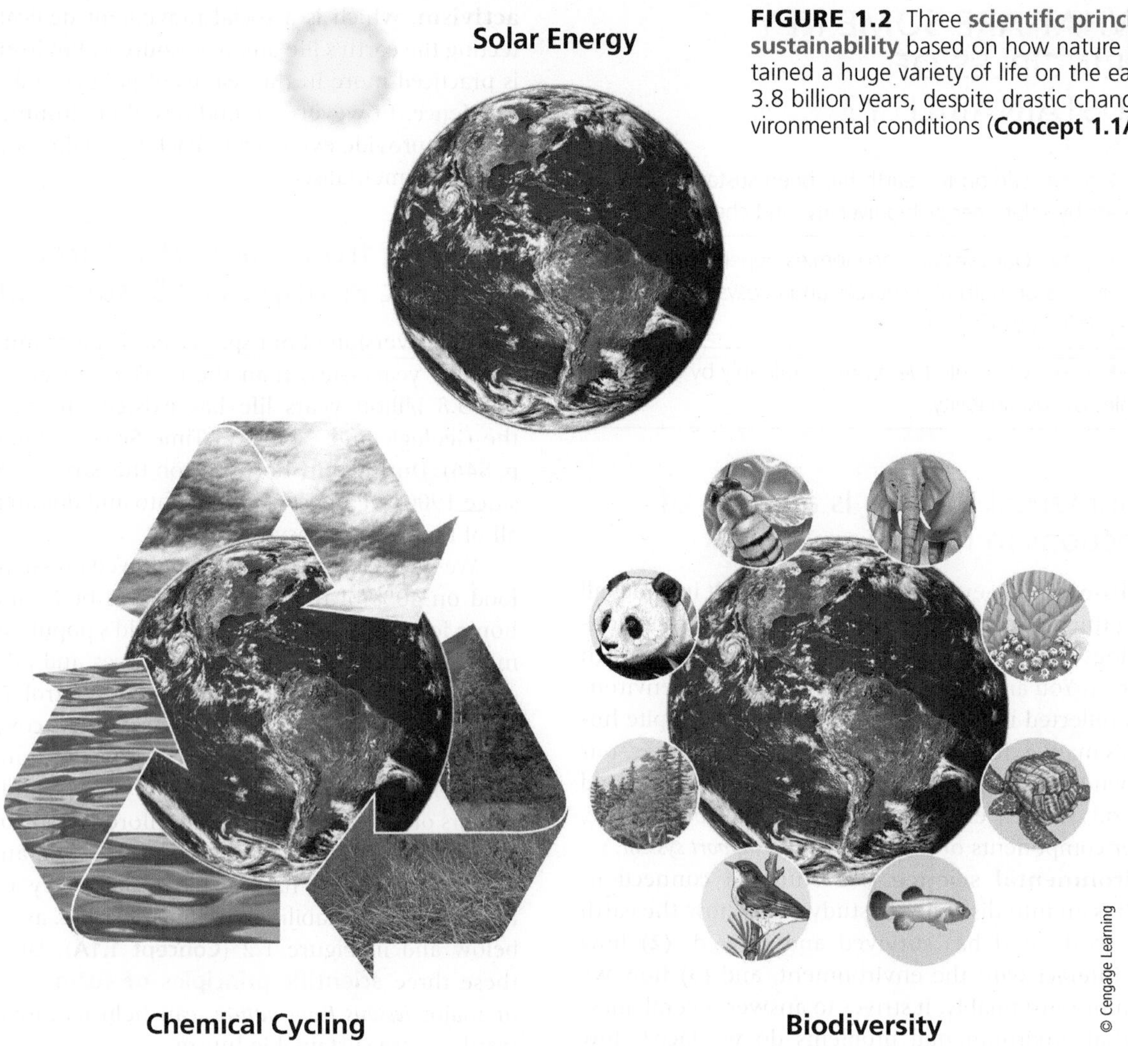

FIGURE 1.2 Three **scientific principles of sustainability** based on how nature has sustained a huge variety of life on the earth for 3.8 billion years, despite drastic changes in environmental conditions (**Concept 1.1A**).

chemicals they need to survive. This means that the wastes and decayed bodies of organisms become nutrients or raw materials for other organisms. In nature, **waste = useful resources**.

Key Components of Sustainability

Sustainability, the integrating theme of this book, has several key components that we use as subthemes. One is **natural capital**—the natural resources and ecosystem services that keep humans and other species alive and that support human economies (Figure 1.3).

Natural resources are materials and energy provided by nature that are essential or useful to humans. They fall into three categories: *inexhaustible resources, renewable resources,* and *nonrenewable (exhaustible) resources* (Figure 1.4). Solar energy is an **inexhaustible resource** because it is expected to last for at least 5 billion years until the death of the star we call the sun. A renewable resource is any resource that can be replenished by natural processes within hours to centuries, as long as people do not use the resource faster than natural processes can replace it. Examples include forests, grasslands, fertile topsoil, fishes, clean air, and freshwater. The highest rate at which people can use a renewable resource indefinitely without reducing its available supply is called its **sustainable yield**.

Nonrenewable or **exhaustible resources** exist in a fixed amount, or *stock*, in the earth's crust. They take millions to billions of years to form through geological processes. On the much shorter human time scale, we can use these resources faster than nature can replace them. Examples of nonrenewable resources include fossil fuel energy resources (such as oil, coal, and natural gas), metallic mineral resources (such as copper and aluminum), and nonmetallic mineral resources (such as salt and sand). As we deplete such resources, sometimes we can find substitutes.

Ecosystem services are natural services provided by healthy ecosystems that support life and human economies at no monetary cost to us (Figure 1.3). For example, forests help purify air and water, reduce soil erosion, regulate climate, and recycle nutrients. Thus, our lives and economies are sustained by energy from the sun and by natural resources and ecosystem services (natural capital) provided by the earth (**Concept 1.1B**).

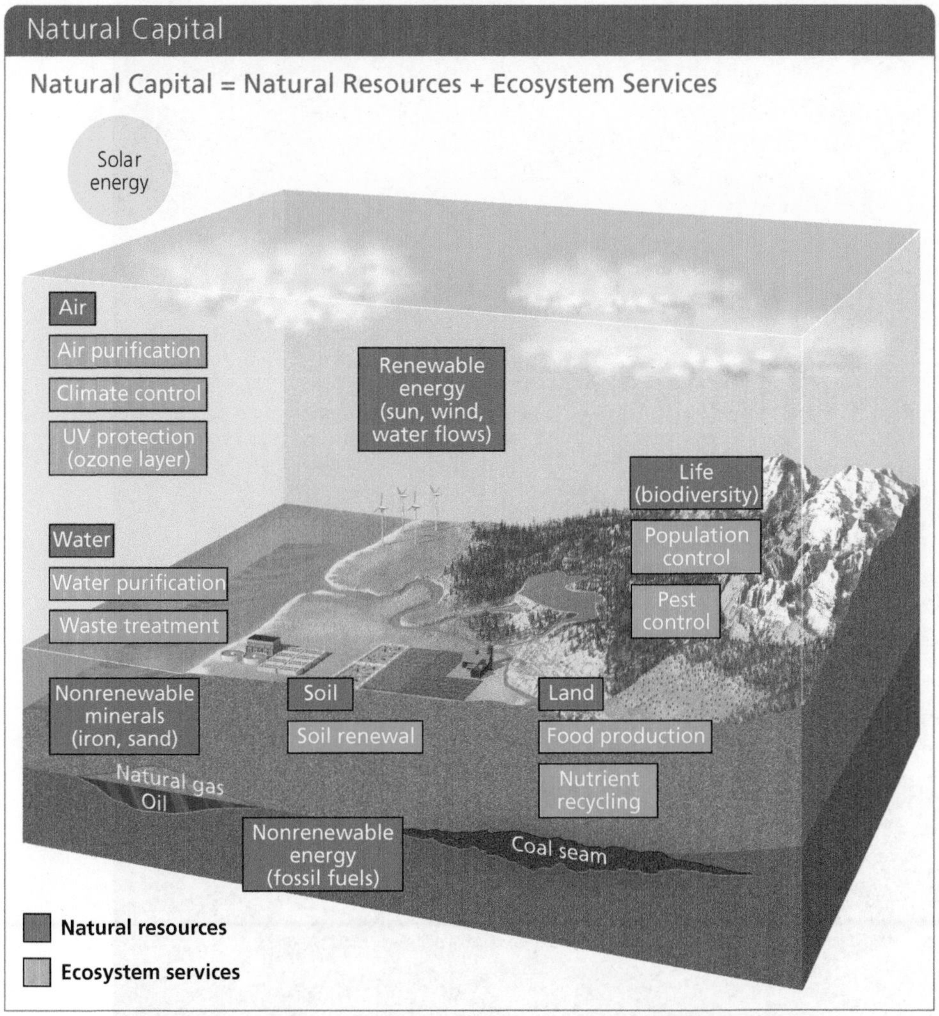

FIGURE 1.3 Natural capital consists of natural resources (blue) and ecosystem services (orange) that support and sustain the earth's life and human economies (**Concept 1.1B**).

Inexhaustible
Solar energy
Wind energy
Geothermal energy

Renewable
Trees
Topsoil
Freshwater

Nonrenewable (Exhaustible)
Fossil fuels (oil, natural gas, coal)
Iron and copper

FIGURE 1.4 We depend on a combination of inexhaustible, renewable, and exhaustible (nonrenewable) natural resources.

FIGURE 1.5 Small remaining area of once diverse Amazon rain forest surrounded by vast soybean fields in the Brazilian state of Mato Grosso.

A vital ecosystem service is **nutrient cycling**, which is a **scientific principle of sustainability**. For example, without nutrient cycling in topsoil there would be no land plants, no pollinators (another ecosystem service), and no humans or other land animals. This would also disrupt the ecosystem services that purify air and water.

A second component of sustainability—and another subtheme of this text—is that human activities can *degrade natural capital*. We do this by using renewable resources faster than nature can restore them and by overloading the earth's normally renewable air, water, and soil with pollution and wastes. For example, people in many parts of the world are replacing forests with crop plantations (Figure 1.5) that require large inputs of energy, water, fertilizer, and pesticides. We also add pollutants to the air and dump chemicals and wastes into rivers, lakes, and oceans faster than they can be cleansed through natural processes. Many of the plastics and other synthetic materials people use poison wildlife and disrupt nutrient cycles because they cannot be broken down and used as nutrients by other organisms.

This leads us to a third component of sustainability: creating *solutions* to the environmental problems we face. For example, a solution to the loss of forests (see chapter-opening photo) is to stop burning or cutting down mature forests. This cannot be done unless citizens become educated about the ecosystem services forests provide and governments pass laws to protect forests.

Conflicts can arise when environmental protection has a negative economic effect on groups of people or certain industries. Dealing with such conflicts often involves both sides making compromises or *trade-offs*. For example, a timber company might be persuaded to plant and harvest trees in an area that it had already cleared or degraded instead of clearing an undisturbed forest area. In return, the government may subsidize (pay part of the cost) of planting the new trees.

Each of us can play an important role in learning how to live more sustainably. Thus, *individuals matter*—another sustainability subtheme of this book.

Three Additional Principles of Sustainability

Economics, politics, and ethics can provide us with three additional **principles of sustainability** (Figure 1.6):

- **Full-cost pricing** (from economics): Some economists urge us to find ways to include the harmful environmental and health costs of producing and using goods and services in their market prices. This practice, called *full-cost pricing*, would give consumers information about the harmful environmental impacts of products.
- **Win-win solutions** (from political science): Political scientists often look for *win-win solutions* to environmental problems based on cooperation and compromise that will benefit the largest number of people as well as the environment.
- **Responsibility to future generations** (from ethics): Ethics is a branch of philosophy devoted to studying ideas about what is right or wrong. According to environmental ethicists, we should leave the planet's life-support systems in a condition that is as good as or better than it is now as our responsibility to future generations.

These six **principles of sustainability** (see inside back cover of book) can serve as guidelines to help us move toward a future that is more sustainable ecologically, economically, and socially. This includes using biomimicry as a major tool for learning how to live more sustainably (**Core Case Study** and Individuals Matter 1.1).

Countries Differ in Their Resource Use and Environmental Impact

The United Nations (UN) classifies the world's countries as economically more developed or less developed, based primarily on their average income per person. **More-developed countries**—industrialized nations with high average incomes per person—include the United States, Japan, Canada, Australia, and Germany and most other European countries. These countries, with 17% of the world's population, use about 70% of the earth's natural resources. The United States, with only 4.3% of the world's population, uses about 30% of the world's resources.

All other nations are classified as **less-developed countries**, most of them in Africa, Asia, and Latin America. Some are *middle-income, moderately developed countries*

FIGURE 1.6 Three **principles of sustainability** based on economics, political science, and ethics can help us make a transition to a more environmentally and economically sustainable future.

Left: Minerva Studio/Shutterstock.com. Center: mikeledray/Shutterstock.com. Right: iStockphoto.com/ Kali Nine LLC

INDIVIDUALS MATTER 1.1

ecowatch.com

Janine Benyus: Using Nature to Inspire Sustainable Design and Living

Janine Benyus has had a lifelong interest in learning how nature works and how to live more sustainably. She realized that 99% of the species that have lived on the earth became extinct because they could not adapt to changing environmental conditions. She views the surviving species as examples of *natural genius* that we can learn from.

Benyus says that when we need to solve a problem or design a product, we should ask: Has nature done this and how did it do it? We should also think about what nature does not do as a clue to what we should not do, she argues. For example, nature does not produce waste materials or chemicals that cannot be broken down and recycled.

Benyus has set up the nonprofit Biomimicry Institute that has developed a curriculum for K–12 and university students and has a 2-year program to train biomimicry professionals. She has also established a network called Biomimicry 3.8, named for the 3.8 billion years during which organisms have developed their genius for surviving. It is a network of scientists, engineers, architects, and designers who share examples of successful biomimicry through an online database called AskNature.org.

such as China, India, Brazil, Thailand, and Mexico. Others are *low-income, least-developed countries* such as Nigeria, Bangladesh, Congo, and Haiti. (For a map showing high-, upper-middle-, lower-middle-, and low-income countries see Supplement 4, Figure 4, p. S19). The less-developed countries, with 83% of the world's population, use about 30% of the world's natural resources.

1.2 HOW ARE OUR ECOLOGICAL FOOTPRINTS AFFECTING THE EARTH?

CONCEPT 1.2A Humans dominate the earth with the power to sustain, add to, or degrade the natural capital that supports all life and human economies.

CONCEPT 1.2B As our ecological footprints grow, we deplete and degrade more of the earth's natural capital that sustains us.

Good News: Many People Have a Better Quality of Life

As the world's dominant animal, we have an awesome power to degrade or sustain our life-support system. We decide whether forests are preserved or cut down and engineer the flows of rivers. Our activities affect the temperature of the atmosphere and the temperature and acidity of the ocean. We can also contribute to the extinction of species. At the same time, our creativity, economic growth, scientific research, grassroots political pressure by citizens, and regulatory laws have improved the quality of life for many of the earth's people, especially in the United States and in most other more-developed countries.

We have developed an astounding array of useful materials and products. We have learned how to use wood, fossil fuels, the sun, wind, flowing water, the nuclei of certain atoms, and the earth's heat (geothermal energy) to supply us with enormous amounts of energy. We have created artificial environments in the form of buildings and cities. We have invented computers to extend our brains, robots to do much of our work, and electronic networks to enable instantaneous global communication.

Globally, life spans are increasing, infant mortality is decreasing, education is on the rise, some diseases are being conquered, and the population growth rate has slowed. Although one out of seven people live in extreme poverty, we have witnessed the greatest reduction in poverty in human history. The food supply is generally more abundant and safer, air and water are getting cleaner in many parts of the world, and exposure to toxic chemicals is more avoidable. We have protected some endangered species and ecosystems and restored some grasslands and wetlands, and forests are growing back in some areas that we cleared.

Scientific research and technological advances financed by affluence helped achieve these improvements in life and environmental quality. Education also spurred many citizens to insist that businesses and governments work toward improving environmental quality. We are a globally connected species with growing access to information that could help us to shift to a more sustainable path.

We Are Living Unsustainably

According to a large body of scientific evidence, we are living unsustainably. We waste, deplete, and degrade much of the earth's life-sustaining natural capital—a process known as **environmental degradation**, or **natural capital degradation** (Figure 1.7).

According to research by the Wildlife Conservation Society and the Columbia University Center for International Earth Science Information Network, human activities directly affect about 83% of the earth's land surface (excluding Antarctica) (Figure 1.8). This land is used for urban development, growing crops, energy production, pasture for livestock, mining, timber cutting, and other purposes.

In parts of the world, we are destroying forests and grasslands, withdrawing water from some rivers and underground aquifers faster than nature replenishes them, and harvesting many fish species faster than they can be renewed. We also litter the land and oceans with wastes faster than they can be recycled by the earth's natural chemical cycles. In addition, we add pollutants to the air (including some that are altering the earth's climate), soil, aquifers, rivers, lakes, and oceans.

In many parts of the world, renewable forests are shrinking (Figure 1.5), deserts are expanding, and topsoil is eroding. The lower atmosphere is warming, floating ice and many glaciers are melting at unexpected rates, sea levels are rising, and ocean acidity is increasing. There are more intense floods, droughts, severe weather, and forest fires in many areas. In a number of regions, rivers are running dry, harvests of many species of fish are dropping sharply, and 20% of the world's species-rich coral reefs are gone and others are threatened. Species are becoming extinct at least 100 times faster than in pre-human times. And extinction rates are projected to increase at least another 100-fold during this century, creating a mass extinction caused by human activities.

In 2005 the UN released its *Millennium Ecosystem Assessment*, a 4-year study by 1,360 experts from 95 countries. According to this study, human activities have overused about 60% of the ecosystem services provided by nature (see orange boxes in Figure 1.3), mostly since 1950. According to these researchers, "human activity is putting such a strain on the natural functions of Earth that the ability of the planet's ecosystems to sustain future generations can no longer be taken for granted." They also concluded that there are scientific, economic, and political solutions to these problems that could be implemented within a few decades.

There is much talk about saving the earth, but the earth does not need saving. It has been around for 4.6 billion years, has sustained life for 3.8 billion years, and has

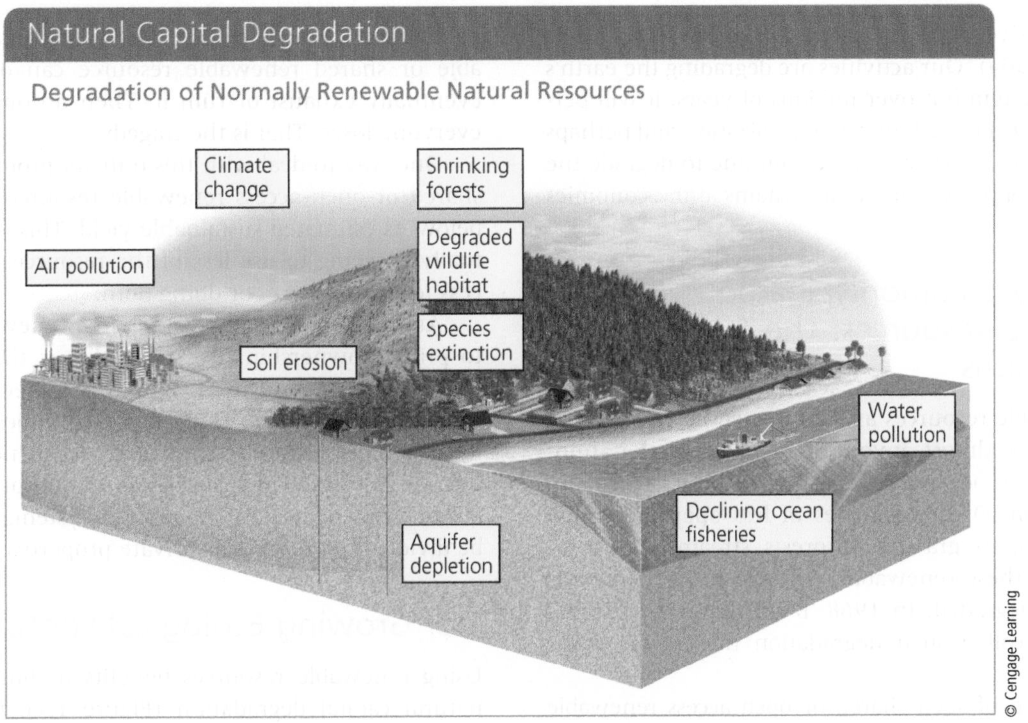

FIGURE 1.7 Natural capital degradation: Degradation of normally renewable natural resources and natural services (Figure 1.3), mostly from population growth and increased resource use per person.

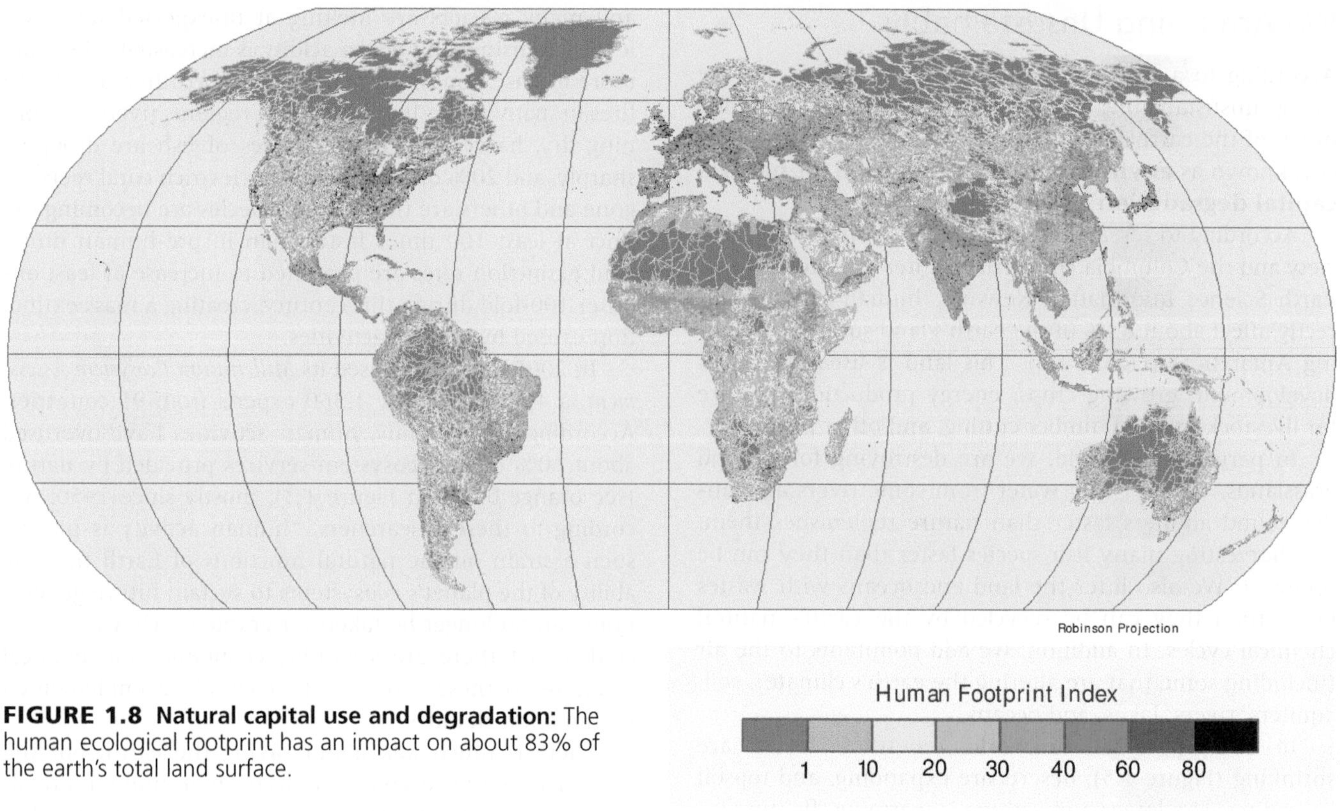

FIGURE 1.8 Natural capital use and degradation: The human ecological footprint has an impact on about 83% of the earth's total land surface.

survived massive changes in environmental conditions (**Core Case Study**). Our activities are degrading the earth's life-support system but over millions of years, it will persist. What needs saving is our own civilization and perhaps the existence of our species if we continue to degrade the earth's life-support system that sustains our economies and us.

Degrading Commonly Shared Renewable Resources: The Tragedy of the Commons

Some renewable resources are not owned by anyone and can be used by almost anyone. Examples are the atmosphere, the open ocean and its fishes, and the earth's life-support system. Other examples of less open, but often *shared resources*, are grasslands, forests, streams, and aquifers. Many of these renewable resources have been environmentally degraded. In 1968, biologist Garrett Hardin (1915–2003) called such degradation the *tragedy of the commons*.

Degradation of such shared or open-access renewable resources occurs because each user reasons, "The little bit that I use or pollute is not enough to matter, and anyway, it's a renewable resource." When the level of use is small, this logic works. Eventually, however, the cumulative effect of large numbers of people trying to exploit a widely available or shared renewable resource can degrade it and eventually exhaust or ruin it. Then no one benefits and everyone loses. That is the tragedy.

One way to deal with this difficult problem is to use a shared or open-access renewable resource at a rate well below its estimated sustainable yield. This is done by mutually agreeing to use less of the resource, regulating access to the resource, or doing both.

Another way is to convert shared renewable resources to private ownership. The reasoning is that if you own something, you are more likely to protect your investment. However, history shows that this does not necessarily happen. In addition, this approach is not possible for open-access resources such as the atmosphere, the open ocean, and our global life-support system, which cannot be divided up and sold as private property.

Our Growing Ecological Footprints

Using renewable resources benefits us but can result in natural capital degradation (Figure 1.7), pollution, and wastes. This harmful environmental impact is called an **ecological footprint**—the amount of biologically productive land and water (biocapacity) needed to supply a population in an area with renewable resources and to

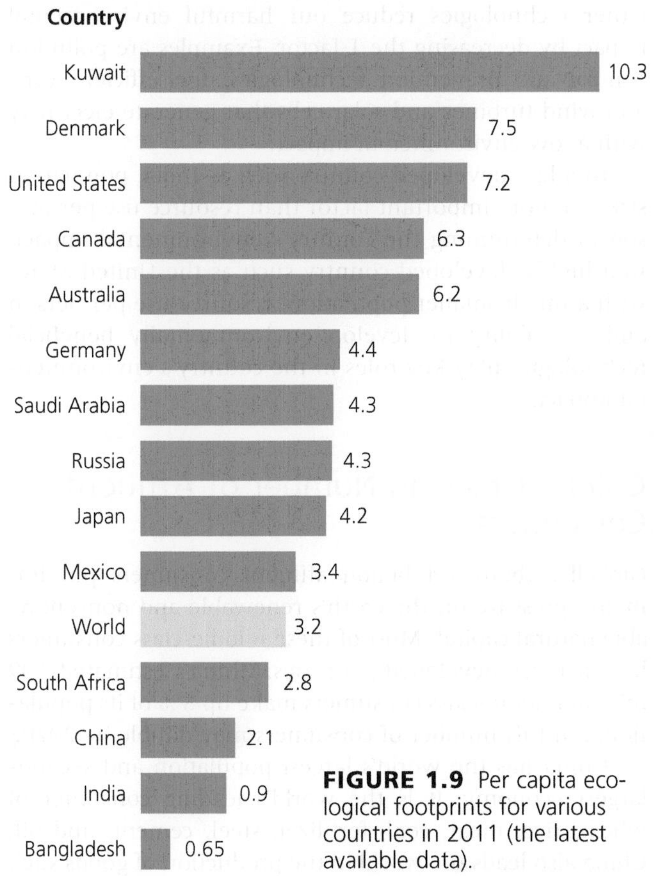

FIGURE 1.9 Per capita ecological footprints for various countries in 2011 (the latest available data).

Compiled by the authors using data from Global Footprint Network and WWF: Living Planet Report 2014.

absorb and recycle the wastes and pollution such resource use produces. Ecologist William Rees and environmental scientist Mathis Wackernagel developed this concept in the 1990s.

This measure of sustainability relates to the earth's **biocapacity**—the ability of its productive ecosystems to regenerate the renewable resources used by a population, city, region, country, or the world, and to absorb the resulting wastes and pollution indefinitely. The largest component of our ecological footprint is the air pollution, climate change, and ocean acidification caused by the burning of fossil fuels—oil, coal, and natural gas—to provide 90% of the commercial energy used in the world and in the United States. The **per capita ecological footprint** is the average ecological footprint of an individual in a given country or area (Figure 1.9).

Scientists estimate how much land and water we need to support an area's people, economy, and average lifestyle and then compare it with how much land and water is available to do this. If the total ecological footprint is larger than its biocapacity, the area is said to have an *ecological deficit*. In other words, its people are living unsustainably by depleting natural capital instead of living off the renewable resources and ecosystem services provided by such capital. Figure 1.10 is a map of ecological debtor and creditor countries (see also the Ecological Footprint Analysis exercise on p. 27).

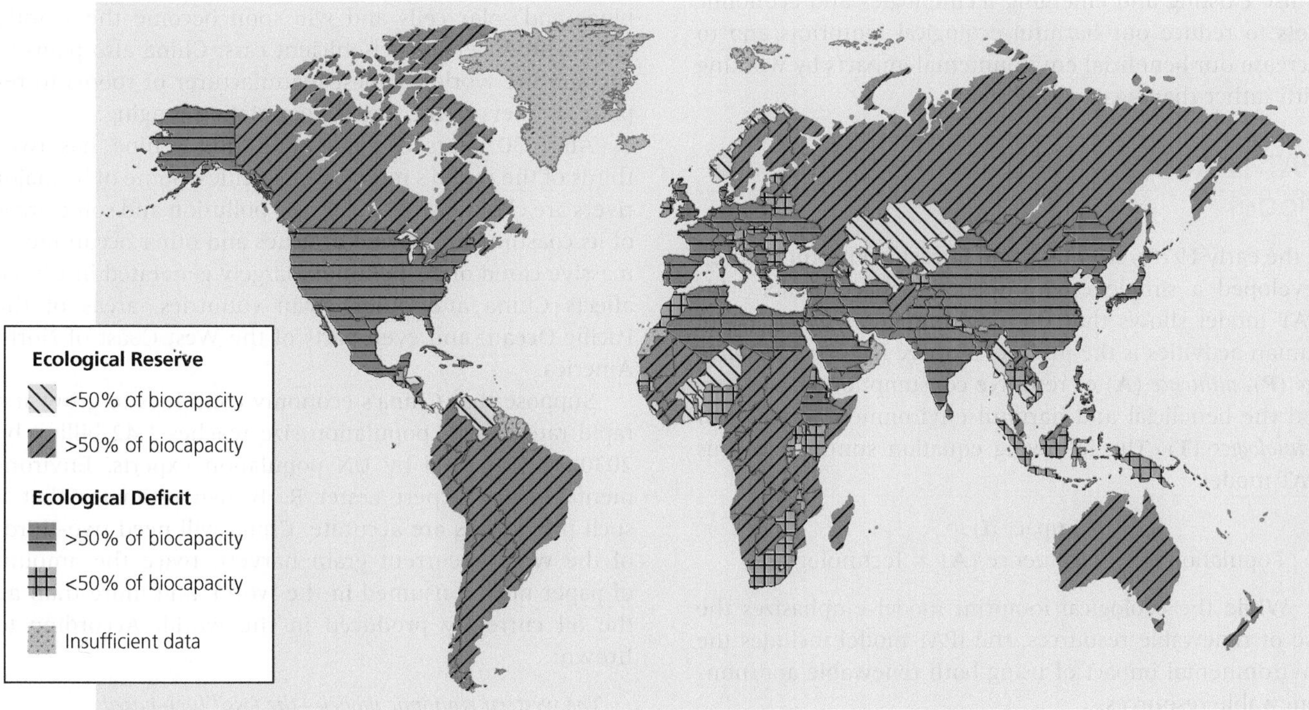

Compiled by the authors using data from Global Footprint Network and WWF: Living Planet Report 2014.

FIGURE 1.10 *Ecological debtors and creditors.* The ecological footprints of some countries exceed their biocapacity, while other countries have ecological reserves. ***Critical thinking:*** Why do you think that the United States is an ecological debtor country?

The World Wide Fund for Nature (WWF) and the Global Footprint Network (headed by Wackernagel) estimate that we would need 1.5 planet Earths to sustain the world's 2012 rate of resource use far into the future. In other words, the world's total ecological footprint in 2012 was 50% higher than the planet's estimated long-term biocapacity. This overdraft of the earth's natural resources and ecosystem services is being passed on to future generations.

1.5 Number of earths needed to sustain the world's 2012 rate of renewable resource use indefinitely

Architect William McDonough and scientist Michael Braungart recognize the need to reduce our harmful environmental impacts but call for us to also create and expand our beneficial environmental impacts. For example, we can replant forests on degraded land, restore degraded wetlands and grasslands, and protect species from becoming extinct.

According to research in ecology and environmental science, everything in nature is connected. Thus, what we do to the earth, we do to ourselves. This is why living more sustainably requires expanding our beneficial environmental impact. Throughout this book, we discuss ways to use existing and emerging technologies and economic tools to reduce our harmful ecological footprints and to increase our beneficial environmental impacts by working with rather than against the earth.

IPAT Is Another Environmental Impact Model

In the early 1970s, scientists Paul Ehrlich and John Holdren developed a simple environmental impact model. This IPAT model shows that the environmental *impact* (**I**) of human activities is the product of three factors: *population size* (**P**), *affluence* (**A**) or resource consumption per person, and the beneficial and harmful environmental effects of *technologies* (**T**). The following equation summarizes this IPAT model:

$$\text{Impact } (\mathbf{I}) = \text{Population } (\mathbf{P}) \times \text{Affluence } (\mathbf{A}) \times \text{Technology } (\mathbf{T})$$

While the ecological footprint model emphasizes the use of renewable resources, the IPAT model includes the environmental impact of using both renewable and nonrenewable resources.

The T factor can be harmful or beneficial. Some forms of technology such as polluting factories, gas-guzzling motor vehicles, and coal-burning power plants increase our harmful environmental impact by raising the T factor. Other technologies reduce our harmful environmental impact by decreasing the T factor. Examples are pollution control and prevention technologies, fuel-efficient cars, and wind turbines and solar cells that generate electricity with a low environmental impact.

In a less-developed country such as India, population size is a more important factor than resource use per person in determining the country's environmental impact. In a highly developed country such as the United States with a much smaller population, resource use per person and the ability to develop environmentally beneficial technologies play key roles in the country's environmental impact.

China's Growing Number of Affluent Consumers

Globally, about 1.4 billion affluent consumers put immense pressure on the earth's renewable and nonrenewable natural capital. Most of these middle-class consumers live in more-developed countries. China's estimated 109 million middle-class consumers make up 8% of its population, but this number of consumers may double by 2020.

China has the world's largest population and second-largest economy. It is the world's leading consumer of wheat, rice, meat, coal, fertilizer, steel, cement, and oil. China also leads the world in the production of goods such as televisions, cell phones, refrigerators, and drones for consumers. It is the world's largest producer of wind turbines and solar cells and will soon become the world's largest producer of fuel-efficient cars. China also plans to become the world's largest manufacturer of robots to replace workers as the world's population is aging.

After 30 years of industrialization, China has two-thirds of the world's most polluted cities. Some of its major rivers are choked with waste and pollution and some areas of its coastline are devoid of fishes and other ocean life. A massive cloud of air pollution, largely generated in China, affects China and other Asian countries, areas of the Pacific Ocean, and even parts of the West Coast of North America.

Suppose that China's economy continues to grow at a rapid rate and its population size reaches 1.42 billion by 2030, as projected by UN population experts. Environmental policy expert Lester R. Brown estimates that if such projections are accurate, China will need two-thirds of the world's current grain harvest, twice the amount of paper now consumed in the world, and more than all the oil currently produced in the world. According to Brown:

> *The western economic model—the fossil fuel–based, automobile-centered, throwaway economy—is not going to work for China . . . or for the other 3 billion people in developing countries who are also dreaming the "American dream."*

Cultural Changes Can Increase or Shrink Our Ecological Footprints

Until about 10,000 to 12,000 years ago, we were mostly *hunter–gatherers* who obtained food by hunting wild animals or scavenging their remains, and gathering wild plants. Our hunter–gatherer ancestors lived in small groups, consumed few resources, had few possessions, and moved as needed to find enough food to survive.

Since then, three major cultural changes have occurred. *First* was the *agricultural revolution*, which began around 10,000 years ago when humans learned how to grow and breed plants and animals for food, clothing, and other purposes and began living in villages instead of frequently moving to find food. They had a more reliable source of food, lived longer, and produced more children who survived to adulthood.

Second was the *industrial–medical revolution*, beginning about 300 years ago when people invented machines for the large-scale production of goods in factories. Many people move from rural villages to cities to work in the factories. This shift involved learning how to get energy from fossil fuels (such as coal and oil) and how to grow large quantities of food in an efficient manner. It also included medical advances that allowed a growing number of people to have longer and healthier lives. *Third*, about 50 years ago the *information–globalization revolution* began when we developed new technologies for gaining rapid access to all kinds of information and resources on a global scale.

Each of these three cultural changes gave us more energy and new technologies with which to alter and control more of the planet's resources to meet our basic needs and increasing wants. They also allowed expansion of the human population, mostly because of larger food supplies and longer life spans. In addition, these cultural changes resulted in greater resource use, pollution, and environmental degradation and allowed us to dominate the planet and expand our ecological footprints (Figures 1.8 and 1.9).

On the other hand, some technological leaps have enabled us to shrink our ecological footprints by reducing our use of energy and matter resources and our production of wastes and pollution. For example, the use of the energy-efficient LED light bulbs and energy-efficient cars and buildings is on the rise.

Many environmental scientists and other analysts see such developments as evidence of an emerging fourth major cultural change: a **sustainability revolution**, in which we could learn to live more sustainably during this century. This involves avoiding degradation and depletion of the natural capital that supports all life and our economies and restoring natural capital that we have degraded (Figure 1.3). Making this shift involves learning how nature has sustained life for over 3.8 billion years and using these lessons from nature to shrink our ecological footprints and grow our beneficial environmental impacts. GOOD NEWS

1.3 WHAT CAUSES ENVIRONMENTAL PROBLEMS AND WHY DO THEY PERSIST?

CONCEPT 1.3A Basic causes of environmental problems are population growth, wasteful and unsustainable resource use, poverty, avoidance of full-cost pricing, increasing isolation from nature, and different environmental worldviews.

CONCEPT 1.3B Our environmental worldviews play a key role in determining whether we live unsustainably or more sustainably.

Basic Causes of Environmental Problems

To deal with the environmental problems we face we must understand their causes. According to a significant number of environmental and social scientists, the major causes of today's environmental problems are:

- population growth
- wasteful and unsustainable resource use
- poverty
- omission of the harmful environmental and health costs of goods and services in market prices
- increasing isolation from nature
- competing environmental worldviews.

We discuss each of these causes in detail in later chapters of this book. Let us begin with a brief overview from them.

The Human Population Is Growing at a Rapid Rate

Exponential growth occurs when a quantity increases at a fixed percentage per unit of time, such as 0.5% or 2% per year. Exponential growth starts slowly but after a few doublings it grows to enormous numbers because each doubling is twice the total of all earlier growth. When we plot the data for an exponentially growing quantity, we get a curve that looks like the letter J.

For an example of the awesome power of exponential growth, consider a simple form of bacterial reproduction in which one bacterium splits into two every 20 minutes. Starting with one bacterium, after 20 minutes, there would be two; after an hour, there would be eight; ten hours later, there would be more than 1,000; and after just 36 hours (assuming that nothing interfered with their reproduction), there would be enough bacteria to form a layer 0.3 meters (1 foot) deep over the entire earth's surface.

The human population has grown exponentially (Figure 1.11) to the current population of 7.3 billion people. In 2015, the rate of growth was 1.20%. Although this rate

Country	Population (millions)
China	1,372
India	1,314
United States	321
Indonesia	256
Brazil	205
Pakistan	199
Nigeria	182
Bangladesh	160
Russia	144
Mexico	127

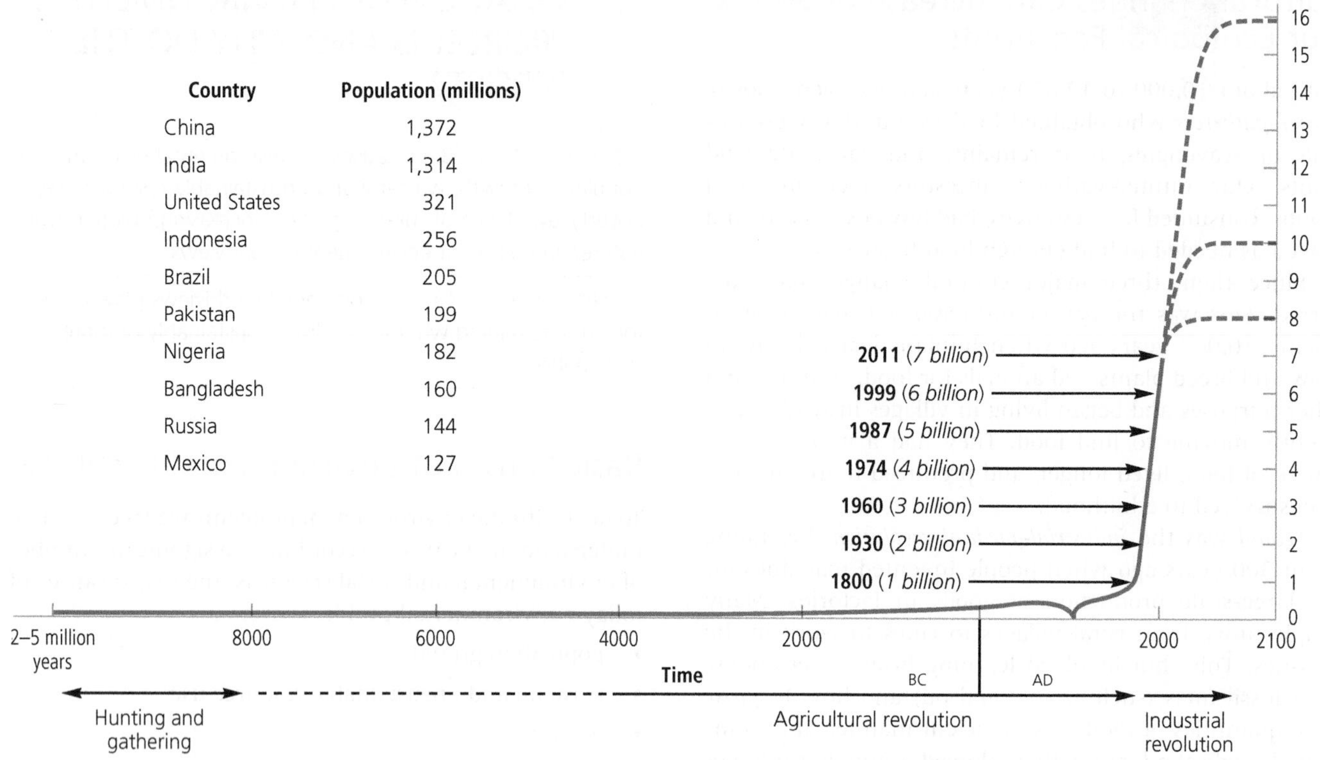

Compiled by the authors using data from the World Bank, United Nations, and Population Reference Bureau. Photo: NASA.

FIGURE 1.11 *Exponential growth:* The J-shaped curve represents past exponential world population growth, with projections to 2100 showing possible population stabilization as the J-shaped curve of growth changes to an S-shaped curve. The top 10 countries (left) represent nearly 60% of the world's total population in 2015. ***Data analysis:*** By what percentage did the world's population increase between 1960 and 2015? (This figure is not to scale.)

of growth seems small, it added 88.9 million people to the world's 7.34 billion people. By 2050, the population could reach 9.8 billion—an addition of 2.5 billion people.

CONSIDER THIS . . .

CONNECTIONS Exponential Growth and Doubling Time: The Rule of 70

The doubling time of the human population or of any exponentially growing quantity can be calculated by using the rule of 70: doubling time (years) = 70/annual growth rate (%). The world's population is growing at about 1.20% per year. At this rate how long will it take to double its size?

No one knows how many people the earth can support indefinitely. However, our large and expanding ecological footprints and the resulting widespread natural capital degradation are disturbing warning signs.

Some analysts call for us to reduce environmental degradation by slowing population growth and level it off at around 8 billion by 2050 instead of 9.8 billion. We examine the possible ways to do this in Chapter 6. Other analysts call for us to shift from environmentally harmful to environmentally beneficial forms of economic growth, which we discuss in Chapter 23.

Affluence and Unsustainable Resource Use

The lifestyles of the world's expanding population of consumers are built on growing affluence, or resource consumption per person, as more people earn higher incomes. As total resource consumption and average resource consumption per person increase, so does environmental degradation, wastes, and pollution from the increase in environmental footprints.

The effects can be dramatic. The WWF and the Global Footprint Network estimate that the United States, with only 4.3% of the world's population, is responsible for about 23% of the global ecological footprint. The average American consumes about 30 times the amount of resources that the average Indian consumes and 100 times the amount consumed by the average person in the world's poorest countries. The WWF has projected that we would need five planet Earths if everyone used renewable resources at the same rate as the average American did in 2012.

On the other hand, affluence can allow for widespread and better education that can lead people to become more concerned about environmental quality. Affluence also makes more money available for developing technologies to reduce pollution, environmental degradation, and resource

FIGURE 1.12 Poor settlers in Peru have cleared and burned this small plot in a tropical rain forest in the Amazon and planted it with seedlings to grow food for their survival.

5 Number of earths needed to sustain the world's population at U.S. consumption rates in 2012

waste along with ways to increase our beneficial environmental impacts.

Poverty Has Harmful Environmental and Health Effects

Poverty is a condition in which people lack enough money to fulfill their basic needs for food, water, shelter, health care, and education. *Bad News*: According to the World Bank, about one of every three people, or 2.6 billion people, struggled to live on less than $2.25 a day in 2014. In addition, 1 billion people living in *extreme poverty* struggled to live on the equivalent of less than $1.25 a day—less than what many people spend for a bottle of water or a cup of coffee. Could you do this? *Good news*: The percentage of the world's population living on less than $1.25 a day decreased from 52% in 1981 to 14% in 2014.

Poverty causes a number of harmful environmental and health effects. The daily lives of the world's poorest people center on getting enough food, water, and cooking and heating fuel to survive. Typically, these individuals are too desperate for short-term survival to worry about long-term environmental quality or sustainability. Thus, collectively, they may be forced to degrade forests (Figure 1.12), topsoil, and grasslands, and deplete fisheries and wildlife populations to stay alive.

Poverty does not always lead to environmental degradation. Some of the poor increase their beneficial environmental impact by planting and nurturing trees and conserving the soil that they depend on as a part of their long-term survival strategy.

CONSIDER THIS . . .

CONNECTIONS Poverty and Population Growth

To many poor people, having more children is a matter of survival. Their children help them gather firewood, haul water, and tend crops and livestock. The children also help take care of their aging parents, most of whom do not have social security, health care, and retirement funds. This daily struggle for survival is largely why populations in some of the poorest countries continue to grow at high rates.

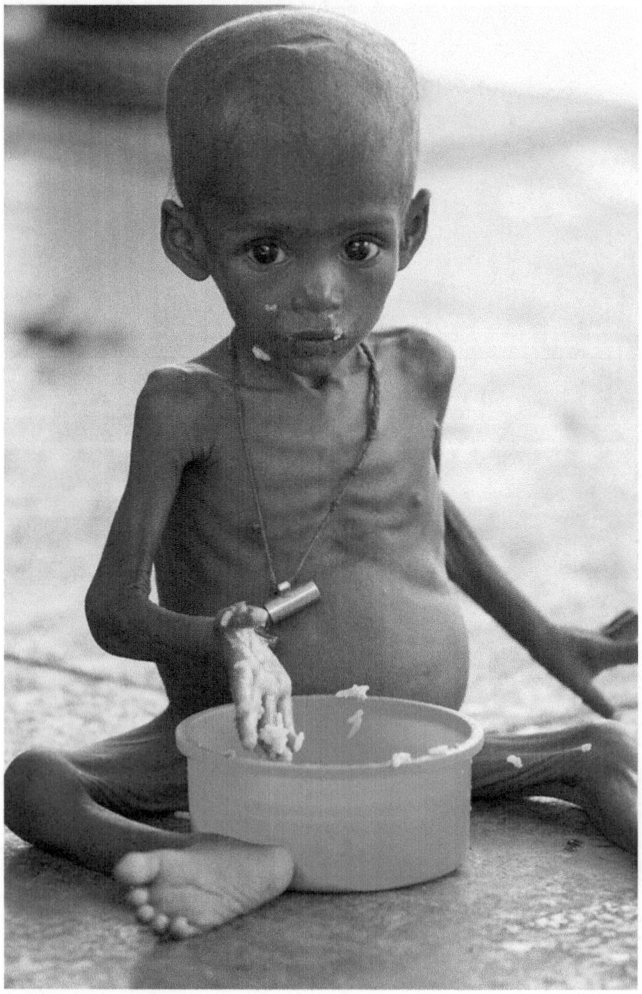

FIGURE 1.13 One of every three children younger than age 5 in less-developed countries, such as this starving child in Bangladesh, suffers from severe malnutrition caused by a lack of calories and protein.

Environmental degradation can have severe health effects on the poor. One problem is life-threatening *malnutrition*, a lack of protein and other nutrients needed for good health (Figure 1.13). Another effect is illness caused by limited access to adequate sanitation facilities and clean drinking water. As a result, about one of every nine of the world's people get water for drinking, washing, and cooking from sources polluted by human and animal feces.

In 2010 the World Health Organization estimated that these factors—mostly related to poverty—cause premature death for about 7 million children under age 5 each year. Some hopeful news is that this number of annual deaths is down from about 10 million in 1990. Even so, every day an average of at least 19,000 young children die prematurely from these causes. This is equivalent to *95 fully loaded 200-passenger airliners crashing every day with no survivors*. The news media rarely cover this ongoing human tragedy.

CONSIDER THIS . . .

THINKING ABOUT The Poor, the Affluent, and Environmental Harm

Some see the rapid population growth in less-developed countries as the primary cause of our environmental problems. Others say that the high rate of resource use per person in more-developed countries is a more important factor. Which factor do you think is more important? Why?

Prices of Goods and Services Rarely Include Their Harmful Environmental and Health Costs

Another basic cause of environmental problems has to do with how the marketplace prices goods and services. Companies providing goods for consumers generally are not required to pay for most of the harmful environmental and health costs of supplying such goods. For example, timber companies pay the cost of clear-cutting forests but do not pay for the resulting environmental degradation and loss of wildlife habitat.

The primary goal of a company is to maximize profits for its owners or stockholders, so it is not inclined to add these costs to its prices voluntarily. Because the prices of goods and services do not include most of their harmful environmental and health costs, consumers have no effective way to know the harm caused by what they buy. This lack of information is a major reason for why we are degrading key components of our life-support system (Figure 1.7).

For example, producing and using gasoline results in air pollution and other problems that damage the environment and people's health. Scientists and economists have estimated that the price of gasoline to U.S. consumers would rise by $3.18 per liter ($12 per gallon) if the estimated short- and long-term harmful environmental and health costs were included in its pump price. Thus, when gas costs $2 per gallon, U.S. consumers are really paying about $14 per gallon, as discussed in more detail in Chapter 23. Consumers pay these hidden costs, but not at the gas pump.

CONSIDER THIS . . .

THINKING ABOUT Real Gasoline Prices

Suppose the price of gasoline included its harmful environmental and health effects and was therefore $14 a gallon. How would this affect your decision on what type of car to buy or whether to go without a car and instead make greater use of walking, bicycling, and mass transit?

Another problem arises when governments (taxpayers) give companies *subsidies* such as tax breaks and payments to assist them with using resources to run their businesses. This helps to create jobs and stimulate economies, but some subsidies encourage the depletion and degradation of natural capital.

Steve Winter/National Geographic Creative

FIGURE 1.14 These ecotourists atop Asian elephants in India's Kaziranga National Park are learning about threatened barasingha deer and other species.

According to environmental economists, we could live more sustainably and increase our beneficial environmental impact by including the harmful environmental and health costs of the goods and services into market prices and placing a monetary value on the natural capital that supports all economies. This would not be an easy economic or political task, but it can be done—and the alternative is to continue degrading natural capital. Such full-cost pricing is the basis for one of the six **principles of sustainability**.

Economists propose two ways to implement full-cost pricing over the next two decades. One is to shift from environmentally harmful government subsidies to environmentally beneficial subsidies that sustain or enhance natural capital. The other is to increase taxes on pollution and wastes that we want less of and reduce taxes on income and wealth that we want more of. We discuss such *subsidy shifts* and *tax shifts* in Chapter 23.

People Are Increasingly Isolated from Nature

Today, more than half of the world's people and three out of four people in more-developed countries live in urban areas. This shift from rural to urban living is continuing at a rapid pace. Urban environments and the increasing use of cell phones, computers, and other electronic devices are isolating people, especially children, from the natural world. Some argue that this has led to a phenomenon known as *nature-deficit disorder*.

Children and adults can gain many benefits from outdoor activities. Research indicates that experiencing nature (see Figure 1.14 and the chapter-opening photo) can lead to better health, reduced stress, improved mental abilities, and increased imagination and creativity. It also can provide a sense of wonder and connection to the earth's life-support system that keeps us alive and supports our economies.

People Have Different Views about Environmental Problems and Their Solutions

One of the reasons why environmental problems persist is that people differ over the nature and seriousness of the world's environmental problems, as well as how to solve them. These disagreements arise mostly because of differing environmental worldviews. Your **environmental worldview** is your set of assumptions and values concerning how the natural world works and how you think you should interact with the environment.

Environmental ethics, the study of varying beliefs about what is right and wrong with how we treat the environment, provides useful tools for examining worldviews. For example, here are some important *ethical questions* relating to the environment:

- Why should we care about the environment?
- Are we the most important species on the planet or are we just another one of the earth's millions of life forms?
- Do we have an obligation to see that our activities do not cause the extinction of other species? If so, should we try to protect all species or only some? How do we decide which to protect?
- Do we have an ethical obligation to pass the natural world on to future generations in a condition that is as good as or better than what we inherited?
- Should every person be entitled to equal protection from environmental hazards regardless of race, gender, age, national origin, income, social class, or any other factor? (This is the central ethical and political issue for what is known as the *environmental justice* movement; see Chapter 24 for more on this topic.)
- Should we seek to live more sustainably, and if so, how?

CONSIDER THIS . . .

THINKING ABOUT Our Responsibilities

How would you answer each of the questions above? Compare your answers with those of your classmates. Record your answers and, at the end of this course, return to these questions to see if your answers have changed.

People with different environmental worldviews can take the same data, be logically consistent with it, and arrive at quite different answers to such questions. This happens because they start with different assumptions and moral, ethical, or religious beliefs. Environmental worldviews are discussed in detail in Chapter 25, but here is a brief introduction.

There are three major categories of environmental worldviews: human-centered, life-centered, and earth-centered. A **human-centered environmental worldview** sees the natural world as a support system for human life. Two variations in this worldview are the *planetary management worldview* and the *stewardship worldview*. According to both of these variations, humans are separate from and in charge of nature; we should manage the earth for our benefit; and if we degrade or deplete a natural resource or ecosystem service, we can use our technological ingenuity to find a substitute. The stewardship worldview also calls for us to be caring and responsible managers, or *stewards*, of the earth for current and future human generations. It also calls for us to encourage environmentally beneficial forms of economic growth and development and discourage environmentally harmful forms.

According to the **life-centered environmental worldview**, all species have value in fulfilling their particular role within the biosphere, regardless of their potential or actual use to humans. Most people with a life-centered worldview believe we have an ethical responsibility to avoid hastening the extinction of species through our activities.

According to the **earth-centered environmental worldview**, we are part of, and dependent on, nature, and the earth's natural capital exists for all species, not just for humans. According to this view, our economic success and the long-term survival of our cultures, our species, and many other species depend on learning how life on the earth has sustained itself for billions of years (Figure 1.2) and integrating such lessons from nature (**Core Case Study** and Science Focus 1.1) into the ways we think and act.

CASE STUDY

The Rise of Environmental Conservation and Protection in the United States

When European colonists arrived in what is now the United States in the early 1600s, they viewed North America as a land with inexhaustible resources and a wilderness to be conquered and managed for human use. As the colonists spread across the continent, they cleared forests to build settlements, plowed up grasslands to plant crops, and mined gold, lead, and other minerals.

In 1864, George Perkins Marsh, a scientist and member of Congress from Vermont, questioned the idea that America's resources were inexhaustible. He used scientific studies and case studies to show how the rise and fall of past civilizations were linked to the use and misuse of their soils, water supplies, and other resources. He was one of the founders of the U.S. conservation movement.

Early in the 20th century, this movement split into two factions with differing views over how to use U.S. public lands owned jointly by all American citizens. The *preservationist view*, led by naturalist John Muir (Figure 1.15),

Some Biomimicry Principles

According to Janine Benyus (Individuals Matter 1.1): "The study of biomimicry reveals that life creates conditions conducive to life." She calls for us to evaluate each of the goods and services we produce and use by asking: Is it something nature would do? Does it help sustain life? Will it last?

Benyus recognizes three levels of biomimicry. The first involves mimicking the characteristics of species such as bumps on a whale's fins or the wing and feather designs of birds that are believed to have enhanced the long-term survival of such species. The second and deeper level involves mimicking the processes that species use to make shells, feathers, and other parts that benefit their long-term survival without using or producing toxins and without using the high-temperature or high-pressure processes we use in manufacturing. The third and deepest level involves mimicking the long-term survival strategies and beneficial environmental effects of natural ecosystems such as forests and coral reefs. Benyus is working with others to use this third level of biomimicry to design more sustainable cities.

Since 1997 scientists, engineers, and others working in the field of biomimicry have identified several principles that have sustained life on the earth for billions of years. They have found that life:

- Runs on sunlight, not fossil fuels
- Does not waste energy
- Uses only what it needs
- Adapts to changing environmental conditions
- Depends on biodiversity for population control and adaptation
- Creates no waste because the matter outputs of one organism are resources for other organisms
- Does not pollute its own environment
- Does not produce chemicals that cannot be recycled by the earth's chemical cycles.

By learning from nature and using such principles, innovative scientists, engineers, and business people are leading a *biomimicry revolution* by creating life-friendly goods and services and profitable businesses that could enrich and sustain life far into the future.

CRITICAL THINKING

Which, if any, of the proposed principles of biomimicry do you follow in your life? How might your lifestyle change if you followed all of these principles? Would you resist or embrace doing this?

wanted wilderness areas on some public lands to be left untouched so they could be preserved indefinitely, an idea that was not enacted into law until 1964. He was largely responsible for establishing Yosemite National Park in 1890. In 1892 he founded the Sierra Club, which to this day is a political force working on behalf of the environment.

The *conservationist view* was promoted by Teddy Roosevelt (Figure 1.16) and Gifford Pinchot. Roosevelt was president of the United States and Pinchot was the first chief of the U.S. Forest Service. They believed all public lands should be managed wisely and scientifically, primarily to provide resources for people. Roosevelt's term of office, 1901–1909, has been called the country's *Golden Age of Conservation*. He established 36 national wildlife reserves and more than tripled the size of the national forest reserves.

Aldo Leopold (Figure 1.17)—wildlife manager, professor, writer, and conservationist—was trained in the conservation view but shifted toward the preservation view. In 1935 he helped found the U.S. Wilderness Society. Through his writings, especially his 1949 book *A Sand*

FIGURE 1.15 As leader of the preservationist movement, John Muir (1838–1914) called for setting aside some of the country's public lands as protected wilderness, an idea that was not enacted into law until 1964.

© Cengage Learning

FIGURE 1.16 Effective protection of forests and wildlife on federal lands did not begin until Theodore (Teddy) Roosevelt (1858–1919) became president.

Robert McCabe/Courtesy of the University of Wisconsin-Madison Archives

FIGURE 1.17 Aldo Leopold (1887–1948) became a leading conservationist and his book, *A Sand County Almanac*, is considered an environmental classic that helped to inspire the modern conservation and environmental movements.

County Almanac, he laid the groundwork for the field of environmental ethics. He argued that the role of the human species should be to protect nature, not conquer it.

Later in the 20th century, the concept of resource conservation was broadened to include preservation of the *quality* of the planet's air, water, soil, and wildlife.

U.S. Fish and Wildlife Service

FIGURE 1.18 Rachel Carson (1907–1964) alerted us to the harmful effects of the widespread use of pesticides. Many environmental historians mark Carson's wake-up call as the beginning of the modern environmental movement in the United States.

A prominent pioneer in that effort was biologist Rachel Carson (Figure 1.18). In 1962 she published *Silent Spring*, which documented the pollution of air, water, and wildlife from the widespread use of pesticides such as DDT. This influential book heightened public awareness of pollution problems and led to the regulation of several dangerous pesticides.

Between 1940 and 1970, the United States underwent rapid economic growth and industrialization. This increased air and water pollution and produced large quantities of solid and hazardous wastes. Air pollution was so bad in many cities that drivers had to use their car headlights during the daytime. Thousands died each year from the harmful effects of air pollution. A stretch of the Cuyahoga River running through Cleveland, Ohio, was so polluted with oil and other flammable pollutants that it caught fire several times. There was a devastating oil spill off the California coast in 1969. Well-known wildlife species such as the American bald eagle, the grizzly bear, the whooping crane, and the peregrine falcon became endangered.

Growing publicity over these problems led the American public to demand government action. When the first Earth Day was held on April 20, 1970, some 20 million people in more than 2,000 U.S. communities and college and university campuses attended rallies to demand improvements in environmental quality. The first Earth Day and the resulting bottom-up political pressure it created led the U.S. government to establish the Environmental

Protection Agency (EPA) in 1970 and to enact most of the U.S. environmental laws now in place during the 1970s, known as the *decade of the environment*. During this period, the United States led the world in environmental awareness, wildlife conservation, and environmental protection.

Since 1970, many grassroots environmental organizations have sprung up to help deal with environmental threats. Interest in environmental issues has grown on many college and university campuses, resulting in the expansion of environmental science and environmental studies courses and programs. In addition, awareness of critical, complex, and largely invisible environmental issues has increased. They include losses in biodiversity, aquifer depletion, ocean warming and acidification, and atmospheric warming and climate change.

In the 1980s, there was a political backlash against environmental laws and regulations led by some corporate leaders and members of Congress. They contended that environmental laws were hindering economic growth and tried to weaken or eliminate many environmental laws passed during the 1970s and do away with the EPA. The 1990s saw increasingly sophisticated "disinformation" campaigns, funded by powerful business interests, that were meant to confuse or mislead the public on important environmental issues—efforts that continue today.

Since the 1980s, environmental leaders and their supporters have had to spend much of their time and resources fighting efforts to keep key environmental laws from being weakened or repealed. Some analysts call for the United States to regain and strengthen its global role in improving environmental quality and in making the shift to more environmentally sustainable societies and economies.

1.4 WHAT IS AN ENVIRONMENTALLY SUSTAINABLE SOCIETY?

CONCEPT 1.4 Living sustainably means living on the earth's natural income without depleting or degrading the natural capital that supplies it.

Protecting Natural Capital and Living on Its Income

An **environmentally sustainable society** protects natural capital and lives on its income. Such a society would meet the current and future basic resource needs of its people. This would be done in a just and equitable manner without compromising the ability of future generations to meet their basic resource needs. This is in keeping with the ethical **principle of sustainability**.

Imagine that you win $1 million in a lottery. Suppose you invest this money (your capital) and earn 10% interest per year. If you live on just the interest income made by your capital, you will have a sustainable annual income of $100,000. You can spend $100,000 each year indefinitely and not deplete your capital. However, if you consistently spend more than your income, you will deplete your capital. Even if you spend just $10,000 more per year while still allowing the interest to accumulate, your money will be gone within 18 years.

The lesson here is an old one: *Protect your capital and live on the income it provides*. Deplete or waste your capital and you will move from a sustainable to an unsustainable lifestyle.

The same lesson applies to our use of the earth's natural capital (Figure 1.3). This natural capital is a global trust fund of natural resources and ecosystem services available to people now and in the future and to the earth's other species. *Living sustainably* means living on **natural income**, which is the renewable resources such as plants, animals, soil, clean air, and clean water, provided by the earth's natural capital. By preserving and replenishing the earth's natural capital that supplies this income, we can reduce our ecological footprints and expanding our beneficial environmental impact (**Concept 1.4**). For example, the earth's elephants are in trouble and some people are working to help protect them (Individuals Matter 1.2).

We Can Live More Sustainably

Living more sustainability means learning to live within limits imposed on all life by the earth and the unbreakable scientific laws that govern our use of matter and energy (discussed in the next chapter). Doing this requires:

- Learning from nature (**Core Case Study** and Science Focus 1.1)
- Protecting natural capital
- Not wasting resources (there is no waste in nature)
- Recycling and reusing nonrenewable resources
- Using renewable resources no faster than nature can replenish them
- Incorporating the harmful health and environmental impacts of producing and using goods and services in their market prices
- Preventing future ecological damage and repairing past damage
- Cooperating with one another to find win-win solutions to the environmental problems we face
- Accepting the ethical responsibility to pass the earth that sustains us on to future generations in a condition as good as or better than what we inherited.

INDIVIDUALS MATTER 1.2

Tuy Sereivathana: Elephant Protector

Courtesy Tom Dusenbery

Since 1970, Cambodia's rain forest cover has dropped from over 70% of the country's land area to 3%, primarily because of population growth, rapid development, illegal logging, and warfare. This severe forest loss forced elephants to search for food and water on farmlands. This set up a conflict between elephants and poor farmers, who killed the elephants to protect their food supply.

Since 1995, Tuy Sereivathana, with a master's degree in forestry, has been on a mission to accomplish two goals. One is to more than double the population of Cambodia's endangered Asian elephants by 2030. The other is to show poor farmers that protecting elephants and other forms of wildlife can help them escape poverty.

Sereivathana has helped farmers set up nighttime lookouts for elephants. He has taught them to scare elephants away by using foghorns and fireworks and to use solar-powered electric fences to mildly shock them. He has also encouraged farmers to stop growing watermelons and bananas, which elephants love, and to grow eggplant and chili peppers that elephants shun.

Since 2005, mostly because of Sereivathana's efforts, no elephants have been killed in Cambodia over conflicts with humans. In 2010 Sereivathana was one of the six recipients of the Goldman Environmental Prize (often dubbed the "Nobel Prize for the environment"). In 2011 he was named a National Geographic Explorer.

Environmental problems are so complex and widespread that it may seem hopeless, but that is not true. There is plenty of reason to hope and to act. For instance, consider these two pieces of good news from the social sciences. *First,* research suggests that it takes only 5–10% of the population of a community, a country, or the world to bring about major social and environmental change. *Second,* this research also shows that such change can occur much faster than most people believe.

Anthropologist Margaret Mead summarized the potential for social change: "Never doubt that a small group of thoughtful, committed citizens can change the world. Indeed, it is the only thing that ever has." Engaged citizens in communities and schools around the world are proving Mead right.

One of our goals in writing this book has been to provide a realistic vision of how we can live more sustainably. We base this vision not on immobilizing fear, gloom, and doom, but on education about how the earth sustains life and on energizing and realistic hope.

BIG IDEAS

- We can ensure a more sustainable future by relying more on energy from the sun and other renewable energy sources, protecting biodiversity through the preservation of natural capital, and avoiding the disruption of the earth's vital chemical cycles.
- A major goal for achieving a more sustainable future is full-cost pricing—the inclusion of harmful environmental and health costs in the market prices of goods and services.
- We will benefit ourselves and future generations if we commit ourselves to finding win-win solutions to environmental problems and to leaving the planet's life-support system in a condition as good as or better than what we now enjoy.

Tying It All Together

Learning from the Earth and Sustainability

We opened this chapter with a **Core Case Study** about learning from nature by understanding how the earth—our only truly sustainable system—has sustained an incredible diversity of life for 3.8 billion years despite drastic and long-lasting changes in the planet's environmental conditions. Part of the answer involves learning how to apply the six **principles of sustainability** (Figures 1.2 and 1.6 and inside back cover of this book) to the design and management of our economic and social systems, and to our individual lifestyles.

We can use such strategies to slow the rapidly expanding losses of biodiversity, to sharply reduce our production of wastes and pollution, to switch to more sustainable sources of energy, and to promote more sustainable forms of agriculture and other uses of land and water. We can also use these principles to sharply reduce poverty and slow human population growth.

Pecold/Shutterstock.com

You and your fellow students have the good fortune to be members of the 21st century's *transition generation* that will play a major role in deciding whether humanity creates a more sustainable future or continues on an unsustainable path toward further environmental degradation and disruption. It is an incredibly exciting and challenging time to be alive as we struggle to develop a more sustainable relationship with the earth that keeps us alive and supports our economies.

Chapter Review

Core Case Study

1. What is **sustainability**? What is **biomimicry**? Explain why learning from the earth is a key to learning how to live more sustainably.

Section 1.1

2. What are the three key concepts for this section? Define **environment**. Distinguish among **environmental science**, **ecology**, and **environmentalism**. What is an **ecosystem**? What are three **scientific principles of sustainability** derived from how the natural world works? Define **solar energy**, **biodiversity**, and **chemical cycling** (or **nutrient cycling**) and explain why they are important to life on the earth.

3. Define **natural capital**. Define **natural resources** and **ecosystem services**, and give two examples of each. Give three examples of how we are degrading natural capital. Explain how finding solutions to environmental problems involves making trade-offs. Explain why individuals matter in dealing with the environmental problems we face. What are three economic, political, and ethical **principles of sustainability**? What is **full-cost pricing** and why is it important? Describe the role of Janine Benyus in promoting the important and growing field of biomimicry.

4. What is a **resource**? Distinguish between an **inexhaustible resource** and a **renewable resource** and give an example of each. What is the **sustainable yield** of a renewable resource? Define and give an example of a **nonrenewable** or **exhaustible resource**. Distinguish between **more-developed countries** and **less-developed countries** and give one example each of a high-income, middle-income, and low-income country.

Section 1.2

5. What is the key concept for this section? How have humans improved the quality of life for many people? How are humans living unsustainably? Define and give three examples of **environmental degradation** (or **natural capital degradation**). About what percentage of the earth's natural or ecosystem services have been degraded by human activities? What is the tragedy of the commons? What are two ways to deal with this effect?

6. What is an **ecological footprint**? What is a **per capita ecological footprint**? Define **biocapacity**. Use the ecological footprint concept to explain how we are living unsustainably. What is the IPAT model for estimating our environmental impact? Explain how three major cultural changes taking place over

the last 10,000 years have increased our overall environmental impact. What would a **sustainability revolution** involve?

Section 1.3

7. What are the two key concepts for this section? Identify six basic causes of the environmental problems that we face. What is **exponential growth**? What is the rule of 70? What is the current size of the human population? About how many people are added each year? How big is the world's population projected to be in 2050? How do Americans, Indians, and the average people in the poorest countries compare in terms of average resource consumption per person? Summarize the potentially harmful and beneficial environmental effects of affluence.

8. What is **poverty** and what are three of its harmful environmental and health effects? About what percentage of the world's people struggle to live on the equivalent of $1.25 a day? About what percentage have to live on $2.25 a day? How are poverty and population growth connected? List three major health problems faced by many of the poor.

9. Explain how excluding the harmful environmental and health costs of production from the prices of goods and services affects the environmental problems we face. What is the connection between government subsidies, resource use, and environmental degradation? What are two ways to include the harmful environmental and health costs of the goods and services in their market prices? Explain how a lack of knowledge about nature and the importance of natural capital, along with our increasing isolation from nature, can intensify the environmental problems we face. What is an **environmental worldview**? What is **environmental ethics**? What are five important ethical questions relating to the environment? Distinguish among the **human-centered**, **life-centered**, and **earth-centered environmental worldviews**. What are three levels of biomimicry? List eight key biomimicry principles.

Section 1.4

10. What is the key concept for this section? What is an **environmentally sustainable society**? What is **natural income** and how is it related to sustainability? Describe Tuy Sereivathana's efforts to prevent elephants from becoming extinct in Cambodia and to reduce the country's poverty. List nine principles for living more sustainably. What are two pieces of good news about making the transition to a more sustainable society? What are this chapter's three big ideas?

Note: Key terms are in bold type. Knowing the meanings of these terms will help you in the course you are taking.

Critical Thinking

1. Why is biomimicry so important? Find an example of something in nature that you think could be mimicked for some beneficial purpose. Explain that purpose and how biomimicry could apply.

2. What do you think are the three most environmentally unsustainable components of your lifestyle? List two ways in which you could apply each of the six **principles of sustainability** (Figures 1.2 and 1.6) to making your lifestyle more environmentally sustainable.

3. For each of the following actions, state one or more of the three **scientific principles of sustainability** that are involved: **(a)** recycling aluminum cans; **(b)** using a rake instead of a leaf blower; **(c)** walking or bicycling to class instead of driving; **(d)** taking your own reusable bags to a store to carry your purchases home; and **(e)** volunteering to help restore a prairie or other degraded ecosystem.

4. Explain why you agree or disagree with the following propositions:
   a. Stabilizing population is not desirable because, without more consumers, economic growth would stop.
   b. The world will never run out of resources because we can use technology to find substitutes and to help us reduce resource waste.
   c. We can shrink our ecological footprints while creating beneficial environmental impacts.

5. Should nations with large ecological footprints reduce their footprints to decrease their harmful environmental impact and leave more resources for nations with smaller footprints and for future generations? Explain.

6. When you read that at least 19,000 children age 5 and younger die each day (13 per minute) from preventable malnutrition and infectious disease, what is your response? How would you address this problem?

7. Explain why you agree or disagree with each of the following statements: **(a)** humans are superior to other forms of life; **(b)** humans are in charge of the earth; **(c)** the value of other forms of life depends only on whether they are useful to humans; **(d)** all forms of life have a right to exist; **(e)** all economic growth is good; **(f)** nature has an almost unlimited storehouse of resources for human use; **(g)** technol-

ogy can solve our environmental problems; **(h)** I don't have any obligation to future generations; and **(i)** I don't have any obligation to other forms of life.

8. What are the basic beliefs of your environmental worldview? Record your answer. At the end of this course, return to your answer to see if your environmental worldview has changed. Are the beliefs included in your environmental worldview consistent with the answers you gave to Question 7 above? Are your actions that affect the environment consistent with your environmental worldview? Explain.

Doing Environmental Science

Estimate your own ecological footprint by using one of the many estimator tools available on the Internet. Is your ecological footprint larger or smaller than you thought it would be, according to this estimate? Why do you think this is so? List three ways in which you could reduce your ecological footprint. Try one of them for a week, and write a report on this change. List three ways you could increase your beneficial environmental impact.

Global Environment Watch Exercise

Go to your MindTap course to access the GREENR database. Use the world maps in Figure 1, p. S14, in Supplement 4 and Figure 1.10 to choose one more-developed country and one less-developed country to compare their ecological footprints. Use the "World Map" link at the top of the page to access information about the countries you have chosen to research. Once on the country page, view the "Quick Facts" panel at the right. Click on the ecological footprint number to view a graph of both the ecological footprint and biocapacity of each country. Using those graphs, determine whether these countries are living sustainably or not. What would be some reasons for these trends?

Ecological Footprint Analysis

If the *ecological footprint per person* of a country or the world is larger than its *biocapacity per person* to replenish its renewable resources and absorb the resulting waste products and pollution, the country or the world is said to have an *ecological deficit*. If the reverse is true, the country or the world has an *ecological credit* or *reserve*. Use the data in the accompanying table to calculate the ecological deficit or credit for the countries listed. (As an example, this value has been calculated and filled in for World.)

1. Which three countries have the largest ecological deficits? For each of these countries, why do you think it has a deficit?

2. Rank the countries with ecological credits in order from highest to lowest credit. For each country, why do you think it has an ecological credit?

3. Rank all of the countries in order from the largest to the smallest per capita ecological footprint.

Place	Per Capita Ecological Footprint (hectares per person)	Per Capita Biocapacity (hectares per person)	Ecological Credit (+) or Deficit (−) (hectares per person)
World	2.6	1.8	−0.8
United States	6.8	3.8	
Canada	7.0	13.0	
Mexico	2.4	1.3	
Brazil	2.5	9.0	
South Africa	2.5	1.2	
United Arab Emirates	8.0	0.7	
Israel	4.6	0.3	
Germany	4.3	1.9	
Russian Federation	4.4	6.6	
India	0.9	0.4	
China	0.5	0.8	
Australia	7.5	15.0	
Bangladesh	0.65	0.35	
Denmark	4.0	4.0	
Japan	3.7	0.7	
United Kingdom	4.0	1.1	

Compiled by the authors using data from World Wide Fund for Nature: *Living Planet Report 2014*.

CHAPTER 2

Science, Matter, Energy, and Systems

> Science is built up of facts, as a house is built of stones; but an accumulation of facts is no more a science than a heap of stones is a house.
>
> HENRI POINCARÉ

Key Questions

2.1 What do scientists do?

2.2 What is matter and what happens when it undergoes change?

2.3 What is energy and what happens when it undergoes change?

2.4 What are systems and how do they respond to change?

Researchers measuring a 3,200-year-old giant sequoia in California's Sequoia National Park.

Michael Nichols/National Geographic Creative

Core Case Study

How Do Scientists Learn about Nature? Experimenting with a Forest

Suppose, a logging company plans to cut down all of the trees on the land behind your house. You are concerned and want to know about the possible harmful environmental effects of this action.

One way to learn about such effects is to conduct a *controlled experiment*, just as environmental scientists do. They begin by identifying key *variables*, such as water loss and soil nutrient content that might change after the trees are cut down. Then they set up two groups. One is the *experimental group*, in which a chosen variable is changed in a known way. The other is the *control group*, in which the chosen variable is not changed. They then compared the results from the two groups.

Botanist F. Herbert Bormann, forest ecologist Gene Likens, and colleagues carried out such a controlled experiment. Their goal was to compare the loss of water and soil nutrients from an area of uncut forest (the *control site*) with one that had been stripped of its trees (the *experimental site*).

The scientists built V-shaped concrete dams across the creeks at the bottoms of each forest in the Hubbard Brook Experimental Forest in New Hampshire (Figure 2.1). The dams were designed so that all surface water leaving each forested valley had to flow across a dam, where they could measure its volume and dissolved nutrient content.

First, the researchers measured the amounts of water and dissolved soil nutrients flowing from an undisturbed forested area in one of the valleys (the control site) (Figure 2.1, left). These measurements showed that the undisturbed mature forest was efficient at storing water and retaining chemical nutrients in its soils.

Next, they set up an experimental forest area in a nearby valley (Figure 2.1, right). They cut down all the trees and shrubs in that valley, left them where they fell, and sprayed the area with herbicides to prevent regrowth of vegetation. Then, for 3 years, they compared the outflow of water and nutrients in this experimental site with data from the control site.

The scientists found that, without plants to help absorb and retain water, the amount of water flowing out of the deforested valley increased 30–40%. This excess water ran over the ground rapidly, eroded soil, and removed dissolved nutrients from the topsoil. Overall, the loss of key soil nutrients from the experimental forest was six to eight times that in the nearby uncut control forest.

In this chapter, you will learn more about how scientists study nature and about the matter and energy that make up the world within and around us.

FIGURE 2.1 This controlled field experiment measured the loss of water and soil nutrients from a forest due to deforestation. The forested valley (left) was the control site; the cutover valley (right) was the experimental site.

2.1 WHAT DO SCIENTISTS DO?

CONCEPT 2.1 Scientists collect data and develop hypotheses, theories, and laws about how nature works.

Scientists Collect Evidence to Learn How Nature Works

Science is a field of study focused on discovering how nature works and using that knowledge to describe what is likely to happen in nature. Science is based on the assumption that events in the natural world follow orderly cause-and-effect patterns. These patterns can be understood through *observations* (by use of our senses and with instruments that expand our senses), *measurements*, and *experimentation*. Figure 2.2 summarizes the **scientific method**, a research process in which scientists identify a problem for study, gather relevant data, propose a hypothesis that explains the data, gather data to test the hypothesis, and modify the hypothesis as needed. Within this process, scientists use many different methods to learn more about how nature works.

There is nothing mysterious about the scientific process. You use it all the time in making decisions. As the famous physicist Albert Einstein put it, "The whole of science is nothing more than a refinement of everyday thinking."

In this chapter's **Core Case Study**, Bormann and Likens used the scientific method to see how clearing forested land can affect its ability to store water and retain soil nutrients. They designed an experiment to collect **data**, or information, to answer their question. From the data collected, they proposed a **scientific hypothesis**—a testable explanation of their data. Bormann and Likens came up with the following hypothesis to explain their data: land cleared of vegetation and exposed to rain and melting snow retains less water and loses soil nutrients. They tested this hypothesis twice. The first set of data measured the amount of soil nutrient nitrogen in the runoff. Later they repeated their controlled experiment to determine the amount of the soil nutrient phosphorus in the runoff. Both experiments confirmed their hypothesis.

The experimenters wrote scientific articles describing their research and submitted it to a scientific journal for publication. Before publishing the articles, the journal editor had them evaluated by other scientists in their fields (their peers in the scientific community). A larger body of experts in the field then evaluated the published articles. These reviews and further research by other scientists supported their results and hypothesis.

Another way to study nature is to develop a **model**, or an approximate physical or mathematical simulation of a system. Scientists use models to study complex systems such as the earth's climate and the forest studied by Bormann and Likens. Data from the research carried out by Bormann and Likens and from other scientists' research were fed into such models, which also supported the results of their research.

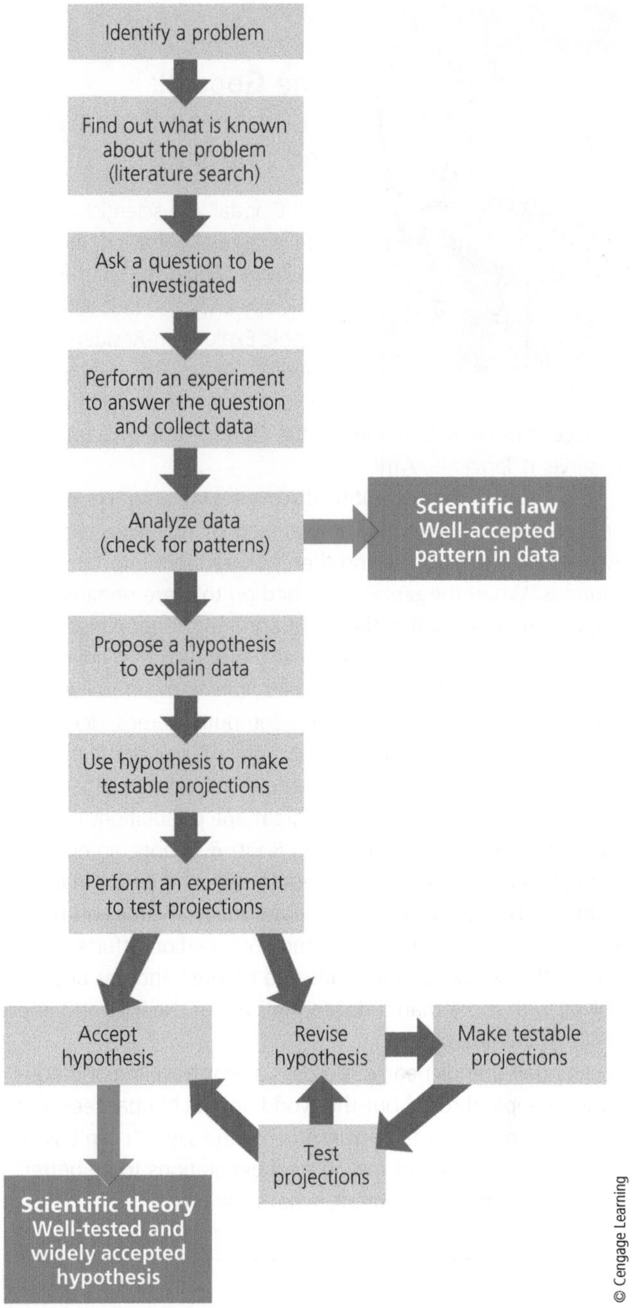

FIGURE 2.2 Scientific method: The general process that scientists use for discovering and testing ideas about how the natural world works.

A well-tested and widely accepted scientific hypothesis or a group of related hypotheses is called a **scientific theory**, which is one of the most important and certain results of science. The research conducted by Bormann and Likens and other scientists led to the scientific theory that trees and other plants hold soil in place and retain water and nutrients needed to support the plants.

INDIVIDUALS MATTER 2.1

JENS SCHLUETER/AFP/Getty Images

Jane Goodall: Chimpanzee Researcher and Protector

Jane Goodall is a scientist who studies animal behavior. She has a PhD from England's Cambridge University and is a National Geographic Explorer-in-Residence Emeritus. At age 26, she began a 50-year career of studying chimpanzee social and family life in the Gombe Stream Game Reserve in Tanzania, Africa.

One of her major scientific discoveries was that chimpanzees make and use tools. She watched chimpanzees modifying twigs or blades of grass and then poking them into termite mounds. When the termites latched on to these primitive tools, the chimpanzees pulled them out and ate the termites. Goodall and several other scientists have also observed that chimpanzees, including captive chimpanzees, can learn simple sign language, do simple arithmetic, play computer games, develop relationships, and worry about and protect one another.

In 1977 she established the Jane Goodall Institute, an organization that works to preserve great ape populations and their habitats. In 1991 Goodall started *Roots & Shoots*, an environmental education program for youth with chapters in more than 130 countries. She has received many awards and prizes for her scientific contributions and conservation efforts. She has written 27 books for adults and children and has been involved with more than a dozen films about the lives and importance of chimpanzees.

Goodall spends nearly 300 days a year traveling and educating people throughout the world about chimpanzees and the need to protect the environment. She says, "I can't slow down. . . . If we're not raising new generations to be better stewards of the environment, what's the point?"

Scientists Are Curious and Skeptical and Demand Evidence

Good scientists are curious about how nature works (Individuals Matter 2.1). Scientists tend to be highly skeptical about new data and hypotheses. They say, "Show me your evidence. Explain the reasoning behind the scientific ideas or hypotheses that you propose to explain your data."

An important part of the scientific process is **peer review**. This involves scientists publishing details of the methods they used, the results of their experiments, and the reasoning behind their hypotheses for other scientists working in the same field (their *peers*) to evaluate. Scientific knowledge advances in this self-correcting way, with scientists questioning and confirming the data and hypotheses of their peers. Sometimes new data and analysis can lead to revised hypotheses (Science Focus 2.1).

Critical Thinking and Creativity Are Important in Science

Scientists use logical reasoning and critical thinking skills (see p. xxviii) to learn about nature. Thinking critically involves three steps:

1. Be skeptical about everything you read or hear.
2. Evaluate evidence and hypotheses using inputs and opinions from a variety of reliable sources.
3. Identify and evaluate your personal assumptions, biases, and beliefs and distinguish between facts and opinions before coming to a conclusion.

Logic and critical thinking are important tools in science, but imagination, creativity, and intuition are also vital. According to Albert Einstein, "There is no completely logical way to a new scientific idea."

Scientific Theories and Laws: The Most Important and Certain Results of Science

We should never take a scientific theory lightly. It has been tested widely, is supported by extensive evidence, and is accepted as being a useful explanation of some phenomenon by most scientists in a particular field or related fields of study. So when you hear someone say, "Oh, that's just a theory," you will know that he or she does not have a clear understanding of what a scientific theory is and how it is an important result of science.

Another important and reliable outcome of science is a **scientific law**, or **law of nature**—a well-tested and widely accepted description of what we find always happening in the same way in nature. An example is the *law of gravity*. After making many thousands of observations and measurements of objects falling from different heights, scientists developed the following scientific law: all objects fall to the earth's surface at predictable speeds. Scientific laws cannot be broken.

Science Can Be Reliable, Unreliable, and Tentative

Reliable science consists of data, hypotheses, models, theories, and laws that are accepted by most of the scientists who are considered experts in the field under study. Scientific results and hypotheses that are presented as reliable without having undergone peer review, or are discarded as a result of peer review or additional research, are considered to be **unreliable science**.

Revisions in a Popular Scientific Hypothesis

For years, the story of Easter Island has been used in textbooks as an example of how humans can seriously degrade their own life-support system and as a warning about what we are doing to our life-support system.

What happened on this small island in the South Pacific is a story about environmental degradation and the collapse of an ancient civilization of Polynesians living there. Over the years, many researchers have studied the island and its remains, including hundreds of huge statues (Figure 2.A).

Some scientists drilled cores of sediment from lakebeds and studied grains of pollen from palm trees and other plants in sediment layers to reconstruct the history of plant life on the island. Based on these data, they hypothesized that as their population grew, the Polynesians began living unsustainably by using the island's palm forest trees faster than they could be renewed.

By studying charcoal remains in the island's layers of soil, scientists hypothesized that when the forests were depleted, there was no firewood for cooking or keeping warm and no wood for building large canoes used to catch fish, shellfish, and other forms of seafood. They also hypothesized that, with the forest cover gone, soils eroded, crop yields plummeted, famine struck, the population dwindled, violence broke out, and the civilization collapsed.

In 2001 anthropologist Terry L. Hunt and archeologist Carl Lippo carried out new research to test the older hypotheses about what happened on Easter Island. They used radiocarbon data and other analyses to propose some new hypotheses. *First*, their research indicated that the Polynesians arrived on the island about 800 years ago, not 2,900 years ago, as had been thought. *Second*, their population size probably never exceeded 3,000, contrary to the earlier estimate of up to 15,000.

Third, the Polynesians did use the island's trees and other vegetation in an unsustainable manner, and visitors reported that by 1722, most of the island's trees were gone. However, one question not answered by the earlier hypothesis was, why did the trees never grow back? Based on new evidence Hunt and Lippo hypothesized that rats, which either came along with the original settlers as stowaways or were brought along as a source of protein for the long voyage, played a key role in the island's permanent deforestation. Over the years, the rats multiplied rapidly into the millions and devoured the seeds that would have regenerated large areas of the forests. According to this new hypothesis, the rats played a key role in the fall of the civilization on Easter Island.

Fourth, the collapse of the island's civilization was not due to famine and warfare. Instead, it likely resulted from epidemics when European visitors unintentionally exposed the islanders to infectious diseases to which they had no immunity. This was followed by invaders who raided the island and took away some islanders as slaves. Later Europeans took over the land, used the remaining islanders for slave labor, and introduced sheep that devastated the island's remaining vegetation.

Hunt and Lippo's research and hypotheses indicate that the Easter Island tragedy may not be as clear an example of the islanders bringing about their own ecological collapse as was once thought. This story is an excellent example of how science works. The gathering of new scientific data and the reevaluation of older data led to revised hypotheses that challenged some of the earlier thinking about the decline of civilization on Easter Island. Scientists are gathering new evidence to test these two versions of what happened on Easter Island. This could lead to some new hypotheses.

shin/Shutterstock.com

FIGURE 2.A These and several hundred other statues were created by an ancient civilization of Polynesians on Easter Island. Some of them are as tall as a five-story building and weigh as much as 89 metric tons (98 tons).

CRITICAL THINKING

Does the new doubt about the original Easter Island hypothesis mean that we should not be concerned about using resources unsustainably on the island in space that we call Earth? Explain.

Preliminary scientific results without adequate testing and peer review are viewed as **tentative science**. Some of these results and hypotheses will be validated and classified as reliable. Others may be discredited and classified as unreliable. This is how scientific knowledge advances.

Science Has Limitations

Environmental science and science in general have several limitations. *First,* scientists cannot prove anything absolutely because there is always some degree of uncertainty in measurements, observations, models, and the resulting hypotheses and theories. Instead, scientists try to establish that a particular scientific theory has a very high *probability* or *certainty* (typically 90–95%) of being useful for understanding some aspect of the natural world.

Scientists do not use the word *proof* in the same way as many nonscientists use it, because it can falsely imply "absolute proof." For example, most scientists would not say: "Science has proven that cigarette smoking causes lung cancer." Instead, they might say: "Overwhelming evidence from thousands of studies indicates that people who smoke regularly for many years have a greatly increased chance of developing lung cancer."

CONSIDER THIS . . .

THINKING ABOUT Scientific Proof

Does the fact that science can never prove anything absolutely mean that its results are not valid or useful? Explain.

A *second* limitation of science is that scientists are human and not always free of bias about their own results and hypotheses. However, the high standards for evidence and peer review uncover or greatly reduce personal bias and falsified results.

A *third* limitation is that many systems in the natural world involve a huge number of variables with complex interactions. This makes it too difficult, costly, and time consuming to test one variable at a time in controlled experiments such as the one described in this chapter's **Core Case Study**. To deal with this, scientists develop *mathematical models* that can take into account the interactions of many variables and run the models on high-speed computers.

A *fourth limitation* of science involves the use of statistical tools. For example, there is no way to measure accurately the number of metric tons of soil eroded annually worldwide. Instead, scientists use statistical sampling and mathematical methods to estimate such numbers.

Despite these limitations, science is the most useful way that we have of learning about how nature works and projecting how it might behave in the future.

2.2 WHAT IS MATTER AND WHAT HAPPENS WHEN IT UNDERGOES CHANGE?

CONCEPT 2.2A Matter consists of elements and compounds, which in turn are made up of atoms, ions, or molecules.

CONCEPT 2.2B Whenever matter undergoes a physical or chemical change, no atoms are created or destroyed (the law of conservation of matter).

Matter Consists of Elements and Compounds

Matter is anything that has mass and takes up space. It can exist in three *physical states*—solid, liquid, and gas—and two *chemical forms*—elements and compounds.

An **element** such as gold or mercury (Figure 2.3) is a type of matter with a unique set of properties and that cannot be broken down into simpler substances by chemical means. Chemists refer to each element with a unique one- or two-letter symbol such as C for carbon and Au for gold. They have arranged the known elements based on their chemical behavior in a chart known as the **periodic table of elements** (see Supplement 3, Figure 1, p. S6). The periodic table contains 118 elements, not all of which occur naturally. See a list of the elements and their symbols that you need to know to understand the material in this book in Supplement 3, Table 1, p. S7.

Most matter consists of **compounds**, combinations of two or more different elements held together in fixed proportions. For example, water is a compound (H_2O) containing the elements hydrogen and oxygen, and sodium chloride (NaCl) contains the elements sodium and chlorine.

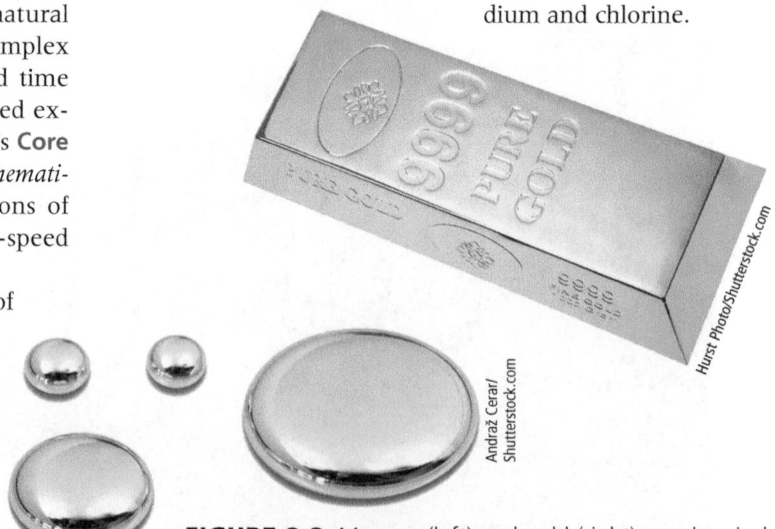

FIGURE 2.3 Mercury (left) and gold (right) are chemical elements. Each has a unique set of properties and cannot be broken down into simpler substances.

Elements and Compounds Are Made of Atoms, Molecules, and Ions

The basic building block of matter is an **atom**—the smallest unit of matter into which an element can be divided and still have its distinctive chemical properties. The idea that all elements are made up of atoms is called the **atomic theory** and is the most widely accepted scientific theory in chemistry.

Atoms are incredibly small. For example, more than 3 million hydrogen atoms could sit side by side on the period at the end of this sentence. If you could view atoms with a supermicroscope, you would find that each different type of atom contains a certain number of three types of *subatomic particles*: **neutrons**, with no electrical charge; **protons**, each with a positive electrical charge (+); and **electrons**, each with a negative electrical charge (−).

Each atom has an extremely small center called the **nucleus**, which contains one or more protons and, in most cases, one or more neutrons. Outside of the nucleus, we find one or more electrons in rapid motion (Figure 2.4).

Each element has a unique **atomic number** equal to the number of protons in the nucleus of its atom. Carbon (C), with 6 protons in its nucleus, has an atomic number of 6, whereas uranium (U), a much larger atom, has 92 protons in its nucleus and thus an atomic number of 92.

Because electrons have so little mass compared to protons and neutrons, most of an atom's mass is concentrated in its nucleus. The mass of an atom is described by its **mass number**, the total number of neutrons and protons in its nucleus. For example, a carbon atom with 6 protons and 6 neutrons in its nucleus (Figure 2.4) has a mass number of 12 ($6 + 6 = 12$) and a uranium atom with 92 protons and 143 neutrons in its nucleus has a mass number of 235 ($92 + 143 = 235$).

Each atom of a particular element has the same number of protons in its nucleus. However, the nuclei of atoms of a particular element can vary in the number of neutrons they contain, and, therefore, in their mass numbers. The forms of an element having the same atomic number but different mass numbers are called **isotopes** of that element. Scientists identify isotopes by attaching their mass numbers to the name or symbol of the element. For example, the three most common isotopes of carbon are carbon-12 (with six protons and six neutrons, Figure 2.4), carbon-13 (with six protons and seven neutrons), and carbon-14 (with six protons and eight neutrons). Carbon-12 makes up about 98.9% of all naturally occurring carbon.

A second building block of matter is a **molecule**, a combination of two or more atoms of the same or different elements held together by *chemical bonds*. Molecules are the basic building blocks of many compounds. Examples are water, hydrogen gas, and methane (the main component of natural gas).

A third building block of some types of matter is an **ion**. It is an atom or a group of atoms with one or more net positive (+) or negative (−) electrical charges from losing or gaining negatively charged electrons. Chemists use a superscript after the symbol of an ion to indicate the number of positive or negative electrical charges. The hydrogen ion (H^+) and sodium ion (Na^+) are examples of positive ions. Examples of negative ions are the hydroxide ion (OH^-) and chloride ion (Cl^-). Another example of a negative ion is the nitrate ion (NO_3^-), a nutrient essential for plant growth. In this chapter's **Core Case Study**, Bormann and Likens measured the loss of nitrate ions (Figure 2.5) from the deforested area (Figure 2.1, p. 30) in their controlled experiment. See Supplement 3, Table 2, p. S7 for a list of the chemical ions used in this book.

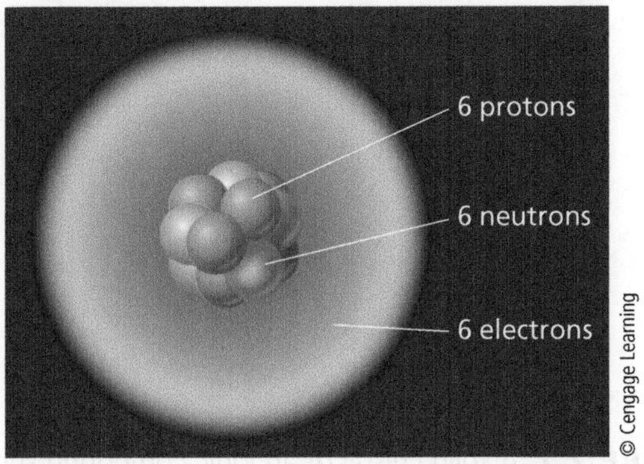

FIGURE 2.4 Simplified model of a carbon-12 atom. It consists of a nucleus containing six protons, each with a positive electrical charge, and six neutrons with no electrical charge. Six negatively charged electrons are found outside its nucleus.

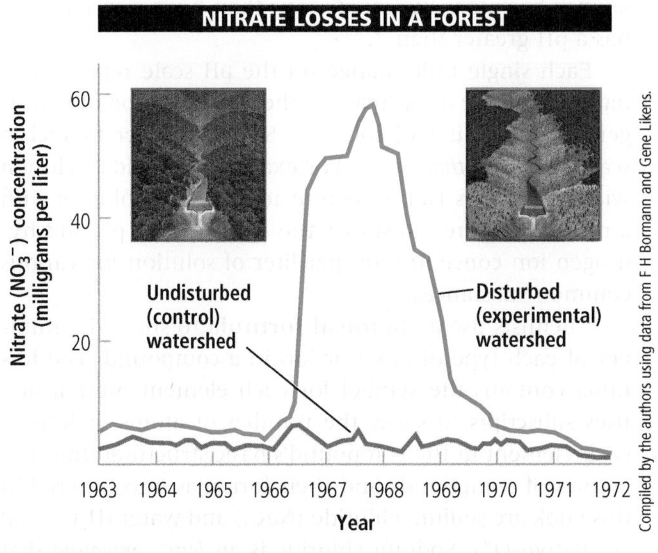

FIGURE 2.5 Loss of nitrate ions (NO_3^-) from a deforested watershed in the Hubbard Brook Experimental Forest (**Core Case Study**, Figure 2.1, p. 30). ***Data analysis:*** By what percent did the nitrate concentration increase between 1965 and the peak concentration between 1967 and 1968?

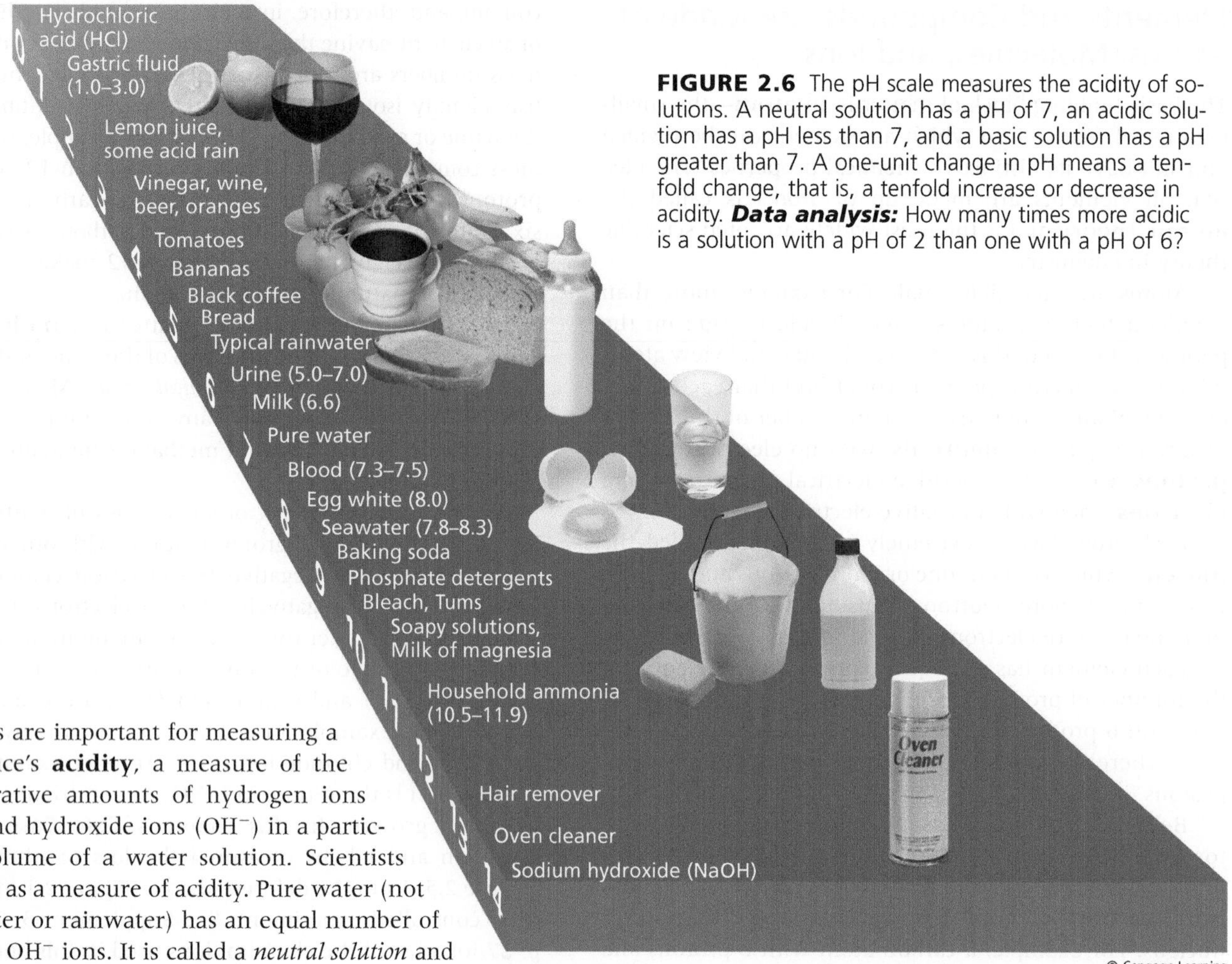

FIGURE 2.6 The pH scale measures the acidity of solutions. A neutral solution has a pH of 7, an acidic solution has a pH less than 7, and a basic solution has a pH greater than 7. A one-unit change in pH means a tenfold change, that is, a tenfold increase or decrease in acidity. ***Data analysis:*** How many times more acidic is a solution with a pH of 2 than one with a pH of 6?

Ions are important for measuring a substance's **acidity**, a measure of the comparative amounts of hydrogen ions (H^+) and hydroxide ions (OH^-) in a particular volume of a water solution. Scientists use **pH** as a measure of acidity. Pure water (not tap water or rainwater) has an equal number of H^+ and OH^- ions. It is called a *neutral solution* and has a pH of 7. An *acidic solution* has more hydrogen ions than hydroxide ions and has a pH less than 7. A *basic solution* has more hydroxide ions than hydrogen ions and has a pH greater than 7.

Each single unit change on the pH scale represents a tenfold increase or decrease in the concentration of hydrogen ions in a liter of solution. Scientists refer to such a scale as a *logarithmic scale*. For example, an acidic solution with a pH of 3 is 10 times more acidic than a solution with a pH of 4. Figure 2.6 shows the approximate pH and hydrogen ion concentration per liter of solution for various common substances.

Chemists use a **chemical formula** to show the number of each type of atom or ion in a compound. The formula contains the symbol for each element present and uses subscripts to show the number of atoms or ions of each element in the compound's basic structural unit. Examples of compounds and their formulas encountered in this book are sodium chloride (NaCl) and water (H_2O, read as "H-two-O"). Sodium chloride is an *ionic compound* that is held together in a three-dimensional array by the attraction between oppositely charged sodium ions (Na^+) and chloride ions (Cl^-) (Figure 2.7).

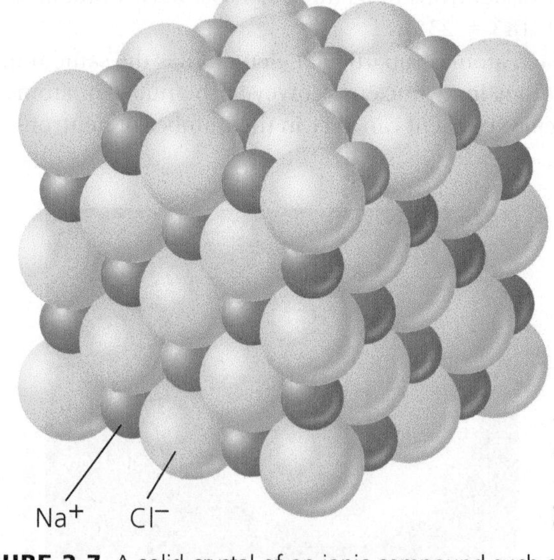

FIGURE 2.7 A solid crystal of an ionic compound such as sodium chloride (NaCl) consists of a three-dimensional array of oppositely charged ions held together by the strong forces of attraction between oppositely charged ions.

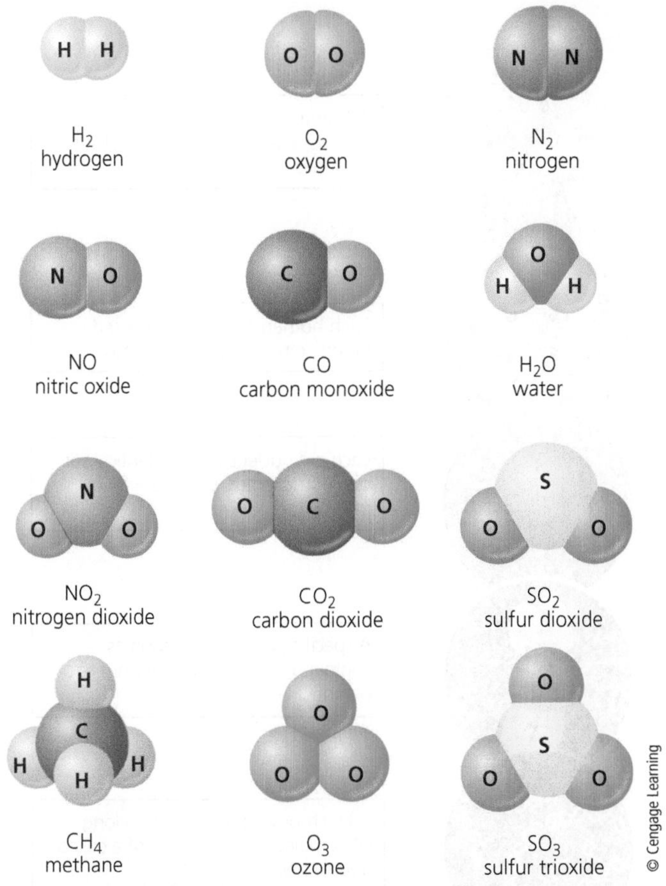

FIGURE 2.8 Chemical formulas and shapes for some covalent compound molecules.

Sodium chloride and many other ionic compounds tend to dissolve in water and break apart into their individual ions (Na^+ and Cl^-).

Other compounds called *covalent compounds* are made up of uncharged atoms. An example is water (H_2O). The bonds between the hydrogen and oxygen atoms in water molecules are called *covalent bonds* and form when the atoms in the molecule share one or more pairs of their electrons. Figure 2.8 shows the chemical formulas and shapes of molecules that are the building blocks for several common covalent compounds.

Organic Compounds Are the Chemicals of Life

Plastics, table sugar, vitamins, aspirin, penicillin, and most of the chemicals in your body are called **organic compounds**, which contain at least two carbon atoms combined with atoms of one or more other elements. The exception is methane (CH_4), with only one carbon atom.

The millions of known organic (carbon-based) compounds include *hydrocarbons*—compounds of carbon and hydrogen atoms—such as methane (CH_4), the main component of natural gas. *Simple carbohydrates (simple sugars)* are organic compounds that contain carbon, hydrogen, and oxygen atoms. An example is glucose ($C_6H_{12}O_6$), which most plants and animals break down in their cells to obtain energy.

Larger and more complex organic compounds, called polymers or *macromolecules*, form when a large number of simple organic molecules (*monomers*) are linked together by chemical bonds, somewhat like rail cars linked in a freight train. Four types of macromolecules—*complex carbohydrates, proteins, nucleic acids,* and *lipids*—are the molecular building blocks of life.

Complex carbohydrates consist of two or more monomers of *simple sugars* (such as glucose, $C_6H_{12}O_6$) linked together. One example is the starches that plants use to store energy and to provide energy for animals that feed on plants. Another is cellulose, the earth's most abundant organic compound, which is found in the cell walls of bark, leaves, stems, and roots.

Proteins are large polymer molecules formed by linking together long chains of monomers called *amino acids*. Living organisms use about 20 different amino acid molecules to build a variety of proteins. Some proteins store energy. Others are components of the immune system and chemical messengers, or *hormones*, that turn various bodily functions of animals on or off. In animals, proteins are also components of hair, skin, muscle, and tendons. In addition, some proteins act as *enzymes* that catalyze or speed up certain chemical reactions.

Nucleic acids are large polymer molecules made by linking large numbers of monomers called *nucleotides*. Each nucleotide consists of a *phosphate group*, a *sugar molecule*, and one of four different *nucleotide bases* (represented by A, G, C, and T, the first letter in each of their names). Two nucleic acids—DNA (**d**eoxyribo**n**ucleic **a**cid) and RNA (**r**ibo**n**ucleic **a**cid)—help build proteins and carry hereditary information used to pass traits from parent to offspring. Bonds called *hydrogen bonds* (see Supplement 3, Figure 3, p. S9) between parts of the nucleotides in DNA hold two DNA strands together like a spiral staircase, forming a double helix (Figure 2.9).

The different molecules of DNA in the millions of species found on the earth are like a vast and diverse genetic library. Each species is a unique book in that library. If the DNA coiled in your body were unwound, it would stretch about 960 million kilometers (600 million miles)—more than six times the distance between the sun and the earth.

Lipids, a fourth building block of life, are a chemically diverse group of large organic compounds that do not dissolve in water. Examples are *fats and oils* for storing energy, *waxes* for structure, and *steroids* for producing hormones.

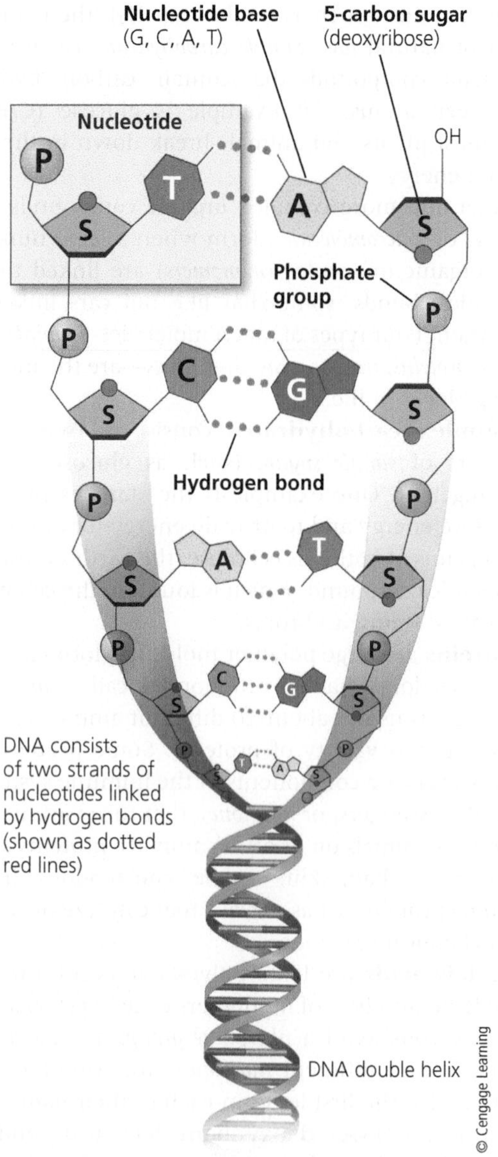

FIGURE 2.9 Portion of a DNA molecule, which is composed of spiral (helical) strands of nucleotides. Each nucleotide contains three units: phosphate (P), a sugar (S), which is deoxyribose), and one of four different nucleotide bases represented by the letters A, G, C, and T.

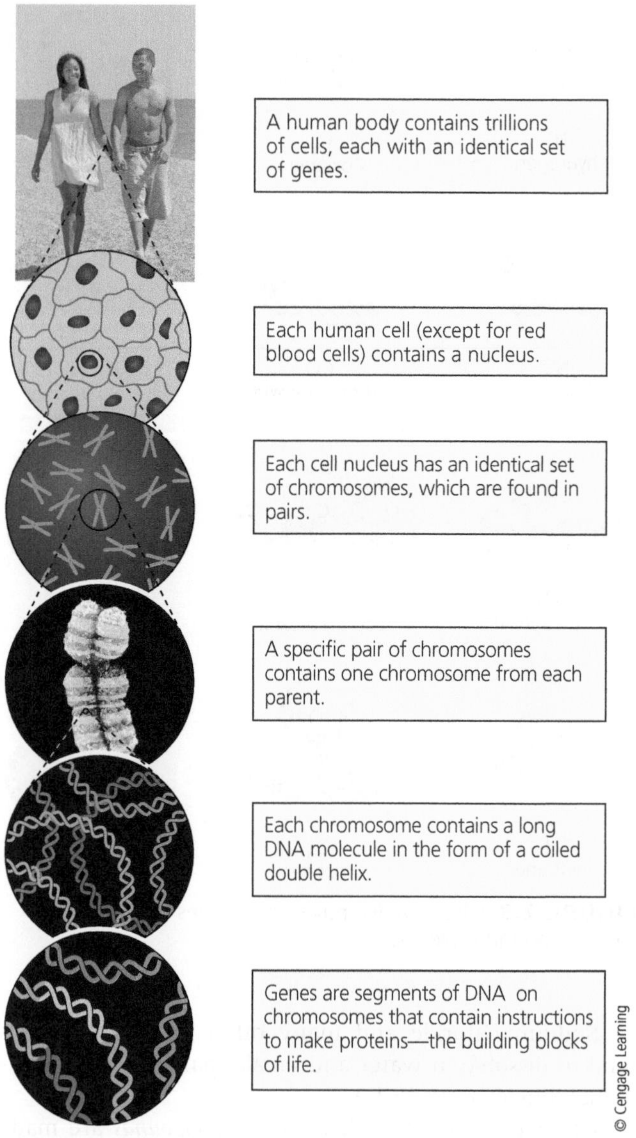

FIGURE 2.10 The relationships among cells, nuclei, chromosomes, DNA, and genes.

Photo: Flashon Studio/Shutterstock.com

Matter Comes to Life through Cells, Genes, and Chromosomes

All organisms are composed of one or more **cells**—the fundamental structural and functional units of life. The idea that all living things are composed of cells is called the **cell theory**. It is the most widely accepted scientific theory in biology.

DNA molecules are made up of sequences of nucleotides called **genes**. Each of these segments of DNA contains instructions, or codes, called *genetic information*. The coded information in each segment of DNA is a **trait** that passes from parents to offspring during reproduction in an animal or plant.

Thousands of genes make up a single **chromosome**, a double helix DNA molecule wrapped around one or more proteins. Genetic information coded in your chromosomal DNA is what makes you different from an oak leaf, a mosquito, and your parents. Figure 20.10 shows the relationships of genetic material to cells.

Matter Can Change

When matter undergoes a **physical change**, there is no change in its chemical composition. A piece of aluminum foil cut into small pieces is still aluminum foil. When solid water (ice) melts and when liquid water boils, the resulting liquid water and water vapor remain as H_2O molecules.

When a **chemical change**, or **chemical reaction**, takes place, there is a change in the chemical composition of the substances involved. Chemists use a *chemical equation* to show how chemicals are rearranged in a chemical reaction. For example, coal is made up almost entirely of the element carbon (C). When coal is burned completely in a power plant, the solid carbon in the coal combines with oxygen gas (O_2) from the atmosphere to form the gaseous compound carbon dioxide (CO_2). Chemists use the following shorthand chemical equation to represent this chemical reaction:

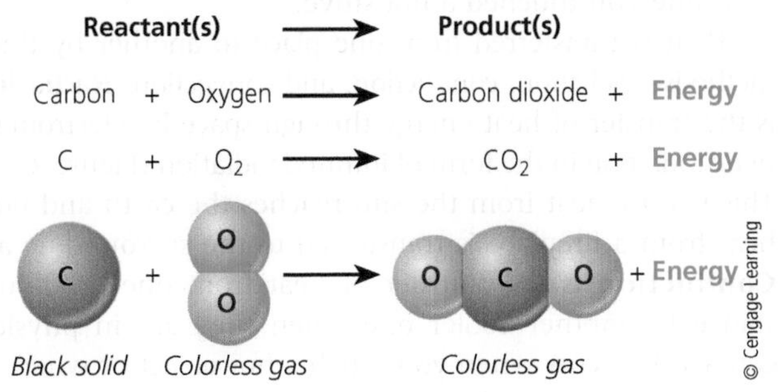

Law of Conservation of Matter

We can change elements and compounds from one physical or chemical form to another. We cannot, however, create or destroy any of the atoms involved in the physical or chemical change. All we can do is to rearrange atoms, ions, or molecules into different spatial patterns (physical changes) or chemical combinations (chemical changes). This finding, based on many thousands of measurements, describes an unbreakable scientific law known as the **law of conservation of matter**: Whenever matter undergoes a physical or chemical change, no atoms are created or destroyed (**Concept 2.2B**).

Chemists obey this scientific law by balancing the equation for a chemical reaction to account for the fact that no atoms are created or destroyed. Passing electricity through water (H_2O) can break it down into hydrogen (H_2) and oxygen (O_2), as represented by the following equation:

H_2O	$\rightarrow$	H_2	+	O_2
2 H atoms		2 H atoms		2 O atoms
1 O atom				

This equation is unbalanced because one atom of oxygen is on the left side of the equation but two oxygen atoms are on the right side. We cannot change the subscripts of any of the formulas to balance this equation because that would change the arrangements of the atoms, leading to different substances. Instead, we must use different numbers of the molecules involved to balance the equation. For example, we could use two water molecules:

$2\ H_2O$	$\rightarrow$	H_2	+	O_2
4 H atoms		2 H atoms		2 O atoms
2 O atoms				

This equation is still unbalanced. Although the numbers of oxygen atoms on both sides of the equation are now equal, the numbers of hydrogen atoms are not. We can correct this problem by recognizing that the reaction must produce two hydrogen molecules:

$2\ H_2O$	$\rightarrow$	$2\ H_2$	+	O_2
4 H atoms		4 H atoms		2 O atoms
2 O atoms				

Now the equation is balanced, and the law of conservation of matter has been observed.

2.3 WHAT IS ENERGY AND WHAT HAPPENS WHEN IT UNDERGOES CHANGE?

CONCEPT 2.3A Whenever energy is converted from one form to another in a physical or chemical change, no energy is created or destroyed (first law of thermodynamics).

CONCEPT 2.3B Whenever energy is converted from one form to another in a physical or chemical change, we end up with lower-quality or less-usable energy than we started with (second law of thermodynamics).

Energy Comes in Many Forms

Suppose you find this book on the floor and you pick it up and put it on your desktop. In doing this, you have to do *work*, or use a certain amount of muscular force to move the book from one place to another. In scientific terms, work is done when any object is moved a certain distance (work = force × distance). When you touch a hot object such as a stove, *heat* (or thermal energy) flows from the stove to your finger. Both of these examples involve **energy**: the ability to do work. Energy quantities are typically expressed in measurement units such as joules, kilojoules (1,000 joules), calories, and kilocalories (1,000 calories). (See Supplement 1, p. S1.)

There are two major types of energy: *moving energy* (called kinetic energy) and *stored energy* (called potential energy). Matter in motion has **kinetic energy**. Examples are flowing water, a car speeding down the highway, electricity (electrons flowing through a wire or other conducting material), and wind (a mass of moving air that we

FIGURE 2.11 Kinetic energy, created by the gaseous molecules in a mass of moving air, turns the blades of these wind turbines. The turbines then convert this kinetic energy to electrical energy, which is another form of kinetic energy.

can use to produce electricity, as shown in Figure 2.11). **Electric power** is the rate at which electric energy is transferred through a wire or other conducting material. It is commonly expressed in units of watts or megawatts (1 million watts) per hour.

In another form of kinetic energy called **electromagnetic radiation**, energy travels from one place to another in the form of *waves* formed from changes in electrical and magnetic fields. There are many different forms of electromagnetic radiation (Figure 2.12). Each form has a different *wavelength*—the distance between successive peaks or troughs in the wave—and *energy content*. Those with short wavelengths have more energy than do those with longer wavelengths. Visible light makes up most of the spectrum of electromagnetic radiation emitted by the sun.

Another form of kinetic energy is **heat**, or **thermal energy**, the total kinetic energy of all moving atoms, ions, or molecules in an object, a body of water, or a volume of gas such as the atmosphere. If the atoms, ions, or molecules in a sample of matter move faster, the matter will become warmer. When two objects at different temperatures make contact with each another, heat flows from the warmer object to the cooler object. You learned this the first time you touched a hot stove.

Heat is transferred from one place to another by three methods—radiation, conduction, and convection. **Radiation** is the transfer of heat energy through space by electromagnetic radiation in the form of infrared radiation (Figure 2.12). This is how heat from the sun reaches the earth and how heat from a fireplace is transferred to the surrounding air. **Conduction** is the transfer of heat from one solid substance to another cooler one when they are in physical contact. It occurs when you touch a hot object or when an electric stove burner heats a pan. **Convection** is the transfer of heat energy within liquids or gases when warmer areas of the liquid or gas rise to cooler areas and cooler liquid or gas takes its place. As a result, heat circulates through the air or liquid such as water being heated in a pan.

The other major type of energy is **potential energy**, which is stored and potentially available for use. Examples of this type of energy include a rock held in your hand, the water in a reservoir behind a dam, the chemical energy stored in the carbon atoms of coal or in the molecules of the food you eat, and **nuclear energy** stored in the strong forces that hold the particles (protons and neutrons) in the nuclei of atoms together.

You can change potential energy to kinetic energy. If you hold this book in your hand, it has potential energy. If you drop it on your foot, the book's potential energy

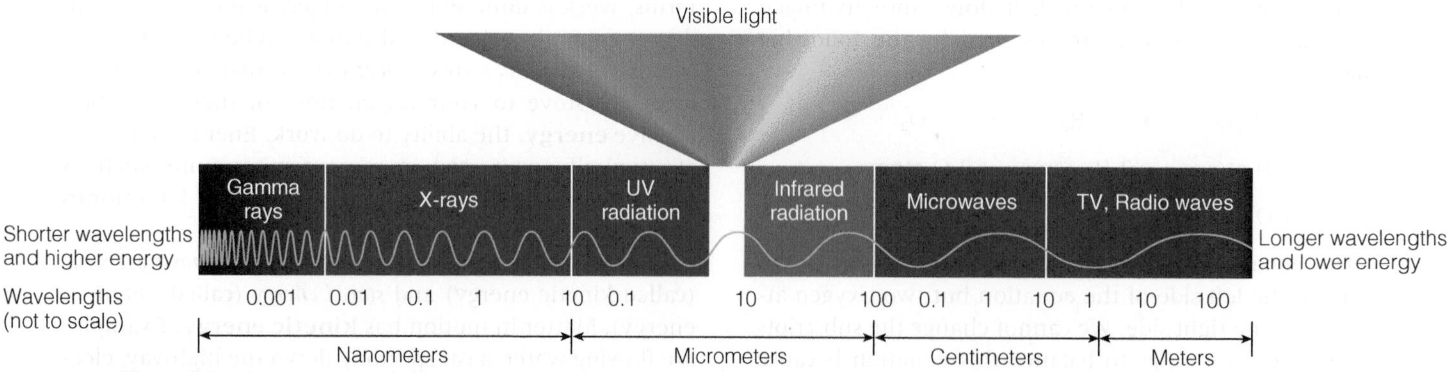

FIGURE 2.12 The *electromagnetic spectrum* consists of a range of electromagnetic waves, which differ in wavelength (the distance between successive peaks or troughs) and energy content.

Songquan Deng/Shutterstock.com

FIGURE 2.13 The water stored in this reservoir behind a dam has potential energy, which becomes kinetic energy when the water flows through channels built into the dam where it spins a turbine and produces electricity—another form of kinetic energy.

changes to kinetic energy during its fall. When a car engine burns gasoline, the potential energy stored in the chemical bonds of the gasoline molecules changes into kinetic energy that propels the car, and into heat that flows into the environment. When water in a reservoir flows through channels in a dam (Figure 2.13), its potential energy becomes kinetic energy used to spin turbines in the dam to produce electricity—yet another form of kinetic energy.

Energy Is Renewable or Nonrenewable

Scientists divide energy resources into two major categories: renewable energy and nonrenewable energy. **Renewable energy** is energy gained from resources that are replenished by natural processes in a relatively short time. Examples are solar energy, wind, moving water, firewood from trees, and heat that comes from the earth's interior (geothermal energy).

Nonrenewable energy is energy from resources that can be depleted and are not replenished by natural processes within a human time scale. Examples are energy produced by the burning of oil, coal, and natural gas, and nuclear energy released when the nuclei of atoms of uranium fuel are split apart.

About 99% of the energy that keeps us warm and supports the plants that we and other organisms eat comes from the sun. This is the basis of the solar energy **principle of sustainability** (see Figure 1.2, p. 6). Without inexhaustible solar energy, the earth would be frozen and life as we know it would not exist.

Commercial energy—energy that is sold in the marketplace—makes up the remaining 1% of the energy we use to supplement the earth's direct input of solar energy. About 90% of the commercial energy used in the world and 90% of that used in the United States comes from the burning of nonrenewable *fossil fuels*—oil, coal, and natural gas. They are called fossil fuels because they were formed over hundreds of thousands to millions of years as layers of the decaying remains of ancient plants and animals were exposed to intense heat and pressure within the earth's crust.

90% Percentage of the commercial energy used in the world and in the United States provided by fossil fuels

Energy Varies in Its Quality

Some types of energy are more useful than others. **Energy quality** is a measure of the capacity of energy to do useful work. **High-quality energy** is concentrated energy that has a high capacity to do useful work. Examples are high-temperature heat, concentrated sunlight, high-speed wind, and the energy released when we burn wood, gasoline, natural gas, or coal.

By contrast, **low-quality energy** is so dispersed that it has little capacity to do useful work. The enormous number of moving molecules in the atmosphere or in an ocean together has such low-quality energy, and such a low temperature, that we cannot use them to move things or to heat things to high temperatures.

Energy Changes Obey Two Scientific Laws

After observing and measuring energy being changed from one form to another in millions of physical and chemical changes, scientists summarized their results in the **first law of thermodynamics**, also known as the **law of conservation of energy**. According to this scientific law, whenever energy is converted from one form to another in a physical or chemical change, no energy is created or destroyed (**Concept 2.3A**).

No matter how hard we try or how clever we are, we cannot get more energy out of a physical or chemical change than we put in. This law is one of nature's basic rules that we cannot violate.

Because energy cannot be created or destroyed, only converted from one form to another, you may be think we will never run out of energy. Think again. If you fill a car's tank with gasoline and drive around all day or run your cell phone battery down, something has been lost. What is it? The answer is *energy quality*, the amount of energy available for performing useful work.

Thousands of experiments have shown that whenever energy is converted from one form to another in a physical or chemical change, we end up with lower-quality or less-usable energy than we started with (**Concept 2.3B**). This is observation is known as the **second law of thermodynamics**. The low-quality energy usually takes the form of heat that flows into the environment. The random motion of air or water molecules further disperses this heat, decreasing its temperature to the point where its energy quality is too low to do much useful work.

In other words, *when energy is changed from one form to another, it always goes from a more useful to a less useful form*. This is another scientific law that cannot be violated. This means we cannot recycle or reuse high-quality energy to perform useful work. Once the high-quality energy in a serving of food, a tank of gasoline, or a chunk coal is released, it is degraded to low-quality heat and dispersed into the environment.

Energy efficiency is a measure of how much work results from each unit of energy that is put into a system. Suppose you turn on a lamp with an incandescent bulb powered by electricity produced by a coal-burning power plant. This electricity is transported by a power line to your house and then through house wires to the light bulb. Because of the second law of thermodynamics, some of the original energy produced by burning the coal is lost as waste heat to the environment in each step of this process. The amount of heat lost in each step depends on the energy efficiency of the technologies used. As a result of these losses, only 5% of the chemical energy in the coal ends up producing the light from the bulb.

Thus, 95% of the money spent for the amount of light in this example was wasted. Some of this energy and money waste was the automatic result of the second law of thermodynamics. The rest was lost mostly because of low energy efficiency of the power plant (35%) and the light bulb (5%). A key to reducing this waste of energy and money is to improve the energy efficiency of the power plant and light bulb or replace them with newer, more energy-efficient technologies. We are still using the energy-wasting power plants but we are shifting from inefficient incandescent light bulbs to much more efficient light-emitting diode (LED) light bulbs.

Scientists estimate that about 84% of the energy used in the United States is either unavoidably wasted because of the second law of thermodynamics (41%) or unnecessarily wasted (43%). Thus, thermodynamics teaches us an important lesson: the cheapest and quickest way to get more energy and cut energy bills is to stop wasting almost half the energy we use. One way to reduce the unnecessary waste of energy and money is to improve the *energy efficiency* of the power plants, automobile engines, and devices powered by electricity such as lights, refrigerators, and air conditioners. This will save money and sharply reduce air pollution, including emissions of climate-changing carbon dioxide.

43% Percentage of the commercial energy used in the United States that is unnecessarily wasted

2.4 WHAT ARE SYSTEMS AND HOW DO THEY RESPOND TO CHANGE?

CONCEPT 2.4 Systems have inputs, flows, and outputs of matter and energy, and feedback can affect their behavior.

Systems and Feedback Loops

A **system** is any set of components that function and interact in some regular way. Examples are a cell, the human body, a forest, an economy, a car, a TV set, and the earth.

Most systems have three key components: **inputs** of matter, energy, and information from the environment; **flows** or **throughputs** of matter, energy, and information within the system; and **outputs** of matter, energy, and information to the environment (Figure 2.14) (**Concept 2.4**). A system can become unsustainable if the throughputs of matter and energy resources exceed the ability of the system's environment to provide the required resource inputs and to absorb or dilute the system's outputs of matter and energy (mostly heat).

When people ask you for feedback they are usually seeking your response to something they said or did. They might feed your response back into their mental process to help them decide whether and how to change what they are saying or doing. Most systems are affected by **feedback**, any process that increases (positive feedback) or decreases (negative feedback) a change to a system (**Concept 2.4**). Such a process, called a **feedback loop**, occurs when an output of matter, energy, or information is fed back into the system as an input and changes the system. A **positive feedback loop** causes a system to change further in the same direction. An example is what happened when researchers removed the vegetation from a valley in the Hubbard Brook Experimental Forest (Figure 2.15) (**Core Case Study**).

When a natural system becomes locked into a positive feedback loop, it can reach an **ecological tipping point**. Beyond this point, the system can change so drastically that it suffers from severe degradation or collapse. Reaching and exceeding a tipping point is somewhat like stretching a rubber band. We can get away with stretching it to several times its original length. At some point, however, we reach an irreversible tipping point where the rubber band breaks. Similarly, if you lean back on the two rear legs of a chair, at some point the chair will tip back and you will land on the floor. Many types of ecological tipping points will be discussed throughout this book.

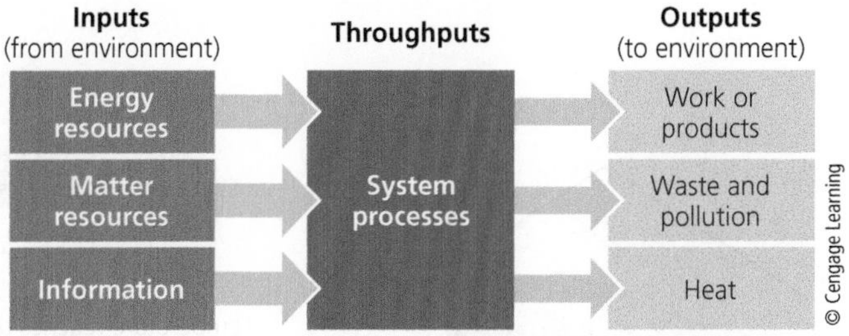

FIGURE 2.14 Simplified model of a system.

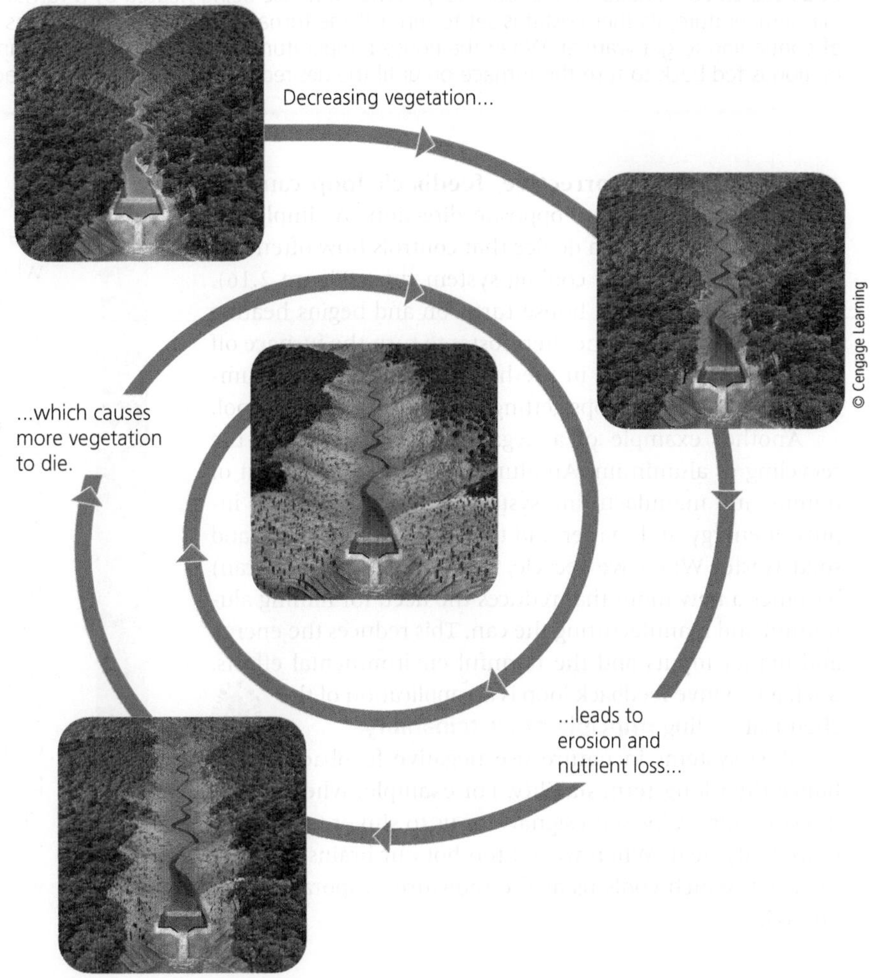

FIGURE 2.15 A *positive feedback loop.* Decreasing vegetation in a valley causes increasing erosion and nutrient losses that in turn cause more vegetation to die, resulting in more erosion and nutrient losses.

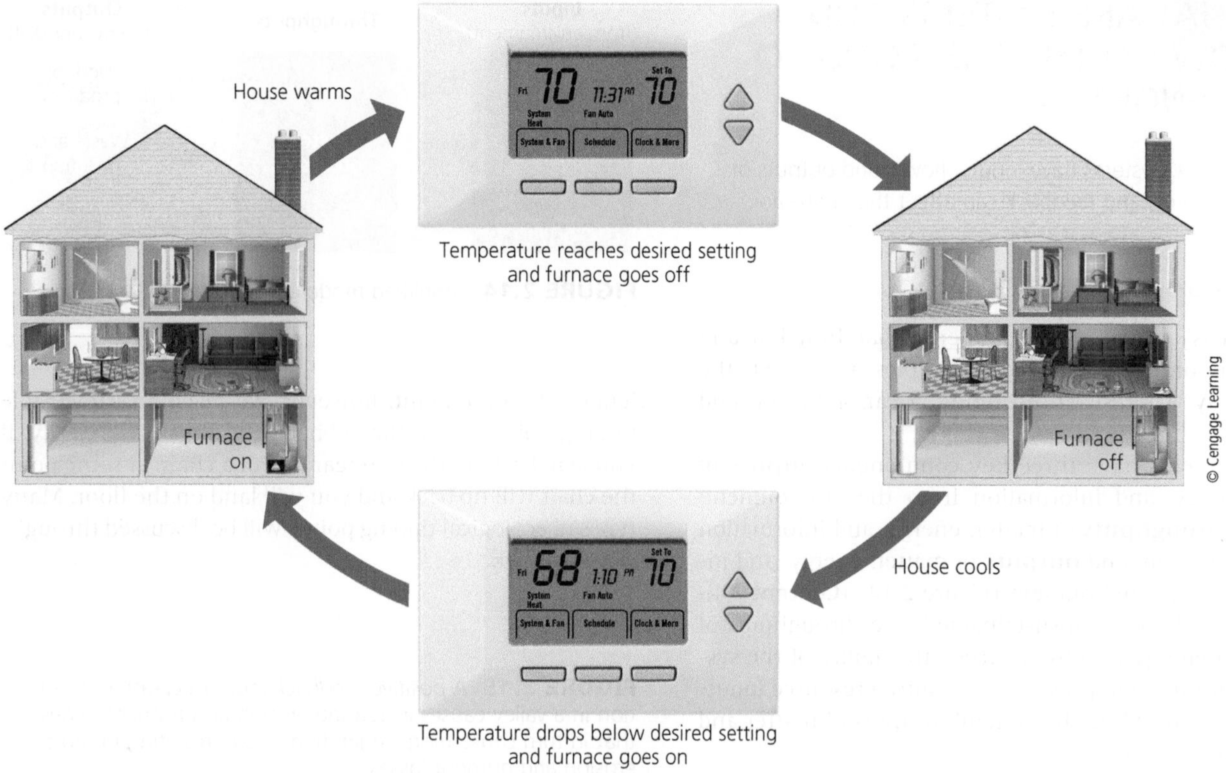

FIGURE 2.16 A *negative feedback loop.* When a house being heated by a furnace gets to a certain temperature, its thermostat is set to turn off the furnace, and the house begins to cool instead of continuing to get warmer. When the house temperature drops below the set point, this information is fed back to turn the furnace on until the desired temperature is reached again.

A **negative**, or **corrective**, **feedback loop** causes a system to change in the opposite direction. A simple example is a thermostat, a device that controls how often and how long a heating or cooling system runs (Figure 2.16). When the furnace in a house turns on and begins heating the house, we can set the thermostat to turn the furnace off when the temperature in the house reaches the set number. The house then stops getting warmer and starts to cool.

Another example of a negative feedback loop is the recycling of aluminum. An aluminum can is an output of mining and manufacturing systems that requires large inputs of energy and matter and that produces pollution and solid waste. When we recycle, the output (the used can) becomes a new input that reduces the need for mining aluminum and manufacturing the can. This reduces the energy and matter inputs and the harmful environmental effects. Such a negative feedback loop is an application of the chemical cycling **principle of sustainability**.

Most systems in nature use negative feedback to enhance their long-term stability. For example, when we get too cold our brains send signals for us to shiver to produce more body heat. When we get too hot our brains cause us to sweat, which cools us as the moisture evaporates from our skin.

BIG IDEAS

- According to the *law of conservation of matter*, no atoms are created or destroyed whenever matter undergoes a physical or chemical change. Thus, we cannot do away with matter; we can only change it from one physical state or chemical form to another.
- According to the *first law of thermodynamics*, or the *law of conservation of energy*, whenever energy is converted from one form to another in a physical or chemical change, no energy is created or destroyed. This means that in causing such changes, we cannot get more energy out than we put in.
- According to the *second law of thermodynamics*, whenever energy is converted from one form to another in a physical or chemical change, we always end up with a lower-quality or less-usable form of energy than we started with. This means that we cannot recycle or reuse high-quality energy.

Tying It All Together

The Hubbard Brook Forest Experiment and Sustainability

In the controlled experiment discussed in this chapter's **Core Case Study**, the clearing of a mature forest degraded some of its natural capital. Specifically, the loss of trees and vegetation altered the ability of the forest to retain and recycle water and other critical plant nutrients—a crucial ecological function based on the chemical cycling **principle of sustainability**.

This clearing of vegetation also violated the solar energy and biodiversity **principles of sustainability**. For example, the cleared forest lost most of its plants that had used solar energy to produce food for the forest's animals, which supplied nutrients to the soil when they died. Thus, the forest lost many of its key nutrients that would normally have been recycled. It also lost much of its life-sustaining biodiversity.

Many of the results of environmental science are based on this sort of experimentation. Throughout this textbook, we explore other examples of how scientists learn about nature and use these results to understand how our actions affect the environment and how we can solve some of our environmental problems.

steve estvanik/Shutterstock.com

Chapter Review

Core Case Study

1. Describe the controlled scientific experiment carried out in the Hubbard Brook Experimental Forest.

Section 2.1

2. What is the key concept for this section? What is **science**? List the steps involved in a scientific process. What are **data**? Distinguish between a **scientific hypothesis** and a **scientific theory**. What is **peer review** and why is it important? What is a **model**? Summarize scientist Jane Goodall's achievements. Summarize the scientific lessons learned from research on the fall of the ancient civilization on Easter Island.

3. Explain why scientific theories and laws are the most important and most certain results of science and why people often use the term *theory* incorrectly. What is a **scientific law (law of nature)**? Explain why we cannot break such laws. Distinguish among **reliable science**, **unreliable science**, and **tentative science**. What are four limitations of science?

Section 2.2

4. What are the two key concepts for this section? What is **matter**? Distinguish between an **element** and a compound and give an example of each. What is the **periodic table of elements**? Define **atom**, **molecule**, and **ion** and give an example of each. What is the **atomic theory**? Distinguish among **protons**, **neutrons**, and **electrons**. What is the **nucleus** of an atom? Distinguish between the **atomic number** and the **mass number** of an element. What are **isotopes**? What is **acidity**? What is **pH**? Define **chemical formula** and give two examples.

5. Define and give two examples of an **organic compound**. What are three types of organic polymers that are important to life? What is a **cell**? What is the cell theory? Define **gene**, **trait**, and **chromosome**.

6. Define and distinguish between a **physical change** and a **chemical change (chemical reaction)** in matter and give an example of each. What is the **law of conservation of matter**?

Section 2.3

7. What are the two key concepts for this section? What is **energy**? Define and distinguish between **kinetic energy** and **potential energy** and give an example of each. What is electric power? Define and give two examples of **electromagnetic radiation**. What is **heat (thermal energy)**? Explain how heat is transferred from one place to another by radiation, conduction, and convection. Define and give two examples of **electromagnetic radiation**. Distinguish between renewable energy and nonrenewable energy and give two examples of each. What percentage of the commercial energy used in the world and what percentage used in the United States is provided by fossil fuels?

8. What is **energy quality**? Distinguish between **high-quality energy** and **low-quality energy** and give an example of each. What is the **first law of thermodynamics (law of conservation of energy)** and why is it important? What is the **second law of thermodynamics** and why is it important? Explain why the second law means that we can never recycle or reuse high-quality energy. What is **energy efficiency**? What percentage of the commercial energy used in the United States is unnecessarily wasted? Why is it important to reduce this waste and how can we do this?

Section 2.4

9. What is the key concept for this section? Define and give an example of a **system**. Distinguish among the **inputs**, **flows (throughputs)**, and **outputs** of a system. What is **feedback**? What is a **feedback loop**? Distinguish between a **positive feedback loop** and a **negative (corrective) feedback loop** in a system, and give an example of each. What is an **ecological tipping point**?

10. What are this chapter's *three big ideas*? Explain how the Hubbard Brook Experimental Forest controlled experiments illustrated the three *scientific principles of sustainability*.

Note: Key terms are in bold type.

Critical Thinking

1. What ecological lesson can we learn from the controlled experiment on the clearing of forests described in the **Core Case Study** that opened this chapter?

2. Suppose you observe that all of the fish in a pond have disappeared. How might you use the scientific process described in the **Core Case Study** and in Figure 2.2 to determine the cause of this fish kill?

3. Respond to the following statements:
   a. Scientists have not absolutely proven that anyone has ever died from smoking cigarettes.
   b. The *natural greenhouse effect*—the warming effect of certain gases such as water vapor and carbon dioxide in the lower atmosphere—is not a reliable idea because it is just a scientific theory.

4. A tree grows and increases its mass. Explain why this is not a violation of the law of conservation of matter.

5. How do the first and second laws of thermodynamics affect our use of energy resources such as fossil fuels, solar energy, and wind energy?

6. Suppose someone wants you to invest money in an automobile engine, claiming that it will produce more energy than is found in the fuel used to run it. What would be your response? Explain.

7. Use the second law of thermodynamics to explain why we can use oil only once as a fuel, or in other words, why we cannot recycle or reuse its high-quality energy.

8. For one day, **(a)** you have the power to revoke the law of conservation of matter, and **(b)** you have the power to violate the first law of thermodynamics. For each of these scenarios, list three ways in which you would use your new power. Explain your choices.

Doing Environmental Science

Find a newspaper or magazine article or a report on the Internet that attempts to discredit a scientific hypothesis because it has not been proven, or a report of a new scientific hypothesis that has the potential to be controversial. Analyze the piece by doing the following: **(1)** determine its source (authors or organization); **(2)** detect an alternative hypothesis, if any, that is offered by the authors; **(3)** determine the primary objective of the authors (for example, to debunk the original hypothesis, to state an alternative hypothesis, or to raise new questions); **(4)** summarize the evidence given by the authors for their position; and **(5)** compare the authors' evidence with the evidence for the original hypothesis. Write a report summarizing your analysis and compare it with those of your classmates.

Global Environment Watch Exercise

Go to your MindTap course to access the GREENR database. Starting on the home page, under "Browse Issues and Topics", click on *Resource Management*, then select *Forests and Deforestation*. Browse the articles listed there and find one that involves a controlled experiment or some other form of scientific research in a forest. Determine what the hypothesis was that the researchers were testing. Summarize their research methods and any conclusions that were reached. Was the research similar in any way to that described in the **Core Case Study**? Explain.

Data Analysis

Consider the graph to the right that compares the losses of calcium from the experimental and control sites in the Hubbard Brook Experimental Forest (**Core Case Study**). Note that this figure is very similar to Figure 2.5, which compares loss of nitrates from the two sites. After studying this graph, answer these questions.

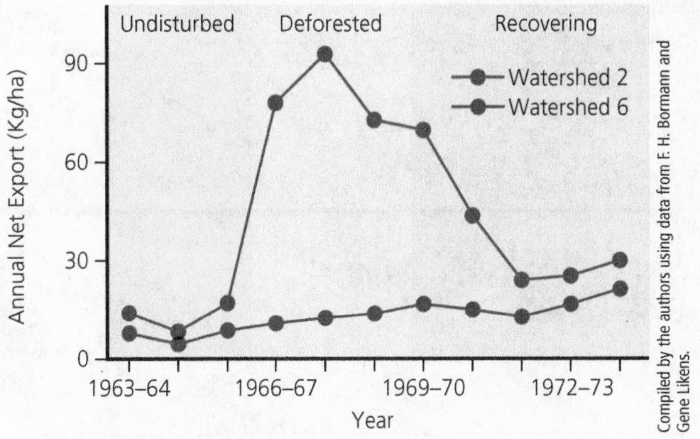

1. In what year did the loss of calcium from the experimental site begin a sharp increase? In what year did it peak? In what year did it level off?
2. In what year were the calcium losses from the two sites closest together? In the span of time between 1963 and 1972, did they ever get that close again?
3. Does this graph support the hypothesis that cutting the trees from a forested area causes the area to lose nutrients more quickly than leaving the trees in place? Explain.

CHAPTER 3

Ecosystems: What Are They and How Do They Work?

First Law of Ecology: Everything is connected to everything else.

BARRY COMMONER

Key Questions

3.1 How does the earth's life-support system work?

3.2 What are the major components of an ecosystem?

3.3 What happens to energy in an ecosystem?

3.4 What happens to matter in an ecosystem?

3.5 How do scientists study ecosystems?

Marine scientist measuring new growth on a table coral.

Brian J. Skerry/National Geographic Creative

Core Case Study

Tropical Rain Forests Are Disappearing

Tropical rain forests are found near the earth's equator and contain an amazing variety of life. These lush forests are warm year round and have high humidity because it rains almost daily. Rain forests cover only 2% of the earth's land but contain up to half of the world's known terrestrial plant and animal species. These properties make rain forests natural laboratories in which to study *ecosystems*—communities of organisms that interact with one another and with the physical environment of matter and energy in which they live.

To date, at least half of the earth's rain forests have been destroyed or degraded by humans cutting down trees, growing crops, grazing cattle, and building settlements (Figure 3.1). The destruction and degradation of these centers of biodiversity is increasing. Ecologists warn that without protection, most of these forests will be gone or severely degraded by the end of this century.

Why should we care that tropical rain forests are disappearing? Scientists give three reasons. *First*, clearing these forests reduces the earth's vital biodiversity by destroying the habitats for many of the earth's species. *Second*, destroying these forests contributes to atmospheric warming and speeds up climate change, which you will learn about in Chapter 19. How does this occur? Eliminating large areas of trees faster than they can grow back decreases the ability of the forests to remove some of the human-generated emissions of carbon dioxide (CO_2), a gas that contributes to atmospheric warming and climate change.

Third, large-scale loss of tropical rain forests can change regional weather patterns. Sometimes such changes can prevent the regrowth of rain forests in cleared or degraded areas. When this *ecological tipping point* is reached, tropical rain forests in such areas become less diverse tropical grasslands.

In this chapter, you will learn how tropical rain forests and other ecosystems work, how human activities are affecting them, and how we can help sustain them.

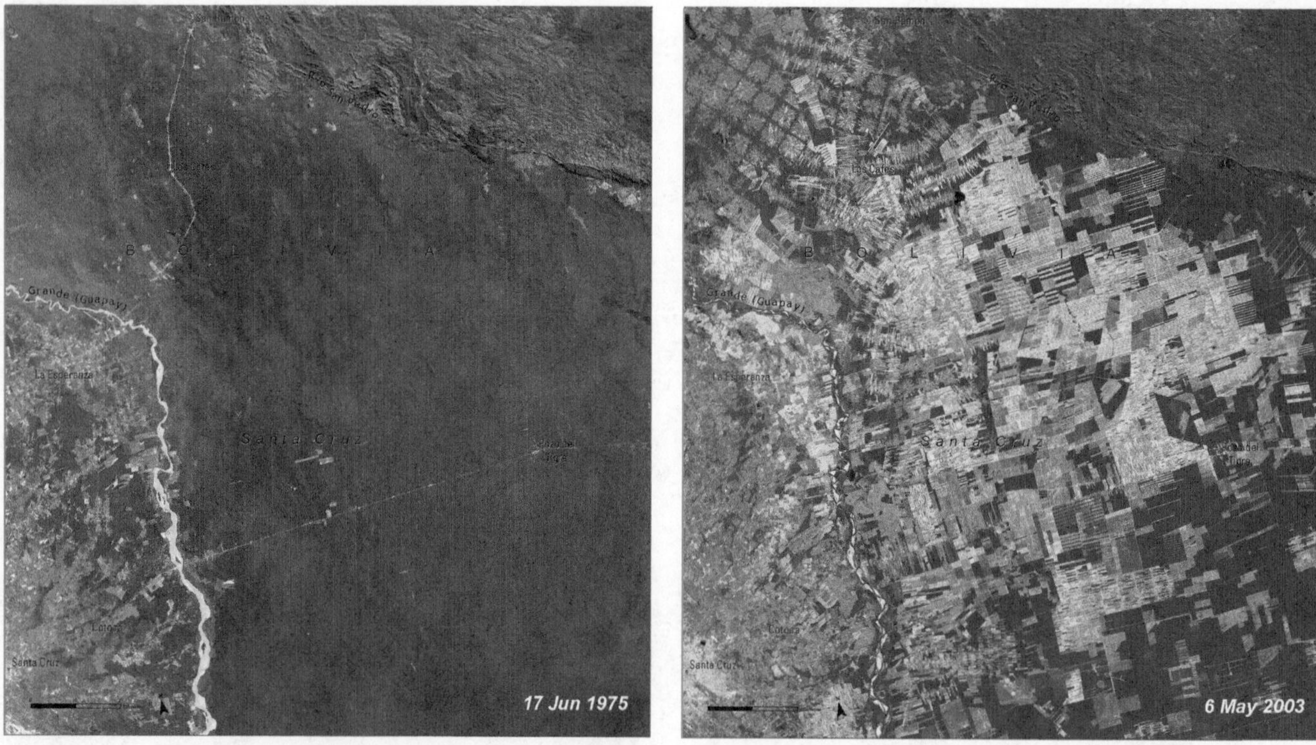

Left: United Nations Environment Programme; Right: United Nations Environment Programme

FIGURE 3.1 Natural capital degradation: Satellite image of the loss of tropical rain forest, cleared for farming, cattle grazing, and settlements, near the Bolivian city of Santa Cruz between June 1975 (left) and May 2003 (right). This is the latest available view of the area but forest degradation has continued since 2003.

3.1 HOW DOES THE EARTH'S LIFE-SUPPORT SYSTEM WORK?

CONCEPT 3.1A The four major components of the earth's life-support system are the atmosphere (air), the hydrosphere (water), the geosphere (rock, soil, and sediment), and the biosphere (living things).

CONCEPT 3.1B Life is sustained by the flow of energy from the sun through the biosphere, the cycling of nutrients within the biosphere, and gravity.

Earth's Life-Support System Has Four Major Components

The earth's life-support system consists of four main systems (Figure 3.2) that interact with one another. They are the atmosphere (air), the hydrosphere (water), the geosphere (rock, soil, and sediment), and the biosphere (living things) (**Concept 3.1A**).

The **atmosphere** is a spherical mass of air surrounding the earth's surface. Its innermost layer, the **troposphere**, extends about 17 kilometers (11 miles) above sea level at the equator and about 7 kilometers (4 miles) above the earth's North and South Poles. The troposphere contains the air we breathe. It is 78% nitrogen (N_2) and 21% oxygen (O_2). The remaining 1% of air is mostly water vapor, carbon dioxide, and methane.

The next layer of the atmosphere is the **stratosphere**. It reaches 17 to 50 kilometers (11–31 miles) above the earth's surface. The layer of the stratosphere closest to the earth's surface contains enough ozone (O_3) gas to filter out about 95% of the sun's harmful *ultraviolet (UV) radiation*. This global sunscreen allows life to exist on the surface of the planet.

The **hydrosphere** includes all of the water on or near the earth's surface. It is found as *water vapor* in the atmosphere, as *liquid water* on the surface and underground, and as *ice*—polar ice, icebergs, glaciers, and ice in frozen soil-layers called *permafrost*. Salty oceans that cover that about 71% of the earth's surface contain 97% of the planet's water and support almost half of the world's species. About 2.5% of the earth's water is freshwater and three-fourths of that is ice.

The **geosphere** contains the earth's rocks, minerals, and soil. It consists of an intensely hot *core*, a thick *mantle* of very hot rock, and a thin outer *crust* of rock and soil. The crust's upper portion contains soil chemicals or nutrients that organisms need to live, grow, and reproduce. It also contains nonrenewable *fossil fuels*—coal, oil, and natural gas—and minerals that we extract and use.

The **biosphere** consists of the parts of the atmosphere, hydrosphere, and geosphere where life is found. If you compare the earth with an apple, the biosphere would be as thick as the apple's skin.

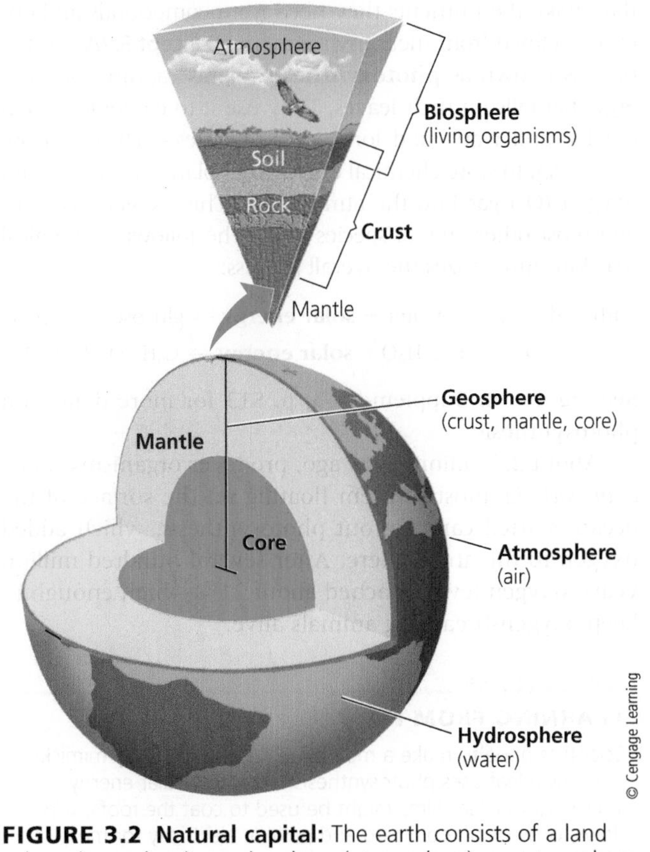

FIGURE 3.2 Natural capital: The earth consists of a land sphere (*geosphere*), an air sphere (*atmosphere*), a water sphere (*hydrosphere*), and a life sphere (*biosphere*) (**Concept 3.1A**).

Three Factors Sustain the Earth's Life

Life on the earth depends on three interconnected factors (**Concept 3.1B**):

1. One-way flow of high-quality energy from the sun. The sun's energy supports plant growth, which provides energy for plants and animals, in keeping with the solar energy **principle of sustainability**. As solar energy interacts with carbon dioxide (CO_2), water vapor, and several other gases in the troposphere, it warms the troposphere—a process known as the **greenhouse effect** (Figure 3.3). Without this natural process, the earth would be too cold to support most of the forms of life we find here today.
2. *Cycling of nutrients* through parts of the biosphere. **Nutrients** are chemicals that organisms need to survive. Because the earth does not get significant inputs of matter from space, its fixed supply of nutrients must be recycled to support life. This is in keeping with the chemical cycling **principle of sustainability**.
3. *Gravity* allows the planet to hold on to its atmosphere and enables the movement and cycling of chemicals through air, water, soil, and organisms.

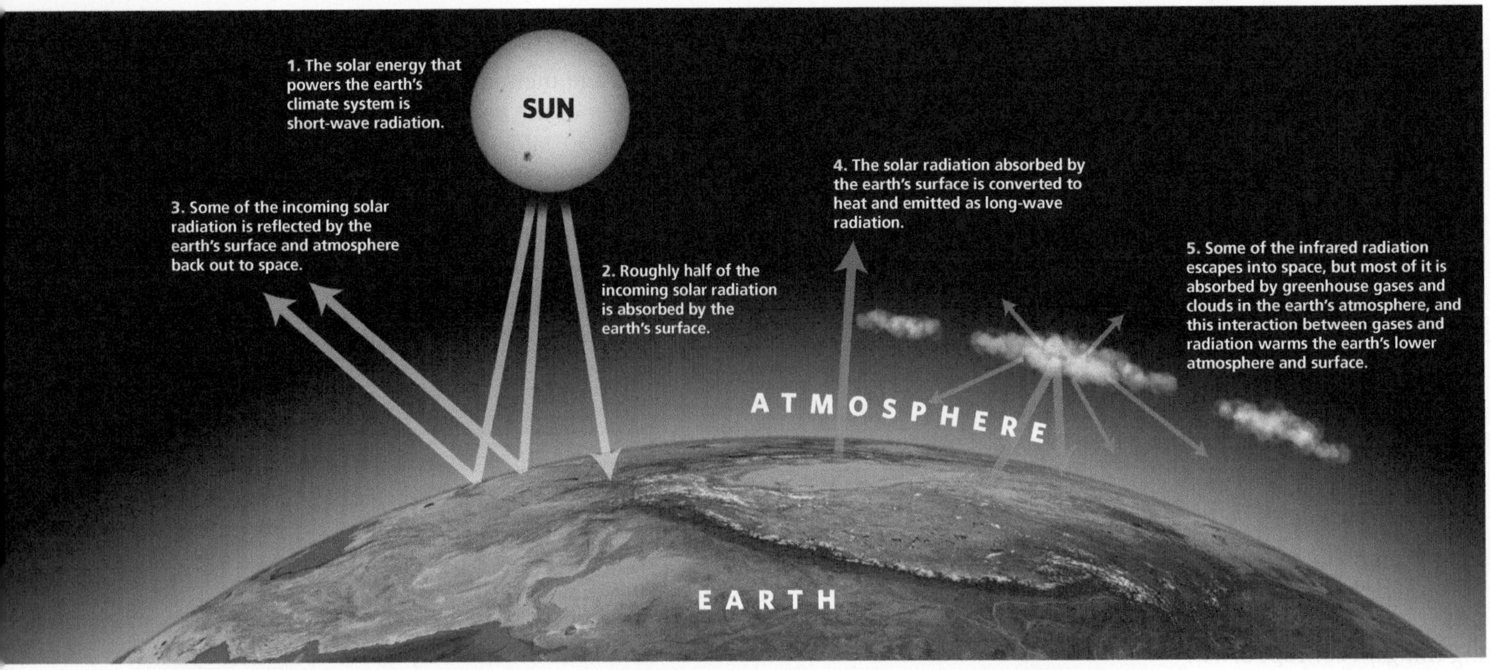
1. The solar energy that powers the earth's climate system is short-wave radiation.
SUN
3. Some of the incoming solar radiation is reflected by the earth's surface and atmosphere back out to space.
2. Roughly half of the incoming solar radiation is absorbed by the earth's surface.
4. The solar radiation absorbed by the earth's surface is converted to heat and emitted as long-wave radiation.
5. Some of the infrared radiation escapes into space, but most of it is absorbed by greenhouse gases and clouds in the earth's atmosphere, and this interaction between gases and radiation warms the earth's lower atmosphere and surface.
ATMOSPHERE
EARTH

National Geographic Visual Atlas of the World. Washington, DC: National Geographic Society, 2008.

FIGURE 3.3 *Greenhouse Earth.* High-quality solar energy flows from the sun to the earth. It is degraded to lower-quality energy (mostly heat) as it interacts with the earth's air, water, soil, and life forms, and eventually returns to space. Certain gases in the earth's atmosphere retain enough of the sun's incoming energy as heat to warm the planet in what is known as the *greenhouse effect.*

3.2 WHAT ARE THE MAJOR COMPONENTS OF AN ECOSYSTEM?

CONCEPT 3.2A Some organisms produce the nutrients they need, others get the nutrients they need by consuming other organisms, and some recycle nutrients back to producers by decomposing the wastes and remains of other organisms.

CONCEPT 3.2B Soil is a renewable resource that provides nutrients that support terrestrial plants and helps purify water and control the earth's climate.

Ecosystems Have Several Important Components

Scientists classify matter into levels of organization ranging from atoms to galaxies. Ecologists study five levels of matter—the *biosphere, ecosystems, communities, populations,* and *organisms*—which are shown and defined in Figure 3.4.

The biosphere and its ecosystems are made up of living (*biotic*) and nonliving (*abiotic*) components (Figure 3.5). Nonliving components include water, air, nutrients, rocks, heat, and solar energy. Living components include plants, animals, and microbes.

Ecologists assign each organism in an ecosystem to a *feeding level,* or **trophic level**, based on its source of nutrients. Organisms are classified as *producers* and *consumers*. **Producers** (also called **autotrophs**) are organisms, such as green plants, that make the nutrients they need from compounds and energy obtained from their environment (**Concept 3.2A**). In the process known as **photosynthesis**, plants capture solar energy that falls on their leaves. They use it to combine carbon dioxide and water and form carbohydrates such as glucose ($C_6H_{12}O_6$), to store chemical energy that plants need and emit oxygen (O_2) gas into the atmosphere. This oxygen keeps us and most other animal species alive. The following chemical reaction summarizes the overall process:

$$\text{carbon dioxide} + \text{water} + \textbf{solar energy} \rightarrow \text{glucose} + \text{oxygen}$$
$$6\,CO_2 + 6\,H_2O + \textbf{solar energy} \rightarrow C_6H_{12}O_6 + 6\,O_2$$

See Figure 10, Supplement 3, p. S13 for more details on photosynthesis.

About 2.8 billion years ago, producer organisms called *cyanobacteria,* most of them floating on the surface of the ocean, started carrying out photosynthesis, which added oxygen to the atmosphere. After several hundred million years, oxygen levels reached about 21%—high enough to keep oxygen-breathing animals alive.

CONSIDER THIS . . .

LEARNING FROM NATURE

Scientists hope to make a molecular-sized solar cell by mimicking how a leaf uses photosynthesis to capture solar energy. These artificial leaf films might be used to coat the roofs, windows, or walls of a building and provide electricity for most homes and other buildings.

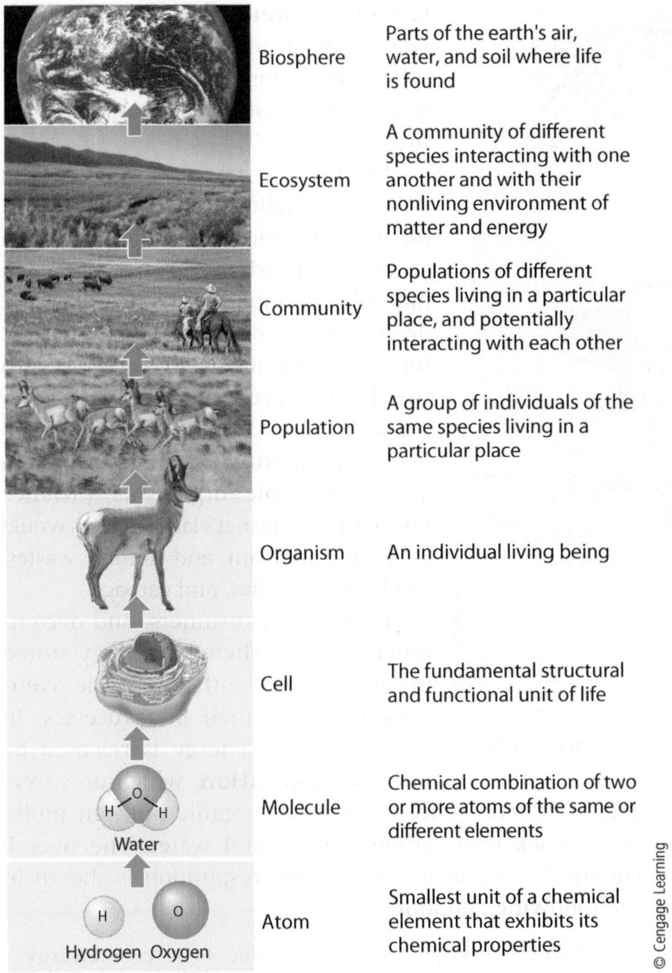
Biosphere
Parts of the earth's air, water, and soil where life is found
Ecosystem
A community of different species interacting with one another and with their nonliving environment of matter and energy
Community
Populations of different species living in a particular place, and potentially interacting with each other
Population
A group of individuals of the same species living in a particular place
Organism
An individual living being
Cell
The fundamental structural and functional unit of life
Molecule
Chemical combination of two or more atoms of the same or different elements
Water
Atom
Smallest unit of a chemical element that exhibits its chemical properties
Hydrogen
Oxygen
© Cengage Learning

FIGURE 3.4 Ecology focuses on the top five of these levels of the organization of matter in nature.

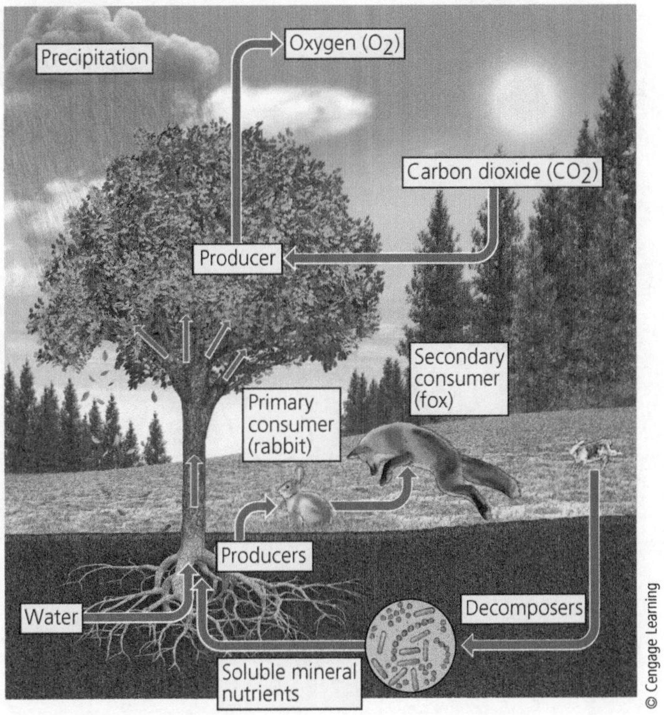
Precipitation
Oxygen (O_2)
Carbon dioxide (CO_2)
Producer
Secondary consumer (fox)
Primary consumer (rabbit)
Producers
Water
Decomposers
Soluble mineral nutrients
© Cengage Learning

FIGURE 3.5 Key living (biotic) and nonliving (abiotic) components of an ecosystem in a field.

Nelik/Shutterstock.com

FIGURE 3.6 This lioness (a carnivore) is feeding on a freshly killed zebra (an herbivore) in Kenya, Africa.

Today, most producers on land are trees and other green plants. In freshwater and ocean ecosystems, algae and aquatic plants growing near shorelines are the major producers. In open water of the oceans, floating and drifting microscopic organisms known as *phytoplankton* are the dominant producers.

Some producer bacteria live in dark and extremely hot water around fissures on the ocean floor. Their source of energy is heat from the earth's interior, or *geothermal energy*. They are an exception to the solar energy principle of sustainability.

The other organisms in an ecosystem are **consumers** (also called **heterotrophs**) that cannot produce the nutrients they need (**Concept 3.2A**). They get their nutrients by feeding on other organisms (producers or other consumers) or on their wastes and remains.

There are several types of consumers. **Primary consumers**, or **herbivores** (plant eaters), are animals that eat mostly green plants. Examples are caterpillars, giraffes, and zooplankton (tiny sea animals that feed on phytoplankton). **Carnivores** (meat eaters) are animals that feed on the flesh of other animals. Some carnivores, including spiders, lions (Figure 3.6), and most small fishes, are **secondary consumers** that feed on the flesh of herbivores. Other carnivores such as tigers, hawks, and killer whales (orcas) are **tertiary** (or higher-level) **consumers** that feed on the flesh of herbivores and other carnivores. Some of these relationships are shown in Figure 3.5. **Omnivores** such as pigs, rats, and humans eat both plants and animals.

CONSIDER THIS . . .

THINKING ABOUT What You Eat

When you ate your most recent meal, were you an herbivore, a carnivore, or an omnivore?

javaman/Shutterstock.com

FIGURE 3.7 The vultures and Marabou storks, eating the carcass of an animal that was killed by another animal, are detritivores.

Decomposers are consumers that get nourishment by releasing nutrients from the wastes or remains of plants and animals. These nutrients return to the soil, water, and air for reuse by producers (**Concept 3.2A**). Most decomposers are bacteria and fungi. Other consumers, called **detritus feeders**, or **detritivores**, feed on the wastes or dead bodies (detritus) of other organisms. Examples are earthworms, soil insects, hyenas, and vultures (Figure 3.7).

Detritivores and decomposers can transform a fallen tree trunk into simple inorganic molecules that plants can absorb as nutrients (Figure 3.8). In natural ecosystems, the wastes and dead bodies of organisms are resources for other organisms in keeping with the chemical cycling **principle of sustainability**. Without decomposers and detritivores, many of which are microscopic organisms (Science Focus 3.1), the planet's land surfaces would be buried in plant and animal wastes, dead animal bodies, and garbage.

Producers, consumers, and decomposers use the chemical energy stored in glucose and other organic compounds to fuel their life processes. In most cells, this energy is released by **aerobic respiration**, which uses oxygen to convert glucose (or other organic nutrient molecules) back into carbon dioxide and water. The overall chemical reaction for the aerobic respiration is shown in the following equation:

glucose + oxygen → carbon dioxide + water + **energy**

$$C_6H_{12}O_6 + 6\ O_2 \rightarrow 6\ CO_2 + 6\ H_2O + \textbf{energy}$$

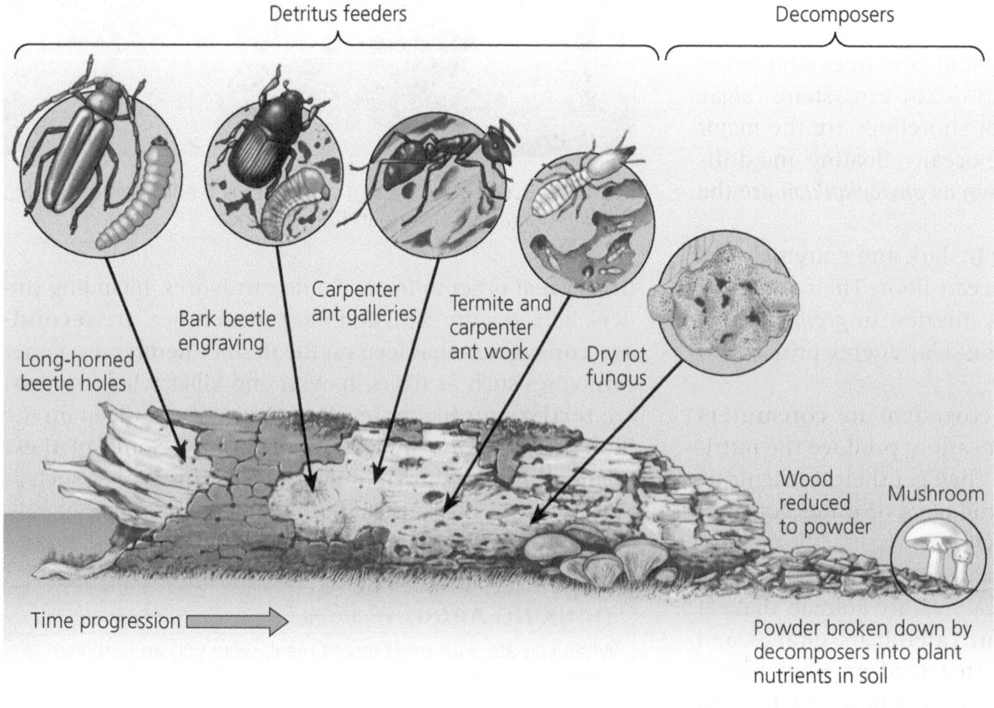

© Cengage Learning

FIGURE 3.8 Various detritivores and decomposers (mostly fungi and bacteria) can "feed on" or digest parts of a log and eventually convert its complex organic chemicals into simpler inorganic nutrients that can be taken up by producers.

Many of the World's Most Important Organisms Are Invisible to Us

They are everywhere. Trillions can be found inside your body, on your skin, in a handful of soil, and in a cup of ocean water. They are *microbes*, or *microorganisms*, catchall terms for many thousands of species of bacteria, protozoa, fungi, and floating phytoplankton. Though most of them are too small to be seen with the naked eye, they are the biological rulers of the earth and play key roles in the earth's life-support system and in our bodies.

Microbes do not get the respect they deserve. Most of us view them primarily as threats to our health in the form of infectious bacteria or fungi that cause athlete's foot and other skin diseases, and protozoa that cause diseases such as malaria. But these harmful microbes are in the minority.

Scientist Philip Tarr and other researchers have identified more than 10,000 species of bacteria, fungi, and other microbes that live in or on our bodies. Many of them provide us with vital services. Bacteria in our intestinal tracts help break down the food we eat, and microbes in our noses help prevent harmful bacteria from reaching our lungs. According to recent research, the greater the diversity of the bacterial zoo in our stomachs, the better our health is likely to be. We are alive largely because of these multitudes of microbes toiling away completely out of sight.

Bacteria and other microbes help purify the water we drink by breaking down plant and animal wastes in the water. Bacteria and fungi (such as yeast) also help to produce foods such as bread, cheese, yogurt, soy sauce, beer, and wine. Bacteria and fungi in the soil decompose organic wastes into nutrients that can be taken up by plants that are then eaten by humans and other plant eaters. Without these tiny creatures, we would go hungry and be up to our necks in waste matter.

Some microorganisms, particularly phytoplankton in the ocean, provide much of the planet's oxygen. They also help regulate the atmosphere's average temperature by removing some of the carbon dioxide produced when we burn coal, natural gas, and gasoline from the atmosphere. Scientists are working on using microbes to develop new medicines and fuels. Genetic engineers are inserting genetic material into existing microorganisms to convert them to microbes that can be used to clean up polluted water and soils.

Some microorganisms assist us in controlling plant diseases and populations of insect species that attack our food crops. By relying more on these microbes for pest control, we could reduce the use of potentially harmful chemical pesticides. In other words, microbes are a vital part of the earth's natural capital.

CRITICAL THINKING

What are two advantages that microbes have over humans for thriving in the world?

To summarize, ecosystems and the biosphere are sustained by the *one-way energy flow* from the sun and the *nutrient cycling* of key materials—in keeping with two of the **scientific principles of sustainability** (Figure 3.9).

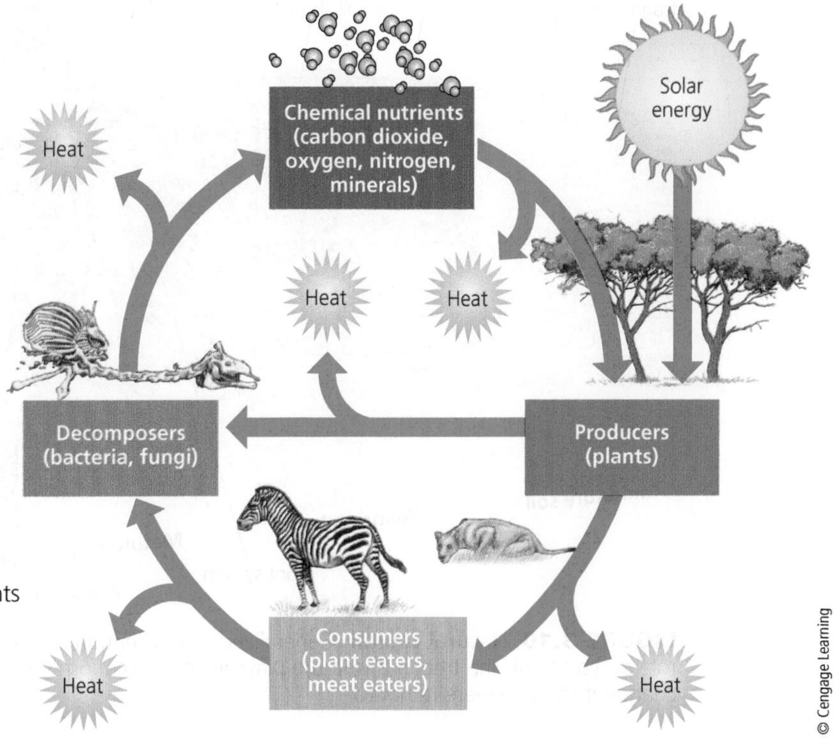

© Cengage Learning

FIGURE 3.9 Natural capital: The main components of an ecosystem are energy, chemicals, and organisms. Nutrient cycling and the flow of energy—first from the sun, then through organisms, and finally into the environment as low-quality heat—link these components.

Soil Is the Foundation of Life on Land

Terrestrial life depends on soil, one of the most important components of the earth's natural capital. The minerals that make up your muscles, bones, and most other parts of your body come almost entirely from soil. Soil also supplies most of the nutrients needed for plant growth and purifies water. Through aerobic respiration, organisms living in soil remove some of the carbon dioxide in the atmosphere and store it as organic carbon compounds, thereby helping to control the earth's climate.

Soil is much more than the dirt that we wash off our hands and clothes. **Soil** is a complex mixture of rock pieces and particles, mineral nutrients, decaying organic matter, water, air, and living organisms that support plant life, which supports animal life (**Concept 3.2B**). Life on land depends on roughly 15 centimeters (6 inches) of topsoil—the earth's living skin.

Soil formation begins when physical, chemical, and biological processes called **weathering** break down bedrock into small pieces. Various forms of plant and animal life begin living on the weathered particles. Their wastes and decaying bodies add organic matter and minerals to the slowly forming soil. Decomposers and detritivores break down fallen leaves and wood (Figure 3.8) and add organic matter and nutrients to the soil. Air (mostly nitrogen and oxygen) and water occupy pores or spaces between soil particles. Over hundreds to thousands of years, various types of life build up distinct layers of mineral and organic matter on a soil's original bedrock.

Most *mature soils* contain several horizontal layers or *horizons*. A cross-sectional view of the horizons of a soil is called a **soil profile** (Figure 3.10, right). The major horizons in a mature soil are **O** (leaf litter), **A** (topsoil), **B** (subsoil), and **C** (weathered parent material), which build up over the parent material. Each layer has a distinct texture, composition, and thickness that vary with the soils formed in different climates and biomes such as deserts, grasslands, and forests (Figure 3.11). Soil forms faster in wet, warm climates.

The roots of most plants and the majority of a soil's organic matter are found in the soil's two upper layers: the O-horizon of leaf litter and the A-horizon of topsoil. In a fertile soil, these two layers teem with bacteria, fungi, earthworms, and numerous small insects, all interacting by feeding on and decomposing one another. The leaf litter and topsoil layers are also habitats for larger animals such as snails, reptiles, amphibians, and burrowing animals such as moles.

Every handful of topsoil contains billions of bacteria and other decomposer organisms. They break down some of the soil's complex organic compounds into a mixture of the partially decomposed bodies dead plants and animals, called *humus*. A fertile soil that produces high crop yields

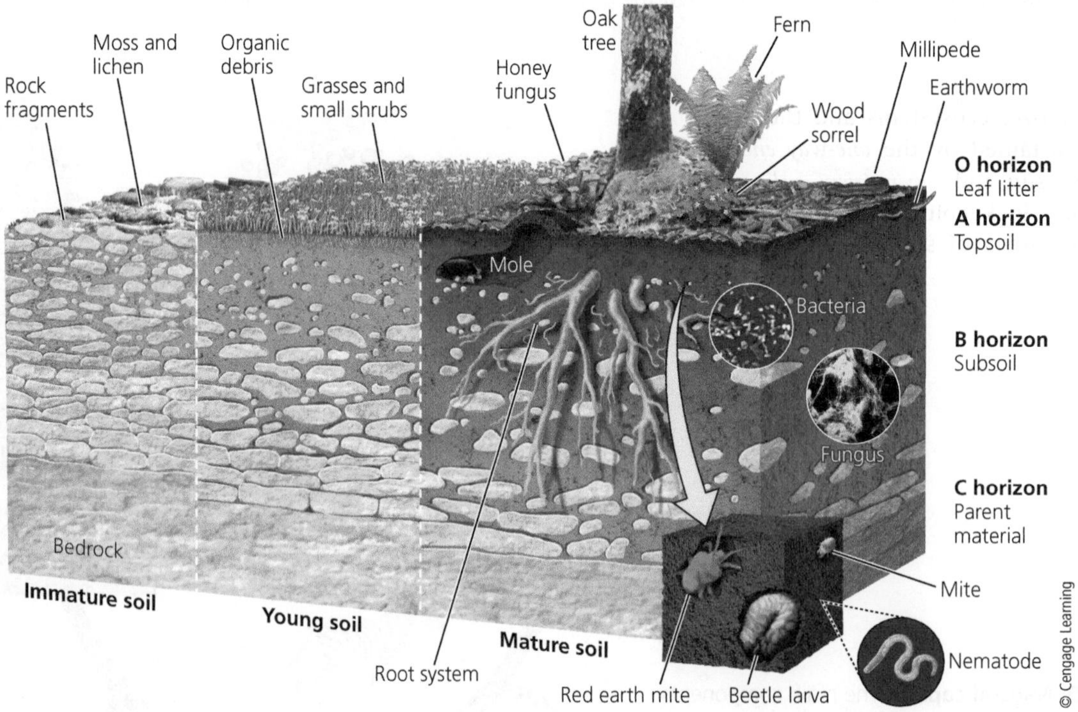

FIGURE 3.10 Natural capital: Generalized soil formation and soil profile. ***Critical thinking:*** What role do you think the tree in this figure plays in soil formation? How might the soil formation process change if the tree were removed?

FIGURE 3.11 Natural capital: Soil profiles of the principal soil types typically found in five types of terrestrial ecosystems.

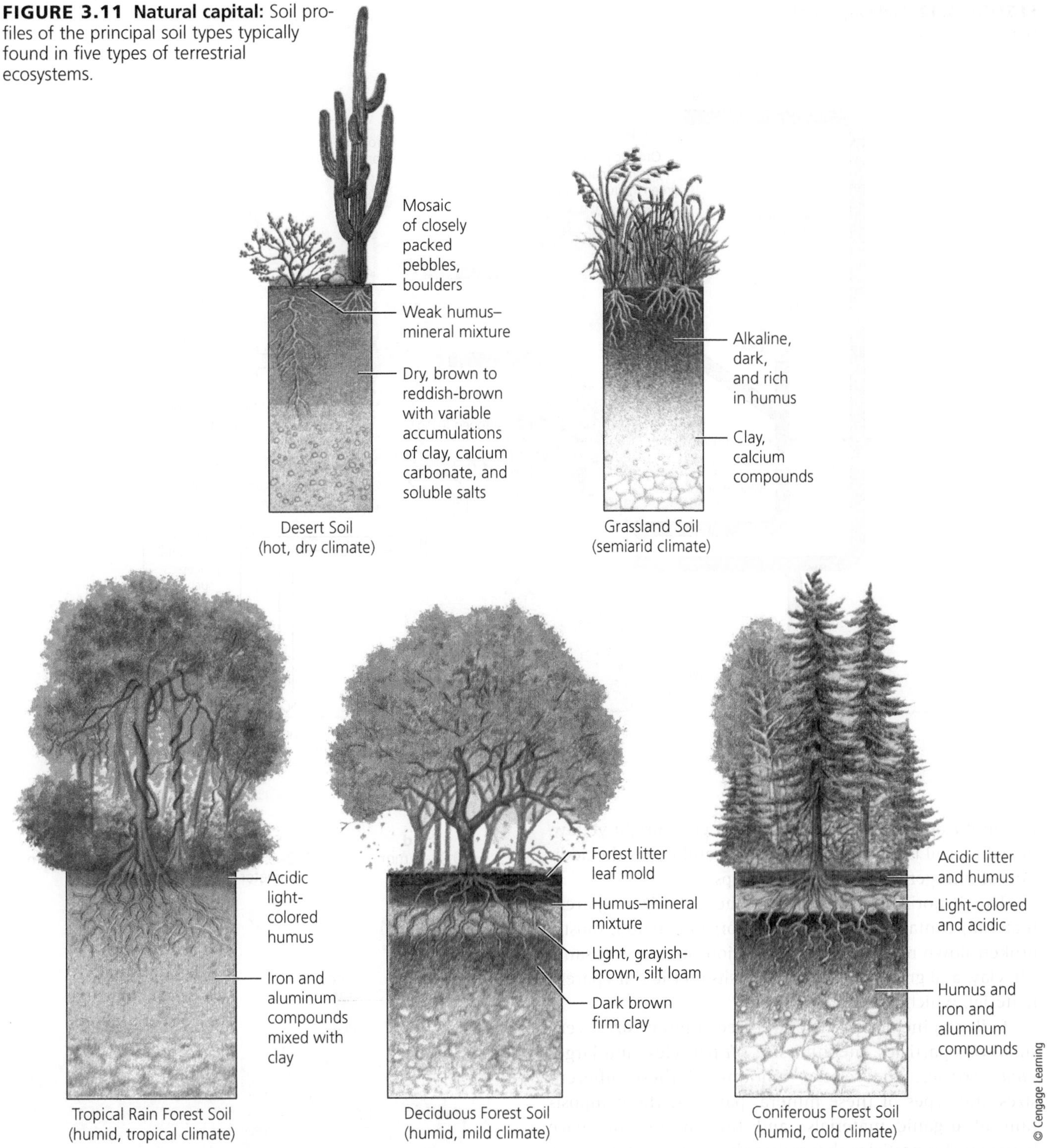

has a thick topsoil layer with a lot of humus mixed with minerals from weathered rock.

Moisture in topsoil dissolves nutrients needed for plant growth. The resulting solution is drawn up by the roots of the plants and transported through their stems into the leaves of plants. This movement of nutrients from the topsoil to plant leaves and to insects and other animals that eat the leaves is part of the chemical cycling process essential for the earth's life. Healthy soils retain more water and help reduce the severity of drought by releasing some of the water into the atmosphere. Figure 3.12 shows the movement of plant nutrients in soils.

The color of topsoil indicates how useful it is for growing crops or other plants. Black or dark brown topsoil is

FIGURE 3.12 Pathways of plant nutrients in soils.

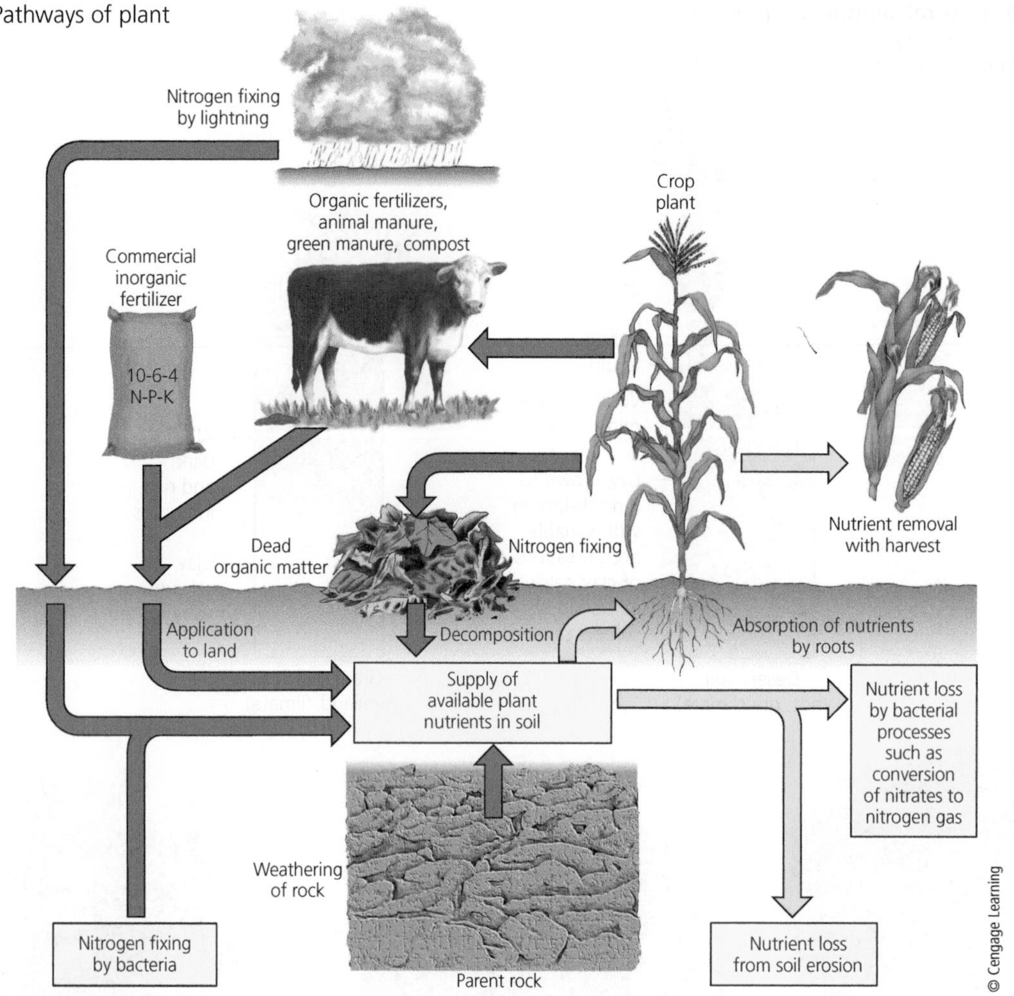

rich in nitrogen and organic matter. A gray, bright yellow, or red topsoil is low in organic matter and needs the addition of nitrogen to support most crops.

The B-horizon (subsoil) and the C-horizon (parent material) contain most of a soil's inorganic matter, mostly broken-down rock, consisting of various mixtures of sand, silt, clay, and gravel. The C-horizon sits on the soil's parent material, which is often bedrock.

Soils can include particles of three different sizes: very small *clay* particles, medium-size *silt* particles, and larger *sand* particles. The relative amounts of these different sizes and types of these mineral particles, the composition of organic materials, and the amount of space between the particles determine the texture of a soil. A soil's texture affects how rapidly water flows through it (Figure 3.13).

To get an idea of a soil's texture, take a small amount of topsoil, moisten it, and rub it between your fingers and thumb. A gritty feel means it contains a lot of sand, which allows water to flow through it rapidly. A sticky feel means high clay content that you can roll into a clump. Very little water penetrates this type of soil. Loam topsoil has a crumbly and spongy texture with many of its particles

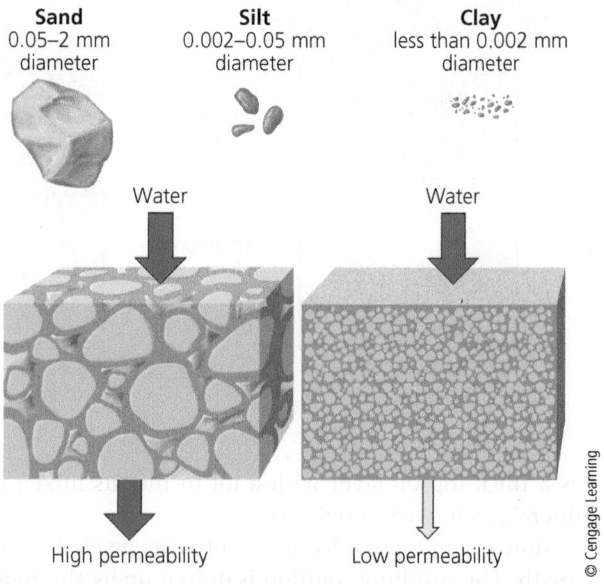

FIGURE 3.13 Natural capital: The size, shape, and degree of clumping of soil particles determine the number and volume of spaces for air and water within a soil. Water can flow more easily through soils with more spaces (left) than through soils with fewer spaces (right).

clumped loosely together. It is best suited for plant growth because it allows water to follow through it at a nourishing trickle.

Soil is a renewable resource but it is renewed very slowly and becomes a nonrenewable resource if we deplete it faster than nature can replenish it. The formation of just 2.5 centimeters (1 inch) of topsoil can take hundreds to thousands of years. Removing plant cover from soil exposes its topsoil to erosion by water and wind. This explains why protecting and renewing topsoil is a key to sustainability. You will learn more about soil erosion and soil conservation in Chapter 12.

3.3 WHAT HAPPENS TO ENERGY IN AN ECOSYSTEM?

CONCEPT 3.3 As energy flows through ecosystems in food chains and food webs, there is a decrease in the high-quality chemical energy available to organisms at each successive feeding level.

Energy Flows through Ecosystems in Food Chains and Food Webs

Chemical energy, stored as nutrients in the bodies and wastes of organisms, flows through ecosystems from one trophic (feeding) level to another. A sequence of organisms with each one serving as a source of nutrients or energy for the next level of organisms is called a **food chain** (Figure 3.14).

Every use and transfer of energy by organisms from one feeding level to another involves a loss of some high-quality energy to the environment as low-quality energy in the form of heat, in accordance with the second law of thermodynamics. A graphic display of the energy loss at each trophic level is called a **pyramid of energy flow**. Figure 3.15 illustrates this energy loss for a food chain, assuming a 90% energy loss for each level of the chain.

The large loss in chemical energy between successive trophic levels explains why food chains and webs rarely have more than four or five trophic levels.

CONSIDER THIS . . .

CONNECTIONS Energy Flow and Feeding People

Energy flow pyramids explain why the earth could support more people if they all ate at a low trophic level by consuming grains, vegetables, and fruits directly rather than passing such crops through another trophic level and eating herbivores such as cattle, pigs, sheep, and chickens. About two-thirds of the world's people survive primarily by eating wheat, rice, and corn at the first trophic level, mostly because they cannot afford to eat much meat.

Ecologists can estimate the *number of organisms* feeding at each trophic level. Here is a hypothetical example: 100,000 blades of grass (producer) might support 30 rabbits (herbivore), which might support 1 fox (carnivore). Ecologists also measure **biomass**—the total mass of organisms in each trophic level—as illustrated by this hypothetical example: 1,000 kilograms (2,200 pounds) of producers might provide 100 kilograms (220 pounds) of food for herbivores, which might provide 10 kilograms (22 pounds) of food for

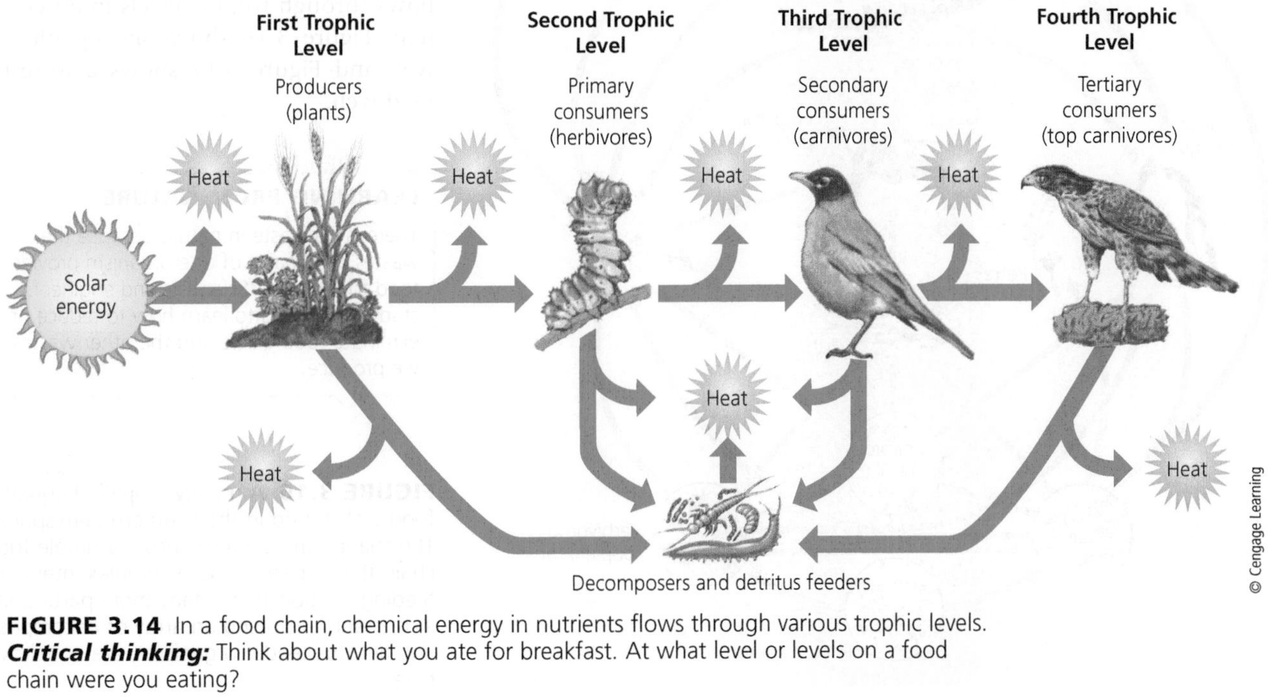

FIGURE 3.14 In a food chain, chemical energy in nutrients flows through various trophic levels. ***Critical thinking:*** Think about what you ate for breakfast. At what level or levels on a food chain were you eating?

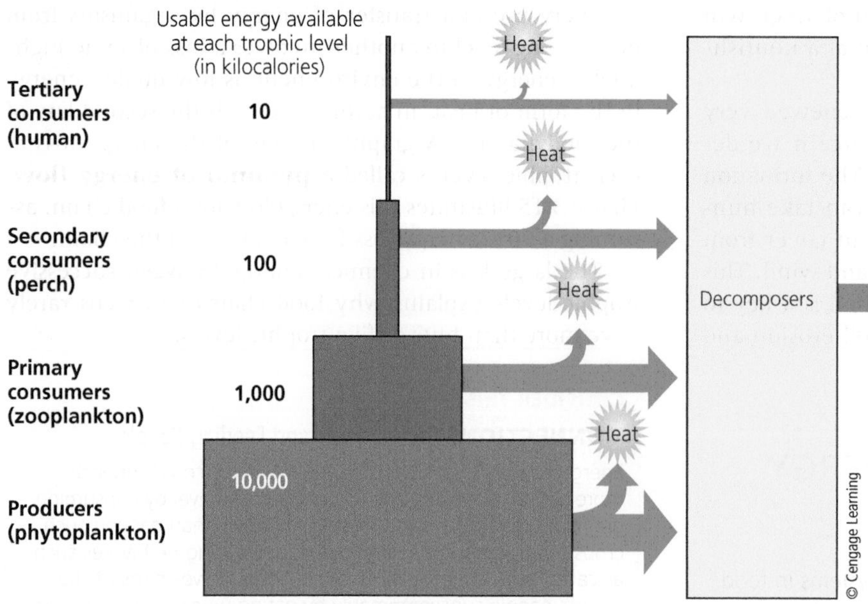

FIGURE 3.15 Generalized *pyramid of energy flow* showing the decrease in usable chemical energy available at each succeeding trophic level in a food chain or food web. This model assumes that with each transfer from one trophic level to another, there is a 90% loss of usable energy to the environment in the form of low-quality heat. (Calories and joules are used to measure energy. 1 kilocalorie = 1,000 calories = 4,184 joules.) ***Critical thinking:*** Why is a vegetarian diet more energy efficient than a meat-based diet?

carnivores, which might supply a top carnivore with 1 kilogram (2.2 pounds) of food.

In natural ecosystems, most consumers feed on more than one type of organism, and most organisms are eaten or decomposed by more than one type of consumer. Because of this, organisms in most ecosystems form a complex network of interconnected food chains called a **food web**. Food chains and food webs show how producers, consumers, and decomposers are connected to one another as energy flows through trophic levels in an ecosystem. Figure 3.16 shows an aquatic food web and Figure 3.17 shows a terrestrial food web.

CONSIDER THIS . . .

LEARNING FROM NATURE

There is no waste in nature because the wastes or remains of one organism provide food for another. Scientists and engineers study food webs to learn how to reduce or eliminate food waste and the other wastes we produce.

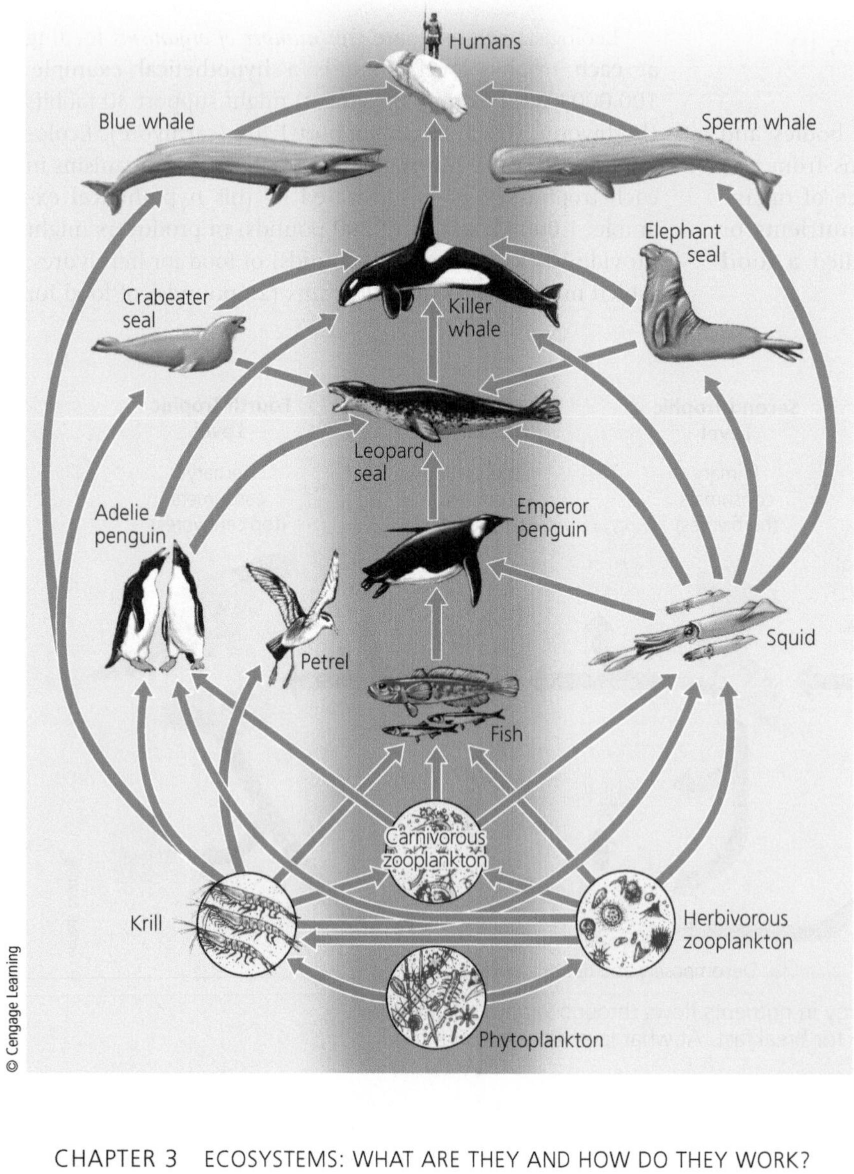

FIGURE 3.16 A greatly simplified aquatic food web found in the southern hemisphere. The shaded middle area shows a simple food chain that is part of these complex interacting feeding relationships. Many more participants in the web, including an array of decomposer and detritus feeder organisms, are not shown here.

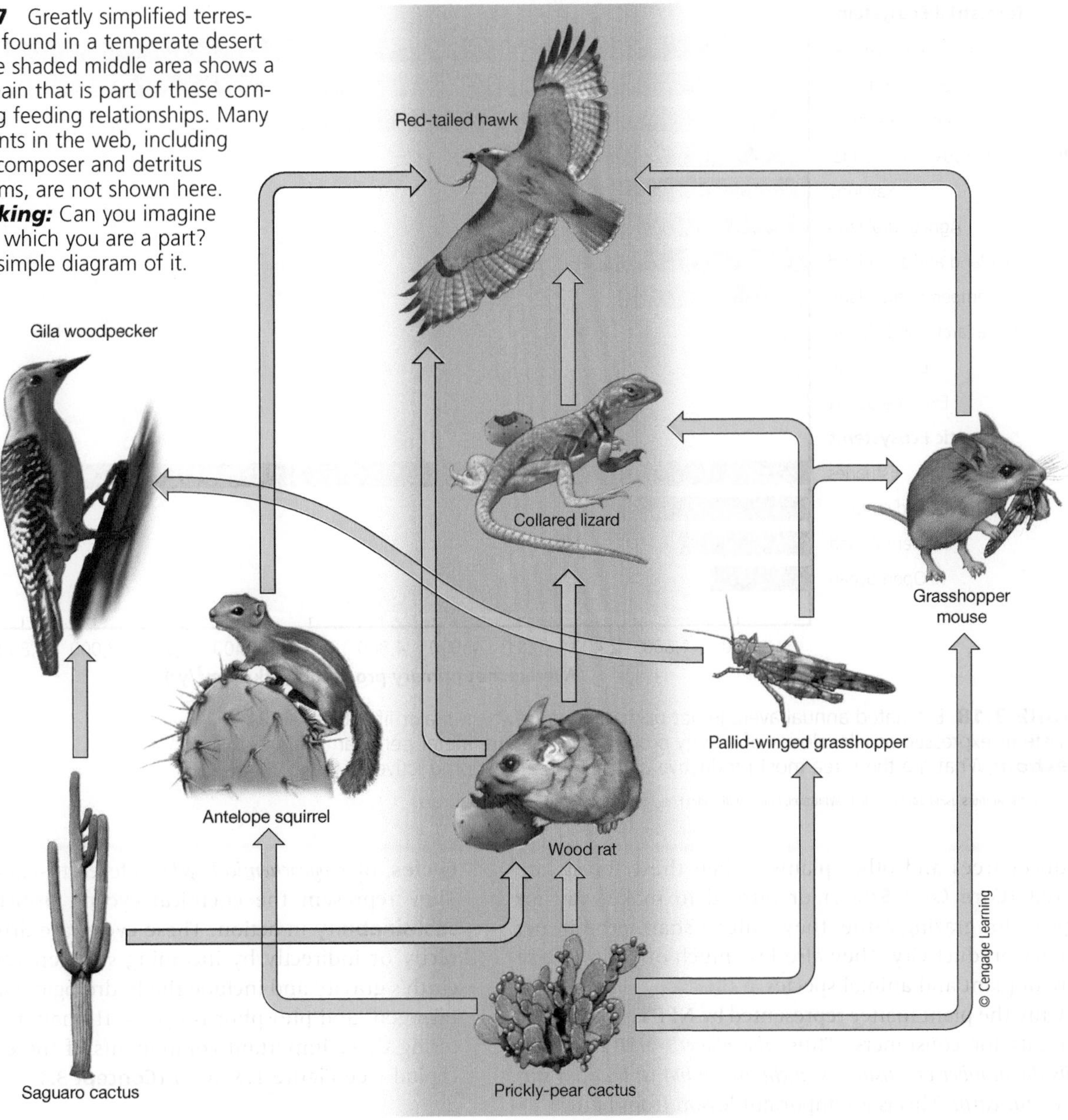

FIGURE 3.17 Greatly simplified terrestrial food web found in a temperate desert ecosystem. The shaded middle area shows a simple food chain that is part of these complex interacting feeding relationships. Many more participants in the web, including an array of decomposer and detritus feeder organisms, are not shown here. ***Critical thinking:*** Can you imagine a food web of which you are a part? Try drawing a simple diagram of it.

Some Ecosystems Produce Plant Matter Faster Than Others Do

Gross primary productivity (GPP) is the *rate* at which an ecosystem's producers (such as plants and phytoplankton) convert solar energy into chemical energy stored in compounds found in their tissues. It is usually measured in terms of energy production per unit area over a given time span, such as kilocalories per square meter per year (kcal/m^2/yr). The map in Figure 5 on p. S20 of Supplement 4 compares gross primary productivity across North America.

To stay alive, grow, and reproduce, producers must use some of their stored chemical energy for their own respiration. **Net primary productivity (NPP)** is the *rate* at which producers use photosynthesis to produce and store chemical energy *minus* the *rate* at which they use some of this stored chemical energy through aerobic respiration. NPP measures how fast producers can make the chemical energy that is stored in their tissues and that is potentially available to other organisms (consumers) in an ecosystem.

Gross primary productivity is similar to the *rate* at which you make money, or the number of dollars you earn per year. *Net primary productivity* is similar to the amount of money earned per year that you can spend after subtracting your expenses such as the costs of transportation, clothes, food, and supplies.

Terrestrial ecosystems and aquatic life zones differ in their NPP as illustrated in Figure 3.18. Despite its low NPP, the open ocean produces more of the earth's biomass per year than any other ecosystem or life zone. This happens because oceans cover 71% of the earth's surface and contain huge numbers of phytoplankton and other producers.

Tropical rain forests have a high net primary productivity because they have a large number and variety of

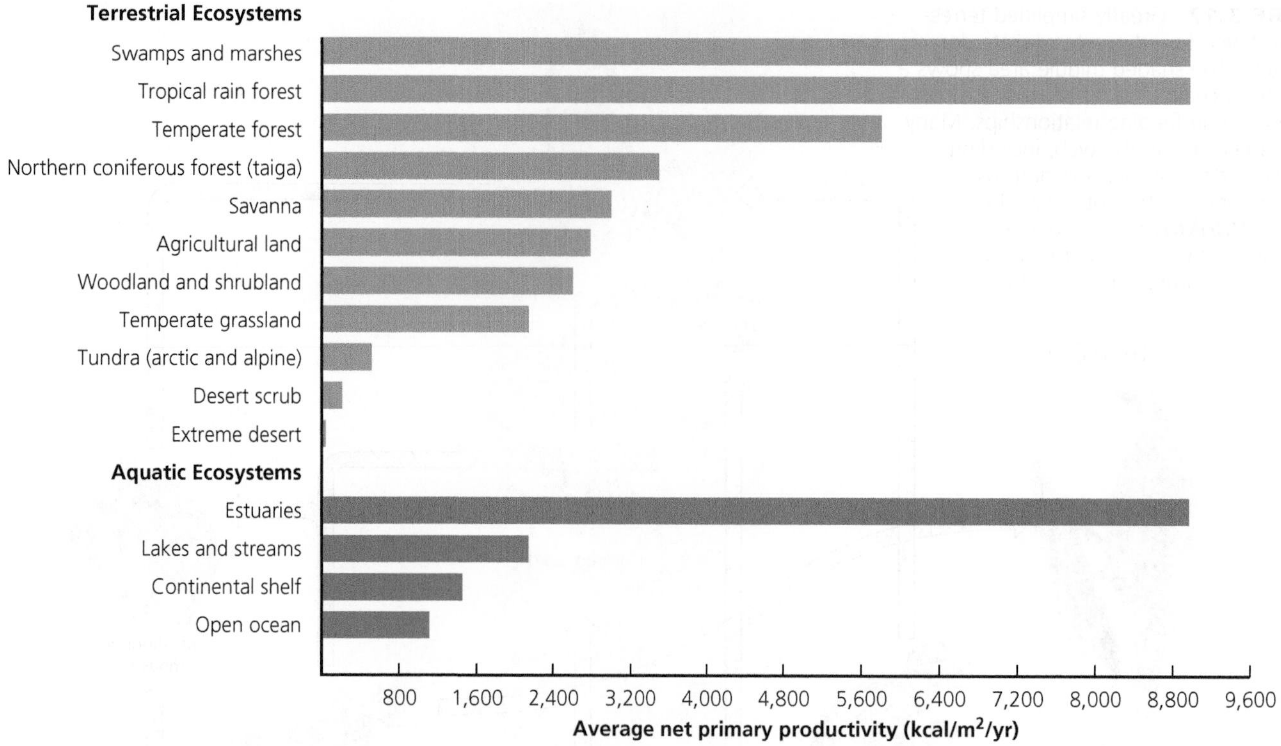

FIGURE 3.18 Estimated annual average *net primary productivity* in major life zones and ecosystems expressed as kilocalories of energy produced per square meter per year (kcal/m^2/yr). ***Question:*** What are the three most productive and the three least productive systems?

(Compiled by the authors using data from R. H. Whittaker, Communities and Ecosystems, 2nd ed., New York: Macmillan, 1975.)

producer trees and other plants. When these forests are cleared (**Core Case Study**) or burned to make way for crops or for grazing cattle, they suffer a sharp drop in net primary productivity. They also lose much of their diverse array of plant and animal species.

Only the plant matter represented by NPP is available as nutrients for consumers. Thus, *the planet's NPP ultimately limits the number of consumers (including humans) that can survive on the earth.* This is an important lesson from nature.

3.4 WHAT HAPPENS TO MATTER IN AN ECOSYSTEM?

CONCEPT 3.4 Matter, in the form of nutrients, cycles within and among ecosystems and the biosphere, and human activities are altering these chemical cycles.

Nutrients Cycle Within and Among Ecosystems

The elements and compounds that make up nutrients move continually through air, water, soil, rock, and living organisms within ecosystems, in cycles called **nutrient cycles**, or *biogeochemical cycles* (life-earth-chemical cycles). They represent the chemical cycling **principle of sustainability** in action. These cycles are driven directly or indirectly by incoming solar energy and by the earth's gravity and include the hydrologic (water), carbon, nitrogen, and phosphorus cycles. Human activities are altering these important components of the earth's natural capital (see Figure 1.3, p. 7) (**Concept 3.4**).

CONSIDER THIS . . .

CONNECTIONS Nutrient Cycles and Life

Nutrient cycles connect past, present, and future forms of life. Some of the carbon atoms in your skin may once have been part of an oak leaf, a dinosaur's skin, or a layer of limestone rock. Your grandmother, George Washington, or a hunter–gatherer who lived 25,000 years ago may have inhaled some of the nitrogen (N_2) molecules you just inhaled.

The Water Cycle Sustains All Life

Water (H_2O) is an amazing substance (Science Focus 3.2) that is necessary for life on the earth. The **hydrologic cycle**, also called the **water cycle**, collects, purifies, and distributes the earth's fixed supply of water, as shown in Figure 3.19.

The sun powers the water cycle. Incoming solar energy causes *evaporation*—the conversion of some of the liquid

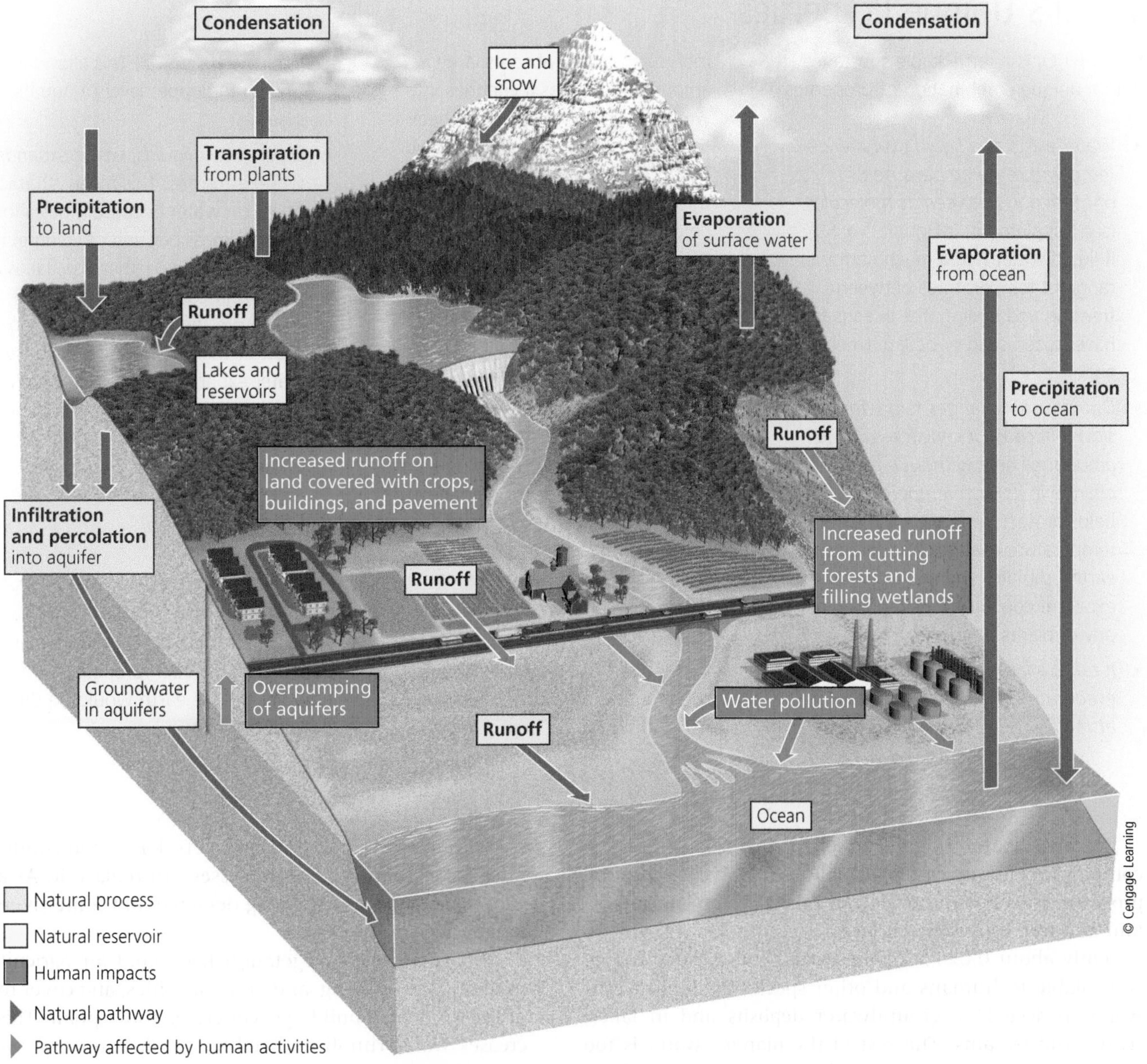

FIGURE 3.19 Natural capital: Simplified model of the *water cycle*, or *hydrologic cycle*, in which water circulates in various physical forms within the biosphere. The red arrows and boxes identify major effects of human activities on this cycle. ***Critical thinking:*** What are three ways in which your lifestyle directly or indirectly affects the hydrologic cycle?

water in the earth's oceans, lakes, rivers, soil, and plants to vapor. This water vapor rises into the atmosphere, where it condenses into droplets. Gravity then draws the water back to the earth's surface as *precipitation* (rain, snow, sleet, and dew).

Most precipitation falling on terrestrial ecosystems becomes **surface runoff**. This water flows into streams, rivers, lakes, wetlands, and oceans, from which it can evaporate to repeat the cycle. Some precipitation seeps into the upper layers of soils and is used by plants, and some evaporates from the soils back into the atmosphere. Some precipitation also sinks through soil into underground layers of rock, sand, and gravel called **aquifers**. This water stored underground is called **groundwater**. Some precipitation is converted to ice that is stored in *glaciers*.

Water's Unique Properties

Water (H_2O) is a remarkable substance with a unique combination of properties:

- *Water exists as a liquid over a wide temperature range because of forces of attraction between its molecules* (see Supplement 3, Figure 3, p. S9). If liquid water had a much narrower range of temperatures between freezing and boiling, the oceans might have frozen solid or boiled away long ago.
- *Liquid water changes temperature slowly because it can store a large amount of heat without a large change in its own temperature.* This helps protect living organisms from temperature changes, moderates the earth's climate, and makes water an excellent coolant for car engines and power plants.
- *It takes a large amount of energy to evaporate water because of the attractive forces between its molecules.* Water absorbs large amounts of heat as it changes into water vapor and releases this heat as the vapor condenses back to liquid water. This helps to distribute heat throughout the world and to determine regional and local climates. It also makes evaporation a cooling process—explaining why you feel cooler when perspiration evaporates from your skin.
- *Liquid water can dissolve a variety of compounds.* It carries dissolved nutrients into the tissues of living organisms, flushes waste products out of those tissues, serves as an all-purpose cleanser, and helps to remove and dilute the water-soluble wastes of civilization. This property also means that water-soluble wastes can easily pollute water.
- *Water filters out wavelengths of the sun's ultraviolet radiation that would harm some aquatic organisms.* (See Figure 2.12, p. 40.) This allows life to exist in the upper layer of aquatic systems.
- *Unlike most liquids, water expands when it freezes.* This means that ice floats on water because it has a lower density (mass per unit of volume) than liquid water has. Otherwise, lakes and streams in cold climates would freeze solid from the bottom up and loose most of their aquatic life. Because water expands on freezing, it can break pipes, crack a car's engine block (if it does not contain antifreeze), break up pavement, and fracture rocks (which helps form soil—see Figure 3.10, p. 56).

CRITICAL THINKING

Pick two of the special properties listed above and, for each property, explain how life on the earth would be different if it did not exist.

Because water is good at dissolving many different compounds, it can easily be polluted. However, natural processes in the water cycle can purify water—an important and free ecosystem service.

Only about 0.024% of the earth's huge water supply is available to humans and other species as liquid freshwater in accessible groundwater deposits and in lakes, rivers, and streams. The rest of the planet's water is too salty, is too deep underground to extract at affordable prices, or is stored as ice.

Humans alter the water cycle in three major ways (see the red arrows and boxes in Figure 3.19). *First,* sometimes we withdraw freshwater from rivers, lakes, and aquifers at rates faster than natural processes can replace it. As a result, some aquifers are being depleted and some rivers no longer flow to the ocean.

0.024%
Percentage of the earth's freshwater supply that is available to humans and other species

Second, we clear vegetation from land for agriculture, mining, road building, and other activities, and cover much of the land with buildings, concrete, and asphalt. This increases water runoff and reduces infiltration that would normally recharge groundwater supplies.

Third, we drain and fill wetlands for farming and urban development. Left undisturbed, wetlands provide the ecosystem service of flood control. They act like sponges to absorb and hold overflows of water from drenching rains or rapidly melting snow.

Carbon Cycles among Living and Nonliving Things

Carbon is the basic building block of the carbohydrates, fats, proteins, DNA, and other organic compounds required for life. Various compounds of carbon circulate

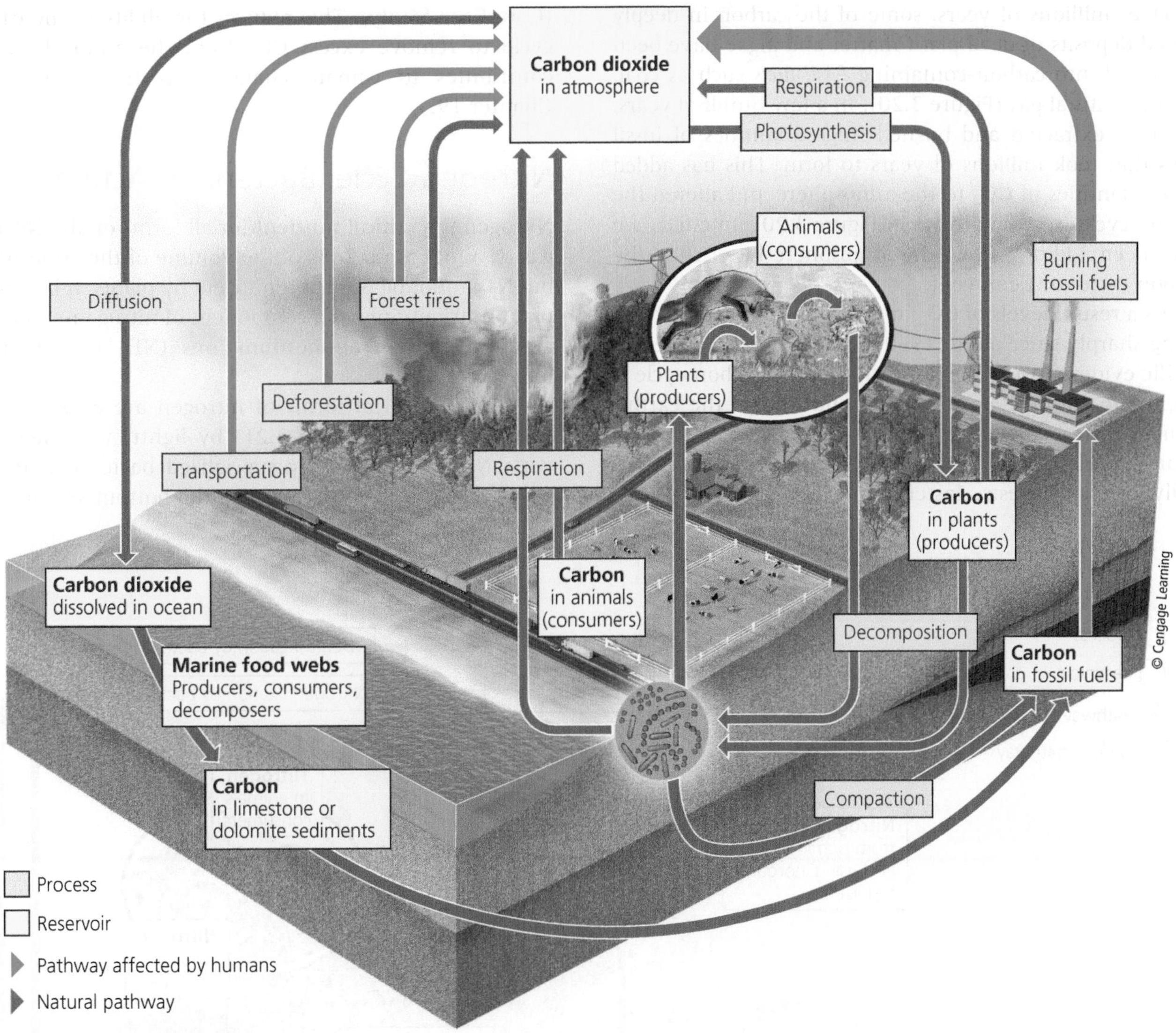

FIGURE 3.20 Natural capital: Simplified model showing the circulation of various chemical forms of carbon in the global *carbon cycle*. Red arrows show major harmful impacts of human activities shown by the red arrows. (Yellow box sizes do not show relative reservoir sizes.) ***Critical thinking:*** What are three ways in which you directly or indirectly affect the carbon cycle?

through the biosphere, the atmosphere, and parts of the hydrosphere, in the **carbon cycle** shown in Figure 3.20.

A key component of the carbon cycle is carbon dioxide (CO_2) gas. It makes up about 0.040% of the volume of the earth's troposphere. Carbon dioxide (along with water vapor in the water cycle) affects the temperature of the atmosphere through the greenhouse effect (Figure 3.3) and thus plays a major role in determining the earth's climate. If the carbon cycle removes too much CO_2 from the atmosphere, the atmosphere will cool, and if it generates too much CO_2, the atmosphere will get warmer. Thus, even slight changes in this cycle caused by natural or human factors can affect the earth's climate, which helps determine the types of life that can exist in various places, as you will learn in Chapter 7.

Carbon is cycled through the biosphere by a combination of *photosynthesis* by producers that removes CO_2 from the air and water, and *aerobic respiration* by producers, consumers, and decomposers that adds CO_2 in the atmosphere. Typically, CO_2 remains in the atmosphere for 100 years or more. Some of the CO_2 in the atmosphere dissolves in the ocean. In the ocean, decomposers release carbon that is stored as insoluble carbonate minerals and rocks in bottom sediment for long periods. Marine sediments are the earth's largest store of carbon, mostly as carbonates.

Over millions of years, some of the carbon in deeply buried deposits of dead plant matter and algae have been converted into carbon-containing *fossil fuels* such as coal, oil, and natural gas (Figure 3.20). In a few hundred years, we have extracted and burned huge quantities of fossil fuels that took millions of years to form. This has added large quantities of CO_2 to the atmosphere and altered the carbon cycle (see red arrows in Figure 3.20). In effect, we have been adding CO_2 to the atmosphere faster than the carbon cycle can remove it.

As a result, levels of CO_2 in the atmosphere have been rising sharply since about 1960. There is considerable scientific evidence that this disruption of the carbon cycle is helping to warm the atmosphere and change the earth's climate. Another way in which we alter the cycle is by clearing carbon-absorbing vegetation from forests, especially tropical forests, faster than they can grow back (**Core Case Study**). This reduces the ability of the carbon cycle to remove excess CO_2 from the atmosphere and contributes to climate change, which we discuss in Chapter 19.

Nitrogen Cycle: Bacteria in Action

Nitrogen is a critical nutrient for all forms of life. Nitrogen gas (N_2) makes up 78% of the volume of the atmosphere, but N_2 cannot be used as a nutrient by plants. It becomes a plant nutrient only as a component of nitrogen-containing ammonia (NH_3), ammonium ions (NH_4^+), and nitrate ions (NO_3^-).

These chemical forms of nitrogen are created in the **nitrogen cycle** (Figure 3.21) by lightning, which converts N_2 to NH_3, and by specialized bacteria in topsoil. Other bacteria in topsoil and in the bottom sediments of

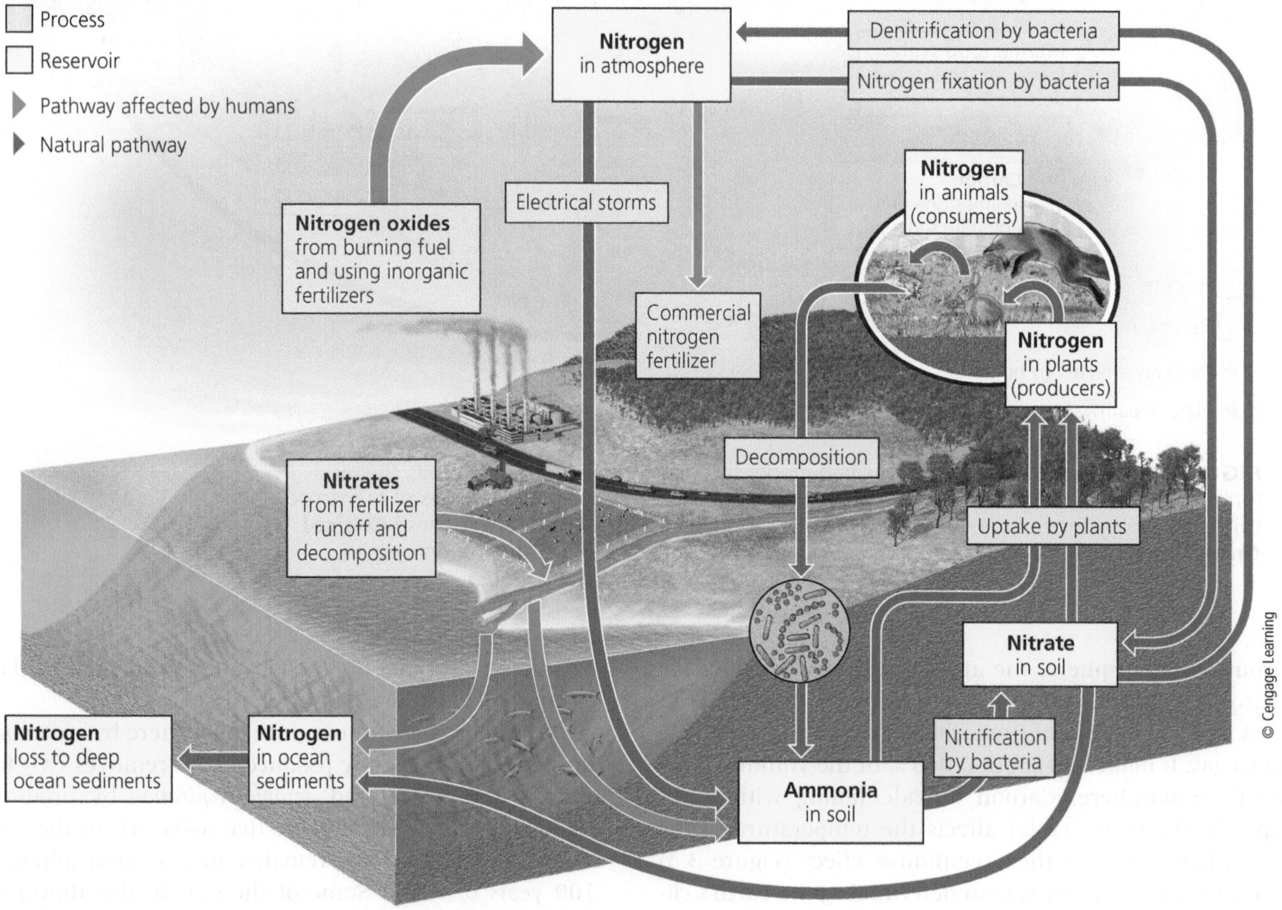

FIGURE 3.21 Natural capital: Simplified model showing the circulation of various chemical forms of nitrogen in the *nitrogen cycle*, with major harmful human impacts shown by the red arrows. (Yellow box sizes do not show relative reservoir sizes.) ***Critical thinking:*** What are two ways in which the carbon cycle and the nitrogen cycle are linked?

aquatic systems convert NH_3 to NH_4^+ and nitrate ions (NO_3^-) that are taken up by the roots of plants. The plants then use these forms of nitrogen to produce various proteins, nucleic acids, and vitamins. Animals that eat plants consume these nitrogen-containing compounds, as do detritus feeders and decomposers. Bacteria in waterlogged soil and bottom sediments of lakes, oceans, swamps, and bogs convert nitrogen compounds into nitrogen gas (N_2). The gas is released to the atmosphere to begin the nitrogen cycle again.

Humans intervene in the nitrogen cycle in several ways (see red arrows in Figure 3.21). When we burn gasoline and other fuels, the resulting high temperatures convert some of the N_2 and O_2 in air to nitric oxide (NO). In the atmosphere, NO can be converted to nitrogen dioxide gas (NO_2) and nitric acid vapor (HNO_3), which can return to the earth's surface as damaging *acid deposition*, commonly called *acid rain*. Acid rain damages buildings, statues, and forests.

We remove large amounts of N_2 from the atmosphere and combine it with H_2 to make ammonia ($N_2 + 3H_2 \rightarrow 2NH_3$) and ammonium ions ($NH_4^+$) used in fertilizers. We add nitrous oxide (N_2O) to the atmosphere through the action of anaerobic bacteria on nitrogen-containing fertilizer or organic animal manure applied to the soil. This greenhouse gas can warm the atmosphere and take part in reactions that deplete stratospheric ozone, which keeps most of the sun's harmful ultraviolet radiation from reaching the earth's surface.

We alter the nitrogen cycle in aquatic ecosystems by adding excess nitrates (NO_3^-) to these systems. The nitrates contaminate bodies of water through agricultural runoff of fertilizers, animal manure, and discharges from municipal sewage treatment systems. This plant nutrient can cause excessive growth of algae that disrupt aquatic systems.

Our nitrogen inputs into the environment have risen sharply and are projected to continue rising (Figure 3.22). According to a 2015 study by a team of scientists, we are disrupting the nitrogen cycle because these human inputs have exceeded the planet's environmental limit for nitrogen.

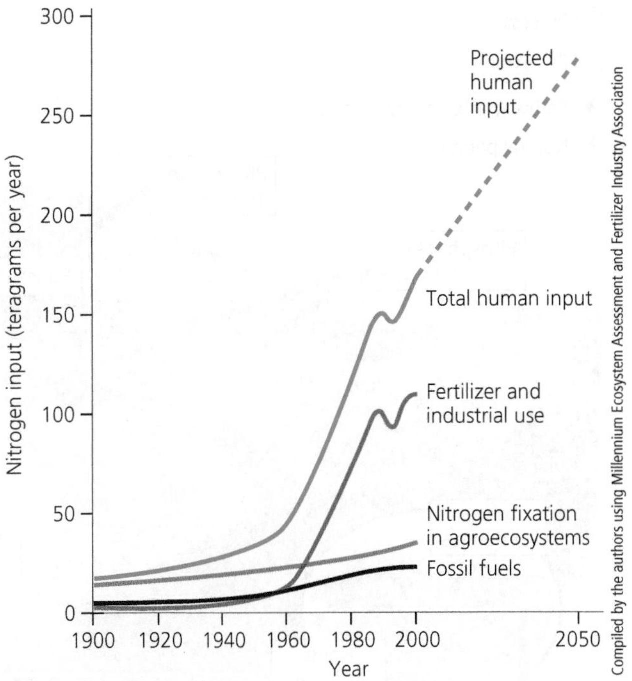

FIGURE 3.22 Global trends in the inputs of nitrogen into the atmosphere from human activities, with projections to 2050. ***Data analysis:*** By what percentage did the overall nitrogen input increase between 1960 and 2000? By what percentage is it projected to increase between 2000 and 2050?

Phosphorus Cycles through Water, Rock, and Food Webs

Phosphorus is another nutrient that supports life. The cyclic movement of phosphorus (P) through water, the earth's crust, and living organisms is called the **phosphorus cycle** (Figure 3.23). The major reservoir for phosphorus in this cycle is phosphate rocks that contain phosphate ions (PO_4^{3-}), which are an important plant nutrient. Phosphorus does not cycle through the atmosphere because few of its compounds exist as a gas. Phosphorus cycles more slowly than water, carbon, and nitrogen.

As water runs over exposed rocks, it slowly erodes away inorganic compounds that contain phosphate ions. Water carries these ions into the soil, where they are absorbed by the roots of plants and by other producers. Phosphate compounds are then transferred by food webs from producers to consumers and eventually to detritus feeders and decomposers.

When phosphate and other phosphorus compounds wash into the ocean, they are deposited as marine sediment and can remain trapped for millions of years. Over time, geological processes can uplift and expose these seafloor deposits, from which phosphate can be eroded and freed up to reenter the phosphorus cycle.

Most soils contain little phosphate, which often limits plant growth on land unless phosphorus (as phosphate salts mined from the earth) is applied to the soil as a fertilizer. Lack of phosphorus also limits the growth of producer populations in many freshwater streams and lakes. This is because phosphate salts are only slightly soluble in water and do not release many phosphate ions to producers in aquatic systems.

Human activities, including the removal of large amounts of phosphate from the earth to make fertilizer,

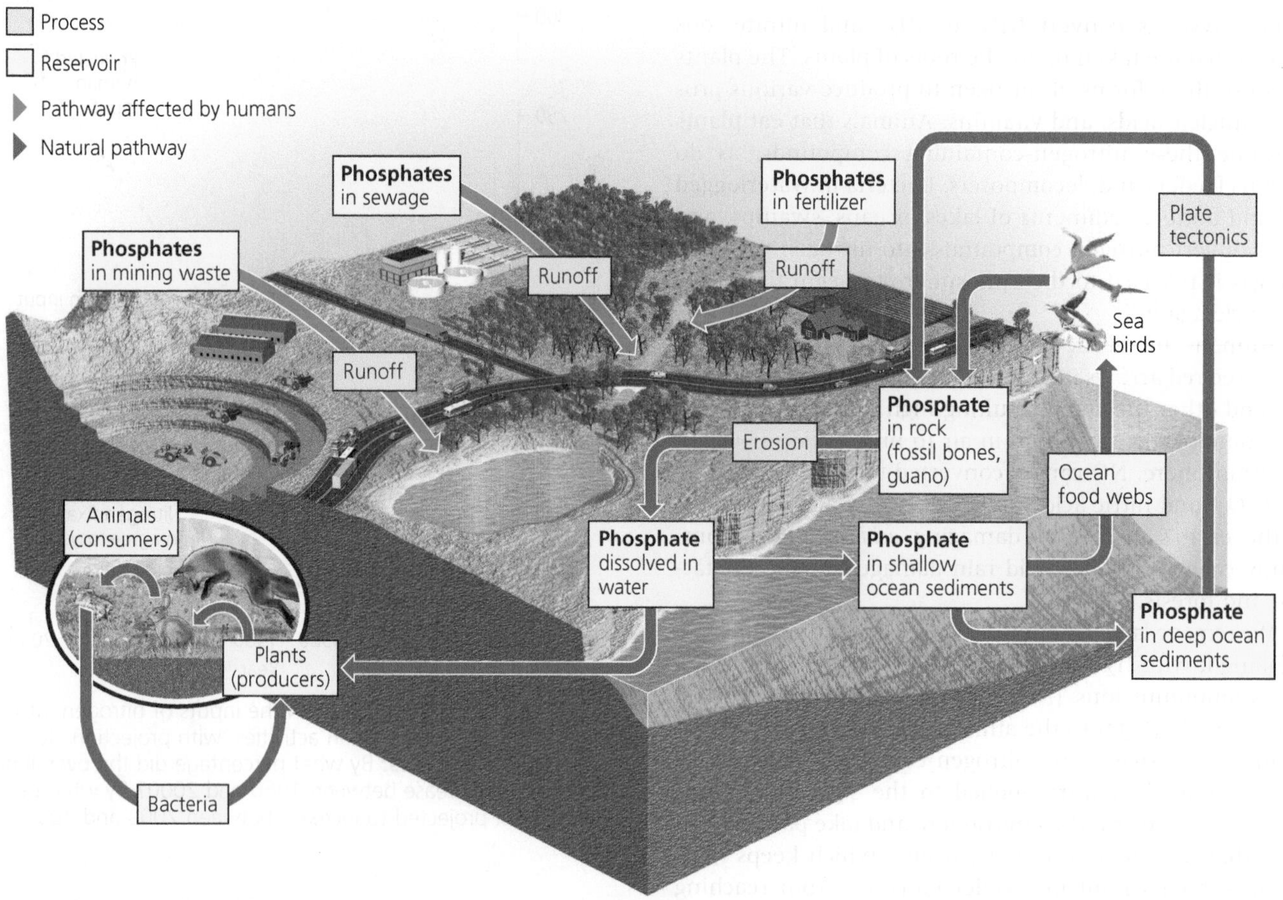

FIGURE 3.23 Natural capital: Simplified model showing the circulation of various chemical forms of phosphorus (mostly phosphates) in the *phosphorus cycle*, with major harmful human impacts shown by the red arrows. (Yellow box sizes do not show relative reservoir sizes.) ***Critical thinking:*** What are two ways in which the phosphorus cycle and the nitrogen cycle are linked? What are two ways in which the phosphorus cycle and the carbon cycle are linked?

disrupt the phosphorus cycle (see red arrows in Figure 3.23). By clearing tropical forests (**Core Case Study**), we reduce phosphate levels in tropical soils. Eroded topsoil and fertilizer washed from fertilized crop fields, lawns, and golf courses carry large quantities of phosphate ions into streams, lakes, and oceans. There they stimulate the growth of producers such as algae and various aquatic plants, which can upset chemical cycling and other processes in bodies of water. According to a number of scientific studies, we are disrupting the phosphorus cycle because our inputs of phosphorus into the environment (primarily for use as fertilizer) have exceeded the planet's environmental limit for phosphorus (Science Focus 3.3).

CONSIDER THIS . . .

LEARNING FROM NATURE

Scientists study the water, carbon, nitrogen, and phosphorus cycles to help us learn how to recycle the wastes we create.

3.5 HOW DO SCIENTISTS STUDY ECOSYSTEMS?

CONCEPT 3.5 Scientists use field research, laboratory research, and mathematical and other models to learn about ecosystems and how much stress they can take.

Some Scientists Study Nature Directly

Scientists use field research, laboratory research, and mathematical and other models to learn about ecosystems (**Concept 3.5**). *Field research,* sometimes called "muddy-boots biology," involves going into forests (see Chapter 2 opening photo, pp. 28–29), oceans (see opening photo this chapter), and other natural settings to study the structure of ecosystems and to learn what happens in them. Most of what we know about ecosystems has come from such research (Individuals Matter 3.1). **GREEN CAREER: Ecologist**

SCIENCE FOCUS 3.3

Planetary Boundaries

For most of the past 10,000–12,000 years, humans have been living in an era called the *Holocene*. During this era, we have enjoyed a favorable climate and other environmental conditions. This general stability allowed the human population to grow, develop agriculture, and take over a large share of the earth's land and other resources (see Figure 1.8, p. 12).

Most geologists contend that we are still living in the Holocene era, but some scientists disagree. According to them, when the Industrial Revolution began (around 1750) we entered an era called the *Anthropocene* (the era of man). In this new era, our ecological footprints have expanded significantly and are changing and stressing the earth's life-support system, especially since 1950.

In 2015 an international team of 18 leading researchers in their fields, led by Will Steffen and Johan Rockstrom of the Stockholm Resilience Centre, published a paper estimating how close we are to exceeding nine major *planetary boundaries*, or *ecological tipping points*, because of human activities. They warn that exceeding them could change how the planet operates and could trigger abrupt and long-lasting or irreversible environmental changes. This could seriously degrade the earth's life-support system and our economies.

The researchers estimated that we have exceeded four of these planetary boundaries. They are **(1)** *disruption of the nitrogen and phosphorus cycles*, mostly from greatly increased use of fertilizers to produce food; **(2)** *biodiversity loss* from replacing biologically diverse forests and grasslands with simplified fields of single crops; **(3)** *land system change* from agriculture and urban development; and **(4)** *climate change* from disrupting the carbon cycle, mostly by overloading it with carbon dioxide produced by the burning of fossil fuels.

There is an urgent need for more research to fill in the missing data on these planetary boundaries. This would help scientists to further evaluate how close we are to exceeding them and how exceeding them could affect humans, other species, and the earth's life-support systems. Regardless of what we call the era we are living in, such information, combined with taking action to live more sustainably, could help us to avoid exceeding such boundaries by shrinking our ecological footprints while expanding our beneficial environmental impacts.

CRITICAL THINKING

Which two of these boundaries do you think are the most important?

Scientists use a variety of methods to study tropical forests (**Core Case Study**). Some build tall construction cranes to reach the canopies. This, along with climbing trees and installing rope walkways between treetops, helps them identify and observe the diversity of species living or feeding in these treetop habitats.

Ecologists carry out controlled experiments by isolating and changing a variable in part of an area and comparing the results with nearby unchanged areas. You learned about a classic example of this in the Core Case Study of Chapter 2 (p. 30).

Scientists also use aircraft and satellites equipped with sophisticated cameras and other *remote sensing* devices to scan and collect data on the earth's surface. They use *geographic information system (GIS)* software to capture, store, analyze, and display this information. For example, GIS software can convert digital satellite images into global, regional, and local maps. These maps show variations in vegetation, gross primary productivity (see Supplement 4, Figure 5, p. S20), soil erosion, deforestation, air pollution emissions, water usage, drought, flooding, pest outbreaks, and other variables.

Some researchers attach tiny radio transmitters to animals and use global positioning systems (GPSs) to track where and how far animals go. This technology is important for studying endangered species, which you will learn about in Chapter 9. Scientists also study nature by using cell phone cameras and mounting time-lapse cameras or video cameras on stationary objects or small drones. **GREEN CAREERS: GIS Analyst; Remote Sensing Analyst**

Ecologists Do Laboratory Research and Use Models

Ecologists supplement their field research by conducting *laboratory research*. In laboratories, scientists create, set up, and observe models of ecosystems and populations. They create such simplified systems in containers such as culture tubes, bottles, aquariums, and greenhouses, and in indoor and outdoor chambers. In these structures, they control temperature, light, CO_2, humidity, and other variables.

These systems make it easier for scientists to carry out controlled experiments. Laboratory experiments are often faster and less costly than similar experiments in the field. However, scientists must consider how well their scientific observations and measurements in simplified, controlled systems in laboratory conditions reflect what takes place under the more complex and often-changing conditions found in nature.

INDIVIDUALS MATTER 3.1

Thomas E. Lovejoy—Forest Researcher and Biodiversity Educator

Luiz Rampelotto/ZUMAPRESS/Newscom

For several decades, conservation biologist and National Geographic Explorer Thomas E. Lovejoy has played a major role in educating scientists and the public about the need to understand and protect tropical forests. He has carried out research in the Amazon forests of Brazil since 1965, which focused on estimating the minimum area necessary for sustaining biodiversity in national parks and biological reserves in tropical forests. In 1980 he coined the term *biological diversity*.

Lovejoy served as the principal adviser for the popular and widely acclaimed public television series *Nature*. He has also written numerous articles and books on issues related to conserving biodiversity. In addition to teaching environmental science and policy at George Mason University, he has held several prominent posts, including director of the World Wildlife Fund's conservation program and president of the Society for Conservation Biology. In 2012 he was awarded the Blue Planet Prize for his efforts to understand and sustain the earth's biodiversity.

Since the late 1960s, ecologists have developed mathematical models that simulate ecosystems, and they run the models on high-speed supercomputers. The models help them understand large and complex systems, such as lakes, oceans, forests, and the earth's climate, that cannot be adequately studied and modeled in field or laboratory research. **GREEN CAREER: Ecosystem modeler**

Ecologists call for greatly increased research on the condition of the world's ecosystems to see how they are changing. This will help them develop strategies for preventing or slowing their degradation. It will also help us to avoid going beyond ecological tipping points. Exceeding such boundaries could cause severe degradation and even the collapse of some ecosystems (Science Focus 3.3).

BIG IDEAS

- Life is sustained by the flow of energy from the sun through the biosphere, the cycling of nutrients within the biosphere, and gravity.
- Some organisms produce the nutrients they need, others survive by consuming other organisms, and still others live on the wastes and remains of organisms while recycling nutrients that are used again by producer organisms.
- Human activities are altering the flow of energy through food chains and food webs and the cycling of nutrients within ecosystems and the biosphere.

Tying It All Together

Tropical Rain Forests and Sustainability

This chapter began with a discussion of the importance of the world's incredibly diverse tropical rain forests (**Core Case Study**). These ecosystems showcase the functioning of the three **scientific principles of sustainability**, which apply as well to the world's other ecosystems.

First, producers within rain forests rely on *solar energy* to produce a vast amount of biomass through photosynthesis. *Second*, species living in the forests take part in and depend on the *cycling of nutrients* and the flow of energy within the forests and throughout the biosphere. *Third*, tropical rain forests contain a huge and vital part of the earth's *biodiversity*, and interactions among the many species living in these forests help to sustain these complex ecosystems.

We also reported recent research on the possible long-lasting, harmful effects of exceeding key planetary boundaries. In many of the chapters to follow, we will further examine such risks. We will also consider ways in which we can apply the six **principles of sustainability** to avoid exceeding key planetary boundaries, to live more sustainably, and to expand our beneficial environmental impacts.

Aneka/Shutterstock.com

Chapter Review

Core Case Study

1. What are three harmful effects of the clearing and degradation of tropical rain forests?

Section 3.1

2. What are the two key concepts for this section? Define and distinguish among the **atmosphere**, **troposphere**, **stratosphere**, **hydrosphere**, **geosphere**, and **biosphere**. What three interconnected factors sustain life on the earth? Describe the flow of energy to and from the earth. What is the **greenhouse effect** and why is it important?

Section 3.2

3. What are the two key concepts for this section? Define **ecology**. Define **organism**, **population**, **community**, and **ecosystem**, and give an example of each. Distinguish between the living and nonliving components in ecosystems and give two examples of each.

4. What is a **trophic level**? Distinguish among **producers (autotrophs)**, **consumers (heterotrophs)**, **decomposers**, and **detritus feeders (detritivores)**, and give an example of each. Summarize the process of **photosynthesis** and explain how it provides us with food and the oxygen in the air that we breathe. Distinguish among **primary consumers (herbivores)**, **carnivores**, **secondary consumers**, **tertiary consumers**, and **omnivores**, and give an example of each.

5. Explain the importance of microbes. What is **aerobic respiration**? What two processes sustain ecosystems and the biosphere and how are they linked?

6. What is **soil**? Explain how a mature soil forms. What is **weathering**? Why is soil such an important resource? What is a **soil profile**? Describe the four horizons in a mature soil. What does the color of topsoil tell us about its ability to grow crops? What three types of particles are found in soils?

Section 3.3

7. What is the key concept for this section? Define and distinguish between a **food chain** and a **food web**. Explain what happens to energy as it flows through food chains and food webs. What is a **pyramid of energy flow**? Distinguish between **gross primary productivity (GPP)** and **net primary productivity (NPP)**, and explain their importance. What are the two most productive land ecosystems and the two most productive aquatic ecosystems? What percentage of the world's NPP do humans use?

Section 3.4

8. What is the key concept for this section? What happens to matter in an ecosystem? What is a **nutrient cycle**? Explain how nutrient cycles connect past, present, and future life. Describe the **hydrologic cycle**, or **water cycle**. What three major processes are involved in the water cycle? What is **surface runoff**? Define **groundwater**. What is an **aquifer**? What percentage of the earth's water supply is available to humans and other species as liquid freshwater? Summarize the unique properties of water. List three ways that humans are altering the water cycle. Explain how clearing a rain forest can affect local weather and climate.

9. Describe the **carbon**, **nitrogen**, and **phosphorus cycles**, and explain how human activities are affecting each cycle. Summarize Thomas Lovejoy's role in protecting the world's tropical forests and protecting biodiversity.

Section 3.5

10. What is the key concept for this section? List three ways in which scientists study ecosystems. Explain why we need much more basic data about the condition of the world's ecosystems. Distinguish between the Holocene era and the proposed Anthropocene era. What is a planetary boundary (ecological tipping point) and why are such boundaries important? List four boundaries that we may have exceeded. What are this chapter's three big ideas? Explain how tropical rain forests (**Core Case Study**) showcase the functioning of the three **scientific principles of sustainability**.

Note: Key terms are in bold type.

Critical Thinking

1. How would you explain the importance of tropical rain forests (**Core Case Study**) to people who think that such forests have no connection to their lives?

2. Explain **(a)** why the flow of energy through the biosphere depends on the cycling of nutrients, and **(b)** why the cycling of nutrients depends on gravity.

3. Explain why microbes are important. What are two ways in which they benefit your health or lifestyle? Write a brief description of what you think would happen to you if microbes were eliminated from the earth.

4. Make a list of the foods you ate for lunch or dinner today. Trace each type of food back to a particular producer species. Describe the sequence of feeding levels that led to your feeding.

5. Use the second law of thermodynamics (see Chapter 2, p. 42) to explain why many poor people in less-developed countries live on a mostly vegetarian diet.

6. How might your life and the lives of any children or grandchildren you might eventually have be affected if human activities continue to intensify the water cycle?

7. What would happen to an ecosystem if **(a)** all of its decomposers and detritus feeders were eliminated, **(b)** all of its producers were eliminated, and **(c)** all of its insects were eliminated? Could an ecosystem exist with producers and decomposers but no consumers? Explain.

8. If we have exceeded planetary boundaries for the nitrogen and phosphorus cycles, biodiversity loss, land change, and climate change by disrupting the carbon cycle, how might this affect **(a)** you, **(b)** any children you might have, and **(c)** any grandchildren you might have?

Doing Environmental Science

Visit a nearby terrestrial ecosystem and identify its major producers, primary and secondary consumers, detritus feeders, and decomposers. Take notes and describe at least one example of each of these types of organisms. Make a simple sketch showing how these organisms might be related to each other or to other organisms in a food chain or food web. Think of two ways in which this food web or chain could be disrupted. Write a report summarizing your research and conclusions.

Global Environment Watch Exercise

Go to your MindTap course to access the GREENR database. Using the "Basic Search" box at the top of the page, search for *Nitrogen Cycle* and look for information on how humans are affecting the nitrogen cycle. Specifically look for impacts on the atmosphere and on human health from emissions of nitrogen oxides, and look for the harmful ecological effects of the runoff of nitrate fertilizers into rivers and lakes. Make a list of these impacts and use this information to review your daily activities. Find three things that you do regularly that contribute to these impacts.

Data Analysis

Recall that net primary productivity (NPP) is the *rate* at which producers can make the chemical energy that is stored in their tissues and that is potentially available to other organisms (consumers) in an ecosystem. In Figure 3.18, it is expressed as units of energy (kilocalories, or kcal) produced in a given area (square meters, or m^2) per year. Look again at Figure 3.18 and consider the differences in NPP among various ecosystems. Then answer the following questions:

1. What is the approximate NPP of a tropical rain forest in kcal/m^2/yr? Which terrestrial ecosystem produces about one-third of that rate? Which aquatic ecosystem has about the same NPP as a tropical rain forest?
2. Early in the 20th century, large areas of temperate forestland in the United States were cleared to make way for agricultural land. For each unit of this forest area that was cleared and replaced by farmland, by about how much was NPP reduced?
3. Why do you think deserts and grasslands have dramatically lower NPP than swamps and marshes?
4. About how many times higher is NPP in estuaries than it is in lakes and streams? Why do you think this is so?

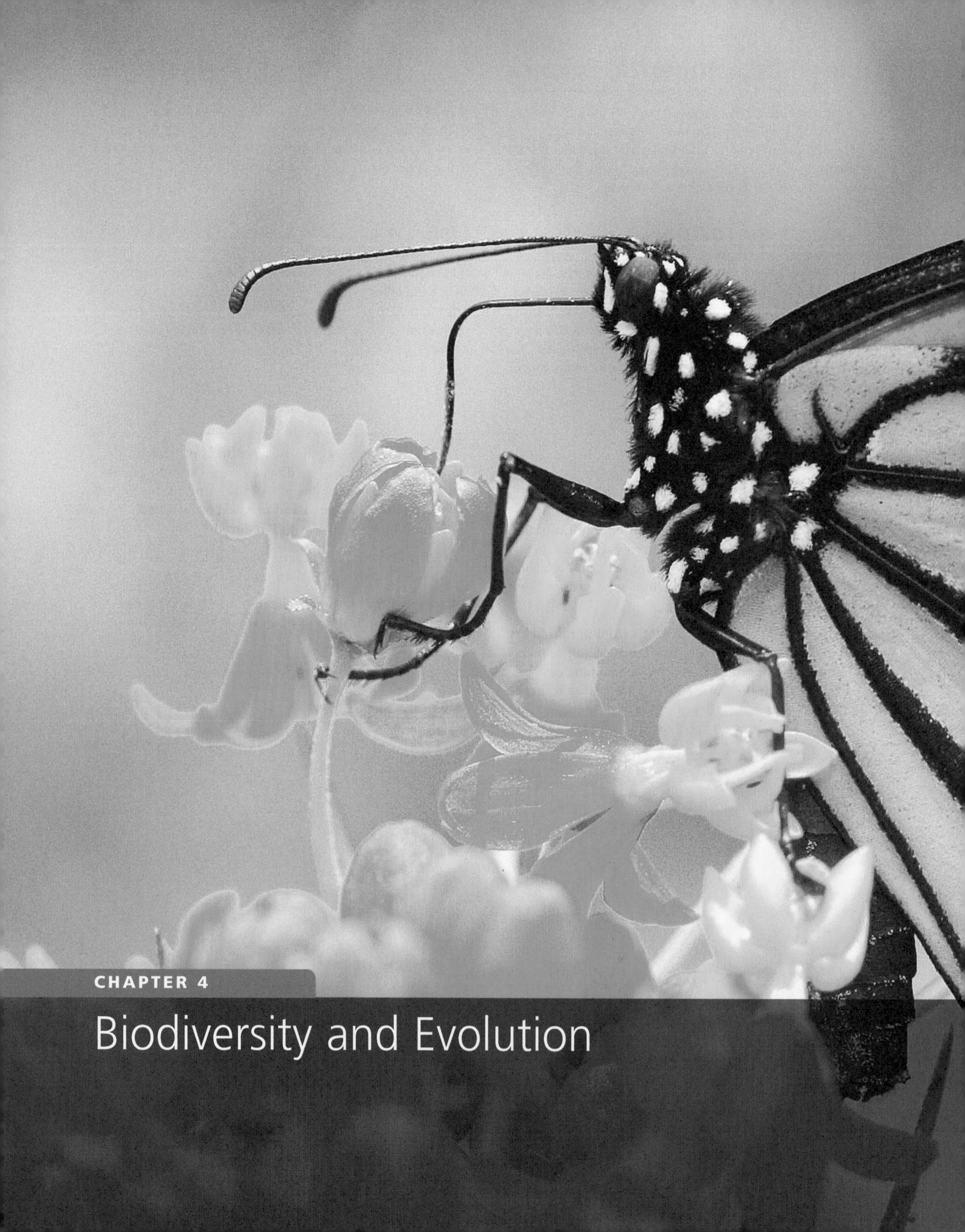

CHAPTER 4

Biodiversity and Evolution

Nothing in biology makes sense except in the light of evolution.
THEODOSIUS DOBZHANSKY

Key Questions

4.1 What are the major types of life on the earth?

4.2 What is biodiversity and why is it important?

4.3 What roles do species play in ecosystems?

4.4 How does the earth's life change over time?

4.5 What factors affect biodiversity?

Monarch butterfly on a flower.

Ariel Bravy/Shutterstock.com

Core Case Study

Why Are Amphibians Vanishing?

FIGURE 4.1 Specimens of some of the nearly 200 amphibian species that have gone extinct since the 1970s.

Joel Sartore/National Geographic Creative

Amphibians—frogs, toads, and salamanders—were among the first vertebrates (animals with backbones) to leave the earth's waters and live on land. Amphibians have adjusted to and survived environmental changes more effectively than many other species, but their world is changing rapidly.

An amphibian lives part of its life in water and part on land. Human activities such as the use of pesticides and other chemicals can pollute the land and water habitats of amphibians. Many of the more than 7,000 known amphibian species have problems adapting to these changes.

Since 1980, populations of hundreds of amphibian species have declined or vanished (Figure 4.1). According to the International Union for Conservation of Nature (IUCN), about 33% of known amphibian species face *extinction*. A 2015 study by biodiversity expert Peter Crane found that 200 frog species have gone extinct since the 1970s, and frogs are going extinct 10,000 times faster than their historical rates.

No single cause can account for the decline of many amphibian species, but scientists have identified a number of factors that affect amphibians at various points in their life cycles. For example, frog eggs lack shells to protect the embryos they contain from water pollutants, and adult frogs ingest the insecticides present in many of the insects they eat. We explore these and other factors later in this chapter.

Why should we care if some amphibian species become extinct? Scientists give three reasons. *First*, amphibians are sensitive *biological indicators* of changes in environmental conditions. These changes include habitat loss, air and water pollution, ultraviolet (UV) radiation, and climate change. The growing threats to the survival of an increasing number of amphibian species indicate that environmental conditions for amphibians and many other species are deteriorating in many parts of the world.

Second, adult amphibians play important roles in biological communities. Adult amphibians eat more insects (including mosquitoes) than many species of birds eat. In some habitats, the extinction of certain amphibian species could lead to population declines or extinction of animals that eat amphibians or their larvae, such as aquatic insects, reptiles, birds, fish, mammals, and other amphibians.

Third, a number of pharmaceutical products come from compounds found in secretions from the skin of certain amphibians. Many of these compounds have been isolated and used as painkillers and antibiotics and in treatments for burns and heart disease. If amphibians vanish, these potential medical benefits vanish with them.

The threat to amphibians is part of a greater threat to the earth's biodiversity. In this chapter, we will learn about biodiversity, how it arose on the earth, why it is important, and how it is threatened. We will also consider possible solutions to these threats.

4.1 WHAT ARE THE MAJOR TYPES OF LIFE ON EARTH?

CONCEPT 4.1 The earth's 7 million to 100 million species vary greatly in their characteristics and ecological roles.

Earth's Organisms Are Many and Varied

Every organism is composed of one or more cells. Based on their cell structure, organisms can be classified as *eukaryotic* or *prokaryotic*. All organisms except bacteria are **eukaryotic**. Their cells are encased in a membrane and have a distinct *nucleus* (a membrane-bounded structure containing genetic material in the form of DNA—see Figure 2.9, p. 38) and several other internal parts enclosed by membranes (Figure 4.2, right). Bacterial cells are **prokaryotic**, enclosed by a membrane but containing no distinct nucleus or other internal parts enclosed by membranes (Figure 4.2, left).

Scientists group organisms into categories based on their greatly varying characteristics (**Concept 4.1**), a process called *taxonomic classification*. The largest category is the *kingdom*, which includes all organisms that have one or several common features. Biologists recognize six kingdoms. Two are different types of *bacteria* (eubacteria and archaebacteria) with single cells that are prokaryotic (Figure 4.2, left). The other four kingdoms are *protists, plants, fungi,* and *animals* with eukaryotic cells (Figure 4.2 right).

Protists are mostly many-celled eukaryotic organisms such as golden brown and yellow-green algae and protozoans. Most *fungi* are many-celled organisms such as mushroom, molds, mildews, and yeasts.

Plants include certain types of algae (including red, brown, and green algae), mosses, ferns, and flowering plants whose flowers produce seeds. Flowering plant species make up about 90% of the plant kingdom. Some flowering plants such as corn and marigolds are **annuals** that live for one growing season, die, and have to be replanted. Others are **perennials**, such as roses and grapes, that can live for two or more seasons before they die and have to be replanted.

Animals are many-celled eukaryotic organisms. Most are **invertebrates**, with no backbones. They include jellyfish, worms, insects, shrimp, snails, clams, and octopuses. Other animals, called **vertebrates**, have backbones. Examples include amphibians (**Core Case Study**), fishes, reptiles (alligators and snakes), birds (robins and eagles), and mammals (humans, whales, elephants, bats, and tigers).

Kingdoms are divided into *phyla*, which are divided into subgroups called *classes*. Classes are subdivided into *orders*, which are further divided into *families*. Families consist of *genera* (singular, genus), and each genus contains one or more *species*. Figure 4.3 shows the detailed taxonomic classification for the current human species: *Homo sapiens sapiens*.

A **species** is a group of living organisms with characteristics that distinguish it from other groups of organisms. In sexually reproducing organisms, individuals must be able to mate with similar individuals and produce fertile offspring in order to be classified as a species.

Estimates of the number of species range from 7 million to 100 million, with a best guess of 7 million to 10 million species (**Concept 4.1**). Biologists have identified about 2 million species.

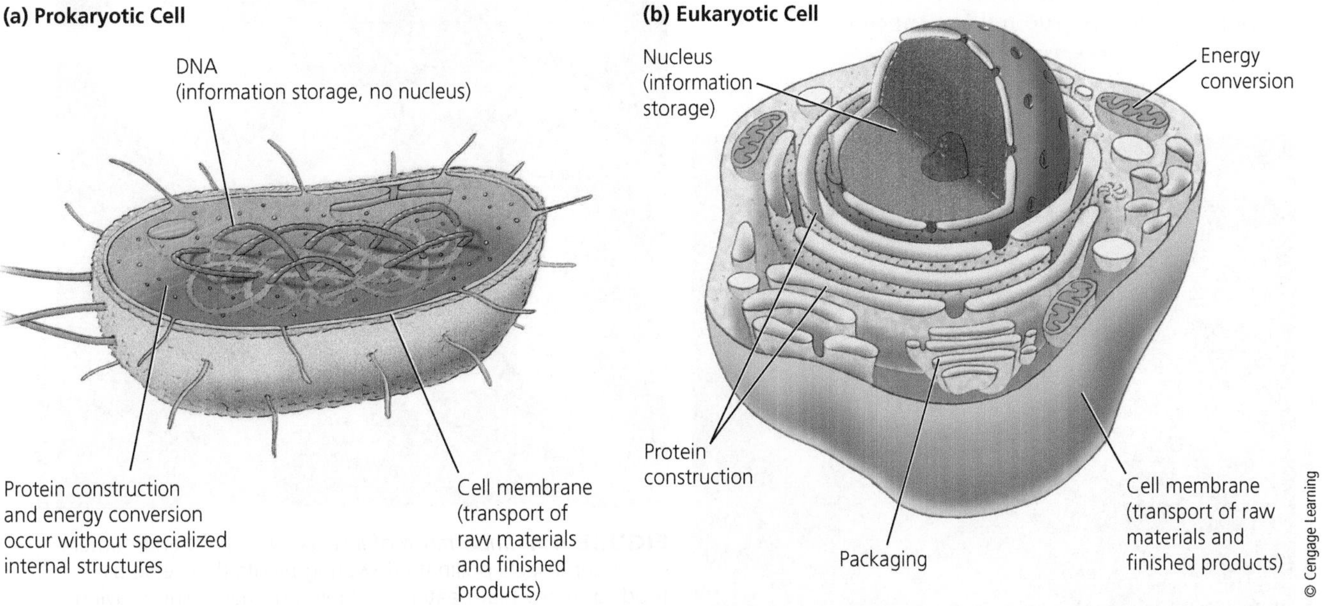

FIGURE 4.2 Comparison of key components of a prokaryotic cell (left) and eukaryotic cell (right).

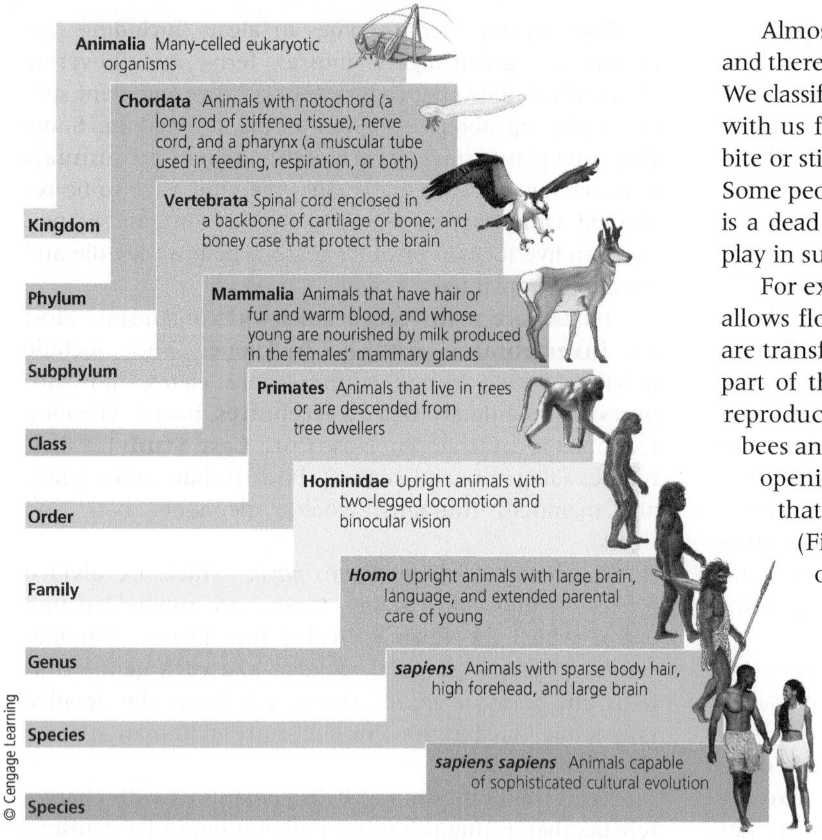

Animalia Many-celled eukaryotic organisms
Chordata Animals with notochord (a long rod of stiffened tissue), nerve cord, and a pharynx (a muscular tube used in feeding, respiration, or both)
Vertebrata Spinal cord enclosed in a backbone of cartilage or bone; and boney case that protect the brain
Mammalia Animals that have hair or fur and warm blood, and whose young are nourished by milk produced in the females' mammary glands
Primates Animals that live in trees or are descended from tree dwellers
Hominidae Upright animals with two-legged locomotion and binocular vision
Homo Upright animals with large brain, language, and extended parental care of young
sapiens Animals with sparse body hair, high forehead, and large brain
sapiens sapiens Animals capable of sophisticated cultural evolution
Kingdom
Phylum
Subphylum
Class
Order
Family
Genus
Species
Species
© Cengage Learning

FIGURE 4.3 How the current human species got its name: *Homo sapiens sapiens.*

Almost half of the world's identified species are insects, and there may be 10–30 million insect species on the planet. We classify many insect species as *pests* because they compete with us for food, spread human diseases such as malaria, bite or sting us, and invade our lawns, gardens, and houses. Some people fear insects and many think the only good bug is a dead bug. They fail to recognize the vital roles insects play in sustaining the earth's life.

For example, *pollination* is a vital ecosystem service that allows flowering plants to reproduce. When pollen grains are transferred from the flower of one plant to a receptive part of the flower of another plant of the same species, reproduction occurs. Many flowering species depend on bees and other insects to pollinate their flowers (chapter-opening photo and Figure 4.4, left). In addition, insects that eat other insects—such as the praying mantis (Figure 4.4, right)—help to control the populations of at least half the species of insects that we call pests. This free pest control service is another vital ecosystem service. In addition, insects make up an increasingly large part of the human food supply in some parts of the world.

Some insect species reproduce at an astounding rate and can rapidly develop new genetic traits such as resistance to pesticides. They also have an exceptional ability to evolve into new species when faced with changing environmental conditions.

Research indicates that some human activities are threatening insect populations such as honeybees. We discuss this environmental problem more fully in Chapter 9.

2 million

The number of species that scientists have identified out of the world's estimated 7 million to 100 million species

Darlyne A. Murawski/National Geographic Creative

Dr. Morley Read/Shutterstock.com

FIGURE 4.4 *Importance of insects:* Bees (left) and numerous other insects pollinate flowering plants that serve as food for many plant eaters, including humans. This praying mantis, which is eating a moth (right), and many other insect species help to control the populations of most of the insect species we classify as pests.

4.2 WHAT IS BIODIVERSITY AND WHY IS IT IMPORTANT?

CONCEPT 4.2 The biodiversity found in genes, species, ecosystems, and ecosystem processes is vital to sustaining the earth's life.

Biodiversity Is the Variety of Life

Biodiversity, or **biological diversity**, is the variety of life on the earth. It has four components, as shown in Figure 4.5.

One is **species diversity**, the number and abundance of the different kinds of species living in an ecosystem. Species diversity has two components, one being **species richness**, the number of different species in ecosystem. The other is **species evenness**, a measure of the comparative abundance of all species in an ecosystem.

For example, a species-rich ecosystem has a large number of different species. However, this tells us nothing about how many members of each species are present. If it has many of one or more species and just a few of others, its species evenness is low. If it has roughly equal numbers of each species, its species evenness is high. For example, if an ecosystem has only three species, its species richness is low. However, if there are roughly equal numbers of each of the three species, it species evenness is high. Species-rich ecosystems such as rain forests tend

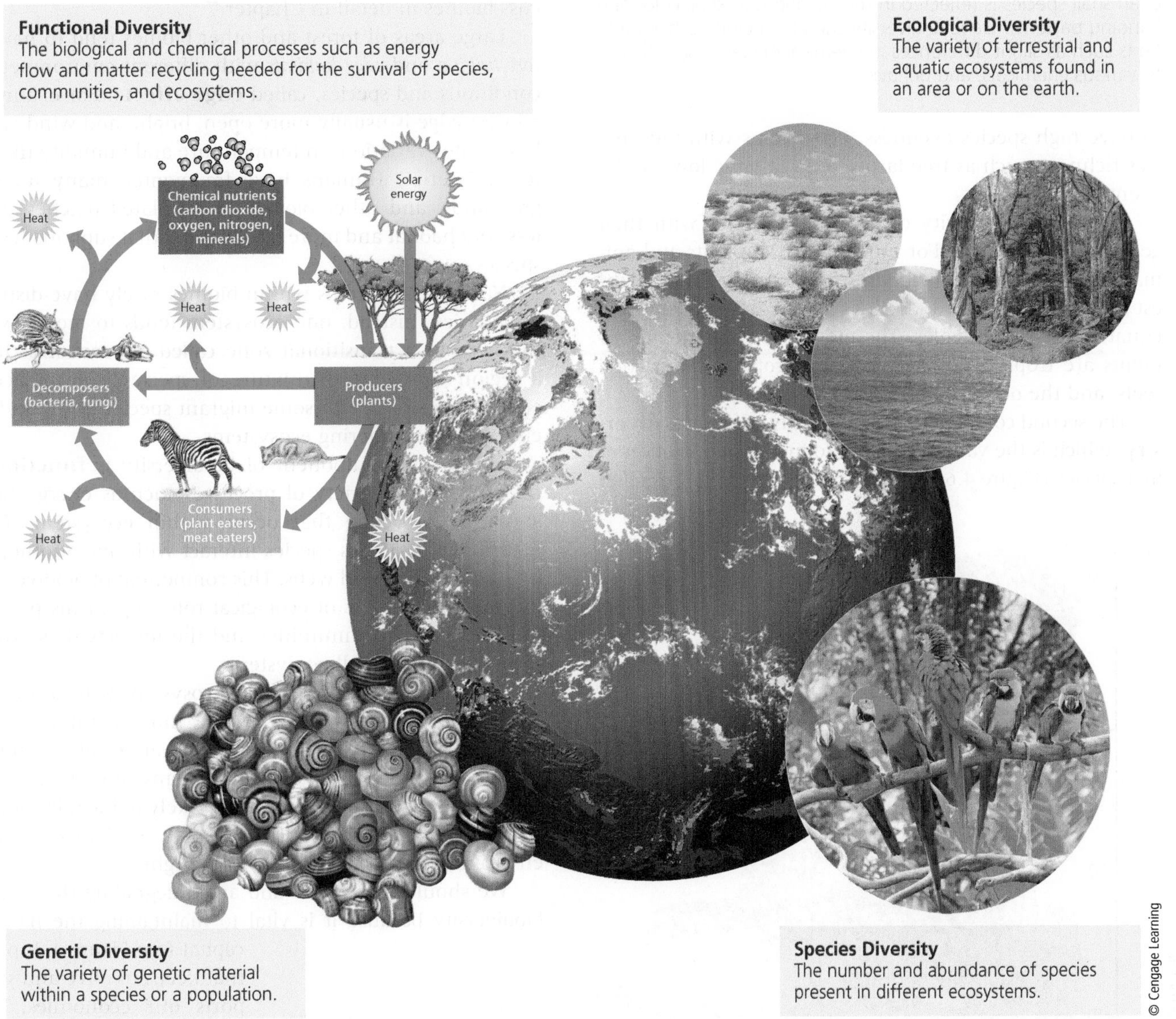

Functional Diversity
The biological and chemical processes such as energy flow and matter recycling needed for the survival of species, communities, and ecosystems.
Ecological Diversity
The variety of terrestrial and aquatic ecosystems found in an area or on the earth.
Chemical nutrients (carbon dioxide, oxygen, nitrogen, minerals)
Solar energy
Heat
Heat
Heat
Heat
Heat
Decomposers (bacteria, fungi)
Producers (plants)
Consumers (plant eaters, meat eaters)
Genetic Diversity
The variety of genetic material within a species or a population.
Species Diversity
The number and abundance of species present in different ecosystems.
© Cengage Learning

FIGURE 4.5 Natural capital: The major components of the earth's *biodiversity*—one of the planet's most important renewable resources and a key component of its natural capital (see Figure 1.3, p. 7).

Right side, top left: Laborant/Shutterstock.com; right side, top right: leungchopan/Shutterstock.com; right side, top center: Elenamiv/Shutterstock.com; bottom right: Juriah Mosin/Shutterstock.com.

© Cengage Learning

FIGURE 4.6 *Genetic diversity* in this population of a Caribbean snail species is reflected in the variations in shell color and banding patterns. Genetic diversity can also include other variations such as slight differences in chemical makeup, sensitivity to various chemicals, and behavior.

to have high species evenness. Ecosystems with low species richness, such as tree farms, tend to have low species evenness.

The species diversity of ecosystems varies with their geographical location. For most terrestrial plants and animals, species diversity (primarily species richness) is highest in the tropics and declines as we move from the equator toward the poles. The most species-rich environments are tropical rain forests, large tropical lakes, coral reefs, and the ocean-bottom zone.

The second component of biodiversity is **genetic diversity**, which is the variety of genes found in a population or in a species (Figure 4.6). Genes contain genetic information that give rise to specific traits, or characteristics, that are passed on to offspring through reproduction. Species have a better chance of surviving and adapting to environmental changes if they have greater genetic diversity.

The third component of biodiversity, **ecosystem diversity**, refers to the earth's diversity of biological communities such as deserts, grasslands, forests, mountains, oceans, lakes, rivers, and wetlands. Biologists classify terrestrial (land) ecosystems into **biomes**—large regions such as forests, deserts, and grasslands characterized by distinct climates and certain prominent species (especially vegetation). Biomes differ in their *community structure* based on the types, relative sizes, and stratification of their plant species (Figure 4.7). Figure 4.8 shows the major biomes across the midsection of the United States. We discuss biomes in detail in Chapter 7.

Large areas of forest and other biomes tend to have a *core habitat* and *edge habitats* with different environmental conditions and species, called **edge effects**. For example, a forest edge is usually more open, bright, and windy and has greater variations in temperature and humidity than a forest interior. Humans have fragmented many forests, grasslands, and other biomes into isolated patches with less core habitat and more edge habitat that supports fewer species.

Natural ecosystems within biomes rarely have distinct boundaries. Instead, one ecosystem tends to merge with the next in a transitional zone called an **ecotone**. It is a region containing a mixture of species from adjacent ecosystems along with some migrant species not found in either of the bordering ecosystems.

The fourth component of biodiversity is **functional diversity**—the variety of processes such as energy flow and matter cycling that occur within ecosystems (see Figure 3.9, p. 55) as species interact with one another in food chains and food webs. This component of biodiversity includes the variety of ecological roles organisms play in their biological communities and the impacts these roles have on their overall ecosystems.

A more biologically diverse ecosystem with a greater variety of producers can produce more plant biomass, which in turn can support a greater variety of consumer species. Biologically diverse ecosystems also tend to be more stable because they are more likely to include species with traits that enable them to adapt to changes in the environment, such as disease or drought.

We should care about and avoid degrading the earth's biodiversity because it is vital to maintaining the natural capital (see Figure 1.3, p. 7) that keeps us alive and supports our economies. We use biodiversity as a source of food, medicine, building materials, and fuel. Biodiversity also provides natural

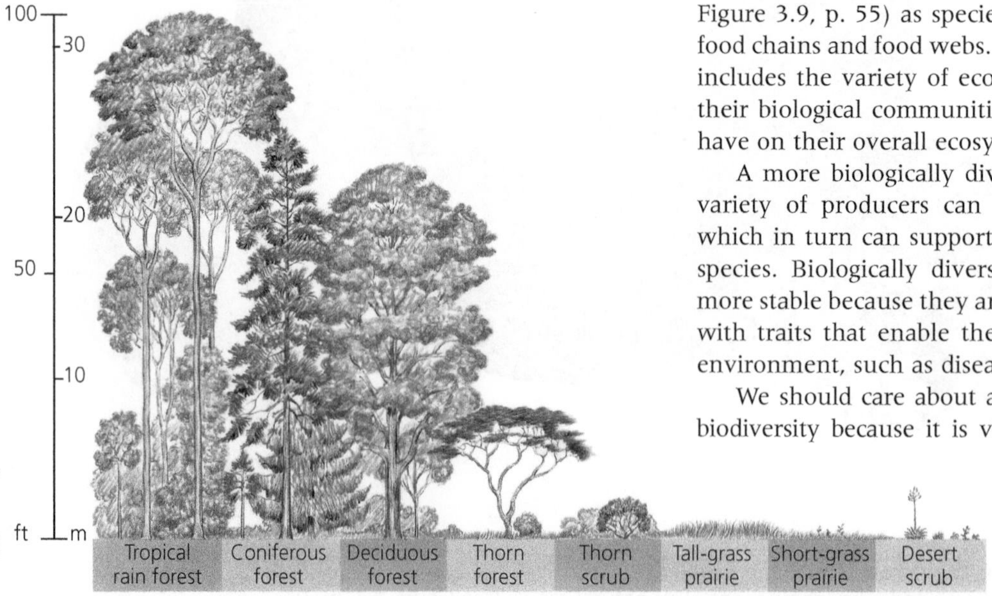

© Cengage Learning

FIGURE 4.7 *Community structure:* Generalized types, relative sizes, and stratification of plant species in communities or ecosystems in major terrestrial biomes.

INDIVIDUALS MATTER 4.1

Edward O. Wilson: Champion of Biodiversity

As a boy growing up in the southeastern United States, Edward O. Wilson became interested in insects at age 9. He has said, "Every kid has a bug period. I never grew out of mine."

Before entering college, Wilson decided he would specialize in the study of ants. He became one of the world's experts on ants and then steadily widened his focus, eventually to include the entire biosphere. One of Wilson's landmark works is *The Diversity of Life*, published in 1992. In that book, he presented the principles and practical issues of biodiversity more completely than anyone had to that point. Today, he is recognized as one of the world's leading experts on biodiversity. He is now deeply involved in writing and lecturing about the need for global conservation efforts and is working on Harvard University's *Encyclopedia of Life*, a database of information on the world's known species.

Wilson has won more than 100 national and international awards and has written 32 books, two of which won the Pulitzer Prize for General Nonfiction. In 2013 he received the National Geographic Society's highest award, the Hubbard Medal. About the importance of biodiversity, he writes: "How can we save Earth's life forms from extinction if we don't even know what most of them are? . . . I like to call Earth a little known planet."

ecosystem services such as air and water purification, renewal of topsoil, decomposition of wastes, and pollination. In addition, the earth's variety of genetic information, species, and ecosystems provide raw materials for the evolution of new species and ecosystem services, as they respond to changing environmental conditions. Biodiversity is an ecological insurance policy.

According to ocean researcher Sylvia Earle, "The bottom line answer to the question about why biodiversity matters is fairly simple. The rest of the living world can get along without us but we can't get along without them." We owe much of what we know about biodiversity to researchers such as Earle and biologist Edward O. Wilson (Individuals Matter 4.1).

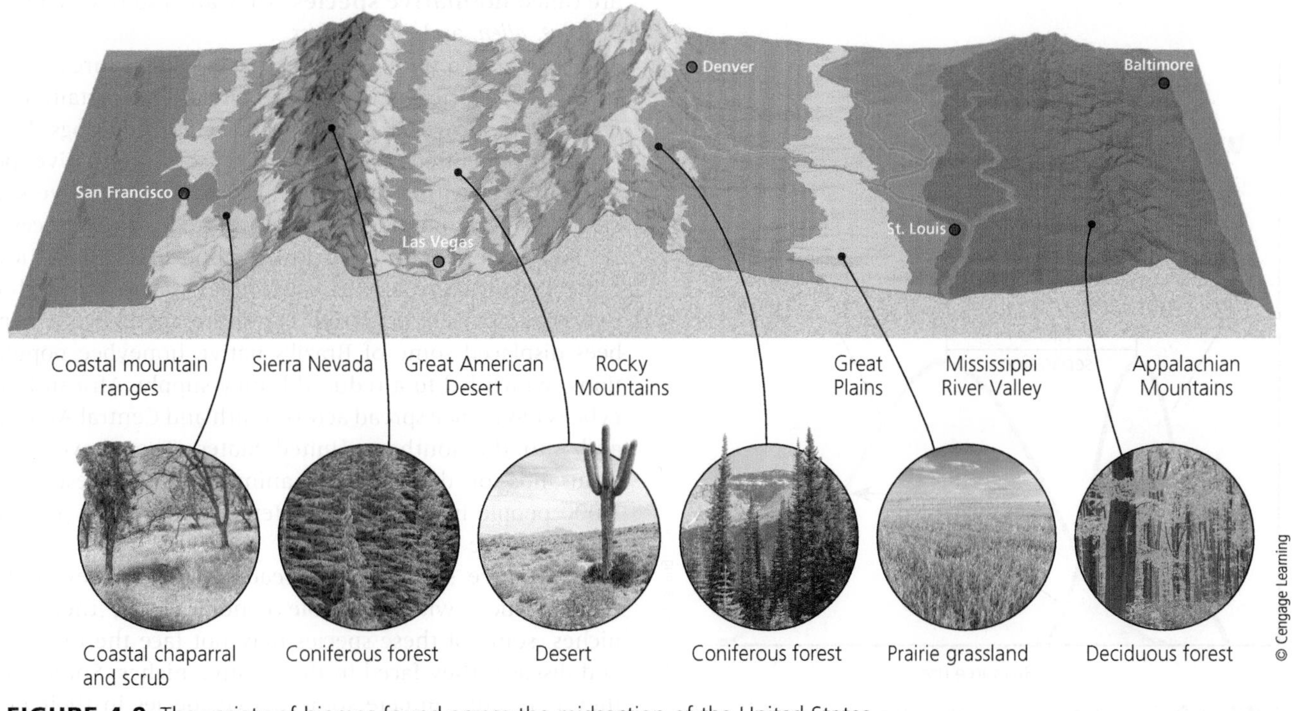

FIGURE 4.8 The variety of biomes found across the midsection of the United States.

First: Zack Frank/Shutterstock.com; second: Robert Crum/Shutterstock.com; third: Joe Belanger/Shutterstock.com; fourth: Protasov AN/Shutterstock.com; fifth: Maya Kruchankova/Shutterstock.com; sixth: Marc von Hacht/Shutterstock.com

4.3 WHAT ROLES DO SPECIES PLAY IN ECOSYSTEMS?

CONCEPT 4.3A Each species plays a specific ecological role called its *niche*.

CONCEPT 4.3B Any given species may play one or more of four important roles—native, nonnative, indicator, or keystone—in a particular ecosystem.

Each Species Plays a Role

Each species plays a role within the ecosystem it inhabits (**Concept 4.3A**). Ecologists describe this role as its **ecological niche**. It is a species' way of life in its ecosystem and includes everything that affects its survival and reproduction, such as how much water and sunlight it needs, how much space it requires, what it feeds on, what feeds on it, how and when it reproduces, and the temperatures and other conditions it can tolerate. The niche of a species differs from its **habitat**—the place, or type of ecosystem, in which a species lives and obtains what it needs to survive.

Ecologists use the niches of species to classify them as *generalists* or *specialists*. **Generalist species**, such as raccoons, have broad niches (Figure 4.9, right curve). They can live in many different places, eat a variety of foods, and often tolerate a wide range of environmental conditions. Other generalist species include flies, cockroaches, rats, coyotes, white-tailed deer, and humans.

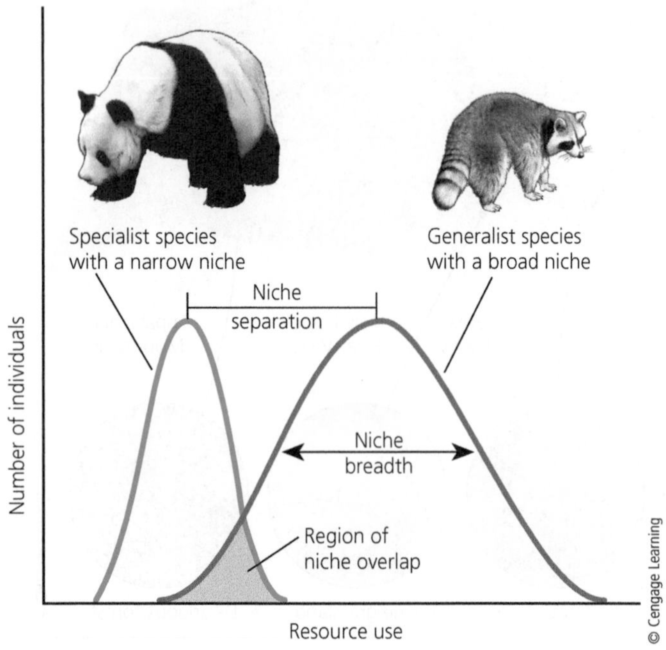

FIGURE 4.9 Specialist species such as the giant panda have a narrow niche (left curve) and generalist species such as the raccoon have a broad niche (right curve).

In contrast, **specialist species**, such as the giant panda, occupy narrow niches (Figure 4.9, left curve). They may be able to live in only one type of habitat, eat only one or a few types of food, or tolerate a narrow range of environmental conditions. For example, different specialist species of shorebirds feed on crustaceans, insects, or other organisms found on sandy beaches and their adjoining coastal wetlands (Figure 4.10).

Because of their narrow niches, specialists are more prone to extinction when environmental conditions change. For example, China's *giant panda* is highly endangered because of a combination of habitat loss, low birth rate, and its specialized diet consisting mostly of bamboo.

Is it better to be a generalist or a specialist? It depends. When environmental conditions undergo little change, as in a tropical rain forest, specialists have an advantage because they have fewer competitors. Under rapidly changing environmental conditions, the more adaptable generalist usually is better off.

There Are Four Major Ecosystem Roles for Species

Niches can be classified further in terms of specific roles that certain species play within ecosystems. Ecologists describe these roles as *native*, *nonnative*, *indicator*, and *keystone*. Any given species may play one or more of these roles in a particular ecosystem (**Concept 4.3B**).

Native species normally live and thrive in a particular ecosystem. Other species that migrate into or that are deliberately or accidentally introduced into an ecosystem are called **nonnative species**. They are also referred to as *invasive*, *alien*, and *exotic species*.

People often think of nonnative species as threatening. In fact, most nonnative species, including certain food crops, trees, flowers, chickens, cattle, fish, and dogs, have certainly benefitted people. However, some nonnative species compete with and reduce an ecosystem's native species, causing unintended and unexpected consequences.

For example, in 1957 Brazil imported wild African honeybees (Figure 4.11) to help increase honey production. The opposite occurred. The more aggressive African bees displaced some of Brazil's native honeybee populations, which led to a reduced honey supply. African honeybees have since spread across South and Central America and into the southern United States. They have killed thousands of domesticated animals and an estimated 1,000 people in the Western Hemisphere, many of them allergic to bee stings.

Nonnative species can spread rapidly, if they find a new location with favorable conditions. In their new niches, some of these species may not face the predators and diseases they faced in their native niches. Sometimes they may out-compete some native species in their new locations. We examine this environmental threat in more detail in Chapter 9.

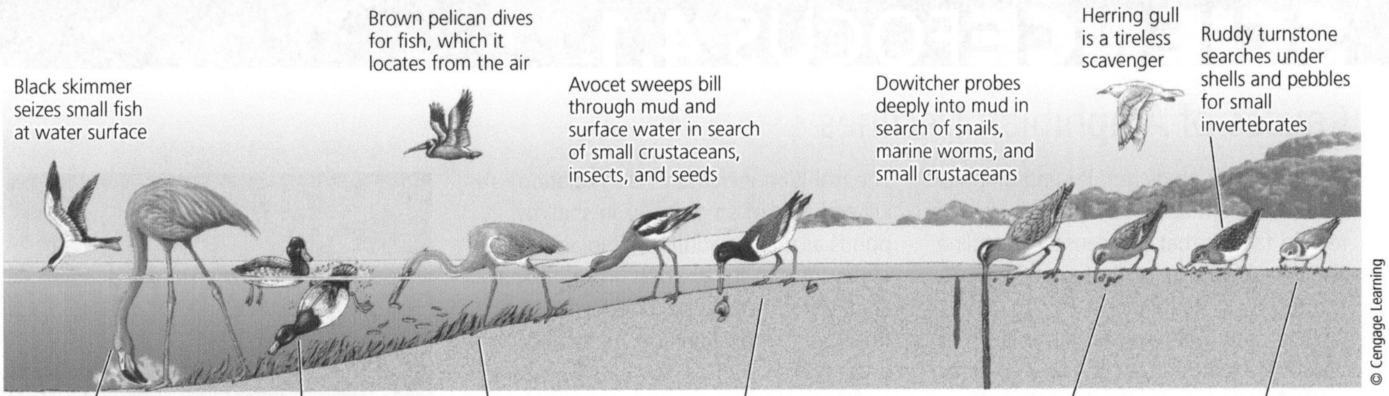
Brown pelican dives for fish, which it locates from the air
Herring gull is a tireless scavenger
Ruddy turnstone searches under shells and pebbles for small invertebrates
Black skimmer seizes small fish at water surface
Avocet sweeps bill through mud and surface water in search of small crustaceans, insects, and seeds
Dowitcher probes deeply into mud in search of snails, marine worms, and small crustaceans
Flamingo feeds on minute organisms in mud
Scaup and other diving ducks feed on mollusks, crustaceans, and aquatic vegetation
Louisiana heron wades into water to seize small fish
Oystercatcher feeds on clams, mussels, and other shellfish into which it pries its narrow beak
Knot (sandpiper) picks up worms and small crustaceans left by receding tide
Piping plover feeds on insects and tiny crustaceans on sandy beaches
© Cengage Learning

FIGURE 4.10 Various bird species in a coastal wetland occupy specialized feeding niches. This specialization reduces competition and allows for sharing of limited resources.

Indicator Species Can Sound the Alarm

Species that provide early warnings of changes in environmental conditions in an ecosystem are called **indicator species**. They are like biological smoke alarms. In this chapter's **Core Case Study**, you learned that some amphibians are classified as indicator species.

Scientists have been working hard to identify some of the possible causes of the declines in amphibian populations (Science Focus 4.1).

Birds are also excellent biological indicators. They are found almost everywhere and are affected quickly by environmental changes such as the loss or fragmentation of their habitats and the introduction of chemical pesticides. We explore this problem more fully in Chapter 9.

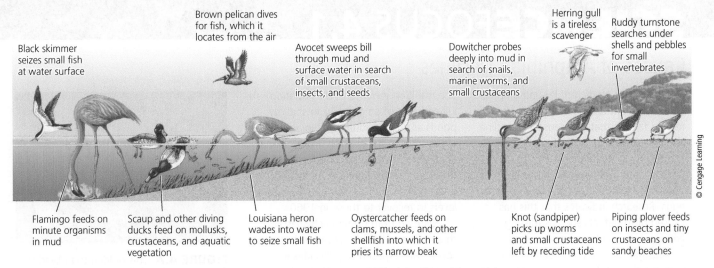

John Lindsay Smith/Shutterstock.com

FIGURE 4.11 Wild African honeybees, popularly known as "killer bees," were imported to Brazil to increase the honey supply but ended up decreasing it and spreading to Central America and parts of the United States.

Keystone Species Play Critical Roles

A keystone is the wedge-shaped stone placed at the top of a stone archway. Remove this stone and the arch collapses. In some communities and ecosystems, ecologists hypothesize that certain species play a similar role. A **keystone species** has a large effect on the types and abundance of other species in an ecosystem. Without the keystone species, the ecosystem would be dramatically different or might cease to exist.

Keystone species play several critical roles in helping to sustain ecosystems. One is the *pollination* of flowering plant species by butterflies, honeybees (Figure 4.4, left), hummingbirds, bats, and other species. In addition, top predator keystone species feed on and help to regulate the populations of other species. Examples are wolves, leopards, lions, some shark species, and the American alligator (see the following Case Study).

The loss of a keystone species in an ecosystem can lead to population crashes and extinctions of other species that depend on them for certain ecosystem services. This is why it so important for scientists to identify keystone species and work to protect them.

CASE STUDY

The American Alligator—A Keystone Species That Almost Went Extinct

The American alligator (Figure 4.12) is a keystone species in subtropical wetland ecosystems in the southeastern United States. Alligators play several important ecological roles. They dig deep depressions known as gator holes. During dry periods, these depressions hold freshwater and serve as refuges for aquatic life. The depressions supply freshwater and food for fishes, insects, snakes, turtles, birds, and other

Causes of Amphibian Declines

Scientists who study reptiles and amphibians have identified natural and human-related factors that can cause the decline and disappearance of these indicator species.

One natural threat is *parasites* such as flatworms that feed on certain amphibian eggs. Research indicates that this has caused birth defects such as missing limbs or extra limbs in some amphibians.

Another natural threat comes from *viral and fungal diseases*. For example, the chytrid fungus infects a frog's skin and causes it to thicken. This reduces the frog's ability to take in water through its skin and leads to death from dehydration. Such diseases can spread easily, because adults of many amphibian species congregate in large numbers to breed.

Habitat loss and fragmentation is another major threat to amphibians. It is mostly a human-caused problem resulting from the clearing of forests and the draining and filling of freshwater wetlands for farming and urban development.

Another human-related problem is *higher levels of UV radiation* from the sun. Ozone (O_3) that forms in the stratosphere protects the earth's life from harmful UV radiation emitted by the sun. During the past few decades, ozone-depleting chemicals released into the troposphere by human activities have drifted into the stratosphere and have destroyed some the stratosphere's protective ozone. The resulting increase in UV radiation can kill embryos of amphibians in shallow ponds as well as adult amphibians basking in the sun for warmth. International action has been taken to reduce the threat of stratospheric ozone depletion, but it will take about 50 years for ozone levels to recover to those in 1960.

Pollution from human activities also threatens amphibians. Frogs and other species are exposed to pesticides in ponds and in the bodies of insects that they eat. This can make them more vulnerable to bacterial, viral, and fungal diseases and to some parasites. Amphibian expert and National Geographic Explorer Tyrone Hayes, a professor of biology at University of California, Berkeley, conducts research on how some pesticides can harm frogs and other animals by disrupting their endocrine systems.

Climate change is also a concern. Amphibians are sensitive to even slight changes in temperature and moisture. Warmer temperatures may lead amphibians to breed too early. Extended dry periods also lead to a decline in amphibian populations by drying up breeding pools that frogs and other amphibians depend on for reproduction and survival through their early stages of life (Figure 4.A).

Overhunting is another human-related threat, especially in areas of Asia and Europe, where frogs are hunted for their leg meat. Another threat is the invasion of amphibian habitats by *nonnative predators and competitors*, such as certain fish species. Some of this immigration is natural, but humans accidentally or deliberately transport many species to amphibian habitats.

According to most amphibian experts, a combination of these factors, which vary from place to place, is responsible for most of the decline and extinctions of amphibian species. This amounts to a biological "fire alarm."

Charles H. Smith/U.S. Fish and Wildlife Service

FIGURE 4.A This golden toad lived in Costa Rica's high-altitude Monteverde Cloud Forest Reserve. The species became extinct in 1989, apparently because its habitat dried up.

CRITICAL THINKING

Of the factors listed above, which three do you think could be most effectively controlled by human efforts?

animals. The large nesting mounds alligators build provide nesting and feeding sites for some herons and egrets. Red-bellied turtles lay their eggs in old gator nests.

When alligators excavate holes and build nesting mounds, they help keep vegetation from invading shorelines and open water. Without this ecosystem service, freshwater ponds and coastal wetlands fill in with shrubs and trees, and dozens of species can disappear from these ecosystems, due to such changes. The alligators also help ensure the presence of game fish such as bass and bream by eating large numbers of gar, a predatory fish that hunts these species.

In the 1930s, hunters began killing American alligators for their exotic meat and their soft belly skin. People used the skin to make expensive shoes, belts, and purses. Other people hunted alligators for sport or out of dislike for the large reptile. By the 1960s, hunters and poachers had wiped out 90% of the alligators in the state of Louisiana, and the alligator population in the Florida Everglades was near extinction.

In 1967 the U.S. government placed the American alligator on the endangered species list. By 1977, because it was protected, its populations had made a strong comeback and the alligator was removed from the endangered

Martha Marks/Shutterstock.com

FIGURE 4.12 *Keystone species:* The American alligator plays an important ecological role in its marsh and swamp habitats in the southeastern United States by helping support many other species.

species list. Today, there are more than a million alligators in Florida, and the state allows property owners to kill alligators that stray onto their land.

To conservation biologists, the comeback of the American alligator is an important success story in wildlife conservation. Recently, however, large and rapidly reproducing Burmese and African pythons released deliberately or accidently by humans have invaded the Florida Everglades. These nonnative invaders feed on young alligators, and could threaten the long-term survival of this keystone species in the Everglades.

CONSIDER THIS . . .

THINKING ABOUT The American Alligator and Biodiversity

What are two ways in which the American alligator supports one or more of the four components of biodiversity (Figure 4.5) within its environment?

4.4 HOW DOES THE EARTH'S LIFE CHANGE OVER TIME?

CONCEPT 4.4A The scientific theory of evolution through natural selection explains how life on the earth changes over time due to changes in the genes of populations.

CONCEPT 4.4B Populations evolve when genes mutate and give some individuals genetic traits that enhance their abilities to survive and to produce offspring with these traits (natural selection).

Evolution Explains How Organisms Change Over Time

How did the earth end up with such an amazing diversity of species? The scientific answer is **biological evolution** (or simply **evolution**)—the process by which the earth's life forms change genetically over time. These changes occur in the genes of populations of organisms from generation to generation (**Concept 4.4A**). According to this scientific theory, species have evolved from earlier, ancestral species through **natural selection**. Through this process, individuals with certain genetic traits are more likely to survive and reproduce under a specific set of environmental conditions. These individuals then pass these traits on to their offspring (**Concept 4.4B**).

A large body of scientific evidence supports this idea. As a result, *biological evolution through natural selection* has become a widely accepted scientific theory. It explains how the earth's life has changed over the past 3.8 billion years and why we have today's diversity of species.

Most of what we know about the history of life on the earth comes from **fossils**—the remains or traces of past organisms. Fossils include mineralized or petrified replicas of skeletons, bones, teeth, shells, leaves, and seeds, or impressions of such items found in rocks (Figure 4.13). Scientists have discovered fossil evidence in successive layers of sedimentary rock such as limestone and sandstone. They have also studied evidence of ancient life contained in ice core samples drilled from glacial ice at the earth's poles and on mountaintops.

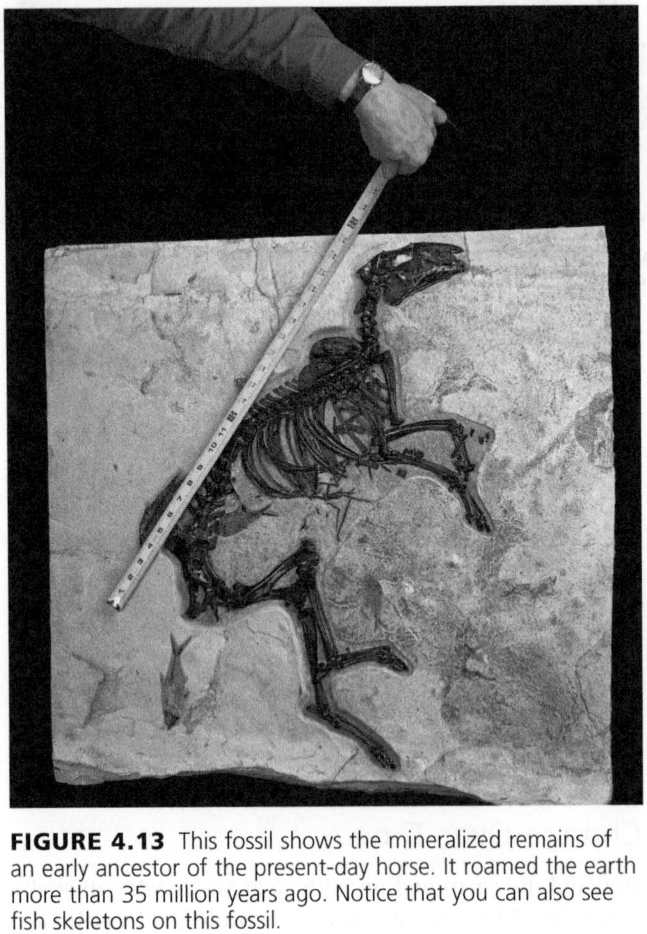

Ira Block/National Geographic Creative

FIGURE 4.13 This fossil shows the mineralized remains of an early ancestor of the present-day horse. It roamed the earth more than 35 million years ago. Notice that you can also see fish skeletons on this fossil.

This total body of evidence is called the *fossil record*. This record is uneven and incomplete because many past forms of life left no fossils and some fossils have decomposed. Scientists estimate that the fossils found so far represent probably only 1% of all species that have ever lived. There are still many unanswered scientific questions about the details of evolution by natural selection, and research continues in this area.

Evolution Depends on Genetic Variability and Natural Selection

The idea that organisms change over time and are descended from a single common ancestor has been discussed since the time of the early Greek philosophers. No one had developed an explanation of how this happened until 1858 when naturalists Charles Darwin (1809–1882) and Alfred Russel Wallace (1823–1913) independently proposed the concept of natural selection as a mechanism for biological evolution. Darwin gathered evidence for this idea and published it in his 1859 book, *On the Origin of Species by Means of Natural Selection*.

Biological evolution by natural selection involves changes in a population's genetic makeup through successive generations (**Concept 4.4A**). Populations—not individuals—evolve by becoming genetically different.

The first step in this process is the development of **genetic variability**: a variety in the genetic makeup of individuals in a population. This occurs primarily through **mutations**, which are changes in the coded genetic instructions in the DNA in a gene. During an organism's lifetime, the DNA in its cells is copied each time one of its cells divides and whenever it reproduces. In a lifetime, this happens millions of times and results in various mutations.

Most mutations result from random changes in the DNA's coded genetic instructions that occur in only a tiny fraction of these millions of divisions. Some mutations also occur from exposure to external agents such as radioactivity, ultraviolet radiation from the sun, and certain natural and human-made chemicals called *mutagens*.

Mutations can occur in any cell, but only those that take place in the genes of reproductive cells are passed on to offspring. Sometimes a mutation can result in a new genetic trait, called a *heritable trait*, which can be passed from one generation to the next. In this way, populations develop genetic differences among their individuals. Some mutations are harmful to offspring and some are beneficial.

The next step in biological evolution is *natural selection*, which explains how populations evolve in response to changes in environmental conditions by changing their genetic makeup. Through natural selection, environmental conditions favor increased survival and reproduction of certain individuals in a population. These favored individuals possess heritable traits that give them

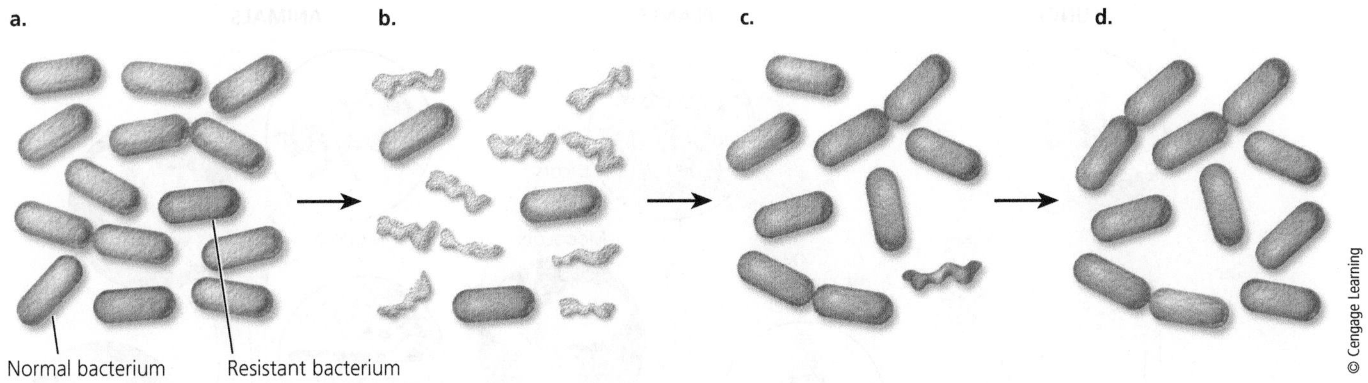

FIGURE 4.14 *Evolution by natural selection:* **(a)** A population of bacteria is exposed to an antibiotic, which **(b)** kills all individuals except those possessing a trait that makes them resistant to the drug; **(c)** the resistant bacteria multiply and eventually **(d)** replace all or most of the nonresistant bacteria.

an advantage over other individuals in the population. Such a trait is called an **adaptation**, or **adaptive trait**. An adaptive trait improves the ability of an individual organism to survive and to reproduce at a higher rate than other individuals in a population can under prevailing environmental conditions.

An example of natural selection at work is *genetic resistance*. It occurs when one or more organisms in a population have genes that can tolerate a chemical (such as a pesticide or antibiotic) that is designed to kill them. The resistant individuals survive and reproduce more rapidly than the members of the population that do not have such genetic traits. Genetic resistance can develop quickly in populations of organisms such as bacteria and insects that produce large numbers of offspring. For example, some disease-causing bacteria have developed genetic resistance to widely used antibacterial drugs, or *antibiotics* (Figure 4.14).

Our own species is an example of evolution by natural selection. If we compressed the earth's 4.6 billion years of geological and biological history into a 24-hour day, the human species arrived about a tenth of 1 second before midnight. In that short time, and especially during the last 100 years, we have dominated most of the earth's land and aquatic systems with a growing ecological footprint (see Figure 1.8, p. 12, and Figure 1.9, p. 13). Evolutionary biologists attribute our ability to dominate the planet to three major adaptations (Figure 4.15):

- *Strong opposable thumbs* that allowed humans to grip and use tools better than the few other animals that have thumbs
- *The ability to walk upright,* which gave humans agility and freed up their hands for many uses
- *A complex brain,* which allowed humans to develop many skills, including the ability to talk, read, and write in order to transmit complex ideas

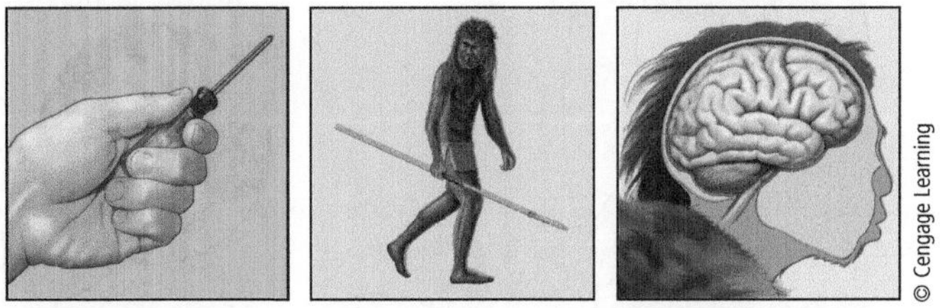

FIGURE 4.15 *Homo sapiens sapiens:* We have had three advantages over other mammals that helped us to become the earth's dominant species within an eye blink of time in the 3.8-billion-year history of life on the earth.

To summarize the process of biological evolution by natural selection: Genes mutate, individuals are selected, and populations that are better adapted to survive and reproduce under existing environmental conditions evolve (**Concept 4.4B**).

Evolutionary biologists study patterns of evolution by examining the similarities and differences among species based on their physical and genetic characteristics. They use this information to develop branching **evolutionary tree**, or **phylogenetic tree**, diagrams that depict the hypothetical evolution of various species from common ancestors. They use fossil, DNA, and other evidence to hypothesize the evolutionary pathways and connections among species. Figure 4.16 is a phylogenetic tree diagram tracing the hypothesized development of the earth's six kingdoms of life over 3.8 billion years.

On an evolutionary timescale, as new species arise, they have new genetic traits that can enhance their survival as long as environmental conditions do not change dramatically. The older species from which they originated and the new species evolve and branch out along different lines or lineages of species that can be recorded in phylogenetic tree diagrams (Figure 4.16).

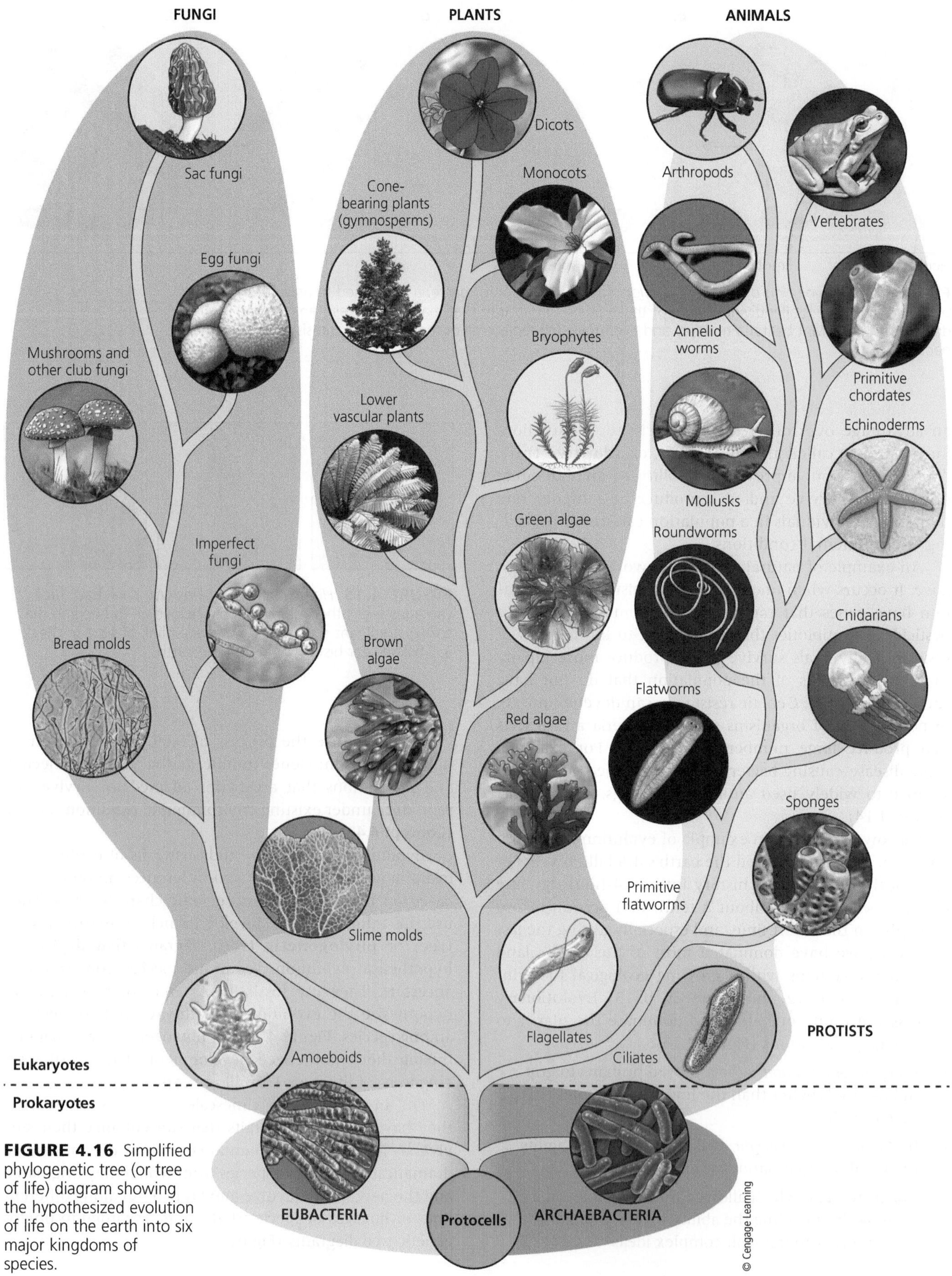

FIGURE 4.16 Simplified phylogenetic tree (or tree of life) diagram showing the hypothesized evolution of life on the earth into six major kingdoms of species.

Limits to Adaptation through Natural Selection

In the not-too-distant future, will adaptations to new environmental conditions through natural selection protect us from harm? Will adaptations allow the skin of our descendants to become more resistant to the harmful effects of UV radiation, enable our lungs to cope with air pollutants, and improve the ability of our livers to detoxify pollutants in our bodies?

Scientists in this field say *not likely* because of two limitations on adaptation through natural selection. *First*, a change in environmental conditions leads to adaptation only for genetic traits already present in a population's gene pool, or if they arise from mutations—which occur randomly.

Second, even if a beneficial heritable trait is present in a population, the population's ability to adapt may be limited by its reproductive capacity. Populations of genetically diverse species that reproduce quickly can often adapt to a change in environmental conditions in a short time (days to years). Examples are dandelions, mosquitoes, rats, bacteria, and cockroaches. By contrast, species that cannot produce large numbers of offspring rapidly—such as elephants, tigers, sharks, and humans—take thousands or even millions of years to adapt through natural selection.

Myths about Evolution through Natural Selection

There are a number of misconceptions about biological evolution through natural selection. Here are five common myths:

- *Survival of the fittest means survival of the strongest*. To biologists, *fitness* is a measure of reproductive success, not strength. Thus, the fittest individuals are those that leave the most descendants, not those that are physically the strongest.
- *Evolution explains the origin of life*. It does not. However, it does explain how species have evolved after life came into being around 3.8 billion years ago.
- *Humans evolved from apes or monkeys*. Fossil and other evidence shows that humans, apes, and monkeys evolved along different paths from a common ancestor that lived 5 million to 8 million years ago.
- *Evolution by natural selection is part of a grand plan in nature in which species are to become more perfectly adapted*. There is no evidence of such a plan. Instead, evidence indicates that the forces of natural selection and random mutations can push evolution along any number of paths.
- *Evolution by natural selection is not important because it is just a theory*. This reveals a misunderstanding of what a scientific theory is. A scientific hypothesis becomes a scientific theory only after thorough testing and evaluation by the experts in a field. Numerous polls show that evolution by natural selection is widely accepted by over 95% of biologists because it is the best explanation for the earth's biodiversity and how populations of different species have adapted to changes in the earth's environmental conditions over billions of years.

4.5 WHAT FACTORS AFFECT BIODIVERSITY?

CONCEPT 4.5A As environmental conditions change, the balance between the formation of new species and the extinction of existing species determines the earth's biodiversity.

CONCEPT 4.5B Human activities are decreasing biodiversity by causing the extinction of many species and by destroying or degrading habitats needed for the development of new species through natural selection.

How Do New Species Arise?

Under certain circumstances, natural selection can lead to an entirely new species. In this process, called **speciation**, one species evolves into two or more different species. For sexually reproducing organisms, a new species forms when a separated population of a species evolves to the point where its members can no longer interbreed and produce fertile offspring with members of another population of its species that did not change or that evolved differently.

Speciation, especially among sexually reproducing species, happens in two phases: first geographic isolation, and then reproductive isolation. **Geographic isolation** occurs when different groups of the same population of a species become physically isolated from one another for a long time. Part of a population may migrate in search of food and then begin living as a separate population in an area with different environmental conditions. Winds and flowing water may carry a few individuals far away where they establish a new population. A flooding stream, a new road, a hurricane, an earthquake, or a volcanic eruption, as well as long-term geological processes (Science Focus 4.2), can also separate populations. Human activities, such as the creation of large reservoirs behind dams and the clearing of forests, can also create physical barriers for certain species. The separated populations can develop quite different genetic characteristics because they are no longer exchanging genes.

In **reproductive isolation**, mutation and change by natural selection operate independently in the gene pools of geographically isolated populations. If this process continues long enough, members of isolated populations of sexually reproducing species can become different in genetic makeup. Then they cannot produce live, fertile offspring if they are rejoined and attempt to interbreed. When that is the case, speciation has occurred and one species has become two (Figure 4.17).

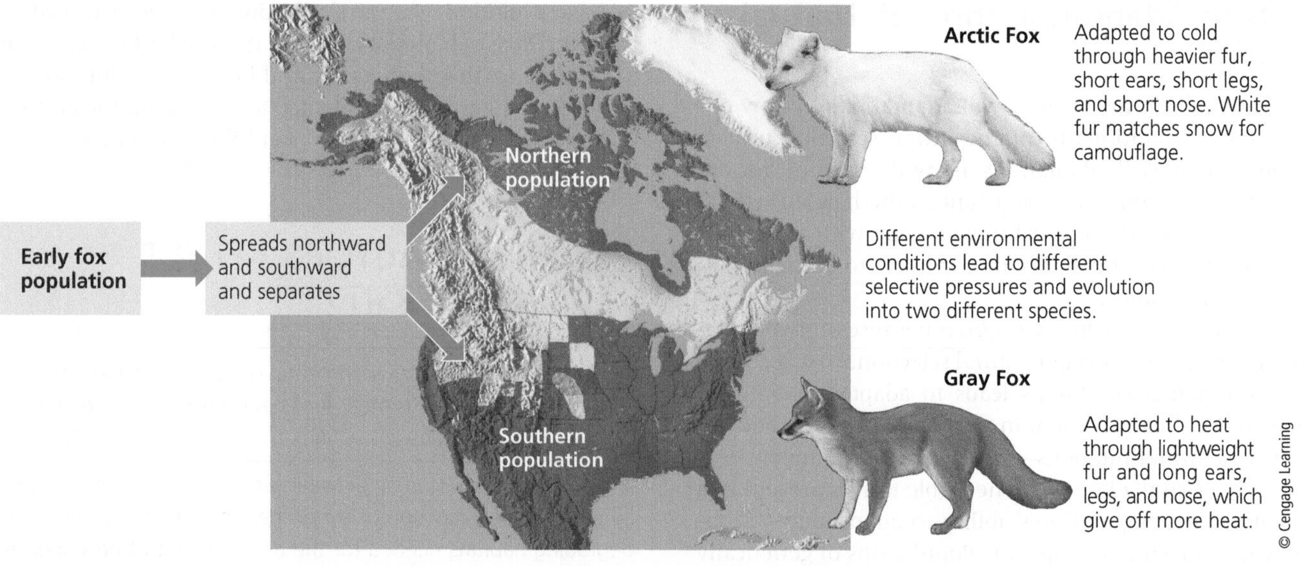
Arctic Fox
Adapted to cold through heavier fur, short ears, short legs, and short nose. White fur matches snow for camouflage.
Northern population
Early fox population
Spreads northward and southward and separates
Different environmental conditions lead to different selective pressures and evolution into two different species.
Gray Fox
Southern population
Adapted to heat through lightweight fur and long ears, legs, and nose, which give off more heat.
© Cengage Learning

FIGURE 4.17 *Geographic isolation* can lead to reproductive isolation, divergence of gene pools, and speciation.

SCIENCEFOCUS 4.2

Geological Processes Affect Biodiversity

The earth's surface has changed dramatically over its long history. Scientists discovered that huge flows of molten rock within the earth's interior have broken its surface into a number of gigantic solid plates, called *tectonic plates*. For hundreds of millions of years, these plates have drifted slowly on the planet's mantle (Figure 4.B).

Rock and fossil evidence indicates that 200–250 million years ago, all of the earth's present-day continents were connected in a supercontinent called Pangaea (Figure 4.B, left). About 175 million years ago, Pangaea began splitting apart as the earth's tectonic plates moved. Eventually tectonic movement resulted in the present-day locations of the continents (Figure 4.B, right).

The drifting of tectonic plates has had two important effects on the evolution and distribution of life on the earth. *First*, the locations of continents and oceanic basins have greatly influenced the earth's climate, which plays a key role in where plants and animals can live. *Second*, the breakup, movement, and joining of continents has allowed species to move and adapt to new environments. This led to the formation of large number of new species through speciation.

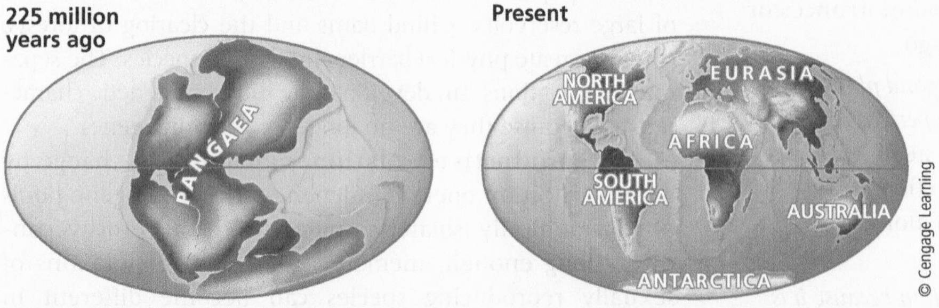
225 million years ago
PANGAEA
Present
NORTH AMERICA
EURASIA
AFRICA
SOUTH AMERICA
AUSTRALIA
ANTARCTICA
© Cengage Learning

FIGURE 4.B Over millions of years, the earth's continents have moved very slowly on several gigantic tectonic plates. ***Critical thinking:*** How might an area of land splitting apart cause the extinction of a species?

Sometimes tectonic plates that are grinding along next to one another shift quickly. Such sudden movement of tectonic plates can cause *earthquakes*. This can affect biological evolution by causing fissures in the earth's crust that, on rare occasions, can separate and isolate populations of species. Over long periods, this can lead to the formation of new species as each isolated population changes genetically in response to new environmental conditions.

Volcanic eruptions that occur along the boundaries of tectonic plates can also affect extinction and speciation by destroying habitats and reducing, isolating, or wiping out populations of species. We discuss these geological processes in detail in Chapter 14.

CRITICAL THINKING

The earth's tectonic plates, including the one you are riding on, typically move at about the rate at which your fingernails grow. If they stopped moving, how might this affect the future biodiversity of the planet?

Artificial Selection, Genetic Engineering, and Synthetic Biology

Scientists use **artificial selection** to change the genetic characteristics of populations with similar genes. First, they select one or more desirable genetic traits that already exist in the population of a plant or animal such as a type of wheat, dog, or fruit. Then they use *selective breeding*, or *crossbreeding*, to control which members of a population have the opportunity to reproduce to increase the numbers of individuals in a population with the desired traits (Figure 4.18).

Artificial selection is limited to crossbreeding between genetic varieties of the same species or between species that are genetically similar to one another, and is not a form of speciation. Most of the grains, fruits, and vegetables we eat are produced by artificial selection. Artificial selection has also given us food crops with higher yields, cows that give more milk, trees that grow faster, and many different varieties of dogs and cats. However, traditional crossbreeding is a slow process.

Scientists have learned how to speed this process of change by manipulating genes to get desirable genetic traits or eliminate undesirable ones in a plant or animal. They do this by transferring segments of DNA with the desired trait from one species to another through a process called **genetic engineering**. In this process, also known as *gene splicing*, scientists alter an organism's genetic material by adding, deleting, or changing segments of its DNA to produce desirable traits or to eliminate undesirable ones. Scientists have used genetic engineering to develop modified crop plants, new drugs, pest-resistant plants, and animals that grow rapidly.

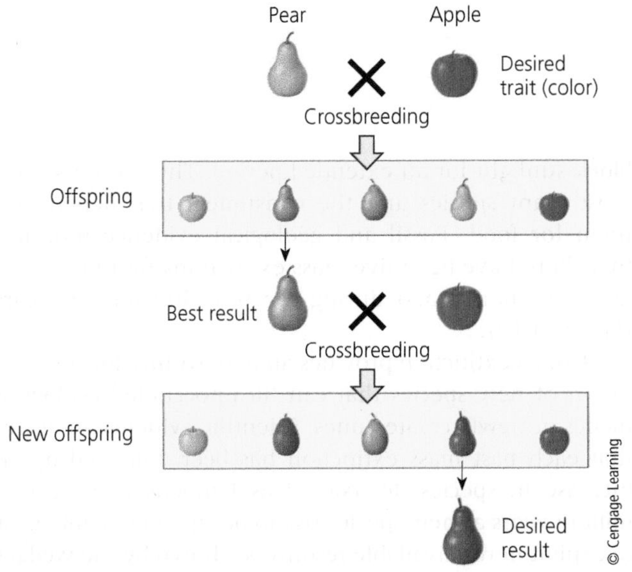

FIGURE 4.18 Artificial selection involves the crossbreeding of species that are close to one another genetically. In this example, similar fruits are being crossbred to yield a pear with a certain color.

There are five steps in this process:

1. Identify a gene with the desired trait in the DNA found in the nucleus of a cell from the donor organism.
2. Extract a small circular DNA molecule, called a *plasmid*, from a bacterial cell.
3. Insert the desired gene from step 1 into the plasmid to form a *recombinant DNA plasmid*.
4. Insert the recombinant DNA plasmid into a cell of another bacterium, which rapidly divides and reproduces large numbers of bacterial cells with the desired DNA trait.
5. Transfer the genetically modified bacterial cells to a plant or animal that is to be genetically modified.

The result is a **genetically modified organism (GMO)** with its genetic information modified in a way not found in natural organisms. Genetic engineering enables scientists to transfer genes between different species that would not interbreed in nature. For example, scientists can put genes from a cold-water fish species into a tomato plant to give it properties that help it resist cold weather. Recently, scientists have learned how to treat certain genetic diseases by altering or replacing the genes that cause them. Genetic engineering has revolutionized agriculture and medicine. However, it is a controversial technology, as we discuss in Chapter 12.

Another emerging field is synthetic biology, as discussed in Science Focus 4.3.

Extinction Eliminates Species

Another factor affecting the number and types of species on the earth is **extinction**. Extinction occurs when an entire species ceases to exist. When environmental conditions change dramatically, a population of a species faces three possible futures: (1) *adapt* to the new conditions through natural selection, (2) *migrate* (if possible) to another area with more favorable conditions, or (3) *become extinct*.

Species found in only one area, called **endemic species**, are especially vulnerable to extinction. They exist on islands and in other isolated areas. For example, many species in tropical rain forests have highly specialized roles and are vulnerable to extinction. These organisms are unlikely to be able to migrate or adapt to rapidly changing environmental conditions. Many of these endangered species are amphibians (**Core Case Study**).

Fossils and other scientific evidence indicate that all species eventually become extinct. In fact, the evidence indicates that 99.9% of all species that have existed on the earth have gone extinct. Throughout most of the earth's long history, species have disappeared at a low rate, called the **background extinction rate**. Based on the fossil record and analysis of ice cores, biologists

Synthetic Biology

Synthetic biology, or **biological engineering**, is a new and rapidly growing field. This technology enables scientists to make new sequences of DNA and use such genetic information to design and create artificial cells, tissues, body parts, and organisms not found in nature.

This process starts with a computer code of an organism's entire genetic sequence (genome). Then engineers insert new sequences of the four nucleotide bases, adenine (A), cytosine (C), guanine (G), and thymine (T) (see Figure 2.9, p. 38), to create a new and different genetic sequence or genome. Next, they transplant the new genome into the cell of a bacterium to transform it into a different, human-created species of bacteria. In May 2010 J. Craig Venter and his colleagues announced that they had created an entirely new living and self-replicating bacterium by using a computer to come up with a new genome.

This was a dramatic demonstration of the ability of synthetic biology to bypass the long process of evolution by natural selection and create new forms of life in a short time. This technology uses science and engineering to alter the planet's life by reducing the cell to a machine that can assemble forms of life like products in a factory.

Proponents of this new technology want to use it to create bacteria that can produce hydrogen gas or liquid fuels from sunlight to help us phase out the use of fossil fuels. They also view it as a way to destroy harmful substances in our bodies and to create new vaccines to prevent diseases and drugs to combat parasitic diseases such as malaria. It could also be used to develop instructions for three-dimensional printers to print human body parts and products such as bioplastics, packaging, car parts, and clothes.

Synthetic biology could also be used to create bacteria and algae that break down oil, industrial wastes, toxic heavy metals, pesticides, and radioactive waste in contaminated soil and water. We are a long way from achieving such goals but the potential is there. If used properly and ethically, this new technology could help us live more sustainably.

The problem is that like any technology, synthetic biology can be used for good or bad. It could also be used to create biological weapons such as deadly bacteria that spread new diseases, to destroy existing oil deposits, or to interfere with the chemical cycles that keep us alive. It might also end up hindering the ability of decomposers to breakdown and recycle wastes or adding new pollutants to soil and water.

To some analysts, we have used our increasing power over the earth's natural processes to take over most of the planet's surface and degrade our own life-support system. Adding this new form of power over the earth's life and processes could worsen and speed up our harmful and unsustainable environmental impacts. According to naturalist and writer Diane Ackerman, "We're at a dangerous age in our evolution as a species: clever, headstrong, impulsive, and far better at tampering with nature than understanding it." This is why many scientists call for increased monitoring and regulation of this new technology before it gets out of control.

CRITICAL THINKING

Explain why you are for or against the widespread use of synthetic biology. How would you control its use?

estimate that the average annual background extinction rate has been about 0.0001% of all species per year, which amounts to 1 species lost for every million species on the earth per year. At this rate, if there were 10 million species on the earth, about 10 of them, on average, would go extinct every year.

Evidence indicates that life on the earth has been sharply reduced by several periods of **mass extinction** during which there is a significant rise in extinction rates, well above the background rate. In such a catastrophic, widespread, and often global event, large groups of species (20–95% of all species) are wiped out. The causes of such extinctions are unknown, but possible events include gigantic volcanic eruptions and collisions with giant meteors and asteroids. Such events would trigger drastic environmental changes on a global scale, including massive release of debris into the atmosphere that would block sunlight for an extended period. This would kill off most plant species and the consumers that depend on them for food. Fossil and geological evidence indicates that there have been five mass extinctions (at intervals of 20–60 million years) during the past 500 million years (Figure 4.19).

A mass extinction provides an opportunity for the evolution of new species that can fill unoccupied ecological niches or newly created ones. Scientific evidence indicates that each past mass extinction has been followed by an increase in species diversity. This happens over several million years as new species rise to occupy new habitats or to exploit newly available resources (shown by the wedges in Figure 4.19) following each mass extinction.

As environmental conditions change, the degree of balance between speciation and extinction determines the earth's biodiversity (**Concept 4.5A**). The existence of

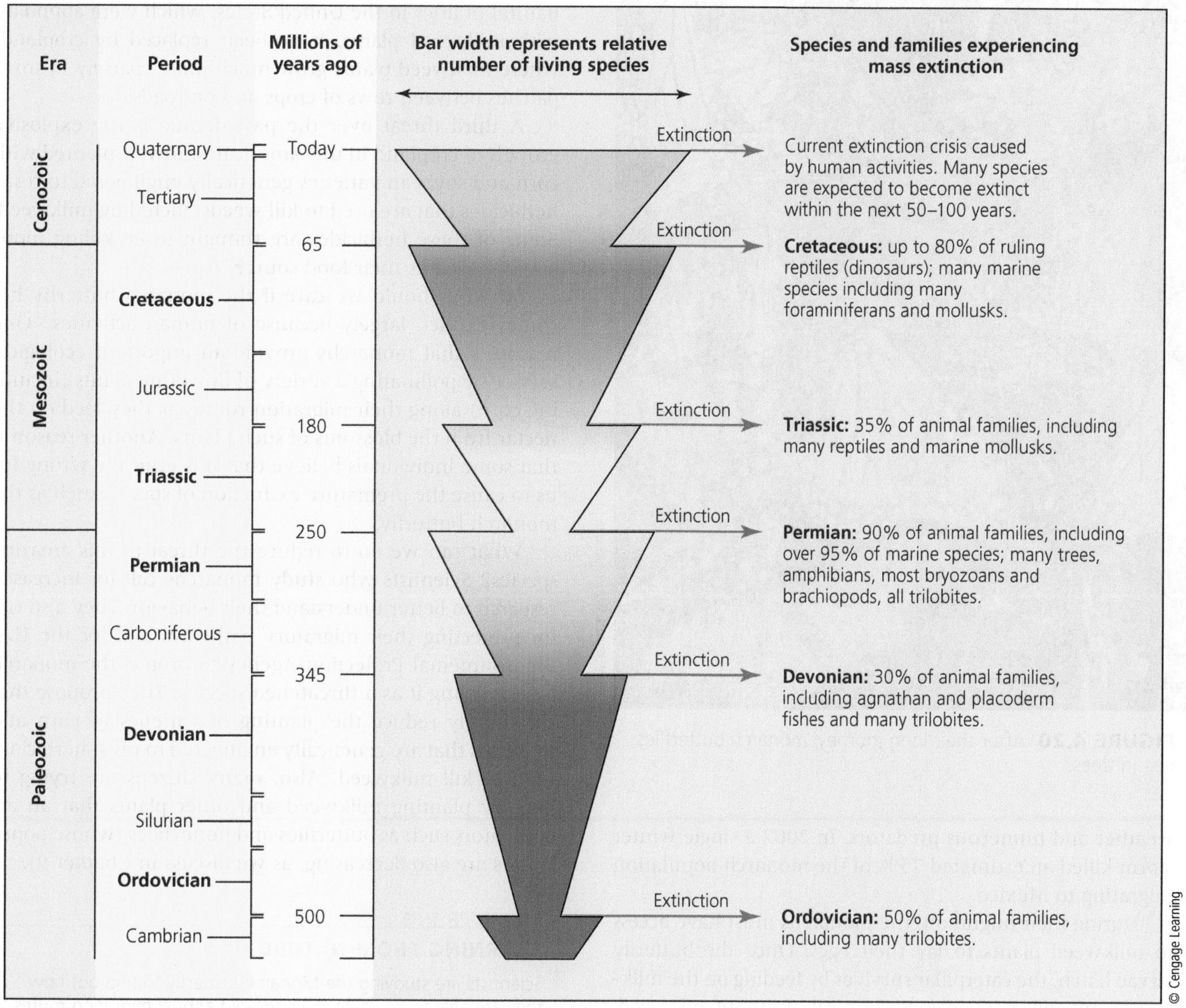
Era
Period
Millions of years ago
Bar width represents relative number of living species
Species and families experiencing mass extinction
Cenozoic
Mesozoic
Paleozoic
Quaternary
Tertiary
Cretaceous
Jurassic
Triassic
Permian
Carboniferous
Devonian
Silurian
Ordovician
Cambrian
Today
65
180
250
345
500
Extinction
Current extinction crisis caused by human activities. Many species are expected to become extinct within the next 50–100 years.
Extinction
Cretaceous: up to 80% of ruling reptiles (dinosaurs); many marine species including many foraminiferans and mollusks.
Extinction
Triassic: 35% of animal families, including many reptiles and marine mollusks.
Extinction
Permian: 90% of animal families, including over 95% of marine species; many trees, amphibians, most bryozoans and brachiopods, all trilobites.
Extinction
Devonian: 30% of animal families, including agnathan and placoderm fishes and many trilobites.
Extinction
Ordovician: 50% of animal families, including many trilobites.
© Cengage Learning

FIGURE 4.19 Scientific evidence indicates that the earth has experienced five mass extinctions over the past 500 million years and that human activities have initiated a new sixth mass extinction.

millions of species today means that speciation, on average, has kept ahead of extinction. However, evidence indicates that the global extinction rate is rising dramatically, as we discuss more fully in Chapter 9. Many scientists see this is as evidence that we are experiencing the beginning of a new sixth mass extinction (Figure 4.19).

There is also considerable evidence that much of the current rise in the extinction rate and the resulting loss of biodiversity are primarily due to human activities (**Concept 4.5B**). As our ecological footprint spreads over the planet (see Figure 1.8, p. 12), so does extinction. Research indicates that the largest cause of the rising rate of species extinctions is the loss, fragmentation, and degradation of habitats. We examine this issue further in Chapter 9. The Case Study that follows discusses the threat of extinction for the monarch butterfly because of human activities.

CASE STUDY

The Threatened Monarch Butterfly

The beautiful North American monarch butterfly (see chapter-opening photo) is in serious trouble. This species is known for its annual 3,200- to 4,800-kilometer (2,000- to 3,000-mile) migration from the northern parts of Canada and the United States to a small number of tropical forest areas in central Mexico. They arrive during winter on a predictable schedule and later return to their North American summer ranges. Another monarch population in the Midwestern United States makes a shorter annual journey to coastal northern California and then returns to the midwest.

During their annual round-trip journeys, these two populations of monarchs face serious threats from bad

FIGURE 4.20 After their long journey, monarch butterflies rest in trees.

weather and numerous predators. In 2002 a single winter storm killed an estimated 75% of the monarch population migrating to Mexico.

During their migration, the monarchs must have access to milkweed plants to lay their eggs. Once the butterfly larvae hatch, the caterpillar survives by feeding on the milkweed plant and later becomes a butterfly. Without milkweed, the monarch cannot reproduce and faces extinction.

Once the monarchs reach their winter forest destinations in Mexico and California, they cling to trees by the millions (Figure 4.20) as they rest. Each year, biologists estimate the monarch's population size by measuring the total areas of forest they occupy at these destinations.

The overall estimated monarch population varies from year to year, mostly because of changes in weather and other natural conditions. However, the U.S. Fish and Wildlife Service estimates that since 1975, this overall population has dropped by nearly a billion. Since 1996, the number of monarchs migrating to their winter range in Mexico has dropped by more than 90%.

The monarchs now face three serious threats from human activities in addition to their historic threats from nature. One threat is the steady loss of their forest habitat in Mexico, due to logging (most of it illegal), and loss of their northern California habitat due to coastal development. A second threat is reduced access to milkweed plants essential for their survival during their migration. Almost all of the natural prairies in the United States, which were abundant with milkweed plants, have been replaced by croplands where milkweed plants grow much more sparsely in small patches between rows of crops and on roadsides.

A third threat over the past decade is the explosive growth of cropland in the American Midwest planted with corn and soybean varieties genetically engineered to resist herbicides that are used to kill weeds, including milkweed. Some of these herbicides are thought to be killing monarchs as well as their food source.

So why should we care if the monarch butterfly becomes extinct, largely because of human activities? One reason is that monarchs provide an important ecological service by pollinating a variety of flowering plants (including corn) along their migration routes as they feed on the nectar from the blossoms of such plants. Another reason is that some individuals believe that it is ethically wrong for us to cause the premature extinction of species such as the monarch butterfly.

What can we do to reduce the threat to this amazing species? Scientists who study monarchs call for increased research to better understand their behavior. They also call for protecting their migratory pathways and for the U.S. Environmental Protection Agency to protect the monarch by classifying it as a threatened species. They propose that we sharply reduce the planting of varieties of corn and soybeans that are genetically engineered to resist herbicides used to kill milkweed. Also, many citizens are trying to help by planting milkweed and other plants that attract pollinators such as butterflies and honeybees (whose populations are also decreasing, as we discuss in Chapter 9).

CONSIDER THIS . . .

LEARNING FROM NATURE

Scientists are studying the Monarch butterfly to find out how they are able to navigate their age-old annual migration routes and arrive at the same places in Mexico and California on the same day of each year.

BIG IDEAS

- Each species plays a specific ecological role, called its *niche*, in the ecosystems where it is found.
- As environmental conditions change, the genes in some individuals mutate and give those individuals genetic traits that enhance their abilities to survive and to produce offspring with these traits.
- The degree of balance between speciation and extinction in response to changing environmental conditions determines the earth's biodiversity, which helps to sustain the earth's life and our economies.

Tying It All Together

Amphibians and Sustainability

This chapter's **Core Case Study** describes the increasing losses of amphibian species and explains why these species are ecologically important. In this chapter, we studied the importance of biodiversity—the numbers and varieties of species found in different parts of the world, along with genetic, ecosystem, and functional diversity.

We examined the variety of roles played by species in ecosystems—their niches. For example, we saw that some species, including many amphibians, are indicator species that warn us about threats to biodiversity, ecosystems, and the biosphere. Others such as the American alligator are keystone species that play vital roles in sustaining the ecosystems where they live.

Robert King/Shutterstock.com

We also studied the scientific theory of biological evolution through natural selection, which explains how life on the earth changes over time due to changes in the genes of populations and how new species can arise. We learned that the earth's biodiversity is the result of a balance between the formation of new species (speciation) and extinction of species due to changing environmental conditions.

The ecosystems where amphibians and other species live are functioning examples of the three **scientific principles of sustainability** in action. These species depend on solar energy, the cycling of nutrients, and biodiversity. Disruptions in any of these forms of natural capital can result in degradation of these species' populations and their ecosystems.

Chapter Review

Core Case Study

1. Describe the threats to many of the world's amphibian species and explain why we should avoid hastening the extinction of amphibian species through our activities.

Section 4.1

2. What is the key concept for this section? Distinguish between **eukaryotic** and **prokaryotic** organisms in terms of their cell structures. What are the six kingdoms of life? Distinguish between **annual** and **perennial** plants and between **invertebrate** and **vertebrate** animals.

3. What is a **species**? About how many species exist on the earth? How many of these species have we identified? Why are insects such important species?

Section 4.2

4. What is the key concept for this section? Define **biodiversity (biological diversity)** and list and describe its four major components. Why is biodiversity important? What is **species diversity**? Distinguish between **species richness** and **species evenness**. What is **genetic diversity**? What is **ecosystem diversity**? Define and give three examples of **biomes**. What are **edge effects**? What is an **ecotone**? What is **functional diversity**? Summarize the scientific contributions of Edward O. Wilson.

Section 4.3

5. What are the two key concepts for this section? Define and distinguish between **ecological niche** and a **habitat**. Distinguish between **generalist species** and **specialist species** and give an example of each.

6. Define and distinguish among **native**, **nonnative**, **indicator**, and **keystone species** and give an example of each. List six factors that threaten many species of frogs and other amphibians with extinction. Describe the role of the American alligator as a keystone species.

Section 4.4

7. What are the two key concepts for this section? Define **biological evolution (evolution)** and **natural selection** and explain how they are related. What is the scientific theory of biological evolution through natural selection? What are **fossils** and how do scientists use them to understand evolution? What is a **mutation** and what role do mutations play in evolution through natural selection? What is **genetic variability**? What is an **adaptation**, or **adaptive trait**? Explain how harmful bacteria can become genetically resistant to antibiotics. What three genetic adaptations have helped humans to become such a dominant species? What is an **evolutionary tree** or **phylogenetic tree**? How does it help describe evolution? What are two limitations on evolution through natural selection? What are five common myths about evolution through natural selection?

Section 4.5

8. What are the two key concepts for this section? Define **speciation**. Distinguish between **geographic isolation** and **reproductive isolation** and explain how they can lead to the formation of a new species. Explain how geological processes can affect biodiversity. Define and distinguish between **artificial selection** and **genetic engineering**, give an example of each, and explain how they differ from evolution by natural selection. What is a **genetically modified organism (GMO)**? Define **synthetic biology** or **biological engineering**, explain how it differs from evolution by natural selection, and point out some of its potential benefits and dangers.

9. What is **extinction**? What is an **endemic species** and why are such species vulnerable to extinction? Define and distinguish between the **background extinction rate** and a **mass extinction**. How many mass extinctions has the earth experienced? What is one of the leading causes of the rising rate of extinction? Explain why the monarch butterfly is threatened with extinction.

10. What are this chapter's *three big ideas*? How are ecosystems where amphibians and other species live functioning examples of the three **scientific principles of sustainability**?

Note: Key terms are in bold type.

Critical Thinking

1. What might happen to humans and a number of other species if most or all amphibian species (**Core Case Study**) were to go extinct?

2. How might a reduction in species diversity affect the other three components of biodiversity?

3. Is the human species a keystone species? Explain. If humans become extinct, what are three species that might also become extinct and what are three species whose populations might grow?

4. Why should we care about saving the monarch butterfly from extinction caused by human activities?

5. How would you respond to someone who tells you:
   a. We should not believe in biological evolution because it is "just a theory."
   b. We should not worry about air pollution because natural selection will enable humans to develop lungs that can detoxify pollutants.

5. How would you respond to someone who says that because extinction is a natural process, we should not worry about the loss of biodiversity when species become extinct largely because of our activities?

6. List three aspects of your lifestyle that could be contributing to some of the losses of the earth's biodiversity. For each of these, what are some ways to avoid making this contribution?

7. Congratulations! You are in charge of the future evolution of life on the earth. What are the three most important things that you would do? Explain.

Doing Environmental Science

Study an ecosystem of your choice, such as a meadow, a patch of forest, a garden, or an area of wetland. (If you cannot do this physically, do so virtually by reading about an ecosystem online or in a library.) Determine and list five major plant species and five major animal species in your ecosystem. Which, if any, of these species are indicator species and which of them, if any, are keystone species? Explain how you arrived at these hypotheses. Then design an experiment to test each of your hypotheses, assuming you would have unlimited means to carry them out.

Global Environment Watch Exercise

Go to your MindTap course to access the GREENR database. Using the "Basic Search" box at the top of the page, search for *Amphibians* to find out more about the current state of various amphibian species with regard to threats to their existence (**Core Case Study**). What actions are being taken by various nations and organizations to protect amphibians? Write a short summary report on your research.

Data Analysis

The following table is a sample of a very large body of data reported by J. P. Collins, M. L. Crump, and T. E. Lovejoy III in their book *Extinction in Our Times—Global Amphibian Decline*. It compares various areas of the world in terms of the number of amphibian species found and the number of amphibian species that were endemic, or unique to each area. Scientists like to know these percentages because endemic species tend to be more vulnerable to extinction than do non-endemic species. Study the table and then answer the following questions.

1. Fill in the fourth column by calculating the percentage of amphibian species that are endemic to each area. (Percentage endemic = number of species divided by number of endemic species.)
2. Which two areas have the highest numbers of endemic species? Name the two areas with the highest percentages of endemic species.
3. Which two areas have the lowest numbers of endemic species? Which two areas have the lowest percentages of endemic species?
4. Which two areas have the highest percentages of non-endemic species?

Area	Number of Species	Number of Endemic Species	Percentage Endemic
Pacific/Cascades/Sierra Nevada Mountains of North America	52	43	
Southern Appalachian Mountains of the United States	101	37	
Southern Coastal Plain of the United States	68	27	
Southern Sierra Madre of Mexico	118	74	
Highlands of Western Central America	126	70	
Highlands of Costa Rica and Western Panama	133	68	
Tropical Southern Andes Mountains of Bolivia and Peru	132	101	
Upper Amazon Basin of Southern Peru	102	22	

CHAPTER 5

Species Interactions, Ecological Succession, and Population Control

In looking at nature, never forget that every single organic being around us may be said to be striving to increase its numbers.

CHARLES DARWIN

Key Questions

5.1 How do species interact?

5.2 How do communities and ecosystems respond to changing environmental conditions?

5.3 What limits the growth of populations?

A clownfish gains protection by living among sea anemones and helps protect the anemones from some of their predators.

Morrison/Dreamstime.com

Core Case Study

The Southern Sea Otter: A Species in Recovery

Southern sea otters (Figure 5.1, left) live in giant kelp forests (Figure 5.1, right) in shallow waters along parts of the Pacific coast of North America. Most of the remaining members of this endangered species are found off the California coast between the cities of Santa Cruz and Santa Barbara.

Fast and agile swimmers, the otters dive to the ocean bottom looking for shellfish and food. They swim on their backs on the ocean surface and use their bellies as a table to eat their prey (Figure 5.1, left). Each day, a sea otter consumes 20–35% of its weight in clams, mussels, crabs, sea urchins, abalone, and 40 other species of bottom-dwelling organisms. Their incredibly dense fur traps air bubbles and keeps them warm.

At one time, an estimated 13,000 to 20,000 southern sea otters lived in California's coastal waters. By the early 1900s, they had been hunted almost to extinction by fur traders who killed them for their luxurious fur. Commercial fishers also killed otters, viewing them as competitors in the hunt for valuable abalone and other shellfish.

The otter population grew from a low of 50 in 1938 to 1,850 in 1977 when the U.S. Fish and Wildlife listed the species as endangered. Since then, it has continued to make a slow recovery to 3,054 individuals in 2015, which is below the size needed for it to be removed from the endangered species list.

Why should we care about the southern sea otters of California? One reason is ethical: Many people believe it is wrong to allow human activities to cause the extinction of a species. Another reason is that people love to look at these appealing and highly intelligent animals as they play in the water. As a result, otters help to generate millions of dollars a year in tourism revenues. A third reason—and a key reason in our study of environmental science—is that biologists classify the southern sea otter as a *keystone species* (see p. 83). Scientists hypothesize that in the absence of southern sea otters, sea urchins and other kelp-eating species would probably destroy the Pacific coast kelp forests and much of the rich biodiversity they support.

Biodiversity is an important part of the earth's natural capital and is the focus of one of the three **scientific principles of sustainability**. In this chapter, we look at how species interact and help control one another's population sizes. We also explore how communities, ecosystems, and populations of species respond to changes in environmental conditions.

FIGURE 5.1 An endangered southern sea otter in Monterey Bay, California (USA) uses a stone to crack the shells of the clams that it feeds on (left). It lives in a bed of seaweed called *giant kelp* (right).

Left: Kirsten Wahlquist/Dreamstime.com. Right: Paul Whitted/Shutterstock.com.

5.1 HOW DO SPECIES INTERACT?

CONCEPT 5.1 Five types of interactions among species—interspecific competition, predation, parasitism, mutualism, and commensalism—affect the resource use and population sizes of species.

Competition for Resources

Ecologists have identified five basic types of interactions among species as they share limited resources such as food, shelter, and space. These types of interactions are called *interspecific competition, predation, parasitism, mutualism,* and *commensalism.* These interactions affect the population sizes of the species in an ecosystem and their use of resources (**Concept 5.1**).

Competition is the most common interaction among species. It occurs when members of one or more species interact to use the same limited resources such as food, water, light, and space. Competition between different species is called **interspecific competition**. It plays a larger role in most ecosystems than *intraspecific competition*—competition among members of the same species.

When two species compete with one another for the same resources, their niches (p. 82) overlap. The greater this overlap, the more they compete for key resources. For example, if species A takes over the largest share of one or more key resources, then competing species B must move to another area (if possible) or suffer a population decline.

Given enough time for natural selection to occur, populations can develop adaptations that enable them to reduce or avoid competition with other species. An example is **resource partitioning**, which occurs when different species competing for similar scarce resources evolve specialized traits that allow them to share the same resources. This can involve using parts of the resources or using the resources at different times or in different ways. Figure 5.2 shows resource partitioning by insect-eating bird species. Adaptations allow the birds to reduce competition by feeding in different portions of certain spruce trees and by feeding on different insect species.

Another example of resource partitioning through natural selection involves birds called *honeycreepers* that live in the U.S. state of Hawaii (Figure 5.3). Figure 4.10 (p. 83) shows how the evolution of specialized feeding niches has reduced competition for resources among bird species in a coastal wetland.

Predation

In **predation**, a member of one species is a **predator** that feeds directly on all or part of a member of another species, the **prey**. A brown bear (the predator) and a salmon (the prey) are engaged in a **predator–prey relationship** (Figure 5.4). Such a relationship (between a lion and a zebra) is also shown in Figure 3.6, p. 53. This type of species interaction has a strong effect on population sizes and other factors in many ecosystems.

FIGURE 5.2 *Sharing the wealth:* Resource partitioning among five species of insect-eating warblers in the spruce forests of the U.S. state of Maine. Each species spends at least half its feeding time in its associated yellow-highlighted areas of these spruce trees.

FIGURE 5.3 *Specialist species of honeycreepers:* Through natural selection, different species of honeycreepers have shared resources by evolving specialized beaks to take advantage of certain types of food such as insects, seeds, fruits, and nectar from certain flowers. ***Question:*** Look at each bird's beak and take a guess at what sort of food that bird might eat.

FIGURE 5.4 *Predator–prey relationship:* This brown bear (the predator) in the U.S. state of Alaska has captured and will feed on this salmon (the prey).

CONSIDER THIS . . .

CONNECTIONS Grizzly Bears and Moths

During the summer months, the grizzly bears of the Greater Yellowstone ecosystem in the western United States eat huge amounts of army cutworm moths, which huddle in masses high on remote mountain slopes. In this predator–prey interaction, one grizzly bear can dig out and lap up as many as 40,000 cutworm moths in a day. Consisting of 50–70% fat, the moths offer a nutrient that the bear can store in its fatty tissues and draw on during its winter hibernation.

In a giant kelp forest ecosystem, sea urchins prey on kelp, a type of seaweed (Science Focus 5.1). As a keystone species, southern sea otters (**Core Case Study**) prey on the sea urchins and prevent them from destroying the kelp forests. An adult southern sea otter can eat as many as 1,500 sea urchins a day.

Predators use a variety of ways to help them capture prey. *Herbivores* can walk, swim, or fly to the plants they feed on. Many *carnivores,* such as cheetahs, use speed to chase down and kill prey, such as zebras. Eagles and hawks can fly and have keen eyesight to find prey. Some predators such as female African lions work in groups to capture large or fast-running prey.

Other predators use *camouflage* to hide in plain sight and ambush their prey. For example, praying mantises (see Figure 4.4, right, p. 78) sit on flowers or plants of a color similar to their own and ambush visiting insects. White ermines (a type of weasel), snowy owls, and arctic foxes (Figure 5.5) hunt their prey in snow-covered areas. People camouflage themselves to hunt wild game and use camouflaged traps to capture wild animals. Some predators use *chemical warfare* to attack their prey. For example, some spiders and poisonous snakes use venom to paralyze their prey and to deter their predators.

Prey species have evolved many ways to avoid predators. Some can run, swim, or fly fast and some have highly developed senses of sight, sound, or smell that alert them to the presence of predators. Other adaptations include protective shells (abalone and turtles), thick bark (giant sequoia trees), spines (porcupines), and thorns (cacti and rose bushes).

Other prey species use camouflage to blend into their surroundings. Some insect species have shapes that look like twigs (Figure 5.6a), or bird droppings on leaves. A leaf insect can be almost invisible against its background (Figure 5.6b), as can an arctic hare in its white winter fur.

Prey species also use *chemical warfare*. Some discourage predators by containing or emitting chemicals that are *poisonous* (oleander plants), *irritating* (stinging nettles and bombardier beetles, Figure 5.6c), *foul smelling* (skunks and stinkbugs), or *bad tasting* (buttercups and monarch butterflies, Figure 5.6d). When attacked, some species of squid and octopus emit clouds of black ink, allowing them to escape by confusing their predators.

Many bad-tasting, bad-smelling, toxic, or stinging prey species flash a warning coloration that eating them is risky.

SCIENCE FOCUS 5.1

Threats to Kelp Forests

A kelp forest contains large concentrations of seaweed called *giant kelp*. Anchored to the ocean floor, its long blades grow toward the sunlit surface waters (Figure 5.1, right). Under good conditions, the blades can grow 0.6 meter (2 feet) in a day and the plant can grow as tall as a 10-story building. The blades are flexible and can survive all but the most violent storms and waves.

Kelp forests support many marine plants and animals and are one of the most biologically diverse marine ecosystems. These forests also reduce shore erosion by blunting the force of incoming waves and trapping some of the outgoing sand.

Sea urchins (Figure 5.A) prey on kelp plants. Large populations of these predators can rapidly devastate a kelp forest because they eat the bases of young kelp plants. Scientific studies by biologists, including James Estes of the University of California at Santa Cruz, indicate that the southern sea otter is a keystone species that helps to sustain kelp forests by controlling populations of sea urchins.

Polluted water running off the land also threatens kelp forests. The pollutants in this runoff include pesticides and herbicides that can kill kelp plants and other species and upset the food webs in these aquatic forests. Another runoff pollutant is fertilizer. Its plant nutrients (mostly nitrates) can cause excessive growth of algae and other aquatic plants. This growth blocks some of the sunlight needed to support the growth of giant kelp.

Some scientists warn that the warming of the world's oceans is a growing threat to kelp forests, which require cool water. If coastal waters get warmer during this century, as projected by climate models, many or most of California's coastal kelp forests could disappear.

Kokhanchikov/Shutterstock.com

FIGURE 5.A The purple sea urchin inhabits the coastal waters of the U.S. state of California and feeds on kelp.

CRITICAL THINKING

List three ways in which we could reduce the degradation of giant kelp forest ecosystems.

Examples are the brilliantly colored, foul-tasting monarch butterflies (Figure 5.6d) and poisonous frogs (Figure 5.6e). When a bird eats a monarch butterfly, it usually vomits and learns to avoid monarchs.

Some butterfly species gain protection by looking and acting like other, more dangerous species, a protective device known as *mimicry*. The nonpoisonous viceroy butterfly (Figure 5.6f) mimics the monarch butterfly. Other prey species use *behavioral strategies* to avoid predation. Some attempt to scare off predators by puffing up (blowfish), spreading their wings (peacocks), or mimicking a predator (Figure 5.6h). Some moths have wings that look like the eyes of much larger animals (Figure 5.6g). Other prey species gain some protection by living in large groups such as schools of fish and herds of antelope.

Biologist Edward O. Wilson proposed two criteria for evaluating the dangers posed by various brightly colored animal species. *First*, if they are small and strikingly beautiful, they are probably poisonous. *Second*, if they are strikingly beautiful and easy to catch, they are probably deadly.

Paul Nicklen/National Geographic Creative

FIGURE 5.5 A white arctic fox hunts its prey by blending into its snowy background to avoid being detected.

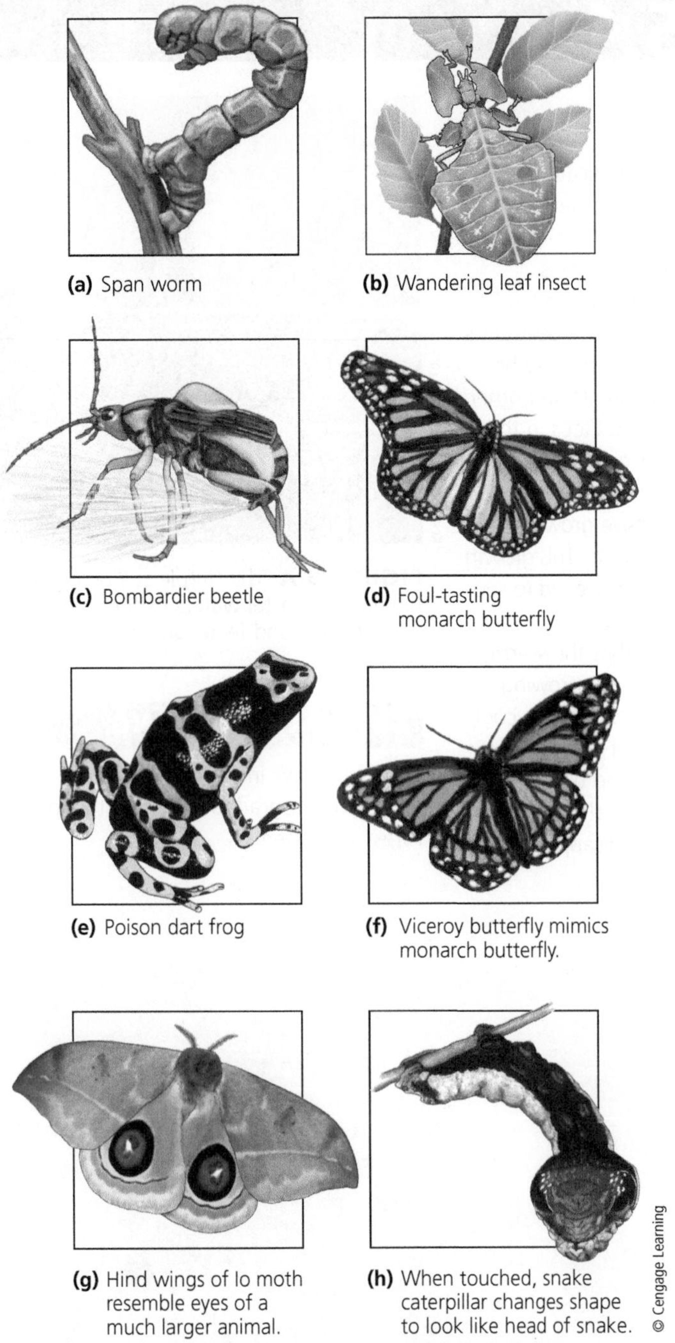

FIGURE 5.6 These prey species have developed specialized ways to avoid their predators: (a, b) *camouflage*, (c, d, e) *chemical warfare*, (d, e, f) *warning coloration*, (f) *mimicry*, (g) *deceptive looks*, and (h) *deceptive behavior*.

Coevolution

At the individual level, members of predator species benefit from their predation and members of prey species are harmed. At the population level, predation plays a role in natural selection. Animal predators tend to kill the sick, weak, aged, and least fit members of a prey population because they are the easiest to catch. Individuals with better defenses against predation thus tend to survive longer and leave more offspring with adaptations that can help them avoid predation. Over time, as a prey species develops traits that make it more difficult to catch, its predators face selection pressures that favor traits increasing their ability to catch their prey. Then the prey species must get better at eluding the more effective predators.

This back-and-forth adaptation is called **coevolution**, a natural selection process in which changes in the gene pool of one species leads to changes in the gene pool of another species. It can play an important role in controlling population growth of predator and prey species. When populations of two species interact as predator and prey over a long time, genetic changes occur in both species that help them to become more competitive or to avoid or reduce competition.

For example, coevolution can be observed between bats and certain species of moths they feed on. Bats prey on certain species of moths that they hunt at night using echolocation. They emit pulses of high-frequency sound that bounce off their prey. Then they capture the returning echoes that tell them where their prey is located. Over time, certain moth species have evolved ears that are sensitive to the sound frequencies that bats use to find them. When they hear these frequencies, they drop to the ground or fly evasively. Some bat species evolved ways to counter this defense by changing the frequency of their sound pulses. In turn, some moths evolved their own high-frequency clicks to jam the bats' echolocation systems. Some bat species then adapted by turning off their echolocation systems and using the moths' clicks to locate their prey.

CONSIDER THIS . . .

LEARNING FROM NATURE

Bats and dolphins use echolocation to navigate and locate prey in the darkness of night and in the ocean's murky water. Scientists are studying how they do this to improve our sonar systems, sonic imaging tools for detecting underground mineral deposits, and medical ultrasound imaging systems.

Parasitism, Mutualism, and Commensalism

Parasitism occurs when one species (the *parasite*) lives in or on another organism (the *host*). The parasite benefits by extracting nutrients from the host. A parasite weakens its host but rarely kills it, since doing so would eliminate the source of its benefits. For example, tapeworms are parasites that live part of their life cycle inside their hosts. Others such as mistletoe plants and blood-sucking sea lampreys (Figure 5.7) attach themselves to the outsides of their hosts. Some parasites (such as fleas and ticks) move from one host to another whereas others (such as certain protozoa) spend their adult lives within a single host. Parasites help keep their host populations in check.

FIGURE 5.7 *Parasitism:* This blood-sucking, parasitic sea lamprey has attached itself to an adult lake trout from one of the Great Lakes (USA, Canada).

In **mutualism**, two species interact in ways that benefit both by providing each with food, shelter, or some other resource. An example is pollination of flowering plants by species such as honeybees, hummingbirds, and butterflies (see Chapter 4 opening photo, pp. 74–75) that feed on the nectar of flowers. Figure 5.8 shows an example of a mutualistic relationship that combines *nutrition* and *protection*. It involves birds that ride on the backs or heads of large animals such as elephants, rhinoceroses, and impalas. The birds remove and eat parasites and pests (such as ticks and flies) from the animals' bodies and often make noises warning the animals when predators are approaching.

Another example of mutualism involves clownfish, which usually live within sea anemones (see chapter-opening photo), whose tentacles sting and paralyze most fish that touch them. The clownfish, which are not harmed by the tentacles, gain protection from predators and feed on the waste matter left from the anemones' meals. The sea anemones benefit because the clownfish protect them from some of their predators and parasites.

Mutualism might appear to be a form of cooperation between species. However, each species is concerned only for its own survival.

Commensalism is an interaction that benefits one species but has little, if any, beneficial or harmful effect on the other. One example involves plants called *epiphytes* (air plants), which attach themselves to the trunks or branches of trees (Figure 5.9) in tropical and subtropical forests. The plants gain access to sunlight, water from the humid air and rain, and nutrients falling from the tree's upper leaves

FIGURE 5.8 *Mutualism:* Oxpeckers feed on parasitic ticks that infest animals such as this impala and warn of approaching predators.

FIGURE 5.9 *Commensalism:* This pitcher plant is attached to a branch of a tree without penetrating or harming the tree. This carnivorous plant feeds on insects that become trapped inside it.

and limbs, but their presence apparently does not harm the tree. Similarly, birds benefit by nesting in trees, generally without harming them.

5.2 HOW DO COMMUNITIES AND ECOSYSTEMS RESPOND TO CHANGING ENVIRONMENTAL CONDITIONS?

CONCEPT 5.2 The species composition of a community or ecosystem can change in response to changing environmental conditions through a process called *ecological succession*.

Ecological Succession Creates and Changes Ecosystems

The types and numbers of species in biological communities and ecosystems change in response to changing environmental conditions. The normally gradual change in species composition in a given terrestrial area or aquatic system is called **ecological succession** (**Concept 5.2**). Ecologists recognize two major types of ecological succession, depending on the conditions present at the beginning of the process.

Primary ecological succession involves the gradual establishment of communities of different species in lifeless areas. This type of succession begins where there is no soil in a terrestrial ecosystem or no bottom sediment in an aquatic ecosystem. Examples include bare rock exposed by a retreating glacier (Figure 5.10), newly cooled lava from a volcanic eruption, an abandoned highway or parking lot, and a newly created shallow pond or lake (Figure 5.11). Primary succession usually takes hundreds to thousands of years because of the need to build up fertile soil or aquatic sediments to provide the nutrients needed to establish a plant community.

Species such as lichens and mosses that quickly colonize the newly exposed rocks are called **pioneer species**. They often have seeds or spores that can travel long distances and quickly spread over the exposed rock (Figure 5.10). As lichens grow and spread, they release

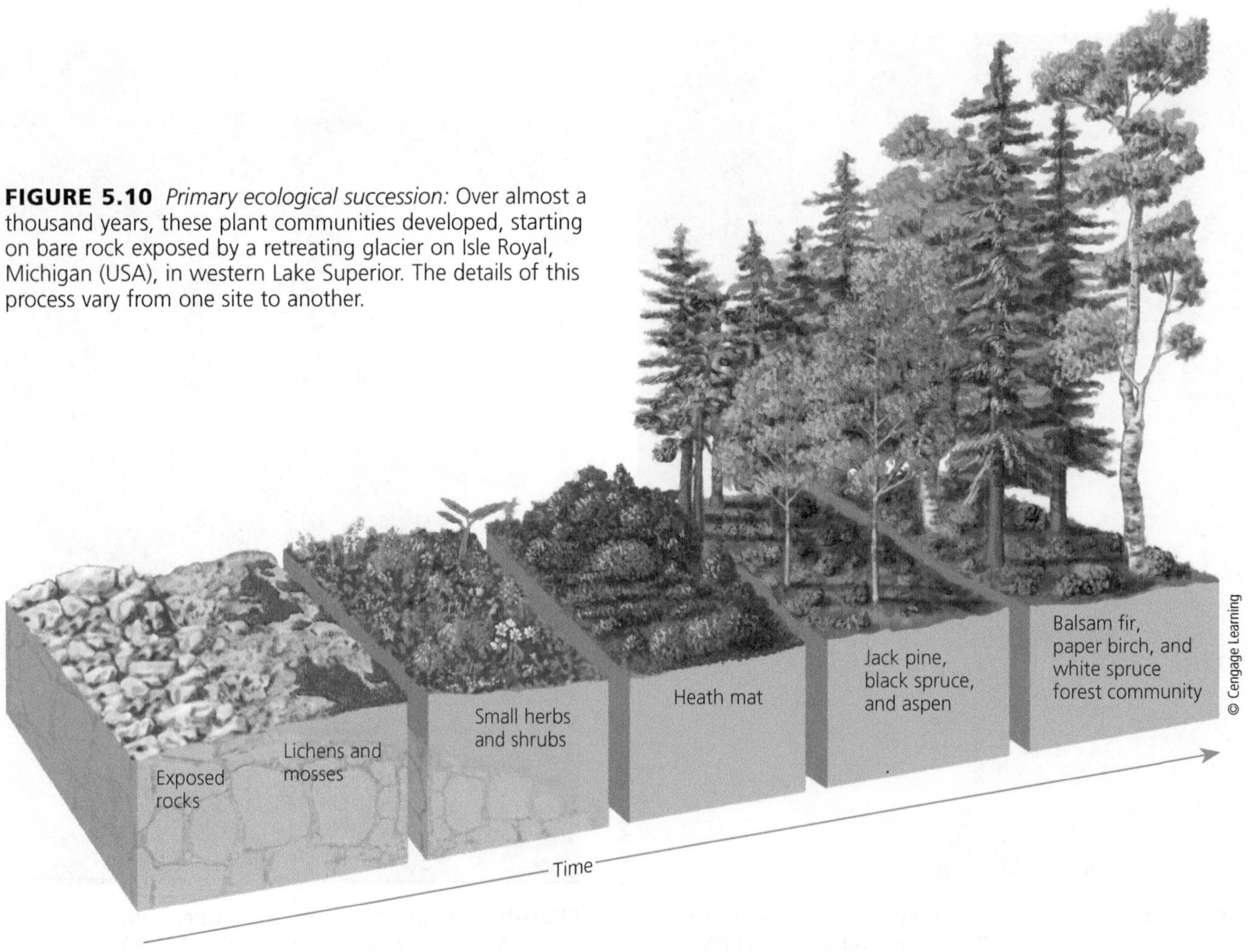

FIGURE 5.10 *Primary ecological succession:* Over almost a thousand years, these plant communities developed, starting on bare rock exposed by a retreating glacier on Isle Royal, Michigan (USA), in western Lake Superior. The details of this process vary from one site to another.

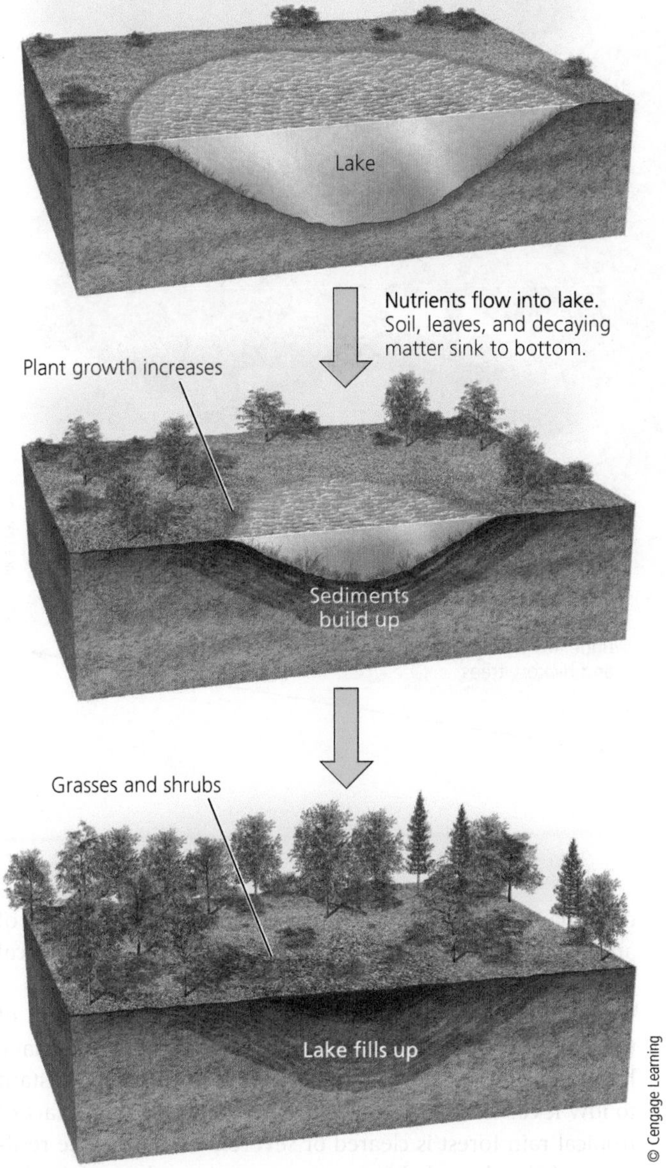

FIGURE 5.11 *Primary ecological succession* in a lake basin in which sediments and plants have been gouged out by a glacier. When the glacier melts, the lake basin begins accumulating sediments and plant and animal life. Over hundreds to thousands of years, the lake can fill with sediments and become a terrestrial habitat.

acids that can breakdown the rock and start the soil formation process (see Figure 3.10, p. 56). As the soil slowly forms, small plants, insects, and worms invade and add more nutrients that build up the soil. Each successive wave of new organisms changes the environmental conditions in ways that provide more nutrients, habitats, and favorable environmental conditions for future arrivals.

The other, more common type of ecological succession is **secondary ecological succession**, in which a series of terrestrial communities or ecosystems with different species develop in places containing soil or bottom sediment. This type of succession begins in an area where an ecosystem has been disturbed, removed, or destroyed, but some soil or bottom sediment remains. Candidates for secondary succession include abandoned farmland (Figure 5.12), burned or cut forests, heavily polluted streams, and flooded land. Because some soil or sediment is present, new vegetation can begin to grow, usually within a few weeks. On land, growth begins with the germination of seeds already in the soil and seeds imported by wind or in the droppings of birds and other animals.

Ecological succession is an important ecosystem service that can enrich the biodiversity of communities and ecosystems by increasing species diversity and interactions among species. Such interactions enhance sustainability by promoting population control and increasing the complexity of food webs. Primary and secondary ecological succession are examples of *natural ecological restoration*.

Ecologists have identified three factors that affect how and at what rate ecological succession occurs. One is *facilitation*, in which one set of species makes an area suitable for species with different niche requirements, and often less suitable for itself. For example, as lichens and mosses gradually build up soil on a rock in primary succession, herbs and grasses can move in and crowd out the lichens and mosses (Figure 5.10).

A second factor is *inhibition*, in which some species hinder the establishment and growth of other species. For example, needles dropping off some pine trees make the soil beneath the trees too acidic for most other plants to grow there. A third factor is *tolerance*, in which plants in the late stages of succession succeed because they are not in direct competition with other plants for key resources. Shade-tolerant plants, for example, can live in shady forests because they do not need as much sunlight as the trees above them do (Figure 5.12).

Is There a Balance of Nature?

According to the traditional view, ecological succession proceeds in an orderly sequence along an expected path until a certain stable type of *climax community* (Figures 5.10 and 5.12), which is assumed to be in balance with its environment, occupies an area. This equilibrium model of succession is what ecologists once meant when they talked about the *balance of nature*.

Over the last several decades, many ecologists have changed their views about balance and equilibrium in nature based on ecological research. There is a general tendency for succession to lead to more complex, diverse, and presumably more resilient ecosystems that can withstand changes in environmental conditions if the changes are

FIGURE 5.12 *Secondary ecological succession:* Natural restoration of disturbed land on an abandoned farm field in the U.S. state of North Carolina. It took 150 to 200 years after the farmland was abandoned for the area to become covered with a mature oak and hickory forest.

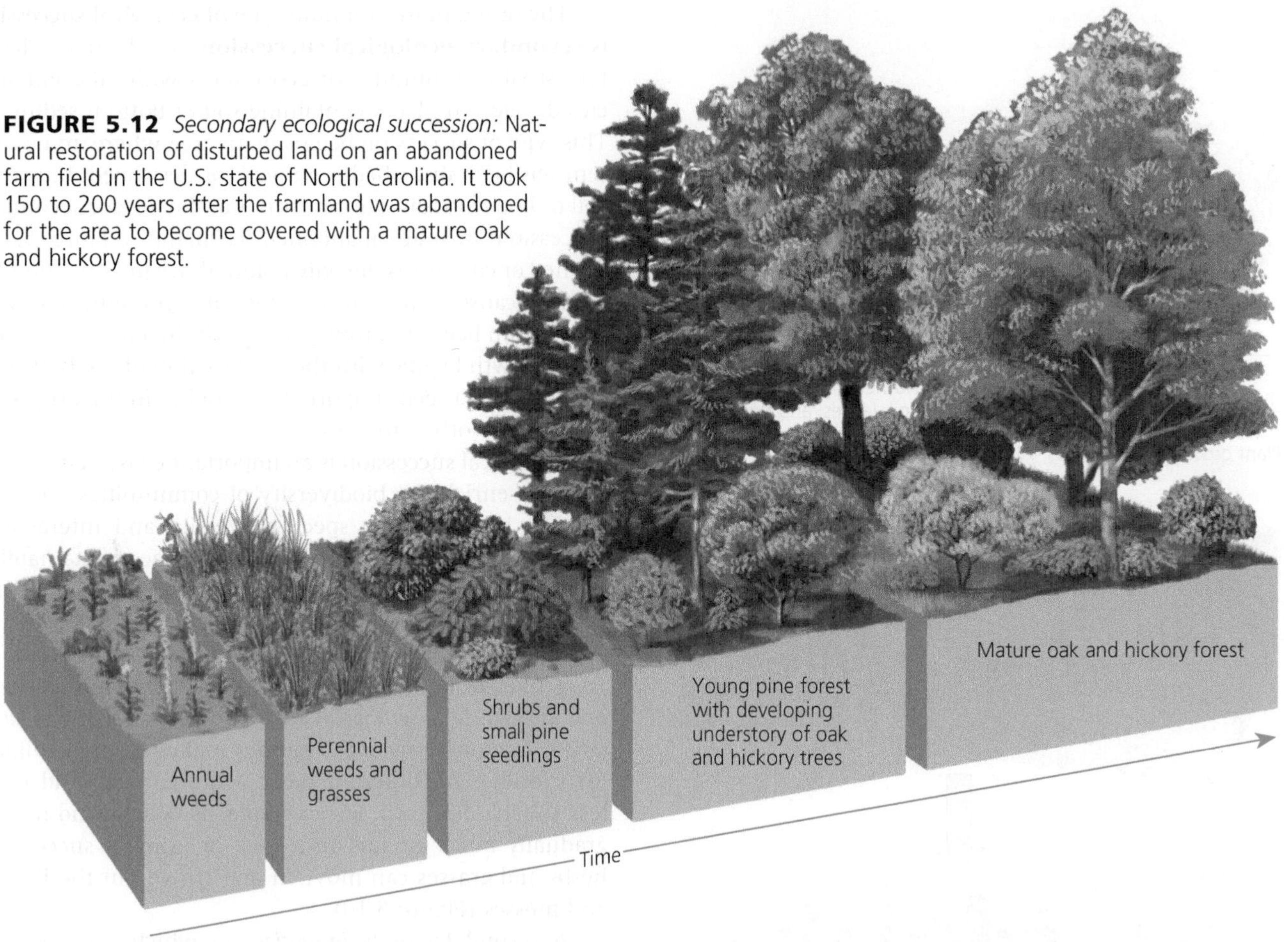

not too large or too sudden. However, the current scientific view is that we cannot predict a given course of succession or view it as inevitable progress toward an ideally adapted climax plant community or ecosystem. Rather, ecological succession reflects the ongoing struggle by different species for enough light, water, nutrients, food, space, and other key resources.

Living Systems Are Sustained through Constant Change

All living systems, from a cell to the biosphere, constantly change in response to changing environmental conditions. Living systems have complex processes that interact to provide some degree of stability, or sustainability. This stability, or the capacity to withstand external stress and disturbance, is maintained by constant change in response to changing environmental conditions. In a mature tropical rain forest, some trees die and others take their places. However, unless the forest is cut, burned, or otherwise destroyed, you would still recognize it as a tropical rain forest 50 or 100 years from now.

Ecologists distinguish between two aspects of stability or sustainability in ecosystems. One, called **inertia**, or **persistence**, is the ability of an ecosystem to survive moderate disturbances. A second factor, **resilience**, is the ability of an ecosystem to be restored through secondary ecological succession after a severe disturbance.

Evidence suggests that some ecosystems have one of these properties but not the other. Tropical rain forests have high species diversity and high inertia and thus are resistant to low levels of change or damage. But once a large tract of tropical rain forest is cleared or severely damaged, the resilience of the degraded forest ecosystem may be so low that the degradation reaches an ecological tipping point. Beyond that point, the forest might not be restored by secondary ecological succession. One reason is that most of the nutrients in a tropical rain forest are stored in its vegetation, not in the topsoil. Once the nutrient-rich vegetation is gone, frequent rains on a large cleared area of land can remove most of the remaining soil nutrients and thus prevent the return of a tropical rain forest to such an area.

By contrast, grasslands are much less diverse than most forests. Thus, they have low inertia and can burn easily. Because most of their plant matter is stored in underground roots, these ecosystems have high resilience and can recover quickly after a fire because their root systems produce new grasses. Grassland can be destroyed only if its roots are plowed up and something else is planted in its place, or if it is severely overgrazed by livestock or other herbivores.

FIGURE 5.13 A population, or *school*, of Anthias fish on coral in Australia's Great Barrier Reef.

iStockphoto.com/Rich Carey

5.3 WHAT LIMITS THE GROWTH OF POPULATIONS?

CONCEPT 5.3 No population can grow indefinitely because of limitations on resources and because of competition among species for those resources.

Populations Can Grow, Shrink, or Remain Stable

A **population** is a group of interbreeding individuals of the same species (Figure 5.13). **Population size** is the number of individual organisms in a population at a given time. The size of a population may increase, decrease, go up and down in cycles, or remain roughly the same in response to changing environmental conditions.

Scientists use sampling techniques to estimate the sizes of large populations of species such as oak trees that are spread over a large area and squirrels that move around and are hard to count. Typically, they count the number of individuals in one or more small sample areas and use this information to estimate the number of individuals in a larger area.

Populations of different species vary in their distribution over their habitats, or *dispersion*, as shown in Figure 5.14. Most populations live together in *clumps* or *groups* such as packs of wolves, schools of fish (Figure 5.13), and flocks of birds. Southern sea otters (**Core Case Study**), for example, are usually found in groups known as rafts or pods ranging in size from a few to several hundred animals.

Living in groups allows organisms to cluster where resources are available. Group living also provides some protection from predators, and gives some predator species a better chance of getting a meal.

Four variables—*births*, *deaths*, *immigration*, and *emigration*—govern changes in population size. A population increases through birth and immigration (arrival of individuals from outside the population). Populations decrease through death and emigration (departure of individuals from the population):

$$\text{Population change} = \text{Individuals added} - \text{Individuals lost}$$

$$\text{Population change} = (\text{Births} + \text{Immigration}) - (\text{Deaths} + \text{Emigration})$$

a. Clumped (elephants) **b.** Uniform (creosote bush) **c.** Random (dandelions)

FIGURE 5.14 Three general habitat *dispersion patterns* for individuals in a population.

Left: EcoPrint/Shutterstock.com. Center: kenkistler/Shutterstock.com. Right: Nataly Lukhanina/Shutterstock.com.

Several Factors Can Limit Population Size

Each population in an ecosystem has a **range of tolerance**—a range of variations in its physical and chemical environment under which it can survive. For example, a trout population may do best within a narrow band of temperatures (*optimum level* or *range*), but a few individuals can survive above and below that band (Figure 5.15). If the water becomes too hot or too cold, none of the trout can survive.

Individuals within a population may also have slightly different tolerance ranges for temperature, other physical factors, or chemical factors. These occur because of small differences in their genetic makeup, health, and age. Such differences allow for evolution through natural selection. The individuals that have a wider tolerance for change in some factor such as temperature are more likely to survive such a change and produce offspring that can tolerate it.

Various physical or chemical factors can determine the number of organisms in a population and how fast a population grows or declines. Sometimes one or more factors, known as **limiting factors**, are more important than other factors in regulating population growth.

On land, precipitation often is the limiting factor. Low precipitation levels in desert ecosystems limit desert plant growth. Lack of key soil nutrients limits the growth of plants, which in turn limits populations of animals that eat plants, and animals that feed on such plant-eating animals.

Limiting physical factors for populations in *aquatic systems* include water temperature (Figure 5.15) and water depth and clarity (allowing for more or less sunlight). Other important factors are nutrient availability, acidity, salinity, and the level of oxygen gas in the water (dissolved oxygen content).

An additional factor that can limit the sizes of some populations is **population density**, the number of individuals in a population found within a defined area or volume. It is a measure of how crowded the members of a population are.

Density-dependent factors are variables that become more important as a population's density increases. For example, in a dense population, parasites and diseases can spread more easily, resulting in higher death rates. On the other hand, a higher population density helps sexually reproducing individuals to find mates more easily to produce offspring. Other factors such as flood, fires, landslides, drought, and climate change are considered *density-independent factors*, because any effects they have on a population's size are not related to its density.

No Population Can Grow Indefinitely: J-Curves and S-Curves

Some species have an incredible ability to increase their numbers and grow exponentially (see p. 15). Plotting these numbers against time yields a J-shaped curve of exponential growth when a population increases by a fixed percentage each year (Figure 5.16, left). Members of such populations typically reproduce at an early age, have many offspring each time they reproduce, and reproduce many times with short intervals between generations.

Examples are bacteria and many insect species. For example, with no controls on its population growth, a species of bacteria that can reproduce every 20 minutes would generate enough offspring to form a 0.3-meter-deep (1-foot-deep) layer over the surface of the entire earth in only 36 hours. Such exponential growth occurs in nature

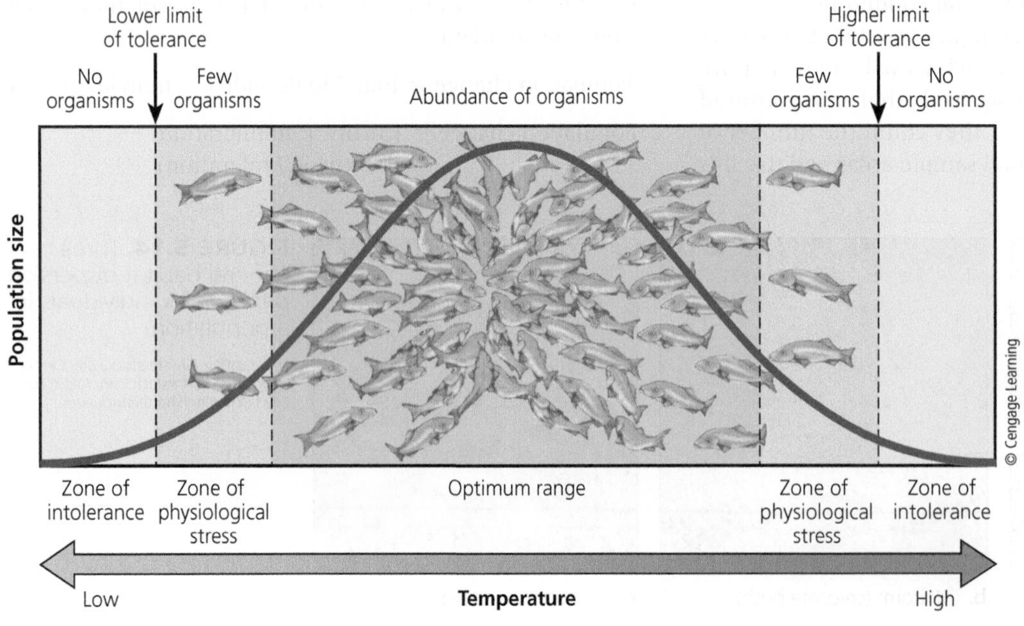

FIGURE 5.15 Range of tolerance for a population of trout to changes in water temperature.

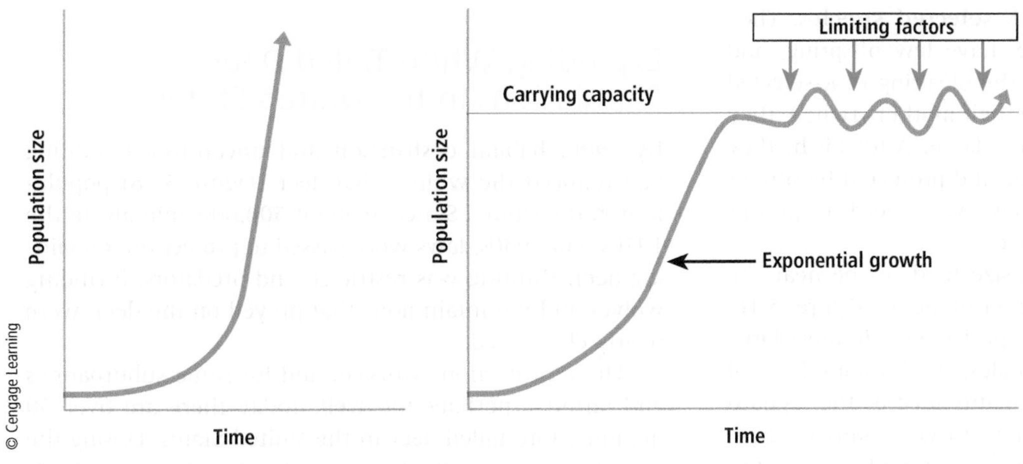

FIGURE 5.16 Populations of species can undergo *exponential growth* represented by a J-shaped curve (left) when resource supplies are plentiful. As resource supplies become limited, a population undergoes *logistic growth*, represented by an S-shaped curve (right), when the size of the population approaches the carrying capacity of its habitat.

when species with a high reproductive potential have few predators, plenty of food and other resources, and little competition from other species for such resources.

However, *there are always limits to population growth in nature*. Research reveals that a rapidly growing population of any species eventually reaches some size limit imposed by limiting factors. These factors include sunlight, water, temperature, space, or nutrients, or exposure to predators or infectious diseases (**Concept 5.3**). **Environmental resistance** is the sum of all such factors in a habitat. Limiting factors largely determine an area's **carrying capacity**, the maximum population of a given species that a particular habitat can sustain indefinitely. The carrying capacity for a population is not fixed and can rise or fall as environmental conditions change the factors that limit the population's growth.

As a population approaches the carrying capacity of its habitat, the J-shaped curve of its exponential growth (Figure 5.16, left) is converted to an S-shaped curve of *logistic growth*, or growth that often fluctuates around the carrying capacity of its habitat (Figure 5.16, right). The population sizes of some species often fluctuate above and below their carrying capacity as shown in the right graph in Figure 5.16.

Some populations do not make a smooth transition from exponential growth to logistic growth. Instead, they use up their resource supplies and temporarily *overshoot*, or exceed, the carrying capacity of their environment. In such cases, the population suffers a sharp decline, called a *dieback*, or **population crash**, unless part of the population can switch to new resources or move to an area that has more resources. Such a crash occurred when reindeer were introduced onto a small island in the Bering Sea in the early 1900s (Figure 5.17).

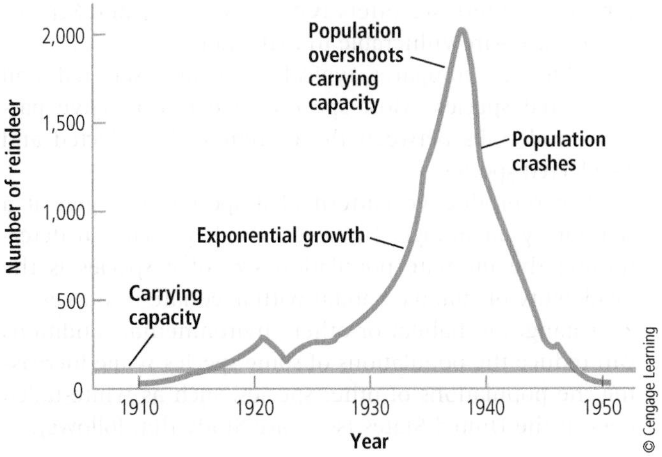

FIGURE 5.17 Exponential growth, overshoot, and population crash of a population of reindeer introduced onto the small Bering Sea island of St. Paul in 1910. ***Data analysis:*** By what percentage did the population of reindeer grow between 1923 and 1940?

Reproductive Patterns

Species vary in their reproductive patterns. Species with a capacity for a high rate of population growth (*r*) (Figure 5.16, left) are called ***r*-selected species**. These species tend to have short life spans and produce many, usually small offspring and give them little or no parental care. As a result, many of the offspring die at an early age. To overcome such losses, *r*-selected species produce large numbers of offspring so that a few will likely survive and have many offspring to sustain the species. Examples of *r*-selected species include algae, bacteria, frogs, most insects, and many fish.

Such species tend to be *opportunists*. They reproduce and disperse rapidly when conditions are favorable or when a disturbance such as a fire or clear-cutting of a forest opens up a new habitat or niches for invasion. Once established, their populations may crash because of unfavorable changes in environmental conditions or invasion by more competitive species. This explains why many opportunist species go through irregular and unstable boom-and-bust cycles in their population sizes.

At the other extreme are ***K*-selected species**. They tend to reproduce later in life, have few offspring, and have long life spans. Typically, the offspring of *K*-selected mammal species develop inside their mothers (where they are safe) and are born relatively large. After birth, they mature slowly and are cared for and protected by one or both parents. In some cases, they live in herds or groups until they reach reproductive age.

Such a species' population size tends to be near the carrying capacity (*K*) of its environment (Figure 5.16, right). Examples of *K*-selected species include most large mammals such as elephants, whales, and humans, birds of prey, and large and long-lived plants such as the saguaro cactus, and most tropical rain forest trees. Many of these species—especially those with low reproductive rates, such as elephants, sharks, giant redwood trees, and California's southern sea otters (**Core Case Study** and Science Focus 5.2)—are vulnerable to extinction.

Table 5.1 compares typical traits of *r*-selected and *K*-selected species. Most species have reproductive patterns and traits between the extremes of *r*-selected and *K*-selected species.

The reproductive pattern of a species may give it a temporary advantage. However, the key factor in determining the ultimate population size of a species is the availability of suitable habitat with adequate resources.

Changes in habitat or other environmental conditions can reduce the populations of some species while increasing the populations of other species, such as white-tailed deer in the United States (see Case Study that follows).

TABLE 5.1 Typical traits of *r*-selected and *K*-selected species

Trait	*r*-Selected Species	*K*-Selected Species
Reproductive potential	High	Low
Population growth rate	Fast	Slow
Time to reproductive maturity	Short	Long
Number of reproductive cycles	Many	Few
Number of offspring	Many	Few
Size of offspring	Small	Larger
Degree of parental care	Low	High
Life span	Short	Long
Population size	Variable with crashes	Stable, near carrying capacity
Role in environment	Usually prey	Usually predators

CASE STUDY

Exploding White-Tailed Deer Populations in the United States

By 1900, habitat destruction and uncontrolled hunting had reduced the white-tailed deer (Figure 5.18) population in the United States to about 500,000 animals. In the 1920s and 1930s, laws were passed to protect the remaining deer. Hunting was restricted and predators, including wolves and mountain lions that preyed on the deer, were nearly eliminated.

These protections worked, and for some suburbanites and farmers, perhaps too well. Today there are over 30 million white-tailed deer in the United States. During the last 50 years, suburbs have expanded and many Americans have moved into the wooded habitat of deer. The gardens and landscaping around their homes provide deer with flowers, shrubs, garden crops, and other plants they like to eat.

Deer prefer to live in the edge areas of forests and woodlots for security and go to nearby fields, orchards, lawns, and gardens for food. A suburban neighborhood can be an all-you-can-eat paradise for white-tailed deer, and their populations in such areas have soared.

In woodlands, deer are consuming native groundcover vegetation, which has allowed nonnative weed species to take over and upset ecosystem food webs. The deer also help to spread Lyme disease (carried by deer ticks) to humans. In addition, each year about 1 million deer–vehicle collisions injure up to 10,000 Americans and kill at

Roy Toft/National Geographic Creative

FIGURE 5.18 White-tailed deer populations in the United States have been growing.

The Future of California's Southern Sea Otters

The population of southern sea otters (**Core Case Study**) has fluctuated in response to changes in environmental conditions (Figure 5.B). One change is a rise in populations of the orcas (killer whales) that feed on them. Scientists hypothesize that orcas started feeding more on southern sea otters when populations of their normal prey, sea lions and seals, began declining. In addition, between 2010 and 2012 the number of sea otters killed or injured by sharks increased, possibly because warmer ocean water brought some sharks closer to the shore.

Another factor affecting sea otters may be parasites that breed in the intestines of cats. Scientists hypothesize that some southern sea otters are dying because cat owners flush feces-laden cat litter down their toilets or dump it in storm drains that empty into coastal waters where parasites from the litter can infect otters.

Toxic algae blooms also threaten otters. The algae thrive on urea, a nitrogen-containing ingredient in fertilizer that washes into coastal waters. Other pollutants released by human activities are PCBs and other fat-soluble toxic chemicals. These chemicals can kill otters by accumulating to high levels in the tissues of the shellfish that otters eat. Because southern sea otters feed at high trophic levels and live close to the shore, they are vulnerable to these and other pollutants in coastal waters.

Other threats to otters include oil spills from ships. The entire California southern sea otter population could be wiped out by a large oil spill from a single tanker off the central west coast or by the rupture of an offshore oil well, should drilling for oil be allowed off this coast. Some sea otters die when they are trapped in underwater nets and traps for shellfish. Others are killed by boat strikes and gunshots.

The factors listed here, mostly resulting from human activities, together with a low reproductive rate and a rising mortality rate, have hindered the ability of the endangered southern sea otter to rebuild its population (Figure 5.B). In 2012, the National Geographic Society funded a project led by Nicole Thometz, a biologist at the University of California, Santa Cruz, to learn more about why juvenile sea otters, in particular, were suffering a high mortality rate. Such information could be used to help biologists to refine recovery plans for the southern sea otter.

Since 2012, the sea otter population has increased, possibly because of an increase in the population of sea urchins, their preferred prey. In 2015, the sea otter population was 3,054, the highest it has been since 1985. If the sea otter population exceeds 3,090 for 3 consecutive years, it may be removed from the endangered species list. If this happens, the otters will still be protected under a California state law.

FIGURE 5.B Changes in the population size of southern sea otters off the coast of the U.S. state of California, 1983–2015.
(Compiled by the authors using data from U.S. Geological Survey.)

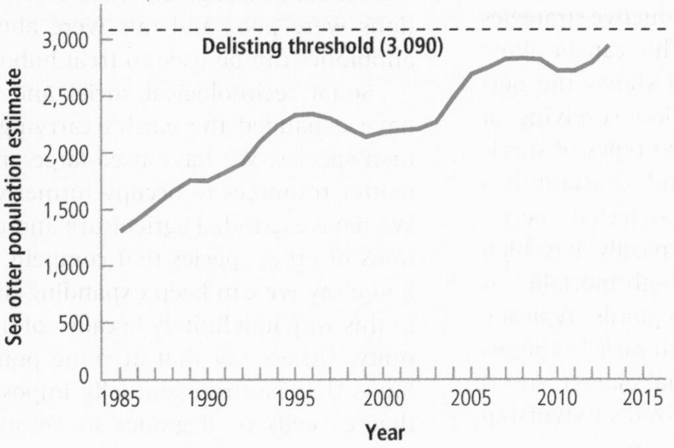

CRITICAL THINKING

How would you design a controlled experiment to test the hypothesis that cat litter flushed down toilets might be killing southern sea otters?

least 200—the highest human death toll from encounters with any wild animal in the United States.

There are no easy solutions to the deer population problem in the suburbs. Changes in hunting regulations that allow for the killing of more female deer have cut down the overall deer population. However, this has had a limited effect on deer populations in suburban areas because it is too dangerous to allow widespread hunting with guns in such populated communities. Some areas have hired experienced and licensed archers who use bows and arrows to help reduce deer numbers. To protect nearby residents the archers hunt from elevated tree stands and only shoot their arrows downward.

Some communities spray the scent of deer predators or of rotting deer meat in edge areas to scare off deer. Others scare off deer by using electronic equipment that emits high-frequency sounds that humans cannot hear. Some homeowners surround their gardens and yards with high, black plastic mesh fencing.

Deer can be trapped and moved from one area to another, but this is expensive and must be repeated whenever they move back into an area. In addition, there are questions concerning where to move the deer and how to pay for such programs.

Darts loaded with contraceptives can be shot into female deer to hold down their birth rates, but this is expensive and must be repeated every year. One possibility is an experimental, single-shot contraceptive vaccine that lasts for several years. Another approach is to trap dominant males and use chemical injections to sterilize them. Both of these approaches are costly and will require years of testing.

Meanwhile, suburbanites can expect deer to chow down on their shrubs, flowers, and garden plants unless they can protect their properties with fences, repellents, or other methods. Suburban dwellers could also stop planting trees, shrubs, and flowers that attract deer around their homes.

CONSIDER THIS . . .

THINKING ABOUT White-Tailed Deer

Some people blame the white-tailed deer for invading farms and suburban yards and gardens to eat food that humans have made easily available to them. Others say humans are mostly to blame because they have invaded deer territory, eliminated most of the predators that kept deer populations under control, and provided the deer with plenty to eat in their lawns, gardens, and crop fields. Which view do you hold? Why? Do you see a solution to this problem?

Species Vary in Their Life Spans

Individuals of species with different reproductive strategies tend to have different *life expectancies*. This can be illustrated by a **survivorship curve**, which shows the percentages of the members of a population surviving at different ages. There are three generalized types of survivorship curves: late loss, early loss, and constant loss (Figure 5.19). A *late loss* population (*K*-selected species such as elephants and rhinoceroses) typically has high survivorship to a certain age, and then high mortality. A *constant loss* population (such as many songbirds) typically has a constant death rate at all ages. For an *early loss* population (many *r*-selected species and annual plants), survivorship is low early in life. These generalized survivorship curves only approximate the realities of nature.

CONSIDER THIS . . .

THINKING ABOUT Survivorship Curves

Which type of survivorship curve applies to the human species?

Humans Are Not Exempt from Nature's Population Controls

Humans are not exempt from population crashes. In 1845 Ireland experienced such a crash after a fungus destroyed its potato crop. About 1 million people died from hunger or diseases related to malnutrition. Millions more migrated to other countries, sharply reducing the Irish population.

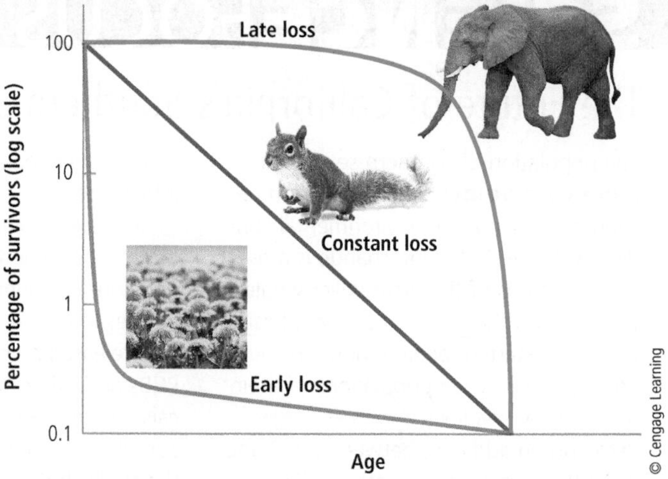

FIGURE 5.19 Survivorship curves for populations of different species, obtained by showing the percentages of the members of a population surviving at different ages.

Top: Gualtiero boffi/Shutterstock.com. Center: IrinaK/Shutterstock.com. Bottom: ultimathule/Shutterstock.com.

During the 14th century, *bubonic plague* spread through densely populated European cities and killed at least 25 million people—one-third of the European population. The bacterium that causes this disease normally lives in rodents. It was transferred to humans by fleas that fed on infected rodents and then bit humans. The disease spread like wildfire through crowded cities, where sanitary conditions were poor and rats were abundant. Today several antibiotics can be used to treat bubonic plague.

So far, technological, social, and other cultural changes have expanded the earth's carrying capacity for the human species. We have used large amounts of energy and matter resources to occupy formerly uninhabitable areas. We have expanded agriculture and controlled the populations of other species that compete with us for resources. Some say we can keep expanding our ecological footprint in this way indefinitely because of our technological ingenuity. Others say that at some point, we will reach the limits that nature eventually imposes on any population that exceeds or degrades its resource base. We discuss these issues in Chapter 6.

BIG IDEAS

- Certain interactions among species affect their use of resources and their population sizes.
- The species composition and population sizes of a community or ecosystem can change in response to changing environmental conditions through a process called *ecological succession*.
- No population can escape natural limiting factors and grow indefinitely.

Tying It All Together

Southern Sea Otters and Sustainability

The sea otters of California are part of a complex ecosystem made up of large underwater kelp forests, bottom-dwelling creatures, and other species that depend on one another for survival. The sea otters act as a keystone species, mostly by feeding on sea urchins and keeping them from destroying the kelp.

fred goldstein/Shutterstock.com

In this chapter, we focused on how biodiversity promotes sustainability, provides a variety of species to restore damaged ecosystems through ecological succession, and limits the sizes of populations. Populations of most plants and animals depend, directly or indirectly, on solar energy, and all populations play roles in the cycling of nutrients in the ecosystems where they live. In addition, the biodiversity in different terrestrial and aquatic ecosystems provides alternative paths for energy flow and nutrient cycling, better opportunities for natural selection as environmental conditions change, and natural population control mechanisms. When we disrupt these paths, we violate the three **scientific principles of sustainability**.

Chapter Review

Core Case Study

1. Explain how southern sea otters act as a keystone species in their environment. Explain why we should care about protecting this species from extinction.

Section 5.1

2. What is the key concept for this section? Define and give an example of **interspecific competition**. How is it different from intraspecific competition? Define and give an example of **resource partitioning** and explain how it can increase species diversity. Define **predation**. Distinguish between a **predator** species and a **prey** species and give an example of each. What is a **predator–prey relationship** and why is it important?

3. Describe three threats to kelp forests and explain why they should be preserved. List three ways in which predators can increase their chances of feeding on their prey and three ways in which prey species can avoid their predators. Define and give an example of **coevolution**.

4. Define **parasitism**, **mutualism**, and **commensalism** and give an example of each. Explain how each of these species interactions, along with predation, can affect the population sizes of species in ecosystems.

Section 5.2

5. What is the key concept for this section? What is **ecological succession**? Distinguish between **primary ecological succession** and **secondary ecological succession** and give an example of each. Define and give an example of a **pioneer species**. Describe three factors that affect how and at what rate succession occurs.

6. Explain why ecological succession does not follow a predictable path and does not necessarily end with a stable climax community. What is the current thinking among ecologists on the concept of a balance of nature? In terms of the stability of ecosystems, distinguish between **inertia (persistence)** and **resilience** and give an example of each.

Section 5.3

7. What is the key concept for this section? Define **population**. Define **population size** and explain how it is estimated, Why do most populations live in clumps? List four variables that govern changes in population size. Write an equation showing how these variables interact. Define **range of tolerance**. Define **limiting factor** and give three examples. Define **population density** and explain how some limiting factors can become more important as a population's density increases.

8. Distinguish between the exponential and logistic growth of a population and describe the nature of their growth curves. Define **environmental resistance**. What is the **carrying capacity** of an environment? Define and give an example of a **population crash**.

9. Describe two different reproductive strategies for species. Distinguish between ***r*-selected species** and ***K*-selected species** and give an example of each. What factors have hindered the recovery of the southern sea otter? Describe the effects of the exploding population of white-tailed deer in the United States and list some possible solutions to this problem. Define **survivorship curve**, describe three types of curves, and for each, give an example of a species that fits that pattern. Explain why humans are not exempt from nature's population controls.

10. What are this chapter's *three big ideas*? Explain how the interactions among plant and animal species in any ecosystem are related to the three **scientific principles of sustainability**.

Note: Key terms are in bold type.

Critical Thinking

1. What difference would it make if the southern sea otter (**Core Case Study**) became extinct primarily because of human activities? What are three things we could do to help prevent the extinction of this species?

2. Use the second law of thermodynamics (Chapter 2, p. 42) and the concept of food chains and food webs to explain why predators are generally less abundant than their prey.

3. How would you reply to someone who argues that we should not worry about the effects that human activities have on natural systems because ecological succession will repair whatever damage we do?

4. How would you reply to someone who contends that efforts to preserve species and ecosystems are not worthwhile because nature is largely unpredictable?

5. What is the reproductive strategy of most species of insect pests and harmful bacteria? Why does this make it difficult for us to control their populations?

6. If the earth's climate continues to change due to atmospheric warming during this century, as most climate scientists project it will, is this likely to favor *r*-selected or *K*-selected species? Explain.

7. List two factors that may limit human population growth in the future. Do you think that we are close to reaching those limits? Explain.

8. If the human species were to suffer a population crash, name three species that might move in to occupy part of our ecological niche. What are three species that would likely decline as a result? Explain why these other species would decline.

Doing Environmental Science

Visit a nearby land area, such as a partially cleared or burned forest, grassland, or an abandoned crop field, and record signs of secondary ecological succession. Take notes on your observations and formulate a hypothesis about what sort of disturbance led to this succession. Include your thoughts about whether this disturbance was natural or caused by humans. Study the area carefully to see whether you can find patches that are at different stages of succession and record your thoughts about what sorts of disturbances have caused these differences. You might want to research the topic of ecological succession in such an area.

Global Environment Watch Exercise

Go to your MindTap course to access the GREENR database. Using the "Basic Search" function at the top of the page, search for *kelp forests* (also sometimes called *kelp beds*), and use the results to find sources of information about how a warmer ocean resulting from climate change might affect California's coastal kelp forests on which the southern sea otters depend (**Core Case Study**). Write a report on what you found. Try to include information on current effects of warmer water on the kelp beds as well as projections about future effects. Also, summarize any information you might find on possible ways to prevent harm to these kelp forests.

Data Analysis

The graph below shows changes in the size of an Emperor penguin population in terms of numbers of breeding pairs on the island of Terre Adelie in the Antarctic. Scientists used this data along with data on the penguins' shrinking ice habitat to project a general decline in the island's Emperor penguin population, to the point where they will be endangered in 2100. Use the graph to answer the following questions.

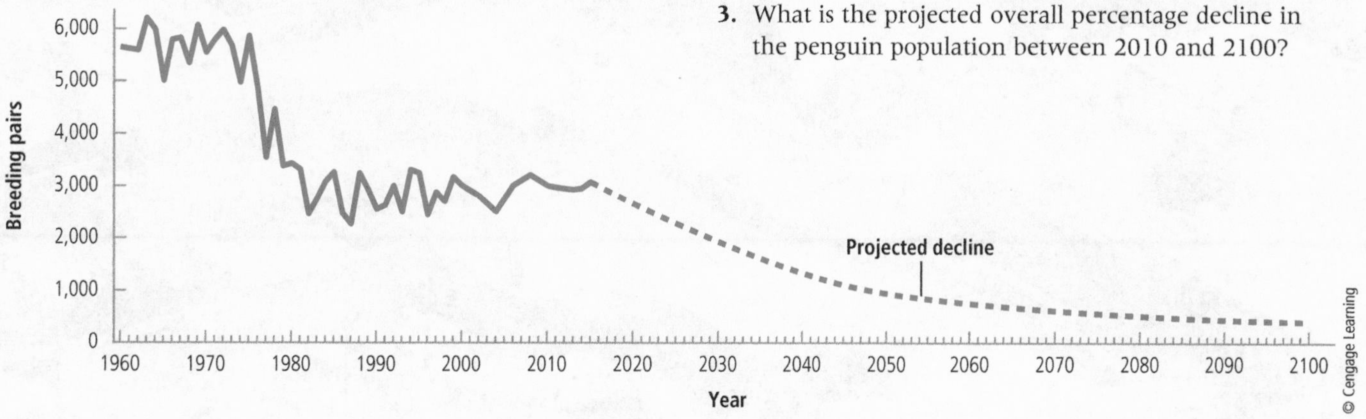

1. If the penguin population fluctuates around the carrying capacity, what was the approximate carrying capacity of the island for the penguin population from 1960 to 1975? What was the approximate carrying capacity of the island for the penguin population from 1980 to 2010?
2. What was the overall percentage decline in the penguin population from 1975 to 2010?
3. What is the projected overall percentage decline in the penguin population between 2010 and 2100?

CENGAGE brain.com For access to MindTap and additional study materials visit www.cengagebrain.com.

CHAPTER 6

The Human Population

> Either we limit our population growth or the natural world will do it for us.
>
> SIR DAVID ATTENBOROUGH

Key Questions

6.1 How many people can the earth support?

6.2 What factors influence the size of the human population?

6.3 How does a population's age structure affect its growth or decline?

6.4 How can we slow human population growth?

Slum in Jaipur, India.

Zanskar/iStock/Getty Images Plus/Getty Images

Core Case Study

Planet Earth: Population 7.3 Billion

It took about 200,000 years—from the time that the latest version of our species *Homo sapiens sapiens* evolved to the 1920s—for our population to reach an estimated 2 billion. It took less than 50 years to add the second 2 billion people (by about 1974), and 25 years to add the third 2 billion (by 1999). Sixteen years later, in 2015, the earth had 7.3 billion people. Figure 1.11 (p. 16) lists the world's 10 most populous countries. In 2015 the top three, in order, were China with 1.37 billion people (Figure 6.1), India with 1.31 billion people, and the United States with 321 million people.

Does it matter that there are now 7.3 billion people on the earth—almost 3 times as many as there were in 1950? Does it matter that each day, 241,000 more people show up for dinner and many of them will go hungry? Some say it does not matter, and they contend that we can develop new technologies that could easily support billions more people.

Many scientists disagree and argue that the current exponential growth of the human population (see Figure 1.11, p. 16) is unsustainable. The reason is that as our population grows, we use more of the earth's natural resources and our ecological footprints expand and degrade the natural capital that keeps us alive and supports our lifestyles and economies.

According to *demographers*, or population experts, three major factors account for the rapid rise of the human population. *First*, the emergence of early and modern agriculture about 10,000 years ago increased food production. *Second*, additional technologies helped humans expand into almost all of the planet's climate zones and habitats (see Figure 1.8, p. 12). *Third*, death rates dropped sharply with improved sanitation and health care and the development of antibiotics and vaccines to control infectious diseases.

What is a sustainable level for the human population? Population experts have made low, medium, and high projections of the human population size by the end of this century, as shown in Figure 1.11, p. 16. No one knows whether any of these population sizes are sustainable or for how long.

In this chapter, we examine population growth trends, the environmental impacts of the growing population, and proposals for dealing with human population growth and decline. ●

MACDUFF EVERTON/National Geographic Creative

FIGURE 6.1 This crowded street is located in China, where almost one-fifth of the world's people live.

6.1 HOW MANY PEOPLE CAN THE EARTH SUPPORT?

CONCEPT 6.1 The rapid growth of the human population and its impact on natural capital raises questions about how long the human population can keep growing.

Human Population Growth

For most of history, the human population grew slowly (see Figure 1.11, p. 16, left part of curve). However, it has grown rapidly for the last 200 years, resulting in the characteristic J-curve of exponential growth (Figure 1.11, right part of curve).

Demographers recognize three important trends related to the current size, growth rate, and distribution of the human population. *First,* the rate of population growth decreased in most years since 1965, slowing to 1.2%, but the world's population is still growing (Figure 6.2). This growth rate may not seem very high, but in 2015 it added about 88 million people to the population—an average of 241,000 people every day. In 2015 China had 19% of the world's population and was growing at a rate of 0.5% a year, India had 18% and was growing at 1.3% a year, and the United States with 4.4% was growing at 0.8% a year.

Second, human population growth is unevenly distributed and this pattern is expected to continue (Figure 6.3). About 2% of the 88 million new arrivals on the planet in 2015 were added to the world's more-developed countries. The other 98% were added to the world's less-developed countries.

At least 95% of the 2.5 billion people projected to be added to the world's population between 2015 and 2050 will be born into the less-developed countries. Most of these countries are not equipped to deal with the pressures of rapid population growth.

The *third* important trend is large numbers of people moving from rural areas to *urban areas,* or cities and their surrounding suburbs. In 2015, 53% of the world's people lived in urban areas, and this percentage is increasing. Most of these urban dwellers live in less-developed countries where resources for dealing with rapidly growing populations are limited. Scientists and other analysts have long pondered the question: How long can the human population continue to grow while sidestepping many of

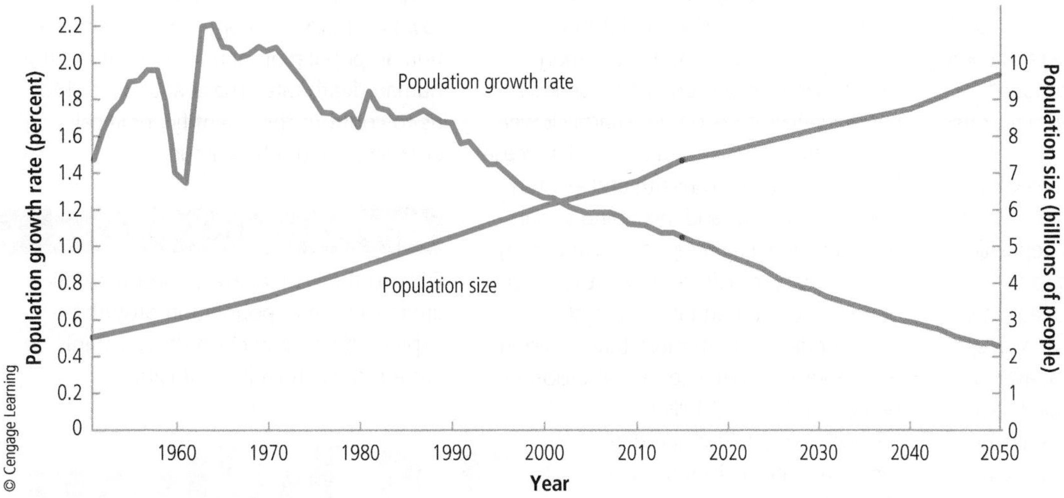

FIGURE 6.2 Global human population size compared with population growth rate, 1950–2015, with projection to 2050 (in blue). ***Critical thinking:*** Why do you think that while the annual growth *rate* of world population has generally dropped since the 1960s, the population has continued to grow?

(Compiled by the authors using data from United Nations Population Division, U.S. Census Bureau, and Population Reference Bureau.)

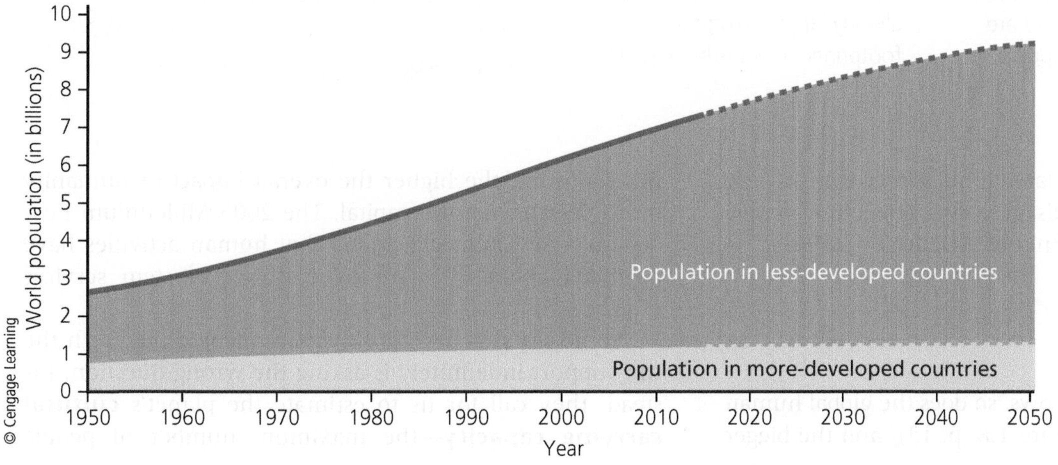

FIGURE 6.3 Most of the world's population growth between 1950 and 2015 took place in the world's less-developed countries. This gap is projected to increase between 2015 and 2050.

(Compiled by the authors using data from United Nations Population Division and Population Reference Bureau.)

How Long Can the Human Population Keep Growing?

Are there physical limits to human population growth and economic growth on a finite planet? Some say yes. Others say no. This debate has been going on since 1798 when Thomas Malthus, a British economist, hypothesized that the human population tends to grow exponentially, while food supplies tend to increase more slowly at a linear rate. However, food production has grown at an exponential rate instead of at a linear rate because of technological advances in industrialized food production.

One current view is that we have already exceeded some of those limits, with too many people collectively degrading the earth's life-support system (see Science Focus 3.3, p. 69). To some analysts, the key problem is the large and rapidly growing number of people in less-developed countries, which have 83% of the world's population. To others, the key factor is *overconsumption* in affluent, more-developed countries with high rates of resource use per person.

Another view of population growth is that technology has allowed us to overcome the environmental limits that all populations of other species face. According to this view, technological advances have increased the earth's carrying capacity for the human species. Some analysts point out that average life expectancy in most of the world has been steadily rising despite warnings from some environmental scientists that we are seriously degrading our life-support system.

These analysts argue that because of our technological ingenuity, there are few, if any, limits to human population growth and resource use per person. They believe that we can continue increasing economic growth and avoid serious damage to our life-support systems by making technological advances in areas such as food production and medicine, and by finding substitutes for resources that we are depleting. They see no need to slow the world's population growth or resource consumption.

Proponents of slowing or stopping population growth point out that currently, we are failing to provide the basic necessities for about 1 billion people—one of every seven on the planet—who struggle to survive on the equivalent of about $1.25 per day. This raises a serious question: How will we meet the basic needs of the additional 2.5 billion people projected to be added mostly to less-developed countries between 2015 and 2050?

Proponents of slowing population growth warn of two potentially serious consequences if we do not sharply lower birth rates. First, death rates could increase because of declining health and environmental conditions and increasing social disruption in some areas, as is happening today in parts of Africa. A worst-case scenario for such a trend is a crash of the human population from more than 7 billion to a more sustainable level of 4 billion or perhaps as low as 2 billion. Second, resource use and degradation of normally renewable resources may intensify as more consumers increase their already large ecological footprints in more-developed countries and in rapidly developing countries such as China, India, and Brazil.

So far, advances in food production and health care have prevented widespread population declines. But there is extensive and growing evidence that we are steadily depleting and degrading much of the earth's irreplaceable natural capital (see Figure 1.3, p. 7). We can get away with this for a while, because the earth's life-support system is resilient. However, such disturbances could reach various *tipping points* beyond which there could be damaging and long-lasting change. In 2015 a team of researchers estimated that we have likely exceeded four key tipping points, or *planetary boundaries* (see Science Focus 3.3).

No one knows how close we are to environmental limits that, many scientists say, eventually will control the size of the human population primarily by raising the human death rate. These analysts call for us to confront this scientific, political, economic, and ethical issue.

CRITICAL THINKING

Do you think there are environmental limits to human population growth? Explain. If so, how close do you think we are to such limits? Explain.

2.5 billion

Projected increase in the world's population between 2015 and 2050

the factors that sooner or later limit the growth of any population? These experts disagree over how many people the earth can support indefinitely (Science Focus 6.1).

Human Population Growth and Natural Capital

As the human population grows, so does the global human ecological footprint (see Figure 1.8, p. 12), and the bigger this footprint, the higher the overall impact of humanity on the earth's natural capital. The 2005 Millennium Ecosystem Assessment concluded that human activities have degraded about 60% of the earth's ecosystem services (Figure 6.4).

Some say that asking how many people the earth the can support indefinitely is asking the wrong question. Instead, they call for us to estimate the planet's **cultural carrying capacity**—the maximum number of people

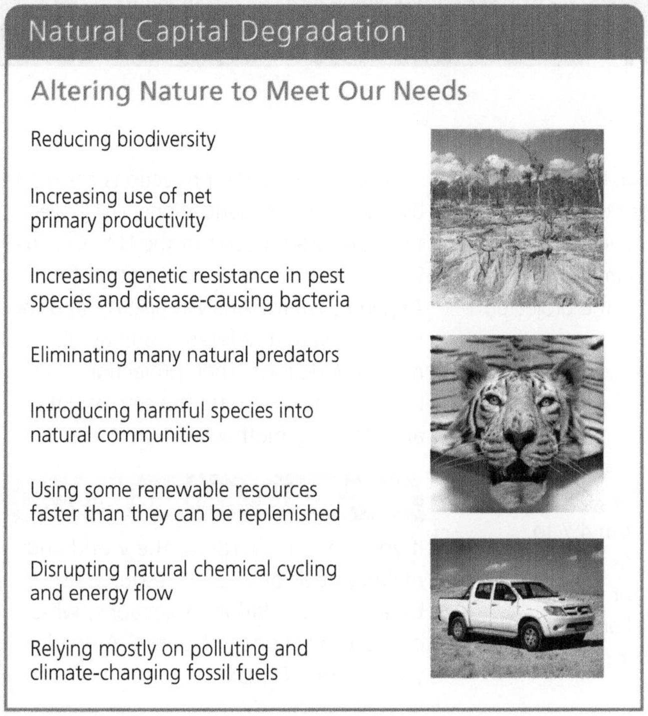

Natural Capital Degradation
Altering Nature to Meet Our Needs
Reducing biodiversity
Increasing use of net primary productivity
Increasing genetic resistance in pest species and disease-causing bacteria
Eliminating many natural predators
Introducing harmful species into natural communities
Using some renewable resources faster than they can be replenished
Disrupting natural chemical cycling and energy flow
Relying mostly on polluting and climate-changing fossil fuels

FIGURE 6.4 Humans have altered the natural systems that sustain their lives and economies in at least eight major ways to meet the increasing needs and wants of their growing population (**Concept 6.1**). ***Critical thinking:*** In your daily living, do you think you contribute directly or indirectly to any of these harmful environmental impacts? Which ones? Explain.

Top: Dirk Ercken/Shutterstock.com. Center: Fulcanelli/Shutterstock.com. Bottom: Werner Stoffberg/Shutterstock.com.

who could live in reasonable freedom and comfort indefinitely, without decreasing the ability of the earth to sustain future generations.

6.2 WHAT FACTORS INFLUENCE THE SIZE OF THE HUMAN POPULATION?

CONCEPT 6.2A Population size increases through births and immigration, and decreases through deaths and emigration.

CONCEPT 6.2B Many factors affect birth rates and death rates and the size of the human population, but the key factor is the average number of children born to the women in the population (*total fertility rate*).

The Human Population Can Grow, Decline, or Stabilize

The basics of global population change are simple. When there are more births than deaths, the human population increases; when there are more deaths than births, the population decreases. When the number of births equals the number of deaths, population size does not change.

Instead of using the total numbers of births and deaths per year, demographers use the **crude birth rate** (the number of live births per 1,000 people in a population in a given year) and the **crude death rate** (the number of deaths per 1,000 people in a population in a given year).

The human population in a particular area grows or declines through the interplay of three factors: *births (fertility)*, deaths *(mortality)*, and *migration*. We can calculate the **population change** of an area by subtracting the number of people leaving a population (through death and emigration) from the number entering it (through birth and immigration) during a year (**Concept 6.2A**):

Population change = (Births + Immigration) − (Deaths + Emigration)

See Figure 9, p. S24, in Supplement 4 for a map comparing generalized rates of population growth among countries and regions in 2014. See Figure 8, pp. S22–S23, in Supplement 4 for more detailed population data for high-, middle-, and low-income countries.

Fertility Rates

Demographers distinguish between two types of fertility rates. One is the **replacement-level fertility rate**: the average number of children that couples in a population must bear to replace themselves. It is slightly higher than two children per couple (typically 2.1) because some children die before reaching their reproductive years, especially in the world's poorest countries.

If we were to reach a global replacement-level fertility rate of 2.1 tomorrow, would it bring an immediate halt to population growth? No, because there is a large number of potential mothers under age 15 who will be moving into their reproductive years.

The second type of fertility rate is the **total fertility rate (TFR)**. It is the average number of children born to the women of childbearing age in a population (**Concept 6.2B**).

Between 1955 and 2015, the global TFR dropped from 5.0 to 2.5. Those who support slowing the world's population growth view this as good news. However, to eventually halt population growth, the global TFR must drop to and remain at the fertility replacement level of 2.1. (See Figure 10, p. S25, in Supplement 4 for a map showing how TFRs vary globally.)

With a TFR of 4.7, Africa's population is growing more than twice as fast as any other continent and is projected to more than double from 1.2 billion in 2015 to 2.5 billion in 2050. Africa is also the world's poorest continent.

Estimates of any population's future numbers can vary considerably, depending mostly on TFR projections. Demographers also have to make assumptions about death rates, migration, and a number of other variables. If their assumptions are wrong, their population forecasts can be inaccurate (Science Focus 6.2).

Projecting Population Change

Estimates of the human population size in 2050 range from 7.8 billion to 10.8 billion people—a difference of 3 billion. The range of estimates varies because many factors affect birth rates and TFRs.

First, demographers have to determine the reliability of current population estimates. While many more-developed countries such as the United States have reliable estimates of their population size, most countries do not. Some countries deliberately inflate or deflate the numbers for economic or political purposes.

Second, demographers make assumptions about trends in fertility. They might assume that fertility is declining by a certain percentage per year. If this estimate is off by a few percentage points, the resulting percentage increase in population can be magnified over a number of years and be quite different from the projected population size increase.

For example, United Nation (UN) demographers assumed that Kenya's fertility rate would decline. Based on that, in 2002 they projected that Kenya's total population would be 44 million by 2050. In reality, the fertility rate rose from 4.7 to 4.8 children per woman. As a result, the UN revised its projection for Kenya's population in 2050 to 81.4 million, which was 85% higher than its earlier projection.

Third, population projections are made by a variety of organizations. UN projections are often cited but the U.S. Census Bureau, the International Institute for Applied Systems Analysis (IIASA), and the U.S. Population Reference Bureau also make projections. Their projections vary because they use differing sets of data and differing methods (Figure 6.A).

CRITICAL THINKING

If you were in charge of the world and making decisions about resource use based on population projections, which of the projections in Figure 6.A would you rely on? Explain.

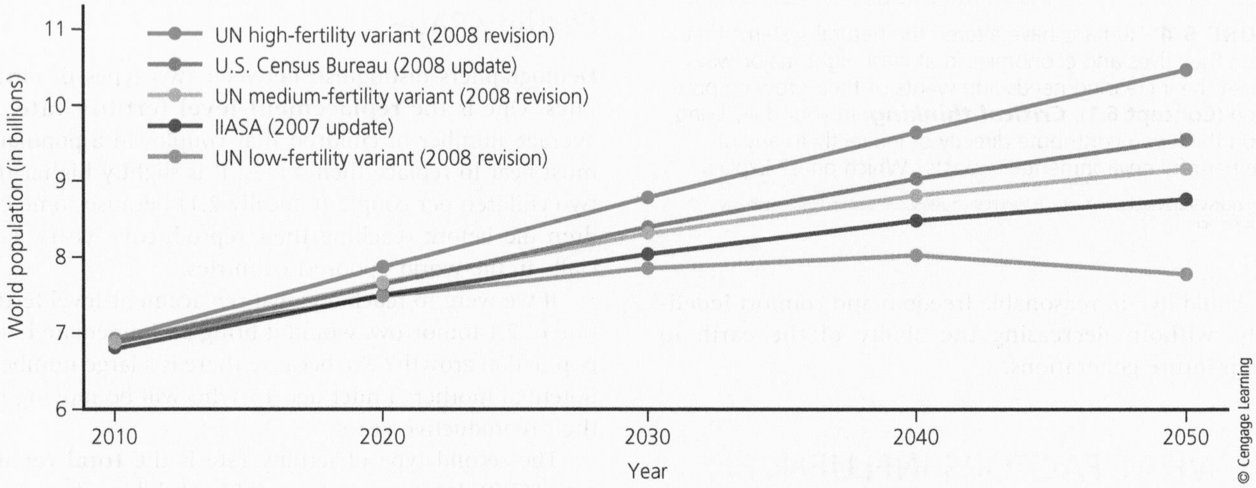

FIGURE 6.A World population projections to 2050 from three different organizations: the UN, the U.S. Census Bureau, and IIASA. Note that the uppermost, middle, and lowermost curves of these five projections are all from the UN, each assuming a different level of fertility. ***Data analysis:*** What are the ranges (differences between the lowest and highest) in these projections for 2030, 2040, and 2050?

(Compiled by the authors using data from United Nations, U.S. Census Bureau, and IIASA.)

CASE STUDY

The U.S. Population—Third Largest and Growing

The population of the United States grew from 76 million in 1900 to 321 million in 2015. This happened despite oscillations in the country's TFR (Figure 6.5) and population growth rate. It took the country 139 years to reach a population of 100 million people, 52 years to add another 100 million (by 1967), and 39 years to add the third 100 million (by 2006).

During the period of high birth rates between 1946 and 1964, known as the *baby boom*, 79 million people were added to the U.S. population. At the peak of the baby boom in 1957, the average TFR was 3.7 children per woman. In most years since 1972, it has been at or below 2.1 children per woman—1.9 in 2015, compared to a global TFR of 2.5.

The drop in the TFR has slowed the rate of population growth in the United States, but the country's population is still growing. In 2015 about 2.5 million people were added to the U.S. population, according to the U.S. Census

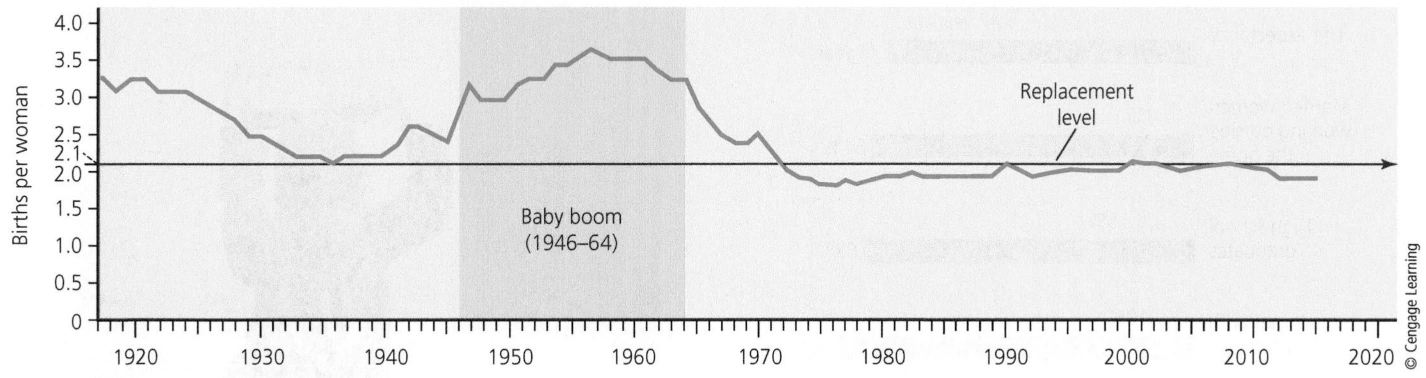

FIGURE 6.5 Total fertility rates for the United States between 1917 and 2015. ***Critical thinking:*** The U.S. fertility rate has declined and remained at or below replacement levels since 1972. So why is the population of the United States still increasing?

(Compiled by the authors using data from Population Reference Bureau and U.S. Census Bureau.)

Bureau. About 1.5 million (60% of the total) were added because there were more births than deaths, and about 1 million (40% of the total) were legal immigrants.

Since 1820, the United States has admitted almost twice as many legal immigrants and refugees as all other countries combined. The number of legal immigrants (including refugees) has varied during different periods because of changes in immigration laws and rates of economic growth (Figure 6.6).

Since 1965, nearly 59 million people have legally immigrated to the United States, most of them from Latin America and Asia, with the government giving preferences for those with technical training or with family members of U.S. citizens. A 2015 study by the U.S. Census Bureau noted that in 2013, China surpassed Mexico as the largest source of new U.S. immigrants. A 2015 study by the Pew Research Center found that between 2009 and 2014, more legal and illegal immigrants in the United States returned to Mexico (where economic conditions are improving) than migrated to the United States.

According to population experts, the country's influx of immigrants has made the country more culturally diverse and has increased economic growth as these citizens worked and started businesses. An estimated 11.7 illegal immigrants also live in the United States. There is controversy over whether to deport those who can be found or to allow these individuals to meet strict criteria for becoming U.S. citizens.

In addition to the fourfold increase in population since 1900, some amazing changes in lifestyles took place in the United States during the 20th century (Figure 6.7), which led to Americans living longer. Along with this came dramatic increases in per capita resource use and much larger total and per capita ecological footprints.

The U.S. Census Bureau projects that between 2015 and 2050, the U.S. population will likely grow from 321 million to 398 million—an increase of 77 million people. Because of a high per-person rate of resource use and the resulting waste and pollution, each addition to the U.S. population has an enormous environmental impact (see Figure 6, p. S20, in Supplement 4).

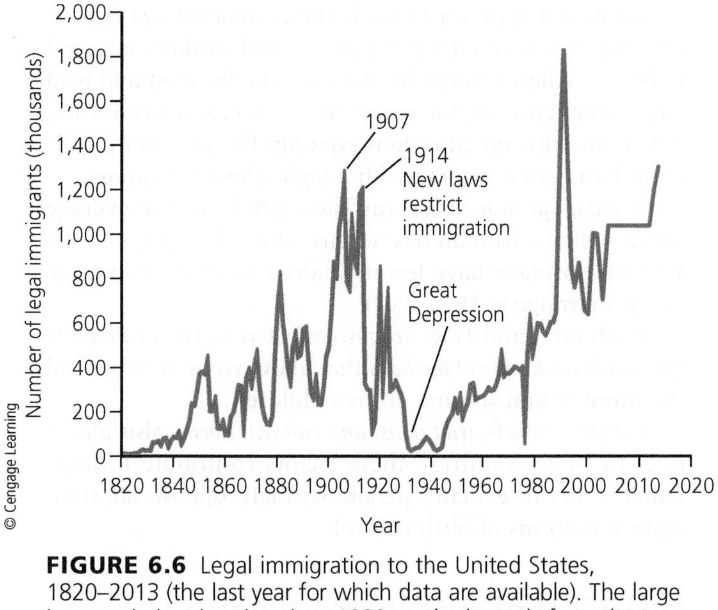

FIGURE 6.6 Legal immigration to the United States, 1820–2013 (the last year for which data are available). The large increase in immigration since 1989 resulted mostly from the Immigration Reform and Control Act of 1986, which granted legal status to certain illegal immigrants who could show they had been living in the country prior to January 1, 1982.

(Compiled by the authors using data from U.S. Immigration and Naturalization Service, the Immigration Policy Institute, and the Pew Hispanic Center.)

77 million

Projected increase in the U.S. population between 2015 and 2050

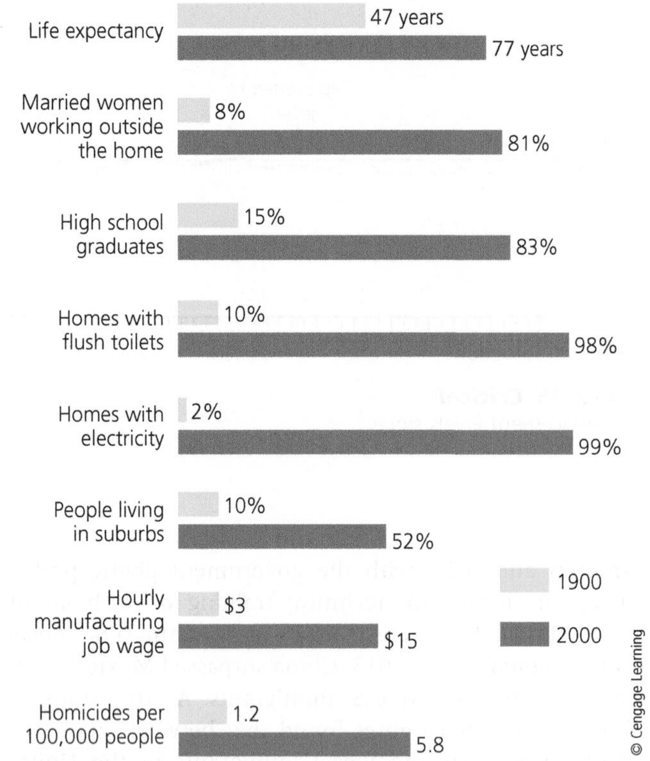

FIGURE 6.7 Some major changes took place in the United States between 1900 and 2000. ***Critical thinking:*** Which two of these changes do you think had the biggest impacts on the U.S. ecological footprint?

(Compiled by the authors using data from U.S. Census Bureau and Department of Commerce.)

FIGURE 6.8 This young boy spends much of his day carrying bricks.

Factors That Affect Birth and Fertility Rates

Many factors affect a country's average birth rate and TFR. One is the *importance of children as a part of the labor force*, especially in less-developed countries. Many poor couples in those countries struggle to survive on less than $2.25 a day. Some of these couples have a large number of children to help them haul drinking water, gather wood for heating and cooking, and grow or find food. Worldwide, 1 of every 10 children between ages 5 and 17 work to help their families survive (Figure 6.8).

Another economic factor is the *cost of raising and educating children*. Birth and fertility rates tend to be lower in more-developed countries, where raising children is much more costly because they do not enter the labor force until they are in their late teens or twenties. In the United States, the average cost of raising a child born in 2013 to age 18 ranged from $169,000 to $390,000 depending on household income (latest data available).

The *availability of pension systems* can influence the number of children couples have, especially poor people in less-developed countries. Pensions reduce a couple's need to have several children to replace those that die at an early age and to help support them in old age.

Urbanization also plays a role. People living in urban areas usually have better access to family planning services and tend to have fewer children than do those living in the rural areas of less-developed countries.

Another important factor is the *educational and employment opportunities available for women*. Total fertility rates tend to be low when women have access to education and paid employment outside the home. In less-developed countries, a woman with no education typically has two more children than does a woman with a high school education.

Average age at marriage (or, more precisely, the average age at which a woman has her first child) also plays a role. Women normally have fewer children when their average age at marriage is 25 or older.

Birth rates and TFRs are also affected by the *availability of reliable birth control methods* that allow women to control the number and spacing of their children.

Religious beliefs, traditions, and cultural norms also play a role. In some countries, these factors contribute to large families, because many people strongly oppose abortion and some forms of birth control.

Factors That Affect Death Rates

The rapid growth of the world's population over the past 100 years is largely the result of declining death rates, especially in less-developed countries. More people in some of these countries live longer, and fewer infants die because of larger food supplies, improvements in food distribution, better nutrition, improved sanitation, safer water supplies, and medical advances such as immunizations and antibiotics.

A useful indicator of the overall health of people in a country or region is **life expectancy**: the average number of years a person born in a particular year can be expected to live. Between 1955 and 2015, average global life expectancy increased from 48 years to 71 years. In 2015 Japan had the world's longest life expectancy of 83 years. Between 1900 and 2015, the average U.S. life expectancy rose from 47 years to 79 years. Research indicates that poverty, which reduces the average life span by 7 to 10 years, is the single most important factor affecting life expectancy.

Another important indicator of the overall health of a population is its **infant mortality rate**, the number of babies out of every 1,000 born who die before their first birthday. It is viewed as one of the best measures of a society's quality of life because it indicates the general level of nutrition and health care. (See Figure 11, p. S26, in Supplement 4 for a map comparing generalized infant mortality rates among the world's countries.) A high infant mortality rate usually indicates insufficient food (*undernutrition*), poor nutrition (*malnutrition*; see Figure 1.13, p. 18), and a high incidence of infectious disease. Infant mortality also affects the TFR. In areas with low infant mortality rates, women tend to have fewer children because fewer of their children die at an early age.

Infant mortality rates in most countries have declined dramatically since 1965 (Figure 6.9). Even so, every year more than 4 million infants die of *preventable* causes during their first year of life, according to UN population experts. Most of these deaths occur in less-developed countries.

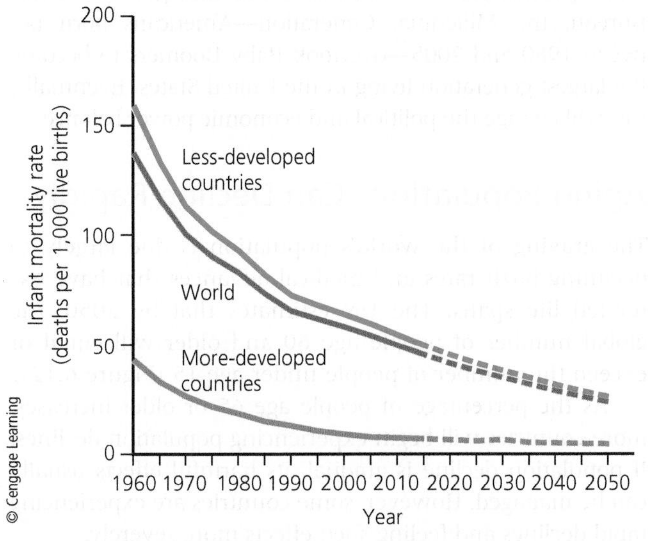

FIGURE 6.9 Comparison of infant mortality rates in more-developed countries and less-developed countries, 1950–2015, with projections to 2050 based on medium population projections.

(Compiled by the authors using data from United Nations Population Division and Population Reference Bureau.)

This average of nearly 11,000 mostly unnecessary infant deaths per day is equivalent to 55 jet airliners, each loaded with 200 infants, crashing *every day* with no survivors.

Between 1900 and 2015, the U.S. infant mortality rate dropped from 165 to 6. This sharp decline was a major factor in the marked increase in U.S. average life expectancy during this period. However, 47 other nations had lower infant mortality rates than the United States in 2015. They include Canada, Cuba, Israel, Japan, South Korea, and most European nations.

Migration

A third factor in population change is **migration**: the movement of people into (*immigration*) and out of (*emigration*) specific geographic areas. Most people who migrate to another area within their country or to another country are seeking jobs and economic improvement. Others are driven by religious persecution, ethnic conflicts, political oppression, or war. There are also *environmental refugees*—people who have to leave their homes and sometimes their countries because of water or food shortages, soil erosion, or some other form of environmental degradation.

6.3 HOW DOES A POPULATION'S AGE STRUCTURE AFFECT ITS GROWTH OR DECLINE?

CONCEPT 6.3 The numbers of males and females in young, middle, and older age groups determine how fast a population grows or declines.

Age Structure

The **age structure** of a population is the numbers or percentages of males and females in young, middle, and older age groups in that population (**Concept 6.3**). In addition to total fertility rates, age structure is an important factor in determining whether the population of a country increases or decreases.

Population experts construct a population *age-structure diagram* by plotting the percentages or numbers of males and females in the total population in each of three age categories: *pre-reproductive* (ages 0–14), consisting of individuals normally too young to have children; *reproductive* (ages 15–44), consisting of those normally able to have children; and *post-reproductive* (ages 45 and older), with individuals normally too old to have children. Figure 6.10 presents generalized age-structure diagrams for countries with rapid, slow, zero, and negative population growth rates.

A country with a large percentage of people younger than age 15 (represented by a wide base in Figure 6.10, far left) will experience rapid population growth unless death rates rise sharply. Because of this *demographic momentum*,

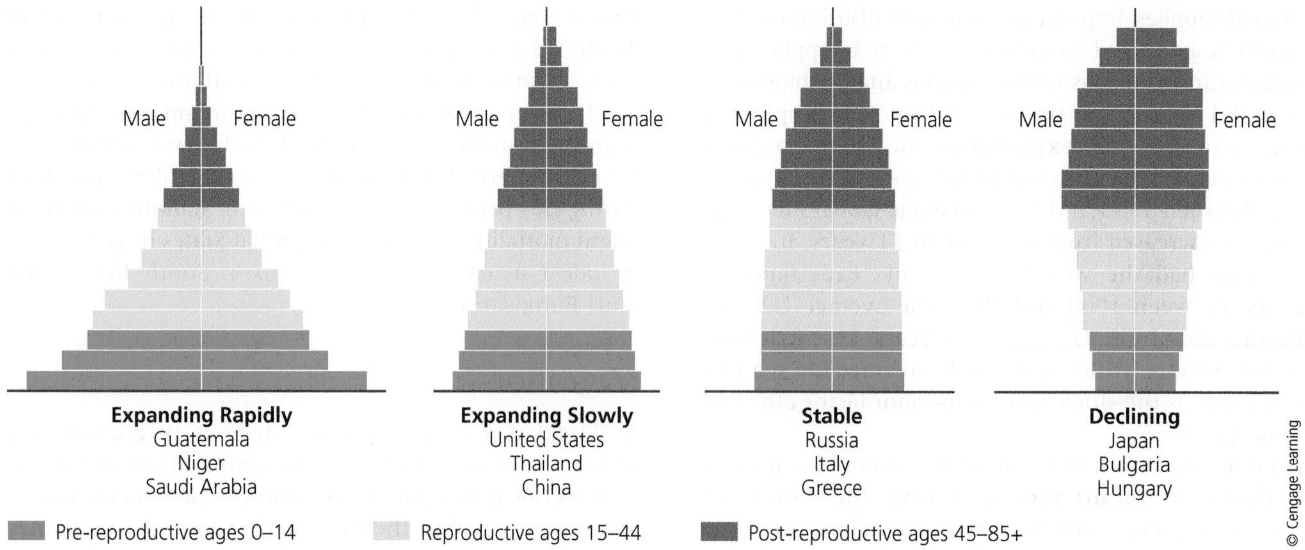

FIGURE 6.10 Generalized population age-structure diagrams for countries with rapid (1.5–3%), slow (0.3–1.4%), zero (0–0.2%), and negative (declining) population growth rates. ***Question:*** Which of these diagrams best represents the country where you live?

(Compiled by the authors using data from U.S. Census Bureau and Population Reference Bureau.)

the number of births in such a country will rise for several decades. This will occur even if women each have an average of only one or two children because of the large number of girls entering their prime reproductive years. Most future human population growth will take place in less-developed countries because of their typically youthful age structure and rapid population growth rates.

The global population of seniors—people who are 65 and older—is projected to triple between 2015 and 2050, when one of every six people will be a senior. (See the Case Study that follows.). An aging population combined with a lower fertility rate results in fewer working-age adults having to support a large number of seniors. For example, in China and the United States between 2010 and 2050, the working-age population is projected to decline sharply. This could lead to a shortage of workers and friction between the younger and older generations in these countries.

CASE STUDY

The American Baby Boom

Changes in the distribution of a country's age groups have long-lasting economic and social impacts. For example, the American baby boom (Figure 6.5) added 79 million people to the U.S. population between 1946 and 1964. Over time, this group looks like a bulge as it moves up through the country's age structure, as shown in Figure 6.11.

For decades, the baby-boom generation has strongly influenced the U.S. economy because it makes up about 25% of the U.S. population. Baby boomers created the youth market in their teens and twenties and are now creating the late-middle-age (ages 50 to 60) and senior markets. In addition to having this economic impact, the large baby-boom generation plays an increasingly important role in deciding who is elected to public office and what laws are passed or weakened.

Since 2011, when the first baby boomers began turning 65, the number of Americans older than age 65 has grown at the rate of about 10,000 a day and will do so through 2030. This process has been called the *graying of America*. As the number of working adults declines in proportion to the number of seniors, there may be political pressure from baby boomers to increase tax revenues to help support the growing senior population. However, in 2015, according to the Census Bureau, the Millennial Generation—Americans born between 1980 and 2005—overtook Baby Boomers to become the largest generation living in the United States. Eventually, this will change the political and economic power balance.

Aging Populations Can Decline Rapidly

The graying of the world's population is due largely to declining birth rates and medical advances that have extended life spans. The UN estimates that by 2050, the global number of people age 60 and older will equal or exceed the number of people under age 15 (Figure 6.12).

As the percentage of people age 65 or older increases, more countries will begin experiencing population declines. If population decline is gradual, its harmful effects usually can be managed. However, some countries are experiencing rapid declines and feeling such effects more severely.

Japan has the world's highest percentage of people age 65 or over and the world's lowest percentage of people under age 15. In 2015 Japan's population was 127 million. By 2050 its population is projected to be 97 million, a 24% drop. As its population declines, there will be fewer adults

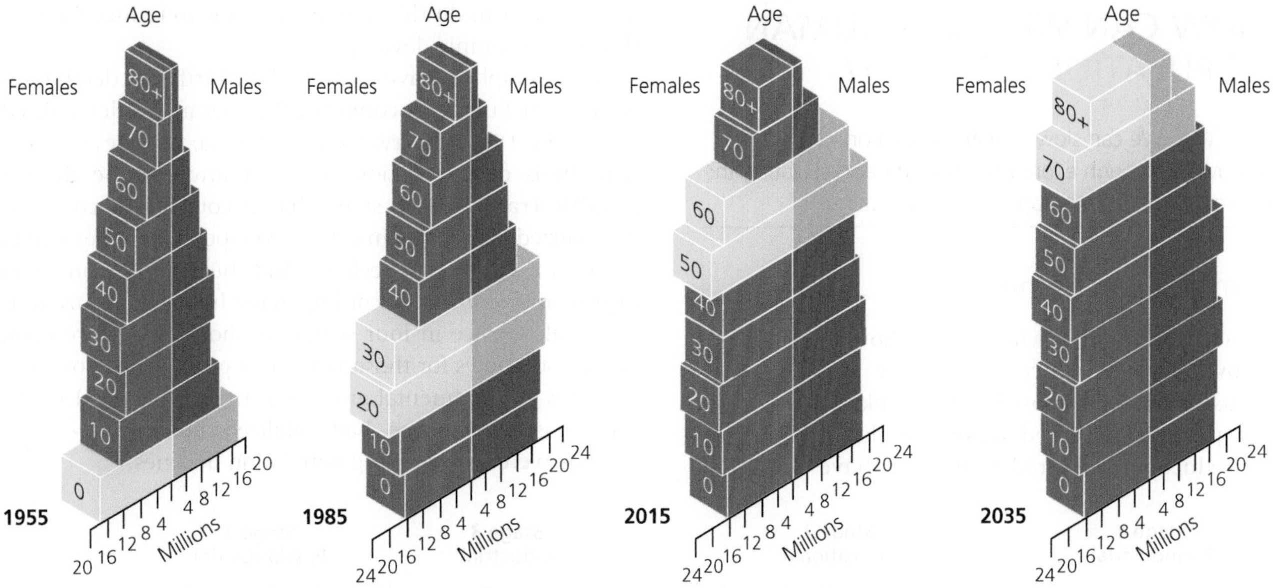

© Cengage Learning

FIGURE 6.11 Age-structure diagrams tracking the baby-boom generation in the United States, 1955, 1985, 2015, and 2035 (projected). ***Critical thinking:*** How might the projected age structure in 2035 affect you?

(Compiled by the authors using data from U.S. Census Bureau and Population Reference Bureau.)

working and paying taxes to support an increasingly elderly population. Because Japan discourages immigration, this could threaten its economic future. In recent years, Japan has been feeling the effects of a declining population. For example, houses in some suburbs have been abandoned and cannot be sold because of a lack of buyers. They could be demolished, but who will pay the costs—the owners who have abandoned them, or the government?

Figure 6.13 lists some of the problems associated with rapid population decline. Countries with rapidly declining populations, in addition to Japan, include Germany, Italy, Bulgaria, Hungary, Romania, Cuba, and Portugal. Other countries facing population declines in the not-too-distant future are China, South Korea, Iran, Russia, Spain, Singapore, and the Netherlands. Population declines are difficult to reverse.

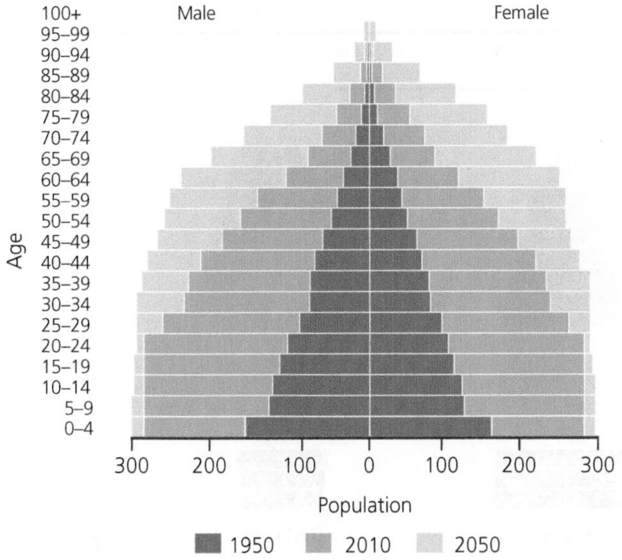

© Cengage Learning

FIGURE 6.12 The world's age structure 1950, 2010, and 2050 (projected). ***Critical thinking:*** How might the projected age structure in 2050 affect you?

(Compiled by the authors using data from U.S. Census Bureau and United Nations Population Division.)

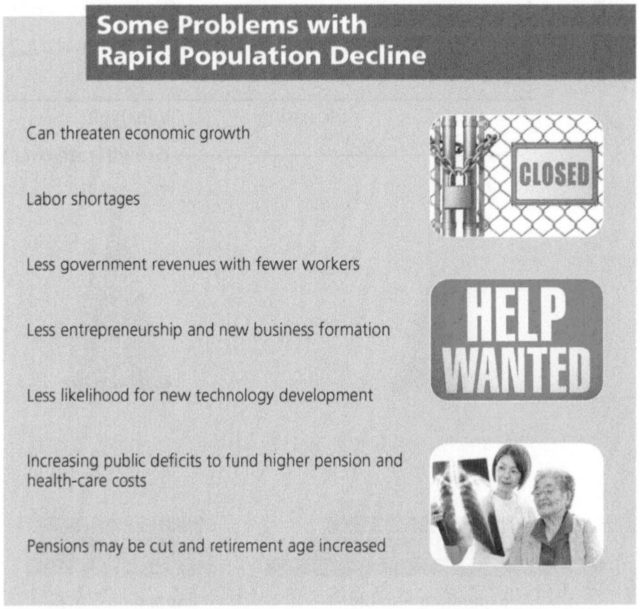

© Cengage Learning

FIGURE 6.13 Rapid population decline can cause several problems. ***Critical thinking:*** Which two of these problems do you think are the most important?

Top: Slavoljub Pantelic/Shutterstock.com. Center: Iofoto/Shutterstock.com. Bottom:sunabesyou/Shutterstock.com.

6.4 HOW CAN WE SLOW HUMAN POPULATION GROWTH?

CONCEPT 6.4 We can slow human population growth by reducing poverty through economic development, elevating the status of women, and encouraging family planning.

Economic Development

There is controversy over whether we should slow population growth (Science Focus 6.1). Some analysts argue that because population growth can be linked to environmental degradation, we need to slow population growth in order to reduce such degradation. They have suggested several ways to do this, one of which is to reduce poverty through economic development.

Demographers have examined the birth and death rates of western European countries that became industrialized during the 19th century. Using such data, they developed a hypothesis on population change known as the **demographic transition**. It states that as countries become industrialized and economically developed, their per capita incomes rise, poverty declines, and their populations tend to grow more slowly. According to the hypothesis, this transition takes place in four stages, as shown in Figure 6.14. Some good news for those who view population growth as a serious environmental problem is that by 2015, 31 countries, mostly in Europe, had stabilized their populations or were experiencing population declines. GOOD NEWS

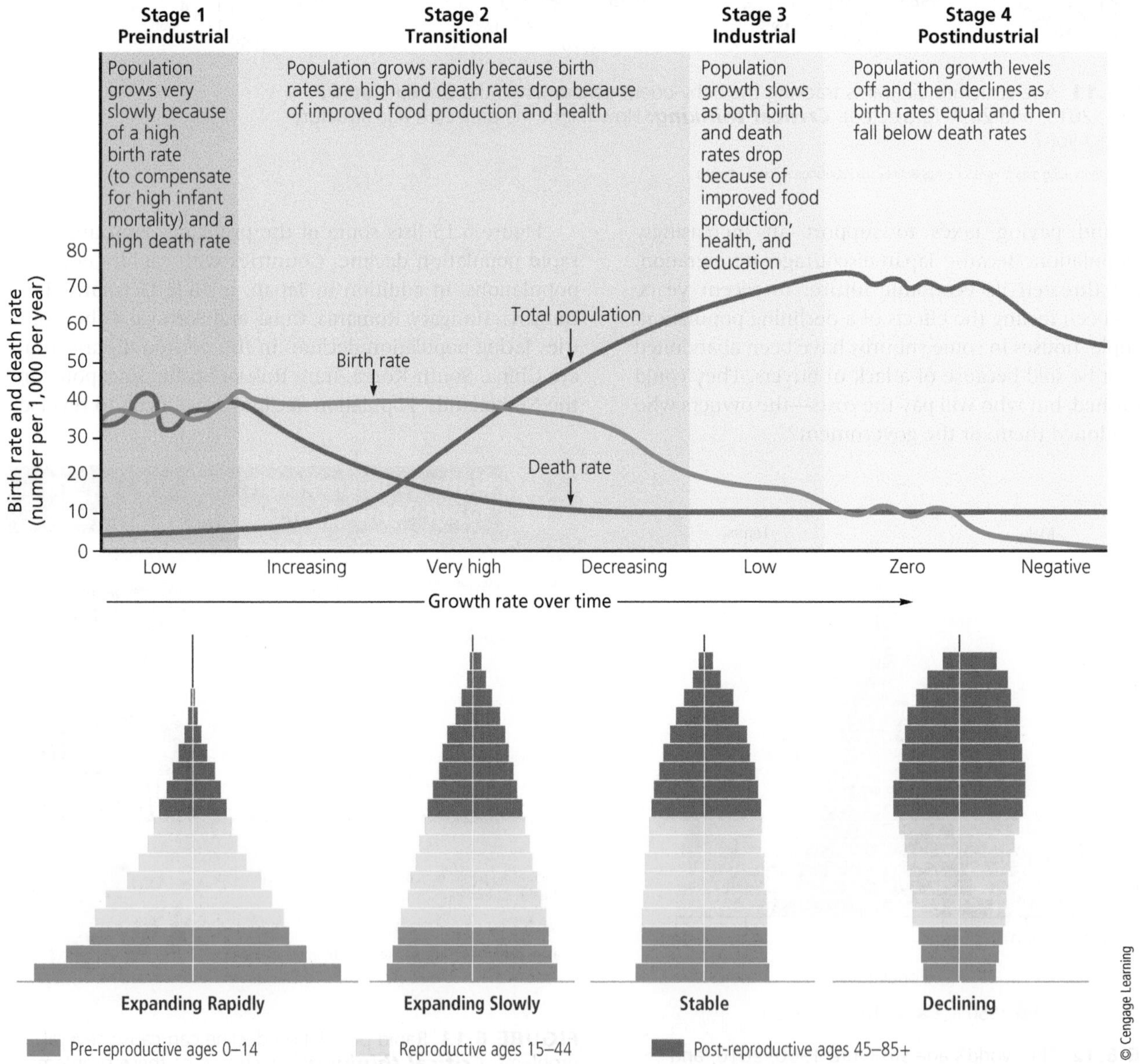

FIGURE 6.14 The *demographic transition*, which a country can experience as it becomes industrialized and more economically developed, can take place in four stages. ***Question:*** At what stage is the country where you live?

Some analysts believe that most of the world's less-developed countries will make a demographic transition over the next few decades. They hypothesize that the transition will occur because newer technologies will help them to develop economically and to reduce poverty.

Other analysts fear that rapid population growth, extreme poverty, war, increasing environmental degradation, and resource depletion could leave some countries with high population growth rates (2.5% to 3%) stuck in stage 2 of the demographic transition. Such countries include Afghanistan, Iraq, Guatemala, Pakistan, Yemen, and Nigeria and a number of other African countries. This highlights the need to reduce poverty as a key to improving human health and stabilizing the population.

Educating and Empowering Women

A number of studies show that women tend to have fewer children if they are educated, can control their own fertility, earn an income of their own, and live in societies that do not suppress their rights.

Only about 30% of the world's girls are enrolled in secondary education, so widespread education of girls is important for their future and for trying to slow population growth. In most societies women have fewer rights and educational and economic opportunities than men have.

Women do almost all of the world's domestic work and child care for little or no pay. They provide more unpaid health care (within their families) than do all of the world's organized health-care services combined. In rural areas of Africa, Latin America, and Asia, women do 60–80% of the work associated with growing food, hauling water, and gathering and hauling wood (Figure 6.15) and animal dung for use as fuel. As one Brazilian woman observed, "For poor women, the only holiday is when you are asleep."

Iv Nikolny/Shutterstock.com

FIGURE 6.15 This woman in Nepal is bringing home firewood. Typically, she spends 2 hours a day, two or three times a week, on this task.

Women make up slightly more than 50% of the world's poor and 66% of the world's 800 million illiterate adults. In many societies, boys are more likely to get an education than are girls because of traditional views of a woman's role in a household. Studies show that one of the major factors for slowing population growth is the widespread education of girls.

Poor women who cannot read often have an average of five to seven children, compared with two or fewer children in societies where most women can read. This highlights the need for all children to get at least an elementary school education.

A growing number of women in less-developed countries are taking charge of their lives and reproductive behavior. As this number grows, such change driven by individual women will play an important role in stabilizing populations. This change will also improve human health, reduce poverty and environmental degradation, and allow more access to basic human rights.

Family Planning

Family planning programs provide education and clinical services that can help couples to choose how many children to have and when to have them. Such programs vary from culture to culture, but most of them provide information on birth spacing, birth control, and health care for pregnant women and infants.

According to studies by the UN Population Division and other population agencies, family planning has been a major factor in reducing the number of unintended pregnancies, births, and abortions. In addition, family planning has reduced rates of infant mortality and the number of mothers and fetuses dying during pregnancy. According to the UN, had there not been the sharp drop in TFRs since the 1970s, with all else being equal, the world's population today would be about 8.5 billion instead of 7.3 billion (**Core Case Study**). Family planning has played an important role in countries that have stabilized their populations.

Family planning also has financial benefits. Studies show that each dollar spent on family planning in countries such as Thailand, Egypt, and Bangladesh saves $10 to $16 in health, education, and social service costs by preventing unwanted births.

Despite these successes, certain problems have hindered progress in some countries. There are three major problems. *First*, according to the UN Population Fund and the Guttmacher Institute, about 40% of all pregnancies in less-developed countries were unplanned and about half of these pregnancies end with abortion. So ensuring access to voluntary contraception would play a key role in stabilizing the populations and reducing the number of abortions in such countries.

Second, according to the same sources, an estimated 225 million women, primarily in 69 of the world's poorest countries, lack access to family planning services. Meeting these current unmet needs for family planning and contraception could prevent 54 million unintended pregnancies, 26 million induced abortions (16 million of them unsafe), 1.1 million infant deaths, and 79,000 pregnancy-related deaths of women per year. This could reduce the projected global population size by more than 1 billion people, at an average cost of $25 per couple per year.

Third, largely because of cultural traditions, male domination, and poverty, one in every three girls in less-developed countries is married before age 18 and one in nine is married before age 14. This occurs despite laws against child marriage. For a poor family, marrying off a daughter can relieve financial pressure.

Some analysts call for expanding family planning programs to educate men about the importance of having fewer children and taking more responsibility for raising them. Proponents also call for greatly increased research in order to develop more effective birth control methods for men.

The experiences of countries such as Japan, Thailand, Bangladesh, South Korea, Taiwan, and China show that a country can achieve or come close to replacement-level fertility within a decade or two. The real population story of the past 50 years has been the sharp reduction in the *rate* of population growth (Figure 6.2) resulting from the reduction of poverty through economic development, empowerment of women, and the promotion of family planning. However, the global population is still growing fast enough to add 2 to 3 billion more people during this century (see Figure 1.11, p. 16).

CASE STUDY

Population Growth in India

For six decades, India has tried to control its population growth with only modest success. The world's first national family planning program began in India in 1952, when its population was nearly 400 million. In 2015, after 63 years of population control efforts, India had 1.31 billion people—the world's second largest population and a TFR of 2.3. Much of this increase occurred because of India's declining death rates.

In 1952 India added 5 million people to its population. In 2015 it added 14 million people—more than any other country. Figure 6.16 shows changes in India's age structure between 2010 and 2035 (projected). The UN projects that by 2029, India will be the world's most populous country, and that by 2050, it will have a population of 1.7 billion.

India has the world's fourth largest economy and a rapidly growing middle class of more than 100 million people—a number nearly equal to a third of the U.S. population. However, the country faces serious poverty, malnutrition, and environmental problems that could worsen as its population continues to grow rapidly. About one-fourth of all people in India's cities live in slums, and prosperity and progress have not touched hundreds of millions of Indians who live in rural villages. According to the World Bank, about 30% of India's population—one-third of the world's extremely poor people—live in extreme poverty on less than $1.25 per day (Figure 6.17). In India, 300 million people—a number almost equal to the entire U.S. population—do not have electricity.

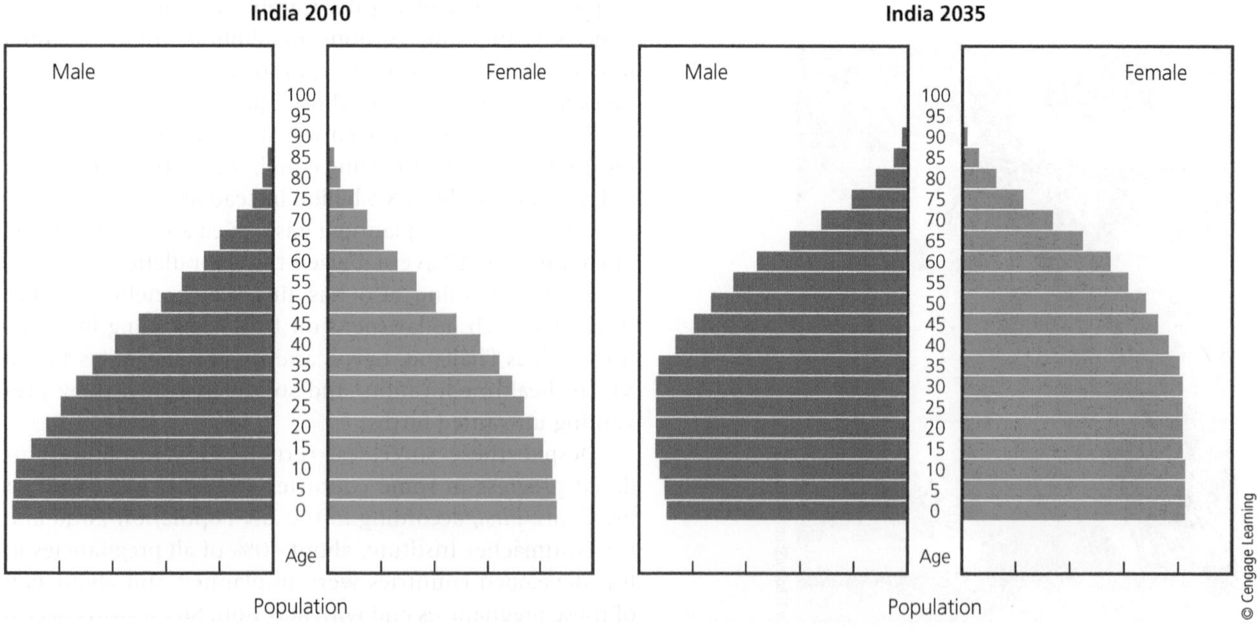

FIGURE 6.16 Age structure changes in India: 2010 and 2035 (projected). ***Critical thinking:*** How might the projected age structure in 2035 affect India's ability to reduce poverty?

(Compiled by the authors using data from U.S. Census Bureau and United Nations Population Division.)

Samrat35/Dreamstime.com

FIGURE 6.17 Homeless people in Kolkata, India.

For decades, the Indian government has provided family planning services throughout the country and has strongly promoted a smaller average family size. Even so, Indian women have an average of 2.3 children.

Three factors help to account for larger families in India. *First,* most poor couples believe they need several children to work and care for them in their old age. *Second,* the strong cultural preference in India for male children means that some couples keep having children until they produce one or more boys. And *third,* although 90% of Indian couples have access to at least one modern birth control method, only about 47% actually use one.

India also faces critical resource and environmental problems. With 18% of the world's people, India has just 2.3% of the world's land resources and 2% of its forests. About half the country's cropland has been degraded by soil erosion and overgrazing. In addition, more than two-thirds of its water is seriously polluted, sanitation services often are inadequate, and many of its major cities suffer from serious air pollution.

India's rapid economic growth is expected to accelerate over the next few decades. This will help many people in India escape poverty, but it will also increase pressure on India's and the earth's natural capital as rates of per capita resource use rise. On the other hand, economic growth may help India to slow its population growth by accelerating its demographic transition.

CASE STUDY

Slowing Population Growth in China

China is the world's most populous country, with 1.37 billion people in 2015. According to the Earth Policy Institute, China's population is projected to peak at 1.4 billion in 2026 and then to begin a slow decline.

In the 1960s, China's large population was growing so rapidly that there was a serious threat of mass starvation. To avoid this, government officials took measures that eventually led to the establishment of the world's most extensive, intrusive, and strict family planning and birth control program.

The centerpiece of the program, established in 1978, was the promotion of one-child families. Married couples pledging to have no more than one child received better housing, more food, free health care, salary bonuses, and preferential job opportunities for their child. Couples who broke their pledge lost such benefits. The government also provided contraceptives, sterilizations, and abortions for married couples.

Since this government-controlled program began in 1978, China has made impressive efforts to feed its people and bring its population growth under control. Between 1978 and 2015, the country reduced its TFR from 3.0 to 1.7. China's one-child policy played a role in this drop in the country's TFR. However, the TFR had already been falling a decade earlier, from 5.9 in 1968 to 3.0 in 1978 when the one-child policy was implemented. This earlier drop and its continuation after 1978 was strongly influenced by increased education and employment opportunities for young women, according to Nobel Laureate Amartya Sen. Currently, China's population is growing more slowly than the U.S. population, even with legal immigration included. Although China has avoided mass starvation, its strict population control program has been accused of violating human rights.

Because of the cultural preference for sons, many Chinese women abort female fetuses. This has reduced the female population and means that about 30 million Chinese men are unable to find anyone to marry.

Since 1980, China has undergone rapid industrialization and economic growth. According to the Earth Policy Institute, between 1990 and 2010 this process reduced the number of people living in extreme poverty by almost 500 million. It has also helped at least 300 million Chinese to become middle-class consumers. However, poverty still plagues millions of people living in China's villages and cities (Figure 6.18).

Over time, China's rapidly growing middle class will consume more total resources. This will put a strain on China's and the earth's natural capital. Like India, China faces serious soil erosion, overgrazing, water pollution, and air pollution problems.

Because of its one-child policy, in recent years the average age of China's population has been increasing at one of the fastest rates ever recorded (Figure 6.19). The UN projects that the number of Chinese over 65 will increase from 130 million in 2015 to 243 million in 2030. This increase is roughly equal to the population of Indonesia, the world's fourth most populous country in 2015. The UN also estimates that by 2030, the country is likely to have too few young workers (ages 15 to 64) to support its rapidly aging population. To help deal with this problem, China plans to become the world's largest manufacturer of industrial robots to be used for manufacturing. It will also sell such robots to other countries.

Because of these concerns, in 2015 the Chinese government abandoned its one-child policy and replaced it with a two-child policy. Married couples can apply to the government for permission to have two children. However, the Chinese people have gotten used to small families. In addition, many married couples may have only one child because of the high cost of raising a second child and the greatly increased educational and employment opportunities for young women.

BIG IDEAS

- The human population is growing rapidly and may soon bump up against environmental limits.
- The combination of population growth and the increasing rate of resource use per person is expanding the overall human ecological footprint and putting a strain on the earth's natural capital.
- We can slow human population growth by reducing poverty, elevating the status of women, and encouraging family planning.

Nikada/iStock/Getty Images Plus/Getty Images

FIGURE 6.18 Old and new housing in heavily populated Shanghai, China, in 2015.

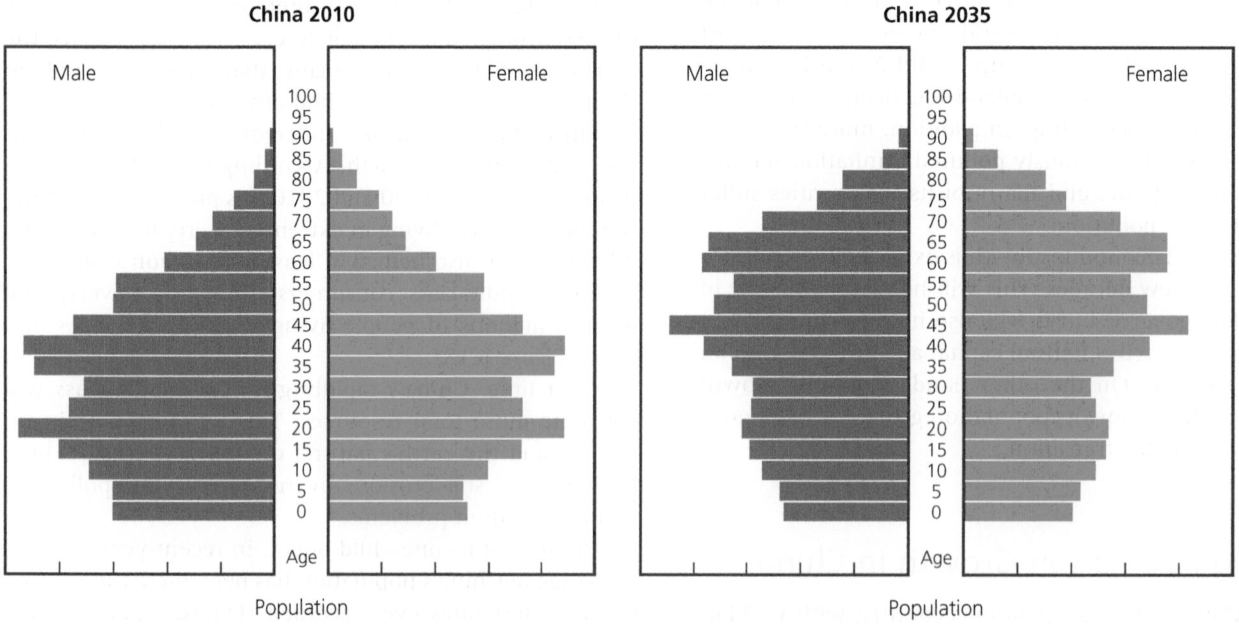

© Cengage Learning

FIGURE 6.19 Age structure in China: 2010 and 2035 (projected). ***Critical thinking:*** How might the projected age structure in 2035 affect China's economy?

(Compiled by the authors using data from U.S. Census Bureau and United Nations Population Division.)

Tying It All Together

World Population Growth and Sustainability

This chapter began with a discussion of the fact that the world's human population has now passed 7.3 billion (**Core Case Study**). We noted that this is a result of exponential population growth and that many environmental scientists believe such growth to be unsustainable in the long run. We briefly considered some of the environmental problems brought on by exponential human population growth. We looked at factors that influence the growth of populations, as well as at how some countries have made progress in controlling population growth.

JeremyRichards/Shutterstock.com

In the first six chapters of this book, you have learned how ecosystems and species have been sustained throughout the earth's history, in keeping with the three **scientific principles of sustainability**, by nature's reliance on solar energy, nutrient cycling, and biodiversity (see inside back cover of this book). These three principles can guide us in dealing with the problems brought on by population growth and decline. By greatly increasing our use of solar, wind, and other renewable-energy technologies, we can cut pollution and emissions of climate-changing gases that are increasing as the population and resource use per person grow. By reusing and recycling more materials, we can cut resource waste and reduce our ecological footprints. By focusing on preserving biodiversity, we can help sustain the life-support system on which we and all other species depend.

Chapter Review

Core Case Study

1. Summarize the story of how human population growth has surpassed 7.3 billion and explain why this is significant to many environmental scientists (**Core Case Study**). List three factors that account for the rapid growth of the human population over the past 200 years.

Section 6.1

2. What is the key concept for this section? What is the range of estimates for the size of the human population in 2050? Summarize the three major population growth trends recognized by demographers. About how many people were added to the world's population in 2015? What are the world's three most populous countries? List eight major ways in which we have altered natural systems to meet our needs. Define **cultural carrying capacity**. Summarize the debate over whether and how long the human population can keep growing.

Section 6.2

3. What are the two key concepts for this section? Define and distinguish between **crude birth rate** and **crude death rate**. List three variables that affect the growth and decline of human populations. Explain how a given area's **population change** is calculated.

4. Define and distinguish between the **replacement-level fertility rate** and the **total fertility rate (TFR)**. How has the global TFR changed since 1955? List three major factors that demographers have to consider in making population projections.

5. Summarize the story of population growth in the United States and explain why it is high compared to population growth in most other more-developed countries. About how much of the annual U.S. population growth is due to immigration? List six changes in lifestyles that have taken place in the United States during the 20th century, leading to a rise in per capita resource use. What is the end effect of such changes in terms of the U.S. ecological footprint?

6. List eight factors that affect birth rates and fertility rates. Explain why there are more boys than girls in some countries. Define **life expectancy** and **infant mortality rate** and explain how they affect the population size of a country. What is **migration**? What are environmental refugees?

Section 6.3

7. What is the key concept for this section? What is the **age structure** of a population? Explain how age structure affects population growth and economic growth. What is demographic momentum? Describe the baby boom in the United States and some of its economic and social effects. What are some problems related to rapid population decline due to an aging population?

Section 6.4

8. What is the key concept for this section? What is the **demographic transition** and what are its four stages? What factors could hinder some less-developed countries in making this transition?

9. Explain how education, reduction of poverty, and empowerment of women can help countries to slow their population growth. What is **family planning** and how can it help to stabilize populations? Describe India's efforts to control its population growth. Describe China's population control program and compare it with that of India.

10. What are this chapter's *three big ideas*? Summarize the story of human population growth and explain how the three **scientific principles of sustainability** can guide us in dealing with the problems that stem from population growth and decline.

Note: Key terms are in bold type.

Critical Thinking

1. Do you think that the global population of 7.3 billion (**Core Case Study**) is too large? Explain. If your answer was *yes*, what do you think should be done to slow human population growth? If your answer was *no*, do you believe that there is a population size that would be too big? Explain.

2. If you could say hello to a new person every second without taking a break and working around the clock, how many people could you greet in a day? How many in a year? How long would it take you to greet the 88 million people who were added to the world's population in 2015? At this same rate, how many years would it take you to greet all 7.3 billion people on the planet?

3. Of the three major environmental worldviews summarized on p. 20, which do you think underlies each of the two major positions on whether the world is overpopulated, as described in Science Focus 6.1?

4. Should everyone have the right to have as many children as they want? Explain. Is your belief on this issue consistent with your environmental worldview?

5. Is it rational for a poor couple in a less-developed country such as India to have four or five children? Explain.

6. Do you think that projected increases in the earth's population size and economic growth are sustainable? Explain. If not, how is this likely to affect your life?

7. Some people think the most important environmental goal is to sharply reduce the rate of population growth in less-developed countries, where at least 95% of the world's population growth is expected to take place between 2015 and 2050. Others argue that the most serious environmental problems stem from high levels of resource consumption per person in more-developed countries. What is your view on this issue? Explain.

8. Experts have identified population growth as one of the major causes of the environmental problems we face. The population of the United States is growing faster than that of any other more-developed country. This fact is rarely discussed and the U.S. government has no official policy for slowing U.S. population growth. Why do think this is so? Do you think there should be such a policy? If so, explain your thinking and list three steps you would take as a leader to slow U.S. population growth. If not, explain your thinking.

Doing Environmental Science

Prepare an age-structure diagram for your community. You will need to estimate how many people belong in each age category (see p. 127). To do this, interview a randomly drawn sample of the population to find out their ages and then divide your sample into age groups. (Be sure to interview equal numbers of males and females.) Then find out the total population of your community and apply the percentages for each age group from your sample to the whole population in order to make your estimates. Create your diagram and then use it to project future population trends. Write a report in which you discuss some economic, social, and environmental effects that might result from these trends.

Global Environment Watch Exercise

Go to your MindTap course to access the GREENR database. Using the "Basic Search" box at the top of the page, search for *aging population*. Choose an article from those that appear. Read the article, take notes, and summarize its main idea. List three major arguments or pieces of evidence cited in the article to support this main idea. Write a report summarizing your findings and answering the following questions: Is there anything in the article that contradicts anything you read in Chapter 6? Is there anything in the article that adds to the knowledge you gained from reading Chapter 6? Explain.

Data Analysis

The chart below shows selected population data for two different countries, A and B. Study the chart and answer the questions that follow.

	Country A	Country B
Population (millions)	144	82
Crude birth rate (number of live births per 1,000 people per year)	43	8
Crude death rate (number of deaths per 1,000 people per year)	18	10
Infant mortality rate (number of babies per 1,000 born who die in first year of life)	100	3.8
Total fertility rate (average number of children born to women during their childbearing years)	5.9	1.3
% of population under 15 years old	45	14
% of population older than 65 years	3	19
Average life expectancy at birth	47	79
% urban	44	75

1. Calculate the rates of natural increase (due to births and deaths, not counting immigration) for the populations of country A and country B. Based on these calculations and the data in the table, for each of the countries, suggest whether it is a more-developed country or a less-developed country and explain the reasons for your answers.
2. Describe where each of the two countries might be in the stages of demographic transition (Figure 6.14). Discuss factors that could hinder either country from progressing to later stages in the demographic transition.
3. Explain how the percentages of people under age 15 in each country could affect its per capita and total ecological footprints.

CHAPTER 7

Climate and Biodiversity

> When we try to pick out anything by itself, we find it hitched to everything else in the universe.
>
> JOHN MUIR

Key Questions

7.1 What factors influence weather?

7.2 What factors influence climate?

7.3 How does climate affect the nature and location of biomes?

Giraffes on tropical grassland (savanna) in Africa.

Oleg Znamensky/Shutterstock.com

Core Case Study

African Savanna

The earth has a great diversity of species and *habitats*, or places where these species can live. Some species live in *terrestrial*, or land, habitats such as grasslands (see chapter-opening photo), forests, and deserts. These three major types of terrestrial ecosystems are called *biomes*. They represent one of the four components of biodiversity (Figure 4.5, p. 79), which is the basis for one of the three **scientific principles of sustainability**.

Why do grasslands grow on some areas of the earth's land while forests and deserts form in other areas? The answer lies largely in differences in *climate*, the average weather conditions in a given region over at least three decades to thousands of years. Differences in climate result mostly from long-term differences in weather, based primarily on average annual precipitation and temperature. These differences lead to three major types of climate—*tropical* (areas near the equator, receiving the most intense sunlight), *polar* (areas near the earth's poles, receiving the least intense sunlight), and *temperate* (areas between the tropical and polar regions).

Throughout these regions, we find different types of ecosystems, vegetation, and animals adapted to the various climate conditions. For example, in tropical areas, we find a type of grassland called a *savanna*. This biome typically contains scattered trees and usually has warm temperatures year-round with alternating dry and wet seasons. Savannas in East Africa are home to *grazing* (primarily grass-eating) and *browsing* (twig- and leaf-nibbling) hoofed animals. They include wildebeests, gazelles, antelopes, zebras, elephants (Figure 7.1), and giraffes (chapter-opening photo), as well as their predators such as lions, hyenas, and humans.

Archeological evidence indicates that our species emerged from African savannas. Early humans lived largely in trees but eventually came down to the ground and learned to walk upright. This freed them to use their hands for using tools such as clubs and spears. Much later, they developed bows and arrows and other weapons that enhanced their abilities to hunt animals for food and clothing made from animal hides.

After the last ice age, about 10,000 years ago, the earth's climate warmed and humans began their transition from hunter–gatherers to farmers growing food on the savanna and on other grasslands. Later, they cleared patches of forest to expand farmland and created villages and eventually towns and cities.

Today, vast areas of African savanna have been plowed up and converted to cropland or used for grazing livestock. Towns are also expanding there, and this trend will continue as the human population in Africa—the continent with the world's fastest population growth—increases. As a result, populations of elephants, lions, and other animals that roamed the savannas for millions of years have dwindled. Many of these animals face extinction in the next few decades because of the loss of their habitats and because people kill them for food and their valuable parts such as the ivory tusks of elephants.

In this chapter, we distinguish between weather and climate and examine the role that climate plays in the location and formation of the major terrestrial ecosystems. We also begin the study of human impacts on these important ecosystems.

Amy Nichole Harris/Fotolia LLC

FIGURE 7.1 Elephants on a tropical African savanna.

7.1 WHAT FACTORS INFLUENCE WEATHER?

CONCEPT 7.1 Key factors that influence weather are moving masses of warm and cold air, changes in atmospheric pressure, and occasional shifts in major winds.

Weather Is Affected by Moving Masses of Warm or Cold Air

Weather is the set of physical conditions of the lower atmosphere, including temperature, precipitation, humidity, wind speed, cloud cover, and other factors, in a given area over a period of hours to days. The most important factors in the weather in any area are atmospheric temperature and precipitation.

Meteorologists use equipment mounted on weather balloons, aircraft, ships, and satellites, as well as radar and stationary sensors, to obtain data on weather variables. They feed these data into computer models to draw weather maps for various parts of the world. Other computer models project upcoming weather conditions based on probabilities that air masses, winds, and other factors will change in certain ways.

Much of the weather we experience results from interactions between the leading edges of moving masses of warm air and cold air (**Concept 7.1**). Weather changes when one air mass replaces or meets another. The most dramatic changes in weather occur along a **front,** the boundary between two air masses with different temperatures and densities.

A **warm front** is the boundary between an advancing warm air mass and the cooler one it is replacing (Figure 7.2, left). Because warm air is less dense (weighs less per unit of volume) than cool air, an advancing warm air mass rises up over a mass of cool air. As the warm air rises, its moisture begins condensing into droplets, forming layers of clouds at different altitudes. Gradually, the clouds thicken, descend to a lower altitude, and often release their moisture as rainfall.

A **cold front** (Figure 7.2, right) is the leading edge of an advancing mass of cold air. Because cold air is denser than warm air, an advancing cold front stays close to the ground and wedges beneath less dense warmer air. It pushes this warm, moist air up, which produces rapidly moving, towering clouds called *thunderheads*. As it passes through, it can cause high surface winds and thunderstorms, followed by cooler temperatures and a clear sky.

Weather Is Affected by Changes in Atmospheric Pressure and Wind Patterns

Atmospheric pressure results from molecules of gases in the atmosphere (mostly nitrogen and oxygen) zipping around at very high speeds and bouncing off everything they encounter. Atmospheric pressure is greater near the earth's surface because the molecules in the atmosphere are squeezed together under the weight of the air above them.

An air mass with high pressure, called a **high,** contains cool, dense air that descends slowly toward the earth's surface and becomes warmer. Because of this warming, water molecules in the air do not form droplets—a process called *condensation*. Thus clouds, which are made of droplets, usually do not form in the presence of a high. Fair weather with clear skies follows as long as this high remains over the area.

A low-pressure air mass, called a **low,** contains low-density, warm air at its center. This air rises, expands, and cools. When its temperature drops below a certain level, called the *dew point,* moisture in the air condenses and forms clouds. The condensation process usually requires that the air contain suspended tiny particles of dust, smoke, sea salts, or volcanic ash, called *condensation nuclei,* around

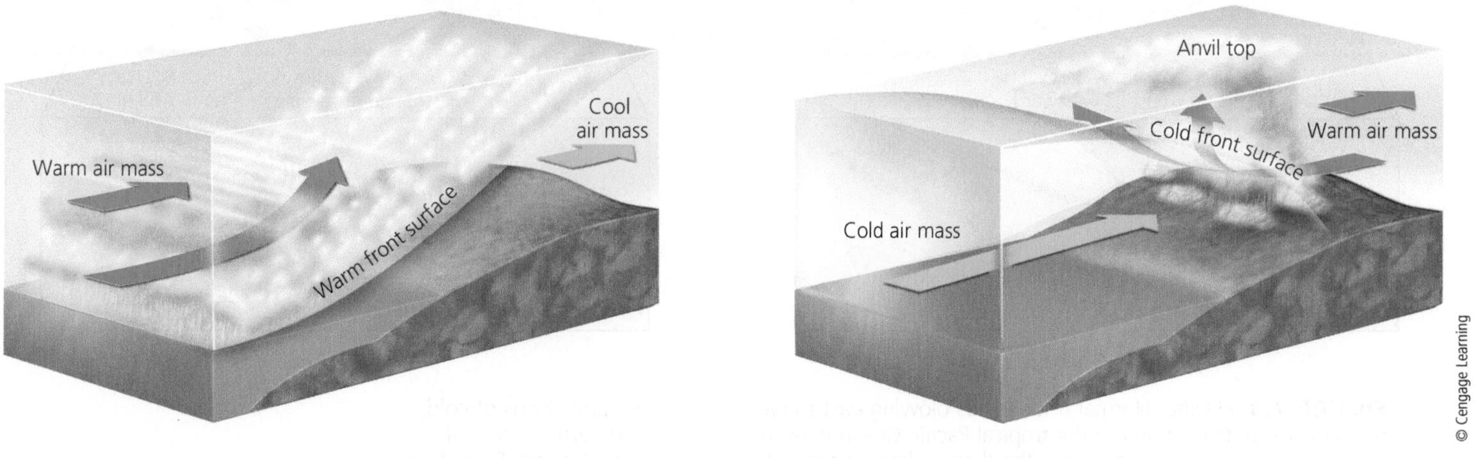

FIGURE 7.2 *Weather fronts:* A warm front (left) occurs when a moving mass of warm air meets and rises up over a mass of denser cool air. A cold front (right) forms when a moving mass of cold air wedges beneath a mass of less dense warm air.

which water droplets can form. If the droplets in the clouds coalesce into larger drops or snowflakes heavy enough to fall from the sky, precipitation occurs. Thus, a low tends to produce cloudy and sometimes stormy weather.

Movement of these air masses is influenced strongly by *jet streams*—powerful winds that circle the globe near the top of the troposphere (see p. 51). They are like fast-flowing rivers of air moving west to east, one in each hemisphere somewhere above and below the equator. They form because of the temperature difference between the equator and the poles, which causes air to move. As the air moves away from the equator, north and south, it is deflected by the earth's rotation and flows generally west to east. The greater the temperature difference, the faster the flow of these winds. Jet streams can influence weather by moving moist air masses from one area to another (**Concept 7.1**). We examine these air flow patterns in more depth later in this chapter.

Every few years, normal wind patterns in the Pacific Ocean (Figure 7.3, left) are disrupted and this affects weather around much of the globe. This change in wind patterns is called the *El Niño–Southern Oscillation*, or *ENSO* (Figure 7.3, right).

In an ENSO, often called simply *El Niño*, winds that usually blow more-or-less constantly from east to west weaken or reverse direction. This allows the warmer waters of the western Pacific to move toward the coast of South America. A horizontal zone of gradual temperature change called the *thermocline*, separating warm and cold waters, sinks in the eastern Pacific. These changes result in drier weather in some areas and wetter weather in other areas. A strong ENSO can alter weather conditions over at least two-thirds of the globe (Figure 7.4)—especially on the coasts of the Pacific and Indian Oceans.

An ENSO is a 1- to 2-year natural weather event. Although it is not a climate event, it can raise the earth's average temperature by as much as 0.25°C (0.45°F) for a year or two. As a result, it can affect the climate by temporarily increasing the earth's average temperature. ENSOs can be extreme in their effects. One such *super ENSO* occurred in 1997 and 1998. This 2-year period of extreme weather, including severe storms, flooding, and temperature extremes, caused $4.5 billion in damages and 23,000 deaths.

La Niña, the reverse of El Niño, cools some coastal surface waters. This natural weather event also occurs every few years and it typically leads to more Atlantic Ocean hurricanes, colder winters in Canada and the northeastern United States, and warmer and drier winters in the southeastern and southwestern United States. It also usually leads to wetter winters in the Pacific Northwest, torrential rains in Southeast Asia, and sometimes more wildfires in Florida. Scientists do not know the exact causes of these weather events or when they are likely to occur, but they do know how to detect and monitor them.

Tornadoes and Tropical Cyclones Are Violent Weather Extremes

Sometimes we experience *weather extremes*. Two examples are violent storms called *tornadoes* (which form over land)

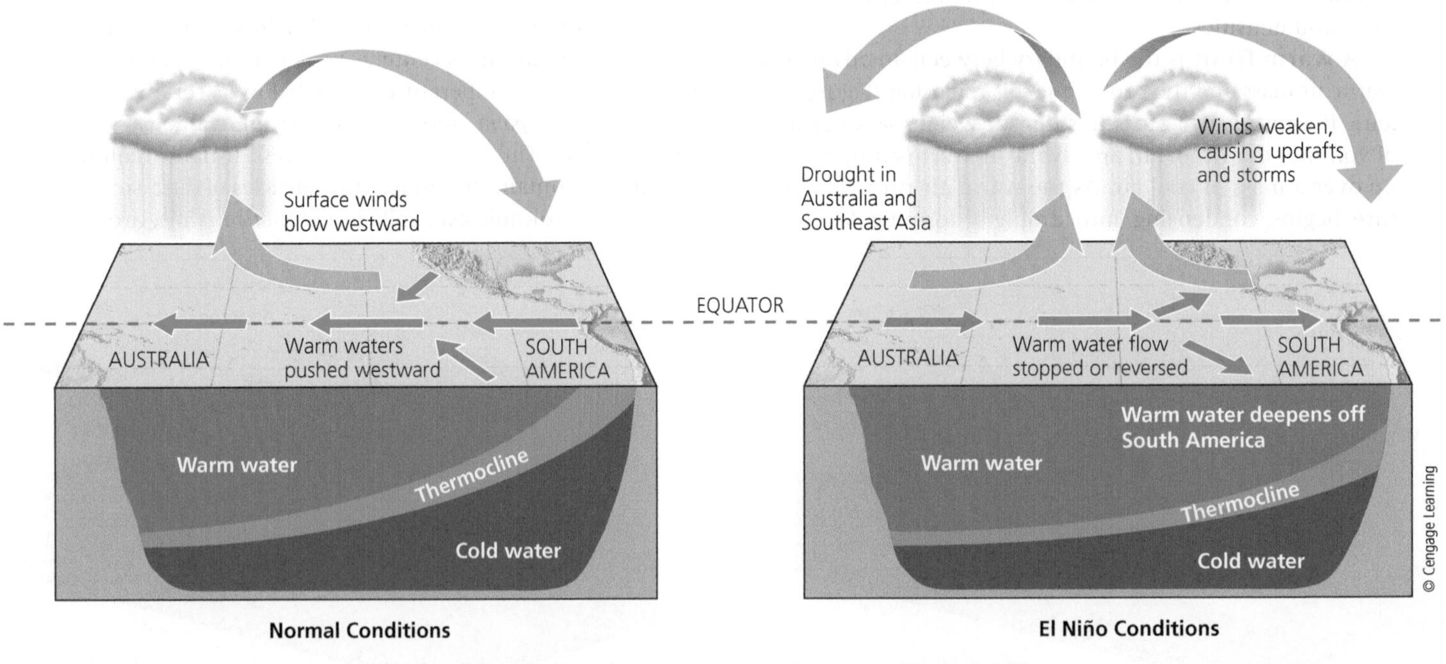

FIGURE 7.3 El Niño: Normal trade winds blowing east to west cause shore upwellings of cold, nutrient-rich bottom water in the tropical Pacific Ocean near the coast of Peru (left). A zone of gradual temperature change called the thermocline separates the warm and cold water. Every few years, a shift in trade winds known as the El Niño–Southern Oscillation (ENSO) disrupts this pattern (right) for 1 to 2 years.

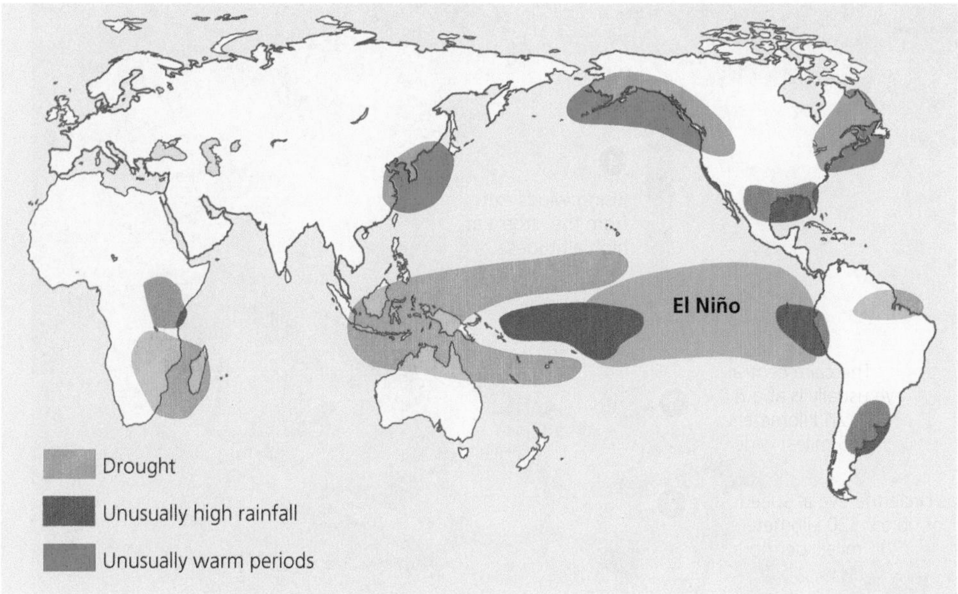

FIGURE 7.4 Typical global weather effects of an El Niño–Southern Oscillation. ***Question:*** How might an ENSO affect the weather where you live or go to school?

(Compiled by the authors using data from United Nations Food and Agriculture Organization and U.S. Weather Service.)

and *tropical cyclones* (which form over warm ocean waters and sometimes pass over coastal land areas).

Tornadoes, or *twisters*, are swirling, funnel-shaped clouds that form over land. They can destroy houses and cause other serious damage in areas where they touch down. The United States is the world's most tornado-prone country, followed by Australia.

Tornadoes in the plains of the Midwestern United States often occur when a large, dry, cold front moving southward from Canada runs into a large mass of warm humid air moving northward from the Gulf of Mexico. As the large warm front moves rapidly over the denser cold-air mass, it rises swiftly and forms strong vertical convection currents that suck air upward (Figure 7.5). Scientists hypothesize that the interaction of the cooler air nearer the ground and the rapidly rising warmer air above causes a spinning, vertically rising air mass, or *vortex*. Most tornadoes in the American Midwest occur in the spring and

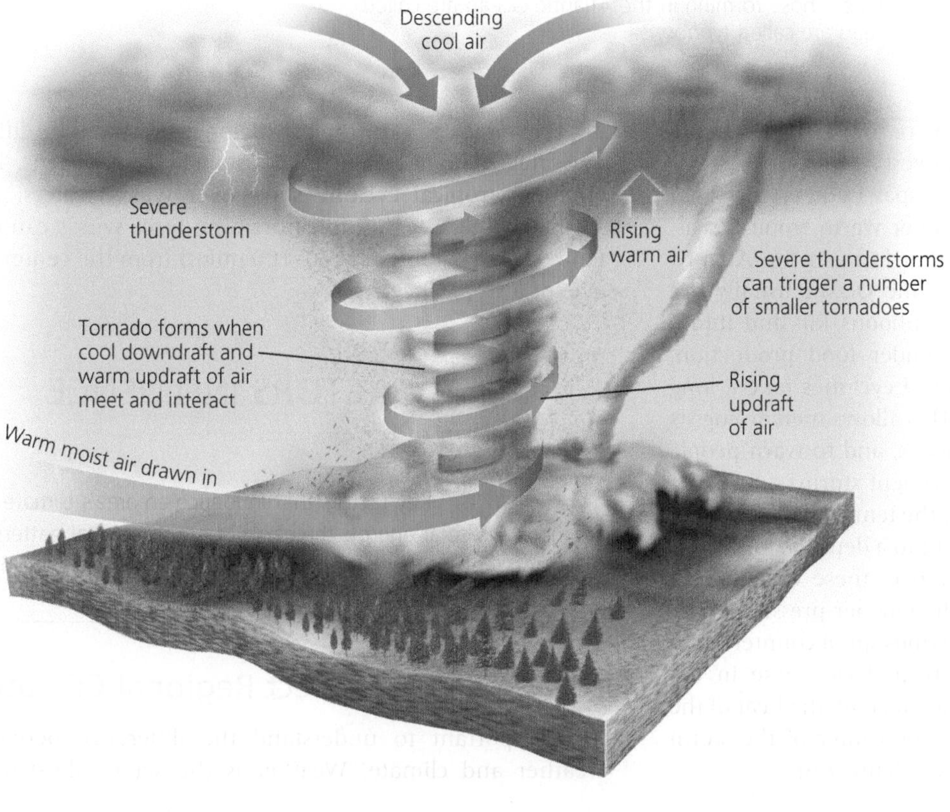

FIGURE 7.5 Formation of a tornado, or twister. The most active tornado season in the United States is usually March through August.

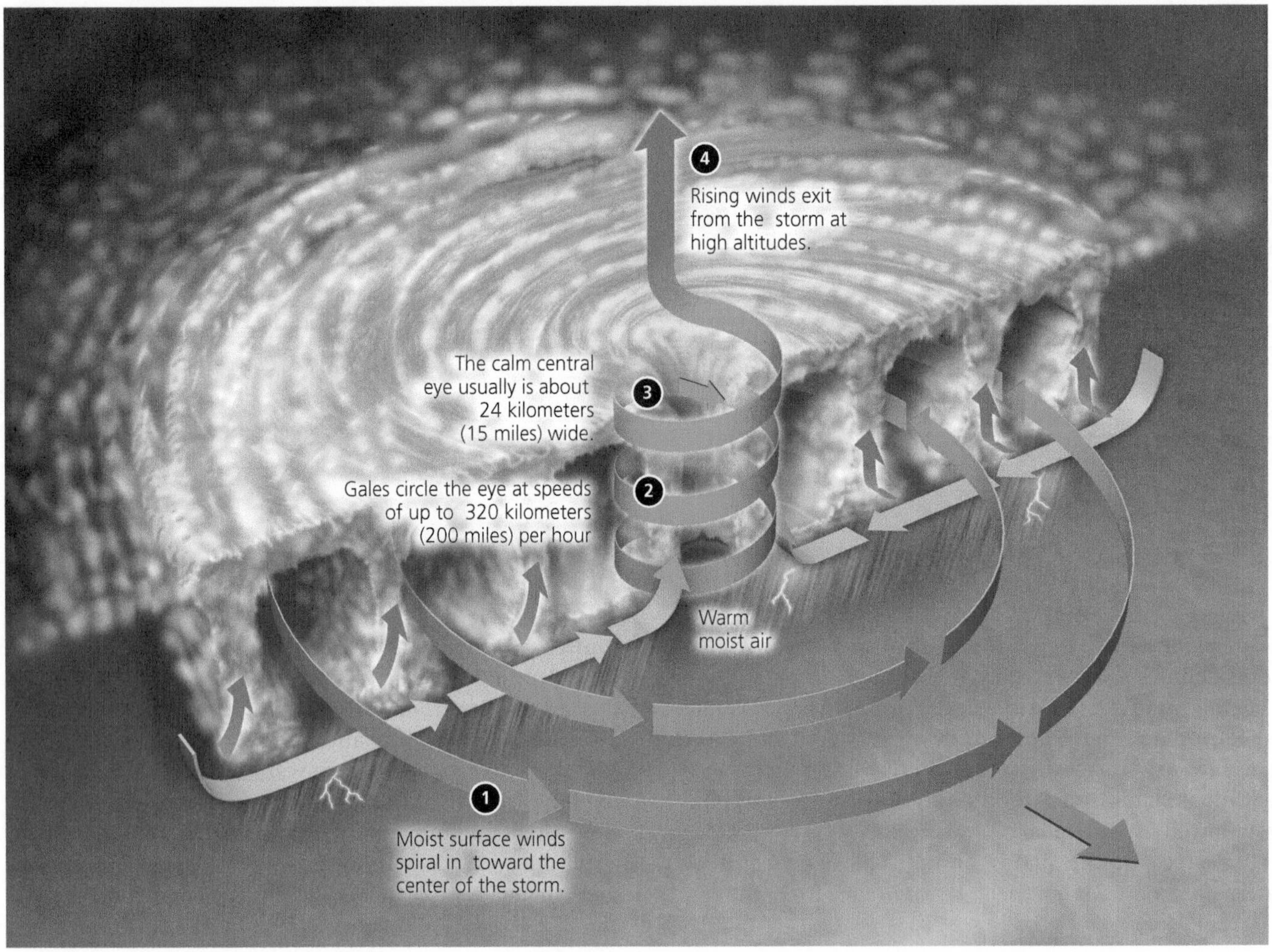
4
Rising winds exit from the storm at high altitudes.
The calm central eye usually is about 24 kilometers (15 miles) wide.
3
Gales circle the eye at speeds of up to 320 kilometers (200 miles) per hour
2
Warm moist air
1
Moist surface winds spiral in toward the center of the storm.
© Cengage Learning

FIGURE 7.6 Formation of a tropical cyclone. Those forming in the Atlantic Ocean are called *hurricanes*. Those forming in the Pacific Ocean are called *typhoons*.

summer when cold fronts from the north penetrate deeply into the Great Plains and the Midwest.

Tropical cyclones (Figure 7.6) are spawned by the formation of low-pressure cells of air over warm tropical seas. *Hurricanes* are tropical cyclones that form in the Atlantic Ocean. Those forming in the Pacific Ocean usually are called *typhoons*. Hurricanes and typhoons kill and injure people, damage property, and hinder food production. Unlike tornadoes, however, tropical cyclones take a long time to form and gain strength. This allows meteorologists to track their paths and wind speeds, and to warn people in areas likely to be hit by these violent storms.

For a tropical cyclone to form, the temperature of ocean water has to be at least 27°C (80°F) to a depth of 46 meters (150 feet). Areas of low pressure over these warm ocean waters draw in air from surrounding higher-pressure areas. The earth's rotation makes these winds spiral counterclockwise in the northern hemisphere and clockwise in the southern hemisphere. Moist air, warmed by the heat of the ocean, rises in a vortex through the center of the storm until it becomes a tropical cyclone (Figure 7.6).

The intensities of tropical cyclones are rated in different categories, based on their sustained wind speeds. The longer a tropical cyclone stays over warm waters, the stronger it gets. Significant hurricane-force winds can extend 64–161 kilometers (40–100 miles) from the center, or eye, of a tropical cyclone.

7.2 WHAT FACTORS INFLUENCE CLIMATE?

CONCEPT 7.2 Key factors that influence an area's climate are incoming solar energy, the earth's rotation, global patterns of air and water movement, gases in the atmosphere, and the earth's surface features.

Several Factors Affect Regional Climates

It is important to understand the difference between weather and climate. Weather is the set of short-term

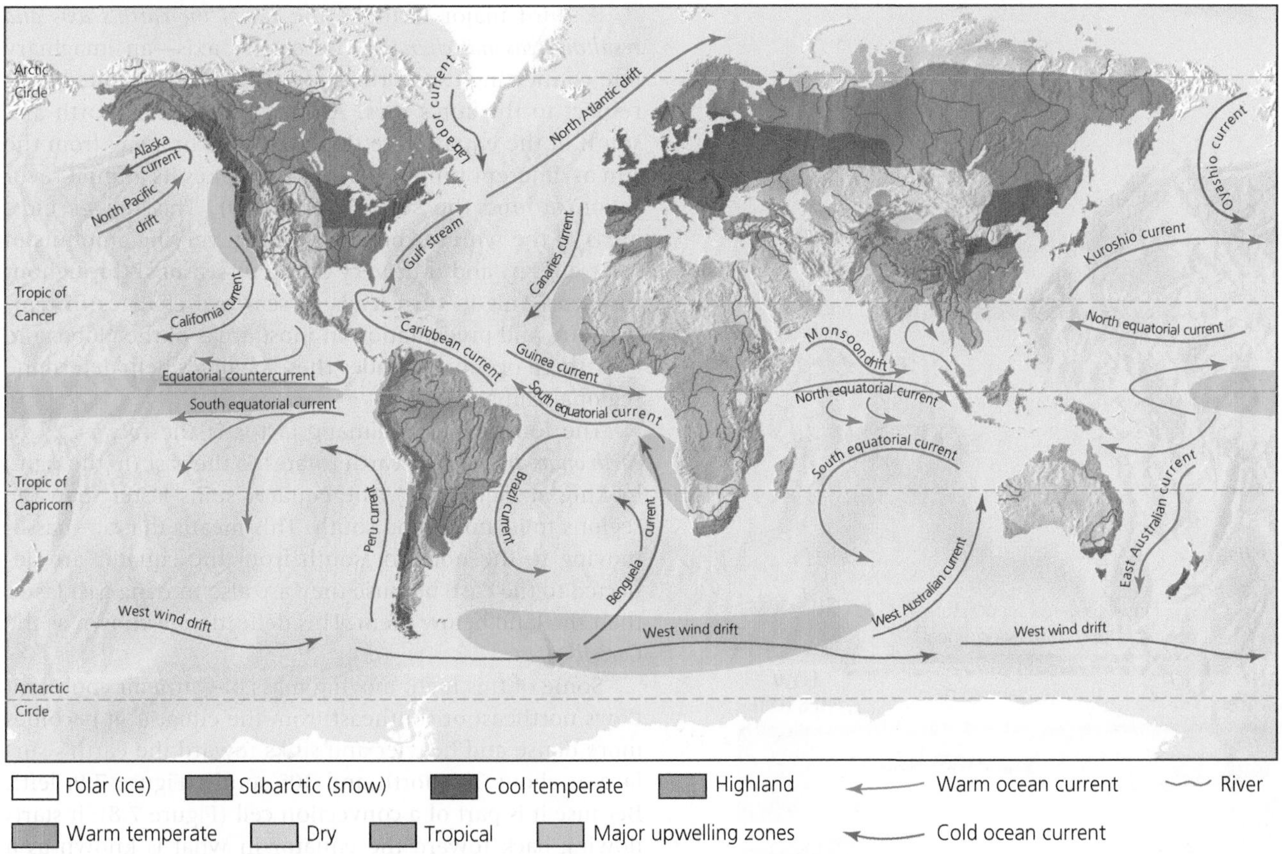

FIGURE 7.7 Natural capital: This generalized map of the earth's current climate zones also shows the major ocean currents and upwelling areas (where currents bring nutrients from the ocean bottom to the surface). ***Question:*** Based on this map, what is the general type of climate where you live?

atmospheric conditions over hours to days to years, whereas **climate** is the general pattern of atmospheric conditions in a given area over periods ranging from at least three decades to thousands of years. Weather often fluctuates daily, from one season to another, and from one year to the next. However, climate tends to change slowly because it is the average of long-term atmospheric conditions.

Climate varies among the earth's different regions primarily because of global air circulation and **ocean currents,** or mass movements of ocean water. Global winds and ocean currents distribute heat and precipitation unevenly between the tropics and other parts of the world. Scientists have described the various regions of the earth according to their climates (Figure 7.7).

Several major factors help determine regional climates. The first is the *cyclical movement of air driven by solar energy*. It is a form of **convection,** the movement of fluid matter (such as gas or water) caused when the warmer and less dense part of a body of such matter rises while the cooler, denser part of the fluid sinks due to gravity. In the atmosphere, convection occurs when the sun warms the air and causes some of it to rise, while cooler air sinks in a cyclical pattern called a **convection cell.**

For example, the air over an ocean is heated when the sun evaporates water. This transfers moisture and heat from the ocean to the atmosphere, especially near the hot equator. This warm, moist air rises, then cools and releases heat and moisture as precipitation (Figure 7.8, right side and top, center). Then the cooler, denser, and drier air sinks, warms up, and absorbs moisture as it flows across the earth's surface (Figure 7.8, left side and bottom) to begin the cycle again.

The second major climatic factor is the *uneven heating of the earth's surface by the sun*. Air is heated much more at the equator, where the sun's rays strike directly, than at the poles, where sunlight strikes at an angle and spreads out over a much greater area (Figure 7.9, left). Thus, solar heating varies with **latitude**—the location between the equator and one of the poles. Latitudes are designated by degrees (X°) north or south. The equator is at 0°, the poles are at 90° north and 90° south, and areas between range from 0° to 90°.

The input of solar energy in a given area, called **insolation**, varies with latitude. This partly explains why tropical regions are hot, polar regions are cold, and temperate regions generally alternate between warm and cool temperatures (Figure 7.9, right).

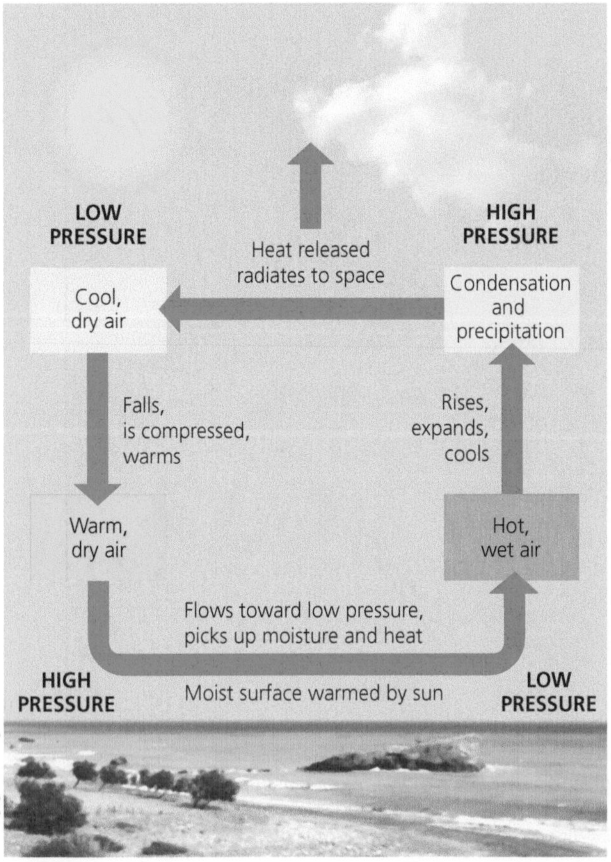

FIGURE 7.8 *Convection cells* play a key role in transferring energy (heat) and moisture through the atmosphere from place to place on the planet.

A third major factor is the *tilt of the earth's axis and resulting seasonal changes*. The earth's axis—an imaginary line connecting the north and south poles—is tilted with respect to the sun's rays. As a result, regions north and south of the equator are tipped toward or away from the sun at different times, as the earth makes its annual revolution around the sun (Figure 7.10). This means most areas of the world experience widely varying amounts of solar energy, and thus very different seasons, throughout the year. This in turn leads to seasonal changes in temperature and precipitation in most areas of the globe, and over three or more decades these changes help determine regional climates.

The fourth major climatic factor is the *rotation of the earth on its axis*. As the earth rotates to the east (to the right, looking at Figure 7.9), the equator spins faster than the regions to its north and south. This means that air masses moving to the north or south from the equator are deflected to the east, because they are also moving east faster than the land below them. This deflection is known as the *Coriolis effect*.

Some of this high, moving mass of warm air cools as it flows northeast or southeast from the equator. It becomes more dense and heavier and sinks toward the earth's surface at about 30° north and 30° south (Figure 7.9, left). Because it is part of a convection cell (Figure 7.8), it starts flowing back toward the equator in what is known as a *Hadley cell*. Because of the Coriolis effect, this air moving

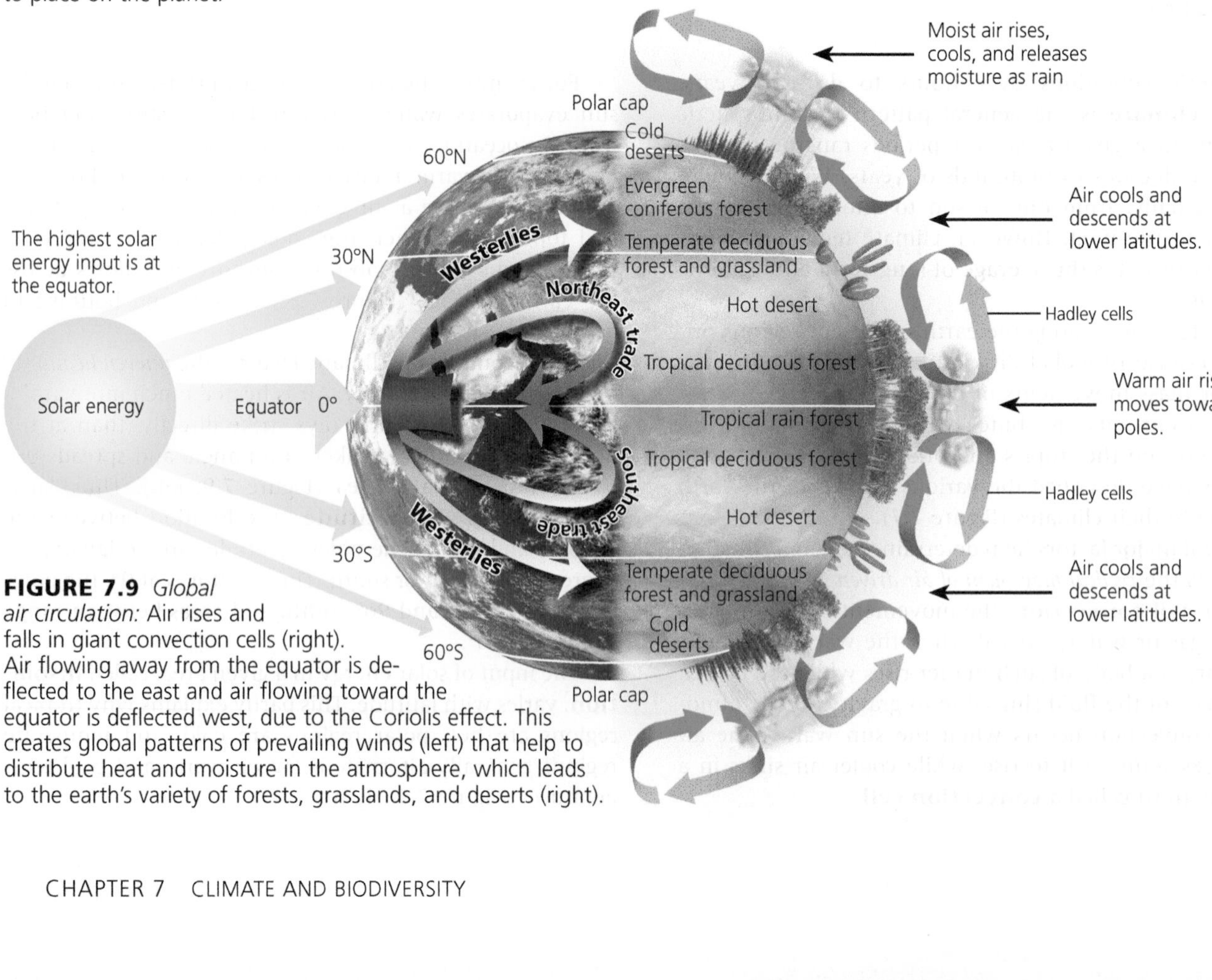

FIGURE 7.9 *Global air circulation:* Air rises and falls in giant convection cells (right). Air flowing away from the equator is deflected to the east and air flowing toward the equator is deflected west, due to the Coriolis effect. This creates global patterns of prevailing winds (left) that help to distribute heat and moisture in the atmosphere, which leads to the earth's variety of forests, grasslands, and deserts (right).

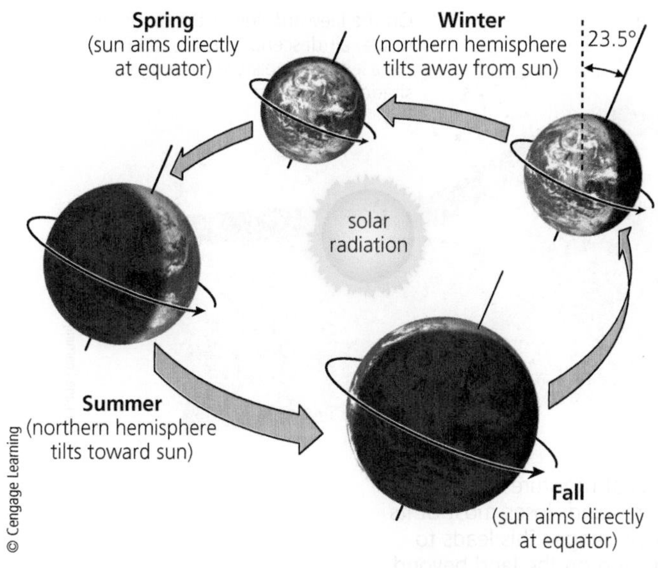

FIGURE 7.10 The earth's axis is tilted about 23.5° with respect to the plane of the earth's path around the sun. The resulting variations in solar energy reaching the northern and southern hemispheres throughout a year result in seasons. ***Critical thinking:*** How might your life be different if the earth's axis was not tilted?

toward the equator curls in a westerly direction. In the northern hemisphere, it thus flows southwest from northeast. In the southern hemisphere, it flows northwest from southeast.

These winds are known as the *northeast trade winds* (north of the equator) and the *southeast trade winds* (south of the equator). They were named long ago when sailing ships used them to move goods in trade between the continents. They are examples of *prevailing winds*—major surface winds that blow almost continuously.

The warm air that does not descend in the Hadley cells at 30° north and 30° south continues moving toward the poles and curving to the east due to the Coriolis effect. These prevailing winds that blow generally from the west in temperate regions of the globe are known as *westerlies* (Figure 7.9, left).

This complex movement of air results in six huge regions between the equator and the poles in which warm air rises and cools, then falls and heats up again in great rolling patterns (Figure 7.9, right). The two nearest the equator are the Hadley cells. These convection cells and the resulting prevailing winds distribute heat and moisture over the earth's surface, thus helping to determine regional climates (**Concept 7.2**).

Finally, a fifth major factor determining regional climates is *ocean currents* (Figure 7.7). They help to redistribute heat from the sun, thereby influencing climate and vegetation, especially near coastal areas. This solar heat, along with differences in water density (mass per unit volume), creates warm and cold ocean currents. They are driven by prevailing winds and the earth's rotation

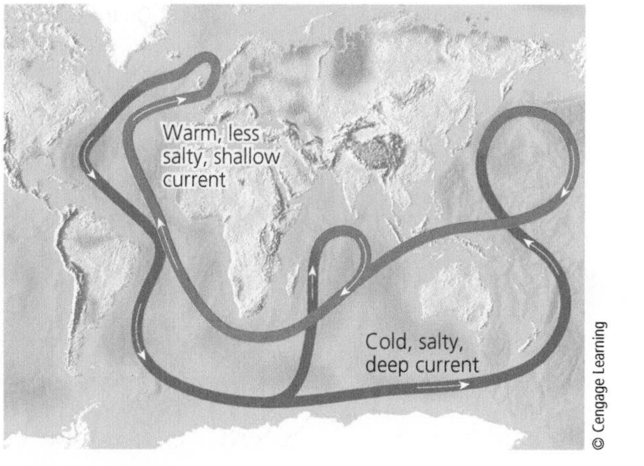

FIGURE 7.11 A connected loop of deep and shallow ocean currents transports warm and cool water to various parts of the earth.

(the Coriolis effect), and continental coastlines change their directions. As a result, between the continents, they flow in roughly circular patterns, called **gyres,** which move clockwise in the northern hemisphere and counterclockwise in the southern hemisphere.

Water also moves vertically in the oceans as denser water sinks while less dense water rises. This creates a connected loop of deep and shallow ocean currents (which are separate from those shown in Figure 7.7). This loop acts somewhat like a giant conveyer belt that moves heat from the surface to the deep sea and transfers warm and cold water between the tropics and the poles (Figure 7.11).

Greenhouse Gases Warm the Lower Atmosphere

As energy flows from the sun to the earth, some of it is reflected by the earth's surface back into the atmosphere. Molecules of certain gases in the atmosphere, including water vapor (H_2O), carbon dioxide (CO_2), methane (CH_4), and nitrous oxide (N_2O), absorb some of this solar energy and release a portion of it as infrared radiation (heat) that warms the lower atmosphere and the earth's surface. These gases, called **greenhouse gases,** play a role in determining the lower atmosphere's average temperatures and thus the earth's climates.

The earth's surface also absorbs much of the solar energy that strikes it and transforms it into longer-wavelength infrared radiation, which then rises into the lower atmosphere. Some of this heat escapes into space, but some is absorbed by molecules of greenhouse gases and emitted into the lower atmosphere as even longer-wavelength infrared radiation (see Figure 2.12, p. 40). Some of this released energy radiates into space, and some adds to the warming of the lower atmosphere and the earth's surface. Together, these processes result in a natural

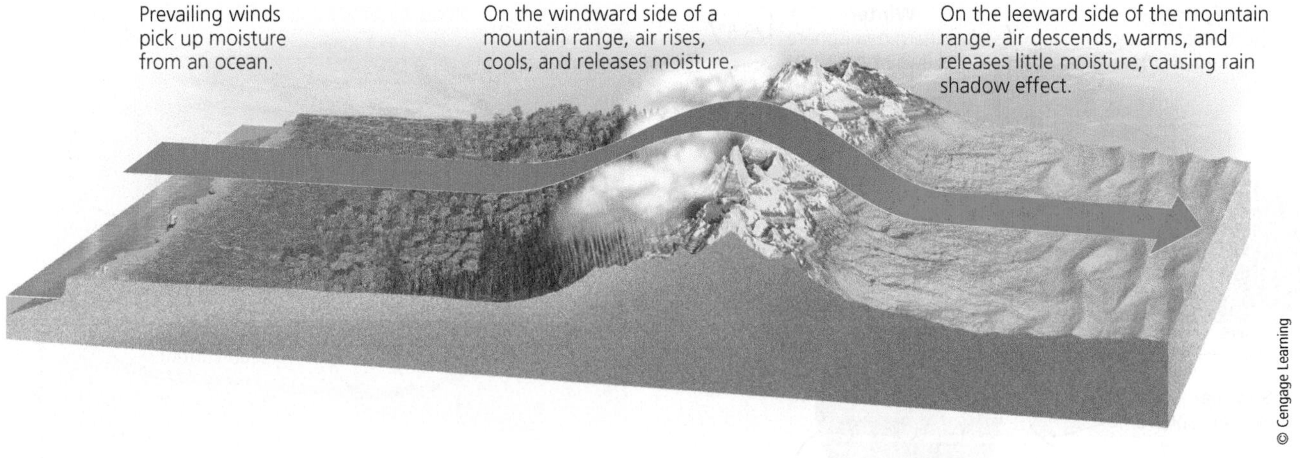
Prevailing winds pick up moisture from an ocean.
On the windward side of a mountain range, air rises, cools, and releases moisture.
On the leeward side of the mountain range, air descends, warms, and releases little moisture, causing rain shadow effect.
© Cengage Learning

FIGURE 7.12 The *rain shadow effect* is a reduction of rainfall and loss of moisture from the landscape on the leeward side of a mountain. Warm, moist air in onshore winds loses most of its moisture as rain and snow that fall on the windward slopes of a mountain range. This leads to semiarid and arid conditions on the leeward side of the mountain range and on the land beyond.

warming of the troposphere, called the **greenhouse effect** (see Figure 3.3, p. 52). Without this natural warming effect, the earth would be a very cold and mostly lifeless planet.

Human activities such as the production and burning of fossil fuels, clearing of forests, and growing of crops release large quantities of the greenhouse gases carbon dioxide, methane, and nitrous oxide into the atmosphere. According to a considerable body of scientific evidence, we are emitting greenhouse gases into the atmosphere faster than they can be removed by the earth's carbon and nitrogen cycles (see Figures 3.20, p. 65, and 3.21, p. 66).

Climate research and climate models indicate that these emissions have played a key role in warming the earth over the last 50 years and thus helping to change its climate. In other words, human activities are enhancing the earth's natural greenhouse effect. If the earth's average atmospheric temperature continues to rise as projected, this will alter temperature and precipitation patterns, raise average sea levels, and shift areas where we can grow crops and where many types of plants and animals (including humans) can live. We discuss this issue more fully in Chapter 19.

The Earth's Surface Features Affect Local Climates

Various topographic features of the earth's surface can create local climatic conditions that differ from the general climate in some regions. For example, mountains interrupt the flow of prevailing surface winds and the movement of storms. When moist air from an ocean blows inland and reaches a mountain range, it is forced upward. As the air rises, it cools, expands, and loses most of its moisture as rain and snow that fall on the windward slope of the mountain.

As shown in Figure 7.12, when the drier air mass passes over the mountaintops, it flows down the leeward slopes (facing away from the wind) and warms up. This warmer air can hold more moisture, but it typically does not release much of it. This tends to dry out plants and soil below. This process is called the **rain shadow effect.** Over many decades, it results in *semiarid* or *arid* conditions on the leeward side of a high mountain range. Sometimes this effect leads to the formation of deserts such as Death Valley, a part of the Mojave Desert, which lies within the U.S. states of California, Nevada, Utah, and Arizona.

Cities also create distinct microclimates based on their weather averaged over three decades or more. Bricks, concrete, asphalt, and other building materials absorb and hold heat, and buildings block wind. Motor vehicles and the heating and cooling systems of buildings release large quantities of heat and pollutants. As a result, cities on average tend to have more haze and smog, higher temperatures, and lower wind speeds than the surrounding countryside. These factors make cities *heat islands.*

7.3 HOW DOES CLIMATE AFFECT THE NATURE AND LOCATION OF BIOMES?

CONCEPT 7.3 Desert, grassland, and forest biomes can be tropical, temperate, or cold depending on their climate and location.

Climate Helps to Determine Where Terrestrial Organisms Can Live

Differences in climate (Figure 7.7) help to explain why one area of the earth's land surface is a desert, another a

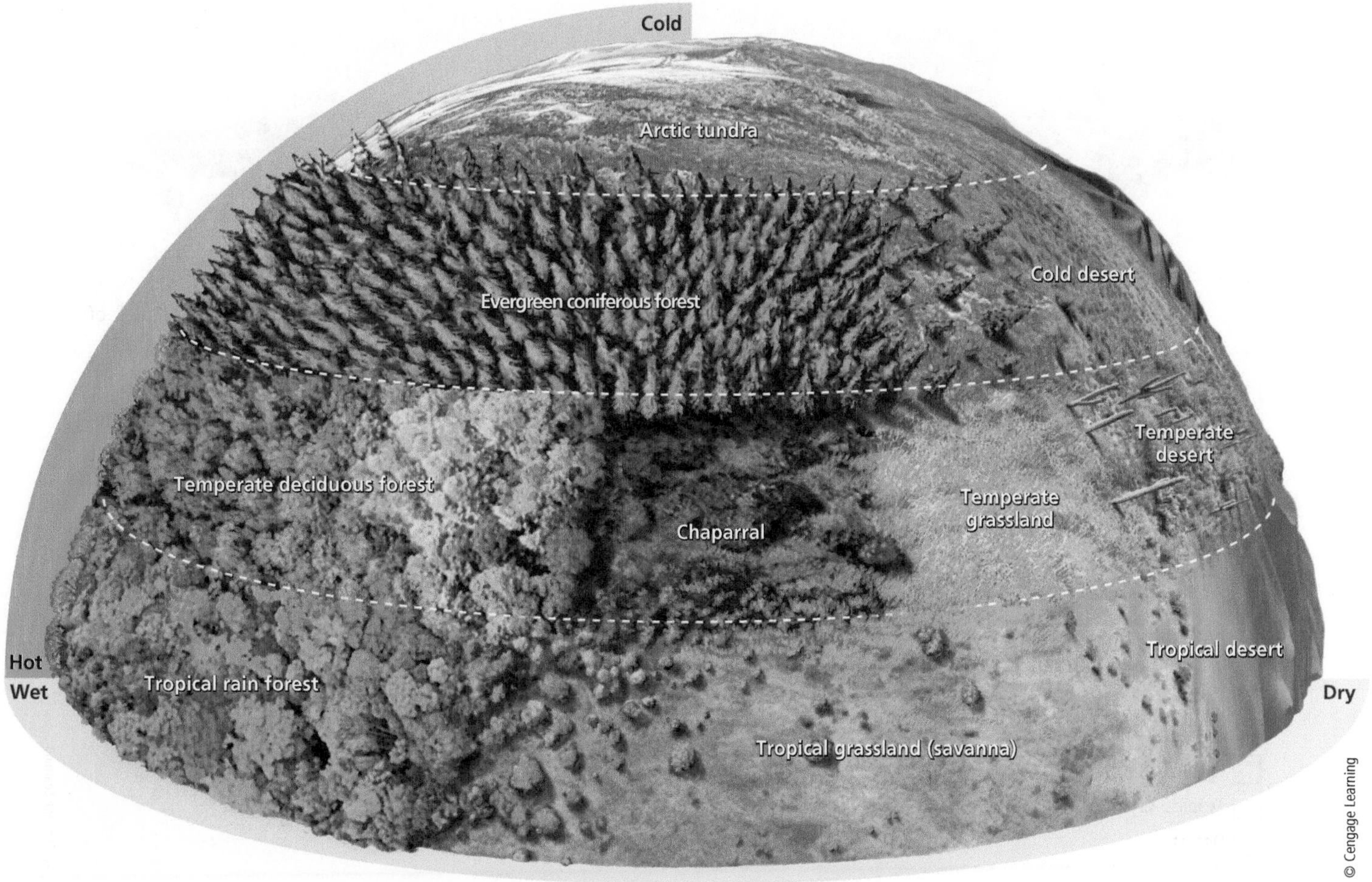

FIGURE 7.13 Natural capital: Average precipitation and average temperature, acting together as limiting factors over a long time, help to determine the type of desert, grassland, or forest in any particular area, and thus the types of plants, animals, and decomposers found in that area (assuming it has not been disturbed by human activities).

grassland, and another a forest. (See Figure 2, p. S16, Supplement 4.) Different climates based on long-term average annual precipitation and temperatures, global air circulation patterns, and ocean currents, lead to the formation of tropical (hot), temperate (moderate), and polar (cold) deserts, grasslands, and forests, as summarized in Figure 7.13 (**Concept 7.3**).

Figure 7.14 shows how scientists have divided the world into **biomes**—large terrestrial regions, each characterized by a certain type of climate and dominant forms of plant life. The variety of biomes and aquatic systems is one of the four components of the earth's biodiversity (see Figure 4.5, p. 79)—a vital part of the earth's natural capital.

By comparing Figure 7.14 with Figure 7.7, you can see how the world's major biomes vary with climate. Figure 4.8 (p. 81) shows how major biomes along the midsection of the United States are related to different climates.

On maps such as the one in Figure 7.14, biomes are shown with sharp boundaries, and each biome is covered with one general type of vegetation. In reality, biomes are not uniform. They consist of a variety of areas, each with somewhat different biological communities but with similarities typical of the biome. These areas occur because of the irregular distribution of the resources needed by plants and animals and because human activities have removed or altered the natural vegetation in many areas. There are also differences in vegetation along the transition zone or ecotone (see p. 80) between any two different ecosystems or biomes.

CONSIDER THIS . . .

THINKING ABOUT Biomes, Climate, and Human Activities

Use Figure 7.7 to determine the general type of climate where you live and Figure 7.14 to determine the general type of biome that should exist where you live. Then use Figure 1.8, p. 12, or Figure 6, p. S20, in Supplement 4 to determine how human ecological footprints have affected the biome where you live.

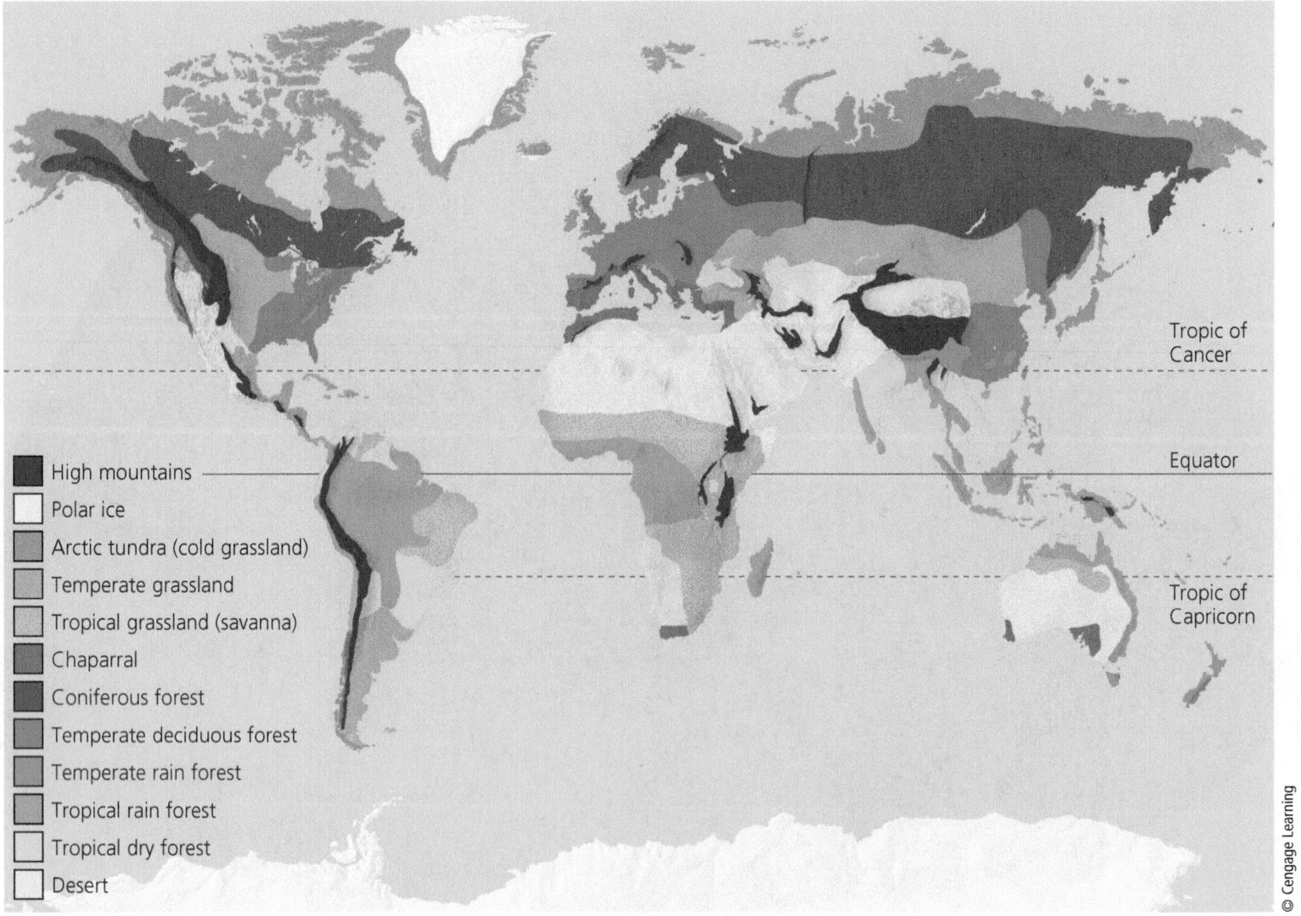

FIGURE 7.14 Natural capital: The earth's major *biomes* result primarily from differences in climate.

Three Types of Deserts

In a *desert*, annual precipitation is low and often scattered unevenly throughout the year. During the day, the baking sun warms the ground and evaporates water from plant leaves and from the soil. At night, most of the heat stored in the ground radiates quickly into the atmosphere. This explains why in a desert, you might roast during the day but shiver at night.

A combination of low rainfall and varying average temperatures over many decades creates a variety of desert types—tropical, temperate, and cold (Figures 7.13 and 7.14 and **Concept 7.3**). Tropical deserts (Figure 7.15, top photo) such as the Sahara and the Namib of Africa are hot and dry most of the year (Figure 7.15, top graph). They have few plants and a hard, windblown surface strewn with rocks and sand.

In *temperate deserts* (Figure 7.15, center photo), daytime temperatures are high in summer and low in winter and there is more precipitation than in tropical deserts (Figure 7.15, center graph). The sparse vegetation consists mostly of widely dispersed, drought-resistant shrubs and cacti or other succulents adapted to the dry conditions and temperature variations.

In *cold deserts* such as the Gobi Desert in Mongolia, vegetation is sparse (Figure 7.15, bottom photo). Winters are cold, summers are warm or hot, and precipitation is low (Figure 7.15, bottom graph). In all types of deserts, plants and animals have evolved adaptations that help them to stay cool and to get enough water to survive (Science Focus 7.1).

Desert ecosystems are vulnerable to disruption because they have slow plant growth, low species diversity, slow nutrient cycling due to lack of humus (see Figure 3.10, p. 56), low bacterial activity in the soils, and very little water. It can take decades to centuries for their soils to recover from disturbances such as off-road vehicle traffic, which can also destroy the habitats for a variety of animal species that live underground. The lack of vegetation, especially in tropical and polar deserts, also makes them vulnerable to heavy wind erosion from sandstorms.

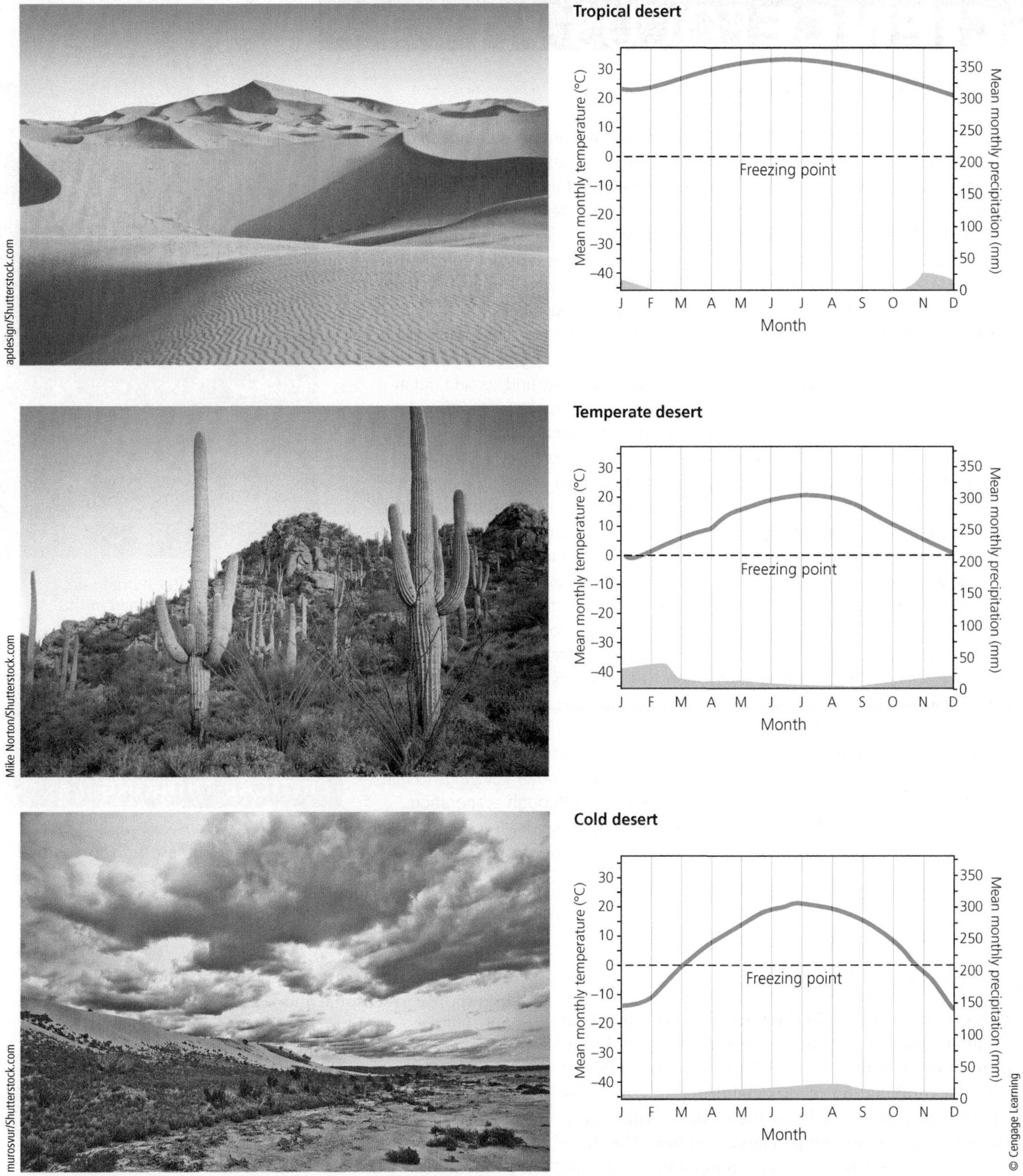

FIGURE 7.15 These climate graphs track the typical variations in annual temperature (red) and precipitation (blue) in tropical (savanna), temperate, and cold deserts. Top photo: a *tropical desert* in Morocco. Center photo: a *temperate desert* in southeastern California, with saguaro cactus, a prominent species in this ecosystem. Bottom photo: a *cold desert*, Mongolia's Gobi Desert. ***Data analysis:*** Which month of the year has the highest temperature and which month has the lowest rainfall for each of the three types of deserts?

Staying Alive in the Desert

Adaptations for survival in the desert have two themes: *beat the heat* and *every drop of water counts*.

Desert plants have evolved a number of strategies based on such adaptations. During long hot and dry spells, plants such as mesquite and creosote drop their leaves to survive in a dormant state. *Succulent* (fleshy) *plants* such as the saguaro ("sah-WAH-ro") cactus (Figure 7.A and Figure 7.15, middle photo) have no leaves that can lose water to the atmosphere through *transpiration*. They also store water and synthesize food in their expandable, fleshy tissue and they reduce water loss by opening their pores only at night to take up carbon dioxide (CO_2). The spines of these and many other desert plants guard them from being eaten by herbivores seeking the precious water they hold.

Some desert plants use deep roots to tap into groundwater. Others such as prickly pear and saguaro cacti use widely spread shallow roots to collect water after brief showers and store it in their spongy tissues.

Some desert plants conserve water by having wax-coated leaves that reduce water loss. Others such as annual wildflowers and grasses store much of their biomass in seeds that remain inactive, sometimes for years, until they receive enough water to germinate. Shortly after a rain, these seeds germinate, grow, and carpet some deserts with dazzling arrays of colorful flowers (Figure 7.A) that last for up to a few weeks.

Most desert animals are small. Some beat the heat by hiding in cool burrows or rocky crevices by day and coming out at night or in the early morning. Others become dormant during periods of extreme heat or drought. Some larger animals such as camels can drink massive quantities of water when it is available and store it in their fat for use as needed. In addition, the camel's thick fur helps it keep cool because the air spaces in the fur insulate the camel's skin against the outside heat. In addition, camels do not sweat, which reduces their water loss through evaporation. Kangaroo rats never drink water. They get the water they need by breaking down fats in seeds that they consume.

Insects and reptiles such as rattlesnakes have thick outer coverings to minimize water loss through evaporation, and their wastes are dry feces and a dried concentrate of urine. Many spiders and insects get their water from dew or from the food they eat.

Anton Foltin/Shutterstock.com

FIGURE 7.A After a brief rain, these wildflowers bloomed in this temperate desert in Picacho Peak State Park in the U.S. state of Arizona.

CRITICAL THINKING

What are three steps you would take to survive in the open desert if you had to?

Three Types of Grasslands

Grasslands occur primarily in the interiors of continents in areas that are too moist for deserts to form and too dry for forests to grow (Figures 7.13 and 7.14). Grasslands persist because of a combination of seasonal drought, grazing by large herbivores, and occasional fires—all of which keep shrubs and trees from growing in large numbers. The three main types of grassland—tropical, temperate, and cold (arctic tundra)—result from long-term combinations of low average precipitation and varying average temperatures (Figure 7.16) (**Concept 7.3**).

One major type of *tropical grassland* is *savanna* (**Core Case Study** and Science Focus 7.2). It contains widely scattered clumps of trees and usually has warm temperatures year-round with alternating dry and wet seasons (Figure 7.16, top graph). Herds of grazing and browsing animals migrate across the savanna to find water and food in response to seasonal and year-to-year variations in rainfall (Figure 7.16, blue areas in top graph) and food availability. Savanna plants, like those in deserts, are adapted to survive drought and extreme heat. Many have deep roots that tap into groundwater.

CONSIDER THIS . . .

CONNECTIONS Savanna Grassland Niches and Feeding Habits

As an example of differing niches, some large herbivores have evolved specialized eating habits that minimize competition among species for the vegetation found on the savanna (**Core Case Study**). For example, giraffes eat leaves and shoots from the tops of trees, elephants eat leaves and branches farther down, wildebeests prefer short grasses, and zebras graze on longer grasses and stems.

erichon/Shutterstock.com

Tropical grassland (savanna)

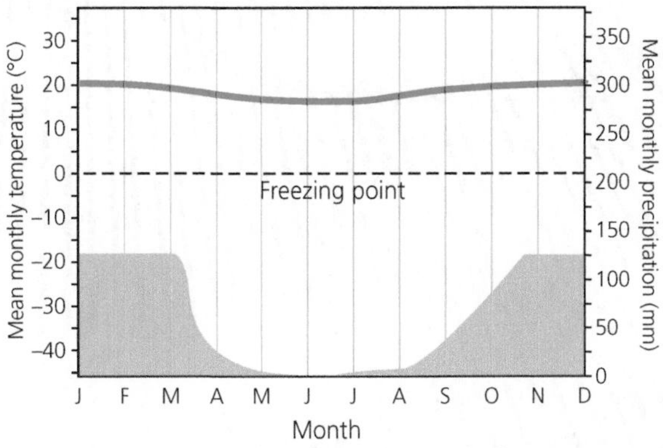

biletskiy/Shutterstock.com

Temperate grassland (prairie)

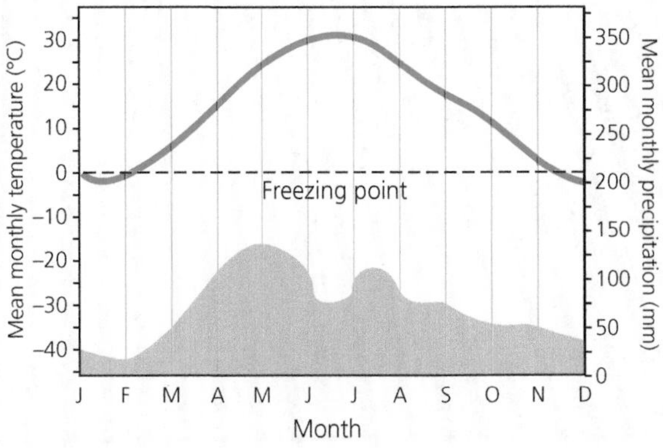

Fexel/Shutterstock.com

Cold grassland (arctic tundra)

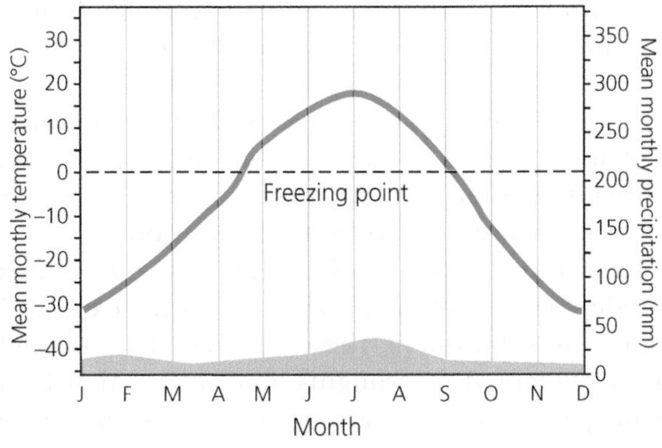

FIGURE 7.16 These climate graphs track the typical variations in annual temperature (red) and precipitation (blue) in tropical, temperate, and cold (arctic tundra) grasslands. Top photo: *savanna (tropical grassland)* in Kenya, Africa, with zebras grazing (**Core Case Study**). Center photo: *prairie (temperate grassland)* in the U.S. state of Illinois. Bottom photo: *arctic tundra (cold grassland)* in Iceland in fall. ***Data analysis:*** Which month of the year has the highest temperature and which month has the lowest rainfall for each of the three types of grassland?

EPA Documerica/National Archives and Records Administration

FIGURE 7.17 Natural capital degradation: This intensively cultivated cropland is an example of the replacement of biologically diverse temperate grasslands (such as in the center photo of Figure 7.16) with a monoculture crop.

In a *temperate grassland,* winters can be bitterly cold, summers are hot and dry, and annual precipitation is sparse and falls unevenly throughout the year (Figure 7.16, center graph). Because the aboveground parts of most of the grasses die and decompose each year, organic matter accumulates to produce deep, fertile topsoil (Figure 3.11, p. 57). This topsoil is held in place by a thick network of the grasses' intertwined roots unless the topsoil is plowed up, which exposes it to high winds found in these biomes. This biome's grasses are adapted to droughts and to fires that burn the plant parts above the ground but do not harm the roots, from which new grass can grow.

In the midwestern and western areas of the United States, we find two types of temperate grasslands depending primarily on average rainfall: *short-grass prairies* (Figure 7.16, center photo) and the *tallgrass prairies* (which get more rain). In all prairies, winds blow almost continuously and evaporation is rapid, often leading to fires in the summer and fall. This combination of winds and fires helps to maintain such grasslands by hindering tree growth. Many of the world's natural temperate grasslands have been converted to farmland, because their fertile soils are useful for growing crops (Figure 7.17) and grazing cattle.

CONSIDER THIS . . .

THINKING ABOUT Prairies

Some people say the widespread destruction and degradation of prairies is justified because they believe that these grasslands are being underutilized and should be put to use for humans. To others, the belief that prairies are not useful in their natural state reflects our lack of understanding of how nature works and of our separation from the rest of nature. Which view do you support? Why?

Cold grasslands, or *arctic tundra,* lie south of the arctic polar ice cap (Figures 7.13 and 7.14). During most of the year, these treeless plains are bitterly cold (Figure 7.16, bottom graph), swept by frigid winds, and covered with ice and snow. Winters are long with few hours of daylight, and the scant precipitation falls primarily as snow.

Under the snow, this biome is carpeted with a thick, spongy mat of low-growing plants, primarily grasses, mosses,

Revisiting the Savanna: Elephants as a Keystone Species

As in all biomes, the African savanna (**Core Case Study**) has food webs, which define its character and keep it functioning (Figure 7.B). Like many food webs, savanna food webs often include one or more keystone species that play a major role in maintaining the structure and functioning of the ecosystem.

Ecologists view elephants as a keystone species in the African savanna. They eat woody shrubs and young trees. This helps keep the savanna from being overgrown by these woody plants and prevents the grasses, which form the foundation of the food web (Figure 7.B), from dying out. If they were to die out, antelopes, zebras, and other grass-eaters would leave the savanna in search of food and with them would go the carnivores such as lions and hyenas that feast on these grass-eaters. Elephants also dig for water during drought periods, creating or enlarging waterholes that are used by other animals. Without African elephants, savanna food webs would collapse and the savanna would become shrubland.

FIGURE 7.C One reason African elephants are being threatened with extinction is their financially valuable ivory tusks.

Conservation scientists classify the African elephant as *vulnerable* to extinction. In 1979 there were an estimated 1.3 million wild African elephants. Today, an estimated 400,000 remain in the wild. This sharp decline is due mostly to the illegal killing of elephants for their valuable ivory tusks (Figure 7.C, left). Since 1990 there has been an international ban on the sale of ivory, and in some areas elephants are protected as threatened or endangered species; however, the illegal killing of elephants for their valuable ivory continues (Figure 7.C, right).

Another major threat to elephants is the loss and fragmentation of their habitats as human populations have expanded and taken over more land. Elephants are eating or trampling the crops of settlers who have moved into elephant habitat areas, and this has caused the killing of some elephants by farmers. If the multiple threats are not curtailed, elephants may disappear from the savanna within your lifetime.

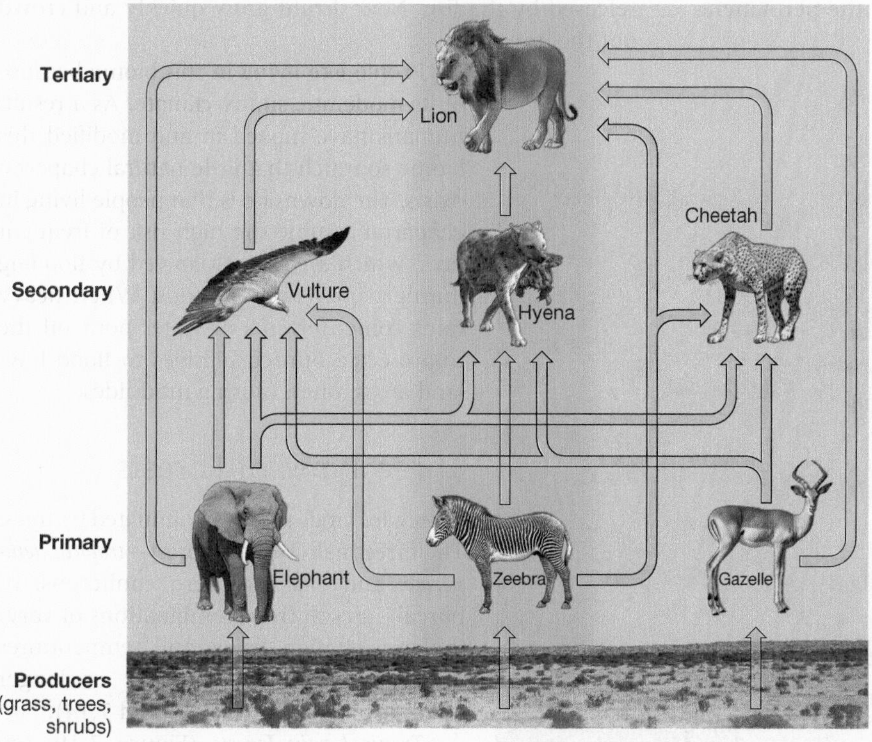

FIGURE 7.B A savanna food web.

CRITICAL THINKING

Do you think African governments would be justified in setting aside large areas of elephant habitat and prohibiting development there? Explain your reasoning. What are other alternatives for preserving African elephants on the savanna?

lichens, and dwarf shrubs. Trees and tall plants cannot survive in the cold and windy tundra because they would lose too much of their heat. Most of the annual growth of the tundra's plants occurs during the short 7- to 8-week summer, when there is daylight almost around the clock.

One outcome of the extreme cold is the formation of **permafrost,** underground soil in which captured water stays frozen for more than 2 consecutive years. During the brief summer, the permafrost layer keeps melted snow and ice from draining into the ground. Thus, shallow lakes, marshes, bogs, ponds, and other seasonal wetlands form when snow and frozen surface soil melt. Hordes of mosquitoes, black flies, and other insects thrive in these shallow surface pools. They serve as food for large colonies of migratory birds (especially waterfowl) that migrate from the south to nest and breed in the tundra's summer bogs and ponds.

Animals in this biome survive the intense winter cold through adaptations such as thick coats of fur (arctic wolf, arctic fox, and musk oxen) or feathers (snowy owl) and living underground (arctic lemming). In the summer, caribou (often called reindeer) and other types of deer migrate to the tundra to graze on its vegetation.

Tundra is vulnerable to disruption. Because of the short growing season, tundra soil and vegetation recover very slowly from damage or disturbance. Human activities in the arctic tundra—primarily on and around oil and natural gas drilling sites, pipelines, mines, and military bases—leave scars that persist for centuries.

Another type of tundra, called *alpine tundra,* occurs above the limit of tree growth but below the permanent snow line on high mountains. The vegetation is similar to that found in arctic tundra, but it receives more sunlight than arctic vegetation gets. During the brief summer, alpine tundra can be covered with an array of beautiful wildflowers.

Chaparral—a Dry Temperate Biome

In many coastal regions that border on deserts, we find a biome known as *temperate shrubland* or *chaparral* (Spanish for *thicket*). Because it is close to the sea, it has a slightly longer winter rainy season than the bordering desert has and experiences fogs during the spring and fall seasons. Chaparral is found along coastal areas of southern California, the Mediterranean Sea, central Chile, southern Australia, and southwestern South Africa.

This biome consists mostly of dense growths of low-growing evergreen shrubs and occasional small trees with leathery leaves (Figure 7.18). Its animal species include mule deer, chipmunks, jackrabbits, lizards, and a variety of birds. The soil is thin and not very fertile. During the long, hot, and dry summers, chaparral vegetation dries out. In the late summer and fall, fires started by lightning or human activities spread swiftly.

Research reveals that chaparral is adapted to and maintained by occasional fires. Many of the shrubs store food reserves in their fire-resistant roots and have seeds that sprout only after a hot fire. With the first rain, annual grasses and wildflowers spring up and use nutrients released by the fire. New shrubs grow quickly and crowd out the grasses.

People like living in this biome because of its moderate, sunny climate. As a result, humans have moved in and modified this biome so much that little natural chaparral exists. The downside is that people living in chaparral assume the high risk of frequent fires, which are often followed by flooding during winter rainy seasons. When heavy rains come, torrents of water pour off the unprotected burned hillsides to flood lowland areas, often causing mudslides.

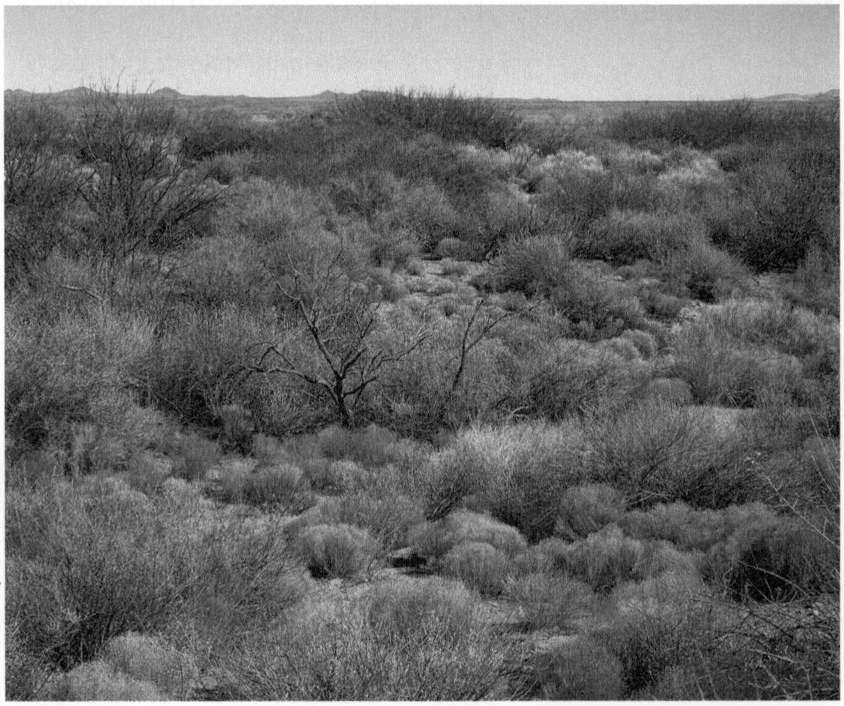

Minden Pictures/Superstock

FIGURE 7.18 Lowland chaparral in the Rio Grande River Valley, New Mexico (USA).

Three Types of Forests

Forests are lands that are dominated by trees. The three main types of forest—*tropical, temperate,* and *cold* (northern coniferous, or boreal)—result from combinations of varying precipitation levels and temperatures averaged over three decades or longer (**Concept 7.3**) (Figures 7.13 and 7.14).

Tropical rain forests (Figure 7.19, top photo) are found near the equator (Figure 7.14), where hot, moisture-laden air rises and dumps its moisture. These

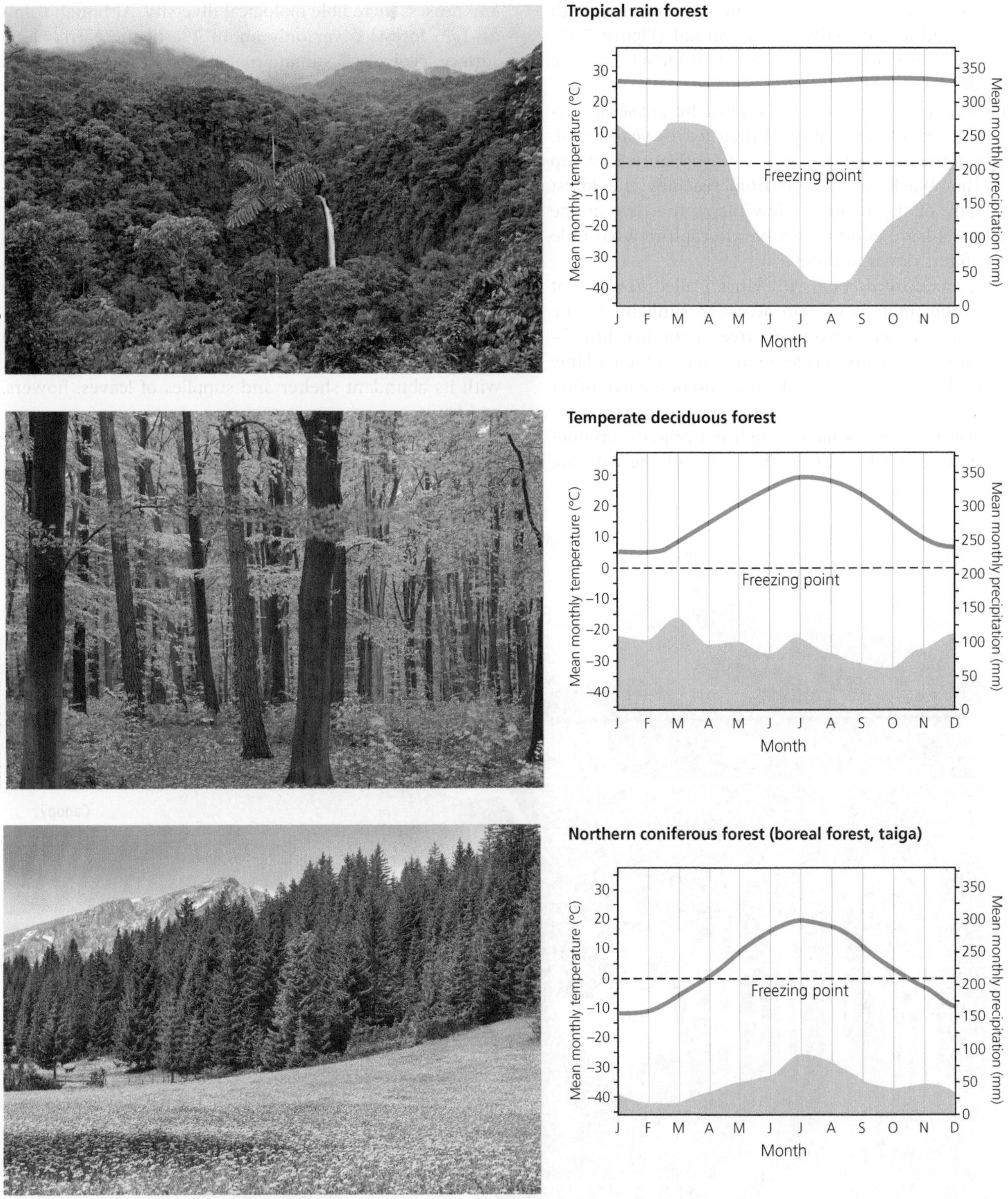

FIGURE 7.19 These climate graphs track the typical variations in annual temperature (red) and precipitation (blue) in tropical, temperate, and cold (northern coniferous, or boreal) forests. Top photo: the closed canopy of a *tropical rain forest* in Costa Rica. Middle photo: a *temperate deciduous forest* in autumn. Bottom photo: a *northern coniferous forest* in Canada. ***Data analysis:*** Which month of the year has the highest temperature and which month has the lowest rainfall for each of the three types of forest?

lush forests have year-round, warm temperatures, high humidity, and almost daily heavy rainfall (Figure 7.19, top graph). This warm and wet climate is ideal for a wide variety of plants and animals.

Tropical rain forests are dominated by *broadleaf evergreen plants*, which keep most of their leaves year-round. The tops of the trees form a dense *canopy* (Figure 7.19, top photo) that blocks most light from reaching the forest floor. Many of the relatively few plants that live at the ground level have enormous leaves to capture what little sunlight filters down to them.

Some trees are draped with vines (called *lianas*) that reach for the treetops to gain access to sunlight. In the canopy, the vines grow from one tree to another, providing walkways for many species living there. When a large tree is cut down, its network of lianas can pull down other trees.

Tropical rain forests have a high net primary productivity (see Figure 3.18, p. 62). They are teeming with life and possess incredible biological diversity. Although tropical rain forests cover only about 2% of the earth's land surface, ecologists estimate that they contain at least 50% of the known terrestrial plant and animal species. A single tree in these forests may support several thousand different insect species. Plants from tropical rain forests are a source of a variety of chemicals, many of which have been used as blueprints for making most of the world's prescription drugs.

Rain forest species occupy a variety of specialized niches in distinct layers, which contribute to their high species diversity. Vegetation layers are structured, for the most part, according to the plants' needs for sunlight, as shown in Figure 7.20. Much of the animal life, particularly insects, bats, and birds, lives in the sunny canopy layer, with its abundant shelter and supplies of leaves, flowers, and fruits.

Dropped leaves, fallen trees, and dead animals decompose quickly in tropical rain forests because of the warm,

FIGURE 7.20 Specialized plant and animal niches are *stratified*, or arranged roughly in layers, in a tropical rain forest. Filling such specialized niches enables many species to avoid or minimize competition for resources and results in the coexistence of a great variety of species.

moist conditions and the hordes of decomposers. About 90% of the nutrients released by this rapid decomposition are quickly taken up and stored by trees, vines, and other plants. Nutrients that are not taken up are soon leached from the thin topsoil by the frequent rainfall and little plant litter builds up on the ground. The resulting lack of fertile soil (see Figure 3.11, p. 57) helps explain why rain forests are not good places to clear and grow crops or graze cattle on a sustainable basis.

At least half of the world's tropical rain forests have been destroyed or disturbed by human activities such as farming, and the pace of this destruction and degradation is increasing (see Chapter 3 Core Case Study, p. 50). Ecologists warn that without strong protective measures, most of these forests, along with their rich species biodiversity and highly valuable ecosystem services, could be gone by the end of this century.

The second major type of forest is the *temperate forest*, the most common of which is the *temperate deciduous forest* (Figure 7.19, middle photo). Such forests typically see warm summers, cold winters, and abundant precipitation—rain in summer and snow in winter months (Figure 7.19, middle graph). They are dominated by a few species of *broadleaf deciduous trees* such as oak, hickory, maple, aspen, and birch. Animal species living in these forests include predators such as wolves, foxes, and wildcats. They feed on herbivores such as white-tailed deer, squirrels, rabbits, and mice. Warblers, robins, and other bird species live in these forests during the spring and summer, mating and raising their young.

In these forests, most of the trees' leaves, after developing their vibrant colors in the fall (Figure 7.19, middle photo), drop off the trees. This allows the trees to survive the cold winters by becoming dormant. Each spring, the trees sprout new leaves and spend their summers growing and producing until the cold weather returns.

Because they have cooler temperatures and fewer decomposers than tropical forests have, temperate forests also have a slower rate of decomposition. As a result, they accumulate a thick layer of slowly decaying leaf litter, which becomes a storehouse of soil nutrients (Figure 3.11, p. 57).

On a global basis, temperate forests have been degraded by various human activities, especially logging and urban expansion, more than any other terrestrial biome. However, within 100 to 200 years, forests of this type that have been cleared can return through secondary ecological succession (see Figure 5.12, p. 108).

Another type of temperate forest, the *coastal coniferous forests* or *temperate rain forests* (Figure 7.21), are found in scattered coastal temperate areas with ample rainfall and moisture from dense ocean fogs. These forests contain thick stands of large cone-bearing, or *coniferous*, trees that keep most of their leaves (or needles) year-round. Most of these species have small, needle-shaped, wax-coated leaves that can withstand the intense cold and drought of winter. Examples are Sitka spruce, Douglas fir, giant sequoia (see Chapter 2 opening photo, pp. 28–29), and redwoods that once dominated undisturbed areas of these biomes along the coast of North America, from Canada to Northern California in the United States.

In this biome, the ocean moderates the temperature so winters are mild and summers are cool. The trees in these moist forests depend on frequent rains and moisture from summer fogs. Most of the trees are evergreen because the abundance of water means that they have no need to shed their leaves. Tree trunks and the ground are frequently covered with mosses and ferns in this cool and moist

CrackerClips Stock Media/Shutterstock.com

FIGURE 7.21 Temperate rain forest in Olympic National Park in the U.S. state of Washington.

environment. As in tropical rain forests, little light reaches the forest floor.

Many of the redwood, Douglas fir, and western cedar forests have been cleared for their valuable timber and there is constant pressure to cut what remains. This threatens species such as the spotted owl and marbled murrelet that depend on these ecosystems. Clear-cutting also loads streams in these ecosystems with eroded sediment and threatens species such as salmon that depend on clear streams for laying their eggs.

The third major forest type is the *cold*, or *northern coniferous forests* (Figure 7.19, bottom photo), also called *boreal forests* or *taigas* ("TIE-guhs"). They are found south of arctic tundra in northern regions across North America, Asia, and Europe (Figure 7.14) and above certain altitudes in the Sierra Nevada and Rocky Mountain ranges of the United States. In the subarctic, cold, and moist climate of the northernmost boreal forests, winters are long and extremely cold, with winter sunlight available only 6 to 8 hours per day. Summers are short, with cool to warm temperatures (Figure 7.19, bottom graph), and the sun shines as long as 19 hours a day during midsummer.

Most boreal forests are dominated by a few species of *coniferous evergreen trees* or *conifers* such as spruce, fir, cedar, hemlock, and pine. Plant diversity is low because few species can survive the winters when soil moisture is frozen.

Beneath the stands of trees in these forests is a deep layer of partially decomposed conifer needles. Decomposition is slow because of low temperatures, the waxy coating on the needles, and high soil acidity. The decomposing conifer needles make the thin, nutrient-poor topsoil acidic (Figure 3.11, p. 57), which prevents most other plants (except certain shrubs) from growing on the forest floor.

Year-round wildlife in this biome includes bears, wolves, moose, lynx, and many burrowing rodent species. Caribou spend winter in the taiga and summer in the arctic tundra (Figure 7.16, bottom). During the brief summer, warblers and other insect-eating birds feed on flies, mosquitoes, and caterpillars.

Mountains Play Important Ecological Roles

Some of the world's most spectacular environments are high on *mountains* (Figure 7.22), steep or high-elevation lands that cover about one-fourth of the earth's land surface.

Mountains are places where dramatic changes take place over very short distances. In fact, climate and vegetation vary according to *elevation*, or height above sea level, just as they do with latitude (Figure 7.23). If you climb a tall mountain, from its base to its summit, you can observe changes in plant life similar to those you would encounter in traveling from a temperate region to the earth's northern polar region.

About 1.2 billion people (16% of the world's population) live in mountain ranges or in their foothills, and 4 billion people (55% of the world's population) depend on mountain systems for all or some of their water. Because of the steep slopes, mountain soils are easily eroded when the vegetation holding them in place is removed by natural disturbances such as landslides and avalanches, or by human activities such as timber cutting and agriculture. Many mountains are *islands of biodiversity* surrounded by a sea of lower-elevation landscapes transformed by human activities.

Pichugin Dmitry/Shutterstock.com

FIGURE 7.22 Mountains, such as these surrounding a lake, provide vital ecosystem services.

FIGURE 7.23 Climate and vegetation represented by different biomes change with elevation as well as with latitude.

Mountains play an important ecological role. They contain a large portion of the world's forests, which are habitats for much of the planet's terrestrial biodiversity. They often are habitats for *endemic species*—those that are found nowhere else on earth. They also serve as sanctuaries for animals that are capable of migrating to higher altitudes and surviving in such environments. Every year, more of these animals are driven from lowland areas to mountain habitats by human activities and by the warming climate.

CONSIDER THIS . . .

CONNECTIONS Mountains and Climate

Mountains help regulate the earth's climate. Many mountaintops are covered with glacial ice and snow that reflect some solar radiation back into space, which helps to cool the earth. However, many mountain glaciers are melting, primarily because the atmosphere has warmed over recent decades. Whereas glaciers reflect solar energy, the darker rocks exposed by melting glaciers absorb that energy. This helps to warm the atmosphere above them, which melts more ice and warms the atmosphere more—in an escalating positive feedback loop.

Mountains play a critical role in the hydrologic cycle (see Figure 3.19, p. 63) by serving as major storehouses of water. During winter, precipitation is stored as ice and snow. In the warmer weather of spring and summer, much of this snow and ice melts, releasing water to streams for use by wildlife and by humans for drinking and for irrigating crops. As the earth's atmosphere has warmed in recent decades, some mountaintop snow packs and glaciers have been melting earlier in the spring each year. This is leading to lower food production in certain areas, because much of the water needed throughout the summer to irrigate crops is released too early in the season and too quickly.

Despite the ecological, economic, and cultural importance of mountain ecosystems, governments and many environmental organizations have not focused on protecting these areas. However, conservationist, mountain explorer, and National Geographic Explorer Gregg Treinish is trying to change this. He founded the nonprofit Adventurers and Scientists for Conservation (ASC). It connects outdoor adventurers who volunteer to collect data during their travels with researchers who are focused on identifying the effects of climate change on mountain ecosystems. Treinish, who was National Geographic Adventurer of the Year in 2008–2009, leads his own research expeditions to many of the world's rugged mountain regions.

Humans Have Disturbed Much of the Earth's Land

About 60% of the world's major terrestrial ecosystems are being degraded or used unsustainably, as the human ecological footprint gets bigger and spreads across the globe (see Figure 1.8, p. 12), according to the 2005 Millennium Ecosystem Assessment and later updates of such research. Figure 7.24 summarizes the most harmful human impacts on the world's deserts, grasslands, forests, and mountains.

How long can we keep eating away at these terrestrial forms of natural capital without threatening our economies and the long-term survival of our own and many other species? No one knows. But there are increasing signs that we need to come to grips with this vital issue.

Natural Capital Degradation
Major Human Impacts on Terrestrial Ecosystems
Deserts
Grasslands
Forests
Mountains
TO Fabulous
LAS VEGAS
NEVADA
Large desert cities
Destruction of soil and underground habitat by off-road vehicles
Depletion of groundwater
Land disturbance and pollution from mineral extraction
Conversion to cropland
Release of CO_2 to atmosphere from burning grassland
Overgrazing by livestock
Oil production and off-road vehicles in arctic tundra
Clearing for agriculture, livestock grazing, timber, and urban development
Conversion of diverse forests to tree plantations
Damage from off-road vehicles
Pollution of forest streams
Agriculture
Timber and mineral extraction
Hydroelectric dams and reservoirs
Air pollution blowing in from urban areas and power plants
Soil damage from off-road vehicles

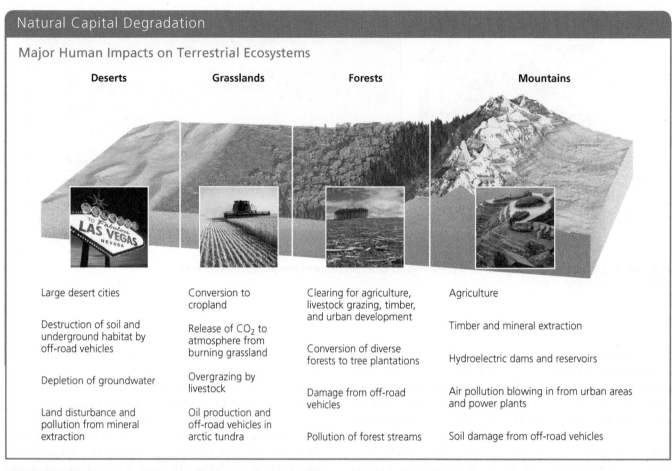

FIGURE 7.24 Human activities have had major impacts on the world's deserts, grasslands, forests, and mountains, as summarized here. ***Critical thinking:*** For each of these biomes, which two of the impacts listed do you think are the most harmful? Explain.

Left: somchaij/Shutterstock.com. Left center: Orientaly/Shutterstock.com. Right center: Timothy Epp/Dreamstime.com. Right: Vasik Olga/Shutterstock.com.

Many environmental scientists call for a global effort to better understand the nature and state of the world's major terrestrial ecosystems and biomes and to use such scientific data to protect the world's remaining wild areas from harmful forms of development. In addition, they call for us to restore many of the land areas that have been degraded, especially in areas that are rich in biodiversity. However, such efforts are highly controversial because of the timber, mineral, fossil fuel, and other resources found on or under many of the earth's remaining wild land areas. These issues are discussed in Chapter 10.

BIG IDEAS

- Differences in climate, based mostly on long-term differences in average temperature and precipitation, largely determine the types and locations of the earth's deserts, grasslands, and forests.
- The earth's terrestrial ecosystems provide important economic and ecosystem services.
- Human activities are degrading and disrupting many of the ecosystem and economic services provided by the earth's terrestrial ecosystems.

Tying It All Together

Tropical African Savanna and Sustainability

This chapter's Core Case Study began with questions about the earth's diversity of terrestrial ecosystems and how they form. We examined the difference between weather and climate and how climate is a major factor in the formation and distribution of these biomes—the world's deserts, grasslands, and forests—as well as the life forms that live in those systems. In particular, we focused on the savanna, a grassland biome that is threatened by expansion of the human population.

Anharris/iStock/Getty Images Plus/Getty Images

The relationships among weather, climate, and biomes and their living inhabitants are in keeping with the three **scientific principles of sustainability**. The earth's dynamic climate system helps to distribute heat from solar energy and to recycle the earth's nutrients. This in turn helps to generate and support the biodiversity found in the earth's various biomes.

Scientists have made progress in understanding the ecology of the world's terrestrial systems, as well as how the vital ecosystem and economic services they provide are being degraded and disrupted. One of the major lessons from their research is: *in nature, everything is connected*. According to these scientists, we urgently need more research on the components and workings of the world's biomes, on how they are interconnected, and on which connections are in the greatest danger of being disrupted by human activities. With such information, we will have a clearer picture of what we can do to help sustain the natural capital on which we and all other species depend.

Chapter Review

Core Case Study

1. Describe the African savanna (**Core Case Study**) and explain why it serves as an example of how differences in climate lead to the formation of different types of ecosystems.

Section 7.1

2. What is the key concept for this section? Define **weather**. Define **front** and distinguish between a **warm front** and a **cold front**. What is **atmospheric pressure**? Define and distinguish between a **high** and a **low**. What are El Niño and La Niña? Summarize their effects. Explain how tornadoes form and describe their effects. What are tropical cyclones and what is the difference between hurricanes and typhoons? How do these storms form?

Section 7.2

3. What is the key concept for this section? Define **climate** and distinguish between weather and climate. Define **ocean currents**. Define **convection** and **convection cell**. Explain how uneven solar heating of the earth affects climate. Define **latitude** and explain how latitudes are designated. Define **insolation** and explain how it is related to latitude. Explain how the tilt of the earth's axis and resulting seasonal changes affect climates.
4. How does the rotation of the earth on its axis affect climates? What is the Coriolis effect? Explain how prevailing winds form and how they affect climate. Explain how ocean currents affect climate. What is a **gyre**?
5. Define and give four examples of a **greenhouse gas**. What is the **greenhouse effect** and why is it important to the earth's life and climate? What is the **rain shadow effect** and how can it lead to the formation of deserts? Why do cities tend to have more haze and smog, higher temperatures, and lower wind speeds than the surrounding countryside?

Section 7.3

6. What is the key concept for this section? Explain how different combinations of annual precipitation and temperatures averaged over several decades, along with global air circulation patterns and ocean currents, lead to the formation of deserts, grasslands,

and forests. What is a **biome**? Explain why there are three major types of each of the major biomes (deserts, grasslands, and forests). Explain why biomes are not uniform. Describe how climate and vegetation vary with latitude and elevation.

7. Explain how the three major types of deserts differ in their climate and vegetation. Why are desert ecosystems vulnerable to long-term damage? How do desert plants and animals survive?

8. Explain how the three major types of grasslands differ in their climate and vegetation. Explain how savanna animals survive seasonal variations in rainfall (**Core Case Study**). Why is the elephant an important component of the African savanna? Why have many of the world's temperate grasslands disappeared? Describe arctic tundra and define **permafrost**. What is chaparral and what are the risks of living there?

9. Explain how the three major types of forests differ in their climate and vegetation. Why is biodiversity so high in tropical rain forests? Why do most soils in tropical rain forests hold few plant nutrients? Why do temperate deciduous forests typically have a thick layer of decaying litter? What are coastal coniferous or temperate rain forests? How do most species of coniferous evergreen trees survive the cold winters in boreal forests? What important ecological roles do mountains play? Summarize the ways in which human activities have affected the world's deserts, grasslands, forests, and mountains.

10. What are this chapter's *three big ideas*? Summarize the connections between climate and terrestrial ecosystems, and explain how these connections are in keeping with the three **scientific principles of sustainability**.

Note: Key terms are in bold type.

Critical Thinking

1. Why is the African savanna (**Core Case Study**) a good example of the three **scientific principles of sustainability** in action? For each of these principles, give an example of how it applies to the African savanna and explain how it is being violated by human activities that now affect the savanna.

2. For each of the following, decide whether it represents a likely trend in weather or in climate: **(a)** an increase in the number of thunderstorms in your area from one summer to the next; **(b)** a decrease of 20% in the depth of a mountain snowpack between 1975 and 2015; **(c)** a rise in the average winter temperatures in a particular area over a decade; and **(d)** an increase in the earth's average global temperature since 1980.

3. Review the five major climatic factors explained in Section 7.2 and explain how each of them has helped to define the climate in the area where you live or go to school.

4. Why do deserts and arctic tundra support a much smaller number and variety of animals than do tropical forests? Why do most animals in a tropical rain forest live in its trees?

5. How might the distribution of the world's forests, grasslands, and deserts shown in Figure 7.14 differ if the prevailing winds shown in Figure 7.9 did not exist?

6. Which biomes are best suited for **(a)** raising crops and **(b)** grazing livestock? Use the three **scientific principles of sustainability** to come up with three guidelines for growing crops and grazing livestock more sustainably in these biomes.

7. What type of biome do you live in? (If you live in a developed area, what type of biome was the area before it was developed?) List three ways in which your lifestyle could be contributing to the degradation of this biome. What are three lifestyle changes that you could make in order to reduce your contribution?

8. You are a defense attorney arguing in court for sparing a large area of tropical rain forest from being cut down and used to produce food. Give your three best arguments for the defense of this ecosystem.

Doing Environmental Science

Find a natural terrestrial ecosystem near where you live or go to school. Study and write a description of the system, including its dominant vegetation and any animal life that you are aware of. Also, note how any human disturbances have changed the system. Return to the system after a month or two and note any changes, based on your earlier notes. Compare your notes with those of your classmates.

Global Environment Watch Exercise

Go to your MindTap course to access the GREENR database. Starting on the home page, under "Browse Issues and Topics", click on *Environment and Ecology*, then select *Forests and Deforestation*. Search within this portal for *tropical rain forests* to find information on **(a)** trends in the global rate of destruction of these forests; **(b)** what areas of the world are seeing rising rates of destruction and what areas are seeing falling rates; and **(c)** what is being done to protect them in various areas. Write a report on your findings.

Data Analysis

In this chapter, you learned how long-term variations in average temperatures and average precipitation play a major role in determining the types of deserts, forests, and grasslands found in different parts of the world. Below are typical annual climate graphs for a tropical grassland (savanna) in Africa (**Core Case Study**) and a temperate grassland in the Midwestern United States.

1. In what month (or months) does the most precipitation fall in each of these areas?
2. What are the driest months in each of these areas?
3. What is the coldest month in the tropical grassland?
4. What is the warmest month in the temperate grassland?

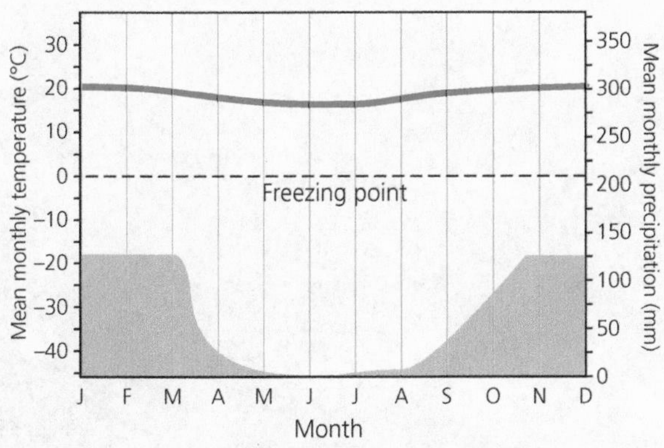

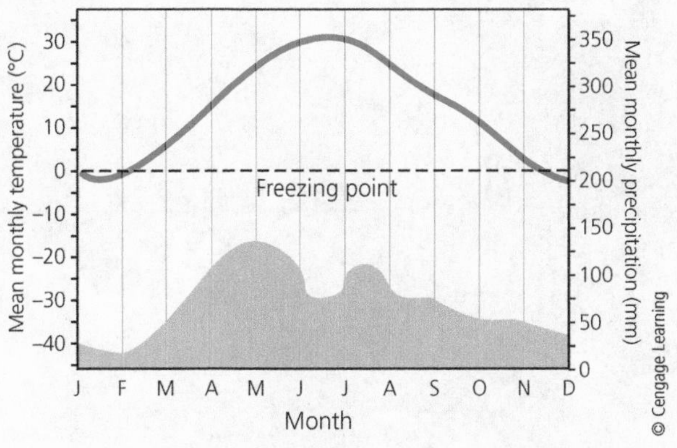

CHAPTER 8

Aquatic Biodiversity

The sea, once it casts its spell, holds one in its net of wonders forever.

JACQUES-YVES COUSTEAU

Coral reef in Egypt's Red Sea.

Vlad61/Shutterstock.com

Key Questions

8.1 What is the general nature of aquatic systems?

8.2 Why are marine aquatic systems important?

8.3 How have human activities affected marine ecosystems?

8.4 Why are freshwater ecosystems important?

8.5 How have human activities affected freshwater ecosystems?

Core Case Study

Why Should We Care about Coral Reefs?

Coral reefs form in clear, warm coastal waters in tropical areas. These stunningly beautiful natural wonders (see chapter-opening photo) are among the world's oldest, most diverse, and most productive ecosystems.

Coral reefs are formed by massive colonies of tiny animals called *polyps* (close relatives of jellyfish). They slowly build reefs by secreting a protective crust of limestone (calcium carbonate) around their soft bodies. When the polyps die, their empty crusts remain behind as part of a platform for more reef growth. The resulting elaborate network of crevices, ledges, and holes serves as calcium carbonate "condominiums" for a variety of marine animals.

Coral reefs are the result of a mutually beneficial relationship between polyps and tiny single-celled algae called *zooxanthellae* ("zoh-ZAN-thel-ee") that live in the tissues of the polyps. In this example of mutualism (see p. 105), the algae provide the polyps with food and oxygen through photosynthesis, and help the corals produce calcium carbonate. Algae also give the reefs their stunning coloration. The polyps, in turn, provide the algae with a well-protected home and some of their nutrients.

Although shallow and deep-water coral reefs occupy only about 0.2% of the ocean floor, they provide important ecosystem and economic services. They act as natural barriers that help to protect 15% of the world's coastlines from flooding and erosion caused by battering waves and storms. They also provide habitats, food, and spawning grounds for one-quarter to one-third of the organisms that live in the ocean, and they produce about one-tenth of the global fish catch. Through tourism and fishing, they provide goods and services worth about $40 billion a year.

Coral reefs are vulnerable to damage because they grow slowly and are disrupted easily. Runoff of soil and other materials from the land can cloud the water and block the sunlight that the algae in shallow reefs need for photosynthesis. Also, the water in which shallow reefs live must have a temperature of 18–30°C (64–86°F) and cannot be too acidic. This explains why a major long-term threat to coral reefs is *climate change*, which could raise the water temperature above tolerable limits in most reef areas. The closely related problem of *ocean acidification* could make it harder for polyps to build reefs and could even dissolve some of their calcium carbonate formations. (We discuss this problem further in this and later chapters.)

One result of stresses such as pollution and rising ocean water temperatures is *coral bleaching* (Figure 8.1), which can cause the colorful algae, upon which corals depend for food, to die off. Without food, the coral polyps die, leaving behind a white skeleton of calcium carbonate. Studies by the Global Coral Reef Monitoring Network and other scientific groups estimate that since the 1950s, some 45% to 53% of the world's shallow coral reefs have been destroyed or degraded. Another 25% to 33% could be lost within 20 to 40 years. These centers of biodiversity are by far the most threatened marine ecosystems.

In this chapter, we explore the nature of aquatic ecosystems, and we begin to examine the effects of human activities on these vital forms of natural capital.

iStockphoto.com/Rainer von Brandis

FIGURE 8.1 This bleached coral has lost most of its algae because of changes in the environment such as warming of the waters and deposition of sediments.

8.1 WHAT IS THE GENERAL NATURE OF AQUATIC SYSTEMS?

CONCEPT 8.1A Saltwater and freshwater aquatic life zones cover almost three-fourths of the earth's surface, with oceans dominating the planet.

CONCEPT 8.1B The key factors determining biodiversity in aquatic systems are temperature, dissolved oxygen content, availability of food, and access to light and nutrients necessary for photosynthesis.

Most of the Earth Is Covered with Water

Saltwater covers about 71% of the earth's surface, and freshwater occupies roughly another 2%. Although the *global ocean* is a single and continuous body of water, geographers divide it into five large areas—the Atlantic, Pacific, Indian, Arctic, and Southern Oceans—separated by the continents. The largest ocean is the Pacific, which contains more than half of the earth's water and covers one-third of the earth's surface. Together, the oceans hold almost 98% of the earth's water (**Concept 8.1A**). Each of us is connected to, and utterly dependent on, the earth's global ocean through the water cycle (see Figure 3.19, p. 63).

73%
Percentage of the earth that is covered with water

The aquatic equivalents of biomes are called **aquatic life zones**—saltwater and freshwater portions of the biosphere that can support life. The distribution of many aquatic organisms is determined largely by the water's **salinity**—the amounts of various salts such as sodium chloride (NaCl) dissolved in a given volume of water. It is measured in *parts per thousand (ppt)* or the number of units of salt dissolved in 1,000 units of water. The salinity of ocean water averages 35 ppt. The salinity of freshwater that we use for drinking and other purposes is 0.5 ppt or less.

Aquatic life zones are classified into two major types: **saltwater** or **marine life zones** (oceans and their bays, estuaries, coastal wetlands, shorelines, coral reefs, and mangrove forests) and **freshwater life zones** (lakes, rivers, streams, and inland wetlands). Some systems such as estuaries are a mix of saltwater and freshwater, but scientists classify them as marine systems for purposes of discussion.

Aquatic Species Drift, Swim, Crawl, and Cling

Saltwater and freshwater life zones contain several major types of organisms. One type consists of **plankton,** which can be divided into three groups. The first group consists of drifting organisms called *phytoplankton,* which includes many types of algae. These tiny aquatic plants and even smaller *ultraplankton*—the second group of plankton—are the producers that make up the base of most aquatic food chains and webs (see Figure 3.16, p. 60). Through photosynthesis, they produce about half of the earth's oxygen, on which we depend for survival.

The third group is made up of drifting animals called *zooplankton,* which feed on phytoplankton and on other zooplankton (see Figure 3.16, p. 60). The members of this group range in size from single-celled protozoa to large invertebrates such as jellyfish (Figure 8.2).

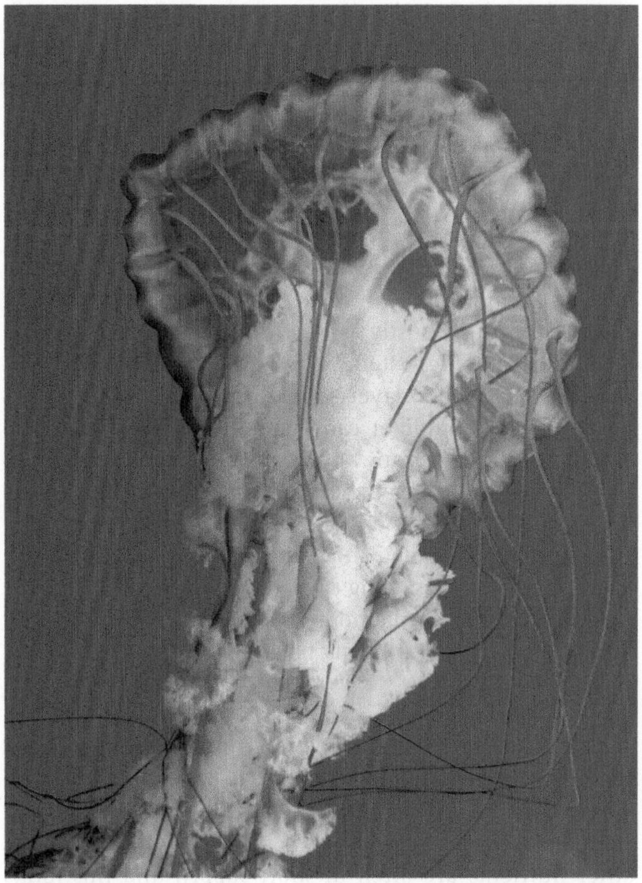

Keng Po Leung/Dreamstime.com

FIGURE 8.2 Jellyfish are drifting zooplankton that use their long tentacles with stinging cells to stun or kill their prey.

Jefferychin/Shutterstock.com

FIGURE 8.3 Starfish on a coral reef. The starfish is also called a sea star because it is not a fish.

A second major type of aquatic organism is **nekton,** strongly swimming consumers such as fish, turtles, and whales. The third type, **benthos,** consists of bottom-dwellers such as oysters and sea stars (Figure 8.3), which anchor themselves to ocean-bottom structures; clams and worms, which burrow into the sand or mud; and lobsters and crabs, which walk about on the sea floor. A fourth major type is **decomposers** (mostly bacteria), which break down organic compounds in the dead bodies and wastes of aquatic organisms into nutrients that aquatic primary producers can use.

Key factors determining the types and numbers of organisms found in different areas of the ocean are *temperature, dissolved oxygen content, availability of food,* and *availability of light and nutrients required for photosynthesis,* such as carbon (as dissolved CO_2 gas), nitrogen (as NO_3^-), and phosphorus (mostly as PO_4^{3-}) (**Concept 8.1B**).

In deep aquatic systems, photosynthesis is largely confined to the upper layer—the *euphotic* or *photic zone*—through which sunlight can penetrate. The depth of the euphotic zone in oceans and deep lakes is reduced when the water is clouded by excessive growth of algae. This is called an *algal bloom* and it results from nutrient overloads. This cloudiness is called **turbidity.** It is also caused by soil and other sediments being carried by wind, rain and melting snow from cleared land into adjoining bodies of water. This is one of the problems plaguing shallow coral reefs (**Core Case Study**).

In shallow systems such as small open streams, lake edges, and ocean shorelines, ample supplies of nutrients for primary producers are usually available, which tends to make these areas high in biodiversity. By contrast, in most areas of the open ocean, nitrates, phosphates, iron, and other nutrients are often in short supply, and this limits net primary productivity (NPP) (see Figure 3.18, p. 62) and the diversity of species.

8.2 WHY ARE MARINE AQUATIC SYSTEMS IMPORTANT?

CONCEPT 8.2 Saltwater ecosystems provide major ecosystem and economic services and are irreplaceable reservoirs of biodiversity.

Oceans Provide Vital Ecosystem and Economic Services

Oceans provide enormously valuable ecosystem and economic services (Figure 8.4) that help keep us and other species alive and support our economies. They produce more than half of the oxygen we breathe and, as a vital part of the water cycle, provide most of the rain that sustains our water supply. Every year, we remove almost 100 million tonnes (110 million tons) of seafood from the oceans.

As land dwellers, we have a distorted and limited view of the oceans that cover most of the earth's surface. We know more about the surface of the moon than we know about the earth's oceans. According to aquatic scientists, the scientific investigation of poorly understood marine and freshwater aquatic systems could yield immense ecological and economic benefits.

Natural Capital

Marine Ecosystems

Ecosystem Services	Economic Services
Oxygen supplied through photosynthesis	Food
Water purification	Energy from waves and tides
Climate moderation	Pharmaceuticals
CO_2 absorption	Harbors and transportation routes
Nutrient cycling	Recreation and tourism
Reduced storm damage (mangroves, barrier islands, coastal wetlands)	Employment
Biodiversity: species and habitats	Minerals

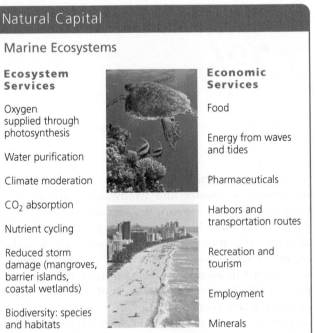

© Cengage Learning

FIGURE 8.4 Marine systems provide a number of important ecosystem and economic services (**Concept 8.2**). ***Critical thinking:*** Which two ecosystem services and which two economic services do you think are the most important? Why?

Top: Willyam Bradberry/Shutterstock.com. Bottom: James A. Harris/Shutterstock.com.

CONSIDER THIS . . .

LEARNING FROM NATURE

Engineers are learning how whales use sound waves to communicate over long distances underwater to improve our underwater communication technologies.

Marine aquatic systems are enormous reservoirs of biodiversity. Marine life is found in three major *life zones*: the coastal zone, the open sea, and the ocean bottom (Figure 8.5).

The **coastal zone** is the warm, nutrient-rich, shallow water that extends from the high-tide mark on land to the gently sloping, shallow edge of the *continental shelf* (the submerged part of the continents). It makes up less than 10% of the world's ocean area, but it contains 90% of all marine species and most large commercial fisheries. This zone's aquatic systems include estuaries, coastal marshes, mangrove forests, and coral reefs.

Estuaries and Coastal Wetlands Are Highly Productive

An **estuary** is an aquatic zone where a river meets the sea (Figure 8.6). It is a partially enclosed body of water where seawater mixes with the river's freshwater, as well as with nutrients and pollutants in runoff from the land.

Estuaries are associated with **coastal wetlands**—coastal land areas covered with water all or part of the

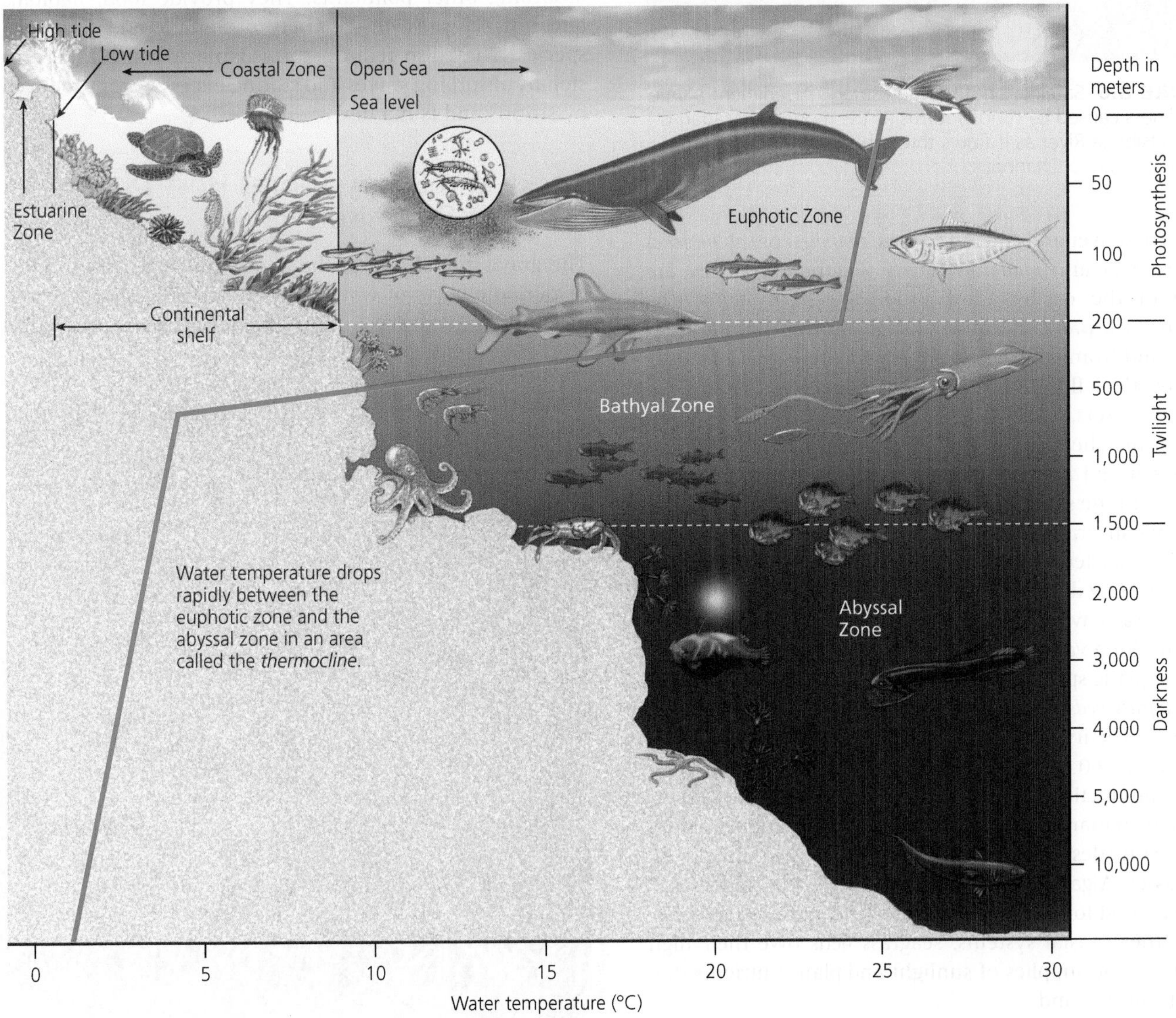

FIGURE 8.5 Major life zones and vertical zones (not drawn to scale) in an ocean. Actual depths of zones may vary. Available light determines the euphotic, bathyal, and abyssal zones. Temperature zones also vary with depth, shown here by the red line. ***Critical thinking:*** How is an ocean similar to a rain forest? (*Hint:* See Figure 7.20, p. 158.)

NASA

FIGURE 8.6 Satellite photo of an *estuary*. A sediment plume (cloudiness caused by runoff) forms at the mouth of Madagascar's Betsiboka River as it flows through the estuary and into the Mozambique Channel.

year. These wetlands include *coastal marshes*, or *salt marshes* (Figure 8.7) and *mangrove forests* (Figure 8.8). They are some of the earth's most productive ecosystems (see Figure 3.18, p. 62) because of high nutrient inputs from rivers and from adjoining land, rapid circulation of nutrients by tidal flows, and ample sunlight penetrating the shallow waters.

Salt marshes exhibit the edge effect (see p. 80) where terrestrial and aquatic plant communities meet. In areas away from the saltiest parts of a marsh, plant life is diverse with marsh grasses as well as some land plants and the animals that feed on them. Mangrove forests around the world host 69 different species of trees and shrubs that can grow in salty water. They provide habitat, food, and nursery sites for a variety of fishes, shellfish, crabs, snakes, and other aquatic species.

Seagrass beds (Figure 8.9), another component of marine biodiversity, occur in shallow coastal waters and host as many as 60 species of grasses and other plants. These highly productive and physically complex systems support a variety of marine species. A few species, including manatees, sea turtles, and dugongs (sea cows), feed directly on seagrasses. Algae attach themselves to the grass blades and serve as food for various mollusks, fish, and other animals. Like other coastal systems, seagrass beds owe their high NPP to ample supplies of sunlight and plant nutrients that flow from the land.

These coastal aquatic systems provide important ecosystem and economic services. They help to maintain water quality in tropical coastal zones by filtering toxic pollutants, excess plant nutrients, and sediments, and by

Jcdesign/Shutterstock.com

FIGURE 8.7 Coastal marsh in the U.S. state of South Carolina.

absorbing other pollutants. They provide food, habitats, and nursery sites for a variety of aquatic and terrestrial species. They also reduce storm damage and coastal erosion by absorbing waves and storing excess water produced by storms and tsunamis.

Rocky and Sandy Shores Host Different Types of Organisms

The gravitational pull of the moon and sun causes **tides,** or periodic flows of water onto and off the shore, to rise and fall about every 6 hours in most coastal areas. The

Manit Larpluechai/Dreamstime.com

FIGURE 8.8 Mangrove forest on the coast of Thailand. Mangroves have roots that curve up from the mud and water to obtain oxygen from the air, working somewhat like a snorkel.

James Forte/National Geographic Creative

FIGURE 8.9 Seagrass beds, such as this one near the coast of San Clemente Island, California, support a variety of marine species.

area of shoreline between low and high tides is called the **intertidal zone.** Organisms living in this zone must be able to avoid being swept away or crushed by waves. They need to survive when immersed during high tides and left high and dry (and much hotter) at low tides. They must also survive changing levels of salinity when heavy rains dilute saltwater. To deal with such stresses, most intertidal organisms hide in protective shells, dig in, or hold on tight to something.

CONSIDER THIS . . .

LEARNING FROM NATURE

The blue mussel produces a nontoxic, biodegradable glue to cling to underwater rocks in oceans. Scientists have mimicked this process to produce nontoxic glues that can be used under and above water.

On some coasts, steep *rocky shores* are pounded by waves (Figure 8.10, top). The numerous pools and other habitats in these intertidal zones contain a great variety of species. Each occupies a different niche to deal with daily and seasonal changes in environmental conditions such as temperature, water flows, and salinity.

Other coasts have gently sloping *barrier beaches,* or *sandy shores,* that support other types of marine organisms (Figure 8.10, bottom). Most of them keep hidden from view and survive by burrowing, digging, and tunneling in the sand. These beaches and their adjoining coastal wetlands are also home to a variety of shorebirds that have evolved in specialized niches to feed on crustaceans, insects, and other organisms (see Figure 4.10, p. 83).

Many of these same species also live on *barrier islands*—low, narrow, sandy islands that form offshore, parallel to coastlines. Undisturbed barrier beaches generally have one or more rows of sand dunes in which the sand is held in place by the roots of grasses and other plants. These dunes are the first line of defense against the ravages of the sea. Real estate developers frequently remove the protective dunes or cover them with buildings and roads. Large storms can then flood and even sweep away seaside construction and severely erode the sandy beaches.

CASE STUDY

Revisiting Coral Reefs—Amazing Centers of Biodiversity

Coral reefs (**Core Case Study** and chapter-opening photo) are some of the world's oldest and most diverse and productive ecosystems. They are the marine equivalents of tropical rain forests, with complex interactions among their diverse populations of species.

Worldwide, coral reefs are being damaged and destroyed at an alarming rate by a variety of human activities. The newest growing threat is **ocean acidification**—the rising levels of acidity in ocean waters. This is occurring because the oceans absorb about 25% of the CO_2 emitted into the atmosphere by human activities, especially the burning of fossil fuels. The CO_2 reacts with ocean water to form a weak acid (carbonic acid, H_2CO_3). This reaction decreases the levels of carbonate ions (CO_3^{2-}) necessary for the formation of coral reefs and the shells and skeletons of many marine organisms. This makes it harder for these species to thrive and reproduce. At some point, this rising acidity could slowly dissolve corals and the shells and skeletons of some marine species.

Ocean acidification and other forms of degradation could have devastating effects on the biodiversity and food webs of coral reefs. This will in turn degrade the ecosystem services that reefs provide. It will also have a severe impact on the approximately 500 million people who depend on coral reefs for food or for income from fishing and tourism. We discuss the threat from ocean acidification in more detail in Chapter 11.

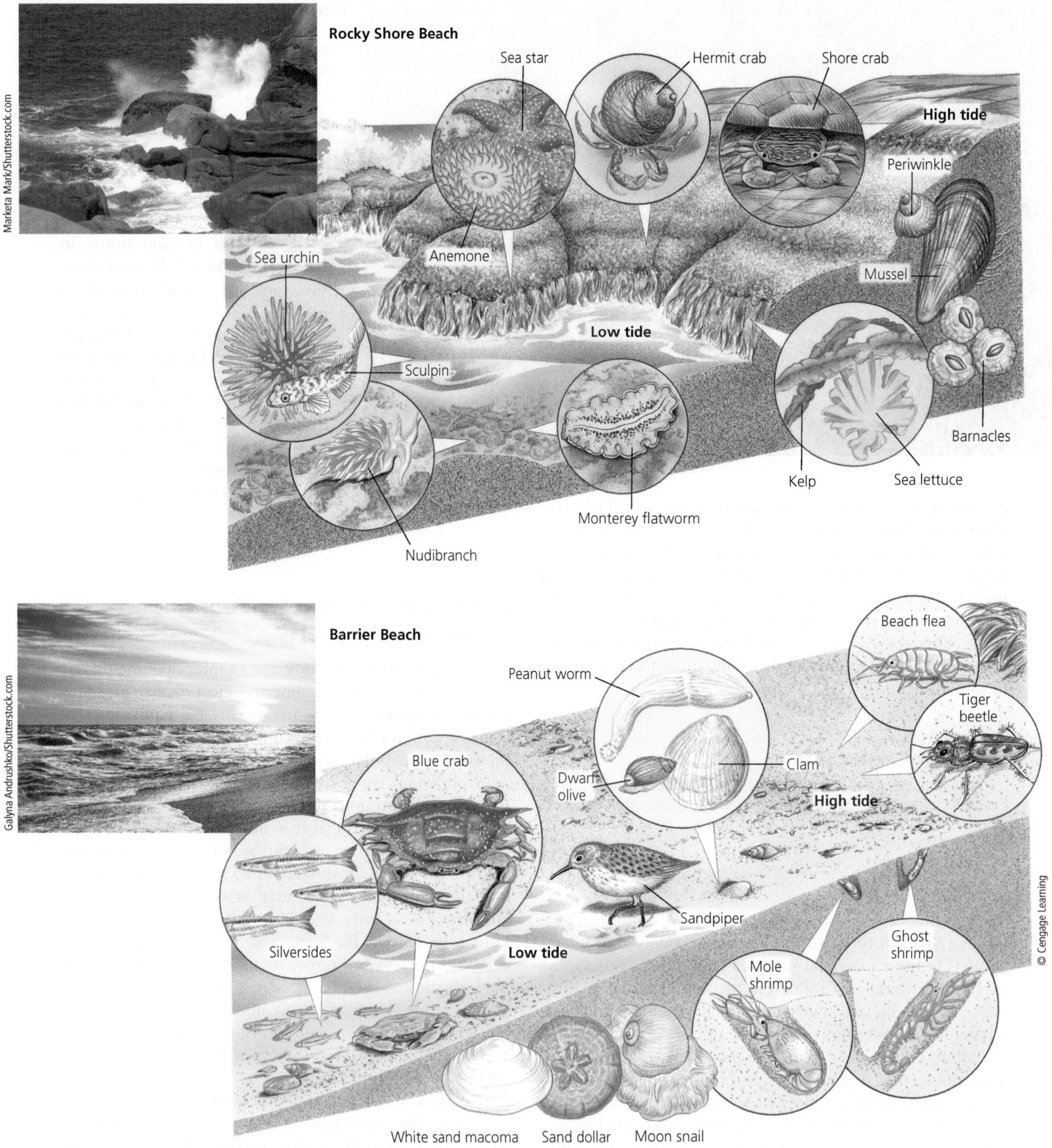

FIGURE 8.10 *Living between the tides:* Some organisms with specialized niches are found in various zones on rocky shore beaches (top) and barrier or sandy beaches (bottom). Organisms are not drawn to scale.

In 2013 Richard Vevers and teams of marine scientists launched the Catlin Seaview Survey and the Underwater Earth Project. These researchers are using sophisticated underwater cameras to create three-dimensional digital images of the world's major coral reefs. These projects aim to provide a baseline of the health of the world's coral reefs and to identify areas that need emergency protection to keep the reefs from dying (**Core Case Study**).

The Open Sea and the Ocean Floor Host a Variety of Species

The sharp increase in water depth at the edge of the continental shelf separates the coastal zone from the vast volume of the ocean called the **open sea.** This aquatic life zone is divided into three *vertical zones* (Figure 8.5), or layers, primarily based on the degree of penetration of sunlight. Temperatures also change with depth (Figure 8.5, red line) and scientists use them to define zones of varying species diversity in these layers.

The *euphotic zone* is the brightly lit upper zone, where drifting phytoplankton carry out about 40% of the world's photosynthetic activity. Large, fast-swimming predatory fishes such as swordfish, sharks, and bluefin tuna populate the euphotic zone.

Nutrient levels are low and levels of dissolved oxygen are high in the euphotic zone. The exception to this is those areas called *upwelling zones*. An **upwelling,** or upward movement of ocean water, brings cool and nutrient-rich water from the bottom of the ocean to the warmer surface. There it supports large populations of phytoplankton, zooplankton, fish, and fish-eating seabirds. Strong upwellings occur along the steep western coasts of some continents when winds blowing along the coasts push surface water away from the land. This draws water up from the ocean bottom (Figure 8.11 and Figure 7.3, p. 142). Figure 7.7 (p. 145) shows the oceans' major upwelling zones.

The second major open sea zone is the *bathyal zone*, the dimly lit middle zone that receives little sunlight and therefore does not contain photosynthesizing producers. Zooplankton and smaller fishes, many of which migrate to feed on the surface at night, are found in these waters.

The deepest open sea zone, called the *abyssal zone*, is dark and cold. There is no sunlight to support photosynthesis, and this water has little dissolved oxygen. Nevertheless, the deep ocean floor is teeming with life because it contains enough nutrients to support a large number of species. Most of this zone's organisms get their food from showers of dead and decaying organisms—called *marine snow*—drifting down from the upper zones. Some abyssal-zone organisms, including many types of worms, are *deposit feeders*, which take mud into their guts and extract nutrients from it. Others such as oysters, clams, and sponges are *filter feeders*, which pass water through or over their bodies and extract nutrients from it.

Net primary productivity (NPP) is quite low in the open sea, except in upwelling areas. However, because the open sea covers so much of the earth's surface, it makes the largest contribution to the earth's overall NPP. In fact, scientists have learned that the open sea contains more biodiversity than they thought a few years ago (Science Focus 8.1).

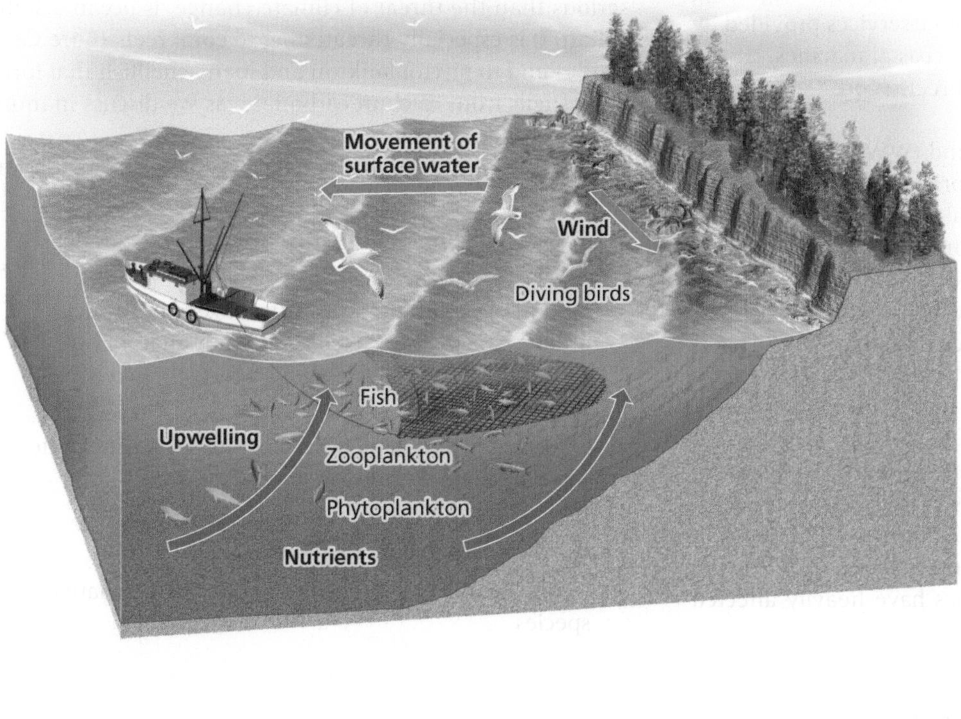

FIGURE 8.11 A shore upwelling occurs when deep, cool, nutrient-rich waters are drawn up to replace surface water moved away from a steep coast by wind flowing along the coast toward the equator.

We Are Still Learning about the Ocean's Biodiversity

Scientists have long assumed that open-ocean waters contained few microbial life forms. But recent research has challenged that assumption and greatly increased our knowledge of the ocean's genetic diversity.

A team of scientists led by J. Craig Venter took 2 years to conduct a census (an estimated count based on sampling) of ocean microbes. They sailed around the world, stopping every 320 kilometers (200 miles) to pump seawater through extremely fine filters, from which they gathered data on bacteria, viruses, and other microbes. It was the most thorough of such censuses ever conducted.

Using a supercomputer, they counted genetic coding for 6 million new proteins—double the number that had previously been known. They also reported that they were discovering new genes and proteins at the same rate at the end of their voyage as they had at the start of it. This indicated that there is still much more of this biodiversity to discover.

This means that the ocean contains a much higher diversity of microbial life than had previously been thought. Ocean-water microbes play an important role in the absorption of carbon by the ocean, as well as in the ocean food web. Venter has led more expeditions to continue the sampling in other areas.

CRITICAL THINKING

Why was the *rate* of discovery of new genes and proteins important to Venter and his colleagues? Explain.

8.3 HOW HAVE HUMAN ACTIVITIES AFFECTED MARINE ECOSYSTEMS?

CONCEPT 8.3 Human activities threaten aquatic biodiversity and disrupt ecosystem and economic services provided by marine ecosystems.

Human Activities Are Disrupting and Degrading Marine Ecosystems

Certain human activities are disrupting and degrading many of the ecosystem and economic services provided by marine aquatic systems, especially coastal marshes, shorelines, mangrove forests, and coral reefs (**Core Case Study**) (**Concept 8.3**).

In 2015 the World Wildlife Fund (WWF) reported that more than 35% of the world's original mangrove forest area had been lost to agricultural and urban expansion, marinas, roadways, and other forms of coastal development. According to the International Union for the Conservation of Nature (IUCN), more than one of every six species of mangrove are in danger of extinction. In addition, since 1980 about 29% of the world's seagrass beds have been lost to pollution and other disturbances.

The U.S. National Center for Ecological Analysis and Synthesis (NCEAS) used computer models to analyze and provide the first-ever comprehensive map of the effects of 17 different types of human activities on the world's oceans. In this 4-year study, an international team of scientists found that human activities have heavily affected 41% of the world's ocean area.

Harmful human activities increase with the number of people living on or near coasts. In 2014 about 45% of the world's population and more than half of the U.S. population lived along or near coasts, and these percentages are rising.

The biggest threat to marine systems, according to many marine scientists, is climate change (which we explore fully in Chapter 19). Because land-based glaciers in Greenland and other parts of the world are slowly melting, sea levels are rising. The rise in sea levels projected for this century would destroy many shallow coral reefs and flood coastal marshes and other coastal ecosystems, as well as many coastal cities.

A second threat, which some scientists view as more serious than the threat of climate change, is ocean acidification. It is especially threatening to coral reefs (**Core Case Study**) and to phytoplankton and many shellfish that form their shells from calcium carbonate, as we discuss in more detail in Chapter 11.

Other major threats to marine systems from human activities include the following:

- Coastal development, which destroys or degrades coastal habitats
- Runoff of pollutants such as fertilizers, pesticides, and livestock wastes (see Case Study that follows) and pollution from cruise ships and oil tanker spills
- Overfishing and depletion of commercial fish species populations
- Destruction of ocean bottom habitats by fishing trawlers dragging weighted nets
- Invasive species that deplete populations of native species

Natural Capital Degradation

Major Human Impacts on Marine Ecosystems and Coral Reefs

Marine Ecosystems

Coral Reefs

Half of coastal wetlands lost to agriculture and urban development

Over one-fifth of mangrove forests lost to agriculture, aquaculture, and development

Beaches eroding due to development and rising sea levels

Ocean-bottom habitats degraded by dredging and trawler fishing

At least 20% of coral reefs severely damaged and 25–33% more threatened

Ocean warming

Rising ocean acidity

Rising sea levels

Soil erosion

Algae growth from fertilizer runoff

Bleaching

Increased UV exposure

Damage from anchors and from fishing and diving

© Cengage Learning

FIGURE 8.12 Human activities have major harmful impacts on all marine ecosystems (left) and particularly on coral reefs (right) (**Concept 8.3**). *Critical thinking:* Which two of the threats to marine ecosystems do you think are the most serious? Why? Which two of the threats to coral reefs do you think are the most serious? Why?

Top left: Jorg Hackemann/Shutterstock.com. Top right: Rich Carey/Shutterstock.com. Bottom left: Piotr Marcinski/Shutterstock.com. Bottom right: Rostislav Ageev/Shutterstock.com.

CASE STUDY

The Chesapeake Bay—An Estuary in Trouble

Since 1960, the Chesapeake Bay (Figure 8.13 inset map on right)—the largest estuary in the United States—has been in trouble from water pollution, mostly because of human activities. One problem is population growth. Between 1940 and 2014, the number of people living in the Chesapeake Bay area grew from 3.7 million to 17.9 million, and is projected to reach 20 million by 2030, according to estimates by the Chesapeake Bay Program.

The estuary receives wastes from point and nonpoint sources scattered throughout its huge drainage basin, which lies in parts of six states and the District of Columbia (Figure 8.13). The shallow bay has become a huge pollution sink because only 1% of the waste entering it is flushed into the Atlantic Ocean. Phosphate and nitrate levels have been high in many parts of the bay, causing algal blooms that deplete the oxygen dissolved in the waters, making them unsuitable for most forms of aquatic life. Commercial harvests of the bay's once-abundant oysters and crabs, as well as several important fish species, fell sharply after the 1960s because of a combination of pollution, overfishing, and disease.

Point sources, primarily sewage treatment plants and industrial plants, add large amounts of phosphates to the bay. However, nonpoint sources—mostly runoff of fertilizer and animal wastes from agricultural land—accounted for 58% of the phosphates, 42% of the nitrates, and 58% of sediments entering the bay in 2012. The bay also receives inputs of pesticides from direct spraying over water and as runoff from nearby croplands. And discharges from sewage treatment plants into the bay often contain pharmaceutical chemicals that have entered the sewage treatment system after use by humans. Most treatment systems do not remove many of these chemicals, some of which can harm aquatic species. Finally, runoff of sediment,

Figure 8.12 shows some of the effects of these human impacts on marine systems in general (left) and on coral reefs in particular (right). Scientists are working to learn more about the little-understood marine ecosystems, our effects on them, and the ways in which we can seek to preserve them (Individuals Matter 8.1). We examine human impacts on aquatic life zones more closely in Chapters 11 and 19.

CONSIDER THIS . . .

THINKING ABOUT Coral Reef Destruction

How might the loss of most of the world's remaining tropical coral reefs (**Core Case Study**) affect your life and the lives of any children or grandchildren you might have? What are two things you could do to help reduce this loss?

INDIVIDUALS MATTER 8.1

Enric Sala: Working to Protect Ocean Ecosystems

MAURICIO HANDLER

Marine biologist Enric Sala has made a career of working to protect undisturbed marine ecosystems. He travels to remote areas with the goal of learning what marine ecosystems were like before human activities disrupted them. In 2008, he launched National Geographic's Pristine Seas project to find, survey, and help protect the last wild places in the ocean. Pristine Seas aims for an ocean where representative examples of all major ecosystems are protected, so that they can be healthier, more productive, and more resilient to the impacts of ocean warming and acidification.

After exploring the Southern Line Islands undisturbed coral reef system in the South Pacific, Sala reported that "all the scientific data confirm that humans are the most important factor in determining the health of coral reefs." He says that reefs are killed by "a combination of the local impact of human activities such as fishing and pollution with the global impact of human-induced climate change."

Sala suggests that in order to allow coral reefs and other systems to survive and function, we need to "take out less and throw in less." One way to accomplish this is to establish large areas of protected ocean habitat, free of human activities of any kind. Sala was instrumental in establishing such *marine protected areas*. To date, Pristine Seas has helped to create nine of the largest marine reserves on the planet, covering an area of over 3 million square km—more than six times the area of Spain. Over the next three years, Sala's team will target 12 more locations, aiming to inspire the protection of a total of 20 of the wildest places in the ocean. For his outstanding scientific work, Sala has been named a National Geographic Explorer.

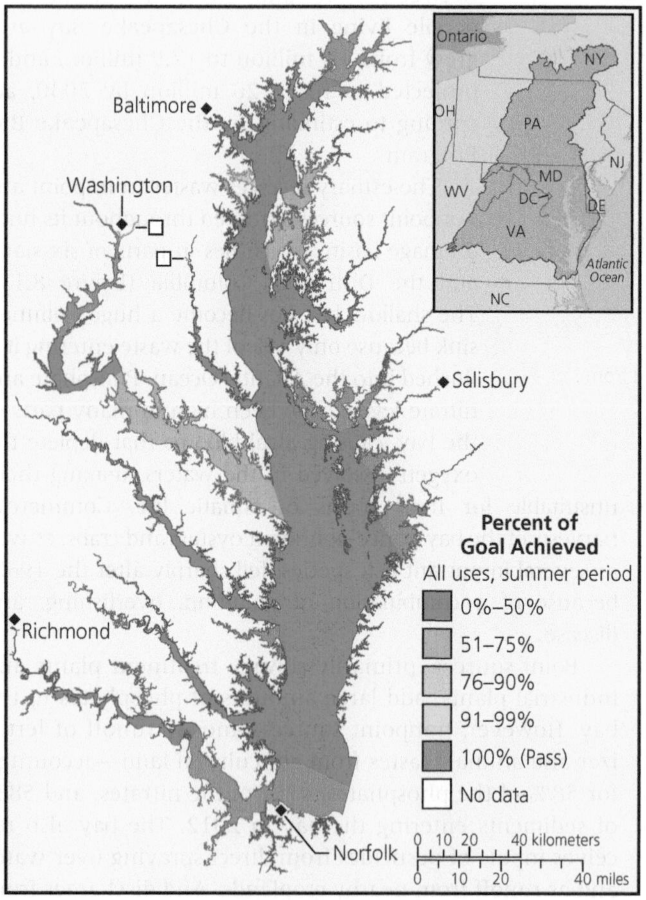

FIGURE 8.13 The Chesapeake Bay has established goals for restoration of the bay waters as measured by dissolved oxygen (DO) content. As of 2012 (latest available data), the goals were achieved in parts of the estuary but not in others, as reflected on this map.

(Compiled by the authors using data from Chesapeake Bay Program.)

mostly from soil erosion, harms the bay's seagrasses on which crabs and young fish depend.

A century ago, oysters were so abundant that they filtered and cleaned the Chesapeake's entire volume of water every 3 days. This important form of natural capital provided by these *keystone species* (see p. 83) helped reduce excess nutrients and algal blooms. Now the oyster population has been reduced to the point where this filtration process takes a year.

In 1983 the United States implemented the Chesapeake Bay Program. In this ambitious attempt at *integrated coastal management,* citizens' groups, communities, and state and federal governments worked together to reduce pollution in the bay. One strategy was to set land-use regulations to reduce agricultural and urban runoff in the bay's drainage area. Other strategies included banning phosphate detergents, upgrading sewage treatment plants, and monitoring industrial discharges more closely. Watershed management plans have been established and sensitive land areas have been protected through donations or purchases by conservation organizations. Some adjoining wetlands have been restored and large areas of the bay were replanted with seagrasses to help filter out excessive nutrients and other pollutants.

In 2012 the Chesapeake Bay Program had met its goals in some areas of the bay, but not in others (Figure 8.13, main map). Results have been mixed, mostly because of a growing population, continued development, and a drop in state and federal funding.

However, progress has been made with the help of a U.S. Environmental Protection Agency (EPA) plan that puts limits on the amount of agricultural nutrients entering the bay from the six Chesapeake Bay states and the

District of Columbia. In 2011 a team of scientists led by Rebecca R. Murphy reported encouraging news. After analyzing 60 years of water-quality data, the scientists found that the sizes of oxygen-depleted zones that had been growing every summer in the bay had leveled off in the 1980s and had been declining since then. This meant that efforts to reduce inputs of pollutants were having a positive effect on the bay.

Also, crab populations in the Chesapeake Bay have rebounded due to a set of measures put in place in 2008 by the states of Maryland and Virginia. While the bay's blue crab population was on the verge of collapse in 2003, it has now grown to sustainable levels. In 2015 Chesapeake Bay Program researchers announced that the population of female blue crabs was no longer considered to be depleted.

In 2014, however, this progress and future progress was threatened by a lawsuit filed by the attorneys general of 21 states, most of them not in the Chesapeake Bay region, to strike down the EPA's Chesapeake cleanup plan. The lawsuit contends that the plan represents an overreach of power by the EPA, violates the language of the Clean Water Act, and raises serious concerns about states' rights. The plaintiffs fear that it could lead to similar water cleanup regulations in watersheds in other major farm states such as Mississippi and Kansas. This suit—backed by the American Farm Bureau, a powerful agricultural interest group that has filed but lost similar lawsuits against the EPA—alleges that the EPA water pollution regulations will hinder agricultural production in the affected farm states.

William Baker, President of the Chesapeake Bay Foundation, calls the suit by outsider states "absolutely bizarre." He calls the cleanup program "one of the great examples of people working together across federal, state, local business, and agriculture, and it's working."

8.4 WHY ARE FRESHWATER ECOSYSTEMS IMPORTANT?

CONCEPT 8.4 Freshwater ecosystems provide major ecosystem and economic services and are irreplaceable reservoirs of biodiversity.

Water Stands in Some Freshwater Systems and Flows in Others

Precipitation that does not sink into the ground or evaporate becomes **surface water**—freshwater that flows or is stored in bodies of water on the earth's surface. *Freshwater aquatic life zones* include *standing (lentic)* bodies of freshwater such as lakes, ponds, and inland wetlands, and *flowing (lotic)* systems such as streams and rivers.

Surface water that flows into such bodies of water is called **runoff.** A **watershed,** or **drainage basin,** is the land area that delivers runoff, sediment, and dissolved substances to a stream, lake, or wetland. Although freshwater systems cover less than 2.2% of the earth's surface, they provide a number of important ecosystem and economic services (Figure 8.14).

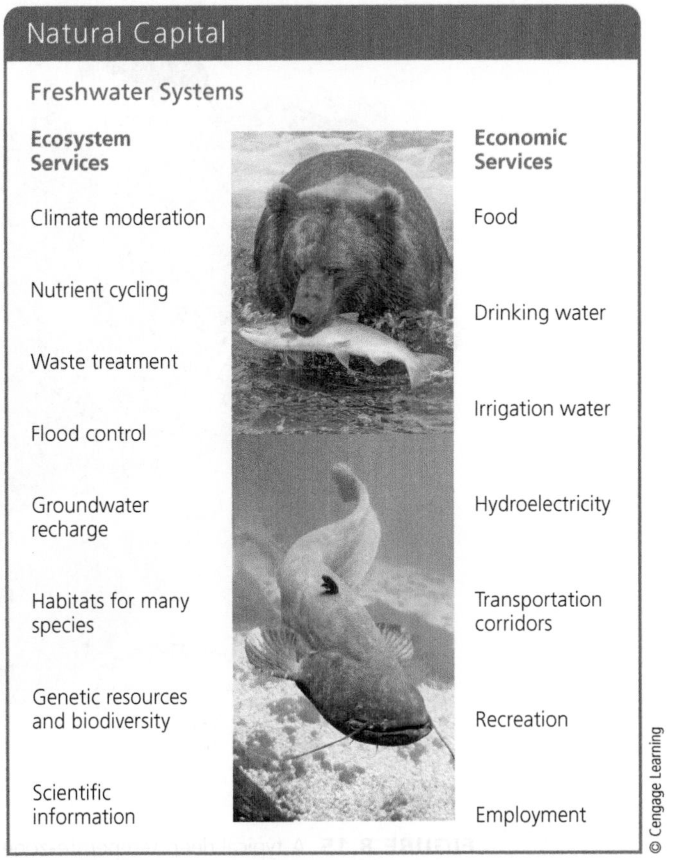

Natural Capital
Freshwater Systems
Ecosystem Services
Climate moderation
Nutrient cycling
Waste treatment
Flood control
Groundwater recharge
Habitats for many species
Genetic resources and biodiversity
Scientific information
Economic Services
Food
Drinking water
Irrigation water
Hydroelectricity
Transportation corridors
Recreation
Employment
© Cengage Learning

FIGURE 8.14 Freshwater systems provide many important ecosystem and economic services (**Concept 8.4**). ***Critical thinking:*** Which two ecosystem services and which two economic services do you think are the most important? Why?

Top: Galyna Andrushko/Shutterstock.com. Bottom: Kletr/Shutterstock.com.

Lakes are large natural bodies of standing freshwater formed when precipitation, runoff, streams, rivers, and groundwater seepage fill depressions in the earth's surface. Causes of such depressions include glaciation, displacement of the earth's crust, and volcanic activity. A lake's watershed supplies it with water from rainfall, melting snow, and streams.

Freshwater lakes vary in size, depth, and nutrient content. Deep lakes normally consist of four distinct life zones that are defined by their depth and distance from shore (Figure 8.15). The top layer, called the *littoral zone,* is near the shore and consists of the shallow sunlit waters to the depth at which rooted plants stop growing. It has a high level of biodiversity because of ample sunlight and inputs of nutrients from the surrounding land. Species living in the littoral zone include many rooted plants; animals such as turtles, frogs, and crayfish; and fish such as bass, perch, and carp.

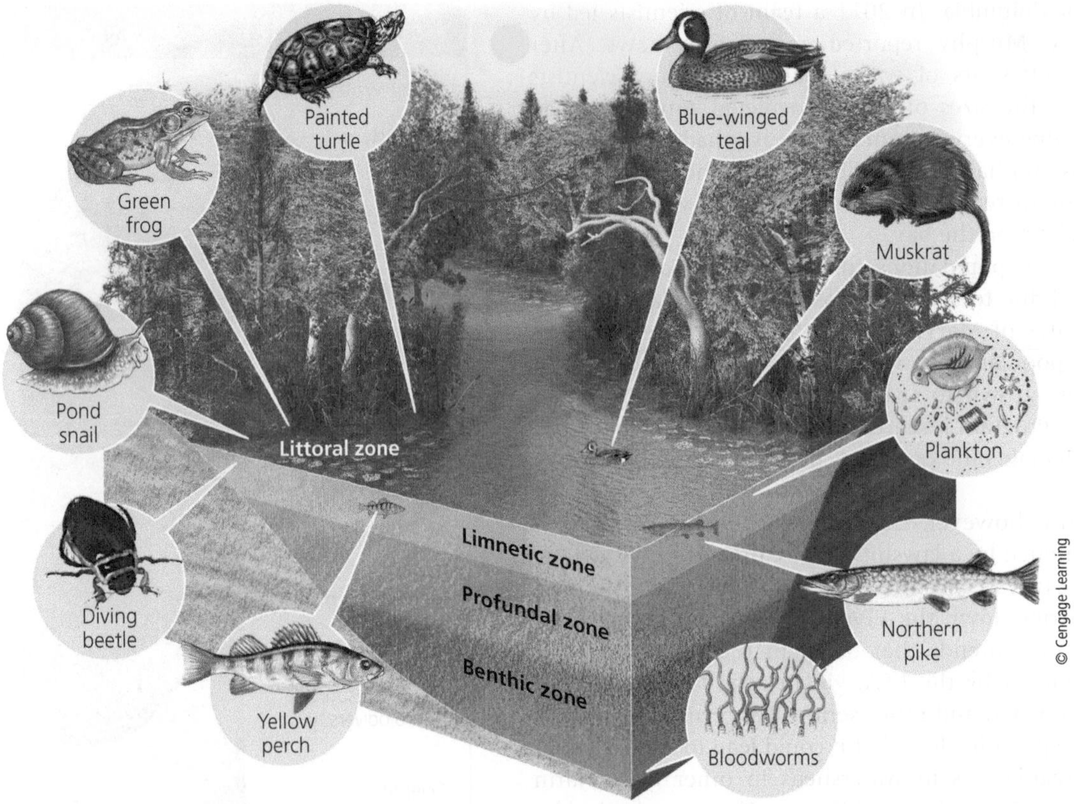

FIGURE 8.15 A typical deep temperate-zone lake has distinct zones of life. ***Critical thinking:*** How are deep lakes similar to tropical rain forests? (*Hint:* See Figure 7.20, p. 158.)

The next layer is the *limnetic zone*, the open, sunlit surface layer away from the shore that extends to the depth penetrated by sunlight. This is the main photosynthetic zone of the lake, the layer that produces the food and oxygen that support most of the lake's consumers. Its most abundant organisms are phytoplankton and zooplankton. Some large species of fish spend most of their time in this zone, with occasional visits to the littoral zone to feed and reproduce.

The *profundal zone* is the volume of deeper water lying between the limnetic zone and the lake bottom. It is too dark for photosynthesis. Without sunlight and plants, oxygen levels are often low. Fishes adapted to the lake's cooler and darker water, such as perch, are found in this zone.

The bottom of the lake is called the *benthic zone*, inhabited mostly by decomposers, detritus feeders, and some bottom-feeding species of fish such as catfish. The benthic zone is nourished mainly by dead matter that falls from the littoral and limnetic zones and by sediment washing into the lake.

Some Lakes Have More Nutrients than Others

Ecologists classify lakes according to their nutrient content and primary productivity. Lakes that have a small supply of plant nutrients are called **oligotrophic lakes.** This type of lake (Figure 8.16) is often deep and can have steep banks. Glaciers and mountain streams supply water to many of these lakes, which usually have crystal-clear water and small populations of phytoplankton and fish species, such as smallmouth bass and trout. Because of their low levels of nutrients, these lakes have a low net primary productivity (NPP).

Over time, sediments, organic material, and inorganic nutrients wash into most oligotrophic lakes, and plants grow and decompose to form bottom sediments. A lake with a large supply of nutrients is called a **eutrophic lake** (Figure 8.17). Such lakes typically are shallow and have murky brown or green water. Because of their high levels of nutrients, these lakes have a high NPP. Most lakes fall somewhere between the two extremes of nutrient enrichment.

Human inputs of nutrients through the atmosphere and from urban and agricultural areas within a lake's watershed can accelerate the eutrophication of the lake. This process, called **cultural eutrophication,** often puts excessive nutrients into lakes.

Freshwater Streams and Rivers Carry Large Volumes of Water

In drainage basins, water accumulates in small streams that join to form rivers. Collectively, streams and rivers

tusharkoley/Shutterstock.com

FIGURE 8.16 Trillium Lake in the U.S. state of Oregon with a view of Mount Hood.

carry huge amounts of water from highlands to lakes and oceans. They drain an estimated 75% of the earth's land surface. Many streams begin in mountainous or hilly areas, which collect and release water falling to the earth's surface as precipitation. The downward flow of surface water and groundwater from mountain highlands to the sea typically takes place in three aquatic life zones characterized by different environmental conditions: the *source zone*, the *transition zone*, and the *floodplain zone* (Figure 8.18). Rivers and streams can differ from this generalized model.

In the narrow *source zone* (Figure 8.18, left), headwater streams are usually shallow, cold, clear, and swiftly flowing (Figure 8.18, inset photo). As this water tumbles over rocks, waterfalls, and rapids, it dissolves large amounts of oxygen from the air. Most of these streams are not very productive because of a lack of nutrients and primary producers. Their nutrients come primarily from organic matter, mostly leaves, branches, and the bodies of living and dead insects that fall into the stream from nearby land.

The source zone is populated by cold-water fish species that need lots of dissolved oxygen. Fishes in this habitat, such as trout and minnows, tend to have streamlined and muscular bodies that allow them to swim in the rapid, strong currents. Other animals such as riffle beetles have compact, hard, or flattened bodies that allow them to live among or under stones in fast-flowing headwater streams. Most of the plants in this zone are algae and mosses attached to rocks and other surfaces underwater.

In the *transition zone* (Figure 8.18, center), headwater streams merge to form wider, deeper, and warmer streams that flow down gentler slopes with fewer obstacles. They can be more turbid (containing suspended sediments) and slower flowing than headwater streams, and they

Nicholas Rjabow/Dreamstime.com

FIGURE 8.17 This eutrophic lake has received large flows of plant nutrients. As a result, its surface is covered with mats of algae.

tend to have less dissolved oxygen. The warmer water and other conditions in this zone support more producers, as well as cool-water and warm-water fish species (such as black bass) with slightly lower oxygen requirements.

As streams flow downhill, they shape the land through which they pass. Over millions of years, the friction of moving water has leveled mountains and cut deep canyons. Sand, gravel, and soil carried by streams and rivers have been deposited as sediment in low-lying areas. In these *floodplain zones* (Figure 8.18, right), streams join into wider and deeper rivers that flow across broad, flat valleys. Water in this zone usually has higher temperatures and less dissolved oxygen than water in the two higher zones. The slow-moving rivers sometimes support large populations of producers such as algae and cyanobacteria, as well as rooted aquatic plants along the shores.

Because of increased erosion and runoff over a larger area, water in the floodplain zone often is muddy and contains high concentrations of silt. These murky waters support distinctive varieties of fishes, including carp and catfish. At its mouth, a river may divide into many channels as it flows through its **delta**—an area at the mouth of a river built up by deposited sediment, usually containing coastal wetlands and estuaries (Figure 8.7).

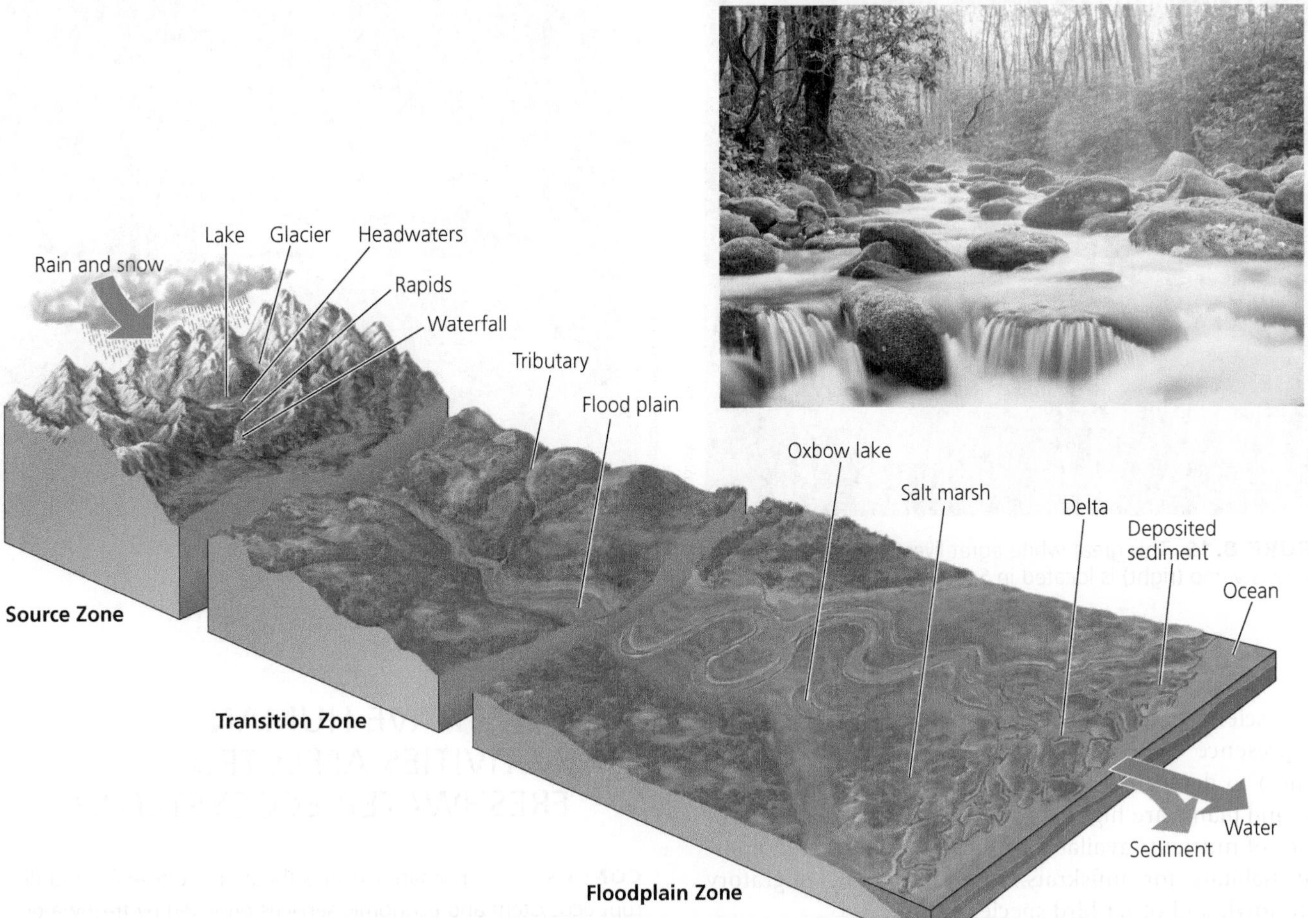

kurdistan/Shutterstock.com

© Cengage Learning

FIGURE 8.18 There are three zones in the downhill flow of water: the *source zone*, which contains *headwater* streams found in highlands and mountains (see inset photo); the *transition zone*, which contains wider, lower-elevation streams; and the *floodplain zone*, which contains rivers that empty into larger rivers or into the ocean. ***Critical thinking:*** How might the building of many dams and reservoirs along a river's path to the ocean affect its sediment input into the ocean and change the river's delta?

Coastal deltas and wetlands, as well as inland wetlands and floodplains, are important parts of the earth's natural capital (see Figure 1.3, p. 7). They absorb and slow the velocity of floodwaters from coastal storms, hurricanes, and tsunamis and provide habitats for a diversity of marine life.

CONSIDER THIS . . .

CONNECTIONS Stream Water Quality and Watershed Land

Streams receive most of their nutrients from bordering land ecosystems. Such nutrients come from falling leaves, animal feces, insects, and other forms of biomass washed into streams during heavy rainstorms or by melting snow. Chemicals and other substances flowing off the land can also pollute streams. Thus, the levels and types of nutrients and pollutants in a stream depend on what is happening in the stream's watershed.

Freshwater Inland Wetlands Are Vital Sponges

Inland wetlands are lands located away from coastal areas that are covered with freshwater all or part of the time—excluding lakes, reservoirs, and streams. They include *marshes* (Figure 8.19, left), *swamps* (Figure 8.19, right), and *prairie potholes* (depressions carved out by ancient glaciers). Other examples are *floodplains*, which receive excess water from streams or rivers during heavy rains and floods.

Some wetlands are covered with water year-round. Others, called *seasonal wetlands*, remain under water or are soggy for only a short time each year. The latter include prairie potholes, floodplain wetlands, and arctic tundra (see Figure 7.16, bottom, p. 153). Some can stay dry for years before water covers them again. In such

FloridaStock/Shutterstock.com

Jayne Chapman/Shutterstock.com

FIGURE 8.19 This great white egret lives in an inland marsh in the Florida Everglades (left). This cypress swamp (right) is located in South Carolina.

cases, scientists must use the composition of the soil or the presence of certain plants (such as cattails and bulrushes) to determine that a particular area is a wetland. Wetland plants are highly productive because of an abundance of nutrients available to them. Wetlands are important habitats for muskrats, otters, beavers, migratory waterfowl, and other bird species.

Inland wetlands provide a number of free ecosystem and economic services. They take part in:

- filtering and degrading toxic wastes and pollutants;
- reducing flooding and erosion by absorbing storm water and releasing it slowly, and by absorbing overflows from streams and lakes;
- sustaining stream flows during dry periods;
- recharging groundwater aquifers;
- maintaining biodiversity by providing habitats for a variety of species;
- supplying valuable products such as fishes and shellfish, blueberries, cranberries, and wild rice; and
- providing recreation for birdwatchers, nature photographers, boaters, anglers, and waterfowl hunters.

CONSIDER THIS . . .

THINKING ABOUT Inland Wetlands

Of the ecosystem and economic services listed above, which two do you think are the most important? Why? List two ways in which our daily activities directly or indirectly degrade inland wetlands.

8.5 HOW HAVE HUMAN ACTIVITIES AFFECTED FRESHWATER ECOSYSTEMS?

CONCEPT 8.5 Human activities threaten biodiversity and disrupt ecosystem and economic services provided by freshwater lakes, rivers, and wetlands.

Human Activities Are Disrupting and Degrading Freshwater Systems

Human activities are disrupting and degrading many of the ecosystem and economic services provided by freshwater rivers, lakes, and wetlands (**Concept 8.5**) in four major ways. *First,* dams and canals restrict the flows of about 40% of the world's 237 largest rivers. This alters or destroys terrestrial and aquatic wildlife habitats along these rivers and in their coastal deltas and estuaries. By reducing the flow of sediments to river deltas, these structures also lead to degraded coastal wetlands and greater damage from coastal storms (see Case Study that follows).

Second, flood control levees and dikes built along rivers disconnect the rivers from their floodplains, destroy aquatic habitats, and alter or degrade the functions of adjoining wetlands.

Third, cities and farms add pollutants and excess plant nutrients to nearby streams, rivers, and lakes. For example, runoff of nutrients into a lake (Figure 8.17) causes explosions in the populations of algae and cyanobacteria, which deplete the lake's dissolved oxygen. When these organisms die and sink to the lake bottom, decomposers go to work and further deplete the oxygen in deeper waters.

Fishes and other species may then die off, which can mean a major loss in biodiversity.

Fourth, many inland wetlands have been drained or filled to grow crops or have been covered with concrete, asphalt, and buildings. More than half of the inland wetlands estimated to have existed in the continental United States during the 1600s no longer exist. About 80% of these lost wetlands were drained to grow crops. The rest were lost to mining, logging, oil and gas extraction, highway construction, and urban development. The heavily farmed U.S. state of Iowa has lost about 99% of its original inland wetlands.

This loss of natural capital has been an important factor in increasing flood damage in parts of the United States. Many other countries have suffered similar losses. For example, 80% of all inland wetlands in Germany and France have been destroyed.

When we look further into human impacts on aquatic systems in Chapter 11, we will also explore possible solutions to environmental problems that result from these impacts, as well as ways to help sustain aquatic biodiversity. This area of study will offer great opportunities to young scientists and other professionals in the years to come. (See Individuals Matter 8.2.)

CASE STUDY

River Deltas and Coastal Wetlands—Vital Components of Natural Capital Now in Jeopardy

Coastal river deltas, mangrove forests, and coastal wetlands provide considerable natural protection against flood and wave damage from coastal storms, hurricanes, typhoons, and tsunamis. They weaken the force of waves and absorb excess storm water like sponges.

When we remove or degrade these ecosystems, any damage from a natural disaster such as a hurricane is intensified. As a result, flooding in places such as New Orleans, Louisiana (USA), and Venice, Italy, is a largely self-inflicted *unnatural* disaster. For example, Louisiana, which contains about 40% of all coastal wetlands in the lower 48 states, has lost more than a fifth of such wetlands since 1950 to oil and gas wells and other forms of coastal development.

We have built dams, levees, and hydroelectric power plants on many of the world's rivers to control water flows and to generate electricity. This helps to reduce flooding along rivers, but it also reduces flood protection provided by the coastal deltas and wetlands. Because river sediments are deposited in the reservoirs behind dams, the river deltas do not get their normal inputs of sediment to build them back up, and they *subside*, or sink into the sea.

As a result, 24 of the world's 33 major river deltas are sinking rather than rising, according to a study led by geologist James Syvitski. The study found that 85% of the world's sinking deltas have experienced severe flooding in recent years, and that global delta flooding is likely to increase by 50% by the end of this century. This is because of dams and other human-made structures that reduce the flow of silt. It is also due partly to the projected rise in sea levels resulting from climate change. This poses a serious threat to the roughly 500 million people in the world who live on river deltas.

For example, the Mississippi River once delivered huge amounts of sediments to its delta each year. But the multiple dams, levees, and canals built in this river system funnel much of this sediment load through the wetlands and out into the Gulf of Mexico. Instead of building up delta lands, this causes them to subside. As many of the river delta's freshwater wetlands have been lost to this subsidence, saltwater from the Gulf has intruded and killed many plants that depended on the river water, further degrading this coastal aquatic system.

Subsidence helps to explain why the city of New Orleans, Louisiana (Figure 8.20), has long been 3 meters (10 feet) below sea level. Dams and levees were built to help protect the city from flooding. However, in 2005 the

National Oceanic and Atmospheric Administration (NOAA)

FIGURE 8.20 Much of the U.S. city of New Orleans, Louisiana, was flooded by the storm surge that accompanied Hurricane Katrina, which made landfall just east of the city on August 29, 2005.

INDIVIDUALS MATTER 8.2

Alexandra Cousteau: Environmental Advocate, Filmmaker, and National Geographic Explorer

© Bill Zelman

Alexandra Cousteau is proud of her heritage as granddaughter of Captain Jacques-Yves Cousteau and daughter of Philippe Cousteau. Her father and grandfather were legendary underwater explorers who brought the mysteries and wonders of the oceans into living rooms around the world with their films and books.

The focus of Cousteau's work is advocating the importance of conservation and sustainable management of water in order to preserve a healthy planet. She seeks to make water one of the defining issues of this century, stating: "We live on a water planet, which means we're all downstream from one another. Where water comes from, where it goes, and its quality is intricately connected to our quality of life."

Cousteau is utilizing tools not even imagined by her grandfather—those of social networking and other modes of mobile communication. She believes that environmental advocates can use such new media tools and technology to inform people about how their actions affect our water. For example, she imagines a day in the future when knowing the quality and quantity of our water is as easy as checking the weather on our smart phones.

Cousteau's nonprofit Blue Legacy International harnesses technology to tell the stories of our water planet and provides film and digital resources to allow others to explore and understand water issues.

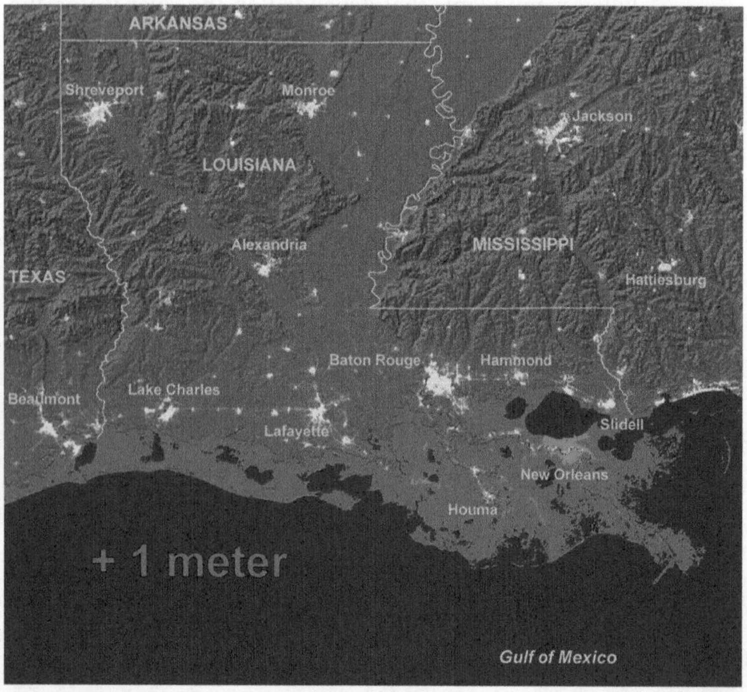

Used by permission from Jonathan Overpeck and Jeremy Weiss, University of Arizona

FIGURE 8.21 The areas in red represent projected coastal flooding that would result from a 1-meter (3-foot) rise in sea level due to projected climate change by the end of this century.

powerful winds and waves from Hurricane Katrina overwhelmed these defenses. They have been rebuilt, but subsidence will put New Orleans further below sea level in the future. Add to this the reduced protection from degraded coastal wetlands, and you have a recipe for a major and possibly more damaging disaster if the area is struck by another major hurricane.

To make matters worse, global sea levels have risen almost 0.3 meters (1 foot) since 1900 and are projected to rise another 0.3–0.9 meter (1–3 feet) by the end of this century. This is because climate change is warming the ocean, causing its waters to expand. Also, the melting of glaciers and other land-based ice is adding to the ocean's volume. Such a rise in sea level would put many of the world's coastal areas, including New Orleans and most of Louisiana's present-day coast, under water (Figure 8.21).

BIG IDEAS

- Saltwater and freshwater aquatic life zones cover almost three-fourths of the earth's surface, and oceans dominate the planet.
- The earth's aquatic systems provide important ecosystem and economic services.
- Certain human activities threaten biodiversity and disrupt ecosystem and economic services provided by aquatic systems.

Tying It All Together

Coral Reefs and Sustainability

This chapter's Core Case Study pointed out the ecological and economic importance of the world's incredibly diverse coral reefs. They are living examples of the three **scientific principles of sustainability** in action. They thrive on solar energy, play key roles in the cycling of carbon and other nutrients, and sustain a great deal of aquatic biodiversity.

In this chapter, we have also seen that coral reefs and other aquatic systems are being severely stressed by a variety of human activities. Research shows that when such harmful human activities are reduced, some coral reefs and other endangered aquatic systems can recover quickly.

As with terrestrial systems, scientists have made a start in understanding the ecology of the world's aquatic systems and how humans are degrading and disrupting the ecosystem and economic services they provide. However, we still know far too little about these vital parts of the earth's life-support system. Scientists argue that we urgently need more research on the components of aquatic life zones, on how they are interconnected, and on which systems are in the greatest danger of being disrupted.

We can take the lessons on how life in aquatic ecosystems has sustained itself for billions of years and use these **scientific principles of sustainability** to help sustain our own systems and the ecosystems on which we depend. By relying more on solar energy and less on fossil fuels, we could drastically cut pollution of aquatic systems and CO_2 emissions that are causing ocean warming and ocean acidification. By reusing and recycling more of the materials and chemicals we use, we could further reduce pollution and disruption of the chemical cycling within aquatic systems. And by learning more about aquatic biodiversity and its importance, we could go a long way toward preserving it and sustaining its valuable ecosystem services.

JonMilnes/Shutterstock.com

Chapter Review

Core Case Study

1. What are **coral reefs** and why should we care about them? What is coral bleaching? What are the major threats to coral reefs?

Section 8.1

2. What are the two key concepts for this section? What percentage of the earth's surface is covered with water? What is an **aquatic life zone**? What is **salinity** and how is it measured? Distinguish between a **saltwater (marine) life zone** and a **freshwater life zone**, and give two examples of each. Define **plankton** and describe three types of plankton. Distinguish among **nekton**, **benthos**, and **decomposers** and give an example of each. List four factors that determine the types and numbers of organisms found in the three layers of aquatic life zones. What is **turbidity** and how does it occur? Describe one of its harmful impacts.

Section 8.2

3. What is the key concept for this section? What major ecosystem and economic services are provided by marine systems? What are the three major life zones in an ocean? Define **coastal zone**. Distinguish between an **estuary** and a **coastal wetland** and explain why each has high net primary productivity. Explain the ecological and economic importance of coastal marshes, mangrove forests, and seagrass beds.
4. What is the **intertidal zone**? Distinguish between rocky and sandy shores and describe some of the organisms often found on each type of shoreline. What

is a barrier island? What is **ocean acidification** and why is it a threat to coral reefs? Define **open sea** and describe its three major zones. What is an **upwelling** and how does it affect ocean life? Why does the open sea have a low NPP? What have scientists recently learned about the ocean's biodiversity?

Section 8.3

5. What is the key concept for this section? What are the two biggest threats to marine systems? List five human activities that pose major threats to marine systems and eight human activities that threaten coral reefs. Explain why the Chesapeake Bay is an estuary in trouble. Summarize the story of what is being done to address its problems.

Section 8.4

6. What is the key concept for this section? Define **surface water**, **runoff**, and **watershed (drainage basin)**. What major ecosystem and economic services do freshwater systems provide? What is a **lake**? What four life zones are found in deep lakes? Distinguish between **oligotrophic** and **eutrophic lakes**. What is **cultural eutrophication**?

7. Describe the three zones that downward-flowing water passes through as it flows from highlands to lower elevations. What are the characteristic life forms of each zone? What is the connection between water quality and watershed land?

8. Give three examples of **inland wetlands** and explain the ecological and economic importance of such wetlands.

Section 8.5

9. What is the key concept for this section? What are four major ways in which human activities are disrupting and degrading freshwater systems? Describe losses of inland wetlands in the United States in terms of the area of wetlands lost and the resulting loss of ecosystem and economic services. Explain how the building of dams and other structures on rivers can affect the river deltas and associated coastal wetlands. How do these effects in turn threaten human coastal communities?

10. What are this chapter's *three big ideas*? How do coral reefs showcase the three **scientific principles of sustainability**? How can we use these three principles to help sustain the earth's vital aquatic ecosystems?

Note: Key terms are in bold type.

Critical Thinking

1. What are three steps that governments and private interests could take to protect the world's remaining coral reefs (**Core Case Study**)?

2. Can you think of any ways in which you might be contributing to the degradation of a nearby or distant aquatic ecosystem? Describe the system and how your actions might be affecting it. What are three things you could do to reduce your impact?

3. You are a defense attorney arguing in court for protecting a coral reef (**Core Case Study**) from harmful human activities. Give your three most important arguments for the defense of this ecosystem.

4. How would you respond to someone who argues that we should use the deep portions of the world's oceans to deposit our radioactive and other hazardous wastes because the deep oceans are vast and are located far away from human habitats? Give reasons for your response.

5. From the list of threats to marine ecosystems listed in Figure 8.12, pick the three that you think are the most serious. For each of them, if it continues to degrade ocean ecosystems during your lifetime, how might this affect you? Can you think of ways in which you might be contributing to each problem? What could you do to reduce your impact?

6. Suppose a developer builds a housing complex overlooking a coastal marsh (Figure 8.7) and the result is pollution and degradation of the marsh. Describe the effects of such a development on the wildlife in the marsh, assuming at least one species is eliminated as a result.

7. Suppose you have a friend who owns property that includes a freshwater wetland and the friend tells you she is planning to fill the wetland to make more room for her lawn and garden. What would you say to this friend?

8. Congratulations! You are in charge of the world. What are the three most important features of your plan to help sustain the earth's aquatic biodiversity?

Doing Environmental Science

Find an aquatic ecosystem near where you live or go to school, such as a lake or wetland. Study and write a description of the system, including its dominant vegetation and any animal life that you are aware of. Also, note how any human disturbances have changed the system. Return to the system after a month or two and note any changes, based on your earlier notes. Compare your notes with those of your classmates.

Global Environment Watch Exercise

Go to your MindTap course to access the GREENR database. Starting on the home page, under "Browse Issues and Topics", click on *Environment and Ecology*, then select *Coral Reefs* and use the topic portal to find information on **(a)** trends in the global rate of coral reef destruction; **(b)** what areas of the world are seeing rising rates of coral reef destruction and what areas are seeing falling rates; and **(c)** what is being done to protect coral reefs in various areas. Write a report on your findings.

Data Analysis

Some 45–53% of the world's shallow coral reefs have been destroyed or severely damaged (**Core Case Study**). A number of factors have played a role in this serious loss of aquatic biodiversity, including ocean warming, sediment from coastal soil erosion, excessive algal growth from fertilizer runoff, coral bleaching, rising sea levels, ocean acidification, overfishing, and damage from hurricanes.

In 2005 scientists Nadia Bood, Melanie McField, and Rich Aronson conducted research to evaluate the recovery of coral reefs in Belize from the combined effects of mass bleaching and Hurricane Mitch in 1998. Some of these reefs are in protected waters where no fishing is allowed. The researchers speculated that reefs in waters where no fishing is allowed should recover faster than reefs in waters where fishing is allowed. The graph to the left shows some of the data they collected from three highly protected (unfished) sites and three unprotected (fished) sites to evaluate their hypothesis. Study this graph and then answer the following questions.

Effects of restricting fishing on the recovery of unfished and fished coral reefs damaged by the combined effects of mass bleaching and Hurricane Mitch in 1998.

(Compiled by the authors with data from Melanie McField, et al., *Status of Caribbean Coral Reefs after Bleaching and Hurricanes in 2005*, NOAA, 2008. Report available at www.coris.noaa.gov/activities/caribbean_rpt/.)

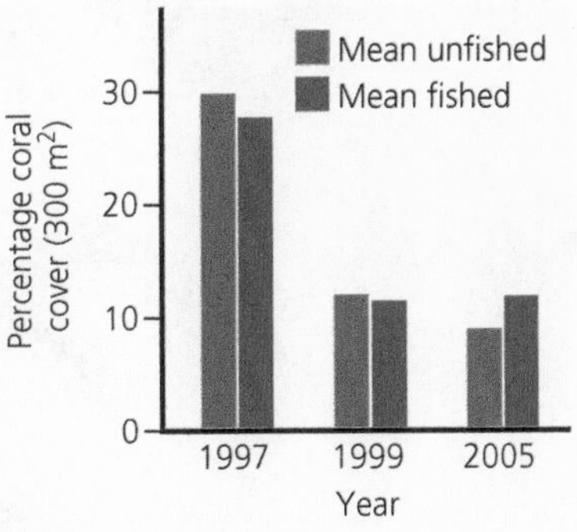

1. By about what percentage did the mean coral cover drop in the protected (unfished) reefs between 1997 and 1999?
2. By about what percentage did the mean coral cover drop in the protected (unfished) reefs between 1997 and 2005?
3. By about what percentage did the coral cover drop in the unprotected (fished) reefs between 1997 and 1999?
4. By about what percentage did the coral cover change in the unprotected (fished) reefs between 1997 and 2005?
5. Do these data support the hypothesis that coral reef recovery should occur faster in areas where fishing is prohibited? Explain.

CENGAGE brain.com For access to MindTap and additional study materials visit www.cengagebrain.com.

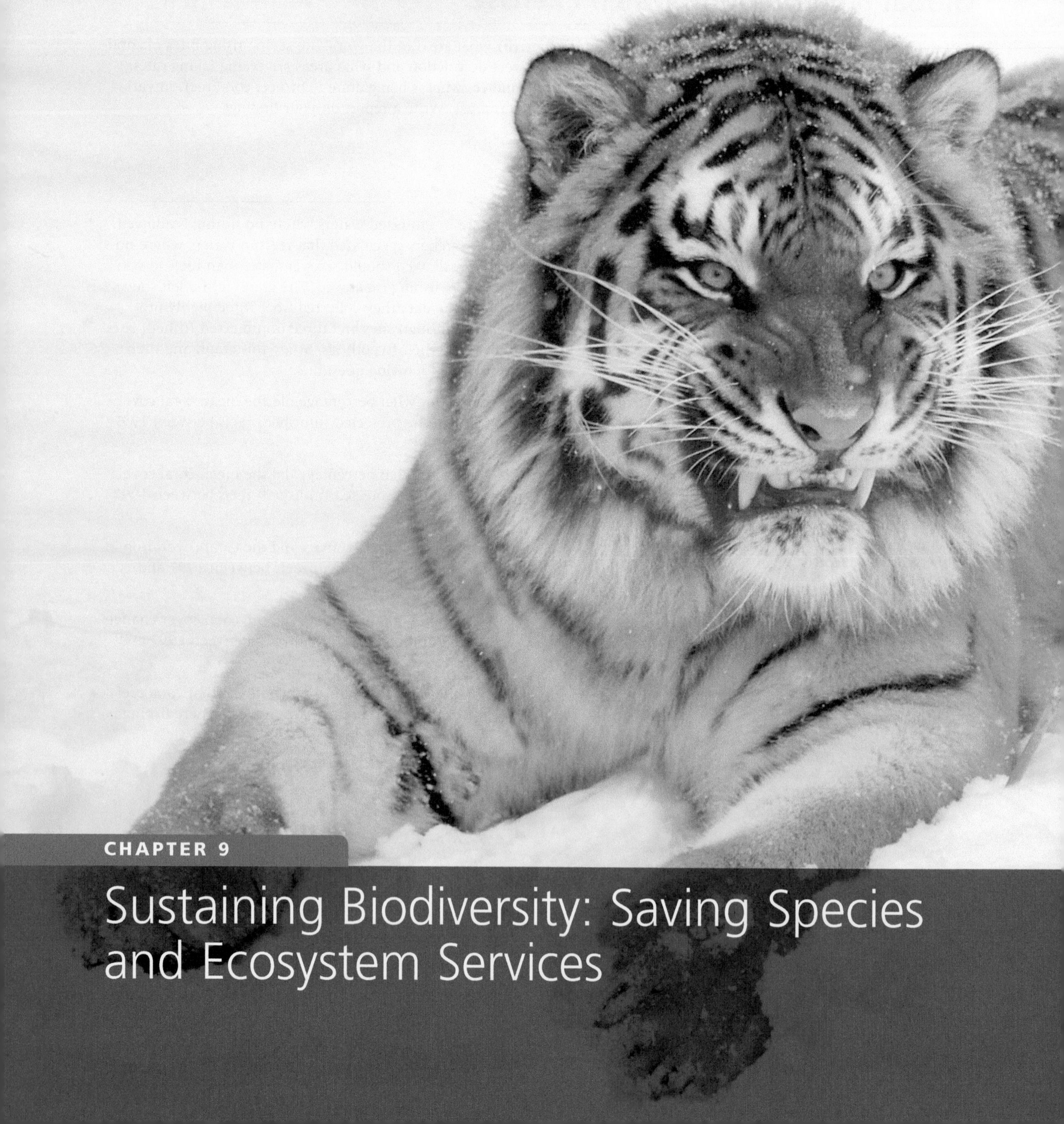

CHAPTER 9

Sustaining Biodiversity: Saving Species and Ecosystem Services

> The last word in ignorance is the person who says of an animal or plant: "What good is it?" . . . If the land mechanism as a whole is good, then every part of it is good, whether we understand it or not.
>
> **ALDO LEOPOLD**

Endangered wild Siberian tiger.

Volodymyr Burdiak/Shutterstock.com

Key Questions

9.1 What role do humans play in the loss of species and ecosystem services?

9.2 Why should we try to sustain wild species and the ecosystem services they provide?

9.3 How do humans accelerate species extinction and degradation of ecosystem services?

9.4 How can we sustain wild species and the ecosystem services they provide?

Core Case Study

Where Have All the Honeybees Gone?

In meadows, forests, farm fields, and gardens around the world, industrious honeybees (Figure 9.1) flit from one flowering plant to another. They are collecting liquid flower nectar to take back to their hives. The bees also collect pollen grains, which stick to hairs on their legs. They feed young honeybees the protein-rich pollen, and the adults feed on the honey made from the collected nectar and stored in the hive.

Honeybees provide one of nature's most important ecosystem services: *pollination*. It involves a transfer of pollen stuck on their bodies from the male to female reproductive organs of the same flower or among different flowers. This fertilization enables the flower to produce seeds and fruit. Honeybees pollinate many plant species and some of our most important food crops, including many vegetables, fruits, and tree nuts such as almonds. European honeybees pollinate about 71% of vegetable and fruit crops that provide 90% of the world's food and a third of the U.S. food supply.

Darlyne A. Murawski/National Geographic Creative

FIGURE 9.1 European honeybee drawing nectar from a flower.

Many U.S. growers rent European honeybees from commercial beekeepers that truck about 2.7 million hives to farms across the country to pollinate different crops during the weeks when they are in bloom. Nature relies on the earth's free pollination service provided by a diversity of bees and other wild pollinators (such as butterflies, hummingbirds, and bats). In contrast, farmers practicing industrialized agriculture on vast croplands and orchards rely mostly on renting this single bee species to pollinate their crops.

However, European honeybees have been in decline since the 1980s. Since 2006, massive numbers of these bees in the United States and in parts of Europe have been disappearing from their colonies. This phenomenon, called **colony collapse disorder (CCD),** occurs when all the bees abandon a colony. Most CCDs used to occur during winter, but since 2010 they have been occurring year-round. Between 2008 and 2015, 23% to 43% of the European honeybee colonies in the United States suffered from CCD. This was above the historical loss rates of 10% to 15%. There has also been a sharp drop in the populations of other pollinators, including monarch butterflies (see Case Study, Chapter 4, p. 93).

Some farmers believe that we need the industrialized honeybee pollination system to grow enough food. Others see such heavy dependence on a single bee species as a potentially dangerous violation of the earth's biodiversity **principle of sustainability** (see Inside Back Cover). They warn that this dependence could put food supplies at risk if the population of European honeybees continues to decline. If this occurs, food prices and hunger will rise. They call for more reliance on the free crop pollination services provided by variety of wild bee species and other pollinators.

This is a classic case of how the decline of a species can threaten vital ecosystem and economic services. Scientists project that during this century, human activities—especially those that contribute to habitat loss and climate change—are likely to play a key role in the extinction of one-fifth to one-half of the world's known plant and animal species. Many scientists view this threat as one of the most serious and long-lasting environmental and economic problems we face. In this chapter, we discuss the causes of this problem and possible ways to deal with it.

9.1 WHAT ROLE DO HUMANS PLAY IN THE LOSS OF SPECIES AND ECOSYSTEM SERVICES?

CONCEPT 9.1 Species are becoming extinct at least 1,000 times faster than the historical rate and by the end of this century, the extinction rate is projected to be 10,000 times higher.

Extinctions Are Natural but Sometimes They Increase Sharply

When a species can no longer be found anywhere on the earth, it has suffered **biological extinction.** A species can become extinct when it cannot adapt and successfully reproduce under new environmental conditions, or a catastrophic environmental event may wipe out its members.

Extinction is a natural process and has occurred at a low rate throughout most of the earth's history. This natural rate is known as the **background extinction rate.** Scientists estimate that the background rate typically amounts to a loss of about 1 species per year for every 1 million species living on the earth. This amounts to 10 natural extinctions a year if the earth has 10 million species.

The extinction of many species in a relatively short period of geologic time is called a **mass extinction.** Geologic, fossil, and other records indicate that the earth has experienced five mass extinctions, during which 50–90% of the species present at that time went extinct (see Figure 4.19, p. 93, and Supplement 6, p. S46) over thousands of years. The largest mass extinction took place some 250 million years ago and wiped about 90% of the world's existing species.

The causes of past mass extinctions are poorly understood but probably involved global changes in environmental conditions. Examples are sustained and significant global warming or cooling, large changes in sea levels and ocean water acidity, and catastrophes such as multiple large-scale volcanic eruptions and large asteroids or comets hitting the planet.

Mass extinctions devastate life on the earth. But they also provide opportunities for new life forms to emerge, diversify, and fill empty ecological niches. Scientific evidence indicates that after each mass extinction, the earth's overall biodiversity returned to equal or higher levels. However, each recovery took several million years. The existence of millions of species today means speciation, on average, has kept ahead of extinction. It also demonstrates the biodiversity **principle of sustainability** as a factor in the long-term sustainability of life on the earth.

Scientific evidence indicates that extinction rates have increased as the human population has grown and spread over most of the globe. In this expansion, humans have destroyed and degraded habitats, consumed huge quantities of resources, and created large and growing ecological footprints (see Figure 1.8, p. 12). The extinction of one species can lead to the extinction of other species that depend on it for food or ecosystem services, and biodiversity researchers project that the rate of extinction will continue to increase. In the words of biodiversity expert Edward O. Wilson (see Individuals Matter 4.1, p. 81), "The natural world is everywhere disappearing before our eyes—cut to pieces, mowed down, plowed under, gobbled up, replaced by human artifacts."

Scientists estimate that the current annual extinction rate is 1,000 times the natural background extinction rate (Science Focus 9.1) (**Concept 9.1**). Assuming there are 10 million species on the earth, this means that today we are losing an estimated 10,000 species per year, compared to the background extinction rate of 10 species per year.

Since 1970, the World Wildlife Fund (WWF) has published a *living planet index (LPI)* based on assessing the sizes of 10,000 populations of over 3,000 vertebrate species. Between 1970 and 2010, the LPI declined by 52%. Population declines were greater **(1)** in tropical areas (56%) than in temperate areas (36%), **(2)** in freshwater systems (76%) than in terrestrial (39%) and marine systems (39%), and **(3)** in low-income countries (58%) than in middle-income countries (18%). In high-income countries, the LPI rose by 10%.

Biodiversity researchers project that during this century, the extinction rate is likely to rise to at least 10,000 times the background rate—mostly because of habitat loss and degradation, climate change, ocean acidification, and other effects of human activities (**Concept 9.1**). At this rate, if there were 10 million species on the earth, 100,000 species would be expected to disappear each year—an average of about 274 species per day or about 11 every hour. By the end of this century, most of the big carnivorous cats, including cheetahs, tigers (see chapter-opening photo), and lions might possibly exist only in zoos and small wildlife sanctuaries. Most elephants, rhinoceroses, gorillas, chimpanzees, and orangutans will likely disappear from the wild.

So why does this matter? According to several biodiversity researchers, including Edward O. Wilson and Stuart Pimm, at this extinction rate, an estimated 20% to 50% of the world's 2 million identified animal and plant species could vanish from the wild by the end of this century. Many other species that have not been identified will also disappear. If these estimates are correct, the earth is entering a *sixth mass extinction* caused primarily by human activities. And this mass extinction is projected to take place within a human lifetime instead of over many thousands of years like the previous five mass extinctions (see Supplement 6, p. S46). As conservation ecologist Thomas E. Lovejoy (Individuals Matter 3.1, p. 70) puts it: "The sixth great extinction has started and the issue is how far do we let it go."

Estimating Extinction Rates

Louis Agassi Fuertes/National Geographic Creative

FIGURE 9.A Painting of the last pair of North American passenger pigeons, which once were the world's most abundant bird species. They became extinct in the wild in 1912 mostly because of habitat loss and overhunting.

Scientists who try to catalog extinctions, estimate past extinction rates, and project future extinction rates face three problems. *First*, because the natural extinction of a species typically takes a long time, it is difficult to document. *Second*, we have identified only about 2 million of the world's estimated 7 million to 10 million and perhaps as many as 100 million species. *Third*, scientists know little about the ecological roles of most of the species that have been identified.

One approach to estimating future extinction rates is to study records documenting past rates at which easily observable mammals and birds (Figure 9.A) have become extinct. Most of these extinctions have occurred since humans began to dominate the planet about 10,000 years ago. This information can be compared with fossil records of extinctions that occurred before that time.

Another approach is to observe how reductions in habitat area affect extinction rates. The *species–area relationship*, studied by Edward O. Wilson (see Individuals Matter 4.1, p. 81) and Robert MacArthur, suggests that, on average, a 90% loss of land habitat in a given area can cause the extinction of about 50% of the species living in that area. Thus, we can base extinction rate estimates on the rates of habitat destruction and degradation, which are increasing around the world.

Scientists also use mathematical models to estimate the risk of a particular species becoming endangered or extinct within a certain period and run them on computers. These models include factors such as trends in population size, past and projected changes in habitat availability, interactions with other species, and genetic factors.

Researchers are working hard to get more and better data and to improve the models they use in order to make better estimates of extinction rates and to project the effects of such extinctions on vital ecosystem services such as pollination (**Core Case Study**). These scientists contend that our need for better data and models should not delay our acting now to keep from hastening extinctions and the accompanying losses of ecosystem services through human activities.

CRITICAL THINKING

Does the fact that extinction rates can only be estimated make them unreliable? Why or why not?

20–50%
Percentage of the earth's known species that could disappear this century primarily because of human activities

Such large-scale loss of species would likely impair some of the earth's vital ecosystem services such as air and water purification, natural pest control, and pollination (**Core Case Study**). According to the Millennium Ecosystem Assessment, 15 of 24 of the earth's major ecosystem services are in decline. Conservation scientists view this potential massive loss of biodiversity and ecosystem services within the span of a human lifetime as one of the most important and long-lasting environmental and economic problems humanity faces. By saving as many species as possible from extinction—especially keystone species (see p. 83)—we could increase our beneficial environmental impact and help sustain and enrich our own lives and economies.

Wilson, Pimm, and other extinction experts consider a projected extinction rate of 10,000 times the background extinction rate to be low, for several reasons:

- The rate of extinction and the resulting threats to ecosystem services are likely to increase sharply during the next 50–100 years because of the harmful environmental impacts of the growing human population and its growing per capita use of resources.
- The current and projected extinction rates in the world's **biodiversity hotspots**—areas that are highly endangered centers of biodiversity—are much higher than the global average.

- We are eliminating, degrading, fragmenting, and simplifying many biologically diverse environments—including tropical forests, coral reefs, wetlands, and estuaries—that serve as potential sites for the emergence of new species. Thus, in addition to greatly increasing the rate of extinction, we may be hindering the long-term recovery of biodiversity by eliminating or degrading places where new species can evolve. In other words, we are also creating a *speciation crisis*.

Biologists Philip Levin, Donald Levin, and others warn that, while our activities are likely to reduce the speciation rates for some species, they are likely to increase the speciation rates for rapidly reproducing species. Examples are weeds, rats, and insects such as cockroaches. Rapidly expanding populations of such species could crowd and compete with various other species, further accelerating their extinction and threatening key ecosystem services.

Endangered and Threatened Species Are Ecological Smoke Alarms

Biologists classify species that are heading toward biological extinction as either *endangered* or *threatened*. An **endangered species** has so few individual survivors that the species could soon become extinct. A **threatened species** (also known as a *vulnerable* species) has enough remaining individuals to survive in the short term, but because of declining numbers, it is likely to become endangered in the near future.

Each year the International Union for Conservation of Nature (IUCN) evaluates the status of a large number of species and its findings in its annual Red List. Figure 9.2 shows 4 of the 22,784 species listed in 2015 as being critically endangered, endangered, or threatened. Between 1996 and 2015, the total number of species in these three categories increased by 96%. The actual number of species in trouble is very likely much higher. Some species have characteristics that increase their chances of becoming extinct (Figure 9.3).

Characteristic	Examples
Low reproductive rate (K-strategist)	Blue whale, giant panda, rhinoceros
Specialized niche	Blue whale, giant panda, Everglades kite
Narrow distribution	Elephant seal, desert pupfish
Feeds at high trophic level	Bengal tiger, bald eagle, grizzly bear
Fixed migratory patterns	Blue whale, whooping crane, sea turtle
Rare	African violet, some orchids
Commercially valuable	Snow leopard, tiger, elephant, rhinoceros, rare plants and birds
Require large territories	California condor, grizzly bear, Florida panther

FIGURE 9.3 Certain characteristics put a species in greater danger of becoming extinct.

a. Mexican gray wolf: About 42 in the forests of Arizona and New Mexico

b. California condor: 225 in the southwestern United States (up from 9 in 1986)

c. Whooping crane: 437 in North America

d. Sumatran tiger: 400–600 on the Indonesian island of Sumatra

FIGURE 9.2 Endangered natural capital: These four critically endangered species are threatened with extinction, largely because of human activities. The number below each photo indicates the estimated total number of individuals of that species remaining in the wild.

As biodiversity expert Edward O. Wilson puts it, "The first animal species to go are the big, the slow, the tasty, and those with valuable parts such as tusks and skins."

Species can also become *regionally extinct* in the areas where they are normally found. A species can also become *functionally extinct* when its populations crash to the point where its interactions with other species are lost or greatly diminished. Important ecosystem services that depend on these interactions might also then be lost or diminished, and this is often difficult to detect until it is too late.

For example, the American alligator is a keystone species in its marsh and swamp habitats of the southeastern United States. (See Case Study, Chapter 4, p. 83.) When its numbers dwindled in the 1960s, certain ecosystem services, such as the building of gator nests, diminished, and bird species that depended on these nesting sites declined. After the alligator was placed on the U.S. endangered species list, it made a strong comeback and its ecosystems recovered. GOOD NEWS

9.2 WHY SHOULD WE TRY TO SUSTAIN WILD SPECIES AND THE ECOSYSTEM SERVICES THEY PROVIDE?

CONCEPT 9.2 We should avoid hastening the extinction of wild species because of the ecosystem and economic services they provide, because it can take millions of years for nature to recover from large-scale extinctions, and because many people believe that species have a right to exist regardless of their usefulness to us.

Species Are a Vital Part of the Earth's Natural Capital

According to the WWF, only about 61,000 orangutans (Figure 9.4) remain in the wild. Most of them are in the tropical forests of Borneo, Asia's largest island. These

Seatraveler/Dreamstime.com

FIGURE 9.4 Natural capital degradation: These endangered orangutans depend on a rapidly disappearing tropical forest habitat in Borneo. ***Critical thinking:*** What difference will it make if human activities hasten the extinction of the orangutan?

highly intelligent animals are disappearing at an estimated rate of 1,000–2,000 per year. A key reason is that much of their tropical forest habitat is being cleared for plantations that grow oil palms. They are a source of palm oil—the world's most widely used vegetable oil. It is an ingredient in numerous products such as cookies, cosmetics, and biodiesel fuel for motor vehicles. Another reason they are disappearing is smuggling. An illegally smuggled, live orangutan sells on the black market for up to $10,000.

The effects of these human-related activities are compounded by the fact that orangutans have the lowest birth rate of all the mammals. They give birth to one young about once every 8 years. As a result, even small decreases in their numbers can be devastating to these primates. Without urgent protective action, the endangered orangutan may disappear in the wild within the next two decades.

Cheryl Knott—a biological anthropologist, National Geographic Explorer, and Associate Professor of Anthropology at Boston University—has been studying endangered orangutans in a national park in Borneo. Her research goals are to understand how fluctuations in food availability in the orangutans' environment shape their behavioral and physiological adaptations.

Does it matter that orangutans—or any species, for that matter—may disappear in the wild largely due to human activities? New species eventually evolve to take the places of species lost through background and mass extinctions, so why should we care if we greatly speed up the global extinction rate over the next 50–100 years? According to biologists, there are four major reasons why we should prevent our activities from causing or hastening the extinction of other species.

First, the world's species provide vital *ecosystem services* (see Figure 1.3, p. 7) that help to keep us alive and support our economies (**Concept 9.2**). For example, we depend on honeybees (**Core Case Study**) and other insects for pollination of many food crops. We also depend on certain species of birds, amphibians (see Chapter 4, Core Case Study, p. 76), and spiders for natural control of insect pests. Aquatic species that live in streams can help purify flowing water. Trees produce oxygen that organisms need to survive. Earthworms aerate topsoil, which helps improve soil health. By eliminating a species or sharply reducing its population—especially a species such as the orangutan that plays a keystone role—we can speed up the extinction of other species. This can upset ecosystems and degrade their important ecosystem services.

A *second* major reason for preventing extinctions caused or hastened by human activities is that many species contribute to *economic services* that we depend on (**Concept 9.2**). Various plant species provide economic value as food crops, wood for fuel, lumber for construction, paper from trees, and substances for medicines. *Bioprospectors* search tropical forests and other ecosystems to find plants and animals that scientists can use to make medicinal drugs (Figure 9.5)—an example of *learning from nature*. According to a United Nations University report, 62% of all cancer drugs have been derived from the discoveries of bioprospectors. Less than 0.5% of the world's known plant species have been examined for their medicinal properties.

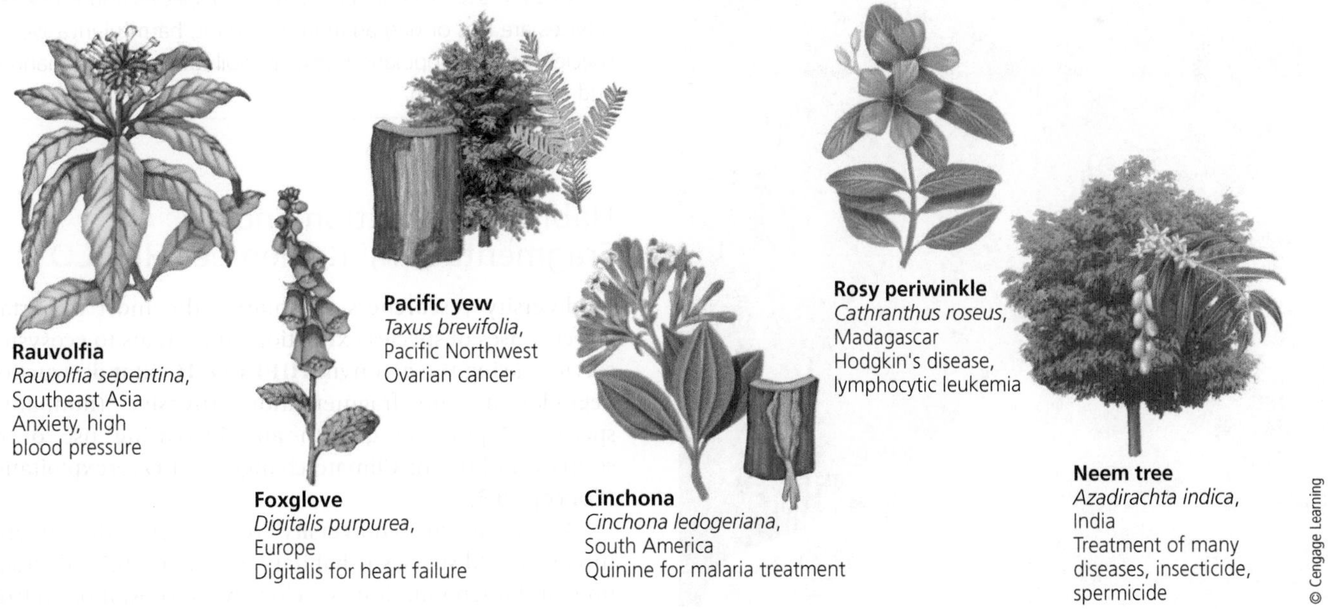

FIGURE 9.5 Natural capital: These plant species are examples of *nature's pharmacy*. Once the active ingredients in the plants have been identified, scientists can usually produce them synthetically. The active ingredients in 9 of the 10 leading prescription drugs originally came from wild organisms.

CONSIDER THIS . . .

LEARNING FROM NATURE

Scientist Richard Wrangham is identifying medicinal compounds useful for humans by observing which plants chimpanzees eat to heal themselves.

Preserving species and their habitats also provides economic benefits in the form of *ecotourism*. This rapidly growing industry specializes in environmentally responsible travel to natural areas. Ecotourism promotes conservation, low environmental impact, respect for local cultures, and support for local economies. Travelers who sign up for ecotours have the chance to see endangered species such as orangutans and parrots (Figure 9.6) in the wild. Some tours also allow travelers to help with data collection or other scientific work. Revenues from ecotourism generate more than $1 million per minute in tourist expenditures worldwide. **GREEN CAREER: Ecotourism guide**

A *third* major reason for not hastening extinctions through our activities is that extinction can hinder speciation. It will take 5 million to 10 million years for natural speciation to replace the species we are likely to wipe out during this century.

Fourth, many people believe that wild species, such as orangutans, have a right to exist, regardless of their usefulness to us (**Concept 9.2**). According to this worldview, we have an ethical responsibility to protect the earth's species from becoming extinct because of our activities, and to prevent degradation of the world's ecosystem services.

ROY TOFT/National Geographic Creative

FIGURE 9.6 Many species of wildlife such as this endangered hyacinth macaw in Mato Grosso, Brazil, are sources of beauty and pleasure. Habitat loss and illegal capture in the wild by pet traders endanger this species.

This ethical viewpoint raises a number of challenging questions. Since we cannot save all species from the harmful consequences of our actions, we have to make choices about which ones to protect. Should we protect more animal species than plant species and, if so, which ones should we protect? Some people support protecting familiar and appealing species such as elephants, whales, tigers, giant pandas, and orangutans (Figure 9.4), but care much less about protecting plants that serve as the base of the food supply for other species (**Core Case Study**). Others might think little about getting rid of species that most people fear or hate, such as mosquitoes, cockroaches, disease-causing bacteria, snakes, sharks, and bats.

In summary, a flourishing diversity of life on the earth is essential for sustaining the planetary life-support system on which the human species and other species depend. To biologist Edward O. Wilson, carelessly and rapidly eliminating species that make up an essential part of the world's biodiversity is like burning millions of books that we have never read.

9.3 HOW DO HUMANS ACCELERATE SPECIES EXTINCTION AND DEGRADATION OF ECOSYSTEM SERVICES?

CONCEPT 9.3 The greatest threats to species and ecosystem services are loss or degradation of habitat, harmful invasive species, human population growth, pollution, climate change, and overexploitation.

Habitat Destruction and Fragmentation]: Remember HIPPCO

Biodiversity researchers summarize the most important direct causes of species extinction and threats to ecosystem services using the acronym **HIPPCO**: **H**abitat destruction, degradation, and fragmentation; **I**nvasive (nonnative) species; **P**opulation growth and increasing use of resources; **P**ollution; **C**limate change; and **O**verexploitation (**Concept 9.3**).

According to biodiversity researchers, the greatest threat to wild species is habitat loss (Figure 9.7), degradation, and fragmentation. Specifically, deforestation in tropical areas (see Figure 3.1, p. 50) is the greatest threat to species and to the ecosystem services they provide. The next largest threat is the destruction and degradation of coastal wetlands and coral reefs (see Chapter 8, Core Case

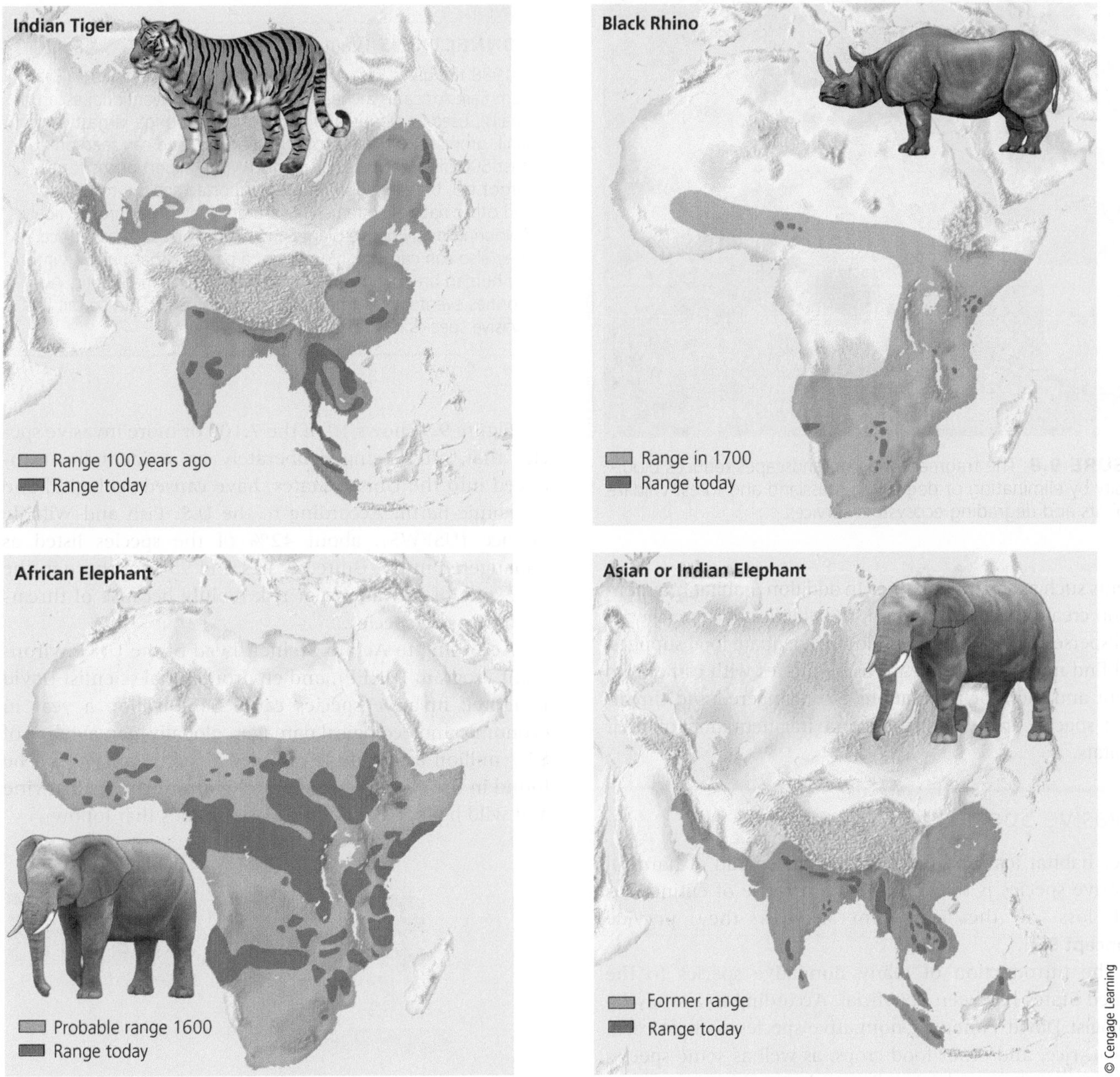

FIGURE 9.7 Natural capital degradation: These maps reveal the reductions in the ranges of four wildlife species, mostly as the result of severe habitat loss and fragmentation and illegal hunting for some of their valuable body parts. ***Critical thinking:*** Would you support expanding these ranges even though this would reduce the land available for human habitation and farming? Explain.

(Compiled by the authors using data from International Union for Conservation of Nature and World Wildlife Fund.)

Study, p. 168), the plowing of grasslands for planting of crops (see Figure 7.17, p. 154), and the pollution of streams, lakes, and oceans.

Island species—many of them found nowhere else on earth—are especially vulnerable to extinction. If their habitats are destroyed, degraded, or fragmented into patches, they have nowhere else to go. This is why the Hawaiian Islands are America's "extinction capital"—with 63% of its species at risk.

Habitat fragmentation occurs when a large, intact area of habitat such as a forest or natural grassland is divided into smaller, isolated patches or *habitat islands* (Figure 9.8). Roads, logging operations, crop fields, and urban developments divide forests and natural grasslands. This reduces tree cover in forests and blocks animal migration routes. Fragmentation can divide populations of a species into increasingly isolated small groups that are more vulnerable to predators, competitor species, disease, and catastrophic

FIGURE 9.8 The fragmentation of landscapes reduces biodiversity by eliminating or degrading grassland and forest wildlife habitats and degrading ecosystem services.

events such as storms and fires. In addition, habitat fragmentation creates barriers that limit the abilities of some species to disperse and colonize areas, locate adequate food supplies, and find mates. Scientists are using drones with cameras to count and monitor populations of endangered and threatened species and degradation and fragmentation of their habitats.

Invasive Species

After habitat loss and degradation, the spread of harmful invasive species is the second largest cause of extinctions and loss of the ecosystem services they provide (**Concept 9.3**).

The introduction of many nonnative species to the United States has been beneficial. According to a study by ecologist David Pimentel, nonnative species such as corn, wheat, rice, and other food crops, as well as some species of cattle, poultry, and other livestock, provide more than 98% of the U.S. food supply. Other deliberately introduced species have helped control pests. In the 1600s, English settlers brought highly beneficial European honeybees (**Core Case Study**) to North America to provide honey and we now use them to pollinate most major crops.

GOOD NEWS

A problem can occur when an introduced species does not face the natural predators, competitors, parasites, viruses, bacteria, or fungi that controlled its populations in its native habitat. This can allow some nonnative species to outcompete populations of many native species for food, disrupt ecosystem services, transmit new diseases, and lead to economic losses. When this happens the nonnative species are viewed as harmful *invasive species*. Invasive species rarely cause the global extinction of other species, but they can cause population declines and local and regional extinctions of some native species.

CONSIDER THIS . . .

CONNECTIONS Giant Snails and Meningitis

In 1988 the giant East African land snail was imported to Brazil from East Africa as a cheap substitute for conventional escargot (snails), used as a source of food. It is the world's largest land snail, growing to the size of a human fist, and can feed on at least 500 different types of plants. When export prices for escargot fell, breeders dumped the imported snails into forests and other natural systems. Since then, they have spread widely, devouring many native plants and food crops such as lettuce. They also can carry rat lungworm, a parasite that burrows into the human brain and causes potentially lethal meningitis. Authorities eventually banned imports of the snail, but so far, this invasive species has been unstoppable.

Figure 9.9 shows 10 of the 7,100 or more invasive species that, after being deliberately or accidentally introduced into the United States, have caused ecological and economic harm. According to the U.S. Fish and Wildlife Service (USFWS), about 42% of the species listed as endangered in the United States and 95% of those in the U.S. state of Hawaii are at risk mainly because of threats from invasive species.

According to Achim Steiner, head of the UN Environment Program (UNEP), and environmental scientist David Pimentel, invasive species cause $1.4 trillion a year in economic and ecological damages, globally—an average of $2.7 million a minute. Examples of such damage can be found in the effects of deliberately introducing kudzu vine and wild boars (see the two Case Studies that follow).

$162 million

Estimated hourly global cost of invasive species

CASE STUDY

The Kudzu Vine and Kudzu Bugs

Some invasive species, such as *kudzu vine*, have been deliberately introduced into ecosystems. In the 1930s, this plant was imported from Japan and planted in the southeastern United States in an effort to control soil erosion.

Kudzu does control erosion, but it grows so rapidly that it engulfs hillsides, gardens, trees, stream banks, cars (Figure 9.10), and anything else in its path. Dig it up or burn it, and it still keeps spreading. It can grow in sunlight or shade and is very difficult to kill, even with herbicides that can contaminate water supplies. Scientists have found a common fungus that can kill kudzu within a few hours, but they need to investigate any harmful side effects it may have.

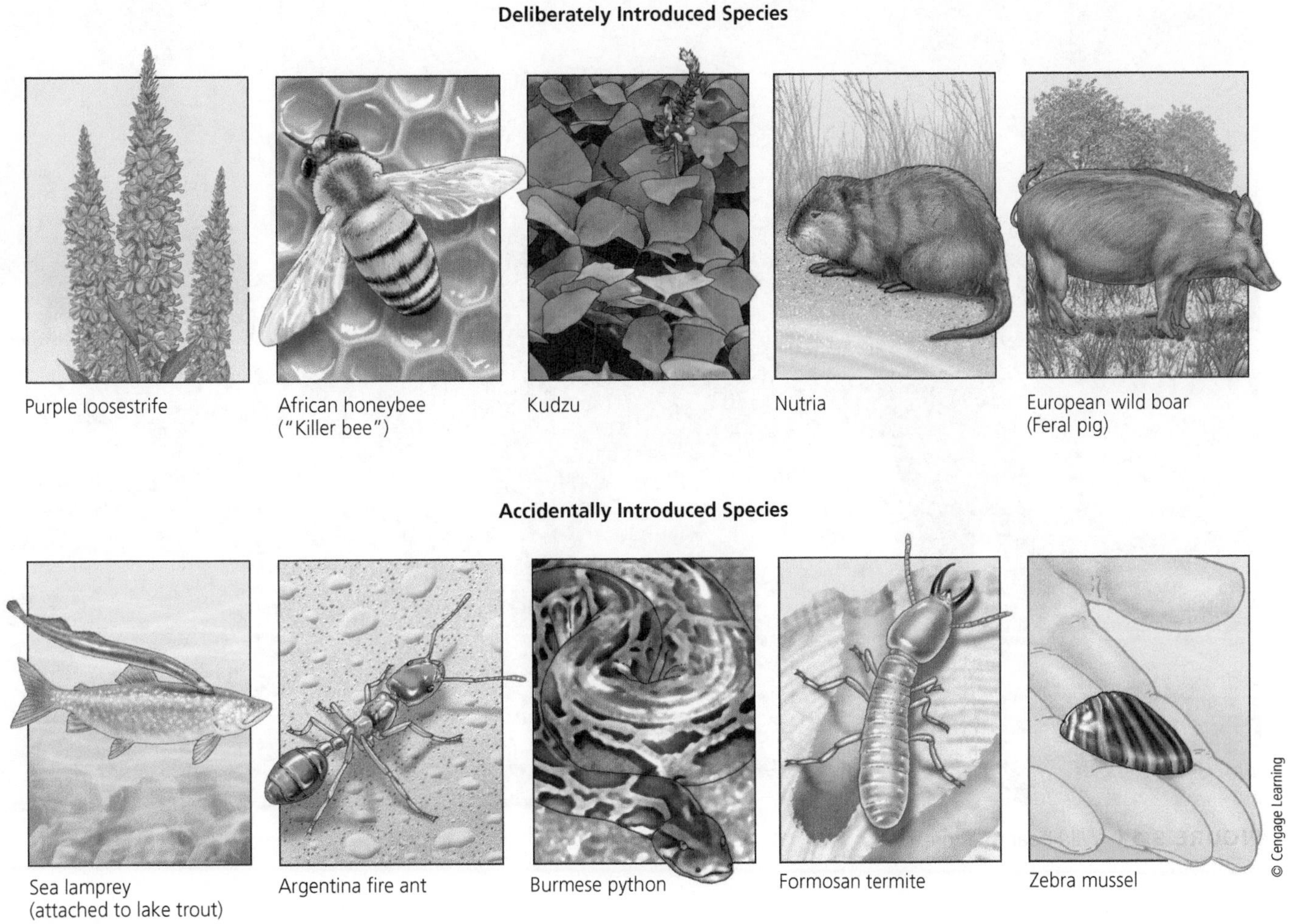

FIGURE 9.9 Some of the estimated 7,100 harmful invasive species that have been deliberately or accidentally introduced into the United States.

FIGURE 9.10 Kudzu has grown over this car in the U.S. state of Georgia.

Nicknamed "the vine that ate the South," kudzu has spread throughout much of the southeastern United States. As the climate gets warmer, it could spread to the north.

Kudzu is considered a menace in the United States. However, for thousands of years Asians have used a powdered form of kudzu in herbal remedies to treat a range of ailments such as fever, inflammation, flu, dysentery, hangovers, and the effects of insect and snake bites.

Almost every part of the kudzu plant is edible, making it an inexpensive and readily available source of nutrition. Because it can grow rapidly where other plants cannot and is drought tolerant, it has helped people survive droughts and famines and restore severely degraded land.

Because ingesting small amounts of kudzu powder can lessen one's desire for alcohol, it could be used to reduce alcoholism and binge drinking. Although kudzu can engulf and kill trees, it might eventually help to save some of them. Researchers at the Georgia Institute of Technology have found that kudzu could replace trees as a source of fiber for making paper. It is also being evaluated as a raw material for producing biofuel.

Eduard Kyslynskyy/Shutterstock.com

FIGURE 9.11 Wild boar in a forest.

The brown, pea-sized Kudzu bug is another invasive species that was imported into the United States from Japan. It breeds in and feeds on patches of kudzu, and it can help to reduce the spread of the vine. However, it spreads even more rapidly than the kudzu vine. It also feeds on soybeans and thus could pose a major threat to soy crops.

Some pesticides can kill this bug, but might end up boosting their numbers by promoting genetic resistance to the pesticides. Researchers hope to change this bug through genetic engineering in such a way that it will stop eating soybeans. They are also evaluating the use of a wasp whose larvae attack kudzu bug embryos. However, so far, scientists see no way to eradicate this rapidly spreading invader species.

CASE STUDY

Wild Boar Invasions

The wild boar (Figure 9.11, also known as wild hogs or feral pigs) is widely distributed over the earth's land surface. Humans have introduced different versions of this species to numerous countries so that they can be hunted for sport and as a source of game meat.

In the early 1900s, the Eurasian wild boar was introduced to the U.S. states of New York and North Carolina and kept in privately owned fenced hunting reserves. Some escaped from the reserves and others were moved to new areas and released into the wild for hunting.

The wild boar has multiplied rapidly. Its population in the United States is estimated to be 5.5 million to 8 million—up from about 2 million in 1990. Wild boars have established populations in 36 states and have been spotted in 11 other states. In order, the three states with the most wild boars are Texas (with an estimated 3 million), Florida, and California.

Wild boars have many of the ideal qualities for a successful and destructive invasive species. They are big, strong, intelligent, hard to kill or trap, and vicious when trapped. As adults, they typically weigh around 90 kilograms (200 pounds), run up to 40 kilometers per hour (25 miles per hour), jump as high as 0.9 meters (3 feet), and climb out of traps with walls as high as 1.8 meters (6 feet). In 2004 a legendary wild boar known as Hogzilla was shot. It was about 2.4 meters (8 feet) long, weighed about 360 kilograms (800 pounds), and had sharp tusks nearly 46 centimeters (18 inches) long.

Wild boars prefer forests but can live almost anywhere. They prefer plants and roots but can eat pretty much anything, including quail, the eggs of endangered sea turtles, and baby lambs, goats, calves, and deer. They are nocturnal animals that come out at night to forage for food.

They use their long, plow-like snouts and strong necks to dig up land to a depth of 0.9 meters (3 feet), upturning large rocks as they go. They devour crops and uproot pastures, lawns, and forest floors. In such areas, one result is soil erosion that muddies streams and destroys habitats for many animals, including ground-nesting birds, voles, and salamanders. By destroying native vegetation, boars can alter forest food webs and open the door to invasive plant species.

Wild boars breed at a high rate and do not have enough natural predators to control their dispersed and rapidly growing populations. They are among the most destructive invasive species in the United States and each year cause about $1.5 billion in damages and control costs.

Efforts to control wild boar populations include shooting and trapping. Researchers are also trying to develop poisons and birth control chemicals to use on the boars. After several decades of such efforts, the boars have been eliminated from several small islands. However, eliminating them in the continental United States and on other continents seems to be impossible.

Some Accidentally Introduced Species Can Disrupt Ecosystems

Many unwanted nonnative invaders arrive from other continents as stowaways on aircraft, in the ballast water of tankers and cargo ships, and as hitchhikers on imported products such as wooden packing crates. Cars and trucks can also spread the seeds of nonnative plant species embedded in their tire treads. Many tourists return home with living plants that can multiply and become invasive. Some of these plants might also contain insects that can invade new areas, multiply rapidly, and threaten crops.

Florida is the global capital for invasive species. Some of its many troublesome invasive species include Burmese pythons, African pythons, and several species of boa constrictors, all of which have invaded the Florida Everglades. About 1 million of these snakes, imported from Africa and Asia, have been sold as pets. Some buyers, after learning that these reptiles do not make good pets, let them go in the wetlands of the Everglades.

The Burmese python (Figure 9.12) can live 20 to 25 years and grow as long as 5 meters (16 feet). It can weigh as much 77 kilograms (170 pounds) and be as big around as a telephone pole. These snakes have huge appetites. They feed at night, eating a variety of birds and mammals (rabbits, raccoons, and white-tailed deer). Occasionally they will eat other reptiles, including the American alligator—a keystone species in the Everglades ecosystem (see Chapter 4, Case Study, p. 83). Pythons seize their prey with their sharp teeth, wrap around them, and squeeze them to death, before feeding on them. They have also been known to eat cats and dogs, small farm animals, and geese. Research indicates that predation by these snakes is altering the food webs and ecosystem services of the Everglades.

According to wildlife scientists, the Burmese python population in Florida's wetlands cannot be controlled. They are hard to find and kill or capture and they reproduce rapidly. Trapping and moving the snakes from one area to another has not worked because they are able to return to the areas where they were captured. Another concern is that the Burmese python could spread to other swampy wetlands in the southern half of the United States.

Invasive species are a serious ecological and economic threat, but the situation is not hopeless. A rule of thumb is that only 1 of every 100 species that invade an area is able to establish a self-sustaining population and reduce the biodiversity of the ecosystem it has invaded. In addition, scientists have found that some invaders end up increasing the biodiversity of the areas they have moved into by creating new habitats and niches for other species.

Dan Callister/Alamy Stock photo

FIGURE 9.12 Employees of the South Florida Water Management District hunted and captured this invasive Burmese python in the Florida Everglades.

Prevention Is the Best Way to Reduce Threats from Invasive Species

Once a harmful nonnative species becomes established in an ecosystem, removing it is almost impossible. Americans pay more than $160 billion a year to eradicate or control an increasing number of invasive species—without much success. Clearly, the best way to limit the harmful impacts of these organisms is to prevent them from being introduced into ecosystems.

Scientists suggest several ways to do this, including the following measures:

- Funding an intensive research program to identify the major characteristics of successful invaders, the types of ecosystems that are vulnerable to invaders, and the natural predators, parasites, bacteria, and viruses that could be used to control populations of established invaders.
- Increasing ground surveys and drone and satellite observations to track invasive plant and animal species, and developing better models for predicting how they spread and what harmful effects they could have.
- Identifying major harmful invader species and establishing international treaties banning their transfer from one country to another and increasing inspection of imported goods to enforce such bans.
- Educating the public about the effects of releasing exotic plants and pets into the environment near where they live.

Figure 9.13 shows some of the things you can do to help prevent or slow the spread of harmful invasive species.

What Can You Do?

Controlling Invasive Species

- Do not buy wild plants and animals or remove them from natural areas.
- Do not release wild pets in natural areas.
- Do not dump aquarium contents or unused fishing bait into waterways or storm drains.
- When camping, use only local firewood.
- Brush or clean pet dogs, hiking boots, mountain bikes, canoes, boats, motors, fishing tackle, and other gear before entering or leaving wild areas.

FIGURE 9.13 Individuals matter: Some ways to prevent or slow the spread of harmful invasive species. ***Critical thinking:*** Which two of these actions do you think are the most important ones to take? Why? Which of these actions do you plan to take?

Population Growth, High Rates of Resource Use, Pollution, and Climate Change

Past and projected *human population growth* and rising rates of *resource use per person* have greatly expanded the human ecological footprint (see Figure 1.8, p. 12). People have eliminated, degraded, and fragmented vast areas of wildlife habitat (Figure 9.8) as they have spread out over the planet, and they use resources at increasing rates. This has caused the extinction of many species (**Concept 9.3**).

Pollution from human activities also threatens some species with extinction. Pollutants can contaminate soil, water, and air. They may kill organisms outright or decrease their ability to reproduce or perform their ecological roles.

According to the USFWS, each year pesticides kill about one-fifth of the European honeybee colonies that pollinate almost a third of all U.S. food crops (**Core Case Study** and Science Focus 9.2). The USFWS estimates that pesticides also kill more than 67 million birds and 6 to 14 million fish each year. They also threaten about 20% of the country's endangered and threatened species.

During the 1950s and 1960s, populations of fish-eating birds such as ospreys, brown pelicans, and bald eagles plummeted because of the widespread use of a pesticide called DDT. The concentration of a chemical derived from the DDT did not break down and was biologically magnified in the fatty tissues of the birds as it moved up through their food chains (Figure 9.14). Concentrated amounts of the chemical in the fatty tissues of these top predator birds decreased their ability to produce calcium in the eggshells they laid. This led to eggshells so thin that they cracked before hatching and reduced the ability of the species to reproduce successfully. Also hard hit in those years were predatory birds such as the prairie falcon, sparrow hawk, and peregrine falcon that help control populations of rabbits, ground squirrels, and other crop eaters. Since the United States banned DDT in 1972, most of these bird species have made a comeback. GOOD NEWS

According to a study by Conservation International, the habitat loss and disruption of food webs associated with projected *climate change* could drive a quarter to half of all known land animals and plants to extinction by the end of this century. After reviewing 131 studies, ecologist and evolutionary biologist Mark Urban published a 2015 paper projecting that climate change will accelerate the sixth extinction, with a major loss of biodiversity and ecosystem services. South America, Australia, and New Zealand will experience the most extinctions because they have many species found nowhere else in the world and they rely on habitats that are not found anywhere else.

The hardest-hit species will be those that have limited ranges or low tolerance for temperature changes. For example, studies indicate that the polar bear is threatened

FIGURE 9.14 *Bioaccumulation and biomagnification:* DDT is a fat-soluble chemical that can accumulate in the fatty tissues of animals. In a food chain or food web, the accumulated DDT is biologically magnified in the bodies of animals at each higher trophic level, as it was in the case of a food chain in the U.S. state of New York, illustrated here. ***Critical thinking:*** How does this story demonstrate the value of pollution prevention?

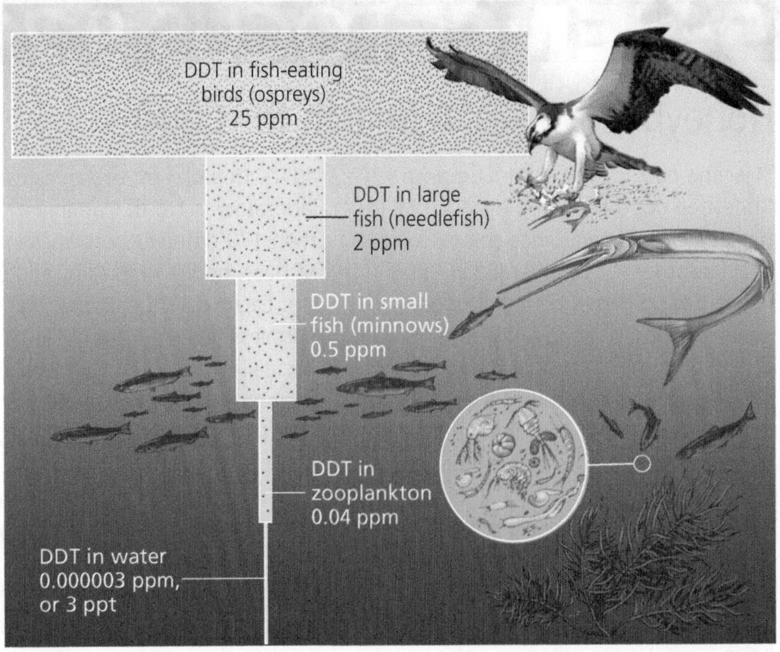

© Cengage Learning

because of higher temperatures and melting sea ice in its polar habitat. The shrinkage of floating ice is making it harder for polar bears to find seals, their favorite prey, especially in the southern parts of the polar bear's range in the Arctic, which has warmed the most (Figure 9.15).

According to the IUCN and the U.S. Geological Survey, the world's total polar bear population is likely to decline by 30–35% by 2050 due to loss of habitat and prey. By the end of this century, polar bears might be found only in zoos.

Vladimir Seliverstov/Dreamstime.com

FIGURE 9.15 On floating ice in the Arctic sea, this polar bear has killed a bearded seal, one of its major sources of food. ***Critical thinking:*** Do you think it matters that the polar bear may become extinct in the wild during this century primarily because of human activities? Explain.

Honeybee Losses: A Search for Causes

Over the past 50 years, the European honeybee population in the United States has been cut in half. Since colony collapse disorder (CCD) emerged in 2006 (**Core Case Study**), commercial beekeepers in the United States have lost 25–50% of their hives on average each year. Scientific research has found several possible reasons for this decline. They include *parasites*, *viruses*, *pesticides*, *stress*, and *poor nutrition*.

Parasites such as the varroa mite feed on the blood of adult honeybees and their larvae. This weakens the immune systems and shortens the life spans of honeybees. Varroa mites may also cause deformities in developing larvae. These parasites have killed millions of honeybees since first appearing in the United States in 1987—probably among bees imported from South America.

A number of *viruses* are known to affect the winter survival of European honeybees. Recently, scientists discovered that the tobacco ringspot virus could be infecting honeybees that feed on pollen containing the virus. The virus is thought to attack the bees' nervous systems. In addition, the virus has been detected in varroa mites, which may help spread the virus as they feed on honeybees.

As honeybees forage for nectar, they come in contact with harmful *pesticides* that they sometimes carry back to the hives. Research indicates that widely used insecticides called *neonicotinoids*—derived from nicotine that tobacco plants use as a natural insecticide—may play a role in CCD. Evidence suggests that they can disrupt the nervous systems of bees and can decrease their ability to find their way back to their hives. These chemicals can also disrupt the immune systems of bees and make them vulnerable to the harmful effects of other threats.

Pesticides in the air can also end up in the stored pollen in hives. A researcher at a U.S. Department of Agriculture (USDA) laboratory in North Carolina found more than 170 different pesticides in samples of bees, honeycomb wax, and stored pollen. Each pesticide exposes the bees to a harmless dose, but exposure to such a cocktail of pesticides can weaken bees' immune systems and make them vulnerable to deadly parasites, viruses, and fungi.

Stress from being transported long distances around the United States (Figure 9.B) can also play a role. Overworking and overstressing honeybees by moving them around the country can weaken their immune systems and make them more vulnerable to death from parasites, viruses, fungi, and pesticides.

Another factor is *poor nutrition*. In natural ecosystems, honeybees gather nectar and pollen from a variety of flowering plants, but industrial worker honeybees feed mostly on pollen or nectar from one crop or a small number of crops that may lack the nutrients they need. In winter, bees in hives where most of the honey has been removed for sale are often fed sugar or high fructose corn syrup that provide calories but not enough protein for good health.

The growing consensus among bee researchers is that CCD occurs because of a combination of these factors. These annual bee deaths raise the costs for beekeepers and farmers who use their services and could put many of them out of business if the problem continues. This could lead to higher food prices.

CRITICAL THINKING

Can you think of some ways in which commercial beekeepers could lessen one or more of the threats described here? Explain.

Cristi111/Dreamstime.com

FIGURE 9.B European honeybee hive boxes in an acacia orchard. Each year, commercial beekeepers rent and deliver several million hives by truck to farmers throughout the United States.

The Illegal Killing, Capturing, and Selling of Wild Species

Some protected animals are illegally killed (poached) for their valuable parts or are captured and sold live to collectors. Organized crime has moved into illegal wildlife smuggling because of the huge profits involved. At least two-thirds of all live animals illegally smuggled around the world die in transit. Few of the smugglers are caught or punished.

To poachers, a highly endangered, live eastern mountain gorilla (of which there are about 880 left in the wild) is worth $150,000, and the pelt of an endangered giant panda (1,864 left in the wild in China) can bring $100,000.

A single poached black rhinoceros horn can be worth as much as $250,000 in Asia's black market. Four of the five rhino species, including the northern white rhino (Figure 9.16), are critically endangered, mostly because so many have been illegally killed for their horns. A rhino's horn is composed of keratin, the same protein that makes up your fingernails. Powdered rhino horn has long been used in traditional medicines for a variety of ailments in many Asian countries, including China, India, and Vietnam. It is also alleged to be a male aphrodisiac, but there is no verifiable evidence for this claim.

The illegal killing of elephants, especially African savanna elephants (see Figure 7.1, p. 140), for their valuable ivory tusks has increased in recent years, despite an international ban on the trade of ivory. An adult male elephant's two tusks can be worth $375,000 on the black market. China has the largest market for illegal ivory, followed by the United States. Elephant numbers have fallen from an estimated 10 million in 1913 to around 500,000 today, and elephants are being killed at a rate of 30,000 a year, according to WildAid.

CONSIDER THIS . . .

CONNECTIONS Drones, Elephants, and Poachers

Researchers are using small drones with cameras connected to smart phones to track and monitor wildlife species such as endangered elephants and rhinos in Africa, tigers in Nepal, and orangutans in Sumatra. Drones with infrared cameras can find illegal poachers at night, expose their locations to wildlife rangers, and deter them by using bright strobe lights.

Since 1900 the overall number of the world's wild tigers has declined by 99%, mostly because of a 90% loss of tiger habitat, caused primarily by rapid human population growth, and poaching. More than half of the remaining 3,200 tigers are in India, which is doing more than other countries to protect them by establishing tiger reserves and working hard to protect the tigers in such reserves from poaching (Figure 9.17).

Photoshot Holdings Ltd/Alamy Stock Photo

FIGURE 9.16 A poacher in South Africa killed this critically endangered northern white rhinoceros for its two horns. This species is now extinct in the wild. ***Question:*** What would you say if you could talk to the poacher who killed this animal?

FIGURE 9.17 India has done more than any country to protect its remaining highly endangered tiger population. These poachers were caught while trying to sell a tiger skin in Madhya Pradesh State, India.

The Indian, or Bengal, tiger is at risk because a coat made from its fur can sell for as much as $100,000 in Tokyo. The bones and penis of a single tiger can fetch as much as $70,000 in China, the world's biggest market for such illegal items. According to the World Wildlife Fund, without emergency action to curtail poaching and preserve tiger habitat, few if any tigers, including the Sumatran tiger (Figure 9.2d), will be left in the wild by 2022.

In India, conservation biologist and National Geographic Explorer Krithi Karanth is studying conflicts between the rapidly growing number of humans and the declining populations of wildlife such as tigers and Asian elephants. As habitats for these animals shrink, they damage farmers' crops while trying to find food. Karanth has visited more than 10,000 households across 4,000 villages in India, and has enlisted 500 "citizen scientists" to help her interview villagers and collect data. Her goals are to document the disappearance of wildlife, and the conflicts between humans and wildlife, and to find effective ways to reduce such conflicts.

CONSIDER THIS . . .

THINKING ABOUT Tigers

Would it matter to you if all of the world's wild tigers were to disappear? Explain. List two steps you could take to help protect the world's remaining wild tigers from extinction.

Around the globe, the legal and illegal trade in wild species for use as pets is a huge and very profitable business. However, many owners of exotic wild pets do not know that, for every live animal captured and sold in the legal and illegal pet market, many others are killed or die in transit.

According to the IUCN, more than 60 bird species, mostly parrots (Figure 9.6), are endangered or threatened because of the wild bird trade (see Case Study that follows). In response, the United States passed the Wild Bird Conservation Act in 1992. This act makes it illegal to import parrots into the United States. Any parrot purchased today in the United States must be from a domestic breeder.

Buyers of wild animals might also be unaware that some imported exotic animals carry diseases such as Hantavirus, Ebola virus, Asian bird flu, herpes B virus (carried by most adult macaques), and salmonella (from pets such as hamsters, turtles, and iguanas). These diseases can spread from pets to their owners and then to other people.

Other wild species whose populations are depleted because of the pet trade include many amphibians (see Chapter 4, Core Case Study, p. 76), various reptiles, and tropical fishes taken mostly from the coral reefs of Indonesia and the Philippines. Some divers catch tropical fish by using plastic squeeze bottles of poisonous cyanide to stun them. For each fish caught alive, many more die and the cyanide solution kills the polyps, the tiny animals that create coral reefs.

Some exotic plants, especially orchids and cacti (see Figure 7.15, center, p. 151), are endangered because they are removed, and sold, often illegally, to collectors to decorate houses, offices, and landscapes. A mature crested saguaro cactus can earn a cactus rustler as much as $15,000. An orchid collector might pay $5,000 for a single rare orchid.

CASE STUDY

A Disturbing Message from the Birds

Approximately 70% of the world's 10,000 or more known bird species are declining in numbers, and much of this decline is related to human activities, summarized by HIPPCO. According to the IUCN 2014 Red List of Endangered Species, roughly one of every eight (13%) of all bird species is threatened with extinction, mostly by habitat loss, degradation, and fragmentation (the H in HIPPCO)—primarily in tropical forests.

According to the 2014 *State of the Birds* study, almost one-third of the 800 or more bird species in the United States are endangered (Figure 9.2b and c), threatened, or in decline, mostly because of habitat loss and degradation, invasive species, and climate change. About one-third of all endangered and threatened bird species in the United States live in Hawaii.

Sharp declines in bird populations have occurred among songbird species that migrate long distances. These birds nest deep in North American woods in the summer

INDIVIDUALS MATTER 9.1

Juliana Machado Ferreira: Conservation Biologist and National Geographic Explorer

© REBECCA DROBIS

Every year, poachers illegally remove about 38 million wild animals from their natural habitats in Brazil. Some of these animals stay in Brazil and others end up in the United States, Europe, and other parts of the world. Juliana Machado Ferreira is a conservation biologist with a PhD in genetics who fights this illegal removal of wildlife in her native country of Brazil.

She founded FREELAND Brasil to help combat this highly profitable illegal trade. Many people in Brazil keep parrots, macaws, songbirds, monkeys, and other wild animals in their homes as pets and believe that it is harmless cultural tradition. Her organization educates the public about the harmful ecological effects of removing birds and other species from the wild for the amusement of the people who take them.

The group is also doing research on the ecological effects of removing animals from the wild, rescuing mistreated and abandoned animals, and trying to return some of them to the wild. Ferreira has used her knowledge of genetics to develop molecular markers that can help identify the origins of illegal birds seized by police so that the birds can be returned to the areas where they lived. Ferreira is also trying to get the Brazilian government to pass and enforce strict laws against illegal wildlife trafficking.

In 2014 she was selected as a National Geographic Explorer. When asked what people can do to help save wild species, she says: "Do not regard wild animals as pets."

---

and spend their winters in Central or South America or on the Caribbean Islands. The primary causes of these population declines appear to be habitat loss and fragmentation of the birds' breeding habitats in North America and Central and South America. In addition, the populations of 40% of the world's water birds are in decline, mostly because of the global loss of wetlands.

After habitat loss, the intentional or accidental introduction of nonnative species such as bird-eating rats is the second greatest danger, affecting about 28% of the world's threatened birds. Other invasive species (the I in HIPPCO) include snakes (such as the brown tree snake) and mongooses, which kill hundreds of millions of birds every year. In the United States, feral cats and pet cats kill at least 1.4 billion birds each year, according to a study by Peter Mara of the Smithsonian Conservation Biology Institute.

Population growth, the first P in HIPPCO, also threatens some bird species, as more people spread out over the landscape each year and increase their use of timber, food, and other resources, which results in destruction or disturbance of bird habitats. According to bird expert Daniel Klem, Jr., about 600 million birds die each year from collisions with windows in the United States and Canada.

Pollution, the second P in HIPPCO, is another major threat to birds. Countless birds are exposed to oil spills, insecticides, herbicides, and toxic lead from shotgun pellets that fall into wetlands and from lead sinkers left by anglers. (However, the use of lead in shotgun pellets and sinkers in the United States is being phased out.) A 2015 study by researchers led by Leanne M. Flair found that persistent chemicals known as polychlorinated biphenyls (PCBs) may be hindering the ability of some common migratory songbirds to migrate, thereby playing a role in their decline.

Another rapidly growing threat to birds is climate change, the C in HIPPCO. A study done for the WWF found that the effects of climate change, such as heat waves and flooding, are causing declines of some bird populations in every part of the globe. Such losses are expected to increase sharply during this century.

Overexploitation (the O in HIPPCO) is also a major threat to bird populations. Fifty-two of the world's 388 parrot species are threatened, partly because so many parrots are captured for sale as pets, often illegally and usually to buyers in Europe and the United States. According to the USFWS, collectors of exotic birds might pay $10,000 or more for an endangered hyacinth macaw (Figure 9.6) smuggled out of Brazil (Individuals Matter 9.1). However, during its lifetime, a single hyacinth macaw left in the wild could attract an estimated $165,000 in ecotourism revenues.

At least 23 species of seabirds face extinction, largely due to the harmful effects of industrialized fishing. For example, many diving birds drown after becoming hooked on baited lines or trapped in huge nets that are set out by fishing boats.

Biodiversity scientists view this decline of bird species with alarm. One reason is that birds are excellent *indicator*

INDIVIDUALS MATTER 9.2

Çağan Hakkı Sekercioğlu: Protector of Birds and National Geographic Explorer

Rebecca Hale/National Geographic

Çağan Sekercioğlu, assistant professor in the University of Utah Department of Biology, is a bird expert, a tropical biologist, an accomplished wildlife photographer, and a National Geographic Explorer. He has seen over 64% of the planet's known bird species in 75 countries, developed a global database on bird ecology, and become an expert on the causes and consequences of bird extinctions around the world.

In 2007 Sekercioğlu founded KuzeyDoğa. It is an award-winning ecological research and community-based conservation organization devoted to conserving and protecting the wildlife of northeastern Turkey. He also developed Turkey's first protected wildlife corridor, which would stretch across the eastern half of the country, according to his plan. In 2011 he was named Turkey's Scientist of the Year.

Based on his extensive research Sekercioğlu estimates that the percentage of the world's known bird species that are endangered could approximately double from 13% in 2013 to 25% by the end of this century. He says, "My ultimate goal is to prevent extinctions and consequent collapses of critical ecosystem processes while making sure that human communities benefit from conservation as much as the wildlife they help conserve. . . . I don't see conservation as people versus nature, I see it as a collaboration."

species because they live in every climate and biome, respond quickly to environmental changes in their habitats, and are relatively easy to track and count. To these scientists, the decline of many bird species indicates widespread environmental degradation.

A second reason for alarm is that birds perform critically important economic and ecosystem services throughout the world. For example, many birds play specialized roles in pollination and seed dispersal, especially in tropical areas. Extinctions of these bird species could lead to extinctions of plants that depend on the birds for pollination. Then, some specialized animals that feed mostly on these plants might also become extinct. Such a *cascade of extinctions,* in turn, could affect our own food supplies and well-being. Biodiversity scientists (Individuals Matter 9.2) urge us to listen more carefully to what birds are telling us about the state of the environment, for the birds' sake, as well as for ours.

CONSIDER THIS . . .

CONNECTIONS African Vultures and Poachers

Detritus feeders such as vultures circle above animals such as elephants and rhinos that have been killed by poachers for their valuable ivory tusks and horns. This can help wildlife protection rangers in Africa locate poachers. To prevent this, poachers in parts of Africa have been killing thousands of vultures by poisoning the carcasses of dead elephants and rhinos. This is endangering some vulture species and preventing them from playing their important role in the chemical cycling of nutrients needed by plants.

Rising Demand for Bushmeat Threatens Some African Species

For centuries, indigenous people in much of West and Central Africa have sustainably hunted wildlife for *bushmeat* as a source of food. In the last three decades, bushmeat hunting in some areas has skyrocketed. Some hunters provide the bushmeat as a food source for rapidly growing populations. Others sell exotic meats from gorillas (Figure 9.18) and other species to restaurants in major cities. Logging roads in once-inaccessible forests have made hunting these animals much easier. As a result, some forests in areas such as Africa's Congo basin are being stripped of many of their antelopes (the most commonly hunted bushmeat animal), monkeys, apes, elephants, hippos, and other wild animals.

Bushmeat hunting has driven at least one species—Miss Waldron's red colobus monkey—to complete extinction. It is also a factor in the reduction of some populations of orangutans (Figure 9.4), gorillas, chimpanzees, elephants, and hippopotami. Another problem is that butchering and eating some forms of bushmeat has helped to spread fatal diseases such as HIV/AIDS and the Ebola virus from wild animals to humans.

The U.S. Agency for International Development (USAID) is trying to reduce unsustainable hunting for bushmeat in some areas of Africa by introducing alternative sources of food, including farmed fish. They are also showing villagers how to breed large rodents such as cane rats as a source of protein.

FIGURE 9.18 *Bushmeat* such as this severed head of an endangered lowland gorilla in the Congo is consumed as a source of protein by local people in parts of West and Central Africa. It is also sold in national and international marketplaces and served in some restaurants, where wealthy patrons regard gorilla meat as a source of status and power. ***Critical thinking:*** How, if at all, is this different from killing a cow for food?

9.4 HOW CAN WE SUSTAIN WILD SPECIES AND THE ECOSYSTEM SERVICES THEY PROVIDE?

CONCEPT 9.4 We can reduce species extinction and sustain ecosystem services by establishing and enforcing national environmental laws and international treaties and by creating and protecting wildlife sanctuaries.

Treaties and Laws Can Help Protect Species

Some governments are working to reduce species extinction and sustain ecosystem services (see the Case Study that follows) by establishing and enforcing international treaties and conventions, as well as national environmental laws (**Concept 9.4**).

One of the most far reaching of international agreements is the 1975 *Convention on International Trade in Endangered Species of Wild Fauna and Flora (CITES)*. This treaty, signed by 181 countries, bans the hunting, capturing, and selling of threatened or endangered species. It lists 931 species that are in danger of extinction and that cannot be commercially traded as live specimens or for their parts or products. It restricts the international trade of roughly 5,600 animal species and 30,000 plant species that are at risk of becoming threatened.

CITES has helped reduce the international trade of many threatened animals, including elephants, crocodiles, cheetahs, and chimpanzees. But the effects of this treaty have been limited because enforcement varies from country to country and convicted violators often pay only small fines. In addition, member countries can exempt themselves from protecting any listed species, and much of the highly profitable illegal trade in wildlife and wildlife products goes on in countries that have not signed the treaty.

Another important treaty is the *Convention on Biological Diversity (CBD)*, ratified or accepted by 196 countries (but as of 2015, not by the United States). CBD legally commits participating governments to reduce the global rate of biodiversity loss and to share the benefits from use of the world's genetic resources. It also includes efforts to prevent or control the spread of harmful invasive species.

This convention is a landmark in international law because it focuses on ecosystems rather than on individual species. It also links biodiversity protection to issues such as the traditional rights of indigenous peoples. However, implementation has been slow because some key countries, including the United States, have not ratified it. The law also contains no severe penalties or other enforcement mechanisms.

CASE STUDY

The U.S. Endangered Species Act

Some countries have strong laws that help sustain wild species and ecosystem services. The United States enacted the **Endangered Species Act (ESA)** in 1973 and has amended it several times. The act is designed to identify and protect endangered species in the United States and abroad (**Concept 9.4**). The ESA creates recovery programs for the species it lists. Its goal is to help each species' numbers recover to levels where legal protection is no longer needed. When that happens, a species can be taken off the list, or delisted.

Under the ESA, the National Marine Fisheries Service (NMFS) is responsible for identifying and listing endangered and threatened ocean species, while the U.S. Fish and Wildlife Service (USFWS) identifies and lists all other endangered and threatened species. Any decision by either agency to list or delist a species must be based on biological factors alone, without consideration of economic or political factors. However, economic factors can be used in deciding whether and how to protect endangered habitat and in developing recovery plans for listed species. The ESA also forbids federal agencies (except the U.S. Department of Defense) to carry out, fund, or authorize projects that would jeopardize any endangered or threatened species or destroy or modify its critical habitat.

For offenses committed on private lands, fines as high as $100,000 and 1 year in prison can be imposed to ensure protection of the habitats of endangered species. Although this provision has rarely been used, it has been controversial because at least 90% of the listed species spend part of their life cycles on private lands. Since 1982, the ESA has been amended to give private landowners various economic incentives to save endangered species living on their lands. The ESA also makes it illegal for Americans to sell or buy any product made from an endangered or threatened species or to hunt, kill, collect, or injure such species in the United States.

This act is probably the most successful and far-reaching environmental law adopted by any nation. Between 1973 and December 2015, the number of U.S. species on the official endangered and threatened species lists increased from 92 to 1,591, with 1,160 (73%) having active recovery plans. According to a study by the Nature Conservancy, 33% of the country's species are at risk of extinction, and 15% of all species are at high risk—far more than the current number listed.

According to a study by the CBD, 90% of the ESA-protected species are recovering at the rate projected in their recovery plans and 99% of the listed species have been saved from extinction. In addition, between 2003 and 2014, the cumulative area designated as critical habitats increased almost tenfold. Successful recovery plans include those for the American alligator (see p. 83), gray wolf, peregrine falcon, bald eagle (Figure 9.19), humpback whale, and brown pelican. GOOD NEWS

The ESA also requires that all commercial shipments of wildlife and wildlife products enter or leave the country through one of 17 designated airports and ocean ports. The 120 full-time USFWS inspectors can inspect only a small fraction of the more than 200 million wild animals brought legally into the United States annually. Each year, tens of millions of wild animals are also brought in illegally, but few illegal shipments of endangered or threatened animals or plants are confiscated. In addition, many violators are not prosecuted and convicted violators often pay only a small fine.

FIGURE 9.19 The American bald eagle has been removed from the U.S. endangered species list. This eagle is about to catch a fish with its powerful talons.

Since 1995, there have been numerous efforts to weaken the ESA and to reduce its meager annual budget. Opponents of the act contend that it puts the rights and welfare of endangered plants and animals above those of people. Some critics would do away with this act entirely. They call it an expensive failure because, as of spring 2015, only 30 species had recovered enough to be removed from the endangered list (Figure 9.19). Ten species also became extinct while on the list, although eight of those species were very likely extinct when they were listed.

Most biologists insist that the act is one of the world's most successful environmental laws, for several reasons. *First,* species are listed only when they are in serious danger of extinction. ESA supporters argue that this is similar to a hospital emergency department set up to take only the most desperate cases, often with little hope for recovery. Such a facility could not be expected to save all or even most of its patients.

Second, according to federal data, the conditions of more than half of the listed species are stable or improving, 90% are recovering at rates specified by their recovery plans, and 99% of the protected species are still surviving. A hospital emergency department having similar results would be considered an astounding success story.

Third, it takes many decades for a species to reach the point where it is in danger of extinction. Thus, it takes many decades to bring a species back to the point where it can be removed from the endangered list.

Fourth, the 2014 budget for protecting endangered species amounted to an average expenditure of about 61 cents per U.S. citizen. To ESA supporters, it is amazing that the federal agencies responsible for enforcing the act have managed to stabilize or improve the conditions of 99% of the listed species on such a small budget.

A national poll conducted by the CBD and Public Policy Polling in 2013 found that two out of three Americans polled want the ESA strengthened or left alone. However, in 2014 some members of Congress were attempting to weaken the law. They were supporting legislation that would require state and congressional approval for adding any new species to the protected list, which would essentially make it impossible to add any new species to the protected list. The proposed legislation would also delist any species after 5 years—a way to phase out this popular law without voting to repeal it. And it would allow state governors to decide whether and how their states would comply with ESA regulations.

Some ESA supporters agree that the act can be improved. They cite a U.S. National Academy of Sciences study that recommended several changes in the way the law is being implemented:

- Greatly increase the meager funding for implementing the act.
- Put greater emphasis on developing recovery plans more quickly.
- When a species is first listed, establish the core of its habitat as critical for its survival and give that area maximum protection.
- Provide more technical and financial assistance to landowners who want to help protect endangered species on their property.

Most biologists and wildlife conservationists believe that the United States also needs a new law that emphasizes protecting and sustaining biological diversity and ecosystem services rather than focusing mostly on saving individual species. (We examine this idea further in Chapter 10.)

Wildlife Refuges and Other Protected Areas

In 1903 President Theodore Roosevelt established the first U.S. federal wildlife refuge at Pelican Island, Florida, to help protect the brown pelican and other birds from extinction (Figure 9.20). This approach has worked well. In 2009 the brown pelican was removed from the U.S. Endangered Species list. By 2015 there were more than 560 refuges in the National Wildlife Refuge System. Each year, more than 47 million Americans visit these refuges to hunt, fish, hike, and watch birds and other wildlife. GOOD NEWS

More than three-fourths of the refuges serve as wetland sanctuaries that are vital for protecting migratory waterfowl. At least one-fourth of all U.S. endangered and threatened species have habitats in the refuge system, and some refuges have been set aside specifically for certain endangered species (**Concept 9.4**). Such areas have helped Florida's key deer, the brown pelican, and the trumpeter swan to recover.

FIGURE 9.20 The Pelican Island National Wildlife Refuge in Florida was America's first National Wildlife Refuge.

George Gentry/U.S. Fish and Wildlife Service; Inset: Chuck Wagner/Shutterstock.com

There is also bad news about the U.S. refuge system. According to a General Accounting Office study, activities that are harmful to wildlife, such as mining, oil drilling, and use of off-road vehicles, are legally allowed in nearly 60% of the nation's wildlife refuges. Biodiversity researchers urge the U.S. government to set aside more refuges and to increase the long-underfunded budget for the refuge system.

Elsewhere in the world, reserves and refuges have also been successful, and public awareness has played a big role in their success. Dereck and Beverly Joubert are National Geographic Explorers and award-winning filmmakers who, for more than 30 years, have been studying, filming, and writing about threatened lions, leopards, and cheetahs, and other big-cat predators in Africa. They hope to heighten public awareness of the plight of these animals. Their efforts have contributed to the establishment of protected reserves for big cats and other African wildlife in Botswana, Tanzania, and Kenya.

National Geographic is funding several other efforts to preserve wild species, including that of Maia Raymundo, who is studying a critically endangered species of fruit bat in the Philippines, threatened by hunting and high rates of deforestation. Conservation biologists are alarmed about the steep decline of many bat species, which provide vital pollination and insect control services. Some populations of these fruit bats have decreased by as much as 98% in large areas of their range. Raymundo's goal is to identify and protect critical habitat areas for the endangered bats.

Seed Banks, Botanical Gardens, and Wildlife Farms

Seed banks are refrigerated, low-humidity storage environments used to preserve genetic information and the seeds of endangered plant species. More than 1,000 seed banks around the world collectively hold about 3 million samples.

Some species cannot be preserved in seed banks. Banks also vary in quality, are expensive to operate, and are difficult to protect against destruction by fire or other mishaps. However, the Svalbard Global Seed Vault, an underground facility on a remote island in the Arctic, will eventually contain 100 million of the world's seeds. It is not vulnerable to fires or storms.

The world's 1,600 *botanical gardens* contain living plants that represent almost one-third of the world's known plant species. But they contain only about 3% of the world's rare and threatened plant species and have limited space and funding to preserve most of those species. Similarly, an *arboretum* is land set aside for protecting, studying, and displaying various species of trees and shrubs. There are hundreds of arboreta around the world.

We can take pressure off some endangered or threatened species by raising individuals of these species on *farms* for commercial sale. In Florida, American alligators are raised on farms for their meat and hides. Butterfly farms established to raise and protect endangered species flourish in Papua New Guinea, where many butterfly species are threatened by development activities. These farms are also used to educate visitors about the need to protect butterfly species.

Zoos and Aquariums

Zoos, aquariums, game parks, and animal research centers preserve some individuals of critically endangered animal species. The long-term goal is to reintroduce the species into protected wild habitats.

Two techniques for preserving endangered terrestrial species are egg pulling and captive breeding. *Egg pulling* involves collecting wild eggs laid by critically endangered bird species and then hatching them in zoos or research centers. In *captive breeding*, some or all of the wild individuals of a critically endangered species are collected for breeding in captivity, with the aim of reintroducing the offspring into the wild. Captive breeding has been used to save the peregrine falcon and the California condor (Figure 9.2b).

Other techniques for increasing the populations of captive species include *artificial insemination*, which involves inserting semen into a female's reproductive system. Another technique is *embryo transfer*, the surgical implantation of eggs of one species into a surrogate mother of another species. Also used are incubators, and *cross fostering*, when offspring are raised by parents of a similar species. Scientists also match individuals for mating by using DNA analysis along with computer databases that hold information on family lineages of endangered zoo animals—a computer dating service for zoo animals. This helps increase genetic diversity.

The ultimate goal of captive breeding programs is to build populations to a level where they can be reintroduced into the wild. Successes include the black-footed ferret, the golden lion tamarin (a highly endangered monkey species), the Arabian oryx, and the California condor (Figure 9.2b). For reintroduction to be successful, individuals raised in captivity must be able to survive in the wild. They must have suitable habitat and they must be protected from overhunting, poaching, pollution, and other environmental hazards. These challenges can make reintroduction difficult.

One problem for captive breeding programs is that a captive population of an endangered species must typically number 100 to 500 individuals in order to avoid extinction resulting from accidents, diseases, or the loss of genetic diversity through inbreeding. Recent genetic research indicates that 10,000 or more individuals are needed for an endangered species to maintain its capacity for biological evolution. Zoos and research centers do not have the funding or space to house such large populations.

SISSE BRIMBERG/National Geographic Creative

FIGURE 9.21 The Monterey Bay Aquarium in Monterey, California (USA), contains this tidewater pool, which is used to train rescued sea otter pups to survive in the wild.

Public aquariums (Figure 9.21) that exhibit unusual and attractive species of fish and marine animals such as seals and dolphins help to educate the public about the need to protect such species. Some carry out research on how to save endangered species. However, mostly because of limited funds, public aquariums have not served as effective gene banks for endangered marine species, especially marine mammals that need large volumes of water.

While zoos, aquariums, and botanical gardens perform valuable services, they cannot by themselves solve the growing problem of species extinction.

The Precautionary Principle

Biodiversity scientists call for us to take precautionary action to avoid hastening species extinction and disrupting essential ecosystem services. This approach is based on the **precautionary principle:** When substantial preliminary evidence indicates that an activity can harm human health or the environment, we should take precautionary measures to prevent or reduce such harm even if some of the cause-and-effect relationships have not been fully established scientifically. It is based on the commonsense idea behind many adages, including "Better safe than sorry" and "Look before you leap."

Scientists use the precautionary principle to argue for both the preservation of species and protection of entire ecosystems and their ecosystem services. Implementing this principle puts the emphasis on preventing species extinction instead of waiting until a species is nearly extinct before taking emergency action that can be too late. The precautionary principle is also used as a strategy for dealing with other challenges such as preventing exposure to harmful chemicals in the air we breathe, the water we drink, and the food we eat. We discuss the pros and cons of using this principle to prevent pollution in Chapter 17.

Protecting Species and Ecosystem Services Raises Difficult Questions

Efforts to prevent the extinction of wild species and the accompanying losses of ecosystem services require the use of financial and human resources that are limited. This raises some challenging questions:

- Should we focus on protecting species or should we focus more on protecting ecosystems and the ecosystem services they provide?
- How do we allocate limited resources between these two priorities?
- How do we decide which species should get the most attention in our efforts to protect as many species as possible? For example, should we focus on protecting the most threatened species or on protecting keystone species?
- Protecting species that are appealing to humans, such as the giant panda, orangutans (Figure 9.4), and tigers, can increase public awareness of the need for wildlife conservation. Is this more important than focusing on the ecological importance of species when deciding which ones to protect?
- How do we determine which habitat areas are the most critical to protect?
- How do we allocate limited resources among such biodiversity hotspots?

Conservation biologists struggle with these questions all the time. Regardless of the answers, each of us can help in the efforts to protect species from extinction due largely to human activities. Figure 9.22 lists some ways in which you can help protect species and increase your beneficial environmental impact.

What Can You Do?

Protecting Species

- Do not buy furs, ivory products, or other items made from endangered or threatened animal species.
- Do not buy wood or wood products from tropical or old-growth forests.
- Do not buy pet animals or plants taken from the wild.
- Tell friends and relatives what you're doing about this problem.

FIGURE 9.22 Individuals matter: You can help prevent the extinction of species. ***Critical thinking:*** Which two of these actions do you believe are the most important ones to take? Why?

BIG IDEAS

- We are hastening the extinction of wild species and degrading the ecosystem services they provide by destroying and degrading natural habitats, introducing harmful invasive species, and increasing human population growth, pollution, climate change, and overexploitation.
- We should avoid causing or hastening the extinction of wild species because of the ecosystem and economic services they provide and because their existence should not depend primarily on their usefulness to us.
- We can work to prevent the extinction of species and to protect overall biodiversity and ecosystem services by establishing and enforcing environmental laws and treaties and by creating and protecting wildlife sanctuaries.

Tying It All Together

Honeybees and Sustainability

Malwina Szweda/Shutterstock.com

In this chapter, we learned about the human activities that are hastening the extinction of many species and about how we might curtail those activities. We learned that as many as half of the world's known wild species could go extinct during this century, largely because of human activities that threaten many species and some of the vital ecosystem services they provide. For example, populations of honeybees, vital for pollinating crops that supply much of our food, have been declining for a variety of reasons (**Core Case Study**), many of them related to human activities. One of the key reasons for such problems is that most people are unaware of the highly valuable ecosystem and economic services provided by the earth's species.

Acting to prevent the extinction of species from human activities implements two of the three **scientific principles of sustainability**. It preserves not only the earth's biodiversity, but also the vital ecosystem services that sustain us, including chemical cycling. It also implements the ethical **principle of sustainability** that call for us to leave the earth in a condition that is as good as or better than what we inherited (see Inside Back Cover).

Chapter Review

Core Case Study

1. What economic and ecological services do honeybees provide? How are human activities contributing to the decline of many populations of European honeybees? What is **colony collapse disorder (CCD)**?

Section 9.1

2. What is the key concept for this section? Define and distinguish between **biological extinction** and **mass extinction**. What is the **background extinction rate**? What percentage of the world's identified species is likely to become extinct primarily because of human activities during this century? How many of the earth's 24 major ecosystem services are in decline? Give three reasons why many extinction experts believe that projected extinction rates are probably on the low side. Explain how scientists estimate extinction rates and describe the challenges they face in doing so. Distinguish between **endangered species** and **threatened species** and give an example of each. List four characteristics that make some species especially vulnerable to extinction.

Section 9.2

3. What is the key concept for this section? What are four reasons for trying to avoid hastening the extinction of wild species? Describe two economic and two ecological benefits of species diversity. Explain how saving other species and the ecosystem services they provide can help us to save our own species and our cultures and economies.

Section 9.3

4. What is the key concept for this section? What is **HIPPCO**? What is the greatest threat to wild species? What is **habitat fragmentation**? Describe the major effects of habitat loss and fragmentation. Why are island species especially vulnerable to extinction? What are habitat islands?

5. Give two examples of the benefits that have been gained by the introduction of nonnative species. What are the pros and cons of planting the kudzu vine? Describe the harmful environmental impacts of the wild boar and Burmese pythons. Explain why prevention is the best way to reduce threats from invasive species and list four ways to implement this strategy.

6. Summarize the roles of population growth, overconsumption, pollution, and climate change in the extinction of wild species. Explain how concentrations of pesticides such as DDT can accumulate to high

levels in food webs. List possible causes of the decline of European honeybee populations in the United States. Give three examples of species that are threatened by poaching. Why are wild tigers likely to disappear from the wild within a few decades? What is the connection between infectious diseases in humans and the pet trade?

7. List the major threats to the world's bird populations and give two reasons for protecting bird species from extinction. Describe the threat to some forms of wildlife from the increased hunting for bushmeat.

Section 9.4

8. What is the key concept for this section? Name two international treaties that are used to help protect species. What is the U.S. Endangered Species Act? How successful has it been, and why is it controversial?

9. Summarize the roles and limitations of wildlife refuges, seed banks, botanical gardens, wildlife farms, zoos, and aquariums in protecting some species. Describe the role of captive breeding in efforts to prevent species extinction and give an example of success in returning a nearly extinct species to the wild. What is the **precautionary principle** and how can it be used to prevent the extinction of species? What are three important questions related to protecting biodiversity?

10. What are this chapter's *three big ideas*? Explain how preventing the extinction of species from human activities implements three of the six **principles of sustainability**.

Note: Key terms are in bold type.

Critical Thinking

1. What are three aspects of your lifestyle that might directly or indirectly contribute to declines in European honeybee populations and the endangerment of other pollinator species (**Core Case Study**)?

2. Give your response to the following statement: "Eventually, all species become extinct. So it does not really matter that the world's remaining tiger species or a tropical forest plant are endangered mostly because of human activities." Be honest about your reaction, and give arguments to support your position.

3. Do you accept the ethical position that each species has the right to survive without human interference, regardless of whether it serves any useful purpose for humans? Explain. Would you extend this right to the *Anopheles* mosquito, which transmits malaria, and to harmful infectious bacteria? Explain. If your answer is no, where would you draw the line?

4. Wildlife ecologist and environmental philosopher Aldo Leopold wrote this with respect to preventing the extinction of wild species: "To keep every cog and wheel is the first precaution of intelligent tinkering." Explain how this statement relates to the material in this chapter.

5. What would you do if wild boar invaded and tore up your yard or garden? Explain your reasoning behind your course of action. How might your actions affect other species or the ecosystem you are dealing with?

6. How do you think your daily habits might contribute directly or indirectly to the extinction of some bird species? What are three things that you think should be done to reduce the rate of extinction of bird species?

7. Which of the following statements best describes your feelings toward wildlife?
   a. As long as it stays in its space, wildlife is okay.
   b. As long as I do not need its space, wildlife is okay.
   c. I have the right to use wildlife habitat to meet my own needs.
   d. When you have seen one redwood tree, elephant, or some other form of wildlife, you have seen them all, so preserve a few of each species in a zoo or wildlife park and do not worry about protecting the rest.
   e. Wildlife should be protected in its current ranges.

8. How might your life change if human activities contribute to the projected extinction of 20–50% of the world's identified species during this century? How might this affect the lives of any children or grandchildren you eventually might have?

Doing Environmental Science

Identify examples of habitat destruction or degradation in the area in which you live or go to school. Try to determine and record any harmful effects that these activities have had on the populations of one wild plant and one animal species. (Name each of these species and describe how they have been affected.) Do some research on the Internet and/or in a school library on *wildlife management plans,* and then develop a management plan for restoring the habitats and species you have studied. Try to determine whether trade-offs are necessary with regard to the human activities you have observed, and account for these trade-offs in your management plan. Compare your plan with those of your classmates.

Global Environment Watch Exercise

Go to your MindTap course to access the GREENR database. Starting on the home page, under "Browse Issues and Topics", click on *Environment and Ecology*, then select *Extinction*, and scroll to statistics on the portal page. Click on "Known Causes of Animal Extinction since 1600." You will find four general categories of causes. Thinking about history from 1600 through today, how do you think humans have changed their impact on species in each of these categories? Has the impact increased or decreased over this period? Give specific examples of changes in this timeframe to support your answers.

Data Analysis

Examine the following data released by the World Resources Institute and answer these questions:

1. Complete the table by filling in the last column. For example, to calculate this value for Costa Rica, divide the number of threatened breeding bird species by the total number of known breeding bird species and multiply the answer by 100 to get the percentage.
2. Arrange the countries from largest to smallest according to total land area. Does there appear to be any correlation between the size of country and the percentage of threatened breeding bird species? Explain your reasoning.

Country	Total Land Area in Square Kilometers (Square Miles)	Protected Area as Percent of Total Land Area (2003)	Total Number of Known Breeding Bird Species (1992–2002)	Number of Threatened Breeding Bird Species (2002)	Threatened Breeding Bird Species as Percent of Total Number of Known Breeding Bird Species
Afghanistan	647,668 (250,000)	0.3	181	11	
Cambodia	181,088 (69,900)	23.7	183	19	
China	9,599,445 (3,705,386)	7.8	218	74	
Costa Rica	51,114 (19,730)	23.4	279	13	
Haiti	27,756 (10,714)	0.3	62	14	
India	3,288,570 (1,269,388)	5.2	458	72	
Rwanda	26,344 (10,169)	7.7	200	9	
United States	9,633,915 (3,718,691)	15.8	508	55	

Compiled by the authors using data from World Resources Institute, Earth Trends, Biodiversity and Protected Areas, Country Profiles.

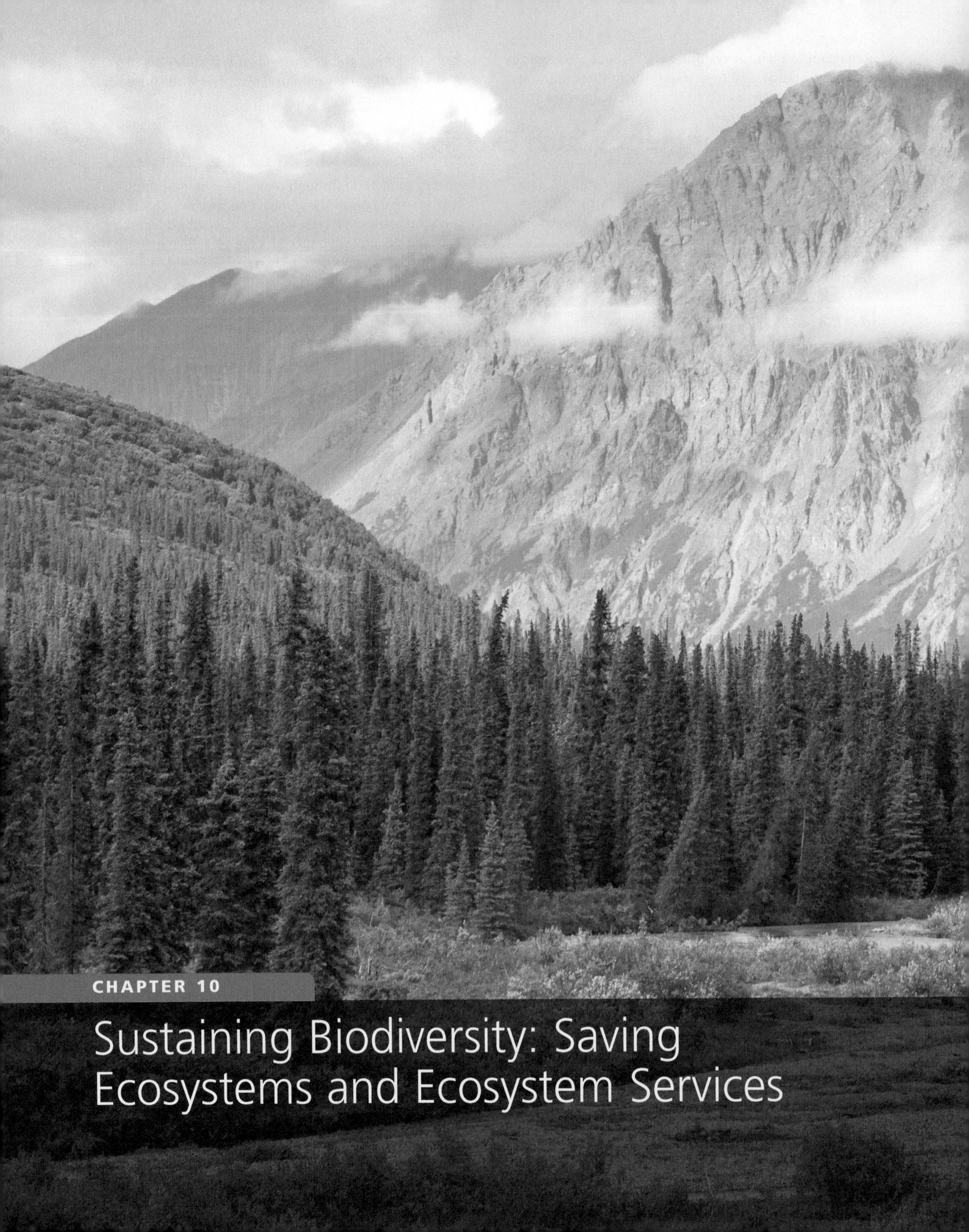

CHAPTER 10

Sustaining Biodiversity: Saving Ecosystems and Ecosystem Services

> Taken at the level of the entire globe, biological diversity can be considered the single measure of how humanity is affecting the environment.
>
> **THOMAS E. LOVEJOY**

Denali National Park, Alaska (USA).

Jerryway/Dreamstime.com

Key Questions

10.1 What are the major threats to forest ecosystems?

10.2 How can we manage and sustain forests?

10.3 How can we manage and sustain grasslands?

10.4 How can we sustain terrestrial biodiversity and ecosystem services?

Core Case Study

Costa Rica—A Global Conservation Leader

Tropical forests once covered Central America's Costa Rica, a country smaller in area than the U.S. state of West Virginia. Between 1963 and 1983, politically powerful ranching families cleared much of the country's forests in order to graze cattle.

Despite such widespread forest loss, Costa Rica is a superpower of biodiversity, with an estimated 500,000 plant and animal species. A single park in Costa Rica is home to more bird species than are found in all of North America.

This oasis of biodiversity results mostly from two factors. One is the country's tropical geographic location. It lies between two oceans and has coastal and mountainous regions with a variety of microclimates and habitats for wildlife. The other is the government's strong commitment to land conservation.

In the mid-1970s, Costa Rica established a system of nature reserves and national parks (Figure 10.1). By 2014, this system included more than 27% of its land—6% of it reserved for indigenous peoples. Costa Rica has increased its beneficial environmental impact by devoting a larger proportion of its land to biodiversity conservation than any other country—an example of the biodiversity **principle of sustainability** in action.

To reduce *deforestation*—the clearing and loss of forests—the government eliminated subsidies for converting forestland to rangeland. Instead, it pays landowners to maintain or restore forests and manage them under 10- to 15-year agreements.

The strategy worked. Costa Rica has gone from having one of the world's highest deforestation rates to having one of the lowest. Over three decades, forests went from covering 20% of its land to covering 50% of it.

Ecologists warn that human population growth, economic development, and poverty put increasing pressure on the earth's ecosystems and on the ecosystem services they provide. According to a recent joint report by two United Nations environmental bodies: "Unless radical and creative action is taken to conserve the earth's biodiversity, many local and regional ecosystems that help to support human lives and livelihoods are at risk of collapsing."

This chapter is devoted to helping you understand the threats to the earth's forests, grasslands, and other storehouses of terrestrial biodiversity. We will also look at ways to sustain these vital ecosystems and the ecosystem services they provide.

gary yim/Shutterstock.com

FIGURE 10.1 La Fortuna Falls is located in a tropical rain forest in Costa Rica's Arenal Volcano National Park.

10.1 WHAT ARE THE MAJOR THREATS TO FOREST ECOSYSTEMS?

CONCEPT 10.1A Forest ecosystems provide ecosystem services far greater in economic value than the value of wood and other raw materials they provide.

CONCEPT 10.1B Unsustainable cutting and burning of forests and climate change are the chief threats to forest ecosystems.

Forests Provide Important Economic and Ecosystem Services

Forests provide highly valuable economic and ecosystem services (Figure 10.2 and **Concept 10.1A**). Through photosynthesis, forests remove CO_2 from the atmosphere and store it in organic compounds as part of the global carbon cycle (see Figure 3.20, p. 65). By performing this ecosystem service, forests help stabilize average atmospheric temperatures and moderate the earth's climate. Forests also produce oxygen, purify water, and reduce runoff and flooding by storing water and releasing it slowly.

Natural Capital

Forests

Ecosystem Services	Economic Services
Support energy flow and chemical cycling	Fuelwood
Reduce soil erosion	Lumber
Absorb and release water	Pulp to make paper
Purify water and air	Mining
Influence local and regional climate	Livestock grazing
Store atmospheric carbon	Recreation
Provide numerous wildlife habitats	Jobs

FIGURE 10.2 Forests provide important ecosystem and economic services (**Concept 10.1A**). ***Critical thinking:*** Which two ecosystem services and which two economic services do you think are the most important?

Photo: Val Thoermer/Shutterstock.com

Forests support biodiversity by providing habitats for about two-thirds of the earth's terrestrial species. They are also home for more than 300 million people and about 1 billion people living in extreme poverty depend on forests for their survival.

Along with highly valuable ecosystem services, forests provide us with raw materials, especially wood. More than half of the wood removed from the earth's forests is used as *biofuel* for cooking and heating. The remainder of the harvest, called *industrial wood,* is used primarily to make lumber and paper.

Forests play a role in maintaining human health. Traditional medicines, used by 80% of the world's people, are derived mostly from plant species native to forests. Certain chemicals found in tropical forest plants are used as blueprints for making most prescription drugs (Figure 9.5, p. 197). Forests also remove air pollutants, including 13% of U.S. emissions of climate-changing carbon dioxide (CO_2) from the burning of fossil fuels.

Scientists and economists have estimated the economic value of major ecosystem services that the world's forests and other ecosystems provide (Science Focus 10.1).

Forests Vary in Age and Structure

Natural and planted forests occupy about 31% of the earth's land surface (excluding Greenland and Antarctica). In order, the five countries with the largest forest areas are Russia, Brazil, Canada, the United States, and China.

Scientists divide forests into major types, based on their age and structure. Forest structure refers to the distribution of vegetation, both horizontally and vertically, as well as the type, size, and shape of that vegetation. An **old-growth forest,** or **primary forest,** is an uncut or regrown forest that has not been seriously disturbed by human activities or natural disasters for 200 years or more (Figure 10.3). Old-growth forests are reservoirs of

FIGURE 10.3 Old-growth forest in Poland.

Putting a Price Tag on Nature's Ecosystem Services

Currently, forests and other ecosystems are valued mostly for their economic services (Figure 10.2, right). Ecologists and ecological economists have estimated the monetary value of the ecosystem services that forests and other ecosystems provide. This information can be used to apply the full-cost pricing **principle of sustainability** to forest ecosystem services and increase our beneficial environmental impact.

In 2014 a team of ecologists, economists, and geographers, led by ecological economist Robert Costanza, estimated the monetary worth of 17 ecosystem services provided by 16 of the earth's biomes. Examples are waste treatment, erosion control, climate regulation, nutrient cycling, habitat, food production, and recreation.

Their conservative estimate was that the global monetary value of these services is at least \$125 trillion per year—much more than the \$76 trillion that the entire world spent on goods and services in 2014. This means that every year, the earth provides you and every other person in the world with ecosystem services worth at least \$17,123, on average. On a global basis, the top five ecosystem services are waste treatment (\$22.5 trillion per year), recreation (\$20.6 trillion), erosion control (\$16.2 trillion), food production (\$14.8 trillion), and nutrient cycling (\$11.1 trillion).

The researchers also estimated that since 1997 the world has been losing ecosystem services with an estimated value of about \$20.2 trillion a year. This annual loss is more than the \$16.5 trillion gross national product (GNP) of the United States in 2014.

According to their research, the world's forests provide us with ecosystem services worth at least \$15.6 trillion per year. This is hundreds of times more than the economic value of lumber, paper, and other wood products that forests provide us (**Concept 10.1A**). About 71% of the total value of ecosystem services from forests comes from climate regulation, genetic resources, and recreation.

We can draw four important conclusions from this and related studies: **(1)** the earth's ecosystem services are essential for all humans and their economies; **(2)** the economic value of these services is huge; **(3)** these ecosystem services will be an ongoing source of ecological income, as long as the ecosystems that provide these services are used sustainably; and **(4)** we need to use *full-cost pricing* to include the huge economic value of these irreplaceable ecosystem services in the prices of goods and services provided by forests.

CRITICAL THINKING

Some people believe that we should not try to put economic values on the world's irreplaceable ecosystem services because their value is infinite. Do you agree with this view? Explain. What is the alternative?

\$125 trillion

Conservative estimate of the annual value of nature's ecosystem services

biodiversity because they provide ecological niches for a multitude of wildlife species (see Figure 7.20, p. 158).

A **second-growth forest** is a stand of trees resulting from secondary ecological succession (see Figure 5.12, p. 108). They develop after the trees in an area have been removed by human activities, such as clear-cutting for timber or conversion to cropland, or by natural forces such as fire or hurricanes.

A **tree plantation,** also called a **tree farm** or **commercial forest** (Figure 10.4), is a managed forest containing only one or two species of trees that are all the same age. They are often grown on land that was cleared of an old-growth or second-growth forest and are usually harvested by clear-cutting as soon as they become commercially valuable. The land is then replanted and clear-cut again in a regular cycle (see drawing, Figure 10.4). When managed carefully, plantations can produce wood at a rapid rate and could supply most of the wood used for industrial purposes, including papermaking and construction. This would help protect the world's remaining old-growth and second-growth forests, as long as they are not cleared to make room for tree plantations.

Use of tree plantations is increasing, and they now occupy about 7% of the world's forest area. In order, the six countries with the largest areas of tree plantations are China, India, the United States, Russia, Canada, and Sweden.

The downside of tree plantations is that, with only one or two tree species, they are less biologically diverse and less sustainable than old-growth and second-growth forests. Tree plantations also do not provide the amount of wildlife habitat and many of the ecosystem services that diverse natural forests do. In addition, repeated cutting and replanting of trees can eventually deplete the nutrients in the plantation's topsoil. This can hinder the regrowth of any type of forest on such land.

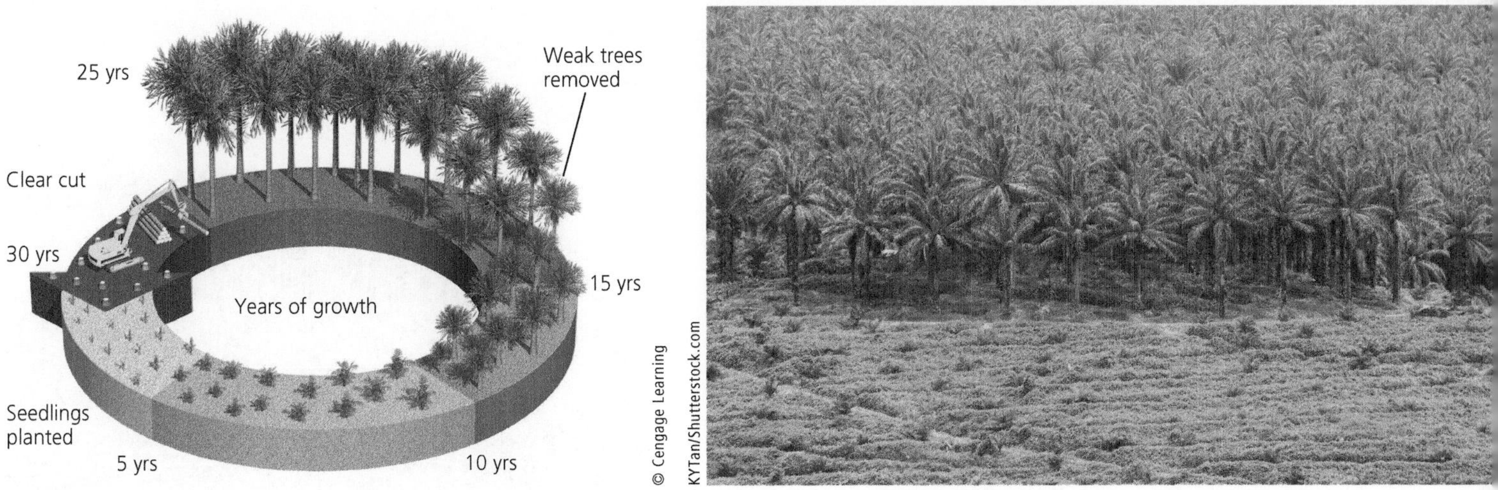

FIGURE 10.4 Oil palm tree plantation. A large area of diverse tropical forest was cleared and planted with this monoculture of oil palm trees.

Ways to Harvest Trees

Because of the immense economic value of forests, the harvesting of wood for timber and to make paper is one of the world's major industries. The first step in harvesting trees is to build roads for access and timber removal. Even carefully designed logging roads can have harmful effects (Figure 10.5), including topsoil erosion, sediment runoff into waterways, habitat loss, and loss of biodiversity. Logging roads also expose forests to invasion by disease-causing organisms and nonnative pests, and to disturbances from human activities such as mining, farming, and ranching.

Loggers use a variety of methods to harvest trees. With *selective cutting*, loggers cut intermediate-aged or mature trees singly or in small groups, leaving the forest largely intact (Figure 10.6a). This allows a forest to produce economically valuable trees on a sustainable basis. However, loggers often remove all the trees from an area in what is called a *clear-cut* (Figure 10.6b and Figure 10.7). Clear-cutting is the most efficient and sometimes the most cost-effective way to harvest trees. It also provides profits in the shortest time for landowners and timber companies. However, clear-cutting can harm or destroy an ecosystem by causing forest soil erosion, sediment pollution of nearby waterways, and losses in biodiversity. Clear-cutting also contributes to atmospheric warming and the resulting climate change by releasing stored carbon dioxide (CO_2) into the atmosphere and reducing the uptake of CO_2 by forests.

A variation of clear-cutting that allows a more sustainable timber yield without widespread destruction is *strip cutting* (Figure 10.6c). It involves clear-cutting a strip of trees along the contour of the land within a corridor narrow enough to allow natural forest regeneration within a few years. After regeneration, loggers cut another strip next to the first, and so on.

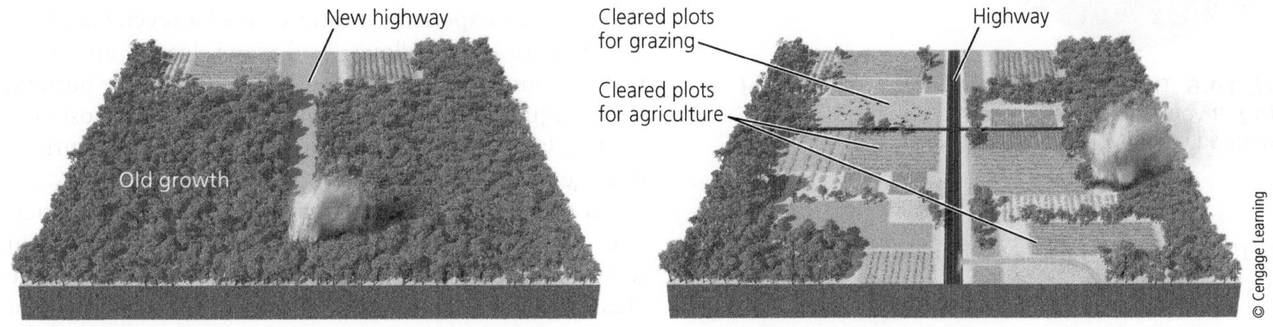

FIGURE 10.5 Natural capital degradation: Building roads into previously inaccessible forests is the first step in harvesting timber, but it also paves the way to fragmentation, destruction, and degradation of forest ecosystems.

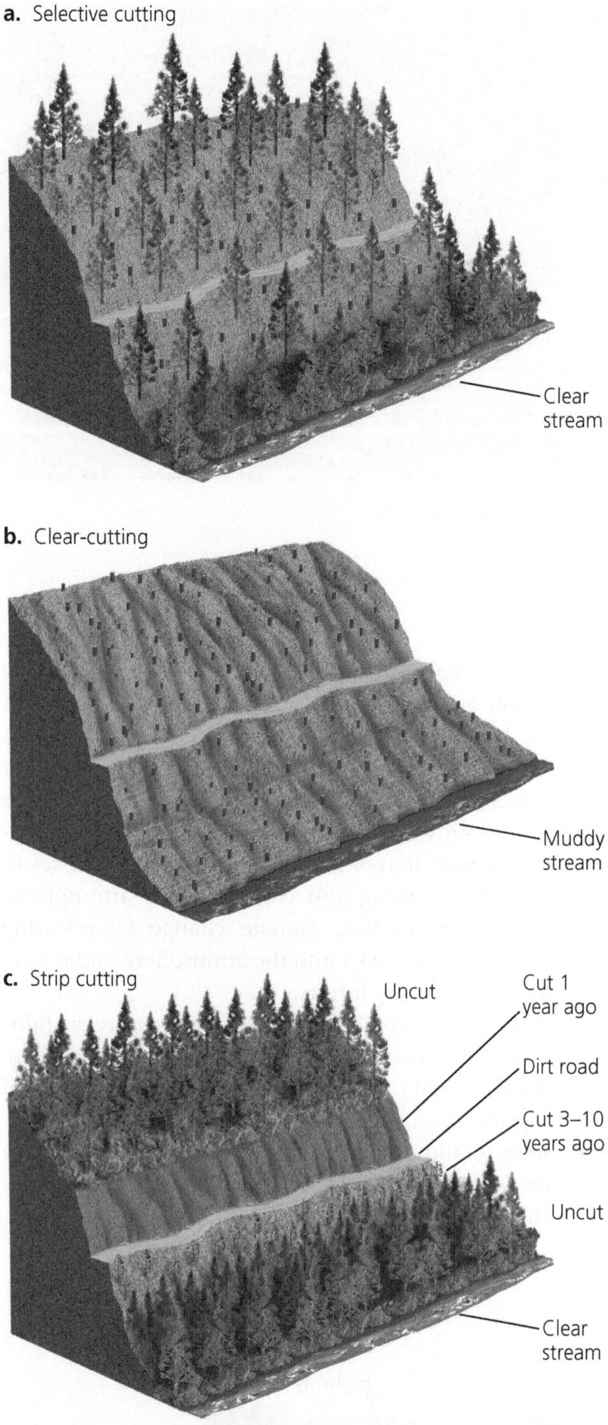

FIGURE 10.6 Three major ways to harvest trees. ***Critical thinking:*** If you were cutting trees in a forest you owned, which method would you choose and why?

Fires and Insects Affect Forest Ecosystems

Two types of fires can affect forest ecosystems. *Surface fires* (Figure 10.8, left) usually burn only undergrowth and leaf litter on the forest floor. They kill seedlings and small trees,

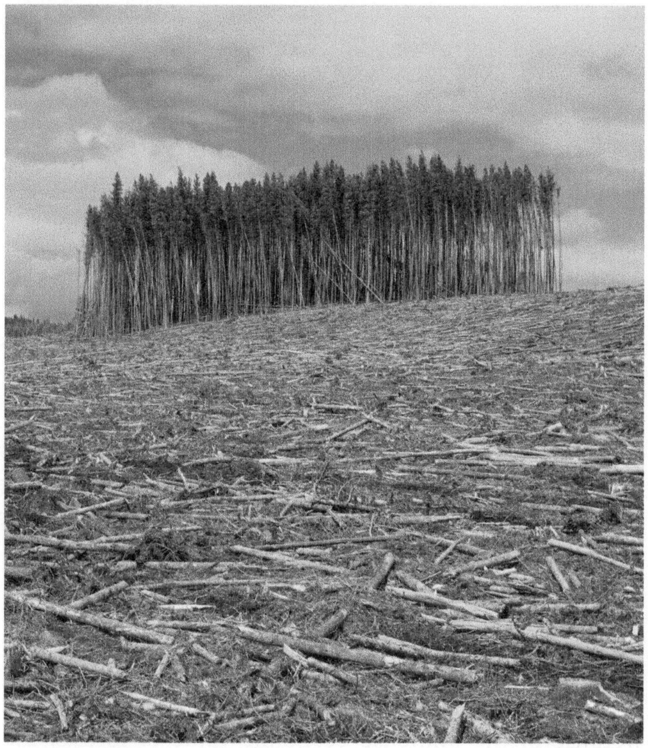

FIGURE 10.7 Clear-cut forest.

but spare most mature trees and allow most wild animals to escape.

Occasional surface fires have several ecological benefits. They:

- burn away flammable material such as dry brush to help prevent fires that are more destructive;
- free valuable plant nutrients trapped in slowly decomposing litter and undergrowth;
- release seeds from the cones of tree species such as lodgepole pines and stimulate the germination of other seeds such as those of the giant sequoia; and
- help control destructive insects and tree diseases.

Another type of fire, called a *crown fire* (Figure 10.8, right), is an extremely hot fire that leaps from treetop to treetop, burning whole trees. They usually occur in forests that have not experienced surface fires for several decades. The absence of fire allows dead wood, leaves, and flammable ground litter to accumulate. These rapidly burning fires can jump to the forest canopy and destroy most vegetation, kill wildlife, increase topsoil erosion, and burn or damage buildings and homes.

In recent years, wildfires have increased in frequency, size, and intensity in the western United States and in many other areas of the world. A key factor is climate change, which is creating hotter and drier conditions that lead to longer fire seasons. Another factor is suppression of smaller fires, which leads to dense and overgrown forests with a large accumulation of flammable ground litter. Fire

David J Moorhead, University of Georgia, Bugwood.org

age fotostock/Alamy Stock Photo

FIGURE 10.8 Surface fires (left) usually burn only undergrowth and leaf litter. They can help to prevent more destructive crown fires (right).

damage has also increased because of the rapid growth in the number of homes and communities in forested areas. According to the Intergovernmental Panel on Climate Change (IPCC), each 1°C (1.8°F) increase in the global average temperature is likely to increase the area of forest burned in the western United States by a factor of 2 to 5.

Insects have always threatened forests. However, in recent years pine bark beetles, which are about the size of a grain of rice, have devastated large areas of conifer pine forests in the western areas of Canada and the United States (Figure 10.9). The primary reason for this population explosion of pine bark beetles and several other insect pests is a warmer climate with winters that do not get cold enough to kill off the insects and control their populations. The warmer winters also allow the beetles to spread to forests at higher elevations and latitudes.

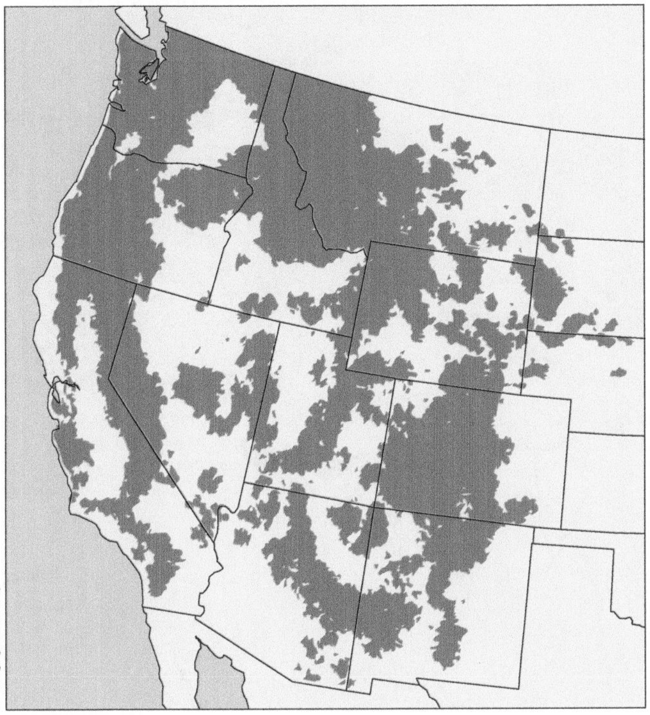

© Cengage Learning

FIGURE 10.9 Forest destruction by pine bark beetles in the western United States, 2000–2014. (U.S. Forest Service.)

Almost Half of the World's Old-Growth Forests Have Been Cut Down

Deforestation is the temporary or permanent removal of large expanses of forest for agriculture, settlements, or other uses. Surveys by the World Resources Institute (WRI) indicate that during the past 8,000 years, deforestation has eliminated almost half of the earth's old-growth forest cover. Most of this loss occurred in the last 65 years.

These forest losses are concentrated mostly in less-developed countries, especially those in the tropical areas of Latin America, Indonesia, and Africa. However, scientists are also concerned about the increased clearing of the northern boreal forests (see Figure 7.19, bottom, p. 157) of Alaska, Canada, Scandinavia, and Russia that make up about one-fourth of the world's forested area.

According to the WRI, if current deforestation rates continue, about 40% of the world's remaining intact forests will be logged or converted to other uses within two decades if not sooner. Clearing large areas of forests, especially old-growth forests, has important short-term economic benefits (Figure 10.2, right column), but it also has a number of long-term harmful environmental effects (Figure 10.10), including severe erosion and loss of topsoil (Figure 10.11) (**Concept 10.1B**).

Natural Capital Degradation

Deforestation

- Water pollution and soil degradation from erosion
- Acceleration of flooding
- Local extinction of specialist species
- Habitat loss for native and migrating species
- Release of CO_2 and loss of CO_2 absorption

FIGURE 10.10 Deforestation has some harmful environmental effects that can reduce biodiversity and degrade the ecosystem services provided by forests (Figure 10.4, left).

Some countries, such as China, Costa Rica (**Core Case Study**), and the United States (see the following Case Study), have increased their forest cover. Some of these increases resulted from natural reforestation by secondary ecological succession on cleared forest areas and abandoned croplands (see Figure 5.12, p. 108).

Forest cover has also increased because of the spread of commercial tree plantations and a global program, sponsored by the United Nations Environment Programme (UNEP), to plant billions of trees throughout much of the world—many of them in tree plantations. China leads the world in new forest cover, mostly due to its plantations of fast-growing trees. In 2000, China established the world's largest *payments for ecosystem services* (PES) program, in response to extensive flooding from deforestation. It pays individuals to stop logging forests and to reforest already logged lands (Figure 10.12). It worked. By 2010, the program had cut the country's deforestation rate in half and sharply reduced flooding. It also helped participants work their way out of poverty. GOOD NEWS

CASE STUDY

Many Cleared Forests in the United States Have Grown Back

Forests cover about 30% of the U.S. land area. They provide habitats for more than 80% of the country's wildlife species and contain about two-thirds of the nation's surface water.

FIGURE 10.11 Severe soil erosion and desertification caused by the clearing an area of tropical forest followed by livestock overgrazing.

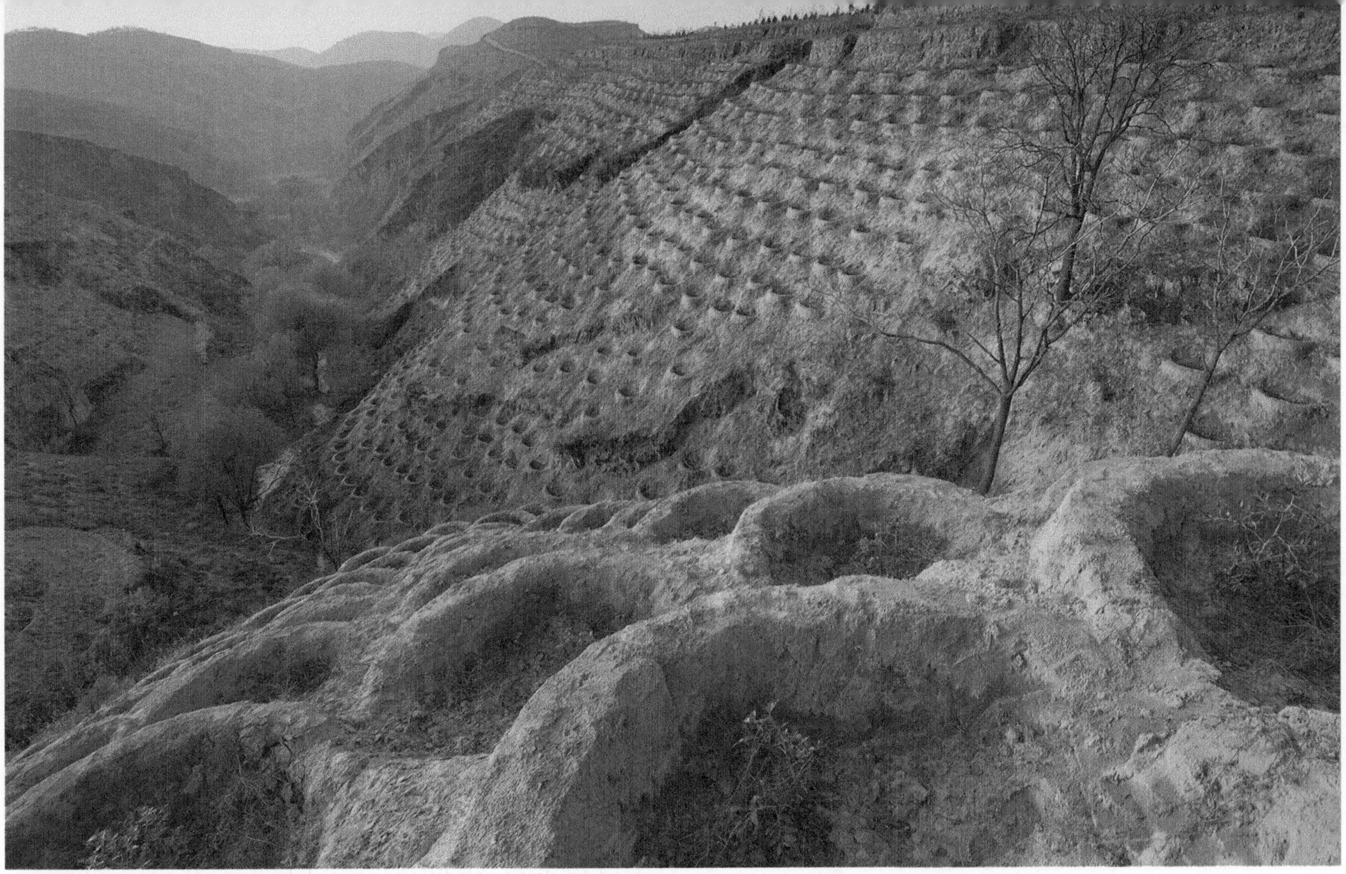

FIGURE 10.12 Tree seedlings planted to reduce erosion and reforest hills in China's Loess Plateau.

Today, forests in the United States (including tree plantations) cover more area than they did in 1920. The primary reason for this is that many of the old-growth forests cleared or partially cleared between 1620 and 1920 have grown back through secondary ecological succession (Figure 10.13).

There are now diverse second-growth (and in some cases third-growth) forests in every region of the United States except in much of the West. Environmental writer Bill McKibben has cited this forest regrowth in the United States—especially in the East—as "the great environmental success story of the United States, and in some ways, the whole world."

Protected forests make up about 40% of the country's total forest area, mostly in the *National Forest System*, which consists of 155 national forests managed by the U.S. Forest Service but owned jointly by the citizens of the United States. However, since the mid-1960s, a large area of the nation's remaining old-growth and diverse second-growth forests have been cut down and replaced with biologically simplified tree plantations.

A new and growing threat is the clearing of hardwood forests in the southern United States to produce wood pellets about as long as a pencil. The pellets are exported to Europe where they are burned in wood-burning power plants that used to burn coal. Critics say that this practice is destroying diverse U.S. forests.

Recent research indicates that old-growth forests have been declining in California (pp. 2–3) and in several other parts of the world. For example, California has lost half of its biggest trees since the 1930s. Large old trees play important keystone ecological roles. They serve as nesting sites and shelter for a variety of species and provide food for numerous animals. They resist damage from wildfires, help reduce climate change by removing CO_2 from the atmosphere, and help build topsoil.

Reasons for these declines are selective harvesting, clear-cutting, and drought. An important and growing factor is a warmer climate, which reduces the snowpacks that provide water for the trees and causes them to lose more water via transpiration. This drying of the trees is contributing to more severe wildfires.

Tropical Forests Are Disappearing Rapidly

Tropical forests (see Figure 7.19, top, p. 157) cover about 6% of the earth's land area—roughly the area of the continental United States. Climate and biological data indicate that mature tropical forests once covered at least twice the area they do today. Most of this loss of half of the world's tropical forests has taken place since 1950 (see Chapter 3, Core Case Study, p. 50). According to the Earth Policy Institute and Global Forest Watch, between 2000 and 2014,

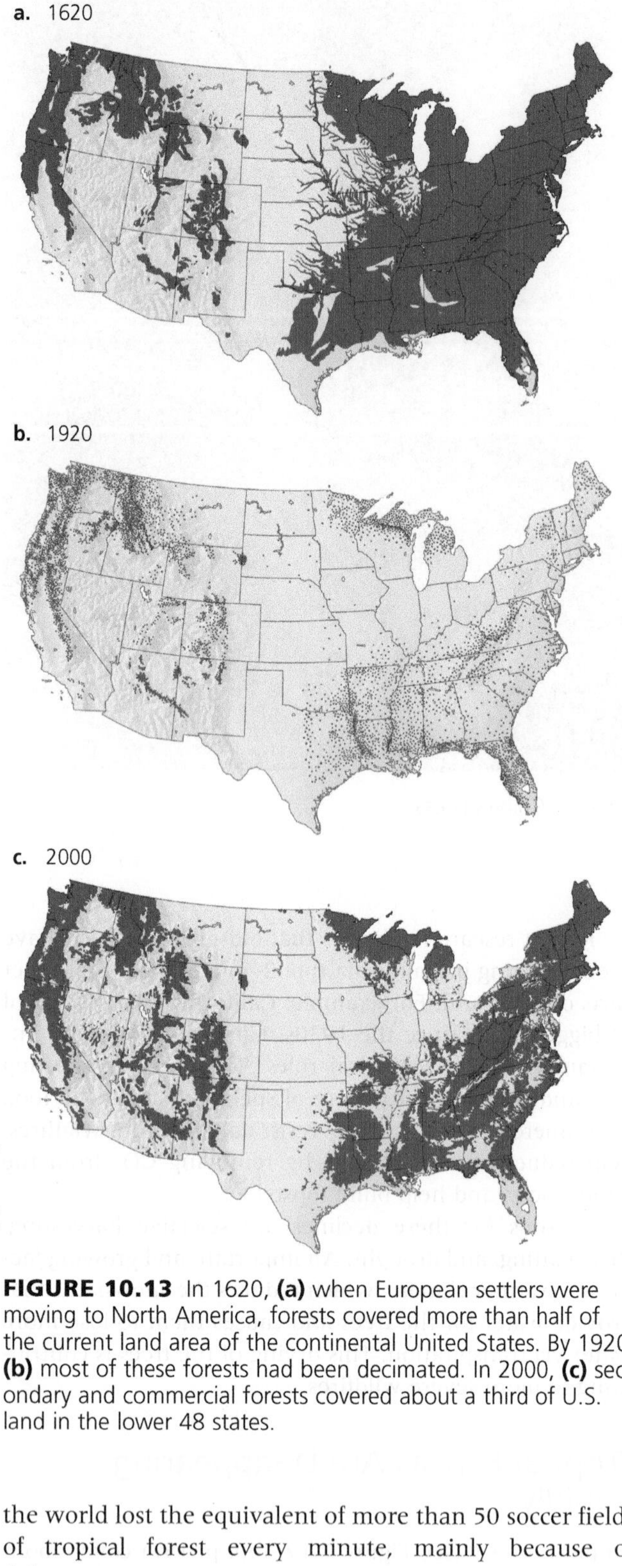

FIGURE 10.13 In 1620, **(a)** when European settlers were moving to North America, forests covered more than half of the current land area of the continental United States. By 1920, **(b)** most of these forests had been decimated. In 2000, **(c)** secondary and commercial forests covered about a third of U.S. land in the lower 48 states.

the world lost the equivalent of more than 50 soccer fields of tropical forest every minute, mainly because of deforestation.

In 2014, the WRI and more than 40 other organizations, including Google and the UNEP, created Global Forest Watch. This free online monitoring and mapping system provides near-real-time reliable data about what is happening to the world's forests. Satellite scans and ground-level surveys indicate that large areas of tropical forests are being cut rapidly in parts of Africa, Southeast Asia, and South America—especially in Brazil's vast Amazon Basin, which has more than 40% of the world's remaining tropical forests that are important living libraries of biodiversity.

Currently, these forests absorb and store about one-third of the world's terrestrial carbon emissions as part of the carbon cycle. Thus, in reducing these forests, we reduce their carbon absorption and contribute to climate change. Burning and clearing tropical forests also adds carbon dioxide to the atmosphere, accounting for 10–15% of global greenhouse gas emissions.

In 2014 the huge Amazon Basin lost an area of forest roughly the size of the U.S. state of Delaware—mostly for raising soybeans in plantations (see Figure 1.5, p. 8) and grazing cattle. About 20% of the tropical forest in the Amazon Basin has been cleared and nearly that much more has been degraded by human activities. We are reducing tropical forests at a time when we need more, not fewer, trees to absorb some of the massive amounts of CO_2 that we are adding to the atmosphere.

Water evaporating from trees and vegetation in tropical rain forests plays a major role in determining the amount of rainfall there. Removing large areas of trees can lead to drier conditions that dehydrate the topsoil by exposing it to sunlight and allowing it to be blown away. This makes it difficult for a forest to grow back in the area. Such forest areas are often replaced by tropical grassland or savanna. Scientists project that if current burning and deforestation rates continue, 20–30% of the Amazon Basin will be turned into tropical grassland or savanna within the next 50 years, and most of it could become savanna by 2080. Currently, tropical forests absorb and store one-third of the world's terrestrial emissions of carbon dioxide as part of the carbon cycle. Converting these forests to grassland will accelerate atmospheric warming and climate change and affect people and economies throughout the world.

Studies indicate that at least half of the world's known species of terrestrial plants, animals, and insects live in tropical forests. Because of their specialized niches (see Figure 7.20, p. 158), many of these species are highly vulnerable to extinction when their forest habitats are destroyed or degraded. The UN Food and Agriculture Organization (FAO) warns that at the current global rate of tropical deforestation, as much as half of the world's remaining old-growth tropical forests will be gone or severely degraded by the end of this century (**Concept 10.1B**).

Indonesia leads the world in tropical deforestation, cutting and burning large areas of rainforest and planting vast oil palm plantations (Figure 10.4). This adds large amounts of CO_2 to the atmosphere. Palm oil harvested from the fruit of the trees is used to make biodiesel fuel for cars (especially in Europe), as well as thousands of other products, including detergents, food products, and cosmetics. Producing palm oil is very profitable because of high yields and a failure to include the harmful environmental effects of palm oil production in its market price. Indonesia has the world's third largest area of undisturbed tropical forest. However, according to the UNEP, 98% of its undisturbed tropical forest will be destroyed by 2021, unless strict conservation measures are implemented and enforced.

Tropical deforestation results from a number of underlying and direct causes. Underlying causes, such as pressures from population growth and poverty, push subsistence farmers and the landless poor into tropical forests, where they cut or burn trees for firewood or try to grow enough food to survive (see Figure 1.12, p. 17). Government subsidies can accelerate other direct causes such as large-scale logging and livestock overgrazing (Figure 10.11) by reducing the costs of these enterprises.

The major direct causes of tropical deforestation vary by area. Tropical forests in Brazil and other South American countries are cleared or burned primarily for cattle grazing and large soybean plantations (see Figure 1.5, p. 8). In Indonesia, Malaysia, and other areas of Southeast Asia, large plantations of oil palm trees (Figure 10.4) are replacing diverse tropical forests. In Africa, the primary direct cause of deforestation is people clearing plots for small-scale farming and harvesting wood for fuel.

Global trade has increased tropical forest loss and degradation. According to tropical forest researcher William Laurance, China is the destination for more than half of the world's timber shipments and much of the paper pulp shipped internationally. Much of this timber comes from the unsustainable cutting of tropical forests. To gain access to this timber, China is funding numerous road-building projects into tropical forests, which opens them to numerous threats (Figure 10.5).

The degradation of a tropical forest usually begins when a road is built deep into the forest interior for logging and settlement. Loggers then use selective cutting (Figure 10.6a) to remove the largest and best trees. When loggers cut one big tree, many other trees often fall with it because of their shallow roots and the network of vines connecting the trees in the forest's canopy. For every selectively logged tree, up to 30 other trees are damaged, which increases erosion and fire risk and reduces biodiversity.

Burning is widely used to clear forest areas for agriculture, settlement, and other purposes. Healthy rain forests do not burn naturally, but roads, settlements, and other developments fragment them. The resulting patches of forest dry out and easily ignite. According to research by forest scientists, widespread fires in the Amazon basin are changing weather patterns by raising temperatures and reducing rainfall. The resulting droughts dry out the forests and make them more likely to burn.

International corporations do much of this logging and often sell the land to ranchers who burn the remaining timber to clear the land for cattle grazing. Within a few years, the land is overgrazed. Then the ranchers sell the degraded land to large-scale farm operators who plow it up to plant large crops such as soybeans (see Figure 1.5, p. 8), or to settlers for small-scale farming (see Figure 1.12, p. 17). After a few years of crop growing and erosion from rain, the topsoil is depleted of nutrients. Then the farmers and settlers move on to newly cleared land to repeat this environmentally destructive deforestation cycle. Biologist Edward O. Wilson (see Individuals Matter 4.1, p. 81) says that "Destroying a rainforest for economic gain is like burning a Renaissance painting to cook a meal."

CONSIDER THIS . . .

THINKING ABOUT Tropical Forests

Why should you care if most of the world's remaining tropical forests are burned, cleared, and converted to tropical grassland within your lifetime? What are three ways in which this might affect your life or the lives of any children and grandchildren that you might have?

10.2 HOW CAN WE MANAGE AND SUSTAIN FORESTS?

CONCEPT 10.2 We can sustain forests by emphasizing the economic value of their ecosystem services, halting government subsidies that hasten their destruction, protecting old-growth forests, harvesting trees no faster than they are replenished, and planting trees to reestablish forests.

Managing Forests More Sustainably

Three widely used approaches to managing forests are maximum sustainable yield, ecosystem-based management, and adaptive management.

The maximum sustainable yield (MSY) approach involves harvesting the maximum amount of trees without reducing future yields. Trees in a forest or plantation grow fastest at an intermediate age. Then they slow down and eventually reach a maximum size based on the carrying capacity of their environment, represented by an S-curve of population growth (see Figure 5.16, p. 111, right). The goal of maximum sustainable yield management is to harvest trees of an intermediate size at the midpoint of the growth curve between planting and the area's carrying capacity. This approach sounds good, but there are difficulties. Calculating the MSY is almost impossible because it is difficult to know an area's carrying capacity and the point at which the MSY has been reached. In addition, changes in environmental conditions over several decades of forest growth can change an area's carrying capacity.

The goal of *ecosystem-based management* is to harvest a renewable resource such as trees in ways that minimize

Certifying Sustainably Grown Timber and Products Such as the Paper Used in This Book

Collins Pine owns and manages a large area of productive timberland in the northeastern part of the U.S. state of California. Since 1940, the company has used selective cutting to help maintain the ecological and economic sustainability of its timberland.

Since 1993, Scientific Certification Systems (SCS) of Oakland, California, has evaluated the company's timber production. SCS. It is part of the nonprofit Forest Stewardship Council® (FSC®), which was formed to develop environmentally sound and responsible forestry practices for use in certifying timber and timber products.

Each year, SCS evaluates Collins Pine's landholdings. It has consistently found that their cutting of trees has not exceeded long-term forest regeneration; roads and harvesting systems have not caused unreasonable ecological damage; topsoil has not been damaged; and downed wood (boles) and standing dead trees (snags) are left to provide wildlife habitat. Year after year, SCS has found that the company is a good employer and a good steward of its land and water resources.

The FSC reported that, by 2013, it had certified about 6% of the world's forest area in 80 countries and about 7% of U.S. forests. The FSC also certifies 5,400 manufacturers and distributors of wood products. The paper used in this book was produced with the use of responsibly managed timber, as certified by the FSC symbol, and contains recycled paper fibers in the body and cover stocks. Figure 10.A shows the FSC certification used for this textbook.

CRITICAL THINKING

Should governments provide subsidies or tax breaks to encourage the use of sustainably grown timber and sustainably produced forest products? Explain.

the harmful impacts of harvesting on an ecosystem and the ecological services it provides. This can be a useful approach. However, it is often limited because of a lack of knowledge about how ecosystems in different areas work.

Adaptive management involves using available knowledge to harvest forests or other resources, evaluating the results, and modifying the approach or using a different approach as needed. This approach recognizes that there will be failures because of inadequate ecological knowledge and that we can learn from such failures.

Figure 10.14 lists specific ways to grow and harvest trees more sustainably, regardless of what management approach is used (**Concept 10.2**). One tool is the certification of sustainably grown timber and of sustainably produced forest products. This helps inform consumers about products made from sustainably grown wood. The nonprofit Forest Stewardship Council (FSC) oversees the certification of forestry operations that meet certain responsible forestry forest standards (Science Focus 10.2).

Loggers could be encouraged or required to make greater use of more sustainable selective cutting (Figure 10.6a) and strip cutting (Figure 10.6c) to harvest trees instead of clear-cutting (Figure 10.6b). To reduce damage to neighboring trees when cutting and removing individual trees, loggers can first cut the canopy vines (lianas) that connect them.

Solutions

More Sustainable Forestry

- Include ecosystem services of forests in estimates of their economic value
- Identify and protect highly diverse forest areas
- Stop logging in old-growth forests
- Stop clear-cutting on steep slopes
- Reduce road-building in forests and rely more on selective and strip cutting
- Leave most standing dead trees and larger fallen trees for wildlife habitat and nutrient cycling
- Put tree plantations only on deforested and degraded land
- Certify timber grown by sustainable methods

FIGURE 10.14 Ways to grow and harvest trees more sustainably (**Concept 10.2**). ***Critical thinking:*** Which three of these methods of more sustainable forestry do you think are the best methods? Why?

Many economists urge governments to begin making a shift to more sustainable forest management. They recommend phasing out government subsidies and tax breaks that encourage forest degradation and deforestation and replacing them with forest-sustaining subsidies and tax breaks. This would likely lead to higher prices on unsustainably produced timber and wood products, in keeping with the full-cost pricing **principle of sustainability**. Costa Rica (**Core Case Study**) is taking the lead in using this approach. Governments can also encourage tree-planting programs to help restore degraded forests. **GREEN CAREER: Sustainable forestry**

Improving Management of Forest Fires

In the 1940s, the U.S. Forest Service and the National Advertising Council launched the Smokey Bear campaign to educate the public about the dangers of forest fires. This campaign has helped prevent many forest fires, saved many lives, and prevented billions of dollars in losses of trees, wildlife, and human-built structures. However, this campaign has convinced much of the public that all forest fires are bad and should be prevented or put out, which is not the case. Ecologists warn that trying to prevent all forest fires can actually make forests more vulnerable to fires in the long run. It increases the likelihood of destructive crown fires (Figure 10.8, right) due to the accumulation of highly flammable underbrush in some forests.

Instead, ecologists and forest fire experts recommend several strategies limiting the harmful effects of forest fires:

- Use carefully planned and controlled fires, called *prescribed burns*, to remove flammable small trees and underbrush in the highest-risk forest areas.
- Allow some fires on public lands to burn underbrush and smaller trees, as long as the fires do not threaten human-built structures or human lives.
- Protect houses and other buildings in fire-prone areas by thinning trees and other vegetation in a zone around them and eliminating the use of highly flammable construction materials such as wood shingles.
- Use drones, equipped with infrared sensors, to detect forest fires and monitor progress in fighting them.

A fourth approach is to thin forest areas vulnerable to fire by clearing away small fire-prone trees and underbrush under careful environmental controls. It can include the use of prescribed burns to remove flammable debris produced by this process. This can help prevent the loss of economically valuable timber and wildlife habitats. It can also reduce the rapidly rising costs of fighting an increasing number of large forest fires.

However, many forest fire scientists warn that such thinning should not remove economically valuable medium-size and large trees, for two reasons. First, these are the most fire-resistant trees. Second, their removal encourages dense growth of more flammable young trees and underbrush and leaves behind highly flammable slash. In addition, a study by U.S. Forest Service researchers found that thinning forests without using prescribed burning to remove the accumulated brush and deadwood (slash) from the thinning can greatly increase rather than decrease fire damage. Forest scientists also oppose removing dead and dying trees (snags) because they play important ecological roles in providing habitats for a variety of animals.

Despite such warnings from scientists, the U.S. Congress under lobbying pressure from timber companies passed the 2003 Healthy Forests Restoration Act in response to devastating fires in California in 2002. It allows timber companies to cut down economically valuable medium-size and large trees (primarily to provide fire breaks) and snags in most of the country's 155 national forests. This is allowed in return for their clearing away smaller, more fire-prone trees and underbrush. However, these operations typically do not include prescribed burns as part of the thinning process. This law also exempts most thinning projects from environmental reviews, which are otherwise required by forest protection laws in the national forests.

Reducing the Demand for Harvested Trees

According to the Worldwatch Institute and to forestry analysts, *up to 60% of the wood consumed in the United States is wasted unnecessarily*. This waste results from inefficient use of construction materials, excessive packaging, overuse of junk mail, inadequate paper recycling, and the failure to reuse or find substitutes for wooden shipping containers.

One way to reduce demand for harvested trees is to produce tree-free paper. Pulp from trees is typically used to make paper, but paper manufacturers can also use fibers from nontree sources. For example, China uses rice straw and other agricultural residues to make some of its paper. Most of the small amount of tree-free paper produced in the United States is made from the fibers of a rapidly growing woody annual plant called *kenaf* (pronounced "kuh-NAHF," Figure 10.15). Kenaf and other nontree fibers such as hemp yield more paper pulp per area of land than tree farms do without using pesticides and herbicides.

According to the USDA, kenaf is "the best option for tree-free papermaking in the United States" and could replace wood-based paper within 20–30 years. However, while timber companies successfully lobby for government subsidies to grow and harvest trees to make paper, there are no major lobbying efforts or subsidies for producing paper from kenaf and other alternatives to trees.

Another way to reduce the demand for tree cutting is to reduce the use of throwaway paper products made from

U.S. Department of Agriculture

FIGURE 10.15 Solutions: The pressure to cut trees to make paper could be greatly reduced by planting and harvesting a fast-growing plant known as kenaf.

PASCAL MAITRE/National Geographic Creative

FIGURE 10.16 Truckload of charcoal in Somalia.

trees. Instead, we can instead choose reusable plates, cups, cloth napkins and handkerchiefs, and cloth bags.

Humans have always used trees for fuel, but the demand for wood as fuel is becoming unsustainable in many areas (see the following Case Study).

CASE STUDY

Deforestation and the Fuelwood Crisis

More than 2 billion people in less-developed countries use fuelwood (see Figure 6.15, p. 131) and charcoal (Figure 10.16) made from wood for heating and cooking. Most of these countries suffer from fuelwood shortages because people cut trees for fuelwood and forest products 10 to 20 times faster than new trees are being planted. The FAO warns that, by 2050, the demand for fuelwood could easily be 50% greater than the amount that can be sustainably supplied.

For example, Haiti, a country with 11 million people, was once a tropical paradise, 60% of it covered with forests. Now it is an ecological and economic disaster. Largely because its trees were cut for fuelwood and to make charcoal, less than 2% of its land is now covered with trees (Figure 10.17). With the trees gone, soils have eroded away in many areas, making it much more difficult to grow crops.

The U.S. Agency for International Development funded the planting of 60 million trees over more than two decades in Haiti. However, local people cut most of them down for firewood and to make charcoal before they could grow into mature trees. This unsustainable use of natural capital and failure to follow the biodiversity and chemical cycling **principles of sustainability** have played a role in the downward spiral of environmental degradation, poverty, disease, social injustice, crime, and violence in Haiti. The situation is likely to get worse because between 2015 and 2050 Haiti's population is projected to grow by 55% from 11 million to 17 million.

One way to reduce the severity of the fuelwood crisis in less-developed countries is to establish small plantations of fast-growing fuelwood trees and shrubs around farms and in community woodlots. Providing villagers with affordable and more fuel-efficient wood stoves and solar-powered ovens is another way to reduce cutting down trees for fuelwood. Another option is stoves that burn renewable biomass, such as sun-dried roots of various gourds and squash plants, or methane produced from crop and animal wastes. In addition to reducing unsustainable deforestation, these options would greatly reduce the large number of deaths caused by indoor air pollution from open fires and poorly designed stoves.

James P. Blair/National Geographic Creative

FIGURE 10.17 Natural capital degradation: Haiti's deforested brown landscape (left) contrasts sharply with the heavily forested green landscape of its neighboring country, the Dominican Republic.

Scientists are also looking for ways to produce charcoal for heating and cooking without cutting down trees. For example, Professor Amy Smith of the Massachusetts Institute of Technology is developing a way to make charcoal from the fibers in a waste product called bagasse, which is left over from sugar cane processing in Haiti and many other countries.

Countries such as South Korea, Nepal, and Senegal have used such methods to reduce fuelwood shortages, while sustaining biodiversity through reforestation and efforts to reduce topsoil erosion. Indeed, the mountainous country of South Korea is a global model for its successful reforestation following severe deforestation during the war between North and South Korea, which ended in 1953. Today, forests cover almost two-thirds of the country, and tree plantations near villages supply fuelwood on a sustainable basis. GOOD NEWS

Reducing Tropical Deforestation

Analysts have suggested various ways to protect tropical forests and use them more sustainably (Figure 10.18).

At the international level, *debt-for-nature swaps* can make it financially attractive for countries to protect their tropical forests and use them more sustainably, a concept pioneered by conservation biologist Thomas R. Lovejoy (see Individuals Matter 3.1, p. 70). Under the terms of such swaps, participating countries agree to set aside and

Solutions

Sustaining Tropical Forests

Prevention

- Protect the most diverse and endangered areas
- Educate settlers about sustainable agriculture and forestry
- Subsidize only sustainable forest use
- Protect forests through debt-for-nature swaps and conservation concessions
- Certify sustainably grown timber
- Reduce poverty and slow population growth

Restoration

- Encourage regrowth through secondary succession
- Rehabilitate degraded areas
- Concentrate farming and ranching in already-cleared areas

FIGURE 10.18 Ways to protect tropical forests and to use them more sustainably (**Concept 10.2**). ***Critical thinking:*** Which three of these solutions do you think are the best ones? Why?

Top: Stillfx/Shutterstock.com. Center: Manfred Mielke/USDA Forest Service Bugwood.org.

INDIVIDUALS MATTER 10.1

Florence Reed—Preserving Tropical Forests and Overcoming Poverty

Photo by Bill Kizorek, courtesy of Sustainable Harvest International

When Florence Reed worked as a Peace Corps Volunteer in Panama she observed poor farmers moving into tropical forests. Once there, they cut and burned small plots of forest to grow enough food to survive. After a few years, they had depleted the soil of nutrients and had to move to another area and clear and burn more trees.

Reed decided to dedicate her life to helping poor farmers learn how to grow nutritious food on the same land year after year and to raise their incomes without having to clear and burn more forest. In 1997, she founded Sustainable Harvest International, a nonprofit organization dedicated to her goal.

Sustainable Harvest International helps farmers learn and use sustainable agricultural practices, including seed saving, organic vegetable farming, tree planting, water conservation, and growing a diversity of crops on their plots. Sustainable Harvest International also teaches farm families how to increase their income by selling surplus crops.

Reed works only with farmers who ask for help. She encourages farmers to cooperate and share their resources and knowledge, and Sustainable Harvest International works with them until they learn to use the farming techniques on their own—typically within five years.

By 2015, Sustainable Harvest International had worked with more than 2,500 families in more than 100 farming communities. Each of those families then goes on to teach many more. Since its inception in 1997, the families Sustainable Harvest International partners with have planted 3.8 million trees, restoring 6,110 hectares (15,100 acres) of forest devastated by slash-and-burn agriculture. In addition, they've converted 6,845 hectares (16,913 acres) to sustainable agriculture. The income of a typical family working with Sustainable Harvest International increases by at least 23%. Lastly, by the time families graduate from the program, they are no longer exposed to harmful synthetic pesticides and their diets are much more varied and nutritious.

---

protect forest reserves in return for foreign aid or debt relief. In a similar strategy, called *conservation concessions*, governments or private conservation organizations pay nations for agreeing to preserve their natural resources such as forests.

National governments can also take steps to reduce deforestation (**Core Case Study**). Between 2005 and 2013, Brazil cut its deforestation rate by 80% by cracking down on illegal logging and setting aside a large conservation reserve in the Amazon Basin. Governments can also end subsidies that fund the construction of logging roads and instead subsidize sustainable forestry and the planting of trees in reforestation programs.

Consumers can reduce the demand for unsustainable and illegal logging in tropical forests by buying only wood and wood products that have been certified as sustainably produced by the FSC and other organizations, including the Rainforest Alliance and the Sustainability Action Network.

People who live in tropical forests, many of them poor farmers trying to feed their families, also want to help sustain tropical forests. Many of them are looking for ways to grow the food they need without having to cut and burn trees, and several organizations are assisting them (Individuals Matter 10.1).

Another important strategy is *reforestation*—the replanting of forests, especially on degraded and abandoned land. Throughout the world, reforestation promotes biodiversity, conserves topsoil, and reduces flooding. It also provides firewood and helps slow climate change by removing CO_2 from the atmosphere. The late Wangari Maathai (Figure 10.19, left) promoted tree planting (Figure 10.19, right) in her native country of Kenya and throughout the world in what became the Green Belt Movement. The main goal of this movement was to organize poor women in rural Kenya to plant and protect trees in order to replenish forests, provide fuelwood, and reduce soil erosion and stream pollution. This helped women to escape poverty by earning money for their reforestation work. Forty other African countries have adopted the program.

In 2004 Wangari Maathai became the first African woman and the first environmentalist to be awarded Nobel Peace Prize for her "contribution to sustainable development, democracy, and peace." After learning that she had won the prize, she planted a tree and urged everyone in the world to plant one. In 2006 she launched a joint campaign with the UNEP to plant a billion trees around the world. By 2014, almost 14 billion trees had been planted in 193 countries.

FIGURE 10.19 In 1977 Wangari Maathai (1940–2011) (left) founded the internationally acclaimed Green Belt Movement, which promoted the planting trees in Kenya (right).

10.3 HOW CAN WE MANAGE AND SUSTAIN GRASSLANDS?

CONCEPT 10.3 We can sustain the productivity of grasslands by controlling the numbers and distribution of grazing livestock and by restoring degraded grasslands.

Some Rangelands Are Overgrazed

Grasslands cover about on-fourth of the earth's land surface. They provide important ecosystem services, including soil formation, erosion control, chemical cycling, storage of atmospheric carbon dioxide in biomass, and maintenance of biodiversity.

After forests, grasslands are the ecosystems most widely used and altered by human activities. **Rangelands** are unfenced natural grasslands in temperate and tropical climates that supply *forage*, or vegetation for grazing (grass-eating) and browsing (shrub-eating) animals. Cattle, sheep, and goats graze on about 42% of the world's natural grasslands. This could increase to 70% by the end of this century, according to the UN Millennium Ecosystem Assessment. Livestock also graze in **pastures,** which are managed grasslands or fenced meadows often planted with domesticated grasses or other forage crops such as alfalfa and clover.

Blades of rangeland grass grow from the base, not at the tip. As long as only the upper portion of the blade is eaten and the lower portion remains in the ground, rangeland grass is a renewable resource that can be grazed again and again. Moderate levels of grazing are healthy for grasslands, because removal of mature vegetation stimulates rapid regrowth and encourages greater plant diversity.

Overgrazing occurs when too many animals graze an area for too long and damage or kill the grasses (Figure 10.20, left) and their roots and exceed the area's

FIGURE 10.20 Natural capital degradation: To the left of the fence is overgrazed rangeland. The land to the right of the fence is lightly grazed.

carrying capacity for grazing. Overgrazing reduces grass cover, exposes the topsoil to erosion by water and wind, and compacts the soil, which lessens its capacity to hold water. Overgrazing also promotes the invasion of rangeland by species such as sagebrush, mesquite, cactus, and cheatgrass, which cattle will not eat. The FAO has estimated that overgrazing by livestock has reduced productivity on as much as 20% of the world's rangeland.

Managing Rangelands More Sustainably

Managing rangelands more sustainably and preventing overgrazing typically involves controlling how many animals are allowed to graze in a given area and for how long, in order to avoid exceeding the area's carrying capacity for grazing animals (**Concept 10.3**). One method for doing this is called *rotational grazing*, in which small groups of cattle are confined by portable fencing to one area for a few days and then moved to a new location.

Cattle prefer to graze around ponds and other natural water sources, especially along streams or rivers lined by strips of vegetation known as *riparian zones*. Overgrazing can destroy the vegetation in such areas (Figure 10.21, left). Ranchers can protect overgrazed land through rotational grazing and by fencing off damaged areas, which eventually leads to their natural restoration by ecological succession (Figure 10.21, right). Ranchers can also move cattle around by providing supplemental feed at selected sites and by strategically locating watering ponds or tanks and salt blocks.

The least expensive way to deal with degradation of rangelands is to prevent it, by using methods such as those described in Science Focus 10.3.

10.4 HOW CAN WE SUSTAIN TERRESTRIAL BIODIVERSITY AND ECOSYSTEM SERVICES?

CONCEPT 10.4A We can establish and protect wilderness areas, parks, and nature preserves.

CONCEPT 10.4B We can identify and protect biological hotspots that are highly threatened centers of biodiversity.

CONCEPT 10.4C We can protect important ecosystem services, restore damaged ecosystems, and share areas that we dominate with other species.

Strategies for Sustaining Terrestrial Biodiversity

Since the 1960s, a number of strategies have been used to help sustain terrestrial biodiversity. The main strategies are:

- Protecting species from extinction, as discussed in Chapter 9.

FIGURE 10.21 Natural capital restoration: In the mid-1980s, cattle had degraded the vegetation and soil on this stream bank along the San Pedro River in the U.S. state of Arizona (left). Within 10 years, the area was restored through secondary ecological succession (right) after grazing and off-road vehicle use were banned (**Concept 10.3**).

Left: U.S. Bureau of Land Management. Right: U.S. Bureau of Land Management.

SCIENCEFOCUS 10.3

Holistic Range Management: Learning from Nature

Biologist and rancher Alan Savory has developed an ecological approach to more sustainable rangeland management. He based it on understanding how huge moving herds of wild grazing animals have lived off grasslands for thousands of years without seriously overgrazing them.

Some rangeland managers call for sharply reducing the number of grazing livestock to reduce overgrazing and conversion of grassland to desert (desertification). Savory contends that reducing the number of grazing livestock would increase overgrazing and desertification because grassland needs grazing and short-term trampling by a large number of grazing animals to stimulate its growth.

Short-term trampling by a large moving herd aerates the soil, increases grass seed germination by pressing seeds into the soil, and promotes a diversity of grasses. It also increases nutrient recycling and soil fertility by pressing decaying and dying grasses into the soil.

Savory calls this ecologically based approach *holistic herd management*, arguing that it is more sustainable and effective than traditional grazing. He confines a large herd of grazing animals to a small area for a few days so that they can trample the soil in each area and not feed long enough to overgraze the grass. He rotates the herd every few days over a large area. It is rotational grazing with a much larger number of animals than the conventional approach. This example of *learning from nature* is based on how nature has sustained grasslands for thousands of years. More research needs to be done to evaluate the effectiveness and costs of this approach.

CRITICAL THINKING

Design an experiment to evaluate the effectiveness of this grazing method compared to conventional rotational grazing.

- Setting aside wilderness areas that are protected from harmful human activities.
- Establishing parks and nature reserves in which people and nature can interact with some restrictions.
- Identifying and protecting *biodiversity hotspots* that contain a high diversity of species and that are under severe threat of extinction from human activities.
- Shifting new development to lands that have already been cleared or degraded.
- Avoiding the destruction of forests and grasslands by increasing crop productivity on existing cropland.
- Protecting important ecosystem services.
- Rehabilitating and partially restoring damaged ecosystems.
- Sharing areas that we dominate with other species.

According to most ecologists and conservation biologists, the best way to preserve terrestrial biodiversity is to create a worldwide network of areas that are strictly or partially protected from harmful human activities. In 2014 less than 13% of the earth's land area (excluding Antarctica) was protected to some degree in more than 177,000 wilderness areas, parks, nature reserves, and wildlife refuges (Figure 10.22). However, no more than 6% of the earth's land is strictly protected from potentially harmful human activities. In other words, we have reserved 94% of the earth's land for human use.

6% Percentage of the earth's land that is strictly protected from potentially harmful human activities

To some, protecting more than 6% of the earth's land is a stunning conservation achievement. Others argue that it is not nearly enough to ward off a decline in the world's vital biodiversity in the face of human population growth, increasing resource use, and long-lasting climate change. Conservation biologists call for strictly protecting at least 20% of the earth's land area in a global system of biodiversity reserves. The system would include multiple examples of all the earth's biomes. The International Union for the Conservation of Nature (IUCN) estimates that making this investment to protect and sustain a vital part of our life-support system would cost roughly $23 billion a year—more than four times the amount spent on such protection today.

Many developers and resource extractors oppose protecting even 13% of the earth's land (the current amount), arguing that these protected areas might contain valuable resources that would provide economic benefits. In contrast, ecologists and conservation biologists view protected areas as islands of biodiversity and ecosystem services that help to sustain all life and economies indefinitely and that serve as centers of future evolution.

FIGURE 10.22 Caribou migrating from Alaska's Arctic National wildlife Refuge to their Yukon wintering grounds.

Establishing Wilderness Areas

One way to protect existing wildlands from human exploitation is to designate them as **wilderness.** Designated wilderness areas are essentially undisturbed by humans and are protected by federal law from harmful human activities (**Concept 10.4A**). For example, forestry, road and trail development, mining, and building construction are not allowed. Theodore Roosevelt (Figure 1.16, p. 22), the first U.S. president to set aside protected areas, summarized his thoughts on what to do with wilderness: "Leave it as it is. You cannot improve it."

Most developers and resource extractors oppose establishing protected wilderness areas because they contain resources that could provide short-term economic benefits. Ecologists and conservation biologists take a longer view. To them such areas provide a long-term "ecological insurance policy" for humans and other species. They are islands of biodiversity and ecosystem services that will be necessary to support life and human economies in the future. In addition, they are needed as *centers for future evolution* in response to changes in environmental conditions.

In 1964 the U.S. Congress passed the Wilderness Act, which allowed the government to protect undeveloped tracts of U.S. public land from development as part of the National Wilderness Preservation System (Figure 10.23). The country's area of protected wilderness grew by nearly twelvefold between 1964 and 2012. Even so, only 5% of all U.S. land is protected as wilderness—more than 54% of it in Alaska. Only about 2.7% of the land of the lower 48 states is protected as wilderness, most of it in the West.

As the human population and its ecological footprint expand, it will be increasingly difficult and expensive to protect existing wilderness areas and to establish new ones. In addition, climate change is projected to threaten the biodiversity and composition of many existing wilderness areas.

Establishing Parks and Other Nature Reserves

According to the IUCN, there are more than 6,600 major national parks located in more than 120 countries (see chapter-opening photo). These are areas where people can enjoy and interact with nature under certain restrictions.

However, most of these parks are too small to sustain many large animal species. In addition, most of them receive little protection and are often called "paper parks." Many parks also suffer from invasions by harmful nonnative species that can outcompete and reduce the populations of native species. Some national parks are so popular that large numbers of visitors are degrading the natural features that make them attractive (see the Case Study that follows).

Parks in less-developed countries have the greatest biodiversity of all the world's parks, but only about 1% of these parklands are protected. Local people in many of

tusharkoley/Shutterstock.com

FIGURE 10.23 Diablo Lake lies in a wilderness area of North Cascades National Park in the U.S. state of Washington.

these countries enter the parks illegally in search of wood, game animals, and other natural products that they need for their daily survival. Loggers and miners also operate illegally in many of these parks, as do wildlife poachers who kill animals to obtain and sell items such as rhino horns (see Figure 9.16, p. 207), elephant tusks, and furs. Park services in most of the less-developed countries have too little money and too few personnel to fight these invasions, either by force or through education.

CASE STUDY

Stresses on U.S. Public Parks

The U.S. National Park System, established in 1912, includes 59 major national parks, sometimes called the country's crown jewels that are owned jointly by all U.S. citizens (see chapter-opening photo). The U.S. national park system also has 339 monuments, recreational areas, battlefields, historic sites, and other areas. States, counties, and cities also operate public parks.

In 1872 Congress set aside public land for Yellowstone National Park—the world's first national park. Historian, conservationist, and writer Wallace Stegner called it "the best idea America ever had."

Popularity threatens many parks. Between 1960 and 2014, the number of recreational visitors to U.S. national parks more than tripled, reaching about 293 million. In order, the three most visited places in the National Park System in 2014 were the Golden Gate National Recreation Area, the Blue Ridge Parkway, and the Great Smoky Mountain National Park.

In some U.S. parks and other public lands, dirt bikes, dune buggies, jet skis, snowmobiles, and other off-road vehicles destroy or damage vegetation, disturb wildlife, and negatively affect the park experience for many visitors. Some visitors expect parks to have grocery stores, laundries, bars, and other such conveniences. Cell phone towers now degrade the pristine nature of some parks.

A number of parks also suffer damage from the migration or deliberate introduction of nonnative species. European wild boars (see Figure 9.11, p. 202), imported into the state of North Carolina in 1912 for hunting, threaten vegetation in parts of the popular Great Smoky Mountains National Park. Nonnative mountain goats in Washington State's Olympic National Park trample and destroy the root systems of native vegetation and accelerate soil erosion.

Native species—some of them threatened or endangered—are killed in, or illegally removed from, almost half of all U.S. national parks. However, the endangered gray wolf was successfully reintroduced into Yellowstone National Park after a 50-year absence (Science Focus 10.4).

Many U.S. national parks have become threatened islands of biodiversity surrounded by commercial development. Their wildlife and recreational value are threatened by nearby activities such as mining, logging, livestock grazing, coal-fired power plants, and urban development. The National Park Service reports that air pollution, mainly from coal-fired power plants and dense vehicle traffic, impairs scenic views more than 90% of the time in many U.S. national parks.

The National Park Service estimated that in 2014, the national parks had at least an $11.5 billion backlog for long overdue maintenance and repairs to trails, buildings, and other park facilities. Some analysts say that some of the funds needed for such purposes could come from private concessionaires who provide campgrounds, restaurants, hotels, and other services for park visitors. They pay the government franchise fees averaging only about 6–7% of their gross receipts, and many large concessionaires with long-term contracts pay as little as 0.75%. Analysts say these percentages could reasonably be increased to around 20%.

Since the 1930s, there have been efforts to sell U.S. National Parks and other public lands to private owners and developers. These pressures are increasing, as discussed in Chapter 25.

Designing and Managing Nature Reserves

In establishing nature reserves, the size and design of the reserve is important. Research by Thomas E. Lovejoy (see Individuals Matter 3.1, p. 70) and other scientists indicates that large nature reserves typically sustain more species and provide greater habitat diversity than do small reserves. Their research also indicates that in some areas, several well-placed medium-size reserves may better protect a variety of habitats and sustain more biodiversity than a single large reserve with the same area.

Establishing protected *habitat corridors* between isolated reserves can benefit more species and allow migration by vertebrates that need large ranges. Corridors will also allow some species to move to areas that are more favorable if climate change alters their existing areas.

On the other hand, corridors can threaten isolated populations by allowing movement of fire, disease, and pest and invasive species between reserves. They can also increase exposure of migrating species to natural predators, human hunters, and pollution. Some research suggests that the benefits of corridors outweigh their potential harmful effects, especially as the climate changes.

Conservation biologists call for using the *buffer zone concept*, whenever possible, to design and manage nature reserves. Establishing a buffer zone means strictly protecting an inner core of a reserve, usually by establishing one or more buffer zones in which local people can extract resources sustainably without harming the inner core (see the Case Study that follows). By 2015, the United Nations had used this concept to create a global network of 631 *biosphere reserves* in 119 countries. However, most biosphere reserves fall short of these design ideals and receive too little funding for their protection and management. The Case Study that follows describes Costa Rica's approach to sustaining its remarkable biodiversity.

CASE STUDY

Identifying and Protecting Biodiversity in Costa Rica

For several decades, Costa Rica (**Core Case Study**) has been using government and private research agencies to identify the plants and animals that make it one of the world's most biologically diverse countries (Figure 10.24). The government consolidated the country's parks and reserves into several large conservation areas, or *megareserves*, with the goal of protecting and sustaining 80% of the country's biodiversity (Figure 10.25). GOOD NEWS

Each reserve contains a protected inner core surrounded by two buffer zones that local and indigenous people can use for sustainable logging, crop farming, cattle grazing, hunting, fishing, and ecotourism. Instead of shutting local people out of reserve areas, this approach enlists local people as partners in protecting a reserve from activities such as illegal logging and poaching. It is an application of the biodiversity and win-win **principles of sustainability**.

In addition to its ecological benefits, this strategy has paid off financially. Today, Costa Rica's largest source of income is its $3-billion-a-year travel and tourism industry, almost two-thirds of which involves ecotourism. In 2013 this industry provided 12% of the country's total employment and 12% of its economic activity.

There are potential threats to Costa Rica's conservation efforts. One serious threat is the clearing of forests to grow

SCIENCEFOCUS 10.4

Reintroducing the Gray Wolf to Yellowstone National Park

In the 1800s, at least 350,000 gray wolves (Figure 10.B) roamed over 75% of America's lower 48 states—especially in the West. The wolves preyed on bison, elk, caribou, and deer. Between 1850 and 1900, most of them were shot, trapped, or poisoned by ranchers, hunters, and government employees. This drove the gray wolf to near extinction in the lower 48 states.

Ecologists recognize the important role that this keystone predator species once played in the Yellowstone National Park region. The wolves culled herds of bison, elk, moose, and mule deer, and kept down coyote populations. By leaving some of their kills partially uneaten, they provided meat for scavengers such as ravens, bald eagles, ermines, grizzly bears, and foxes.

When the number of gray wolves declined, herds of plant-browsing elk, moose, and mule deer expanded and overbrowsed the willow and aspen trees growing near streams and rivers. This led to increased soil erosion and declining populations of other wildlife species such as beaver, which eat willow and aspen. This in turn affected species that depend on wetlands created by dam-building beavers.

In 1974 the gray wolf was listed as an endangered species in the lower 48 states. In 1987 the U.S. Fish and Wildlife Service (USFWS) proposed reintroducing gray wolves into Yellowstone National Park to try to help stabilize the ecosystem. The proposal brought angry protests from ranchers who feared the wolves would leave the park and attack large numbers of their cattle and sheep and from hunters who feared the wolves would kill too many big-game animals. Mining and logging companies objected, fearing that the government would halt their operations on wolf-populated federal lands.

Federal wildlife officials caught gray wolves in Canada and northwest Montana and in 1996 relocated 41 of them in Yellowstone National Park. Scientists estimate that the long-term carrying capacity of the park is 110 to 150 gray wolves. As of December 2014, the park had 104 wolves in 11 packs.

The reintroduction of this keystone species has turned the park into a living ecological laboratory. Wildlife ecologist Robert Crabtree and other scientists have been using radio collars to track some of the wolves and are studying the ecological effects of reintroducing the wolves. Their research indicates that the return of this keystone predator has decreased populations of elk, the wolves' primary food source. The leftovers of elk killed by wolves have also been an important food source for scavengers such as bald eagles and ravens.

The wolves' presence, with a projected decline in elk numbers, was supposed to promote the regrowth of young aspen trees that elk feed on and had depleted. However, a study led by U.S. Geological Survey scientist Matthew Kauffman indicated that the aspen were not recovering despite a 60% decline in elk numbers. Declining populations of elk were also supposed to allow for the return of willow trees along streams. Research indicates that willows have only partly recovered.

The wolves have cut in half the Yellowstone population of coyotes—the top predators in the absence of wolves. This has reduced coyote attacks on cattle from area ranches and has led to larger populations of small animals such as ground squirrels, mice, and gophers, which are hunted by coyotes, eagles, and hawks.

Volodymyr Burdiak/Shutterstock.com

FIGURE 10.B After becoming almost extinct in much of the western United States, the *gray wolf* was listed and protected as an endangered species in 1974.

Overall, this experiment has had some important ecological benefits for the Yellowstone ecosystem, but more research is needed. The focus has been on the gray wolf, but other factors such as drought and the rise of populations of bears and cougars may play a role in the observed ecological changes and need to be examined. Some scientists hypothesize that the long-term absence of wolves led to a number of changes in plant and animal numbers and diversity that are difficult to reverse.

The wolf reintroduction has also produced economic benefits for the region. One of the main attractions of the park for many visitors is the hope of spotting wolves chasing their prey across its vast meadows.

CRITICAL THINKING

If the gray wolf population in the park were to reach its estimated carrying capacity of 110 to 150 wolves, would you support a program to kill wolves to maintain this population level? Explain. Can you think of other alternatives?

Vladimir Melnik/Shutterstock.com

FIGURE 10.24 This scarlet macaw parrot is one of the more than half a million species found in Costa Rica.

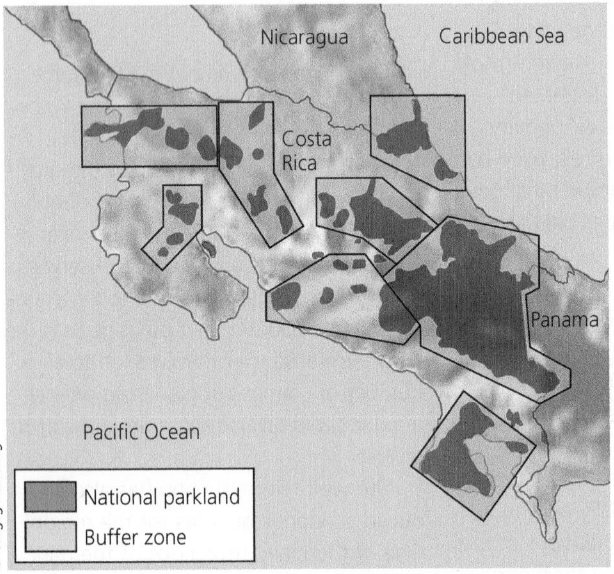

FIGURE 10.25 Solutions: Costa Rica has created several *megareserves*. Green areas are protected natural parklands and yellow areas are the surrounding buffer zones.

pineapples in plantations for export to China. Ecotourism helps to fund parks and conservation efforts and reduces exploitation of conservation areas by providing income for local people in visited areas, but excessive numbers of ecotourists can degrade sensitive areas.

The Ecosystems Approach for Sustaining Terrestrial Biodiversity

Most wildlife biologists and conservationists believe that the best way to keep from hastening the extinction of wild species through human activities is to protect threatened habitats and ecosystem services. This *ecosystems approach* would generally employ the following five-point plan:

1. Map the world's terrestrial ecosystems and create an inventory of the species contained in each of them, along with the ecosystem services they provide.
2. Identify terrestrial ecosystems that are resilient and can recover if not overwhelmed by harmful human activities, along with ecosystems that are fragile and need protection.
3. Protect the most endangered terrestrial ecosystems and species, with emphasis on protecting plant biodiversity and ecosystem services.
4. Restore as many degraded ecosystems as possible.
5. Make development biodiversity-friendly by providing significant financial incentives (such as tax breaks and subsidies) and technical help to private landowners who agree to help protect endangered ecosystems.

Protecting Biodiversity Hotspots

The ecosystems approach calls for identifying and taking emergency action to protect the earth's **biodiversity hotspots**—areas that are rich in highly endangered species found nowhere else and that are also threatened by human activities. These areas have suffered serious ecological disruption, mainly due to rapid population growth and the resulting pressure on natural resources and ecosystem services (**Concept 10.4B**). Environmental scientist Norman Myers first proposed this idea in 1988.

Figure 10.26 shows 34 terrestrial biodiversity hotspots biologists have identified. (For a map of hotspots in the United States, see Figure 7, p. S21, in Supplement 4.) According to the IUCN, these areas cover only about 2.3% of the earth's land surface, but contain an estimated 50% of the world's flowering plant species and 42% of all terrestrial vertebrates (mammals, birds, reptiles, and amphibians).

These hotspots are home for the majority of the world's endangered or critically endangered species, as well as for 1.2 billion people. With this approach, we can conserve nearly half of the world's terrestrial plant and animal species by preserving only about 2.3% of the earth's land surface. However, only 5% of the total area of these hotspots is truly protected with government funding and law enforcement, as described in the Case Study that follows.

CASE STUDY

Madagascar: An Endangered Biodiversity Hotspot

Madagascar, the world's fourth largest island, lies in the Indian Ocean off the east coast of Africa. Most of its

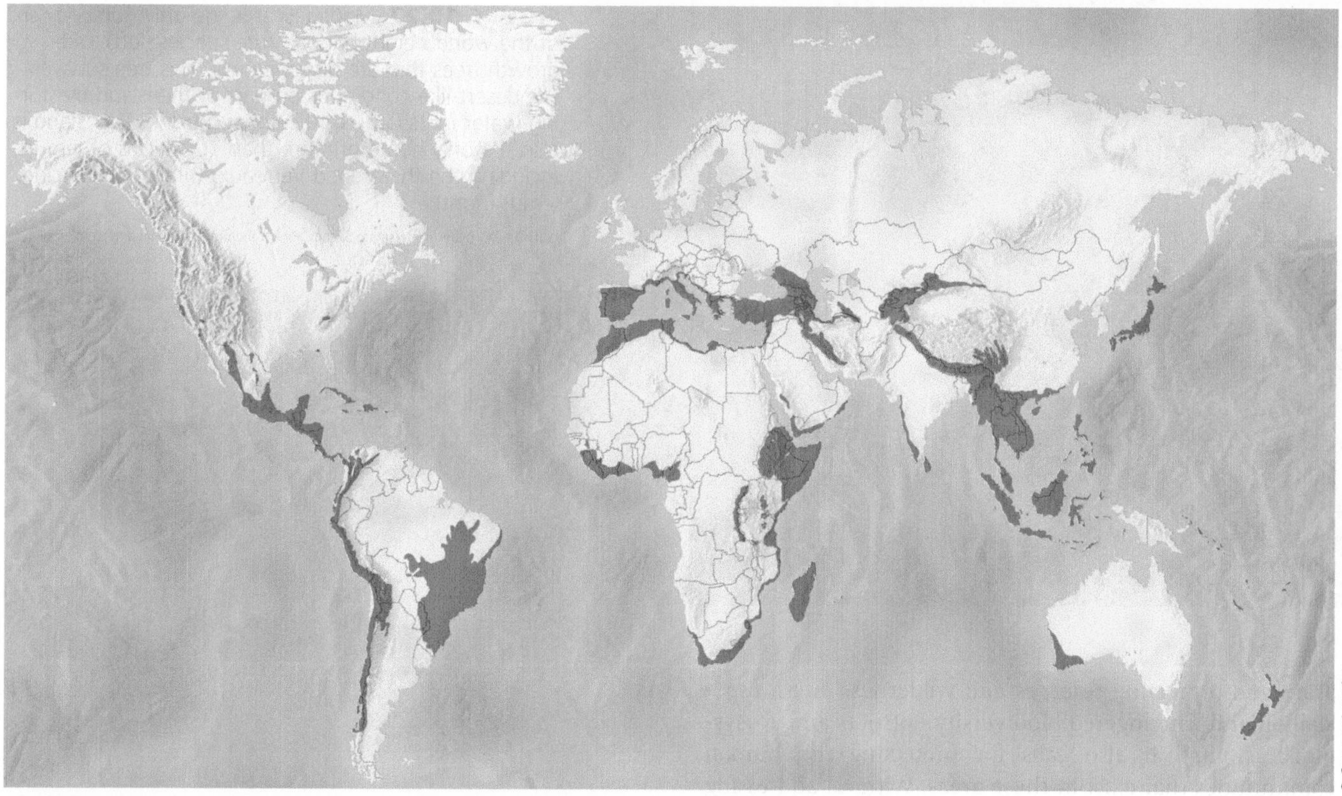

FIGURE 10.26 *Endangered natural capital:* Biologists have identified these 34 biodiversity hotspots. Compare this map with the global map of the human ecological footprint, shown in Figure 1.8, p. 12. ***Critical thinking:*** Why do you think so many hotspots are located near coastal areas?

(Compiled by the authors using data from the Center for Applied Biodiversity Science at Conservation International.)

numerous species have evolved in near isolation from mainland Africa and all other land areas for at least 40 million years. As a result, roughly 90% of the more than 200,000 plant and animal species (Figure 10.27) found in this Texas-size biodiversity hotspot are found nowhere else on the earth.

Many of Madagascar's plant and animal species are among the world's most endangered, primarily because of habitat loss. People have cut down or burned more than 90% of Madagascar's original forests to get firewood and lumber and to make way for small farms, large rice plantations, and cattle grazing.

Only about 17% of the island's original vegetation remains, which means most of its topsoil is exposed. Hence, Madagascar is one of the world's most eroded countries. Huge quantities of its topsoil have run off its hills, flowing as sediment in its rivers and emptying into its coastal waters (see Figure 8.6, p. 172). This explains why Madagascar is one of the world's most threatened biodiversity hotspots.

Since 1984, the government, conservation organizations, and scientists worldwide have mounted efforts to slow the country's rapid loss of biodiversity. Such efforts are hampered by Madagascar's rapid population growth. Between 1994 and 2014, its population grew from 12 million to 23 million and is projected to grow to 53 million by 2050. The country is also very poor, with 90% of its population struggling to survive on the equivalent of less than $2.25 per day. This puts pressure on its dwindling forest resources.

Despite the efforts to preserve Madagascar's biodiversity, less than 3% of its land area is officially protected. To reduce the rapid losses of biodiversity, the country will need to slow its population growth drastically and teach many of its people how to make a living from reforestation, ecotourism, and more sustainable uses of its forests, wildlife, and soil resources.

Protecting Ecosystem Services

Another way to sustain the earth's biodiversity is to identify and protect areas in which vital ecosystem services (see the orange boxed labels in Figure 1.3, p. 7) are being impaired. This ecosystem services approach recognizes that most of the world's ecosystems are already dominated or influenced by human activities and that such pressures are increasing as the human population, urbanization, resource use, and the human ecological footprint all expand. Proponents of this approach recognize that setting

FIGURE 10.27 Madagascar is the only home for six of the world's eight baobab tree species (left), old-growth trees that are disappearing. This tree survives the desert-like conditions on part of the island by storing water in its large bottle-shaped trunk. The island is also the only home of more than 70 species of lemurs, including the threatened Verreaux's sifaka, or dancing lemur (right).

Left: David Thyberg/Shutterstock.com. Right: Richlindie/Dreamstime.com.

aside and protecting reserves and wilderness areas, especially highly endangered biodiversity hotspots and ecosystems, is vital. It also calls for protecting the human communities that exist in these areas. Without addressing such issues as poverty, population growth, urbanization, and resource use, ecosystem services will continue to decline.

Proponents of this strategy call for identifying highly stressed *life raft ecosystems*. These are generally areas with high poverty levels and where most people depend on ecosystem services for survival. This dependence has degraded ecosystems severely enough to threaten humans and other species. In such areas, residents, public officials, and conservation scientists need to work together to protect human communities along with the natural biodiversity and ecosystem services that support all life and economies. Instead of pitting people against nature, this approach applies the win-win **principle of sustainability**. Environmental scientist Gretchen Daily is working to develop tools for evaluating the value of ecosystem services (Individuals Matter 10.2).

Restoring Damaged Ecosystems

Almost every natural place on the earth has been affected to some degree by human activities, most often in harmful ways. However, it is possible to partially reverse much of the damage through **ecological restoration,** the process of repairing damage humans have caused to various ecosystems. Examples include replanting forests (see the Case Study that follows), reintroducing keystone native species (Science Focus 10.4), removing harmful invasive species, freeing river flows by removing dams, and restoring grasslands, coral reefs, wetlands, and stream banks (Figure 10.21, right). This is an important way to expand our beneficial environmental impact and leave the earth in better shape than we found it.

By studying how natural ecosystems recover, scientists are learning how to employ and enhance ecological succession processes by using a variety of approaches, including the following four:

- *Restoration*: returning a degraded habitat or ecosystem to a condition as close as possible to its original one.
- *Rehabilitation*: turning a degraded ecosystem into a functional or useful ecosystem without trying to restore it to its original condition. Examples include removing pollutants from abandoned mining or industrial sites and replanting trees to reduce soil erosion in clear-cut forests.
- *Replacement*: replacing a degraded ecosystem with another type of ecosystem. For example, a degraded forest might be replaced by a productive pasture or tree plantation.
- *Creating artificial ecosystems*: for example, artificial wetlands have been created in some areas to help reduce flooding and to treat sewage.

Researchers have suggested the following four-step strategy for carrying out most forms of ecological restoration and rehabilitation.

1. Identify the causes of the degradation, such as pollution, farming, overgrazing, mining, or invasive species.

INDIVIDUALS MATTER 10.2

Gretchen Daily—Shining a Light on Ecosystem Services

Courtesy of Gretchen Daily

Gretchen Daily, Professor of Environmental Science at Stanford University, is alarmed at the growing threats to natural capital—earth's lands, waters, and the biodiversity, or machinery of nature, that they embody. She is considered one of the world's experts on natural capital and the flow of "ecosystem services" that it generates.

In 2006, Daily cofounded the *Natural Capital Project* with Stanford University, World Wildlife Fund, and The Nature Conservancy. It has three major goals. *First*, develop methods for measuring the economic and other values—to human health and well-being—of ecosystem services in a credible and practical way. *Second*, find ways to integrate these values into major decisions made by governments and businesses. And, *third*, take success to scale by tailoring and replicating models that work across world regions or industry sectors.

The group has developed a software tool called InVEST (for Integrated Valuation of Ecosystem Services and Trade-offs). It helps users compare how different choices—in where and how to develop or conserve land, can affect the ecosystem service benefits provided, by forests or wetlands, for example. It assigns long-term economic values to services such as flood control, water purification, and climate stability and projects how these services are likely to grow or shrink with various factors such as clear-cutting, selective logging, restoration, and preservation.

This tool is being used in over 100 countries, helping landowners and investors evaluate the effects of these and other factors in deciding how to use a piece of forested land. After heavy deforestation through the 1990s, China has invested over $100 billion in reforestation, for example, and is now reaping financial and well-being benefits of hydropower production efficiency, a more secure water supply, and flood control.

2. Stop the degradation by eliminating or sharply reducing these factors.
3. Reintroduce keystone species to help restore natural ecological processes, as was done with gray wolves in the Yellowstone ecosystem (Science Focus 10.4).
4. Protect the area from further degradation to allow natural recovery (Figure 10.21, right).

By following this general plan, conservationist and National Geographic Explorer Sean Gerrity is working with his 35-person team in the U.S. state of Montana to create American Prairie Reserve, the largest nature reserve in the continental United States, a refuge for people and wildlife preserved forever as a part of America's heritage. Their goal is to restore the wildlife and ecosystem services that were common to this unique area of North America's grasslands for more than 11,000 years.

CASE STUDY

Ecological Restoration of a Tropical Dry Forest in Costa Rica

Costa Rica (**Core Case Study**) is the site of one of the world's largest ecological restoration projects. In the lowlands of its Guanacaste National Park, a tropical dry forest was burned, degraded, and fragmented for conversion to cattle ranches and farms. Now it is being restored and reconnected to a rain forest on nearby mountain slopes. GOOD NEWS

Daniel Janzen, professor of conservation biology at the University of Pennsylvania and a leader in the field of restoration ecology, used his own MacArthur Foundation grant money to purchase the Guanacaste forestland for designation as a national park. He has also raised more than $10 million for restoring the park.

Janzen recognizes that ecological restoration and protection of the park will fail unless the people living in the surrounding area believe they will benefit from such efforts. His vision is to see that the nearly 40,000 people who live near the park play an essential role in the restoration of the forest, a concept he calls *biocultural restoration*.

In the park, local farmers are paid to remove nonnative species and to plant tree seeds and seedlings started in Janzen's laboratory. Local grade school, high school, and university students and citizens' groups study the park's ecology during field trips. The park's location near the Pan American Highway makes it an ideal area for ecotourism, which stimulates the local economy.

This project also serves as a training ground in tropical forest restoration for scientists from around the world. Research scientists working on the project give guest classroom lectures and lead field trips. Janzen believes that education, awareness, and involvement—not guards and fences—are the best ways to protect largely intact ecosystems from unsustainable use. This is an application of the biodiversity and win-win **principles of sustainability**.

Sharing Areas We Dominate with Other Species

We dominate most of the world's ecosystems, which is a cause of species extinction and of the degradation and loss of ecosystem services. For this reason, efforts to protect terrestrial biodiversity must focus on finding ways for humans to share some of the spaces we dominate with other species. Ecologist Michael L. Rosenzweig calls this approach **reconciliation ecology**. It focuses on establishing and maintaining new habitats to conserve species diversity in places where people live, work, or play. This is one way for us to increase our beneficial environmental impact.

By encouraging sustainable forms of ecotourism, people can protect local wildlife and ecosystems and provide economic resources for their communities. In the Central American country of Belize, conservation biologist Robert Horwich helped establish a local sanctuary for the black howler monkey. He convinced local farmers to set aside strips of forest to serve as habitats and corridors through which these monkeys can travel. The reserve, run by a local women's cooperative, has attracted ecotourists and biologists. Local residents receive income for housing and guiding these visitors.

Without proper controls, ecotourism can lead to degradation of popular sites if they are overrun by visitors or are degraded by the construction of nearby hotels and other tourist facilities. However, when managed properly, ecotourism can be a useful form of reconciliation ecology.

Reconciliation ecology is also a way to protect vital ecosystem services. Some people are learning how to protect insect pollinators, such as butterflies and honeybees (see Chapter 9, Core Case Study, p. 192), that are vulnerable to pesticides and habitat loss in their communities. Neighborhoods and municipal governments are doing so by reducing or eliminating the use of pesticides on their lawns, fields, golf courses, and parks. Neighbors can also work together to plant gardens of flowering plants as a source of food for bees, butterflies, and other pollinators. According to honeybee experts, people trying to help bees in this way should avoid using glyphosate herbicides and plants that contain neonicotinoid insecticides.

People have also worked together to protect bluebirds within human-dominated habitats. In such areas, bluebird populations have declined because most of their nesting trees have been cut down. Specially designed boxes have provided artificial nesting places for bluebirds. Their widespread use has allowed populations of this species to grow.

GOOD NEWS

These and many other examples of people working together on projects to restore degraded ecosystems are applications of the biodiversity and win-win **principles of sustainability**. Figure 10.28 lists some ways in which you can help sustain and expand the earth's terrestrial biodiversity.

Protecting the earth's vital biodiversity and increasing our beneficial environmental impact will not be implemented without bottom-up political pressure on elected officials from individual citizens and groups. It also will require cooperation among scientists, engineers, and key people in government and the private sector. Individuals also need to "vote with their wallets" by buying only products and services that do not harm terrestrial biodiversity.

What Can You Do?

Sustaining Terrestrial Biodiversity

- Plant trees and take care of them
- Recycle paper and buy recycled paper products
- Buy sustainably produced wood and wood products and wood substitutes such as recycled plastic furniture and decking
- Help restore a degraded forest or grassland
- Landscape your yard with a diversity of native plants

FIGURE 10.28 Individuals matter: Ways to help to sustain terrestrial biodiversity. ***Critical thinking:*** Which two of these actions do you think are the most important ones to take? Why? Which of these things do you already do?

BIG IDEAS

- The economic value of the ecosystem services provided by the world's ecosystems is far greater than the value of raw materials obtained from those systems.
- We can manage forests, grasslands, and nature reserves more effectively by protecting more land and by preventing overuse and degradation of these areas and the renewable resources they contain.
- We can sustain terrestrial biodiversity and ecosystem services and increase our beneficial environmental impact by protecting severely threatened areas and ecosystem services, restoring damaged ecosystems, and sharing with other species much of the land that we dominate.

Tying It All Together

Sustaining Costa Rica's Biodiversity

Eduardo Rivero/Shutterstock.com

In this chapter, you learned how human activities are destroying or degrading much of the earth's terrestrial biodiversity. You learned the importance of preserving what remains of diverse and highly endangered biodiversity hotspots and of sustaining the earth's ecosystem services. You also saw how to reduce this destruction and degradation by using the earth's resources more sustainably and by employing restoration ecology and reconciliation ecology. The **Core Case Study** introduced much of this by reporting on what Costa Rica is doing to protect and restore its precious biodiversity.

Preserving terrestrial biodiversity and the ecosystem services it provides involves applying the three **scientific principles of sustainability**. First, it means respecting biodiversity and understanding the value of sustaining it. In addition, if we rely less on fossil fuels and more on direct solar energy and its indirect forms, such as wind and flowing water, we will generate less pollution. We will also interfere less with chemical cycling and other forms of natural capital that sustain biodiversity and our own lives and economies.

Applying the three additional **principles of sustainability** will also help preserve biodiversity. By placing economic value on ecosystem services, we would acknowledge their importance by helping implement *full-cost pricing*. Working together to find *win-win solutions* to problems of environmental degradation benefits the earth and its people. Our actions can be guided by an ethical responsibility to sustain biodiversity and ecosystem services for current and future generations.

Chapter Review

Core Case Study

1. Summarize the story of Costa Rica's efforts to preserve its rich biodiversity.

Section 10.1

2. What are the two key concepts for this section? What major ecological and economic benefits do forests provide? Describe the efforts of scientists and economists to put a price tag on the major ecosystem services provided by forests and other ecosystems. Distinguish among **old-growth (primary) forests**, **second-growth forests**, and **tree plantations** (**tree farms** or **commercial forests**). What are the pros and cons of tree plantations?

3. Explain how building roads into previously inaccessible forests can harm the forests. Distinguish among selective cutting, clear-cutting, and strip cutting in the harvesting of trees. What are the advantages and disadvantages of clear-cutting forests? What are two types of forest fires? What are some ecological benefits of occasional surface fires? What factors could increase the number of crown fires? How can insects such as pine bark beetles affect forests?

4. What is **deforestation** and what parts of the world are experiencing the greatest forest losses? What percentage of the world's old-growth forests has been eliminated? List some major harmful environmental effects of deforestation. Describe China's reforestation program. Summarize the story of reforestation in the United States. What ecological roles do large trees play and why have they declined in parts of the world? Explain how increased reliance on tree plantations can reduce overall forest biodiversity and degrade forest topsoil. Summarize the trends in tropical deforestation. Explain how widespread tropical deforestation can convert a tropical forest to tropical grassland (savanna). What percentage of the world's remaining old-growth tropical forests is likely to be gone or degraded by the end of this century? Describe the loss of rainforest in Indonesia. What are the key underlying or direct causes and the indirect causes of tropical deforestation? Describe the typical cycle of use and destruction for an area of tropical forest.

Section 10.2

5. What is the key concept for this section? List three methods for managing forests and list

their limitations. What is certified sustainably grown timber? List four ways to manage forests more sustainably. What are four ways to reduce the harm caused by forest fires to forests and to people? What are three ways to reduce the need to harvest trees? Describe the global fuelwood crisis and solutions to this problem. Describe South Korea's reforestation program. What are five ways to protect tropical forests and use them more sustainably? Explain how Sustainable Harvest International helps poor farmers in tropical forests. Summarize the story of Wangari Maathai and the Green Belt Movement.

Section 10.3

6. What is the key concept for this section? Distinguish between **rangelands** and **pastures**. What is **overgrazing** and what are its harmful environmental effects? What are three ways to reduce overgrazing and use rangelands more sustainably? Describe Alan Savory's holistic range management system.

Section 10.4

7. What are the three key concepts for this section? What are seven strategies for sustaining terrestrial biodiversity? What percentage of the world's land has been set aside and protected as nature reserves, and what percentage should be protected, according to conservation biologists? What is **wilderness** and why is it important, according to conservation biologists? Summarize the history of wilderness protection in the United States. What are the major environmental threats to national parks in the world and in the United States? Describe some of the ecological effects of reintroducing the gray wolf to Yellowstone National Park. Why do ecologists support the establishment of large nature reserves and habitat corridors to connect nature reserves? What is the buffer zone concept? How has Costa Rica applied this approach?
8. What is a **biodiversity hotspot** and why is it important to protect such areas? Explain why Madagascar is an endangered biodiversity hotspot. Explain the importance of protecting ecosystem services and list three ways to do this. What is a life raft ecosystem? Describe Gretchen Daily's efforts to place an economic value on ecosystem services.
9. Define **ecological restoration**. What are four approaches to restoration? Summarize the science-based, four-step strategy for carrying out ecological restoration and rehabilitation. Describe the ecological restoration of Guanacaste National Park in Costa Rica. Define and give three examples of **reconciliation ecology**.
10. What are this chapter's three big ideas? Explain the relationship between preserving biodiversity, as it is done in Costa Rica, and the six **principles of sustainability**.

Note: Key terms are in bold type.

Critical Thinking

1. Why do you think Costa Rica (**Core Case Study**) has set aside a much larger percentage of its land for biodiversity conservation than the United States has? Should the United States reserve more of its land for this purpose? Explain.
2. If we fail to protect a much larger percentage of the world's remaining old-growth forests and tropical rain forests, what are three harmful effects that this failure is likely to have on any children and grandchildren you eventually might have?
3. In the early 1990s, Miguel Sanchez, a subsistence farmer in Costa Rica, was offered $600,000 by a hotel developer for a piece of land that he and his family had been using sustainably for many years. An area under rapid development surrounded the land, which contained an old-growth rain forest and a black sand beach. Sanchez refused the offer. Explain how Sanchez's decision was an application of the ethical **principle of sustainability**. What would you have done if you were Sanchez? Explain.
4. Halting the destruction and degradation of tropical rain forests is a key to preserving the world's biodiversity and slowing global climate change. Since this will benefit the entire world during this and future generations, should the United States and the world's other more-developed nations pay tropical, less-developed countries to preserve their remaining tropical forests, as Norway and the United Kingdom have done? Explain. Do you think that the long-term economic and ecological benefits of doing this would outweigh the short-term economic costs? Explain.
5. Are you in favor of establishing more wilderness areas in the United States (or in the country where you live)? Explain. What might be some disadvantages of doing this?
6. You are a defense attorney arguing in court for preserving an old-growth forest that developers want to clear for a suburban development. Give your three strongest arguments for preserving this ecosystem. How would you counter the argument that preserving the forest would harm the economy by causing a loss of jobs in the timber industry?
7. Do you support or oppose the U.S. 2003 Healthy Forests Restoration Act? Explain.

8. It would cost about $76 billion a year to sustain the earth's terrestrial biodiversity. Do you think we should spend this money? How might a decision not to make this investment affect you and any children or grandchildren you might have?

Doing Environmental Science

Pick an area near where you live or go to school that hosts a variety of plants and animals. It could be a yard, an abandoned lot, a park, a forest, or some part of your campus. Visit this area at least three times and make a survey of the plants and animals that you find there, including any trees, shrubs, groundcover plants, insects, reptiles, amphibians, birds, and mammals. Also, take a small sample of the topsoil and find out what organisms are living there. (Be careful to get permission from whoever owns or manages the land before doing any digging.) Using guidebooks and other resources to help identify different species, record your findings, and categorize them into the general types of organisms listed above. Then do some research to find out about the ecosystem services that some or all of these organisms provide. Try to find and record five of these services. Finally, do some research to find a range of values that economists have assigned to these ecosystem services at the global level. Write a report summarizing your findings.

Global Environment Watch Exercise

Go to your MindTap course to access the GREENR database. Starting on the home page, under "Browse Issues and Topics", click on *Environment and Ecology*, then select *Forests and Deforestation* to enter the portal. Go to the Statistics heading and click "View All." On this page, click on "Share of Tropical Deforestation." Choose one of these countries and research the deforestation in this country further (tip: use the World Map feature). Write a report on your findings and include possible solutions for this deforestation problem. Solutions may include those legislated by governments, as well as those being tried by private individuals or companies.

Ecological Footprint Analysis

The table below compares five countries in terms of rain forest area and losses. Study the table and then answer the questions that follow.

Country	Area of Tropical Rain Forest (square kilometers)	Area of Deforestation per Year (square kilometers)	Annual Rate of Tropical Forest Loss
A	1,800,000	50,000	
B	55,000	3,000	
C	22,000	6,000	
D	530,000	12,000	
E	80,000	700	

1. What is the annual rate of tropical rain forest loss, as a percentage of total forest area, in each of the five countries? Answer by filling in the blank column in the table.

2. What is the annual rate of tropical deforestation collectively in all of the countries represented in the table?

3. According to the table, and assuming the rates of deforestation remain constant, which country's tropical rain forest will be destroyed first?

4. Assuming the rate of deforestation in country C remains constant, how many years will it take for all of its tropical rain forests to be destroyed?

5. Assuming that a hectare (1.0 hectare = 0.01 square kilometer) of tropical rain forest absorbs 0.85 metric tons (1 metric ton = 2,200 pounds) of carbon dioxide per year, what would be the total annual growth in the carbon footprint (carbon emitted but not absorbed by vegetation because of deforestation) in metric tons of carbon dioxide per year for each of the five countries in the table?

CHAPTER 11

Sustaining Aquatic Biodiversity and Ecosystem Services

With every drop of water you drink, with every breath you take, you are connected to the sea, no matter where on Earth you live.

SYLVIA EARLE

Key Questions

11.1 What are the major threats to aquatic biodiversity and ecosystem services?

11.2 How can we protect and sustain marine biodiversity?

11.3 How should we manage and sustain marine fisheries?

11.4 How should we protect and sustain wetlands?

11.5 How should we protect and sustain freshwater lakes, rivers, and fisheries?

11.6 What should be our priorities for sustaining aquatic biodiversity?

Endangered green sea turtle swimming over a coral reef.

George Grall/National Geographic Creative

The Great Jellyfish Invasion

Jellyfish, which are not fish, were the first invertebrate animals in the earth's oceans. They dominated the oceans for at least 500 million years in the Precambrian period, long before fish evolved (see Supplement 6, p. S46).

A jellyfish (see Figure 8.2, p. 169) is mostly made of water and has no brain, blood, head, heart, bones, or protective shell. The bell-shaped bodies of jellyfish are filled with a jelly-like substance. A network of nerves at the base of their dangling tentacles can detect warmth, food, odors, and vibration.

Jellyfish move by drifting on water currents and by squirting pulsating jets of water from their bodies. They use tentacles dangling from their bodies to sting or stun prey, to draw food into their mouth on the underside of their body, and to defend themselves against predators such as other jellyfish, tuna, sharks, swordfish, and sea turtles.

Jellyfish eat almost anything that floats their way. Most are carnivorous and typically feed on zooplankton, fish eggs, small fish, shrimp, and other jellyfish. They are caught for food in 15 countries, and are considered a delicacy in countries such as China and Japan.

Jellyfish sizes vary widely. Some are as small as a mosquito. The largest jellyfish is the lion's mane (Figure 11.1). It has a body up to 2.4 meters (8 feet) wide, has tentacles as long as 30 meters (100 feet), and can weigh as much as 150 kilograms (350 pounds).

The sting of a jellyfish can cause itching or a burning sensation that lasts for several days. Most stings occur when people accidentally brush up against a jellyfish. However, a sting from the Portuguese man-of-war, the Australian box jelly, or the tiny Irukandji jellyfish can kill a human within minutes if untreated. Each year, the stings of lethal jellyfish kill around 40 people on average.

Jellyfish are often found in large swarms or *blooms* of thousands, even millions of individuals. In recent years, the numbers of these blooms observed by scientists and fishers have been rising. Often they are as long as five to six city blocks, and one *megabloom* was about 300 meters (1,000 feet) long.

These blooms have a number of harmful economic effects. They cause beach closings, disrupt commercial fishing operations by clogging or tearing holes in nets, wipe out coastal fish farms, and clog ship engines. They can also close down coal-burning and nuclear power plants by blocking their cooling water intakes.

According to Chinese oceanographer Wei Hao and other marine scientists, the startling growth of jellyfish populations threatens to upset marine food webs and ecosystem services and turn some of the world's most productive ocean areas into jellyfish empires. Once jellyfish take over a marine ecosystem, they might dominate it for millions of years, as they did in the Precambrian period. Later in this chapter, we discuss why jellyfish populations have been increasing at an alarming rate.

In this chapter, we examine the effects of human activities on aquatic biodiversity. We also explore ways to prevent or lessen these effects in order to help sustain aquatic life.

wizdata/Fotolia LLC

FIGURE 11.1 The lion's mane is the largest jellyfish species. It can grow to 30 meters (100 feet) in length.

11.1 WHAT ARE THE MAJOR THREATS TO AQUATIC BIODIVERSITY AND ECOSYSTEM SERVICES?

CONCEPT 11.1 Aquatic species and the ecosystem and economic services they provide are threatened by habitat loss, invasive species, pollution, climate change, and overexploitation—all made worse by the growth of the human population and resource use.

We Have Much to Learn about Aquatic Biodiversity

We live on a water planet with 71% of its surface covered by salty ocean water to an average depth of 3.7 kilometers (2.3 miles). Yet, we have explored less than 5% of the earth's interconnected oceans and know relatively little about marine biodiversity and its many functions. We also have limited knowledge about freshwater biodiversity.

Scientists have observed three general patterns related to marine biodiversity. *First,* the greatest marine biodiversity occurs around coral reefs, in estuaries, and on the deep-ocean floor. *Second,* biodiversity is greater near the coasts than in the open sea because of the larger variety of producers and habitats in coastal areas. *Third,* biodiversity is generally greater in the bottom region of the ocean than in the surface region because of the larger variety of habitats and food sources on the ocean bottom.

The deepest part of ocean, where sunlight does not penetrate (see Figure 8.5, p. 171), is the planet's least explored environment but this is changing. More than 2,400 scientists from 80 countries are working on a 10-year project to catalog the species in the deep ocean zone. They have used remotely operated deep-sea vehicles to identify more than 17,000 species living in this zone and are adding a few thousand new species every year.

Why should we care about sustaining life in the oceans? What difference will it make if coral reefs, sharks, or whales disappear? There are a number of economic, health-related, and ecological reasons:

- Worldwide, about 850 million jobs in fishing and tourism depend on the oceans.
- About 40% of the world's people get 15–20% of their animal protein and essential nutrition from seafood.
- Oceans generate more than half of the oxygen we breathe.
- Oceans help slow atmospheric warming and climate change by absorbing about 25% of the carbon dioxide produced by human activities.
- Oceans absorb 90% of the excess heat we add to the atmosphere.
- Natural barriers such as coral reefs, mangrove forests, and sea-grass beds reduce the impacts from tsunamis and major storms.

These economic and ecosystem services are provided by the diversity of species living and interacting in the oceans. This explains why learning about and sustaining marine biodiversity should be one of our top priorities. Freshwater systems, which occupy only 1% of the earth's surface, also provide important economic and ecological services (see Figure 8.14, p. 179).

Human Activities Are Destroying and Degrading Aquatic Habitats

Human activities have destroyed or degraded much of the world's coastal wetlands, coral reefs, mangroves, and ocean bottom. They also have disrupted many freshwater ecosystems. In a 2015 study, ecologist Douglas J. McCauley and a team of other scientists reviewed data about the state of the oceans from hundreds of sources and concluded that "the oceans are facing a major extinction event." They also said that "the impacts are accelerating but they're not so bad we can't reverse them."

As with terrestrial biodiversity, the greatest threats to aquatic biodiversity and ecosystem services can be remembered with the aid of the acronym HIPPCO, with H standing for *habitat loss and degradation* (**Concept 11.1**). Scientists reported in 2006 that coastal habitats were disappearing at rates 2 to 10 times higher than the rate of tropical forest loss.

Shallow, warm-water coral reefs are centers of aquatic biological diversity (Figure 11.2 and Chapter 8, Core Case Study, p. 168). These crown jewels of the ocean occupy only 0.1% of the world's oceans, but are home for about 25% of world's marine fish species.

Since the 1950s, about half of the world's shallow, warm-water coral reefs (90% in the Indian Ocean and Caribbean) have been destroyed (Figure 11.3) or degraded, and another 25–33% could be lost within 20 to 40 years, according to the Global Coral Reef Monitoring Network. Threats include coastal development, overfishing, pollution, warmer ocean water, and ocean acidification (Science Focus 11.1).

Today, shallow coral reefs, on average, are exposed to the warmest and most acidic ocean waters of the past 400,000 years—a double threat from the excess carbon dioxide that we have been adding to the atmosphere, mostly from burning fossil fuels. Warmer ocean waters can cause shallow tropical corals to expel their colorful algae and leave behind white coral—a process called **coral bleaching** (see Figure 8.1, p. 168). It can weaken and sometimes kill corals. Marine biologist Malin L. Pinsky said: "If you cranked up the aquarium heater and dumped some acid in the water, your fish would not be very happy. In effect, that's what we are doing to the oceans."

Vlad61/Shutterstock.com

FIGURE 11.2 Coral reefs are endangered centers of aquatic biodiversity.

Richard Whitcombe/Shutterstock.com

FIGURE 11.3 Dead coral reef.

CONSIDER THIS . . .

CONNECTIONS Sunscreens and Coral Reefs

Certain ingredients in many sunscreens have been shown to promote the growth of a harmful virus within the algae that live in coral reefs. When people dive down to see the reefs, these chemicals can wash off. This can kill the algae and promote coral bleaching.

If given enough time, many species of corals can adapt to changes in environmental conditions such as warmer and perhaps more acidic water. However, corals are threatened with rapidly rising water temperatures, acidity, and sea levels during this century, which will not give them enough time for adaptation.

According to fossil and other evidence, coral reefs were devastated in each of the earth's five mass extinctions that took place over the last half-billion years (see Figure 4.19, p. 93). Evidence indicates that, in each mass extinction, high levels of dissolved CO_2 and prolonged ocean warming and acidification played a role in dissolving the calcium carbonate that coral polyps use to build the reefs. With each mass extinction, the corals disappeared and did not come back for 4 million to 10 million years. If we have triggered a sixth mass extinction as some scientists say, many of the coral species that are currently centers of marine biodiversity are likely to disappear again for millions of years.

CONSIDER THIS . . .

LEARNING FROM NATURE

Some reefs have been resilient enough to recover from coral bleaching, and scientists are researching how this occurs. Scientists are also investigating how shallow reefs in certain naturally acidic waters have survived.

United Nations Environment Programme (UNEP) scientists reported that a fifth of the world's ecologically and economically important mangrove forests (see Figure 8.8 p. 172) have been lost since 1980. They continue to be destroyed for firewood, coastal construction, and shrimp farming. Another study revealed that 58% of the world's coastal sea-grass beds (see Figure 8.9, p. 173) have been degraded or destroyed, mostly by dredging and coastal development.

Sea-bottom habitats are faring no better. Each year, like giant submerged bulldozers, thousands of trawler fishing boats drag up to 60-meter-wide (200-foot-wide) nets weighted down with chains and steel plates over the ocean floor to harvest a few species of bottom fish and shellfish (Figure 11.4). This destroys large areas of deep, cold-water coral reefs and other ocean-bottom habitats, which could take decades or centuries to recover. According to marine scientist Elliot Norse, "Bottom trawling is probably the largest human-caused disturbance to the biosphere." An increase in seabed mining also threatens biologically diverse ocean-bottom habitats.

Habitat disruption is also a problem in freshwater aquatic zones. The main causes are the building of dams and excessive withdrawal of river water for irrigation and urban water supplies. These activities destroy aquatic habitats, decrease water flows, and disrupt freshwater biodiversity. Globally, the extinction rate for freshwater species is 5 times the rate for terrestrial species, according to the latest IUCN Red List.

Ocean acidification (Science Focus 11.1) could kill off much of the oceans' phytoplankton that are the base of marine food webs. According to a study by marine scientist Ove Hoegh-Guldberg and 16 other scientists, unless we take action soon to significantly reduce CO_2 emissions, the oceans may become too acidic and too warm for most of the world's coral reefs to survive this century.

Invasive Species Can Degrade Aquatic Biodiversity

Another threat to aquatic biodiversity is the deliberate or accidental introduction of hundreds of harmful

SCIENCEFOCUS 11.1

Ocean Acidification: The Other CO_2 Problem

By burning increasingly large amounts of carbon-containing fossil fuels, especially since 1950, we have added carbon dioxide (CO_2) to the atmosphere faster than it can be removed by the carbon cycle (see Figure 3.20, p. 65)

Extensive research indicates that this increase in the atmospheric concentration of CO_2 has played a key role in the observed increase in the atmosphere's average temperature, especially since 1980, and has changed the earth's climate. Research and climate models indicate that continuing to increase CO_2 levels in the atmosphere will likely lead to severe disruption of the earth's climate during this century, as we discuss in Chapter 19.

Another serious environmental problem related to CO_2 emissions is *ocean acidification*, a change in ocean chemistry. The oceans have helped reduce atmospheric warming and climate change by absorbing about 25% of the excess CO_2 that human activities have added to the atmosphere. When this absorbed CO_2 combines with ocean water, it forms carbonic acid (H_2CO_3), a weak acid also found in carbonated drinks. This increases the level of hydrogen ions (H^+) in the water and makes the water less basic (with a lower pH; see Figure 2.6, p. 36). This also decreases the level of carbonate ions (CO_3^{2-}) in the water because these ions react with hydrogen ions (H^+) to form bicarbonate ions (HCO_3^-).

The problem is that many aquatic species—including phytoplankton, corals, sea snails, crabs, and oysters—use carbonate ions to produce calcium carbonate ($CaCO_3$), the main component of their shells and bones. In less basic waters, carbonate ion concentrations drop (Figure 11.A) and shell-building species and coral reefs grow more slowly. When the hydrogen ion concentration of seawater gets high enough, their calcium carbonate begins to dissolve. Species that survive will have damaged or weaker shells and bones.

According to a 2013 study by more than 540 of the world's leading experts on ocean acidification, the average acidity of ocean water has risen 30% (actually a 30% decrease in average basicity) since 1800. It has risen 15% since the 1990s, with the largest increase occurring in deep cold waters near the poles, especially in the Arctic Sea, and along the West Coast of the United States and is projected to keep acidifying throughout this century. According to the report, the oceans are acidifying "faster than at any time during the last 300 million years."

Evidence indicates that this change in ocean chemistry is contributing to coral bleaching, to holes in the shells of tiny sea snails in the U.S. Pacific Ocean, and to die-offs at oyster farms off the coast of Washington and Oregon. More research is needed to evaluate the ecological impacts of a warmer and more acidic ocean and to determine whether some of the affected organisms can adapt to such changes.

Because the world's oceans are a key component of our life-support system, a survival rule should be *do not change the chemistry of the oceans*—a rule that we are violating. According to the world's key ocean acidification scientific experts, we are altering the chemistry of the entire ocean ecosystem from the tropics to the poles with little idea of the consequences.

Ocean acidification is a serious threat to ocean life, terrestrial life, and the human species and economies. If we fail to act rapidly to reduce this serious threat, marine biologists project that ocean food webs will shift dramatically, as corals and other calcifying organisms die off and as green algae and jellyfish, which thrive in acidic and warm waters, become more dominant (**Core Case Study**).

CRITICAL THINKING

How might widespread losses of some forms of marine aquatic life due to ocean acidification affect life on land? How might it affect your life? (*Hint:* Think food webs.)

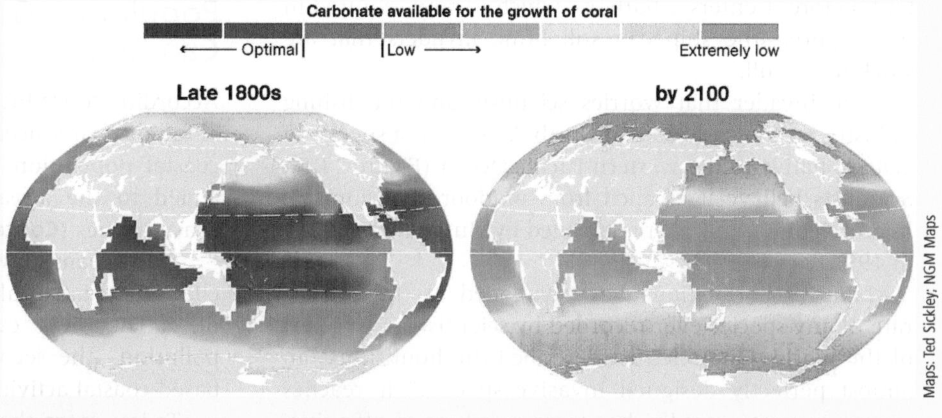

FIGURE 11.A Calcium carbonate levels in ocean waters, calculated from historical data (left), and projected for 2100 (right). Colors shifting from blue to red indicate where waters are becoming less basic. In the late 1800s, when CO_2 began to pile up rapidly in the atmosphere, tropical corals were not yet affected by ocean acidification. However, today carbonate levels have dropped substantially near the Poles, and by 2100, they may be too low even in the tropics for most coral reefs to survive. (Sources: Andrew G. Dickson, Scripps Institution of Oceanography, U.C. San Diego, and Sarah Cooley, Woods Hole Oceanographic Institution. Used by permission from National Geographic.)

Courtesy of Peter J. Auster/National Undersea Research Center

FIGURE 11.4 Natural capital degradation: An area of ocean bottom before (left) and after (right) a trawler net scraped it like a gigantic bulldozer. ***Critical thinking:*** What land activities are comparable to this?

invasive species (see Figure 9.9, p. 201)—the I in HIPPCO—into coastal waters, wetlands, and lakes throughout the world (**Concept 11.1**). According to the U.S. Fish and Wildlife Service, bioinvaders are blamed for about two-thirds of all fish extinctions in the United States since 1900 and have caused huge economic losses.

Many of the more than 1,450 different aquatic invader species in the United States arrived in the ballast water stored in tanks in large cargo ships to keep them stable. The ships take in ballast water from one harbor, along with whatever microorganisms and tiny fish species it contains, and dump it into another—an environmentally and economically harmful effect of globalized trade. Even when ballast water is flushed from an oceangoing ship's tank before it enters a harbor—which is now required in many ports—the ship can still bring invaders that are stuck to its hull.

One invader that worries scientists and the fishing industry on the east coast of North America is a species of *lionfish*, native to the western Pacific Ocean (Figure 11.5). Scientists believe it escaped from outdoor aquariums in Miami, Florida, that were damaged by Hurricane Andrew in 1992.

Lionfish populations have exploded at the highest rate of any species ever recorded by scientists in this part of the world. One scientist described the lionfish as "an almost perfectly designed invasive species." It reaches sexual maturity rapidly, has large numbers of offspring, and is protected by venomous spines. It competes with popular reef fish species such as grouper and snapper, taking their food and eating their young. One ray of hope for controlling this population is that the lionfish tastes good. Scientists are hoping to see a growing market for lionfish as seafood, but it is difficult and costly to prepare.

CONSIDER THIS . . .

CONNECTIONS Lionfish and Coral Reef Destruction

Researchers have found that lionfish eat at least 50 species of prey fish, including parrotfish, that normally consume enough algae around coral reefs to keep the algae from overgrowing and killing the corals. Scientists warn that, where lionfish are now the dominant species such as in the Bahamas, unchecked algae could overwhelm and destroy some coral reefs.

In addition to threatening native species, invasive species can disrupt and degrade whole ecosystems and their ecosystem services. This is the focus of study for a growing number of researchers (Science Focus 11.2).

Population Growth and Pollution Can Reduce Aquatic Biodiversity

According to UNEP, about 80% of the world's people live along or near seacoasts, mostly in large coastal cities. This coastal population growth—the first P in HIPPCO—has added to the already intense pressure on the world's coastal zones (**Concept 11.1**).

The oceans have become a sewer for much of the chemical, biological, and solid wastes that humans produce. The UNEP estimates that about 80% of all ocean pollution—the second P in HIPPCO—comes from land-based coastal activities.

Today, more than 400 oxygen-depleted zones ("dead zones") have formed in coastal areas around the world, and the number is increasing. They form when high levels of plant nutrients from fertilizers and soil erosion flow from the land into rivers that empty into coastal waters. These inputs support large algal blooms. When these algae die, they sink to the bottom, where bacteria begin to decompose them. Because the decomposition requires

SCIENCEFOCUS 11.2

How Invasive Carp Have Muddied Some Waters

Lake Wingra lies within the city of Madison, Wisconsin, surrounded mostly by a forest preserve. The lake contains a number of invasive plant and fish species, including purple loosestrife (see Figure 9.9, p. 201) and common carp. The carp were introduced in the late 1800s and since then have made up as much as half of the fish biomass in the lake. They devour algae called *chara*, which would normally cover the lake bottom and stabilize its sediments. Consequently, fish movements and winds stir these sediments, which accounts for much of the water's excessive *turbidity*, or cloudiness.

Knowing this, Dr. Richard Lathrup, a limnologist (lake scientist) who worked with Wisconsin's Department of Natural Resources, hypothesized that removing the carp would help to restore the natural ecosystem of Lake Wingra. Lathrop speculated that if the carp were removed, the bottom sediments would settle and become stabilized, allowing the water to clear. Clearer water would in turn allow native plants to receive more sunlight and become reestablished on the lake bottom, replacing purple loosestrife and other invasive plants that now dominate its shallow shoreline waters.

Lathrop and his colleagues installed a thick, heavy vinyl curtain around a 1-hectare (2.5-acre), square-shaped perimeter that extended out from the shore. This barrier hung from buoys on the surface to the bottom of the lake, isolating the volume of water within it. The researchers then removed all of the carp from this study area and began observing results. Within 1 month, the waters within the barrier were noticeably clearer, and within a year, the difference in clarity was dramatic (Figure 11.B) and native plants once again grew in the shallow shoreline waters.

Lathrop notes that removing and keeping carp out of Lake Wingra would be a daunting task, perhaps impossible, but his controlled scientific experiment clearly shows the effects that an invasive species can have on an aquatic ecosystem. And it reminds us that preventing the introduction of invasive species in the first place is the best and least expensive way to avoid such effects.

FIGURE 11.B *Lake Wingra* in Madison, Wisconsin (USA) became clouded with sediment partly because of the introduction of invasive species such as the common carp. Removal of carp in the experimental area shown here resulted in a dramatic improvement in the clarity of the water.

CRITICAL THINKING

What are two other results of this controlled experiment that you might expect? (*Hint:* Think food webs.)

oxygen, levels of this dissolved gas in the water become depleted. Marine organisms either suffocate due to lack of dissolved oxygen or leave the area if they can. It is another example of how human activities can lead to changes in ocean chemistry and biodiversity.

Toxic pollutants from industrial and urban areas can kill some forms of aquatic life by poisoning them. Some highly toxic mercury in ocean waters comes from natural sources. However, toxic mercury released into the atmosphere by coal-burning plants can be taken up by ocean producers and magnified to high levels in ocean food webs. Because of this input from natural and human sources, top predator fish such as sharks, tilefish, swordfish, king mackerel, and white tuna can contain high levels of toxic mercury.

FIGURE 11.5 The *common lionfish* has invaded the eastern coastal waters of North America, where it has few, if any, predators.

Doris Alcorn/U.S. National Maritime Fisheries

FIGURE 11.6 This Hawaiian monk seal was slowly starving to death before a discarded piece of plastic was removed from its snout.

Partially decomposed particles of plastic items dumped from ships and garbage barges, and left as litter on beaches, kill up to 1 million seabirds and 100,000 mammals annually (Figure 11.6). These animals mistake the plastic particles for plankton or small fish. Certain compounds in these plastics can be concentrated in some types of seafood that people eat.

Climate Change Is a Growing Threat

Climate change—the C in HIPPCO—threatens aquatic biodiversity (**Concept 11.1**) and ecosystem services. Greenhouse gas emissions and heat, mostly from the burning of fossil fuels, have played an important role in warming the atmosphere and changing the earth's climate. For decades, the earth's oceans have absorbed about 90% of this excess heat. If they had not, the earth's atmosphere would be much warmer and the climate would be changing much more rapidly. As energy expert John Abraham puts it: "The ocean is doing us a favor by grabbing about 90% of our heat output. But it is not going to do it forever."

As a result, the ocean has been getting warmer. The surface warms the most, but vertical currents (see Figure 7.11, p. 147) transfer some of this heat to deep water. This ocean heating affects marine food webs and makes some marine habitats unlivable for some species by exceeding the range of temperatures they can tolerate (see Figure 5.15, p. 110). Unless they can migrate to cooler water, they can face extinction. Measurements indicate that the ability of the ocean to absorb heat is slowing. Eventually some of this heat will flow back into the atmosphere and accelerate atmospheric warming and climate change.

Ocean warming has caused some marine species to migrate from the equator toward the poles to cooler waters. This can threaten some of these species because their new habitats can be less hospitable, having lower dissolved oxygen levels. A warmer ocean could also boost populations of some species such as the coral-eating thorn-of-crown starfish that poses a threat to Australia's Great Barrier Reef.

The ocean has removed and dissolved about 27% of the excess CO_2 we have added to the atmosphere. Because of ocean currents, most of this CO_2 ends up in the southern hemisphere oceans and is forced into deep ocean waters by whirlpools driven by winds. As the ocean warms, it holds less CO_2. The combination of rising ocean temperatures could release huge amounts of CO_2 into the atmosphere and trigger rapid climate change.

A big threat from a warmer ocean is a rising sea level due to thermal expansion as ocean water warms and the partial melting of land-based ice in glaciers and ice sheets as atmospheric temperatures rise. In 2014 the Intergovernmental Panel on Climate Change (IPCC) estimated that the average global sea level is likely to rise by 40–60 centimeters (1.3–2 feet) by the end of this century—about 10 times the rise that occurred in the 20th century. Recent research projects a rise of 1.1–1.2 meters (3–4 feet) by 2100.

Such a rise in sea level would destroy shallow coral reefs, swamp some low-lying islands, drown many coastal wetlands, and put many coastal areas such as a large part of the U.S. Gulf and East Coasts underwater (see Figure 8.21, p. 186). In addition, some Pacific island nations could lose more than half of their protective coastal mangrove forests by 2100, according to a UNEP study.

CONSIDER THIS . . .

CONNECTIONS Protecting Mangroves and Dealing with Climate Change

Protecting mangrove forests and restoring them in areas where they have been destroyed are important ways to reduce the impacts of rising sea levels and storm surges, because mangroves forests can slow storm-driven waves. These ecosystem services will become more important if tropical storms become more intense because of climate change. Protecting and restoring these natural coastal barriers is much cheaper and more effective than building concrete sea walls or moving threatened coastal towns and cities inland.

Warmer and more acidic ocean water is also stressing phytoplankton, the foundation of marine food webs (see Figure 3.16, p. 60). These tiny life forms produce half of the earth's oxygen and absorb a great deal of the CO_2 emitted by human activities. A team of scientists led by marine ecologist Boris Worm found that global phytoplankton populations have declined by 40% since the 1950s, probably because of warmer and more acidic ocean waters.

Overfishing and Overharvesting: Gone Fishing, Fish Gone

Fish and fish products provide 20% of the world's animal protein for billions of people. A **fishery** is a concentration of a wild aquatic species suitable for commercial harvesting in a given ocean area or inland body of water.

Today, 4.4 million fishing boats hunt for and harvest fish from the world's oceans. These industrial fishing fleets use global satellite positioning equipment, sonar fish-finding devices, huge nets, long fishing lines, spotter planes and drones, and refrigerated factory ships that can process and freeze their enormous catches. These highly efficient fleets supply the growing demand for seafood, but critics say that they are overfishing many species (Figure 11.7), reducing marine biodiversity, and degrading important marine ecosystem services. Figure 11.8 shows the major methods used for the commercial harvesting of various marine fishes and shellfish.

For example, trawlers have destroyed vast areas of ocean-bottom habitat (Figure 11.4). In addition, their nets often capture endangered sea turtles (Figure 11.9), causing them to drown.

Another fishing method, *purse-seine fishing* (Figure 11.8), is used to catch surface-dwelling species such as tuna, mackerel, anchovies, and herring, which tend to feed in schools near the surface or in shallow waters. After a spotter plane locates a school, the fishing vessel encloses it with a large purse-seine net. Some of these nets have killed large numbers of dolphins that swim on the surface above schools of tuna.

Some fishing vessels also use *long-lining*, which involves putting out lines up to 100 kilometers (60 miles) long, hung with thousands of baited hooks to catch swordfish, tuna, sharks, and ocean-bottom species such as halibut and cod. Long lines also hook and kill large numbers of sea turtles, dolphins, and seabirds each year.

With *drift-net fishing*, fish are caught by drifting nets that can hang as deep as 15 meters (50 feet) below the surface and extend to 64 kilometers (40 miles) long. These nets trap and kill large quantities of unwanted fish, called **bycatch,** along with marine mammals and sea turtles. Nearly one-third of the world's annual fish catch by weight consists of bycatch species that are mostly thrown overboard dead or dying. This adds to the depletion of these species and puts stress on some of the species that feed on them.

A **fishprint** is the area of ocean needed to sustain the fish consumption of an average person, a nation, or the world based on the weight of fish they consume annually. It helps scientists and government officials distinguish between sustainable and unsustainable levels of fishing and evaluate the effects of policy change. The world's fishing nations are harvesting more fish than their populations can sustain in the long run. According to the Woods Hole Oceanographic Institute, 57% of the world's commercial fisheries have been fully exploited and another 30% have been overfished.

87% Percentage of the world's commercial fisheries that have been fully exploited or overfished

In most cases, overfishing leads to *commercial extinction*, which occurs when it is no longer profitable to continue harvesting the affected species. Overfishing can temporarily deplete a species, as long as depleted areas and fisheries are allowed to recover. However, as industrialized fishing fleets take more and more of the world's available fish and shellfish, recovery times for severely depleted populations are increasing and can be two decades or more. Some depleted fisheries may not recover as jellyfish and other invasive species move in and take over their food webs (**Core Case Study**).

In the late 1950s, fishing fleets began overfishing the 500-year-old northwest Atlantic cod fishery off the coast of Newfoundland, Canada. They used bottom trawlers to

MARK EDWARDS/Still Pictures/Aurora Photos

FIGURE 11.7 The threatened Atlantic bluefin tuna is being overfished because of its high market value.

zaferkizilkaya/Shutterstock.com

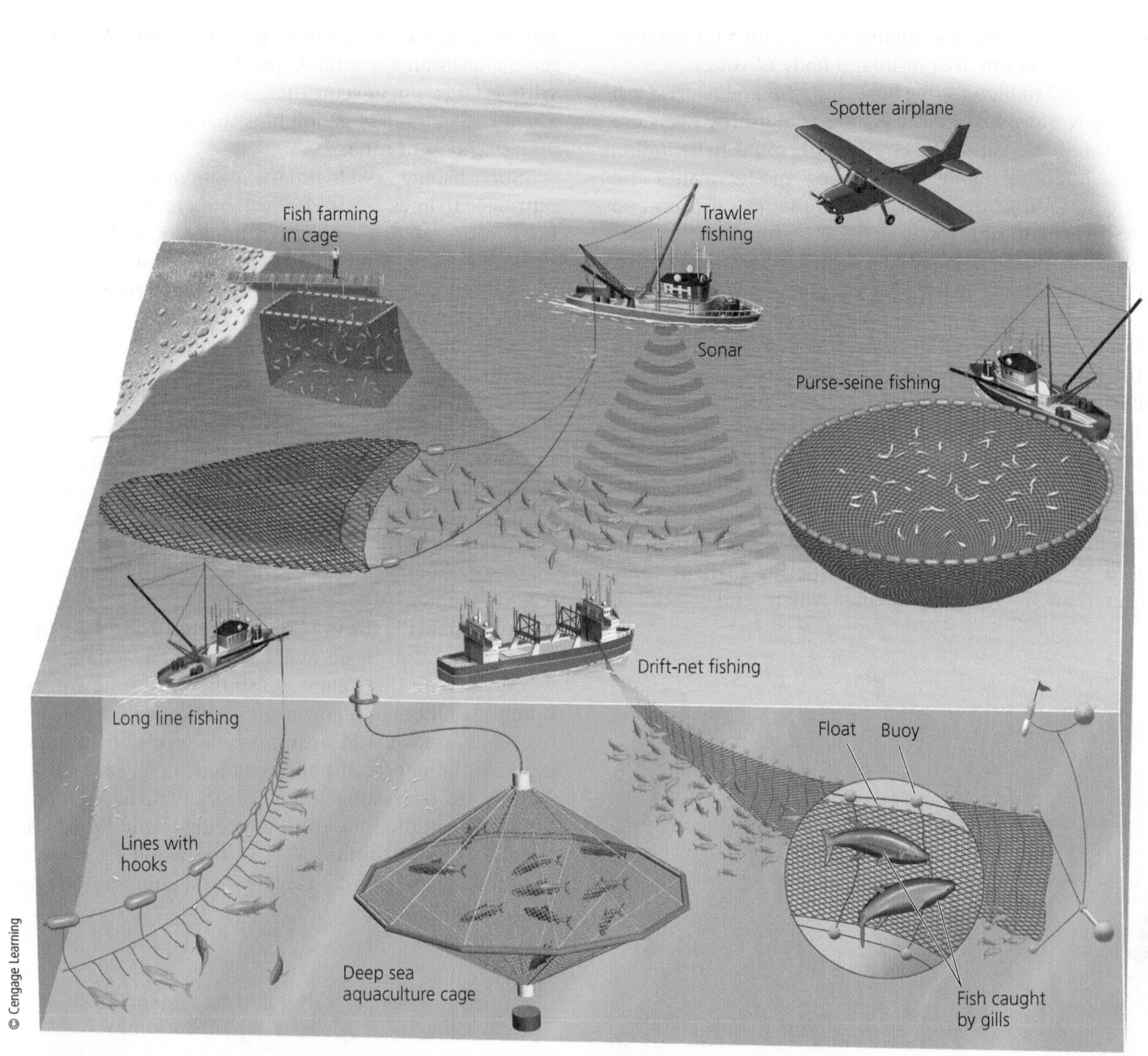

FIGURE 11.8 Major commercial fishing methods used to harvest various marine species, along with some methods used to raise fish through aquaculture.

capture larger shares of the stock, reflected in the sharp rise in the graph in Figure 11.10. This extreme overexploitation of the fishery led to a steady decline in the fish catch throughout the 1970s. After a slight recovery in the 1980s, the fishery collapsed and in 1992 it was shut down. This put at least 35,000 fishers and fish processors out of work in more than 500 coastal communities. The fishery was reopened on a limited basis in 1998, but in 2003 it was closed indefinitely.

One result of the increasingly efficient global hunt for fish is that larger individuals of commercially valuable wild species—including cod, marlin, swordfish, and tuna—are becoming scarce. A study conducted by scientists at Dalhousie University in Canada found that 90% or more of these and other large, predatory, open-ocean fishes disappeared between 1950 and 2006, a trend that is increasing. Another effect of overfishing is that when larger predatory species dwindle, rapidly reproducing invasive species such as jellyfish (**Core Case Study**) can take over and disrupt ocean food webs.

As commercially valuable species are overfished, the fishing industry has turned to other species such as sharks, which are now being overfished (see the Case Study that follows). In addition, as large species are overfished, the fishing industry has been working its way down to lower trophic levels of marine food webs by shifting to smaller marine species such as anchovies, herring, sardines, and shrimp-like krill—known as forage fish. Ninety percent of

Mark Foley/State Archives of Florida, Florida Memory

FIGURE 11.9 This endangered green sea turtle died after being caught in a fishing net.

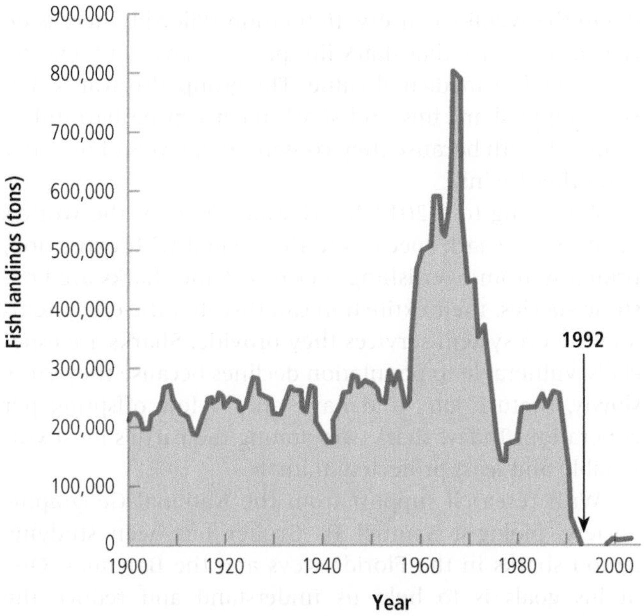

FIGURE 11.10 Natural capital degradation: Collapse of Newfoundland's Atlantic cod fishery. ***Data analysis:*** By roughly what percentage did the catch of Atlantic cod drop between the peak catch in 1960 and 1970? (Compiled by the authors using data from Millennium Ecosystem Assessment.)

this catch is converted to fishmeal and fish oil, and fed to farmed fish. Scientists warn that this reduces the food supply for larger fish species and makes it harder for them to rebound from overfishing. The result will be further disruption of marine ecosystems and their ecosystem services. Overfishing can also make affected species more vulnerable to the stresses of ocean warming and ocean acidification. Marine mammals such as whales are also threatened by overfishing (see the second Case Study that follows).

Industrialized fishing—humanity's last large-scale hunting and gathering operation—requires huge amounts of energy to propel the ships, fuel spotter planes, and freeze the catches. According to a 2014 study by Robert Parker, the species that require the largest energy input are, in order, shrimp, lobster, Pacific albacore tuna, scallops, and skipjack tuna.

CASE STUDY

Why Should We Protect Sharks?

Sharks have roamed the world's oceans for at least 400 million years. Certain shark species are keystone species in their ecosystems. Sharks, like any good predator feeding at or near the tops of their food webs, probably remove injured and sick animals. Without this ecosystem service, the oceans would teem with dead and dying fish

FIGURE 11.11 The threatened whale shark (left), which feeds on plankton, is the ocean's largest fish and is quite friendly to humans. The scalloped hammerhead shark (right) is endangered.

Left: Colin Parker/National Geographic My Shot/National Geographic Creative Right: Frantisek hojdysz/Fotolia LLC

and marine mammals. Shark activity also influences the distribution and feeding habits of other species, which helps maintain balance in marine ecosystems.

The world's more than 480 known species of sharks vary widely in size. They range from the 15-centimeter (6-inch) long dwarf dog shark to the whale shark (Figure 11.11, left), which can grow to the length of a city bus and weigh as much as two full-grown African elephants.

Many people, influenced by movies, popular novels, and media coverage of shark attacks, think of sharks as vicious monsters. In reality, the three largest species—the whale shark (Figure 11.12, left), basking shark, and megamouth shark—are gentle giants. These sharks swim through the water with their mouths open, filtering out and swallowing huge quantities of zooplankton and small fish.

Media reports on shark attacks greatly exaggerate the dangers to humans from sharks. Every year, members of a few species, including the great white, bull, tiger, and hammerhead sharks (Figure 11.11, right), injure 60 to 80 people, and typically kill 6 to 10 people worldwide. Sometimes these sharks are thought to mistake swimmers and people on surfboards or paddle boards for their usual prey of sea lions and other marine mammals. According to shark attack files at the Florida Museum of History, you are much more likely to be killed by a falling coconut or by falling out of bed than by a shark. In 2014 more people in the world were killed taking selfies (12) than by shark attacks (8).

Each year human activities kill more than 200 million sharks. As many as 73 million sharks die each year after being caught for their valuable fins, a practice called *shark finning*. After capture, their fins are cut off. Then the sharks are thrown into the water and in the worst cases, they bleed to death or drown because they can no longer swim. Sharks are also killed for their livers, meat, hides, and jaws, and because we fear or hate them. Each year, an estimated 100 million sharks die when fishing lines and nets trap them.

Harvested shark fins are widely used in Asia as an ingredient in expensive soup ($100 or more a bowl) and as an alleged pharmaceutical cure-all. The large dorsal fin of a whale shark (Figure 11.11, left) can be worth up to $10,000 in Hong Kong or Taiwan and is often hung outside Asian restaurants to advertise shark fin soup. According to the wildlife conservation group WildAid, there is no reliable evidence that shark fins provide flavor or have any nutritional or medicinal value. The group also warns that consuming shark fins and shark meat can be harmful to human health because they contain high levels of mercury and other toxins.

According to a 2014 IUCN study, 25% of the world's open-ocean shark species are threatened with extinction, primarily from overfishing. Because some sharks are keystone species, their extinction can threaten the ecosystems and the ecosystem services they provide. Sharks are especially vulnerable to population declines because they grow slowly, mature late, and have only a few offspring per generation. Today, sharks are among the earth's most vulnerable and least protected animals.

With research support from the National Geographic Society, biologist Samuel H. Gruber has been studying lemon sharks in the Florida Keys and the Bahamas. One of his goals is to help us understand and reduce the slaughter of these sharks, which he calls "fantastic and amazing creatures." Protecting sharks is one way for us to live more sustainably by increasing our beneficial environmental impacts.

CONSIDER THIS . . .

LEARNING FROM NATURE

Sharks easily glide through the water because tiny grooves in their skin form continuous channels for water flow. Scientists are studying this to design ship hulls that will save energy and money by moving through the water with less resistance.

CASE STUDY

Protecting Whales: A Success Story . . . So Far

Overharvesting threatens some marine mammals with extinction. The most prominent examples are whales, or *cetaceans*, that range in size from the 0.9-meter (3-foot) porpoise to the giant 15- to 30-meter (50- to 100-foot) highly endangered blue whale. Cetaceans are divided into two major groups: *toothed whales* and *baleen whales* (Figure 11.12).

Toothed whales, such as the porpoise, sperm whale, and killer whale (orca), bite and chew their food and feed mostly on squid, octopus, and other marine animals. *Baleen whales*, such as the blue, gray, humpback, minke, and fin, are filter feeders. Attached to their upper jaws are plates made of baleen, or whalebone, which they use to filter plankton, especially tiny shrimp-like krill (Figure 3.16, p. 60), from the seawater.

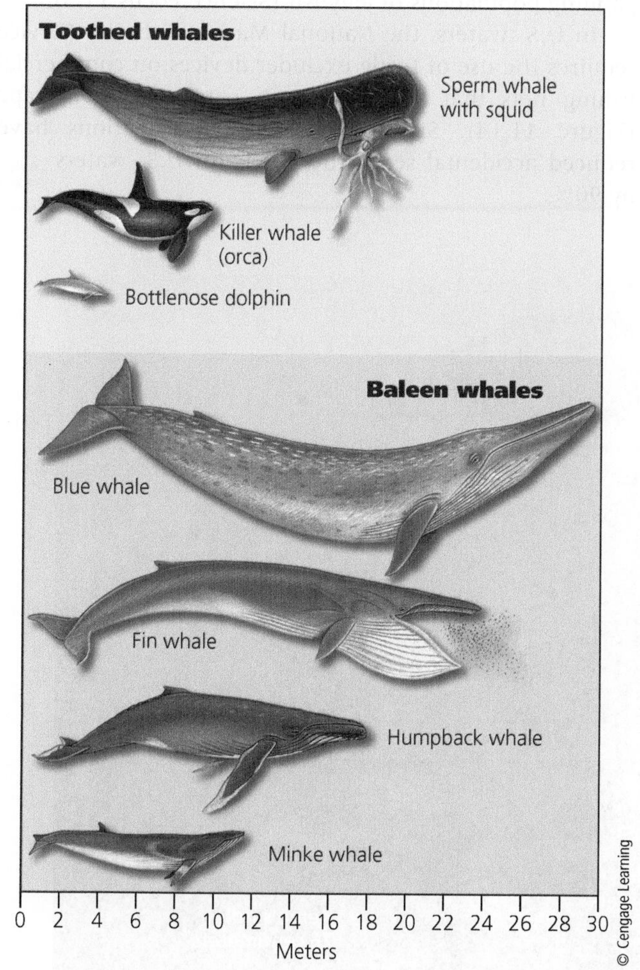

FIGURE 11.12 Cetaceans are classified as either toothed whales or baleen whales.

Whales are easy to kill because of their large size and because of their need to come to the surface to breathe. Whale hunters became efficient at hunting and killing whales using radar, spotters in airplanes, fast ships, and harpoon guns. Whale harvesting, mostly in international waters, has followed the classic pattern of a tragedy of the commons (p. 12). In the 20th century, about 3 million whales were killed by whalers from 46 countries, led by Norway, Japan, the U.S.S.R., and the United Kingdom. Between 1925 and 1975, this overharvesting drove 8 of the 11 major species to commercial extinction and the blue whale close to biological extinction.

The endangered blue whale is the earth's largest animal. Its heart is the size of a small car, its tongue weighs as much as an adult elephant, and some of its blood vessels are big enough for you to swim through. Also, it is one of the fastest-swimming animals in the sea.

In 1946 the International Whaling Commission (IWC) was established to regulate the whaling industry by setting annual quotas to prevent overharvesting. However, IWC quotas often were based on insufficient data or were ignored by whaling countries. Without enforcement powers, the IWC was not able to stop the decline of most commercially hunted whale species.

In 1970 the United States stopped all commercial whaling and banned all imports of whale products. Under pressure from conservationists and governments of many nonwhaling nations, the IWC imposed a moratorium on commercial whaling starting in 1986. It worked. The estimated number of whales killed commercially worldwide dropped from 42,480 in 1970 to about 2,000 in 2014.

Most whaling today is done by Japan, Norway, and Iceland. Since the 1986 ban on commercial whaling, these nations and some tropical island nations have pressured the IWC to lift the ban and reverse the international ban on buying and selling whale products. They contend that the ban on whaling is emotionally motivated and interferes with their cultural diet traditions, which they say are no different from the western cultural tradition of killing cows for beef. Many conservationists dispute this claim and contend that whales are intelligent and highly social mammals that should be protected for ethical and ecological reasons.

Whaling proponents also point out that populations of minke, humpback, and several other whale species have rebounded enough since the moratorium on whaling to be removed from the ban. Others question IWC estimates of the allegedly recovered whale species, noting the inaccuracy of such estimates in the past.

Extinction of Aquatic Species Is a Growing Threat

Beyond the *commercial extinction* of a number of fish species, many marine and freshwater fish species are threatened with *biological extinction* mostly because of a combination of habitat loss, overfishing, pollution, and ocean acidification.

Among the most threatened of all marine species are sea turtles (see chapter-opening photo) that have been roaming the oceans for more than 110 million years—about 550 times longer than our species has been around. Today six of the seven species of sea turtles (Figure 11.13) are in danger of becoming extinct, mostly because human activities taking place during the last 100 years have reduced the world's number of sea turtles by about 95%.

Sea turtles spend most of their lives traveling throughout the world's oceans, but adult females normally return to the beaches where they were born to lay their eggs. They come ashore at night and use their back flippers to dig nests on sand beaches and coastal dunes. Each female lays a clutch of around 100 to 110 eggs, buries them, and returns to the ocean. After the baby turtles hatch, they dig their way out of the nest, often at night, to scamper toward the water. During this dangerous trip, birds and other predators eat many of them. Only about one of every thousand sea turtle hatchlings survives to adulthood.

Sea turtles are threatened by trawler fishing (Figure 11.4), which destroys many of the coral gardens that have served as their feeding grounds. The turtles are also hunted for their skin and their eggs are taken for food. They often drown after becoming entangled in fishing nets (Figure 11.9) and lines, as well as in lobster and crab traps.

Beachgoers and motor vehicles sometimes crush their nests. Artificial lights can disorient newly hatched baby turtles, which try to find their way to the ocean by moving toward moonlight reflected from the ocean's surface. Going in the wrong direction increases their chances of ending up as food for predators.

Pollution of ocean water is another threat. Sea turtles can mistake discarded plastic bags for jellyfish and choke to death trying to eat them. In addition, the threat of rising sea levels from climate change during this century will flood many sea turtle nesting habitats and change ocean currents, which could disrupt their migration routes.

Several sea turtle species eat jellyfish (**Core Case Study**) and help control their populations. Continuing population declines of these endangered sea turtles could promote the takeover of parts of the sea by rapidly expanding populations of jellyfish (Science Focus 11.3).

In U.S. waters, the National Marine Fisheries Service requires the use of turtle excluder devices on commercial fishing nets that allow captured sea turtles to escape (Figure 11.14). Since 1990, fishing regulations have reduced accidental sea turtle deaths in U.S. waters by 90%. GOOD NEWS

FIGURE 11.13 There are seven species of sea turtles, six of them threatened with extinction. All but the flatback are found in U.S. waters.

SCIENCEFOCUS 11.3

Why Are Jellyfish Populations Increasing?

Jellyfish look fragile but have a number of survival traits. They can reproduce rapidly, survive without food for long periods of time by shrinking and waiting until conditions improve, eat many different things, and feed in murky waters. They can also survive in low-oxygen, warm, and acidic waters.

Scientists have identified several possible causes for the increased number and size of jellyfish swarms in many areas of the world's oceans (**Core Case Study**). One likely factor is the decline in populations of their natural predators such as tuna, sharks, swordfish, and endangered sea turtles, mostly due to overfishing. Overfishing also increases food sources for jellyfish because it reduces populations of fish such anchovies and sardines that eat much of the same food as jellyfish. As humans overfish commercially valuable fish species, jellyfish can take over, eat fish eggs and small fish, and hinder or prevent the recovery of overfished species.

Human activities play a key role in ocean warming and ocean acidification, which are favorable for jellyfish. Another environmental change caused by human activities is the overfertilization of coastal waters by huge inputs of the plant nutrients nitrogen and phosphorus. They come from the runoff of fertilizers from large areas of land into rivers that flow into coastal waters. The excess plant nutrients create massive blooms of algae and other plankton, which die, sink, and are decomposed. This process, called *cultural eutrophication*, depletes dissolved oxygen in ocean bottom waters and kills or drives away marine organisms that require oxygen. Jellyfish populations can erupt in these so-called "dead zones" such as the massive one that forms each year in the Gulf of Mexico. At least 400 other dead zones, where almost nothing but jellyfish survive, form in various parts of the world.

Jellyfish have also benefited from the rapid increase in global trade involving ships carrying goods throughout the world. Some species of jellyfish can survive in the ballast water of ships, and polyps of various jellyfish species can attach to ship hulls and end up inhabiting new areas of the ocean throughout the world.

More research is needed, but the rise of jellyfish in the world's ocean waters is probably the result of a mixture of these factors. Think about it. We are doing things that favor jellyfish: depleting populations of their major predators, depleting populations of fish species that compete with them for food, warming and making the oceans more acidic, creating numerous large oxygen-depleted zones, and using ships to spread them throughout the world.

CRITICAL THINKING

What are two ways in which exploding jellyfish populations might affect your life or the lives of any children and grandchildren you might eventually have?

© NOAA

FIGURE 11.14 This endangered loggerhead turtle is escaping a fishing net equipped with a turtle excluder device.

11.2 HOW CAN WE PROTECT AND SUSTAIN MARINE BIODIVERSITY?

CONCEPT 11.2 We can help to sustain marine biodiversity by using laws and economic incentives to protect species, setting aside marine reserves to protect ecosystems and ecosystem services, and using community-based integrated coastal management.

We Can Protect Marine Species with Laws and Treaties

Protecting marine biodiversity is difficult for several reasons:

- The human ecological footprint and the fishprint are expanding so rapidly that it is difficult to monitor their impacts.
- Much of the damage to the oceans and other bodies of water is not visible to most people.
- Many people incorrectly view the seas as an inexhaustible resource that can absorb an almost

infinite amount of waste and pollution and still produce all the seafood we want.

- Most of the world's ocean area lies outside the legal jurisdiction of any country. Thus, much of it is an open-access resource, subject to overexploitation. This is a classic example of the tragedy of the commons (see Chapter 1, p. 12).

Nevertheless, there are several ways to protect and sustain marine biodiversity, thereby increasing our beneficial environmental impact. One involves *passing and enforcing laws* (**Concept 11.2**). For example, we can protect endangered and threatened aquatic species, as discussed in Chapter 9.

National and international laws and treaties that help protect marine species include the 1975 Convention on International Trade in Endangered Species (CITES), the 1979 Global Treaty on Migratory Species, the U.S. Marine Mammal Protection Act of 1972, the U.S. Endangered Species Act of 1973 (ESA; see pp. 211–213), the U.S. Whale Conservation and Protection Act of 1976, and the 1995 International Convention on Biological Diversity. The ESA and several international agreements have been used to identify and protect endangered and threatened marine species, including whales, seals, sea lions, and sea turtles. However, it is hard to get all nations to comply with some of these agreements. Even when agreements and regulations are enforced, the resulting fines and punishments for violators are often inadequate.

We Can Establish Marine Sanctuaries

By international law, a country's offshore fishing zone extends to 370 kilometers (200 nautical miles) from its shores. Foreign fishing vessels can take certain quotas of fish within such zones, called *exclusive economic zones*, but only with a government's permission. Ocean areas beyond the legal jurisdiction of any country are known as the *high seas*, and laws and treaties pertaining to them are difficult to monitor and enforce.

For example, illegal poachers catch an estimated one of every five fish sold in stores and served in restaurants. This illegal catch is difficult to monitor and control. However, in 2015 a new satellite tracking system, funded by the Pew Charitable Trusts, was launched to help crack down on this illegal activity. GOOD NEWS

The United Nations Law of the Sea treaty, which went into effect in 1984, has been signed by 162 countries but not by the United States. Under this treaty, the world's coastal nations have jurisdiction over 36% of the ocean surface and 90% of the world's fish stocks. Instead of using this treaty to protect their fishing grounds, many governments have promoted overfishing by subsidizing fishing fleets and failing to establish and enforce stricter regulation of fish catches in their coastal waters.

We can establish *protected marine sanctuaries*. Since 1986, the IUCN has helped several nations to establish a global system of *marine protected areas* (MPAs)—areas of ocean partially protected from human activities. According to the U.S. National Ocean Service, there are more than 5,800 MPAs worldwide. MPAs cover 2.8% of the world's ocean surface and their numbers are growing.

However, most MPAs allow dredging, trawler fishing (Figure 11.4), and other ecologically harmful resource extraction activities. Many of them are too small to be effective in protecting larger species. However, since 2007, the U.S. state of California has been establishing the nation's most extensive network of MPAs in which fishing is banned or strictly limited. In 2011 Costa Rica (see Chapter 10, Core Case Study, p. 222) expanded one of its MPAs to help protect a number of marine species, including the critically endangered leatherback sea turtle and the endangered scalloped hammerhead shark (Figure 11.11, right). The United States has nearly 1,800 MPAs, which receive varying degrees of protection. GOOD NEWS

Marine Reserves: An Ecosystem Approach to Marine Sustainability

Scientists and policy makers call for protecting and sustaining entire marine ecosystems and their ecosystem services within a global network of fully protected *marine reserves*, some of which already exist. These areas are declared off-limits to commercial fishing, dredging, mining, and waste disposal in order to enable their ecosystems to recover and flourish.

When patrolled and protected, marine reserves work and they work quickly. Studies show that in fully protected marine reserves, on average, commercially valuable fish populations double, average fish size grows by almost a third, fish reproduction triples, and species diversity increases by almost one-fourth. These improvements can happen within 2 to 4 years after strict protection begins. Reserves also benefit nearby fisheries because fish move into and out of the reserves. Currents also carry fish larvae produced inside reserves to adjacent fishing grounds, thus bolstering fish populations there. In addition, marine reserves increase the ability of protected marine ecosystems to respond to the growing stresses of ocean warming and ocean acidification. GOOD NEWS In 2014 the United States created the world's largest marine reserve by expanding its Kingman Reef Reserve around a couple of remote Pacific Islands (Figure 11.15). This marine reserve is more than twice the size of the U.S. state of Texas. Deep-sea mining and tuna fishing are banned in these waters.

Despite the importance of protected reserves, only 1.2% of the world's oceans are fully protected compared to 5% of the world's land. In addition, many of the existing reserves are fully protected only on paper because of shortages of funding and a need for more trained staff to manage and monitor them. In other words, at least 98.8%

Brian J. Skerry/National Geographic Creative

FIGURE 11.15 Gray reef sharks in the recently expanded Kingman Reef Reserve in the Pacific Ocean.

of the world's oceans are not effectively protected from harmful human activities.

98.8%
Percentage of the world's oceans that lack effective protection from harmful human activities

Many marine scientists want to increase our beneficial environmental impact by setting aside 10–30% of the world's oceans as fully protected marine reserves. Sylvia Earle, one of the world's leading marine scientists, is spearheading this effort (Individuals Matter 11.1).

Marine scientists also call for establishing protected corridors to connect the global network of marine reserves, especially those in coastal waters. This would help species move to different habitats in response to the effects of ocean warming, acidification, and many forms of ocean pollution.

CONSIDER THIS . . .

THINKING ABOUT Marine Reserves

Do you support setting aside at least 30% of the world's oceans as fully protected marine reserves? Explain. How would this affect your life?

Restoration Helps Protect Marine Biodiversity, but Prevention Is the Key

A dramatic example of marine system restoration is Japan's attempt to restore its largest coral reef—90% of it dead—by seeding it with new corals. Divers drill holes into the dead reefs and insert ceramic discs holding sprigs of fledgling coral. Figure 11.16 shows how protection has helped to restore coral reefs near Kanton Island, an atoll located in the South Pacific.

Many scientists support efforts to restore aquatic systems, but they warn that these projects could fail if the problems that caused their degradation are not addressed. They call for more emphasis on *preventing aquatic ecosystem degradation*, which is far less expensive and more effective than restoration efforts.

For example, a study by IUCN and scientists from the Nature Conservancy concluded that the world's shallow coral reefs and mangrove forests could survive currently projected climate change if we reduce other threats such as overfishing and pollution. However, while some shallow coral species may be able to adapt to warmer temperatures, they may not have enough time to do this unless we act now to slow down the rate of ocean warming.

To deal with problems of pollution and overfishing, marine scientists call for countries and coastal communities to closely monitor and regulate fishing and coastal land development and greatly reduce pollution from land-based activities. Coastal residents should also think about

INDIVIDUALS MATTER 11.1

Sylvia Earle—Advocate for the Oceans

Tyrone Turner/National Geographic Creative

Sylvia Earle is one of the world's most respected oceanographers and is a National Geographic Explorer. She has taken a leading role in helping us to understand and protect the world's oceans. *Time* magazine named her the first Hero for the Planet and the U.S. Library of Congress calls her "a living legend."

Earle has led more than 100 ocean research expeditions and has spent more than 7,000 hours underwater, either diving or descending in research submarines to study ocean life. She has focused her research on the ecology and conservation of marine ecosystems, with an emphasis on developing deep-sea exploration technology.

She is the author of more than 175 publications and has been a participant in numerous radio and television productions. During her long career, Earle has also served as the Chief Scientist of the U.S. National Oceanic and Atmospheric Administration (NOAA). She also founded three companies that develop submarines and other devices for deep-sea exploration and research. She has received more than 100 major international and national honors, including a place in the National Women's Hall of Fame.

Earle is currently leading a campaign called Mission Blue to finance research and to ignite public support for a global network of marine protected areas, which she calls "hope spots." Her goal is to help save and restore the oceans, which she calls "the blue heart of the planet." She says, "There is still time, but not a lot, to turn things around. . . . This mostly blue planet has kept us alive. It's time for us to return the favor."

Brian J. Skerry/National Geographic Creative

FIGURE 11.16 Recovery of a coral reef in a protected area near Kanton Island in the South Pacific.

the chemicals they put on their lawns and the kinds of waste they generate, much of which ends up in coastal waters.

One strategy emerging in some coastal communities is *integrated coastal management*—a community-based effort to develop and use coastal resources more sustainably (**Concept 11.2**). The overall aim of such programs is for fishers, business owners, developers, scientists, citizens, and politicians to identify shared problems and goals in their use of marine resources. The idea is to develop workable, cost-effective, and adaptable solutions that will help to preserve biodiversity, ecosystem services, and environmental quality, while also meeting economic and social goals.

This requires participants to seek reasonable short-term trade-offs that can lead to long-term ecological and economic benefits—an example of applying the win-win **principle of sustainability**. For example, fishers might have to give up harvesting various fish species in certain areas until stocks recover enough to restore biodiversity in those areas. This might help them to secure a more sustainable future for their businesses.

Australia manages its huge Great Barrier Reef Marine Park in this way, and more than 100 integrated coastal management programs are being developed throughout the world. Another example of such management in the United States is the Chesapeake Bay Program (see Chapter 8, Case Study, p. 177).

11.3 HOW SHOULD WE MANAGE AND SUSTAIN MARINE FISHERIES?

CONCEPT 11.3 Sustaining marine fisheries will require improved monitoring of fish and shellfish populations, cooperative fisheries management among communities and nations, reduction of fishing subsidies, and careful consumer choices in buying seafood.

We Need to Estimate and Monitor Fishery Populations

The first step in protecting and sustaining the world's marine fisheries is to make the best possible estimates of their fish and shellfish populations (**Concept 11.3**). The traditional approach has used a *maximum sustained yield (MSY)* model to project the maximum number of individuals that can be harvested annually from fish or shellfish stocks without causing a population drop. However, the MSY concept has not worked very well because of the difficulty in estimating the populations and growth rates of fish and shellfish stocks. In addition, harvesting a particular species at its estimated maximum sustainable level can affect the populations of other marine species.

In recent years, some fishery biologists and managers have begun using the *optimum sustained yield (OSY)* concept, which attempts to take into account interactions among species. Similarly, another approach is *multispecies management* of a number of interacting species, which takes into account their competitive and predator–prey interactions. An even more ambitious approach is to develop complex computer models for managing multispecies fisheries in *large marine systems*. However, it is a political challenge to get groups of nations to cooperate in planning and managing such large systems.

There are uncertainties built into using any of these approaches because the biology of marine species and their interactions is poorly understood, and because of limited data on changing ocean conditions. As a result, many fishery and environmental scientists are increasingly interested in using the *precautionary principle* for managing fisheries and large marine systems. This means sharply reducing fish harvests and closing some overfished areas until they recover and until we have more information about what levels of fishing they can sustain.

Some Communities Cooperate to Regulate Fish Harvests

An obvious step to take in protecting marine biodiversity—and therefore fisheries—is to regulate fishing. Traditionally, many coastal fishing communities have developed allotment and enforcement systems for controlling fish catches in which each fisher gets a share of the total allowable catch. Such *catch-share systems* have sustained fisheries and jobs in many communities for hundreds and sometimes thousands of years.

However, the influx of large state-of-the-art fishing boats and international fishing fleets has weakened the ability of many coastal communities to regulate and sustain local fisheries. Community management systems have often been replaced by *co-management*, in which coastal communities and the government work together to manage fisheries. Currently, more than 700 of the world's fisheries are co-managed.

With this approach, a central government typically sets quotas for various species and divides the quotas among communities. The government may also limit fishing seasons and regulate the types of fishing gear that can be used to harvest a particular species. Each community then allocates and enforces its quota among its members based on its own rules. When it works, community-based co-management illustrates that we can prevent overfishing and the tragedy of the commons.

Government Subsidies Can Encourage Overfishing

Governments around the world give more than $30 billion per year in subsidies to fishers to help them keep their businesses running, according to fishery experts U. R. Sumaila and Daniel Pauly. Some marine scientists estimate that, each year, $10 billion to $14 billion of these subsidies is spent to encourage overfishing and expansion of the fishing industry. The result is too many boats chasing too few fish. Some argue that such subsidies are not a wise investment because they promote overfishing of targeted fish stocks, which causes economic losses of about $50 billion a year, according to the World Bank.

CONSIDER THIS . . .

THINKING ABOUT Fishing Subsidies

Do you think that government fishing subsidies that promote unsustainable fishing should be eliminated or drastically reduced? Explain. Would your answer be different if your livelihood depended on commercial fishing?

Consumers Can Choose Sustainably Produced Seafood

An important component of sustaining aquatic biodiversity and ecosystem services is bottom-up pressure from consumers who demand *sustainably produced seafood* (**Concept 11.3**). We can purchase only sustainably harvested seafood or sustainably farmed seafood. One way to help consumers make such choices is to label sustainably caught and raised fresh and frozen fish and shellfish. The London-based Marine Stewardship Council (MSC) was created in 1999 to support sustainable fishing and to certify sustainably produced seafood. Only certified fisheries are allowed to use the MSC's "Fish Forever" eco-label, which certifies that the fish were caught by fishers who used environmentally sound and socially responsible practices. Another approach is to certify and label products of sustainable *aquaculture*, or fish-farming operations.

An important way for consumers of seafood raised by aquaculture to help sustain aquatic biodiversity is to choose plant-eating species of fish, such as tilapia, catfish, and carp. Carnivorous species, such as salmon and shrimp, raised through aquaculture are often fed fishmeal made from wild-caught fish and some of these species are being overfished.

Individuals can also help reduce the waste of seafood. Between 2009 and 2013, about 45% of all edible seafood in the United States ended up in the trash, mostly because consumers and restaurants did not cook it before it spoiled. Figure 11.17 summarizes actions that individuals, organizations, and governments can take to help sustain global fisheries, marine biodiversity, and ecosystem services.

Solutions

Managing Fisheries

Fishery Regulations

Set low catch limits

Improve monitoring and enforcement

Economic Approaches

Reduce or eliminate fishing subsidies

Certify sustainable fisheries

Protection

Establish no-fishing areas

Establish more marine protected areas

Consumer Information

Label sustainably harvested fish

Publicize overfished and threatened species

Bycatch

Use nets that allow escape of smaller fish

Use net escape devices for seabirds and sea turtles

Aquaculture

Restrict locations of fish farms

Improve pollution control

Nonnative Invasions

Kill or filter organisms from ship ballast water

Clean aquatic recreation gear

FIGURE 11.17 Ways to manage fisheries more sustainably and protect marine biodiversity. ***Critical thinking:*** Which four of these solutions do you think are the best ones? Why?

11.4 HOW SHOULD WE PROTECT AND SUSTAIN WETLANDS?

CONCEPT 11.4 We can maintain the ecosystem and economic services of wetlands by protecting remaining wetlands and restoring degraded wetlands.

Coastal and Inland Wetlands Are Disappearing

Coastal wetlands and marshes (Figure 8.7, p. 172) and inland wetlands support aquatic biodiversity and provide vital economic and ecosystem services. Their ecosystem services include feeding downstream waters, reducing flooding by storing storm water, reducing storm damage by absorbing waves (coastal wetlands), recharging groundwater supplies, reducing pollution, preventing erosion, and providing fish and wildlife habitat.

Despite their ecological value, the United States has lost more than half of its coastal and inland wetlands since 1900. Other countries have lost even more, and the rate of loss of wetlands throughout the world is accelerating. China, for example, has lost about 60% of its original coastal wetlands, New Zealand has lost 92%, and Italy has lost 95%.

The U.S. state of Louisiana has the largest area of coastal wetlands in the lower 48 states but is losing them faster than any other state. One cause of such losses is subsidence (sinking) of coastal land near the Mississippi River delta. Because the river is heavily dammed, sediments that naturally replenish the delta do not make it to the Gulf of Mexico, so the delta is shrinking and the land subsiding. Other threats to coastal wetlands are oil spills and a rising sea level due to climate change.

For centuries, people have drained, filled in, or covered over swamps, marshes, and other wetlands to create rice fields or other cropland, to accommodate expanding cities and suburbs, and to build roads. Wetlands have also been destroyed to extract minerals, oil, and natural gas, and to eliminate breeding grounds for insects that cause diseases such as malaria. To make matters worse, coastal wetlands in many parts of the world will probably be under water before the end of this century because of rising sea levels.

Preserving and Restoring Wetlands

Some laws protect wetlands. In the United States, zoning laws have been used to steer development away from wetlands. The U.S. government requires a federal permit to fill in wetlands occupying more than 1.2 hectares (3.0 acres) or to deposit dredged material in wetlands. According to the U.S. Fish and Wildlife Service, this law has helped to cut the average annual wetland loss by 80% since 1969.

However, there are continuing attempts by land developers to weaken such wetlands protection. Only about 6% of the country's remaining inland wetlands are federally protected, and state and local wetland protection is inconsistent and generally weak because of intense pressure from coastal developers and landowners.

94% Percentage of U.S. inland wetlands that are not protected by federal law against development and degradation

The stated goal of current U.S. federal policy is *zero net loss* in the functioning and value of coastal and inland wetlands. A policy known as *mitigation banking* allows destruction of existing wetlands as long as an equal or greater area of the same type of wetland is created, enhanced, or restored. However, a study by the National Academy of Sciences found that at least half of the attempts to create new wetlands failed to replace lost ones. Furthermore, most of the created wetlands did not provide the ecosystem services of natural wetlands, even decades after completion. The study also found that wetland creation and restorations often fail to meet the standards set for them and are not adequately monitored.

Creating and restoring wetlands has become a profitable business. Private investment bankers make money by buying wetland areas and restoring or upgrading them or creating new wetland. This creates wetlands banks or credits that the bankers sell to developers. This approach is a small step toward full-cost pricing because it puts a monetary value on the biodiversity and ecosystem services of wetlands that are sold by the bankers.

However, it is difficult to restore, enhance, or create wetlands (see the Case Study that follows). Thus, most U.S. wetland banking systems require replacing each area of destroyed wetland with twice the same area of restored, enhanced, or created wetland (Figure 11.18) as a built-in ecological insurance policy.

Ecologists urge using mitigation banking only as a last resort. They also call for making sure that new replacement wetlands are created and evaluated *before* existing wetlands are destroyed. This example of applying the precautionary principle is usually the reverse of what is actually done.

CONSIDER THIS . . .

THINKING ABOUT Wetlands Mitigation

Should a new wetland be created and evaluated before anyone is allowed to destroy the wetland it is supposed to replace? Explain.

CASE STUDY

Can We Restore the Florida Everglades?

South Florida's Everglades was once a 100-kilometer-wide (62-mile-wide), knee-deep sheet of water flowing slowly south from Lake Okeechobee to Florida Bay (Figure 11.19, red dashed lines). As this shallow body of water—known as the "River of Grass"—trickled south, it created a vast network of wetlands with a variety of wildlife habitats.

To help preserve the wilderness in the lower end of the Everglades system, in 1947 the U.S. government established Everglades National Park. However, this protection effort did not work—as conservationists had predicted—because of a massive water distribution and land development project to the north. Between 1962 and 1971, the U.S. Army Corps of Engineers transformed the wandering 166-kilometer-long (103-mile-long) Kissimmee River into a mostly straight 84-kilometer (52-mile) canal flowing into Lake Okeechobee (Figure 11.19, black dashed line). The

FIGURE 11.18 This human-created wetland is located near Orlando, Florida (USA).

canal provided flood control by speeding the flow of water, but it drained large wetlands north of Lake Okeechobee, which farmers then converted to grazing land.

This and other projects have provided south Florida's rapidly growing population with a reliable water supply and flood protection. However, much of the original Everglades has been drained, paved over, polluted by agricultural runoff, and invaded by a number of plant and animal species. The Everglades is now less than half its original size and much of it has dried out, leaving large areas vulnerable to summer wildfires.

The Everglades National Park is known for its astonishing biodiversity, and each year more than a million people from all over the world visit the park. However, its biodiversity has been decreasing, mostly because of habitat loss, pollution, and invasive species. About 90% of the wading birds in Everglades National Park have vanished and populations of many remaining wading bird species have dropped sharply. In addition, populations of vertebrates, from deer to turtles, are down 75–95%.

By the 1970s, state and federal officials recognized that this huge plumbing project was reducing populations of native plants and wildlife—a major source of tourism revenues for Florida. It was also cutting the water supply for the 5.8 million residents of south Florida. In 1990 Florida's state government and the U.S. government agreed on a plan for the world's largest ecological restoration project, known as the Comprehensive Everglades Restoration Plan. The U.S. Army Corps of Engineers is supposed to carry out this joint federal and state plan to partially restore the Everglades.

The project has several ambitious goals, including restoration of the curving flow of more than half of the Kissimmee River; removal of 400 kilometers (248 miles) of canals and levees that block natural water flows south of Lake Okeechobee; conversion of large areas of farmland to marshes; the creation of 18 large reservoirs and underground water storage areas to store water for the lower Everglades and for south Florida's population; and building a canal–reservoir system for catching the water now flowing out to sea and pumping it back into the Everglades.

Will this huge ecological restoration project work? It depends not only on the abilities of scientists and engineers but also on prolonged political and economic support from citizens, the state's powerful sugarcane and agricultural industries, and elected state and federal

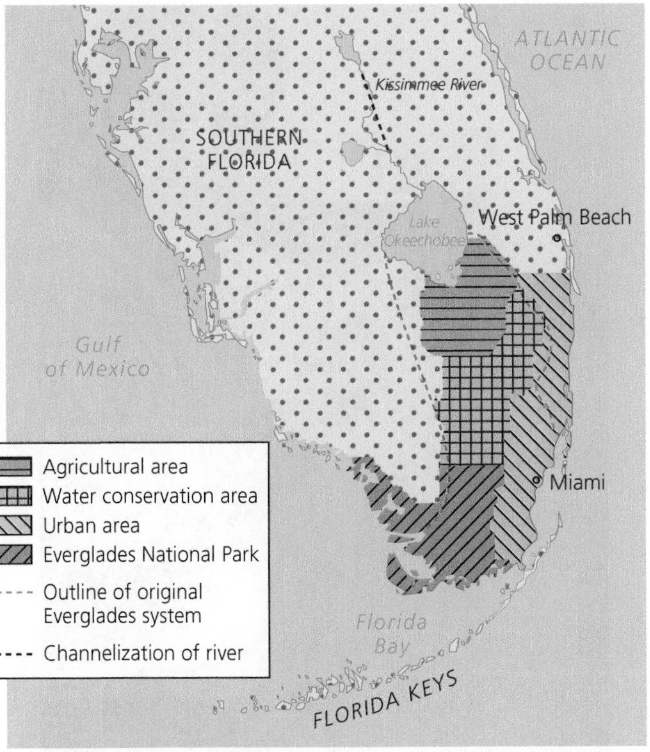

FIGURE 11.19 Florida's Everglades is the site of the world's largest ecological restoration project—an attempt to undo and redo an engineering project that has been destroying this vast wetland and threatening water supplies for south Florida's rapidly growing population.

officials. Some restrictions on phosphorus discharges from sugarcane plantations have been relaxed, which could worsen pollution problems. The project has also had cost overruns and funding shortages and is considerably behind schedule.

CONSIDER THIS . . .

THINKING ABOUT Everglades Restoration

Do you support carrying out the proposed plan for partially restoring the Florida Everglades, including making the federal government (taxpayers) responsible for half of the funding? Explain.

11.5 HOW SHOULD WE PROTECT AND SUSTAIN FRESHWATER LAKES, RIVERS, AND FISHERIES?

CONCEPT 11.5 Freshwater ecosystems are strongly affected by human activities on adjacent lands, and protection of these ecosystems must include protection of their watersheds.

Freshwater Ecosystems Are in Jeopardy

The ecosystem and economic services provided by many of the world's freshwater lakes, rivers, and fisheries are severely threatened by human activities (**Concept 11.5**). Currently, 40% of the freshwater fish species in North America are vulnerable, threatened, or endangered, according to a joint study by U.S., Canadian, and Mexican scientists.

Many of the world's freshwater wetlands have been destroyed. Aquatic species have been crowded out of at least half of the world's freshwater habitat areas and 40% of the world's rivers have been dammed or otherwise engineered. Invasive species, pollution, and climate change threaten the ecosystems of many lakes, rivers, and wetlands. Many freshwater fish stocks are overharvested and during this century, increasing human population pressure and climate change will intensify these threats.

Sustaining and restoring the biodiversity and ecosystem services provided by freshwater lakes and rivers is a complex and challenging task, as shown by the following Case Study.

CASE STUDY

Can the Great Lakes Survive Repeated Invasions by Alien Species?

Invasions by nonnative species are a major threat to the biodiversity and ecological functioning of many lakes, as illustrated by what has happened to the five Great Lakes, located on the border between the United States and Canada.

Collectively, the Great Lakes are the world's largest body of freshwater. Since the 1920s, these lakes have been invaded by at least 180 nonnative species and the number keeps rising. Many of the alien invaders arrive on the hulls of, or in bilge-water discharges of, oceangoing ships that have been entering the Great Lakes through the St. Lawrence Seaway since 1959.

One of the biggest threats, the *sea lamprey*, reached the westernmost Great Lakes as early as 1920. This parasite attaches itself to almost any kind of fish and kills the victim by sucking out its blood (Figure 5.7, p. 105). Over the years, it has depleted populations of many important sport fish species such as lake trout. The United States and Canada keep the lamprey population down by applying a chemical that kills lamprey larvae where they spawn in streams that feed the lakes—at a cost of about $15 million a year.

In 1986 larvae of the *zebra mussel* (Figure 9.9, p. 201) arrived in ballast water discharged from a European ship near Detroit, Michigan. This thumbnail-sized mollusk reproduces rapidly. It has displaced other mussel species and thus depleted the food supply for some other Great Lakes aquatic species. The mussels have also clogged irrigation pipes, shut down water intake pipes for power plants and city water supplies, fouled beaches, and jammed ships' rudders. They have grown in thick masses on many boat hulls, piers, and other exposed aquatic surfaces (Figure 11.20). It

FIGURE 11.20 These *zebra mussels* are attached to a water current meter in Lake Michigan.

FIGURE 11.21 *Asian carp* may be the next major invasive species to threaten the Great Lakes.

has also spread to freshwater communities in parts of southern Canada and 18 eastern U.S. states. Damages and attempts to control this mussel cost the two countries about \$1 billion a year—an average of \$114,000 per hour.

Sometimes, native species can help control a harmful invasive species. For example, populations of zebra mussels are declining in some parts of the Great Lakes because a native sponge growing on their shells is preventing them from opening up their shells to breathe. However, it is not clear whether the sponges will effectively control the invasive mussels and what harmful ecological effects the sponges might have.

In 1989 a larger and potentially more destructive species, the *quagga mussel*, invaded the Great Lakes, probably discharged in the ballast water of a Russian freighter. It can survive at greater depths and tolerate more extreme temperatures than the zebra mussel can. By 2009, scientists reported that quagga mussels had rapidly replaced many other bottom-dwellers in Lake Michigan. This has reduced the food supply for many fish and other species, thus leading to a major disruption of the lake's food web. There is concern that quagga mussels may spread by river transport and eventually colonize eastern U.S. ecosystems such as the Chesapeake Bay (see Case Study, p. 177) and waterways in parts of Florida.

The *Asian carp* (Figure 11.21) is the most recent threat to the Great Lakes system. In the 1970s, catfish farmers in the southern United States imported two species of Asian carp to help remove suspended matter and algae from their aquaculture farm ponds. Heavy flooding during the 1990s caused many of these ponds to overflow, which allowed some of the carp to enter the Mississippi River. After working their way up the Mississippi River system, these invaders are now close to entering Lake Michigan, if they have not already done so. Joel Brammeier, president of the Alliance for the Great Lakes, warned that "if Asian carp get into Lake Michigan, there is no stopping them."

These highly prolific fish can quickly grow as long as 1.2 meters (4 feet) and weigh up to 50 kilograms (110 pounds). They can eat as much as 20% of their own body weight in plankton every day, which can disrupt lake food webs. When startled they jump clear of the water, and several boaters have been hit and injured by jumping carp. These fish have no natural predators in the rivers they have invaded or in the Great Lakes.

Federal, state, and local agencies are developing plans to prevent the Asian carp from reaching and spreading throughout the Great Lakes, which would threaten the lakes' $7 billion-a-year fishing industry. Proposed measures include electric barriers to limit access through rivers flowing into the Great Lakes and laws to prevent the transporting of fish across state borders. In addition, some chemical companies are developing carp-specific poisons.

Managing River Basins Is Complex and Controversial

Rivers and streams provide important ecosystem services (Figure 11.22), but overfishing, pollution, dams, and water withdrawal for irrigation are disrupting these services. Currently, these ecosystem services are given little or no monetary value when the costs and benefits of dam and reservoir projects are assessed. According to environmental economists, attaching even crudely estimated monetary values to these ecosystem services—an application of the full-cost pricing **principle of sustainability**—would help to sustain them.

An example of such disruption and loss of freshwater biodiversity is what happened in the Columbia River, which runs through parts of southwestern Canada and the northwestern United States. It has 119 dams, 19 of which are major generators of inexpensive hydroelectric power. It also supplies water for major urban areas and large irrigation projects.

The Columbia River dams have benefited many people, but have also sharply reduced populations of wild salmon. These migratory fish hatch in the upper reaches of the streams and rivers that form the headwaters of the Columbia River, migrate to the ocean where they spend most of their adult lives, and then swim upstream to return to the place where they were hatched to spawn and die. Dams interrupt their life cycle by interfering with the migration of young fish downstream, and blocking the return of mature fish attempting to swim upstream to their spawning grounds.

Since the dams were built, the Columbia River's wild Pacific salmon population has dropped by 94% and nine of the Pacific Northwest salmon species are listed as endangered or threatened. Since 1980, the U.S. federal government has spent more than $3 billion in efforts to save the salmon, with little success.

Natural Capital

Ecosystem Services of Rivers

- Deliver nutrients to sea to help sustain coastal fisheries
- Deposit silt that maintains deltas
- Purify water
- Renew and renourish wetlands
- Provide habitats for wildlife

FIGURE 11.22 Rivers and streams provide some important ecosystem services. ***Critical thinking:*** Which two of these services do you think are the most important? Why?

Protecting Watersheds

Sustaining freshwater aquatic systems begins with understanding that land and water are connected. For example, lakes and streams receive many of their nutrients from the ecosystems of bordering land. Such nutrient inputs come from falling leaves, animal feces, and pollutants generated by people, all of which are washed into bodies of water by rainstorms and melting snow. Therefore, to protect a stream or lake from excessive inputs of nutrients and pollutants, we must protect its watershed (**Concept 11.5**).

As with marine systems, freshwater ecosystems can be protected through laws, economic incentives, and restoration efforts. For example, restoring and sustaining the ecosystem and economic services of rivers will probably require taking down some dams and restoring river flows. In addition, some scientists and politicians call for protecting all remaining free-flowing rivers by prohibiting the construction of new dams.

In 1968 the U.S. Congress passed the National Wild and Scenic Rivers Act to establish protection of rivers with outstanding wildlife, geological, scenic, recreational, historical, or cultural values. The law classified *wild rivers* as those that are relatively inaccessible (except by trail), and *scenic rivers* as rivers of great scenic value that are free of dams, mostly undeveloped, and accessible in only a few places by roads. These rivers are now protected from widening, straightening, dredging, filling, and damming. However, the Wild and Scenic Rivers System keeps only 3% of U.S. rivers free flowing and protects less than 1% of the country's total river length.

Protecting Freshwater Fisheries

Sustainable management of freshwater fisheries involves supporting populations of commercial and sport fish species, preventing such species from being overfished, and reducing or eliminating populations of harmful invasive species. The traditional approach to managing freshwater fish species is to regulate the time and length of fishing seasons and the number and size of fish that can be taken.

Other techniques include building reservoirs, ponds, and stocking them with fish and protecting and creating fish spawning sites. In addition, some fishery managers try to protect fish habitats from sediment buildup and other forms of pollution. They also work to prevent or reduce large human inputs of plant nutrients that spur the excessive growth of aquatic plants.

Some fishery managers seek to control predators, parasites, and diseases by improving habitats, breeding genetically resistant fish varieties, and using antibiotics and disinfectants. Hatcheries can be used to restock ponds, lakes, and streams with prized species such as trout, and entire river basins can be managed to protect valued species such as salmon.

11.6 WHAT SHOULD BE OUR PRIORITIES FOR SUSTAINING AQUATIC BIODIVERSITY?

CONCEPT 11.6 Sustaining the world's aquatic biodiversity requires mapping it, protecting aquatic hotspots, creating large and fully protected marine reserves, protecting freshwater ecosystems, and restoring degraded coastal and inland wetlands.

Using an Ecosystem Approach to Sustain Aquatic Biodiversity and Ecosystem Services

Edward O. Wilson (see Individuals Matter 4.1, p. 81) and other biodiversity experts have proposed the following priorities for an ecosystem approach to sustaining aquatic biodiversity and ecosystem services (**Concept 11.6**):

- Map and inventory the world's aquatic biodiversity.
- Identify and preserve the world's aquatic biodiversity hotspots and areas where deteriorating ecosystem services threaten people and many forms of aquatic life.
- Create large and fully protected marine reserves to allow damaged marine ecosystems to recover and to allow fish stocks to be replenished.
- Protect and restore the world's lakes and river systems, which are among the world's most threatened ecosystems, but emphasize pollution prevention because ecological restorations are expensive and have a high failure rate.
- Initiate worldwide ecological restoration projects in systems such as coral reefs and inland and coastal wetlands.
- Find ways to raise the incomes of people who live on or near protected waters so that they can become partners in the protection and sustainable use of aquatic ecosystems.

There is growing evidence that the current harmful effects of human activities on aquatic biodiversity and ecosystem services could be reversed over the next two decades. Doing this will require implementing an ecosystem approach to sustaining both terrestrial and aquatic ecosystems. According to Edward O. Wilson, such a conservation strategy would cost about $30 billion per year—an amount that could be provided by a tax of one penny per cup of coffee consumed in the world each year.

CONSIDER THIS . . .

THINKING ABOUT The Cost of Sustaining Ecosystems

Would you be willing to pay an extra penny for each cup of coffee you buy to help pay for sustaining ecosystems and biodiversity? Can you think of other things for which you would pay a little more for this effort?

This strategy for protecting the earth's vital biodiversity and ecosystem services will not be implemented without bottom-up political pressure on elected officials from individual citizens and groups. It will also require cooperation among scientists, engineers, and business and government leaders in applying the win-win **principle of sustainability**.

A key part of this strategy will be for individuals to "vote with their wallets" by trying to buy only products and services that do not have harmful impacts on terrestrial and aquatic biodiversity. For example, we can eat lower on the aquatic food chain by choosing plant-eating species such as tilapia, carp, and catfish instead of carnivorous species such as salmon, shrimp, and sea bass. The highly respected Monterey Bay Aquarium publishes and regularly updates a simple sustainable seafood guide called *Seafood Watch* that can be viewed at their website or with a free mobile phone app.

According to Sylvia Earle (Individuals Matter 11.1), failure to act now "means that in 50 years there may be no coral reefs and no commercial fishing, because the fish will simply be gone. . . . Imagine what that means to our life-support system."

BIG IDEAS

- The world's aquatic systems provide important economic and ecosystem services, and scientific investigation of these poorly understood ecosystems could lead to immense ecological and economic benefits.
- Aquatic ecosystems and fisheries are being severely degraded by human activities that lead to aquatic habitat disruption and loss of biodiversity.
- We can sustain aquatic biodiversity by establishing protected sanctuaries, managing coastal development, reducing water pollution, and preventing overfishing.

Tying It All Together

Invading Jellyfish and Aquatic Sustainability

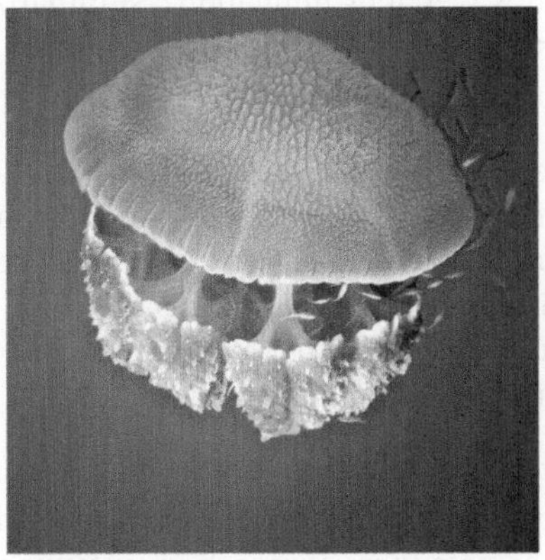
Olga Khoroshunova/Dreamstime.com

This chapter began with a look at populations of a number of jellyfish species that are exploding, disrupting marine food webs, and taking over parts of the ocean (**Core Case Study**). Throughout the chapter, we examined how loss of aquatic habitats, invasive species, population pressures, climate change, ocean acidification, and overexploitation are harming many marine and freshwater aquatic species. We looked at how many of the world's fisheries are being depleted. We discussed why jellyfish populations are exploding and why this is a serious threat.

We also explored possible solutions to these problems, including the jellyfish invasions. We know that when areas of the oceans are left undisturbed, marine ecosystems tend to recover their natural functions, and fish populations can rebound quickly. In addition, the best approach to sustaining freshwater biodiversity is to use an ecosystem approach.

We can achieve greater success in sustaining aquatic biodiversity by applying the three **scientific principles of sustainability**. This means reducing inputs of sediments and excess nutrients, which cloud water, lessen the input of solar energy, and upset aquatic food webs and the natural cycling of nutrients in aquatic systems. It also means valuing aquatic biodiversity and putting a high priority on preserving the biodiversity and ecosystem services of aquatic systems. Applying the full-cost pricing, win-win, and ethical **principles of sustainability** can help us to achieve these goals.

Chapter Review

Core Case Study

1. What are jellyfish? What problems are they causing?

Section 11.1

2. What is the key concept for this section? How much do we know about the habitats and species that make up the earth's aquatic biodiversity? What are three general patterns of marine biodiversity? Give six reasons for caring about marine biodiversity. What are the six major threats to aquatic diversity and ecosystems as summarized by the acronym HIPPCO? How have coral reefs been threatened? What is **coral bleaching** and what causes it? How can sunscreen lotion harm coral reefs? What is the state of the world's mangrove forests, sea-grass beds, and ocean-bottom habitats? What is **ocean acidification**? What causes it, what are its projected harmful effects on ocean life, and how could we reduce these threats?
3. Describe the harmful effects of the marine invasive species known as the lionfish. What are two harmful effects on aquatic systems resulting from the growth of the human population in coastal areas? Give two examples of how pollution is affecting aquatic systems. What are "dead zones" and how do they form? What is the threat from toxic mercury? What are three ways in which projected climate change could threaten aquatic biodiversity?
4. Define **fishery**. What are three major harmful effects of overfishing? Describe the effects of trawler fishing, purse-seine fishing, long-lining, and drift-net fishing. What is **bycatch**? What is a **fishprint**? What percentage of the world's commercial fisheries are fully exploited, overexploited, or depleted? Summarize the story of the collapse of the Atlantic cod fishery. What is happening to the world's large, predatory fishes and why does it matter? What are the harmful effects of the fishing industry's shifting to lower trophic levels of marine food webs? Summarize the arguments for protecting sharks. Summarize efforts to protect whales from overharvesting and extinction. Describe the status of sea turtles and explain how human activities are threatening their existence. Why are jellyfish populations growing and what difference does this make? How could we prevent them from taking over much of the ocean?

Section 11.2

5. What is the key concept for this section? List three reasons why it is difficult to protect marine biodiversity. How have laws and treaties been used to help sustain

aquatic species? What is the main obstacle to enforcement of international agreements? How can economic incentives help to sustain aquatic biodiversity? Give an example of how this can happen.

6. Explain how marine protected areas (MPAs) and marine reserves can help sustain aquatic biodiversity and ecosystem services. What percentage of the world's oceans is strictly protected from harmful human activities in marine reserves? What percentage should be strictly protected, according to marine scientists? Summarize the contributions of Sylvia Earle to the protection of aquatic biodiversity. Give two examples of how marine systems can be restored. Describe the roles of fishing communities and individual consumers in regulating fishing and coastal development. Describe Japan's attempt to restore its largest coral reef. What is integrated coastal management?

Section 11.3

7. What is the key concept for this section? Describe three ways of estimating the sizes of fish populations and list their limitations. How can the precautionary principle be applied in managing fisheries and large marine systems? What are catch-share and co-management systems and how can they help to sustain fisheries? How can government subsidies encourage overfishing? Explain how consumer choices can help to sustain fisheries, aquatic biodiversity, and ecosystem services. List five ways to manage global fisheries more sustainably.

Section 11.4

8. What is the key concept for this section? What percentage of the U.S. coastal and inland wetlands has been destroyed since 1900? What are the major threats to wetlands and their ecosystem services? How does the United States attempt to reduce wetland losses? What is wetlands mitigation? Summarize the story of efforts to restore the Florida Everglades.

Section 11.5

9. What is the key concept for this section? Describe the major threats to the world's rivers and other freshwater systems. Explain how invasions by nonnative species are threatening the Great Lakes. What are some ways to help sustain river systems? What are three ways to protect freshwater habitats and fisheries?

Section 11.6

10. What is the key concept for this section? List six priorities for applying the ecosystem approach to sustaining aquatic biodiversity. What are this chapter's three big ideas? How can we apply the three **scientific principles of sustainability** in efforts to control populations of jellyfish (**Core Case Study**) and thus help sustain aquatic biodiversity and ecosystem services?

Note: Key terms are in bold type.

Critical Thinking

1. Why should you be concerned about jellyfish populations taking over large areas of the world's oceans? Why are jellyfish considered to be indicator species (see p. 83)?

2. What do you think are the three greatest threats to aquatic biodiversity? For each of them, explain your thinking.

3. Overall, why are aquatic species more vulnerable to extinction hastened by human activities than terrestrial species are?

4. How might continued overfishing of marine species affect your life? How could it affect the lives of any children or grandchildren you might have? What are three things you could do to help prevent overfishing?

5. Should fishers who harvest fish from a country's publicly owned waters be required to pay the government fees for the fish they catch? Explain. If your livelihood depended on commercial fishing, would you be for or against such fees?

6. Why do you think no-fishing marine reserves recover their biodiversity faster and more effectively than do protected areas where fishing is allowed but restricted? Explain.

7. Some scientists consider ocean acidification to be one of the most serious environmental and economic threats that the world faces. How do you contribute to ocean acidification in your daily life? What would you do to help reduce the threat of ocean acidification?

8. How might your life and the lives of any children or grandchildren you might have be affected if we fail to control the spread of jellyfish populations? What are three things you could do to help prevent this from happening?

Doing Environmental Science

Pick a coastal area, river, stream, lake, or wetland near where you live and research and write a brief account of its history. Then survey and take notes on its present condition. Has its condition improved or deteriorated during the last 10 years? What governmental or private efforts are being used to protect this aquatic system? Write a report summarizing your findings. Based on your report along with your ecological knowledge of this system, write up some recommendations to policy makers for protecting it. Try presenting your recommendations to one or more local policy makers.

Global Environment Watch Exercise

Go to your MindTap course to access the GREENR database. Starting on the home page, under "Browse Issues and Topics", click on *Environment and Ecology,* then select *Coral Reefs*. Use this topic portal to do the following: **(a)** identify two cases of coral reef degradation—one where the reef continues to be degraded and one where a degraded reef is being restored; **(b)** for each case, list the main causes for this degradation; **(c)** for each case, describe any efforts to protect or restore the reef; and **(d)** for each case, give your best estimation of whether the reef will survive and explain your reasoning.

Ecological Footprint Analysis

A fishprint provides a measure of a country's fish harvest in terms of area. The unit of area used in fishprint analysis is the global hectare (gha), a unit weighted to reflect the relative ecological productivity of the area fished. When compared with the fishing area's *sustainable biocapacity* (its ability to provide a stable supply of fish year after year, expressed in terms of yield per area), its fishprint indicates whether the country's annual fishing harvest is sustainable. The fishprint and biocapacity are calculated using the following formulas:

Fishprint in (gha) = metric tons of fish harvested per year/productivity in metric tons per hectare × weighting factor

Biocapacity in (gha) = sustained yield of fish in metric tons per year/ productivity in metric tons per hectare × weighting factor

The following graph shows the earth's total fishprint and biocapacity. Study it and answer the questions that follow.

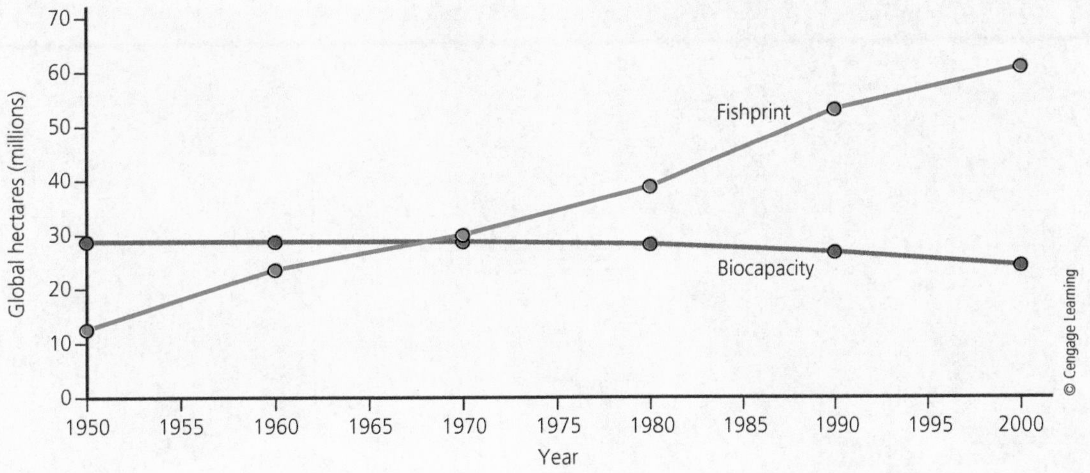

1. Based on the graph,
   a. What was the status of the global fisheries with respect to sustainability in 2000?
   b. In what year did the global fishprint begin to exceed the biological capacity of the world's oceans?
   c. By how much did the global fishprint exceed the biological capacity of the world's oceans in 2000?
2. Assume a country harvests 18 million metric tons of fish annually from an ocean area with an average productivity of 1.3 metric tons per hectare and a weighting factor of 2.68. What is the annual fishprint of that country?
3. If biologists determine that this country's sustained yield of fish is 17 million metric tons per year,
   a. What is the country's sustainable biological capacity?
   b. Is the county's annual fishing harvest sustainable?
   c. To what extent, as a percentage, is the country undershooting or overshooting its biological capacity?

CENGAGE brain.com For access to MindTap and additional study materials visit www.cengagebrain.com.

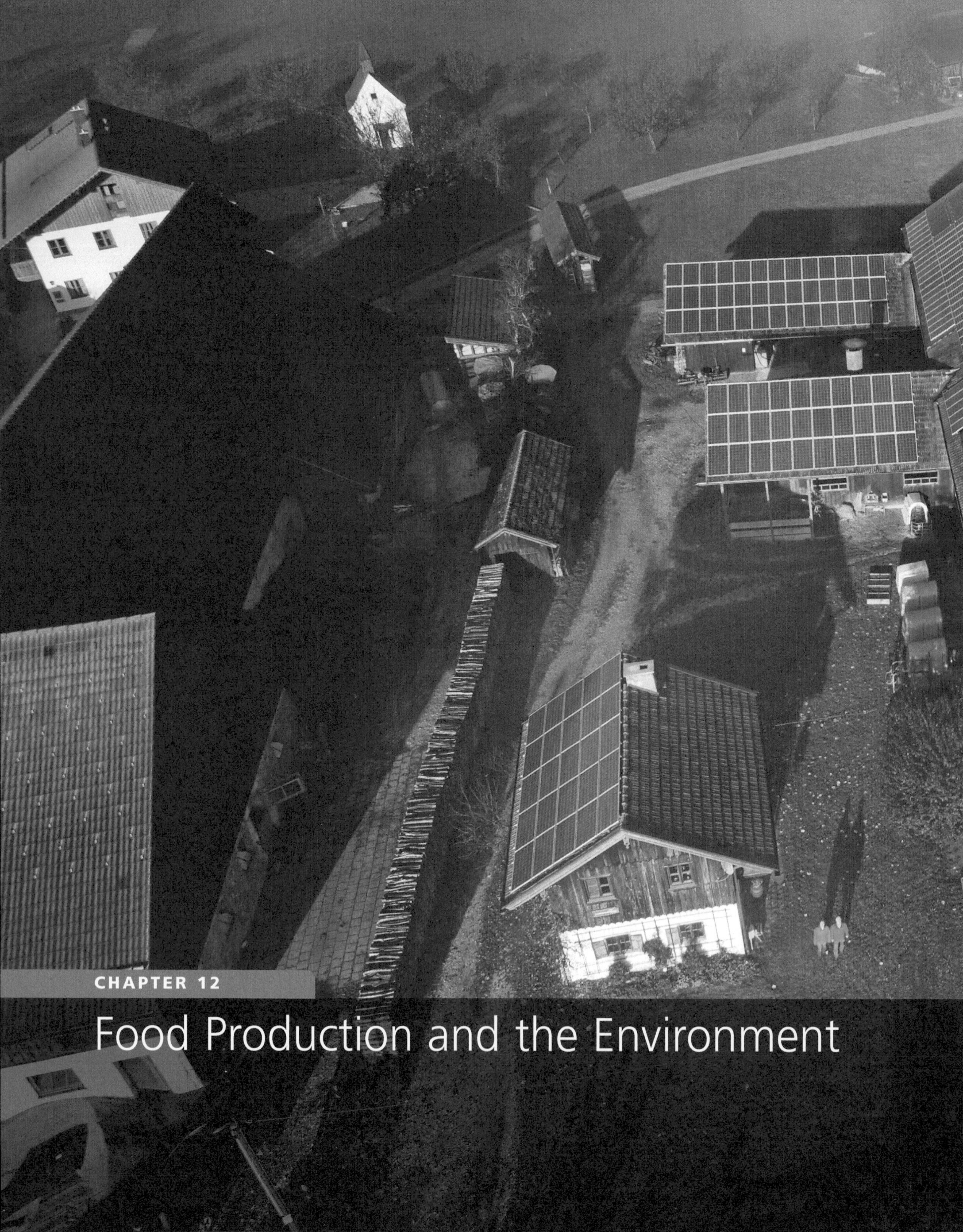

CHAPTER 12

Food Production and the Environment

> There are two spiritual dangers in not owning a farm. One is the danger of supposing that breakfast comes from the grocery, and the other that heat comes from the furnace.
>
> ALDO LEOPOLD

Key Questions

12.1 Why is good nutrition important?

12.2 How is food produced?

12.3 What are the environmental effects of industrialized food

12.4 How can we protect crops from pests more sustainably?

12.5 How can we produce food more sustainably?

12.6 How can we improve food

This farm in Bavaria, Germany uses solar cells on its roofs to provide its electricity.

Michael Melford/National Geographic Creative

Core Case Study

Growing Power—An Urban Food Oasis

A *food desert* is an urban area where people have little or no easy access to grocery stores or other sources of nutritious food. In the United States, an estimated 23.5 million people, including 6.5 million children, live in such urban neighborhoods. They tend to rely on convenience stores and fast-food restaurants that mainly offer high-calorie, highly processed foods that can lead to higher risks of obesity, diabetes, and heart disease.

Will Allen (Figure 12.1) was one of six children of a sharecropper and grew up on a farm in Maryland. He left the farm life for college and a professional basketball career, followed by a successful corporate marketing career. In 1993 Allen had decided to return to his roots. He bought the last working farm within the city limits of Milwaukee, Wisconsin, and, in time, created a food oasis in a food desert.

On this small urban plot, Allen developed Growing Power, Inc. As an ecologically based farm, it is a showcase for forms of agriculture that apply all three **scientific principles of sustainability** (see inside back cover of book). It is powered partly by solar electricity and solar hot water systems and makes use of several greenhouses to capture solar energy for growing food year-round. The farm produces an amazing diversity of crops with about 150 varieties of organic produce. It also produces organically raised chickens, turkeys, goats, fish, and honeybees. And the farm's nutrients are recycled in creative ways. For example, wastes from the farmed fish are used as nutrients to raise some of the crops.

The farm's products are sold locally at Growing Power farm stands throughout the region and to restaurants. Allen also worked with the city of Milwaukee to establish the Farm-to-City Market Basket program through which people can sign up for weekly deliveries of organic produce at modest prices.

In addition, Growing Power runs an educational program in which school children visit the farm and learn about where their food comes from. Allen also trains about 1,000 people every year who want to learn organic farming methods. The farm has also partnered with the city of Milwaukee to create 150 green jobs for unemployed and low-income workers, building greenhouses, and growing food organically. Growing Power has expanded, opening another urban farm in a neighborhood of Chicago, Illinois, and setting up satellite training sites in five other states.

For his creative and energetic efforts, Allen has won several prestigious awards. However, he is most proud of the fact that his urban farm helps to feed more than 10,000 people every year and puts people to work raising good food.

In this chapter, we look at different ways to produce food, the environmental effects of food production, and how we can produce food more sustainably.

Courtesy of Growing Power, Inc.

FIGURE 12.1 Will Allen founded Growing Power—a world-renowned urban farm in Milwaukee, Wisconsin.

12.1 WHY IS GOOD NUTRITION IMPORTANT?

CONCEPT 12.1A Many people in less-developed countries have health problems from not getting enough food, while many people in more-developed countries have health problems from eating too much food.

CONCEPT 12.1B The greatest obstacles to providing enough food for everyone are poverty, war, bad weather, and climate change.

Many People Suffer from Lasting Hunger and Malnutrition

What foods do you think you should eat in order to live a healthy and enjoyable life? What do you really eat? How do those lists compare?

If you find that what you should eat is not what you are eating, you are not alone. Many of today's health problems are related to eating too much of foods that are not nutritious and not enough of those that are. To maintain good health and resist disease, individuals need fairly large amounts of *macronutrients* (such as carbohydrates, proteins, and fats) and smaller amounts of *micronutrients*—vitamins, such as A, B, C, and E, and minerals, such as iron, iodine, and calcium.

One of every eight people on the planet—nearly 800 million in all—are not getting enough to eat. People who cannot grow or buy enough food to meet their basic energy needs suffer from **chronic undernutrition,** or **hunger,** which threatens their ability to lead healthy and productive lives (Figure 12.2) (**Concept 12.1A**).

Most of the world's hungry people live in low-income, less-developed countries and typically can afford only a low-protein, high-carbohydrate, vegetarian diet consisting mainly of grains such as wheat, rice, and corn (Figure 12.3). In other words, they live low on the food chain (see Figures 3.14 and 3.15, pp. 59 and 60). In most cases, people on such limited diets suffer from **chronic malnutrition,** a

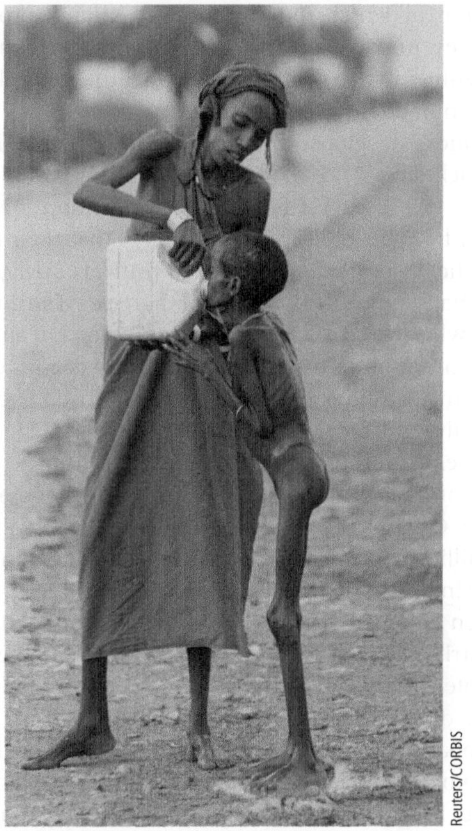

FIGURE 12.2 This woman is feeding her child, who was starving because of famine caused by a civil war that has been going on in Somalia, Africa, most of the time since 1991.

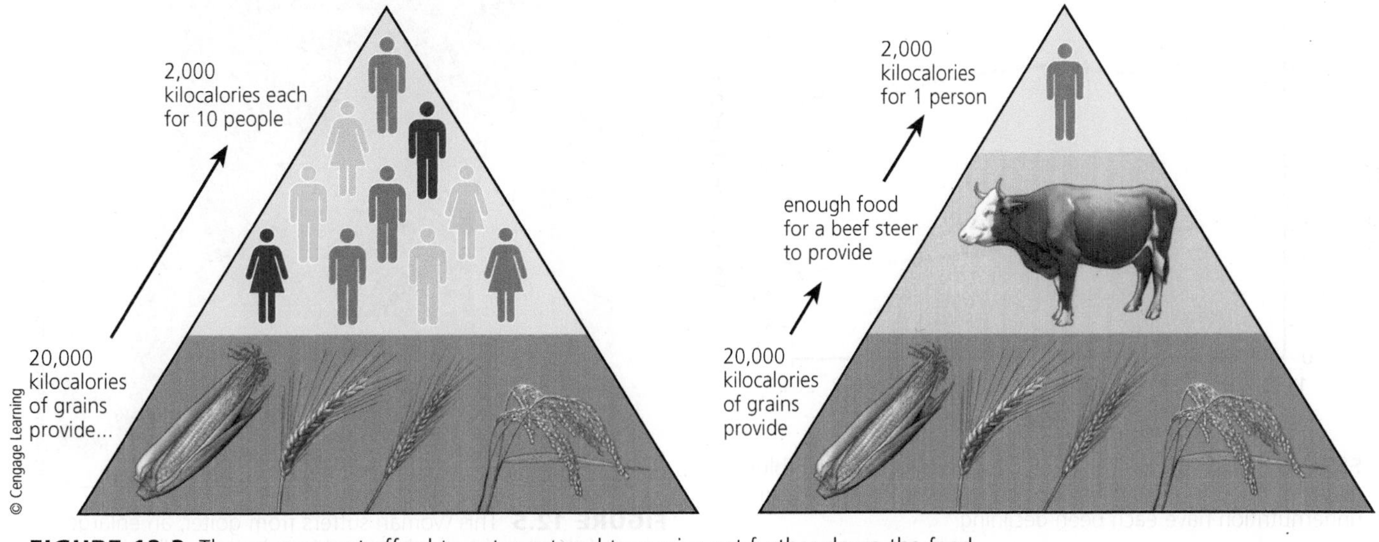

FIGURE 12.3 The poor cannot afford to eat meat and to survive eat further down the food chain on a diet of grain.

condition in which they do not get enough protein and other key nutrients. This can weaken them, make them more vulnerable to disease, and hinder the normal physical and mental development of children.

In more-developed countries, people living in food deserts, including 23.5 million Americans according to the Centers for Disease Control and Prevention (**Core Case Study**), have a similar problem, except that their diet typically is heavy on cheap food loaded with fats, sugar, and salt.

Perhaps the worst form of food shortage is **famine,** which occurs when there is a severe shortage of food in an area and which can result in mass starvation, many deaths, economic chaos, and social disruption. Famines are usually caused by crop failures from drought, flooding, war (Figure 12.2), and other catastrophic events (**Concept 12.1B**).

According to the United Nations (UN) Food and Agriculture Organization (FAO), in 2015 there were about 795 million chronically undernourished or malnourished people in the world. The UN estimates that at least 3 million children younger than age 5 died from chronic hunger and malnutrition in 2013 (the latest year for which the data are available). That was an average of nearly 6 children per minute dying from these causes.

Globally, the number and percentage of people suffering from chronic hunger has been declining since 1992 (Figure 12.4). In 2015 these numbers reached a 25-year low, but there is still a long way to go. In some areas of the world, the news is not good. In Africa, twice as many countries are in a hunger crisis as in 1992. One of every four people living south of the Sahara Desert (sub-Saharan Africa) is undernourished.

A Closer Look at Micronutrients

About 2 billion people, most of them in less-developed countries, suffer from a deficiency of one or more vitamins and minerals, usually *vitamin A, zinc, iron,* and *iodine* (**Concept 12.1A**). According to the World Health Organization (WHO), at least 250,000 children younger than age 6, most of them in less-developed countries, go blind every year from a lack of vitamin A. Within a year, more than half of them die. Providing children with adequate vitamin A and zinc could save an estimated 145,000 lives per year.

Having too little *iron* (Fe)—a component of the hemoglobin that transports oxygen in the blood—is a condition called *anemia.* It causes fatigue, makes infection more likely, and increases a woman's chances of dying from hemorrhage in childbirth. According to the WHO, one of every five people in the world—mostly women and children in less-developed countries—suffers from iron deficiency.

The chemical element *iodine* (I) is essential for proper functioning of the thyroid gland, which produces hormones that control the body's rate of metabolism. Chronic lack of iodine can cause stunted growth, mental retardation, and goiter—a severely swollen thyroid gland that can lead to deafness (Figure 12.5). According to the UN, some 600 million people (almost twice the current U.S. population) suffer from goiter, most of them in less-developed countries. In addition, 26 million children suffer irreversible brain damage every year from lack of iodine.

Health Problems from Eating Too Much

Overnutrition occurs when food energy intake exceeds energy use and causes excess body fat. Too many calories, too little exercise, or both can cause overnutrition.

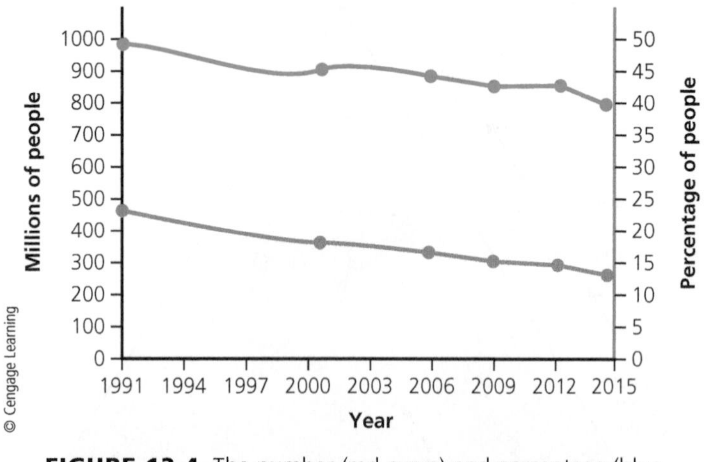

FIGURE 12.4 The number (red curve) and percentage (blue curve) of people in less-developed countries who suffer from undernutrition have each been declining.

(Compiled by the authors using data from U.S. Department of Agriculture, UN Food and Agriculture Organization, and Earth Policy Institute.)

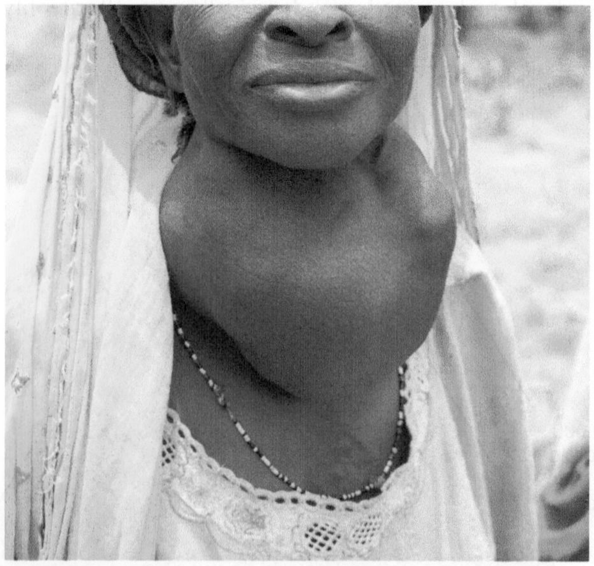

FIGURE 12.5 This woman suffers from goiter, an enlargement of the thyroid gland, caused by a lack of iodine in her diet.

...nderfed and underweight and those ... imilar health ... ptibility to dis- ... d life quality

... to the WHO, ... blems because ... eat and at least ... pulation) have ... ating too much ... active lifestyles, ...

Globally, the num... nd obese people increased from 857 million in 1980 to 2.1 billion in 2013, according to a study led by Christopher Murray. In order, the countries with the most overweight and obese people are the United States, China, India, Russia, and Brazil. According to a 2014 report by the McKinsey Global Institute, the resulting health-care and lost-productivity costs are about $2 trillion a year—more than the combined annual global costs of war, terrorism, and armed violence.

In the United States, according to 2015 statistics from the U.S. Centers for Disease Control and Prevention (CDC), about 72% of all adults over age 20 are obese (38%) or overweight (34%) and 33% of all children ages 2 to 19 are overweight or obese. A 2013 study by Columbia University and the Robert Wood Johnson Foundation found that obesity plays an important role in nearly one in five deaths in the United States from heart disease, stroke, type 2 diabetes, and some forms of cancer.

Poverty Is the Root Cause of Hunger and Malnutrition

Poverty prevents people from growing or buying enough nutritious food to meet their basic needs. This prevents them from having **food security,** or daily access to enough nutritious food to live healthy lives. This is not surprising given that about half of the world's people are trying to survive on the equivalent of $2.25 a day, and one out of six people struggle to get by on $1.25 a day. Other obstacles to food security are war, corruption, bad weather (such as prolonged drought, flooding, and heat waves), and climate change (**Concept 12.1B**).

Each day, there are about 244,000 more people at the world's dinner tables, and by 2050, there will likely be at least 2.5 billion more people to feed. Most of these newcomers will be born in the major cities of less-developed countries (see pp. 118–119). A critical question is, how will we feed the projected 9.8 billion people in 2050 without causing serious harm to the environment? We explore possible answers to this question throughout this chapter.

12.2 HOW IS FOOD PRODUCED?

CONCEPT 12.2 We have used high-input industrialized agriculture and lower-input traditional agriculture to greatly increase food supplies.

Food Production Has Increased Dramatically

Three systems supply most of the world's food, using about 40% of the world's land. *Croplands* produce grains—primarily rice, wheat, and corn—that provide about 77% of the world's food. The rest is provided by *rangelands, pastures,* and *feedlots* that produce meat and meat products and *fisheries* and *aquaculture* (fish farming), which supply fish and shellfish.

These three systems depend on a small number of plant and animal species. Of the estimated 30,000 edible plant species, 14 supply about 90% of the world's food calories. At least half the world's people survive primarily by eating three grain crops—*rice, wheat,* and *corn*—because they cannot afford meat. Only a few species of mammals and fish provide most of the world's meat and seafood. (See Figure 5 in Supplement 5, p. S44, for a closer look at the loss of food diversity in the United States.)

Such food specialization puts us in a vulnerable position, if any of the small number of crop strains, livestock breeds, and fish and shellfish species we depend on become depleted because of disease, environmental degradation, climate change, or other factors. Food specialization violates the biodiversity **principle of sustainability,** which calls for depending on a variety of food sources as an ecological insurance policy against changing environmental conditions.

Despite such genetic vulnerability, since 1960, there has been a staggering increase in global food production from all three of the major food production systems (**Concept 12.2**). Three major technological advances have been especially important: **(1)** the development of **irrigation,** a mix of methods by which water is supplied to crops by artificial means; **(2) synthetic fertilizers**—manufactured chemicals that contain nutrients such as nitrogen, phosphorus, potassium, calcium, and several others; and **(3) synthetic pesticides**—chemicals manufactured to kill or control populations of organisms that interfere with crop production.

Industrialized Crop Production Relies on High-Input Monocultures

There are two major types of agriculture: industrialized and traditional. **Industrialized agriculture,** or **high-input agriculture,** uses heavy equipment (Figure 12.6) along with large amounts of financial capital, fossil fuels,

Brenda Carson/Shutterstock.com

FIGURE 12.6 This farmer, harvesting a wheat crop in the Midwestern United States, relies on expensive heavy equipment and uses large amounts of seed, manufactured inorganic fertilizer and pesticides, and fossil fuels to produce the crop.

water, commercial inorganic fertilizers, and pesticides to produce single crops, or *monocultures*. The major goal of industrialized agriculture is to steadily increase each crop's **yield**—the amount of food produced per unit of land. Industrialized agriculture is practiced on 25% of all cropland, mostly in more-developed countries, and produces about 80% of the world's food (**Concept 12.2**).

Plantation agriculture is a form of industrialized agriculture used primarily in less-developed tropical countries. It involves growing cash crops such as bananas, coffee, vegetables, soybeans (mostly to feed livestock; see Figure 1.5, p. 8), sugarcane (to produce sugar and ethanol fuel), and palm oil (to produce cooking oil and biodiesel fuel). These crops are grown on large monoculture plantations, mostly for export to more-developed countries.

Traditional Agriculture Often Relies on Low-Input Polyculture

Traditional, low-input agriculture provides about 20% of the world's food crops on about 75% of its cultivated land, mostly in less-developed countries. It takes two basic forms. **Traditional subsistence agriculture** combines energy from the sun with the labor of humans and draft animals to produce enough crops for a farm family's survival, with little left over to sell or store as a reserve for hard times. In **traditional intensive agriculture,** farmers try to obtain higher crop yields by increasing their inputs of human and draft animal labor, animal manure for fertilizer, and water. If weather cooperates, farmers can produce enough food to feed their families and have some left over to sell for income.

Some traditional farmers focus on cultivating a single crop, but many grow several crops on the same plot simultaneously, a practice known as **polyculture.** This method relies on solar energy and natural fertilizers such as animal manure. The various crops mature at different times. This provides food year-round and keeps the topsoil covered to reduce erosion from wind and water. Polyculture also lessens the need for fertilizer and water because root systems at different depths in the soil capture nutrients and moisture efficiently. Also, weeds have trouble competing with the multitude and density of crop plants, and this crop diversity reduces the chance of losing most or all of the year's food supply to pests, bad weather, and other misfortunes.

One type of polyculture is known as *slash-and-burn agriculture* (Figure 1.12, p. 17). This type of subsistence agriculture involves burning and clearing small plots in tropical forests, growing a variety of crops for a few years until the soil is depleted of nutrients, and then shifting to other plots to begin the process again. In parts of South America and Africa, some traditional farmers grow as many as 20 different crops together on small cleared plots.

Organic Agriculture Is on the Rise

In recent years there has been increased use of **organic agriculture** in which crops are grown without the use of synthetic pesticides, synthetic inorganic fertilizers, or genetically engineered seed varieties. To be classified as *organically grown*, animals must be raised on 100% organic feed without the use of antibiotics or growth hormones. Growing Power (**Core Case Study**) has become a well-known model for such food production. Figure 12.7 compares organic agriculture with industrialized agriculture.

In the United States, by law, a label of *100 percent organic* (or *USDA Certified Organic*) means that a product is produced only by organic methods and contains all organic ingredients. About 13,000 of the 2.2 million farms in the United States are USDA-certified organic. Products labeled *organic* must contain at least 95% organic ingredients. Those labeled *made with organic ingredients* must contain at least 70% organic ingredients. The word *natural* has no requirement for organic ingredients and is primarily used as an advertising ploy.

Partly because organic farming is more labor intensive, organically grown foods tend to cost more than conventionally produced food. However, Will Allen and other Growing Power farmers (**Core Case Study**) are learning how to use sustainable farming methods to get higher yields of a variety of organic crops at affordable prices.

A Closer Look at Industrialized Crop Production

Farmers have two ways to produce more food: farming more land or getting higher yields from existing cropland. Since 1950, about 88% of the dramatic increase in global grain production has come from using high-input industrialized agriculture to increase crop yields.

This process, called the **green revolution,** involves three steps. *First,* develop and plant monocultures of selectively bred or genetically engineered high-yield varieties of key crops such as rice, wheat, and corn. *Second,* produce high yields by using large inputs of water, synthetic inorganic fertilizers, and pesticides. *Third,* increase the number of crops grown per year on a plot of land. Between 1950 and 1970, in what was called the *first green revolution,* this

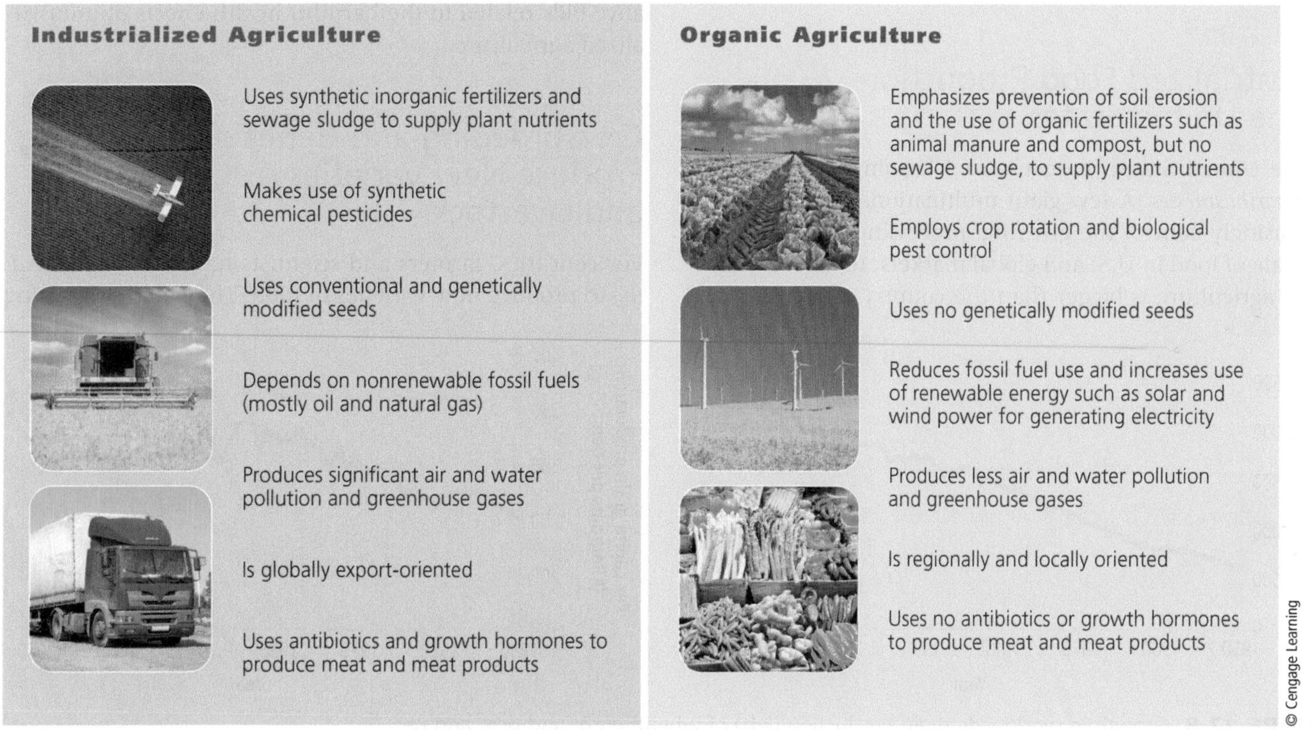
Industrialized Agriculture
Uses synthetic inorganic fertilizers and sewage sludge to supply plant nutrients
Makes use of synthetic chemical pesticides
Uses conventional and genetically modified seeds
Depends on nonrenewable fossil fuels (mostly oil and natural gas)
Produces significant air and water pollution and greenhouse gases
Is globally export-oriented
Uses antibiotics and growth hormones to produce meat and meat products
Organic Agriculture
Emphasizes prevention of soil erosion and the use of organic fertilizers such as animal manure and compost, but no sewage sludge, to supply plant nutrients
Employs crop rotation and biological pest control
Uses no genetically modified seeds
Reduces fossil fuel use and increases use of renewable energy such as solar and wind power for generating electricity
Produces less air and water pollution and greenhouse gases
Is regionally and locally oriented
Uses no antibiotics or growth hormones to produce meat and meat products
© Cengage Learning

FIGURE 12.7 Major differences between industrialized agriculture and organic agriculture.

Left top: BBrown/Shutterstock.com. Left center: Zorandim/Shutterstock.com. Left bottom: Art Konovalov/Shutterstock.com. Right top: Noam Armonn/Shutterstock.com. Right center: Varina and Jay Patel/Shutterstock.com. Right bottom: Adisa/Shutterstock.com.

high-input approach dramatically raised crop yields in most of the world's more-developed countries, especially the United States (see the Case Study that follows).

In the *second green revolution*, which began in 1967, fast-growing varieties of rice and wheat, specially bred for tropical and subtropical climates, have been introduced into middle-income, less-developed countries such as India, China, and Brazil. Producing more food on less land in such countries has helped to protect biodiversity by preserving large areas of forests, grasslands, and wetlands that might otherwise be used for farming.

Largely because of the two green revolutions, between 1950 and 2014, world grain production (Figure 12.8, left) increased by 312% and per capita grain production (Figure 12.8, right) grew by 37%. Today, the world's three largest grain-producing countries—China, India, and the United States—produce almost half of the world's grains. But their rate of growth in crop yields has slowed from an average of 2.2% per decade before 1990 to 1.2% on average since then.

An important factor in expanded industrialized crop production has been the use of **farm subsidies,** or government payments and tax breaks intended to help farmers stay in business and increase their yields. In the United States, most subsidies go to corporate farming operations for raising corn, wheat, soybeans, and cotton on an industrial scale. U.S. government records show that in recent years, 74% of all subsidies went to just 10% of all U.S. farmers.

CASE STUDY

Industrialized Food Production in the United States

In the United States, industrialized farming has evolved into *agribusinesses*. A few giant multinational corporations increasingly control the growing, processing, distribution, and sale of food in U.S. and global markets. In total annual sales, agriculture is bigger than the country's automotive, steel, and housing industries combined. Yet, because of advances in technology, the numbers of U.S. farms and farmers have dropped as production has risen. As a result, the average U.S. farmer now feeds 129 people compared to 19 people in the 1940s.

1% Percentage of the U.S. workforce who are farmers—down from 18% in 1910

Since 1960, U.S. industrialized agriculture has more than doubled the yields of key crops such as wheat, corn, and soybeans that are grown for profit without the need for cultivating more land. Such yield increases have saved large areas of U.S. forests, grasslands, and wetlands from being converted to farmland.

Because of the efficiency of U.S. agriculture, Americans spend the lowest percentage of disposable income in the world—an average of 9%—on food. By contrast, low-income people in less-developed countries typically spend 50–70% of their income on food, according to the FAO.

However, because of a number of *hidden costs* related to food production and consumption, most American consumers are not aware that their actual food costs are much higher than the market prices they pay. Such costs include taxes to pay for farm subsidies and the costs of pollution and environmental degradation, and higher health insurance bills related to the harmful health effects of industrialized agriculture.

Crossbreeding and Genetic Engineering Produce New Varieties of Crops and Livestock

For centuries, farmers and scientists have used *crossbreeding* to produce new varieties of food. They do this by using

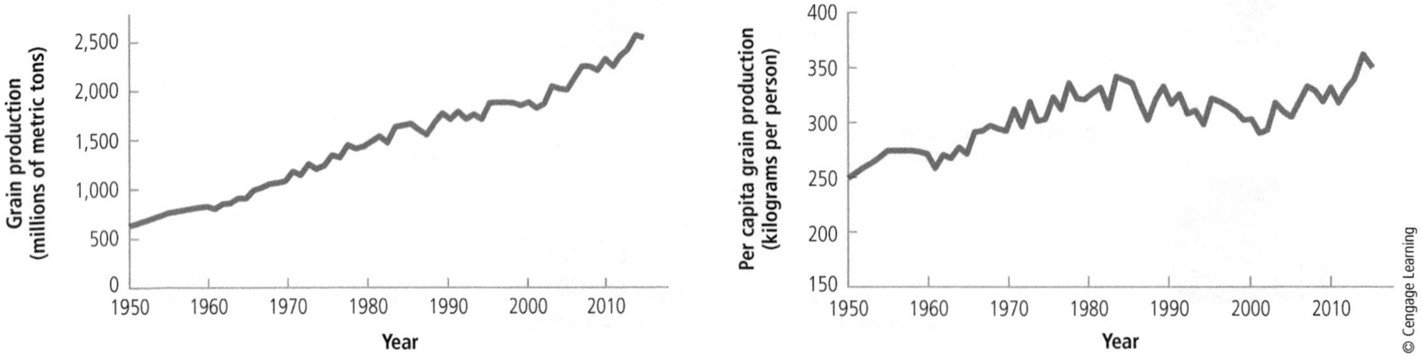

FIGURE 12.8 Growth in worldwide grain production (left) of wheat, corn, and rice, and per capita grain production (right) between 1950 and 2015. ***Critical thinking:*** Why do you think grain production per capita has grown less consistently than total grain production?

(Compiled by the authors using data from U.S. Department of Agriculture, Worldwatch Institute, UN Food and Agriculture Organization, and Earth Policy Institute.)

artificial selection to develop genetically improved varieties of crops (see Figure 4.18, p. 91) and livestock animals. Such selective breeding in this first *gene revolution* has yielded amazing results. For example, ancient ears of corn were about the size of your little finger, and wild tomatoes were once the size of grapes, but most of the large varieties used now were selectively bred. GOOD NEWS

Traditional crossbreeding is a slow process. It often takes 15 years or more to produce a commercially valuable new crop variety and it can combine traits only from species that are genetically similar. Typically, resulting varieties remain useful for only 5 to 10 years before pests and diseases reduce their yields. Important advances are still being made with this method.

Today, a *second gene revolution* is taking place. Scientists are using **genetic engineering** to develop genetically modified strains of crops and livestock animals. They use a process called *gene splicing* to alter an organism's genetic material by adding, deleting, or changing segments of its DNA (see Figure 2.9, p. 38). The goal of this process is to develop genetically modified strains of crops and livestock animals by adding desirable traits or eliminating undesirable traits, often by transferring genes between species that would normally not interbreed in nature. The resulting organisms are called **genetically modified organisms (GMOs).** Developing a new crop variety through genetic engineering (p. 91) takes about half as long as traditional crossbreeding and usually costs less.

Globally, 28 countries planted genetically modified (GM) crops in 2015, with four crops—corn, soybeans, cotton, and canola—leading the way. Three countries—the United States, Brazil, and Argentina—accounted for more than 75% of the total acreage of GM crops in 2015, with the United States accounting for 40%. The European Union severely restricts the use of GM seeds and the importation of products containing GMOs.

According to the U.S. Department of Agriculture (USDA), at least 80% of the food products on U.S. supermarket shelves contain some form of GM food or ingredients and that proportion is growing. The Environmental Working Group estimates that each year, the average American adult consumes 88 kilograms (193 pounds) of genetically modified foods.

Bioengineers hope to develop new GMO varieties of crops that are resistant to heat, cold, drought, insect pests, parasites, and other environmental threats. They also hope to develop crop plants that can grow faster and survive with little or no irrigation and with less use of fertilizer and pesticides. Another possibility is GM foods that last longer in fields before harvest and on food shelves after harvesting. This would help cut food waste.

Despite such potential benefits, some scientists warn that genetic engineering is not free of drawbacks. They point out that we know little about the long-term environmental and health effects of GMOs, which we examine later in this chapter.

Meat Consumption Has Grown Steadily

Meat and animal products such as eggs and milk are sources of high-quality protein and represent the world's second major food-producing system. Between 1950 and 2014, global meat production grew more than sixfold, according to the FAO (Figure 12.9, left).

Since 1974, the total global consumption of meat and meat products has more than doubled. Globally, the three most widely consumed meats are, in order, pork, poultry (chicken and turkey), and beef. The FAO projects that by 2024 poultry will be the largest source of meat. Total meat consumption is likely to double again by 2050 as incomes rise in rapidly developing countries. For example, meat consumption in China increased ninefold between 1978 and 2013, while it leveled off in the United States (Figure 12.9, right).

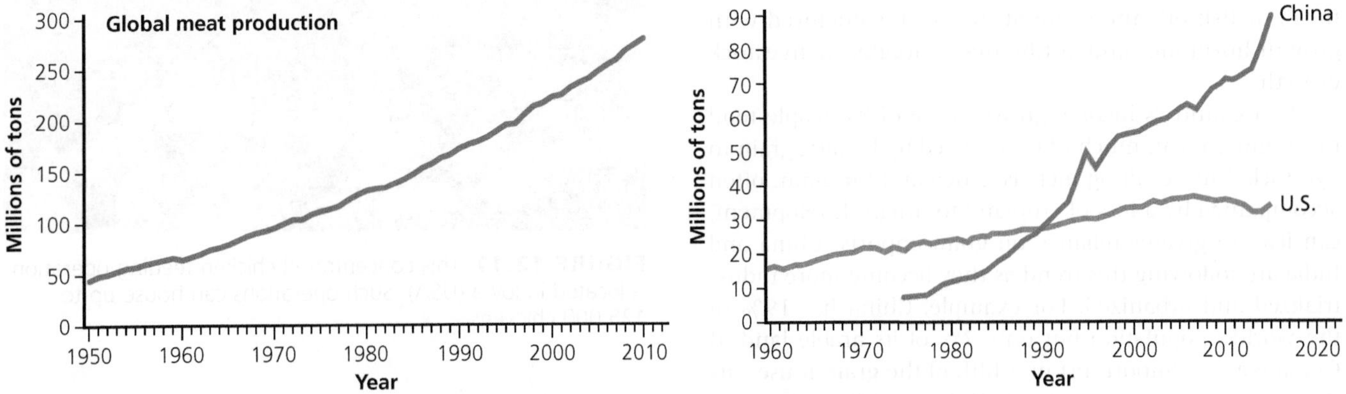

FIGURE 12.9 Global meat production grew more than sixfold between 1950 and 2015 (left). In China, meat consumption has grown ninefold since 1975 while growing more slowly in the United States. ***Data analysis:*** Using the right-hand graph, estimate the average annual rate of growth in consumption between 1980 and 2015 for China and for the United States.

(Compiled by the authors using data from U.S. Department of Agriculture, Worldwatch Institute, UN Food and Agriculture Organization, and Earth Policy Institute.)

PETE MCBRIDE/National Geographic Creative

FIGURE 12.10 *Industrialized beef production:* On this cattle feedlot in Arizona, thousands of cattle are fattened on grain for a few months before being slaughtered.

About half of the world's meat comes from livestock grazing on grass in unfenced rangelands and enclosed pastures. The other half is produced through an industrialized factory farm system. It involves raising large numbers of animals bred to gain weight quickly, mostly in *feedlots* (Figure 12.10) or in crowded pens and cages in huge buildings called *concentrated animal feeding operations (CAFOs)*, also called *factory farms* (Figure 12.11). In these facilities, the animals are fed grain, soybeans, fishmeal, or fish oil, and some of this feed is doctored with growth hormones and antibiotics to accelerate livestock growth.

As a country's income grows, more of its people tend to eat more meat, much of it produced by feeding grain to livestock. The resulting increased demand for grain, often accompanied by a loss of cropland to urban development, can lead to greater reliance on grain imports. China and India are following this trend as they become more industrialized and urbanized. For example, China has 19% of the world's population but only 7% of its arable land. If China were to import just one-fifth of the grain it uses, its imports would equal roughly half the world's entire grain exports. This is one reason why China as well as South Korea, Saudi Arabia, India, and Egypt are buying or leasing cropland in a number of other countries, especially in Africa.

Scott Sinklier/AgStock Images/Terra/Corbis

FIGURE 12.11 This concentrated chicken feeding operation is located in Iowa (USA). Such operations can house up to 125,000 chickens.

Fish and Shellfish Production Are Growing Rapidly

The world's third major food-producing system consists of fisheries and aquaculture. A **fishery** is a concentration of

Puwanai/Shutterstock.com

FIGURE 12.12 *Aquaculture:* Shrimp farms on the southern coast of Thailand.

a particular aquatic species suitable for commercial harvesting in a given ocean area or inland body of water. Industrial fishing fleets use a variety of methods (Figure 11.8, p. 262) to harvest most of the world's marine catch of wild fish. Fish and shellfish are also produced through **aquaculture** or **fish farming** (Figure 12.12). It is the practice of raising fish in freshwater ponds, lakes, reservoirs, and rice paddies, and in underwater cages in coastal and deeper ocean waters.

Between 1950 and 2015, global seafood production of wild and farmed fish increased ninefold (**Concept 12.2**). In 2015 aquaculture accounted for nearly half of the world's fish and shellfish production and the rest were caught mostly by industrial fishing fleets.

According to the FAO, about 87% of the world's ocean fisheries are being harvested at full capacity (57%) or are overfished (30%). Some fishery scientists warn that failure to reduce overfishing and ocean acidification and to slow climate change could severely threaten food security for roughly 3 billion people who now depend on fish for at least 20% of their animal protein.

Between 1980 and 2014, the amount of fish and shellfish produced globally through aquaculture grew almost 12-fold while the global wild catch leveled off and declined (Figure 12.13). Asia accounted for 88% (with China alone accounting for 60%) of the world's aquaculture production in 2014.

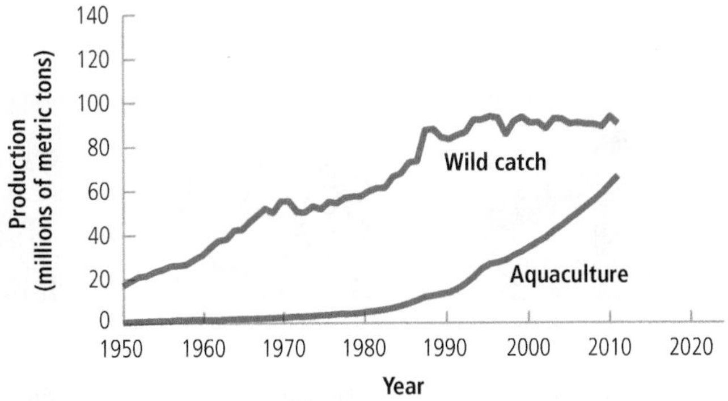

FIGURE 12.13 World seafood production, including both wild catch (marine and inland) and aquaculture, grew between 1950 and 2012, with the wild catch generally leveling off since 1996 and aquaculture production rising sharply since 1990. ***Data analysis:*** In about what year did aquaculture surpass the 1980 wild catch?

(Compiled by the authors using data from UN Food and Agriculture Organization, Worldwatch Institute, and Earth Policy Institute.)

Most of the world's aquaculture involves raising species that feed on algae or other plants—mainly carp in China, catfish in the United States, and tilapia and shellfish in a number of countries. However, the farming of meat-eating species such as shrimp and salmon is growing rapidly, especially in more-developed countries. Such species are often fed fishmeal and fish oil produced from other fish and their wastes.

Industrialized Food Production Requires Huge Inputs of Energy

Industrialized food production has been made possible by use of fossil fuels. Energy—mostly from oil and natural gas—is required to run farm machinery and fishing vessels, to pump irrigation water for crops, and to produce synthetic pesticides and synthetic inorganic fertilizers. Fossil fuels are also used to process food and transport it long distances within and between countries. Agriculture uses about 17% of all the energy used in the United States, more than any other industry.

When we consider the energy used to grow, store, process, package, transport, refrigerate, and cook all plant and animal food, it takes about 10 units of fossil fuel energy to put 1 unit of food energy on the table in the United States. Also, according to a study led by ecological economist Peter Tyedmers, the world's fishing fleets use about 12.5 units of energy to put 1 unit of food energy from seafood on the table. In other words, today's food production systems operate with a large net energy loss.

On the other hand, the amount of energy per calorie used to produce crops in the United States has declined by about 50% since the 1970s. One factor in this decline is that the amount of energy used to produce synthetic nitrogen fertilizer has dropped sharply. Another is the rising use of conservation tillage, which substantially reduces energy use and the harmful environmental effects of plowing.

12.3 WHAT ARE THE ENVIRONMENTAL EFFECTS OF INDUSTRIALIZED FOOD PRODUCTION?

CONCEPT 12.3 Future food production may be limited by soil erosion and degradation, desertification, irrigation water shortages, air and water pollution, climate change, and loss of biodiversity.

Producing Food Has Major Environmental Impacts

Although industrialized agriculture has provided many benefits, many analysts point out that it has greater overall harmful environmental impacts (Figure 12.14) than any other human activity. These environmental effects may limit future food production (**Concept 12.3**).

Natural Capital Degradation

Food Production

Biodiversity Loss	Soil	Water	Air Pollution	Human Health
Conversion of grasslands, forests, and wetlands to crops or rangeland	Erosion	Aquifer depletion	Emissions of greenhouse gases CO_2 from fossil fuel use, N_2O from inorganic fertilizer use, and methane (CH_4) from cattle	Nitrates in drinking water (blue baby)
Fish kills from pesticide runoff	Loss of fertility	Increased runoff, sediment pollution, and flooding from cleared land	Other air pollutants from fossil fuel use and pesticide sprays	Pesticide residues in water, food, and air
Killing of wild predators to protect livestock	Salinization	Pollution from pesticides		Livestock wastes in drinking and swimming water
Loss of agrobiodiversity replaced by monoculture strains	Waterlogging	Algal blooms and fish kills caused by runoff of fertilizers and farm wastes		Bacterial contamination of meat
	Desertification			

FIGURE 12.14 Food production has a number of harmful environmental effects (**Concept 12.3**). ***Critical thinking:*** Which item in each of these categories do you think is the most harmful?

Left: Orientaly/Shutterstock.com. Left center: pacopi/Shutterstock.com. Center: Tim McCabe/USDA Natural Resources Conservation Service. Right center: Mikhail Malyshev/Shutterstock.com. Right: BBrown/Shutterstock.com.

FIGURE 12.15 Natural capital degradation: Flowing water from rainfall is the leading cause of topsoil erosion as seen on this farm in the U.S. state of Tennessee (left). Severe water erosion can become gully erosion, which has damaged this cropland in western Iowa (right).

Left: Tim McCabe/USDA Natural Resources Conservation Service. Right: © USDA Natural Resources Conservation Service.

According to a study by 27 experts assembled by the United Nations Environment Programme (UNEP), agriculture uses massive amounts of the world's resources and has a high environmental impact. It accounts for about 70% of the freshwater removed from aquifers and surface waters, worldwide. It also uses about 38% of the world's ice-free land, emits about 25% of the world's greenhouse gas emissions, and produces about 60% of all water pollution. As a result, many analysts view today's industrialized agriculture as environmentally and economically unsustainable. Let's look closer at the reasons for this conclusion.

Topsoil Erosion Is a Serious Problem

Topsoil, the fertile top layer of many soils (Figure 3.10, p. 56), is a potentially renewable resource and provides vital ecosystem services. Soil is literally the foundation of life on land (see p. 56). Specifically, the fertile top layer, or *topsoil,* is a vital component of natural capital, because it stores the water and nutrients that plants need and recycles these nutrients endlessly as long as they are not removed faster than natural processes replenish them. Thus, sustainable agriculture begins with promoting healthy soils.

One major problem related to agriculture is **soil erosion**—the movement of soil components, especially surface litter and topsoil (Figure 3.10) from one place to another by the actions of wind and water. Some topsoil erosion is natural, but much of it is caused by human activities, including agriculture.

Flowing water causes erosion of three types. *Sheet erosion* occurs on level land and removes thin layers, or sheets, of topsoil (Figure 12.15, left). *Rill erosion* is caused by tiny streams (rivulets) of water that carve small channels, or rills, in topsoil. *Gully erosion* is caused by larger streams of water removing enough soil to create gullies (Figure 12.15, right). The second largest cause of erosion is wind, which loosens and blows topsoil particles away, especially in areas with a dry climate and relatively flat land not covered by vegetation (Figure 12.16).

In undisturbed, vegetated ecosystems, the roots of plants help to anchor topsoil and to reduce erosion. However, topsoil can erode when soil-holding grasses, trees, and other vegetation are removed through activities such as farming (see Figure 7.17, p. 154), deforestation (see Figure 10.7, p. 226), and overgrazing (see Figure 10.11, p. 228).

Lynn Betts/USDA Natural Resources Conservation Service

FIGURE 12.16 Wind is an important cause of topsoil erosion in dry areas that are not covered by vegetation such as this bare crop field in the U.S. state of Iowa.

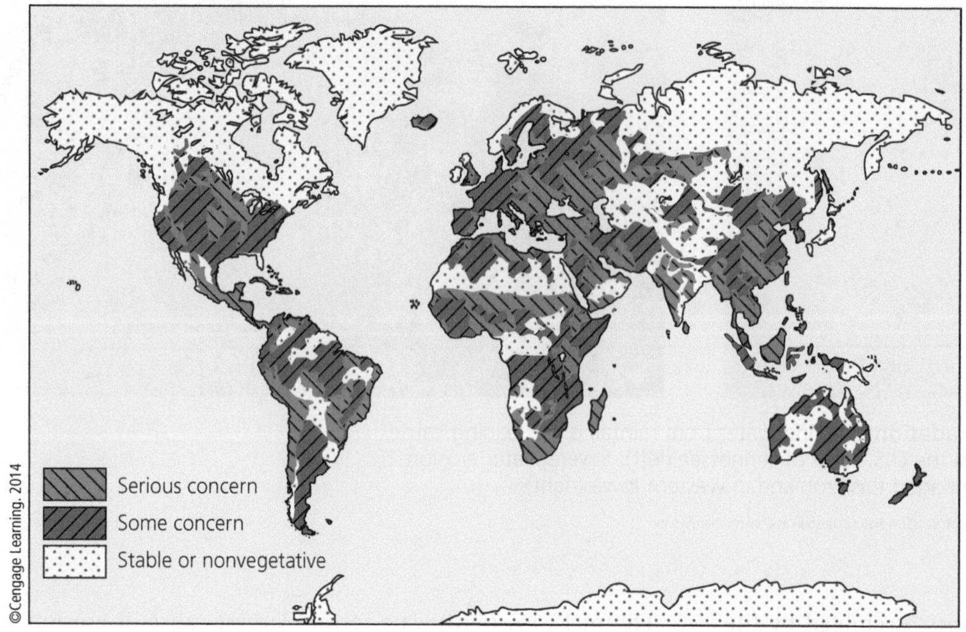

FIGURE 12.17 Natural capital degradation: Topsoil erosion is a serious problem in some parts of the world. ***Critical thinking:*** Can you see any geographical pattern associated with this problem?

(Compiled by the authors using data from the U.N. Environment Programme and the World Resources Institute.)

A joint survey by the UNEP and the World Resources Institute indicated that topsoil is eroding faster than it forms on about one-third of the world's cropland (Figure 12.17). Since 1970, partly because of soil erosion, about one-third of the world's cropland has been abandoned as too degraded for growing crops, according to a 2015 study by Duncan Cameron and other researchers.

The rise of industrialized agriculture has exposed large areas of cropland to severe topsoil erosion, which causes three major harmful effects:

- Loss of soil fertility through depletion of plant nutrients in topsoil.
- Water pollution in surface waters where eroded topsoil ends up as sediment and can cause eutrophication (Figure 8.17, p. 182) by overloading the water with plant nutrients.
- Release of carbon stored in the soil to into the atmosphere as CO_2, which contributes to atmospheric warming and climate change.

Soil pollution is also a problem in some parts of the world. Some of the chemicals emitted into the atmosphere by industrial and power plants and by motor vehicles can pollute soil and irrigation water. Pesticides can also contaminate soil. In 2014 China's environment ministry reported that an estimated 19% of China's arable (farmable) land is contaminated, especially with toxic metals such as cadmium, arsenic, and nickel, and that about 2.5% of the country's cropland is too contaminated to grow food safely.

Drought and Human Activities Are Degrading Drylands

Drylands in regions with arid and semiarid climates occupy about 40% of the world's land area and are home to some 2 billion people. A major threat to food security in some of these areas is **desertification**—the process in which the productive potential of topsoil falls by 10% or more because of a combination of prolonged drought and human activities that expose topsoil to erosion.

Desertification can be *moderate* (with a 10–25% drop in productivity), *severe* (with a drop of 25–50%), or *very severe* (with a drop of more than 50%, usually resulting in large gullies and sand dunes). Only in extreme cases does desertification lead to what we call desert.

Over thousands of years, the earth's deserts have expanded and contracted, primarily because of climate change. However, human use of the land has increased desertification in some parts of the world. Such uses can involve clearing trees, plowing excessively, and overgrazing (Figure 12.18), which have left much topsoil bare and unprotected.

In the 1930s, much of the topsoil in several dry and windy regions of the Midwestern United States was lost because of a combination of poor cultivation practices and prolonged drought. The resulting severe wind erosion led to crop failures and to the formation of a barren *dust bowl.* Thousands of farmers had to abandon their degraded land and move to other parts of the United States.

Researchers at the Earth Policy Institute have warned that overgrazing, overplowing, and deforestation are creating two new dust bowls. One is in the central African Sahel, a vast savanna-like area south of the Sahara Desert. The other straddles northern China and southern Mongolia. The researchers reported that about 90% of China's grasslands are degraded and suffering from desertification. Since 1950, more than 24,000 Chinese villages have been abandoned to spreading sands. And according to the Indian Space Research Organization, 24% of India's land is threatened by desertification.

Dick Ercken/Shutterstock.com

FIGURE 12.18 Severe soil erosion due to overgrazing by cattle has led to desertification.

Another way to degrade topsoil is to deplete it of key plant nutrients, either by repeatedly using it to grow crops (Figure 3.12, p. 58) or by allowing livestock to overgraze the land. Climate change could also contribute to this problem. Researchers have not tied any specific drought to climate change, but it is likely to prolong droughts and increase their severity, which would expand desertification and make some areas unfit for farming.

Excessive Irrigation Can Pollute Soil and Water

Irrigation accounts for about 70% of the water that humanity uses. Currently, the 16% of the world's cropland that is irrigated produces about 44% of the world's food.

However, irrigation has a downside. Most irrigation water is a dilute solution of various salts, such as sodium chloride, that are picked up as the water flows over or through soil and rocks. Irrigation water that is not absorbed into the topsoil evaporates and leaves behind a thin crust of dissolved mineral salts in the topsoil. Repeated applications of irrigation water in dry climates can lead to an accumulation of salts in the upper soil layers—a soil degradation process called **soil salinization.** It stunts crop growth, lowers crop yields, and can eventually kill plants and ruin the land.

The FAO estimates that severe soil salinization has reduced yields on at least 10% of the world's irrigated cropland, and that by 2020, 30% of the world's arable (farmable) land will be salty. The most severe salinization occurs in China, India, Egypt, Pakistan, Mexico, Australia, and Iraq. Salinization affects almost one-fourth of irrigated cropland in the United States, especially in western states (Figure 12.19).

Another problem with irrigation is **waterlogging,** in which water accumulates underground and gradually raises the water table. This can occur when farmers apply large amounts of irrigation water in an effort to reduce salinization by leaching salts deeper into the soil. Waterlogging lowers the productivity of crop plants and kills them after prolonged exposure because it deprives plants of the oxygen they need to survive. At least 10% of the world's irrigated land suffers from this worsening problem, according to the FAO.

USDA Natural Resources Conservation Service

FIGURE 12.19 Natural capital degradation: White alkaline salts have displaced crops that once grew on this heavily irrigated land in the U.S. state of Colorado.

Perhaps the biggest problem resulting from excessive irrigation in agriculture is that it has contributed to depletion of groundwater and surface water supplies in many areas of the world. We discuss this in Chapter 13.

Industrialized Crop Production Contributes to Pollution and Climate Change

Eroded topsoil flows as sediments into streams, lakes, and wetlands where it can smother fish and shellfish and clog irrigation ditches, boat channels, reservoirs, and lakes. This problem gets worse when the eroded sediment contains pesticide residues that can be ingested by aquatic organisms and in some cases biomagnified within food webs (see Figure 9.14, p. 205).

Farmers contribute to pollution through *overfertilizing* their fields. Globally, the use of fertilizers has grown 45-fold since 1940. Whether using natural fertilizers such as animal manure or synthetic fertilizers, excess fertilizers on a farm field can run off into waterways and contribute to eutrophication (see Figure 8.17, p. 182). Nitrates in fertilizer can also percolate down through the soil into aquifers where they can contaminate groundwater used for drinking. According to the FAO, fully one-third of all water pollution from the runoff of nitrogen and phosphorus is due to excessive use of synthetic fertilizers.

Agricultural activities cause a great deal of air pollution and account for more than a quarter of all human-generated emissions of carbon dioxide (CO_2). This is helping to warm the atmosphere and contributing to climate change, which is making some areas unsuitable for growing crops.

Producing Food and Biofuel Reduces Biodiversity

Natural biodiversity and some ecosystem services are threatened when forests are cleared and when croplands replace grasslands to produce food and biofuels (**Concept 12.3**).

For example, one of the fastest-growing threats to world's biodiversity is happening in Brazil. Large areas of tropical forest in its Amazon Basin and in cerrado—a huge tropical grassland region south of the Amazon Basin—are being lost. This land is being burned or cleared for cattle ranches, large plantations of soybeans grown for cattle feed (Figure 1.5, p. 8), and sugarcane used for making ethanol fuel for cars. Another example is in Indonesia, where tropical forests are burned to make way for plantations of oil palm trees (Figure 10.4, p. 225). The trees produce palm oil, which is increasingly used to produce biodiesel fuel for cars. Biodiversity is threatened in these and many other areas because tropical forests and grasslands host many more varieties of organisms than agricultural land can.

A related problem is the increasing loss of **agrobiodiversity**—the genetic variety of animal and plant species used on farms to produce food. Scientists estimate that since 1900, the world has lost 75% of the genetic diversity of agricultural crops. For example, India once planted 30,000 varieties of rice. Now more than 75% of its rice production comes in only 10 varieties, and soon most of its production might come from just one or two varieties. In the United States, about 97% of the food plant varieties available to farmers in the 1940s no longer exist, except perhaps in small amounts in seed banks and occasional home gardens (see Figure 5, Supplement 5, p. S44, in Supplement 5).

Traditionally, farmers have saved seeds from year to year to save money and to have the ability to grow food in times of famine. Families in India and most other less-developed countries still do this, but in the United States, this tradition is disappearing as more farmers plant seeds for genetically engineered crops. Companies that sell these seeds have patents on them, forbid users to save them, and have successfully sued a number of farmers who saved such seeds. In effect, farmers pay seed companies to use or lease their seeds for a year. In 2015, seed companies made $15.3 billion selling these seeds. Ten multinational companies have patents on 65% of the commercial seeds used to plant major crops, and four companies control 43% of the seed supply.

In losing agrobiodiversity, ecologists warn that farming practices are rapidly shrinking the world's genetic "library" of plant varieties, which are critical for increasing food yields. This failure to preserve agrobiodiversity is a serious violation of the biodiversity **principle of sustainability** that could reduce the sustainability of food production (**Concept 12.3**).

Efforts are being made to save endangered varieties of crops and wild plant species important to the world's food supply. Individual plants and seeds from endangered crop varieties and wild plant species are stored in about 1,400 refrigerated seed banks. They are also stored in agricultural research centers and botanical gardens scattered around the world. However, power failures, fires, storms, wars, and unintentional disposal of seeds can cause irreversible losses of these stored plants and seeds. The world's most secure seed bank is the underground Svalbard Global Seed Vault, the so-called "doomsday seed vault," which was carved into the Arctic permafrost on a frozen Norwegian arctic island near the North Pole (Figure 12.20). It is being stocked with duplicates of much of the world's seed collections.

However, the seeds of many plants cannot be stored successfully in seed banks. Because stored seeds do not remain alive indefinitely, they must be planted and germinated periodically, and new seeds must be collected for storage. Unless this is done regularly, seed banks become seed morgues.

Jim Richardson/National Geographic Creative

FIGURE 12.20 The Svalbard Global Seed Vault is the most secure seed bank in the world.

There Is Controversy over Genetically Engineered Foods

Although genetic engineering could help to improve food security, controversy has arisen over the use of this technology. Perhaps the most worrisome problem for many is that, while many of us are consuming genetically modified (GM) foods daily, we know little about their long-term health effects. For example, one type of GMO makes use of *bacillus thuringiensis*, a natural soil bacterium with a gene, known as *Bt*, that produces a chemical toxic to some insects. This gene has been inserted into corn plants which then incorporate the Bt toxin in their leaves, making them resistant to damage by certain insects. The *Global Citizens' Report on the State of GMOs* summarized findings indicating that GM crops with Bt toxins could threaten human health by triggering an inflammatory response leading to diseases such as diabetes and heart disease.

Another problem is that promised yield increases from the use of GM crops have not materialized. Studies have shown small increases during the first years of such use with generally plateauing or falling yields thereafter.

In the soil and in other parts of the biosphere, GM crops could threaten biodiversity. Some GM crops are designed to enable the use of herbicides, which could be part of the reason for declining populations of Monarch butterflies (see Case Study, p. 93) and other pollinators. With declines in pollinators will come declines in the plant communities that depend on them and the animals that depend on those plants. The end effect could be a cascade of biodiversity losses.

Critics also point out that if GM crops or seeds released into the environment caused some harmful genetic or ecological effects, as some scientists project, those organisms could not be recalled. Also, genes in plant pollen from genetically engineered crops have been known to spread to non–genetically engineered species. This could result in hybrids with wild crop varieties, which could reduce the natural genetic biodiversity of the wild strains. This could in turn reduce the gene pool from which new species can evolve or be engineered—a violation of the biodiversity **principle of sustainability**.

With some GM crops, it is necessary to use herbicides, which can put farmers on a costly treadmill, using larger amounts or more toxic herbicides as weeds become resistant to them. One University of Minnesota study found that farmers who cultivated GM crop varieties earned less money over a 14-year period than those who continued to grow non-GM crops.

Some 64 countries require that food labels identify genetically modified food content. Polls consistently indicate that around 90% of U.S. consumers want to have such information clearly listed on food labels. In 2016, Congress passed a law requiring GMO content labels. However, it allows food manufacturers three choices: a symbol, a label, or a digital bar code enabling buyers to read the labels on their smart phones. Many consumers are opposed to having to use a smart phone to scan a bar code to get this information because one-third of the U.S. population—many of them low-income, minority, and elderly—do not have smart phones. Also, most shoppers are not likely to take the time to scan bar codes for all of the items they buy.

The Ecological Society of America and various critics of genetically engineered crops call for more controlled field experiments and testing, to better understand the long-term ecological and health risks of using GM crops. They also want stricter regulation of this rapidly growing technology.

CONSIDER THIS . . .

CONNECTIONS GM Crops and Organic Food Prices

The spread of GM crop genes by wind carrying pollen from field to field threatens the production of certified organic crops, which must be grown without such genes to be classified as organic. Because organic farmers have to perform expensive tests to detect GMOs or take costly planting measures to prevent the spread of GMOs to their fields from nearby crop fields, they have sometimes had to raise the prices of their produce.

Figure 12.21 summarizes the major potential benefits and drawbacks of GMOs.

There Are Limits to Expanding Green Revolutions

So far, several factors have limited the success of the green revolutions and may limit them more in the future (**Concept 12.3**). For example, without large inputs of water and synthetic inorganic fertilizers and pesticides, most green revolution and genetically engineered crop varieties produce yields that are no higher (and are sometimes lower) than those from traditional strains. These high inputs also cost too much for most subsistence farmers in less-developed countries.

Trade-Offs

Genetically Modified Crops and Foods

Potential Benefits

- May need less fertilizer, pesticides, and water
- Can be resistant to insects, disease, frost, and drought
- Can grow faster and could raise yields
- May tolerate higher levels of herbicides
- Could have longer shelf life

Possible Drawbacks

- Have unpredictable genetic and ecological effects
- May put toxins in food
- Could repel or harm pollinators
- Can promote pesticide-resistant insects, herbicide-resistant weeds, and plant diseases
- Could disrupt seed market and reduce biodiversity

FIGURE 12.21 Use of genetically modified crops and foods has advantages and disadvantages. ***Critical thinking:*** Which two advantages and which two disadvantages do you think are the most important? Why?

Top: Lenar Musin/Shutterstock.com. Bottom: oksix/Shutterstock.com.

Scientists point out that where such inputs do increase yields, there comes a point where yields stop increasing because of the inability of crop plants to take up nutrients from additional fertilizer and irrigation water. This helps to explain the slowdown in the rate of growth in global yields for most major crops since 1990.

Can we expand the green revolutions by irrigating more cropland? Since 1978, the amount of irrigated land per person has been declining, and it is projected to fall much more by 2050. One reason for this is population growth, which is projected to add 2.5 billion more people between 2015 and 2050. Other factors are limited availability of irrigation water, soil salinization, and the fact that most of the world's conventional farmers do not have enough money to irrigate their crops.

Climate change is expected to affect crop yields during this century. Several studies, including one from the International Rice Research Institute, indicate that for every 1°C increase in the global average temperature, wheat, rice, and corn yields will all shrink by 10%. Also, mountain glaciers that provide irrigation for many millions of people in China, India, and South America are melting and this will lessen the area of crops that can be irrigated. During this century, fertile croplands in coastal areas, including many of the major rice-growing floodplains and river deltas in Asia, are likely to be flooded by rising sea levels resulting from climate change. Food production could also drop sharply in some major food-producing areas because of longer and more intense droughts and heat waves, also likely resulting from climate change.

Can we increase the food supply by cultivating more land? Farming already uses about 38% of the world's ice-free land surface for croplands and pastures. Clearing more tropical forests and irrigating arid land could more than double the area of the world's cropland. However, massive clearing of forests would decrease biodiversity, speed up climate change and its harmful effects, and increase soil erosion. Also, much of this land has poor soil fertility, steep slopes, or both, and cultivating such land would be expensive and probably not ecologically sustainable.

Crop yields could be increased with the use of conventional or GM crops that are more tolerant of drought, which are still being tested. Commercial fertilizers have played a role in green revolutions, but their use in more-developed countries has reached a level of diminishing returns in terms of increased crop yields. However, there are parts of the world, especially in Africa, where additional fertilizer could boost crop production.

Organic Farming Has Some Drawbacks

Organic crop production is not without its critics. One potential problem is the leaching of nitrate into groundwater from composted manure used as a natural fertilizer. If it is not applied correctly or is applied too densely, this leaching could be a problem. The same is true (to a greater extent as some studies indicate) with synthetic fertilizers applied to conventional crops.

Large-scale composting generates greenhouse gases, especially methane and nitrous oxide. The same is true of synthetic fertilizer use. Different studies report varying comparative amounts of emissions and again, they depend on how the natural or synthetic fertilizers are applied. This is a topic that needs further research.

Another drawback is that, in place of herbicides, some organic farmers resort to plowing to control weeds. This damages the soil and can lead to soil erosion and loss of soil nutrients. The same problems occur on many conventional farms.

To deal with this issue, conventional and organic farmers have developed **no-till farming** (also known as *conservation tillage farming*), a technique in which farmers do not plow the soil and inject seeds into untilled soil using special machines. Weeds are controlled by use of herbicides or some other method that does not disturb the soil. Researchers at the Rodale Institute in Kutztown, Pennsylvania (USA), have developed an organic no-till system involving a dense cover crop and a roller-crimper that flattens and kills the cover crop plants to enhance their mulching effect. This is a way to prevent weed growth while using no-till agriculture without herbicides.

Figure 12.22 summarizes the benefits and drawbacks of organic crop production.

Trade-Offs

Organic Farming

Advantages	Disadvantages
Reduces soil erosion	More use of manure can cause more surface and groundwater pollution
Retains more water in soil during drought years	Large-scale composting can generate greenhouse gases
Improves soil fertility	Lower crop yields in large-scale production
Uses less energy and emits less CO_2	Can require plowing for weed control leading to more erosion
Eliminates water pollution from pesticides and synthetic fertilizers	Higher cost can lead to higher prices
Benefits birds, bats, bees, and other wildlife	

FIGURE 12.22 Organic farming has advantages and disadvantages. ***Critical thinking:*** Which single advantage and which single disadvantage do you think are the most important? Why?

Top: Robert Kneschke/Shutterstock.com Bottom: Marbury67/Dreamstime.com

Industrialized Meat Production Harms the Environment

Proponents of industrialized meat production point out that it has increased meat supplies, reduced overgrazing, and kept food prices down. But feedlots and concentrated animal feeding operations (CAFOs) produce widespread harmful health and environmental effects. Analysts point out that meat produced by industrialized agriculture is artificially cheap because most of its harmful environmental and health costs are not included in the market prices of meat and meat products, a violation of the full-cost pricing **principle of sustainability**.

CAFOs and feedlots are truly industrial-scale livestock operations. A large feedlot contains 1,000 or more cattle (Figure 12.10) and large CAFOs can house 2,500 or more hogs or 125,000 or more chickens or turkeys (Figure 12.11).

A major problem with feedlots and CAFOs is that huge amounts of water are used to irrigate the grain crops that feed the livestock. According to waterfootprint.org, producing a quarter-pound hamburger requires 1,700 liters (450 gallons) of water—the equivalent of 15 to 20 showers for the average person. It takes more water to raise a given amount of beef than it takes to raise the same quantities of pork, chicken, eggs, and milk combined.

Much of this irrigation water is used inefficiently and is helping to deplete groundwater supplies in many areas of the world. Large volumes of water are also used to wash away livestock wastes. Much of this wastewater then flows into streams and other waterways and pollutes those aquatic ecosystems.

According to the USDA, animal waste produced by the American meat industry amounts to about 67 times the amount of waste produced by the country's human population. Ideally, manure from CAFOs should be returned to the soil as a nutrient-rich fertilizer in keeping with the chemical cycling **principle of sustainability**. However, it is often so contaminated with residues of antibiotics and pesticides that it is unfit for use as a fertilizer.

Despite potential contamination, up to half of the manure slurry from CFOs in the United States is applied to nearby fields and creates severe odor problems for people living nearby. Much of the other half of CAFO animal waste is pumped into large lagoons, which eventually leak and pollute nearby surface and groundwater, produce foul odors, and emit large quantities of climate-changing greenhouse gases into the atmosphere.

Industrialized meat production uses large amounts of energy (mostly from oil), which is another source of air pollution and greenhouse gases. Also, as part of their digestion process, cattle and dairy cows release methane (CH_4), a greenhouse gas with about 25 times the warming potential of CO_2 per molecule. According to the FAO study, *Livestock's Long Shadow*, industrialized livestock production generates about 18% of the world's greenhouse gases—more than all of the world's cars, trucks, buses, and planes combined. The entire meat production process contributes 35–40% of the global annual emissions of methane and 65% of all emissions of nitrous oxide (N_2O), a gas with about 300 times the atmospheric warming capacity of CO_2 per molecule.

Another growing problem is the use of antibiotics in industrialized livestock production facilities. The U.S. Food and Drug Administration (FDA) and the Union of Concerned Scientists have estimated that about 80% of all antibiotics used in the United States (and 50% of those in the world) are added to animal feed. This is done to try to prevent the spread of diseases in feedlots and CAFOs and to promote the growth of the animals. This heavy antibiotics use plays a role in the rise of genetic resistance among many disease-causing bacteria (see Figure 4.14, p. 87), which is resulting in new, more genetically resistant infectious disease organisms, some of which can infect humans.

Figure 12.23 summarizes the advantages and disadvantages of using feedlots and CAFOs.

When livestock are grass-fed, environmental impacts can still be high, especially when forests are cut down or burned to make way for grazing land, as is done in Brazil's Amazon forests. According to an FAO report, overgrazing and erosion by livestock has degraded about 20% of the world's grasslands and pastures. The same report estimated that rangeland grazing and industrialized livestock production has caused about 55% of all topsoil erosion and sediment pollution.

Trade-Offs

Feedlots and CAFOs

Advantages	Disadvantages
Increased meat production	Animals unnaturally confined and crowded
Higher profits	Large inputs of grain, fishmeal, water, and fossil fuels
Less land use	Greenhouse gas (CO_2 and CH_4) emissions
Reduced overgrazing	Concentration of animal wastes that can pollute water
Reduced soil erosion	Use of antibiotics can increase genetic resistance to microbes in humans
Protection of biodiversity	

FIGURE 12.23 Use of animal feedlots and confined animal feeding operations has advantages and disadvantages. ***Critical thinking:*** Which single advantage and which single disadvantage do you think are the most important? Why?

Top: Mikhail Malyshev/Shutterstock.com. Bottom: Maria Dryfhout/Shutterstock.com.

In addition, grass-fed cows emit more of the powerful greenhouse gas methane than do grain-fed cows. Thus, expanding grass-fed production could increase the agricultural contributions to deforestation and climate change.

CONSIDER THIS . . .

CONNECTIONS Meat Production and Aquatic Biodiversity

Synthetic fertilizers are used in the Midwestern United States to produce corn for animal feed and ethanol fuel for cars. Much of this fertilizer washes from cropland and into the Mississippi River, which empties into the Gulf of Mexico. The added nitrate and phosphate nutrients overfertilize the gulf's coastal waters. Each year this leads to oxygen depletion in the gulf's waters, which threatens one-fifth of the nation's seafood yield. In other words, growing corn in the Midwest, largely to feed cattle and fuel cars, degrades aquatic biodiversity and seafood production in the Gulf of Mexico.

Aquaculture Can Harm Aquatic Ecosystems

Aquaculture produces about 42% of the world's seafood and is growing rapidly. The World Bank projected that by 2030, aquaculture will produce 62% of all seafood. However, some analysts warn that the harmful environmental effects of aquaculture could limit its future production potential (**Concept 12.3**).

A major environmental problem associated with aquaculture is that about a third of the wild fish caught from the oceans are used to make the fishmeal and fish oil that are fed to farmed fish and livestock. This is contributing to the depletion of many populations of wild fish that are crucial to marine food webs. Aquaculture now uses 70% of all fishmeal and nearly 90% of all fish oil. To satisfy the growing demand for these products, some countries are scooping up thousands of tons of krill from Antarctic waters every year. Krill form the base of the Antarctic food web (Figure 3.16, p. 60) and thus all species, including endangered penguins and whales, depend on them.

Another problem is that some of the fishmeal and fish oil fed to farm-raised fish is contaminated with long-lived toxins such as PCBs and dioxins that are picked up from the ocean floor. This can contaminate farm-raised fish and people who eat such fish. Some fish farms that raise carnivorous fish such as salmon and tuna also produce large amounts of wastes that pollute aquatic ecosystems and fisheries.

Another source of pollution is the use of pesticides and antibiotics on fish farms. In a study by the Norwegian Institute for Water Research, toxic pesticides were found at unhealthy levels in 40–67% of samples tested. Also, farmed species such as shrimp, salmon, tilapia, and trout have been found to contain five different antibiotics. Use of antibiotics on fish farms presents the same problems as it does in CAFOs and feedlots.

Trade-Offs

Aquaculture

Advantages	Disadvantages
High efficiency	Large inputs of land, grain, and fishmeal
High yield	Large waste output
Reduces over-harvesting of fisheries	Loss of mangrove forests and estuaries
Jobs and profits	Dense populations vulnerable to disease

© Cengage Learning

FIGURE 12.24 Use of aquaculture has advantages and disadvantages. ***Critical thinking:*** Which single advantage and which single disadvantage do you think are the most important? Why?

Top: Vladislav Gajic/Shutterstock.com. Bottom: Nordling/Shutterstock.com.

Aquaculture can also destroy or degrade aquatic ecosystems, particularly mangrove forests (Figure 8.8, p. 172). Since aquaculture began growing rapidly in 1990, more than 500,000 hectares (1.2 million acres) of highly productive mangrove forest have been cleared for shrimp farms, according to the Earth Policy Institute. This represents a loss of biodiversity and valuable ecosystem services such as natural flood control in these sensitive coastal areas that are expected to experience severe flooding because of rising sea level caused by climate change.

Figure 12.24 lists the major benefits and drawbacks of aquaculture.

12.4 HOW CAN WE PROTECT CROPS FROM PESTS MORE SUSTAINABLY?

CONCEPT 12.4 We can sharply cut pesticide use without decreasing crop yields by using a mix of cultivation techniques, biological pest controls, and small amounts of selected chemical pesticides as a last resort (integrated pest management).

Nature Controls the Populations of Most Pests

A **pest** is any species that interferes with human welfare by competing with us for food, invading our homes, lawns, or gardens, destroying building materials, spreading disease, invading ecosystems, or simply being a nuisance. Worldwide, only about 100 species of plants (weeds), animals (mostly insects), fungi, and microbes cause most of the damage to the crops we grow.

In natural ecosystems and in many polyculture crop fields, *natural enemies* (predators, parasites, and disease organisms) control the populations of most potential pest species. This free ecosystem service is an important part of the earth's natural capital. For example, biologists estimate that the world's 40,000 known species of spiders kill far more crop-eating insects every year than humans kill with insecticides. Most spiders, including the wolf spider (Figure 12.25), do not harm humans.

When we clear forests and grasslands, plant monoculture crops, and douse fields with chemicals that kill pests, we upset many of these natural population checks and balances that are in keeping with the biodiversity **principle of sustainability**. Then we must devise ways to protect our monoculture crops, tree plantations, lawns, and golf courses from pests that nature has helped to control at no charge.

Synthetic Pesticides Can Help Control Pest Populations

We have developed a variety of **synthetic pesticides**—chemicals used to kill or control organisms we consider to be pests. They are classified as *insecticides* (insect killers), *herbicides* (weed killers), *fungicides* (fungus killers), and *rodenticides* (rat and mouse killers).

Cathy Keifer/Shutterstock.com

FIGURE 12.25 Natural capital: This ferocious-looking wolf spider is eating a grasshopper. It is one of many important insect predators that can be killed by some pesticides.

We did not invent the use of chemicals to repel or kill other species. For nearly 225 million years, plants have been producing chemicals to ward off, deceive, or poison the insects and herbivores that feed on them. Scientists have used such chemicals to create *biopesticides* to kill some pests.

CONSIDER THIS . . .

LEARNING FROM NATURE

In the 1600s, farmers used nicotine sulfate, extracted from tobacco leaves, as an insecticide. Eventually, other *first-generation pesticides*—mainly natural chemicals taken from plants—were developed. Farmers were copying nature's solutions to deal with their pest problems.

A major pest control revolution began in 1939, when entomologist Paul Müller discovered DDT (dichlorodiphenyltrichloroethane)—the first of the so-called *second-generation pesticides* produced in the laboratory. It soon became the world's most-used pesticide, and since then, chemists have created hundreds of other pesticides.

Some second-generation pesticides have turned out to be highly hazardous for birds and other forms of wildlife. In 1962 biologist Rachel Carson (Figure 1.18, p. 22) published her famous book *Silent Spring*, sounding a warning that eventually led to strict controls on the use of DDT and several other widely used pesticides.

Since 1950, synthetic pesticide use has increased more than 50-fold and most of today's pesticides are 10 to 100 times more toxic to pests than those used in the 1950s. Some synthetic pesticides, called *broad-spectrum agents*, can be toxic to beneficial species as well as to pests. Examples are organochlorine compounds (such as DDT), organophosphates (such as malathion and parathion), carbamates, pyrethroids, and neonicotinoids (which have increasingly been linked to the serious decline in honeybee populations). Others, called *selective*, or *narrow-spectrum, agents*, are each effective against a narrowly defined group of organisms. One example is glyphosate, used on corn and soybean crops that are genetically modified to withstand its toxic effects. This widely used herbicide efficiently kills weeds without hurting the crops.

50-fold

The increase in synthetic pesticide use since 1950

Pesticides vary in their *persistence*, the length of time they remain deadly in the environment. Some, such as DDT and related compounds, remain in the environment for years and can be biologically magnified in food chains and food webs (Figure 9.14, p. 205). Others, such as organophosphates, are active for days or weeks and are not biologically magnified but can be highly toxic to humans.

About one-fourth of the pesticides used in the United States are aimed at ridding houses, gardens, lawns, parks, and golf courses of insects and other pests. According to the U.S. Environmental Protection Agency (EPA), the amount of synthetic pesticides used on the average U.S. homeowner's lawn is 10 times the amount (per unit of land area) typically used on U.S. croplands.

Benefits of Synthetic Pesticides

Use of synthetic pesticides has its advantages and disadvantages. Proponents contend that the benefits of pesticides (Figure 12.26, left) outweigh their harmful effects (Figure 12.26, right). They point to the following benefits:

- *They have saved human lives.* Since 1945, DDT and other insecticides probably have prevented the premature deaths of at least 7 million people (some say as many as 500 million) from insect-transmitted diseases such as malaria (carried by the *Anopheles* mosquito), bubonic plague (carried by rat fleas), and typhus (carried by body lice and fleas).
- *They can increase food supplies* by reducing food losses due to pests, for some crops.
- *They can help farmers to control soil erosion and build soil fertility.* In conventional no-till farming, farmers apply herbicides instead of weeding the soil by plowing. This dramatically reduces soil erosion and soil nutrient depletion.
- *They can help farmers to reduce costs and increase profits.* The costs of using pesticides can be regained, at least in the near term, through higher crop yields. Glyphosate

Trade-Offs

Synthetic Pesticides

Advantages	Disadvantages
Expand food supplies	Promote genetic resistance
Raise profits	Can kill pests' natural enemies and harm wildlife and people
Work fast	Can pollute air, water, and land
Are safe if used properly	Are expensive for farmers

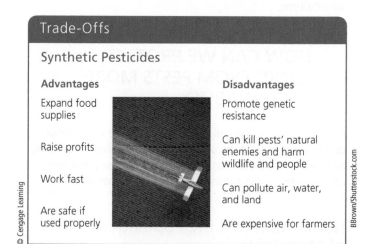

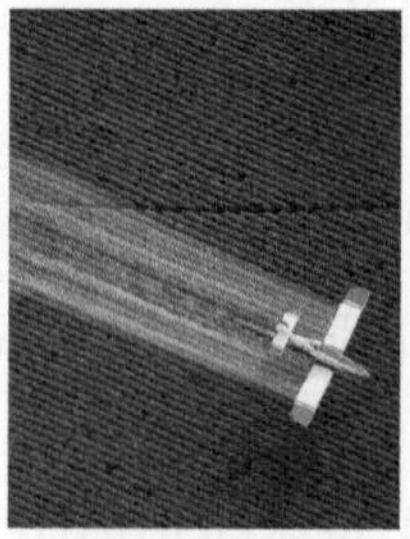

© Cengage Learning

BBrown/Shutterstock.com

FIGURE 12.26 Use of synthetic pesticides has advantages as well as disadvantages. ***Critical thinking:*** Which single advantage and which single disadvantage do you think are the most important? Why?

and other herbicides help farmers kill weeds efficiently, sparing them the costs of mechanical weeding.

- *They work fast.* Pesticides control most pests quickly, have a long shelf life, and are easily shipped and applied.
- *Newer pesticides are safer to use and more effective than many older ones.* Greater use is being made of biopesticides derived from plants, which are generally safer to use and less damaging to the environment than are many older pesticides.

Problems with Synthetic Pesticides

Opponents of widespread use of synthetic pesticides contend that the harmful effects of these chemicals (Figure 12.26, right) outweigh their benefits (Figure 12.26, left). They cite several problems.

- *They accelerate the development of genetic resistance to pesticides in pest organisms* (Figure 12.27). Insects breed rapidly, and within 5 to 10 years (much sooner in tropical areas), they can develop immunity to widely used pesticides through natural selection. In the same way, weeds develop resistance to herbicides. Since 1945, about 1,000 species of insects and rodents (mostly rats) have developed genetic resistance to one or more pesticides. By 2015, the International Survey of Herbicide-Resistant Weeds had identified more than 450 species of *superweeds*—weeds resistant to glyphosate and other herbicides.
- *They can put farmers on a financial treadmill.* Because of genetic resistance, farmers can find themselves having to pay more and more for a chemical pest control program that can become less and less effective as pests develop resistance to the pesticides.
- *Some insecticides kill natural predators and parasites that help to control the pest populations.* About 100 of the 300 most destructive insect pests in the United States were minor pests until widespread use of insecticides wiped out many of their natural predators. (See the Case Study that follows.)
- *Some pesticides harm wildlife.* According to the USDA and the U.S. Fish and Wildlife Service, each year, some of the pesticides applied to crops poison honeybee colonies on which we depend for pollination of many food crops (see Chapter 9, Core Case Study, p. 192, and Science Focus 9.2, p. 206). According to a study by the Center for Biological Diversity, pesticides menace about a third of all endangered and threatened species in the United States.
- *Pesticides that are applied inefficiently can pollute the environment.* According to the USDA, about 98–99.9% of the insecticides and more than 95% of the herbicides applied by aerial spraying or ground spraying do not reach the target pests. They end up in the air, surface water, groundwater, bottom sediments, food, and nontarget organisms, including humans.
- *Some pesticides threaten human health.* The WHO and UNEP have estimated that pesticides annually poison at least 3 million agricultural workers in less-developed countries. The EPA estimates that each year, pesticides poison 10,000 to 20,000 farmworkers in the United States. Household pesticides such as ant and roach sprays sicken another 2.5 million people per year. According to studies by the National Academy of

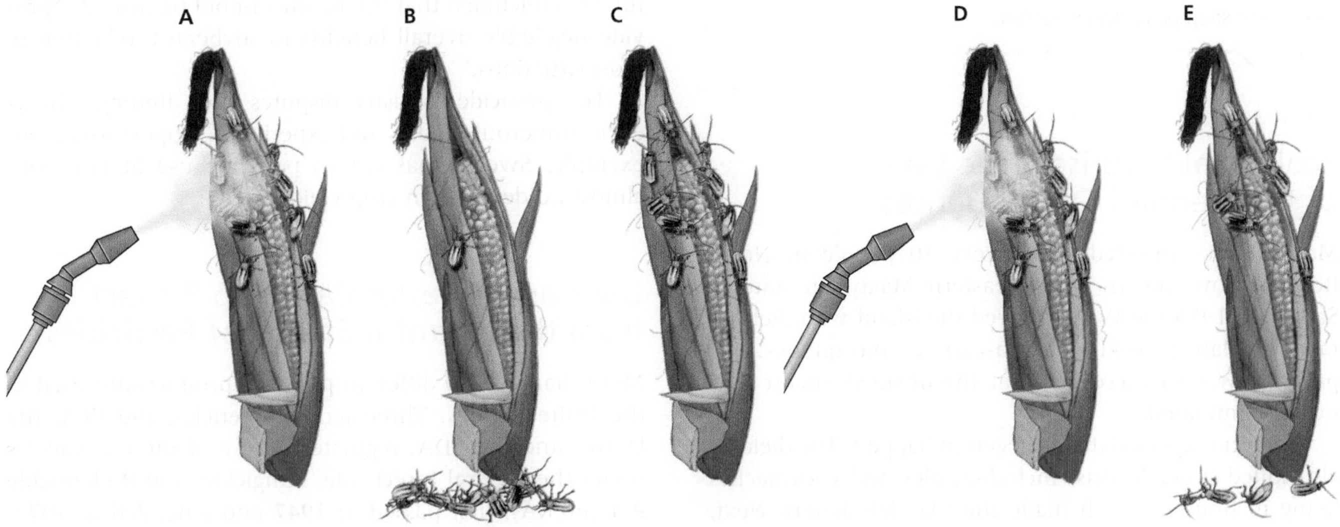

FIGURE 12.27 When a pesticide is sprayed on a crop **(a)**, a few pest insects resist it and survive **(b)**. The survivors reproduce and pass on their trait for resistance to the pesticide **(c)**. When the crop is sprayed again **(d)**, more insects resist and survive it and continue reproducing **(e)**. The pesticide has now become ineffective and the farmer must look for a stronger pesticide.

What Can You Do?

Reducing Exposure to Pesticides

- Grow some of your food using organic methods
- Buy certified organic food
- Wash and scrub all fresh fruits and vegetables
- Eat less meat, no meat, or certified organically produced meat
- Before cooking, trim the fat from meat

FIGURE 12.28 Individuals matter: You can reduce your exposure to pesticides. ***Critical thinking:*** Which three of these steps do you think are the most important ones to take? Why?

Sciences, pesticide residues in food cause an estimated 4,000–20,000 cases of cancer per year in the United States. In 2015 the WHO classified glyphosate as "probably carcinogenic to humans." Glyphosate-based herbicides have been shown to cause DNA damage, infertility, low sperm count, and prostrate or testicular cancer in rats. The pesticide industry disputes these claims, arguing that if used as directed, pesticides do not remain in the environment at levels high enough to cause serious environmental or health problems. Figure 12.28 lists some ways to reduce your exposure to synthetic pesticides.

CONSIDER THIS . . .

CONNECTIONS Pesticides and Food Choices

The Environmental Working Group (EWG) produces an annual list of fruits and vegetables that tend to have the highest pesticide residues. In 2015 these foods (EWG's "dirty dozen") were apples, peaches, nectarines, strawberries, grapes, celery, spinach, sweet bell peppers, cucumbers, cherry tomatoes, imported snap peas, and potatoes.

CASE STUDY

Ecological Surprises: The Law of Unintended Consequences

Malaria once infected 9 of every 10 people in North Borneo, now known as the eastern Malaysian state of Sabah. In 1955 the WHO sprayed the island with dieldrin (a DDT relative) to kill malaria-carrying mosquitoes. The program was so successful that the dreaded disease was nearly eliminated.

Then unexpected things began to happen. The dieldrin also killed other insects, including flies and cockroaches living in houses, which made the islanders happy. Next, small insect-eating lizards living in the houses died after gorging themselves on dieldrin-contaminated insects. Then cats began dying after feeding on the lizards. In the absence of cats, rats flourished in and around the villages. When the residents became threatened by sylvatic plague carried by rat fleas, the WHO parachuted healthy cats onto the island to help control the rats. Operation Cat Drop worked.

But then the villagers' roofs began to fall in. The dieldrin had killed wasps and other insects that fed on a type of caterpillar that was not affected by the insecticide. With most of its predators eliminated, the caterpillar population exploded, munching its way through its favorite food: the leaves used in thatch roofs.

Ultimately, this story ended well. Both malaria and the unexpected effects of the spraying program were brought under control. Nevertheless, this chain of unintended and unforeseen events reminds us that whenever we intervene in nature and affect organisms that interact with one another, we need to ask, "Now what will happen?"

Pesticide Use Has Not Consistently Reduced U.S. Crop Losses to Pests

Largely because of genetic resistance and the loss of many natural predators, synthetic pesticides have not always succeeded in reducing U.S. crop losses. David Pimentel, an expert on insect ecology, evaluated data from more than 300 agricultural scientists and economists. He found that between 1942 and 1997, estimated crop losses from insects almost doubled from 7% to 13%, despite a 10-fold increase in the use of synthetic insecticides. He also estimated that alternative pest management practices could cut the use of synthetic pesticides by half on 40 major U.S. crops without reducing crop yields (**Concept 12.4**).

In 2014 the EPA evaluated the use of three controversial neonicotinoids for insect control on U.S. soybean crops. It found that there was no difference in soybean yields between treated and untreated crops. The report further concluded that the neonicotinoid treatments "provide negligible overall benefits to soybean production in most situations."

The pesticide industry disputes such findings. However, numerous studies and experience support them. For example, Sweden has cut its pesticide use in half with almost no decrease in crop yields.

Laws and Treaties Can Help Protect Us from the Harmful Effects of Pesticides

More than 20,000 different pesticide products are used in the United States. Three federal agencies, the EPA, the USDA, and the FDA, regulate the use of these pesticides under the Federal Insecticide, Fungicide, and Rodenticide Act (FIFRA), first passed in 1947 and amended in 1972. Critics argue that that FIFRA has not been well enforced, and the EPA says that the U.S. Congress has not provided them with enough funds to carry out the complex and lengthy process of evaluating pesticides for toxicity.

As of 2015, the United States was continuing to use at least five highly toxic pesticides that were banned in many other countries. Neonicotinoids were banned in the European Union (EU) in 2013. Paraquat, a pesticide linked to Parkinson's disease, has been banned in the EU and China. The EU is also phasing out 1,3-D, or Telone, a chemical used to sterilize soil. It poses a cancer risk to humans and animals. Glyphosate is banned in the Netherlands, Ontario, Canada, and Sri Lanka, and Brazil is considering a ban. The popular herbicide atrazine has been banned in the EU for over 11 years. It persists in water and has been found in about 88% of all U.S. drinking water, according to an analysis of USDA data by the Pesticide Action Network. It has also been linked to hormone and immune system disorders and birth defects.

In 1996 Congress passed the Food Quality Protection Act, mostly because of growing scientific evidence and citizen pressure concerning the effects of small amounts of pesticides on children. This act requires the EPA to reduce the allowed levels of pesticide residues in food by a factor of 10 when there is inadequate information on the potentially harmful effects on children.

Between 1972 and 2015, the EPA used FIFRA to ban or severely restrict the use of 64 active pesticide ingredients, including DDT and most other chlorinated hydrocarbon insecticides. However, according to studies by the National Academy of Sciences, federal laws regulating pesticide use generally are inadequate and poorly enforced. A 2015 study by the U.S. General Accounting Office found that the FDA tests less that one-tenth of 1% of all imported fruits and vegetables. Also, the FDA does not test foods for some pesticide residues that are strictly regulated by the EPA.

Although laws within countries protect citizens to some extent, banned or unregistered pesticides may be manufactured in one country and exported to other countries. For example, U.S. pesticide companies make and export to other countries pesticides that have been banned or severely restricted—or never evaluated—in the United States. Other industrial countries also export banned or unapproved pesticides.

However, in what environmental scientists call a *circle of poison*, or the *boomerang effect*, residues of synthetic pesticides that have been banned in one country but exported to other countries can return to the exporting countries on imported food. Winds can also carry persistent pesticides from one country to another.

In 2000 more than 100 countries developed an international agreement to ban or phase out the use of 12 especially hazardous persistent organic pollutants (POPs)—9 of them persistent hydrocarbon pesticides such as DDT and other chemically similar pesticides. By 2015, the initial list of 12 chemicals had been expanded to 25. In 2004 the POPs treaty went into effect—a legal application of the precautionary principle (see p. 215). By 2015, it had been signed or ratified by 179 countries, not including the United States.

CONSIDER THIS . . .

THINKING ABOUT Exporting Pesticides

Should companies be allowed to export synthetic pesticides that have been banned, severely restricted, or not approved for use in their home countries? Explain.

There Are Alternatives to Synthetic Pesticides

Experience and research indicate that **(1)** pesticides are not necessary to feed the world, **(2)** pesticides are not rigorously tested for safety, **(3)** GMOs increase reliance on herbicides, and **(4)** there are alternatives to conventional synthetic pesticides. Many scientists urge us to reduce the use of pesticides by turning to biological, ecological, and cultivation tools for controlling pests (**Concept 12.4**).

One important *biological control* involves using natural predators (Figure 12.25), parasites (Figure 12.29), and disease-causing bacteria and viruses to help regulate pest populations. This approach is usually nontoxic to other species and less costly than applying pesticides. However, some biological control agents are difficult to mass-produce and are often slower acting and more difficult to apply than synthetic pesticides are. Sometimes the agents can multiply and become pests themselves.

Two other examples of biological controls are sex attractants (called pheromones) and hormones, chemicals produced by organisms to control their developmental processes. Pheromones can be used to lure pests into traps or to attract their natural predators into crop fields. They

FIGURE 12.29 Natural capital: In this example of biological pest control, a wasp is parasitizing a gypsy moth caterpillar.

have little chance of causing genetic resistance and are not harmful to nontarget species. Scientists have learned how to identify and use hormones that disrupt an insect's normal life cycle, thereby preventing it from reaching maturity and reproducing. Both of these methods are costly and can take weeks to be effective.

In using *ecological controls,* farmers work with nature instead of against it. For example, by practicing polyculture, they use plant diversity to provide habitats for the predators of pest species right on their crop fields. This is a simple application of the biodiversity **principle of sustainability**.

Cultivation controls can be another useful way to work with nature to control pests. For example, farmers can plant different crops on the same plots from year to year and adjust planting times so that major insect pests either starve or are eaten by their natural predators. Some farmers are using large machines to vacuum up harmful bugs.

Finally, some scientists call for using genetic engineering to speed up the development of pest- and disease-resistant crop strains. But controversy persists over whether the projected advantages of using GM plants outweigh their projected disadvantages (Figure 12.21).

Integrated Pest Management Is a Component of More Sustainable Agriculture

Many pest control experts and farmers believe the best way to control crop pests is through **integrated pest management (IPM),** a program in which each crop and its pests are evaluated as parts of an ecosystem (**Concept 12.4**). The overall aim of IPM is to reduce crop damage to an economically tolerable level with minimal use of synthetic pesticides.

When farmers using IPM detect an economically damaging level of pests in any field, they start with biological methods (natural predators, parasites, and disease organisms) and cultivation controls (such as altering planting times and using large machines to vacuum up harmful pests). They apply small amounts of synthetic pesticides—preferably biopesticides—only when insect or weed populations reach a threshold where the potential cost of pest damage to crops outweighs the cost of applying the pesticide.

IPM has a good record of accomplishment. In Sweden and Denmark, farmers have used it to cut their synthetic pesticide use by more than half. In Cuba, where organic farming is used almost exclusively, farmers make extensive use of IPM. In Brazil, IPM has reduced pesticide use on soybeans by as much as 90%. In Japan, many farmers save money by using ducks for pest control in rice paddies. The ducks' droppings also provide nutrients for the rice plants.

According to the U.S. National Academy of Sciences, a well-designed IPM program can reduce synthetic pesticide use and pest control costs by 50–65%, without reducing crop yields and food quality. IPM can also reduce inputs of fertilizer and irrigation water and slow the development of genetic resistance because of reduced use of pesticides. IPM is an important application of the biodiversity **principle of sustainability** and expands society's beneficial environmental impact.

Despite its important benefits, IPM—like any other form of pest control—has some drawbacks. It requires expert knowledge about each pest situation and takes more time than relying strictly on synthetic pesticides. Methods developed for a crop in one area might not apply to areas with even slightly different growing conditions. Initial costs may be higher, although long-term costs typically are lower than the use of conventional pesticides.

Several UN agencies and the World Bank have joined to establish an IPM facility. Its goal is to promote the use of IPM by disseminating information and establishing networks among researchers, farmers, and agricultural extension agents involved in IPM. Widespread use of IPM has been hindered in the United States and other countries by government subsidies that support use of pesticides, as well as by opposition from pesticide manufacturers, and a shortage of IPM experts. **GREEN CAREER: Integrated pest management**

12.5 HOW CAN WE PRODUCE FOOD MORE SUSTAINABLY?

CONCEPT 12.5 We can produce food more sustainably by using resources more efficiently, sharply decreasing the harmful environmental effects of industrialized food production, and eliminating government subsidies that promote such harmful impacts.

Conserve Topsoil

Land used for food production must have fertile topsoil, which takes hundreds of years to form. Thus, sharply reducing topsoil erosion is a key component of more sustainable agriculture and an important way to increase society's beneficial environmental impact.

Soil conservation involves using a variety of methods to reduce topsoil erosion and restore soil fertility, mostly by keeping the land covered with vegetation. For example, *terracing* involves converting steeply sloped land into a series of broad, nearly level terraces that run across the land's contours (Figure 12.30a). Each terrace retains water for crops and reduces topsoil erosion by controlling runoff.

On less steeply sloped land, *contour planting* (Figure 12.30b) can be used to reduce topsoil erosion. It involves plowing and planting crops in rows across the slope of the land

a. Lim Yong Hian/Shutterstock.com

b. Ron Nichols/USDA Natural Resources Conservation Service

c. inga spence/Alamy Stock Photo

d. Fedorov Oleksiy/Shutterstock.com

FIGURE 12.30 Soil conservation methods include **(a)** terracing; **(b)** contour planting and strip cropping; **(c)** intercropping; and **(d)** windbreaks between crop fields.

rather than up and down. Each row acts as a small dam to help hold topsoil by slowing runoff. Similarly, *strip-cropping* (Figure 12.30b) helps to reduce erosion and to restore soil fertility with alternating strips of a row crop (such as corn or cotton) and another crop that completely covers the soil, called a *cover crop* (such as alfalfa, clover, oats, or rye). The cover crop traps topsoil that erodes from the row crop and catches and reduces water runoff.

Alley cropping, or *agroforestry* (Figure 12.30c), is another way to slow erosion and to maintain soil fertility. One or more crops, usually legumes or other crops that add nitrogen to the soil, are planted together in alleys between orchard trees or fruit-bearing shrubs, which provide shade. This reduces water loss by evaporation and helps retain and slowly release soil moisture.

Farmers can also establish *windbreaks*, or *shelterbelts*, of trees around crop fields to reduce wind erosion (Figure 12.30d). The trees retain soil moisture, supply wood for fuel, and provide habitats for birds and insects that help with pest control and pollination.

Another way to greatly reduce topsoil erosion is to eliminate or minimize the plowing and tilling of topsoil and to leave crop residues on the ground. Such conservation-tillage farming uses special tillers and planting machines that inject seeds and fertilizer directly through crop residues on the ground into minimally disturbed topsoil. Weeds are controlled with herbicides.

SCIENCEFOCUS 12.1

Hydroponics: Growing Crops without Soil

Plants need sunlight, carbon dioxide (from the air), and mineral nutrients such as nitrogen and phosphorus. Traditionally, farmers have obtained these nutrients from soil. **Hydroponics** involves growing plants by exposing their roots to a nutrient-rich water solution instead of soil, usually inside of a greenhouse (Figure 12.A).

Indoor hydroponic farming yields a number of benefits. Crops can be grown year-round under controlled conditions, regardless of weather conditions. In dense urban areas, crops can be grown on rooftops, underground with artificial lighting (as is now done in Tokyo, Japan), and on floating barges, thus requiring much less land. Because the nutrient and water solution is recycled, farmers can cut their use of fertilizers and water. This reduces water pollution from the runoff of fertilizers into streams or other waterways. Also, in a well-controlled greenhouse environment, there is little or no need for pesticides.

With these advantages, some scientists say we could use hydroponics to produce a larger amount of the world's food without causing most of the serious harmful environmental effects of industrialized food production. Such systems are expensive to build, although they typically are less expensive to operate than conventional systems, in the end.

CRITICAL THINKING

What are some possible drawbacks to the use of hydroponics?

Khoo Si Lin/Shutterstock.com

FIGURE 12.A These salad greens are being grown hydroponically (without soil) in a greenhouse. The plant roots are immersed in a trough and exposed to nutrients dissolved in running water that can be reused.

This type of farming increases crop yields and greatly reduces soil erosion and water pollution from sediment and fertilizer runoff. It also helps farmers survive prolonged drought by helping to keep more moisture in the soil. However, one drawback is that the greater use of herbicides promotes the growth of herbicide-resistant weeds that force farmers to use larger doses of weed killers or, in some cases, to return to plowing. However, organic no-till methods being developed at the Rodale Institute (p. 301) could be herbicide-free and will not require plowing.

According to government surveys, farmers used conservation tillage methods on about 35% of U.S. cropland. Worldwide, it is used on only about 10% of all cropland, but its use is increasing.

Another way to conserve topsoil is to grow plants without using soil. Some producers are raising crops in greenhouses, using a system called *hydroponics* (Science Focus 12.1). At the Growing Power farm (**Core Case Study**), Will Allen has developed such a system for raising salad greens and fish together. Wastewater from the fish tanks flows into hydroponic troughs where it nourishes the plants. The plant roots filter the water, which is then returned to the fish runs. This closed-loop, chemical-free *aquaponic system* conserves soil, water, and energy while supporting more than 100,000 tilapia and perch, which are sold in local markets along with the salad greens.

Still another way to conserve topsoil is to retire the estimated one-tenth of the world's highly erodible cropland. The goal would be to identify *erosion hotspots*, withdraw these areas from cultivation, and plant them with grasses or trees, at least until their topsoil has been renewed.

Some countries, such as the United States, have paid farmers to set aside considerable areas of cropland for conservation purposes. Under the 1985 Food Security Act (Farm Act), more than 400,000 farmers participating in the Conservation Reserve Program received subsidy payments for taking highly erodible land out of production and replanting it with grass or trees for 10 to 15 years. Since 1985, these efforts have cut total topsoil losses on U.S. cropland by 40%.

Restore Soil Fertility

Another way to protect soil is to restore some of the lost plant nutrients that have been washed, blown, or leached out of topsoil, or that have been removed by repeated crop harvesting. To do this, farmers can use **organic fertilizer** derived from plant and animal materials or synthetic inorganic fertilizer made of inorganic compounds that contain *nitrogen*, *phosphorus*, and *potassium* along with trace amounts of other plant nutrients.

There are several types of organic fertilizers. One is *animal manure*: the dung and urine of cattle, horses, poultry, and other farm animals. It improves topsoil structure, adds organic nitrogen, and stimulates the growth of beneficial soil bacteria and fungi. Another type, called *green manure*, consists of freshly cut or growing green vegetation that is plowed into the topsoil to increase the organic matter and humus available to the next crop. A third type is *compost,* produced when microorganisms break down organic matter such as leaves, crop residues, food wastes, paper, and wood in the presence of oxygen.

The Growing Power farm (**Core Case Study**) depends greatly on its large piles of compost. Will Allen invites local grocers and restaurant owners to send their food wastes to add to the pile. To make this compost, Allen uses millions of red wiggler worms, which reproduce rapidly and eat their own weight in food wastes every day, converting it to plant nutrients. Also, the process of composting generates a considerable amount of heat, which is used to help warm some of the farm's greenhouses during cold months.

Yet another form of organic fertilizer is *biochar*—a form of charcoal made from woody materials that are often discarded through a process called *pyrolysis,* in which the materials are heated at low temperatures in containers that limit oxygen input until the materials become charcoal. This ancient process can provide heat and power and the biochar can then be buried to enrich soil. It has the added benefit of removing carbon from the atmosphere, thereby cutting back on atmospheric warming and climate change. Biochar can be produced from wood wastes, but when large areas of trees are cleared to produce biochar, as some companies are now doing, the use of biochar becomes unsustainable.

We degrade soils when we plant crops such as corn and cotton on the same land several years in a row, a practice that can deplete nutrients—especially nitrogen—in the topsoil. One way to reduce such losses is through **crop rotation,** in which a farmer plants a series of different crops in the same area from season to season. For example, where a nitrogen-depleting crop is grown one year, the farmer can plant the same area the following year with a crop such as legumes, which add nitrogen to the soil.

Many farmers, especially those in more-developed countries, rely on synthetic inorganic fertilizers. The use of these products accounts for about 25% of the world's crop yield. While these fertilizers can replace depleted inorganic nutrients, they do not replace organic matter. Completely restoring topsoil nutrients requires both inorganic and organic fertilizers. Many scientists are encouraging farmers, especially those in less-developed countries, to make greater use of green manure and biochar as a more sustainable way to restore soil fertility.

Reduce Soil Salinization and Desertification

We know how to prevent and deal with soil salinization, as summarized in Figure 12.31. The problem is that most of these solutions are costly.

Reducing desertification is not easy. We cannot control the timing and location of prolonged droughts caused by changes in weather and climate patterns. But we can reduce population growth, overgrazing, deforestation, and destructive forms of planting and irrigation in dryland areas, which have left much land vulnerable to topsoil erosion and thus desertification. We can also work to decrease the human contribution to projected climate change, which could increase the severity of droughts in large areas of the world during this century.

It is possible to restore land suffering from desertification by planting trees and other plants that anchor topsoil and hold water. We can also grow trees and crops together (alley cropping, Figure 12.30c), and establish windbreaks around farm fields (Figure 12.30d).

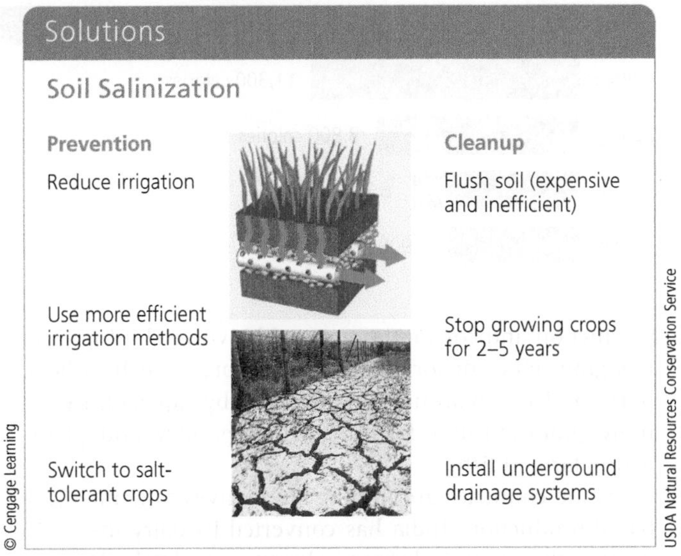
Solutions
Soil Salinization
Prevention
Reduce irrigation
Use more efficient irrigation methods
Switch to salt-tolerant crops
Cleanup
Flush soil (expensive and inefficient)
Stop growing crops for 2–5 years
Install underground drainage systems
© Cengage Learning
USDA Natural Resources Conservation Service

FIGURE 12.31 Ways to prevent and ways to clean up soil salinization (**Concept 12.6**). ***Critical thinking:*** Which two of these solutions do you think are the best ones? Why?

CONSIDER THIS . . .

LEARNING FROM NATURE

Scientists are studying how termites build rich soil around their mounds on otherwise dry and degraded land. By carrying vegetation particles to their mounds to fertilize their food stores (microbes and fungi), they concentrate nitrogen, phosphorus, and other nutrients. Researchers hope to use such methods to slow desertification in certain areas.

Produce and Consume Meat and Dairy Products More Sustainably

Meat production has a huge environmental impact and meat consumption is the largest factor in the growing ecological footprints of individuals in affluent nations.

Some types of meat are produced more efficiently than others (Figure 12.32). For example, producing a pound of beef requires more than three times the amount of grain needed to produce a pound of pork. It takes more than four times the amount of grain used to produce a pound of chicken or turkey. A more sustainable form of meat

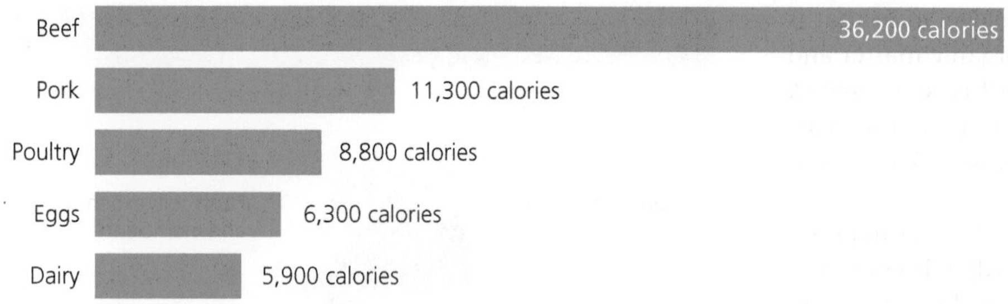

FIGURE 12.32 Amount of feed required to produce 1,000 calories of meat for human consumption, comparing five different foods. ***Data analysis:*** For each of the others (pork, poultry, eggs, and dairy), how many calories could be produced with the grain required to produce 1,000 calories of beef?

(Compiled by the authors using data from U.S. Department of Agriculture.)

production and consumption would involve shifting from less grain-efficient forms of animal protein, such as beef, pork, and carnivorous fish produced by aquaculture, to more grain-efficient forms, such as poultry and plant-eating farmed fish.

GOOD NEWS

Some countries are finding alternatives to grain-based production. India has converted its dairy industry, the world's largest, almost completely to crop residues such as rice straw and corn stalks. This saves energy and prevents pollution and greenhouse gas emissions related to raising grain. India's dairy industry has continued to grow with this conversion.

Insects are another form of protein. Would you consider trying a beetle salad, a caterpillar stew, or a handful of crunchy fried ants for a snack? If you're thinking *yuck*, you're not alone. Yet, at least 2,000 species of insects provide nutrients to more than 2 billion people, according to the FAO. In countries such as Thailand, Australia, and Mexico, insects are deep-fried with spices, made into flours and tasty sauces, baked into breads, and cooked in stews. Insects can be rich in protein and fiber and healthy fats. They can provide vital micronutrients such as calcium, iron, B vitamins, and zinc.

A growing number of people have one or two meatless days per week. Others go further and eliminate most or all meat from their diets, replacing it with a balanced vegetarian diet of fruits, vegetables, and protein-rich foods such as peas, beans, and lentils. According to one estimate, if all Americans picked one day per week to have no meat, the reduction in greenhouse gas emissions would be equivalent to taking 30 to 40 million cars off the road for a year.

CONSIDER THIS . . .

THINKING ABOUT Meat Consumption

If you do not do this already, would you be willing to live lower on the food chain by eating much less meat, or no meat at all? Explain.

Practice More Sustainable Aquaculture

Various organizations have established guidelines, standards, and certifications to encourage more sustainable aquaculture and fishing practices. The Aquaculture Stewardship Council (ASC), for example, has developed aquaculture sustainability standards, although it has certified only about 4.6% of the world's aquaculture production operations. The Marine Stewardship Council (MSC) performs a similar program for wild-caught fisheries. Like the organic certification, the programs have associated labels for food products that help consumers purchase more sustainable options.

Scientists and producers are working on ways to make aquaculture more sustainable and to reduce its harmful environmental effects. One such approach is *open-ocean aquaculture*, which involves raising large carnivorous fish in large underwater pens—some as large as a high school gymnasium. They are located far offshore (Figure 11.8, p. 262), where rapid currents can sweep away fish wastes and dilute them. Some farmed fish can escape from such operations and breed with wild fish, and this approach is costly. However, the environmental impact of raising fish far offshore is smaller than that of raising fish near shore and much smaller than that of industrialized commercial fishing.

Other fish farmers are reducing coastal damage from aquaculture by raising shrimp and fish species in inland facilities using freshwater ponds and tanks. In such *recirculating aquaculture systems*, the water used to raise the fish is continually recycled. Similarly, the Growing Power aquaponic system (**Core Case Study**) captures its fish wastes and converts them to fertilizer used to grow salad greens. This eliminates fish waste pollution of aquatic systems and reduces the need for antibiotics and other chemicals used to combat disease among farmed fish. It also eliminates the problem of farmed fish escaping into natural aquatic systems.

Making aquaculture more sustainable will require some fundamental changes for producers and consumers. One change is for more consumers to choose fish species such as carp, tilapia, and catfish that eat algae and other vegetation rather than fish oil and fishmeal produced from other fish. Raising carnivorous fishes such as salmon, trout, tuna, grouper, and cod contributes to overfishing and population crashes of species and is unsustainable. Aquaculture producers can avoid this problem by raising herbivorous fishes as long as they do not try to increase yields by feeding fishmeal to such species, as many of them are doing.

Fish farmers can also emphasize *polyaquaculture*, which has been part of aquaculture for centuries, especially in Southeast Asia. Polyaquaculture operations raise fish or shrimp along with algae, seaweeds, and shellfish in coastal lagoons, ponds, and tanks. The wastes of the fish or shrimp feed the other species. Polyaquaculture applies the chemical cycling and biodiversity **principles of sustainability**. **GREEN CAREER: Sustainable aquaculture**

Figure 12.33 lists some ways to make aquaculture more sustainable and to reduce its harmful environmental effects.

Expand Organic Agriculture

One component of more sustainable food production is organic agriculture. The Rodale Institute has been conducting a side-by-side comparison of organic and industrialized farming systems for more than 30 years.

Since its launch in 1981, the Farming Systems Trial (FST) has found evidence of some major benefits of organic farming, including the following:

- *Organic farming builds soil organic matter.* Synthetic fertilizers leach, or pass through the soil, more quickly than nutrients from manures, composts, or cover crops. Levels of beneficial soil fungi are also higher in organic soils. Thus on the FST organic plots, the nutrients remain available to the plants much more than on the conventional plots.
- *Organic systems reduce erosion and water pollution.* Because of the leaching action, fertilizer chemicals end up in surface water and groundwater. Most of the precipitation on organic fields soaks into the soil rather than running off, more so than on the conventional fields.
- *Organic farming uses less fossil fuel energy.* The FST organic systems used 33–50% less energy than the conventional systems did. Producing nitrogen fertilizer, not used at all on organic crops, is the single greatest energy input in industrial farming.
- *Organic farming cuts greenhouse gas emissions.* The FST's conventional systems produced 40% more greenhouse gases, mostly because they involved the production and use of synthetic fertilizers. Also, in the organic system, carbon content in the soil increased, while soils in the conventional plots lost carbon.
- *Organic yields match conventional yields.* During its first 3 years while the FST organic system was being developed, it had lower yields than the conventional system. Since then, organic corn, wheat, and soybean yields have matched conventional yields.
- *Organic systems are more weed-tolerant.* Organic corn and soybean crops tolerate much higher levels of weed competition than do conventional crops, while producing equivalent yields without the use of herbicides.
- *Organic crops compare favorably in years of drought.* Organic corn yields were 31% higher in the FST than conventional yields in years of drought. By comparison, genetically engineered "drought-tolerant" varieties saw yield increases of 6.7–13.3% over conventional varieties.
- *Organic farming can be more profitable.* The FST's average net return for the organic systems was $558/acre/year, compared with $190/acre/year for the conventional systems.

Solutions

More Sustainable Aquaculture

- Protect mangrove forests and estuaries
- Improve management of wastes
- Reduce escape of aquaculture species into the wild
- Set up self-sustaining polyaquaculture systems that combine aquatic plants, fish, and shellfish
- Certify sustainable forms of aquaculture

FIGURE 12.33 Ways to make aquaculture more sustainable and reduce its harmful effects. ***Critical thinking:*** Which two of these solutions do you think are the best ones? Why?

Some studies have supported the FST findings, while others have not. Some studies have indicated that yields of organic crops in more-developed countries can be as much as 20% lower than yields of conventionally raised crops. This is largely because most agriculture in these countries is done on an industrial scale. However, in less-developed countries, it is done on a smaller scale. A study by the UNEP surveyed 114 small-scale farms in 24 African countries and found that yields more than doubled where organic farming practices had been used. In some East African operations, the yield increased by 128% with these changes. GOOD NEWS

Another major benefit of organic farming is that by eating USDA 100% certified organic foods, consumers reduce their exposure to pesticide residues and to bacteria-resistant antibiotics that can be found in conventional foods. A 2014 study led by Carlo Leifert concluded that organically grown fruits, vegetables, and grains also contain substantially higher levels of cancer-fighting antioxidants than their conventionally grown counterparts.

One hindrance to expanded use of organic farming, compared with industrialized agriculture, is that it requires more human labor to use methods such as integrated pest management and crop rotation. However, this could be seen as a benefit, especially in less-developed countries, because organic farming creates jobs. A UN report concluded that organic farming provides over 30% more jobs per hectare than nonorganic farming.

Shift to More Sustainable Food Production

Modern industrialized food production yields large amounts of food at affordable prices. However, to a growing number of analysts, it is unsustainable because of its harmful environmental and health costs (Figure 12.14) and because it violates the three **scientific principles of sustainability**. It relies heavily on nonrenewable fossil fuels and thus adds greenhouse gases and other air

INDIVIDUALS MATTER 12.1

David Tilman—Polyculture Researcher

Todd Reubold/University of Minnesota

One of the world's most prominent ecologists and agricultural experts is David Tilman, a professor at the University of Minnesota. Since 1981, he has conducted more than 150 long-term controlled experiments on a university-owned grassland. For example, he applied certain mixes of fertilizers and water on experimental plots and observed how the plants responded. He then compared the results from the experimental plots with those from control plots that did not receive the experimental treatments. He also used varying mixes of plant species in such experiments, focusing on the benefits of polyculture.

With ecologist Peter Reich, Tilman found that carefully controlled polyculture plots with 16 different species of plants consistently outproduced plots with 9, 4, or only 1 type of plant species. Such research explains why some analysts argue for greatly expanding the use of polyculture to produce food more sustainably.

Tilman's findings also support the scientific idea that biodiversity can make ecosystems more stable and sustainable. A diverse forest, for example, suffers damage during pest infestations, but there are enough varying species to withstand the damage. Some species are wiped out, but not the whole forest. A vast field of corn or wheat, on the other hand, is highly vulnerable to heat waves, drought, disease, and pests, and it can be destroyed during extreme events.

For his important research efforts, Tilman has received several awards. In 2010 he was awarded the prestigious Heineken Prize for Environmental Sciences for his important contributions to the science of ecology.

---

pollutants to the atmosphere and contributes to climate change. It also reduces biodiversity and agrobiodiversity and interferes with the cycling of plant nutrients. These harmful effects are hidden from consumers because most of them are not included in the market prices of food—a violation of the full-cost pricing **principle of sustainability**.

In addition to expanding the use of organic agriculture, a more sustainable food production system would have several components (Figure 12.34). One component would be to rely less on conventional monoculture and more on organic polyculture. Research shows that, on average, low-input polyculture produces higher yields per unit of land than does high-input industrialized monoculture. It also uses less energy and fewer resources and provides more food security for small landowners.

Ecologist David Tilman (Individuals Matter 12.1) has been instrumental in demonstrating the benefits of polyculture. The Growing Power farm (**Core Case Study**) practices polyculture by growing a variety of crops in several greenhouses.

Of particular interest to some scientists is the idea of using polyculture to grow *perennial crops*—crops that grow back year after year on their own (Science Focus 12.2).

Another key to developing more sustainable agriculture is to shift from using fossil fuels to relying more on renewable energy for food production—an important application of the solar energy **principle of sustainability** that has been well demonstrated by the Growing Power farm (**Core Case Study**). To produce the electricity and fuels needed for food production, farmers can make greater use of renewable solar energy (see chapter-opening photo), wind, flowing water, and biofuels produced

Solutions

More Sustainable Food Production

More	Less
High-yield polyculture	Soil erosion
Organic fertilizers	Soil salinization
Biological pest control	Water pollution
Integrated pest management	Aquifer depletion
Efficient irrigation	Overgrazing
Perennial crops	Overfishing
Crop rotation	Loss of biodiversity and agrobiodiversity
Water-efficient crops	Fossil fuel use
Soil conservation	Greenhouse gas emissions
Subsidies for sustainable farming	Subsidies for unsustainable farming

FIGURE 12.34 More sustainable, low-input food production has a number of major components. ***Critical thinking:*** For each list in this diagram (left and right), which two items do you think are the most important? Why?

Top: Marko5/Shutterstock.com. Center: Anhong/Dreamstime.com. Bottom: pacopi/Shutterstock.com

SCIENCEFOCUS 12.2

Perennial Polyculture and the Land Institute

Some scientists call for greater reliance on polycultures of perennial crops as a component of more sustainable agriculture. Such crops can live for many years without having to be replanted and are better adapted to regional soil and climate conditions than are most annual crops.

More than three decades ago, plant geneticist Wes Jackson co-founded the Land Institute in the U.S. state of Kansas. One of the institute's goals has been to grow a diverse mixture of edible perennial plants to supplement traditional annual monoculture crops and to help reduce the latter's harmful environmental effects. Examples in this polyculture mix include perennial grasses, plants that add nitrogen to the soil, sunflowers, grain crops, and plants that provide natural insecticides. Some of these plants could also be used as a source of renewable biofuel for motor vehicles.

Researchers are trying to improve yields of different varieties of these perennial crops, which are not all as high as yields of annual crops. However, in the U.S. state of Washington, researchers have bred perennial wheat varieties that have a 70% higher yield than today's commercially grown annual wheat varieties.

The Land Institute's approach, called *natural systems agriculture*, copies nature by growing a diversity of perennial crops using organic methods. It has a number of environmental benefits. Because there is no need to till the soil and replant perennials each year, this approach produces much less topsoil erosion and water pollution. It also reduces the need for irrigation because the deep roots of such perennials retain more water than do the shorter roots of annuals (Figure 12.B). There is little or no need for chemical fertilizers and pesticides, and thus little or no pollution from these sources. Perennial polycultures also remove and store more carbon from the atmosphere, and growing them requires less energy than does growing crops in conventional monocultures.

This approach has worked well in some areas of the world. For example, in Malawi, Africa, farmers have greatly raised their crop yields by planting rows of perennial pigeon peas between rows of corn. The pea plants have doubled the carbon and nitrogen content in the soils, while increasing soil water retention. These legume plants also provide a welcome source of protein for the farm families.

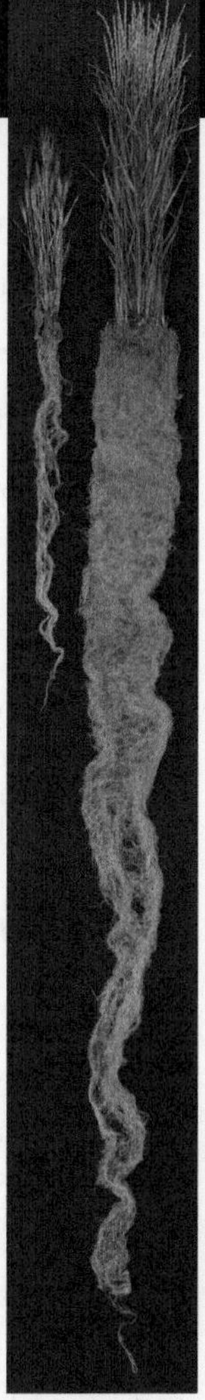

Photo Courtesy of The Land Institute

FIGURE 12.B The roots of an annual wheat crop plant (left) are much shorter than those of big bluestem (right), a tallgrass prairie perennial plant.

Some scientists note that adapting perennials to the grand scale of industrialized crop production will be challenging and time-consuming. However, Wes Jackson calls for governments to promote this and other forms of more sustainable agriculture in order to gain their many benefits. He reminds us that "if our agriculture is not sustainable, then our food supply is not sustainable."

CRITICAL THINKING

Why do you think large seed companies generally oppose this form of more sustainable agriculture?

from farm wastes in tanks called *biogas digesters*. Most proponents of more sustainable agriculture call for using more environmentally sustainable forms of both high-yield polyculture and high-yield monoculture (**Concept 12.5**).

Agricultural experts such as Jonathan Foley argue that industrialized farming could play a major role in shifting to more sustainable food production. Foley notes that farmers are already finding ways to apply pesticides and fertilizers in smaller amounts and more precisely, using computerized tractors and remote sensing and GPS technology. Fertilizers can also be mixed and tailored to different soil conditions to help minimize runoff into waterways, and irrigation can be done much more efficiently than it typically is. Such methods could increase production on the current total area of farmland by 50–60%, according to Foley.

Proponents of more sustainable food production systems say that education is an important key to making a shift toward a more sustainable food production system. They seek to inform people, especially young consumers, about where their food really comes from, how it is produced, and what the environmentally harmful effects of food production are. They also call for economic policies that reward more sustainable agriculture. A major part of such policies would be to shift subsidies from unsustainable to more sustainable food production processes. Simply put, food should be good for people and for the planet. This requires people having *food literacy* by knowing where there food comes from, how it is produced, and its impact on human health and the environment.

Figure 12.35 lists ways in which you can promote more sustainable food production.

What Can You Do?

More Sustainable Food Production

- Eat less meat, no meat, or organically certified meat
- Choose sustainably produced herbivorous fish
- Use organic farming to grow some of your food
- Buy certified organic food
- Eat locally grown food
- Compost food wastes
- Cut food waste

FIGURE 12.35 Individuals matter: Ways to promote more sustainable food production (**Concept 12.6**). ***Critical thinking:*** Which three of these actions do you think are the most important? Why?

12.6 HOW CAN WE IMPROVE FOOD SECURITY?

CONCEPT 12.6 We can improve food security by reducing poverty and chronic malnutrition, producing food more sustainably, relying more on locally grown food, and cutting food waste.

We Can Use Government Policies to Improve Food Production and Security

Agriculture is a financially risky business. Whether farmers have a good or bad year depends on factors over which they have little control, including weather, crop prices, pests and diseases, interest rates on loans, and global food markets.

Governments use two main approaches to influence food production in hopes of strengthening food security. *First,* they can *control food prices* by placing a legally mandated upper limit on them in order to keep them artificially low. This helps consumers, but farmers may find it harder to make a living.

Second, they can *provide subsidies* by giving farmers price supports, tax breaks, and other financial support to help them stay in business and to encourage them to increase food production. In the United States, most subsidies go to industrialized food production, often in support of environmentally harmful practices. For example, by subsidizing corn and soybeans, the government pushes prices on those grains down and indirectly supports CAFOs and feedlots, which depend on corn and soy for their feed, and their harmful environmental and health effects.

Some opponents of subsidies call for ending them. They point to New Zealand, which ended farm subsidies in 1984. After the shock wore off, innovation took over and production of some foods such as milk quadrupled. Brazil has also ended most of its farm subsidies. Some analysts call for replacing traditional subsidies for farmers with subsidies that promote more environmentally sustainable farming practices.

Similarly, government subsidies to fishing fleets can promote overfishing and the reduction of aquatic biodiversity. For example, several governments give the highly destructive bottom-trawling industry (Figure 11.4, p. 258) millions of dollars per year in subsidies, which is the main reason fishers who use this environmentally destructive practice can stay in business. Many analysts call for replacing those harmful subsidies with subsidies that would sharply reduce overfishing and promote more sustainable fishing and aquaculture.

Some analysts call for all countries to establish sustainable agriculture laws to guide people in producing food more sustainably. Such laws could set standards for sustainable production, cut subsidies that promote unsustainable practices, regulate industrial farm pollution, and set goals for agricultural greenhouse gas reductions. Some have argued for levying more taxes on pesticide and fertilizer use and methane emissions. The proceeds could be used to subsidize organic farming, integrated pest management, and sustainable food production on farms.

Other Government and Private Programs Are Increasing Food Security

Government and private programs aimed at reducing poverty can improve food security (**Concept 12.1B**). For example, some programs provide small loans at low interest rates to poor people to help them start businesses or buy land to grow their own food.

Some analysts urge governments to establish special programs focused on saving children from the harmful health effects of poverty. Studies by the United Nations Children's Fund (UNICEF) indicate that one-half to two-thirds of nutrition-related childhood deaths could be prevented at an average annual cost of $5 to $10 per child. This involves simple measures such as immunizing more children against childhood diseases, preventing dehydration due to diarrhea by giving infants a mixture of sugar and salt in their water, and combatting blindness by giving children an inexpensive vitamin A capsule twice a year.

Some farmers and plant breeders are working on preserving a diverse gene pool as another way to improve food security. For example, an organization called the Global Crop Diversity Trust is seeking to prevent the disappearance of 100,000 varieties of food crops. The trust is working with about 50 seed banks around the world to cultivate and store seeds from endangered varieties of many food plant species.

In the quest for food security, some critics recognize the potential benefits of GM crops (Figure 12.21, left). However, they point out that most of the GM crops developed so far have provided very few of these benefits while they present potentially serious drawbacks (Figure 12.21, right).

Still, many scientists think that GM crops hold great promise. A survey by the Pew Research Center and the American Association for the Advancement of Science

(AAAS) indicated that 88% of AAAS scientists polled think it is safe to eat GM foods, while just 37% of the general public agreed with this.

Many private, mostly nonprofit, organizations are also working to help individuals, communities, and nations to improve their food security and produce food more sustainably. For example, Growing Power's Will Allen (**Core Case Study**) argues that instead of trying to transfer complex technologies such as genetic engineering to less-developed countries, we could be helping them to develop simple, sustainable, local food production and distribution systems. He argues this would give them more control over their food security.

Sustainable agriculturalists and National Geographic Explorers Cid Simones and Paola Segura work with small farmers to show them how to grow food more sustainably on small plots in the tropical forests of Brazil. They train one family at a time. In return, each family must teach five other families and thus help to spread more sustainable farming methods.

We Can Grow and Buy More Food Locally and Cut Food Waste

One way to increase food security is to grow more of our food locally or regionally, ideally with USDA 100% certified organic farming practices. A growing number of consumers are becoming *locavores*, who try to buy as much of their food as possible from local and regional producers in farmers' markets, which provide access to fresher seasonal foods, many of them grown organically.

In addition, many people participate in *community-supported agriculture (CSA)* programs. In these programs, people buy shares of a local farmer's crops and receive a box of fruits or vegetables on a regular basis during the growing season. Growing Power (**Core Case Study**) runs such a program for inner-city residents. For many of these people, the organically grown food they get from the urban farm greatly improves their diets and increases their chances of living longer and healthier lives.

By buying locally, people support local economies and farm families. Buying locally reduces fossil fuel energy costs for food producers, as well as the greenhouse gas emissions from storing and transporting food products over long distances. There are limits to this benefit, however. Food scientists point out that the largest share of carbon footprint for most foods is in production. Thus, for example, an apple grown through high-input agriculture and trucked across North America could have a larger footprint than an apple grown through low-input farming and sent on a ship from South America.

An increase in the demand for locally grown food could result in more small, diversified farms that produce organic, minimally processed food from plants and animals. Such *eco-farming* could be one of this century's challenging new careers for many young people. **GREEN CAREER: Small-scale sustainable agriculture**

Sustainable agriculture entrepreneurs and ordinary citizens who live in urban areas could grow more of their own food, as the Growing Power farm has shown (**Core Case Study**). According to the USDA, approximately 15% of the world's food is grown in urban areas, and this percentage could easily be doubled. Increasingly, people are sharing garden space, labor, and produce in community gardens (Figure 12.36) in vacant lots. People are planting gardens and raising chickens in backyards, growing dwarf fruit trees in large containers of soil, and raising vegetables in containers on rooftops, balconies, and patios. One study estimates that converting 10% of American lawns into food-producing gardens would supply one-third of the country's fresh produce.

Many urban schools, colleges, and universities are benefitting greatly from having gardens on school grounds. Not only do the students have a ready source of fresh produce, but they also learn about where their food comes from and how to grow their own food more sustainably.

In the future, much of our food might be grown in cities within specially designed high-rise buildings. Such a building would have rooftop solar panels for generating electricity, and it could capture and recycle rainwater for irrigating its wide diversity of crops. Sloped glass facing south would bring in sunlight, and excess heat collected in this way could be stored in tanks containing water or sand underneath the building for use as needed. This approach would put into practice the three **scientific principles of sustainability**.

Finally, people can sharply cut food waste (**Concept 12.6**). Jonathan Foley reported in 2014 that 25% of the world's food calories are lost or wasted. In poor countries with unreliable food storage and transportation, much is lost before it gets to consumers. In wealthy countries, much waste occurs in restaurants, homes, and supermarkets. According to studies by the EPA and the Natural Resources Defense Council, Americans throw away 30–40% of the country's food supply each year while 49 million Americans experience chronic hunger. According to the USDA, the foods most often thrown away in the United States are, in order,

FIGURE 12.36 Community gardens like this one are helping people without much land to grow their own food.

perishable items such as vegetables and fruits, roots and tubers, fish and seafood, dairy products, and meat.

Food policy expert Lester Brown has discussed our food security challenges in terms of a food equation that is out of balance because its demand side is growing and its supply side is shrinking (Figure 12.37, top). He suggests that we can meet these challenges by working on both the demand side and the supply side to find solutions that will be difficult but achievable (Figure 12.37, bottom). His argument provides an excellent summary of how food production, currently unsustainable, could be made more sustainable, thereby leading us toward global food security.

BIG IDEAS

- About 795 million people have health problems because they do not get enough to eat and 2.1 billion people face health problems from eating too much.
- Modern industrialized agriculture has a greater harmful impact on the environment than any other human activity.
- More sustainable forms of food production could greatly reduce the harmful environmental and health impacts of industrialized food production systems.

Trade-Offs

Challenges

Demand Side	Supply Sides
Growing population	Soil erosion
People moving up the food chain	Depletion of aquifers
	Stagnant grain yields
Turning food into biofuel	Rising temperature

Solutions

Demand Side	Supply Sides
Stabilize population	Conserve soil
Eradicate poverty	Use water efficiently
Reduce excessive meat consumption	Find ways to increase yields
Eliminate biofuel subsidies	Stabilize climate

© Cengage Learning

FIGURE 12.37 *The Food Equation:* Demand and supply are currently out of balance (upper part of figure) and growing more so, due to unsustainable food production methods. We can rebalance the food equation with workable solutions (lower part of figure).

(Compiled by the authors using data from Brown, Lester R. *Plan B 4.0*, Norton, 2009.)

Tying It All Together

Growing Power and Sustainability

Baloncici/Shutterstock.com

This chapter began with a look at how Growing Power, the ecologically based urban farm (**Core Case Study**), is providing a diversity of good food to people living in a food desert. Its founder Will Allen, in demonstrating how organic food can be grown more sustainably at affordable prices, is showing how to make the transition to more sustainable food production while applying the three **scientific principles of sustainability**. Modern industrialized agriculture, aquaculture, and other forms of industrialized food production violate all of these principles.

Making the transition to more sustainable food production means relying more on solar (see chapter-opening photo) and other forms of renewable energy and less on fossil fuels. It also means sustaining chemical cycling by conserving topsoil and returning crop residues and animal wastes to the soil. It involves helping to sustain natural, agricultural, and aquatic biodiversity by relying on a greater variety of crop and animal strains and seafood, produced by certified organic methods and sold locally in grocery stores and farmers' markets. Controlling pest populations through broader use of conventional and perennial polyculture and integrated pest management will also help to sustain biodiversity.

Such efforts will be enhanced if we can slow the growth of the human population and stop our wasteful use of food and other resources. Governments could help these efforts by replacing environmentally harmful agricultural and fishing subsidies and tax breaks with more environmentally beneficial ones. Finally, the transition to more sustainable food production would be accelerated for the benefit of the environment as well as current and future generations if we could find ways to include the harmful environmental and health costs of food production in the market prices of food, in keeping with the economic, political, and ethical **principles of sustainability**.

Chapter Review

Core Case Study

1. What is a food desert? Summarize the benefits that the Growing Power farm has brought to its community. How does the farm showcase the three **scientific principles of sustainability**?

Section 12.1

2. What are the two key concepts for this section? What is a basic requirement for maintaining good health and resisting disease? Distinguish between **chronic undernutrition (hunger)** and **chronic malnutrition** and describe their harmful effects. What is **famine**? Describe the effects of diet deficiencies in vitamin A, iron, and iodine. What is **overnutrition** and what are its harmful effects? About how many people in the world suffer from hunger and what percentage of them are overweight or obese? What is the biggest contributor to hunger and malnutrition? Define **food security**.

Section 12.2

3. What is the key concept for this section? What three systems supply most of the world's food? Define and distinguish among **irrigation**, **synthetic fertilizers**, and **synthetic pesticides**. Define **industrialized agriculture (high-input agriculture)**. Define **yield**. What is **plantation agriculture**? Distinguish between **traditional subsistence agriculture** and **traditional intensive agriculture**. Define **polyculture** and summarize its benefits. Define **organic agriculture** and compare its main components with those of conventional industrialized agriculture. What is a **green revolution**? Define **farm subsidies**. Summarize the story of industrialized food production in the United States. What is an example of a hidden cost of food production?

4. Define **genetic engineering** and distinguish between it and crossbreeding through artificial selection. Describe the second gene revolution based on genetic engineering. What is a **genetically modified organism (GMO)**? Summarize the growth of industrialized meat production. What are feedlots and CAFOs? What is a **fishery**? What is **aquaculture (fish farming)**? Explain why industrialized food production requires large inputs of energy. Why does it result in a net energy loss?

Section 12.3

5. What is the key concept for this section? List two major benefits of high-yield industrialized agriculture. What is **topsoil** and why is it one of our most important resources? What is **soil erosion** and what are its two major harmful environmental effects? What is **desertification** and what are its harmful environmental effects? Define **soil salinization** and **waterlogging** and explain why they are harmful. What is soil pollution and what are two of its causes?

6. Summarize industrialized agriculture's contribution to current and projected climate change. Explain how synthetic fertilizer use has increased and list two effects of overfertilization. Explain how industrialized food production systems have caused losses in biodiversity. What is **agrobiodiversity** and how is it being affected by industrialized food production? List the advantages and disadvantages of using genetic engineering in food production. What factors can limit green revolutions? Compare the benefits and drawbacks of organic agriculture. Define **no-till farming**. Compare the benefits and harmful effects of industrialized meat production. Explain the connection between feeding livestock and the formation of ocean dead zones. Compare the benefits and harmful effects of aquaculture.

Section 12.4

7. What is the key concept for this section? What is a **pest**? Define and give two examples of a **pesticide**. What are biopesticides? Summarize the advantages and disadvantages of using synthetic pesticides. Describe the use of laws and treaties to help protect U.S. citizens from the harmful effects of pesticides. What are three general classes of alternatives to conventional pesticides? Give an example of each. Define **integrated pest management (IPM)** and list its advantages.

Section 12.5

8. What is the key concept for this section? What is **soil conservation**? Describe six ways to reduce topsoil erosion. What is **hydroponics**? Define **organic fertilizers** and name and describe three types of organic fertilizers. What is biochar? Define **crop rotation** and explain how it can help restore topsoil fertility. What are some ways to prevent and some ways to clean up soil salinization? How can we reduce desertification? What are some ways to make meat production and consumption more sustainable? Describe three ways to make aquaculture more sustainable. List eight benefits of organic agriculture. What are four important components of a more sustainable food production system? List the advantages of relying more on organic polyculture and perennial crops. What are three important ways in which individual consumers can help to promote more sustainable food production?

Section 12.6

9. What is the key concept for this section? What are the two main approaches used by governments to influence food production? How have governments used subsidies to influence food production and what have been some of their effects? What are two other ways in which organizations are improving food security? Explain three of the benefits of buying locally grown food. When it is not a good choice? How can urban farming help to increase food security? According to Lester Brown, how can we meet the challenges of improving food security?

10. What are this chapter's three big ideas? Explain how making the transition to more sustainable food production such as that promoted by the Growing Power farm (**Core Case Study**) will involve applying the six **principles of sustainability**.

Note: Key terms are in bold type.

Critical Thinking

1. Suppose you got a job with Growing Power, Inc. (**Core Case Study**) and were given the assignment to turn an abandoned suburban shopping center and its large parking lot into an organic farm. Write up a plan for how you would accomplish this.

2. Do you think that the advantages of organic agriculture outweigh its disadvantages? Explain. Do you eat or grow organic foods? If so, explain your reasoning for making this choice. If not, explain your reasoning for some of the food choices you do make.

3. Food producers can now produce more than enough food to feed everyone on the planet a healthy diet. Given this fact, why do you think that nearly a billion people are chronically undernourished or malnourished? Assume you are in charge of solving this problem, and write a plan for how you will accomplish it.

4. Explain why you support or oppose greatly increasing the use of **(a)** genetically modified foods and **(b)** organic perennial polyculture.

5. Suppose you work for a farmer and are given the assignment of deciding whether to use no-till agriculture on the farmer's fields or to continue using conventional plowing and weed control methods. Compare the advantages and disadvantages of each and decide how you will advise your boss. Write up a report and provide evidence to support your arguments.

6. You are the head of a major agricultural agency in the area where you live. Weigh the advantages and disadvantages of using synthetic pesticides and explain why you would support or oppose the increased use of such pesticides as a way to help farmers raise their yields. What are the alternatives? Give evidence to support your claims.

7. If the mosquito population in the area where you live were proven to be carrying malaria or some dangerous virus, would you want to spray DDT in your yard, inside your home, or all through the local area to reduce this risk? Explain. What are the alternatives?

8. According to physicist Albert Einstein, "Nothing will benefit human health and increase the chances of survival of life on Earth as much as the evolution to a vegetarian diet." Explain your interpretation of this statement. Are you willing to eat less meat or no meat? Explain.

Doing Environmental Science

For 1 week, weigh the food that is purchased in your home and the food that is thrown out. Also, keep track of the types of food you eat, using categories such as fruits, vegetables, meats, dairy, and other more specific categories if you wish. Record and compare these numbers and other data from day to day. Develop a plan for cutting your household food waste in half. Consider making a similar study for your school cafeteria and reporting the results and your recommendations to school officials.

Global Environment Watch Exercise

Go to your MindTap course to access the GREENR database. Starting on the home page, under "Browse Issues and Topics", click on *Resource Management,* then select *Soil Erosion.* Use this portal to searchfor information on causes of soil erosion and how it affects soil fertility. Write a report on your findings. If you were to overhear a group of farmers complaining about how much money they must spend on fertilizers, what suggestions would you give them for saving money? Include your answer to this question, along with your reasoning, in your report.

Ecological Footprint Analysis

The following table gives the world's fish harvest and population data.

1. Use the world fish harvest and population data in the table to calculate the per capita fish consumption for 1990–2012 in kilograms per person. (*Hints:* 1 million metric tons equals 1 billion kilograms; the human population data are expressed in billions; and per capita consumption can be calculated directly by dividing the total amount consumed by a population figure for any year.)
2. Did per capita fish consumption generally increase or decrease between 1990 and 2012?
3. In what years did per capita fish consumption decrease?

WORLD FISH HARVEST

Years	Fish Catch (million metric tons)	Aquaculture (million metric tons)	Total (million metric tons)	World Population (in billions)	Per Capita Fish Consumption (kilograms/person)
1990	84.8	13.1	97.9	5.27	
1991	83.7	13.7	97.4	5.36	
1992	85.2	15.4	100.6	5.44	
1993	86.6	17.8	104.4	5.52	
1994	92.1	20.8	112.9	5.60	
1995	92.4	24.4	116.8	5.68	
1996	93.8	26.6	120.4	5.76	
1997	94.3	28.6	122.9	5.84	
1998	87.6	30.5	118.1	5.92	
1999	93.7	33.4	127.1	6.00	
2000	95.5	35.5	131.0	6.07	
2001	92.8	37.8	130.6	6.15	
2002	93.0	40.0	133.0	6.22	
2003	90.2	42.3	132.5	6.31	
2004	94.6	45.9	140.5	6.39	
2005	94.2	48.5	142.7	6.46	
2006	92.0	51.7	143.7	6.54	
2007	90.1	52.1	142.2	6.61	
2008	89.7	52.5	142.3	6.69	
2009	90.0	55.7	145.7	6.82	
2010	89.0	59.0	148.0	6.90	
2011	93.5	62.7	156.2	7.00	
2012	90.2	66.5	156.7	7.05	

Compiled by the authors using data from UN Food and Agriculture Organization and Earth Policy Institute.

CHAPTER 13

Water Resources

Through the cycling of water, across space and time, we are linked to all life. . . . Water's gift is life. No water, no life.

SANDRA POSTEL

Key Questions

13.1 Will we have enough usable water?

13.2 Is groundwater a sustainable resource?

13.3 How can we increase freshwater supplies?

13.4 Can water transfers expand water supplies?

13.5 How can we use freshwater more sustainably?

13.6 How can we reduce the threat of flooding?

Glen Canyon Dam and Lake Mead on the Colorado River, Arizona USA

Pytyczech/Dreamstime.com

Core Case Study

The Colorado River Story

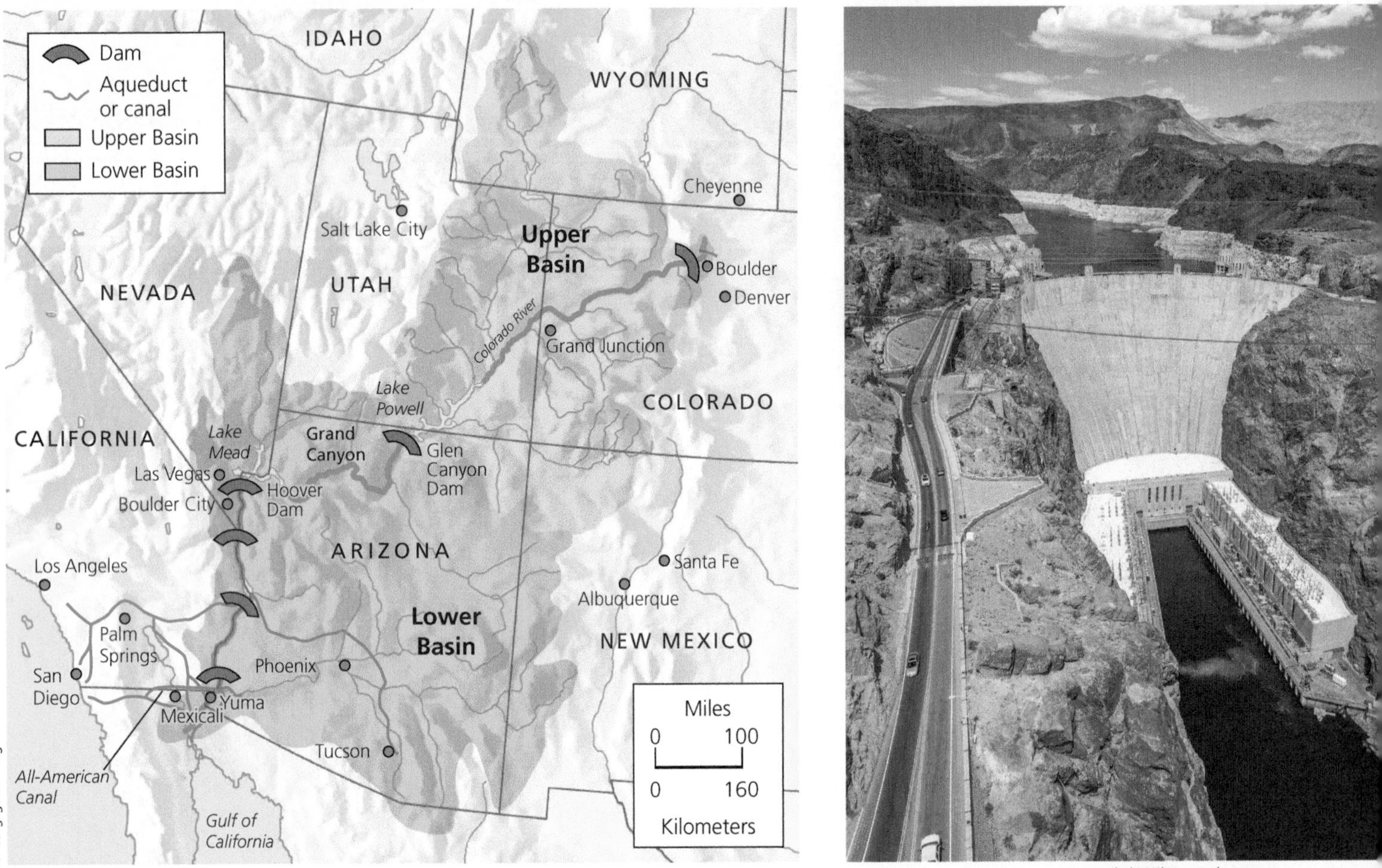

© Cengage Learning

milosk50/Shutterstock.com

FIGURE 13.1 The *Colorado River basin:* The area drained by this river system is more than one-twelfth of the land area of the lower 48 states. This map shows 6 of the river's 14 dams. The photo shows the Hoover Dam and Reservoir of Nevada and Arizona.

The Colorado River, the major river of the arid southwestern United States, flows 2,300 kilometers (1,400 miles) through seven states to the Gulf of California (Figure 13.1, map). Most of its water comes from snowmelt in the Rocky Mountains. During the past 100 years, this once free-flowing river has been tamed by a gigantic plumbing system consisting of 14 major dams and reservoirs (Figure 13.1, photo) and canals that carry water to farmers, ranchers, industries, and cities.

This system of dams and reservoirs provides electricity from its hydroelectric plants to roughly 40 million people in seven states—about one of every eight people in the United States. It supplies irrigation water that is used to produce about 15% of the nation's crops and livestock. It also provides 90% of the drinking water for Las Vegas, Nevada, and large amounts of the water used in Phoenix, Arizona, and San Diego and Los Angeles, California. Take away the Colorado River's dam-and-reservoir system, and these cities would become largely uninhabitable desert areas. In California's Imperial Valley, vast fields of vegetables would eventually give way to cactus and mesquite plants.

So much water is withdrawn from this river to grow crops and support cities in a desert-like climate that very little of it reaches the sea. To make matters worse, since 1999, the system has experienced severe **drought,** a prolonged period, usually a season or more, in which precipitation is lower than normal and evaporation is higher than normal. As a result, in 2015 Lake Mead (chapter-opening photo) sank to a record low water level.

This overuse of the Colorado River illustrates the challenges faced by governments and people living in arid and semiarid regions with shared river systems. There are many of these arid, water-short areas scattered around the globe. In such areas, population growth and economic growth are putting increasing demands on limited or decreasing supplies of surface water.

To many analysts, emerging shortages of water for drinking and irrigation in several parts of the world represent one of the major environmental challenges of this century.

13.1 WILL WE HAVE ENOUGH USABLE WATER?

CONCEPT 13.1A We are using available freshwater unsustainably by extracting it faster than nature can replace it, and by wasting, polluting, and underpricing this irreplaceable natural resource.

CONCEPT 13.1B Freshwater supplies are not evenly distributed, and 1 of every 10 people on the planet does not have adequate access to clean water.

Freshwater Is an Irreplaceable Resource That We Are Managing Poorly

We live on a planet that is unique in our solar system because of a precious layer of water—most of it saltwater—covering about 71% of its surface. Look in the mirror. What you see is about 60% water, most of it inside your cells.

Water is an amazing chemical with unique properties that help to keep us and other species alive (see Science Focus 3.2, p. 64). You could survive for several weeks without food, but for only a few days without **freshwater,** or water that contains very low levels of dissolved salts. We have no substitute for this vital form of natural capital (**Concept 13.1A**).

It takes huge amounts of water to supply food and most of the other things that we use to meet our daily needs and wants. Water also plays a key role in determining the earth's climates and in removing and diluting some of the pollutants and wastes that we produce. And over eons, it has sculpted the planet's surface, creating valleys, canyons, and other land features.

Freshwater is one of the earth's most important forms of natural capital, but despite its importance, it is also one of our most poorly managed resources. We use it inefficiently and pollute it, and we do not value it highly enough. As a result, it is available at too low a cost to billions of consumers, and this encourages waste and pollution of this resource, for which we have no substitute **(Concept 13.1A).**

Access to freshwater is a *global health issue*. The World Health Organization (WHO) has estimated that each day, an average of more than 4,100 people die from waterborne infectious diseases because they do not have access to safe drinking water.

Access to freshwater is also an *economic issue* because water is vital for producing food and energy and for reducing poverty. According to the WHO, just 52% of the world's people have water piped to their homes. The rest have to find and carry it from distant sources or wells. This daily task usually falls to women and children (Figure 13.2).

Shiv Ji Joshi/National Geographic Creative

FIGURE 13.2 Each day these women carry water to their village in a dry area of India.

Water is also a *national and global security issue* because of increasing tensions within and between some nations over access to limited freshwater resources that they share.

Finally, water is an *environmental issue* because excessive withdrawal of freshwater from rivers and aquifers has resulted in falling water tables, dwindling river flows (**Core Case Study**), shrinking lakes, and disappearing wetlands. This, in combination with water pollution in many areas of the world, has degraded water quality, reduced fish populations, hastened the extinction of some aquatic species, and degraded aquatic ecosystem services (see Figures 8.4, p. 170, 8.12, p. 177, and 8.14, p. 179).

Most of the Earth's Freshwater Is Not Available to Us

Only *0.024%* of the planet's enormous water supply is readily available to us as liquid freshwater stored in accessible underground deposits and in lakes, rivers, and streams. The rest is in the salty oceans (about 96.5% of the earth's volume of liquid water), in frozen polar ice caps and glaciers (1.7%), and underground.

Fortunately, the world's freshwater supply is continually recycled, purified, and distributed in the earth's *hydrologic cycle* (see Figure 3.19, p. 63). However, this vital ecosystem service begins to fail when we overload it with water pollutants or withdraw freshwater from underground and surface water supplies faster than natural processes replenish it.

In addition, research indicates that atmospheric warming is altering the water cycle by evaporating more water into the atmosphere. As a result, wet places will get wetter with more frequent and heavier flooding and dry places will get drier with more intense drought.

We have paid little attention to our effects on the water cycle mostly because we have thought of the earth's freshwater as a free and infinite resource. As a result, we have placed little or no economic value on the irreplaceable ecosystem services that water provides (**Concept 13.1A**), a serious violation of the full-cost pricing **principle of sustainability** (see Inside Back Cover). On a global basis, we have plenty of freshwater, but it is not distributed evenly (**Concept 13.1B**). Differences in average annual precipitation and economic resources divide the world's countries and people into water *haves* and *have-nots*. For example, Canada, with only 0.5% of the world's population, has 20% of its liquid freshwater, while China, with 19% of the world's people, has only 6.5% of the supply.

Groundwater and Surface Water Are Critical Resources

Some precipitation soaks into the ground and sinks downward through spaces in soil, gravel, and rock until an impenetrable layer of rock or clay stops it. The freshwater in these underground spaces is called **groundwater**—a key component of the earth's natural capital (Figure 13.3).

The spaces in soil and rock close to the earth's surface hold little moisture. However, below a certain depth, in the **zone of saturation,** these spaces are completely filled with freshwater. The top of this groundwater zone is the **water table.** It falls in dry weather, or when we remove groundwater from this zone faster than nature can replenish it, and it rises in wet weather.

Deeper down are geological layers called **aquifers,** caverns and porous layers of sand, gravel or rock through which groundwater flows. Some aquifers contain caverns with rivers of groundwater flowing through them. However, most aquifers are like large, elongated sponges where groundwater seeps through porous layers of sand, gravel, or rock—typically moving only a meter or so (about 3 feet) per year and rarely more than 0.3 meter (1 foot) per day. Watertight layers of rock or clay below such aquifers keep the freshwater from escaping deeper into the earth.

We use pumps to bring this groundwater to the surface for irrigating crops and supplying households and industries. Most aquifers are replenished, or *recharged,* naturally by precipitation that sinks downward through exposed soil and rock. Others are recharged from the side from nearby lakes, rivers, and streams (Figure 13.3).

According to the U.S. Geological Survey (USGS), groundwater makes up 95% of all freshwater available to us and other forms of life. However, most aquifers recharge slowly. Because so much of the earth's urban area landscapes have been built on or paved over, freshwater can no longer penetrate the ground to recharge aquifers below such areas. And in dry areas of the world, there is little precipitation available to recharge aquifers.

Some aquifers, called *deep aquifers,* were filled with water by glaciers that melted thousands of years ago. For this reason, they are also called *fossil aquifers*. Most of them, because of geological factors, cannot be recharged or will require many thousands of years to recharge. Many are located under coastal seabeds and were formed long ago when sea levels were lower and these areas were part of the land mass. Deep aquifers are nonrenewable, at least on the human timescale.

Another crucial resource is **surface water,** the freshwater from rain and melted snow that flows or is stored in lakes, reservoirs, wetlands, streams, and rivers. Precipitation that does not soak into the ground or return to the atmosphere by evaporation is called **surface runoff.** The land from which surface runoff drains into a particular stream, lake, wetland, or other body of water is called its **watershed,** or **drainage basin.** The drainage basin for the Colorado River is shown in yellow and green on the map in Figure 13.1 (**Core Case Study**).

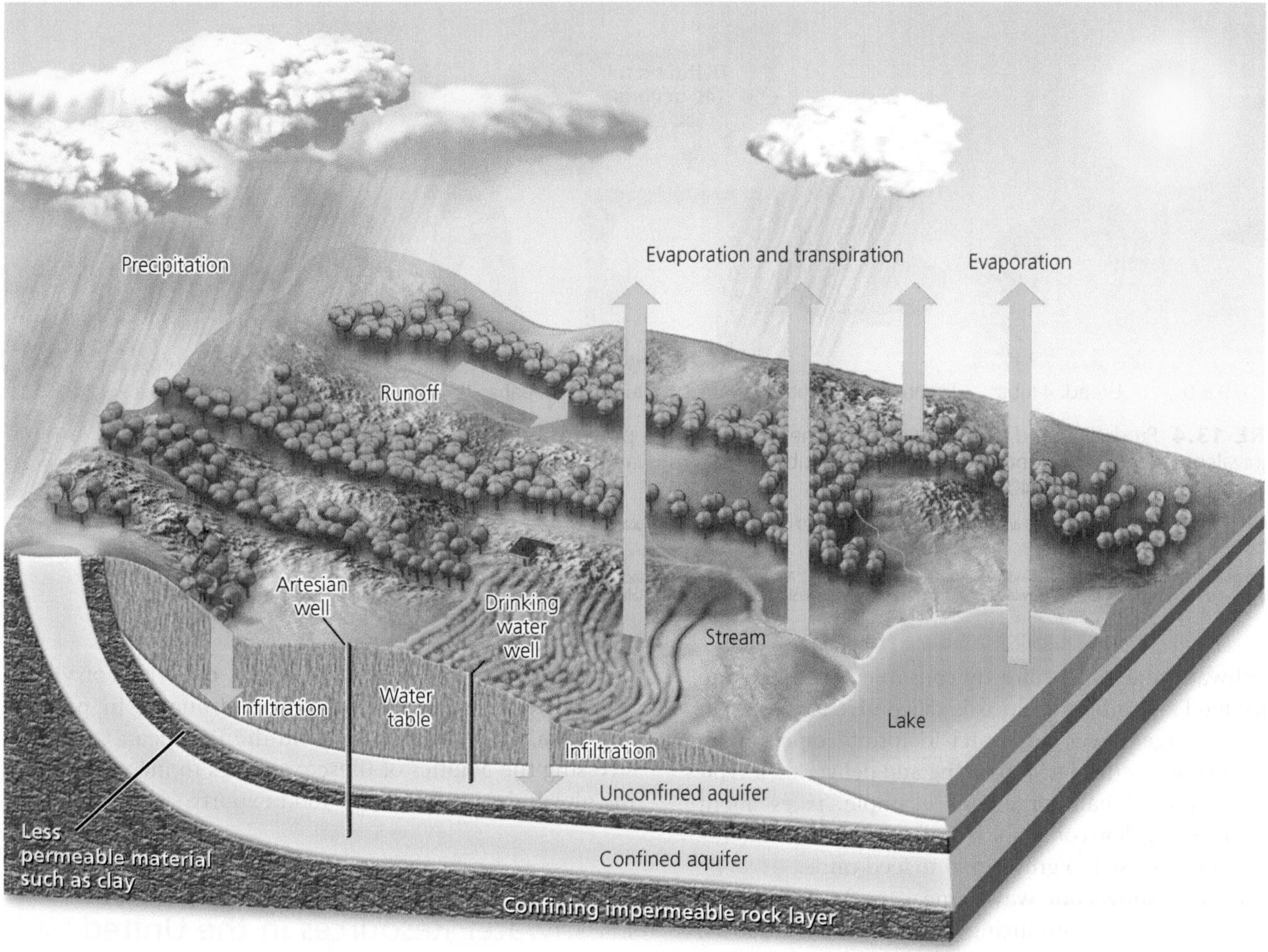

FIGURE 13.3 Natural capital: Much of the water that falls in precipitation seeps into the ground to become groundwater, stored in aquifers.

CONSIDER THIS . . .

CONNECTIONS Groundwater and Surface Water

There is usually a connection between surface water and groundwater because much groundwater flows into rivers, lakes, estuaries, and wetlands. Thus, if we remove groundwater in a particular location faster than it is replenished, nearby streams, lakes, and wetlands can dry up. This process degrades aquatic biodiversity and other ecosystem services.

We Are Using Increasing Amounts of the World's Reliable Runoff

According to *hydrologists*, scientists who study water and its properties and movement, two-thirds of the annual surface runoff of freshwater into rivers and streams is lost in seasonal floods and is not available for human use. The remaining one-third is **reliable surface runoff,** which we can generally count on as a source of freshwater from year to year. **GREEN CAREER: Hydrologist**

During the last century, the human population tripled, global water withdrawals increased sevenfold, and per capita withdrawals quadrupled. As a result, we now withdraw an estimated 34% of the world's reliable runoff. This is a global average. In the arid American Southwest, up to 70% of the reliable runoff is withdrawn for human purposes, mostly for irrigation (**Core Case Study**). Some water experts project that because of population growth, rising rates of water use per person, longer dry periods in some areas, and unnecessary water waste, we are likely to be withdrawing up to 90% of the world's reliable freshwater runoff by 2025.

Worldwide, we use 70% of the freshwater we withdraw each year from rivers, lakes, and aquifers to irrigate cropland and raise livestock. In arid regions, up to 90% of the regional water supply is used for food production. Industry uses roughly another 20% of the water withdrawn globally each year, and cities and residences use the remaining 10%. Our **water footprint** is a rough measure of the volume of freshwater that we use or pollute, directly and indirectly, to stay alive and to support our lifestyles. (See the Case Study that follows for information on U.S. water use.)

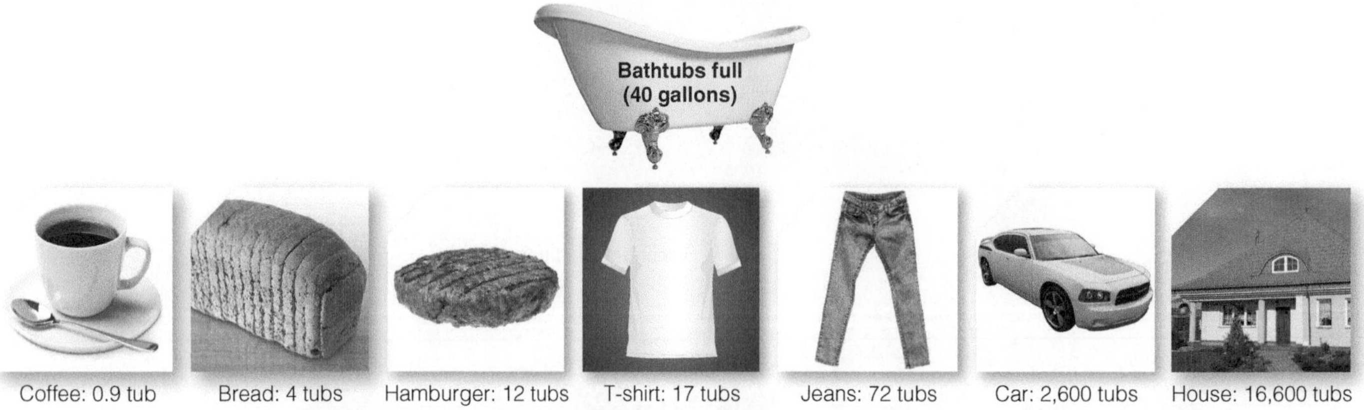

FIGURE 13.4 Producing and delivering a single one of each of the products listed here requires the equivalent of nearly one and usually many bathtubs full of freshwater, called *virtual water*. *Note:* 1 bathtub = 151 liters (40 gallons).

(Compiled by the authors using data from UN Food and Agriculture Organization, UNESCO-IHE Institute for Water Education, World Water Council, and Water Footprint Network.)

Bathtub: Baloncici/Shutterstock.com. Coffee: Aleksandra Nadeina/Shutterstock.com. Bread: Alexander Kalina/Shutterstock.com. Hamburger: Joe Belanger/Shutterstock.com. T-shirt: grmarc/Shutterstock.com. Jeans: Eyes wide/Shutterstock.com. Car: L Barnwell/Shutterstock.com. House: Rafal Olechowski/Shutterstock.com

Freshwater that is not directly consumed but is used to produce food and other products is called **virtual water.** It makes up a large part of our water footprints, especially in more-developed countries. Producing and delivering a typical quarter-pound hamburger, for example, takes about 2,400 liters (630 gallons or about 16 bathtubs) of freshwater—most of which is used to grow grain to feed cattle.

Figure 13.4 shows one way to measure the amounts of virtual water used for producing and delivering products. These values can vary depending on how much of the supply chain is included, but they give us a rough estimate of the size of our water footprints.

Because of global trade, the virtual water used to produce and transport products such as coffee and wheat (also called *embedded water*) is often withdrawn as groundwater or surface water in another part of the world. Thus, water can be imported in the form of products, often from countries that are short of water.

The three largest water footprints in the world belong to India, the United States, and China, in that order. Each of these is at least twice the size of any other country's footprint outside of the top three. Large exporters of virtual water—mostly in the form of wheat, corn, soybeans, and other foods—are the European Union, the United States, Canada, Brazil, India, and Australia. Indeed, Brazil's supply of freshwater per person is more than 8 times the U.S. supply per person, 14 times China's supply, and 29 times India's supply. Brazil is becoming one of the world's largest exporters of virtual water. However, prolonged severe droughts in parts of Australia, the United States, and the European Union are stressing the abilities of these countries to meet the growing global demand for their food exports.

CASE STUDY

Freshwater Resources in the United States

According to the USGS, the major uses of groundwater and surface freshwater in the United States are the cooling of electric power plants, irrigation, public water supplies, industry, and livestock production (Figure 13.5, left).

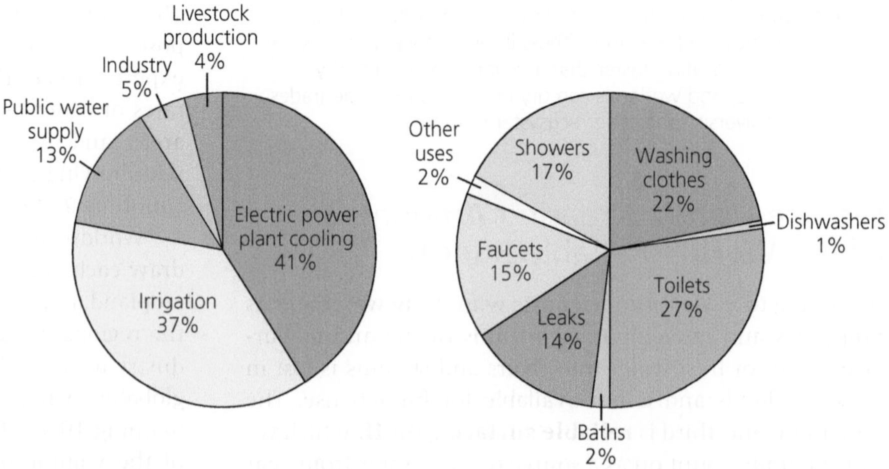

FIGURE 13.5 Comparison of primary uses of water in the United States (left) and uses of water in a typical U.S. household (right). ***Data analysis:*** Which three categories, added together (right), are smaller than the amount of water lost in leaks?

(Compiled by the authors using data from U.S. Geological Survey, World Resources Institute, and American Water Works Association.)

Every day, the average American directly uses between 300 and 377 liters (80 to 100 gallons) of freshwater—enough water to fill at least two typical bathtubs full of water. (A typical bathtub can contain about 151 liters or 40 gallons of water.) Household water is used mostly for flushing toilets, washing clothes, taking showers, and running faucets, or is lost through leaking pipes, faucets, and other fixtures (Figure 13.5, right).

The United States has more than enough renewable freshwater to meet its needs. However, it is unevenly distributed and much of it is contaminated by agricultural and industrial practices. The eastern states usually have ample precipitation, whereas many western and southwestern states have little (Figure 13.6).

In the eastern United States, most water is used for manufacturing and for cooling power plants (with most of the water heated and returned to its source). In many parts of this area, the most serious water problems are flooding, occasional water shortages because of drought, and pollution.

In the arid and semiarid regions of the western half of the United States (**Core Case Study**), irrigation counts for as much as 85% of freshwater use. Much of it is lost to evaporation and a great deal of it is used to grow thirsty crops. The major water problem is a shortage of freshwater runoff caused by low precipitation (Figure 13.6), high evaporation, and recurring prolonged drought.

Groundwater is one of the most precious of all U.S. resources. About half of all Americans (and 95% of all rural residents) rely on it for drinking water. It makes up about half of all irrigation water, feeds about 40% of the country's streams and rivers, and provides about one-third of the water used by U.S. industries.

Water tables in many water-short areas, especially in the dry western states, are dropping quickly as farmers and rapidly growing urban areas draw down many aquifers faster than they can be recharged. In 2010 the USGS estimated total U.S. water withdrawals to be at the lowest level since 1970. However, in 2014 the U.S. Government Accountability Office found that even with per capita water use dropping, water managers from 40 of the 50 states expected water shortages in some areas by 2025 or before. Their projections factored in drought, population growth, urban sprawl, and rising consumption of meat and other water-intensive products.

The U.S. Department of the Interior has mapped out *water hotspots* in 17 western states (Figure 13.7). In these areas, competition for scarce freshwater to support growing urban areas, irrigation, recreation, and wildlife could trigger intense political and legal conflicts between states and between rural and urban areas within states. In addition, Columbia University climate researchers led by Richard Seager used well-tested climate models to project that the southwestern United States is very likely to have long periods of extreme drought throughout most of the rest of this century. According to current research, atmospheric warming does not cause drought. However, it makes a drought worse because warmer temperatures dry out the soil, reducing evaporation of soil moisture, which normally helps reduce drought conditions.

The Colorado River system (Figure 13.1) will be directly affected by such drought. There are three major problems associated with the use of freshwater from this river (**Core Case Study**). *First*, the Colorado River basin includes some

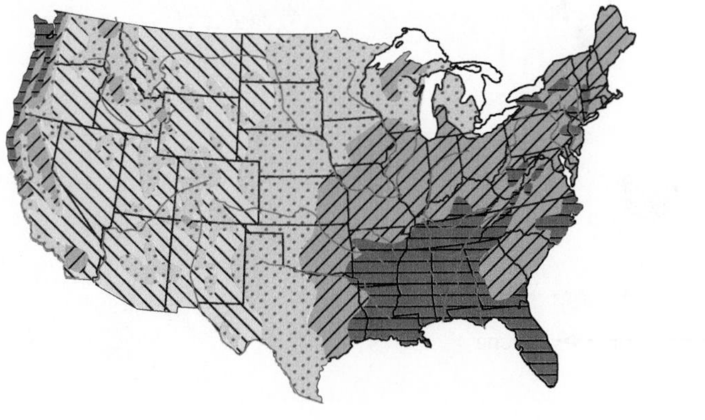

FIGURE 13.6 Long-term average annual precipitation and major rivers in the continental United States.

(Compiled by the authors using data from U.S. Water Resources Council and U.S. Geological Survey.)

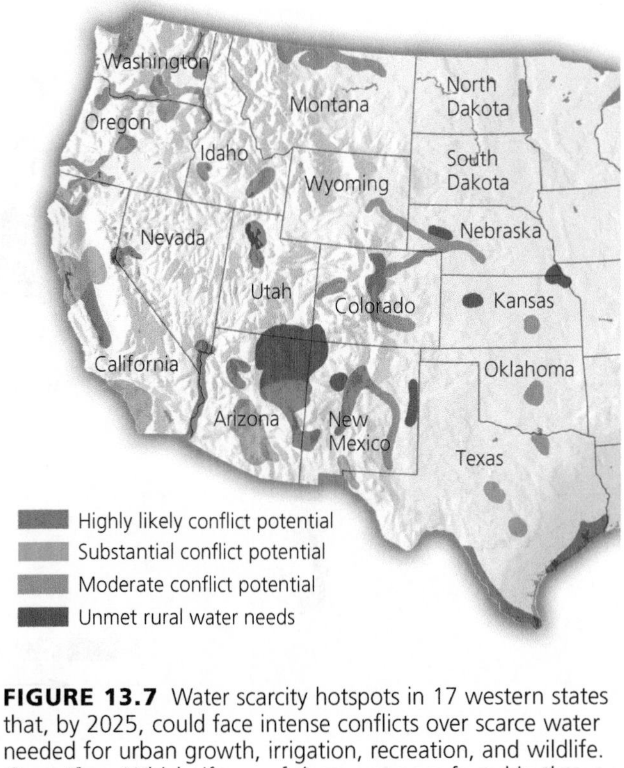

FIGURE 13.7 Water scarcity hotspots in 17 western states that, by 2025, could face intense conflicts over scarce water needed for urban growth, irrigation, recreation, and wildlife. ***Question:*** Which, if any, of these areas are found in the Colorado River basin (**Core Case Study**)?

(Compiled by the authors using data from U.S. Department of the Interior and U.S. Geological Survey.)

of the driest lands in the United States and Mexico. *Second,* long-standing legal agreements between Mexico and the affected western states allocated more freshwater for human use than the river can supply, even in rare years when there is no drought. These pacts allocated no water for protecting aquatic and terrestrial wildlife. *Third,* since 1960, because of drought, damming, and heavy withdrawals, the river has rarely flowed all the way to the Gulf of California and this has severely degraded the river's aquatic ecosystem (which we discuss further later in this chapter).

Freshwater Shortages Will Grow

Freshwater scarcity stress is a measure based on the amount of freshwater available compared to the amount used for human purposes. Like the Colorado River (**Core Case Study**), many of the world's major river systems are highly stressed (Figure 13.8). They include the Nile, Jordan, Yangtze, and Ganges Rivers, whose flows regularly dwindle to almost nothing in some locations.

More than 30 countries—most of them in the Middle East and Africa—now face stress from freshwater scarcity, according to the UN. By 2050, some 60 countries, many of them in Asia, with three-fourths of the world's population, are likely to be suffering from such freshwater scarcity stress. The Chinese government has reported that two-thirds of China's 600 major cities face freshwater shortages.

Currently, about 30% of the earth's land area—a total area roughly 5 times the size of the United States—experiences severe drought. By 2059, as much as 45% of the earth's land surface could experience an even higher level of drought, called *extreme drought,* due to natural cycles and projected climate change, according to a study by climate researcher David Rind and his colleagues.

In 276 of the world's water basins, two or more countries share the available freshwater supplies. However, not all of these countries participate in water-sharing agreements. As a result, international conflicts over water are likely to occur as populations grow, as demand for water increases, and as supplies shrink in many parts of the world.

In 2015 the United Nations (UN) and the WHO reported that about 783 million people—about 2.4 times the U.S. population—did not have regular access to enough clean water for drinking, cooking, and washing, mostly due to poverty (**Concept 13.1B**). The report also noted that more than 2 billion people had gained access to clean water between 1990 and 2012. GOOD NEWS

783 million

Number of people without regular access to clean water

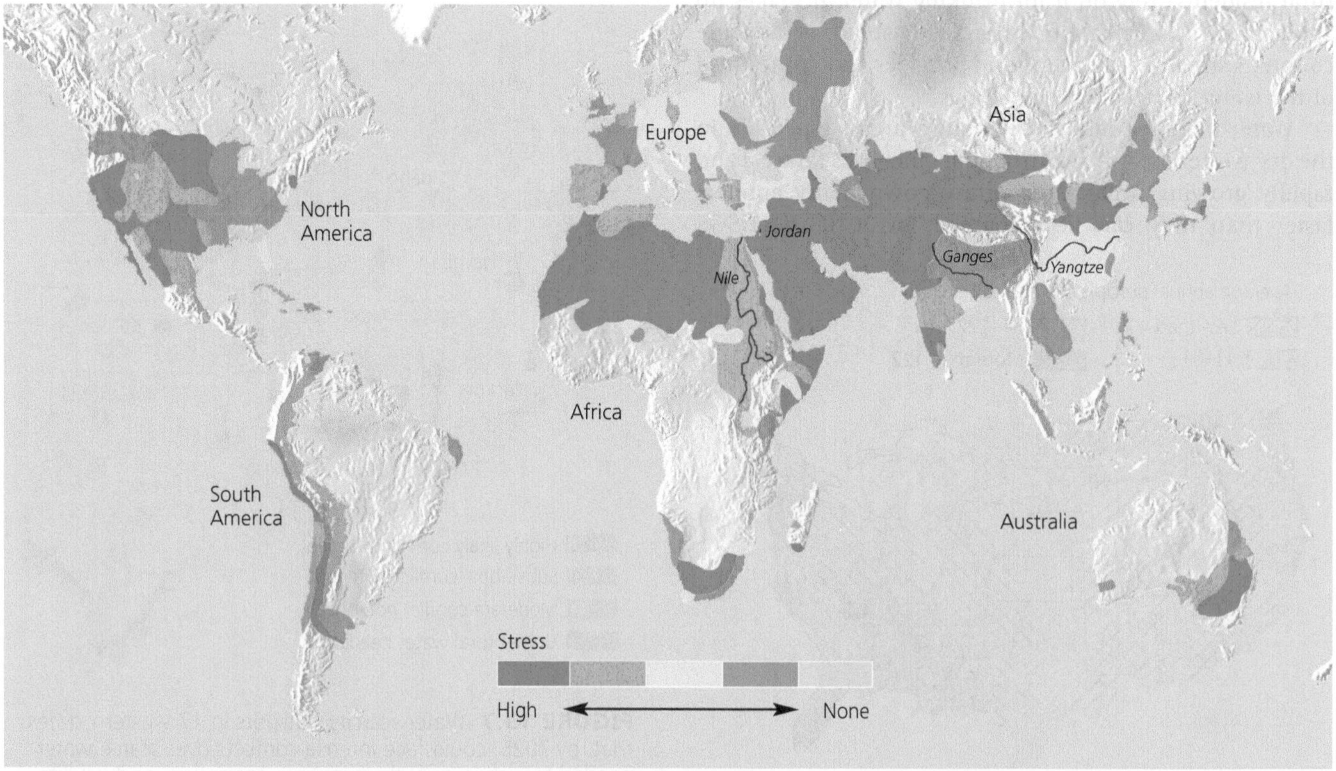

FIGURE 13.8 Natural capital degradation: The world's major river basins differ in their degree of freshwater *scarcity stress* (**Concept 13.1B**).

(Compiled by the authors using data from World Commission on Water Use for the 21st Century, UN Food and Agriculture Organization, and World Water Council.)

Using Satellites to Monitor Groundwater Supplies

Since 2002, hydrologist Jay S. Famiglietti and his colleagues have been using two twin satellites (Figure 13.A) to measure small variations in the planet's gravitational pull that tell them about changes in ice cover, snow cover, surface water, soil moisture, and groundwater supplies. The two satellites, each the size of a small car, travel in the same orbit, one about 217 kilometers (135 miles) behind the other. They constantly beam microwaves toward each other and can detect changes of less than the diameter of a human hair in this distance between the satellites. When the leading satellite speeds up, increasing the distance, it means the mass of the earth beneath the satellite has increased, tugging a bit harder on the satellite. Various forms of water content on and under the surface of the earth can cause such an increase.

By comparing thousands of these satellite measurements to data from ground measurements and computer modeling, Famiglietti and his team have learned what various changes in the data mean. In 2015 the team reported new data indicating that in 21 of the world's 37 largest aquifers, more water was withdrawn between 2003 and 2013 than had been replaced by snow and rainfall, and some were being depleted rapidly. These aquifers provide freshwater for hundreds of millions of people, and 13 of them were showing little or no recharge over that 10-year period.

For example, in California's agricultural Central Valley, aquifers were drawn down by more than enough to fill Lake Powell, the second-largest reservoir in the United States. According to the scientists, the satellite data indicate that large aquifers in several other areas, including North Africa, northeastern China, and northern India, are also being overpumped. These are all areas where water shortages are worsening every year.

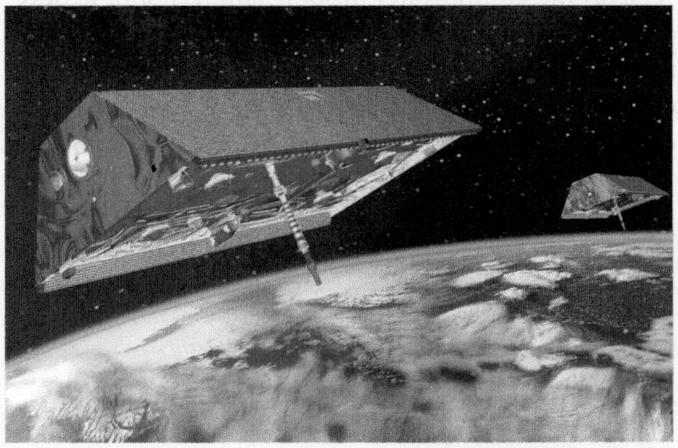

NASA

FIGURE 13.A These twin satellites measure slight variations in the earth's gravitational field that scientists use to detect and measure changes in large bodies of groundwater.

CRITICAL THINKING

If you were a scientist on this team, how would you double-check the data you got from the satellites? (*Hint:* think of other ways to take such measurements.)

Many analysts view the likelihood of expanding water shortages as one of our most serious environmental, health, and economic challenges. Scientists have found a number of ways to obtain information that will help in meeting this challenge, including use of satellites (Science Focus 13.1). In the following section, we will explore some approaches to dealing with water shortages.

CONSIDER THIS . . .

CONNECTIONS Virtual Water, Grain, and Hunger

Some water-short countries are reducing their irrigation water needs by importing grain, thereby freeing up more of their own water supplies for industrial and urban development. The result is a competition for the world's grain, which includes indirect competition for water (virtual water used to grow grain). This competition could increase grain prices, lead to food shortages, and increase hunger among the poor.

13.2 IS GROUNDWATER A SUSTAINABLE RESOURCE?

CONCEPT 13.2 Groundwater used to supply cities and grow food is being pumped from many aquifers faster than it is renewed by precipitation.

Groundwater Withdrawals are Unsustainable in Some Areas

Aquifers provide drinking water for nearly half of the world's people. Most aquifers are renewable resources unless the groundwater they contain becomes contaminated or is removed faster than it is replenished. Relying more on groundwater has advantages and disadvantages (Figure 13.9).

Trade-Offs

Withdrawing Groundwater

Advantages	Disadvantages
Useful for drinking and irrigation	Aquifer depletion from overpumping
Exists almost everywhere	Sinking of land (subsidence) from overpumping
Renewable if not overpumped or contaminated	Some deeper aquifers are nonrenewable
Cheaper to extract than most surface waters	Pollution of aquifers lasts decades or centuries

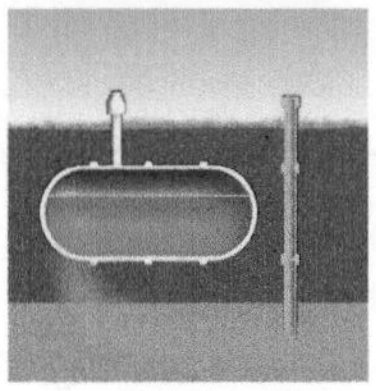

FIGURE 13.9 Withdrawing groundwater from aquifers has advantages and disadvantages. ***Critical thinking:*** Which two advantages and which two disadvantages do you think are the most important? Why?

Top: Ulrich Mueller/Shutterstock.com

Test wells and satellite data (Science Focus 13.1) indicate that water tables are falling in many areas of the world. One reason is that the rate at which water is being pumped out of most of the world's aquifers (mostly to irrigate crops) is greater than the rate of natural recharge from rainfall and snowmelt (**Concept 13.2**). The world's three largest grain producers—China, the United States, and India—as well as Mexico, Saudi Arabia, Iran, Iraq, Egypt, Pakistan, Spain, and other countries are overpumping many of their aquifers.

Every day, the world withdraws enough freshwater from aquifers to fill a convoy of large tanker trucks that could stretch 480,000 kilometers (300,000 miles)—well beyond the distance to the moon. According to the World Bank, in 2012 more than 400 million people were consuming grain produced through this unsustainable use of groundwater. This number is growing.

The widespread drilling of wells by farmers, especially in India and China, has accelerated aquifer overpumping. As water tables fall, farmers drill deeper wells and buy larger pumps to bring more water to the surface. This process eventually depletes the groundwater in some aquifers or at least removes all the water that can be pumped at an affordable cost.

In Saudi Arabia, freshwater has been pumped from a deep, nonrenewable aquifer to irrigate crops such as wheat (Figure 13.10). It also is used to fill fountains and

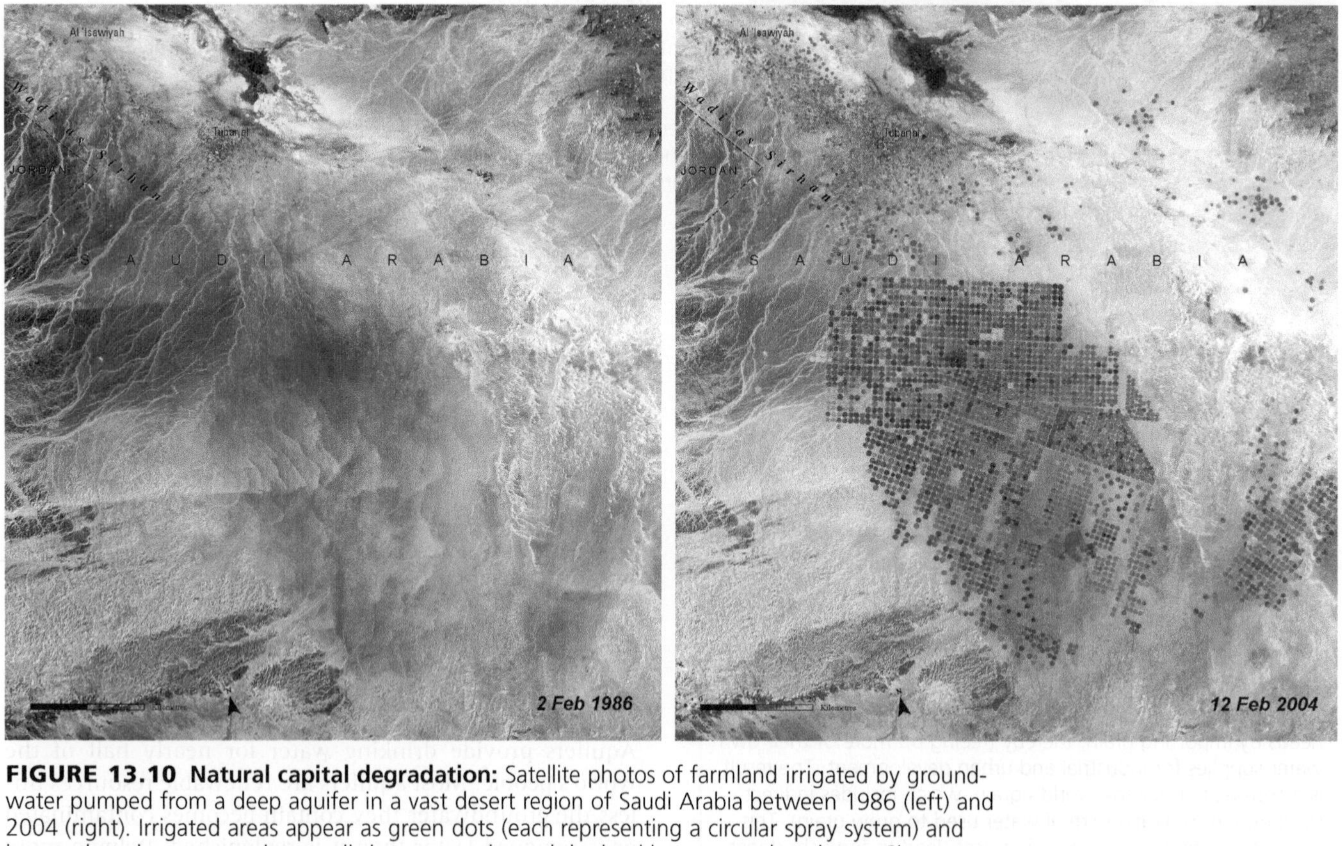

FIGURE 13.10 Natural capital degradation: Satellite photos of farmland irrigated by groundwater pumped from a deep aquifer in a vast desert region of Saudi Arabia between 1986 (left) and 2004 (right). Irrigated areas appear as green dots (each representing a circular spray system) and brown dots show areas where wells have gone dry and the land has returned to desert. Since 2004, many more wells have gone dry.

Left: U.N. Environment Programme and U.S. Geological Survey. Right: U.N. Environment Programme and U.S. Geological Survey.

swimming pools, which lose a great deal of water through evaporation into the dry desert air. In 2008 Saudi Arabia announced that it had largely depleted its major deep aquifer. The country stopped producing wheat in 2016 and will continue to import grain (virtual water) to help feed its 32 million people.

In the United States, aquifer depletion is a growing problem, especially in the vast Ogallala Aquifer (see Case Study that follows).

CASE STUDY

Overpumping the Ogallala Aquifer

In the United States, groundwater is being withdrawn from aquifers, on average, four times faster than it is replenished, according to the USGS. One of the most serious overdrafts of groundwater is in the lower half of the Ogallala Aquifer, one of the world's largest known aquifers, which lies under eight Midwestern states from southern South Dakota to Texas (Figure 13.11).

The Ogallala Aquifer supplies about one-third of all the groundwater used in the United States and turned the Great Plains into one of world's most productive irrigated agricultural regions (Figure 13.12). The Ogallala is essentially a one-time deposit of liquid natural capital with a slow rate of recharge. *Hydrogeologists* (scientists who study groundwater and its movements) estimate that since 1960, we have withdrawn between a third and half of this water and that if it were to be depleted, it could take 6,000 years to recharge naturally.

In parts of the southern half of the Ogallala, groundwater is being pumped out 10–40 times faster than the slow natural recharge rate. This, along with reduced access to Colorado River water (**Core Case Study**) and population growth, has led to the shrinkage of irrigated croplands in Texas, Arizona, Colorado, and California. It has also led to increased competition for water among farmers, ranchers, and growing urban areas.

Government *subsidies*—payments or tax breaks designed to increase crop production—have encouraged farmers to continue growing water-thirsty crops in dry areas, which has accelerated depletion of the Ogallala Aquifer. In particular, corn—a very thirsty crop—has been planted widely on fields watered by the Ogallala.

The aquifer also supports biodiversity. In various places, groundwater from the Ogallala flows out of the ground onto land or onto lake bottoms through exit points called *springs*. In some cases, springs feed wetlands, which are vital habitats for many species, especially birds. When the water tables fall, many of these aquatic oases of biodiversity dry out.

FIGURE 13.11 Natural capital degradation: Areas of greatest aquifer depletion from groundwater overdraft in the continental United States. The blowup section (right) shows where water levels in the Ogallala Aquifer have dropped sharply at its southern end beneath parts of Kansas, Oklahoma, Texas, and New Mexico. ***Critical thinking:*** Should the amount of water that farmers can withdraw from the Ogallala Aquifer be restricted? How would you enforce this? How might this affect food production?

(Compiled by the authors using data from U.S. Water Resources Council and U.S. Geological Survey.)

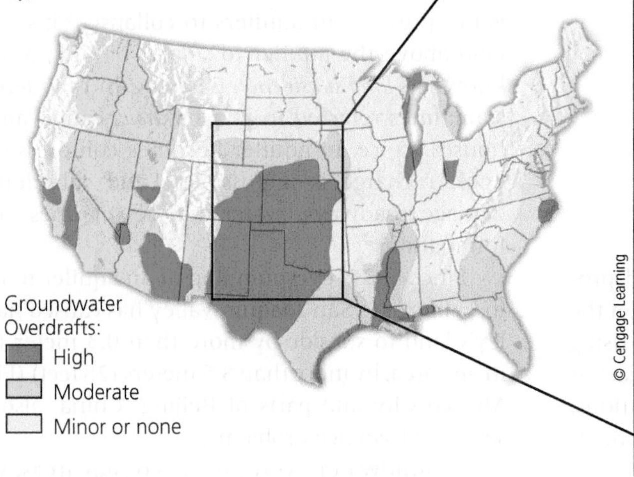

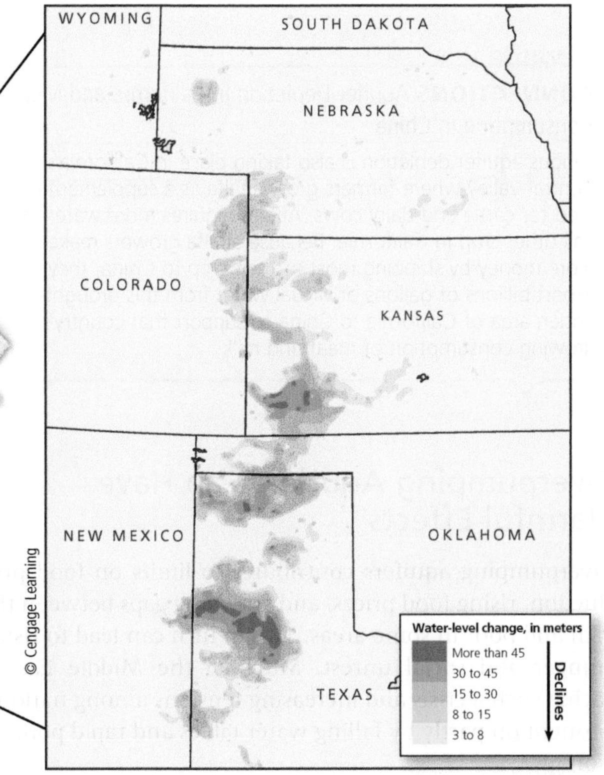

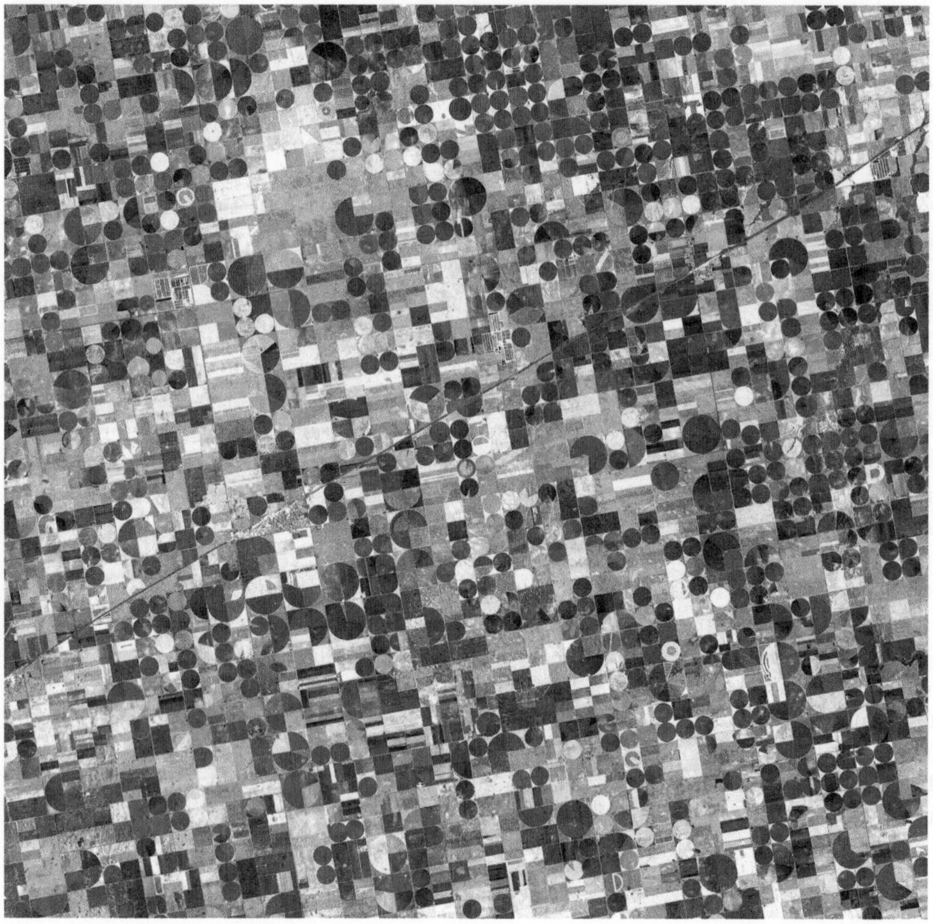

NASA/GSFC/METI/ERSDAC/JAROS, and U.S./Japan ASTER Science Team

FIGURE 13.12 Satellite photo of crop fields in the U.S. state of Kansas. Center-pivot irrigation uses long, suspended pipes that swing around a central point in each field. Dark green circles are irrigated fields of corn, light green circles are sorghum, and light yellow circles are wheat. Brown areas are fields that have been recently harvested and plowed under. The water used to irrigate these crops is pumped from the Ogallala Aquifer.

CONSIDER THIS . . .

CONNECTIONS Aquifer Depletion in California and Meat Consumption in China

Serious aquifer depletion is also taking place in California's Central Valley where farmers grow alfalfa as a supplemental feed for cattle and dairy cows. Alfalfa requires more water than any other crop in California. Because alfalfa growers make more money by shipping most of their crop to China, they export billions of gallons of virtual water from this drought-ridden area of California to China to support that country's growing consumption of meat and milk.

Overpumping Aquifers Can Have Harmful Effects

Overpumping aquifers contributes to limits on food production, rising food prices, and widening gaps between the rich and poor in some areas. This in turn can lead to rising hunger and social unrest. Much of the Middle East is facing such a crisis and increasing tensions among nations, brought on partly by falling water tables and rapid population growth.

Also, as water tables drop, the energy and financial costs of pumping the water from lower depths rise sharply because farmers must drill deeper wells, buy larger pumps, and use more electricity to run the pumps. Poor farmers cannot afford such costs and often lose their land. This forces them to work for richer farmers or to migrate to cities that are crowded with people struggling to survive.

Withdrawing large amounts of groundwater sometimes causes the sand and rock that is held in place by water pressure in aquifers to collapse. This can cause the land above the aquifer to *subside* or sink, a phenomenon known as *land subsidence*. Extreme and sudden subsidence, sometimes referred to as a *sinkhole*, can swallow cars and houses. Once an aquifer becomes compressed by subsidence, recharge is impossible. Land subsidence can also damage roadways, water and sewer lines, and building foundations.

Since 1925, overpumping of an aquifer to irrigate crops in California's San Joaquin Valley has caused half of the valley's land to subside by more than 0.3 meter (1 foot) and, in one area, by more than 8.5 meters (28 feet) (Figure 13.13). Mexico City and parts of Beijing, China, also suffer from severe subsidence problems.

Groundwater overdrafts in coastal areas, where many of the world's largest cities and industries are found, can pull saltwater into freshwater aquifers. The resulting

Dick Ireland/U.S. Geological Survey

FIGURE 13.13 This pole shows subsidence from overpumping of an aquifer for irrigation in California's San Joaquin Central Valley between 1925 and 1977. In 1925 this area's land surface was near the top of the pole. Since 1977, this problem has gotten worse.

contaminated groundwater is undrinkable and unfit for irrigation. This problem is especially serious in coastal areas of the U.S. states of California, Texas, Florida, Georgia, South Carolina, and New Jersey, as well as in coastal areas of Turkey, Thailand, and the Philippines.

Figure 13.14 lists ways to prevent or slow the problem of aquifer depletion by using this potentially renewable resource more sustainably. The challenge is to educate people about the dangers of depleting vital underground supplies of water that they cannot see.

Deep Aquifers Might Be Tapped

With global shortages of freshwater looming, scientists are evaluating deep aquifers as future sources of freshwater. Preliminary results suggest that some of these aquifers hold enough freshwater to support billions of people for centuries.

There are five major problems related to tapping these ancient deposits of freshwater. *First,* they are nonrenewable on a human time scale. *Second,* little is known about the geological and ecological impacts of pumping large amounts of freshwater from deep aquifers, especially those located under seabeds. *Third,* no international treaties govern access to deep aquifers that flow beneath more than one country. Without such treaties, wars could break out over this resource. *Fourth,* the costs of tapping deep aquifers are unknown and could be high. *Fifth,* recent research indicates that much of this water is salty and contaminated with arsenic and uranium.

Recent research indicates that the supply of freshwater available from renewable aquifers not too far underground is smaller than previous estimates. Thus, our current unsustainable use of many of these aquifers is a serious environmental problem that threatens this vital source of freshwater.

Solutions

Groundwater Depletion

Prevention	Control
Use water more efficiently	Raise price of water to discourage waste
Subsidize water conservation	Tax water pumped from wells near surface water
Limit number of wells	Build rain gardens in urban areas
Stop growing water-intensive crops in dry areas	Use permeable paving material on streets, sidewalks, and driveways

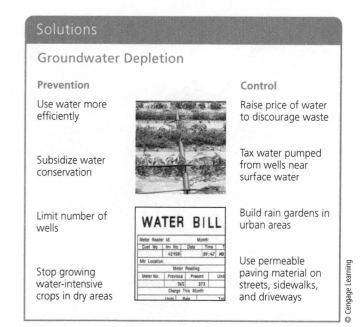

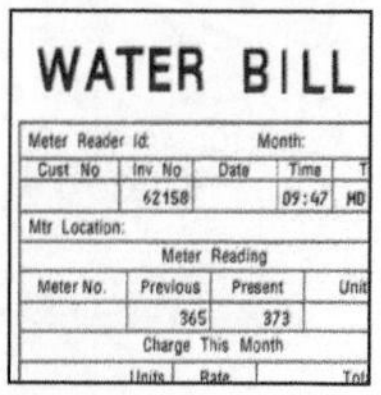

© Cengage Learning

FIGURE 13.14 Ways to prevent or slow groundwater depletion by using freshwater more sustainably. ***Critical thinking:*** Which two of these solutions do you think are the most important? Why?

Top: Anhong/Dreamstime.com. Bottom: Banol2007/Dreamstime.com.

13.3 HOW CAN WE INCREASE FRESHWATER SUPPLIES?

CONCEPT 13.3A Large dam-and-reservoir systems and water transfer projects have greatly expanded water supplies in some areas, but have also disrupted ecosystems and displaced people.

CONCEPT 13.3B We can convert salty ocean water to freshwater, but the energy and other costs are high, and the resulting salty brine must be disposed of without harming aquatic or terrestrial ecosystems.

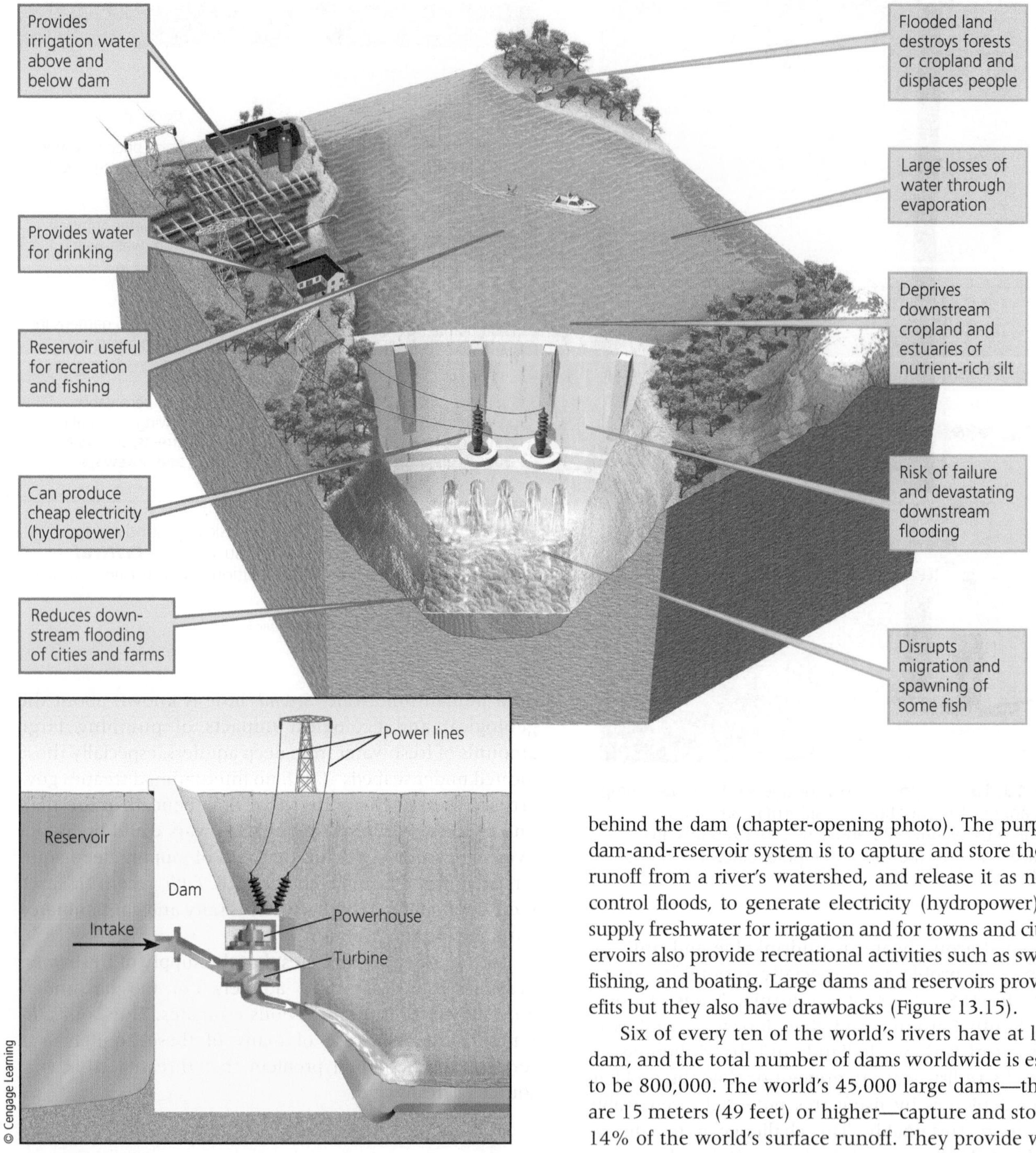
Provides irrigation water above and below dam
Provides water for drinking
Reservoir useful for recreation and fishing
Can produce cheap electricity (hydropower)
Reduces downstream flooding of cities and farms
Flooded land destroys forests or cropland and displaces people
Large losses of water through evaporation
Deprives downstream cropland and estuaries of nutrient-rich silt
Risk of failure and devastating downstream flooding
Disrupts migration and spawning of some fish
Power lines
Reservoir
Dam
Intake
Powerhouse
Turbine
© Cengage Learning

FIGURE 13.15 Trade-offs: Large dam and reservoir systems have advantages (green) and disadvantages (orange) (**Concept 13.3**). ***Critical thinking:*** Which single advantage and which single disadvantage do you think are the most important? Why?

Large Dams Provide Benefits and Create Problems

A **dam** is a structure built across a river to control its flow. Usually, dammed water creates an artificial lake, or **reservoir,** behind the dam (chapter-opening photo). The purpose of a dam-and-reservoir system is to capture and store the surface runoff from a river's watershed, and release it as needed to control floods, to generate electricity (hydropower), and to supply freshwater for irrigation and for towns and cities. Reservoirs also provide recreational activities such as swimming, fishing, and boating. Large dams and reservoirs provide benefits but they also have drawbacks (Figure 13.15).

Six of every ten of the world's rivers have at least one dam, and the total number of dams worldwide is estimated to be 800,000. The world's 45,000 large dams—those that are 15 meters (49 feet) or higher—capture and store about 14% of the world's surface runoff. They provide water for almost half of all irrigated cropland and supply more than half the electricity used in 65 countries. The United States has about 75,000 dams, according to the U.S. Army Corps of Engineers. They capture and store about half of the country's entire river flow.

Dams have increased the annual reliable runoff available for our uses by nearly 33%. As a result, the world's reservoirs now hold 3–6 times more freshwater than the total amount flowing at any moment in all of the world's natural rivers. On the downside, this engineering approach to river management has displaced 40 million to 80 million people from their homes and impaired some of the important ecosystem services that rivers provide (see Figure 8.14, left, p. 179) (**Concept 13.3A**).

A study by the World Wildlife Fund (WWF) estimated that about one out of five of the world's freshwater fish and plant species are either extinct or endangered, primarily because dams and water withdrawals have sharply decreased certain river flows. The study found that only 21 of the planet's 177 longest rivers consistently run all the way to the sea before running dry. As a result, aquatic habitat along rivers and at their mouths has been severely degraded (see Case Study that follows).

Within 50 years, reservoirs behind dams typically fill up with sediments (mud and silt), which makes them useless for storing water or producing electricity. In the Colorado River system (**Core Case Study**), the equivalent of roughly 20,000 dump-truck loads of silt are deposited on the bottoms of the Lake Powell and Lake Mead reservoirs every day. Eventually, these two reservoirs will be too full of silt to function as designed. According to American Rivers, almost 1,000 U.S. dams were removed between 1976 and 2014. More dams will be removed as they fill with silt, because about 85% of all U.S. dam-and-reservoir systems will be 50 years old or more by 2025.

If climate change occurs as projected, it will intensify shortages of water in many parts of the world. For example, mountain snows that feed the Colorado River system (**Core Case Study**) will melt faster and earlier, making less freshwater available to the river system when it is needed for irrigation during hot and dry summer months.

If some of the Colorado River's largest reservoirs keep dropping dramatically or become filled with silt during this century, the region will experience costly water and economic disruptions. For example, by 2013, the water level in Lake Mead had dropped below the Hoover Dam's intake pipes. The city of Las Vegas has been forced to spend more than $800 million to build lower intake pipes in order to maintain hydroelectric production.

Also likely are political and legal battles over who will get how much of the region's greatly diminished freshwater supply. Agricultural production would drop sharply and the region's major desert cities would be challenged to survive. A report from the U.S. Bureau of Reclamation concluded that over the next 50 years, the Colorado River will not be able to meet the projected water demands of Arizona, New Mexico, and California.

Nearly 3 billion people in South America, China, India, and other parts of Asia—that is, nearly half the world's population—depend on river flows fed by mountain glaciers, which act like aquatic savings accounts. They store precipitation as ice and snow in wet periods and release it slowly during dry seasons for use on farms and in cities. In 2015, according to the World Glacier Monitoring Service, many of these mountain glaciers had been shrinking for 24 consecutive years, mostly due to a warming atmosphere.

CASE STUDY

How Dams Can Kill an Estuary

Since 1905, the amount of water flowing to the mouth of the Colorado River (**Core Case Study**) has dropped dramatically. In most years since 1960, the river has dwindled to a small, sluggish stream by the time it reaches the Gulf of California (Figure 13.16).

The Colorado River once emptied into a vast *delta*, the wetland area at the mouth of a river containing the river's estuary. The delta covered an area of more than 800,000 hectares (2 million acres)—the size of the state of Rhode Island. It hosted forests, lagoons, and marshes rich in plant and animal life and supported a thriving coastal fishery for hundreds of years.

Since the damming of the Colorado River—within one human lifetime—this biologically diverse delta ecosystem has collapsed and is now covered mostly by mud flats and desert. All but one-tenth of the river's flow was diverted for use in seven U.S. states. Most of the remaining 10% is assigned to farms and to the growing cities of Mexicali and Tijuana in Mexico. The delta's wildlife are now mostly gone and its coastal fishery that fed many generations of area residents is disappearing.

Historically, about 80% of the water withdrawn from the Colorado has been used to irrigate crops and raise cattle. That is because the government paid for the dams and reservoirs and has supplied many farmers and ranchers with water at low prices. These government subsidies

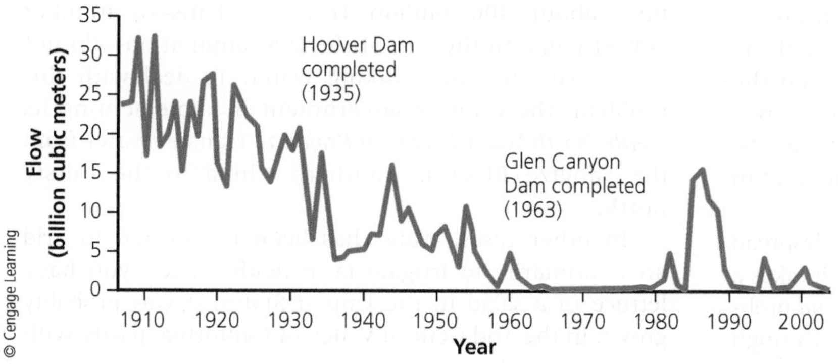

FIGURE 13.16 The measured flow of the Colorado River (**Core Case Study**) at its mouth has dropped sharply since 1905 as a result of multiple dams, water withdrawals for agriculture and urban areas, and prolonged drought. ***Data analysis:*** How much higher than the highest flow after 1935 (when the Hoover Dam was built) was the highest flow before 1935?

(Compiled by the authors using data from U.S. Geological Survey.)

have led to inefficient use of irrigation water for growing thirsty crops such as rice, cotton, almonds, and alfalfa.

In 2014 the floodgates of the Morelos Dam near Yuma, Arizona, were opened for 2 months to send Colorado River water through the delta to the Gulf of California for the first time in years. Researchers are evaluating the effects of this experiment, but short-term results were dramatic, according to National Geographic Fellow and water policy expert Sandra Postel. Thousands of trees began to grow along the river's banks and groundwater in the delta area was partially recharged for the first time in many years.

Water experts call for the seven states using the Colorado River to enact and enforce strict water conservation measures. They also call for phasing out state and federal government subsidies for agriculture in this region, shifting water-thirsty crops to less arid areas, and banning or severely restricting the watering of golf courses and lawns in the desert areas of the Colorado River basin. They suggest that the best way to implement such solutions is to sharply raise the historically low price of the river's freshwater over the next decade—another application of the full-cost pricing **principle of sustainability**.

CONSIDER THIS . . .

THINKING ABOUT The Colorado River

What are three steps you would take to deal with the problems of the Colorado River system?

Removing Salt from Seawater to Provide Freshwater

Desalination is the process of removing dissolved salts from ocean water or from brackish (slightly salty) water in aquifers or lakes. It is another way to increase supplies of freshwater.

The two most widely used methods for desalinating water are distillation and reverse osmosis. *Distillation* involves heating saltwater until it evaporates (leaving behind salts in solid form) and condenses as freshwater. *Reverse osmosis* (or *microfiltration*) uses high pressure to force saltwater through a membrane filter with pores small enough to remove the salt and other impurities.

According to the International Desalination Association, there are more than 17,000 desalination plants operating in 150 countries. Most of them arid nations of the Middle East, North Africa, the Caribbean Sea, and the Mediterranean Sea. In 2013 the number of U.S. plants was 324 and growing. Desalination supplies less than 1% of the demand for freshwater in the United States and in the world.

There are three major problems with the widespread use of desalination. *First* is the high cost, because it takes a lot of energy to remove salt from seawater. A *second* problem is that pumping large volumes of seawater through pipes requires the use of chemicals to sterilize the water and to keep down algae growth. This kills many marine organisms and requires large inputs of energy and money. *Third,* desalination produces huge quantities of salty wastewater that must be disposed of. Dumping it into nearby coastal ocean waters increases the salinity of those waters, which can threaten food resources and aquatic life, especially near coral reefs, marshes, and mangrove forests. Disposing of it on land could contaminate groundwater and surface water (**Concept 13.3B**).

Currently, desalination is practical only for water-short countries and cities that can afford its high cost. However, scientists and engineers are working to develop better and more affordable desalination technologies (Science Focus 13.2).

CONSIDER THIS . . .

LEARNING FROM NATURE

Scientists are trying to develop more efficient and affordable ways to desalinate seawater by mimicking how our kidneys take salt out of water and how fish in the sea survive in saltwater.

13.4 CAN WATER TRANSFERS EXPAND WATER SUPPLIES?

CONCEPT 13.4 Transferring water from one place to another has greatly increased water supplies in some areas but has also disrupted ecosystems.

Water Transfers Have Benefits and Drawbacks

In some heavily populated dry areas of the world, governments have tried to solve water shortage problems by transferring water from water-rich areas to water-poor areas. For example, in northern China, rapidly growing cities, including Beijing with 21 million people, have helped to deplete underlying aquifers. According to the Chinese Academy of Sciences, two-thirds of China's 669 major cities have water shortages. In addition, about 300 million rural residents—a number almost equal to the size of the U.S. population—do not have access to safe drinking water. To deal with this problem, the Chinese government is implementing its *South–North Water Diversion Project* to transfer water from the Yangtze River in southern China to the thirsty north.

In other cases, water has been transferred to arid areas primarily to irrigate farm fields. When you have lettuce in a salad in the United States, it was probably grown in the arid Central Valley of California, partly with

SCIENCEFOCUS 13.2

The Search for Better Desalination Technology

Reverse osmosis (Figure 13.B, left) is the favored desalination technology because it requires much less energy than distillation, but it is still energy intensive. In this process, high pressure is applied to seawater in order to squeeze the freshwater out of it. The membrane that filters out the salt must be strong enough to withstand such pressure. Seawater must be pretreated and treated again after desalination to make it pure enough for drinking and for irrigating crops.

Much of the scientific research in this field is aimed at improving the membrane and the pre- and post-treatment processes to make desalination more energy efficient. Scientists are working to develop new, more efficient and affordable membranes that can separate freshwater from saltwater under lower pressure, which would require less energy. One promising material that might serve this purpose is one-atom-thick graphene (see Chapter 14, Case Study, p. 369). Such technological advances have brought the cost of desalination down, but not enough yet to make it affordable or useful for large-scale irrigation or to meet much of the world's demand for drinking water.

At the University of Texas, doctoral student Kyle Knust has invented the Waterchip—a small device that removes salt from saltwater using an electrical current. Water flows down a Y-shaped channel and where it splits, an electrode emits a charge that separates water from salt. Knust says this small-scale device could be scaled up to produce larger amounts of desalinated water using half the energy of an osmosis plant. A team of scientists at the Massachusetts Institute of Technology (MIT), led by Martin Z. Bazant, is evaluating the use of an electric shock to separate saltwater and freshwater.

Scientists are considering ways to use solar and wind energy—applying one of the three **scientific principles of sustainability** (see Inside Back Cover)—in combination with conventional power sources to help bring down the cost of desalinating seawater. In 2012 Saudi Arabia completed the world's largest solar-powered desalination project. It uses concentrated solar energy to power new filtration technology at a plant that will meet the daily water needs of 100,000 people.

Two Australian companies, Energetech and H2AU, have joined forces to build an experimental desalination plant that uses the power generated by ocean waves to drive reverse-osmosis desalination. This approach produces no air pollution and uses renewable energy. Some scientists argue for building fleets of such floating desalination plants. They could operate out of sight from coastal areas and transfer the water to shore through seabed pipelines or in food-grade shuttle tankers. Because of their distance from shore, the ships could draw water from depths below where most marine organisms are found. The resulting brine could be returned to the ocean and diluted far away from coastal waters.

These methods would cut the costs of desalination, but they would still be high. Analysts expect desalination to be used more widely in the future, as water shortages become worse. However, there is still a lot of research to do before desalination can become an affordable major source of freshwater. **GREEN CAREER: Desalination engineer**

CRITICAL THINKING

Do you think that improvements in desalination will justify highly inefficient uses of water, such as maintaining swimming pools, fountains, and golf courses in desert areas? Explain.

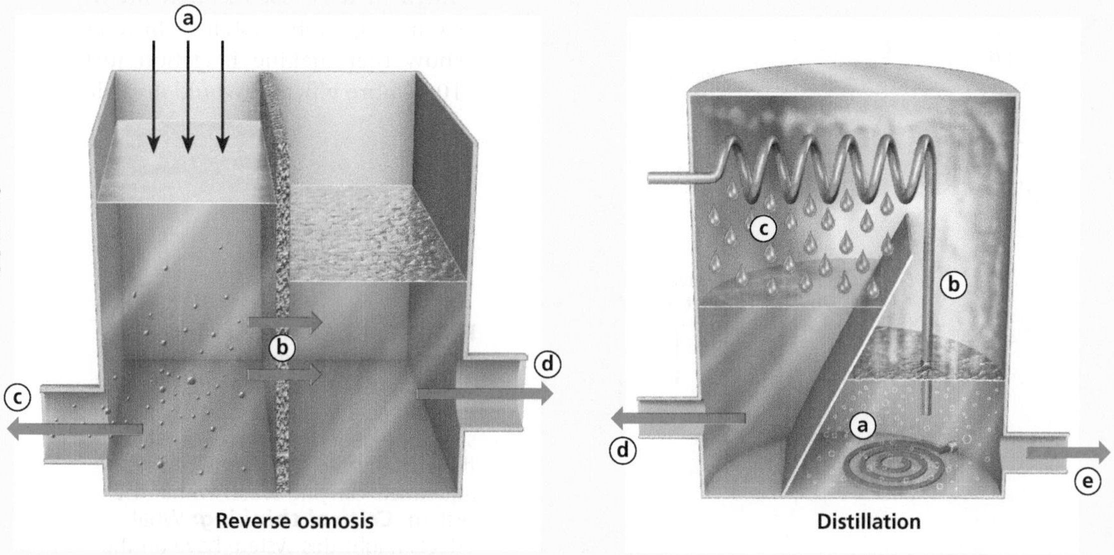

FIGURE 13.B *Desalination:* Reverse osmosis (left) involves applying high pressure **(a)** to force sea water from one chamber into another through a semipermeable membrane **(b)** that separates the salt **(c)**, producing freshwater **(d)**. Distillation (right) involves heating sea water **(a)** to produce steam **(b)**, which is then condensed **(c)** and collected as freshwater **(d)**, while brine is also collected **(e)** for processing.

the use of irrigation water from snow melting off the tops of the High Sierra Mountains of northeastern California. The California State Water Project (Figure 13.17) is one of the world's largest freshwater transfer projects. It uses a maze of giant dams, pumps, and lined canals, or *aqueducts* (photo in Figure 13.17), to transfer freshwater from the mountains to heavily populated cities and agricultural regions in water-poor central and southern California.

These massive water transfers have yielded many benefits. Some 440 million Chinese will be served by the South–North Water Diversion Project. California's Central Valley supplies half of the United States' fruits and vegetables, and the cities of San Diego and Los Angeles have grown and flourished because of the water transfer.

However, water transfers also have high environmental, economic, and social costs. They usually involve large water losses, through evaporation and leaks in the water-transfer systems. They also degrade ecosystems in areas from which the water is taken (**Concept 13.4**). China's water transfer—moving 23 trillion liters (6 trillion gallons of water) per year—will be expensive. It will displace more than 350,000 villagers who will have to move from lands they have farmed for generations. And scientists worry that removing huge volumes of water from the Yangtze River could severely damage its ecosystem, which has been suffering from its worst drought in 50 years.

In California, sending water south has degraded the Sacramento River and reduced the flushing action that helps to cleanse the San Francisco Bay of pollutants. As a result, the bay has suffered from pollution and the flow of freshwater to its coastal marshes and other ecosystems has dropped, putting stress on wildlife species that depend on these ecosystems. Water was also diverted from streams that flow into Mono Lake, an important feeding stop for migratory birds. This lake experienced an 11-meter (35-foot) drop in its water level before the diversions were stopped. For a while, the lake's entire ecosystem was in jeopardy.

Federal and state governments typically subsidize water transfers. In the California project, subsides have promoted inefficient use of large volumes of water to irrigate thirsty crops such as lettuce, alfalfa, and almonds in desert-like areas. In central California, agriculture consumes three-fourths of the water that is transferred, and much of it is lost through inefficient irrigation systems. Studies show that making irrigation just 10% more efficient would provide

Sarahleen/National Geographic Creative

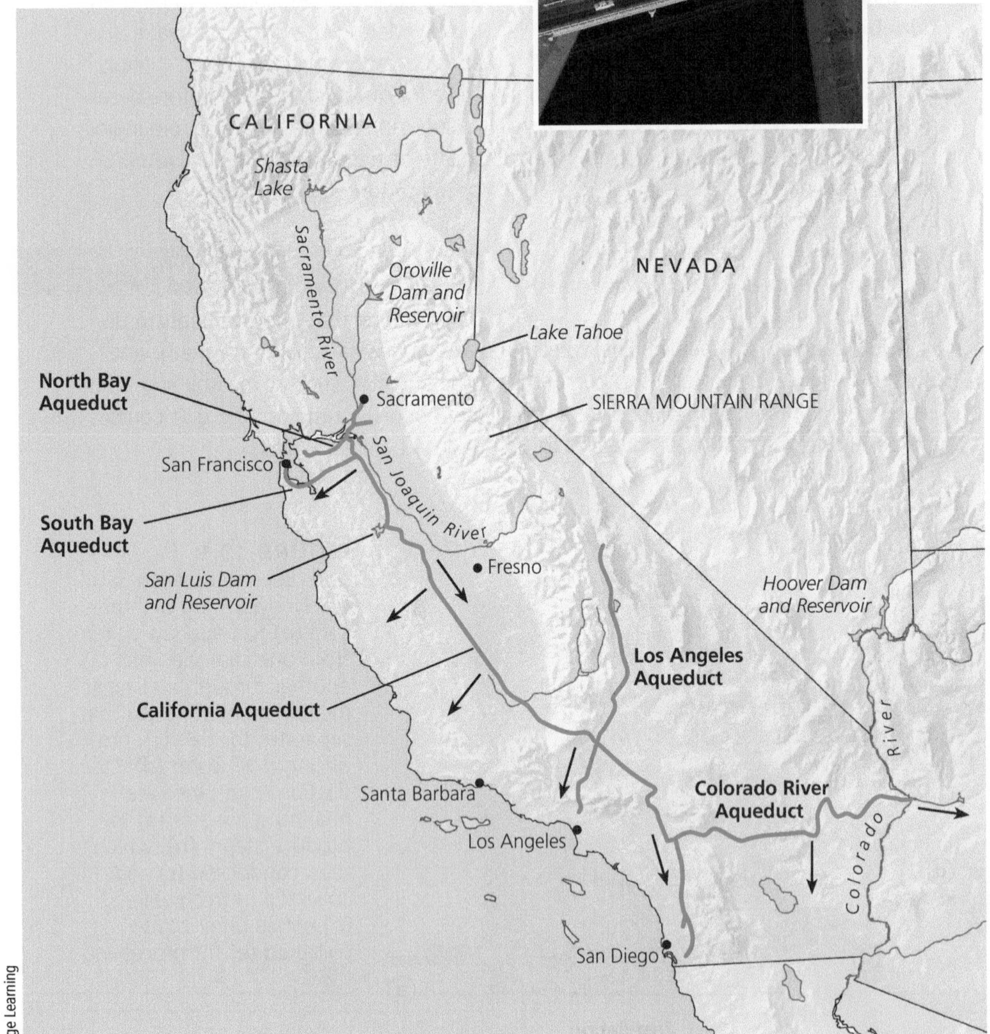

© Cengage Learning

FIGURE 13.17 The California State Water Project transfers huge volumes of freshwater from one watershed to another. The arrows on the map show the general direction of water flow. The photo shows one of the aqueducts carrying water within the system. ***Critical thinking:*** What effects might this system have on the areas from which the water is taken?

all the water necessary for domestic and industrial uses in southern California.

According to several studies, climate change will make matters worse in many areas where water is being removed for transfers. In southern China, climate change could intensify and prolong the drought and create a need for an even larger and more expensive transfer of water. California depends on *snowpacks*, bodies of densely packed, slowly melting snow in the High Sierra Mountains, for more than 60% of its freshwater, according to the Sierra Nevada Conservancy. Projected atmospheric warming could shrink the snowpacks by as much as 40% by 2050 and by as much as 90% by the end of this century. This will sharply reduce the amount of freshwater available for northern residents and ecosystems, as well for the transfer of water to central and southern California.

There are many other examples around the world of water transfers that have resulted in environmental degradation (see the following Case Study).

CASE STUDY

The Aral Sea Disaster: An Example of Unintended Consequences

The shrinking of the Aral Sea (Figure 13.18) is the result of a water transfer project in central Asia. Starting in 1960, enormous amounts of irrigation water were diverted from the two rivers that supply water to the Aral Sea. The goal was to create one of the world's largest irrigated areas, mostly for raising cotton and rice. The irrigation canal, the world's longest, stretches more than 1,300 kilometers (800 miles)—roughly the distance between the two U.S. cities of Boston, Massachusetts, and Chicago, Illinois.

This project, coupled with drought and high evaporation rates due to the area's hot and dry climate, has caused a regional ecological and economic disaster. Since 1961, the sea's salinity has risen sevenfold and the average level of its water has dropped by an amount roughly equal to the height of a six-story building. The Southern Aral Sea has lost 90% of its volume of water and most of its lake bottom is now a white salt desert (Figure 13.18, right photo). Water withdrawals reduced the two rivers feeding the sea to mere trickles.

About 85% of the area's wetlands have been eliminated and about half the local bird and mammal species have disappeared. The sea's greatly increased salt concentration—three times saltier than ocean water—has caused the presumed local extinction of 26 of the area's 32 native fish species. This has devastated the area's fishing industry, which once provided work for more than 60,000 people. Fishing villages and boats once located on the sea's coastline now sit abandoned in a salty desert.

Winds pick up the sand and salty dust and blow it onto fields as far as 500 kilometers (310 miles) away. As the salt spreads, it pollutes water and kills wildlife, crops, and other vegetation. Aral Sea dust settling on glaciers in the Himalayas is causing them to melt at a faster-than-normal rate.

The shrinkage of the Aral Sea has also altered the area's climate. The shrunken sea no longer acts as a thermal buffer to moderate the heat of summer and the extreme cold of winter. Now there is less rain, summers are hotter and drier, winters are colder, and the growing season is shorter.

1976

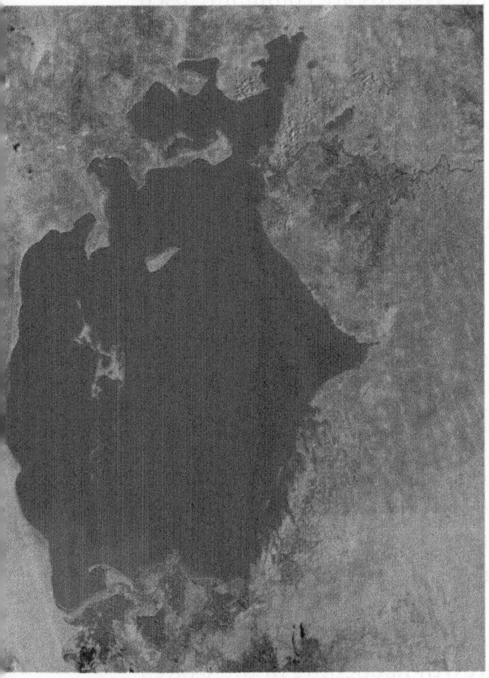

2015

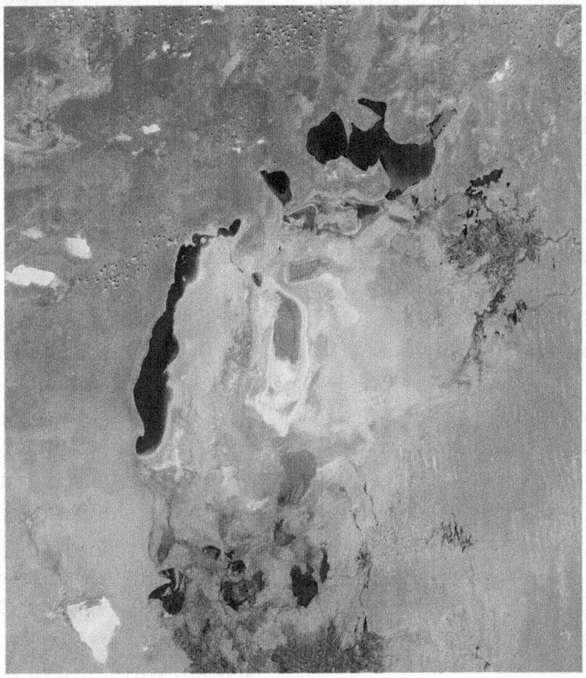

NASA

FIGURE 13.18 Natural capital degradation: The *Aral Sea*, straddling the borders of Kazakhstan and Uzbekistan, was one of the world's largest saline lakes. These satellite photos show the sea in 1976 (left) and in 2015 (right). ***Question:*** What do you think should be done to help prevent further shrinkage of the Aral Sea?

The combination of such climate change and severe salinization has reduced crop yields by 20–50% on almost one-third of the area's cropland—the opposite of the project's intended effects.

Since 1999, the UN, the World Bank, and the five countries surrounding the lake have worked to improve irrigation efficiency. They have also partially replaced thirsty crops with other crops that require less irrigation water. Because of a dike built to block the flow of water from the Northern Aral Sea into the southern sea, the level of the northern sea has risen by 2 meters (7 feet), its salinity has dropped, dissolved oxygen levels are up, and it supports a healthy fishery.

However, the formerly much larger southern sea is still shrinking. By 2012, its eastern lobe was essentially gone (Figure 13.18, right photo). The European Space Agency projects that the rest of the Southern Aral Sea could dry up completely by 2020.

13.5 HOW CAN WE USE FRESHWATER MORE SUSTAINABLY?

CONCEPT 13.5 We can use freshwater more sustainably by cutting water waste, raising water prices, slowing population growth, and protecting aquifers, forests, and other ecosystems that store freshwater.

Cutting Water Waste Would Have Many Benefits

According to water resource expert Mohamed El-Ashry of the World Resources Institute, about 66% of the freshwater used in the world and about 50% of the freshwater used in the United States is lost through evaporation, leaks, and inefficient use. El-Ashry estimates that it is economically and technically feasible to reduce such losses to 15%, thereby meeting most of the world's future freshwater needs.

Why do we have such large losses of freshwater? According to water resource experts, there are two major reasons. First, the cost of freshwater to most users is low due mostly to government subsidies—a violation of the full-cost pricing **principle of sustainability**. This gives users little or no financial incentive to invest in water-saving technologies.

Higher prices for freshwater encourage water conservation but make it difficult for low-income farmers and city dwellers to buy enough water to meet their needs. When South Africa raised water prices, it dealt with this problem by establishing *lifeline rates,* which give each household a set amount of free or low-priced water to meet basic needs. When users exceed this amount, they pay increasingly higher prices as their water use increases. This is a *user-pays* approach.

The second major cause of unnecessary waste of freshwater is a lack of government subsidies for improving the efficiency of water use. Withdrawing some of the subsidies that encourage inefficient water use and replacing them with subsidies for more efficient water use would sharply reduce water losses. Understandably, farmers and industries that receive subsidies that keep water prices low have vigorously opposed efforts to eliminate or reduce them.

We Can Improve Efficiency in Irrigation

Since 1980, the amount of food that can be grown per drop of water has roughly doubled. In addition, since the 1970s, the amount of water used per person in the United States has dropped by about 33%, after rising for decades. Most of these water savings have come from improvements to irrigation efficiency in the United States and other more-developed countries. GOOD NEWS

However, there is still a long way to go, especially in less-developed countries. Only about 60% of the world's irrigation water reaches crops, which means that most irrigation systems are highly inefficient. The most inefficient system, commonly used in less-developed countries, is *flood irrigation*. With this method, water is pumped from a groundwater or surface water source through unlined ditches where it flows by gravity to the crops being watered (Figure 13.19, left). This method delivers far more water than is needed for crop growth, and typically, about 45% of it is lost through evaporation, seepage, and runoff.

Another inefficient system is the traditional spray irrigation system, a widely used tool of industrialized crop production. It sprays huge volumes of water onto large fields, and as much as 40% of this water is lost to evaporation, especially in dry and windy areas, according to the U.S. Geological Survey. These systems are commonly used in the Midwestern United States and have helped to draw down the Ogallala Aquifer (Case Study, p. 333).

More efficient irrigation technologies greatly reduce water losses by delivering water more precisely to crops—a *more crop per drop* strategy. For example, a *center-pivot, low-pressure sprinkler* (Figure 13.19, right), which uses pumps to spray water on a crop, allow about 80% of the water to reach crops. *Low-energy, precision application (LEPA) sprinklers,* another form of center-pivot irrigation, put 90–95% of the water where crops need it.

Drip, or *trickle irrigation*, also called *micro-irrigation* (Figure 13.19, center), is the most efficient way to deliver small amounts of water precisely to crops. It consists of a network of perforated plastic tubing installed at or below the ground level. Small pinholes in the tubing deliver drops of water at a slow and steady rate, close to the roots of individual plants. These systems drastically reduce water waste because 90–95% of the water input reaches the crops.

Since the early 1990s, the global area of cropland on which drip irrigation is used has increased more than sixfold, with most of this growth happening in the United

FIGURE 13.19 Traditional irrigation methods rely on gravity and flowing water (left). Newer systems such as center-pivot, low-pressure sprinkler irrigation (right) and drip irrigation (center) are far more efficient.

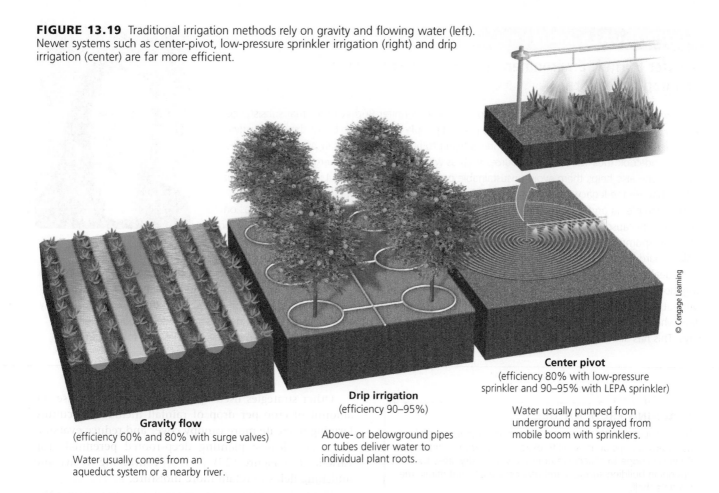

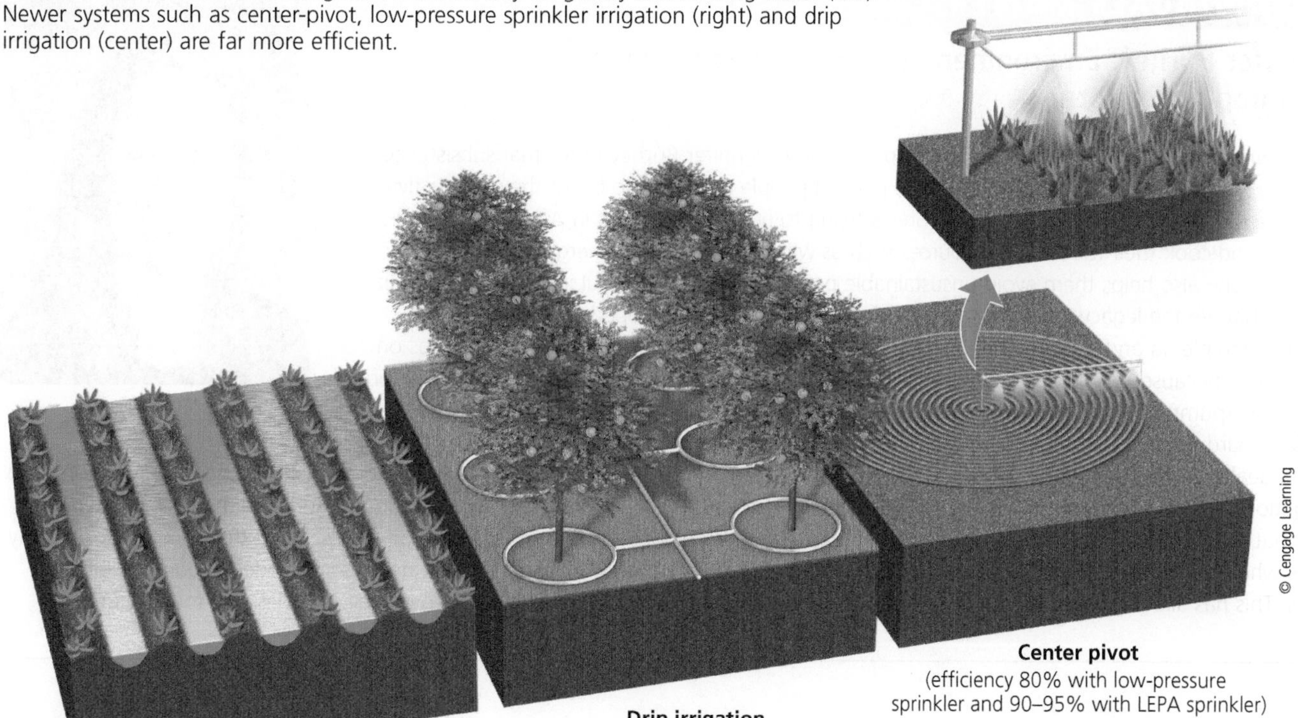

© Cengage Learning
Center pivot
(efficiency 80% with low-pressure sprinkler and 90–95% with LEPA sprinkler)
Water usually pumped from underground and sprayed from mobile boom with sprinklers.
Drip irrigation
(efficiency 90–95%)
Above- or belowground pipes or tubes deliver water to individual plant roots.
Gravity flow
(efficiency 60% and 80% with surge valves)
Water usually comes from an aqueduct system or a nearby river.

States, China, and India. Still, drip irrigation is used on less than 4% of the irrigated crop fields in the world and in the United States, largely because most drip irrigation systems are costly. This percentage rises to 13% in the U.S. state of California, 66% in Israel, and 90% in Cyprus. If freshwater were priced closer to the value of the ecosystem services it provides, and if government subsidies for inefficient use of water were reduced or eliminated, drip irrigation could be used to irrigate most of the world's crops.

According to the UN, reducing the current global withdrawal of water for irrigation by just 10% would save enough water to grow crops and meet the estimated additional water demands of the earth's cities and industries through 2025.

Poor Farmers Conserve Water Using Low-Tech Methods

Many of the world's poor farmers use low-cost, traditional irrigation technologies that are far more sustainable than most large-scale irrigation systems. For example, millions of farmers in Bangladesh and other countries where water tables are high use human-powered treadle pumps to bring groundwater up to the earth's surface and into irrigation ditches (Figure 13.20). These wooden devices are inexpensive and easy to build from local materials. One such pump developed by the nonprofit International Development Enterprises (IDE) uses 60–70% less water than a conventional gravity-flow system to irrigate the same amount of cropland at one-tenth the cost of conventional drip systems.

Other farmers in some less-developed countries use buckets, small tanks with holes, or simple plastic tubing systems for drip irrigation. One ingenious system makes use of solar energy to drive drip irrigation (Individuals Matter 13.1).

Rainwater harvesting is another simple and inexpensive way to provide water. It involves using pipes from rooftops and channels dug in the ground to direct rainwater that would otherwise run off the land. It can be stored in underground or aboveground storage tanks (cisterns), ponds, and plastic barrels for use during dry seasons. This is especially useful in countries such as India, where much of the rain comes in a short monsoon season.

In dry mountainous coastal areas, such as in Peru, some communities are capturing water from fog that rolls in off the ocean on most days. On the seaward hillsides, they erect large flat nets on which the fog condenses. The resulting water drops roll off the nets into troughs that channel the water into holding tanks.

INDIVIDUALS MATTER 13.1

Jennifer Burney: Environmental Scientist and National Geographic Explorer

UC San Diego

Environmental scientist and National Geographic Explorer Jennifer Burney notes that subsistence farmers represent the majority of the world's poorest people and need to boost their productivity for better standards of living and health. She is trying to help such farmers in Africa to grow, distribute, and cook their food using resources such as water, fertilizer, and energy as efficiently as possible. She also helps them avoid unsustainable practices such as wasteful irrigation and fertilizer runoff that are the legacy of large-scale industrial farming in the developed world.

For example, in arid sub-Saharan Africa, farmers must depend on rainfall for raising crops on small plots because only 20% of the rainfall flows into streams and aquifers while the rest evaporates. Overpumping can quickly deplete the groundwater. These factors, worsened by drought, make it hard for farmers to feed their families.

To deal with this problem, Burney has helped farmers to connect two technologies—solar energy systems and drip irrigation. Drip irrigation systems sip water and drip it directly onto plant roots instead of pumping and dumping it. Solar-powered pumps work without the need for batteries or fuel. On sunny days, when crops need water more, the solar panels speed the pumping; on cloudy days when there is less evaporation, the pumping slows down. Thus, only the amount of water that is needed is pumped on most days. This has allowed farmers to grow fruits and vegetables on a larger scale and to improve their incomes and food security.

CONSIDER THIS . . .

LEARNING FROM NATURE

The Namibian beetle survives in Africa's arid Namib Desert by using tiny bumps on its shell to extract water from night fog. Engineers hope to collect water in the world's dry areas by designing building surfaces and other materials that mimic the beetle's shell.

Other strategies used by poor farmers to increase the amount of crop per drop of rainfall include polyculture farming to create more canopy cover and reduce evaporative water losses; planting deep-rooted perennial crop varieties (see Figure 12.B, p. 315); controlling weeds; and mulching fields to retain more moisture.

Figure 13.21 summarizes several ways to reduce water losses in crop irrigation. Since 1950, Israel has used many of these techniques to slash irrigation water losses by 84% while irrigating 44% more land. Israel now treats and reuses 30% of its municipal sewage water for crop production and plans to increase this to 80% by 2025. The government also gradually eliminated most water subsidies to raise Israel's price of irrigation water, which is now one of the highest in the world. GOOD NEWS

We Can Cut Freshwater Losses in Industries and Homes

Producers of chemicals, paper, oil, coal, primary metals, and processed foods consume almost 90% of the freshwater used by industries in the United States. Some of these industries recapture, purify, and recycle water to reduce their water use and water treatment costs. For example, more than 95% of the water used to make steel can be recycled. Even so, most industrial processes could be redesigned to use much less water. **GREEN CAREER: Water conservation specialist** GOOD NEWS

Courtesy of International Development Enterprises

FIGURE 13.20 Solutions: In areas of Bangladesh and India, where water tables are high, many small-scale farmers use treadle pumps to supply irrigation water to their fields.

Solutions

Reducing Irrigation Water Losses

- Avoid growing thirsty crops in dry areas
- Import water-intensive crops and meat
- Encourage organic farming and polyculture to retain soil moisture
- Monitor soil moisture to add water only when necessary
- Expand use of drip irrigation and other efficient methods
- Irrigate at night to reduce evaporation
- Line canals that bring water to irrigation ditches
- Irrigate with treated wastewater

FIGURE 13.21 Ways to reduce freshwater losses in irrigation. ***Critical thinking:*** Which two of these solutions do you think are the best ones? Why?

Flushing toilets with freshwater (most of it clean enough to drink) is the single largest use of domestic freshwater in the United States and accounts for about 27% of home water use. Since 1992, U.S. government standards have required that new toilets use no more than 6.1 liters (1.6 gallons) of water per flush. Even at this rate, just two flushes of such a toilet use more than the daily amount of water available for all uses to many of the world's poor people living in arid regions.

Other water-saving appliances are widely available. Low-flow showerheads can save large amounts of water by cutting the flow of a shower in half. Front-loading clothes washers use 30% less water than top-loading machines use. According to the American Water Works Association, if the typical American household were to stop all water leaks and use these devices, along with low-flow toilets and faucets, it could cut its daily water use by nearly a third. According to UN studies, 30–60% of the water supplied in nearly all of the world's major cities in less-developed countries is lost, primarily through leakage from water mains, pipes, pumps, and valves. Water experts say that fixing these leaks should be a high priority for water-short countries, because it would increase water supplies and cost much less than building dams or importing water.

Even in advanced industrialized countries such as the United States, these losses to leakage average 10–30%. However, leakage losses have been reduced to about 3% in Copenhagen, Denmark, and to 5% in Fukuoka, Japan. In 1 year, a faucet leaking water at the rate of 1 drop per second can waste 10,000 liters (2,650 gallons). The customer's water bill goes up and so does the energy bill if the water is leaking from a hot water faucet. Not detecting and fixing water leaks from faucets, pipes, and toilets is equivalent to burning money.

Many homeowners and businesses in water-short areas are using drip irrigation on their properties to cut water losses. Some are also using smart sprinkler systems with moisture sensors that have cut water used for watering lawns by up to 40%. GOOD NEWS Others are copying nature by replacing green lawns with a mix of native plants that need little or no watering (Figure 13.22). Such water-thrifty landscaping saves money by reducing water use by 30–85% and by sharply reducing labor, fertilizer, and fuel requirements. It also can help landowners to reduce polluted runoff, air pollution, and yard wastes.

In some more-developed countries, people who live in arid areas maintain green lawns by watering them heavily. Some communities and housing developments in water-short areas have even passed ordinances that require green lawns and prohibit the planting of native vegetation in place of lawns.

Water used in homes can be reused and recycled. About 50–75% of a typical household's *gray water*—used water from bathtubs, showers, sinks, dishwashers, and clothes washers—could be recovered and stored. This water can be reused to irrigate lawns and nonedible plants, to flush toilets, and to wash cars. Such efforts mimic the way nature recycles water, and thus they follow the chemical cycling **principle of sustainability**.

The relatively low cost of water in most communities is one of the major causes of excessive water use and waste in homes and industries. About one-fifth of all U.S. public water systems do not use water meters and charge a single low annual rate for almost unlimited use of high-quality freshwater.

When the U.S. city of Boulder, Colorado, introduced water meters, water use per person dropped by 40%. In some cities in Brazil, people buy *smart cards*, each of which contains a certain number of water credits that entitle their owners to measured amounts of freshwater. Brazilian officials say this approach saves water and typically reduces household water bills by 40%.

Figure 13.23 summarizes various ways to use water more efficiently in industries, homes, and businesses (**Concept 13.3**).

We Can Use Less Water to Remove Wastes

Currently, we use large amounts of freshwater to flush away industrial, animal, and household wastes. According to the UN Food and Agriculture Organization (FAO), if current growth trends in population and water use continue, within 40 years, we will need the world's entire reliable flow of river water just to dilute and transport the wastes we produce each year.

We could save much of this freshwater by recycling and reusing gray water from homes and businesses for flushing wastes and cleaning equipment. In Singapore, all sewage water is treated at reclamation plants for reuse by

karolinapatryk/Thinkstock

FIGURE 13.22 This yard in a dry area of the southwestern United States uses a mix of plants that are native to the arid environment and require little watering.

industry. U.S. cities such as Las Vegas, Nevada, and Los Angeles, California, are also beginning to clean up and reuse some of their wastewater. However, only about 7% of the water in the United States is recycled, cleaned up, and reused. Sharply raising this percentage would be a way to apply the chemical cycling **principle of sustainability**.

Another way to keep freshwater out of the waste stream is to rely more on waterless composting toilets. These devices convert human fecal matter to a small amount of dry and odorless soil-like humus material that can be removed from a composting chamber every year or so and returned to the soil as fertilizer. One of the authors (Miller) used a composting toilet for over a decade with no problems, while living and working deep in the woods in a small passive solar home and office used for evaluating solutions to water, energy, and other environmental problems (see p. xxxiii).

As water shortages grow in many parts of the world, people are using methods discussed here to use water more sustainably. Their experiences can be instructive to people who want to avoid water shortages in the first place. (See the Case Study that follows.)

CASE STUDY

How Californians Are Dealing with Water Woes

In 2015 the state of California had been experiencing drought for 4 years. Due to climate change, the state's climate in the future is projected to be hotter and drier than it was during the 20th century. It is also likely to see more extreme weather events, including large flooding episodes, as well as more intense droughts.

In some areas of California, the effects of drought include dwindling aquatic ecosystems and municipal water supplies, increasingly frequent wildfires, crop losses, and parched lawns. In 2014 NASA satellite data (Science Focus 13.1)

Solutions

Reducing Water Losses

- Redesign manufacturing processes to use less water
- Recycle water in industry
- Fix water leaks
- Landscape yards with plants that require little water
- Use drip irrigation on gardens and lawns
- Use water-saving showerheads, faucets, appliances, and toilets (or waterless composting toilets)
- Collect and reuse gray water in and around houses, apartments, and office buildings
- Raise water prices and use meters, especially in dry urban areas

© Cengage Learning

FIGURE 13.23 Ways to reduce freshwater losses in industries, homes, and businesses (**Concept 13.6**). ***Critical thinking:*** Which three of these solutions do you think are the best ones? Why?

showed the state's major river basins to be well below normal (Figure 13.24). In 2015 the Sierra Nevada snowpack, which has provided a third to half of the state's water supply in the past, was just 5% of its historical average size. As of late 2015, the drought across 97% of the state was classified as *severe* or worse.

According to the Natural Resources Defense Council (NRDC), agriculture uses 78% to 80% of California's available water in most years. Due to water restrictions imposed by the state in 2013, along with complex water rights laws, some Central Valley farmers have lost their surface water irrigation sources. In turn, many have drilled deeper wells into the valley's aquifers that require decades to recharge. Hydrologists see this overpumping of the aquifers as unsustainable and dangerous in the long term.

The strategies that Californians are using to deal with drought could teach us much about how to use freshwater more sustainably. First, many water supply agencies raised their water rates to encourage conservation. One common approach involves an increasing-step rate structure. Users pay a certain rate for a set amount of water. Once they surpass that amount, their rate goes up. There are typically three or four such thresholds within the structure, so the largest users pay much more per volume of water than do those who use water at the lowest level.

Second, California residents are conserving water in a number of ways. According to the Public Policy Institute of California, at least 40% of residential water use is for watering lawns. The next three largest uses are for swimming pools, toilets, and showers. Many Californians are replacing their grass lawns with water-saving ground cover or native vegetation adapted to dry conditions. Others are installing more efficient toilets and showerheads and are showering and washing clothes less frequently. A smaller number of people are fixing water leaks and draining their pools. In 2015 California's urban residents reached the 25% water use reduction goal set by the state government.

Third, there is growing interest in *conjunctive use,* also referred to as *water banking*—finding ways to save water

FIGURE 13.24 One of California's major reservoirs, Lake Oroville, dropped dramatically between 2011 (left) and 2014 (right), due largely to drought.

Justin Sullivan/Getty Images

for future use. Climate models indicate that much of California's future precipitation is likely to come in shorter high-volume bursts that could be captured and stored. One approach is to locate an empty aquifer, of which there are several in California. Excess water is then channeled to the overlying land area where it sinks into the ground to recharge the aquifer. This water does not evaporate and can be withdrawn in the future as needed.

The problem is that water does not travel predictably underground and often flows to other areas. Thus it cannot always be withdrawn from where it was deposited. Also, withdrawing water from the deposit area can pull water from other areas, lowering the water tables and causing water supply problems in those areas. Thus, conjunctive use is not a perfect solution.

Another strategy promoted by many is for farmers to shift from producing thirsty crops such as alfalfa, lettuce, and almonds, to producing less water-intensive crops. The Public Policy Institute of California estimated that the amount of water used to grow almonds in 2013 was larger than that used by all homes and businesses in San Francisco and Los Angeles combined. In addition, California is the leading dairy state and dairy products are among the most water-intensive. It has been estimated that the average American consumes 1,132 liters (300 gallons) of California water every week by eating foods produced in the state.

Desalination is another option that many are promoting (Science Focus 13.2). In 2015 the largest desalination plant in the western hemisphere was opened in Carlsbad, north of San Diego. It was designed to supply 300,000 state residents with freshwater. However, because of current high costs and potentially harmful environmental effects, desalination is a controversial and limited solution to water shortages.

Yet another strategy is the recycling of water—restoring wastewater to drinking water quality. Orange County's Groundwater Replenishment System takes sewer and other wastewater and processes it to the point where it exceeds all state and federal drinking water standards. It meets the needs of about 600,000 people. Water can also be reused through gray water systems. Many homes in California are equipped with new *purple pipes*—special pipes built in for this purpose and distinctively color-coded.

We Can Use Water More Sustainably

More sustainable water use would include a variety of strategies (Figures 13.14, 13.21, and 13.23) aimed not only at conserving water and using it efficiently, but also at protecting water supplies and the ecosystems that sustain them (**Concept 13.5**). Such strategies would have to be applied at local and regional levels, as well as national and international levels.

However, to be successful, these strategies would also have to be applied at the personal level. Each of us can reduce our water footprints by using less freshwater and using it more efficiently (Figure 13.25).

What Can You Do?

Water Use and Waste

- Use water-saving toilets, showerheads, and faucets
- Take short showers instead of baths
- Turn off sink faucets while brushing teeth, shaving, or washing
- Wash only full loads of clothes or use the lowest possible water-level setting for smaller loads
- Repair water leaks
- Wash your car from a bucket of soapy water, use gray water, and use the hose for rinsing only
- If you use a commercial car wash, try to find one that recycles its water
- Replace your lawn with native plants that need little if any watering
- Water lawns and gardens only in the early morning or evening and use gray water
- Use drip irrigation and mulch for gardens and flowerbeds

FIGURE 13.25 Individuals matter: You can reduce your use and waste of freshwater. ***Question:*** Which of these steps have you taken? Which would you like to take?

13.6 HOW CAN WE REDUCE THE THREAT OF FLOODING?

CONCEPT 13.6 We can lessen the threat of flooding by protecting more wetlands and natural vegetation in watersheds, and by not building in areas subject to frequent flooding.

Some Areas Get Too Much Water from Flooding

Some areas have too little freshwater, but others sometimes have too much because of natural flooding by streams, caused mostly by heavy rain or rapidly melting snow. A flood happens when freshwater in a stream overflows its normal channel and spills into the adjacent area, called the **floodplain.**

Human activities contribute to flooding in several ways. People settle on floodplains to take advantage of their many assets. They include fertile soil on flat land suitable for crops, ample freshwater for irrigation, and availability of nearby rivers for transportation and recreation. In efforts to reduce the threat of flooding on floodplains, rivers have been narrowed and straightened (or *channelized*), surrounded by protective dikes and *levees* (long mounds of earth along their banks), and dammed to create reservoirs that store and release water as needed. However, such measures can lead to greatly increased flood damage when heavy snowmelt or prolonged rains overwhelm them.

Floods provide several benefits. They have created some of the world's most productive farmland by depositing nutrient-rich silt on floodplains. They also help recharge groundwater and refill wetlands that are commonly found on floodplains, thereby supporting biodiversity and aquatic ecosystem services.

At the same time, floods kill thousands of people every year and cost tens of billions of dollars in property damage (see the Case Study that follows). Floods are usually considered natural disasters, but since the 1960s, human activities have contributed to a sharp rise in flood deaths and damages, meaning that such disasters are partly human-made.

One such human activity is the *removal of water-absorbing vegetation*, especially on hillsides (Figure 13.26). Once the trees on a hillside have been cut for timber, fuelwood, livestock grazing, or farming, freshwater from precipitation rushes down the barren slopes, erodes precious topsoil, and can increase flooding and pollution in local streams. Such deforestation can also make landslides and mudflows more likely. A 3,000-year-old Chinese proverb says, "To protect your rivers, protect your mountains."

The second human activity that increases the severity of flooding is the *draining of wetlands* that naturally absorb floodwaters. These areas are then often covered with pavement and buildings that greatly increase runoff, which contributes to flooding and pollution of surface waters. When Hurricane Katrina struck the Gulf Coast of the United States in August 2005 and flooded the city of New Orleans, Louisiana, the damage was intensified because of the degradation or removal of coastal wetlands. These wetlands had historically helped to absorb water and buffer this low-lying land from storm surges. For this reason, Louisiana officials are now working to restore some coastal wetlands.

Another human-related factor that will likely increase flooding is a rise in sea levels, projected to occur during this century (mostly the result of climate change related to human activities). Climate change models project that, by 2075, as many as 150 million people living in the world's largest coastal cities—a number nearly equal to half of the current U.S. population—could be flooded out by rising sea levels.

CASE STUDY

Living Dangerously on Floodplains in Bangladesh

Bangladesh is one of the world's most densely populated countries. In 2015 its 160 million people were packed into an area roughly the size of the U.S. state of Wisconsin (which has a population of less than 6 million). And the

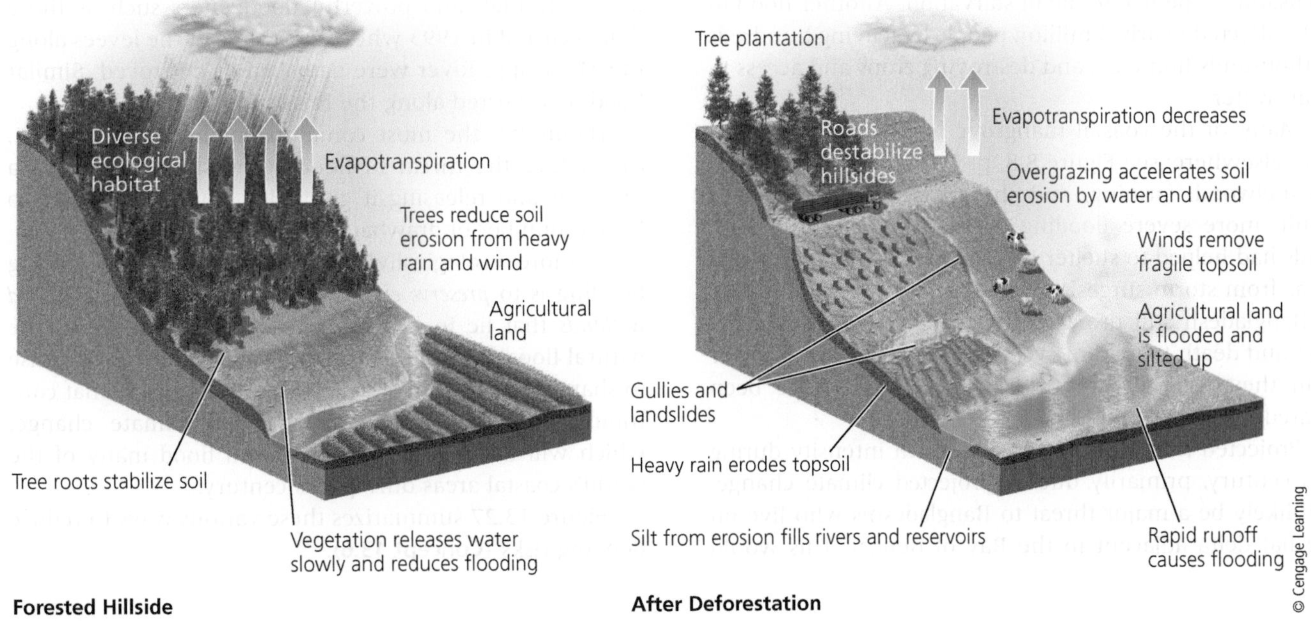

FIGURE 13.26 Natural capital degradation: A hillside before and after deforestation.
Question: How might a drought in this area make these effects even worse?

country's population is projected to increase to 202 million by 2050. Bangladesh is a very flat country, only slightly above sea level, and it is one of the world's poorest countries.

The people of Bangladesh depend on moderate annual flooding during the summer monsoon season to grow rice and help maintain soil fertility in their country's delta basin region, which is fed by numerous river systems. The annual floods also deposit eroded Himalayan soil on the country's crop fields. Bangladeshis have adapted to moderate flooding. Most of the houses have flat thatch roofs on which families can take refuge with their belongings in case of rising waters. The roofs can be detached from the walls, if necessary, and floated like rafts. After the waters have subsided, the roof can be reattached to the walls of the house. However, great floods can overwhelm such defenses.

In the past, great floods occurred every 50 years or so. However, between 1987 and 2015 there were seven severe floods, each covering a third or more of the country with water. Bangladesh's flooding problems begin in the Himalayan watershed, where rapid population growth and unsustainable farming have resulted in deforestation. Monsoon rains now run more quickly off the barren Himalayan foothills, carrying vital topsoil with them (Figure 13.26, right).

This increased runoff of topsoil, combined with heavier-than-normal monsoon rains, has led to more severe flooding along Himalayan rivers, as well as downstream in Bangladesh's delta areas. In 1998 a disastrous flood covered two-thirds of Bangladesh's land area, in some places for 2 months, drowning at least 2,000 people and leaving 30 million homeless. It also destroyed more than one-fourth of the country's crops, which caused thousands of people to die of starvation. Another flood in 2014 affected nearly 3 million people by leaving hundreds of thousands homeless and destroying crops and access to clean water.

Many of the coastal mangrove forests in Bangladesh (and elsewhere; see Figure 8.8, p. 172) have been cleared for fuelwood, farming, and shrimp farming ponds. The result: more severe flooding because these coastal wetlands had helped to shelter Bangladesh's low-lying coastal areas from storm surges, cyclones, and tsunamis. In areas of Bangladesh still protected by mangrove forests, damages and death tolls from cyclones have been much lower than they were in areas where the forests have been cleared.

Projected rises in sea level and storm intensity during this century, primarily due to projected climate change, will likely be a major threat to Bangladeshis who live on the flat delta adjacent to the Bay of Bengal. This would create millions of environmental refugees with no place to go in this already densely populated country.

Bangladesh is one of the few less-developed nations that is implementing plans to adapt to projected rising sea levels. This includes using varieties of rice and other crops that can better tolerate flooding, saltwater, and drought. GOOD NEWS People are also planting small vegetable gardens in bare patches between houses to help reduce their dependence on rice. In addition, they are building ponds to collect monsoon rainwater and a network of earthen embankments to help protect against high tides and storm surges. Bangladesh has been praised in recent years for its work on disaster preparedness, including construction of storm shelters and improved evacuation procedures. Such measures have resulted in declining death tolls and property damage in the face of more frequent storms and flooding.

We Can Reduce Flood Risks

Many scientists argue that we could improve flood control by relying less on engineered devices such as dams and levees and more on nature's systems such as wetlands and forests in watersheds.

One engineering approach is the channelizing of streams, which does reduce upstream flooding. However, it also eliminates the aquatic habitats that lie along a meandering stream by taking the water from those systems and sending it in a faster flow straight down a channel. It also reduces groundwater recharge and often leads to downstream flooding.

Similarly, levees or floodwalls along the banks of a river contain and speed up stream flow and can lead to flooding downstream. They also do not protect against unusually high and powerful floodwaters such as those that occurred in 1993 when two-thirds of the levees along the Mississippi River were damaged or destroyed. Similar flooding occurred along the Mississippi in 2011.

Damming, the most common engineering approach, can reduce the threat of flooding by storing water in a reservoir and releasing it gradually. However, dams also have a number of drawbacks (Figure 13.15).

A more ecologically oriented approach to reducing flooding is to *preserve existing wetlands* and *restore degraded wetlands* that lie in floodplains to take advantage of the natural flood control they provide. We would also be wise to sharply reduce emissions of greenhouse gases that contribute to atmospheric warming and climate change, which will likely raise sea levels and flood many of the world's coastal areas during this century.

Figure 13.27 summarizes these various ways to reduce flooding risks (**Concept 13.6**).

Solutions

Reducing Flood Damage

Prevention	Control
Preserve forests in watersheds	Strengthen and deepen streams (channelization)
Preserve and restore wetlands on floodplains	Build levees or floodwalls along streams
Tax development on floodplains	Build dams
Increase use of floodplains for sustainable agriculture and forestry	

FIGURE 13.27 Methods for reducing the harmful effects of flooding (**Concept 13.7**). ***Critical thinking:*** Which two of these solutions do you think are the best ones? Why?

Top: allensima/Shutterstock.com. Bottom: Zeljko Radojko/Shutterstock.com.

BIG IDEAS

- One of the major global environmental problems is the growing shortage of freshwater in many parts of the world.
- We can expand water supplies in water-short areas in a number of ways, but the most important ways are to reduce overall water use and to use water much more efficiently.
- We can use water more sustainably by reducing water use, using water more efficiently, cutting water losses, raising water prices, and protecting aquifers, forests, and other ecosystems that store and release water.

Tying It All Together

The Colorado River and Sustainability

The **Core Case Study** that opens this chapter discusses the problems and tensions that can occur when a large number of U.S. states share a limited river water resource in a water-short region. Such problems are representative of those faced by many other dry regions of the world, especially areas where the population is growing rapidly and water resources are dwindling for various reasons.

Large dams, river diversions, levees, and other big engineering schemes have helped to provide much of the world with electricity, food from irrigated crops, drinking water, and flood control. However, they have also degraded the aquatic natural capital necessary for long-term economic and ecological sustainability by seriously disrupting rivers, streams, wetlands, aquifers, and other aquatic systems.

The three **scientific principles of sustainability** can guide us in using water more sustainably during this century. Scientists hope to use solar energy to desalinate water and expand freshwater supplies. Recycling more water and reducing water waste will help reduce water losses. Preserving biodiversity by avoiding disruption of aquatic systems and their bordering terrestrial systems is a key factor in maintaining water supplies and water quality.

Atm2003/Shutterstock.com

Chapter Review

Core Case Study

1. Summarize the importance of the Colorado River basin in the United States and how human activities are stressing this system. Define **drought** and explain how it has affected the Colorado River system.

Section 13.1

2. What are the two key concepts for this section? Define **freshwater**. Explain why access to water is a health issue, an economic issue, a national and global security issue, and an environmental issue. What percentage of the earth's freshwater is available to us? Explain how water is recycled by the hydrologic cycle and how human activities can interfere with this cycle. Define **groundwater**, **zone of saturation**, **water table**, and **aquifer**, and explain how aquifers are recharged. What are deep aquifers and why are they considered nonrenewable? Define and distinguish between **surface water** and **surface runoff**. What is a **watershed (drainage basin)**?

3. What is **reliable surface runoff**? What percentage of the world's reliable runoff are we using and what percentage are we likely to be using by 2025? How is most of the world's water used? Define **water footprint** and **virtual water** and give two examples of each. Describe the availability and use of freshwater resources in the United States and the water shortages that could occur during this century. What are three major problems resulting from the way people are using water from the Colorado River basin?

4. How many countries face water scarcity today and how many could face water scarcity by 2050? What percentage of the earth's land suffers from severe drought today and how might this change by 2059? How many people in the world lack regular access to clean water today and how high might this number grow by 2025? Why do many analysts view the likelihood of steadily worsening water shortages as one of the world's most serious environmental problems? Explain the connection between water shortages and hunger. How do scientists use satellites to measure changes in water supplies?

Section 13.2

5. What is the key concept for this section? What are the advantages and disadvantages of withdrawing groundwater? Summarize the problem of groundwater depletion in the world and in the United States, especially in the Ogallala Aquifer. List three problems that result from the overpumping of aquifers. List some ways to prevent or slow groundwater depletion. What is the potential for using deep aquifers to expand water supplies?

Section 13.3

6. What are the two key concepts for this section? What is a **dam**? What is a **reservoir**? What are the advantages and disadvantages of using large dams and reservoirs? How do dams affect aquatic wildlife? What has happened to water flows in the Colorado River (**Core Case Study**) since 1960? Explain how the damming of this river has affected its delta. What other problems are likely to further shrink this supply of water? List three possible solutions to the supply problems in the Colorado River basin.

7. Define **desalination** and distinguish between distillation and reverse osmosis as methods for desalinating water. What are three limitations of desalination? What are scientists doing to try to deal with these problems?

Section 13.4

8. What is the key concept for this section? What is a water transfer? Describe two large water-transfer programs, explain how they came about, and summarize the controversy around each of these programs. Summarize the story of the Aral Sea water-transfer project and its disastrous consequences.

Section 13.5

9. What is the key concept for this section? What percentage of available freshwater is lost through inefficient use and other causes in the world and in the United States? What are two major reasons for those losses? Describe three major irrigation methods and list ways to reduce water losses in irrigation. What are three ways in which people in less-developed countries conserve water? List four ways to reduce water waste in industries and homes and three ways to use less water to remove wastes. What are five ways in which Californians are dealing with their severe drought and water shortages? List four ways in which you can reduce your use and waste of water.

Section 13.6

10. What is the key concept for this section? What is a **floodplain** and why do people like to live on floodplains? What are the benefits and harms of flooding? List two human activities that increase the risk of flooding. Describe the flooding risks that many people in Bangladesh face and what they are doing about it. List and compare two engineering approaches to flood control and two ecologically oriented approaches. What are this chapter's *three big ideas*? Explain how the **scientific principles of sustainability** can guide us in using water more sustainably during this century.

Note: Key terms are in bold type.

Critical Thinking

1. What do you think are the three most important priorities for dealing with the water resource problems of the Colorado River basin, as discussed in the **Core Case Study** that opens this chapter? Explain your choices.

2. List three ways in which human activities are affecting the water cycle. How might these changes to the water cycle affect your lifestyle? How might your lifestyle be contributing to these effects?

3. Explain how our current use of the earth's water resources can be viewed as a good example of the tragedy of the commons (see Chapter 1, p. 12).

4. Many argue that government freshwater subsidies promote the expansion of productive farmland, stimulate local economies, and help to keep food and electricity prices low. Do you think this is reason enough for governments to continue providing subsidies to farmers and cities? Explain.

5. Explain why you are for or against **(a)** raising the price of water while providing lower lifeline rates for poor consumers, **(b)** withdrawing government subsidies that provide farmers with water at low cost, and **(c)** providing government subsidies to farmers for improving irrigation efficiency.

6. Calculate how many liters (and gallons) of water are lost in 1 month by a toilet or faucet that leaks 2 drops of water per second. (One liter of water equals about 3,500 drops and 1 liter equals 0.265 gallon.) How many bathtubs (each containing about 151 liters or 40 gallons) could be filled with this lost water?

7. List the three most important ways in which you could use water more efficiently. Which, if any, of these measures do you already take?

8. List three ways in which human activities increase the harmful effects of flooding. What is the best way to prevent each of these human impacts? Do you think they should be prevented? Why or why not?

Doing Environmental Science

Investigate water use at your school. Try to determine all specific sources of any water losses, taking careful notes and measurements for each of them, and estimate how much water is lost per hour, per day, and per year from each source. Sum these estimated amounts to arrive at an estimate of total water losses for a year at your school. Develop a water conservation plan for your school and submit it to school officials.

Global Environment Watch Exercise

Go to your MindTap course to access the GREENR database. Using the "Basic Search" box at the top of the page, search for *Ogallala Aquifer*. Research articles that quantify how much the aquifer has declined and list areas over the aquifer where the decline is the worst. Look for projections on how much more the aquifer could decline in the future and take notes on this. Find information on the causes of this decline and determine which are the largest causes. Learn what is being done to address each of these causes and write a report explaining the causes, projections, and possible ways to slow the decline of the Ogallala Aquifer.

Ecological Footprint Analysis

The following table is based on data from the Water Footprint Network, a science-based organization that promotes the sustainable use of water through sharing knowledge and building awareness of how water is used. It shows the amounts of *embedded water*, or water required to produce each of the products listed in the first column. Study the table and then answer the questions that follow.

Product	Liters per kilogram (kg) or product	Gallons per pound (lb) or product
Beef	15,400/kg	1,855/lb.
Pork	5,990/kg	722/lb.
Chicken	4,325/kg	521/lb.
Milk	255/glass (250 ml)	68/glass (8 oz.)
Eggs	196/egg	52/egg
Coffee	132/cup	35/cup
Beer	74/glass (250 ml)	20/glass (8 oz.)
Wine	54/glass (250 ml)	14/glass (8 oz.)
Apple	125/average size apple	33/apple
Banana	160/large banana	42/banana
Tomato	50/medium tomato	13/tomato
Rice	2,500/kg	301/lb.
Bread	1,608/kg	426/lb.
Cotton t-shirt	2,495/shirt	661/shirt

Questions:

1. Find a loaf of common wheat bread, count the slices per loaf, and calculate the amount of water used to produce each slice. Calculate the amount of water used to produce one-third pound of beef. Combine these two to arrive at the approximate amount of water used to produce an average restaurant hamburger.
2. In terms of embedded water, how many tomatoes can be produced for each pound of pork? For each pound of beef?
3. If you drank coffee and/or milk today, how many gallons (or liters) of embedded water did you drink?
4. In terms of embedded water, how many kilograms (and pounds) of rice are represented in one t-shirt? How many bananas? How many apples?
5. In terms of embedded water, how many pounds of chicken can be produced for each pound of beef? How many pounds of rice?

CHAPTER 14

Geology and Mineral Resources

Civilization exists by geological consent, subject to change without notice.

WILL DURANT

Open-pit copper mine in Utah. It is almost 5 kilometers (3 miles) wide and 1,200 meters (4,000 feet) deep, and is getting deeper.

Lee Prince/Shutterstock.com

Key Questions

14.1 What are the earth's major geological processes and what are mineral resources?

14.2 How long might supplies of nonrenewable mineral resources last?

14.3 What are the environmental effects of using nonrenewable mineral resources?

14.4 How can we use mineral resources more sustainably?

14.5 What are the earth's major geological hazards?

Core Case Study

The Real Cost of Gold

Mineral resources are extracted from the earth's crust through a variety of processes called **mining**. They are processed into an amazing variety of products that make life easier and provide economic benefits and jobs. However, extracting minerals from the ground and using them to manufacture products results in a number of harmful environmental and health effects.

For example, gold mining often involves digging up massive amounts of rock (Figure 14.1) containing only small concentrations of gold. Many newlyweds would be surprised to know that mining enough gold to make their wedding rings produces roughly enough mining waste to equal the total weight of more than three midsize cars. This waste is usually left piled near the mine site and can pollute the air and nearby surface water.

About 90% of the world's gold mining operations extract the gold by spraying a solution of highly toxic cyanide salts onto piles of crushed rock. The solution reacts with the gold and then drains off the rocks, pulling some gold with it, into settling ponds (Figure 14.1, foreground). After the solution is recirculated a number of times, the gold is removed from the ponds.

Until sunlight breaks down the cyanide, the settling ponds are extremely toxic to birds and mammals that go to them in search of water. These ponds can also leak or overflow, posing threats to underground drinking water supplies and to fish and other organisms in nearby lakes and streams. Special liners in the settling ponds can help prevent leaks, but some have failed. According to the U.S. Environmental Protection Agency (EPA), all such liners are likely to leak, eventually.

In 2000 snow and heavy rains washed out an earthen dam on one end of a cyanide leach pond at a gold mine in Romania. The dam's collapse released large amounts of water laced with cyanide and toxic metals into the Tisza and Danube Rivers, which flow through parts of Romania, Hungary, and Yugoslavia. Several hundred thousand people living along these rivers were told not to fish or to drink or withdraw water from them or from wells along the rivers. Businesses located there were shut down. Thousands of fish and other aquatic animals and plants were killed. This accident and another one that occurred in January 2001 could have been prevented if the mining company had installed a stronger containment dam and a backup collection pond to prevent leakage into nearby surface water.

In addition, in parts of Africa and Latin America, millions of poor miners have illegally cleared areas of tropical forest and dug huge pits to find gold. In these operations, they typically use toxic mercury to extract the gold from the *ore*, or the rock containing the gold. Many of these miners have been poisoned by mercury. Because these illegal mining operations pollute the air and water with mercury and its toxic compounds, they have become a regional and global threat.

In 2015 the world's top five gold-producing countries were, in order, China, Australia, Russia, the United States, and Canada. These countries vary in how they deal with the environmental impacts of gold mining.

In this chapter, we look at the earth's dynamic geological processes, the valuable minerals such as gold that some of these processes produce, and the potential supplies of these resources. We will also study the environmental impacts of extracting and processing these resources, and how people can use these resources more sustainably.

Larry Mayer/Getty Images

FIGURE 14.1 Gold mine in the Black Hills of the U.S. state of South Dakota with cyanide leach piles and settling ponds in foreground.

14.1 WHAT ARE THE EARTH'S MAJOR GEOLOGICAL PROCESSES AND WHAT ARE MINERAL RESOURCES?

CONCEPT 14.1A Dynamic processes within the earth and on its surface produce the mineral resources we depend on.

CONCEPT 14.1B Mineral resources are nonrenewable because it takes millions of years for the earth's rock cycle to produce or renew them.

The Earth Is a Dynamic Planet

Geology is the scientific study of dynamic processes taking place on the earth's surface and in its interior. Scientific evidence indicates that the earth formed about 4.6 billion years ago. As the primitive earth cooled over millions of years, its interior separated into three major concentric zones: the *core*, the *mantle*, and the *crust* (Figure 14.2). They make up the *geosphere* (Figure 3.2, p. 51), which is part of the earth's life-support system.

The **core** is the earth's innermost zone and is composed primarily of iron (Fe). The inner core is extremely hot and has a solid center. It is surrounded by the outer core, a thick layer of *molten rock*, or hot fluid rock, and semisolid material.

Surrounding the core is a thick zone called the **mantle**—a zone made mostly of solid rock that can be soft and pliable at very high temperatures. The outermost part of the mantle is entirely solid rock. Beneath it is the **asthenosphere**—a volume of hot, partly melted rock that flows.

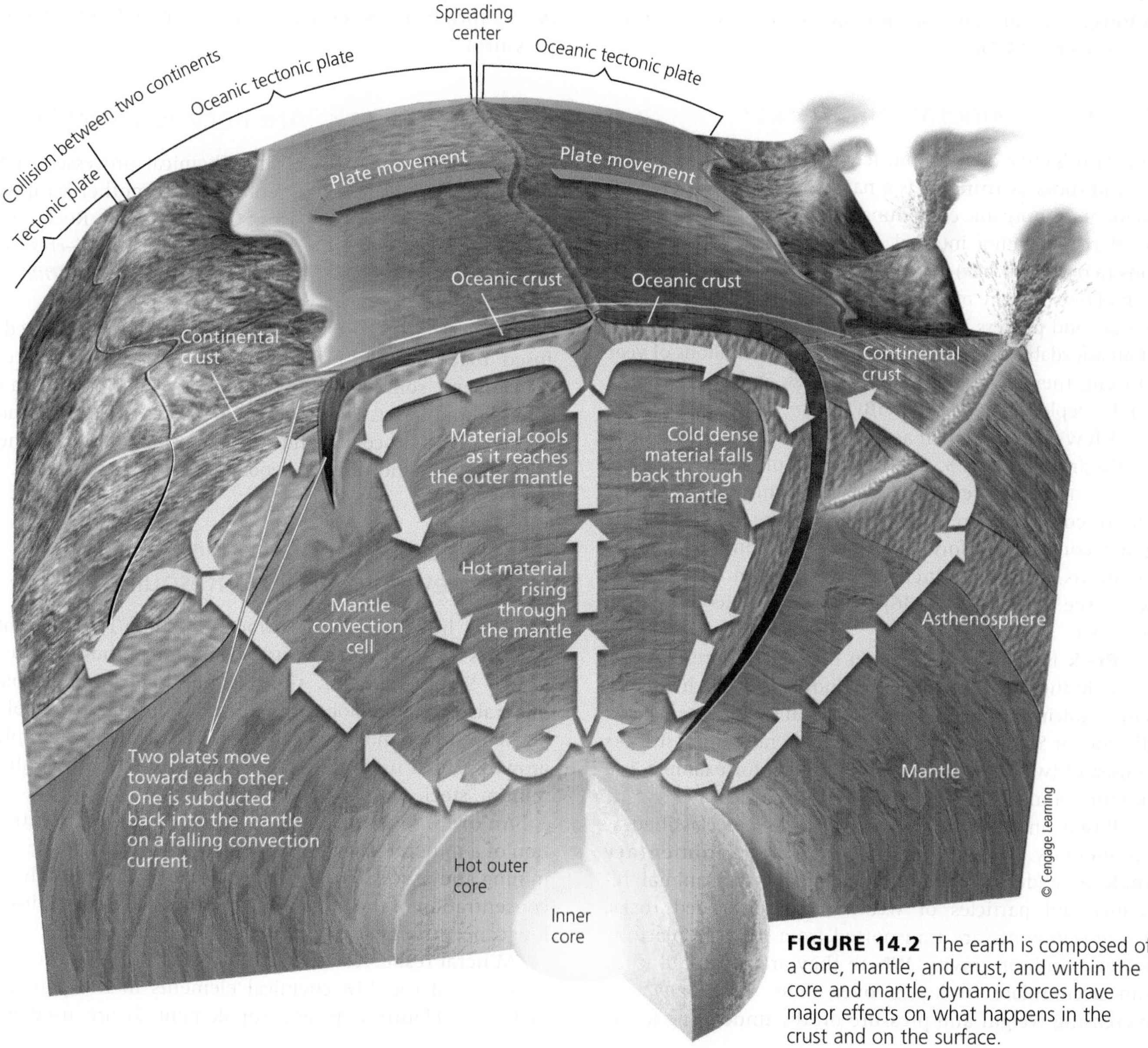

FIGURE 14.2 The earth is composed of a core, mantle, and crust, and within the core and mantle, dynamic forces have major effects on what happens in the crust and on the surface.

Tremendous heat within the core and mantle generates *convection cells,* or *currents.* The innermost material heats, rises, and begins to cool. As it cools, it becomes denser and sinks back toward the core where it is reheated, completing a huge loop of slowly moving material. These loops within the mantle operate like gigantic conveyer belts (Figure 14.2). Some of the molten rock in the asthenosphere flows upward into the crust, where it is called *magma.* When magma erupts onto the earth's surface, it is called *lava.* This cycling moves rock and minerals and transfers heat and energy within the earth and to its surface.

The outermost and thinnest zone of solid material is the earth's **crust.** It consists of the *continental crust,* which underlies the continents (including the continental shelves extending into the oceans), and the *oceanic crust,* which underlies the ocean basins and makes up 71% of the earth's crust. The combination of the crust and the rigid, outermost part of the mantle is called the **lithosphere.** This zone contains the mineral resources that we use (**Concept 14.1A**).

What Are Minerals and Rocks?

The earth's crust beneath our feet consists mostly of minerals and rocks. A **mineral** is a naturally occurring chemical element or inorganic compound that exists as a solid with a regularly repeating internal arrangement of its atoms or ions (a *crystalline solid*). A **mineral resource** is a concentration of one or more minerals in the earth's crust that we can extract and process into raw materials and useful products at an affordable cost. Because minerals take millions of year to form, they are *nonrenewable resources,* and their supplies can be depleted (**Concept 14.1B**).

A few minerals, such as mercury and gold (**Core Case Study** and Figure 2.3, p. 34), consist of a single chemical element. However, most of the more than 2,000 identified mineral resources that we use occur as inorganic compounds formed by various combinations of elements. Examples include salt (sodium chloride, or NaCl; see Figure 2.7, p. 36) and quartz (silicon dioxide, or SiO_2).

Rock is a solid combination of one or more minerals found in the earth's crust. Some kinds of rock, such as limestone (calcium carbonate, or $CaCO_3$) and quartzite (silicon dioxide, or SiO_2), contain only one mineral, but most rocks consist of two or more minerals. Granite, for example, is a mixture of mica, feldspar, and quartz crystals.

Based on the way they form, rocks are classified as sedimentary, igneous, or metamorphic. **Sedimentary rock** is made of *sediments*—dead plant and animal remains and particles of weathered and eroded rocks. These sediments are transported from place to place by water, wind, or gravity. Where they are deposited, they can accumulate in layers over time. Eventually, the increasing weight and pressure on the underlying layers transform the sedimentary layers to rock. Examples include *sandstone* and *shale* (formed from pressure created by deposited layers made mostly of sand or silt), *dolomite* and *limestone* (formed from the compacted shells, skeletons, and other remains of dead aquatic organisms), and *lignite* and *bituminous coal* (derived from compacted plant remains).

Igneous rock forms below or on the earth's surface under intense heat and pressure when magma wells up from the earth's mantle and then cools and hardens. Examples include *granite* (formed underground) and *lava rock* (formed aboveground). Igneous rock forms the bulk of the earth's crust but is usually covered with layers of sedimentary rock.

Metamorphic rock forms when an existing rock is subjected to high temperatures (which may cause it to melt partially), high pressures, chemically active fluids, or a combination of these agents. Examples include *slate* (formed when shale and mudstone are heated) and *marble* (produced when limestone is exposed to heat and pressure).

The Earth's Rocks Are Recycled Slowly

The interaction of physical and chemical processes that change the earth's rocks from one type to another is called the **rock cycle** (Figure 14.3 and **Concept 14.1B**). Rocks are recycled over millions of years by three processes—*erosion, melting,* and *metamorphism*—which produce *sedimentary, igneous,* and *metamorphic* rocks, respectively.

In these processes, rocks are broken down, melted, fused together into new forms by heat and pressure, cooled, and sometimes recrystallized within the earth's interior and crust. The rock cycle is the slowest of the earth's cyclic processes and plays the major role in the formation of concentrated deposits of mineral resources.

We Depend on a Variety of Nonrenewable Mineral Resources

We know how to find and extract more than 100 different minerals from the earth's crust. According to the U.S. Geological Survey (USGS), the quantity of nonrenewable minerals extracted globally increased more than threefold between 1995 and 2014. To support its rapid economic growth, China used 45–55% of the world's iron ore, aluminum, steel, nickel, copper, and zinc in 2014.

An **ore** is rock that contains a large enough concentration of a mineral—often a metal—to make it profitable for mining and processing. A **high-grade ore** contains a high concentration of the mineral. A **low-grade ore** contains a low concentration.

Mineral resources are used for many purposes. Today, about 60 of the 118 chemical elements in the periodic table (see Figure 1, p. S6, Supplement 3) are used for

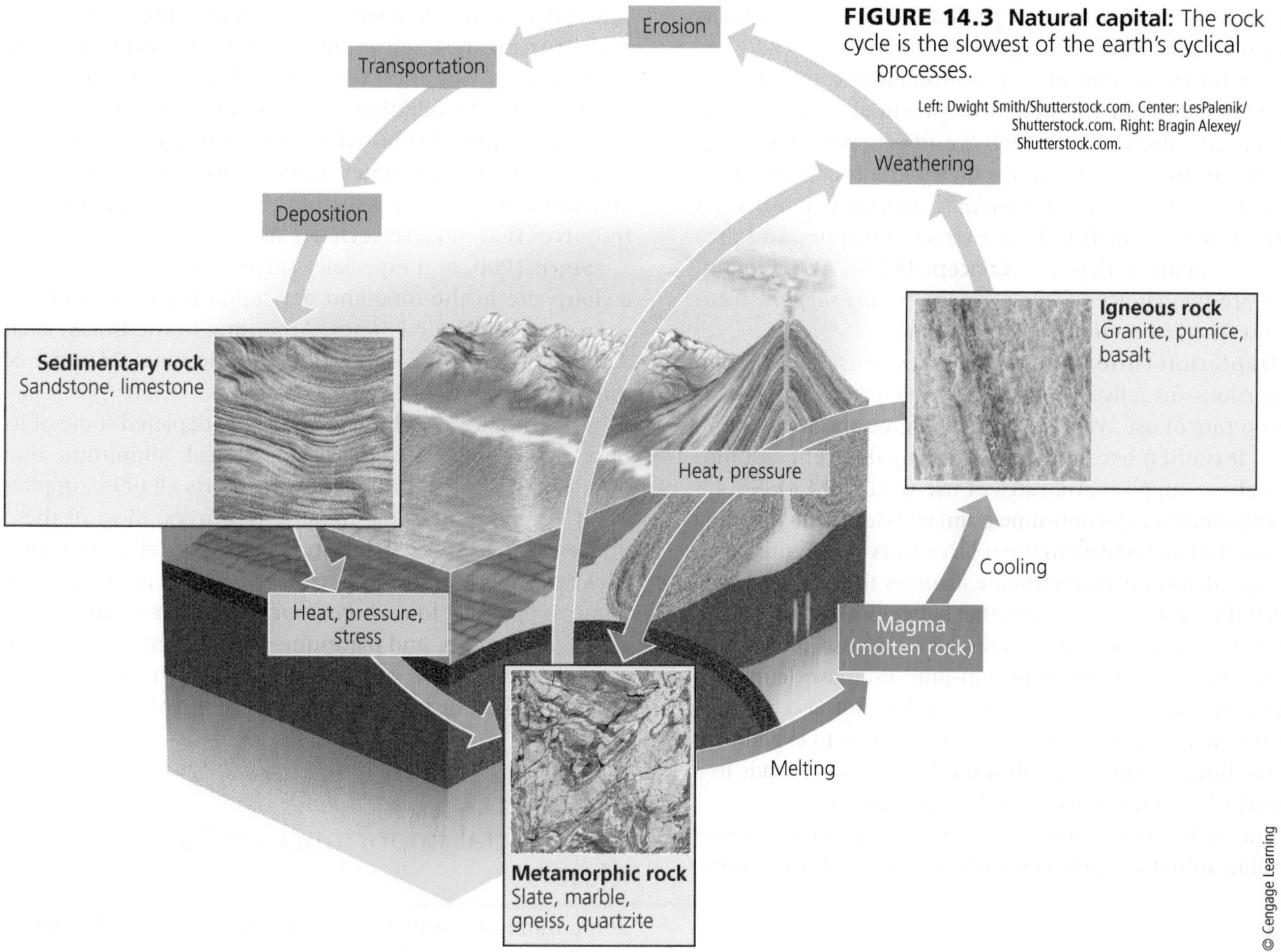

FIGURE 14.3 Natural capital: The rock cycle is the slowest of the earth's cyclical processes.

Left: Dwight Smith/Shutterstock.com. Center: LesPalenik/Shutterstock.com. Right: Bragin Alexey/Shutterstock.com.

making computer chips. *Aluminum* (Al) is used as a structural material in beverage cans, motor vehicles, aircraft, and buildings. *Steel,* an essential material used in buildings, machinery, and motor vehicles, is a mixture (or *alloy*) of iron (Fe) and other elements that gives it certain physical properties. *Manganese* (Mn), *cobalt* (Co), and *chromium* (Cr) are widely used in steel alloys. *Copper* (Cu), a good conductor of electricity, is used to make electrical and communications wiring and plumbing pipes. *Gold* (Au) (**Core Case Study**) is a component of electrical equipment, tooth fillings, jewelry, coins, and some medical implants. *Molybdenum* (Mo) is widely used to harden steel and to make it more resistant to corrosion.

There are several widely used nonmetallic mineral resources. *Sand,* which is mostly silicon dioxide (SiO_2), is used to make glass, bricks, and concrete for the construction of roads and buildings. *Gravel* is used for roadbeds and to make concrete. Another common nonmetallic mineral is *limestone* (mostly calcium carbonate, or $CaCO_3$), which is crushed to make concrete and cement. Still another is *phosphate,* used to make inorganic fertilizers and certain detergents.

14.2 HOW LONG MIGHT SUPPLIES OF NONRENEWABLE MINERAL RESOURCES LAST?

CONCEPT 14.2A Nonrenewable mineral resources exist in finite amounts and can become economically depleted when it costs more than it is worth to find, extract, and process the remaining deposits.

CONCEPT 14.2B There are several ways to extend supplies of mineral resources, but each of them is limited by economic and environmental factors.

Supplies of Nonrenewable Mineral Resources Can Be Economically Depleted

Most published estimates of the supply of a given nonrenewable mineral resource refer to its **reserves:** identified deposits from which we can extract the mineral profitably at current prices. Reserves can be expanded when we find new, profitable deposits or when higher prices or improved mining technologies make it profitable to extract deposits that previously were too expensive to remove.

The future supply of any nonrenewable mineral resource depends on the actual or potential supply of the mineral and the rate at which we use it. We have never completely run out of a nonrenewable mineral resource, but a mineral becomes *economically depleted* when it costs more than it is worth to find, extract, transport, and process the remaining deposits (**Concept 14.2A**). At that point, there are five choices: *recycle or reuse existing supplies, waste less, use less, find a substitute,* or *do without.*

Depletion time is the time it takes to use up a certain proportion—usually 80%—of the reserves of a mineral at a given rate of use. When experts disagree about depletion times, it is often because they are using different assumptions about supplies and rates of use (Figure 14.4).

The shortest depletion-time estimate assumes no recycling or reuse and no increase in the reserve (curve A, Figure 14.4). A longer depletion-time estimate assumes that recycling will stretch the existing reserve and that better mining technology, higher prices, or new discoveries will increase the reserve (curve B). The longest depletion-time estimate (curve C) makes the same assumptions as A and B, but also assumes that people will reuse and reduce consumption to expand the reserve further. Finding a substitute for a resource leads to a new set of depletion curves for the new mineral.

The earth's crust contains abundant deposits of nonrenewable mineral resources such as iron and aluminum.

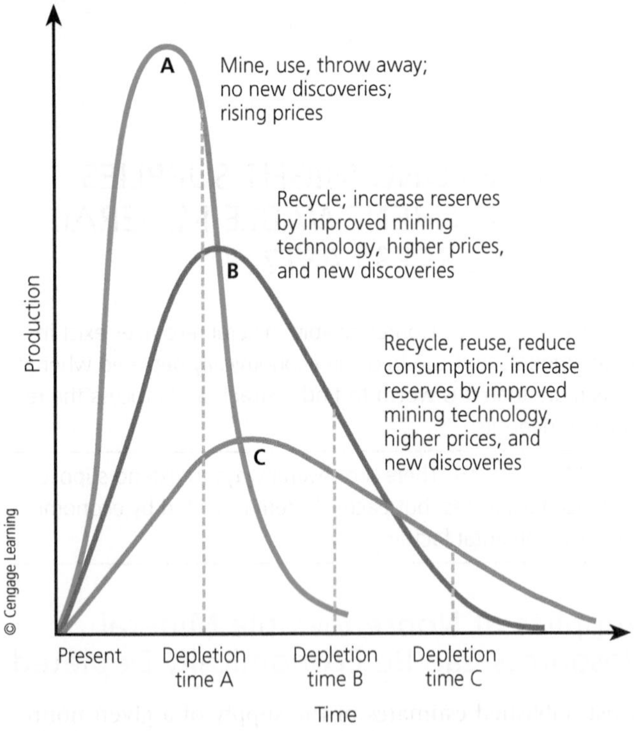

FIGURE 14.4 Natural capital depletion: Each of these *depletion curves* for a mineral resource is based on a different set of assumptions. Dashed vertical lines represent the times at which 80% depletion occurs.

But concentrated deposits of important mineral resources such as manganese, chromium, cobalt, platinum, and *rare earth elements* (see the Case Study that follows) are relatively scarce. In addition, deposits of many mineral resources are not distributed evenly among countries. Five nations—the United States, Canada, Russia, South Africa, and Australia—supply most of the nonrenewable mineral resources that modern societies use.

Since 1900, and especially since 1950, there has been a sharp rise in the total and per capita use of mineral resources in the United States. According to the USGS, each American uses an average of 22 metric tons (24 tons) of mineral resources per year.

The United States has economically depleted some of its once-rich deposits of metals such as lead, aluminum, and iron. Currently, the United States imports all of its supplies of 24 key nonrenewable mineral resources. Most of these imports come from reliable and politically stable countries. However, there are serious concerns about access to adequate supplies of four *strategic metal resources*—manganese, cobalt, chromium, and platinum—that are essential for the country's economic and military strength. The United States has little or no reserves of these metals.

CASE STUDY

The Crucial Importance of Rare Earth Metals

Some mineral resources are familiar, such as gold, copper, aluminum, sand, and gravel. Less well known are the *rare earth metals and oxides,* which are crucial to the technologies that support modern lifestyles and economies.

The 17 rare earth metals, also known as *rare earths,* include scandium, yttrium, and 15 lanthanide chemical elements, including lanthanum (see the Periodic Table, Figure 1 of Supplement 3, p. S6). Because of their superior magnetic strength and other unique properties, these elements and their compounds are important for a number of widely used technologies.

Rare earths are used to make LCD flat screens for computers and television sets, energy-efficient compact fluorescent and LED light bulbs, solar cells, fiber-optic cables, cell phones, and digital cameras. They are also used to manufacture batteries and motors for electric and hybrid-electric cars (Figure 14.5), solar cells, catalytic converters in car exhaust systems, jet engines, and the powerful magnets in wind turbine generators. Rare earths also go into missile guidance systems, jet engines, smart bombs, aircraft electronics, and satellites.

Without affordable supplies of these metals, industrialized nations could not develop the current versions of cleaner energy technology and other high-tech products that will be major sources of economic growth during this century. Many nations also need these metals to maintain their military strength.

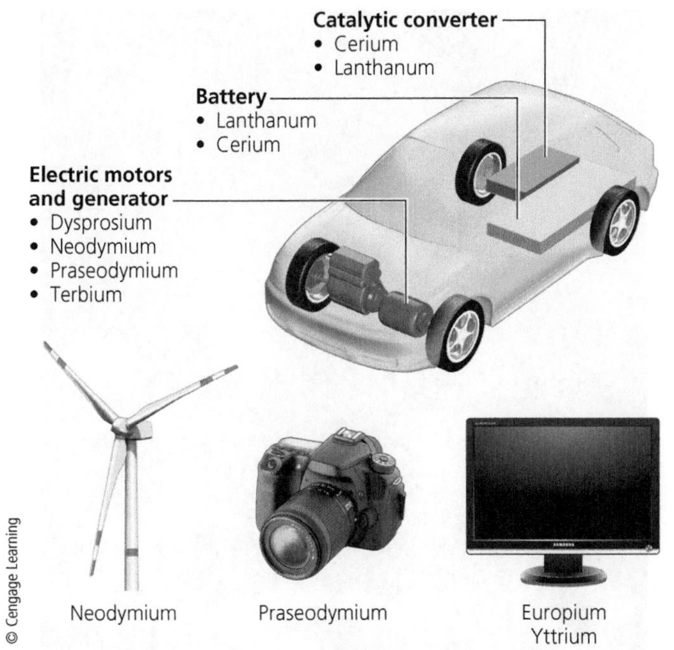

FIGURE 14.5 Rare-earth metals are used to manufacture all-electric and hybrid-electric cars and many other products.

Most rare earth elements are not actually rare, but they are hard to find in concentrations high enough to extract and process at an affordable price. According to the USGS, in 2014 China had roughly 42% of the world's known rare earth reserves, Brazil had the second largest share with 17%, and the United States, with the fifth largest share, had 1.4% of the global reserves.

In 2015 China produced about 90% of the world's rare earth metals and oxides, down from 97% in 2013. Australia and Chile are beginning to increase their shares of global production. China still holds the lead, partly because it does not strictly regulate the environmentally disruptive mining and processing of rare earths. This means that Chinese companies have lower production costs than do companies in countries with stricter regulations. China can thus afford to sell its rare earths at a lower price.

The United States and Japan are heavily dependent on rare earths and their oxides. Japan has no rare earth reserves. In the United States, the only rare earth mine, located in California, was once the world's largest supplier of rare earth metals. However, it closed down because of the expense of meeting pollution regulations, and because China had driven the prices of rare earth metals down to a point where the mine was too costly to operate. In 2015 the company that owns the mine declared bankruptcy.

Market Prices Affect Supplies of Mineral Resources

Geological processes determine the quantity and location of a mineral resource in the earth's crust, but economics determines what part of the known supply is extracted and used. According to standard economic theory, in a competitive market system when a resource becomes scarce, its price rises. Higher prices can encourage exploration for new deposits, stimulate development of better mining technology, and make it profitable to mine lower-grade ores. It can also promote resource conservation and a search for substitutes, but there are limits to these effects (**Concept 14.2B**).

CONSIDER THIS . . .

CONNECTIONS High Metal Prices and Thievery

Resource scarcity can promote theft. For example, copper prices have risen sharply in recent years because of increasing demand. As a result, in many U.S. communities, thieves have been stealing copper to sell it. They strip abandoned houses of copper pipe and wiring and steal outdoor central air conditioning units for their copper coils. They also steal wiring from beneath city streets and copper piping from farm irrigation systems. In 2015 thieves stole copper wiring from New York City's subway system, temporarily shutting down two of the city's busiest lines.

According to some economists, the standard effect of supply and demand on the market prices of mineral resources may no longer apply completely in most more-developed countries. Governments in such countries often use subsidies, tax breaks, and import tariffs to control the supply, demand, and prices of key mineral resources. In the United States, for instance, mining companies get various types of government subsidies, including *depletion allowances*—which allow the companies to deduct the costs of developing and extracting mineral resources from their taxable incomes. These allowances amount to 5–22% of their gross income gained from selling the mineral resources.

Generally, the mining industry maintains that they need subsidies and tax breaks to keep the prices of minerals low for consumers. They also claim that, without subsidies and tax breaks, they might move their operations to other countries where they would not have to pay taxes or comply with strict mining and pollution control regulations.

Can We Expand Reserves by Mining Lower-Grade Ores?

Some analysts contend that we can increase supplies of some minerals by extracting them from lower-grade ores. They point to the development of new earth-moving equipment, improved techniques for removing impurities from ores, and other technological advances in mineral extraction and processing that can make lower-grade ores accessible, sometimes at lower costs.

For example, shortly after World War II, rich deposits of high-grade iron ore in northern Minnesota (USA) were economically depleted. By the 1960s, a new process had been developed for mining *taconite*, a low-grade and plentiful ore that had been viewed as waste rock in the iron mining process. This improvement in mining technology expanded Minnesota's iron reserves and supported a taconite mining industry long after the high-grade iron ore reserves there had been tapped out. Similarly, in 1900 the copper ore mined in the United States was typically 5% copper by weight. Today, it is typically 0.5%, yet copper costs less (when prices are adjusted for inflation).

Several factors can limit the mining of lower-grade ores (**Concept 14.2B**). For example, it requires mining and processing larger volumes of ore, which takes much more energy and costs more. Another factor is the dwindling supplies of freshwater needed for the mining and processing of some minerals, especially in dry areas. A third limiting factor is the growing environmental impacts of land disruption, along with waste material and pollution produced during mining and processing.

One way to improve mining technology and reduce its environmental impact is to use a biological approach, sometimes called *biomining*. Miners use naturally occurring or genetically engineered bacteria to remove desired metals from ores through wells bored into the deposits. This leaves the surrounding environment undisturbed and reduces the air and water pollution associated with removing the metal from metal ores. On the downside, biomining is slow. It can take decades to remove the same amount of material that conventional methods can remove within months or years. So far, biomining methods are economically feasible only for low-grade ores for which conventional techniques are too expensive.

Can We Get More Minerals from the Oceans?

Most of the minerals found in seawater occur in such low concentrations that recovering them takes more energy and money than they are worth. Currently, only magnesium, bromine, and sodium chloride are abundant enough to be extracted profitably from seawater. On the other hand, sediments along the shallow continental shelf and adjacent shorelines contain significant deposits of minerals such as sand, gravel, phosphates, copper, iron, silver, titanium, and diamonds.

Another potential ocean source of some minerals is *hydrothermal ore deposits* that form when superheated, mineral-rich water shoots out of vents in volcanic regions of the ocean floor. When the hot water comes into contact with cold seawater, black particles of various metal sulfides precipitate out and accumulate as chimney-like structures, called *black smokers*, near the hot water vents (Figure 14.6). These deposits are especially rich in minerals such as copper, lead, zinc, silver, gold, and some of the rare earth metals. A variety of more than 300 exotic forms of life—including giant clams, six-foot tubeworms, and eyeless shrimp—live in the dark depths around black smokers.

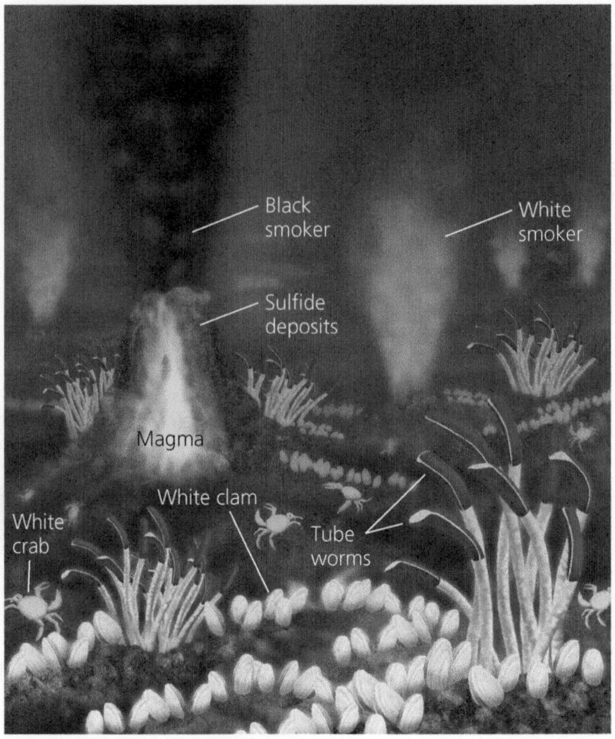

FIGURE 14.6 Natural capital: *Hydrothermal deposits*, or *black smokers*, are rich in various minerals.

Because of the rapidly rising prices of many of these metals, there is growing interest in deep-sea mining. Companies from Australia, the United States, and China have been exploring the possibility of mining black smokers in several areas. In 2012 the U.S. government issued its first-ever approval of large-scale deep-sea mining, proposed for a large area between Hawaii and Mexico. In 2015 the Center for Biological Diversity sued the government to try to prevent the project, arguing that it could damage important habitat for whales, sharks, and sea turtles by destroying seafloor ecosystems.

According to some analysts, seafloor mining is less environmentally harmful than mining on land. Other scientists, however, are concerned because seafloor mining stirs up sediment that can harm or kill organisms that feed by filtering seawater. Supporters of seafloor mining say that the number of potential mining sites, and thus the overall environmental impact, will be small.

Another possible source of metals is the potato-size *manganese nodules* that cover large areas of the Pacific Ocean floor and smaller areas of the Atlantic and Indian Ocean floors. They also contain low concentrations of various rare earth minerals. These modules could be sucked up through vacuum pipes or scooped up by underwater mining machines.

To date, mining on the ocean floor has been hindered by the high costs involved, the potential threat to marine ecosystems, and arguments over rights to the minerals in deep ocean areas that do not belong to any specific country.

14.3 WHAT ARE THE ENVIRONMENTAL EFFECTS OF USING NONRENEWABLE MINERAL RESOURCES?

CONCEPT 14.3 Extracting minerals from the earth's crust and converting them to useful products can disturb the land, erode soils, produce large amounts of solid waste, and pollute the air, water, and soil.

Extracting Minerals Can Have Harmful Environmental Effects

Every metal product has a *life cycle* that includes mining the mineral, processing it, manufacturing the product, and disposal or recycling of the product (Figure 14.7). This process makes use of large amounts of energy and water, and produces pollution and waste at every step of the life cycle (**Concept 14.3**).

The environmental impacts of mining a metal ore are determined partly by the ore's percentage of metal content, or *grade*. The more accessible higher-grade ores are usually exploited first. Mining lower-grade ores takes more money, energy, water, and other resources, and leads to more land disruption, mining waste, and pollution.

Several mining techniques are used to remove mineral deposits. Shallow mineral deposits are removed by **surface mining,** in which vegetation, soil, and rock overlying a mineral deposit are cleared away. This waste material is called **overburden** and is usually deposited in piles called **spoils** (Figure 14.8). Surface mining is used to extract about 90% of the nonfuel mineral resources and 60% of the coal used in the United States.

Different types of surface mining can be used, depending on two factors: the resource being sought and the local topography. In **open-pit mining,** machines are used to dig large pits and remove metal ores containing copper (see chapter-opening photo), gold (**Core Case Study**), or other metals, or sand, gravel, or stone.

LianeM/Shutterstock.com

FIGURE 14.8 Natural capital degradation: This spoils pile in Zielitz, Germany, is made up of waste material from the mining of potassium salts used to make fertilizers.

9 million

Number of people who could sit in Bingham Copper Mine (see chapter-opening photo) if it were a stadium

Strip mining involves extracting mineral deposits that lie in large horizontal beds close to the earth's surface. In **area strip mining,** used on flat terrain, a gigantic earthmover strips away the overburden, and a power shovel—which can be as tall as a 20-story building—removes a mineral resource such as gold (Figure 14.9). The resulting trench is filled with overburden, and a new cut is

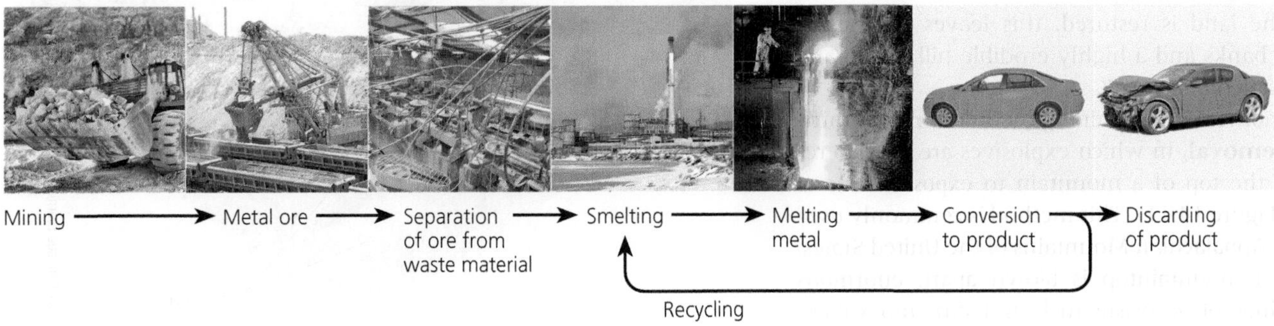

© Cengage Learning

FIGURE 14.7 Each metal product that we use has a *life cycle*.

Left: kaband/Shutterstock.com. Second to left: Andrey N Bannov/Shutterstock.com. Center left: Vladimir Melnik/Shutterstock.com. Center: mares/Shutterstock.com. Center right: Zhu Difeng/Shutterstock.com. Second to right: Michael Shake/Shutterstock.com. Right: Pakhnyushcha/Shutterstock.com.

FIGURE 14.9 Natural capital degradation: Area strip mining for gold in Yukon Territory, Canada.

made parallel to the previous one. This process is repeated over the entire site.

Contour strip mining (Figure 14.10) is used mostly to mine coal and various mineral resources on hilly or mountainous terrain. Huge power shovels and bulldozers cut a series of terraces into the side of a hill. Then, earthmovers remove the overburden, an excavator or power shovel extracts the coal, and the overburden from each new terrace is dumped onto the one below. Unless the land is restored, this leaves a series of spoils banks and a highly erodible hill of soil and rock called a *highwall.*

Another surface mining method is **mountaintop removal,** in which explosives are used to remove the top of a mountain to expose seams of coal (Figure 14.11). This method is commonly used in the Appalachian Mountains of the United States. After a mountaintop is blown apart, enormous machines plow waste rock and dirt into valleys below. This destroys forests, buries mountain streams, and increases the risk of flooding. Wastewater and toxic sludge, produced when the coal is

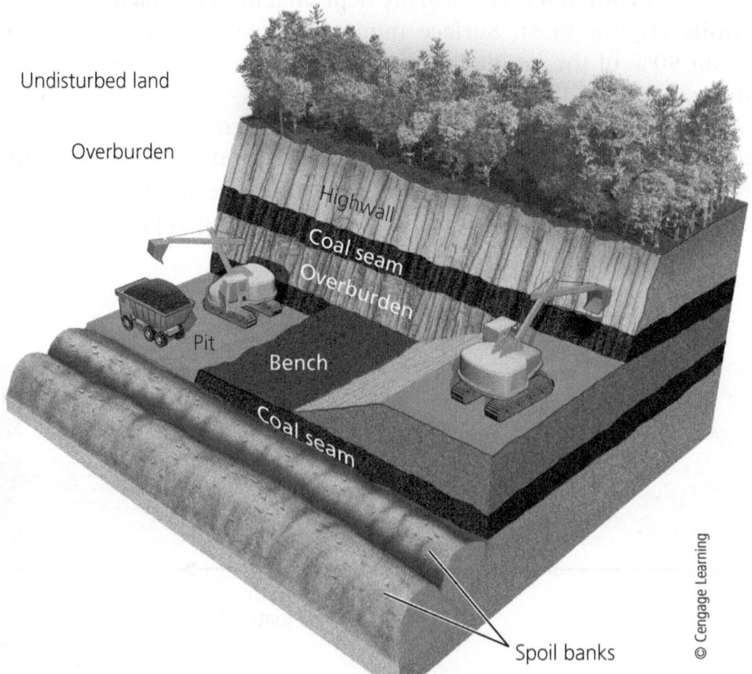

FIGURE 14.10 Natural capital degradation: Contour strip mining is used in hilly or mountainous terrain.

Jim West/Age Fotostock

FIGURE 14.11 Natural capital degradation: Mountaintop removal coal mining near Whitesville, West Virginia.

processed, are often stored behind dams in these valleys. Such dams have been known to overflow or collapse and release toxic substances such as arsenic and mercury.

In the United States, more than 500 mountaintops in West Virginia and other Appalachian states have been removed to extract coal. According to the U.S. Environmental Protection Agency (EPA), the resulting spoils have buried more than 1,100 kilometers (700 miles) of streams—a total roughly equal in length to the distance between the two U.S. cities of New York and Chicago.

The U.S. Department of the Interior estimates that at least 500,000 surface-mined sites dot the U.S. landscape, mostly in the West. Such sites can be cleaned up and restored. The U.S. Surface Mining Control and Reclamation Act of 1977 requires the restoration of surface-mined sites. However, the program is underfunded and many mines have not been reclaimed.

Deep deposits of minerals are removed by **subsurface mining,** in which underground mineral resources are removed through tunnels and shafts (Figure 14.12). This method is used to remove metal ores and coal that are too deep to be extracted by surface mining. Miners dig a deep, vertical shaft and blast open subsurface tunnels and chambers to reach the deposit. Then they use machinery to remove the resource and transport it to the surface.

Subsurface mining disturbs less than one-tenth as much land as surface mining, and usually produces less waste material. However, it can lead to other hazards such as cave-ins, explosions, and fires for miners. Miners often get lung diseases caused by prolonged inhalation of mineral or coal dust in subsurface mines. Another problem is *subsidence*—the collapse of land above some underground mines. It can damage houses, crack sewer lines, break natural gas mains, and disrupt groundwater systems.

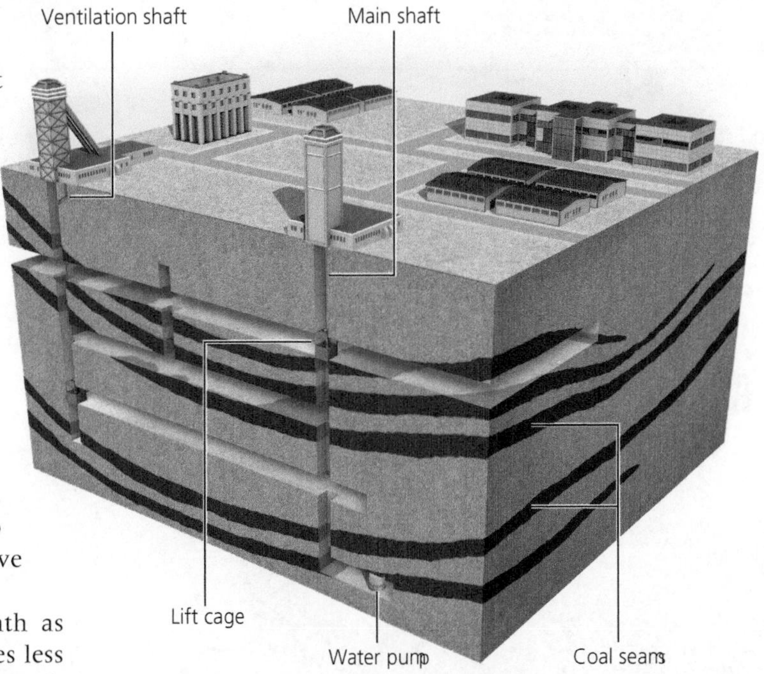

© Cengage Learning

FIGURE 14.12 Subsurface mining of coal.

Surface and subsurface mining operations also produce large amounts of solid waste—three-fourths of all U.S. solid waste—and cause major water and air pollution. For example, *acid mine drainage* occurs when rainwater that seeps through an underground mine or a spoils pile from a surface mine carries sulfuric acid (H_2SO_4) produced when aerobic bacteria act on remaining minerals to nearby streams and groundwater. This is one of the problems often associated with gold mining (**Core Case Study**).

According to the EPA, mining has polluted mountain streams in 40% of the western watersheds in the United States. It accounts for 50% of all the country's emissions of toxic chemicals into the atmosphere. In fact, the mining industry produces more of such toxic emissions than any other U.S. industry.

Where environmental regulations and enforcement are lax, mining is even more harmful to the environment. In China, for instance, the mining and processing of rare earth metals and oxides has stripped land of its vegetation and topsoil. It also has polluted the air, acidified streams, and left toxic and radioactive waste piles.

Removing Metals from Ores Has Harmful Environmental Effects

Ore extracted by mining typically has two components: the ore mineral, containing the desired metal, and waste material. Removing the waste material from ores produces **tailings**—rock wastes that are left in piles or put into ponds where they settle out. Particles of toxic metals in tailings piles can be blown by the wind or washed out by rain, or can leak from holding ponds and contaminate surface water and groundwater.

After the waste material is removed, heat or chemical solvents are used to extract the metals from mineral ores. Heating ores to release metals is called **smelting** (Figure 14.6). Without effective pollution control equipment, a smelter emits large quantities of air pollutants, including sulfur dioxide and suspended toxic particles that damage vegetation and acidify soils in the surrounding area. Smelters also cause water pollution and produce liquid and solid hazardous wastes that require safe disposal. A 2012 study found that lead smelting is the world's second most toxic industry after the recycling of lead-acid batteries.

Using chemicals to extract metals from their ores can also create numerous problems, as noted in the **Core Case Study**. Even on a smaller scale, this is the case. For example, millions of poverty-stricken miners in less-developed countries have gone into tropical forests in search of gold (Figure 14.13). They have cleared trees to get access to gold, and such illegal deforestation has increased rapidly, especially in parts of the Amazon Basin. The miners use toxic mercury to separate gold from its ore. They heat the

Randy Olson/National Geographic Creative

FIGURE 14.13 Illegal gold mining on the banks of the Pra River in Ghana, Africa.

mixture of gold and mercury to vaporize the mercury and leave the gold, causing dangerous air and water pollution. Many of these miners and villagers living near the mines eventually inhale toxic mercury vapor, drink mercury-laden water, or eat fish contaminated with mercury.

14.4 HOW CAN WE USE MINERAL RESOURCES MORE SUSTAINABLY?

CONCEPT 14.4 We can try to find substitutes for scarce resources, reduce resource waste, and recycle and reuse minerals.

Find Substitutes for Scarce Mineral Resources

Some analysts believe that even if supplies of key minerals become too expensive or too scarce due to unsustainable use, human ingenuity will find substitutes (**Concept 14.4**). They point to the current *materials revolution* in which silicon and other materials are replacing some metals for common uses. They also point out the possibilities of finding substitutes for scarce minerals through nanotechnology (Science Focus 14.1), as well as through other emerging technologies.

For example, fiber-optic glass cables that transmit pulses of light are replacing copper and aluminum wires in telephone cables, and nanowires may eventually replace fiber-optic glass cables. High-strength plastics and materials, strengthened by lightweight carbon, hemp, and glass fibers, are beginning to transform the automobile and aerospace industries. These new materials do not need painting (which reduces pollution and costs) and can be molded into any shape. Use of such materials in manufacturing motor vehicles and airplanes could greatly increase vehicle fuel efficiency by reducing vehicle weights. Such new materials are even being used to build bridges. Two such possible breakthrough materials are graphene and phosphorene (see the Case Study that follows).

CONSIDER THIS . . .

LEARNING FROM NATURE

Without using toxic chemicals, spiders rapidly build their webs by producing threads of silk that are capable of capturing insects flying at high speeds. Learning how spiders do this could revolutionize the production of high-strength fibers with a very low environmental impact.

However, resource substitution is not a cure-all. For example, platinum is currently unrivaled as a catalyst and is used in industrial processes to speed up chemical reactions, and chromium is an essential ingredient of stainless steel. We can try to find substitutes for such scarce resources, but this may not always be possible.

CASE STUDY

Graphene and Phosphorene: New Revolutionary Materials

Graphene is made from graphite—a form of carbon that occurs as a mineral in some rocks. Ultrathin graphene consists of a single layer of carbon atoms packed into a two-dimensional hexagonal lattice (somewhat like chicken wire) that can be applied as a transparent film to surfaces (Figure 14.14).

Graphene is one of the world's thinnest and strongest materials and is light, flexible, and stretchable. A single layer of graphene is 150,000 times thinner than a human hair and 100 times stronger than structural steel. A sheet of this amazing material stretched over a coffee mug could support the weight of a car. It is also a better conductor of electricity than copper and conducts heat better than any known material.

The use of graphene could revolutionize the electric car industry by leading to the production of batteries that can be recharged 10 times faster and hold 10 times more power than current car batteries. Graphene composites can also be used to make stronger and lighter plastics, lightweight aircraft and motor vehicles, flexible computer tablets, and TV screens as thin as a magazine. Within 5 years, it might also be used to make flexible, more efficient, less costly solar cells that can be attached to almost anything. Engineers also hope to make advances in desalination by using graphene to make the membrane used in reverse osmosis (see Science Focus 13.2, p. 339).

Researchers are looking into possible harmful effects of graphene production and use. A 2014 study led by Sharon Walker at the University of California–Riverside found graphene oxide in lakes and drinking water storage tanks. This could increase the chances that animals and humans could ingest the chemical, which was found in some early studies to be toxic to mice and human lung cells.

Graphene is made from very high purity and expensive graphite. According to the USGS, in 2013 China controlled about 68% of the world's high-purity graphite production. The United States mines very little natural graphite and imports most of its graphite from Mexico and China, which could restrict U.S. product exports as the use of graphene grows.

Geologists are looking for deposits of graphite in the United States. However, in 2011 a team of Rice University chemists, led by James M. Tour, found ways to make large sheets of high-quality graphene from inexpensive materials found in garbage and from dog feces. If such a process becomes economically feasible, concern over supplies of graphite could vanish, along with any harmful environmental effects of the mining and processing of graphite.

In 2014 a team of researchers at Purdue University was able to isolate a single layer of black phosphorus atoms—a

The Nanotechnology Revolution

Nanotechnology uses science and engineering to manipulate and create materials out of atoms and molecules at the ultra-small scale of less than 100 nanometers. The diameter of the period at the end of this sentence is about a half million nanometers.

At the nanometer level, conventional materials have unconventional and unexpected properties. For example, scientists have learned to link carbon atoms together to form one-atom-thick sheets of carbon called *graphene* (see Case Study, p. 369). These sheets can be shaped into tubes called *carbon nanotubes* that are 60 times stronger than high-grade steel. A nearly invisible thread of this material is strong enough to suspend a pickup truck. Using carbon nanotubes to build cars would make them stronger and safer and would improve gas mileage by making them up to 80% lighter.

Currently, nanomaterials are used in more than 1,300 consumer products and the number is growing. Such products include certain batteries, stain-resistant and wrinkle-free clothes, self-cleaning glass surfaces, self-cleaning sinks and toilets, sunscreens, waterproof coatings for cell phones, some cosmetics, some foods, and food containers that release nanosilver ions to kill bacteria, molds, and fungi.

Nanotechnologists envision innovations such as a supercomputer smaller than a grain of rice, thin and flexible solar cell films that could be attached to or painted onto almost any surface, biocomposite materials that would make our bones and tendons super strong, and nanomolecules specifically designed to seek out and kill cancer cells. Nanotechnology allows us to make materials from the bottom up, using atoms of abundant elements (primarily hydrogen, oxygen, nitrogen, carbon, silicon, and aluminum) as substitutes for scarcer elements, such as copper, cobalt, and tin.

Nanotechnology has many potential environmental benefits. Designing and building products on the molecular level would greatly reduce the need to mine many materials. It would also require less matter and energy and it would reduce waste production. We may be able to use nanoparticles to remove industrial pollutants from contaminated air, soil, and groundwater. Nanofilters might someday be used to desalinate and purify seawater at an affordable cost, thereby helping to increase drinking water supplies. **GREEN CAREER: Environmental nanotechnology**

What's the catch? The main problem is serious concerns about the possible harmful health effects of nanotechnology. Because of the large combined surface area of the huge number of nanoparticles involved in any application, they are more chemically reactive and potentially more toxic to humans and other animals than are conventional materials.

Laboratory studies involving mice and other test animals reveal that nanoparticles can be inhaled deeply into the lungs and absorbed into the bloodstream. This can result in lung damage similar to that caused by mesothelioma, a deadly cancer resulting from the inhalation of asbestos particles. Nanoparticles can also penetrate cell membranes, including those in the brain, and move across the placenta from a mother to her fetus. A 2015 study led by Gary Cherr of the University of California–Davis found that nanoparticles from sunscreens, boat paints, and other consumer products can harm the embryonic development of certain marine organisms.

A panel of experts from the U.S. National Academy of Sciences has said that the U.S. government is not doing enough to evaluate the potential health and environmental risks of using nanomaterials. For example, the U.S. Food and Drug Administration does not maintain a list of the food products and cosmetics that contain nanomaterials. By contrast, the European Union takes a precautionary approach to the use of nanomaterials, requiring that manufacturers demonstrate the safety of their products before they can enter the marketplace.

Many analysts say that, before unleashing nanotechnology more broadly, we should ramp up research on the potential harmful health effects of nanoparticles and develop regulations to control its growing applications until we know more about its possible harmful health effects. Many are also calling for the labeling of all products that contain nanoparticles.

CRITICAL THINKING

Do you think the potential benefits of nanotechnology products outweigh their potentially harmful effects? Explain.

new material known as *phosphorene*. As a semiconductor, it is more efficient than silicon transistors that are used as chips in computers and other electronic devices. Replacing them with phosphorene transistors could make almost any electronic device run much faster while using less power. This could revolutionize computer technology. However, phosphorene must be sealed in a protective coating because it breaks down when exposed to air.

Use Mineral Resources More Sustainably

Figure 14.15 lists several ways to use mineral resources more sustainably. One strategy is to focus on recycling and reuse of nonrenewable mineral resources, especially valuable or scarce metals such as gold, iron, copper, aluminum, and platinum (**Concept 14.4**). Recycling, an application of the chemical cycling **principle of sustainability** (see Inside

INDIVIDUALS MATTER 14.1

Yu-Guo Guo: Designer of Nanotechnology Batteries and National Geographic Explorer

© Jinsong Hu

Yu-Guo Guo is a professor of chemistry and a nanotechnology researcher at the Chinese Academy of Sciences in Beijing. He has invented nanomaterials that can be used to make lithium-ion battery packs smaller, more powerful, and less costly, which makes them more useful for powering electric cars and electric bicycles. This is an important scientific advance, because the battery pack is the most important and expensive part of any electric vehicle.

Guo's innovative use of nanomaterials has greatly increased the power of lithium-ion batteries by enabling electric current to flow more efficiently through what he calls "3-D conducting nanonetworks." With this promising technology, lithium-ion battery packs in electric vehicles can be fully charged in a few minutes. They also have twice the energy storage capacity of today's batteries, and thus will extend the range of electric vehicles by enabling them to run longer. Guo is also interested in developing nanomaterials for use in solar cells and fuel cells that could be used to generate electricity and to power vehicles.

Vincenzo Lombardo/Getty Images

FIGURE 14.14 Graphene, which consists of a single layer of carbon atoms linked together in a hexagonal lattice, is a revolutionary new material.

Solutions

Sustainable Use of Nonrenewable Minerals

- Reuse or recycle metal products whenever possible
- Redesign manufacturing processes to use less mineral resources
- Reduce mining subsidies
- Increase subsidies for reuse, recycling, and finding substitutes

© Cengage Learning

FIGURE 14.15 We can use nonrenewable mineral resources more sustainably (**Concept 14.4**). ***Critical thinking:*** Which two of these solutions do you think are the most important? Why?

Back Cover), has a much lower environmental impact than that of mining and processing metals from ores. For example, recycling aluminum beverage cans and scrap aluminum produces 95% less air pollution and 97% less water pollution, and uses 95% less energy, than mining and processing aluminum ore. We can also extract and recycle valuable gold (**Core Case Study**) from discarded cell phones. Cleaning up and reusing items instead of recycling them has an even lower environmental impact.

Using mineral resources more sustainably is a major challenge in the face of rising demand for many minerals. For example, one way to increase supplies of rare earths is to extract and recycle them from the massive amounts of electronic wastes that are being produced. So far, however, less than 1% of rare earth metals are recovered and recycled.

Another way to use minerals more sustainably is to find substitutes for rare minerals, ideally, substitutes without heavy environmental impacts (**Concept 14.4**). For example, electric car battery makers are beginning to switch from making nickel-metal-hydride batteries, which require the rare earth metal lanthanum, to manufacturing lighter-weight lithium-ion batteries, which researchers are now trying to improve (Individuals Matter 14.1).

Lithium (Li), the world's lightest metal, is a vital component of lithium-ion batteries, which are used in cell phones, iPads, laptop computers, and a growing number of other products. The problem is that some countries, including the United States, do not have large supplies of lithium. The South American countries of Bolivia, Chile, and Argentina have about 80% of the global reserves of lithium. Bolivia alone has about 50% of these reserves, whereas the United States holds only about 3%.

Japan, China, South Korea, and the United Arab Emirates are buying up access to global lithium reserves to ensure their ability to sell lithium or batteries to the rest of the world. Within a few decades, the United States may be heavily dependent on expensive imports of lithium and lithium batteries. However, in 2014 the company Simbol, Inc. began building a plant in California that is designed to extract lithium from brine waste produced by geothermal power plants. If it is successful, this process could lessen some U.S. dependence on imported lithium.

Scientists are also searching for substitutes for rare earth metals that are used to make increasingly important powerful magnets and related devices. In Japan and the United States, researchers are developing a variety of such devices that require no rare earth minerals, are light and compact, and can deliver more power with greater efficiency at a reduced cost.

14.5 WHAT ARE THE EARTH'S MAJOR GEOLOGICAL HAZARDS?

CONCEPT 14.5 Dynamic processes move matter within the earth and on its surface and can cause volcanic eruptions, earthquakes, tsunamis, erosion, and landslides.

The Earth Beneath Your Feet Is Moving

We tend to think of the earth's crust as solid and unmoving. However, the flows of energy and heated material within the earth's convection cells (Figure 14.2) are so powerful that they have caused the lithosphere to break up into a dozen or so huge rigid plates, called **tectonic plates,** which move extremely slowly atop the asthenosphere (Figure 14.16).

These gigantic plates are somewhat like the world's largest and slowest-moving surfboards on which we ride without noticing their movement. Their typical speed is about the rate at which your fingernails grow. Throughout the earth's history, landmasses have split apart and joined together as tectonic plates shifted around, changing the size, shape, and location of the earth's continents (Figure 4.B, p. 90). The slow movement of the continents across Earth's surface is called **continental drift.**

Much of the geological activity at the earth's surface takes place at the boundaries between tectonic plates as they separate, collide, or grind along against each other. The boundary that occurs where two plates move away from each other is called a **divergent boundary** (Figure 14.16). At such boundaries, magma flows up where the plates separate, sometimes hardening and forming new crust and sometimes breaking to the surface and causing volcanic eruptions. Earthquakes can also occur because of divergence of plates, and superheated water can erupt as geysers.

Another type of boundary is the **convergent boundary** (Figure 14.16), where two tectonic plates are colliding. This super-slow-motion collision causes one or both plate edges to buckle and rise, forming mountain ranges. In most cases, one plate slides beneath the other, melting and making new magma that can rise through cracks and form volcanoes near the boundary. The overriding plate is pushed up and made into mountainous terrain.

The third major type of boundary is the **transform plate boundary** (Figure 14.16), where two plates grind along in opposite directions next to each other. The tremendous forces produced at these boundaries can form mountains or deep cracks (Figure 14.17) and cause earthquakes and volcanic eruptions.

Volcanoes Release Molten Rock from the Earth's Interior

An active **volcano** occurs where magma rising in a plume through the lithosphere reaches the earth's surface through a central vent or a long crack, called a *fissure* (Figure 14.18). Magma or molten rock that reaches the earth's surface is called *lava*.

A *volcanic eruption* releases chunks of lava rock, liquid lava, glowing hot ash, and gases (including water vapor, carbon dioxide, and sulfur dioxide) (**Concept 14.5**). Eruptions can be explosive and extremely destructive, causing loss of life and obliterating ecosystems and human communities. They can also be slow and much less destructive with lava gurgling up and spreading slowly across the land or sea floor. It is this slower form of eruption that builds the cone-shaped mountains so commonly associated with volcanoes, as well as layers of rock made of cooled lava on the earth's surface.

While volcanic eruptions can be destructive, they can also form majestic mountain ranges and lakes, and the weathering of lava contributes to fertile soils. Hundreds of volcanoes have erupted on the ocean floor, building cones that have reached the ocean's surface, eventually to form islands that have become suitable for human settlement, such as the Hawaiian Islands.

We can reduce the loss of human life and some of the property damage caused by volcanic eruptions by using historical records and geological measurements to identify high-risk areas, so that people can avoid living in those areas. We also use monitoring devices that warn us when volcanoes are likely to erupt, and in some areas that are prone to volcanic activity, evacuation plans have been developed.

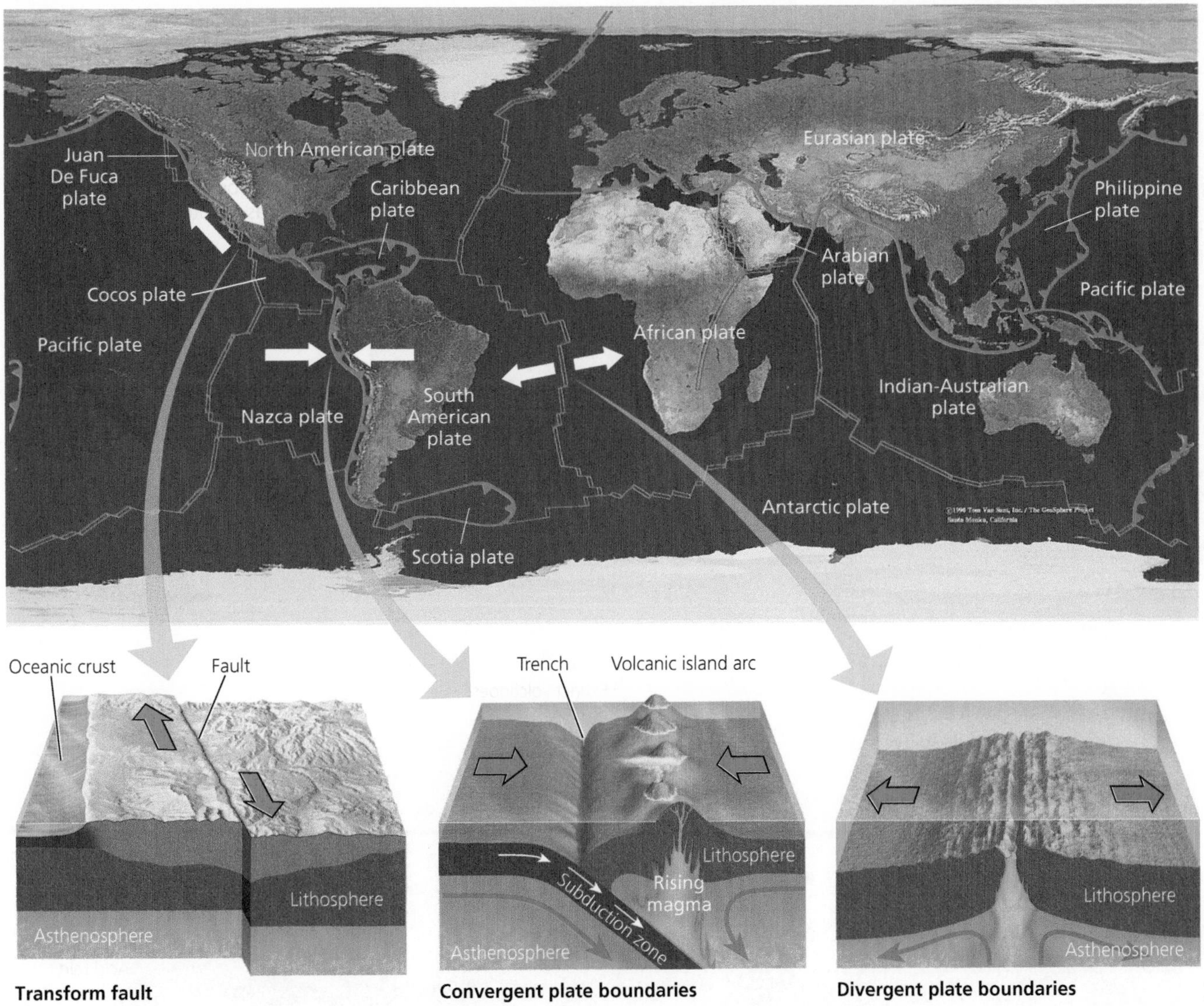

FIGURE 14.16 The earth's crust has been fractured into several major tectonic plates. White arrows indicate examples of where plates are colliding, separating, or grinding along against each other in opposite directions. ***Question:*** Which plate are you riding on?

Earthquakes Are Geological Rock-and-Roll Events

Forces inside the earth's mantle put tremendous stress on rock within the crust. Such stresses can be great enough to cause sudden breakage and shifting of the rock, producing a *fault*, or fracture in the earth's crust (Figure 14.17). When a fault forms, or when there is abrupt movement on an existing fault, energy that has accumulated over time is released in the form of vibrations, called *seismic waves*, which move in all directions through the surrounding rock—an event called an **earthquake** (Figure 14.19 and **Concept 14.5**). Most earthquakes occur at the boundaries of tectonic plates (Figures 14.16 and 14.17).

Seismic waves move upward and outward from the earthquake's *focus* like ripples in a pool of water. Scientists measure the severity of an earthquake by the *magnitude* of its seismic waves. The magnitude is a measure of ground motion (shaking) caused by the earthquake, as indicated by the *amplitude*, or size of the seismic waves when they reach a recording instrument, called a *seismograph*.

Scientists use the *Richter scale*, on which each unit has an amplitude 10 times greater than the next smaller unit. *Seismologists*, or people who study earthquakes, rate earthquakes

FIGURE 14.17 The San Andreas Fault, created by the North American Plate and the Pacific Plate sliding very slowly past each other, runs almost the full length of California (see map). It is responsible for earthquakes of various magnitudes, which have caused rifts on the land surface in some areas (photo).

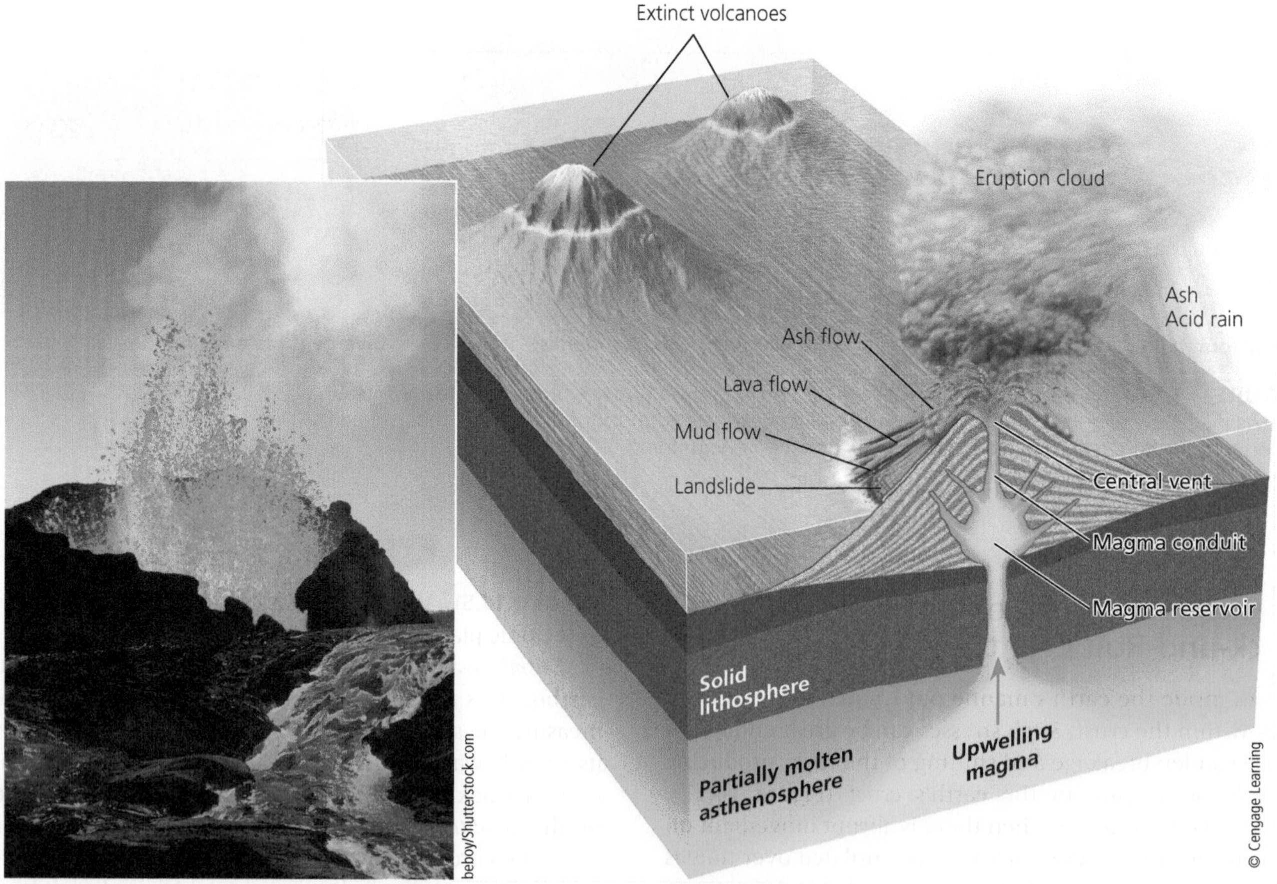

FIGURE 14.18 Sometimes, the internal pressure in a volcano is high enough to cause lava, ash, and gases to be ejected into the atmosphere (photo) or to flow over land, causing considerable damage.

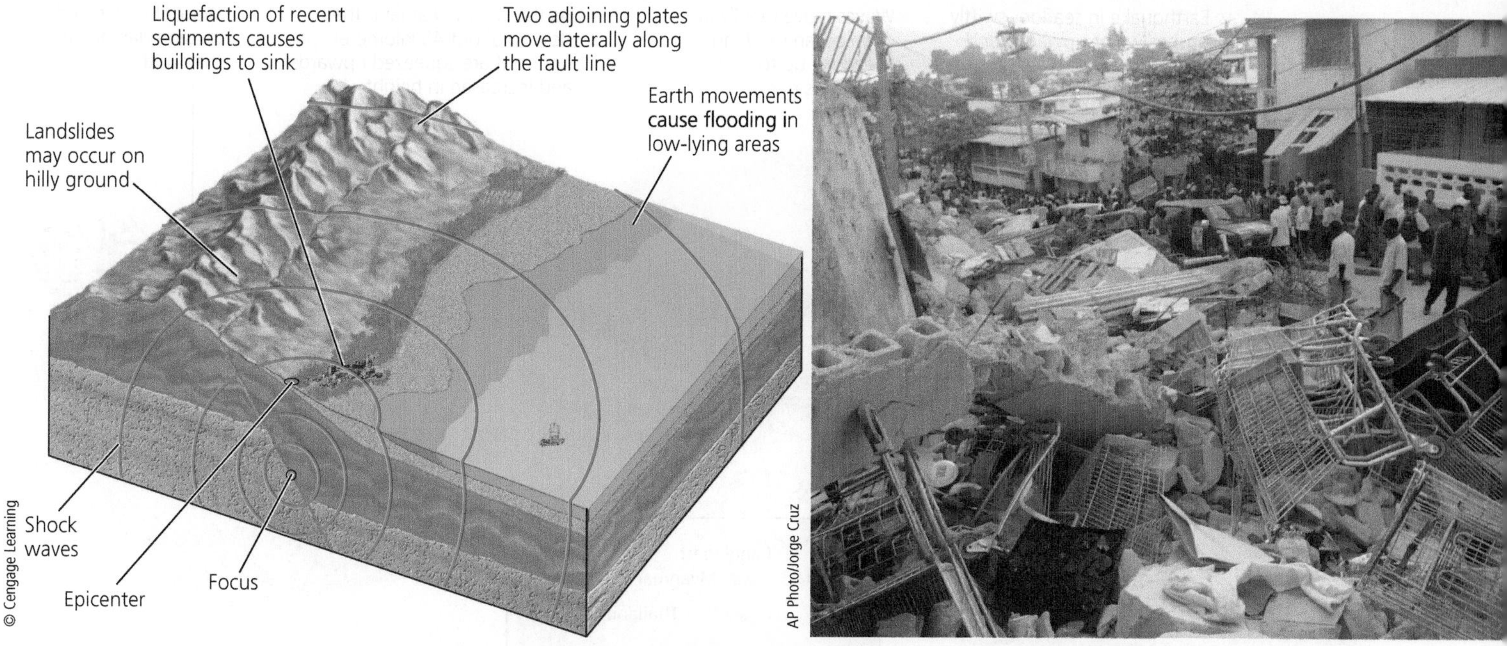
Liquefaction of recent sediments causes buildings to sink
Two adjoining plates move laterally along the fault line
Earth movements cause flooding in low-lying areas
Landslides may occur on hilly ground
Shock waves
Epicenter
Focus
© Cengage Learning
AP Photo/Jorge Cruz

FIGURE 14.19 An *earthquake* (left) is one of nature's most powerful events. The photo shows damage from a 2010 earthquake in Port-au-Prince, Haiti.

as *insignificant* (less than 4.0 on the Richter scale), *minor* (4.0–4.9), *damaging* (5.0–5.9), *destructive* (6.0–6.9), *major* (7.0–7.9), and *great* (over 8.0). The largest recorded earthquake occurred in Chile on May 22, 1960, and measured 9.5 on the Richter scale. Each year, scientists record the magnitudes of more than 1 million earthquakes, most of which are too small to feel.

The primary effects of earthquakes include shaking and sometimes a permanent vertical or horizontal displacement of a part of the crust. These effects can have serious consequences for people and for buildings, bridges, freeway overpasses, dams, and pipelines. A major earthquake is a large rock-and-roll geological event.

One way to reduce the loss of life and property damage from earthquakes is to examine historical records and make geological measurements to locate active fault zones. We can then map high-risk areas and establish building codes that regulate the placement and design of buildings in such areas. Then people can evaluate the risk and factor it into their decisions about where to live. In addition, engineers know how to make homes, large buildings, bridges, and freeways more earthquake resistant, although this is costly.

See Figure 12, p. S27, in Supplement 4, for a map comparing earthquake risks in various areas of the United States, and Figure 13, p. S28, in Supplement 4 for a map of such areas throughout the world.

Earthquakes on the Ocean Floor Can Cause Tsunamis

A **tsunami** is a series of large waves generated when part of the ocean floor suddenly rises or drops (Figure 14.20). Most large tsunamis are caused when certain types of faults in the ocean floor move up or down because of a large underwater earthquake. Other causes are landslides generated by earthquakes and volcanic eruptions (**Concept 14.5**).

Tsunamis are often called *tidal waves,* although they have nothing to do with tides. They can travel across the ocean at the speed of a jet airliner. In deep water the waves are very far apart—sometimes hundreds of kilometers—and their crests are not very high. As a tsunami approaches a coast with its shallower waters, it slows down, its wave crests squeeze closer together, and their heights grow rapidly. It can hit a coast as a series of towering walls of water that can level buildings.

The largest recorded loss of life from a tsunami occurred in December 2004 when a great underwater earthquake in the Indian Ocean with a magnitude of 9.15 caused a tsunami that killed more than 230,000 people and devastated many coastal areas of Indonesia (Figure 14.21 and map in Figure 14.20), Thailand, Sri Lanka, South India, and eastern Africa. It also displaced about 1.7 million people (1.3 million of them in India and Indonesia), and destroyed or damaged about 470,000 buildings and houses. There

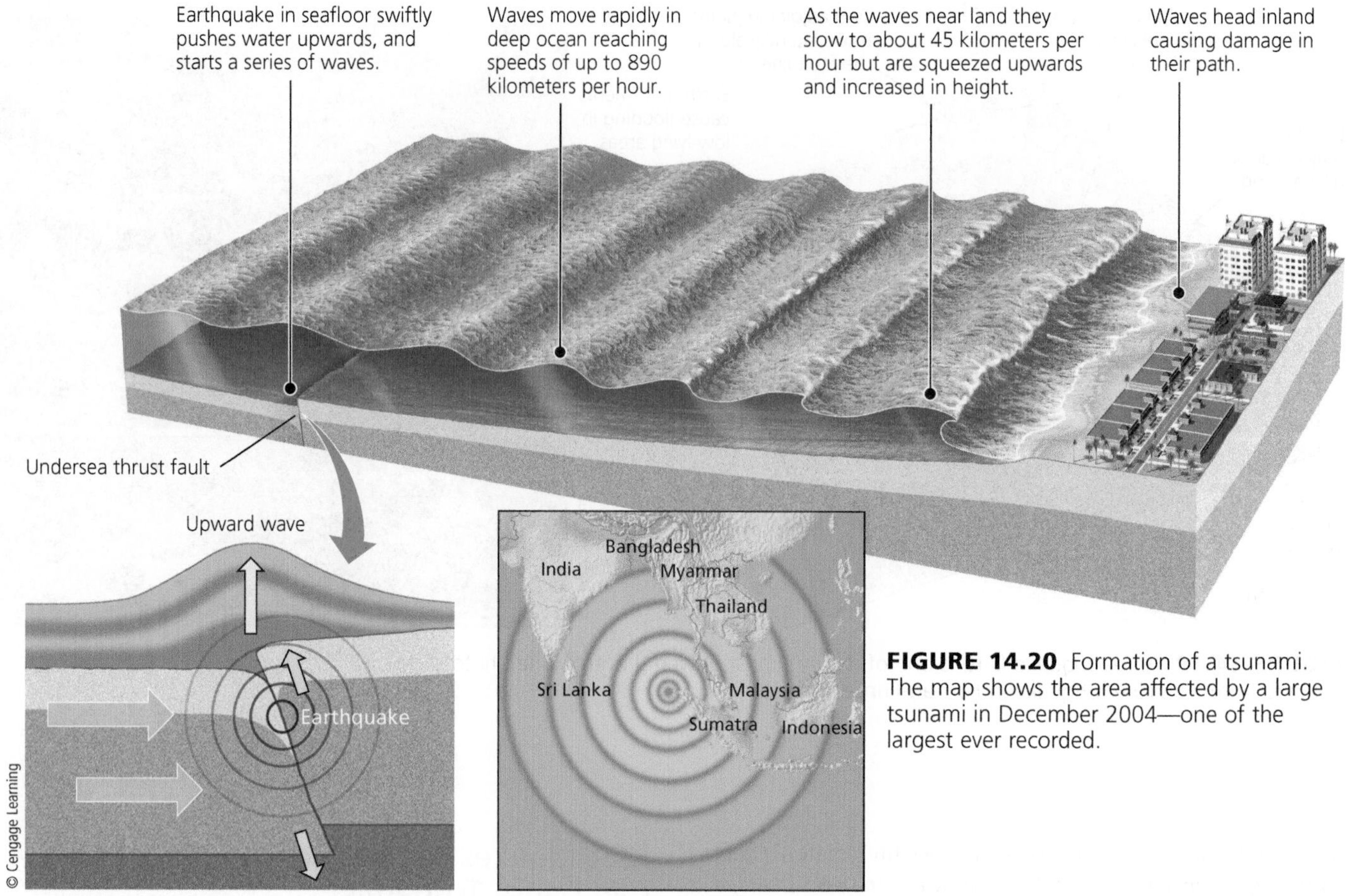
Earthquake in seafloor swiftly pushes water upwards, and starts a series of waves.
Waves move rapidly in deep ocean reaching speeds of up to 890 kilometers per hour.
As the waves near land they slow to about 45 kilometers per hour but are squeezed upwards and increased in height.
Waves head inland causing damage in their path.
Undersea thrust fault
Upward wave
Earthquake
© Cengage Learning
Bangladesh
India
Myanmar
Thailand
Sri Lanka
Malaysia
Sumatra
Indonesia
December 26, 2004, tsunami

FIGURE 14.20 Formation of a tsunami. The map shows the area affected by a large tsunami in December 2004—one of the largest ever recorded.

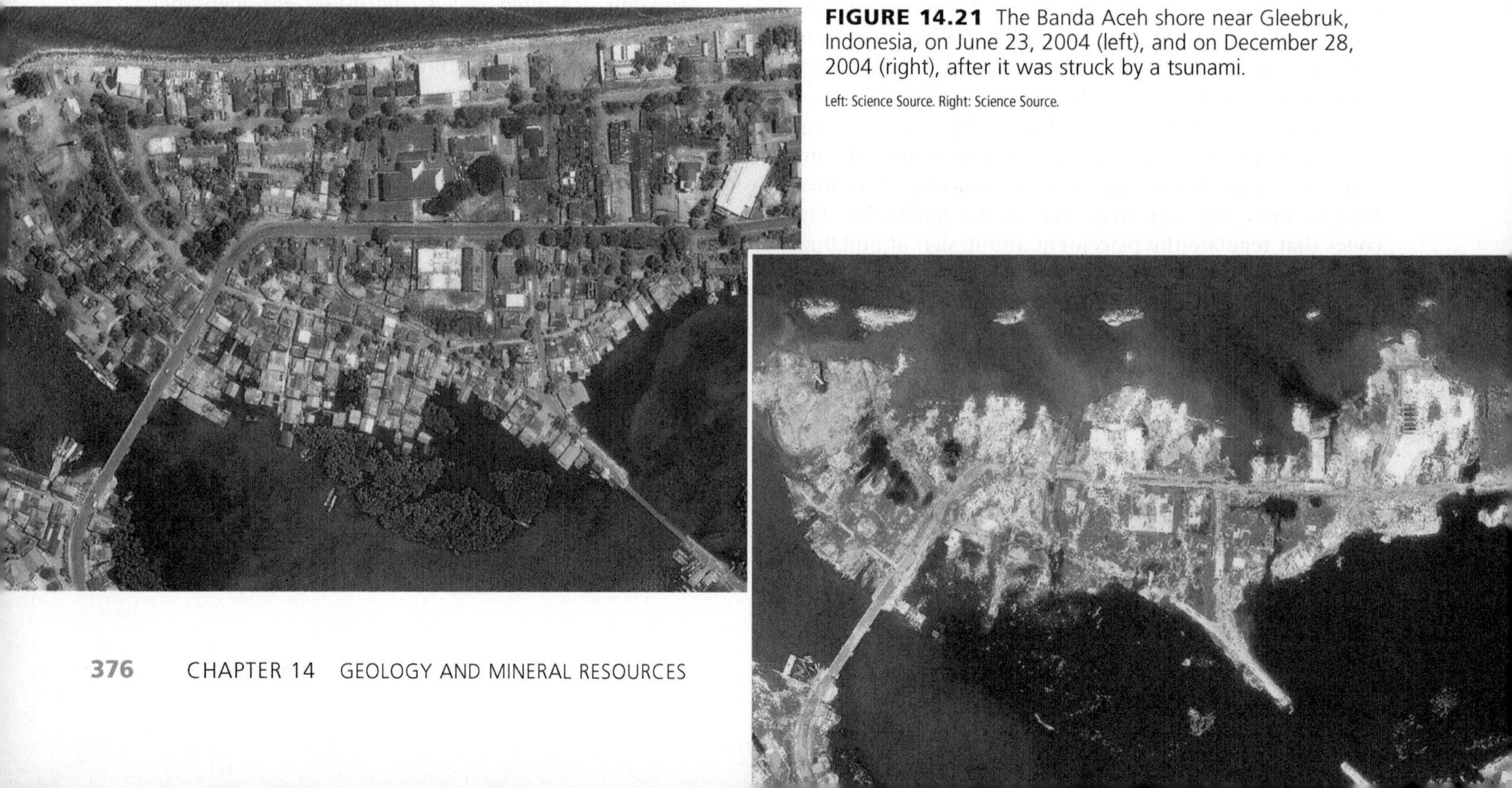

FIGURE 14.21 The Banda Aceh shore near Gleebruk, Indonesia, on June 23, 2004 (left), and on December 28, 2004 (right), after it was struck by a tsunami.

Left: Science Source. Right: Science Source.

were no recording devices in place to provide an early warning of this tsunami.

In 2011 a large tsunami caused by a powerful earthquake off the coast of Japan generated 3-story high waves that killed almost 19,000 people, displaced more than 300,000 people, and destroyed or damaged 125,000 buildings. It also heavily damaged three nuclear reactors, which then released dangerous radioactivity into the surrounding environment.

In some areas, scientists have built networks of ocean buoys and pressure recorders on the ocean floor to collect data that can be relayed to tsunami emergency warning centers. However, these networks are not widespread.

BIG IDEAS

- Dynamic forces that move matter within the earth and on its surface recycle the earth's rocks, form deposits of mineral resources, and cause volcanic eruptions, earthquakes, and tsunamis.
- The available supply of a mineral resource depends on how much of it is in the earth's crust, how fast we use it, the mining technology used to obtain it, its market prices, and the harmful environmental effects of removing and using it.
- We can use mineral resources more sustainably by trying to find substitutes for scarce resources, reducing resource waste, and reusing and recycling nonrenewable minerals.

Tying It All Together

The Real Cost of Gold and Sustainability

In this chapter's Core Case Study, we considered the harmful effects of gold mining as an example of the impacts of our extraction and use of mineral resources. We saw that these effects make gold much more costly, in terms of environmental and human health costs, than is reflected in the price of gold.

In this chapter, we looked at technological developments that could help us to expand supplies of mineral resources and to use them more sustainably. For example, if we develop it safely, we could use nanotechnology to make new materials that could replace scarce mineral resources and greatly reduce the environmental impacts of mining and processing such resources. For example, we might use graphene to produce more efficient and affordable solar cells to generate electricity—an application of the solar energy **principle of sustainability**.

We can also use mineral resources more sustainably by reusing and recycling them, and by reducing unnecessary resource use and waste—applying the chemical cycling **principle of sustainability**. In addition, industries can mimic nature by using a diversity of ways to reduce the harmful environmental impacts of mining and processing mineral resources, thus applying the biodiversity **principle of sustainability**.

Matt Benoit/Shutterstock.com

Chapter Review

Core Case Study

1. Explain why the real cost of gold is more than what most people pay for it. What are some examples of costs not accounted for?

Section 14.1

2. What are the two key concepts for this section? Define **geology**, **core**, **mantle**, **asthenosphere**, **crust**, and **lithosphere**. Define **mineral**, **mineral resource**, and **rock**. Define and distinguish among **sedimentary rock**, **igneous rock**, and **metamorphic rock** and give an example of each. What is the **rock cycle**? Explain its importance. Define **ore** and distinguish between a **high-grade ore** and a **low-grade ore**. List five important nonrenewable mineral resources and their uses.

Section 14.2

3. What are the two key concepts for this section? What are the **reserves** of a mineral resource and how can they be expanded? What two factors determine the future supply of a nonrenewable mineral resource? Explain how the supply of a nonrenewable mineral resource can be economically depleted and list the five choices we have when this occurs. What is **depletion time** and what factors affect it?

4. What five nations supply most of the world's nonrenewable mineral resources? How dependent is the United States on other countries for important nonrenewable mineral resources? Explain the concern over U.S. access to rare earth mineral resources. Describe the conventional view of the relationship between the supply of a mineral resource and its market price. Explain why some economists believe this relationship no longer applies in some countries. Summarize the pros and cons of providing government subsidies and tax breaks for mining companies.

5. Summarize the opportunities and limitations of expanding mineral supplies by mining lower-grade ores. What are the advantages and disadvantages of biomining? Describe the opportunities and possible problems that could result from deep-sea mining.

Section 14.3

6. What is the key concept for this section? Summarize the life cycle of a metal product.

7. What is **surface mining**? Define **overburden** and **spoils**. Define **open-pit mining** and **strip mining**, and distinguish among **area strip mining**, **contour strip mining**, and **mountaintop removal mining**. Describe three harmful environmental effects of surface mining. What is **subsurface mining**? What is acid mine drainage? Define **tailings** and explain why they can be hazardous. What is **smelting** and what are its major harmful environmental effects?

Section 14.4

8. What is the key concept for this section? Give two examples of promising substitutes for key mineral resources. What is **nanotechnology** and what are some of its potential environmental and other benefits? What are some problems that could arise from the widespread use of nanotechnology? Describe the potential of using graphene and phosphorene as new resources. Explain the benefits of recycling and reusing valuable metals. List five ways to use nonrenewable mineral resources more sustainably. What are two examples of research into substitutes for rare earth metals? Explain why uneven distribution of lithium among various countries is a concern.

Section 14.5

9. What is the key concept for this section? What are **tectonic plates**? What is **continental drift**? Define and distinguish among **divergent**, **convergent**, and **transform plate boundaries**. Define **volcano** and describe the nature and major effects of a volcanic eruption. Define **earthquake** and describe its nature and major effects. What is a **tsunami** and what are its major effects?

10. What are this chapter's three big ideas? Explain how we can apply the three **scientific principles of sustainability** to obtain and use gold and other nonrenewable mineral resources in more sustainable ways.

Note: Key terms are in bold type.

Critical Thinking

1. Do you think that the benefits we get from gold—its uses in jewelry, dentistry, electronics, and other uses—are worth the real cost of gold (**Core Case Study**)? If so, explain your reasoning. If not, explain your argument for cutting back on or putting a stop to the mining of gold.

2. You are an igneous rock. Describe what you experience as you move through the rock cycle. Repeat this exercise, assuming you are a sedimentary rock and again assuming you are a metamorphic rock.

3. What are three ways in which you benefit from the rock cycle?

4. Suppose your country's supply of rare earth metals was cut off tomorrow. How would this affect your life? Give at least three examples. How would you adjust to these changes? Explain.

5. Use the second law of thermodynamics (see Chapter 2, p. 42) to analyze the scientific and economic feasibility of each of the following processes:
   a. Extracting certain minerals from seawater
   b. Mining increasingly lower-grade deposits of minerals
   c. Continuing to mine, use, and recycle minerals at increasing rates

6. Suppose you were told that mining deep-ocean mineral resources would mean severely degrading ocean bottom habitats and life forms such as giant tubeworms and giant clams (Figure 14.6). Do you think that such information should be used to prevent ocean bottom mining? Explain.

7. List three ways in which a nanotechnology revolution could benefit you and three ways in which it could harm you. Do you think the benefits outweigh the harms? Explain.

8. What are three ways to reduce the harmful environmental impacts of the mining and processing of nonrenewable mineral resources? What are three aspects of your lifestyle that contribute to these harmful impacts?

Doing Environmental Science

Do research to determine which mineral resources go into the manufacture of each of the following items and how much of each of these resources are required to make each item: **(a)** a cell phone, **(b)** a wide-screen TV, and **(c)** a large pickup truck. Pick three of the lesser-known mineral materials that you have learned about in this exercise and do more research to find out where in the world most of the reserves for that mineral are located. For each of the three minerals you chose, try to find out what kinds of environmental effects have resulted from the mining of the mineral in at least one of the places where it is mined. You might also find out about steps that have been taken to deal with those effects. Write a report summarizing all of your findings.

Global Environment Watch Exercise

Go to your MindTap course to access the GREENR database. Using the "Basic Search" box at the top of the page, search for an article that deals with rare earth metal supplies (**Core Case Study**). Summarize the conclusions expressed in the article. Is there scientific information cited in the article to support the author's conclusions? Give specific examples. Do you think there are any types of supporting scientific data not mentioned in the article that would strengthen the author's conclusions? For example, would you add statistical data to support a point, or data in a graph indicating possible cause-and-effect relationships? Be specific and give reasons for your suggestions.

Data Analysis

Rare earth metals are widely used in a variety of important products (Case Study, p. 362). According to the U.S. Geological Survey, China has about 42% of the world's reserves of rare earth metals. Use this information to answer the following questions.

1. In 2014 China had 55 million metric tons of rare earth metals in its reserves and produced 95,000 metric tons of these metals. At this rate of production, how long will China's rare earth reserves last?

2. In 2014 the global demand for rare earth metals was about 136,000 metric tons. At this annual rate of use, if China were to produce all of the world's rare earth metals, how long would their reserves last?

3. The annual global demand for rare earth metals is projected to rise to at least 182,000 metric tons by 2020. At this rate, if China were to produce all of the world's rare earth metals, how long would its reserves last?

CENGAGE brain.com For access to MindTap and additional study materials visit www.cengagebrain.com.

CHAPTER 15

Nonrenewable Energy

Typical citizens of advanced industrialized nations each consume as much energy in 6 months as typical citizens in less developed countries consume during their entire life.
MAURICE STRONG

Key Questions

15.1 What types of energy resources do we use?

15.2 What are the advantages and disadvantages of using oil?

15.3 What are the advantages and disadvantages of using natural gas?

15.4 What are the advantages and disadvantages of using coal?

15.5 What are the advantages and disadvantages of using nuclear power?

Offshore oil drilling rig.

iStockphoto.com/WitthawatM

Core Case Study

Using Hydrofracking to Produce Oil and Natural Gas

Geologists have known for decades about vast deposits of oil and natural gas that are dispersed and trapped between compressed layers of shale rock formations. These deposits are found deep underground in many areas of the United States, including North Dakota, Texas, and Pennsylvania (see map in Supplement 4, Figure 14, p. S28).

For years, it cost too much to extract such oil (called *tight oil*) and natural gas from shale rock. This changed in the late 1990s when oil and gas producers combined two existing extraction technologies (Figure 15.1). One is **horizontal drilling,** which involves drilling a vertical well deep into the earth, turning the flexible shaft of the drill, and drilling horizontally to gain access to multiple oil and natural gas deposits held tightly between layers in shale rock formations. Usually, wells are drilled vertically for 1.6–2.4 kilometers (1–1.5 miles) or more and then horizontally for up to 1.6 kilometers (1 mile). Two or three horizontally drilled wells can often produce as much oil as 20 vertical wells, which reduces the area of land damaged by drilling operations.

The second technology, called **hydraulic fracturing** (also called hydrofracking or **fracking**), is then used to free the trapped oil and natural gas. High-pressure pumps force a mixture (slurry) of water, sand, and a cocktail of chemicals through holes in the well pipe to fracture the shale rock and create cracks. The sand wedges in the cracks and keeps them open. When the pressure is released a mixture of the oil or natural gas and about half of the slurry flows to the surface through the well pipe (Figure 15.1).

The slurry contains a mix of naturally occurring salts, toxic heavy metals, and radioactive materials leached from the rock, along with some potentially harmful drilling chemicals that oil and natural gas companies are not required to identify. The hazardous slurry is injected into deep underground hazardous waste wells (the most widely used option), sent to sewage treatment plants that often cannot handle the wastes, stored in open air holding ponds that can leak or collapse, or cleaned up and reused in the fracking process—the best but most expensive and least used option. In 2015 the U.S. Environmental Protection Agency (EPA) proposed a rule that would prohibit sending fracking wastewater to sewage treatment plants.

Energy companies drill a well horizontally, frack it several times, and then drill a new well and repeat the process. The use of these two extraction technologies in at least 25 states between 1990 and 2014 has brought about what some call a new era of oil and natural gas production in the United States. It could last as long as market prices of oil and natural gas remain high enough to make these drilling operations profitable. Like any technology, this approach has advantages and disadvantages that we discuss later in this chapter.

In this chapter, we discuss the advantages and disadvantages of using nonrenewable fossil fuels (such as oil, natural gas, and coal) and nuclear power. In the next chapter, we look at the advantages and disadvantages of improving energy efficiency and using a variety of renewable energy resources.

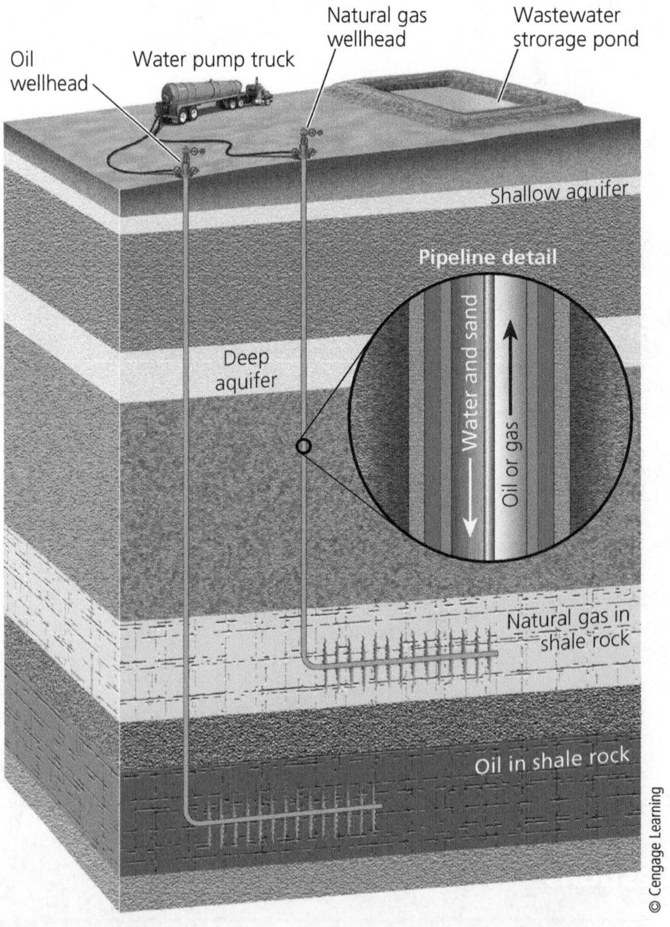

FIGURE 15.1 *Fracking:* Horizontal drilling and hydraulic fracturing are used to release large amounts of oil and natural gas that are tightly held in underground shale rock formations.

15.1 WHAT TYPES OF ENERGY RESOURCES DO WE USE?

CONCEPT 15.1A About 90% of the commercial energy used in the world comes from nonrenewable energy resources (mostly oil, natural gas, and coal) and 10% comes from renewable energy resources.

CONCEPT 15.1B Energy resources vary greatly in their *net energy*—the amount of energy available from a resource minus the amount of energy needed to make it available.

Where Does the Energy We Use Come From?

The energy that heats the earth and makes it livable comes from the sun at no cost to us—in keeping with the solar energy **principle of sustainability**. Without this essentially inexhaustible input of solar energy, the earth's average temperature would be –240°C (–400°F) and life as we know it would not exist.

We use *commercial energy,* energy that is sold in the marketplace, to supplement the sun's life-sustaining energy. We get this supplemental energy from two types of energy resources (**Concept 15.1A**). One type, *nonrenewable energy resources,* includes fossil fuels (oil, coal, and natural gas) and in the nuclei of certain atoms (nuclear energy). We discuss these resources in this chapter. The other, *renewable energy resources,* are wind, flowing water (hydropower), energy from the sun (solar energy), biomass (trees and other plants), and heat in the earth's interior (geothermal energy), which we discuss in Chapter 16.

The world depends on fossil fuels, the product of ancient plant and animal remains buried millions of years ago and subjected to intense heat and pressure. In 2014, 90% of the commercial energy used in the world and in the United States came from nonrenewable resources (mostly fossil fuels) and 10% came from renewable resources (Figure 15.2, right) (**Concept 15.1A**). Electricity, produced mostly from coal, accounted for 18% of the world's commercial energy and 39% of the commercial energy used in the United States in 2014.

90% Percentage of commercial energy used in the world and in the United States that comes from nonrenewable energy (mostly fossil fuels)

Net Energy: It Takes Energy to Get Energy

Producing high-quality energy from any energy resource requires an input of high-quality energy. For example, before oil becomes useful to us, it must be found, pumped up from a deposit beneath the ground or ocean floor, transferred to a refinery, converted to gasoline and other fuels and chemicals, and delivered to consumers. Each of these steps uses high-quality energy, obtained mostly by burning fossil fuels, especially gasoline and diesel fuel produced from oil.

Net energy is the amount of high-quality energy available from a given quantity of an energy resource,

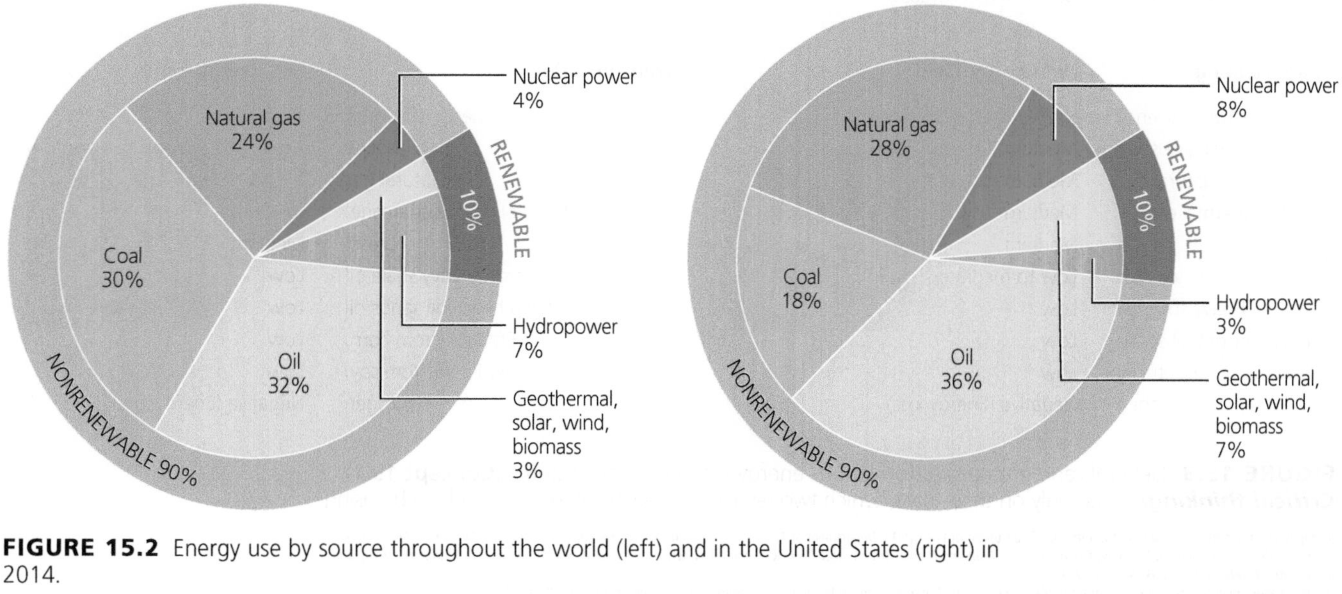

FIGURE 15.2 Energy use by source throughout the world (left) and in the United States (right) in 2014.

(Compiled by the authors using data from British Petroleum, U.S. Energy Information Administration (EIA), and International Energy Agency (IEA).)

minus the high-quality energy needed to make the energy available (**Concept 15.1B**).

Net energy = energy output − energy input

This information can also be expressed as a **net energy ratio (NER),** also known as the **energy returned on investment (EROI).**

Net energy ratio = energy output/energy input

Suppose that it takes about 9 units of high-quality energy to produce 10 units of high-quality energy from an energy resource. Then the net energy is 1 unit of energy ($10 - 9 = 1$) and the net energy ratio is 1.1 ($10/9 = 1.1$), both low values.

Net energy is like the net profit earned by a business after it deducts its expenses. If a business has $1 million in sales and $900,000 in expenses, its net profit is $100,000.

Figure 15.3 shows generalized net energies for energy resources and systems. It is based on several sources of scientific data and classifies estimated net energy as high, medium, low, or negative (negative being a net energy loss).

It is difficult for an energy resource with a low or negative net energy to compete in the marketplace with other energy alternatives with a medium to high net energy unless it receives subsidies and tax breaks from the government (taxpayers) or other outside sources.

For example, electricity produced by nuclear power has a low net energy because large amounts of high-quality energy are needed for each step in the *nuclear power fuel cycle*: to extract and process uranium ore, upgrade it to nuclear fuel, build and operate nuclear power plants, safely store the resulting highly radioactive wastes for thousands of years, and dismantle each highly radioactive plant after its useful life (typically 40–60 years) and safely store its high-level radioactive materials for thousands of years.

The low net energy and the resulting high cost of the entire nuclear fuel cycle is one reason why governments (taxpayers) throughout the world heavily subsidize nuclear-generated electricity to make it available to consumers at an affordable price. Such subsidies hide the true costs of the nuclear power fuel cycle and thus violate the full-cost pricing **principle of sustainability** (see Inside Back Cover).

Another factor that can affect the usefulness of an energy resource is its **energy density,** the amount of energy available per kilogram of the resource. The two energy resources with the highest densities are uranium-235 fuel, used to produce electricity in nuclear power plants, and compressed hydrogen gas (H_2), which when burned does

Electricity	Net Energy Yield
Energy efficiency	High
Hydropower	High
Wind	High
Coal	High
Natural gas	Medium
Geothermal energy	Medium
Solar cells	Low to medium
Nuclear fuel cycle	Low
Hydrogen	Negative (Energy loss)

High-Temperature Industrial Heat	Net Energy Yield
Energy efficiency (cogeneration)	High
Coal	High
Natural gas	Medium
Oil	Medium
Heavy shale oil	Low
Heavy oil from oil sands	Low
Direct solar (concentrated)	Low
Hydrogen	Negative (Energy loss)

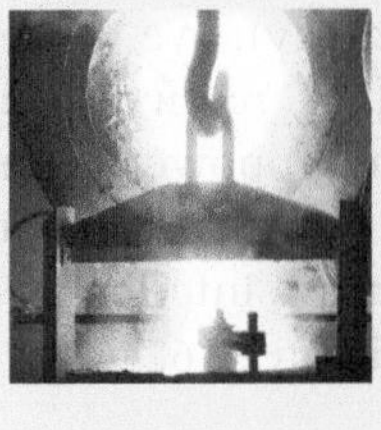

Space Heating	Net Energy Yield
Energy efficiency	High
Passive solar	Medium
Natural gas	Medium
Geothermal energy	Medium
Oil	Medium
Active solar	Low to medium
Heavy shale oil	Low
Heavy oil from oil sands	Low
Electricity	Low
Hydrogen	Negative (Energy loss)

Transportation	Net Energy Yield
Energy efficiency	High
Gasoline	High
Natural gas	Medium
Ethanol (from sugarcane)	Medium
Diesel	Medium
Gasoline from heavy shale oil	Low
Gasoline from heavy oil sands oil	Low
Ethanol (from corn)	Low
Biodiesel (from soy)	Low
Hydrogen	Negative (Energy loss)

FIGURE 15.3 Generalized *net energies* for various energy resources and systems (**Concept 15.1**). ***Critical thinking:*** Based only on these data, which two resources in each category should we be using?

(Compiled by the authors using data from the U.S. Department of Energy; U.S. Department of Agriculture; Colorado Energy Research Institute, *Net Energy Analysis*, 1976; Howard T. Odum and Elisabeth C. Odum, *Energy Basis for Man and Nature*, 3rd ed., New York: McGraw-Hill, 1981, and Charles A. S. Hall and Kent A. Klitgaard, *Energy and the Wealth of Nations*, New York: Springer, 2012.)
Top left: Yegor Korzh/Shutterstock.com. Bottom left: Donald Aitken/National Renewable Energy Laboratory. Top right: Serdar Tibet/Shutterstock.com. Bottom right: Michel Stevelmans/Shutterstock.com.

not emit climate-changing gases or other air pollutants. However, energy density can be misleading because it does not take into account the high-quality energy needed to make the energy resource available for use. For example, the entire nuclear power process of using uranium-235 to produce electricity has a low net energy as described above, and producing hydrogen gas has a net energy loss.

15.2 WHAT ARE THE ADVANTAGES AND DISADVANTAGES OF USING OIL?

CONCEPT 15.2A Conventional crude oil is abundant and has a medium net energy, but causes air and water pollution and releases climate-changing CO_2 to the atmosphere.

CONCEPT 15.2B Unconventional heavy oil from oil shale rock and oil sands exists in potentially large supplies but has a low net energy and a higher environmental impact than conventional oil has.

We Depend Heavily on Oil

Oil is the second most widely used energy resource in the world (Figure 15.2, left) and the most widely used energy in the United States (Figure 15.2, right). We use oil to heat homes, grow food, transport people and goods, make other energy resources available for use, and manufacture most of the things we use every day from plastics to cosmetics to asphalt on roads.

Crude oil, or **petroleum,** is a black, gooey liquid containing a mixture of combustible hydrocarbons (molecules that contain hydrogen and carbon atoms) along with small amounts of sulfur, oxygen, and nitrogen impurities. It is also known as *conventional* or *light crude oil.*

Crude oil was formed from the decayed remains of ancient organisms that were crushed beneath layers of rock and subjected to high temperatures and pressures for millions of years. Deposits of conventional crude oil and natural gas often are trapped beneath layers of nonporous rock within the earth's crust on land or under the seafloor. The crude oil in such deposits is dispersed in the microscopic pores and cracks of these rock formations, somewhat like water saturating a sponge.

Geologists survey landscapes on the ground and from the air to identify rock formations that might have oil deposits beneath them. When they find a promising area, they make a seismic survey of its rock formations. Different types of rocks have different densities and thus reflect shock waves at different speeds. Geologists set off explosives or use machines to pound the earth to send seismic or vibrating shock waves deep underground, and they measure how long it takes the waves to be reflected back. They feed this information into computers to produce *three-dimensional seismic maps* that show the locations and sizes of various underground rock formations, including those containing oil and natural gas deposits.

Once they identify a potential site, geologists drill an exploratory well to learn whether the site has enough oil to be extracted profitably. If it does, a well is drilled and the oil, drawn by gravity out of the rock pores, flows to the bottom of the well and is pumped from there to the surface.

After years of pumping, usually a decade or so, the pressure in a well drops and its rate of crude oil production starts to decline. This point in time is referred to as **peak production** for the well. The same thing can happen to a large oil field when the overall rate of production from its numerous wells begins to decline.

Crude oil from a well cannot be used as it is. It is transported to a refinery by pipeline, truck, rail, or ship (oil tanker). There it is heated in pressurized vessels to separate it into various fuels and other components with different boiling points (Figure 15.4) in a complex process called **refining.** Like all steps in the cycle of oil production and use, refining requires an input of high-quality energy and decreases the net energy of oil. About 2% of the products of refining, called **petrochemicals,** are used as raw materials to make industrial organic chemicals, cleaning fluids, pesticides, plastics, synthetic fibers, paints, medicines, cosmetics, and many other products.

Is the World Running Out of Crude Oil?

We use an astonishing amount of crude oil—the lifeblood of most economies. Laid end to end, the roughly 34 billion barrels of crude oil used worldwide in 2014 would stretch to about 28 million kilometers (18 million miles)—far enough to reach to the moon and back about 37 times. (One barrel of oil contains 159 liters or 42 gallons of oil.)

How much crude oil is there? No one knows, although geologists have estimated the amounts existing in identified oil deposits. **Proven oil reserves** are known deposits from which oil can be extracted profitably at current prices using current technology. Proven oil reserves are not fixed. They grow when we find new deposits or develop new extraction technologies such as horizontal drilling and hydraulic fracking (**Core Case Study**) that make it possible and affordable to produce oil from deposits that were once too costly to tap.

The world is not about to run out of crude oil in the near future. We can produce more conventional light crude oil far offshore from deep ocean seabed deposits and from remote areas such as the Arctic Ocean where drillers face some of the earth's harshest weather, floating icebergs, and other hazards. To be profitable, however, oil from such places would have to sell for at least $100 a barrel because it takes a lot of energy to find and produce these oil deposits. We can also rely more on unconventional heavy oil—a thick type of crude oil that does not flow as easily as light oil from various sources. However, producing oil from these sources has a lower net energy, a higher environmental impact, and higher production costs.

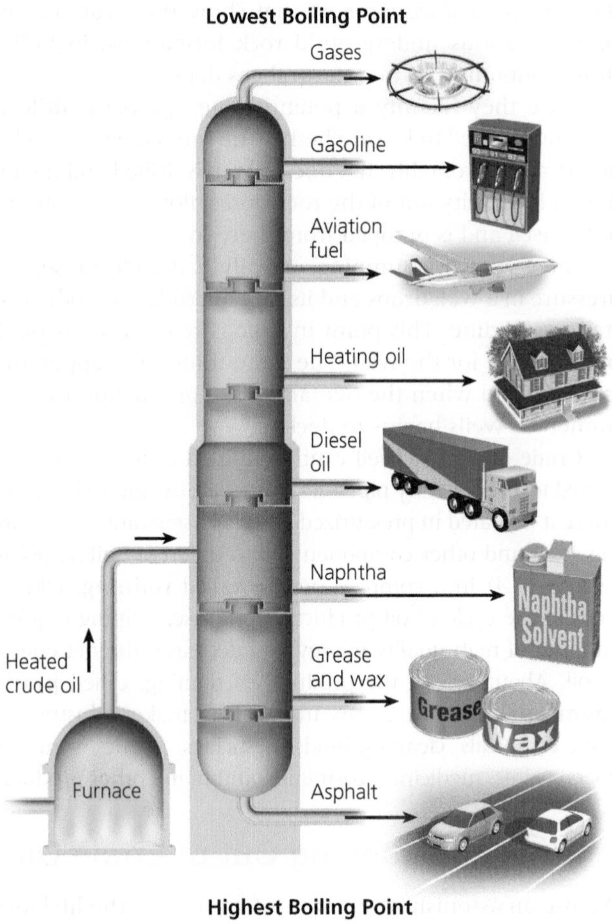

FIGURE 15.4 When crude oil is refined, many of its components are removed at various levels of a distillation column, depending on their boiling points. The most volatile components with the lowest boiling points are removed at the top of the column, which can be as tall as a nine-story building. The photo shows an oil refinery in Texas.

The first oil wells were drilled in the mid-1850s. Oil produced from these shallow wells had a high net energy. It was almost like sticking a metal straw in the ground and standing back as the oil gushed out of the well. Those days are gone.

The net energy of crude oil is still high in places such as Saudi Arabia where huge deposits are located near the earth's surface and yield oil at a cost of $3 to $10 a barrel. However, since 1999, the global net energy for extracting oil has dropped by more than 50%. Producers have had to spend more money and use more energy to dig deeper wells on land and at sea and to extract and transport oil from more remote and challenging areas such as the Arctic. As a result, oil has a medium net energy. There is no global shortage of oil, but there is an increasing shortage of cheap oil as we deplete the world's easy to reach concentrated deposits.

The 12 countries that make up the Organization of Petroleum Exporting Countries (OPEC) have about 81% of the world's proven crude oil reserves, much of it concentrated in large and accessible deposits that are cheap to exploit. Thus they are likely to control most of the world's conventional oil supplies for decades to come. OPEC's member countries are Algeria, Angola, Ecuador, Iran, Iraq, Kuwait, Libya, Nigeria, Qatar, Saudi Arabia, United Arab Emirates, and Venezuela. OPEC nations, especially Saudi Arabia with the world's largest and cheapest proven oil reserves, play a role in global oil prices by agreeing to increase or decrease the amount of oil they produce.

81% Percentage of the world's proven oil reserves held by OPEC's 12 countries

However, the recent increase in U.S. oil production has weakened the ability of OPEC nations to control global oil prices. Between 1970 and 2014, OPEC's share of global oil production dropped from 40% to 30%.

Figure 15.5 shows the three countries with the largest proven reserves, production, and consumption of crude oil in 2014. In 2014 the three largest consumers of oil—the United States, China, and Japan—respectively had only 2.9%, 1.1%, and 0.003% of the world's proven oil reserves.

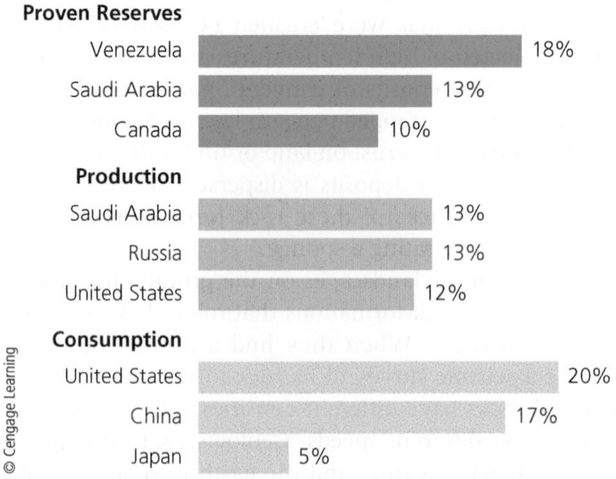

FIGURE 15.5 Countries with the largest proven crude oil reserves, oil production, and oil consumption in 2014.

(Compiled by the authors using data from British Petroleum and the International Energy Agency)

CASE STUDY

Oil Production and Consumption in the United States

The United States gets about 82% of its commercial energy from fossil fuels, with the largest percentage (36%) coming from crude oil (Figure 15.2, right). Since 1982, the United States has had to import some of the oil it uses because its oil consumption has exceeded its domestic production. This gap has been narrowed by rising domestic production of tight oil from shale rock (**Core Case Study**). In 2015 the United States imported about 24% of its crude oil, compared to 60% in 2005. In 2015 the five largest suppliers of imported oil for the United States were Canada, Saudi Arabia, Mexico, Venezuela, and Russia.

Can the United States continue reducing its dependence on oil imports by increasing its oil production? Some say "yes" and project that domestic oil production will increase dramatically over the next few decades—especially from oil found in layers of shale rock (**Core Case Study**).

Other analysts question the long-term availability of oil in layers of shale rock for two reasons. First, shale rock oil fields are not hard to find but drilling a horizontal oil well costs 6 to 10 times more than drilling a vertical well. If oil prices go too low (less than \$50–\$60 a barrel, as happened in recent years), using horizontal drilling and fracking to develop new wells will not be profitable. This explains why, between 2014 and April 2016, the number of rigs drilling for oil in the United States dropped from almost 1,600 to 351.

Second, the output of oil from shale rock drops off about twice as fast as it does in most conventional oil fields. It can drop by as much as a 90% after 2 years of operation. This means that oil companies must drill more profitable wells to maintain production and revenues.

According to the International Energy Agency, oil produced from shale rock in the United States is likely to peak around 2020 and then decline for two to three decades as the richest deposits (called *sweet spots*) are depleted. If this projection is correct, the current U.S. boom in oil production is a temporary bubble, not a long-term source of oil. The long-term problem for the United States is that it uses about 20% of the world's oil production, produces 12% of the world's oil, and has only 2.9% of the world's proven crude oil reserves. (See Figure 14, p. S28, in Supplement 4 for a map of most of North America's proven and potential reserves for crude oil and other fossil fuels.)

Using Crude Oil Has Advantages and Disadvantages

Figure 15.6 lists major advantages and disadvantages of using conventional light oil as an energy resource. A critical problem is that burning oil or any carbon-containing fossil fuel releases the greenhouse gas CO_2 into the atmosphere. According to decades of research and the assessments of 97% of the world's top climate scientists, this plays an important role in warming the atmosphere and changing the world's climate, as discussed in Chapter 19.

Trade-Offs

Conventional Oil

Advantages	Disadvantages
Ample supply for several decades	Water pollution from oil spills and leaks
Net energy is medium but decreasing	Environmental costs not included in market price
Low land disruption	Releases CO_2 and other air pollutants when burned
Efficient distribution system	Vulnerable to international supply interruptions

Photo: Richard Goldberg/Shutterstock.com

© Cengage Learning

FIGURE 15.6 Using conventional light oil as an energy resource has advantages and disadvantages. ***Critical thinking:*** Which single advantage and which single disadvantage do you think are the most important? Why? Do the advantages of relying on conventional light oil outweigh its disadvantages? Explain.

Use of Heavy Oil Has a High Environmental Impact

A potential supply of heavy oil is **shale oil**—oil that is *integrated within* bodies of shale rock, as opposed to *trapped between* layers of shale rock (**Core Case Study**). Producing this shale oil involves mining, crushing, and heating oil shale rock (Figure 15.7, left) to extract a mixture of hydrocarbons called *kerogen* that can be distilled to produce shale oil (Figure 15.7, right). Before the thick shale oil is

U.S. Department of Energy

FIGURE 15.7 Heavy shale oil (right) can be extracted from oil shale rock (left).

FIGURE 15.8 Oil sands surface mining operation in Alberta, Canada, involves clearing boreal forest and strip-mining the land to remove the oil sands.

Christopher Kolaczan/Shutterstock.com

pumped through a pressurized pipeline to a refinery, it must be heated to increase its flow rate and processed to remove sulfur, nitrogen, and other impurities, which reduces its net energy.

About 72% of the world's estimated oil shale rock reserves are buried deep in rock formations located primarily under government-owned land in the U.S. states of Colorado, Wyoming, and Utah. The potential supply is huge but its net energy is low (Figure 15.3). The process also has a large harmful environmental impact, including the production of rock waste, possible leaks of the extracted kerogen to water tables, wastewater, and high water use. If the production price can be significantly lowered or the price of conventional oil rises sharply, and if its harmful environmental effects can be reduced, shale oil could become an important energy source. Otherwise, it will remain in the ground.

A growing source of heavy oil is **oil sands** or **tar sands,** which are a mixture of clay, sand, water, and an organic material called *bitumen*—a thick, sticky, tar-like heavy oil with a high sulfur content. Northeastern Alberta in Canada has three-fourths of the world's proven reserves of oil sands. If we include its conventional light oil and its heavy oil from oil sands, Canada has the world's third largest proven oil reserves. The United States also has large deposits of oil sands in Utah, Wyoming, and Colorado.

In Canada, oil companies produce about half of this oil by drilling vertical wells, pumping steam in to melt the bitumen embedded in the underground sand, and pumping the bitumen to the surface for conversion to heavy synthetic oil. Production of the rest of this oil starts with clear-cutting the forests and strip-mining the land. This process uses large amounts of natural gas to heat water that is used to remove the bitumen and to convert it into a heavy synthetic oil that can flow through a pipeline (Figure 15.8). Some of the bitumen is converted to synthetic oil and diesel fuel used to run the gigantic shovels, trucks, and other machinery.

This process greatly reduces the net energy of oil sands. It takes two to four tons of oil sands and two to five barrels of water to produce one barrel of the heavy synthetic oil. The process also emits large quantities of air pollutants—especially sulfur dioxide, nitrogen oxides, and particulates—and 20% more climate-changing CO_2 than does the production of conventional crude oil, according to a 2015 study by the U.S. Department of Energy.

Producing this heavy oil is expensive, costing $50 or more per barrel. If the global price of oil remains at or below this level (as it did between 2014 and 2016), producers of oil from Canada's oil sands will lose money, and new production will drop. Figure 15.9 lists the major advantages and disadvantages of producing heavy oil from oil sands.

Trade-Offs

Heavy Oil from Oil Sands

Advantages	Disadvantages
Large potential supplies	Low net energy
Easily transported within and between countries	Expensive
Efficient distribution system in place	Releases CO_2 and other air pollutants
	Severe land disruption
	Water pollution and high water use

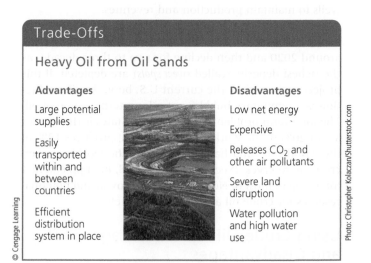

© Cengage Learning

Photo: Christopher Kolaczan/Shutterstock.com

FIGURE 15.9 Using heavy oil from oil sands as an energy resource has advantages and disadvantages (**Concept 15.2**). ***Critical thinking:*** Which single advantage and which single disadvantage do you think are the most important? Why? Do the advantages of relying on heavy oil from oil sands outweigh its disadvantages? Explain.

15.3 WHAT ARE THE ADVANTAGES AND DISADVANTAGES OF USING NATURAL GAS?

CONCEPT 15.3 Conventional natural gas is more plentiful than oil, has a medium net energy and a fairly low production cost, and is a clean-burning fuel, but producing it has created environmental problems.

Natural Gas Is a Versatile and Widely Used Fuel

Natural gas is a mixture of gases of which 50–90% is methane (CH_4). It also contains smaller amounts of heavier gaseous hydrocarbons such as propane (C_3H_8) and butane (C_4H_{10}), and small amounts of highly toxic hydrogen sulfide (H_2S). This versatile fuel has a medium net energy (Figure 15.4) and is widely used for cooking, heating, and industrial purposes. It can also be used as a motor vehicle fuel—especially for fleets of trucks and cars such as taxicabs—and to fuel natural gas turbines used to produce electricity in power plants.

This versatility and the use of horizontal drilling and fracking help to explain why natural gas provided about 28% of U.S. energy in 2014. That year, more than 1,700 natural gas power plants (see map in Figure 17, p. S31, in Supplement 4) produced about 27% of the electricity consumed in the United States. Natural gas power plants are less expensive and take much less time to build than do coal-powered and nuclear power plants. Natural gas also burns cleaner than oil and much cleaner than coal. When burned completely, it emits about 30% less CO_2 than oil and about 50% less than coal for the same amount of energy.

Conventional natural gas is often found in deposits lying above deposits of conventional oil. It also exists in tightly held deposits between layers of shale rock and can be extracted through horizontal drilling and fracking (**Core Case Study**). (See Figure 16, p. S30, in Supplement 4 for a map of U.S. natural gas shale rock deposits.)

When a natural gas deposit is tapped, propane and butane gases can be liquefied under high pressure and removed as **liquefied petroleum gas (LPG).** LPG is stored in pressurized tanks for use mostly in rural areas not served by natural gas pipelines. The rest of the gas (mostly methane) is purified and pumped into pressurized pipelines for distribution across land areas.

Natural gas can also be transported across oceans, by converting it to **liquefied natural gas (LNG)** at a high pressure and at a very low temperature. This highly flammable liquid is transported in refrigerated tanker ships. At its destination port, it is heated and converted back to the gaseous state and then distributed by pipeline. LNG has a lower net energy than natural gas has, because more than a third of its energy content is used to liquefy it, process it, deliver it to users by ship, and convert it back to natural gas.

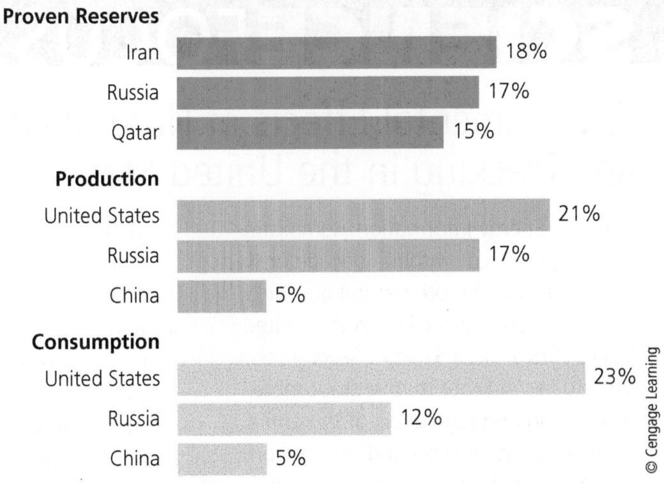

FIGURE 15.10 Countries with the largest proven natural gas reserves, natural gas production, and natural gas consumption in 2014.

(Compiled by the authors using data from British Petroleum and the International Energy Agency)

Figure 15.10 shows the three countries with the largest reserves, production, and consumption of natural gas in 2014. That year, the three largest consumers of natural gas—the United States, Russia, and China—respectively had 5%, 17%, and 1.8% of the world's proven natural gas reserves. (See Figure 3, p. S41, in Supplement 5 for a graph showing global natural gas consumption since 1950.)

Currently, the United States does not have to rely on natural gas imports because of greatly increased production since 1990, mostly due to the growing use of horizontal drilling and fracking in shale rock beds (**Core Case Study**). The country now exports natural gas as LNG to other nations. The increased supply has provided jobs and reduced the price of natural gas. This has benefitted consumers, industries, and utilities, and accelerated a shift from coal to natural gas for generating electricity.

However, natural gas production from shale rock tends to peak and drop off much faster than does production from conventional natural gas wells. In addition, extracting and producing natural gas from shale rock reduces its net energy and, without effective regulation, can increase the harmful environmental impacts of production (Science Focus 15.1). Figure 15.11 lists the advantages and disadvantages of using conventional natural gas as an energy resource.

Can Natural Gas Help to Slow Climate Change?

Some see increased use of natural gas as a way to slow climate change because it emits less CO_2 per unit of energy than coal. Critics cite two problems with this. First, while use of cheap natural gas is reducing the use of coal, its low

SCIENCEFOCUS 15.1

Environmental Effects of Natural Gas Production and Fracking in the United States

The U.S. Energy Information Administration projects that, within a decade or two, at least 100,000 more natural gas wells will be drilled and fracked in the United States (**Core Case Study**). Several scientific studies indicate that without more monitoring and regulation of the entire natural gas production and distribution process, including fracking, greatly increased production of natural gas (and oil) from shale rock (Figure 15.1) could have several harmful environmental effects:

- Fracking requires enormous volumes of water—10 to 100 times more than in a conventional vertical well. In water-short areas, this could reduce available surface water, deplete aquifers, degrade aquatic habitats, and reduce the availability of water for irrigation and other purposes.
- Each fracked well produces huge volumes of hazardous wastewater that flows back to the surface with the released natural gas (or oil). Many scientists warn that, without stricter regulation, the potentially harmful chemicals in these fracking slurries—including arsenic and naturally occurring radioactive elements such as radium—could contaminate groundwater and surface waters, especially from spills and leaking waste storage ponds.
- To prevent leaks of natural gas (or oil) from conventional and fracked wells and from the hazardous wastewater that returns to the surface, steel pipe called *casing* is inserted into the drilled well and cement is pumped into a small space between the steel pipe and the surrounding rock (Figure 15.1). Well-casing failure and poor cementing can release methane (or oil) and contaminants in the wastewater into groundwater and drinking water wells far above the fracking site. When the contaminated water is drawn from a tap, this natural gas can catch fire (Figure 15.A). Experience has shown that cementing the steel well-casing is the weakest link in the process for producing natural gas and oil and in preventing leaks from active and abandoned wells.
- According to recent studies by the National Academy of Sciences and the U.S. Geological Survey, one of the causes of hundreds of small earthquakes in 13 states in recent years has been the shifting of bedrock resulting from the high-pressure injection of fracking wastewater into deep underground hazardous waste wells. Such earthquakes could release hazardous wastewater into aquifers and cause breaks in the steel linings and cement seals of oil and gas well pipes. Oklahoma had so many earthquakes with magnitudes of 3.0 or greater (907 in 2015 compared to 3 from 2010) related to its 411 hazardous waste injection wells that in 2016 the state called for a 40% reduction in the number of injection wells.

In 2015, a four-year EPA study found no evidence that fracking for oil and natural gas has posed a widespread risk to U.S. drinking water, but the EPA cautioned that this conclusion was based on limited data. According to the study, the major threats posed by fracking include potential contamination from well casing and cement seal failures, poor management of the contaminated wastewater resulting from the process, chemical spills at drilling sites, and depletion of water supplies in some areas.

Currently, people who rely on aquifers and streams for their drinking water in areas undergoing rapid increases in shale gas production have little protection from pollution of their air and water supplies that might result from natural gas production. This is because, under political pressure from natural gas suppliers, the U.S. Congress in 2005 excluded natural gas fracking from EPA regulations under seven major federal environmental laws—the Safe Drinking Water Act, Clean Water Act, Clean Air Act, National Environmental Policy Act, Resource Conservation and Recovery Act (which sets standards for the handling of hazardous wastes), Emergency Planning and Community Right to Know Act, and Superfund Act.

MARK THIESSEN/National Geographic Creative

FIGURE 15.A Natural gas fizzing from this faucet in a Pennsylvania home can be lit like a natural gas stove burner. This began happening after an energy company drilled a fracking well in the area, but the company denies responsibility. The homeowners have to keep their windows open year-round to keep the lethal and explosive gas from building up in the house.

Without stricter regulation and monitoring, drilling another 100,000 natural gas wells during the next 10 to 20 years will likely increase the risk of air and water pollution from the natural gas production process. This could cause a public backlash against this technology.

To avoid this, some analysts call for **(1)** getting better data on water quality at drilling sites before and after fracking and on leaks from well-casings, valves, storage tanks, compressors, and pipelines; **(2)** setting higher standards for building and monitoring natural gas wells and monitoring retired wells; **(3)** fixing leaks in the natural gas production, storage, and distribution system; and **(4)** pressuring Congress to revoke all exemptions from federal environmental laws for the natural gas industry. In 2016, the EPA issued regulations designed to curb methane emissions from new oil and natural gas wells by as much as 45% between 2012 and 2025, but the agency imposed no regulations to curb leaks from existing and abandoned wells.

CRITICAL THINKING

How might your life be affected if the four policy changes listed above are not implemented?

Trade-Offs

Conventional Natural Gas

Advantages	Disadvantages
Ample supplies	Low net energy for LNG
Versatile fuel	Production and delivery may emit more CO_2 and CH_4 per unit of energy produced than coal
Medium net energy	Fracking uses and pollutes large volumes of water
Emits less CO_2 and other air pollutants than other fossil fuels when burned	Potential groundwater pollution from fracking

FIGURE 15.11 Using conventional natural gas as an energy resource has advantages and disadvantages. ***Critical thinking:*** Which single advantage and which single disadvantage do you think are the most important? Why? Do you think that the advantages of using conventional natural gas outweigh its disadvantages? Explain.

price could also slow the shift to reliance on energy efficiency and low-carbon solar, wind, and other renewable energy resources that are vital to slowing climate change.

The other problem is that methane (CH_4) is a much more potent greenhouse gas than CO_2 and measurements reveal that the drilling, production, and distribution process for natural gas leaks large quantities of CH_4 into the atmosphere. This raises atmospheric levels of CH_4, which is 34 times more effective per molecule at warming the atmosphere and causing climate change than CO_2. Unless these leaks are plugged, greater reliance on natural gas could hasten rather than slow climate change.

15.4 WHAT ARE THE ADVANTAGES AND DISADVANTAGES OF USING COAL?

CONCEPT 15.4A Conventional coal is plentiful and has a high net energy and a low cost, but using it has a high environmental impact.

CONCEPT 15.4B We can produce gaseous and liquid fuels from coal, but they have a lower net energy and higher environmental impacts than those of conventional coal.

Coal Is a Plentiful but Dirty Fuel

Coal is a solid fossil fuel formed from the remains of land plants that were buried and exposed to intense heat and pressure for 300–400 million years (Figure 15.12).

Coal is burned in power plants (Figure 15.13) that in 2015 generated about 40% of the world's electricity, 33% of the electricity used in the United States (down from 51% in 2003), 65% in China, and 62% in India. Coal is also burned in industrial plants to make steel, cement, and other products. See Figure 3, p. S41, in Supplement 5 for a graph showing global coal consumption since 1950.

Coal is the most abundant fossil fuel. Figure 15.14 shows the three countries with the largest proven reserves, production, and consumption of coal in 2014. That year, the three largest consumers of coal—China, the United States, and India—respectively had 13%, 27%, and 7% of the world's proven coal reserves. Although China is the world's largest consumer of coal, the United States has the largest per capita consumption of coal. See Figure 18, p. S31, in Supplement 4 for a map of the large coal-burning power plants in the United States in 2012 and information on trends in the number of plants since then.

In 2014 China consumed roughly as much coal as the rest of the world combined. Figure 15.15 shows coal consumption in China and the United States since 1980.

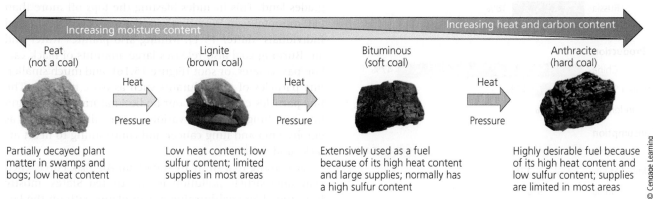

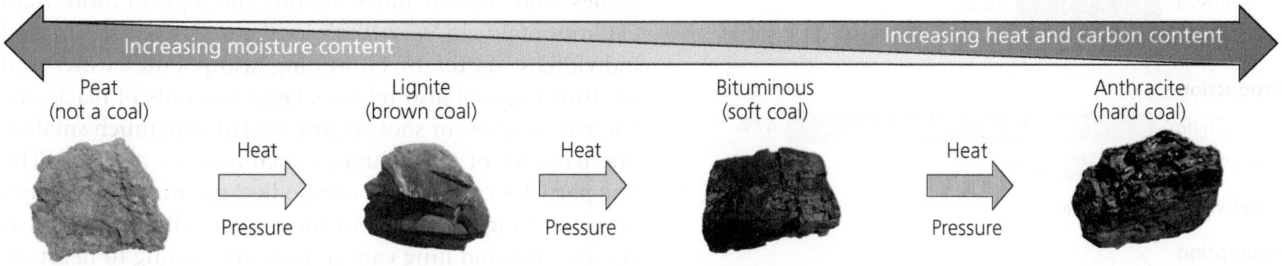
Increasing moisture content
Increasing heat and carbon content
Peat (not a coal)
Lignite (brown coal)
Bituminous (soft coal)
Anthracite (hard coal)
Heat
Pressure
Heat
Pressure
Heat
Pressure
Partially decayed plant matter in swamps and bogs; low heat content
Low heat content; low sulfur content; limited supplies in most areas
Extensively used as a fuel because of its high heat content and large supplies; normally has a high sulfur content
Highly desirable fuel because of its high heat content and low sulfur content; supplies are limited in most areas

FIGURE 15.12 Over millions of years, several different types of coal have formed. Peat is a soil material made of moist, partially decomposed organic matter, similar to coal; it is not classified as a coal, although it is used as a fuel. These different major types of coal vary in the amounts of heat, carbon dioxide, and sulfur dioxide released per unit of mass when they are burned.

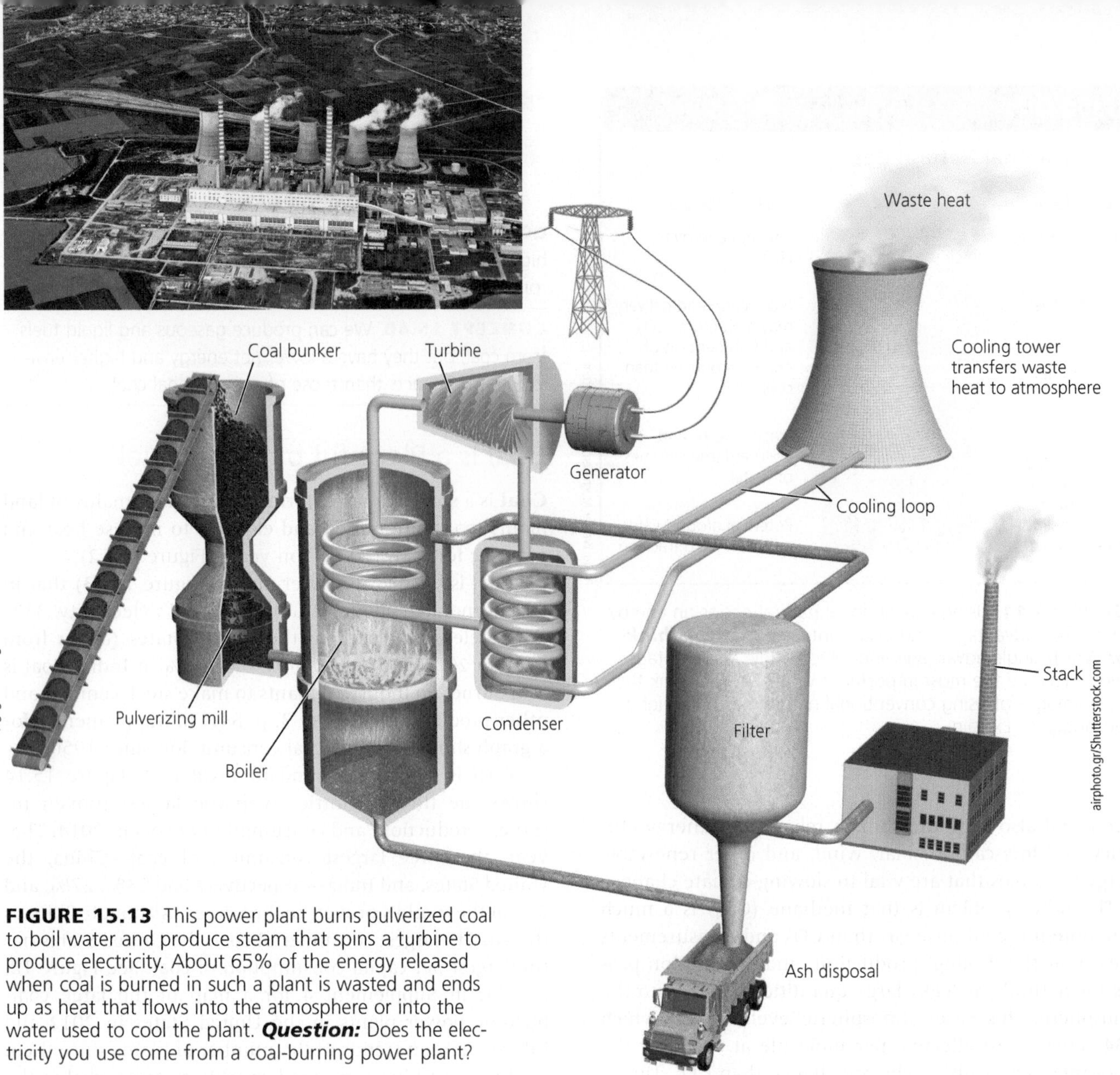

FIGURE 15.13 This power plant burns pulverized coal to boil water and produce steam that spins a turbine to produce electricity. About 65% of the energy released when coal is burned in such a plant is wasted and ends up as heat that flows into the atmosphere or into the water used to cool the plant. ***Question:*** Does the electricity you use come from a coal-burning power plant?

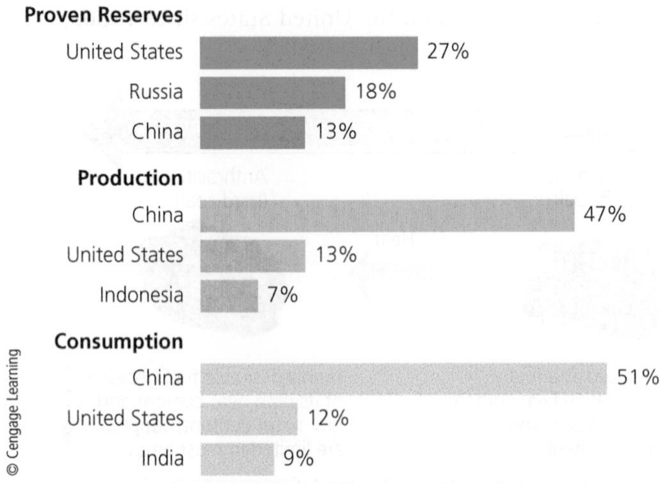

FIGURE 15.14 Countries with the largest proven coal reserves, coal production, and coal consumption in 2014.

(Compiled by the authors using data from British Petroleum and the International Energy Agency)

The problem is that coal is by far the dirtiest of all fossil fuels, starting with the mining of coal, which severely degrades land. This includes blasting the tops off more than 500 mountains in West Virginia (Figure 14.11, p. 367, and Individuals Matter 15.1). Mining also pollutes water and air. Burning coal also releases large amounts of black carbon particulates, or soot (Figure 15.16), and much smaller, fine particles of air pollutants such as toxic mercury. The fine particles can get past our bodies' natural defenses and into our lungs, causing various severe illnesses such as emphysema and lung cancer and contributing to heart attacks and strokes.

According to a study by the Clean Air Task Force, each year fine-particle pollution in the United States, mostly from the older coal-burning power plants without the latest air pollution control technology, kills at least 13,000 people a year. In China, in 2012 outdoor air pollution from the burning of coal contributed to 670,000 premature

INDIVIDUALS MATTER 15.1

Maria Gunnoe: Fighting to Save Mountains

Courtesy of Goldman Environment Prize

In the 1800s, Maria Gunnoe's Cherokee ancestors arrived in what is now Boone County, West Virginia. Her grandfather bought the land where she now lives. In 2000 miners blew up the mountaintop above that land to extract underlying coal (Figure 14.11, p. 367). Gunnoe's land now sits below a 10-story-high pile of mine waste.

With the soil and vegetation gone, rains running off the mountain ridge flooded her land seven times since 2000. They covered her land with toxic sludge, contaminated her water well and the soil she used to grow food, and washed out two small bridges that linked her to the only road out. Gunnoe had to hike to and from the road for years. When she complained to the coal company officials, they called the floods "acts of God."

Gunnoe, a mother of two, refused to leave her land and decided to fight the powerful coal companies and to try to end mountaintop coal mining. Mining companies and miners who worried about losing their jobs viewed her as an enemy. She and her children received death threats and two family dogs were killed. People tried to run her off the road and fired shots around her house. For several years, Gunnoe wore a bulletproof vest when she went outside, but she kept up the fight.

With only a high school education, Gunnoe educated herself about complex mining and water pollution regulations and harmful chemicals found in streams and groundwater. She organized communities and argued in court and testified before Congress contending that the "valley fills" from mountaintop mining violate the federal Clean Water Act by burying streams and destroying aquatic and animal habitats. The EPA agreed and fined coal companies multiple times. She also pressured the president and federal government to ban mountaintop mining because of its harmful health effects.

In 2009 this courageous and inspiring woman won a Goldman Environmental Prize—considered the Nobel Prize equivalent for grassroots environmental leaders around the world.

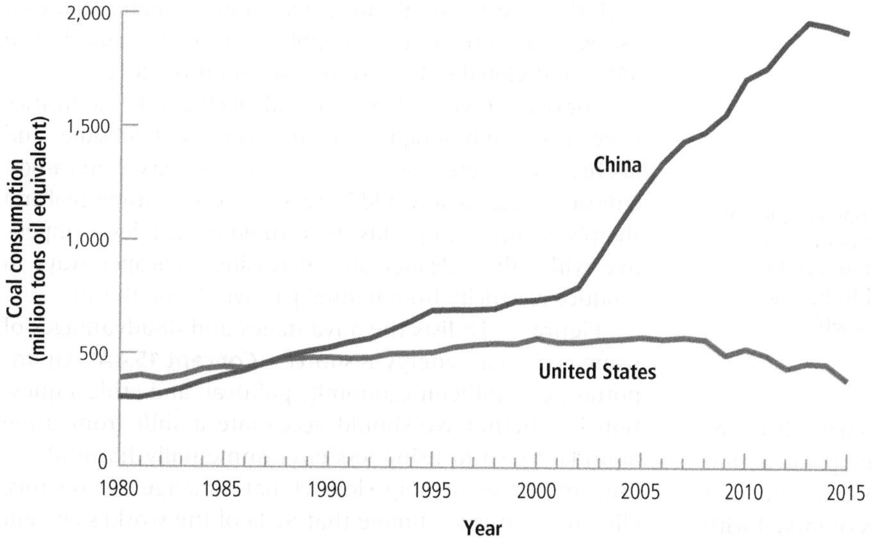

FIGURE 15.15 Coal consumption in the China and the United States, 1980–2014. ***Data analysis:*** In 2012, about how many times more than U.S. coal consumption was that of China?

(Compiled by the authors using data from British Petroleum, International Energy Agency, and Earth Policy Institute)

deaths (an average of 1,835 per day) from strokes, heart attacks, pulmonary obstruction, and lung cancer, according to a study by Teng Fei at Tsinghua University.

Coal-burning power and industrial plants are among the largest emitters of CO_2 (Figure 15.17), which contributes to atmospheric warming and climate change (covered in Chapter 19) and ocean acidification (Science Focus 11.1, p. 257). Because coal is mostly carbon, coal combustion emits about twice as much CO_2 per unit of energy as natural gas and produces about 39% of global CO_2 emissions. China leads the world in such emissions, followed by the United States. Another problem with coal combustion is that it emits trace amounts of radioactive materials as well as toxic mercury into the atmosphere and into lakes, where it can accumulate to high levels in fish consumed by humans.

Because of air pollution laws, many coal-burning plants use scrubbers to remove some of these pollutants before they leave the smokestacks. This reduces air pollution but produces a dust-like material called *coal ash* (Figure 15.13), which can contain dangerous, indestructible chemical elements such as arsenic, lead, mercury, cadmium, and radioactive radium. It must be stored safely, essentially forever. However, political pressure by the U.S. coal industry has kept it from being classified as a hazardous waste and thus it is not subject to much stricter and costly storage regulations. Instead, coal ash is in the same category as household garbage.

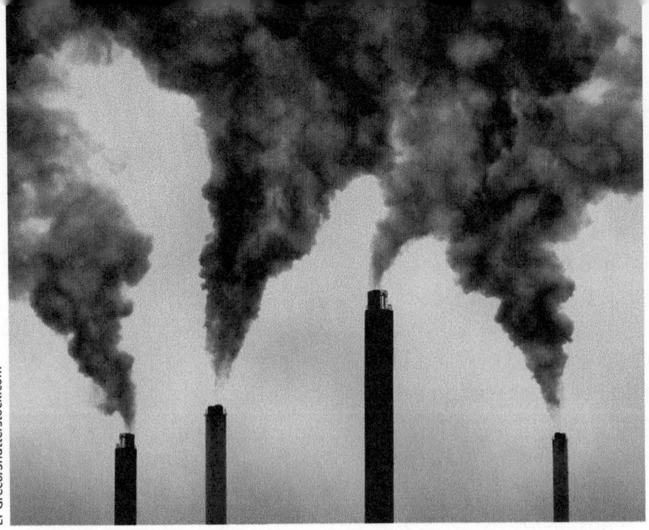

El Greco/Shutterstock.com

FIGURE 15.16 The smokestacks on this coal-burning industrial plant emit large amounts of air pollution because the plant has inadequate air pollution controls.

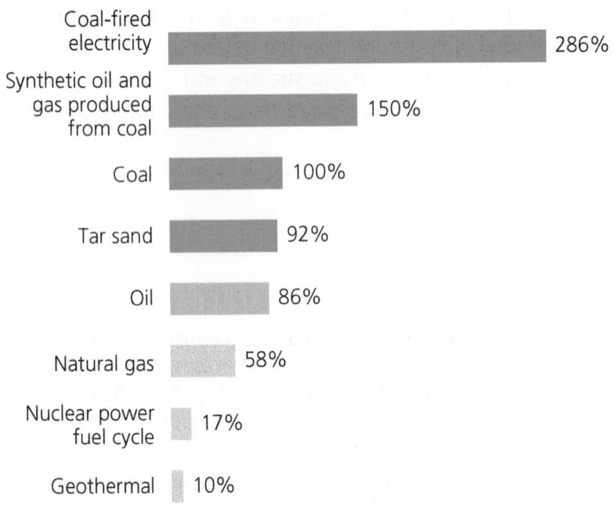

FIGURE 15.17 CO_2 emissions, expressed as percentages of emissions released by burning coal directly, vary with different energy resources. ***Data analysis:*** Which of these produces more CO_2 emissions per kilogram: burning coal to heat a house, or heating with electricity generated by coal?

(Compiled by the authors using data from U.S. Department of Energy.)

In the United States, about 45% of all coal ash is isolated from the environment by being incorporated into materials such as concrete, cement, wallboard, and roof shingles. The other 55% is buried in landfills or mixed with water stored as a slurry in open-air holding ponds. Some of these holding ponds have leaked harmful chemicals into groundwater and nearby rivers, as occurred near Eden, North Carolina, in 2014, and some have collapsed and spilled harmful chemicals into nearby communities, as occurred in Kingston, Tennessee, in 2008.

There is growing pressure to transfer wet coal ash slurry from holding ponds, which are viewed by some as disasters waiting to happen, to double-lined landfills. In China, most coal ash is dumped in open fields and can be blown into the air by wind or washed into streams and rivers by rainfall.

We Are Not Paying the Full Cost of Using Coal

The primary reason that coal is a cheap way to produce electricity is that most of its harmful environmental and health costs are not included in the market price of coal-generated electricity. In 2014 the International Monetary Fund estimated that global coal use causes about $3.2 billion a year in environmental and health damages. According to studies by the Harvard Medical School's Center for Health and the Global Environment and the U.S. National Academy of Sciences, including all such costs would double or triple the price of electricity from coal-fired power plants. This would promote a shift to natural gas and renewable energy resources such as solar and wind whose prices have been falling sharply. As long as these harmful costs are not included in the price of coal burned to provide heat or electricity, it will continue to be widely used in countries with large coal reserves, such as China, the United States, India, Australia, and Indonesia.

There are ways to include such costs as a way to implement the full-cost pricing **principle of sustainability**. They include phasing out subsidies and tax breaks that hide the true costs of coal use, taxing CO_2 emissions from the burning of coal, requiring stricter air pollution controls for coal-burning power and industrial plants, and regulating coal ash as a hazardous waste. If these hidden costs were included in the price of burning coal, the International Monetary Fund estimates that coal use would drop sharply, air pollution deaths would drop 55%, and global CO_2 emissions would drop 20%.

For over 40 years, U.S. coal and electric utility industries have successfully fought to keep government subsidies and tax breaks and prevent coal regulations and taxes on carbon emissions because it would increase the cost of using coal and sharply reduce their profits. This would make it less competitive with other, cleaner and increasingly cheaper ways to produce electricity from natural gas, wind, and the sun.

Figure 15.18 lists the advantages and disadvantages of using coal as an energy resource. (**Concept 15.4A**) An important and difficult economic, political, and ethical question is whether we should accelerate a shift from using abundant coal to using less environmentally harmful energy resources to help slow climate change. To do this, climate scientists estimate that 82% of the world's current coal reserves and 92% of U.S. coal reserves need to be left in the ground. In countries that have large reserves of coal, this is a difficult economic and political challenge.

The Future of Coal

Because of increasing competition from cleaner-burning natural gas, wind power, and solar power, and because of grassroots political opposition, U.S. coal use dropped 18% between 2007 and 2013. In that time, proposals for building 183 new coal plants were scrapped. In addition, several

Trade-Offs

Coal

Advantages	Disadvantages
Ample supplies in many countries	Severe land disturbance and water pollution
Medium to high net energy	Fine particle and toxic mercury emissions threaten human health
Low cost when environmental costs are not included	Emits large amounts of CO_2 and other air pollutants when produced and burned

FIGURE 15.18 Using coal as an energy resource has advantages and disadvantages. ***Critical thinking:*** Which single advantage and which single disadvantage do you think are the most important? Why? Do you think that the advantages of using coal as an energy resource outweigh its disadvantages? Explain.

Trade-Offs

Synthetic Fuels

Advantages	Disadvantages
Large potential supply in many countries	Low to medium net energy
Vehicle fuel	Requires mining 50% more coal with increased land disturbance, water pollution, and water use
Lower air pollution than coal	Higher CO_2 emissions than coal

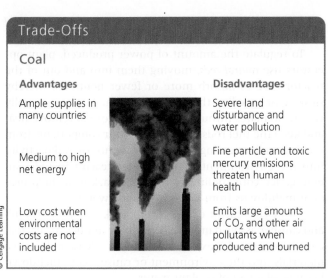

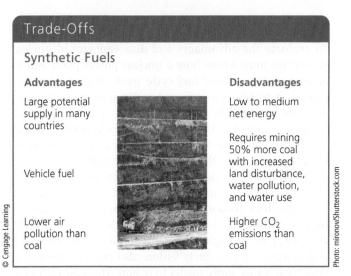

FIGURE 15.19 The use of synthetic natural gas (SNG) and liquid synfuels produced from coal as energy resources has advantages and disadvantages (**Concept 15.2**). ***Critical thinking:*** Which single advantage and which single disadvantage do you think are the most important? Why? Do you think that the advantages of using synfuels produced from coal as an energy source outweigh its disadvantages? Explain.

major U.S. coal producers have declared bankruptcy because of a drop in use of coal to produce electricity. According to the U.S. Energy Information Agency, natural gas should overtake coal as the nation's largest source of electricity by the 2030s, and sooner if the government regulates greenhouse gas emissions from fossil fuel combustion. In response to these changes, U.S. coal producers are exporting increasing amounts of coal to the United Kingdom, the Netherlands, Brazil, South Korea, China, and other countries.

Some countries are going further than the United States in reducing their use of coal. They include Germany, with a 50% drop in coal use since 1965, and France and the United Kingdom, with coal use down 70% since the mid-1990s. In 1997 Denmark banned new coal-fired plants, and it now plans to phase out coal power by 2025.

Does this mean the end of coal? Hardly. There are ample supplies of coal and it will be a cheap source of energy, as long as coal producers have the political and economic power to prevent its harmful health and environmental costs from being included in its market price. And while coal use is dropping in the United States and several European nations, its use is expanding in India and a number of other countries in Asia and Africa.

We Can Convert Coal into Gaseous and Liquid Fuels

We can convert solid coal into **synthetic natural gas (SNG)** by a process called *coal gasification,* which removes sulfur and most other impurities from coal. We can also convert coal into liquid fuels such as methanol and synthetic gasoline through a process called *coal liquefaction*. These fuels, called *synfuels,* are often referred to as cleaner versions of coal.

However, compared to burning coal directly, producing synfuels requires the mining of 50% more coal. Producing and burning synfuels could also add 50% more carbon dioxide to the atmosphere (Figure 15.17). As a result, synfuels have a lower net energy and cost more to produce per unit of energy than does coal production. In addition, it takes larger amounts of water to produce synfuels than to produce coal.

Thus, greatly increasing the use of these synfuels would worsen three of the world's major environmental problems: climate change, ocean acidification caused mostly by CO_2 emissions, and increasing water shortages in many parts of the world (Figure 13.7, p. 329, and Figure 13.8, p. 330). Figure 15.19 lists the advantages and disadvantages of using liquid and gaseous synfuels produced from coal (**Concept 15.4B**).

Another approach to reducing coal's impact on climate change is *carbon capture and sequestration* (CCS). This technology involves removing CO_2 from the smokestacks of coal-burning and industrial power plants and isolating it from the environment by storing it in depleted oil and gas fields and underground coal mines, or in underground salt aquifers. We discuss the pros and cons of CCS in Chapter 19.

15.5 WHAT ARE THE ADVANTAGES AND DISADVANTAGES OF USING NUCLEAR POWER?

CONCEPT 15.5 Nuclear power has a low environmental impact and a very low accident risk, but its use has been limited by a low net energy, high costs, its role in the spread of nuclear weapons technology, fear of accidents, and long-lived radioactive wastes.

How Does a Nuclear Fission Reactor Work?

To evaluate the advantages and disadvantages of nuclear power, we must know how a nuclear power plant and its accompanying nuclear fuel cycle work. A nuclear power plant is a highly complex and costly system designed to perform a relatively simple task: boil water to produce steam that spins a turbine and generates electricity. Nuclear plants cost much more and take longer to build (10 years or more) than any other source of electricity.

What makes nuclear power complex and costly is the use of a controlled nuclear fission reaction to provide the heat. **Nuclear fission** occurs when the nuclei of certain isotopes with large mass numbers (such as uranium-235) are split apart into lighter nuclei when struck by a neutron and release energy. Each fission also releases neutrons, which can cause more nuclei to fission. This cascade of fissions can result in a *chain reaction* that releases an enormous amount of energy in a short time (Figure 15.20).

Nuclear power plants generate heat used to produce electricity through a controlled nuclear fission chain reaction in the core of a *reactor*. A *pressurized light-water reactor* (Figure 15.21) is the most widely used type of reactor.

The fuel for a nuclear reactor is made from uranium ore mined from the earth's crust. After it is mined, the ore must be enriched to increase the concentration of its fissionable uranium-235 from 1% to 3–5%. The enriched uranium-235 is processed into small pellets of uranium dioxide. Each pellet, about the size of an eraser on a pencil, contains the energy equivalent of about a ton of coal. Large numbers of the pellets are packed into closed pipes, called *fuel rods*, which are then grouped together in *fuel assemblies*, to be placed in the core of a reactor.

To regulate the amount of power produced, plant operators use *control rods*, moving them into and out of the reactor core to absorb more or fewer neutrons, thereby slowing or speeding the fission reaction. A *coolant*, usually water, circulates through the reactor's core to remove heat and keep the fuel rods and other reactor components from melting and releasing massive amounts of radioactivity into the environment. A light-water reactor includes an emergency core cooling system as a backup to help prevent meltdowns from a loss of cooling water.

A nuclear reactor cannot explode like an atomic bomb and cause massive damage. The danger in nuclear reactors comes from smaller explosions that can release radioactive materials into the environment or cause a core meltdown because of a loss of coolant water.

A *containment shell* made of thick, steel-reinforced concrete surrounds the reactor core. It is designed to help keep radioactive materials from escaping into the environment, in case there is an internal explosion or a core meltdown. It is also intended to protect the core from external threats such as tornadoes and plane crashes. These essential safety features and the 10 years or more that it typically takes to build a nuclear plant help explain why a new nuclear power plant typically costs $9 billion to $11 billion and why that cost is rising.

In 2014 the world's three leading producers of nuclear power were, in order, the United States, France, and Russia. France generates 75% of its electricity and the United States generates 19% of its electricity using nuclear power. (See Figure 19, p. S32, in Supplement 4 for a map of the 99 large commercial nuclear power reactors in the United States.)

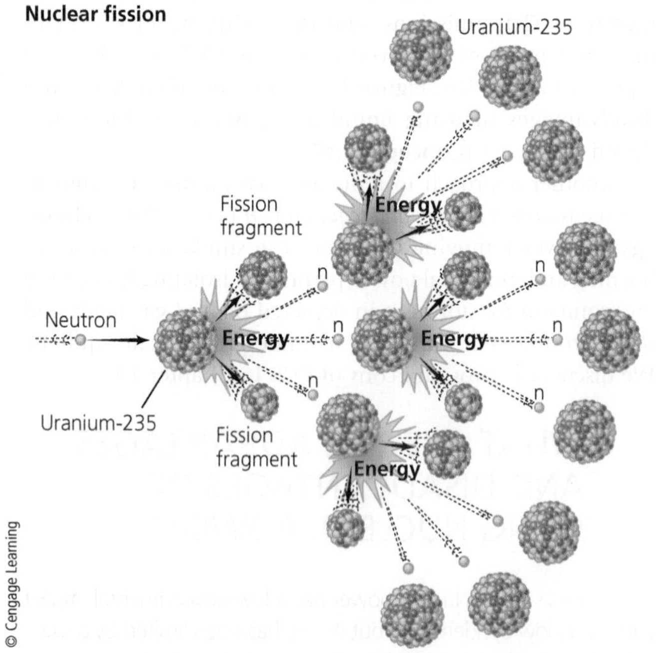

FIGURE 15.20 Nuclear fission is the source of energy for producing electricity in a nuclear power plant.

What Is the Nuclear Fuel Cycle?

Building and running a nuclear power plant is only one part of the **nuclear fuel cycle** (Figure 15.22), which also includes the mining of uranium, processing and enriching the uranium to make fuel, using it in a reactor, safely storing the resulting highly radioactive wastes for thousands of years until their radioactivity falls to safe levels, and retiring the worn-out plant by taking it apart and storing its high- and moderate-level radioactive parts safely for thousands of years.

As long as a reactor is operating safely, the power plant itself has a low environmental impact and a low risk of an accident. However, considering the entire nuclear fuel cycle, the potential environmental impact increases. In evaluating the safety, economic feasibility, net energy, and overall environmental impact of nuclear power, energy experts and economists caution us to look at the entire nuclear fuel cycle, not just the power plant itself. Figure 15.23 lists the major advantages and disadvantages of producing electricity by using the nuclear power fuel cycle (**Concept 15.5**).

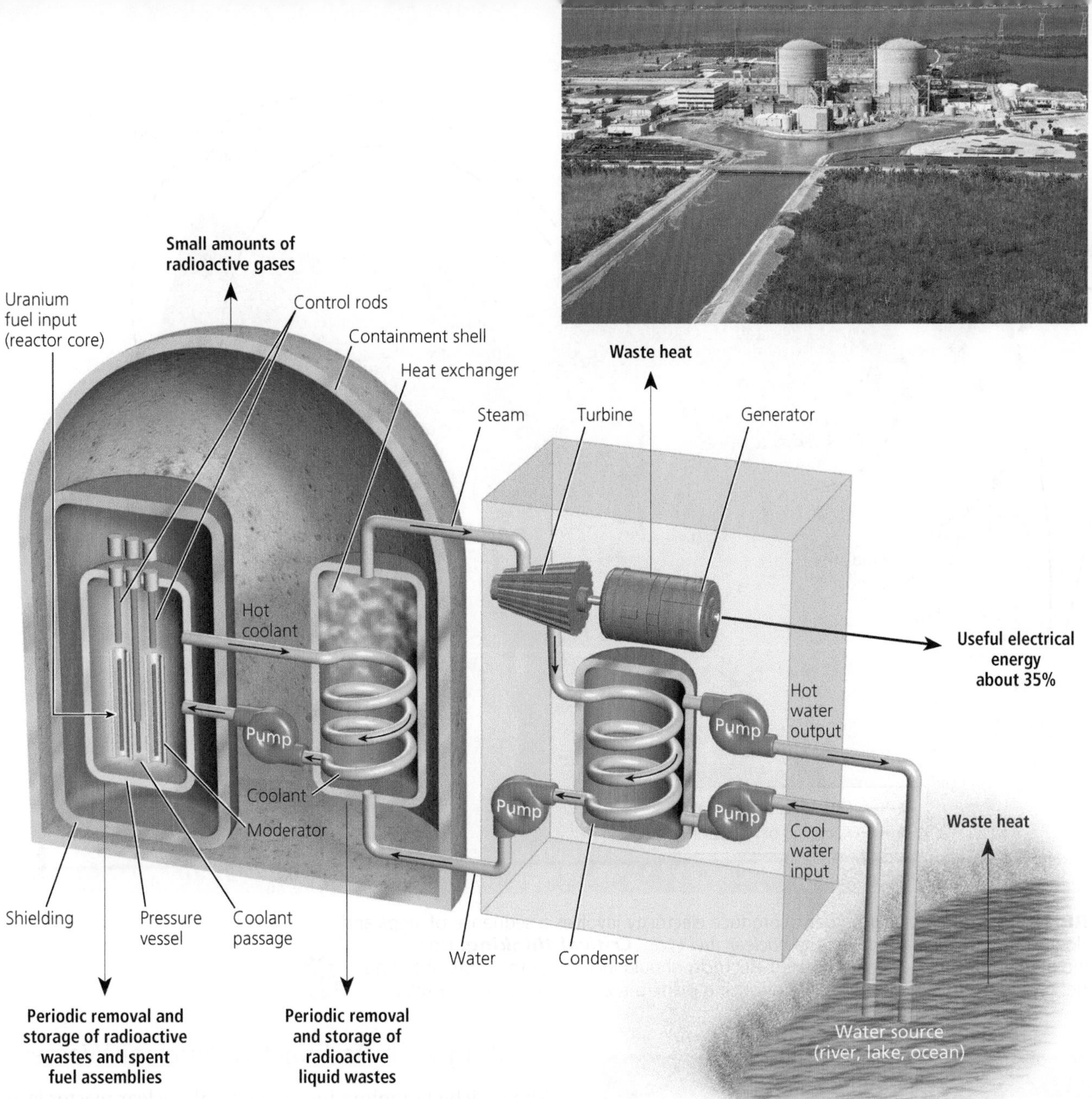

FIGURE 15.21 Nuclear power plant. Intense heat from the nuclear fission of uranium-235 in a chain reaction (Figure 15.20) in the core of this pressurized water-cooled nuclear power plant is transferred to two heat exchangers and converted to steam, which spins a turbine that generates electricity. About 65% of the energy released by the nuclear fission of the plant's uranium fuel is wasted. It ends up as heat that flows into the atmosphere through gigantic cooling towers (shown in the photo) or into water from a nearby source that is used to cool the plant. ***Critical thinking:*** How does this plant differ from the coal-burning plant in Figure 15.13?

The biggest problem with nuclear power is its high cost of building the plant and operating the nuclear fuel cycle, which leads to a low net energy. As a result, nuclear power cannot compete in the marketplace with other energy resources such natural gas, wind, and soon from solar cells unless it is heavily subsidized by governments.

A serious safety and national and global security concern related to commercial nuclear power is the spread of nuclear weapons technology around the world. The United States and 14 other countries have been selling commercial and experimental nuclear reactors and uranium fuel-enrichment and waste reprocessing technology to other countries for decades. Much of this information and equipment can be used to produce bomb-grade uranium and plutonium for use in nuclear weapons. Energy expert John Holdren pointed out that the 60 countries that have nuclear weapons or the knowledge to develop them have gained most of such information by using civilian nuclear power technology. Some critics see this serious threat to global and national security as the single most important reason for not building more nuclear power plants that use uranium-235 or plutonium as a fuel or that produce plutonium-239.

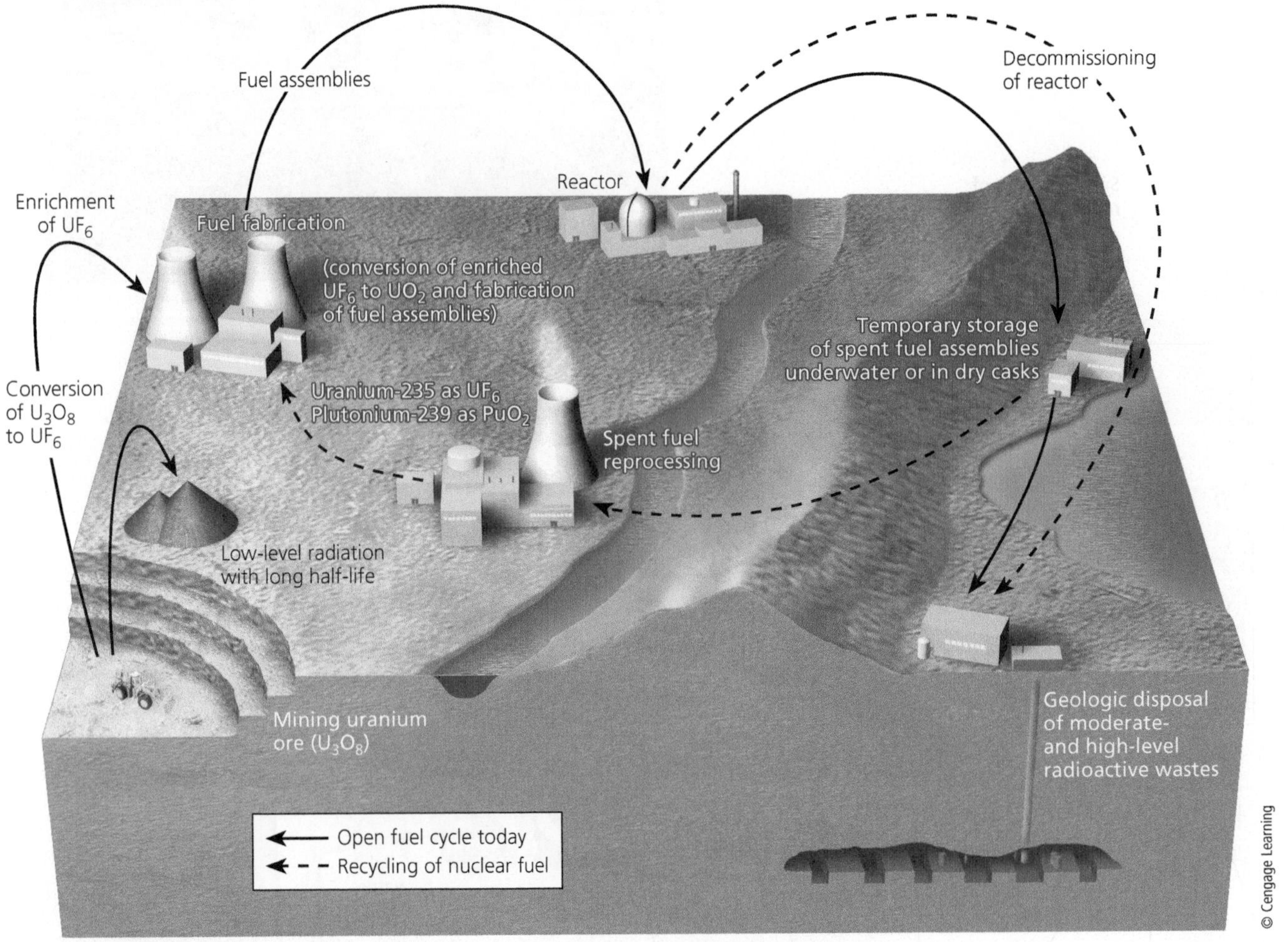

FIGURE 15.22 Using nuclear power to produce electricity involves a sequence of steps and technologies that together are called the *nuclear fuel cycle*. ***Critical thinking:*** Do you think the market price of nuclear-generated electricity should include all the costs of the nuclear fuel cycle, in keeping with the full-cost pricing **principle of sustainability**? Explain.

Trade-Offs

Conventional Nuclear Fuel Cycle

Advantages	Disadvantages
Low environmental impact (without accidents)	Low net energy
	High overall cost
Emits 1/6 as much CO_2 as coal	Produces long-lived, harmful radioactive wastes
Low risk of accidents in modern plants	Promotes spread of nuclear weapons

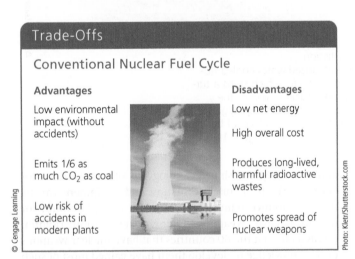

Photo: Klett/Shutterstock.com

© Cengage Learning

FIGURE 15.23 Using the nuclear power fuel cycle (Figure 15.22) to produce electricity has advantages and disadvantages. ***Critical thinking:*** Which single advantage and which single disadvantage do you think are the most important? Why? Do you think that the advantages of using nuclear power outweigh its disadvantages? Explain.

Dealing with Radioactive Nuclear Wastes

The enriched uranium fuel in a typical nuclear reactor lasts for 3–4 years, after which it becomes *spent*, or useless, and must be replaced. The spent-fuel rods are so thermally hot and highly radioactive that they cannot be simply thrown away. Researchers have found that 10 years after being removed from a reactor, a single spent-fuel rod assembly can still emit enough radiation to kill a person standing 1 meter (39 inches) away in less than 3 minutes.

After spent-fuel rod assemblies are removed from reactors, they are stored in *water-filled pools* (Figure 15.24, left). After several years of cooling and decay of some of their radioactivity, they can be transferred to *dry casks* made of heat-resistant metal alloys and concrete and filled with inert helium gas (Figure 15.24, right). These casks are licensed for 20 years and could last for 100 or more years—only a tiny fraction of the thousands of years that the radioactive waste must be safely stored.

Spent nuclear fuel rods can also be processed to remove radioactive plutonium, which can then be used as

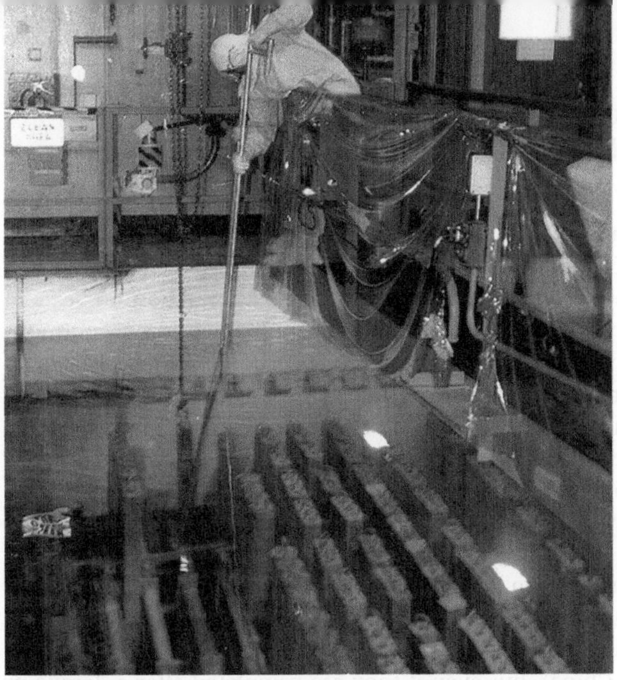

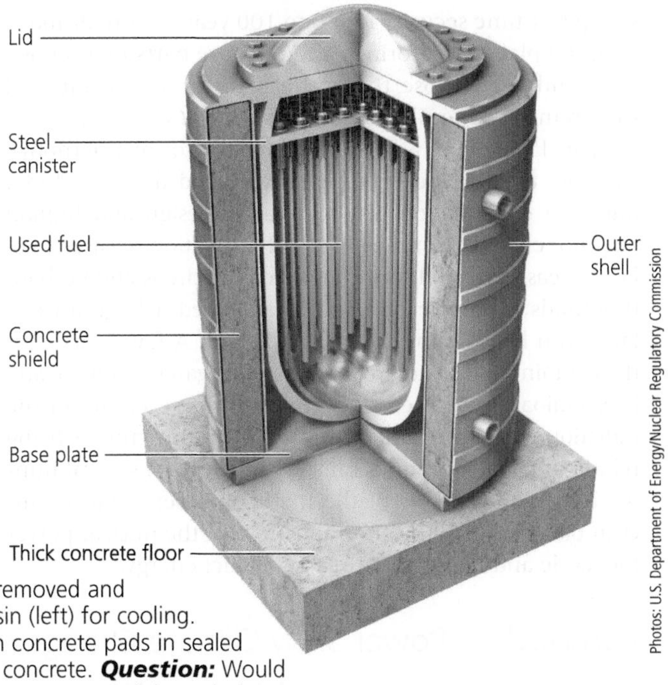

FIGURE 15.24 After 3 or 4 years in a reactor, spent-fuel rods are removed and stored in a deep pool of water contained in a steel-lined concrete basin (left) for cooling. After about 5 years of cooling, the fuel rods can be stored upright on concrete pads in sealed dry-storage casks (right) made of heat-resistant metal alloys and thick concrete. ***Question:*** Would you be willing to live within a block or two of these casks or have them transported through the area where you live in the event that they were transferred to a long-term storage site? Explain.

Photos: U.S. Department of Energy/Nuclear Regulatory Commission

nuclear fuel or for making nuclear weapons, thus closing the nuclear fuel cycle (Figure 15.22). Reprocessing reduces the storage time for the remaining wastes from up to 240,000 years (longer than the current version of the human species has been around) to about 10,000 years.

However, reprocessing is costly and produces bomb-grade plutonium that can be used by nations to make nuclear weapons, as India did in 1974. This is mainly why the U.S. government, after spending billions of dollars, abandoned this fuel recycling approach in 1977. Currently, France, Russia, Japan, India, the United Kingdom, and China reprocess some of their nuclear fuel.

Most scientists and engineers agree in principle that deep burial in an underground repository is the safest and cheapest way to store high-level radioactive wastes for thousands of years. Between 1987 and 2009, the U.S. Department of Energy spent $12 billion on research and testing for a long-term underground nuclear waste storage site in the Yucca Mountain desert region of Nevada. Between 2001 and 2008, the projected cost for completing the repository increased from $58 billion to $96 billion. The project also fell far behind schedule. It was supposed to start accepting wastes in 1998 but this was revised to 2017 and then to 2020. In 2010 this project was abandoned for economic, political, and scientific reasons. These included scientific concerns about earthquake activity. More than 600 earthquakes of magnitude 2.5 and higher occurred in the last 12 years in the repository area.

A government panel is looking for alternative solutions and sites, and some call for completing the Yucca Mountain site. One option being reevaluated is storing the waste in shale rock. Meanwhile, these deadly wastes are building up, with about 78% stored in pools and 22% stored in dry casks (Figure 15.24). There are already enough wastes to fill up the Yucca Mountain repository. After 60 years, no country has come up with a scientific, economic, and politically acceptable solution for storing high-level radioactive wastes for thousands of years.

Meanwhile, as mentioned, the radioactive wastes are being stored indefinitely in pools and dry casks that are not designed to work for more than 20 to 100 years. U.S. electricity consumers have paid over $31 billion (including interest) to a government fund to build a nuclear waste storage site and are still making payments to this fund. Electricity consumers are also paying into a fund to cover the costs of expanding the short-term use of nuclear waste storage casks. Even if all the nuclear power plants in the world were shut down tomorrow, we would still have to find a way to protect ourselves from their high-level radioactive components for thousands of years.

Another costly radioactive waste problem arises when a nuclear reactor reaches the end of its useful life after about 40 to 60 years and must be *decommissioned*. Around the world, 285 of the 390 commercial nuclear reactors operating in 2014 will have to be decommissioned by 2025. In addition, some of the reactors located in coastal areas (including 13 in the United States) could be underwater during this century if the world's sea level continues to rise because of warmer ocean water and melting of land-based ice (glaciers).

Scientists and engineers have proposed three ways to do this: **(1)** remove and store the highly radioactive parts in a permanent, secure repository (which does not yet exist); **(2)** install a physical barrier around the plant and set up full-time security for 30 to 100 years before dismantling the plant and storing its radioactive parts in a repository; and **(3)** enclose the entire plant in a concrete and steel-reinforced tomb, called a *containment structure*.

This last option was used with a reactor at Chernobyl, Ukraine, that exploded and nearly melted down in 1986, due to a combination of poor reactor design and human operator error. The explosion and the radiation released over large areas of several countries killed hundreds and perhaps thousands of people. It also contaminated a large area of land with long-lasting radioactive fallout. A few years after the containment structure was built, it began to crumble and leak radioactive wastes, due to the corrosive nature of the radiation inside the damaged reactor. The structure is being rebuilt at great cost and is unlikely to last even several hundred years. Regardless of the method chosen, retiring nuclear plants adds to the enormous costs of the nuclear power fuel cycle and reduces its already low net energy.

Can Nuclear Power Slow Climate Change?

Nuclear power advocates contend that we could greatly reduce CO_2 emissions that are contributing to climate change by increasing our use of nuclear power. It is often incorrectly reported that nuclear power produces no CO_2 emissions.

While nuclear plants are operating, they do not emit CO_2 but building them and every other step in the nuclear power fuel cycle (Figure 15.22) involves some CO_2 emissions. Such emissions are much lower than those from coal-burning power plants (Figure 15.17) but they still contribute to atmospheric warming and climate change. Nuclear power can help reduce CO_2 emissions, but this would take longer and cost much more than other cheaper and cleaner options such as reducing energy waste and producing electricity from wind and solar energy. According to Mark Lynas, author of several books on climate change, cutting global CO_2 emissions in half between 2015 and 2040 and meeting growing energy needs would require building 12,000 expensive nuclear power reactors (compared to today's 390 reactors, built over the last 60 years)—an average of about one new reactor per day.

Controversy about the Future of Nuclear Power

In the 1950s, researchers predicted that by the year 2000, at least 1,800 nuclear power plants would supply 21% of the world's commercial energy (25% of that in the United States) and most of the world's electricity. After almost 60 years of development, a huge financial investment, and enormous government subsidies, some 390 commercial nuclear reactors (down from 438 reactors in 2005) in 31 countries produced 4% of the world's commercial energy and 15% of its electricity in 2014. In the United States, 99 licensed commercial nuclear power reactors produced about 8% of the country's overall energy and 19% of its electricity in 2014.

Globally in 2015, 67 new nuclear reactors were under construction (24 of them in China) but 49 of them (including 20 in China) were behind schedule, which greatly increases their costs. Another 156 reactors are planned (most of them in China), but even if they are completed after a decade or two, they will not begin to replace the 285 aging reactors that must be decommissioned by 2025. This helps to explain why the production of electricity from nuclear power has stagnated since 2006 (Figure 15.25). Between 2015 and 2050, the world will have to build at least 390 nuclear reactors to produce the same amount of electricity that they produced in 2015. In the United States, electricity from nuclear power has not grown since 2000 and is not expected to grow between 2014 and 2040 because of the costs involved and because electricity can be produced much more quickly and cheaply from natural gas and from wind and solar energy, which emit less CO_2 than the nuclear power fuel cycle emits.

There is controversy over the future of nuclear power. Critics argue that the its two most serious problems are the high cost and low net energy of the nuclear power fuel cycle and its contribution to the spread of technology that can be used to make nuclear weapons. They contend that the nuclear power industry could not exist without high levels of financial support from governments and taxpayers, because of the high cost of ensuring safety and the low net energy of the nuclear fuel cycle.

For example, the U.S. government has provided large research and development subsidies, tax breaks, and loan guarantees to the nuclear industry (with taxpayers accepting the risk of any debt defaults) for more than 50 years. In addition, the government provides

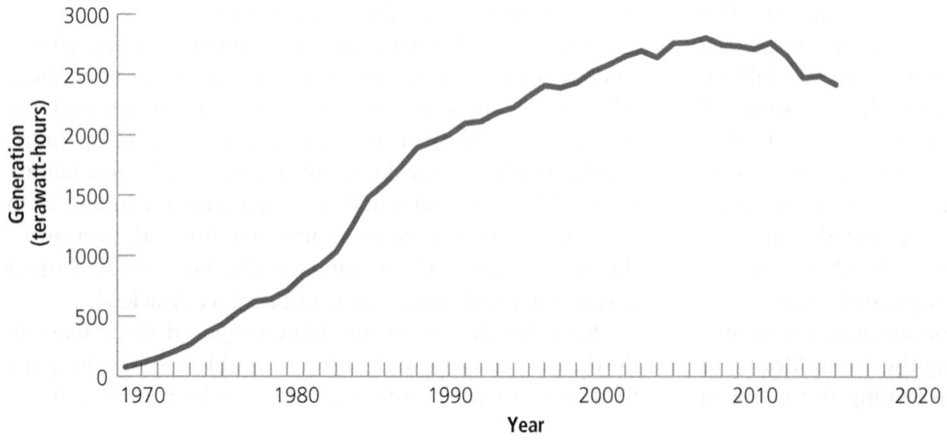

FIGURE 15.25 Global electricity generation from nuclear power, 1970–2014. ***Data analysis:*** By what percentage did electricity produced by nuclear power decrease between 2006 and 2015?

(Compiled by the authors using data from the International Energy Agency, BP, Worldwatch Institute, and Earth Policy Institute)

accident insurance guarantees (under the Price-Anderson Act passed by Congress in 1957), because insurance companies have refused to fully insure any nuclear reactor against the consequences of a catastrophic accident.

According to the nonpartisan Congressional Research Service, since 1948, the U.S. government has spent more than $95 billion (in 2011 dollars) on nuclear energy research and development (R & D)—more than four times the amount spent on R & D for solar, wind, geothermal, biomass, biofuels, and hydropower combined. Many analysts question the need for continuing such taxpayer support for nuclear power, especially since its energy output has not grown for several decades and is unlikely to grow during the next several decades.

CONSIDER THIS . . .

THINKING ABOUT Government Subsidies for Nuclear Power

Do you think the benefits of nuclear power justify high government (taxpayer) subsidies and tax breaks for the nuclear industry? Explain.

Because of the multiple built-in safety features, the risk of exposure to radioactivity from nuclear power plants in the United States and in most other more-developed countries is very low. However, several explosions and partial or complete meltdowns have occurred (see the Case Study that follows). These accidents have dampened public and investor confidence in nuclear power.

Proponents of nuclear power argue that governments should continue funding research, development, and pilot-plant testing of potentially safer and less costly new types of reactors. The nuclear industry claims that hundreds of new *advanced light-water reactors (ALWRs)* could be built in just a few years. ALWRs have built-in safety features designed to make meltdowns and releases of radioactive emissions almost impossible and thus do not need expensive automatic cooling systems. So far, commercial versions of such reactors have not been built or evaluated.

The industry is also evaluating the development of smaller modular light-water reactors—about the size of a school bus—that could be built in a factory, delivered to a site, and installed underground. They need refueling every 5 to 30 years and can quickly be linked together to expand power production as needed. However, because they use fast neutrons instead of slow neutrons to fission uranium-235, the fuel has to be enriched from 1% to 19.9%. This makes it much easier to enrich it to 80–95% for bomb-grade material—worsening the threat of the spread of nuclear weapons. Commercial versions of such rectors have yet to be built and evaluated.

Some scientists call for replacing today's uranium-based reactors with new ones fueled by thorium, a naturally occurring radioactive element found in abundance in the earth's crust throughout the world and that does not need to be enriched like uranium-235. According to scientists, thorium reactors would be much less costly and safer because they cannot melt down. Thorium nuclear plants could be smaller and cheaper per unit of electricity produced than uranium nuclear plants. In addition, the nuclear waste produced by thorium plants could not be used to make nuclear weapons. China and India plan to develop a network of thorium nuclear reactors to produce at least 25% of their electricity by 2050.

Some analysts believe that, in order to be environmentally and economically acceptable, any new-generation nuclear technology should meet the five criteria listed in Figure 15.26. In the United States, even with considerable government subsidies and loan guarantees, most utility companies and money lenders are unwilling to take on the financial risk of building new nuclear plants of any design as long as electricity can be produced more cheaply with the use of natural gas and wind power (and solar cells if solar prices keep falling).

Solutions

- Reactors must be built so that a runaway chain reaction is impossible.
- The reactor fuel and methods of fuel enrichment and fuel reprocessing must be such that they cannot be used to make nuclear weapons.
- Spent fuel and dismantled plants must be easy to dispose of without burdening future generations with harmful radioactive waste.
- Taking its entire fuel cycle into account, nuclear power must generate a net energy high enough so that it does not need government subsidies, tax breaks, or loan guarantees to compete in the open marketplace.
- Its entire fuel cycle must generate fewer greenhouse gas emissions than other energy alternatives.

FIGURE 15.26 Some energy analysts say that any new generation of nuclear power plants should meet these five criteria. ***Critical thinking:*** Do you agree or disagree with these analysts? Explain.

CASE STUDY

The Fukushima Daiichi Nuclear Power Plant Accident in Japan

A nuclear accident occurred on March 11, 2011, at the Fukushima Daiichi Nuclear Power Plant on the northeast coast of Japan. A strong offshore earthquake that caused a severe tsunami (Figure 14.20, p. 376) devastated coastal communities and triggered the nuclear accident. An immense wave of seawater washed over the nuclear plant's protective seawalls and knocked out the circuits and backup diesel generators of the emergency core cooling systems for the plant's three operating reactors. Then, explosions (presumably from the buildup of hydrogen gas produced by the exposed nuclear fuel rods) blew the roofs off three of the reactor buildings and released radioactive materials into the atmosphere and nearby coastal waters.

Evidence indicates that the cores of these three reactors suffered full meltdowns and contaminated a large area with low to moderate levels of radioactivity. In 2013 low-level radioactivity from contaminated groundwater and from one of the plant's 1,000 wastewater storage tanks was leaking into the coastal waters near the plant. In terms of radiation released, this was a local, not a national or global, nuclear accident.

Some 130,000 people were evacuated from the area near the plant and have not been able to return. Although the tsunami killed 15,891 people, no one has died directly from exposure to radioactivity from the nuclear accident. However, eventually, 100 or more people could die from thyroid and other cancers associated with their exposure to radioactivity.

Officials say the costly cleanup and decommissioning of the damaged reactors will take several decades and could cost at least $100 billion. This does not include the costs of decontamination of the surrounding area and compensation for victims of the accident—possibly as high as $400 billion. Officials estimate that it will take three to four decades to remove radioactive materials and wastes from the damaged reactors.

This event damaged the confidence of Japanese citizens in the safety of nuclear power and led the government to shut down all of the country's nuclear reactors but quietly reopened them in 2015. Since then, Japan has relied more on imports of expensive liquefied natural gas (LNG) and cheaper coal to produce electricity. However, Japan has replaced half of the energy it got from nuclear power with energy saved through improvements to energy efficiency.

This serious but not major nuclear power accident (compared to Chernobyl) had five important effects:

- It increased public fear throughout the world about the use of nuclear power to produce electricity.
- It revealed that a single accident can add an estimated $500 billion to the already high cost of nuclear power.
- It led Germany, Switzerland, and Belgium to announce plans for phasing out nuclear power and shifting to increased use of wind and solar energy to produce electricity.
- It increased the exposure of Japanese citizens to air pollution from the use of coal.
- It spurred Japan to reduce its energy use by cutting energy waste and improving energy efficiency.

Is Nuclear Fusion the Answer?

Other proponents of nuclear power hope to develop **nuclear fusion**—in which the nuclei of two isotopes of a light element such as hydrogen are forced together at extremely high temperatures until they fuse to form a heavier nucleus, releasing energy in the process (Figure 15.27). The sun uses nuclear fusion to warm the earth and sustain its life. Some scientists hope that controlled nuclear fusion on the earth will provide an almost limitless source of energy.

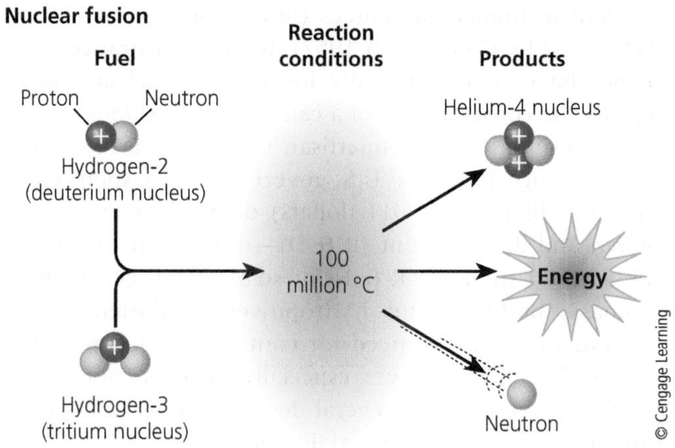

FIGURE 15.27 *Nuclear fusion:* After more than 50 years of research and a $25 billion investment, controlled nuclear fusion is still in the laboratory stage.

With nuclear fusion, there would be no risk of a meltdown or of a release of large amounts of radioactive materials, and little risk of the additional spread of nuclear weapons. Fossil fuels would not be needed to produce electricity, thereby eliminating most of the earth's air pollution and climate-changing CO_2 emissions. Fusion power might also be used to destroy toxic wastes and to supply electricity for desalinating water and for decomposing water to produce clean-burning hydrogen fuel.

However, in the United States, after more than 50 years of research and a $25 billion investment (mostly by the government), controlled nuclear fusion is still in the laboratory stage. None of the approaches tested so far has produced more energy than they used. In 2006 the United States, China, Russia, Japan, South Korea, India, and the European Union agreed to spend at least $12.8 billion in a joint effort to build a large-scale experimental nuclear fusion reactor by 2026 to determine if it can have an acceptable net energy at an affordable cost. By 2014 the estimated cost of this project had more than doubled and it was far behind schedule. Unless there is an unexpected scientific breakthrough, some skeptics say that nuclear fusion is the power of the future—and always will be.

BIG IDEAS

- A key factor to consider in evaluating the long-term usefulness of any energy resource is its net energy.
- Conventional oil, natural gas, and coal are plentiful and have a moderate to high net energy, but use of these fossil fuels, especially coal, has a high environmental impact.
- The nuclear power fuel cycle has a low environmental impact and a low accident risk, but high costs, a low net energy, long-lived radioactive wastes, and its role in spreading nuclear weapons technology have limited its use.

Tying It All Together

Fracking, Nonrenewable Energy, and Sustainability

Alfred Estes, Colorado School of Mines

In the **Core Case Study** that opened this chapter, we looked at how two technologies—horizontal drilling and fracking (see photo)—have greatly increased the production of oil and natural gas in the United States. Later we discussed some problems with fracking and the overall process of extracting oil and natural gas from underground deposits, and we explored ways to improve this process.

We also learned that the long-term usefulness of any energy resource depends on its *net energy*. The net energy of conventional oil, natural gas, and coal is medium to high but will decrease as we use up their easily accessible supplies.

Because of their availability and advantages, we will continue relying on fossil fuels. However, in the long term, this reliance violates the chemical cycling and biodiversity **principles of sustainability**. The technologies we use to obtain energy from fossil fuels disrupt the earth's chemical cycles by diverting huge amounts of water and emitting large quantities of climate-changing greenhouse gases and other air pollutants. Using these technologies also destroys and degrades biodiversity and ecosystem services.

We learned that using nuclear power is costly because the net energy of the nuclear fuel cycle is low, and that its use has spread the ability of countries to produce nuclear weapons.

The challenge is to find ways to reduce the harmful environmental impacts of nonrenewable fossil fuels and nuclear energy. One way to do this is implement the full-cost pricing **principle of sustainability** by including the harmful environmental and health costs of using these resources in their market prices. This could be accomplished with the aid of compromise and trade-offs in the political arena by applying the win-win **principle of sustainability**. Another option is to greatly reduce the enormous amount of energy that we waste and to gradually shift to a mix of less environmentally harmful renewable energy resources, as discussed in the next chapter. Moving in this direction applies the ethical **principle of sustainability** by leaving the planet's life-support systems in a condition as good as or better than what we inherited for future generations.

Chapter Review

Core Case Study

1. What is **horizontal drilling** and **hydraulic fracturing (fracking)** and how are they used to produce oil and natural gas?

Section 15.1

2. What are the two key concepts for this section? What types of commercial energy resources do we use to supplement energy from the sun? Distinguish between nonrenewable and renewable energy resources and give three examples of each type. What percentage of the energy used in the world and in the United States comes from each of these types of energy resources?

3. Define **net energy** and **net energy ratio** and explain why they are important in evaluating energy resources. Explain why some energy resources need help in the form of subsidies to compete in the marketplace, and give an example. What is **energy density**? What two energy resources have the highest energy density? What is a limitation of using energy density to evaluate energy resources?

Section 15.2

4. What are the two key concepts for this section? What is **crude oil (petroleum)** and how are oil deposits detected and removed? What percentages of the commercial energy used in the world and in the United States are provided by conventional crude oil? What is **peak production** for an oil well or oil field? What is **refining**? What are **petrochemicals** and why are they important?

5. What are **proven oil reserves**? Why has the global net energy for extracting oil been dropping? Is there a global shortage of oil? What three countries have the world's largest proven oil reserves? Which three

are the largest producers of oil? Which three are the largest consumers of oil? What percentage of the world's proven oil reserves does the United States and China have? What is OPEC? Summarize the story of oil production and consumption in the United States. What major factors could increase or decrease oil production in the United States? What are the major advantages and disadvantages of using crude oil as an energy resource?

6. What is **shale oil** and how is this heavy oil produced? What are **oil sands**? What is bitumen, and how is it extracted and converted to heavy oil? What are the major advantages and disadvantages of using shale oil and heavy oils produced from oil shale rock and from oil sands?

Section 15.3

7. What is the key concept for this section? Define **natural gas**, **liquefied petroleum gas (LPG)**, and **liquefied natural gas (LNG)**. What three countries have the world's largest proven natural gas reserves? Which three are the largest producers of natural gas? Which three are the largest consumers of natural gas? Why has natural gas production risen sharply in the United States and what two factors could hinder this rise? What are the major advantages and disadvantages of using natural gas as an energy resource? Describe four problems resulting from increased use of fracking to produce natural gas in the United States and four ways to deal with these problems. Discuss whether increased use of natural gas can slow or speed up global climate change.

Section 15.4

8. What are the two key concepts for this section? What is coal, how is it formed, and how do the various types of coal differ? How does a coal-burning power plant work? What three countries have the world's largest proven coal reserves? Which three are the largest producers of coal? Which three are the largest consumers of coal? Summarize the major environmental and health problems caused by the use of coal. What would happen if coal's harmful health and environmental effects were included in its market price? What are the major advantages and disadvantages of using coal as an energy resource? What is **synthetic natural gas (SNG)** and what are the major advantages and disadvantages of using it as an energy resource? What is carbon capture and sequestration (CCS)?

Section 15.5

9. What is the key concept for this section? What is nuclear fission? How does a nuclear fission reactor work and what are its major safety features? Describe the nuclear fuel cycle. What three countries are the three leading users of nuclear power? What percentage of the electricity generated in the United States comes from nuclear power? What is the relationship between nuclear power plants and the spread of nuclear weapons? Explain how highly radioactive spent-fuel rods are stored and what risks this presents. How has the United States dealt with the nuclear waste problem? What can we do with worn-out nuclear power plants? Summarize the arguments over whether or not the widespread use of nuclear power could help to slow projected climate change during this century. Summarize the arguments of experts who disagree over the future of nuclear power. Describe the Fukushima Daiichi nuclear power plant accident and list its five major effects. What is **nuclear fusion** and what is its potential as an energy resource?

10. What are this chapter's *three big ideas*? Explain how we can apply each of the six **principles of sustainability** to reducing the harmful effects of relying on fossil fuels and nuclear energy.

Note: Key terms are in bold type.

Critical Thinking

1. How might greatly increased production of domestic oil and natural gas in the United States over the next two decades (**Core Case Study**) affect the country's future use of coal, nuclear power, and energy from the sun and wind? How might such an increase affect your life during the next 20 years?

2. Should governments give a high priority to considering net energy when deciding what energy resources to support? What are other factors that should be considered? Which factor or factors should get the most weight in deciding what energy resources to use? Explain your thinking.

3. Some analysts argue that in order to continue using oil at the current rate, we must discover and add to global oil reserves the equivalent of two new Saudi Arabian reserves every 7 years. Do you think this is possible? If not, what effects might the failure to find such supplies have on your life and on the lives of any children and grandchildren that you might have?

4. List three things you could do to reduce your dependence on oil and gasoline. Which of these things do you already do or plan to do?

5. Are you for or against the increased use of horizontal drilling and fracking (**Core Case Study**) to produce oil and natural gas in the United States? Explain. What are the alternatives?

6. Are you for or against phasing out the use of coal in the United States by 2050? State your three strongest arguments to explain your position. How might this affect your life and the lives of any children and grandchildren that you might have? What are the alternatives?

7. Are you for or against greatly increasing the use of nuclear power to produce electricity? State your three strongest arguments to explain your position.

8. Are you for or against providing government subsidies and tax breaks to producers of **(a)** oil, **(b)** natural gas, **(c)** coal, and **(d)** electricity from nuclear power? Explain. What are the alternatives?

Doing Environmental Science

Do a survey of energy use at your school, based on the following questions: How is the electricity generated? If it comes from more than one source (such as coal-fired, gas-fired, or nuclear power plants, and wind farms), determine or estimate the percentages of the total for the various sources. Similarly, determine how most buildings are heated, how water is heated, how vehicles are powered, and how the computer network is powered. Write a report summarizing your findings.

Global Environment Watch Exercise

Go to your MindTap course to access the GREENR database. At the top of the page do a "Basic Search" on *fracking.* Use the results to find information on the latest developments in public opposition to, and regulation of, fracking in the United States. Write a report on your findings.

Data Analysis

Use the graph below, comparing U.S. oil consumption and production, to answer the questions that follow.

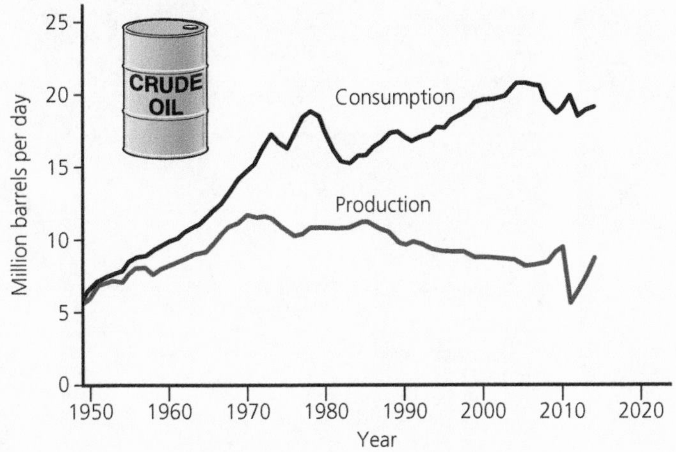

(Compiled by the authors using data from the U.S. Energy Information Agency (EIA))

1. By what percentage did U.S. oil consumption increase between 1982 and 2014?

2. By what percentage did U.S. oil consumption decrease between 2005 and 2011?

3. By what percentage did U.S. oil production decrease between 1985 and 2004?

4. By what percentage did U.S. oil production increase between 2011 and 2014?

What general conclusions can you draw from these data? Compare notes and discuss your conclusions with your classmates.

CENGAGE brain.com For access to MindTap and additional study materials visit www.cengagebrain.com.

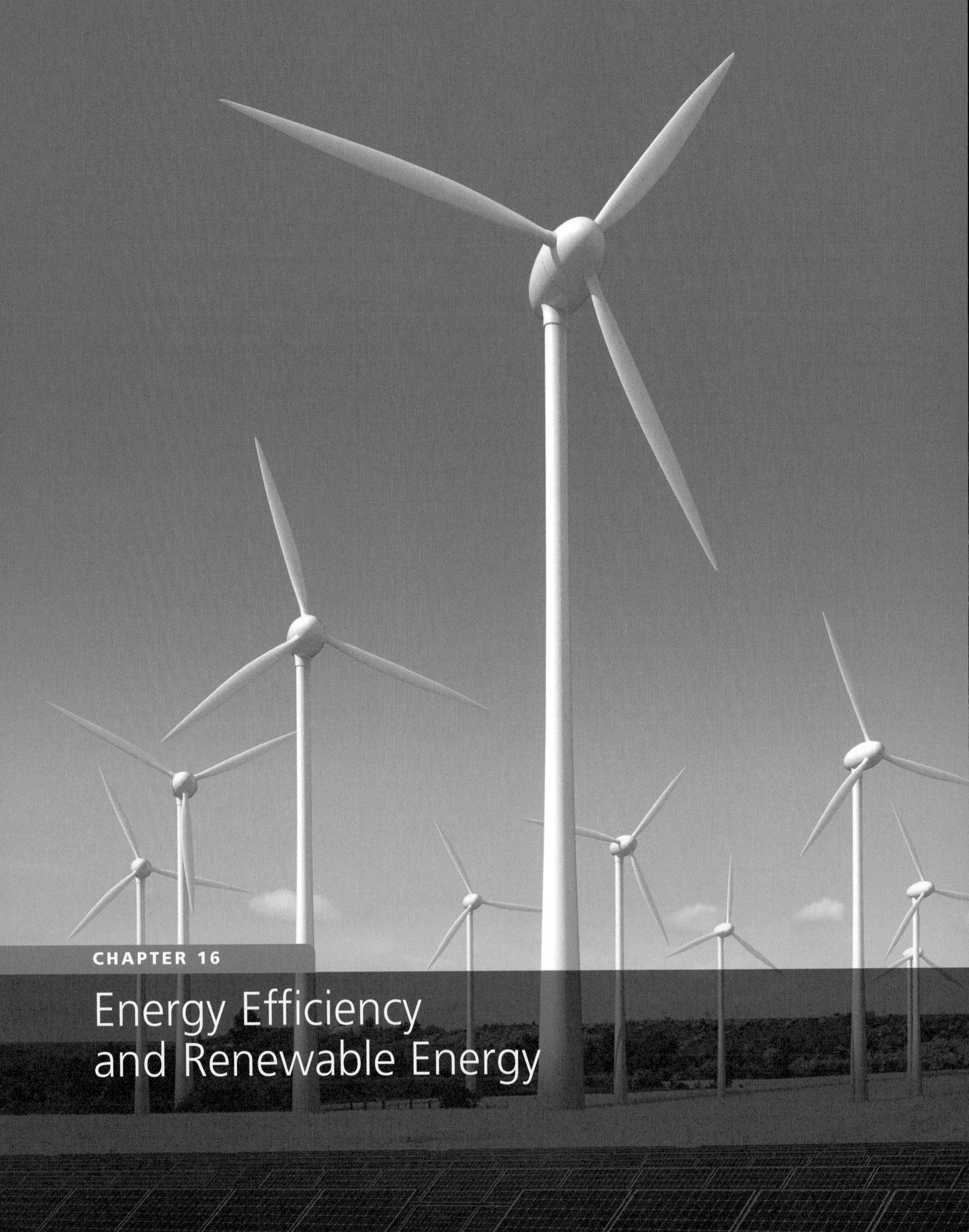

CHAPTER 16

Energy Efficiency and Renewable Energy

> Coal plant smokestacks that dirty the air and alter the climate will be replaced by solar panels on our rooftops and wind turbines turning gracefully in the distance. Welcome to the clean energy era.
>
> LESTER R. BROWN

Key Questions

16.1 Why do we need a new energy transition?

16.2 Why is improving energy efficiency and reducing energy waste an important energy resource?

16.3 What are the advantages and disadvantages of using solar energy?

16.4 What are the advantages and disadvantages of using wind power?

16.5 What are the advantages and disadvantages of using geothermal energy?

16.6 What are the advantages and disadvantages of using biomass as an energy source?

16.7 What are the advantages and disadvantages of using hydropower?

16.8 What are the advantages and disadvantages of using hydrogen as an energy source?

16.9 How can we make the transition to a more sustainable energy future?

Wind turbines and solar cell panels in a rapeseed field.

visdia/Shutterstock.com

Core Case Study

Saving Energy and Money and Reducing Our Environmental Impact

Do you like to burn dollar bills? Do you like wasting 15–30% of the money you spend on electricity, gasoline, and heating and cooling bills? About 43% of the money that Americans spend each year on energy is unnecessarily wasted and provides no useful energy. In addition to burning up mountains of $100 bills, this waste of energy increases our harmful environmental impact.

Some do this by driving cars with engines that waste 80% of the money they spend on gasoline. Only about 20% of the money we spend on gasoline, on average, gets us somewhere. Many also waste energy and money by living in leaky houses and buildings. The left photo in Figure 16.1 is an infrared shot of a home that is poorly insulated and has heat and money leaking out through its doors, windows, walls, and roof. Spending a small amount of money insulating the walls and roof and caulking (sealing) leaks around the windows and doors can drastically reduce this loss of heat and money, as shown in the photo on the right. Many homes in the United States and other countries are so full of leaks that their heat loss in cold weather and heat gain in hot weather are equivalent to having a large, window-sized hole in a wall of the house.

Would you like to have an investment that will earn 25% a year without any risk or taxes, compared to a bank savings account with a 0.1% annual return? Do what the owner of the house in Figure 16.1 did. If you live in a typical home and you could invest $2,000 in insulation and $25 in caulking all the air leaks around doors and windows and in your ceiling and roof, you could reduce your heating and cooling bill by 15–30%. The caulking immediately would start earning you a 100% return on your investment. It would take about 4 years to earn back the money you spent on insulation. After that, you would be earning a 25% annual return on your investment as long as you lived in the house. (Interest rate = 100%/payback period = 100/4 = 25%.) As an added bonus, you could reduce your input of climate-changing CO_2 by about 2.1 metric tons (2.4 tons) a year.

Investing in insulating a house and sealing air leaks is not exciting. But the money you can save is, and you can also reduce your harmful environmental impacts from heating and cooling your house. It is a win-win deal for you and the environment.

In this chapter, we consider the importance of using energy more efficiently and wasting less energy. We also explore ways to make the transition to a more sustainable energy future during this century by shifting from nonrenewable fossil fuels to renewable energy resources such as solar energy, wind power, flowing water, and the earth's internal heat.

Science Source

FIGURE 16.1 A thermogram, or infrared photo, of a house that is poorly insulated and poorly sealed and thus loses heat (reddish areas) and wastes money. ***Critical thinking:*** How do you think the place where you live would compare to this house in terms of heat loss?

16.1 WHY DO WE NEED A NEW ENERGY TRANSITION?

CONCEPT 16.1 The world is in the early stages of a transition from relying on fossil fuels to greater reliance on energy efficiency and renewable energy.

Establishing New Energy Priorities

Shifting to new energy resources is not new. The world has shifted from primary dependence on wood to coal, then from coal to oil, and then to our current dependence on oil, natural gas, and coal as new technologies made these resources more available and affordable. Each of these shifts in key energy resources took about 50 to 60 years. As in the past, it takes an enormous investment in science and engineering, research, technology, and infrastructure to develop and spread the use of new energy resources.

Currently, the world and the United States get about 90% of the commercial energy they use from three carbon-containing fossil fuels—oil, coal, and natural gas (Figure 15.2, p. 383). They have supported tremendous economic growth and improved the lives of many people.

However, we are awakening to the fact that burning fossil fuels, especially coal, is largely responsible for the world's three most serious environmental problems: air pollution, climate change, and ocean acidification. Fossil fuels are affordable because their market prices do not include these and other harmful health and environmental effects.

According to many scientists, energy experts, and energy economists, over the next 50 to 60 years and beyond, we need to make a new energy transition by **(1)** improving energy efficiency and reducing energy waste, **(2)** decreasing our dependence on nonrenewable fossil fuels, and **(3)** relying more on a mix of renewable energy from the sun, wind, the earth's interior heat (geothermal energy), flowing water (hydropower), and biomass (wood and biofuels). We also need to modernize the electrical grids used to transmit electricity produced at fossil fuel and solar power plants and wind farms (see chapter-opening photo).

We can reduce fossil fuel use, but fossil fuels are not going to disappear. In 2065, fossil fuels are projected to provide at least 50% of the world's energy, compared to 90% today, according to the International Energy Agency. The use of coal is likely to decline the most because of its harmful environmental effects and its key role in accelerating climate change and ocean acidification.

Supporters of this restructuring of the global energy system and economy over the next 50 to 60 years and beyond project that it will save money, create profitable business and investment opportunities, and provide jobs. They also project that it will save lives by sharply reducing air pollution, help keep climate change from spiraling out of control and creating ecological and economic chaos, and slow the increase in ocean acidity. Finally, supporters project that this shift will increase our positive environmental impact, and pass the world on to future generations in better shape than we found it.

For four decades, fossil fuel industries and utility companies have been using their economic and political power to slow this energy shift, especially in the United States. Eventually, however, economists expect fossil fuel businesses to fade as cheaper and cleaner energy alternatives emerge, which is how creative capitalism works.

This energy shift is being driven by the availability of perpetual supplies of clean and increasingly cheaper solar and wind energy throughout the world and advances in solar cell and wind turbine technology. This is in contrast to fossil fuels, which are dependent on finite supplies of oil, coal, and natural gas that are not widely distributed, are controlled by a few countries, and are subject to fluctuating prices based on supply and demand.

In this new technology-driven energy economy, an increasing percentage the world's electricity will be produced locally from available sun and wind and regionally from solar cell power plants and wind farms and transmitted to consumers by a modern, interactive, smart electrical grid. Homeowners and businesses with solar panels on their roofs or land will become independent electricity producers. They will be able to heat and cool their homes and businesses, run electrical devices, charge hybrid or electric cars, and sell any excess electricity they produce. The United States will benefit economically, because making such a market-based shift will set off an explosion of increased energy efficiency and renewable energy and battery technological innovations by tapping into the country's renowned ability to innovate.

Like any major societal change, this shift will not be easy. However, to many analysts the harmful environmental, health, and economic effects of not making this shift far outweigh the environmental, health, and economic benefits of making this shift.

This shift is underway and gaining momentum as the cost of electricity produced from the sun and wind continues its rapid fall and investors see a way to make money on two of the world's fastest growing businesses. Between 2009 and 2015, the cost of generating electricity with solar energy fell by 82% and with wind by 61%. Germany (see the Case Study that follows), Sweden, Denmark, and Costa Rica have made the most progress in this shift, and Japan leads the world in reducing energy waste. For example, Costa Rica gets more than 90% of its electricity from renewable resources, mostly hydropower, geothermal energy, and wind. It aims to generate all of its electricity from renewable sources by 2021, mostly by increasing its use of geothermal energy. GOOD NEWS

In the United States, solar and wind power use is increasing and coal use has dropped slightly, mostly because of increased use of affordable domestic supplies of natural gas to produce electricity. However, the United States has yet to commit to making the new energy shift, partly because of almost four decades of opposition by politically and economically powerful fossil fuel and electric utility companies. In addition, increased reliance on using cheap natural gas to produce electricity hinders the shift to wind, solar, and other renewable energy resources.

CASE STUDY

Germany Is a Renewable Energy Superpower

Germany is phasing out nuclear power and between 2006 and 2014, increased the percentage of its electricity produced by renewable energy, mostly from the sun and wind, from 6% to 28%. Its goals for 2050 are to get 60% of its electricity from renewable energy, increase electricity efficiency by 50%, reduce its current emissions of climate-changing CO_2 by 80% to 95%, and sharply reduce its use of coal, which currently provides 40% of its electricity, as a backup energy resource for solar and wind power.

The biggest factor in this shift is Germany's use of a **feed-in-tariff** (also used by Great Britain and 47 other countries). Under a long-term contract, it requires utilities to buy electricity produced by homeowners and businesses from renewable energy resources at a price that guarantees a good return and to feed it into the electrical grid.

German households now own and make money from roughly 80% of the country's solar cell installations. The feed-in-tariff program is financed by all electricity users and costs less than $5 a month per household. It has also created more than 370,000 new renewable energy jobs. In addition, the German government and private investors have subsidized research on technological improvements in solar and wind power and backup energy storage systems that can further lower costs. However, political opposition by Germany's coal industry and some electrical utility companies and the need for costly backup power may slow the country's shift to greater dependence on solar and wind power.

16.2 WHY IS IMPROVING ENERGY EFFICIENCY AND REDUCING ENERGY WASTE AN IMPORTANT ENERGY RESOURCE?

CONCEPT 16.2A Improvements in energy efficiency and reducing energy waste could save at least a third of the energy used in the world and up to 43% of the energy used in the United States.

CONCEPT 16.2B We have a variety of technologies for sharply increasing the energy efficiency of industrial operations, motor vehicles, appliances, and buildings.

We Waste a Lot of Energy and Money

Improving energy efficiency and conserving energy are key strategies in using energy more sustainability. **Energy efficiency** is a measure of how much useful work we can get from each unit of energy we use. Improving energy efficiency means using less energy to provide the same amount of work. No energy-using device operates at 100% efficiency; some energy is always lost as heat, as required by the second law of thermodynamics (p. 42). However, there are ways to improve energy efficiency and waste less energy by using of more fuel-efficient cars, lights, appliances, computers, and industrial process.

We can also cut energy waste by changing our behavior. **Energy conservation** means reducing or eliminating the unnecessary wasting of energy. If you ride your bicycle to school or work rather than driving a car, you are practicing energy conservation. Another way to conserve energy is to turn off lights and electronic devices when you are done using them.

You may be surprised to learn that roughly 84% of all commercial energy used in the United States is wasted (Figure 16.2). About 41% of this energy unavoidably ends up as low-quality waste heat in the environment because of the degradation of energy quality imposed by the second law of thermodynamics. The other 43% is wasted unnecessarily, mostly due to the inefficiency of industrial motors,

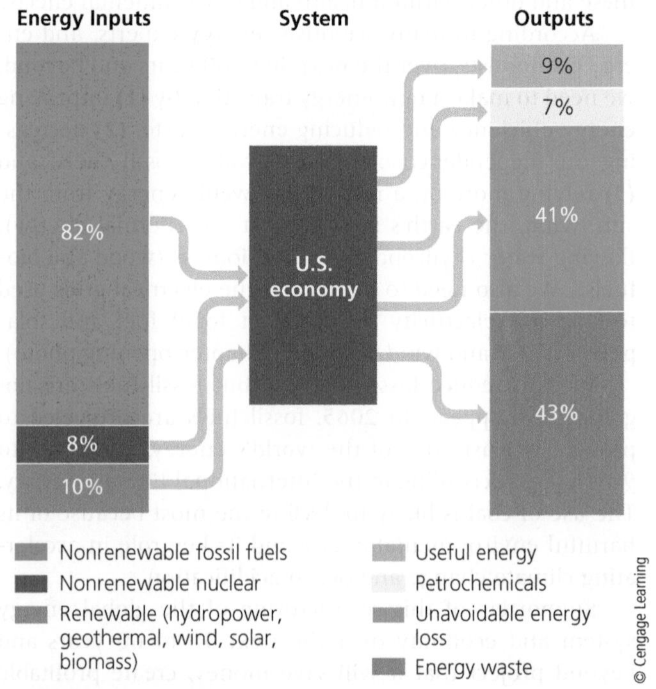

FIGURE 16.2 Flow of commercial energy through the U.S. economy. Only 16% of the country's high-quality energy ends up performing useful tasks. ***Critical thinking:*** What are two examples of unnecessary energy waste?

(Compiled by the authors using data from U.S. Department of Energy.)

motor vehicles, power plants, light bulbs, and numerous other devices (**Concept 16.2A**).

43% Percentage of energy used in the United States that is unnecessarily wasted

Another reason for our costly and wasteful use of energy is that many people live and work in poorly insulated, leaky houses and buildings that require excessive heating during cold weather and excessive cooling during hot weather (see Figure 16.1). In addition, many Americans (and people in other parts of the world) live in ever-expanding suburban areas around large cities where they must depend on cars for getting around. Three of every four Americans commute to work, mostly alone in energy-inefficient vehicles, and only 5% rely on more energy-efficient mass transit. In addition, many people waste energy and money by buying inefficient appliances.

Much of this loss of energy and money is due to heavy reliance on three widely used energy-inefficient technologies:

- Huge *data centers*, filled with racks of electronic servers that process all online information and provide cloud-based data storage for users. Typically, they use only 10% of the electrical energy they consume. The other 90% is lost as heat. Most data centers run 24 hours a day at their maximum capacities, regardless of the demand. They also require large amounts of energy for cooling to keep the servers from overheating.
- The *internal combustion engine*, which propels most motor vehicles, wastes about 80% of the high-quality energy in its gasoline fuel. Thus, only about 20% of the money people spend on gasoline provides them with transportation. The other 80% of the energy released by burning gasoline ends up as waste heat in the atmosphere.
- *Nuclear, coal,* and *natural gas power plants*, which turn their fuel into one-third electricity and two-thirds waste heat that flows into the atmosphere or into waterways. Typically, only about a third of your electricity bill is actually for electricity.

Improvements in energy efficiency and conservation have numerous economic, health, and environmental benefits (Figure 16.3). To most energy analysts, *they are the quickest, cleanest, and usually the cheapest ways to provide more energy, reduce pollution and environmental degradation, and slow climate change and ocean acidification.*

Solutions

Improving Energy Efficiency

- Prolongs fossil fuel supplies
- Reduces oil imports and improves energy security
- Very high net energy
- Low cost
- Reduces pollution and environmental degradation
- Buys time to phase in renewable energy
- Creates local jobs

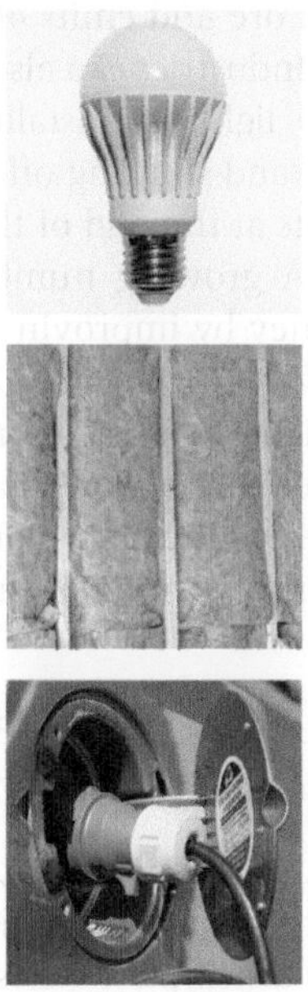

FIGURE 16.3 Improving energy efficiency and conserving energy can have important benefits. ***Critical thinking:*** Which two of these benefits do you think are the most important? Why?

Top: Dmitriy Raykin/Shutterstock.com. Center: V. J. Matthew/Shutterstock.com. Bottom: andrea lehmkuhl/Shutterstock.com.

Improving Energy Efficiency in Industries and Utilities

The industrial sector includes all facilities and equipment used to produce, process, or assemble goods. Industries that use the most energy are those that produce petroleum, chemicals, cement, steel, aluminum, and paper and wood products.

One way for industries and utility companies to save energy is to use **cogeneration** to produce two useful forms of energy—electricity and heat—from the same fuel source. For example, the steam used for generating electricity in a power or industrial plant can be captured and used again to heat the plant or other nearby buildings, rather than being released into the environment as waste heat. The energy efficiency of these systems is 60–80%, compared to 25–35% for coal-fired and nuclear power plants. Denmark leads the world by using cogeneration to produce 53% of its electricity compared to 12% in the United States.

Inefficient motors account for 60% of the electricity used in U.S. industry. Industries can save energy and money by using more energy-efficient variable-speed electric motors that run at the minimum speed needed for each job. In contrast, standard electric motors run at full speed with their output throttled to match the task. This is somewhat like using one foot to push the gas pedal to the floorboard of your car and putting your other foot on the brake pedal to control its speed.

Recycling materials such as steel and other metals can also help industries save energy and money. For example, producing steel from recycled scrap iron uses 75% less high-quality energy than does producing steel from virgin iron ore and emits 40% less CO_2.

Industries can also save energy by using energy-efficient LED lighting; installing smart meters to monitor energy use; and shutting off computers, printers, and nonessential lights at the end of the work day.

A growing number of major corporations are saving money by improving energy efficiency. For example, GOOD NEWS between 1990 and 2014, Dow Chemical Company, which operates 165 manufacturing plants in 37 countries, saved $27 billion in a comprehensive program to improve energy efficiency. Ford Motor Company saves $1 million a year by turning off unused computers.

Building a Smarter and More Energy-Efficient Electrical Grid

In the United States, electricity is delivered to consumers through an electrical grid. This network of transmission and distribution lines carries electricity from power plants, wind farms, and other electricity producers to homes, schools, offices, and other end users.

The U.S. electrical grid system, designed more than 100 years ago, is inefficient and outdated. According to former U.S. energy secretary Bill Richardson, "We're a major superpower with a third-world electrical grid system."

Work is underway to convert and expand the outdated U.S. electrical grid system into a *smart grid*. The new grid would be a digitally controlled, ultra-high-voltage (UHV), and high-capacity system with superefficient transmission lines. It will be less vulnerable to power outages, connect to wind farms and solar power plants throughout the country, be more energy efficient, and help save consumers money.

Such a grid could quickly adjust for a major power loss in one area by automatically rerouting electricity from other parts of the country. It would also enable power companies and consumers to buy electricity produced from wind, solar, and other renewable forms of energy in areas where they are not directly available.

According to the U.S. Department of Energy (DOE), building such a grid would cost the United States up to $800 billion over the next 20 years. However, it would save the U.S. economy $2 trillion during that period. Much of these savings would go to consumers with smart electricity meters that would enable them to track how much electricity they use, when they use it, and what its cost is. They could use this information to limit electricity use during times when rates are higher.

The two fastest growing energy resources in the world and in the United States are solar and wind energy used to produce electricity. However, this growth will be limited unless wind farms and solar cell power plants built in sparsely populated areas or at sea can be connected to a smart grid. A national network of wind farms and solar cell power plants would make the sun and wind reliable sources of electricity around the clock. Without such a grid, the United States will not reap the environmental and economic advantages of relying on the sun and wind to produce much of its electricity.

Making Transportation More Energy-Efficient

In 1975, the U.S. Congress established Corporate Average Fuel Economy (CAFE) standards to improve the average fuel economy of new cars and light trucks, vans, and sport utility vehicles (SUVs) in the United States. Between 1973 and 2013, these standards increased the average fuel economy for such vehicles in the United States from 5 kilometers per liter, or *kpl* (11.9 miles per gallon, or *mpg*) to 10.6 kpl (24.9 mpg). The government goal is for such vehicles to get 23.3 kpl (54.5 mpg) by 2025. Existing fuel economy standards for new vehicles in Europe, Japan, China, and Canada are much higher than this proposed 2025 U.S. standard.

Energy experts project that by 2040, all new cars and light trucks sold in the United States could get more than 43 kpl (100 mpg) using available technology. They call for such government standards as an important way reduce energy waste, cut air pollution, and slow climate change and ocean acidification.

Governments may be pushing for more fuel-efficient vehicles, but consumers do not always buy such vehicles, especially when gasoline prices fall. Most consumers are unaware that gasoline costs them much more than the price they pay at the pump. A number of *hidden gasoline costs* not included in the price of gasoline include government subsidies and tax breaks for oil companies, car manufacturers, and road builders; defense spending to secure access to Middle East oil supplies; costs of pollution control and cleanup; and higher medical bills and health insurance premiums resulting from illnesses caused by air and water pollution from the production and use of motor vehicles. The International Center for Technology Assessment estimated that the hidden costs of gasoline for U.S. consumers amount to $3.18 per liter ($12.00 per gallon).

One way to include more of these hidden costs in the market price is through higher gasoline taxes. This would implement the full-cost pricing **principle of sustainability**. However, higher gas taxes are politically unpopular in the United States. Some analysts call for increasing U.S. gasoline taxes and reducing payroll and income taxes to balance such increases, thereby relieving consumers of any additional financial burden. Another way for governments to encourage higher energy efficiency in transportation is to give consumers significant tax breaks or other economic incentives to encourage them to buy more fuel-efficient vehicles.

Other ways to save energy and money in transportation include building or expanding mass transit systems within cities, constructing high-speed rail lines between cities (as is done in Japan, much of Europe, and China), and carrying more freight by rail instead of in heavy trucks. Another approach is to encourage bicycle use by building bike lanes along highways and city streets.

Switching to Energy-Efficient Vehicles

Energy-efficient vehicles are available. One such vehicle is the gasoline–electric *hybrid car* (Figure 16.4, left). These cars have a small gasoline-powered engine and a battery-powered electric motor used to provide the energy needed for acceleration and hill climbing. The most efficient current models of these cars get a combined city/highway mileage of up to 23 kpl (55 mpg) and emit about 65% less CO_2 per kilometer driven than do comparable conventional cars.

Another option is the *plug-in hybrid electric vehicle* (Figure 16.4, right). These cars can travel 48–97 kilometers (30–60 miles) on electricity alone. Then a small gasoline-powered motor kicks in, recharges the battery, and extends the driving range to 600 kilometers (370 miles) or more. The battery can be plugged into a conventional 110-volt outlet and fully charged in 6 to 8 hours or a much shorter time using a 220-volt outlet. Another option is an all-electric vehicle that runs on a battery only.

According to a DOE study, replacing most of the current U.S. vehicle fleet with plug-in hybrid vehicles over 3 decades would cut U.S. oil consumption by 70–90%, eliminate the need for costly oil imports, save consumers money, and reduce CO_2 emissions by 27%. Recharging the batteries in these cars mostly by electricity generated by clean renewable energy resources such as wind turbines, solar cells, or hydroelectric power would cut U.S. emissions of CO_2 80–90%. This would slow climate change and ocean acidification and save thousands of lives by reducing air pollution from motor vehicles and coal-burning power plants.

The problem for the average consumer is that the prices of hybrid, plug-in hybrid, and all-electric cars are high because of the high cost of their batteries, which have to be replaced every few years. Thus, the key to greatly increasing the use of hybrid, plug-in hybrid, and all-electric motor vehicles is to ramp up research and development of suitable batteries (Science Focus 16.1). Another important factor will be to build a network of recharging stations throughout the country. Israel, Denmark, and several other countries are pioneering the use of stations where motorists can swap dead batteries for fully charged ones in less time than it takes to fill a gas tank. **GREEN CAREER: Plug-in hybrid and all-electric car technology**

Another potential alternative fuel resource is the **hydrogen fuel cell,** which could be used to power electric vehicles. This device uses hydrogen gas (H_2) as a fuel to produce electricity when it reacts with oxygen gas (O_2) in the atmosphere and emits harmless water vapor into the atmosphere. A fuel cell is more efficient than an internal combustion engine, has no moving parts, and requires little maintenance. Their H_2 fuel is usually produced by passing electricity through water or produced from methane stored in a vehicle. Two major problems with fuel cells

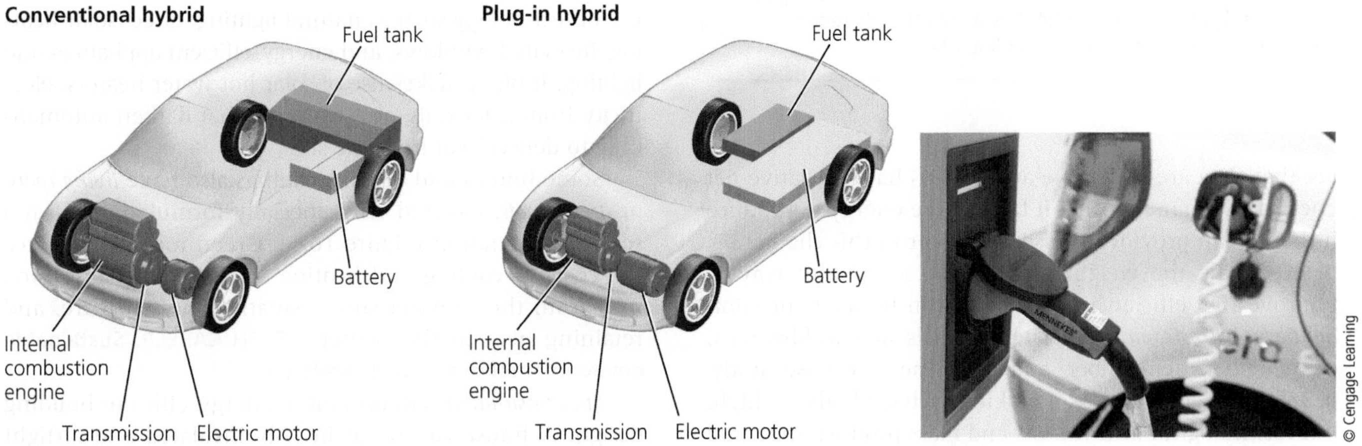

FIGURE 16.4 Solutions: A conventional gasoline–electric hybrid vehicle (left) is powered mostly by a small internal combustion engine with an assist from a strong battery. A plug-in hybrid electric vehicle (right) has a smaller internal combustion engine with a second and more powerful battery that can be plugged into a 110-volt or 220-volt outlet and recharged (see photo). An all-electric vehicle (not shown) runs completely on a rechargeable battery. ***Question:*** Would you buy one of these vehicles? Explain.

Photo: Gyuszko/Dreamstime.com

The Search for Better Batteries

The major obstacle standing in the way of wider use of plug-in hybrid-electric and all-electric vehicles is the need for an affordable, small, lightweight, and easily rechargeable car battery that can store enough energy for long-distances trips.

Lithium-ion batteries are light (because lithium is the lightest solid chemical element) and can pack a lot of energy into a small volume. Many of them are hooked together and used to power hybrid, plug-in hybrid, and all-electric motor vehicles. However, they are expensive, take a long time to recharge, lose their charge even when they are idle, must be replaced every few years, and can catch fire. Lithium-ion batteries are expensive but prices dropped by 50% between 2010 and 2015.

Researchers at the Massachusetts Institute of Technology (MIT) have developed a new type of lithium-ion battery using nanotechnology (Science Focus 14.1, p. 370). It is less expensive and can be charged 40 times faster. In 2014, researchers Joseph DeSimone and Nitash Balsara created a prototype of a nonflammable lithium-ion battery.

Scientists have also developed *super-capacitors*, which are small mechanical batteries consisting of two metal surfaces separated by an electric insulator. They quickly store and release large amounts of electrical energy, thus providing the power needed for quick acceleration. They can be recharged in minutes, can hold a charge much longer than conventional chemical batteries, and do not have to be replaced as frequently as conventional batteries.

If any one or a combination of these or other new battery technologies can be mass-produced at an affordable cost, plug-in hybrid and all-electric vehicles could take over the car and truck market within a few decades, which would greatly reduce air pollution, climate-changing CO_2 emissions, and ocean acidification. **GREEN CAREER: Battery engineer**

CRITICAL THINKING

How might your life change if one or more of the new battery technologies discussed above become a reality?

CONSIDER THIS . . .

LEARNING FROM NATURE

Blue-green algae use sunlight and an enzyme to produce hydrogen from water. Scientists are evaluating this as a way to produce hydrogen fuel for cars and for home heating without the use of costly high-temperature processes or electricity to produce this fuel. If successful, this would sharply reduce emissions of CO_2 and other air pollutants.

are that they are expensive and that H_2 has a negative net energy, which means that it takes more energy to produce it than it can provide, as discussed later in this chapter.

Reducing the weight of a vehicle is another way to improve fuel efficiency. Car bodies can be made of *ultralight* and *ultrastrong* composite materials such as fiberglass, carbon fiber, hemp fiber, and graphene (see Case Study, p. 369). The current cost of making such car bodies is high, but technological innovations and mass production would likely bring these costs down.

Energy conservation can also play a role. Since cars are the biggest energy user for most Americans, shifting to a more fuel-efficient car that gets at least 17 kpl (45 mpg) is one of the best ways to save money and reduce one's harmful environmental impact.

Designing Buildings That Save Energy and Money

Green architecture can help us make a transition to more energy-efficient, resource-efficient, and money-saving buildings over the next few decades. Green architecture makes use of technology such as natural lighting, direct solar heating, insulated windows, and energy-efficient appliances and lighting. It often makes use of solar hot water heaters, electricity from solar cells, and windows that darken automatically to deflect heat from the sun.

Some homes and urban buildings also have *living roofs*, or *green roofs*, covered with specially formulated soil and selected vegetation (Figure 16.5). Green roofs can reduce the costs of cooling and heating a building by absorbing heat from the summer sun, insulating the structure, and retaining heat in the winter. **GREEN CAREER: Sustainable environmental design and architecture**

Superinsulation is important in energy-efficient building design. A house can be so heavily insulated and airtight that heat from direct sunlight, appliances, and human bodies can warm a superinsulated house with little or no need for a backup heating system, even in extremely cold climates.

Superinsulated houses in Sweden use 90% less energy for heating and cooling than do typical American homes of

DIANE COOK/LEN JENSHEL/National Geography Creative

FIGURE 16.5 Green roof on Chicago's City Hall.

the same size. One example of superinsulation is straw-bale construction, in which a house's walls are built of straw bales that are covered inside and out with mud-based adobe bricks (Figure 16.6). Such walls have insulating values of 2–6 times those of conventional walls.

The World Green Building Council and the U.S. Green Building Council's Leadership in Energy and Environmental Design (LEED) have developed standards for certifying that a building meets certain energy efficiency and environmental standards. **GREEN CAREERS: Sustainable environmental design and architecture**

CONSIDER THIS . . .

THINKING ABOUT Energy-Efficient Building Design

What are three ways in which the building in which you live or work could have been designed to cut its waste of energy and money?

Saving Energy and Money in Existing Buildings

Here are ways to reduce energy use in existing buildings and to cut energy and save money on electricity and heating and cooling bills (see **Core Case Study**):

- *Get a home energy audit.*
- *Insulate the building and plug leaks* (Figure 16.1).
- *Use energy-efficient windows.*
- *Seal leaky heating and cooling ducts in attics and unheated basements.*
- *Heat interior spaces more efficiently.* In order, the most energy-efficient ways to heat indoor space are superinsulation (including plugging leaks); a geothermal heat pump that transfers heat stored from underground into a home; passive solar heating; a high-efficiency, conventional heat pump (in warm

FIGURE 16.6 Solutions: Energy-efficient, Victorian-style straw-bale house in Crested Butte, Colorado, during construction (left) and after it was completed (right). ***Question:*** Would you like to live in such a house? Explain.

climates only); and a high-efficiency natural gas furnace.

- *Heat water more efficiently.* One option is a roof-mounted solar hot water heater. Another option is a *tankless instant water heater*. It uses natural gas or liquefied petroleum gas (but not an electric heater, which is inefficient) to deliver hot water only when it is needed rather than keeping water in a large tank hot all the time.
- *Use energy-efficient appliances.* A refrigerator with its freezer in a drawer on the bottom uses about half as much energy as one with the freezer on the top or on the side, which allows dense cold air to flow out quickly when the door is opened. Microwave ovens use 25–50% less electricity than electric stoves do for cooking and 66% less energy than conventional ovens. Front-loading clothes washers use 55% less energy and 30% less water than top-loading models use and cut operating costs in half.
- *Use energy-efficient computers.* According to the EPA, if all computers sold in the United States met its Energy Star requirements, consumers would save $1.8 billion a year in energy costs and reduce greenhouse gas emissions by an amount equal to that of taking about 2 million cars off the road.
- *Use energy-efficient lighting.* The DOE estimates that by switching to LED bulbs over the next 20 years, U.S. consumers could save money and reduce the demand for electricity by an amount equal to the output of 40 new power plants.
- *Stop using the standby mode.* Consumers can reduce their energy use and their monthly power bills by plugging their standby electronic devices into smart power strips that cut off power to a device when it detects that the device has been turned off.

Figure 16.7 lists ways in which you can cut your energy use and save money in your home.

Why Are We Wasting So Much Energy and Money?

Considering its impressive array of economic and environmental benefits (Figure 16.3), why is there so little emphasis on improving energy efficiency and conserving energy? One reason is that energy resources such as fossil fuels and nuclear power are artificially cheap. This is primarily because of the government subsidies and tax breaks they receive and because their market prices do not include the harmful environmental and health costs of producing and using them. This distortion of the energy marketplace violates the full-cost pricing **principle of sustainability**.

Another reason is that there are too few significant government tax breaks, rebates, low-interest and long-term loans, and other economic incentives for consumers and businesses to invest in improving energy efficiency and reducing energy waste. A third reason is that most governments and utility companies have not put a high priority on educating the public about the environmental and economic advantages of improving energy efficiency and conserving energy.

Some critics say an emphasis on improving energy efficiency does not work because of the *rebound effect* in which some people tend to use more energy when they buy energy-efficient devices. For example, some people who buy a more efficient car tend to drive more, which offsets some of their energy and money savings and their reduced environmental impact.

Instead of downplaying efforts to improve energy efficiency, energy experts call for a major program to educate people about the rebound effect and its waste of money and long-lasting harmful health and environmental effects.

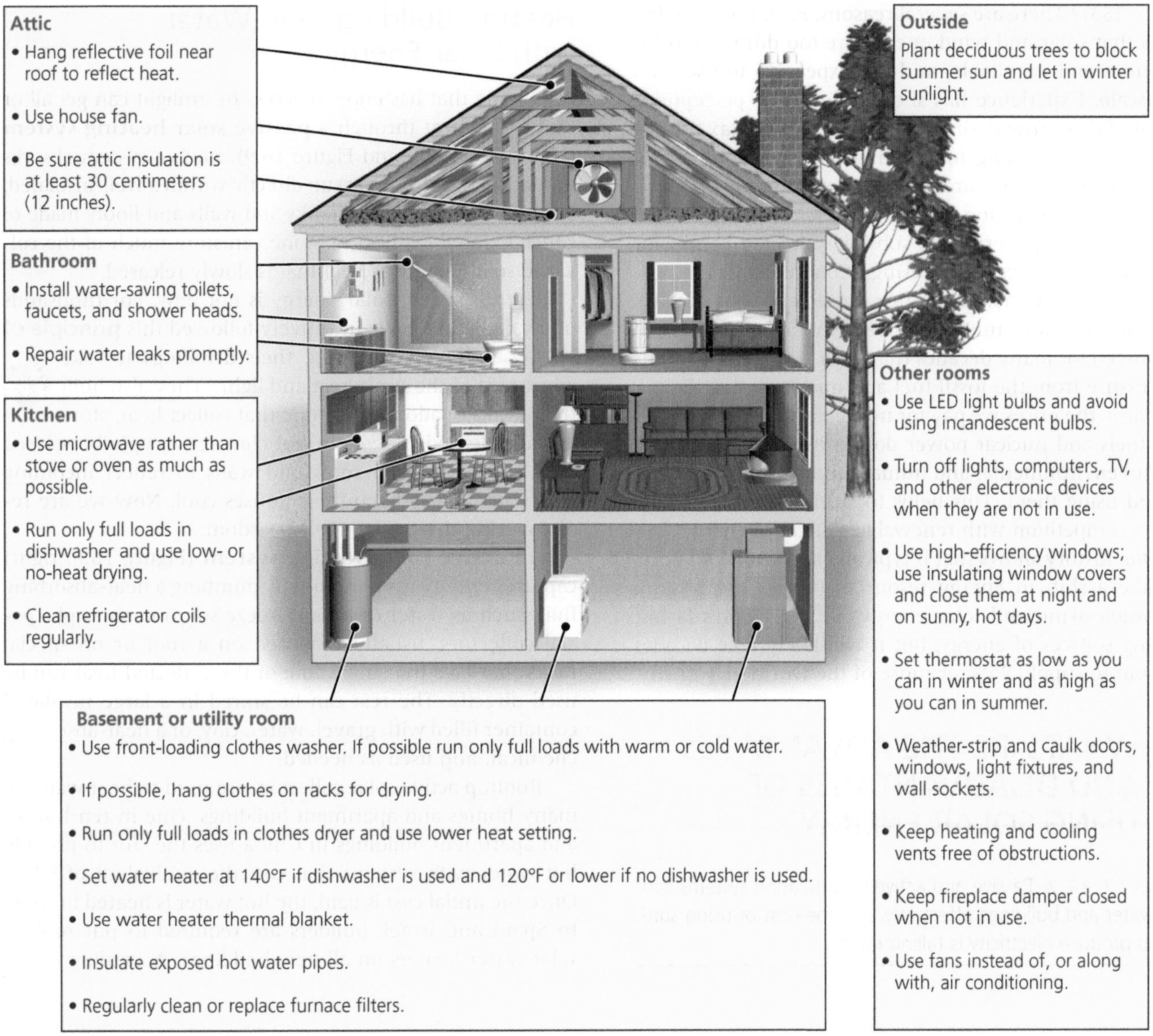
Attic
• Hang reflective foil near roof to reflect heat.
• Use house fan.
• Be sure attic insulation is at least 30 centimeters (12 inches).
Outside
Plant deciduous trees to block summer sun and let in winter sunlight.
Bathroom
• Install water-saving toilets, faucets, and shower heads.
• Repair water leaks promptly.
Kitchen
• Use microwave rather than stove or oven as much as possible.
• Run only full loads in dishwasher and use low- or no-heat drying.
• Clean refrigerator coils regularly.
Other rooms
• Use LED light bulbs and avoid using incandescent bulbs.
• Turn off lights, computers, TV, and other electronic devices when they are not in use.
• Use high-efficiency windows; use insulating window covers and close them at night and on sunny, hot days.
• Set thermostat as low as you can in winter and as high as you can in summer.
• Weather-strip and caulk doors, windows, light fixtures, and wall sockets.
• Keep heating and cooling vents free of obstructions.
• Keep fireplace damper closed when not in use.
• Use fans instead of, or along with, air conditioning.
Basement or utility room
• Use front-loading clothes washer. If possible run only full loads with warm or cold water.
• If possible, hang clothes on racks for drying.
• Run only full loads in clothes dryer and use lower heat setting.
• Set water heater at 140°F if dishwasher is used and 120°F or lower if no dishwasher is used.
• Use water heater thermal blanket.
• Insulate exposed hot water pipes.
• Regularly clean or replace furnace filters.

FIGURE 16.7 Individuals matter: You can save energy and money where you live and reduce your harmful environmental impact. ***Critical thinking:*** Which of these things do you already do? Which additional ones might you do?

Relying More on Renewable Energy

The lesson from one of nature's three **scientific principles of sustainability** is to *rely mostly on solar energy*. Most forms of renewable energy can be traced to the sun, because wind, flowing water, and biomass would not exist, were it not for solar energy. Another form of renewable energy is geothermal energy, or heat from the earth's interior. All of these sources of renewable energy are constantly replenished at no cost to us.

Studies show that with increased and consistent government backing in the form of research and development funds and subsidies and tax breaks, renewable energy could provide 20% of the world's electricity by 2025 and 50% by 2050. In 2012, the National Renewable Energy Laboratory (NREL) projected that, with a crash program, the United States could get 50% of its electricity from renewable energy sources by 2050.

In 2014, 13 countries got more than 30% of their electricity from renewable energy, compared to 13% in the United States. Germany is phasing out nuclear power, gets 14% of its electricity from renewable energy, and plans to increase that to 50% by 2050 (see Case Study, p. 410). In 2015, California legislators approved a plan that would require the state to produce half of its electric power from renewable energy by 2030. And San Diego, California, adopted a legally binding pledge in 2015 to get all of its electricity from renewable energy by 2035.

If renewable energy is so great, why does it provide only 8% of the world's energy (Figure 15.2, left, p. 383) and 5% of the energy used in the United States (Figure 15.2, right p. 383)? There are several reasons. *First*, there are the myths that solar and wind energy are too diffuse, too intermittent and unreliable, and too expensive to use on a large scale. Experience has shown that these perceptions are out of date. *Second*, since 1950, government tax breaks, subsidies, and funding for research and development of renewable energy resources have been much lower than those for fossil fuels and nuclear power.

Third, while government subsidies and tax breaks for renewable have been increasing, Congress must renew them every few years. In contrast, subsidies and tax breaks for fossil fuels and nuclear power have essentially been guaranteed for many decades thanks in large part to political pressure from the fossil fuel and nuclear industries.

Fourth, the prices we pay for nonrenewable energy from fossil fuels and nuclear power do not include most of the harmful environmental and human health costs of producing and using them. This helps to shield them from free-market competition with renewable sources of energy.

Fifth, history shows that it typically takes 50 to 60 years to make a shift in energy resources (**Core Case Study**). Renewable wind and solar energy are the world's fastest growing sources of energy, but it will likely take decades for them to supply 25% or more of the world's electricity.

16.3 WHAT ARE THE ADVANTAGES AND DISADVANTAGES OF USING SOLAR ENERGY?

CONCEPT 16.3 Passive and active solar heating systems can heat water and buildings effectively, and the cost of using sunlight to produce electricity is falling rapidly.

Heating Buildings and Water with Solar Energy

A building that has enough access to sunlight can get all or most of its heat through a **passive solar heating system** (Figure 16.8, left, and Figure 16.9). Such a system absorbs and stores heat from the sun directly within a well-insulated, airtight structure. Water tanks and walls and floors made of concrete, adobe, brick, or stone can store much of the collected solar energy as heat that is slowly released.

Use of passive solar energy is not new. For thousands of years, people have intuitively followed this **principle of sustainability** by orienting their dwellings to take advantage of the sun's heat and light. They also built thick walls of adobe and stone that collect heat, store it, and release it slowly. In hot climates, people have used light colors on their roofs and walls to reflect incoming solar energy and keep their houses cool. Now we are rediscovering this ancient solar wisdom.

An **active solar heating system** (Figure 16.8, right) captures energy from the sun by pumping a heat-absorbing fluid such as water or an antifreeze solution through special collectors, usually mounted on a roof or on special racks that face the sun. Some of the collected heat can be used directly. The rest can be stored in a large insulated container filled with gravel, water, clay, or a heat-absorbing chemical, and used as needed.

Rooftop active solar collectors are used to heat water in many homes and apartment buildings. One in ten houses and apartment buildings in China uses the sun to provide hot water with systems that cost the equivalent of $200. Once the initial cost is paid, the hot water is heated for free. In Spain and Israel, builders are required to put rooftop solar water heaters on all new buildings. According to the

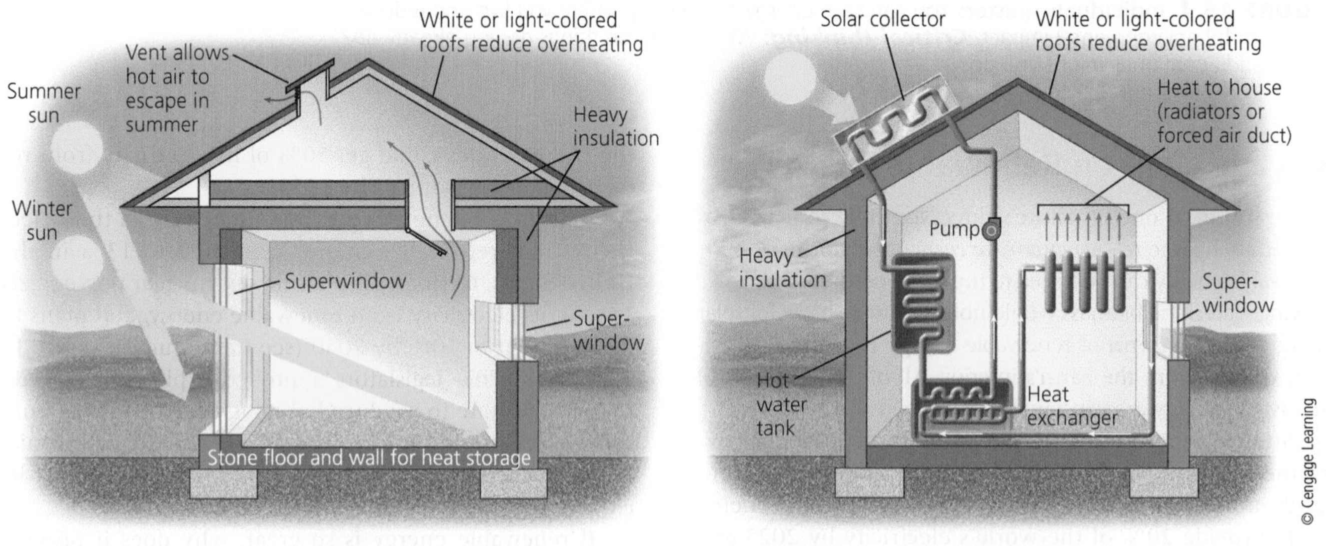

PASSIVE ACTIVE

FIGURE 16.8 Solutions: Passive (left) and active (right) solar home heating systems.

FIGURE 16.9 This passive solar home (right) in Golden, Colorado, collects and stores incoming solar energy to provide much of its heat in a climate with cold winters. Notice the solar hot water heating panels in the yard. Some passive solar houses have sunrooms (see inset photo) to help collect incoming solar energy.

UN Development Programme, solar water heaters could be used to provide half of the world's hot water.

Passive and active solar systems can heat new homes in areas with adequate sunlight, as long as trees or other buildings do not block solar access. (See Figures 20 and 21, pp. S33 and S34, in Supplement 4 for maps showing solar energy availability throughout the world and in the United States, respectively.) Figure 16.10 lists the major advantages and disadvantages of using passive or active solar heating systems for heating buildings.

Cooling Buildings Naturally

Direct solar energy works against us when we want to keep a building cool, but we can use indirect solar energy (mainly wind) to help cool buildings. We can open windows to take advantage of breezes and use fans to keep the air moving. When there is no breeze, superinsulation and high-efficiency windows keep hot air outside.

Other ways to keep buildings cool include blocking the sun with shade trees, broad overhanging eaves, window awnings, or shades. A light-colored roof can reflect up to 90% of the sun's heat (compared to only 10–15% for a dark-colored roof), while a green roof can absorb extra heat. Geothermal heat pumps can pump cool air from underground into a building during summer to cool it as well.

CONSIDER THIS . . .

LEARNING FROM NATURE

Some species of African termites stay cool in a hot climate by building giant mounds that allow air to circulate through them. Engineers have used this design lesson from nature to cool buildings naturally, reduce energy use, and save money.

Trade-Offs

Passive or Active Solar Heating

Advantages	Disadvantages
Medium net energy	Need access to sun 60% of time during daylight
Very low emissions of CO_2 and other air pollutants	Sun can be blocked by trees and other structures
Very low land disturbance	High installation and maintenance costs for active systems
Moderate cost (passive)	Need backup system for cloudy days

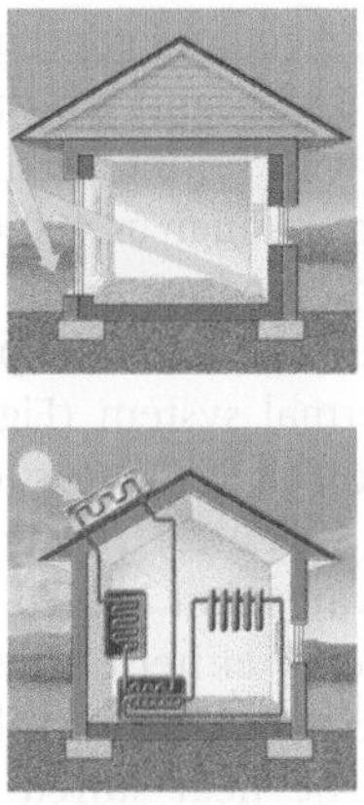

FIGURE 16.10 Heating a house with passive or active solar energy system has advantages and disadvantages (**Concept 16.3**). ***Critical thinking:*** Which single advantage and which single disadvantage do you think are the most important? Why? Do you think that the advantages of using these technologies outweigh their disadvantages?

Concentrating Sunlight to Produce High-Temperature Heat and Electricity

One of the problems with direct solar energy is that it is dispersed. **Solar thermal systems,** also known as *concentrated solar power (CSP)*, use different methods to collect and concentrate solar energy to boil water and produce steam

National Renewable Energy Laboratory (NREL)

FIGURE 16.11 *Solar thermal power:* This solar power plant in California's Mojave Desert uses curved (parabolic) solar collectors to concentrate solar energy to provide enough heat to boil water and produce steam for generating electricity.

Sandia National Laboratories/National Renewable Energy Laboratory

FIGURE 16.12 *Solar thermal power:* In this system in California an array of mirrors tracks the sun and focuses reflected sunlight on a central receiver to boil the water for producing electricity.

for generating electricity. These systems can be used in deserts and other open areas with ample sunlight.

One such system uses rows of curved mirrors, called parabolic troughs, to collect and concentrate sunlight. Each trough focuses incoming sunlight on a pipe that runs through its center and is filled with synthetic oil (Figure 16.11). The oil heats to temperatures as high as 400°C (750°F). That heat is used to boil water and produce steam, which in turn powers a turbine that drives a generator to produce electricity.

Another solar thermal system (Figure 16.12) uses an array of computer-controlled mirrors to track the sun and focus its energy on a central power tower to provide enough heat to boil water that is used to produce electricity. The heat produced by either of these systems can also be used to melt a certain kind of salt stored in a large insulated container. The heat stored in this molten salt system can then be released as needed to produce electricity at night or on cloudy days.

Some analysts see solar thermal power as a growing and important source of the world's electricity. However, because solar thermal systems have a low net energy, they need large government subsidies or tax breaks to be competitive in the marketplace. These systems also require large volumes of cooling water for condensing the steam back to water and for washing the surfaces of the mirrors and parabolic troughs. However, they are usually built in sunny, arid desert areas where water is scarce. Figure 16.13 summarizes the major advantages and disadvantages of using these solar thermal systems.

Solar energy can also be concentrated on a smaller scale. In some sunny areas, people use inexpensive *solar cookers* to

Trade-Offs

Solar Thermal Systems

Advantages	Disadvantages
High potential for growth	Low net energy and high costs
No direct emissions of CO_2 and other air pollutants	Needs backup or storage system on cloudy days
Lower costs with natural gas turbine backup	Requires high water use
Source of new jobs	Can disrupt desert ecosystems

FIGURE 16.13 Using solar energy to generate high-temperature heat and electricity has advantages and disadvantages (**Concept 16.3**). ***Critical thinking:*** Which single advantage and which single disadvantage do you think are the most important? Why? Do you think that the advantages of using these technologies outweigh their disadvantages?

Top: Sandia National Laboratories/National Renewable Energy Laboratory. Bottom: National Renewable Energy Laboratory.

FIGURE 16.14 Solutions: Solar cooker (left) in Costa Rica and simple solar oven (right).

focus and concentrate sunlight for boiling and sterilizing water (Figure 16.14, left) and cooking food (Figure 16.14, right). Inventor Jon Boehner has developed a $6 solar cooker made from a cardboard box. Solar cookers can replace wood and charcoal fires and reduce indoor air pollution, a major killer of many of the world's poor people. They also help to reduce deforestation by decreasing the need for firewood and charcoal (Figure 10.16, p. 234) made from firewood.

Using Solar Cells to Produce Electricity

In 1931, Thomas Edison (inventor of the electric light bulb) told Henry Ford, "I'd put my money on the sun and solar energy. … I hope we don't have to wait until oil and coal run out before we tackle that." Edison's dream is now a reality.

We can convert solar energy directly into electrical energy using **photovoltaic (PV) cells,** commonly called **solar cells.** Most solar cells are thin transparent wafers of purified silicon (Si) or polycrystalline silicon with trace amounts of metals that allow them to conduct electricity.

Solar cells have no moving parts and they operate safely and quietly with no emissions of greenhouse gases and other air pollutants. A typical solar cell has a thickness ranging from less than that of a human hair to that of a sheet of paper. When sunlight strikes solar cells, they produce electricity (a flow of electrons). Many cells wired together in a panel and many panels can be connected to produce electricity for a house or a large solar power plant (Figure 16.15). Such systems can be connected to existing electrical grids or to batteries that store the electrical energy until it is needed.

People can mount arrays of solar cells on rooftops and incorporate them into almost any type of roofing material. Nanotechnology and other emerging technologies will likely allow the manufacturing of solar cells in paper-thin, rigid or flexible sheets (see Figure 14.14, p. 371) that can be printed like newspapers and attached to or embedded in other surfaces such as outdoor walls, windows, drapes, and clothing. Engineers are also developing dirt- and water-repellent coatings to keep solar panels and collectors clean without having to use water. Solar power providers are putting floating arrays of solar cell panels on the surfaces of lakes, reservoirs, ponds, and canals. **GREEN CAREER: Solar-cell technology**

Solar cells have great potential for providing electricity in less-developed countries. Worldwide, 1.3 billion people, most of them in rural villages in such countries, do not have access to electricity. A growing number of these individuals and villages are using rooftop solar panels (Figure 16.16) to power energy-efficient LED lamps (Figure 16.17), which can replace costly and inefficient kerosene lamps that pollute indoor air.

India has more than 300 million mostly rural poor people who are not connected to an electrical grid. Private entrepreneurs in India and Africa are setting up stand-alone *solar-powered microgrids* where a centralized group of solar cell panels are connected by cable to a few dozen homes and local businesses. Customers use cell phones to connect to village smart meters and purchase a certain amount of electricity. The smart meters cut off the power when a user's payment runs out.

Solar cells emit no greenhouse gases, although they are not a carbon-free option, because fossil fuels are used to produce and transport the panels. However, the emissions per unit of electricity produced are much smaller than those generated by using fossil fuels and nuclear power to produce electricity. Conventional solar cells also contain toxic materials that must be recovered when the cells wear out after 20–25 years of use, or when they are replaced by new systems.

One problem with current solar cells is their low energy efficiency. They typically convert only 15–20% of the incoming solar energy into electricity. However, scientists and engineers are rapidly improving the efficiency of solar cells. In 2014, researchers at Germany's Fraunhofer

Ollyy/Shutterstock.com

FIGURE 16.15 *Solar cell power plant:* Huge arrays of solar cells can be connected to produce electricity.

Institute for Solar Energy Systems developed a solar cell with an efficiency of 45%—compared to an efficiency of 35% for fossil fuel and nuclear electric power plants. They plan to scale up this prototype cell for commercial use within a few years. Figure 16.18 lists the major advantages and disadvantages of using solar cells to produce electricity (**Concept 16.3**).

Because of government subsidies and tax breaks for solar cell producers and users and rapidly declining prices, solar cells have become the world's fastest growing way to produce electricity (Figure 16.19). Between 2001 and 2015, the cost per watt of electricity produced by solar cells fell by 83%, and this cost is expected to keep falling.

Some businesses and homeowners are spreading the cost of rooftop solar power systems over decades by including them in their mortgages. Others are leasing solar-cell systems from companies that install and maintain them.

Some communities and neighborhoods are using community solar or shared solar systems to provide electricity for individuals who rent or live in condominiums, or whose access to sunlight is blocked by buildings or trees. Customers buy the power from a centrally located small solar cell power plant. The power is delivered by the local utility and customers share deductions on their monthly bills for any excess power the project sells back to the grid.

Producing electricity from solar cells is expected to grow because solar energy is unlimited and available throughout the world. It is also a technology, not a fuel

Jim Welch/National Renewable Energy Laboratory

FIGURE 16.16 Solutions: A solar-cell panel provides electricity for lighting this hut in rural West Bengal, India. ***Critical thinking:*** Do you think your government should provide aid to poor countries for obtaining solar-cell systems? Explain.

TYRONE TURNER/National Geographic Creative

FIGURE 16.17 Solar powered LED lantern in Mexico.

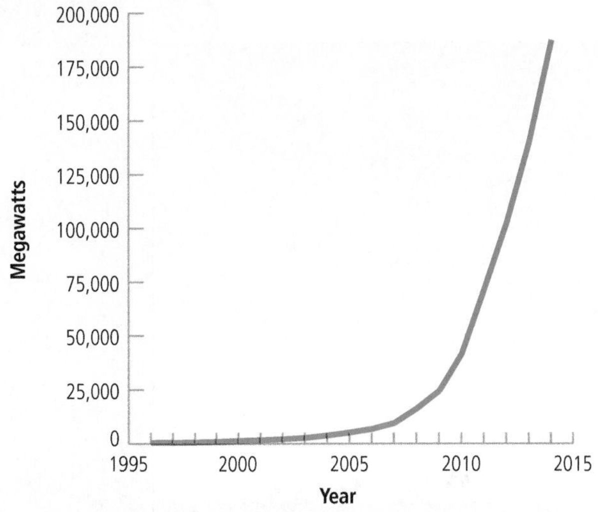

FIGURE 16.19 Global installed electricity capacity of solar cells, 1996–2014. ***Data analysis:*** By what factor and percent did the capacity of installed solar cell increase between 1996 and 2015?

(Compiled by the authors using data from U.S. Energy Information Administration, International Energy Agency, Worldwatch Institute, and Earth Policy Institute.)

Trade-Offs

Solar Cells

Advantages	Disadvantages
	Need access to sun
Little or no direct emissions of CO_2 and other air pollutants	Need electricity storage system or backup
Easy to install, move around, and expand as needed	Low net energy but likely to improve
Competitive cost for newer cells	Solar-cell power plants could disrupt desert ecosystems

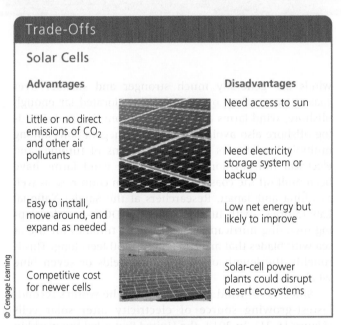

FIGURE 16.18 Using solar cells to produce electricity has advantages and disadvantages (**Concept 16.3**). ***Critical thinking:*** Which single advantage and which single disadvantage do you think are the most important? Why? Do you think that the advantages of using this technology outweigh its disadvantages? Why?

Top: Martin D. Vonka/Shutterstock.com. Bottom: pedrosala/Shutterstock.com.

such as coal or natural gas, the prices of which are controlled by available supplies. Prices for solar cell systems are likely to continue dropping because of technological advances, mass production, and decreased installation costs.

Solar cells cannot produce electricity at night, and storing energy in large batteries for use at night and on cloudy days is expensive. However, researchers at Ohio State University have developed a solar cell panel with a built-in battery that is 25% less expensive and 20% more efficient than conventional batteries. If it can be mass produced, this invention could revolutionize the use of solar energy to produce electricity.

In 2015, solar energy provided only 0.6% of U.S. electricity. However, this new technology is growing exponentially at a rapid rate (Figure 16.19) and prices continue to drop. If pushed hard and supported with government subsidies equivalent to or greater than fossil fuel subsidies, solar energy could supply 5% of U.S. electricity by 2020 and as much as 23% by 2050, according to projections by the NREL.

After 2050, continuing exponential growth and doubling of the large base of solar electricity would allow it to become one of the top sources of electricity for the United States and much of the world. If that happens, it will represent global application of the solar energy **principle of sustainability**. **GREEN CAREER: Solar-cell technology**

16.4 WHAT ARE THE ADVANTAGES AND DISADVANTAGES OF USING WIND POWER?

CONCEPT 16.4 Wind is one of the fastest growing, least expensive, and cleanest ways to produce electricity.

Using Wind to Produce Electricity

Simply put, *wind* is air in motion. Land near the earth's Equator absorbs more solar energy than does land near its

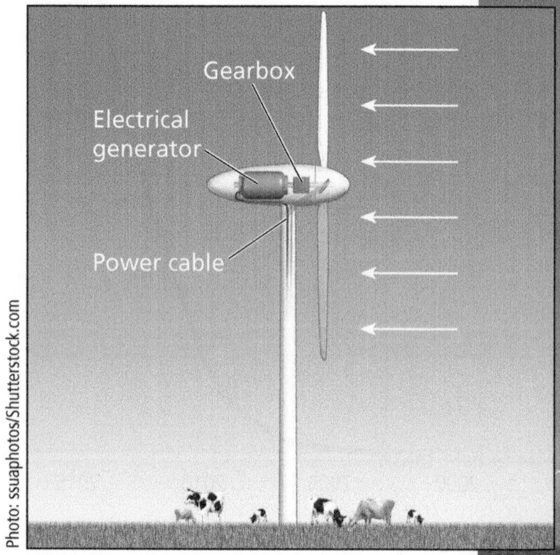

Wind turbine

FIGURE 16.20 Wind turbines convert the kinetic energy in wind to electricity, another form of kinetic energy (moving electrons). Wind power is an indirect form of solar energy.

Poles. This uneven heating of the earth's surface and its atmosphere, combined with the earth's rotation, causes winds to blow (Figure 7.9, p. 146). Because wind is an indirect form of solar energy, using it is a way to apply the solar energy **principle of sustainability**.

The kinetic energy from blowing winds can be captured and converted to electrical energy by devices called *wind turbines*. As a turbine's blades spin, they turn a drive shaft that connects to the blades. The drive shaft then turns an electric generator, which produces electricity (Figure 16.20, left). Groups of wind turbines called *wind farms* transmit electrical energy to electrical grids. Wind farms can operate both on land (chapter-opening photo) and at sea (Figure 16.20, right).

Today's wind turbine towers can be as tall as a 60-story building and have blades as long as 70 meters (230 feet)—the combined length of six school buses. This height allows them to tap into the strong and more constant winds found at higher altitudes on land and at sea (Figure 16.20, right) and to produce more electricity at a lower cost. A typical wind turbine can generate enough electricity to power more than 1,000 homes.

Harvard University researcher Xi Lu estimates that wind power has the potential to produce 40 times the world's current use for electricity. Most of world's wind farms have been built on land in parts of Europe, China, and the United States. However, the frontier for wind energy is offshore wind farms (Figure 16.20, right) because winds are generally much stronger and steadier over coastal waters than over land. When located far enough offshore, wind farms are not visible from the land. Building offshore also avoids the need for negotiations among multiple landowners over the locations of turbines and electrical transmission lines. Offshore wind farms have been built off the coasts of 10 European countries, as well as China and Japan. Researchers at the Sandia National Laboratories—with funding from the DOE—are developing towering hurricane-resistant wind turbines for use at sea with blades that are 200 meters (650 feet) long. This is roughly the length of two football fields or seven blue whales.

Since 1990, wind power has been the world's second-fastest-growing source of electricity after solar cells (Figure 16.21). In 2014, the United States led the world in producing electricity from wind, followed by China and Germany.

In 2013, wind farms in more than 85 countries produced about 3.5% of the world's electricity—enough to provide electricity for more than 500 million people. Experts project that by 2050, this number could grow to 31%. Around the world, more than 400,000 people are employed in the production, installation, and maintenance of wind turbines and such job numbers will grow as wind power continues its rapid expansion (Figure 16.21).

In 2015, wind power produced 45% of Denmark's electricity and the country plans to increase this to 50% by

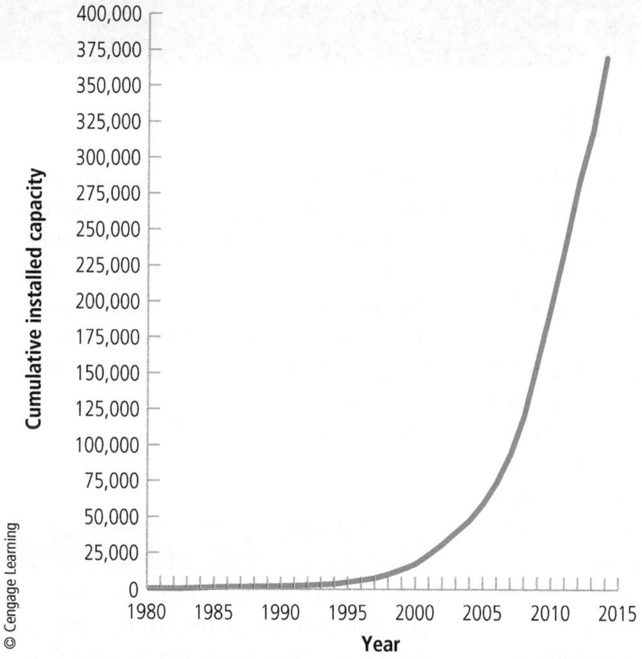

FIGURE 16.21 Global installed capacity for generation of electricity by wind energy, 1980–2014. ***Data analysis:*** In 2014, the world's installed capacity for generating electricity by wind power was about how many times more than it was in 2005?

(Compiled by the authors using data from Global Wind Energy Council, European Wind Energy Association, American Wind Energy Association, Worldwatch Institute, World Wind Energy Association, and Earth Policy Institute.)

2020 and to 85% by 2035. By 2015, wind turbines in the United States were producing 4.7% of the country's electricity, an amount equal to that produced by 64 large nuclear power reactors.

A study published in the *Proceedings of the U.S. National Academy of Sciences* estimated that the United States has enough wind potential to meet 16 to 22 times its current electricity needs. The DOE estimates that wind farms at favorable sites in three states—North Dakota, Kansas, and Texas—could more than meet the electricity needs of the lower 48 states if a modern electrical grid is available to distribute the electricity. In addition, the NREL estimates that winds off the Atlantic and Pacific coasts and the shores of the Great Lakes could generate 4 times the electricity currently used in the lower 48 states—more than enough electricity to replace all of the country's coal-fired power plants. Many states in these regions plan to tap into this vast source of energy and boost their economies.

According to a 2015 study by the DOE, with the continuation of government subsidies, the United States could get 30% of its electricity from wind by 2030. This would create up to 600,000 jobs and lower energy bills. It would also reduce air pollution and slow climate change and ocean acidification by reducing the use of coal to produce electricity.

Wind is abundant, widely distributed, and inexhaustible, and wind power is mostly carbon-free and pollution-free. A wind farm can be built within 9 to 12 months and expanded as needed. And although wind farms can cover large areas of land, the turbines themselves occupy only a small portion of the land.

Many U.S. landowners in favorable wind areas are investing in wind farms. Landowners typically receive $3,000 to $10,000 a year in royalties for each wind turbine placed on their land. The land can still be used for activities such as growing crops or grazing cattle (Figure 16.20, left, and chapter-opening photo). An acre of land in northern Iowa planted in corn can produce about $1,000 worth of ethanol car fuel. The same site used for a single wind turbine can produce $300,000 worth of electricity per year.

Since 1990, prices for electricity produced by wind in the United States (and other countries) have been falling sharply. Wind farms in Texas generate so much electricity that some utility companies are offering free electricity at night. Prices are expected to keep falling because of technological improvements in wind turbine design that increase their energy efficiency and reduce mass-production and maintenance costs. The DOE and the Worldwatch Institute estimate that, if we were to apply the **full-cost pricing principle of sustainability** by including the harmful environmental and health costs of various energy resources in comparative cost estimates, wind energy would be the least costly way to produce electricity.

Like any energy source, wind power has some drawbacks. Land areas with the greatest wind power potential are often sparsely populated and located far from cities. Thus, to take advantage of its huge potential for using wind energy, the United States will have to invest in replacing and expanding its outdated electrical grid with a smart grid system to connect with the country's wind farms.

Because winds can die down, a backup source of power such as natural gas is needed. However, a large number of wind farms in different areas connected to a smart grid could take up the slack when winds die down in any one area without the need for a backup source of energy. This could make wind power a more stable and reliable source of electricity than fossil fuel and nuclear power plants.

Wind turbines can also kill birds and bats—a problem that scientists and wind power developers are working on (Science Focus 16.2).

Figure 16.22 lists the major advantages and disadvantages of using wind to produce electricity. According to many energy analysts, wind power has more benefits and fewer serious drawbacks than any other energy resource, except for reducing energy waste (Figure 16.2) and increasing energy efficiency (Figure 16.3). **GREEN CAREER: Wind-energy engineering**

Making Wind Turbines Safer for Birds and Bats

Wildlife ecologists and bird experts estimate that collisions with wind turbines kill as many as 234,000 birds and 600,000 bats each year in the United States. Such deaths are a legitimate concern.

However, according to studies by the Defenders of Wildlife, the U.S. Fish and Wildlife Service, and the Smithsonian Conservation Biology Institute wind turbines are a minor source of bird and bat deaths compared to other human-related sources and account for only about 0.003% of such deaths. Each year, domestic and feral cats kill 1.4 billion to 3.7 billion birds; collisions with windows 1 billion; cars and trucks 89 million to 340 million; high-tension wires 174 million; and pesticides 72 million. Most of the wind turbines involved in bird and bat deaths were built years ago using outdated designs, and some were built in bird migration corridors.

Developers of new wind farms avoid bird migration corridors, as well as areas with large bat colonies. Newer turbine designs reduce bird and bat deaths considerably by using slower blade rotation speeds along with several other innovations. For example, researchers are also evaluating the use of ultraviolet light to deter birds and bats from turbines. Ultrasonic devices attached to turbine blades scare bats away by emitting high-frequency sounds that we cannot hear. Another approach is to use radar to track large incoming flocks of migrating birds and to shut down the turbines until they pass. In addition, many older wind turbines are being replaced with new designs that sharply reduce bird and bat deaths by providing no perching or nesting surfaces.

CRITICAL THINKING

What would you say to someone who tells you that we should not depend on wind power because wind turbines can kill birds and bats?

Trade-Offs

Wind Power

Advantages

High net energy

Widely available

Low electricity cost

Little or no direct emissions of CO_2 and other air pollutants

Easy to build and expand

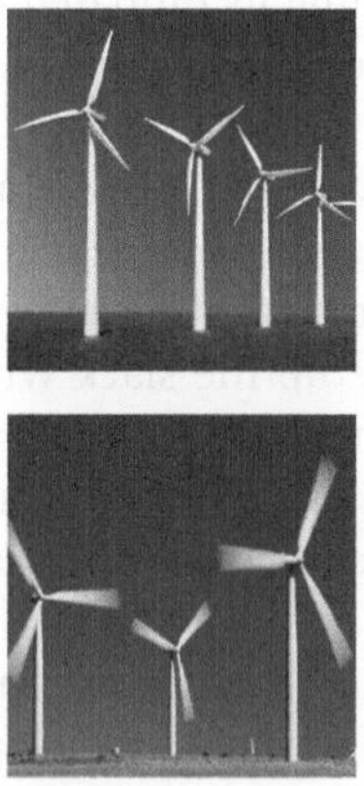

Disadvantages

Needs backup or storage system when winds die down unless connected in a national electrical grid

Visual pollution for some people

Low-level noise bothers some people

Can kill birds if not properly designed and located

FIGURE 16.22 Using wind to produce electricity has advantages and disadvantages (**Concept 16.4**). ***Critical thinking:*** Which single advantage and which single disadvantage do you think are the most important? Why? Do you think the advantages outweigh the disadvantages? Why?

Top: TebNad/Shutterstock.com. Bottom: Yegor Korzh/Shutterstock.com.

16.5 WHAT ARE THE ADVANTAGES AND DISADVANTAGES OF USING GEOTHERMAL ENERGY?

CONCEPT 16.5 Geothermal energy can supply many areas with heat and electricity, and has a generally low environmental impact, but sites where it can be produced economically are limited.

Tapping into the Earth's Internal Heat

Geothermal energy is heat stored in soil, underground rocks, and fluids in the earth's mantle. Geothermal energy is used to heat and cool buildings, and to heat water to produce electricity.

A *geothermal heat pump* system (Figure 16.23) can heat and cool a house almost anywhere in the world. This system makes use of the temperature difference between the earth's surface and underground at a depth of 3–6 meters (10–20 feet), where the temperature typically is 10–20°C (50–60°F) year-round. In winter, a closed loop of buried pipes circulates a fluid, which extracts heat from the ground and carries it to a heat pump, which transfers the heat to a home's heat distribution system. In summer, this system works in reverse, removing heat from a home's interior and storing it below ground.

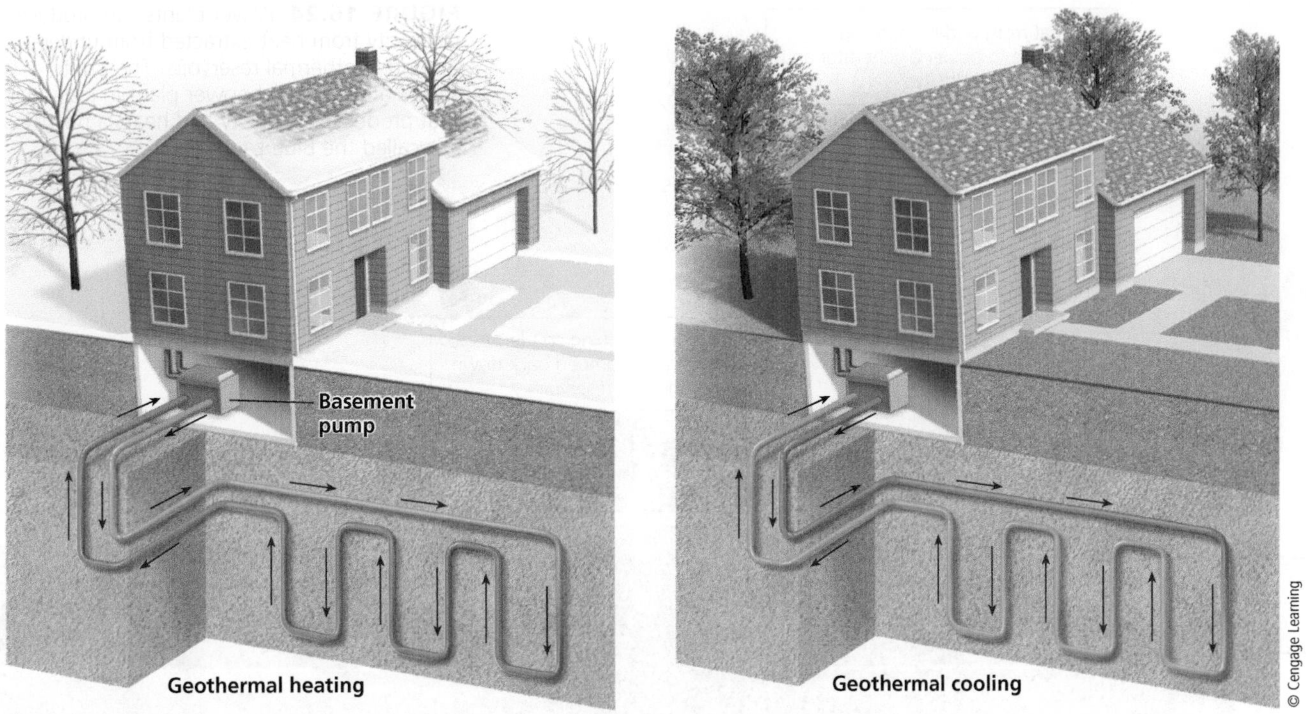

FIGURE 16.23 Natural capital: A geothermal heat pump system can heat or cool a house almost anywhere.

According to the EPA, a geothermal heat pump system is the most energy-efficient, reliable, environmentally clean, and cost-effective way to heat or cool a space. Installation costs can be high but are recouped within 3 to 5 years, after which these systems save energy and money for their owners. Initial costs can be added to a home mortgage to spread the financial burden over two or more decades.

We can also tap into deeper, more concentrated *hydrothermal reservoirs* of geothermal energy (Figure 16.24). Wells are drilled into the reservoirs to extract their dry steam (with a low water content), wet steam (with a high water content), or hot water. The steam or hot water can be used to heat homes and buildings, provide hot water, grow vegetables in greenhouses, raise fish in aquaculture ponds, and spin turbines to produce electricity.

Drilling geothermal wells, like drilling oil and natural gas wells, is expensive and requires a major investment. It is also a risky investment because drilling projects do not always succeed in tapping into concentrated deposits of geothermal energy. Once a successful deposit is found, it can supply geothermal energy for heat or to produce electricity around the clock, as long as heat is not removed from the deposit faster than the earth replaces it—usually at a slow rate. When this happens geothermal energy becomes a nonrenewable resource.

Geothermal energy is used in 24 countries, including the United States, the world's largest producer of geothermal electricity from hydrothermal reservoirs. Most of it is produced in California, Nevada, Utah, and Hawaii (see Figure 25, p. S38, in Supplement 4 for a map of the best geothermal sites in the continental United States). In 2015, there were 134 new geothermal power plants under construction or in development in the United States. When completed, they could triple the country's electrical output from geothermal energy. The U.S. Geothermal Energy Association (GEO) estimates that 90% of the available geothermal energy for producing electricity in the United States and 60% of the potential supply in California has not been tapped.

Figure 16.25 shows the global growth in installed geothermal electricity-generating capacity between 1950 and 2014. Iceland gets almost all of its electricity from renewable hydroelectric (69%) and geothermal (29%) power plants (Figure 16.25, photo) and nearly 95% of its demand for heat and hot water from geothermal energy.

In Peru, a National Geographic Explorer is carrying out research to develop that country's geothermal resources (Individuals Matter 16.1).

Another source of geothermal energy is *hot, dry rock* found 5 kilometers or more (3 miles or more) underground almost everywhere. Water can be injected through deep wells drilled into this rock. Some of this water becomes steam that is brought to the surface and used to spin turbines to generate electricity. According to the U.S. Geological Survey, tapping just 2% of this source of geothermal

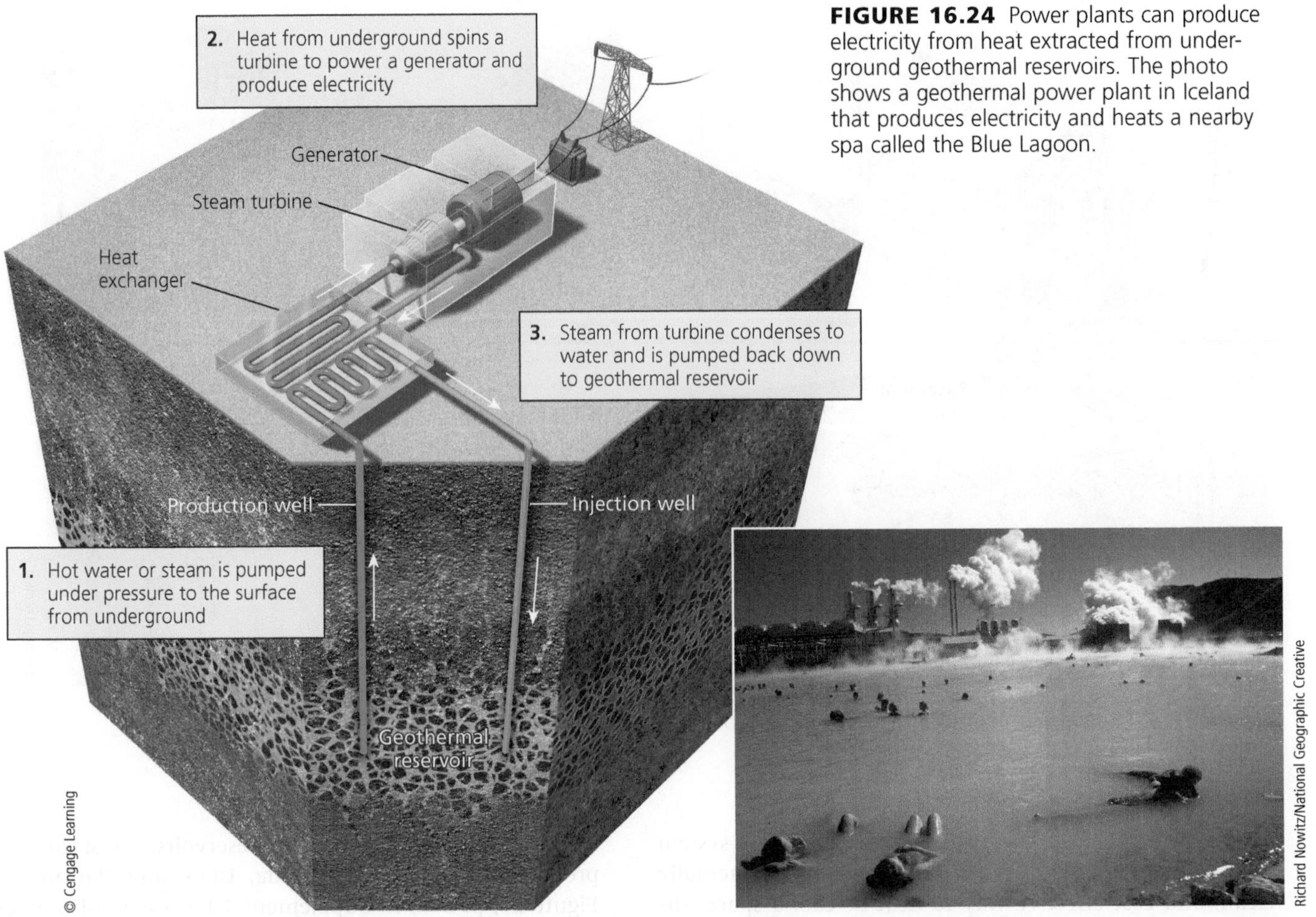

FIGURE 16.24 Power plants can produce electricity from heat extracted from underground geothermal reservoirs. The photo shows a geothermal power plant in Iceland that produces electricity and heats a nearby spa called the Blue Lagoon.

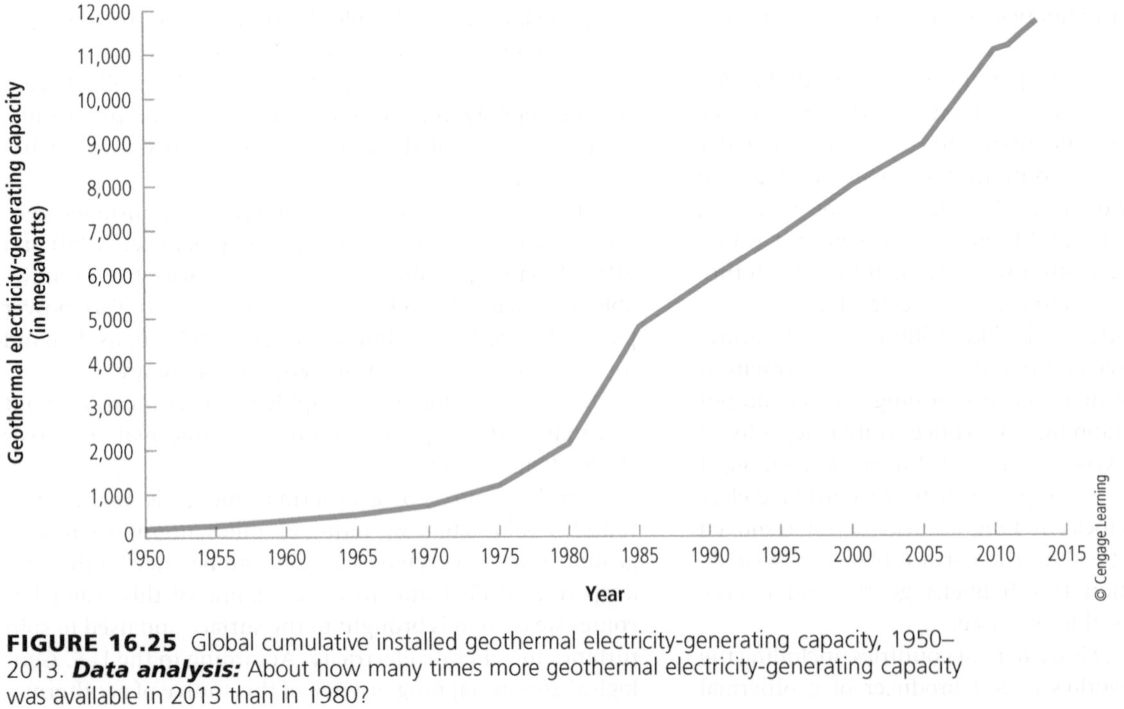

FIGURE 16.25 Global cumulative installed geothermal electricity-generating capacity, 1950–2013. ***Data analysis:*** About how many times more geothermal electricity-generating capacity was available in 2013 than in 1980?

(Compiled by the authors using data from International Energy Agency, Worldwatch Institute, Earth Policy Institute, and BP.)

INDIVIDUALS MATTER 16.1

Andrés Ruzo—Geothermal Energy Sleuth and National Geographic Explorer

© Andres Ruzo

Andrés Ruzo is a geophysicist with a driving passion to learn about geothermal energy and to show how this renewable and clean energy source can help us to solve some of the world's energy problems. As a boy, he spent his summers on the family farm in Nicaragua. Because the farm rests on top of the Casita Volcano, he was able to experience firsthand the power of the earth's heat.

As an undergraduate student at Southern Methodist University (SMU) in Dallas, Texas (USA), because of his boyhood experience, he took a course in volcanology. The course awakened his passion for geology along with a desire to learn more about the earth's heat as a source of energy. This led him to pursue a PhD in geophysics at SMU's Geothermal Laboratory.

Since 2009, Ruzo and his wife and field assistant, Sofia, have been gathering data across Peru to develop the country's first detailed heat flow map—which will help identify areas of geothermal energy potential. Their field work involves lowering temperature measuring equipment down into oil, gas, mining, or water wells. Much of this work was done in the Talara Desert in northwestern Peru, where surface temperatures can exceed 54°C (130°F). Other wells were temperature logged deep in the Amazon rainforest. These data illustrate how thermal energy flows through the upper crust of the earth, and highlights areas where earth's heat can potentially be tapped as a source of energy.

Ruzo believes that geothermal energy is a "sleeping giant" that, if properly harnessed, can be an important renewable source of heat and electricity. He says that his goal in life is "to be a force of positive change in the world."

---

energy in the United States could produce more than 2,000 times the amount of electricity currently used in the country. The limiting factor is its high cost. **GREEN CAREER: Geothermal engineer**

Figure 16.26 lists the major advantages and disadvantages of using geothermal energy (**Concept 16.5**). The two biggest factors limiting the widespread use of geothermal energy are the lack of hydrothermal sites with concentrations of heat high enough to make it affordable and the high cost of drilling the wells and building the plants.

Trade-Offs

Geothermal Energy

Advantages	Disadvantages
Medium net energy and high efficiency at accessible sites	High cost except at concentrated and accessible sources
Lower CO_2 emissions than fossil fuels	Scarcity of suitable sites
Low operating costs at favorable sites	Noise and some CO_2 emissions

© Cengage Learning

FIGURE 16.26 Using geothermal energy for space heating and for producing electricity or high-temperature heat for industrial processes has advantages and disadvantages (**Concept 16.5**). ***Critical thinking:*** Which single advantage and which single disadvantage do you think are the most important? Why? Do you think the advantages of this energy resource outweigh its disadvantages? Why?

Photo: N. Minton/Shutterstock.com

16.6 WHAT ARE THE ADVANTAGES AND DISADVANTAGES OF USING BIOMASS AS AN ENERGY SOURCE?

CONCEPT 16.6A Solid biomass is a potentially renewable resource, but it requires large areas of land and burning it faster than it is replenished produces a net gain in emissions of greenhouse gases.

CONCEPT 16.6B Liquid biofuels derived from biomass can lessen dependence on oil, but devoting large areas of land to biofuel crops can degrade soil and biodiversity and increase greenhouse gas emissions.

Producing Energy by Burning Solid Biomass

Energy can be produced by burning the solid biomass or organic matter found in plants or plant-related material or by converting it to gaseous or liquid biofuels. Examples of biomass fuels include wood, wood pellets, wood wastes, charcoal made from wood, and agricultural wastes such as sugarcane stalks, rice husks, and corncobs.

Biomass is burned mostly for heating and cooking, but can also be used for industrial processes and generating electricity. Biomass used for heating and cooking supply 10% of the world's energy, 35% of the energy used in less-developed countries, and 95% of the energy used in the poorest countries.

Wood is a renewable resource only if it is not harvested faster than it is replenished. The problem is that 2.7 billion people in 77 less-developed countries face a *fuelwood crisis*. They often are forced to meet their fuel needs by harvesting trees faster than new ones can replace them.

One solution is to plant fast-growing trees, shrubs, or perennial grasses in *biomass plantations*. However, repeated cycles of growing and harvesting these plantations can deplete the soil of key nutrients. It can also allow for the spread of nonnative tree species that become invasive. Clearing forests and grasslands to provide fuel also reduces biodiversity and the amount of vegetation that would otherwise capture climate-changing CO_2.

In the southeastern and northwestern United States, virgin and second-growth hardwood forests are being cleared to make wood pellets for fuel. They are mostly exported to European Union countries for use in heating factories and producing electricity. Critics call this an unsustainable practice. The pellet industry denies that they are removing whole trees and says they are using only tree branches and other wood wastes to make the pellets. However, as the volume of wood pellet production has increased, observers are seeing the destruction of large forested areas.

There is also controversy over burning forestry and crop wastes to provide heat and electricity. The supply of such wastes is not as large as some have estimated, and collecting and transporting these widely dispersed wastes to factories and utilities is difficult and expensive. In addition, crop wastes left on fields are valuable soil nutrients, and many argue they should be used as such.

Burning wood and other forms of biomass produces CO_2 and other pollutants such as fine particulates in smoke. In 2014, the EPA proposed phasing in stricter regulations to curb such pollution from new residential wood-burning stoves in the United States, beginning in 2015. Figure 16.27 lists the major advantages and disadvantages of burning solid biomass as a fuel (**Concept 16.6A**).

Using Liquid Biofuels to Power Vehicles

Biomass can also be converted into liquid biofuels for use in motor vehicles. The two most common liquid biofuels are *ethanol* (ethyl alcohol produced from plants and plant wastes) and *biodiesel* (produced from vegetable oils). The biggest producers of liquid biofuel are, in order, the United States (producing ethanol from corn), Brazil (producing ethanol from sugarcane residues), the European Union (producing biodiesel from vegetable oils), and China (producing ethanol from nongrain plant sources).

Trade-Offs

Solid Biomass

Advantages	Disadvantages
Widely available in some areas	Contributes to deforestation
Moderate costs	Clear-cutting can cause soil erosion, water pollution, and loss of wildlife habitat
Medium net energy	Can open ecosystems to invasive species
No net CO_2 increase if harvested, burned, and replanted sustainably	Increases CO_2 emissions if harvested and burned unsustainably
Plantations can help restore degraded lands	

FIGURE 16.27 Burning solid biomass as a fuel has advantages and disadvantages (**Concept 16.6A**). ***Critical thinking:*** Which single advantage and which single disadvantage do you think are the most important? Why? Do you think the advantages outweigh the disadvantages? Why?

Top: Fir4ik/Shutterstock.com. Bottom: Eppic/Dreamstime.com.

Biofuels have three major advantages over gasoline and diesel fuel produced from oil. *First,* biofuel crops can be grown throughout much of the world, which can help more countries reduce their dependence on imported oil. *Second,* if growing new biofuel crops keeps pace with harvesting them, there is no net increase in CO_2 emissions, unless existing grasslands or forests are cleared to plant biofuel crops. *Third,* biofuels are easy to store and transport through existing fuel networks and can be used in motor vehicles at little additional cost.

Since 1975, global ethanol production has increased rapidly, especially in the United States and Brazil. Brazil makes ethanol from *bagasse,* a residue produced when sugarcane is crushed. This sugarcane ethanol has a medium net energy that is 8 times higher than that of ethanol produced from corn. About 45% of Brazil's motor vehicles run on ethanol or ethanol–gasoline mixtures produced from sugarcane grown on only 1% of the country's arable land. This has greatly reduced Brazil's dependence on imported oil.

About 43% of the corn produced in the United States in 2014 was used to make ethanol, which is mixed with gasoline to fuel cars. Studies indicate that corn-based ethanol has a low net energy because of the large-scale use of fossil fuels to produce fertilizers, grow the corn, and convert it to ethanol.

According to a 2015 report by the Environmental Working Group (EWG), producing and burning corn-based ethanol adds at least 20% more greenhouse gases to the atmosphere per unit of energy than does producing and burning gasoline. Growing corn also requires a great deal of water, and ethanol distilleries produce large volumes of wastewater. Another EWG study concluded that the heavily government-subsidized corn-based ethanol program in the United States has taken more than 2 million hectares (5 million acres) of land out of the soil conservation reserve (p. 310), an important topsoil preservation program.

As a result, a number of scientists and energy economists call for withdrawing all government subsidies for corn-based ethanol production and sharply reducing the large amount of ethanol required in U.S. gasoline as mandated by the Energy Independence and Security Act of 2007. Corn producers and ethanol distillers claim that the harmful environmental effects of corn-based ethanol are overblown and that it has many environmental and economic benefits.

An alternative to corn-based ethanol is *cellulosic ethanol*, which is produced from the inedible cellulose that makes up most of the biomass of plants in the form of leaves, stalks, and wood chips. Cellulosic ethanol can be produced from tall and rapidly growing grasses such as switchgrass and miscanthus that do not require nitrogen fertilizers and pesticides. They also do not have to be replanted because they are perennial plants, and they can be grown on degraded and abandoned farmlands.

Ecologist David Tilman (Individuals Matter 12.1, p. 314) estimates that the net energy of cellulosic ethanol is about 5 times that of corn-based ethanol. However, producing cellulosic ethanol is not yet affordable, and growing switchgrass requires even more land than corn. More research is also needed to determine possible environmental impacts.

In Malaysia and Indonesia, large areas of tropical rain forests are being cleared and replaced with plantations of oil palm trees (Figure 10.4, p. 225), which produce a fruit that contains palm oil. After the oil is extracted, about a third of it is exported to Europe to make biodiesel fuel and the rest goes into processed food and cosmetics. The clearing and burning of tropical forests to make space for these palm oil plantations eliminates the vital biodiversity of the forests. It also adds CO_2 to the atmosphere, and the resulting plantations remove far less CO_2 than do the forests they replace.

In a United Nations report on bioenergy, and in another study by R. Zahn and his colleagues, scientists warned that large-scale biofuel crop farming could reduce biodiversity by eliminating more forests, grasslands, and wetlands and increasing soil degradation and erosion. It would also lead to higher food prices if it becomes more profitable to grow corn for biofuel rather than for feeding livestock and people.

Another approach involves using certain strains of algae to produce biofuels. Algae can grow year-round in various aquatic environments. The algae store their energy as natural oils within their cells. This oil can be extracted and refined to make a product very much like gasoline or biodiesel. Researchers estimate that algae can produce 10 times more energy per area of land than plants used to produce cellulosic ethanol can produce. Currently, extracting and refining the oil from algae is too costly. More research is also needed on which types of algae are most suitable and which ways of growing them are the most successful.

Figure 16.28 compares the advantages and disadvantages of using biodiesel and ethanol liquid biofuels.

Trade-Offs

Liquid Biofuels

Advantages	Disadvantages
Reduced CO_2 emissions for some crops	Fuel crops can compete with food crops for land and raise food prices
Medium net energy for biodiesel from oil palms	Fuel crops can be invasive species
Medium net energy for ethanol from sugarcane	Low net energy for corn ethanol and for biodiesel from soybeans
	Higher CO_2 emissions from corn ethanol

FIGURE 16.28 Ethanol and biodiesel biofuels have advantages and disadvantages (**Concept 16.6B**). ***Critical thinking:*** Which single advantage and which single disadvantage do you think are the most important? Why?

16.7 WHAT ARE THE ADVANTAGES AND DISADVANTAGES OF USING HYDROPOWER?

CONCEPT 16.7 We can use water flowing over dams, tidal flows, and ocean waves to generate electricity, but environmental concerns and limited availability of suitable sites may limit the use of these energy resources.

Producing Electricity from Falling and Flowing Water

Hydropower is any technology that uses the kinetic energy of flowing and falling water to produce electricity. It is an indirect form of solar energy because it depends on heat from the sun evaporating surface water. The water is deposited as rain or snow at higher elevations. Through gravity, the water can flow to lower elevations in rivers as part of the earth's solar-powered water cycle (Figure 3.19, p. 63).

The most common approach to harnessing renewable hydropower is to build a high dam across a large river to create a reservoir (see photo on pp. 322–323). Some of the water stored in the reservoir is allowed to flow through large pipes at controlled rates. The flowing water causes blades on a turbine to turn and spin a generator to produce electricity. Electric lines then carry the electricity to where it is needed.

Hydropower is the world's leading renewable energy source and produces about 16% of the world's electricity in 2015. In order, the world's top three producers of hydropower are China, Brazil, and the United States. In 2014, hydropower supplied about 7% of the electricity used in the United States and about half of the electricity used on the West Coast, mostly in Washington and California. Electricity from hydropower has been growing (Figure 16.29).

According to the United Nations, only 13% of the world's potential for hydropower has been developed. Countries with the greatest potential include China, India, and South America, as well as countries in Central Africa and Asia. China has plans to more than double its hydropower output during the next decade and is also building or funding more than 200 hydropower dams around the world.

Hydropower is the least expensive renewable energy resource. Once a dam is up and running, its source of energy—flowing water—is free and is annually renewed by snow and rainfall. Despite their potential, some analysts expect that the use of large-scale hydropower plants will fall slowly over the next several decades, as many existing reservoirs fill with silt and become useless faster than new systems are built. There is also growing concern over emissions of methane, a potent greenhouse gas, from the decomposition of submerged vegetation in hydropower plant reservoirs, especially in warm climates. Scientists at Brazil's National Institute for Space Research estimate that the world's largest dams altogether are the single largest human-caused source of climate-changing methane. The electricity output of hydropower plants may also drop if atmospheric temperatures continue to rise. This will cause melting of mountain glaciers, a primary source of water for these plants.

It is unlikely that large new hydroelectric dams will be built in the United States because most of the best sites already have dams and because of controversy and the high costs involved with building new dams. However, the turbines at many existing U.S. hydropower dams could be modernized and upgraded to increase their output of electricity. Figure 16.30 lists the major advantages and disadvantages of using large-scale hydropower plants to produce electricity (**Concept 16.7**).

Using Tides and Waves to Produce Electricity

Another way to produce electricity from flowing water is to tap into the energy from *ocean tides* and *waves*. In some coastal bays and estuaries, water levels can rise or fall by 6 meters (20 feet) or more between daily high and low tides. Dams can be built across the mouths of such bays and estuaries to capture the energy in these flows for hydropower, but sites with large tidal flows are rare. The only three large tidal energy dams currently operating are in France, Nova Scotia, and South Korea. According to energy

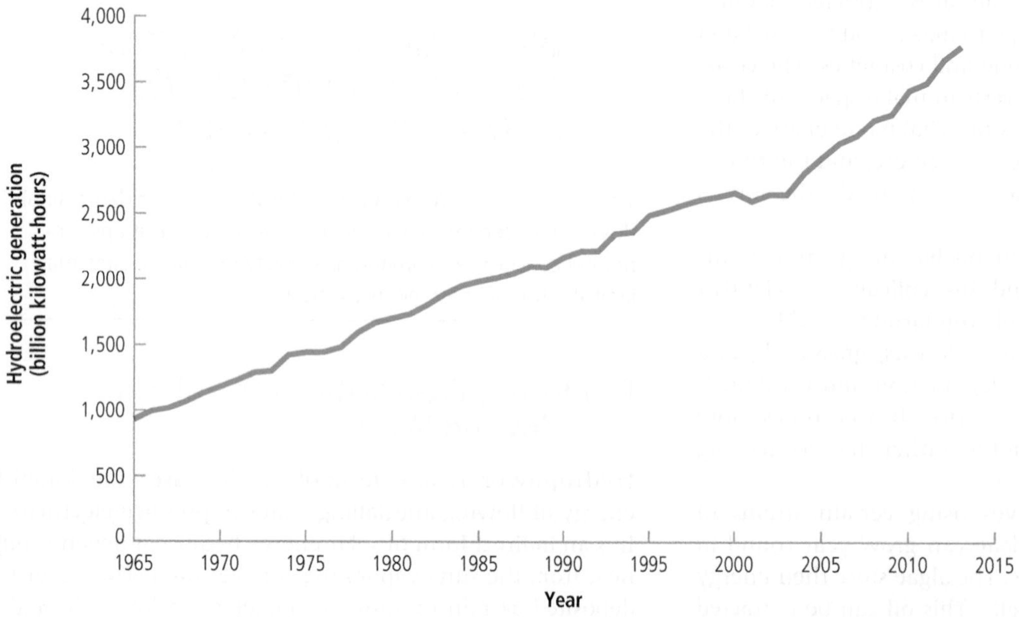

FIGURE 16.29 World hydroelectric generation, 1965–2013. ***Data analysis:*** After 1965, how many years did it take the world to double its hydroelectric capacity?

(Compiled by the authors using data from International Energy Agency, British Petroleum, Worldwatch Institute, and Earth Policy Institute.)

Trade-Offs

Large-Scale Hydropower

Advantages	Disadvantages
High net energy	Large land disturbance and displacement of people
Large untapped potential	High CH_4 emissions from rapid biomass decay in shallow tropical reservoirs
Low-cost electricity	Disrupts downstream aquatic ecosystems
Low emissions of CO_2 and other air pollutants in temperate areas	

FIGURE 16.30 Using large dams and reservoirs to produce electricity has advantages and disadvantages (**Concept 16.7**). ***Critical thinking:*** Which single advantage and which single disadvantage do you think are the most important? Why? Do you think that the advantages of this technology outweigh its disadvantages? Why?

experts, tidal power will make only a minor contribution to the world's electricity production because of the rarity of suitable sites.

For decades, scientists and engineers have been trying to produce electricity by tapping wave energy along seacoasts where there are almost continuous waves. However, production of electricity from tidal and wave systems is limited because of a lack of suitable sites, citizen opposition at some sites, high costs, and equipment damage from saltwater corrosion and storms.

16.8 WHAT ARE THE ADVANTAGES AND DISADVANTAGES OF USING HYDROGEN AS AN ENERGY SOURCE?

CONCEPT 16.8 Hydrogen is a clean energy source as long as it is not produced with the use of fossil fuels, but it has a negative net energy.

Will Hydrogen Save Us?

Hydrogen (H) is the simplest and most abundant chemical element in the universe. The sun produces its energy, which sustains life on the earth, through the nuclear fusion of hydrogen atoms (Figure 15.27, p. 402).

Some scientists say that the fuel of the future is hydrogen gas (H_2). Most of their research has been focused on

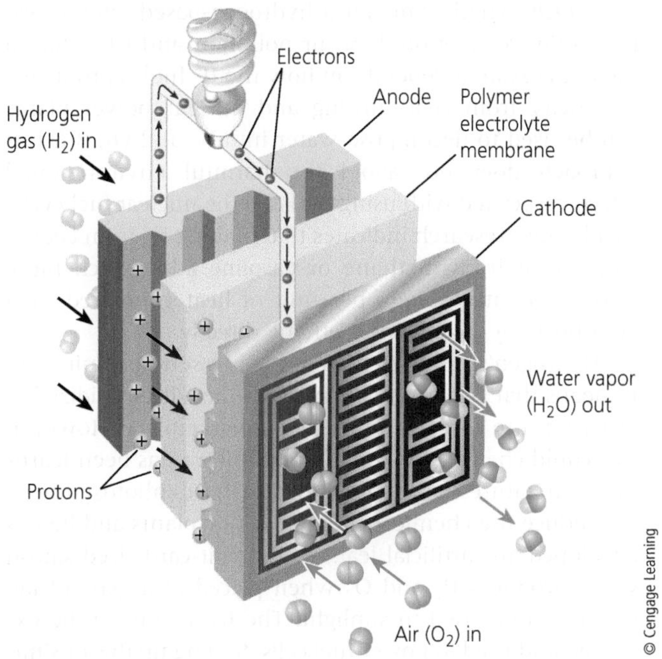

FIGURE 16.31 A fuel cell takes in hydrogen gas and separates the hydrogen atoms' electrons from their protons. The electrons flow through wires to provide electricity, while the protons pass through a membrane and combine with oxygen gas to form water vapor. Note that this process is the reverse of electrolysis, the process of passing electricity through water to produce hydrogen fuel.

using fuel cells (Figure 16.31) that combine H_2 and oxygen gas (O_2) to produce electricity and water vapor ($2\ H_2 + O_2 \rightarrow 2\ H_2O$ + energy), a harmless chemical that is emitted into the atmosphere.

Widespread use of hydrogen as a fuel for running motor vehicles, heating buildings, and producing electricity would eliminate most of the outdoor air pollution that comes from burning fossil fuels. It would also greatly slow climate change and ocean acidification, because production and use of hydrogen do not increase CO_2 emissions unless the H_2 is produced with the use of fossil fuels or nuclear power.

However, there are four major challenges in turning the vision of hydrogen as a major fuel into reality. *First,* there is hardly any hydrogen gas (H_2) in the earth's atmosphere. We can produce H_2 by heating water or passing electricity through it; by stripping it from the methane (CH_4) found in natural gas and from gasoline molecules; and through a chemical reaction involving coal, oxygen, and steam. *Second,* hydrogen has a negative net energy because it takes more high-quality energy to produce H_2 using these methods than we get by burning it. *Third,* although fuel cells are the best way to use H_2, current versions of fuel cells are expensive. However, progress in the development of nanotechnology (see Science Focus 14.1, p. 370) and mass production could lead to less expensive fuel cells.

Fourth, whether or not a hydrogen-based energy system produces less outdoor air pollution and CO_2 than a fossil fuel system depends on how the H_2 fuel is produced. Electricity from coal-burning and nuclear power plants can be used to decompose water into H_2 and O_2. But this approach does not avoid the harmful environmental effects associated with using coal and the nuclear fuel cycle. In addition, research indicates that making H_2 from coal or stripping it from methane or gasoline adds much more CO_2 to the atmosphere per unit of heat generated than does burning the coal or methane directly.

Hydrogen's negative net energy is a serious limitation. It means that this fuel will have to be heavily subsidized in order for it to compete in the open marketplace. However, this could change. Chemist Daniel Nocera has been learning from nature by studying how a leaf uses photosynthesis to produce the chemical energy used by plants and he has developed an "artificial leaf." This credit-card-sized silicon wafer produces H_2 and O_2 when placed in a glass of tap water and exposed to sunlight. The hydrogen can be extracted and used to power fuel cells. Scaling up this or similar processes to produce large amounts of H_2 at an affordable price with an acceptable net energy over the next several decades could represent a tipping point for use of solar energy and hydrogen fuel. Doing so would help implement the solar energy **principle of sustainability** on a global scale. **GREEN CAREER: Fuel cell technology**

Figure 16.32 lists the major advantages and disadvantages of using hydrogen as an energy resource (**Concept 16.8**).

Trade-Offs

Hydrogen

Advantages	Disadvantages
Can be produced from plentiful water at some sites	Negative net energy
No CO_2 emissions if produced with use of renewables	CO_2 emissions if produced from carbon-containing compounds
Good substitute for oil	High costs create need for subsidies
High efficiency in fuel cells	Needs H_2 storage and distribution system

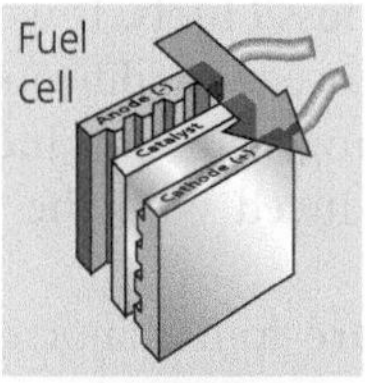

FIGURE 16.32 Using hydrogen as a fuel for vehicles and for providing heat and electricity has advantages and disadvantages (**Concept 16.8**). ***Critical thinking:*** Which single advantage and which single disadvantage do you think are the most important? Why? Do you think that the advantages of hydrogen fuel outweigh its disadvantages? Why?

Photo: LovelaceMedia/Shutterstock.com

16.9 HOW CAN WE MAKE THE TRANSITION TO A MORE SUSTAINABLE ENERGY FUTURE?

CONCEPT 16.9 We can make the transition to a more sustainable energy future by reducing energy waste, improving energy efficiency, using a mix of renewable energy resources, and including the environmental and health costs of energy resources in their market prices.

Shifting to a New Energy Economy

According to its proponents, shifting to a new set of energy resources over the next 50 to 60 years would have numerous environmental, health, and economic benefits.

China (which uses 20% of the world's energy) and the United States (which uses 19% of the world's energy) are the key players in making this shift. China has a long way to go in reducing its heavy dependence on coal and leads the world in climate-changing CO_2 emissions. However, it has launched efforts to make its economy more energy efficient, build a modern smart electrical grid, and install solar hot water heaters on a large scale. China is also building wind farms and solar power plants, supporting research on better batteries and improved solar and wind technologies, and building and selling all-electric cars. It is also making money by producing and selling more wind turbines and solar cell panels than any other country.

The United States is also making efforts to shift to a new energy economy. However, it is falling behind China's efforts and those of countries such as Germany, Sweden, and Denmark. This is mostly because of more than 40 years of successful efforts by powerful fossil fuel and electric utility companies to stop or slow down this shift because it threatens their profits.

This energy and economic transition is underway and is accelerating because it is increasingly driven by market forces. This is the result of the rapidly falling prices of electricity produced by the sun and wind, new energy technologies, and strong and growing public opposition to coal and nuclear power. Investors are moving rapidly into clean energy technologies. According to many scientists and energy economists, the shift to a new energy economy could be further accelerated if citizens, the leaders of emerging renewable energy companies, and energy investors demanded the following from their elected officials:

- Use **full-cost pricing** to include the harmful health and environmental costs of using fossil fuels and all other energy resources in their market prices.
- Tax carbon emissions. This is supported by most economists and many business leaders, and is now done in 40 countries. Use the revenue to reduce taxes on income and wealth and to promote investments and

research in new energy-efficient and renewable energy technologies. (We discuss this further in Chapter 19.)

- Sharply decrease and eventually eliminate government subsidies for fossil fuel industries, which are well-established and profitable businesses.
- Establish a national feed-in-tariff system that guarantees owners of wind farms, solar power plants, and home solar systems a long-term price for energy that they feed into their electrical grids (as is being done in more than 50 countries, many of them in Europe).
- Mandate that a certain percentage (typically 20–40%) of the electricity generated by utility companies be from renewable resources (as is done in 24 countries and in 29 U.S. states).
- Increase government fuel efficiency (CAFE) standards for new vehicles to 43 kilometers per liter (100 miles per gallon) by 2040.

We have the creativity, wealth, and most of the technology to make the transition to a safer, more energy-efficient, and cleaner energy economy within your lifetime. With such a shift, we could greatly increase our beneficial environmental impact. Figure 16.33 lists ways in which you can take part in the transition toward such a future.

What Can You Do?

Shifting to More Sustainable Energy Use

- Walk, bike, or use mass transit or a car pool to get to work or school
- Drive only vehicles that get at least 17 kpl (40 mpg)
- Have an energy audit done in the place where you live
- Superinsulate the place where you live and plug all air leaks
- Use passive solar heating
- For cooling, open windows and use fans
- Use a programmable thermostat and energy-efficient heating and cooling systems, lights, and appliances
- Turn down your water heater's thermostat to 43–49°C (110–120°F) and insulate hot water pipes
- Turn off lights, TVs, computers, and other electronics when they are not in use
- Wash laundry in cold water and air dry it on racks

FIGURE 16.33 Individuals matter: You can make a shift in your own life toward using energy more sustainably. ***Critical thinking:*** Which three of these measures do you think are the most important ones to take? Why? Which of these steps have you already taken and which do you plan to take?

BIG IDEAS

- To make our economies more sustainable, we need to reduce our use of fossil fuels, especially coal, and greatly increase energy efficiency, reduce energy waste, and use a mix of renewable energy resources, especially the sun and wind.
- Making this energy shift will have important economic and environmental benefits.
- Making the transition to a more sustainable energy future will require including the harmful environmental and health costs of all energy resources in their market prices, taxing carbon emissions, and greatly increasing government subsidies and research and development for improving energy efficiency and developing renewable energy resources.

Tying It All Together

Saving Energy and Money and Reducing Our Environmental Impact

In the Core Case Study, we learned that by wasting energy, we waste money and increase our harmful environmental impact. The world and the United States waste so much energy that reducing this waste by increasing energy efficiency and saving energy is the quickest, cleanest, and usually the cheapest way to provide more energy. In doing so, we would also reduce pollution and environmental degradation and slow climate change and ocean acidification.

Over the next 50 to 60 years, we could choose to rely less on fossil fuels, especially coal, and more on increasing energy efficiency and using a mix of solar, wind, and other renewable energy resources. Making this energy shift would have enormous economic, environmental, and health benefits.

Relying more on energy from the sun and wind helps us to implement the solar energy **principle of sustainability**. It also follows the chemical cycling **principle of sustainability** by reducing our excess inputs of CO_2 into the atmosphere, which disrupt the earth's carbon cycle. This energy shift also mimics the earth's biodiversity **principle of sustainability** by reducing the environmental degradation that degrades biodiversity.

Making this shift will require implementing the full-cost pricing **principle of sustainability** by including the harmful health and environmental costs of all energy resources in their market prices. It will also require compromise and trade-offs in the political arena, in keeping with the win-win **principle of sustainability**. By making this energy shift, we would also by implementing the ethical **principle of sustainability** that calls for us to honor future generations by increasing our beneficial environmental impact.

LianeM/Shutterstock.com

Chapter Review

Core Case Study

1. What are two ways in which we waste energy? What are the benefits of reducing this waste?

Section 16.1

2. What is the key concept for this section? Why do we need to make a new energy transition over the next 50 to 60 years? What are the two key components of this energy resource shift? What are the economic and environmental advantages of making this energy transition? List three factors that are driving this shift. How would the United States benefit from it? Describe Germany's progress in making this energy shift.

Section 16.2

3. What are the two concepts for this section? Define and give an example of **energy efficiency**. Explain why improving energy efficiency and reducing energy waste is a major energy resource. What percentage of the energy used in the United States is unnecessarily wasted? Why is so much energy wasted? List three widely used energy-inefficient technologies. What are the major advantages of reducing energy waste? List four ways to save energy and money in industry. What is **cogeneration**? What is a smart electric grid and why is it important? What are U.S. CAFE standards? What are the hidden costs of using gasoline? List four ways to save energy and money in transportation. Distinguish between hybrid, plug-in hybrid, and all-electric vehicles. Explain the importance of developing better and cheaper batteries and list some advances in this area. What are fuel cells and what are their major advantages and disadvantages? What are some green architecture technologies? What is superinsulation? What are 10 ways to improve energy efficiency in existing buildings? List six ways in which you can save energy where you live. What is the rebound effect and how can it be reduced? Give five reasons why we are not making greater use of renewable energy.

Section 16.3

4. What is the key concept for this section? Distinguish between a **passive solar heating system** and an **active solar heating system** and list the major advantages and disadvantages of using such systems for heating buildings. What are three ways to cool houses naturally? What are **solar thermal systems**, how are they used, and what are the major advantages and disadvantages of using them? What is a **solar cell (photovoltaic** or **PV cell)** and what are the major advantages and disadvantages of using such devices to produce electricity?

Section 16.4

5. What is the key concept for this section? What are the advantages of using taller wind turbines? Summarize the global and U.S. potential for wind power. What are the major advantages and disadvantages of using wind to produce electricity?

Section 16.5

6. What is the key concept for this section? What is **geothermal energy** and what are three sources of such energy? What are the major advantages and disadvantages of using geothermal energy as a source of heat and to produce electricity?

Section 16.6

7. What are the two key concepts for this section? What is **biomass** and what are the major advantages and disadvantages of using wood to provide heat and electricity? What are the major advantages and disadvantages of using biodiesel and ethanol to power motor vehicles? Explain how algae and bacteria could be used to produce fuels nearly identical to gasoline and diesel fuel.

Section 16.7

8. What is the key concept for this section? Define **hydropower** and summarize the potential for expanding it. What are the major advantages and disadvantages of using hydropower? What is the potential for using tides and waves to produce electricity?

Section 16.8

9. What is the key concept for this section? What are the major advantages and disadvantages of using hydrogen as a fuel to use in producing electricity and powering vehicles?

Section 16.9

10. What is the key concept for this section? Describe efforts by China and the United States to make the shift to a new energy economy. List six strategies suggested by scientists and energy economists for making the transition to a more sustainable energy future. List five ways you can participate in making the transition to a new energy future. What are this chapter's *three big ideas*? Explain how we would be applying the six **principles of sustainability** by improving energy capacity and shifting to renewable energy resources.

Note: Key terms are in bold type.

Critical Thinking

1. List five ways in which you unnecessarily waste energy during a typical day, and explain how each of these actions violates the three scientific **principle of sustainability**.

2. Congratulations! You have won enough money to build a more sustainable house of your choice. With the goal of maximizing energy efficiency, what type of house would you build? How large would it be? Where would you locate it? What types of materials would you use? What types of materials would you *not* use? How would you heat and cool the house? How would you heat water? What types of lighting, stove, refrigerator, washer, and dryer would you use? Which, if any, of these appliances could you do without?

3. Suppose that a homebuilder installs electric baseboard heat and claims, "It is the cheapest and cleanest way to go." Apply your understanding of the second law of thermodynamics (p. 42) and net energy (Figure 15.3, p. 384) to evaluate this claim. Write a response to the homebuilder summarizing your findings.

4. Do you think that the estimated hidden costs of gasoline should be included in its price at the pump? Explain. Would you favor much higher gasoline taxes to accomplish this if payroll taxes and income taxes were reduced to balance gasoline tax increases, with no net additional cost to consumers? Explain.

5. Suppose that a wind power developer has proposed building a wind farm near where you live. Would you be in favor of the project or opposed to it? Write a letter to your local newspaper or a blog for a website explaining your position and your reasoning. As part of your research, determine how the electricity you use now is generated and where the nearest power plant is located, and include this information in your arguments.

6. Explain why you agree or disagree with the following proposals made by various energy analysts:
   a. Government subsidies for all energy alternatives should be eliminated so that all energy resources can compete on a level playing field in the marketplace.
   b. All government tax breaks and other subsidies for conventional fossil fuels (oil, natural gas, and coal), synthetic natural gas and oil, and nuclear power (fission and fusion) should be eliminated. They should be replaced with subsidies and tax breaks for improving energy efficiency and developing renewable energy alternatives.
   c. Development of renewable energy alternatives should be left to private enterprise and should receive little or no help from the federal government, but the nuclear power and fossil fuels industries should continue to receive large federal government subsidies.
7. What percentages of the U.S. Department of Energy research and development budget would you devote to fossil fuels, nuclear power, renewable energy, and improving energy efficiency? How would you distribute your funds among the various renewable energy options? Explain your thinking.
8. How important is it to make the transition to a new energy future? Do you think it can be done? Explain. How would making such a transition affect your life and the lives of any children or grandchildren that you might have? How would not making such a transition affect your life and the lives of any children or grandchildren that you might have?

Doing Environmental Science

Do a survey of energy use at your school, based on the following questions: How is the electricity generated? How are most of the buildings heated? How is water heated? How are most of the vehicles powered? How is the computer network powered? How could energy efficiency be improved, if at all, in each of these areas? How could your school make use of solar, wind, biomass, and other forms of renewable energy if it does not already do so? Write up a proposal for using energy more efficiently and sustainably at your school and submit it to school officials.

Global Environment Watch Exercise

Go to your MindTap course to access the GREENR database. Starting on the home page, under "Browse Issues and Topics", click on *Energy,* then select *Renewable Energy.* Use this portal to search for information on the renewable energy policy in Germany (you may also use the World Map). What are some of the challenges the German government is facing, and how are they being addressed? What challenges would arise if a U.S. government mandate required a similar transition to renewable energies? Write a report answering these questions and discussing the effects of the "German Experiment" on the United States as it looks to revise its energy policy and address climate change.

Data Analysis

Study the table below and then answer the questions that follow it by filling in the blank columns in the table.

Combined City/Highway Fuel Efficiency for 2015 Models (mpg)

Model	Miles per Gallon (mpg)	Kilometers per Liter (kpl)	Annual Liters (Gallons) of Gasoline	Annual CO_2 Emissions
Toyota Prius Hybrid	55			
Honda CR-Z AV-S7	37			
Chevrolet Cruze A-S6	30			
Ford Fusion AS6	29			
Ford F150 2WD-6 Pickup Truck	20			
Chevrolet Silverado DC15 2WD Pickup Truck	20			
Toyota Tundra Pickup Truck A-S6	16			
Ferrari AM7	13			

Compiled by the authors using data from Fuel Economy Guide 2015, U.S. Environmental Protection Agency.

1. Using Supplement 1 (Measurement Units, p. S1), convert the miles per gallon figures in the table to kilometers per liter (kpl).
2. How many liters (and how many gallons) of gasoline would each car use annually if it was driven 19,300 kilometers (12,000 miles) per year?
3. How many kilograms (and how many pounds) of carbon dioxide would be released into the atmosphere each year by each car, based on the fuel consumption calculated in question 2, assuming that the combustion of gasoline releases 2.3 kilograms of CO_2 per liter (19 pounds per gallon)?

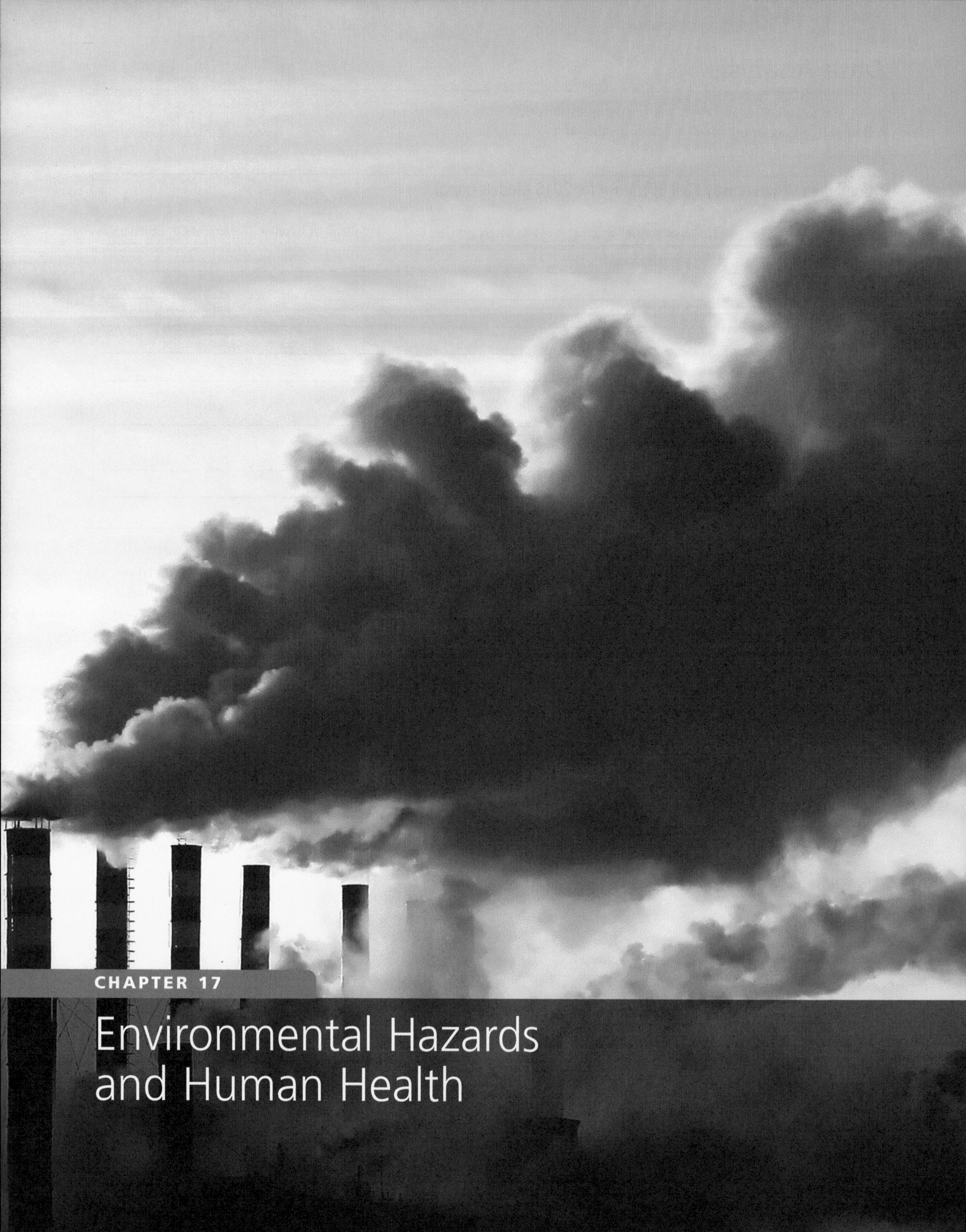

CHAPTER 17

Environmental Hazards and Human Health

The dose makes the poison.

PARACELSUS, 1540

Key Questions

17.1 What major health hazards do we face?

17.2 How do biological hazards threaten human health?

17.3 How do chemical hazards threaten human health?

17.4 How can we evaluate risks from chemical hazards?

17.5 How do we perceive and avoid risks?

Without effective air pollution control, coal-burning factories and power plants release toxic mercury and other air pollutants into the atmosphere.

Dudarev Mikhail/Shutterstock.com

Core Case Study

Mercury's Toxic Effects

Mercury (Hg) and its compounds are all toxic. Research indicates that long-term exposure to high levels of mercury can permanently damage the human nervous system, brain, kidneys, heart, and lungs. Low levels of mercury can cause birth defects and brain damage in fetuses and young children. Pregnant women, nursing mothers and their babies, women of childbearing age, and young children are especially vulnerable to the harmful effects of mercury.

This toxic metal is released into the air from rocks, soil, and volcanoes and by vaporization from the oceans. Such natural sources account for about one-third of the mercury reaching the atmosphere each year. According to the U.S. Environmental Protection Agency (EPA), the remaining two-thirds come from human activities. The largest source of mercury air pollution is thousands of small-scale gold miners in Asia, Africa, and South America (see Chapter 14, p. 368). Other large sources of mercury in the atmosphere include emissions from coal-burning power and industrial plants (chapter-opening photo), cement kilns, smelters, and solid-waste incinerators. Mercury can also be released from household products such as certain types of thermometers, light bulbs, and thermostats when they break or disintegrate.

Because mercury is an element, it cannot be broken down or degraded. Therefore, it accumulates in soil, water, and the tissues of humans and other animals. In the atmosphere, some elemental mercury is converted to more toxic mercury compounds that are deposited in the oceans, in lakes (Figure 17.1), in other aquatic environments, and on land.

Under certain conditions in aquatic systems, bacteria convert inorganic mercury compounds to highly toxic *methylmercury* (CH_3Hg^+). Like DDT (see Figure 9.14, p. 205), methylmercury can be biologically magnified in food chains and food webs. High levels of methylmercury are often found in the tissues of large fishes, such as tuna, swordfish, shark, and marlin, which feed at high trophic levels. However, shrimp and salmon generally have low levels of mercury. A 2015 study led by Kyra St. Pierre found that methylmercury is finding its way from ocean waters into land food webs via lichens that grow on Arctic coastlines and absorb the toxin. Lichens make up 77% of the diet of caribou, which in turn are eaten by humans and other predators and scavengers.

Humans are exposed to mercury in two major ways. People eat contaminated fish and shellfish that are contaminated with methylmercury. This accounts for 75% of all human exposures to mercury. People also inhale vaporized elemental mercury or particles of inorganic mercury salts such as mercury sulfide (HgS) and mercuric chloride ($HgCl_2$)—pollutants that come mostly from many coal-burning power plants and solid-waste incinerators. Studies estimate that 30,000 to 60,000 of the children born each year in the United States are likely to have reduced IQs and possible nervous system damage due to such exposure.

This problem raises three important questions: How do scientists determine the potential harm from exposure to mercury and other chemicals? How serious is the risk of harm from a particular chemical compared to other risks? And what should we do with any evidence of harm?

In this chapter, we will look at how scientists try to answer these and other questions about our exposure to chemicals. We will also examine health threats from disease-causing bacteria, viruses, and protozoa, and from other environmental hazards that kill millions of people every year.

Drake Fleege/Alamy Stock Photo

FIGURE 17.1 Fish are contaminated with mercury in many lakes, including this one in Wisconsin.

17.1 WHAT MAJOR HEALTH HAZARDS DO WE FACE?

CONCEPT 17.1 We face health hazards from biological, chemical, physical, and cultural factors, and from the lifestyle choices we make.

We Face Many Types of Hazards

A **risk** is the probability of suffering harm from a hazard that can cause injury, disease, death, economic loss, or damage. Scientists often state the probability of a risk in terms such as, "The lifetime probability of developing lung cancer from smoking one pack of cigarettes per day is 1 in 250." This means that 1 of every 250 people who smoke a pack of cigarettes every day will likely develop lung cancer over a typical lifetime (usually considered to be 70 years). Probability can also be expressed as a percentage, as in a 30% chance of developing a certain type of cancer. The greater the probability of harm, the greater the risk.

Risk assessment is the process of using statistical methods to estimate how much harm a particular hazard can cause to human health or to the environment. It helps us compare risks and establish priorities for avoiding or managing risks. **Risk management** involves deciding whether and how to reduce a particular risk to a certain level and at what cost. Figure 17.2 summarizes how risks are assessed and managed.

We take risks every day. Examples include choosing to drive or ride in a car without a seatbelt; talking on a phone or texting while driving; eating foods that are high in cholesterol or have too much sugar; drinking too much alcohol; smoking or being in an enclosed space with a smoker; and living in a tornado-, hurricane-, or flood-prone area.

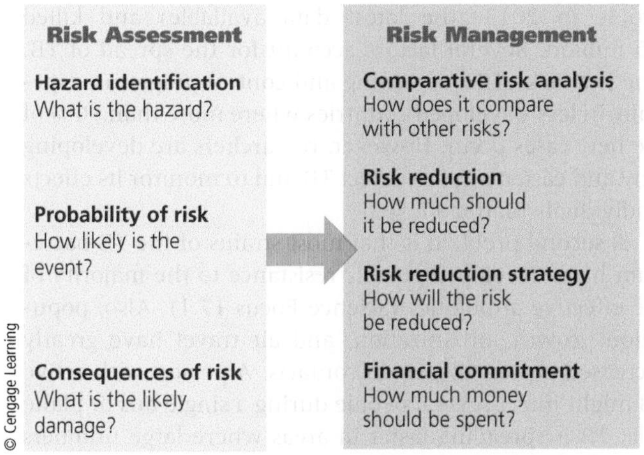
Risk Assessment
Hazard identification
What is the hazard?
Probability of risk
How likely is the event?
Consequences of risk
What is the likely damage?
Risk Management
Comparative risk analysis
How does it compare with other risks?
Risk reduction
How much should it be reduced?
Risk reduction strategy
How will the risk be reduced?
Financial commitment
How much money should be spent?
© Cengage Learning

FIGURE 17.2 Risk assessment and risk management are used to estimate the seriousness of various risks and to help reduce such risks. ***Critical thinking:*** What is an example of how you have applied this process in your daily living?

No one can live a risk-free life, but we can reduce exposure to risks. When assessing risks, it is important to understand how serious the risks are and whether the benefits of certain activities outweigh the risks.

Five major types of hazards threaten human health (**Concept 17.1**):

- *Biological hazards* from more than 1,400 **pathogens,** or organisms that can cause disease in other organisms. Examples are bacteria, viruses, parasites, protozoa, and fungi.
- *Chemical hazards* from harmful chemicals in the air, water, soil, food, and human-made products (**Core Case Study**).
- *Natural hazards* such as fire, earthquakes, volcanic eruptions, floods, tornadoes, and hurricanes.
- *Cultural hazards* in daily life such as unsafe working conditions, criminal assault, and poverty.
- *Lifestyle choices* such as smoking, making poor food choices, and having unsafe sex.

CONSIDER THIS . . .

THINKING ABOUT Health Hazards

Think of a hazard from each of these categories that you may have faced recently. Which one was the most threatening?

17.2 HOW DO BIOLOGICAL HAZARDS THREATEN HUMAN HEALTH?

CONCEPT 17.2 The most serious biological hazards we face are infectious diseases such as flu, acquired immunodeficiency syndrome (AIDS), tuberculosis, diarrheal diseases, and malaria.

Some Diseases Can Spread from Person to Person

An **infectious disease** is a disease caused by a pathogen such as a bacterium, virus, or parasite invading the body and multiplying in its cells and tissues. **Bacteria** are single-cell organisms that are found everywhere and that can multiply rapidly on their own. Most bacteria are harmless or beneficial but some cause diseases such as strep throat or tuberculosis (see Case Study that follows). **Viruses** are pathogens that work by invading a cell and taking over its genetic machinery to copy themselves in order to spread throughout the body. Viruses cause diseases such as flu and AIDS. **Parasites,** organisms that live on or inside other organisms and feed on them, can also cause malaria and other serious infectious diseases.

A **transmissible disease** is an infectious disease that can be transmitted from one person to another. Some transmissible diseases are bacterial diseases such as tuberculosis, many ear infections, and gonorrhea. Others are viral diseases such as the common cold, flu, and AIDS. A **nontransmissible disease** is caused by something other than a living organism and does not spread from one person to another. Transmissible diseases include cardiovascular (heart and blood vessel) diseases, most cancers, asthma, and diabetes.

In 1900, infectious disease was the leading cause of death in the world. Since then, and especially since 1950, the incidences of infectious diseases and the death rates from them have dropped significantly. This has been achieved mostly by a combination of better health care, the use of antibiotics to treat diseases caused by bacteria, and the development of vaccines to prevent the spread of some viral diseases. GOOD NEWS

Despite the declining risk of harm from infectious diseases, they remain serious health threats, especially in less-developed countries. Such diseases can be spread through air, water, food, and body fluids such as feces, urine, blood, and droplets sprayed by sneezing and coughing, and by insects such as mosquitoes. A large-scale outbreak of an infectious disease in an area or a country is called an *epidemic.* A global epidemic is called a *pandemic.* Figure 17.3 shows the annual death tolls from the world's seven deadliest infectious diseases (**Concept 17.2**).

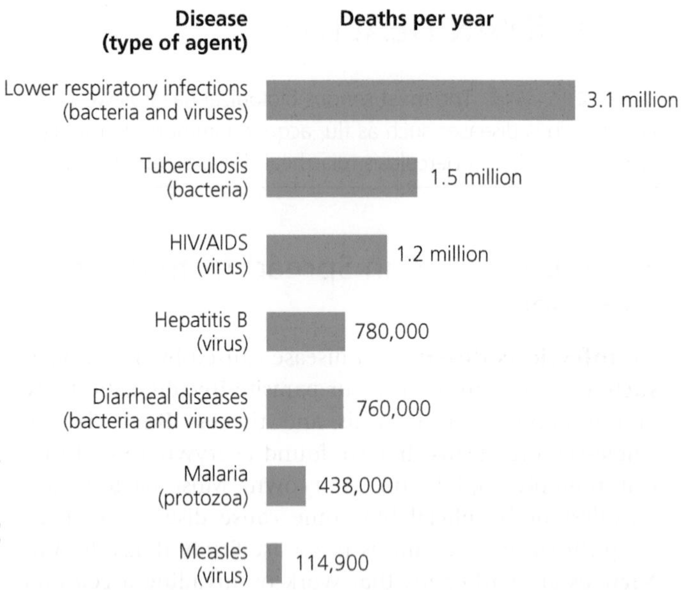

FIGURE 17.3 The World Health Organization has estimated that the world's seven deadliest infectious diseases kill about 8.3 million people per year—most of them poor people in less-developed countries (**Concept 17.2**). ***Data Analysis:*** How many people die from all seven of the infectious diseases every year? Every day?

(Compiled by the authors using data from the World Health Organization (WHO) and the U.S. Centers for Disease Control and Prevention)

CONSIDER THIS . . .

LEARNING FROM NATURE

A shark's skin is covered with tiny bumps that somehow help it to avoid bacterial infections. Scientists are using this information to create antibacterial films with a bumpy structure that could reduce human skin infections.

One reason why infectious disease is still a serious threat is that many disease-carrying bacteria have developed genetic immunity to widely used antibiotics (Science Focus 17.1). Also, many disease-transmitting species of insects such as mosquitoes have become immune to widely used pesticides such as DDT that once helped to control their populations (see Figure 12.27, p. 305).

Another factor that will likely keep infectious diseases high on the list of environmental health threats is climate change. Many scientists warn that warmer temperatures will likely allow some infectious diseases—especially those spread by mosquitoes and other insects that breed more rapidly in warmer climates—to spread more easily from tropical to temperate areas.

CASE STUDY

The Global Threat from Tuberculosis

Tuberculosis (TB) is a bacterial infection that destroys lung tissue and can ultimately lead to death. Since 1990, TB, a highly contagious bacterial infection of the lungs, has been spreading. Many TB-infected people do not appear to be sick, and most of them do not know they are infected. Left untreated, each person with active TB typically infects a number of other people. Without treatment, about half of the people with active TB die from bacterial destruction of their lung tissue (Figure 17.4).

According to the WHO, TB struck about 9 million people in 2013 (the latest data available) and killed 1.5 million. Several factors account for the spread of TB. One is a lack of TB screening and control programs, especially in less-developed countries where more than 90% of the new cases occur. However, researchers are developing new and easier ways to detect TB and to monitor its effects (Individuals Matter 17.1).

A second problem is that most strains of the TB bacterium have developed genetic resistance to the majority of the effective antibiotics (Science Focus 17.1). Also, population growth, urbanization, and air travel have greatly increased person-to-person contacts. A person with active TB might infect several people during a single bus or plane ride. TB is spreading faster in areas where large numbers of poor people crowd together, especially in the rapidly growing slums of less-developed countries.

Slowing the spread of the disease requires early identification and treatment of people with active TB, especially

Genetic Resistance to Antibiotics

Antibiotics are chemicals that can kill bacteria. They have played an important role in the 30-year increase in American life expectancy since 1950.

In 2014, the World Health Organization (WHO) issued a report warning that the age of antibiotics may be ending because many disease-causing bacteria are becoming genetically resistant to the antibiotics that we have long used to fight infectious diseases. The scientific and medical communities are falling behind in their efforts curb the spread of infectious bacterial diseases because of the astounding reproductive rate of bacteria, some of which can grow from a population of 1 to well over 16 million in 24 hours. This allows bacteria to quickly become genetically resistant to an increasing number of antibiotics through natural selection (see Figure 4.14, p. 87). To make matters worse, research indicates that in addition to passing genetic resistance to antibiotics to offspring bacteria can transfer such resistance to one another and to different strains of bacteria.

A major factor in the promotion of such genetic resistance, also called *antibiotic resistance*, is the widespread use of antibiotics on livestock raised in feedlots and CAFOs (see Chapter 12, p. 292). They are used to control disease and to promote growth among dairy and beef cattle, poultry, and hogs that are raised in large numbers in crowded conditions. The U.S. Food and Drug Administration (FDA) has estimated that about 80% of all antibiotics used in the United States and 50% of those used worldwide are added to the feed of healthy livestock. A study in the Proceedings of the National Academy of Sciences projected that such use of antibiotics will increase by about 67% between 2010 and 2030. There is growing concern that residues of some of these antibiotics in meat can play a role in increasing genetic resistance in humans to these antibiotics. According to the Centers for Disease Control and Prevention (CDC), 22% of antibiotic-resistant illness in humans is linked to food, especially food from livestock treated with antibiotics.

Another factor is the overuse of antibiotics for colds, flu, and sore throats, many of which are viral diseases that do not respond to treatment with antibiotics. In many countries, antibiotics are available without a prescription, which promotes excessive and unnecessary use. Yet another factor is the spread of bacteria around the globe by human travel and international trade. The growing use of antibacterial hand soaps and other antibacterial cleansers could also be promoting antibiotic resistance in bacteria.

Because of these factors acting together, every major disease-causing bacterium has now developed strains that resist at least 1 of the roughly 200 antibiotics in use and a growing number of infectious bacteria have developed genetic resistance to more than one antibiotic. We are now seeing the emergence of *superbugs*, bacteria that resist all available antibiotics. The CDC has estimated that each year, at least 2 million Americans get infectious diseases from superbugs, and at least 23,000 die. Also, 1 of every 25 U.S. hospital patients picks up such an infection while in the hospital.

For example, a bacterium known as methicillin-resistant *Staphylococcus aureus*, commonly known as MRSA (or "mersa"), has become resistant to most common antibiotics. MRSA can cause severe pneumonia, a vicious rash, and a quick death if it gets into the bloodstream.

MRSA can be found in hospitals, nursing homes, schools, gyms, and college dormitories. It can be spread through skin contact, unsanitary use of tattoo needles, and contact with poorly laundered clothing and shared items such as towels, bed linens, athletic equipment, and razors. Another worrisome superbug found in hospitals is *Clostridium difficile*, or *C. diff.* It causes severe diarrhea and can live on surfaces such as bed rails and medical equipment. It causes about 250,000 infections and 14,000 deaths per year in the United States, according to the CDC.

Health officials warn that we could be moving into a post-antibiotic era of higher death rates. No major new antibiotics have been developed in recent years, mostly because drug companies lose millions of dollars developing new antibiotics that are used for only a short time to treat infections. They make more money on drugs that users take daily for many years for diseases such as diabetes and high blood pressure. As a result, 15 of the 18 largest pharmaceutical companies no longer make antibiotics.

However, in 2015, researchers led by Kim Lewis discovered a new antibiotic called *teixobactin*, extracted from bacteria that live in dirt. In laboratory mice, it proved to be a powerful drug against tuberculosis, MRSA, and other infections. It works by breaking down a microbe's outer cell walls—an approach that makes it difficult for bacteria to develop resistance to it. It will take years of testing to learn whether teixobactin offers a possible solution to the serious problem of antibiotic resistance.

CRITICAL THINKING

What are three steps that you think we could take to slow the rate at which disease-causing bacteria are developing resistance to antibiotics?

2 million

the annual number of U.S. citizens who get infections that cannot be treated with any known antibiotics

INDIVIDUALS MATTER 17.1

Hayat Sindi: Health Science Entrepreneur

© Hayat Sindi

Growing up in a home of humble means in Saudi Arabia, Hayat Sindi was determined to get an education, become a scientist, and do something for humanity. She was the first Saudi woman to be accepted at Cambridge University. She earned a PhD and taught in the Cambridge's international medical program, and in 2012 she was named a National Geographic Explorer.

As a visiting scholar, Sindi worked with a team of scientists at Harvard University and co-founded a nonprofit company called *Diagnostics for All* to bring low-cost health monitoring to remote, poor areas of the world. The Harvard team sought to develop simple and inexpensive diagnostic tools that could be used to detect certain illnesses and medical problems in remote areas.

One such tool is a small piece of paper the size of a postage stamp, with tiny channels and wells etched into it. Technicians load it with various diagnostic chemicals and then put a drop of a patient's blood, urine, or saliva onto this paper where the chemicals react with the fluid to change its color. Results show up in a minute and can be easily be read to diagnose different medical infections and conditions such as declining liver function, which can result from taking anti-TB drugs. The test can be conducted by a technician with minimal training and requires no electricity, clean water, or special equipment. After the paper is used, it can be burned on the spot to prevent the spread of any infectious agents. Dr. Sindi was named a UNESCO Goodwill Ambassador for science education and has a passion for inspiring women and girls, particularly those in the Middle East, to pursue science.

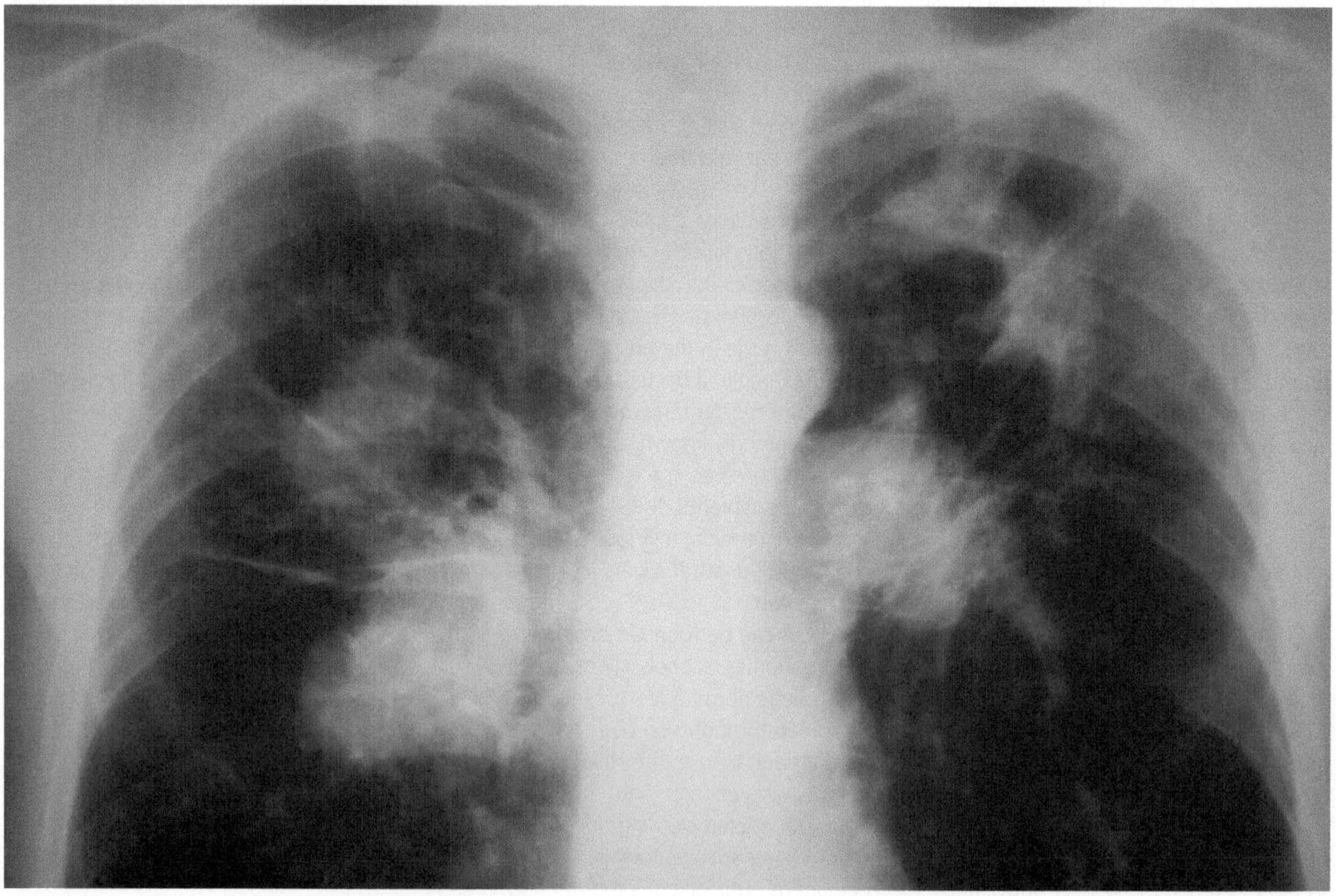

Michael Abbey/Visuals Unlimited/Corbis

FIGURE 17.4 Colorized red areas in this chest X-ray show where TB bacteria have destroyed lung tissue.

those with a chronic cough, which is the primary way in which the disease is spread from person to person. Treatment with a combination of four inexpensive drugs can cure 90% of individuals with active TB, but to be effective, the drugs must be taken every day for 6-9 months. Because symptoms often disappear after a few weeks of treatment, many patients think they are cured and stop taking the drugs. This can allow the disease to recur, possibly in drug-resistant forms, and to spread to other people.

A deadly form of tuberculosis known as *multidrug-resistant TB* is on the rise. About 480,000 new cases occur every year, according to the WHO, and each year the disease kills about 150,000 people. Fewer than half of those cases are cured each year, and only with the best available medical care costing more than $500,000 per person on average. Because this disease cannot be treated effectively with antibiotics, victims must be isolated from the rest of society, some permanently, and they pose a threat to health workers.

CONSIDER THIS . . .

THINKING ABOUT Dealing with Tuberculosis

If you were a public health official, what would you do to try to slow the spread of tuberculosis?

Viral Diseases and Parasites Are Killers

Viruses are not affected by antibiotics and can be deadly. The biggest viral killer is the *influenza* or *flu virus* (**Concept 17.2**), which is transmitted by the body fluids or airborne droplets released when an infected person coughs or sneezes. Influenza often leads to fatal pneumonia. Flu viruses are so easily transmitted that an especially potent flu virus could spread around the world in a pandemic that could kill millions of people in only a few months.

The second biggest viral killer is the *human immunodeficiency virus*, or *HIV* (see the Case Study that follows). On a global scale, according to the Joint United Nations Programme on HIV/AIDS (UNAIDS), HIV infected about 2.0 million people in 2014 (down from 3.1 million in 2000), and the complications resulting from AIDS killed about 1.2 million (down from 1.6 million in 2000). HIV is transmitted by unsafe sex, the sharing of needles by drug users, infected mothers who pass the virus to their babies before or during birth, and exposure to infected blood.

The third largest viral killer is the *hepatitis B virus (HBV)*, which damages the liver. According to the WHO, it kills more than 780,000 people each year. It is spread in the same ways that HIV is spread.

Another deadly virus is Ebola. In 2014, an outbreak of this disease in the West African countries of Guinea, Liberia, and Sierra Leone infected 28,500 people and killed 11,300 of them.

Scientists believe Ebola originated in fruit bats, but it is usually spread to people from other animals and other people. It enters the body primarily through tiny cuts and scrapes in the skin and through mucous membranes of the eye, ear, nose, and mouth. One must contact the bodily fluids of an infected animal or person to get the virus. Within 4 to 10 days of infection, the victim typically develops sudden fever, sore throat, muscle pain, and headache. Advanced symptoms can include coughing, diarrhea, vomiting, chest pain, internal bleeding, and bleeding gums.

According to the WHO, the Ebola virus kills an average of 50% of those it infects within 8 days. No existing drug can cure Ebola, although experimental drugs are being evaluated. A victim's best hope is a strong immune response with intensive supportive care in a hospital, including continual rehydration.

Part of the problem in the 2014 African outbreak was that friends and family members tried to care for Ebola victims under unsterile conditions, and the virus was easily transmitted through casual contacts during such care. The chances of Ebola spreading in the United States and other more-developed countries are slim because hospitals, infection controls, and safe burial procedures are much more readily available than they are in many less-developed countries. More widespread screening of people for the Ebola virus can help reduce its spread (Figure 17.5). However, those who care for patients are at a much higher-than-average risk of getting the disease, no matter where they are.

Another deadly virus is the *West Nile virus*, which is transmitted to humans by the bite of a common mosquito that is infected when it feeds on birds that carry the virus. In the United States, according to the CDC, between 1999 and 2015, the virus caused severe illnesses in nearly 42,500 people and killed about 1,800 people. About 45%

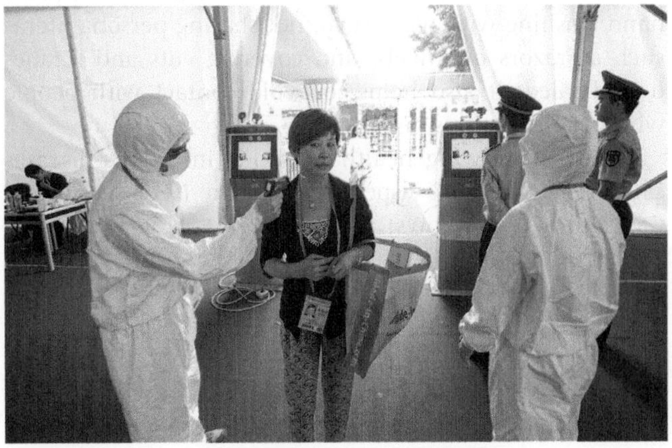

Plavevski/Shutterstock.com

FIGURE 17.5 These health-care workers are screening a woman in China for the Ebola virus. They must wear special suits to avoid all direct contact between their own skin and anyone who might be infected with the virus.

of all infections affect the brain and spinal cord, and such infections account for 93% of all deaths due to West Nile virus.

Between 2010 and 2016, the Zika virus—first discovered in 1947—began spreading rapidly in 42 countries, mostly in Central and South America and the Caribbean. In 2015 it affected more than a million people in Brazil. It is spread by the bite of a mosquito species that also spreads yellow fever and dengue fever. The species is widespread in Latin America and, by 2016, had been found in 30 U.S. states. The disease can spread rapidly in less-developed countries with warm climates, housing without window and door screens, and standing water sources around homes where the mosquitoes can breed.

The virus has little effect on most adults. The main health concern is a link between pregnant women carrying the virus and several birth defects, including babies born with unusually small heads and brains. In 2016, there were no vaccines or treatments for the disease. By July 2016, seven babies with small heads caused by Zika had been born in the U.S. to mothers who had traveled to areas with outbreaks of the virus. Although a small number of infections originated in the United States in 2016, health officials say that there is little risk of a major outbreak in the United States because of the widespread use of window and door screens, air conditioning, and mosquito control programs. Pregnant women or women trying to get pregnant are advised not to travel to countries where the virus exists and is spreading.

Scientists estimate that throughout history, more than half of all infectious diseases were originally transmitted to humans from wild or domesticated animals. The development of such diseases has spurred the growth of the relatively new field of *ecological medicine* (Science Focus 17.2). **GREEN CAREER: Ecological medicine**

Good hygiene can greatly reduce the chances of getting infectious diseases. This includes thorough and frequent hand washing with plain soap, not sharing personal items such as razors or towels, and covering cuts and scrapes until healed. It also helps to avoid contact with people who have infectious diseases.

Another growing health hazard is infectious diseases caused by parasites, especially malaria (see the second Case Study that follows).

CASE STUDY

The Global HIV/AIDS Epidemic

The global spread of AIDS, caused by HIV infection, is a major global health threat. This virus cripples the immune system and leaves the body vulnerable to infections such as TB and rare forms of cancer such as *Kaposi's sarcoma*. A person infected with HIV can live a normal life, but if the infection develops into AIDS, death is likely. An estimated 20% of all people infected with HIV are not aware of the infection and can spread the virus for years before being diagnosed.

Since HIV was identified in 1981, this viral infection has spread around the globe. According to UNAIDS, in 2014, about 36.9 million people worldwide (about 1.2 million in the United States, according to the CDC) were living with HIV. In 2014, there were about 2.0 million new cases of AIDS—half of them in people ages 15 to 49. The CDC estimates that the United States sees 45,000 to 50,000 new cases per year.

Between 1981 and 2014, more than 38 million people died of AIDS-related diseases, according to UNAIDS. According to the CDC, the U.S. death toll for the same period was more than 658,500. In 2014, AIDS killed about 1.2 million people—down from a peak of 2.3 million in 2005. AIDS has reduced the life expectancy of the 750 million people living in sub-Saharan Africa, the area south of the Sahara Desert, from 62 to 47 years, on average, and to 40 years in the seven countries most severely affected by AIDS.

Worldwide, AIDS is the leading cause of death for people of ages 15 to 49. This affects the population age structures in several African countries, including Botswana (Figure 17.6), where 23% of all people between ages 15

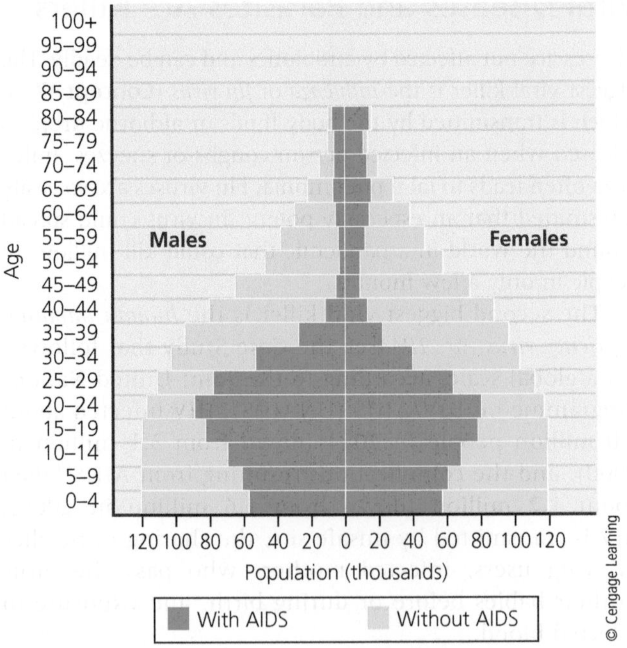

FIGURE 17.6 In Botswana, more than 25% of people ages 15–49 were infected with HIV in 2011. This figure shows two projected age structures for Botswana's population in 2020—one including the possible effects of the AIDS epidemic (red bars), and the other not including those effects (yellow bars). ***Critical thinking:*** How might this affect Botswana's economic development?

(Compiled by the authors using data from the U.S. Census Bureau, UN Population Division, and World Health Organization)

SCIENCEFOCUS 17.2

Ecological Medicine: Tracking Infectious Diseases from Animals to Humans

FIGURE 17.A A deer tick can carry the Lyme disease bacterium from a deer or mouse to a human.

Dariusz Majgier/Shutterstock.com

A number of infectious diseases move from one animal species to another and from wild and domesticated animal species to humans. These infections are on the rise. Studies led by Peter Daszak and other researchers have found that about 60% of the roughly 400 infectious diseases emerging between 1940 and 2014 were transmitted from animals to humans, and the majority of those came from wild animals. Examples of such diseases and their origins include the following:

- HIV—moves from simians (apes or monkeys) to humans (see Case Study, p. 448)
- Lyme disease—moves from wild deer and mice through ticks to humans
- Ebola—thought to have come from bats
- West Nile virus—transmitted from birds via mosquito bites
- Avian flu—a severe flu strain from birds
- Plague—moved from rats to rat fleas to humans
- HBV and dengue fever—moved from apes to humans
- African sleeping sicknesses—moved from wild and domestic grazing animals to humans

Ecological medicine is a new interdisciplinary field devoted to tracking down these unwanted connections between animals and humans. Scientists in this field have identified several human practices that encourage the spread of diseases among animals and people. One is the clearing or fragmenting of forests to make way for settlements, farms, and expanding cities.

For example, in the United States, the push of suburban development into forests has increased the chances of many suburbanites becoming infected with Lyme disease. The bacterium that causes this disease lives in the bodies of deer and white-footed mice and is passed between these two animals and to humans, mostly by certain types of ticks (Figure 17.A), but also by mosquitoes, spiders, fleas, and mites. Left untreated, Lyme disease can cause debilitating arthritis, heart disease, and nervous disorders. Since 1982 when record keeping began, the annual number of Lyme disease cases, estimated in 2014 to be about 300,000, has grown nearly 25-fold, according to the CDC.

The practice of hunting wild game for food has also increased the spread of infectious diseases. In parts of Africa and Asia, local people who kill monkeys and other animals for bush meat (see Figure 9.18, p. 211) come in regular contact with primate blood and can be exposed to a simian strain of HIV, which causes AIDS.

Another important factor in the spread of these diseases is the legal and illegal international trade in wild species. The U.S. Fish and Wildlife Service estimates that more than 200 million wild animals—from kangaroos to iguanas to tropical fish—are legally imported into the United States each year with little quarantining or screening for disease. Most are imported for commercial sale in the pet trade.

Industrialized meat production is another major factor in the spread of food-borne infectious diseases to humans. For example, a deadly form of *E. coli* bacteria sometimes spreads from livestock to humans when people eat meat contaminated by animal manure. *Salmonella* bacteria found on animal hides and in poorly processed, contaminated meat also can cause food-borne disease.

A number of scientists are looking at the connections between climate change and the spread of infectious diseases, especially malaria, meningitis, dengue fever, and West Nile virus. With warmer temperatures, they are concerned that the mosquitoes and other insects that spread these diseases will increase their ranges from tropical areas to temperate areas of the globe.

CRITICAL THINKING

If you were an ecological medicine doctor, where would you put your greatest efforts in researching this problem? Explain.

and 49 were infected with HIV in 2014. The premature deaths of many teachers, health-care workers, farmers, and other adults in these countries has contributed to declines in education, health care, food production, economic development, and political stability. In addition, it has led to the disintegration of many families and large numbers of orphaned children.

The treatment for HIV infection includes a combination of antiviral drugs that can slow the progress of the virus. However, such drugs cost too much to be used widely in the less-developed countries where HIV infections are widespread.

CASE STUDY

Malaria—The Spread of a Deadly Parasite

About 3.2 billion people—44% of the world's population—are at risk of getting malaria (Figure 17.7). Most of them

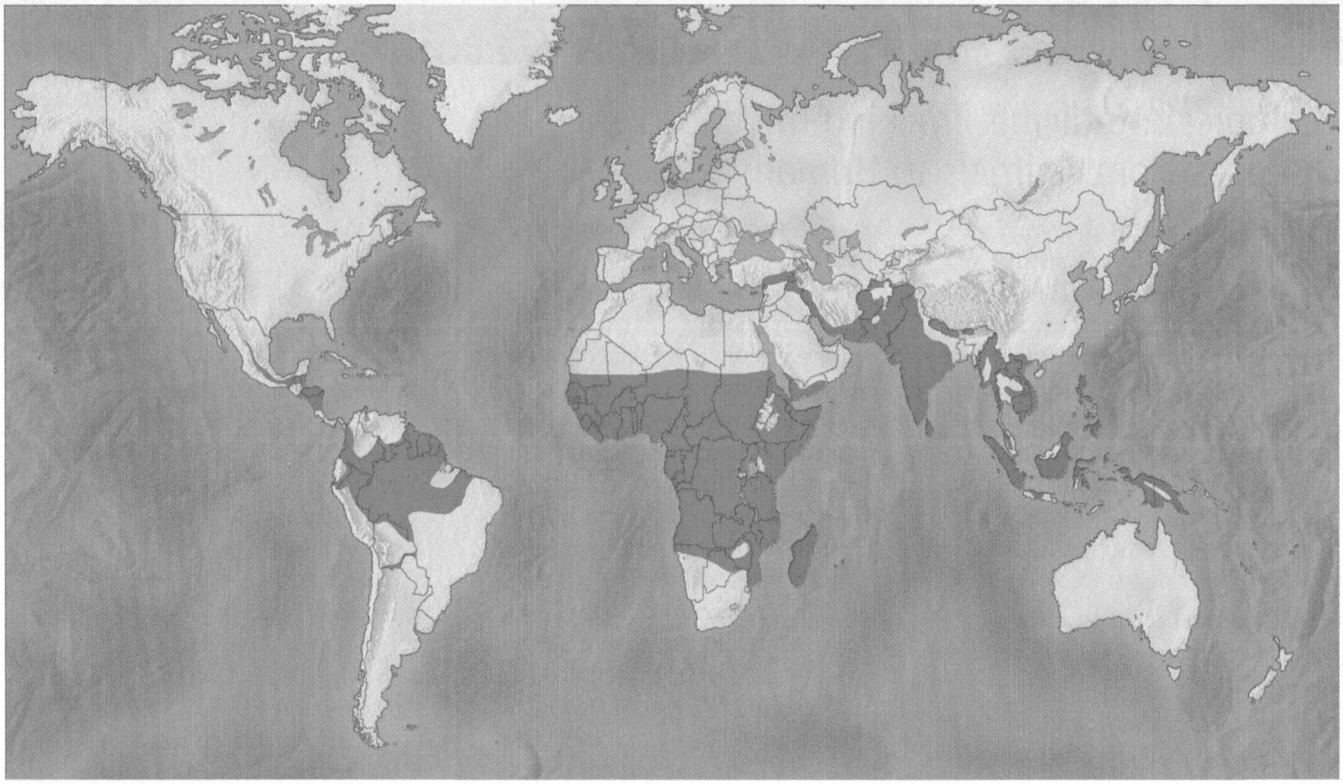

FIGURE 17.7 About 44% of the world's population live in areas in which malaria is prevalent. As the earth warms, malaria may spread to some temperate areas such as the southern half of the United States.

(Compiled by the authors using data from the World Health Organization and U.S. Centers for Disease Control and Prevention.)

live in poor African countries. People traveling to malaria-prone areas are also at risk because there is no vaccine that can prevent this disease.

Malaria is caused by a parasite that is spread by certain mosquito species. The mosquito bites an infected person, picks up the parasite, and passes it to the next person it bites. The parasites then move out of the mosquito, into the human's bloodstream, and multiply in the human's liver. The parasite infects and destroys its victim's red blood cells, causing intense fever, chills, drenching sweats, severe abdominal pain, vomiting, headaches, and increased susceptibility to other diseases. According to the WHO, malaria killed about 438,000 people in 2014—a 63% drop since 2000. However, some experts contend this total could be much higher, because public health records are incomplete in many areas. More than 90% of all malaria victims live in sub-Saharan Africa and most of them are children younger than age 5. Many of the children who survive suffer brain damage or impaired learning ability.

Over the course of human history, malarial protozoa probably have killed more people than all the wars ever fought. The spread of malaria was slowed during the 1950s and 1960s, a time when swamps and marshes where mosquitoes breed were drained or sprayed with insecticides, and drugs were used to kill the parasites in victims' bloodstreams.

Since 1970, however, malaria has come roaring back. Most species of mosquitoes that transmit malaria have become genetically resistant to most insecticides and the *Plasmodium* parasites have become genetically resistant to common antimalarial drugs. During this century, as the average atmospheric temperature rises, populations of malaria-carrying mosquitoes will likely spread from tropical areas to warming temperate areas.

CONSIDER THIS . . .

CONNECTIONS Deforestation and Malaria

The clearing and development of tropical forests has led to the spread of malaria among workers and the settlers who follow them. One study found that a 5% loss of tree cover in one part of Brazil's Amazon forest led to a 50% increase in malaria in that study area. The researchers hypothesized that deforestation creates partially sunlit pools of water that make ideal breeding ponds for malaria-carrying mosquitoes.

Researchers are working to develop new antimalarial drugs and vaccines, as well as biological controls for

Olivier Asselin/Alamy Stock Photo

FIGURE 17.8 This baby in Senegal, Africa, is sleeping under an insecticide-treated mosquito net to reduce the risk of being bitten by malaria-carrying mosquitoes.

Anopheles mosquitoes. During the last 10 years, scientists have made progress in developing a malaria vaccine, but currently no effective vaccine is available.

Another approach is to provide poor people in malarial regions with free or inexpensive insecticide-treated bed nets (Figure 17.8) and window screens. Between 2000 and 2014, the percentage of Africa's population sleeping under mosquito nets increased from 2% to more than 50% and saved 6.2 million lives, according to the WHO. Also, zinc and vitamin A supplements could be given to children to boost their resistance to malaria.

Reducing the Incidence of Infectious Diseases

According to the WHO, the percentage of all deaths worldwide resulting from infectious diseases dropped from 35% to 16% between 1970 and 2015, primarily because a growing number of children were immunized against the major infectious diseases. Between 1990 and 2015, the estimated annual number of children younger than age 5 who died from infectious diseases dropped from nearly 12 million to 4.9 million. GOOD NEWS

CONSIDER THIS . . .

LEARNING FROM NATURE

Water-dwelling tardigrades—tiny segmented animals that grow to only 0.5 millimeters (0.02 inches)—can dry out completely and stay alive for months. Scientists hope to learn how they do this and use this information to store and transport vaccines throughout the world without the need for refrigeration.

Figure 17.9 lists measures that could help prevent or reduce the incidence of infectious diseases—especially in less-developed countries. The WHO has estimated that implementing the solutions listed in Figure 17.9 could save the lives of as many as 4 million children younger than age 5 each year. **GREEN CAREER: Infectious disease prevention**

CONSIDER THIS . . .

CONNECTIONS Drinking Water, Latrines, and Infectious Diseases

More than a third of the world's people—2.6 billion—do not have sanitary bathroom facilities. Nearly 1 billion get their water for drinking, washing, and cooking from sources polluted by animal or human feces. A key to reducing sickness and premature death due to infectious disease is to focus on providing simple latrines and access to safe drinking water.

Solutions

Infectious Diseases

- Increase research on tropical diseases and vaccines
- Reduce poverty and malnutrition
- Improve drinking water quality
- Reduce unnecessary use of antibiotics
- Sharply reduce use of antibiotics on livestock
- Immunize children against major viral diseases
- Provide oral rehydration for diarrhea victims
- Conduct global campaign to reduce HIV/AIDS

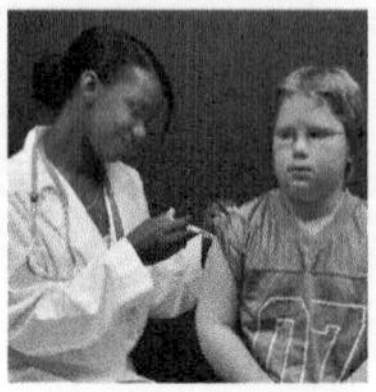

FIGURE 17.9 Ways to prevent or reduce the incidence of infectious diseases, especially in less-developed countries. ***Critical thinking:*** Which three of these approaches do you think are the most important? Why?

17.3 HOW DO CHEMICAL HAZARDS THREATEN HUMAN HEALTH?

CONCEPT 17.3 Certain chemicals in the environment can cause cancers and birth defects and disrupt the human immune, nervous, and endocrine systems.

Some Chemicals Can Cause Cancers, Mutations, and Birth Defects

A **toxic chemical** is an element or compound that can cause temporary or permanent harm or death to humans. The U.S. Environmental Protection Agency (EPA) has listed arsenic, lead, mercury (**Core Case Study**), vinyl chloride (used to make PVC plastics), and polychlorinated biphenyls (PCBs; see the Case Study that follows) as the top five toxic substances in terms of human health.

There are three major types of potentially toxic agents. **Carcinogens** are chemicals, some types of radiation, and certain viruses that can cause or promote *cancer*—a disease in which malignant cells multiply uncontrollably and create tumors that can damage the body and often lead to premature death. Examples of carcinogens are arsenic, benzene, formaldehyde, gamma radiation, PCBs, radon, ultraviolet (UV) radiation, vinyl chloride, and certain chemicals in tobacco smoke.

The second major type of toxic agent, **mutagens,** includes chemicals or forms of radiation that cause or increase the frequency of *mutations*, or changes, in the DNA molecules found in cells. Most mutations cause no harm, but some can lead to cancers and other disorders. For example, nitrous acid (HNO_2), formed by the digestion of nitrite (NO_2^-) preservatives in foods, can cause mutations linked to increases in stomach cancer in people who consume large amounts of processed foods and wine containing such preservatives. Harmful mutations occurring in reproductive cells can be passed on to offspring and to future generations.

Teratogens, a third type of toxic agent, are chemicals that harm or cause birth defects in a fetus or embryo. Ethyl alcohol is a teratogen. Women who drink alcoholic beverages during pregnancy increase their risk of having babies with low birth weight and a number of physical, developmental, behavioral, and mental problems. Other teratogens are mercury (**Core Case Study**), PCBs, lead, formaldehyde, benzene, phthalates, and angel dust.

CASE STUDY

PCBs Are Everywhere—A Legacy from the Past

PCBs are a class of more than 200 chlorine-containing organic compounds that are very stable and nonflammable. They exist as oily liquids or solids but, under certain conditions, they can enter the air as a vapor. Between 1929 and 1977, PCBs were widely used as lubricants, hydraulic fluids, and insulators in electrical transformers and capacitors. They also became ingredients in a variety of products including paints, fire retardants in fabrics, preservatives, adhesives, and pesticides.

The U.S. Congress banned the domestic production of PCBs in 1977 after research showed that they could cause liver cancer and other cancers in test animals. Studies also showed that pregnant women exposed to PCBs gave birth to underweight babies who eventually suffered permanent neurological damage, sharply lower-than-average IQs, and long-term growth problems.

Production of PCBs has also been banned in most other countries, but the potential health threats from these chemicals will be with us for a long time. For decades, PCBs entered the air, water, and soil as they were manufactured, used, and disposed of, as well as through accidental spills and leaks. Because PCBs break down very slowly in the environment, they can travel long distances in the air before landing far away from where they were released. Also, because they are fat-soluble, PCBs can also be biologically magnified in food chains and food webs (Figure 17.10).

As a result, PCBs are now found almost everywhere—in the air, soil, lakes, rivers, fish, birds, most human bodies, and even the bodies of polar bears in the Arctic. According to the EPA, about 70% of all the PCBs made in the United States are still in the environment.

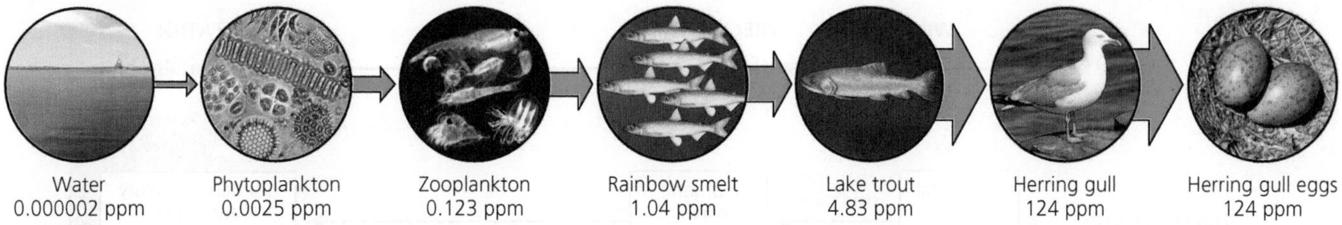

FIGURE 17.10 Biological magnification of polychlorinated biphenyls (PCBs) in an aquatic food chain in the Great Lakes.

Some Chemicals Can Affect Our Immune and Nervous Systems

Since the 1970s, research on wildlife and laboratory animals along with some studies of humans suggest that long-term exposure to some chemicals in the environment can disrupt important body systems, including immune and nervous systems (**Concept 17.3**).

The *immune system* consists of specialized cells and tissues that protect the body against disease and harmful substances by forming *antibodies*, or specialized proteins that render invading agents harmless. Some chemicals such as arsenic, methylmercury (**Core Case Study**), and dioxins can weaken the human immune system and leave the body vulnerable to attacks by allergens and infectious bacteria, viruses, and protozoa.

Some natural and synthetic chemicals in the environment, called *neurotoxins*, can harm the human *nervous system* (the brain, spinal cord, and peripheral nerves). Effects can include behavioral changes, learning disabilities, attention-deficit disorder, paralysis, and death. Examples of neurotoxins are PCBs, arsenic, lead, and certain pesticides.

Methylmercury (**Core Case Study**) is an especially dangerous neurotoxin because it is persists in the environment and, like DDT and PCBs, can be biologically magnified in food chains and food webs (Figure 17.11). According to the Natural Resources Defense Council, predatory fish such as tuna, orange roughy, swordfish, mackerel, grouper, and sharks can have mercury concentrations in their bodies that are 10,000 times higher than the levels in the water around them.

In one study, the EPA found that almost half of the fish tested in 500 lakes and reservoirs across the United States had levels of mercury that exceeded safe levels (Figure 17.1). Similarly, a study by the U.S. Geological Survey of nearly 300 streams across the United States found mercury-contaminated fish in all of the streams surveyed, with one-fourth of the fish exceeding the safe levels determined by the EPA.

The symptoms of mercury poisoning in adults include poor balance and coordination, muscle weakness, tremors, memory problems, insomnia, hearing loss, loss of hair, and loss of peripheral vision. The EPA estimates that about 1 of every 12 women of childbearing age in the United States has enough mercury in her blood to harm a developing fetus. Figure 17.12 lists ways to prevent or reduce human inputs of mercury (**Core Case Study**) into the environment.

Some Chemicals Affect the Endocrine System

The *endocrine system* is a complex network of glands that release tiny amounts of *hormones* into the bloodstreams of humans and other vertebrate animals. Very low levels of these chemical messengers (often measured in parts per billion or parts per trillion) regulate bodily systems that control sexual reproduction, growth, development, learning ability, and behavior. Each hormone has a unique molecular shape that allows it to attach to certain parts of cells called *receptors*, and to transmit a chemical message (Figure 17.13).

Molecules of certain synthetic chemicals have shapes similar to those of natural hormones. This allows them to attach to the molecules of natural hormones and to disrupt the endocrine systems in humans and in some other animals (**Concept 17.3**). These molecules are called *hormonally active agents (HAAs)* or *endocrine disruptors*.

Examples of HAAs include Atrazine and several other widely used herbicides, pesticides, dioxins, lead, and mercury (**Core Case Study**). Some HAAs, including *bisphenol A*, or *BPA* (Science Focus 17.3, p. 456), act as hormone imposters, or *hormone mimics*. They are chemically similar to estrogens (female sex hormones) and can disrupt the endocrine system by attaching to estrogen receptor molecules. Other HAAs, called *hormone blockers*, disrupt the endocrine system by preventing natural hormones such as androgens (male sex hormones) from attaching to their receptors.

Estrogen mimics and hormone blockers can have a number of effects on sexual development and reproduction. Numerous studies involving wild animals, laboratory animals, and humans suggest that the males of species that are exposed to hormonal disruption are generally becoming more feminized.

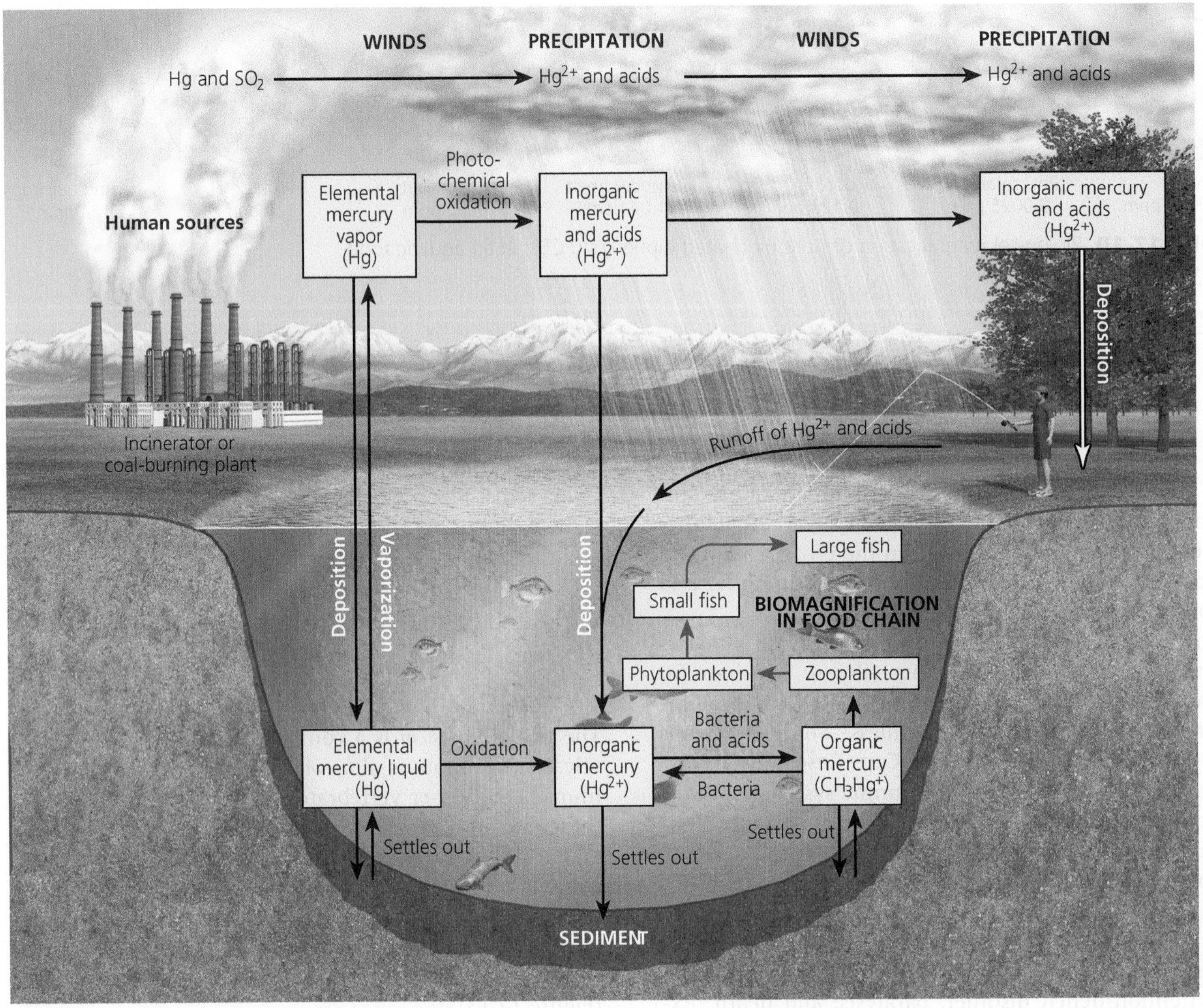

FIGURE 17.11 Movement of different forms of toxic mercury from the atmosphere into an aquatic ecosystem where it is biologically magnified in a food chain. ***Critical thinking:*** What is your most likely exposure to mercury?

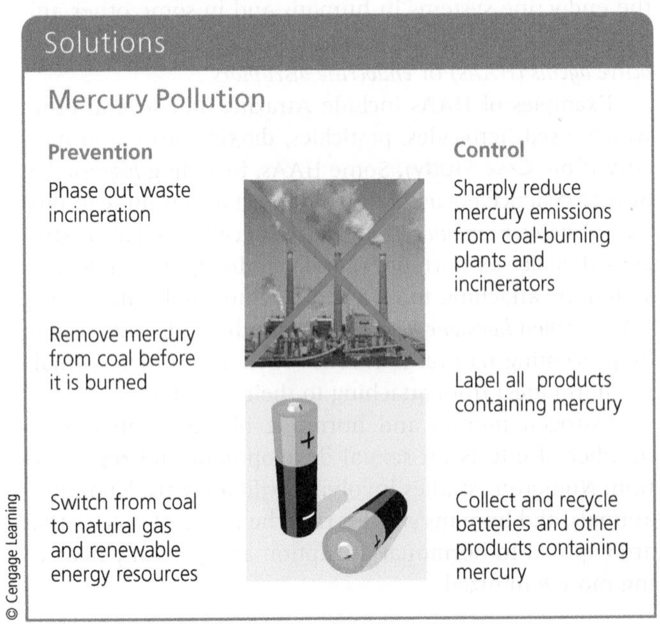

FIGURE 17.12 Ways to prevent or control inputs of mercury (**Core Case Study**) into the environment from human sources—mostly coal-burning power plants and incinerators. ***Critical thinking:*** Which four of these solutions do you think are the most important? Why?

Top: Mark Smith/Shutterstock.com. Bottom: tuulijumala/Shutterstock.com

There is also growing concern about another group of HAAs that affect hormones generated by the thyroid gland. These pollutants, called *thyroid disrupters*, can cause growth, weight, brain, and behavioral disorders. *Perfluorinated chemicals (PFOAs)*, used to make nonstick surfaces on cookware, have been linked to thyroid disease, as well as cancer and birth defects. *Polybrominated diphenyl ethers (PBDEs)*, another set of thyroid disrupters, are flame retardants added to certain fabrics, furniture, plastics, mattresses, and television and computer housings.

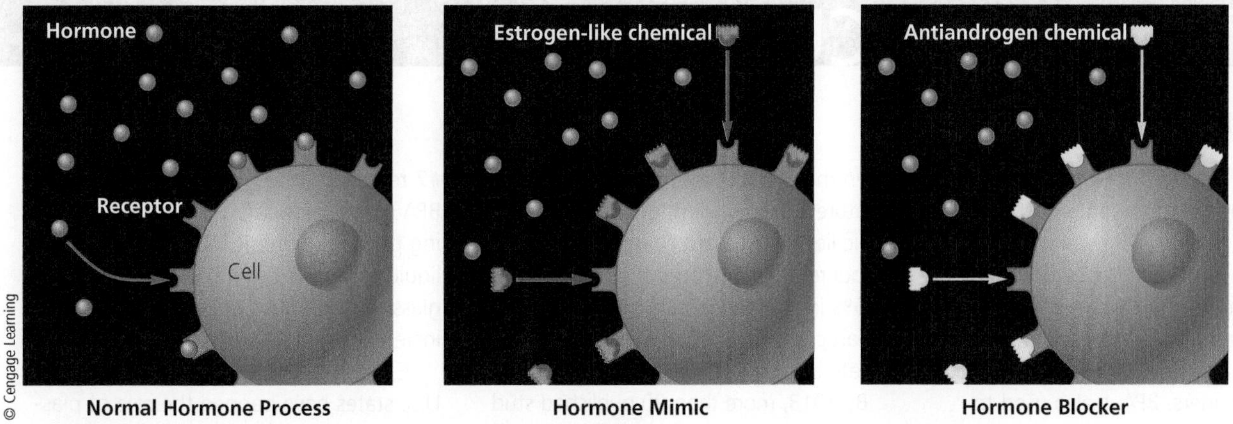

FIGURE 17.13 Each type of hormone has a unique molecular shape that allows it to attach to specially shaped receptors on the surface of, or the inside of, a cell and to transmit its chemical message (left). Molecules of certain pesticides and other synthetic chemicals, called *hormonally active agents* (HAAs, center and right), have shapes similar to those of natural hormones, allowing them to attach to the hormone molecules and disrupt the endocrine systems of humans and other animals.

In 2013, the FDA indicated that the chemicals triclosan and triclocarban, widely used in antibacterial soaps and some deodorants, are likely hormone disrupters and could be contributing to bacterial resistance to antibiotics. The FDA also said that there is no evidence that using these chemicals is any more effective in preventing bacterial infections than is thoroughly washing your hands with plain soap and water. Since 2000, several European countries have restricted the use of triclosan in consumer products.

Scientists are also increasingly concerned about certain HAAs called *phthalates* (pronounced THALL-eights). These chemicals are used to make plastics more flexible and to make cosmetics easier to apply to the skin. They are found in a variety of products, including many detergents, perfumes, cosmetics, baby powders, sunscreens, deodorants, soaps, nail polishes, and shampoos for adults and babies. They are also found in polyvinyl chloride (PVC) plastic products such as soft vinyl toys and vinyl gloves, teething rings, blood storage bags, intravenous drip bags, shower curtains, and some plastic food and drink containers.

Exposure of laboratory animals to high doses of various phthalates has caused birth defects, kidney and liver diseases, immune system suppression, and abnormal sexual development in these animals. Studies have linked exposure of human babies to phthalates with early puberty in girls. A 2014 study led by biologist Heather Patisaul and her colleagues suggested that phthalate exposure might lead to sperm damage in men by interfering with the creation of the male hormone testosterone. The European Union and at least 14 other countries have banned several phthalates. But scientists, government regulators, and manufacturers in the United States are divided on its risks to human health and reproductive systems.

Concerns about BPA, BPS, phthalates, and other HAAs show how difficult it can be to assess the potential harmful health effects from exposure to very low levels of various chemicals. Resolving these uncertainties will take decades of research. Some scientists argue that as a precaution during this period of research, people should sharply reduce their exposure to potentially harmful hormone disrupters (Figure 17.14), especially in products frequently used by pregnant women, infants, young children, and teenagers.

What Can You Do?

Exposure to Hormone Disrupters

- Eat certified organic produce and meats
- Avoid processed, prepackaged, and canned foods
- Use glass and ceramic cookware
- Store food and drinks in glass containers
- Use only natural cleaning and personal care products
- Use natural fabric shower curtains, not vinyl
- Avoid artificial air fresheners, fabric softeners, and dryer sheets
- Use only glass baby bottles and BPA-free, phthalate-free sipping cups, pacifiers, and toys

FIGURE 17.14 Individuals matter: Ways to reduce your exposure to hormone disrupters. ***Critical thinking:*** Which three of these steps do you think are the most important ones to take? Why?

The Controversy over BPA

The estrogen mimic *bisphenol A (BPA)* serves as a hardening agent in certain plastics that are used in a variety of products. They include some baby bottles, sipping cups, and pacifiers, as well as some reusable water bottles, sports drink and juice bottles, microwave dishes, and food storage containers. BPA is also used to make some dental sealants. In addition, individuals can be exposed to BPA when their hands touch the thermal paper used to produce some cash register receipts.

BPA is also used as a liner in most food and beverage cans sold in the United States. This type of liner allows containers to withstand extreme temperatures, keeps canned food from interacting with the metal in the cans, prevents rust in the cans, and helps to preserve the canned food.

A study by the CDC indicated that 93% of Americans older than age 6 had trace levels of BPA in their urine. Although these levels were well below the acceptable level set by the EPA, the EPA level was established in the late 1980s, before much was known about the potential effects of BPA on human health. The CDC study also found that children and adolescents generally had higher urinary BPA levels than adults had.

Research indicates that the BPA in plastics can leach into water or food when the plastic is heated to high temperatures, microwaved, or exposed to acidic liquids. Harvard University Medical School researchers found that there was a 66% increase in BPA levels in the urine of participants who drank from polycarbonate bottles regularly for just 1 week.

By 2013, more than 90 published studies by independent laboratories had found a number of significant adverse effects on test animals from exposure to very low levels of BPA. These effects include brain damage, early puberty, decreased sperm quality, certain cancers, heart disease, type 2 diabetes, liver damage, impaired immune function, impotency in males, and obesity in test animals.

On the other hand, 12 studies funded by the chemical industry found no evidence or only weak evidence of adverse effects from low-level exposure to BPA in test animals. In 2008, the FDA concluded that BPA in food and drink containers was not a health hazard. In 2015, the European Food Safety Authority agreed, concluding that BPA is not appearing in people's body systems at high enough levels to cause harm.

Consumers now have more choices, since most makers of baby bottles, sipping cups, and sports water bottles offer BPA-free alternatives. Many consumers are avoiding plastic containers with a #7 recycling code (which indicates that BPA can be present). People are also using powdered infant formula instead of liquid formula from metal cans, choosing glass bottles, mugs, and food containers instead of those made of plastic.

Canada, the European Union, and six U.S. states have banned the sale of plastic baby bottles that contain BPA. And in 2012, the FDA banned the use of BPA in baby bottles and sipping cups.

Many manufacturers have replaced BPA with bisphenol S (BPS). However, studies indicate that BPS can have effects similar to those of BPA, and was present at detectable levels in the urine of nearly 81% of people tested for it.

There are substitutes for the plastic resins containing BPA or BPS that line most food cans in the United States. However, these replacements are more expensive and the potential health effects of some chemicals they contain need to be evaluated.

CRITICAL THINKING

Should plastics that contain BPA or BPS be banned from use in all children's products? Explain. Should such plastics be banned from use in the liners of canned food containers? Explain.

17.4 HOW CAN WE EVALUATE RISKS FROM CHEMICAL HAZARDS?

CONCEPT 17.4A Scientists use live laboratory animals, case reports of poisonings, and epidemiological studies to estimate the toxicity of chemicals, but these methods have limitations.

CONCEPT 17.4B Many health scientists call for much greater emphasis on pollution prevention to reduce our exposure to potentially harmful chemicals.

Many Factors Determine the Toxicity of Chemicals

We are exposed to small amounts of potentially harmful chemicals every day in the air we breathe, the water we drink, and the food we eat. **Toxicology** is the study of the harmful effects of these and other chemicals on humans and other organisms.

Toxicity is a measure of the ability of a substance to cause injury, illness, or death to a living organism. A basic principle of toxicology is that *any synthetic or natural chemical can be harmful if ingested or inhaled in a large enough quantity*. But the critical question is this: *At what level of exposure to a particular toxic chemical will the chemical cause harm?*

This is a difficult question to answer because of the many variables involved in estimating the effects of human exposure to chemicals. A key factor is the **dose,** the amount of a harmful chemical that a person has ingested, inhaled, or absorbed through the skin at any one time.

The effects of a particular chemical can also depend on the age of the person exposed to it. For example, toxic chemicals usually have a greater effect on fetuses, infants, and children than on adults (see the Case Study that follows). Toxicity also depends on *genetic makeup,* which determines an individual's sensitivity to a particular toxin. People vary widely in their degrees of sensitivity to chemicals (Figure 17.15), and some are sensitive to a number of toxins—a condition known as *multiple chemical sensitivity* (MCS).

Several other variables can affect the level of harm caused by a chemical. One is its *solubility*. Water-soluble toxins can move throughout the environment and get into water supplies, as well as the aqueous solutions that surround the cells in our bodies. Oil- or fat-soluble toxins can penetrate the membranes that surround our cells, because these membranes allow similar oil-soluble chemicals to pass through them. Thus, oil- or fat-soluble toxins can accumulate in body tissues and cells.

Another factor is a substance's *persistence,* or resistance to breaking down. Many chemicals, including DDT and PCBs, were used widely because they are not easily broken down in the environment. This means that they are more likely to remain in the body and have long-lasting harmful health effects.

Biological magnification (see Figure 9.14, p. 205) can also play a role in toxicity. Animals that eat higher on the food chain are more susceptible to the effects of fat-soluble toxic chemicals because of the magnified concentrations of the toxins in their bodies. Examples of chemicals that can be biomagnified include DDT, PCBs (Figure 17.10), and methylmercury (**Core Case Study**).

The health damage resulting from exposure to a chemical is called the **response.** One type of response, an *acute effect,* is an immediate or rapid harmful reaction ranging from dizziness to death. By contrast, a *chronic effect* is a permanent or long-lasting consequence (kidney or liver damage, for example) of exposure to a single dose or to repeated lower doses of a harmful substance.

Natural and synthetic chemicals can be either safe or toxic. Many synthetic chemicals, including many of the medicines we take, are quite safe if used as intended, whereas many natural chemicals such as lead and mercury (**Core Case Study**) are deadly.

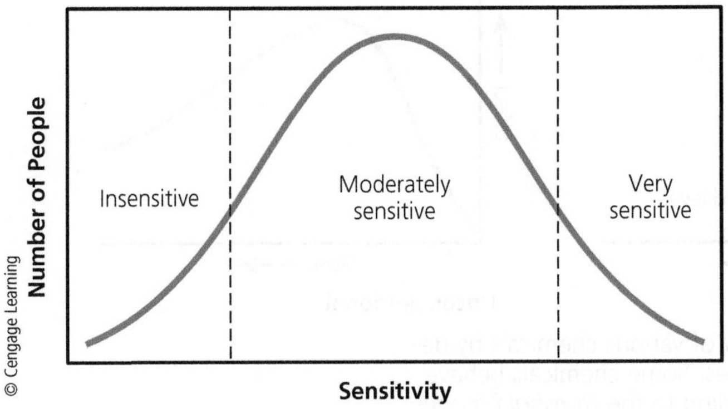

FIGURE 17.15 Individuals in a human population can vary in how sensitive they are to a particular dose of a toxic chemical.

CASE STUDY

Protecting Children from Toxic Chemicals

In one study, the Environmental Working Group analyzed umbilical cord blood from 10 randomly selected newborns in U.S. hospitals. Of the 287 chemicals detected in that study, 180 have been shown to cause cancers in humans or animals, 217 have damaged the nervous systems of test animals, and 208 have caused birth defects or abnormal development in test animals. Scientists do not know what harm, if any, might be caused by the very low concentrations of these chemicals found in the infants' blood.

However, more recent science has caused some experts to suggest that exposure to chemical pollutants in the womb may be related to increasing rates of autism, childhood asthma, and learning disorders. In 2009, researchers for the first time found a connection between the exposure of pregnant women to air pollutants and lower IQ scores in their children as they grew. A team of researchers led by Frederica Perera of Columbia University reported that children exposed to high levels of air pollution before birth scored 4–5 points lower, on average, in IQ tests than did children with less exposure.

Infants and young children are more susceptible to the effects of toxic substances than are adults, for three major reasons. *First,* they generally breathe more air, drink more water, and eat more food per unit of body weight than do adults. *Second,* they are exposed to toxins in dust and soil when they put their fingers, toys, and other objects in their mouths. *Third,* children usually have less well-developed immune systems and body detoxification processes than adults have. Fetuses are also highly vulnerable to trace amounts of toxic chemicals such as methylmercury (**Core Case Study**) that they can receive from their mothers.

The EPA has proposed that in determining any risk, regulators should assume that children have a 10-times higher risk factor than adults have. Some health scientists suggest that to be on the safe side, we should assume that this risk for children is 100 times the risk for adults.

CONSIDER THIS . . .

THINKING ABOUT Toxic Chemical Levels for Children

Do you think environmental regulations should require that the allowed levels of exposure to toxic chemicals for children be 100 times lower than those for adults? Explain your reasoning.

Scientists Use Live Laboratory Animals and Non-Animal Tests to Estimate Toxicity

The most widely used method for determining toxicity is to expose a population of live laboratory animals to measured doses of a specific substance under controlled conditions. Laboratory-bred mice and rats are widely used because, as mammals, their systems function similarly to human systems, to some degree. Also, they are small and can reproduce rapidly under controlled laboratory conditions. Animal tests typically take 2 to 5 years, involve hundreds to thousands of test animals, and can cost as much as $2 million per substance tested. Some of these tests can be painful to the test animals and can kill or harm them. Animal welfare groups want to limit or ban the use of test animals and, at the very least, want to ensure that they are treated in the most humane manner possible.

Scientists estimate the toxicity of a chemical by determining the effects of various doses of the chemical on test organisms and plotting the results in a **dose-response curve** (Figure 17.16). One approach is to determine the *lethal dose*—the dose that will kill an animal. A chemical's *median lethal dose (LD50)* is the dose that can kill 50% of the animals (usually rats and mice) in a test population within a given time period, usually expressed in milligrams of the chemical per kilogram of body weight (mg/kg).

Chemicals vary widely in their toxicity. Some toxins can cause serious harm or death after a single very low dose. For example, a few drops of pure nicotine (found in e-cigarettes) would make you very sick, while a teaspoon of it could kill you. Others such as water or table sugar cause such harm only at dosages so huge that it is nearly impossible to get enough into the body to cause injury or death. Most chemicals fall between these two extremes.

There are three general types of dose-response curves. With the *nonthreshold dose-response model* (Figure 17.16, left), any dosage of a toxic chemical causes harm that increases with the dosage. With the *threshold dose-response model* (Figure 17.16, center), a certain level, or *threshold*, of exposure to the chemical must be reached before any detectable harmful effects occur, presumably because the body can repair the damage caused by low dosages of some substances. With the third type, called the *unconventional model* (Figure 17.16, right), the harmful effects increase with dosage to a certain point and then begin decreasing.

Establishing which of these three models applies at low dosages is extremely difficult and controversial. To be on the safe side, scientists often choose the nonthreshold dose-response model. High dosages are used to reduce the number of test animals needed, obtain results quickly, and lower costs. Use of low dosages would require running tests on millions of laboratory animals for many years, in which case chemical companies and government agencies could not afford to test most chemicals.

For the same reasons, scientists usually use mathematical models to *extrapolate*, or estimate, the effects of low-dose exposures based on the measured results of high-dose exposures. Then they extrapolate these results from test organisms to humans as a way of estimating LD50 values for acute toxicity. Some scientists challenge the validity of extrapolating data from test animals to humans, because human physiology and metabolism often differ from those of the test animals, as well as from person to person. Other scientists say that such tests and models can work well,

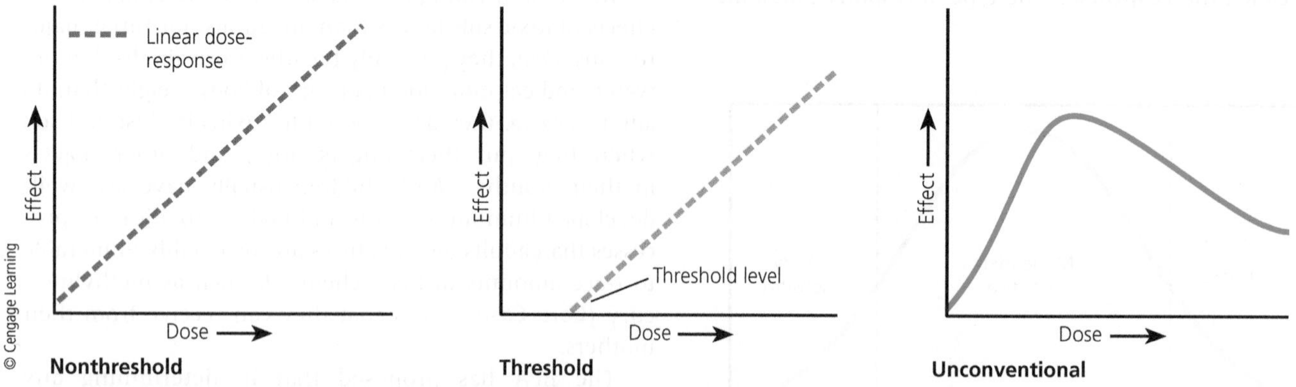

FIGURE 17.16 *Dose-response curves.* Scientists estimate the toxicity of various chemicals by determining how a chemical's harmful effects change as the dose increases. Some chemicals behave according to the *nonthreshold model* (left curve). Others behave according to the *threshold model* (center curve). Still others are unconventional in how they behave (right curve). For all of these graphs, the curves usually vary from being exactly linear, or straight. ***Critical thinking:*** Can you think of commonly used chemicals that fit each of these models? What are they?

especially for revealing cancer risks, when the correct experimental animal is chosen, or when a chemical is toxic to several different test-animal species.

More humane methods for toxicity testing are available and are being used more often to replace testing on live animals. They include making computer simulations and using individual animal cells, instead of whole, live animals. High-speed robot testing devices can now screen the biological activity of more than 1 million compounds a day to help determine their possible toxic effects.

The problems with estimating toxicities in the laboratory get even more complicated (**Concept 17.4A**). In real life, each of us is exposed to a variety of chemicals, some of which can interact in ways that decrease or enhance their individual effects. Toxicologists already have great difficulty in estimating the toxicity of a single substance. Evaluating mixtures of potentially toxic substances, determining how they interact, and deciding which of them are the most harmful can be overwhelming from a scientific and economic standpoint. For example, just studying the interactions among 3 of the 500 most widely used industrial chemicals would take 20.7 million experiments—a physical and financial impossibility.

CONSIDER THIS . . .

THINKING ABOUT Animal Testing

Should laboratory-bred mice, rats, and other animals be used to determine toxicity and other effects of chemicals? Explain.

Other Ways to Estimate the Harmful Effects of Chemicals

Scientists use several other methods to get information about the harmful effects of chemicals on human health. For example, *case reports*, usually made by physicians, provide information about people who have suffered from adverse health effects or died after exposure to a chemical. Most case reports are not reliable for estimating toxicity because the actual dosage and the exposed person's health status are usually unknown. But such reports can provide clues about environmental hazards and suggest the need for laboratory investigations.

Another source of information is *epidemiological studies*, which compare the health of people exposed to a particular chemical (the *experimental group*) with the health of a similar group of people not exposed to the agent (the *control group*). The goal is to determine whether the statistical association between exposure to a toxic chemical and a health problem is strong, moderate, weak, or undetectable.

Four factors can limit the usefulness of epidemiological studies. *First*, in many cases, too few people have been exposed to high enough levels of a toxic agent to detect statistically significant differences. *Second*, the studies usually take a long time. *Third*, closely linking an observed effect with exposure to a particular chemical is difficult because people are exposed to many different toxic agents throughout their lives and can vary in their sensitivity to such chemicals. *Fourth*, we cannot use epidemiological studies to evaluate hazards from new technologies or chemicals to which people have not been exposed.

Are Trace Levels of Toxic Chemicals Harmful?

Almost everyone who lives in a more-developed country is now exposed to potentially harmful chemicals (Figure 17.17) that have built up to trace levels in their blood and in other parts of their bodies. CDC studies have found that the blood of an average American contains traces of 212 different chemicals, including potentially harmful chemicals such as arsenic and BPA.

Should we be concerned about trace amounts of various synthetic chemicals in our air, water, food, and bodies? In most cases, we simply do not know because there are not enough data and because it is so difficult to determine the effects of exposures to low levels of these chemicals (**Concept 17.4A**).

Some scientists view exposures to trace amounts of synthetic chemicals with alarm, especially because of their potential long-term effects on the human immune, nervous, and endocrine systems. Some studies have indicated that exposure to very small amounts of such chemicals at crucial stages of development can have impacts on individuals that can extend to their offspring. Such impacts can affect reproductive and sexual development, neurological systems, and immune system function.

Others scientists view the threats from such exposures as minor. They point out that average life expectancy has been increasing in most countries, especially more-developed countries, for decades. Some scientists contend that the concentrations of such chemicals are so low that they are harmless. All agree that there is a need for much more research on the effects of trace levels of synthetic chemicals on human health.

Why Do We Know So Little about the Harmful Effects of Chemicals?

All methods for estimating toxicity levels and risks have serious limitations (**Concept 17.4A**), but they are all that we have. To minimize harm, scientists and regulators typically set allowed levels of exposure to toxic substances at 1/100th or even 1/1,000th of the estimated harmful levels.

According to risk assessment expert Joseph V. Rodricks, "Toxicologists know a great deal about a few chemicals, a little about many, and next to nothing about most." The U.S. National Academy of Sciences estimates that only 10% of the more than 85,000 registered synthetic

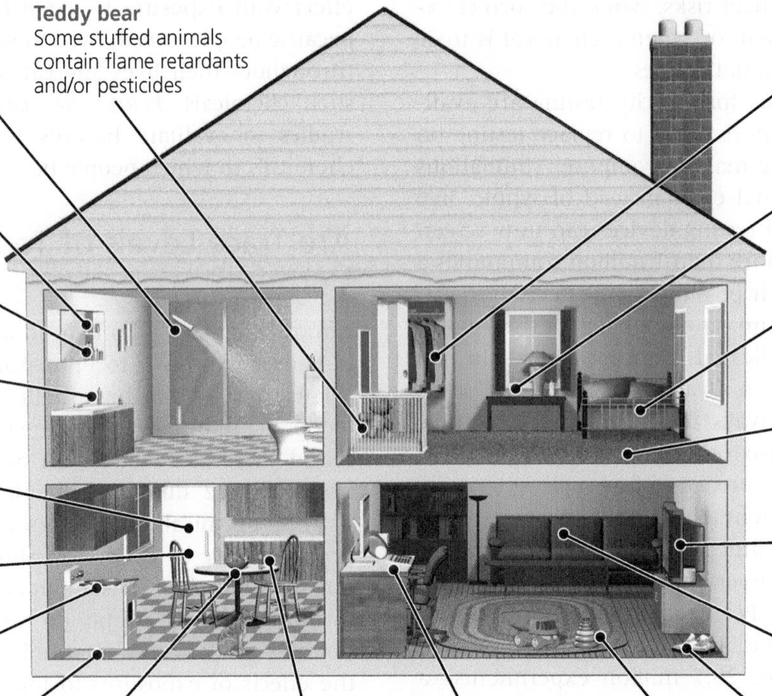
Shampoo
Perfluorochemicals to add shine
Teddy bear
Some stuffed animals contain flame retardants and/or pesticides
Nail polish
Perfluorochemicals and phthalates
Perfume
Phthalates
Hairspray
Phthalates
Food
Some food can contain bisphenol A
Milk
Fat can contain dioxins
Frying pan
Nonstick coating can contain perfluorochemicals
Tile floor
Can contain perfluorochemicals, phthalates, and pesticides
Fruit
Imported fruit may contain pesticides banned in the U.S.
Water bottle
Can contain bisphenol A
Computer
Flame retardant coatings of plastic casing and wiring
Toys
Vinyl toys contain phthalates
Tennis shoes
Can contain phthalates
Clothing
Can contain perfluorochemicals
Baby bottle
Can contain bisphenol A
Mattress
Flame retardants in stuffing
Carpet
Padding and carpet fibers can contain flame retardants, perfluorochemicals, and pesticides
TV
Wiring and plastic casing contain flame retardants
Sofa
Foam padding can contain flame retardants and perfluorochemicals

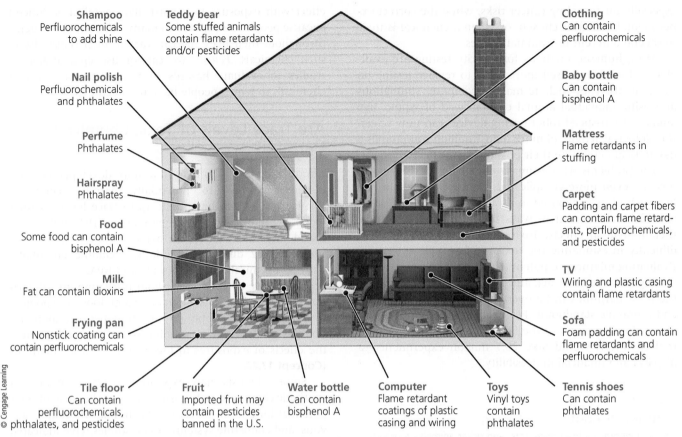

FIGURE 17.17 A number of potentially harmful chemicals are found in many homes. ***Critical thinking:*** Does the fact that we do not know much about the long-term harmful effects of these chemicals make you more likely or less likely to minimize your exposure to them? Why?

(Compiled by the authors using data from the U.S. Environmental Protection Agency, Centers for Disease Control and Prevention, and New York State Department of Health.)

chemicals in commercial use have been thoroughly screened for toxicity. Only 2% have been adequately tested to determine whether they are carcinogens, mutagens, or teratogens. Hardly any of the chemicals in commercial use have been screened for possible damage to the human nervous, endocrine, and immune systems.

Because of insufficient data and the high costs of regulation, federal and state governments do not supervise the use of nearly 99.5% of the commercially available chemicals in the United States. The problem is much worse in less-developed countries.

Pollution Prevention and the Precautionary Principle

Some scientists and health officials, especially those in European Union countries, are pushing for much greater emphasis on *pollution prevention* (**Concept 17.4B**). They say chemicals that are known or suspected to cause significant harm should not be released into the environment. This means looking for harmless or less harmful substitutes for toxic and hazardous chemicals. Another option is to recycle them within production processes to keep them from reaching the environment, as companies such as DuPont and 3M have been doing (see the Case Study that follows).

Pollution prevention is a strategy for implementing the **precautionary principle** (see Chapter 9, p. 215). According to this principle, when there is substantial preliminary evidence that an activity, technology, or chemical substance can harm humans, other organisms, or the environment, decision makers should take measures to prevent or reduce such harm, rather than waiting for more conclusive scientific evidence.

There is controversy over how far we should go in using the precautionary principle. Those who favor a precautionary approach argue that a person or company, proposing to introduce a new chemical or technology should bear the burden of establishing its safety. This would require two major changes in the way we evaluate and manage risks. *First,* we would assume that new chemicals and technologies could be harmful until scientific studies could show otherwise. *Second,* the existing chemicals and technologies that appear to have a strong chance of causing harm would be removed from the market until we could establish their safety. For example, after decades of research revealed the harmful effects of lead, especially on children, lead-based paints and leaded gasoline were phased out in most developed countries.

Manufacturers and businesses contend that widespread application of the precautionary approach and requiring more pollution prevention would make it too expensive and almost impossible to introduce any new chemical or technology. They note that there is always some uncertainty in any scientific assessment of risk.

Proponents of increased reliance on the precautionary principle argue that it would focus the efforts on finding solutions to pollution problems that are based on prevention rather than on cleanup. It also reduces health risks for employees and society, frees businesses from having to deal with pollution regulations, and reduces the threat of lawsuits from injured parties. Proponents also argue that we have an ethical responsibility to reduce serious risks to human health, to the environment, and to future generations, in keeping with the ethical **principle of sustainability** (see the inside back cover of this book).

The European Union is applying the precautionary principle through pollution prevention. The Stockholm Convention of 2000 is an international agreement to ban or phase out the use of 12 of the most notorious *persistent organic pollutants (POPs)*, also called the *dirty dozen*. These highly toxic chemicals have been shown to produce numerous harmful effects, including cancers, birth defects, compromised immune systems, and declining sperm counts and sperm quality in men in a number of countries. The list includes DDT and eight other pesticides, PCBs, and dioxins. In 2009, nine more POPs were added, some of which are widely used in pesticides and in flame-retardants added to clothing, furniture, and other consumer goods. The treaty went into effect in 2004 but has not been formally approved or implemented by the United States.

In 2007, the European Union enacted regulations known as REACH (for *Registration, Evaluation, Authorization, and restriction of CHemicals*). It required the registration of 30,000 untested, unregulated, and potentially harmful chemicals. In the REACH process, the most hazardous substances are not approved for use if safer alternatives exist.

REACH puts more of the burden on industry to show that chemicals are safe. Conventional regulation such as that used in the United States has put the burden on governments to show that they are dangerous. The U.S. chemical regulation structure was enacted in the 1976 Toxic Substances Control Act. At Congressional hearings, experts have testified that this system makes it virtually impossible for the government to limit or ban the use of toxic chemicals. The hearings found that by 2009, the EPA had required testing for only 200 of the more than 85,000 chemicals registered for use and had issued regulations to control fewer than 12 of those chemicals.

In 2011, the EPA took a step toward pollution prevention by issuing a rule to control emissions of mercury (**Core Case Study**) and harmful fine-particle pollution from older coal-burning plants in 28 states. Many eastern states have high depositions of mercury and harmful particles produced by coal-burning power and electric plants in the Midwest and blown eastward by prevailing winds (Figure 17.18). The new air pollution standards could prevent as many as 11,000 premature deaths, 200,000 nonfatal heart attacks, and 2.5 million asthma attacks. However, coal companies are pressuring members of Congress to eliminate this new standard.

Representatives from many nations have developed a UN treaty, known as the Minamata Convention, with the overall goal of reducing global mercury emissions by 15–35% in the next several decades. By January 2016, 128 countries had signed and 22 countries had ratified the treaty. It will go into effect after 50 countries have ratified it. Once in effect, within 5 years, signatory nations must require new coal-fired power plants, industrial plants, and smelters to use the best available mercury emission-control technologies. The treaty also calls for phasing out the use of mercury in many products, including thermometers, light bulbs, switches, batteries, and some cosmetics.

CASE STUDY

Pollution Prevention Pays

The U.S.-based 3M Company makes 60,000 different products in 100 manufacturing plants around the world. In 1975, 3M began a Pollution Prevention Pays (3P) program. Since then, it has reformulated some of its products, redesigned equipment and processes, and reduced its use of hazardous raw materials. It has also recycled and reused more waste materials and sold some of its potentially hazardous but still useful wastes as raw materials to other companies. As of 2015, this program had prevented more than 1.8 million metric tons (2 million tons) of pollutants from reaching the environment and saved the company $1.9 billion.

The 3M 3P program has been successful largely because employees are rewarded if projects they come up with eliminate or reduce a pollutant; reduce the amount of energy, materials, or other resources required in production; or save money through reduced pollution control costs, lower operating costs, or increased sales of new or existing products. Employees at 3M have now completed more than 10,000 3P projects.

Since 1990, a growing number of companies have adopted similar pollution and waste prevention programs that have led to cleaner production. They are learning that, in addition to saving money by preventing pollution and reducing waste production, they have a much easier job of complying with pollution laws and regulations. In addition, they find they can avoid lawsuits based on exposure to harmful chemicals, provide a safer environment for their workers (which can reduce their employee health insurance costs), and improve their public image.

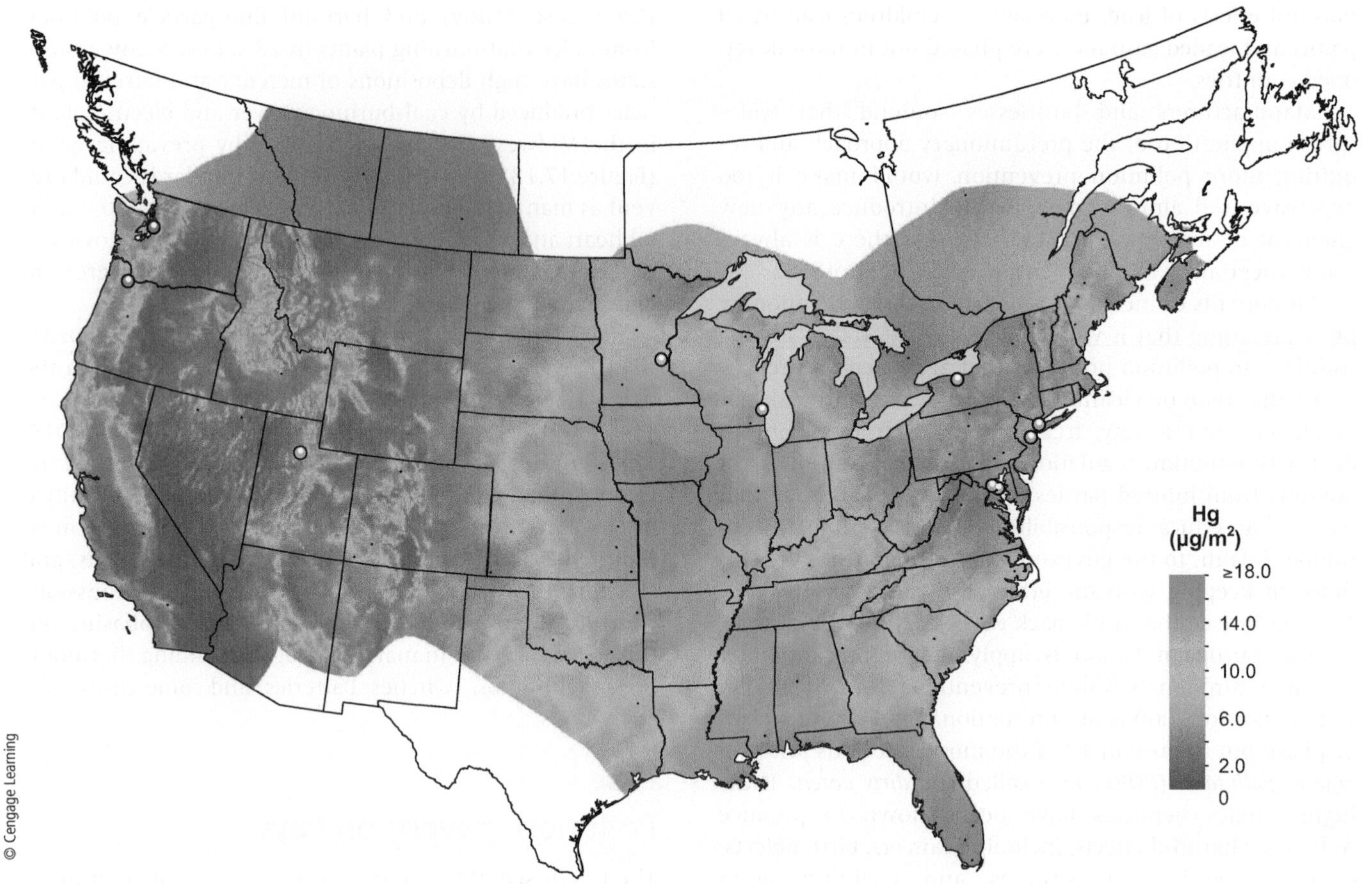

FIGURE 17.18 Atmospheric wet deposition of mercury in the lower 48 states in 2010. ***Critical thinking:*** Why do the highest levels occur mainly in the eastern half of the United States?

(Compiled by the authors using data from the Environmental Protection Agency and the National Atmospheric Deposition Program)

17.5 HOW DO WE PERCEIVE AND AVOID RISKS?

CONCEPT 17.5 We can reduce the major risks we face by becoming informed, thinking critically about risks, and making careful choices.

The Greatest Health Risks Come from Poverty, Gender, and Lifestyle Choices

Risk analysis involves identifying hazards and evaluating their associated risks (*risk assessment*; Figure 17.2, left), ranking risks (*comparative risk analysis*), determining options and making decisions about reducing or eliminating risks (*risk management*; Figure 17.2, right), and informing decision makers and the public about risks (*risk communication*).

Statistical probabilities based on experience, animal testing, and other assessments are used to estimate risks from older technologies and chemicals. To evaluate new technologies and products, risk evaluators use more uncertain statistical probabilities, based on models rather than on actual experience and testing.

In terms of the number of deaths per year (Figure 17.19, left), *the greatest risk by far is poverty*. The high death toll ultimately resulting from poverty is caused by malnutrition, increased susceptibility to normally nonfatal infectious diseases, and often-fatal infectious diseases transmitted by unsafe drinking water (Figure 17.19, right).

Studies indicate that the four greatest risks in terms of shortened life spans are living in poverty, being born male, smoking (see the Case Study that follows), and obesity. Some of the greatest risks of premature death are illnesses that result primarily from lifestyle choices that people make (Figure 17.20) (**Concept 17.1**). For example, overeating can lead to obesity and type-2 diabetes and long-term smoking can cause lung cancer.

According to 2015 statistics from the CDC, about 72% of all U.S. adults over age 20 are obese (38%) or overweight (34%). A 2015 study of the latest data on obesity indicated that 12–14% of all U.S. adults have diabetes, and most of them have type 2 diabetes, which is linked to obesity and inactivity. The study found that another 40% of

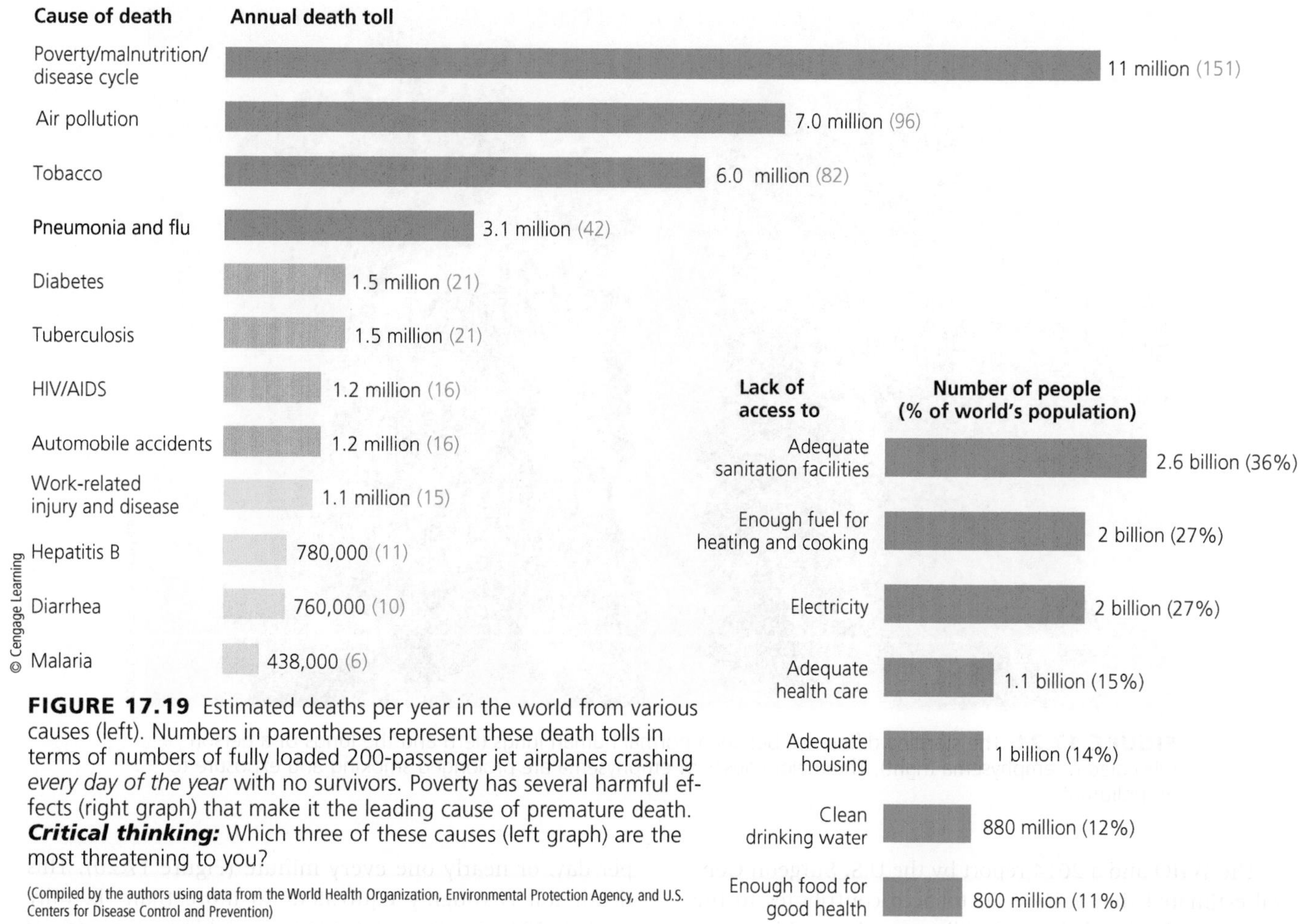

FIGURE 17.19 Estimated deaths per year in the world from various causes (left). Numbers in parentheses represent these death tolls in terms of numbers of fully loaded 200-passenger jet airplanes crashing *every day of the year* with no survivors. Poverty has several harmful effects (right graph) that make it the leading cause of premature death. ***Critical thinking:*** Which three of these causes (left graph) are the most threatening to you?

(Compiled by the authors using data from the World Health Organization, Environmental Protection Agency, and U.S. Centers for Disease Control and Prevention)

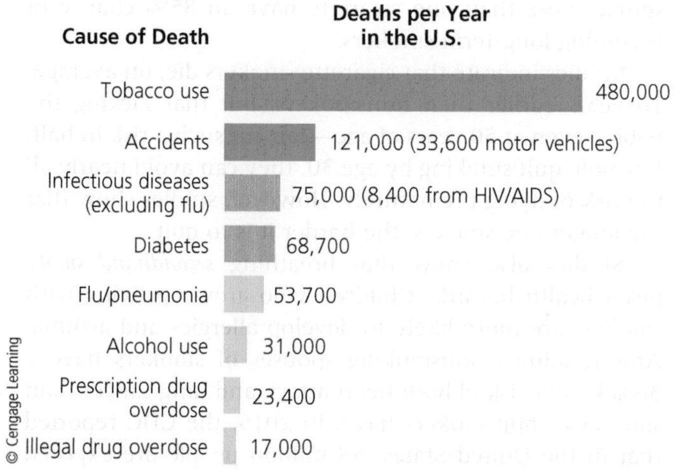

FIGURE 17.20 Leading causes of death in the United States. Some result from lifestyle choices and are preventable. ***Data analysis:*** The number of deaths from tobacco use is how many times the number of deaths from flu/pneumonia?

(Compiled by the authors using data from the U.S. Centers for Disease Control and Prevention.)

all U.S. adults have *prediabetes,* a condition marked by blood sugar levels high enough to put them at risk of developing diabetes.

The best ways to reduce one's risk of such serious health problems are to avoid smoking and exposure to smoke, lose excess weight, reduce consumption of foods containing sugar (especially high fructose corn syrup), eat a variety of fruits and vegetables, exercise regularly, drink little or no alcohol (no more than two drinks in a single day), avoid excess sunlight (which ages skin and can cause skin cancer), and practice safe sex (**Concept 17.5**).

CASE STUDY

Cigarettes and E-Cigarettes

Cigarette smoking is the world's most preventable and largest cause of suffering and premature death among adults. The WHO estimates that smoking contributed to the deaths of 100 million people during the 20th century and could kill 1 billion people during this century unless governments and individuals act to dramatically reduce smoking.

FIGURE 17.21 The startling difference between normal human lungs (left) and the lungs of a person who died of emphysema (right). The major causes of emphysema are prolonged smoking and exposure to air pollutants.

The WHO and a 2014 report by the U.S. Surgeon General estimated that each year, tobacco contributes to the premature deaths of about 6 million people resulting from 25 illnesses, including heart disease, stroke, type 2 diabetes, lung and other cancers, memory impairment, bronchitis, and emphysema (Figure 17.21). This amounts to an average of more than 16,400 deaths every day, or about one every 5 seconds.

In a study led by Rachel A. Whitmer, researchers tracked the health of 21,123 individuals for 30 years. They found that people between the ages of 50 and 60 who had smoked one to two packs of cigarettes daily had a 44% higher chance of getting Alzheimer's disease or vascular dementia (which reduces blood flow to the brain and can erode memory) by age 72.

By 2030, the annual death toll from smoking-related diseases is projected to reach more than 8 million—an average of 21,900 preventable deaths per day—according to the CDC and the WHO. About 80% of these deaths are expected to occur in less-developed countries, especially China, with 350 million smokers. There is little effort to reduce smoking in China, partly because cigarette taxes provide up to 10% of the central government's total annual revenues. According to a 2015 study by Zhengming Chen and a team of other researchers, smoking could lead to 3 million deaths per year in China by 2050.

According to a 2014 CDC report, smoking kills about 480,000 Americans per year—an average of 1,315 deaths per day, or nearly one every minute (Figure 17.20). This death toll is roughly equivalent to more than six fully loaded 200-passenger jet planes crashing *every day of the year* with no survivors. Smoking also causes about 8.6 million illnesses every year in the United States.

The overwhelming scientific consensus is that the nicotine inhaled in tobacco smoke is highly addictive. A British government study showed that adolescents who smoke more than one cigarette have an 85% chance of becoming long-term smokers.

Studies indicate that cigarette smokers die, on average, 10 years earlier than nonsmokers, but that kicking the habit—even at 50 years of age—can cut such a risk in half. If people quit smoking by age 30, they can avoid nearly all the risk of dying prematurely. However, studies show that the longer one smokes, the harder it is to quit.

Studies also show that breathing *secondhand smoke* poses health hazards. Children who grow up living with smokers are more likely to develop allergies and asthma. Among adults, nonsmoking spouses of smokers have a 30% higher risk of both heart attack and lung cancer than spouses of nonsmokers have. In 2015, the CDC reported that in the United States, 58 million people are exposed daily to secondhand smoke, which is responsible for nearly 42,000 deaths per year.

In the United States, the percentage of adults who smoke dropped from more than 50% in the 1950s to 18% in 2013, according to the CDC, and the goal is to reduce

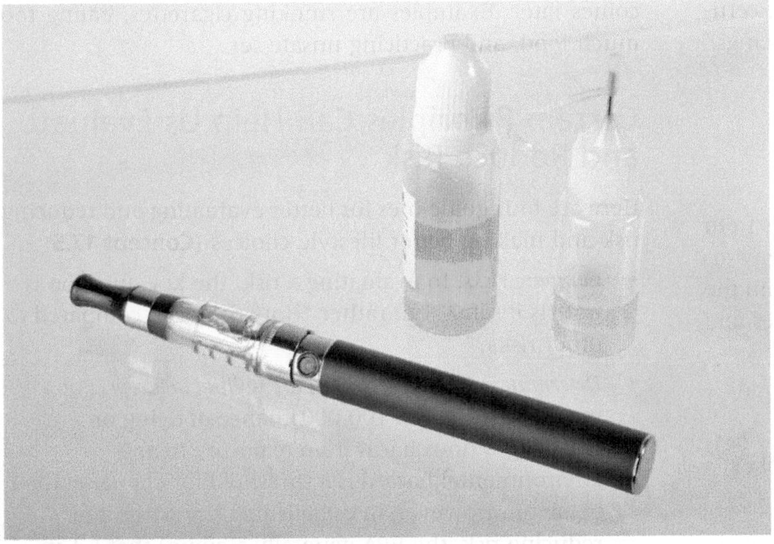

jps/Shutterstock.com

FIGURE 17.22 An e-cigarette that can be refilled with a solution of nicotine (e-juice).

this to less than 10% by 2025. This decline can be attributed to media coverage about the harmful health effects of smoking, sharp increases in cigarette taxes in many states (in New York City, the average price of a pack if cigarettes is $13), mandatory health warnings on cigarette packs, the ban on sales to minors, and bans on smoking in workplaces, bars, restaurants, and public buildings.

A new form of cigarette, called *electronic cigarettes* or *e-cigarettes* (Figure 17.22), is being used increasingly as a substitute for tobacco cigarettes. These devices contain pure nicotine dissolved in a syrupy solvent containing one or more of over 7,000 chemicals to enhance taste and smell. A battery in the device heats the nicotine solution and converts it to a vapor containing liquid particles that the user inhales. Smoking e-cigarettes is called *vaping*. E-cigarettes can be refilled with solutions that vary from 2% to 10% in their concentrations of nicotine.

The CDC reported in 2015 that tobacco smoking among high school students dropped to a historic low of 9.2% of all U.S. students in 2014—down from a high of 36% in 1997. However, in 2014, more than 13% of all U.S. high school students used e-cigarettes—three times the percentage estimated in 2013.

Are e-cigarettes safe? No one knows, because they have not been around long enough to be thoroughly evaluated. E-cigarettes reduce or eliminate the inhalation of tar and numerous other harmful chemicals found in regular cigarette smoke. But they still expose users to highly addictive nicotine, sometimes at levels of up to 5 times as high (10% nicotine) as that found in regular cigarettes (2% nicotine). Smokers may reduce their overall health threat by switching to e-cigarettes. On the other hand, e-cigarettes may lead some adolescents and young adults who have never smoked to become regular cigarette users.

Preliminary research indicates that some e-cigarette vapors contain trace amounts of toxic metals such as cadmium, nickel, lead, and several substances that can cause cancer in test animals. Some of these toxins, not found in regular cigarette smoke, are nanoparticles that are small enough to get past the body's defense systems and travel deep into the lungs. However, it will take much more research for any direct link between e-cigarettes and cancer to be established.

The European Union (EU) has banned the advertising and sales of e-cigarettes and tobacco products to minors, as well as Internet sales of these products. EU regulations also limit the concentration of nicotine in e-cigarettes to 2% and require the disclosure of e-cigarette ingredients. They require that these products have childproof and tamper-proof packaging that carries graphic warnings on the harmful health effects of nicotine. In the United States in 2016, the FDA issued a set of rules that banned the sale of e-cigarettes to anyone under the age of 18. The rules also require package warning labels and make all existing and new e-cigarette products subject to FDA approval.

Estimating Risks from Technologies

The more complex a technological system, and the more people needed to design and run it, the more difficult it is to estimate the risks of using the system. The overall *reliability* of such a system—the probability (expressed as a percentage) that the system will complete a task without failing—is the product of two factors:

$$\text{System reliability (\%)} = \text{Technology reliability (\%)} \times \text{Human reliability (\%)}$$

With careful design, quality control, maintenance, and monitoring, a highly complex system such as a nuclear power plant or a deep-sea oil-drilling rig can achieve a high degree of technological reliability. But human reliability usually is much lower than technological reliability and is almost impossible to predict.

Suppose the estimated technological reliability of a nuclear power plant is 95% (0.95) and human reliability is 75% (0.75). Then the overall system reliability is 71% ($0.95 \times 0.75 = 71\%$). Even if we could make the technology 100% reliable (1.0), the overall system reliability would still be only 75% ($1.0 \times 0.75 = 75\%$).

We can make a system safer by moving more of the potentially fallible elements from the human side to the technological side. However, chance events such as a lightning strike can knock out an automatic control system, and no machine or computer program can completely replace human judgment. Also, the parts in any automated control system are manufactured, assembled, tested, certified, inspected, and maintained by fallible human beings.

Most People Do a Poor Job of Evaluating Risks

Most of us are not good at assessing the relative risks from the hazards that we encounter. Many people deny or shrug off the high-risk chances of death or injury from the voluntary activities they enjoy. Examples of activities and their death rates include the following:

- *Driving* (1 in 3,300 without a seatbelt and 1 in 6,070 with a seatbelt)
- *Smoking* (1 in 250 by age 70 for a pack-a-day smoker)
- *Hang gliding* (1 in 1,250)

Indeed, the most dangerous thing that many people do each day is to drive or ride in a car. Yet some of these same people may be terrified about their chances of being killed by the flu (a 1 in 130,000 chance), a nuclear power plant accident (1 in 200,000), West Nile virus (1 in 1 million), a lightning strike (1 in 3 million), Ebola virus (1 in 4 million), a commercial airplane crash (1 in 9 million), snakebite (1 in 36 million), or shark attack (1 in 281 million).

Five factors can cause people to see a technology or a product as being more or less risky than experts judge it to be. The first factor is *fear*. Research shows that fear causes people to overestimate risks and to worry more about catastrophic risks than they do about common, everyday risks. Studies show that people tend to overestimate numbers of deaths caused by tornadoes, floods, fires, homicides, cancer, and terrorist attacks, and to underestimate death tolls from smoking, flu, diabetes, asthma, heart attack, stroke, and automobile accidents.

The second factor clouding risk evaluation is the *degree of control* individuals have in a given situation. Many people have a greater fear of things over which they do not have personal control. For example, some individuals feel safer driving their own car for long distances (with a 1-in-6,070 risk of dying) than traveling the same distance on a plane (with a risk of 1 in 9 million).

The third factor influencing risk evaluation is whether a risk is catastrophic or chronic. People usually are more frightened by news of catastrophic accidents such as a plane crash than of a cause of death such as smoking, which has a much higher death toll spread out over time.

Fourth, some people have *optimism bias*, the belief that risks that apply to other people do not apply to them. Although people get upset when they see others driving erratically while talking on a cell phone or texting, they may believe that talking on the cell phone or texting does not impair their own driving ability.

A fifth factor affecting risk analysis is that many of the risky things we do are highly pleasurable and give *instant gratification*, while the potential harm from such activities comes later. Examples are smoking cigarettes, eating too much food, and practicing unsafe sex.

Certain Principles Can Help Us Evaluate and Reduce Risk

Here are four guidelines for better evaluating and reducing risk and making better lifestyle choices (**Concept 17.5**):

- *Compare risks.* In evaluating a risk, the key question is not "Is it safe?" but rather "How risky is it compared to other risks?"
- *Determine how much risk you are willing to accept.* For most people, a 1 in 100,000 chance of dying or suffering serious harm from exposure to an environmental hazard is a threshold for changing their behavior. However, in establishing standards and reducing risk, the EPA generally assumes that a 1 in 1 million chance of dying from an environmental hazard is acceptable.
- *Evaluate the actual risk involved.* The news media usually exaggerate the daily risks we face in order to capture our interest and attract more readers, listeners, or television viewers. As a result, most people who are exposed to a daily diet of such exaggerated reports believe that the world is much more dangerous and risk-filled than it really is.
- *Concentrate on evaluating and carefully making important lifestyle choices.* When you worry about a risk, the most important question to ask is, "Do I have any control over this?" There is no point worrying about risks over which we have little or no control. But we do have control over major ways to reduce risks from heart attack, stroke, and certain forms of cancer, by deciding whether to smoke, what to eat, how much alcohol to drink, how much exercise we get, and how safely we drive.

BIG IDEAS

- We face significant hazards from infectious diseases such as flu, AIDS, tuberculosis, diarrheal diseases, and malaria, and from exposure to chemicals that can cause cancers and birth defects, as well as chemicals that can disrupt the human immune, nervous, and endocrine systems.
- Because of the difficulty of evaluating the harm caused by exposure to chemicals, many health scientists call for much greater emphasis on pollution prevention.
- By becoming informed, thinking critically about risks, and making careful choices, we can reduce the major risks we face.

Tying It All Together

Mercury's Toxic Effects and Sustainability

Robert and Jean Pollock/Science Source

In the **Core Case Study** that opens this chapter, we saw that mercury (Hg) and its compounds that occur regularly in the environment can permanently damage the human nervous system, kidneys, and lungs and harm fetuses and cause birth defects. In this chapter, we also learned of many other chemical hazards, as well as biological, physical, cultural, and lifestyle hazards, in the environment. In addition, we saw how difficult it is to evaluate the nature and severity of threats from these various hazards.

One of the important facts discussed in this chapter is that on a global basis, the greatest threat to human health is poverty (often leading to malnutrition and disease), followed by the threats from air pollution, smoking, pneumonia and flu, and HIV/AIDS.

There are some threats that we can do little to avoid, but we can reduce other threats, partly by applying the three **scientific principles of sustainability** (see inside back cover of this book). For example, we can greatly reduce our exposure to mercury and other pollutants by shifting from the use of nonrenewable fossil fuels (especially coal) to wider use of a diversity of renewable energy resources, including solar and wind energy. We can reduce our exposure to harmful chemicals used in the manufacturing of various goods by cutting resource use and waste and by reusing and recycling material resources. We can also mimic biodiversity by using diverse strategies for solving environmental and health problems, and especially for reducing poverty and controlling population growth. In so doing, we also help to preserve the earth's biodiversity and increase our beneficial environmental impact.

Chapter Review

Core Case Study

1. Describe the toxic effects of mercury and its compounds and explain how we are exposed to these toxins.

Section 17.1

2. What is the key concept for this section? Define and distinguish among **risk**, **risk assessment**, and **risk management**. Give an example of how scientists state probabilities. Give an example of a risk from each of the following: biological hazards, chemical hazards, natural hazards, cultural hazards, and lifestyle choices. What is a **pathogen**?

Section 17.2

3. What is the key concept for this section? Define **infectious disease**. Define and distinguish among **bacteria**, **viruses**, and **parasites**, and give examples of diseases that each can cause. Define and distinguish between **transmissible disease** and **nontransmissible disease**, and give an example of each. In terms of death rates, what are the world's four most serious infectious diseases? List five factors that have contributed to genetic resistance in bacteria to commonly used antibiotics. What is MRSA and why is it so threatening?
4. Describe the global threat from tuberculosis and list three factors that have helped it to spread. What is the biggest viral killer and how does it spread? Summarize the threat from the hepatitis B virus. What is the best way to reduce one's chances of getting an infectious disease? What is the focus of ecological medicine and what are some of its findings regarding the spread of diseases? Summarize the health threats from the global HIV/AIDS pandemic.
5. What is malaria and how does it spread? What percentage of the human population is subject to this threat? Explain how deforestation can promote the spread of malaria. List six major ways to reduce the global threat from infectious diseases.

Section 17.3

6. What is the key concept for this section? What is a **toxic chemical**? Define and distinguish among **carcinogens**, **mutagens**, and **teratogens**, and give an example of each. What are PCBs and why are they a threat? Describe the human immune, nervous, and endocrine systems, and for each of these systems, give an example of a chemical that can threaten it. What is a neurotoxin and why is methylmercury (**Core Case Study**) an especially dangerous one? What are six ways to prevent or control environmental inputs of mercury? What are hormonally active agents (HAAs), what risks do they pose, and how can we reduce those risks? Summarize health scientists' concerns about exposure to bisphenol A (BPA) and the controversy over what to do about exposure to this chemical. Summarize the concerns over exposure to phthalates. List six ways to reduce your exposure to HAAs.

Section 17.4

7. What are the two key concepts for this section? Define **toxicology**, **toxicity**, **dose**, and **response**. What are three factors that affect the level of harm caused by a chemical? Give three reasons why children are especially vulnerable to harm from toxic chemicals. Describe how the toxicity of a substance can be estimated by testing laboratory animals, and explain the limitations of this approach. What is a **dose-response curve**? Explain how toxicities are estimated through use of case reports and epidemiological studies, and discuss the limitations of these approaches.

8. Summarize the controversy over the effects of trace levels of chemicals. Why do we know so little about the harmful effects of chemicals? What is the **precautionary principle**? Explain why the use of pollution prevention based on the precautionary principle is controversial. Describe some efforts to apply this principle on national and international levels. What is the Stockholm Convention? What is the Minamata Convention? How did pollution prevention pay off for the 3M Company?

Section 17.5

9. What is the key concept for this section? What is **risk analysis**? In terms of premature deaths, what are the three greatest threats that people face? What are six ways in which poverty can threaten one's health? Describe the health threats from smoking and how we can reduce these threats. Summarize our knowledge of the health effects of using e-cigarettes. How can we reduce the threats resulting from the use of various technologies? What are five factors that can cause people to misjudge risks? List four guidelines for evaluating and reducing risk.

10. What are this chapter's *three big ideas*? Explain how we can lessen the threats of harm from mercury in the environment by applying the three **scientific principles of sustainability**.

Note: Key terms are in bold type.

Critical Thinking

1. Assume that you are a national official with the power to set policy for controlling environmental mercury pollution from human sources (**Core Case Study**). List the goals of your policy and outline a plan for accomplishing those goals. List three or more possible problems that could result from implementing your policy.

2. What are three actions you would take to reduce the global threats to human health and life from each of the following: **(a)** tuberculosis, **(b)** HIV/AIDS, and **(c)** malaria?

3. Explain why you agree or disagree with each of the following statements:
   a. We should not worry much about exposure to toxic chemicals because almost any chemical, at a large enough dosage, can cause some harm.
   b. We should not worry much about exposure to toxic chemicals because, through genetic adaptation, we can develop immunities to such chemicals.
   c. We should not worry much about exposure to toxic chemicals because we can use genetic engineering to reduce our susceptibility to their effects.
   d. We should not worry about exposure to a chemical such as bisphenol A (BPA) because it has not been absolutely proven scientifically that BPA has killed anyone.

4. Should we ban the use of hormone mimics such as BPA in making products to be used by children younger than age 5? Should such a ban be extended to all products? Explain.

5. Workers in a number of industries are exposed to higher levels of various toxic substances than are the general public. Should we reduce the workplace levels allowed for such chemicals? What economic effects might this have?

6. Do you think that electronic cigarettes should be taxed and regulated like conventional cigarettes? Explain.

7. What are the three major risks you face from each of the following: **(a)** your lifestyle, **(b)** where you live,

and **(c)** what you do for a living? Which of these risks are voluntary and which are involuntary? List three steps you could take to reduce each of these risks. Which of these steps do you already take or plan to take?

8. In deciding what to do about risks from chemicals in the area where you live, would you support legislation that requires the use of pollution prevention based on the precautionary principle and on the assumption that chemicals are potentially harmful until shown otherwise? Explain.

Doing Environmental Science

Pick a commonly used and potentially harmful chemical and do some research to learn about **(a)** what it is used for and how widely it is used, **(b)** its potential harm, **(c)** the scientific evidence for such claims, and **(d)** proposed solutions for dealing with this threat. Pick a study area, such as your dorm or apartment building, your block, or your city. In this area, try to determine the level of presence of the chemical you are studying. You could do this by finding four or five examples of items or locations containing the chemical and then estimating the total amount based on your sample. Write a report summarizing your findings.

Global Environment Watch Exercise

Go to your MindTap course to access the GREENR database. At the top of the page do a "Basic Search" on *mercury pollution.* Research the latest developments in studies of the harmful health effects of mercury (**Core Case Study**). Find an example of an effort to prevent or control mercury pollution and write a short report summarizing your findings. Try to find reports of two studies that reach different conclusions about how mercury should be regulated. Summarize the arguments for these conclusions on both sides. Based on what you have found, do you think that mercury pollution should be regulated more strictly in the state or country where you live? Explain your reasoning.

Data Analysis

The graph below shows the effects of AIDS on life expectancy at birth in Botswana, 1950–2000, and projects these effects to 2050. Study the graph and answer the questions that follow.

1. **(a)** By what percentage did life expectancy in Botswana increase between 1950 and 1995? **(b)** By what percentage did life expectancy in Botswana drop between 1995 and 2015?

2. **(a)** By what percentage was life expectancy in Botswana projected to increase between 2015 and 2050? **(b)** By what percentage was life expectancy in Botswana projected to decrease between 1995 and 2050?

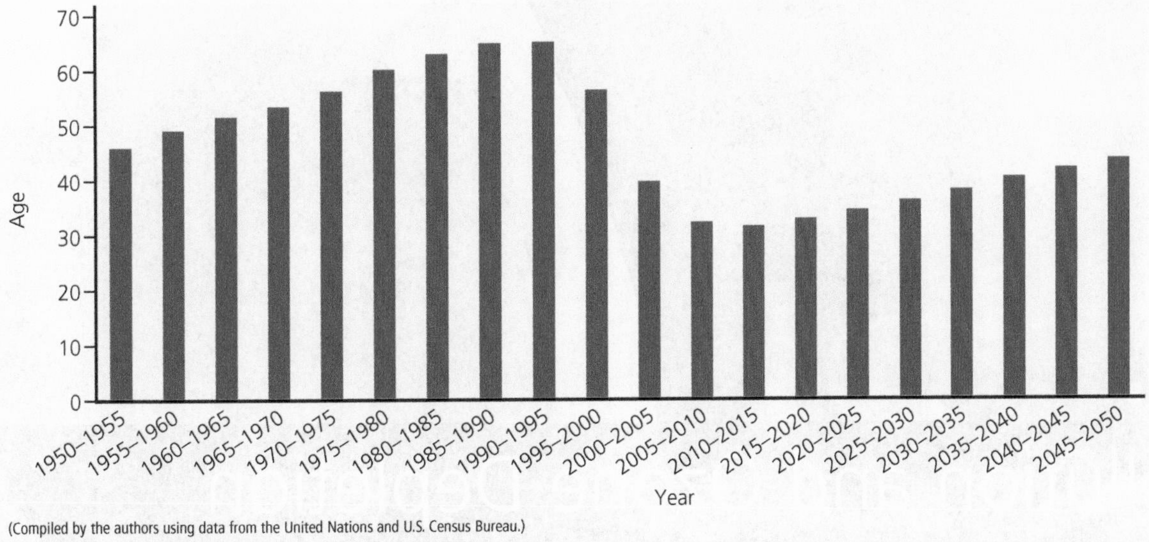

(Compiled by the authors using data from the United Nations and U.S. Census Bureau.)

CHAPTER 18

Air Pollution and Ozone Depletion

> The atmosphere is the key symbol of global interdependence. If we can't solve some of our problems in the face of threats to this global commons, then I can't be very optimistic about the future of the world.
>
> MARGARET MEAD

Key Questions

18.1 What is the nature of the atmosphere?

18.2 What are the major outdoor air pollution problems?

18.3 What is acid deposition and why is it a problem?

18.4 What are the major indoor air pollution problems?

18.5 What are the health effects of air pollution?

18.6 How should we deal with air pollution?

18.7 How have we depleted ozone in the stratosphere and what can we do about it?

Air pollution has become a major health threat, especially in many growing cities such as Beijing, China.

testing/Shutterstock.com

Core Case Study

Los Angeles Air Pollution

In the 1950s, Los Angeles, California (USA), had some of the dirtiest air in the world. Steel and chemical plants, oil refineries, factories, and backyard trash incinerators all pumped smoke and chemicals into the air, creating a dangerous mix of pollutants. More than 2 million motor vehicles also belched pollutants by the ton every day.

This problem had been growing since the 1940s as the city grew rapidly. By 1950, the dirty air was threatening tourism and real estate values. The city's leaders grew concerned and in 1947, Los Angeles (LA) established the nation's first air pollution control district.

By the mid-1950s, scientists had found strong evidence that motor vehicle emissions, when exposed to sunlight, were creating *photochemical smog*—an unhealthy mix of ozone and other chemicals that threatened human health. People began to realize the dangers of air pollution, powerfully symbolized by smog, which was visible and ugly. Concerned parents kept their children home from school on especially bad days, and athletic events were often canceled. Visits to doctors multiplied, and doctors hypothesized possible connections between air pollution and heart disease, respiratory ailments, and lung cancer.

As the science became clearer, public protests became frequent. People showed up wearing gas masks, which became a symbol of LA's pollution woes. Through the 1950s and 1960s, protesters organized anti-smog groups and these grassroots efforts spread across California and beyond, focusing their attention on lawmakers. By the early 1960s, citizens were demanding smog control devices on cars along with other measures to control air pollution. Automakers dragged their feet in response. In 1969, the federal government got involved, taking the car companies to court for working together to delay such technology.

Mounting air pollution problems in LA and across the country, along with a growing public outcry, finally caused the U.S. Congress to pass the Clean Air Act of 1970. This far-reaching law put strict limits on pollution from cars, diesel engines, and coal-burning facilities, forcing automakers and other industries to take measures to control pollution.

Such measures have paid off in the form of cleaner air all over the United States, even while the country's population has grown by 58% since 1970. In LA, ozone levels are 40% of those that existed in 1970, even though the number of cars there has more than doubled. Since 2000, LA has reduced its *fine particle*, or soot, pollution by half. However, LA's smog problem is still among the worst of all U.S. cities (Figure 18.1), according to the American Lung Association. Most observers agree that LA has made great progress, but still has a long way to go in cleaning up its air.

The story of LA air pollution and efforts to deal with it reflects the story of air pollution in general, at least in more-developed countries. In this chapter, we examine that story, as well as the facts about air pollution in less-developed countries—a problem that is no less challenging.

iStockphoto.com/Lee Pettet

FIGURE 18.1 Los Angeles, California, on a smoggy day. Despite major improvements in air quality, the city still sees smoggy days like this one.

18.1 WHAT IS THE NATURE OF THE ATMOSPHERE?

CONCEPT 18.1 The two innermost layers of the atmosphere are the *troposphere*, which supports life, and the *stratosphere*, which contains the protective ozone layer.

The Atmosphere Consists of Several Layers

Life exists under a thin blanket of gases surrounding the earth, called the **atmosphere.** It is divided into several spherical layers defined mostly by temperature differences (Figure 18.2). Our focus in this book is on the atmosphere's two innermost layers: the troposphere and the stratosphere (**Concept 18.1**).

An important atmospheric variable is *density*, the number of gas molecules per unit of air volume. It varies throughout the atmosphere because gravity pulls harder on gas molecules near the earth's surface than it does on molecules high up in the atmosphere. This means that lower layers have more gases (more weight) in them than upper layers do, and are more densely packed with molecules. Thus, the air we breathe at sea level has a higher density than the air we would inhale on top of a high mountain.

Another important atmospheric variable is *atmospheric pressure*—the force, or mass, per unit area of a column of air. This force is caused by the continuous bombardment of a surface such as your skin by the molecules in air. Atmospheric pressure varies with density. It decreases with altitude (see black line in Figure 18.2) because there are fewer gas molecules at higher altitudes. The density and pressure of the atmosphere are important because they play major roles in the weather.

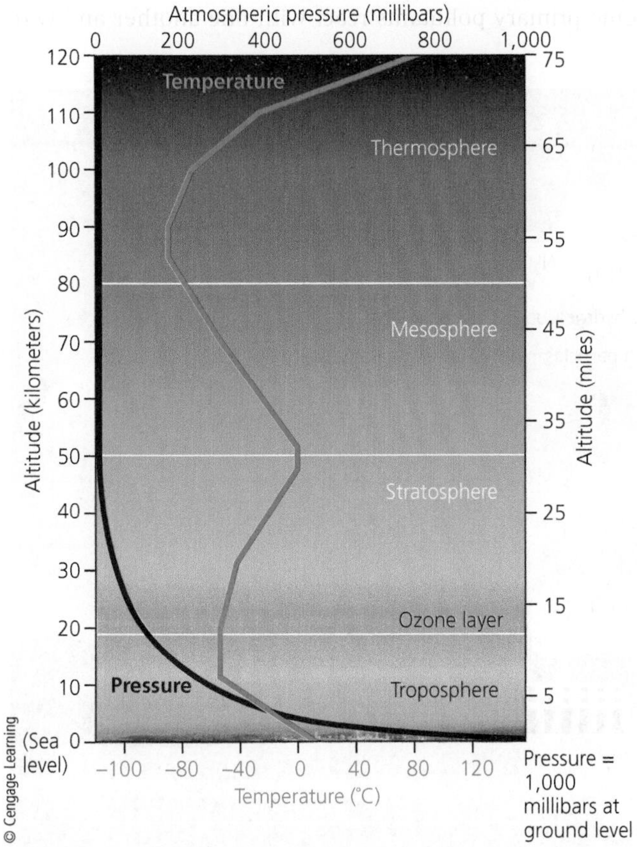

FIGURE 18.2 Natural capital: The earth's atmosphere is a dynamic system that includes four layers. The average temperature of the atmosphere varies with altitude (red line) and with differences in the absorption of incoming solar energy. ***Critical thinking:*** Why do you think most of the planet's air is in the troposphere?

Air Movements and Chemicals in the Troposphere Affect the Earth's Weather and Climate

About 75–80% of the earth's air mass is found in the **troposphere,** the atmospheric layer closest to the earth's surface. This layer extends about 17 kilometers (11 miles) above sea level at the equator and 6 kilometers (4 miles) above sea level over the poles. If the earth were the size of an apple, this lower layer containing the air we breathe would be no thicker than the apple's skin.

Take a deep breath. About 99% of the volume of air you inhaled consists of two gases: nitrogen (78%) and oxygen (21%). The remainder is 0.93% argon (Ar), 0.040% carbon dioxide (CO_2), and smaller amounts of water vapor, dust and soot particles, and other gases, including methane (CH_4), ozone (O_3), and nitrous oxide (N_2O).

Several gases in the troposphere, including H_2O, CO_2, CH_4, and N_2O, are called **greenhouse gases** because they absorb and release energy that warms the troposphere. Rising and falling air currents, winds, and concentrations of CO_2 and other greenhouse gases in the troposphere play major roles in the planet's short-term *weather* and long-term *climate*.

The Stratosphere Is Our Global Sunscreen

The atmosphere's second layer is the **stratosphere,** which extends from about 17 to about 48 kilometers (from 11 to 30 miles) above the earth's surface (Figure 18.2). Although the stratosphere contains less matter than the troposphere, its chemical composition is similar, with two notable exceptions. The stratosphere has a much lower volume of water vapor and a much higher concentration of ozone.

Most of the atmosphere's ozone is concentrated in a portion of the stratosphere called the **ozone layer,** found roughly 17–26 kilometers (11–16 miles) above sea level. Most of the ozone in this layer is produced when oxygen molecules interact with ultraviolet (UV) radiation emitted by the sun.

$$3O_2 + UV \leftrightarrow 2O_3$$

This UV filtering effect of ozone in the lower stratosphere acts as a "global sunscreen" that keeps about 95% of the sun's harmful UV radiation from reaching the earth's surface. The ozone layer allows life to exist on land and helps to protect us from sunburn, skin and eye cancers, cataracts, and damage to our immune systems. It also prevents much of the oxygen in the troposphere from being converted to ground-level ozone, a harmful air pollutant.

18.2 WHAT ARE THE MAJOR OUTDOOR AIR POLLUTION PROBLEMS?

CONCEPT 18.2 Pollutants mix in the air to form harmful *industrial smog*, caused mostly by the burning of coal, and *photochemical smog*, caused by emissions from motor vehicles, industrial facilities, and power plants.

Air Pollution Comes from Natural and Human Sources

Air pollution is the presence of chemicals in the atmosphere in concentrations high enough to harm organisms, ecosystems, or human-made materials, or to alter climate. Almost any chemical in the atmosphere can become a pollutant if it occurs in a high enough concentration. The effects of air pollution range from annoying to lethal.

Air pollutants come from natural and human sources. Natural sources include wind-blown dust, solid and gaseous pollutants from wildfires and volcanic eruptions, and volatile organic chemicals released by some plants. Most natural air pollutants are spread out over the globe and become diluted or are removed by chemical cycles, precipitation, and gravity. However, pollutants emitted by volcanic eruptions and forest fires can temporarily reach harmful levels.

Most human inputs of outdoor air pollutants (Figure 18.3) occur in industrialized and urban areas where people, cars, and factories are concentrated. These pollutants are generated mostly by the burning of fossil fuels in power plants and industrial facilities (*stationary sources*) and in motor vehicles (*mobile sources*).

Scientists classify outdoor air pollutants into two categories. **Primary pollutants** (Figure 18.3, center) are chemicals or substances emitted directly into the air from natural processes and human activities at concentrations high enough to cause harm. While in the atmosphere, some primary pollutants react with one another and with

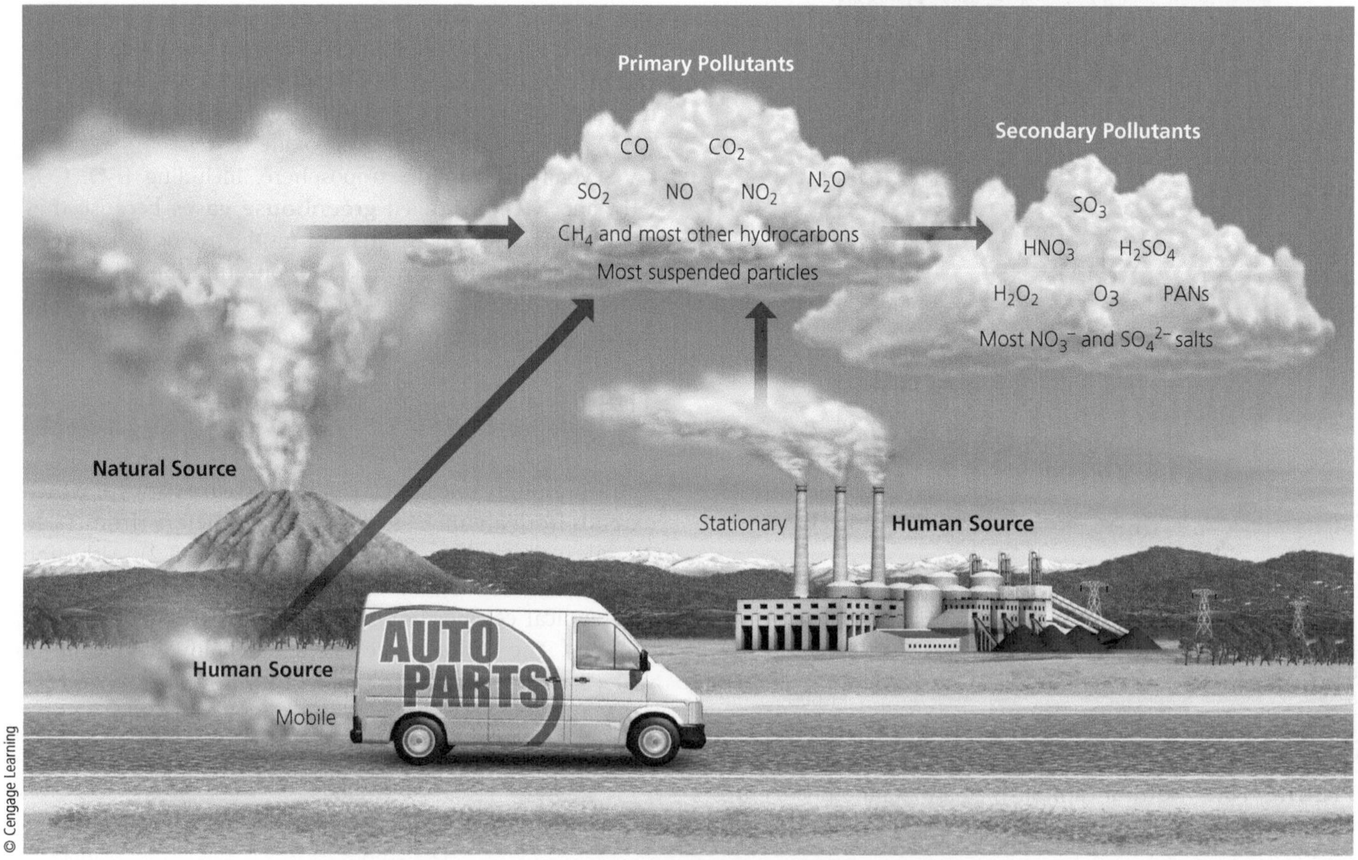

FIGURE 18.3 Human inputs of air pollutants come from *mobile sources* (such as cars) and *stationary sources* (such as industrial, power, and cement plants). Some *primary air pollutants* react with one another and with other chemicals in the air to form *secondary air pollutants*.

SCIENCEFOCUS 18.1

The Atmospheric Brown Clouds

Air pollution is no longer viewed as primarily a localized urban problem. Annual satellite images and studies by the United Nations Environment Programme (UNEP) have found massive, dark brown clouds of pollution—called *atmospheric brown clouds*. They stretch across much of India, Bangladesh, and the industrial heart of China, as well as parts of the western Pacific Ocean (Figure 18.A).

In most years, these clouds cover an area about the size of the continental United States. They contain small particles of dust, smoke, and ash resulting from wind erosion due to drought and from the clearing and burning of forests for planting crops. They also contain particles of soot, or *black carbon*, and of toxic metals such as mercury and lead. These various particles enter the atmosphere from wildfires, the burning of wood and animal dung for heat and cooking, diesel engine exhaust, motor vehicle exhaust, ocean ships burning heavy oil, coal-burning power and industrial plants, metal smelters, and waste incinerators.

These enormous pollution clouds can move across the Asian continent within 3 to 4 days. Satellites have tracked the spread of pollutants from the atmospheric brown clouds over northern China across the Pacific Ocean to the West Coast of the United States. Measurements made by atmospheric scientists show that large portions of the particulate matter, soot, and toxic mercury in the skies above Los Angeles, California (**Core Case Study**), can be traced to China.

Researchers have estimated that the atmospheric brown clouds are directly linked to the deaths of more than 380,000 people a year in China and India. They also affect global weather patterns. In 2015, atmospheric scientists led by Yuan Wang reported on a study finding that pollution from the clouds is strengthening storms over the Pacific Ocean. These storms can push heat from the tropics toward the Arctic region and force cold air to move south. This has probably contributed to abnormally cold weather in North America during winter months in some years.

Long-term studies on other effects of the brown clouds were carried out by an international team of scientists led by V. Ramanathan of the Scripps Institution of Oceanography. Their findings include decreases in the summer monsoon rainfall in some areas, a north-south shift in rainfall patterns in eastern China, accelerated melting of Himalayan glaciers that feed major Asian rivers, and increased levels of ozone in the lower atmosphere in many areas. These effects have helped reduce water supplies and crop yields and have damaged human health.

The researchers also found that soot and some of the other particles that fall onto Himalayan glaciers from the atmospheric brown clouds absorb sunlight and heat the air above those glaciers. This soot also decreases the ability of the glaciers to reflect sunlight back into space. The glaciers then absorb more solar energy and experience increased melting. This adds to the warming of the air above them, which in turn further increases the rate of glacial melting in a runaway positive feedback cycle (see Chapter 2, p. 43). The researchers projected that at the current rate of melting, the Himalayan glaciers could shrink by as much as 75% before 2050, which would pose "a grave danger to the region's water security."

JUSTIN GUARIGLIA/National Geographic Creative

FIGURE 18.A Severe air pollution from the atmospheric brown cloud in Shanghai, China.

CRITICAL THINKING

The Asian economy has provided much of the world, including the United States, with massive amounts of goods in recent decades, which is part of the reason for the industrial pollution in the atmospheric brown clouds. Do you think this means that countries outside of Asia should contribute financial and other resources to help deal with this problem? Explain.

other natural components of air to form new harmful chemicals, called **secondary pollutants** (Figure 18.3, right).

With their high concentrations of cars and factories, urban areas normally have higher outdoor air pollution levels (**Core Case Study**) than rural areas have. However, prevailing winds can spread long-lived primary and secondary air pollutants from urban and industrial areas to the countryside and to other urban areas. Long-lived pollutants entering the atmosphere in India and China now find their way across the Pacific where they affect the West Coast of North America (Science Focus 18.1). In fact, satellite measurements show that long-lived air pollutants from anywhere on the planet can circle the entire globe in about 2 weeks.

Air Pollution Has Harmful Effects

Air pollution connects us all and has affected every place on the planet's surface. Even in arctic regions where very few people live, air pollutants flowing north from Europe, Asia, and North America collect to form *arctic haze.*

Since the 1970s, the quality of outdoor air in most of the more-developed countries has improved, thanks largely to grassroots pressure from citizens in the 1960s and 1970s. This led the governments of the United States and of most European countries to pass and enforce air-pollution-control laws (**Core Case Study**).

On the other hand, in 2015, according to the American Lung Association, 138 million Americans or 43% of the U.S. population lived in areas where air pollution reached dangerous levels during parts of the year. Globally, according to the World Health Organization (WHO), more than 1.1 billion people live in urban areas where outdoor air is unhealthy to breathe. Most of them live in densely populated cities in less-developed countries where air-pollution-control laws do not exist or are poorly enforced.

Prolonged high exposure to air pollutants overloads the body's natural defense mechanisms. Fine and ultrafine particles can get lodged deep in the lungs and contribute to cancer, asthma, heart attack, and stroke. In 2014, the WHO estimated that outdoor and indoor air pollution kills about 7 million people each year. This averages to about 800 deaths every hour. A 2015 study led by Jos Lelieveld at Germany's Max Planck Institute found that outdoor air pollution kills 1.4 million people per year in China and 645,000 per year in India.

Major Outdoor Air Pollutants

Hundreds of different chemicals and substances can pollute outdoor air. Here we focus on six major groups of air pollutants.

CARBON OXIDES. *Carbon monoxide* (CO) is a colorless, odorless, and highly toxic gas that forms during the incomplete combustion of carbon-containing materials (Table 18.1). Major sources are motor vehicle exhaust, the burning of forests and grasslands, the smokestacks of fossil fuel–burning power plants and industries, tobacco smoke, and open fires and inefficient stoves used for cooking or heating.

In the body, CO can combine with hemoglobin in red blood cells, which reduces the ability of blood to transport oxygen to body cells and tissues. Long-term exposure can trigger heart attacks and aggravate lung diseases such as asthma and emphysema. At high levels, CO can cause headache, nausea, drowsiness, confusion, collapse, coma, and death, which is why it is important to have CO detectors in your home.

Carbon dioxide (CO_2) is a colorless, odorless gas. About 93% of the CO_2 in the atmosphere is the result of the natural carbon cycle (see Figure 3.20, p. 65). The rest comes from human activities such as the burning of fossil fuels, which adds CO_2 to the atmosphere, and the removal of forests and grasslands that help remove excess CO_2 from the atmosphere.

TABLE 18.1 Chemical Reactions that Form Major Air Pollutants

Pollutant	Chemical Reaction
Carbon monoxide (CO)	$2C + O_2 \rightarrow 2CO$
Carbon dioxide (CO_2)	$C + O_2 \rightarrow CO_2$
Nitric oxide (NO)	$N_2 + O_2 \rightarrow 2NO$
Nitrogen dioxide (NO_2)	$2NO + O_2 \rightarrow 2NO_2$
Sulfur dioxide (SO_2)	$S + O_2 \rightarrow SO_2$

NITROGEN OXIDES AND NITRIC ACID. *Nitric oxide* (NO) is a colorless gas that forms when nitrogen and oxygen gases react under high temperatures in automobile engines and coal-burning power and industrial plants (Table 18.1). Lightning and certain bacteria in soil and water also produce NO as part of the nitrogen cycle (see Figure 3.21, p. 66).

In the air, NO reacts with oxygen to form *nitrogen dioxide* (NO_2), a reddish-brown gas. Collectively, NO and NO_2 are called *nitrogen oxides* (NO*x*). Some of the NO_2 reacts with water vapor in the air to form *nitric acid* (HNO_3) and nitrate salts (NO_3^-), components of harmful *acid deposition,* which we discuss later in this chapter. Both NO and NO_2 play a role in the formation of *photochemical smog*—a mixture of chemicals formed under the influence of sunlight in cities with heavy traffic (**Core Case Study**). Nitrous oxide (N_2O), a greenhouse gas, is emitted from fertilizers and animal wastes and is produced by the burning of fossil fuels.

At high enough levels, nitrogen oxides can irritate the eyes, nose, and throat, and aggravate lung ailments such as asthma and bronchitis. They can also suppress plant growth, and reduce visibility when they are converted to nitric acid and nitrate salts.

SULFUR DIOXIDE AND SULFURIC ACID. Sulfur dioxide (SO_2) is a colorless gas with an irritating odor. About one-third of the SO_2 in the atmosphere comes from natural sources such as volcanoes. The other two-thirds (and as much as 90% in highly industrialized urban areas) comes from human sources, mostly combustion of sulfur-containing coal in power and industrial plants (Table 18.1), oil refining, and the smelting of sulfide ores.

In the atmosphere, SO_2 can be converted to *aerosols,* which consist of microscopic suspended droplets of sulfuric acid (H_2SO_4) and suspended particles of sulfate (SO_4^{2-}) salts that return to the earth as a component of acid deposition.

O. LOUIS MAZZATENTA/National Geographic Creative

FIGURE 18.4 Sulfuric acid and other air pollutants have damaged this statue in Rome, Italy. The nose and part of the forehead have been restored.

Sulfur dioxide, sulfuric acid droplets, and sulfate particles reduce visibility and aggravate breathing problems. They can damage crops, trees, soils, and aquatic life in lakes. They also corrode metals and damage paint, paper, leather, and the stone used to build walls, statues (Figure 18.4), and monuments.

PARTICULATES. *Suspended particulate matter* (SPM) consists of a variety of solid particles and liquid droplets that are small and light enough to remain suspended in the air for long periods. The U.S. Environmental Protection Agency (EPA) classifies particles as fine, or PM-10 (with diameters less than 10 micrometers, or less than one-fifth the diameter of a human hair); and ultrafine, or PM-2.5 (with diameters less than 2.5 micrometers). About 62% of the SPM in outdoor air comes from natural sources such as dust, wildfires, and sea salt. The other 38% comes from human sources such as coal-burning power and industrial plants (Figure 18.5), motor vehicles, wind erosion from exposed topsoil, and road construction. The EPA has found that fine particles can travel for thousands of kilometers in the atmosphere, while ultrafine particles have been shown to travel for up to 10 kilometers (6 miles) from their sources.

These particles can irritate the nose and throat, damage the lungs, aggravate asthma and bronchitis, and shorten life spans. Toxic particulates such as lead (see the Case Study that follows), cadmium, and polychlorinated biphenyls (PCBs) can cause genetic mutations, reproductive problems, and cancer. Particulates also reduce visibility, corrode metals, and discolor clothing and paints.

JAMES P. BLAIR/National Geographic Creative

FIGURE 18.5 Severe air pollution from an iron and steel factory in Czechoslovakia.

OZONE. One of the major ingredients of photochemical smog is *ozone* (O_3), a colorless and highly reactive gas. It can cause coughing and breathing problems, aggravate lung and heart diseases, reduce resistance to colds and pneumonia, and irritate the eyes, nose, and throat. Ozone also damages plants, rubber in tires, fabrics, and paints.

Ozone in the troposphere near ground level can be harmful at high enough levels, whereas ozone in the stratosphere is beneficial because it protects us from harmful UV radiation. Extensive measurements indicate that human activities have *decreased* the amount of beneficial ozone in the stratosphere and *increased* the amount of harmful ground-level ozone—especially in some urban areas. We examine the issue of decreased stratospheric ozone in the final section of this chapter.

VOLATILE ORGANIC COMPOUNDS (VOCs). Organic compounds that exist as gases in the atmosphere or that evaporate from sources on the earth's surface into the atmosphere are called *volatile organic compounds* (VOCs). Examples are hydrocarbons, emitted by the leaves of many plants, and methane (CH_4), a greenhouse gas that is 25 times more effective per molecule than CO_2 is at warming the atmosphere. About a third of global methane emissions come from natural sources such as plants, wetlands, and termites. The rest come from human sources such as rice paddies, landfills, natural gas wells and pipelines, and from cows (mostly from their belching) raised for meat and dairy production.

Other VOCs are liquids than can evaporate quickly into the atmosphere. Examples are benzene and other industrial solvents, dry-cleaning fluids, and various components of gasoline, plastics, and other products.

An important priority for many public health officials and scientists is to continually improve the monitoring of outdoor air for the presence of dangerous pollutants (Science Focus 18.2).

CASE STUDY

Lead Is a Highly Toxic Pollutant

Lead (Pb) is a soft gray metal used to make various products including lead–acid batteries and bullets, and it was once a common ingredient of gasoline and paints. It is also a particulate pollutant found in air, water, soil, plants, and animals.

Because it is a chemical element, lead does not break down in the environment. This indestructible and potent neurotoxin can harm the nervous system, especially in young children. Severe lead poisoning can leave children to suffer from palsy, partial paralysis, blindness, and mental retardation.

Children under age 6 and unborn fetuses, even with low blood levels of lead, are especially vulnerable to nervous system impairment, lowered IQ (by 2 to 5 points), shortened attention span, hyperactivity, hearing damage, and various behavior disorders. According to many scientists, there is no safe level of lead in children's blood and they call for sharply reducing the currently allowed levels for lead in the air and water.

Since the 1970s, according to the U.S. Centers for Disease Control and Prevention (CDC), the percentage of U.S. children under age 6 with blood lead levels above the safety standard dropped from 85% to less than 1%, which prevented at least 9 million childhood lead poisonings. The primary reason for this drop was that after a decade-long fight with the oil and lead industries, the federal government banned leaded gasoline in 1976. Leaded gasoline was completely phased out by 1986. The government also greatly reduced the allowable levels of lead in paints. This is an excellent example of the effectiveness of pollution prevention.

However, in 2012, the CDC used the latest scientific data to come up with new guidelines for identifying children who have potentially dangerous blood lead levels. These guidelines more than doubled the estimated number of young children at risk from lead poisoning in the United States, raising it to about 535,000. The major source of exposure is peeling lead-based paint and lead-contaminated dust in some 4 million older U.S. homes. Children can inhale or ingest paint particles from these sources when they put dust-covered hands or toys into their mouths. Another source is soils contaminated with lead emitted by motor vehicles before leaded gasoline was banned. Lead can also leach from water pipes and faucets containing lead parts or lead solder (a problem we examine in Chapter 20).

Other sources are older coal-burning power plants that have not been required to meet the emission standards of new plants, as well as lead smelters and waste incinerators. And so far, the U.S. ban on leaded gasoline does not apply to aviation fuel. Although commercial jet fuel does not contain lead, about 75% of all U.S. private aircraft still burn leaded fuel. Lead-free fuel is available, and the Federal Aviation Administration (FAA) hopes to replace all leaded aviation fuel in U.S. private aircraft by 2018. Meanwhile, the estimated 3 million children who live or go to school in areas around the nation's 22,000 airports are exposed to unsafe levels of lead from this source.

CONSIDER THIS . . .

CONNECTIONS Lead and Urban Gardening

Health officials and scientists have urged people who plant urban vegetable gardens to have their garden soils tested for lead. For decades, lead particles fell from the air into urban soils, primarily from the exhaust fumes of vehicles burning leaded gasoline. Soil found to have lead in it can be treated or removed from urban gardens and replaced with uncontaminated soil.

Although the threat from lead has been greatly reduced in the United States, this is not the case in many less-developed countries. While only six smaller, less-developed countries were still using leaded gasoline as of 2012, millions of children are still exposed to lead from other sources. China recently phased out leaded gasoline in less than 3 years. However, in 2014, the National Institutes of Health reported that 40 countries, including China, India, Russia, most of South America, and several African countries, still allow the sale of lead-based paints.

Children and adults in China and several African countries are also exposed to dangerous levels of lead when they work in recycling centers extracting lead and other valuable metals from electronic waste (e-waste)—discarded computers, TV sets, cellphones, and other electronic devices. Much of this e-waste is shipped to such unsafe recycling centers from the United States and other more-developed countries.

The WHO reported in 2015 that globally, lead exposure accounts for about 143,000 deaths and about 600,000 cases of children with intellectual disabilities per year, mostly in less-developed countries. Health scientists have proposed a number of ways to help protect children from lead poisoning (Figure 18.6).

Burning Coal Produces Industrial Smog

Seventy years ago, cities such as London, England, and the U.S. cities of Chicago, Illinois, and Pittsburgh, Pennsylvania, burned large amounts of coal in power plants and factories. They also burned coal to heat homes and often to cook food. People in such cities, especially during winter, were exposed to **industrial smog,** consisting mostly of an unhealthy mix of sulfur dioxide (SO_2), suspended

SCIENCEFOCUS 18.2

Detecting Air Pollutants and Monitoring Air Quality

We can detect the presence of pollutants in the air with the use of chemical instruments and satellites armed with various sensors. The scientists who discovered the components and effects of the atmospheric brown clouds (Science Focus 18.1) used small, unmanned aircraft carrying miniaturized instruments to measure chemical concentrations, temperatures, and other variables within the clouds.

Aerodyne Research in the U.S. city of Boston, Massachusetts, has developed a mobile laboratory that uses sophisticated instruments to make instantaneous measurements of primary and secondary air pollutants from motor vehicles, factories, and other sources. This laboratory can also monitor changes in concentrations of the pollutants throughout a day and under different weather conditions, and it can measure the effectiveness of various air pollution control devices used in cars, trucks, and buses. Scientists are also using nanotechnology (see Science Focus 14.1, p. 370) to try to develop inexpensive nanodetectors for various air pollutants.

In partnership with the EPA, some Google Street View cars are equipped with state-of-the-art sensors that measure atmospheric levels of a number of pollutants, including soot, ozone, and nitrogen oxide gases. If expanded, these data would allow individuals to use Google Earth and Google Maps to monitor air quality on the block where they live.

Another way to detect air pollutants is through biological indicators. For example, a *lichen* is an organism consisting of a fungus and an alga living together, usually in a mutualistic relationship. These hardy pioneer species are good biological indicators of air pollution because they continually absorb air as a source of nourishment. A highly polluted area around an industrial plant might have only gray-green crusty lichens or none at all. An area with moderate air pollution might support only orange crusty lichens (Figure 18.B). In contrast, areas with clean air can support larger varieties of lichens.

Some lichen species are sensitive to specific air-polluting chemicals. Old man's beard and yellow Evernia lichens, for example, can sicken and die in the presence of excessive sulfur dioxide (SO_2), even if the pollutant originates far away. For this reason, scientists discovered SO_2 pollution on Isle Royale, Michigan (USA), in Lake Superior, an island where no car or smokestack has ever intruded. They used Evernia lichens to point the finger northwest toward coal-burning facilities in and around the Canadian city of Thunder Bay, Ontario.

With daily information about air pollution, the EPA has created an air quality indicator called the *Air Quality Index (AQI)* for informing citizens about unsafe levels of pollution in any given area of the country. Scientists collect daily data on the levels of five major pollutants—ground-level ozone, particulates, carbon monoxide, NO_2, and SO_2—using instruments at more than 1,000 locations around the United States. They use these data to compute a daily AQI for each pollutant and an overall AQI for any particular region. AQI values run from 0 to 500, with higher numbers indicating poorer air quality. Values of 200 and over are considered *very unhealthy* or *hazardous* for all people.

CRITICAL THINKING

How do you think the science and technology of air pollution detection and air quality monitoring should be financed? Who should pay for it?

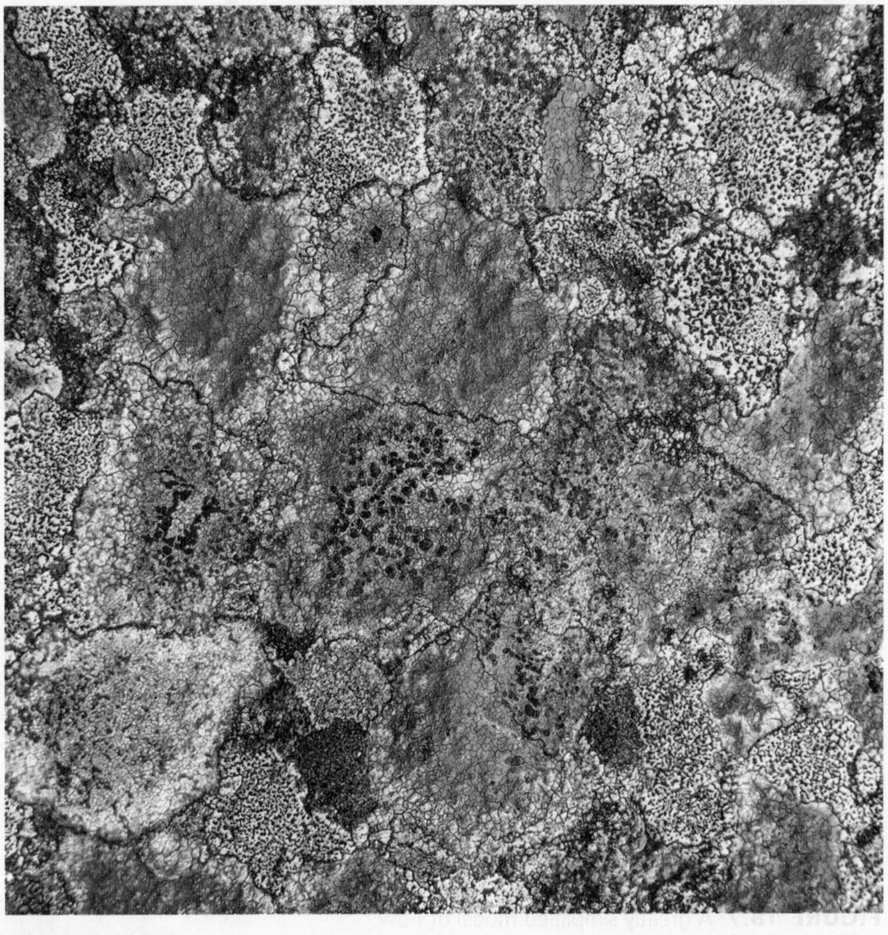

Mr Doomits/Shutterstock.com

FIGURE 18.B Lichens such as these growing on a rock can act as biological indicators of air pollution.

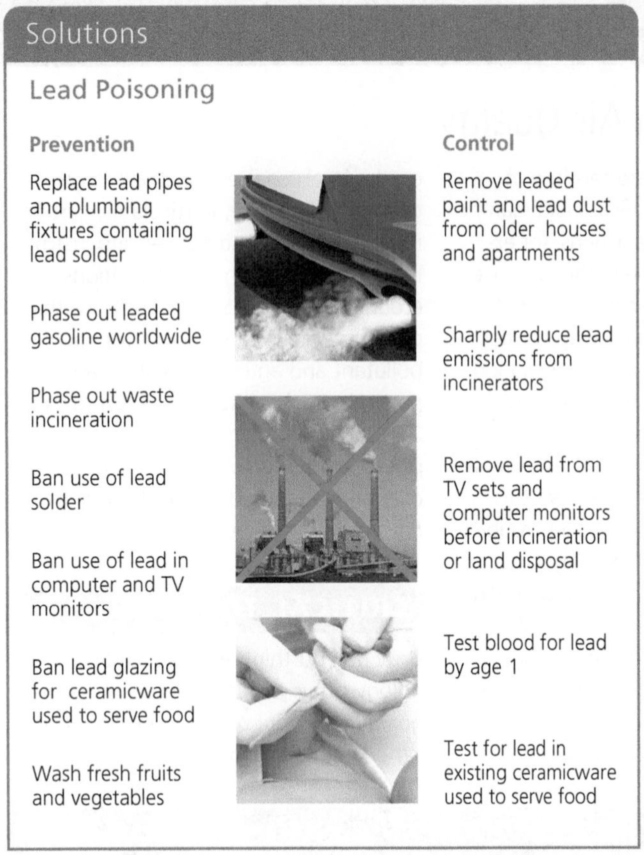

FIGURE 18.6 Ways to help protect children from lead poisoning. ***Critical thinking:*** Which two of these solutions do you think are the best ones? Why?

droplets of sulfuric acid, and a variety of suspended solid particles in outside air (**Concept 18.2**). Those who burned coal inside their homes were often exposed to dangerous levels of indoor air pollutants.

When coal or oil is burned, the sulfur compounds they contain react with oxygen to produce SO_2 gas (Figure 18.7, left), some of which is converted to tiny suspended droplets of sulfuric acid (H_2SO_4). Some of these droplets react with ammonia (NH_3) in the atmosphere to form solid particles of ammonium sulfate, or $(NH_4)_2SO_4$. Also, during combustion of coal and oil, most of the carbon they contain is converted to carbon monoxide (CO) and carbon dioxide (CO_2). Unburned carbon in coal also ends up in the atmosphere as soot or black carbon. Suspended particles of such salts and soot give the resulting smog a gray color (Figure 18.5), which is why it is sometimes called *gray-air smog*.

Today, urban industrial smog is rarely a problem in most of the more-developed countries where coal is burned only in large power and industrial plants with reasonably good air pollution control. However, many of these facilities have tall smokestacks that send the pollutants high into air where prevailing winds carry them downwind to rural areas, and can cause air pollution problems that we deal with later in this chapter.

Industrial smog remains a problem in industrialized urban areas of China, India, Ukraine, Czechoslovakia (Figure 18.5), and other countries where large quantities of coal are still burned in houses, power plants, and factories with inadequate pollution controls. Because of its heavy reliance on coal, China has some of the world's highest levels of industrial smog and 16 of the world's 20 most polluted cities. According to a 2014 Chinese government report, 92% of all Chinese cities did not meet the government's national outdoor air quality standards in 2013 (see chapter-opening photo). Beijing's air quality is among the worst in the world, according to the WHO, mostly from industrial pollution and the burning of coal. China is making some progress in lessening its dependence on coal but has a long way to go.

Sunlight Plus Cars Equals Photochemical Smog

Photochemical smog is a mixture of primary and secondary pollutants formed under the influence of UV radiation from the sun. In greatly simplified terms,

$$\text{VOCs} + NO_x + \text{heat} + \text{sunlight} \rightarrow \begin{array}{l} \text{ground-level ozone}(O_3) \\ + \text{ other photochemical oxidants} \\ + \text{ aldehydes} \\ + \text{ other secondary pollutany} \end{array}$$

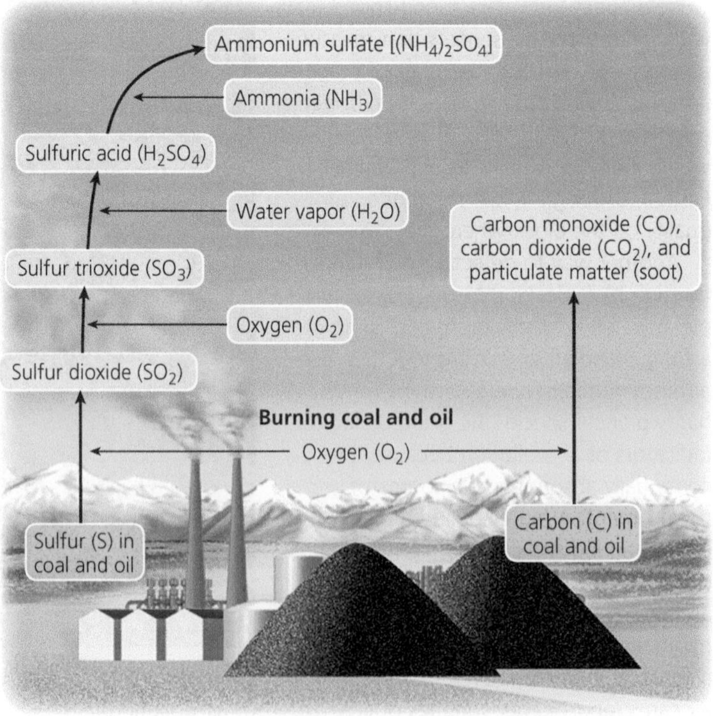

FIGURE 18.7 A greatly simplified model of how pollutants are formed when coal and oil are burned. The result is industrial smog (**Concept 18.2**).

The formation of photochemical smog (Figure 18.8 and **Concept 18.2**) begins when exhaust from morning commuter traffic releases large amounts of NO and VOCs into the air over a city. The NO is converted to reddish-brown NO_2, which is why photochemical smog is sometimes called *brown-air smog* (Figure 18.1). When exposed to ultraviolet radiation from the sun, some of the NO_2 reacts in complex ways with VOCs released by certain trees (such as certain species of oak, sweet gum, and poplar), motor vehicles, and businesses (especially bakeries and dry cleaners).

The resulting photochemical smog is a mixture of ozone, nitric acid, aldehydes, peroxyacyl nitrates (PANs), and other secondary pollutants. Collectively, NO_2, O_3, and PANs in this chemical brew are called *photochemical oxidants* because these damaging chemicals can react with certain compounds in the atmosphere and inside our lungs. Hotter days lead to higher levels of ozone and other components of smog. As traffic increases on a sunny day, photochemical smog (dominated by ozone) usually builds

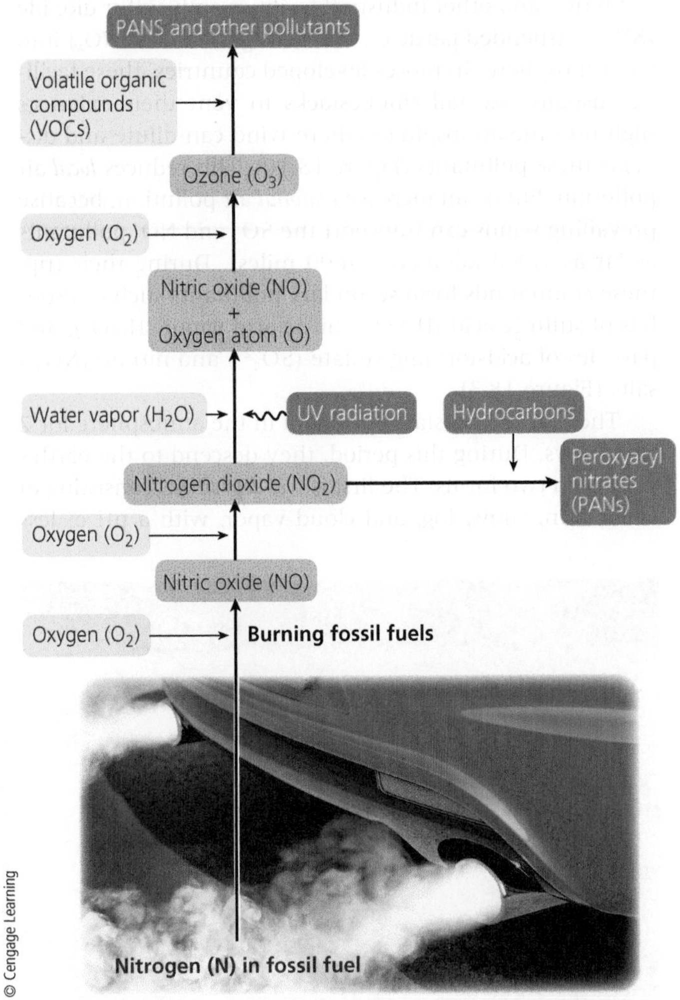

FIGURE 18.8 A greatly simplified model of how the pollutants that make up photochemical smog are formed.

Photo: Ssuaphotos/Shutterstock.com

to peak levels by late morning, irritating some people's eyes and respiratory tracts.

All modern cities have some photochemical smog, but it is much more common in cities with sunny and warm climates, and a large number of motor vehicles. Examples are Los Angeles, California (**Core Case Study** and Figure 18.1), and Salt Lake City, Utah, in the United States; Sydney, Australia; São Paulo, Brazil; Bangkok, Thailand; and Mexico City, Mexico.

CONSIDER THIS . . .

CONNECTIONS Short Driving Trips and Air Pollution

About 60% of the pollution from motor vehicle emissions occurs in the first minutes of operation before pollution control devices are working at top efficiency. Yet 40% of all U.S. car trips take place within 3 kilometers (2 miles) of drivers' homes, and half of the U.S. working population drives 8 kilometers (5 miles) or less to work. Did you drive a car today, and if so, how far did you drive?

Several Factors Affect Levels of Outdoor Air Pollution

Five natural factors help *reduce* outdoor air pollution. First, *gravity* causes particles heavier than air to settle out. Second, *rain and snow* partially cleanse the air of pollutants. Third, *salty sea spray from the oceans* washes out many pollutants from air that flows from land over the oceans. Fourth, *winds* sweep pollutants away and dilute them by mixing them with cleaner air. And fifth, *natural chemical reactions* remove some pollutants. For example, SO_2 can react with O_2 in the atmosphere to form SO_3, which reacts with water vapor to form droplets of H_2SO_4 that fall out of the atmosphere as acidic precipitation.

Six other factors can *increase* outdoor air pollution. First, *urban buildings* slow wind speed and reduce the dilution and removal of pollutants. Second, *hills and mountains* reduce the flow of air in valleys below them and allow pollutant levels to build up at ground level. Third, *high temperatures* promote the chemical reactions leading to the formation of photochemical smog. Fourth, *emissions of volatile organic compounds (VOCs)* from certain trees and plants in urban areas can promote the formation of photochemical smog.

The fifth factor—the so-called *grasshopper effect*—occurs mostly in winter when air pollutants are transported at high altitudes by evaporation and winds from tropical and temperate areas to the earth's polar areas as part of the earth's global air circulation system (see Figure 7.9, p. 146). The grasshopper effect explains why, for decades, pilots have reported seeing a reddish-brown haze over the Arctic. It also helps explain why polar bears, sharks, and native peoples in remote arctic areas have high levels of various harmful pollutants in their bodies.

The sixth factor that increases air pollution has to do with the *vertical movement of air*. During the day, the sun warms the air near the earth's surface. Normally, this warm

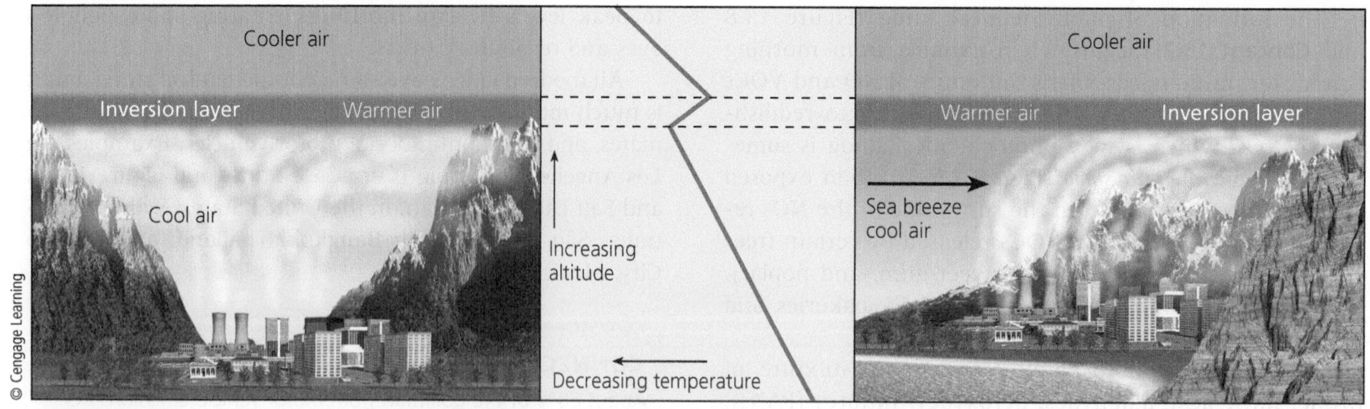

FIGURE 18.9 A *temperature inversion* can take place in either of the two sets of topography and weather conditions shown here. Polluted air can be trapped between mountain ranges and under the inversion layer (left). Or it can be blown by sea breezes and trapped against a mountain range and under the conversion layer (right, see **Core Case Study**).

air and most of the pollutants it contains rise to mix with the cooler air above it and are dispersed. However, under certain atmospheric conditions, a layer of warm air can temporarily lie atop a layer of cooler air nearer the ground. This is called a **temperature inversion.** Because the cooler air near the surface is denser than the warmer air above, it does not rise and mix with the air above. If this condition persists, pollutants can build up to harmful and even lethal concentrations in the trapped layer of cool air near the ground.

Two types of areas are especially susceptible to prolonged temperature inversions. The first is a town or city located in a valley surrounded by mountains where the weather turns cloudy and cold during part of the year (Figure 18.9, left). In such cases, the clouds block much of the winter sunlight that causes air to heat and rise, and the mountains block winds that could disperse the pollutants. As long as these stagnant conditions persist, pollutants in the valley below build up to harmful and even lethal concentrations.

The other type of area vulnerable to temperature inversions is a city with many motor vehicles in an area with a sunny climate, mountains on three sides, and an ocean on the fourth side (Figure 18.9, right). Here, the conditions are ideal for the formation of photochemical smog, worsened by frequent thermal inversions. The surrounding mountains prevent the polluted surface air from being blown away by breezes coming off the sea. This describes several cities, including heavily populated Los Angeles, California (**Core Case Study**), which has prolonged temperature inversions.

18.3 WHAT IS ACID DEPOSITION AND WHY IS IT A PROBLEM?

CONCEPT 18.3 Acid deposition is caused mainly by coal-burning power plants and motor vehicle emissions, and in some regions, it threatens human health, aquatic life and ecosystems, forests, and human-built structures.

Acid Deposition

Most coal-burning power plants, metal ore smelters, oil refineries, and other industrial facilities emit sulfur dioxide (SO_2), suspended particles, and nitrogen oxides (NO_x) into the atmosphere. In more-developed countries, these facilities usually use tall smokestacks to vent their exhausts high into the atmosphere where wind can dilute and disperse these pollutants (Figure 18.10). This reduces *local* air pollution, but it can increase *regional* air pollution, because prevailing winds can transport the SO_2 and NO_x pollutants as far as 1,000 kilometers (600 miles). During their trip, these compounds form secondary pollutants such as droplets of sulfuric acid (H_2SO_4), nitric acid vapor (HNO_3), and particles of acid-forming sulfate (SO_4^{2-}) and nitrate (NO_3^-) salts (Figure 18.3).

These acidic substances remain in the atmosphere for 2 to 14 days. During this period, they descend to the earth's surface in two forms. The first is *wet deposition*, consisting of acidic rain, snow, fog, and cloud vapor, with a pH of less

FIGURE 18.10 Tall smokestacks can reduce local air pollution from burning coal, but they help transfer sulfur dioxide and particulates to downwind areas.

than 5.6—the acidity level of unpolluted rain (Figure 2.6, p. 36). The second is *dry deposition*, consisting of acidic particles. The resulting mixture is called **acid deposition** (Figure 18.11)—often called *acid rain*. Most dry deposition occurs within 2 to 3 days of emission, relatively close to the industrial sources, whereas most wet deposition takes place within 4 to 14 days in more distant downwind areas.

Acid deposition has been occurring since the Industrial Revolution began in the mid-1700s. In 1872, British chemist Robert A. Smith coined the term *acid rain* after observing that rain was eating away stone in the walls of buildings in major industrial areas. Acid deposition is the result of human activities that disrupt the natural nitrogen cycle (see Figure 3.21, p. 66) and sulfur chemical cycle by adding excessive amounts of NO*x* and SO_2 to the atmosphere.

Acid deposition is a regional air pollution problem (**Concept 18.3**) in areas that lie downwind from coal-burning facilities and from urban areas with large numbers of cars (Figure 18.12). In some areas, soils contain *basic* compounds such as calcium carbonate ($CaCO_3$) or limestone that can react with and help neutralize, or *buffer*, some inputs of acids. The areas most sensitive to acid deposition are those with thin, acidic soils that provide no such natural buffering (Figure 18.12, all green and most red areas) and those where the buffering capacity of soils has been depleted by decades of acid deposition.

In the United States, older coal-burning power and industrial plants without adequate pollution controls, especially in the Midwest, emit the largest quantities of SO_2 and other pollutants that cause acid deposition. Because of these emissions and those of other urban industries and motor vehicles, as well as the prevailing west-to-east winds, typical precipitation in the eastern United States is at least 10 times more acidic than natural precipitation is. One of the first experiments to determine this took place in the Hubbard Brook Experimental Forest (see Chapter 2, Core Case Study, p. 30), located in the northeastern United States. There, researchers found that precipitation was several hundred times more acidic than natural rainwater.

Many acid-producing chemicals generated in one country are exported to other countries by prevailing winds. For example, acidic emissions from the United Kingdom and Germany blow south and east into Switzerland and Austria, and north and east into Norway and other neighboring countries. The worst acid deposition occurs in Asia, especially in China, which in 2014 got 66% of its total energy and

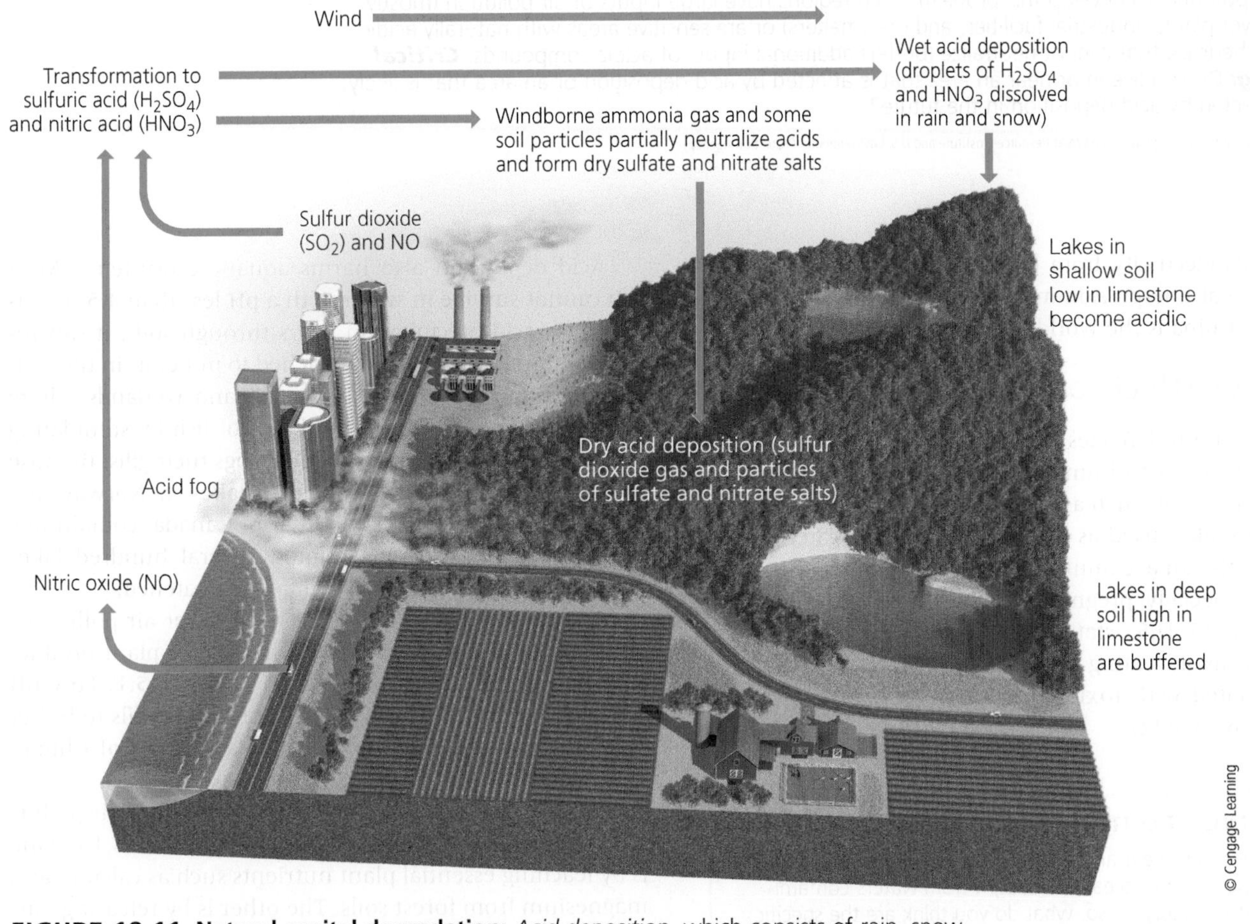

FIGURE 18.11 Natural capital degradation: *Acid deposition*, which consists of rain, snow, dust, or gas with a pH lower than 5.6, is commonly called *acid rain*. ***Critical thinking:*** What are three ways in which your daily activities contribute to acid deposition?

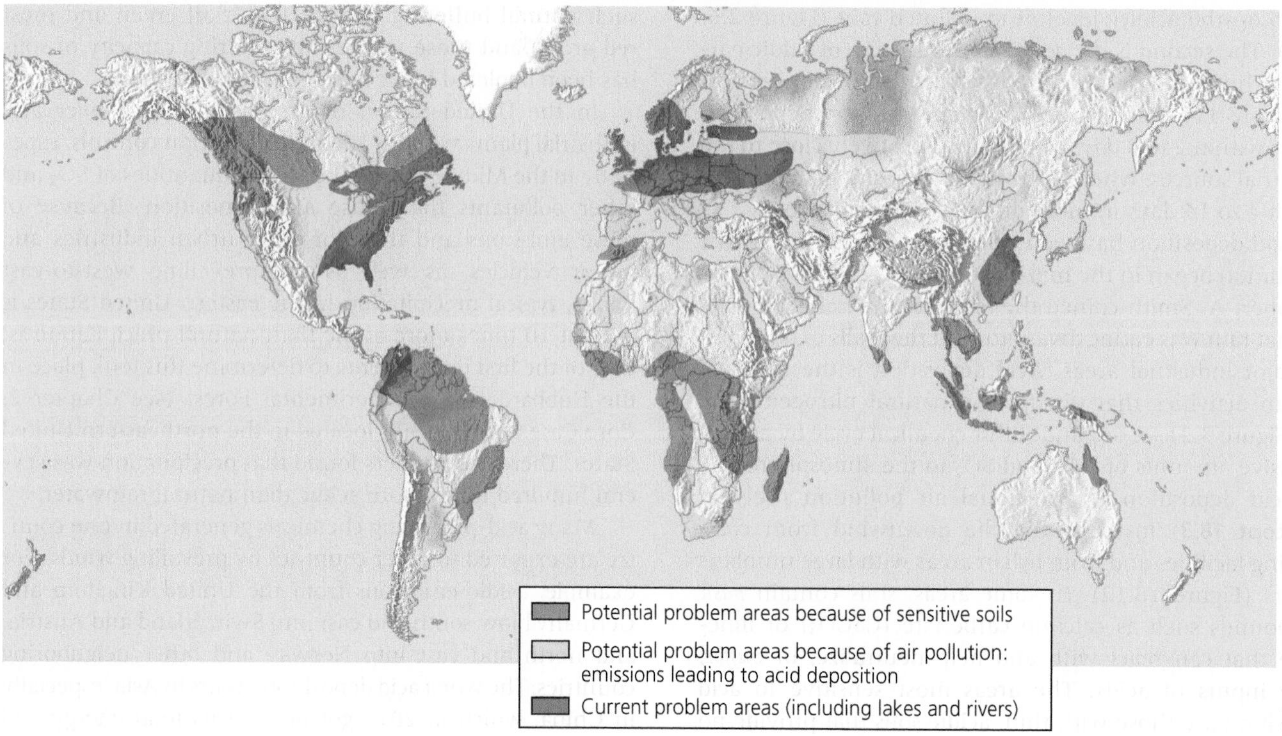

FIGURE 18.12 This map shows regions where acid deposition is now a problem and regions with the potential to develop this problem. Such regions have large inputs of air pollution (mostly from power plants, industrial facilities, and ore smelters) or are sensitive areas with naturally acidic soils and bedrock that cannot neutralize (buffer) additional inputs of acidic compounds. ***Critical thinking:*** Do you live in or near an area that is affected by acid deposition or an area that is likely to be affected by acid deposition in the future?

(Compiled by the authors using data from World Resources Institute and U.S. Environmental Protection Agency.)

67% of its electricity from burning coal, according to the International Energy Administration. According to its government, China is the world's top emitter of SO_2.

Harmful Effects of Acid Deposition

Acid deposition damages statues (Figure 18.4) and buildings, contributes to human respiratory diseases, and can leach toxic metals such as lead and mercury from soils and rocks into lakes used as sources of drinking water. These toxic metals can accumulate in the tissues of fish eaten by people (especially pregnant women) and other animals. Currently, 45 U.S. states have issued warnings telling people to avoid eating fish caught from waters that are contaminated with toxic mercury (see Chapter 17, Core Case Study, p. 442).

CONSIDER THIS . . .

THINKING ABOUT Acid Deposition and Mercury

Do you live in or near an area where government officials have warned people not to eat fish caught from waters contaminated with mercury? If so, what do you think are the specific sources of the mercury pollution and how could this pollution be cleaned up or prevented?

Acid deposition also harms aquatic ecosystems. Most fish cannot survive in water with a pH less than 4.5. In addition, as acid precipitation flows through soils, it can release aluminum ions (Al^{3+}) attached to minerals in the soils and carry them into lakes, streams, and wetlands. There these ions can suffocate many kinds of fish by stimulating excessive mucus formation, which clogs their gills. Because of excess acidity, several thousand lakes in Norway and Sweden, and 1,200 lakes in Ontario, Canada, contain few if any fish. In the United States, several hundred lakes (most in the Northeast) are similarly threatened.

Acid deposition (often along with other air pollutants such as ozone) can harm crops and reduce plant productivity, especially when the soil pH is below 5.1. Low pH reduces plant productivity and the ability of soils to buffer or neutralize acidic inputs. An estimated 30% of China's cropland suffers from excess acidity.

A combination of acid deposition and other air pollutants can also affect forests in two ways (Figure 18.13). One is by leaching essential plant nutrients such as calcium and magnesium from forest soils. The other is by releasing ions of aluminum, lead, cadmium, and mercury from forest soils. These ions are toxic to the trees. These two effects rarely kill trees directly, but they can weaken them and

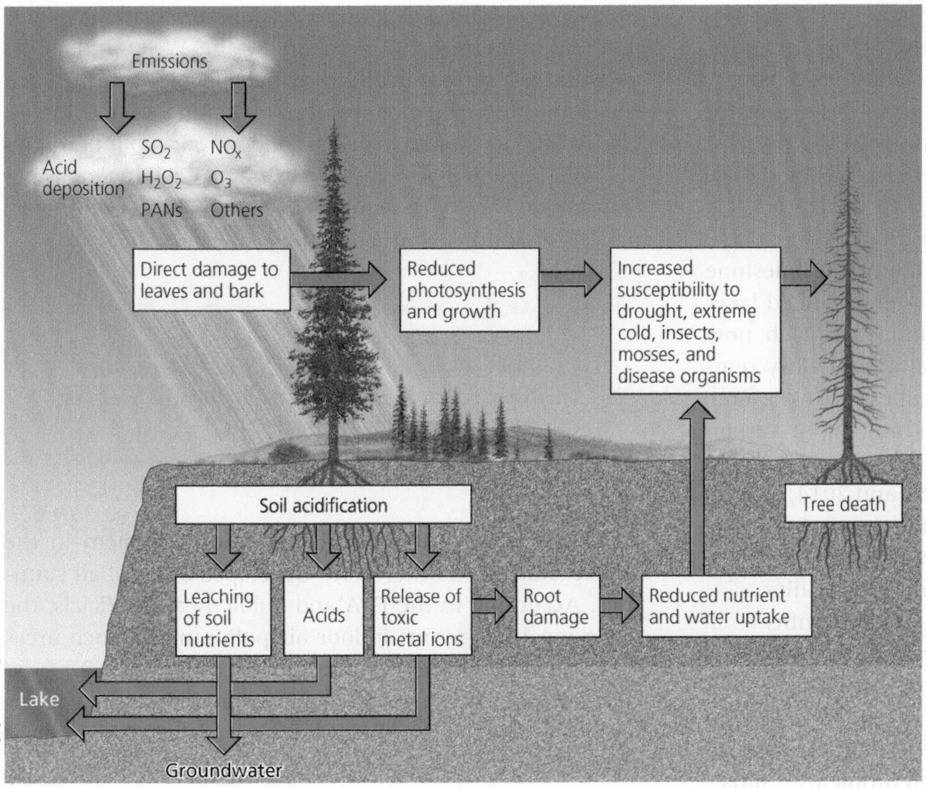

© Cengage Learning

Anticiclo/Shutterstock.com

FIGURE 18.13 Natural capital degradation: Air pollution is one of several interacting stresses that can damage, weaken, or kill trees and pollute surface and groundwater. The inset photo shows trees in a German forest that have died due to exposure to acid deposition and other air pollutants.

leave them vulnerable to stresses such as severe cold, diseases, insect attacks, and drought.

Mountaintop forests are the terrestrial areas hit hardest by acid deposition. These areas tend to have thin soils without much buffering capacity and some of these areas are bathed almost continuously in highly acidic fog and clouds. Some mountaintop forests in the eastern United States, as well as east of Los Angeles, California (**Core Case Study**), are bathed in fog and dews that are as acidic as lemon juice—with about 1,000 times the acidity of unpolluted precipitation.

Since 1994, acid deposition has decreased sharply in the United States, partly because of significant reductions in SO_2 and NO_x emissions from coal-burning facilities under the 1990 amendments to the U.S. Clean Air Act. Even so, soils and surface waters in many areas are still acidic because of the accumulation of acids over decades of acid deposition.

Reducing Acid Deposition

Figure 18.14 summarizes ways to reduce acid deposition. According to most scientific experts on acid deposition, the best solutions are *preventive approaches* that reduce or eliminate emissions of sulfur dioxide (SO_2), nitrogen oxides (NO_x), and particulates.

Although we know how to prevent acid deposition (Figure 18.14, left), implementing these solutions is politically difficult. One problem is that the people and ecosystems it affects often are quite far downwind from the sources of the problem. Also, countries with large supplies of coal (such as China, India, Russia, and the United States) have a strong incentive to use it. In addition, owners of coal-burning power plants resist adding the latest pollution control equipment to their facilities and using low-sulfur coal, arguing that these measures increase the cost of electricity for consumers.

However, in the United States, the use of affordable and cleaner-burning natural gas (see Chapter 15, p. 389)

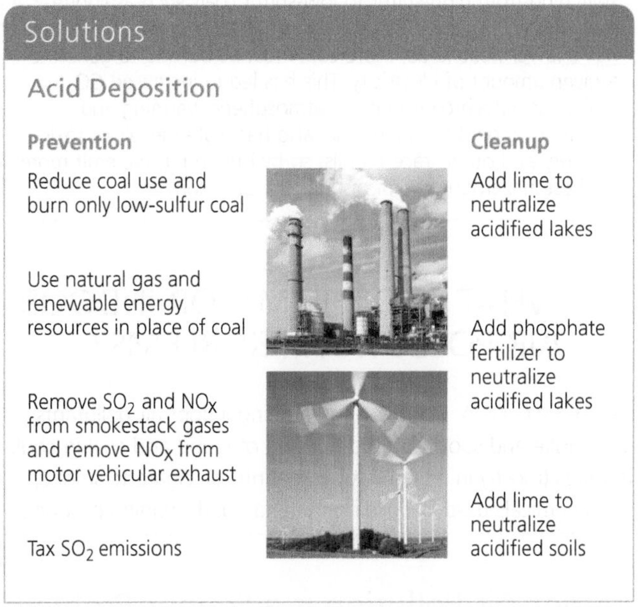
Solutions
Acid Deposition
Prevention
Reduce coal use and burn only low-sulfur coal
Use natural gas and renewable energy resources in place of coal
Remove SO_2 and NO_X from smokestack gases and remove NO_X from motor vehicular exhaust
Tax SO_2 emissions
Cleanup
Add lime to neutralize acidified lakes
Add phosphate fertilizer to neutralize acidified lakes
Add lime to neutralize acidified soils

© Cengage Learning

FIGURE 18.14 Ways to reduce acid deposition and its damage. ***Critical thinking:*** Which two of these solutions do you think are the best ones? Why?

Top: Brittany Courville/Shutterstock.com. Bottom: Yegor Korzhi/Shutterstock.com.

and wind (see Chapter 16, p. 423) for generating electricity is on the rise, and has reduced the use of coal. Environmental scientists point out that including the largely hidden, harmful health and environmental costs of burning coal in its market prices, in keeping with the full-cost pricing **principle of sustainability**, would further reduce coal use, spur the use of cleaner ways to generate electricity, and help prevent acid deposition.

As for technological fixes, large amounts of limestone or ground lime are used to neutralize some acidified lakes and surrounding soils—the only cleanup approach now being used. However, this expensive and temporary remedy usually must be repeated annually. It can also kill some types of plankton and aquatic plants and harm certain wetland plants that need acidic water.

According to the EPA, between 1980 and 2014, air pollution laws in the United States helped to reduce SO_2 emissions from all sources by 80% and nitrogen oxide emissions by 57%. This has helped reduce the acidity of rainfall in parts of the Northeast, Mid-Atlantic, and Midwest regions. However, scientists call for more reductions of these and other harmful emissions from older coal-burning power and industrial plants.

China, the world's largest emitter of SO_2, has one of the world's most serious acid deposition problems. China's SO_2 emissions have declined slightly because of some reduction in coal use since 2011, but the country has a long way to go in curtailing acid deposition.

CONSIDER THIS . . .

CONNECTIONS Low-Sulfur Coal, Atmospheric Warming, and Toxic Mercury

Some U.S. power plants have lowered SO_2 emissions by switching from high-sulfur to low-sulfur coals such as lignite (see Figure 15.12, p. 391). However, because low-sulfur coal has a lower heat value, more coal must be burned to generate a given amount of electricity. This has led to increased CO_2 emissions, which contribute to atmospheric warming and climate change. Low-sulfur coal also has higher levels of toxic mercury and other trace metals, so by burning it, we emit more of these hazardous chemicals into the atmosphere.

18.4 WHAT ARE THE MAJOR INDOOR AIR POLLUTION PROBLEMS?

CONCEPT 18.4 The most threatening indoor air pollutants are smoke and soot from the burning of wood and coal in cooking fires (mostly in less-developed countries), cigarette smoke, and chemicals used in building materials and cleaning products.

Indoor Air Pollution Is a Serious Problem

In less-developed countries, the indoor burning of wood, charcoal, dung, crop residues, coal, and other fuels in open fires (Figure 18.15) and in unvented or poorly vented stoves exposes people to dangerous levels of particulate air pollution. According to a 2014 report by the WHO, indoor air pollution is a serious health threat for people who use such stoves and indoor fires for heating and cooking (**Concept 18.4**). The WHO has estimated that indoor air pollution kills 3.3 million people per year—an average of 9,040 deaths per day. About 77% of those deaths occurred in less-developed nations of Southeast Asia and the Western Pacific.

3.3 million
Annual number of global deaths due to indoor air pollution

Indoor air pollution is also a serious problem in the United States and other more-developed areas of all countries. According to the EPA and public health officials, the three most dangerous indoor air pollutants in such areas are *tobacco smoke* (see Chapter 17, Case Study, p. 463); *formaldehyde* emitted from many building materials and various household products; and *radioactive radon-222 gas*, which can seep into houses from underground rock deposits (see the Case Study that follows).

Formaldehyde (CH_2O) is a colorless, extremely irritating chemical that is considered a carcinogen. It is commonly

Alain Lauga/Shutterstock.com

FIGURE 18.15 Burning wood to cook food inside this dwelling in Nepal exposes this woman and other occupants to dangerous levels of indoor air pollution.

used to make furniture, drapes, carpeting, foam insulation, and other products. It can also be present in plywood, particleboard, paneling, and high-gloss wood used to make flooring and cabinets. According to the EPA and the American Lung Association, 20 to 40 million Americans suffer from chronic breathing problems, dizziness, headaches, sore throats, sinus and eye irritation, and other ailments caused by daily exposure to low levels of formaldehyde emitted from these materials and products. Many manufactured (mobile) homes have been found to have high levels of formaldehyde. The EPA estimates that 1 of every 5,000 people who live for more than 10 years in such homes will likely develop cancer from formaldehyde exposure.

Other common sources of indoor air pollution, according to the EPA, include the following:

- Pesticide residues in the 75% of U.S. homes where pesticides are used indoors at least once a year
- Lead particles brought indoors on shoes and collecting in carpets and furnishings
- Dust mites and cockroach droppings found in some homes, thought to play a role in asthma attacks
- Airborne spores of molds and mildew that can cause headaches, allergic reactions, and asthma attacks
- Candles, almost all of which emit fine-particle soot when burned
- Clothes dryer sheets that emit ammonium salt, linked to asthma
- Gas stoves that emit nitrogen dioxide
- Cleaning products that contain alcohol, chlorine, ammonia, and VOCs
- Air fresheners that emit glycol ethers, which can contribute to fatigue, nausea, and anemia
- Air purifiers that emit ozone

Figure 18.16 summarizes these and other sources of indoor air pollution in a typical home.

Danish and U.S. EPA studies have linked various air pollutants found in buildings to a number of health effects,

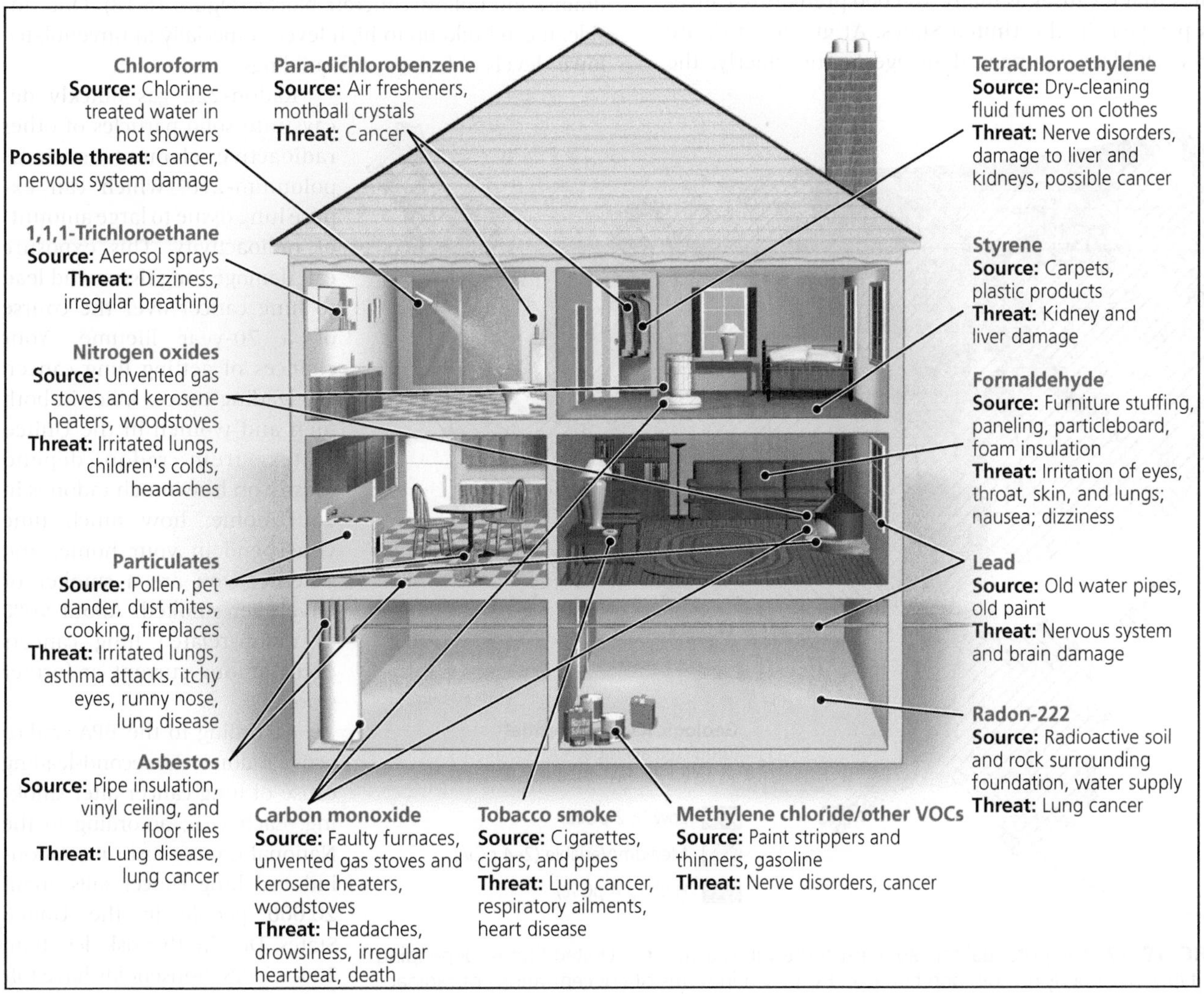

FIGURE 18.16 Numerous indoor air pollutants are found in most modern homes (**Concept 18.4**). ***Question:*** To which of these pollutants are you exposed?

(Compiled by the authors using data from U.S. Environmental Protection Agency.)

a phenomenon known as the *sick-building syndrome*. Such effects include dizziness, headaches, coughing, sneezing, shortness of breath, nausea, burning eyes, sore throats, skin irritation, and respiratory infections. EPA and Labor Department studies indicate that almost one of every five U. S. commercial buildings is exposing employees to such health risks. **GREEN CAREER: Indoor air pollution specialist**

EPA studies have revealed some alarming facts about indoor air pollution in the United States. *First,* levels of several common air pollutants generally are two to five times higher inside U.S. homes and commercial buildings than they are outdoors. In some cases, they are as much as 100 times higher. *Second,* pollution levels inside cars in traffic-clogged urban areas can be up to 18 times higher than outside levels. *Third,* the health risks from exposure to such chemicals are growing because most people in more-developed urban areas spend 70% or more of their time indoors or inside vehicles.

Since 1990, the EPA has placed indoor air pollution at the top of its list of 18 sources of cancer risk. The EPA estimates that it causes as many as 6,000 premature cancer deaths per year in the United States. At greatest risk are smokers, children younger than age 5, the elderly, the sick, pregnant women, people with respiratory or heart problems, and factory workers.

CASE STUDY

Radioactive Radon Gas

Radon-222 is a colorless, odorless, radioactive gas that is produced by the natural radioactive decay of uranium-238, small amounts of which are contained in most rocks and soils. But this isotope is much more concentrated in underground deposits of minerals such as uranium, phosphate, shale, and granite. Figure 18.17 compares the potential geological risk of exposure to radioactive radon across the United States.

When radioactive radon gas from such deposits seeps upward through the soil and is released outdoors, it disperses quickly in the air and decays to harmless levels of radioactivity. However, in buildings above such deposits, radon gas can enter through cracks in a foundation's slab and walls, as well as through well water, openings around sump pumps and drains, and hollow concrete blocks (Figure 18.18). Once inside, it can build up to high levels, especially in unventilated lower levels of homes and buildings.

Radon-222 gas quickly decays into solid particles of other radioactive elements such as polonium-210, which can expose lung tissue to large amounts of radioactivity. This exposure can damage lung tissue and lead to lung cancer over the course of a 70-year lifetime. Your chances of getting lung cancer, the leading cancer killer in both men and women in the United States, from radon depend mostly on how much radon is in your home, how much time you spend in your home, and whether you are a smoker or have ever smoked. About 90% of radon-related lung cancers occur among current or former smokers.

According to the EPA, radioactive radon is the second-leading cause of lung cancer after smoking. Each year, according to the National Cancer Institute, radon-induced lung cancer kills about 20,000 people in the United States. Despite this risk, less than 20% of U.S. households have followed the EPA's recommendation to conduct radon tests, which can be done with inexpensive testing

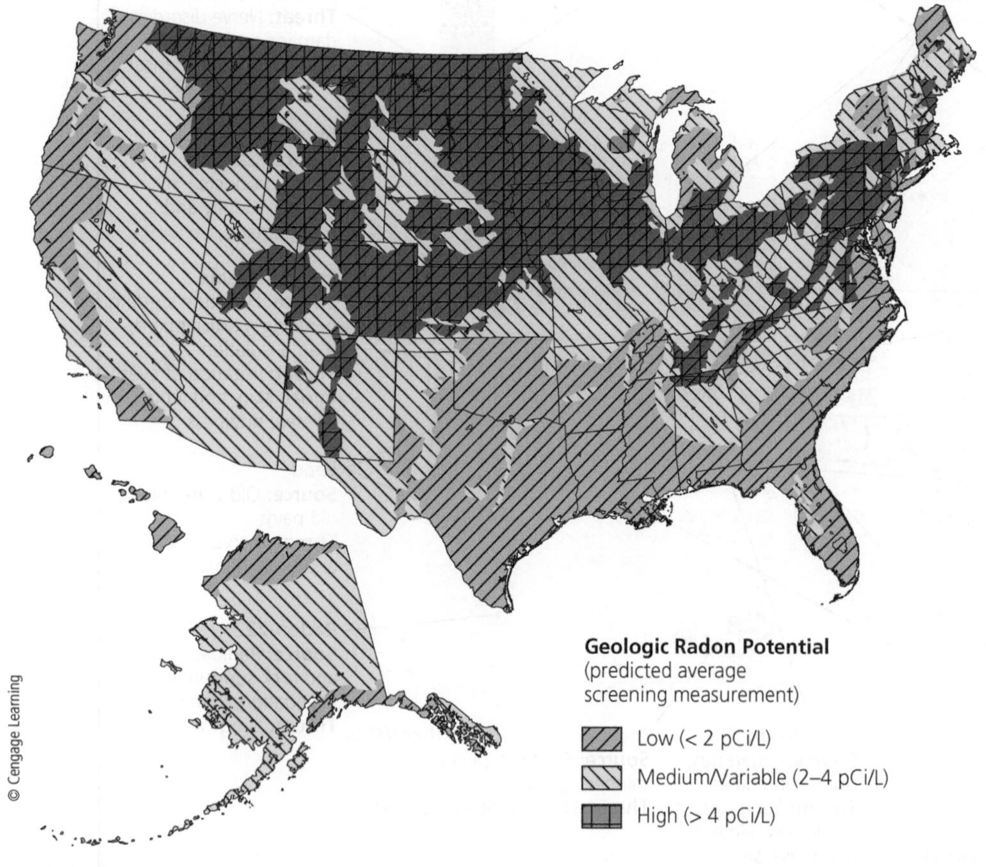

FIGURE 18.17 The potential for radon exposure varies across the United States, depending on the types of underlying soils and bedrock. Expressed in terms of concentrations of radioactive radon in picocuries per liter (pCi/L). ***Question:*** What is the average risk level of exposure to radioactive radon where you live or go to school?

(Compiled by the authors using data from U.S. Geological Survey and U.S. Environmental Protection Agency.)

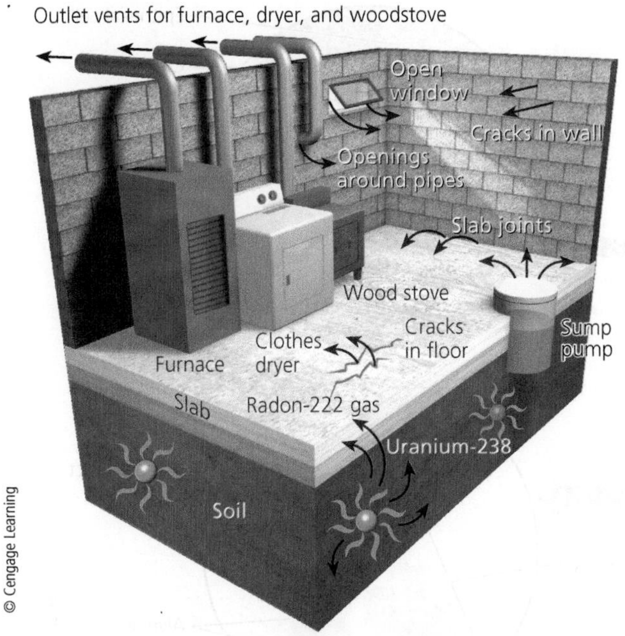

FIGURE 18.18 Ways that radon-222 gas can enter homes and other buildings. ***Question:*** Has anyone tested the indoor air where you live for radon-222?

(Compiled by the authors using data from U.S. Environmental Protection Agency.)

kits. Many schools and day-care centers also have not tested for radon, and only a few states have laws that require radon testing for schools.

When radon is detected, homeowners need to seal all cracks in the foundation's slab and walls. They can also increase ventilation by cracking a window, installing vents in the basement, and using a fan to create cross ventilation.

CONSIDER THIS . . .

THINKING ABOUT Preventing Indoor Air Pollution

What are some steps you could take to prevent indoor air pollution, especially regarding the four most dangerous indoor air pollutants listed above?

18.5 WHAT ARE THE HEALTH EFFECTS OF AIR POLLUTION?

CONCEPT 18.5 Air pollution can contribute to asthma, chronic bronchitis, emphysema, lung cancer, heart attack, and stroke.

Your Body's Natural Air Pollution Defenses Can Be Overwhelmed

Your respiratory system (Figure 18.19) helps to protect you from air pollution in various ways. Hairs in your nose filter out large particles. Sticky mucus in the lining of your upper respiratory tract captures smaller (but not the smallest) particles and dissolves some gaseous pollutants. Hundreds of thousands of tiny, mucus-coated, hair-like structures, called *cilia*, also line your upper respiratory tract. They continually move back and forth and transport mucus and the pollutants it traps to your throat where they are swallowed or expelled through sneezing and coughing.

Prolonged or acute exposure to air pollutants can overload or break down these natural defenses. Fine and ultrafine particulates can lodge deep in the lungs and contribute to lung cancer, asthma, heart attack, and stroke. Years of smoking or breathing polluted air can lead to other lung ailments such as chronic bronchitis and emphysema, (Figure 17.21, p. 464) which leads to acute shortness of breath.

Recent research, including a study done at the University of Southern California, indicates that fine and ultrafine particles in the air can bypass this defense system by moving directly from our nostrils to our brains along neural pathways. Researchers say that once in the brain, these pollutants could be initiating or accelerating degenerative diseases such as Parkinson's and Alzheimer's diseases. Such research is in an early stage, but studies of people who live in areas contaminated with particulate pollution—both autopsies and examinations of living people—show a connection between such pollution and brain disease.

Air Pollution Is a Big Killer

In 2014, the WHO estimated that each year, indoor and outdoor air pollution kills about 7 million people around the world. This averages out to about 800 deaths every hour. According to the WHO, about 28% of these deaths occur in China (see chapter-opening photo) and India, which in 2015 had 4 of the 10 cities in the world with the worst air pollution. Thus, the WHO dubbed air pollution "the world's largest single environmental health risk." The leading direct causes of death related to air pollution are heart attacks, stroke, chronic obstructive pulmonary disease (COPD), and lung cancer.

In 2013, Steven Barrett and other researchers at the Massachusetts Institute of Technology (MIT) estimated that outdoor air pollution, mostly fine-particle pollution, contributes to the deaths of roughly 200,000 Americans every year (Figure 18.20). Typically, these victims die about 10 years earlier than they otherwise might have. About half of these deaths are blamed on car and truck exhaust and the other half on coal-burning power and industrial plants.

According to EPA studies, each year, more than 125,000 Americans get cancer primarily from breathing soot-laden diesel fumes emitted by buses, trucks, tractors, bulldozers and other construction equipment, trains, and ships. A large diesel truck emits as much particulate matter as 150 cars. And according to a study led by Daniel Lack, the world's 100,000 or more diesel-powered oceangoing ships emit almost half as much particulate pollution as do the world's 1 billion motor vehicles.

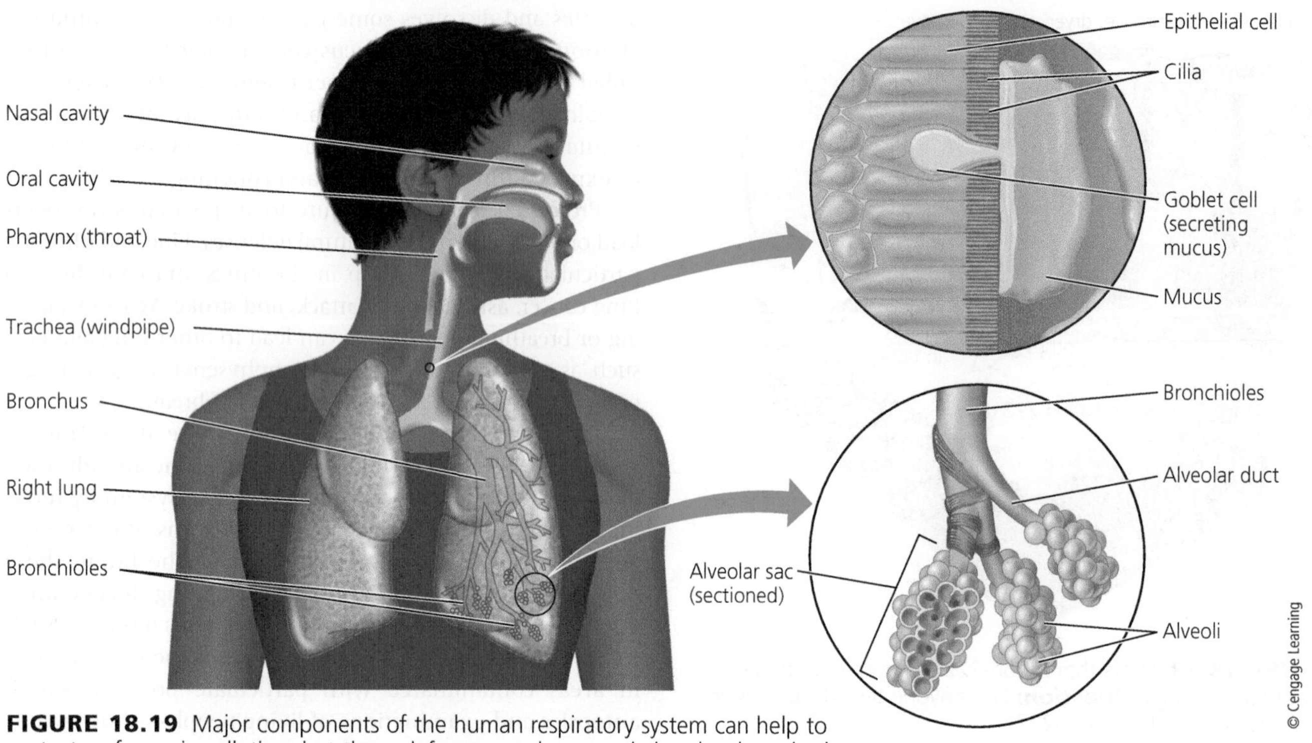

FIGURE 18.19 Major components of the human respiratory system can help to protect us from air pollution, but these defenses can be overwhelmed or breached.

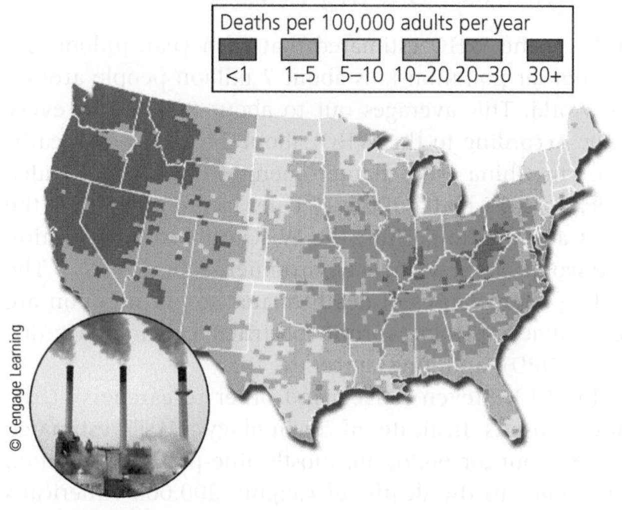

FIGURE 18.20 Distribution of premature deaths from air pollution in the United States, mostly from very small, fine, and ultrafine particles added to the atmosphere by coal-burning power plants. ***Critical thinking:*** Why do the highest death rates occur in the eastern half of the United States? If you live in the United States, what is the risk at your home or where you go to school?

(Compiled by the authors using data from U.S. Environmental Protection Agency.)

Millions more suffer from asthma attacks and other respiratory disorders brought on or aggravated by air pollution (**Concept 18.5**). In 2015, the American Lung Association noted that 46 million Americans, or 14% of the country's population, are exposed daily to particulate pollution levels that exceed EPA safety standards.

18.6 HOW SHOULD WE DEAL WITH AIR POLLUTION?

CONCEPT 18.6 Legal, economic, and technological tools can help us to clean up air pollution, but the best solution is to prevent it.

Laws and Regulations Can Reduce Outdoor Air Pollution

The United States provides an example of how government can reduce air pollution (**Concept 18.6**). The U.S. Congress passed the Clean Air Acts of 1970, 1977, and 1990. With these laws, the federal government established air pollution regulations for key outdoor air pollutants to be enforced by states and major cities.

With the 1970 law, Congress directed the EPA to establish air quality standards for six major outdoor pollutants—carbon monoxide (CO), nitrogen dioxide (NO_2), sulfur dioxide (SO_2), suspended particulate matter (SPM, smaller than PM-10), ozone (O_3), and lead (Pb). One limit, called a *primary standard,* was set to protect human health. Another limit, called a *secondary standard,* was intended to prevent environmental and property damage. Each standard specifies the maximum allowable level for a pollutant, averaged over a specific period. The law has reduced emissions of these pollutants by motor vehicles and industries.

The EPA has also established national emission standards for more than 188 *hazardous air pollutants (HAPs)*—pollutants that can cause serious health and ecological effects. Most of these chemicals are chlorinated hydrocarbons, volatile organic compounds, or compounds of toxic metals. An important public source of information about HAPs is the annual *Toxic Release Inventory* (TRI). The TRI law (passed in 1990 as part of the Pollution Prevention Act) requires more than 20,000 refineries, power plants, mines, chemical manufacturers, and factories to report their releases and waste management methods for 667 toxic chemicals. The TRI, which is available on the Internet, lists this information by community. Since the first TRI report was released in 1988, reported emissions of toxic chemicals have dropped sharply.

According to a 2015 EPA report, the combined emissions of the six major outdoor air pollutants decreased by about 70% between 1980 and 2014, even with significant increases during the same period in gross domestic product, vehicle miles traveled, population, and energy consumption (Figure 18.21).

In 2015, the EPA issued the first federal rules to limit emissions of CO_2 on existing coal-fired power plants beginning in 2022 with full compliance by 2030. According to the EPA, these plants are responsible for nearly 40% of CO_2 emissions in the United States. In addition to slowing climate change, measures to control CO_2 emissions will also result in reduced emissions of other air pollutants. The EPA projects that by 2030, these new regulations will have the effect of cutting nitrogen oxides by 72% and sulfur dioxides by 90%, compared to 2005 levels. The agency also projects that these cuts will prevent 3,600 premature deaths, 1,700 heart attacks, 90,000 asthma attacks, and 300,000 missed work days and school days.

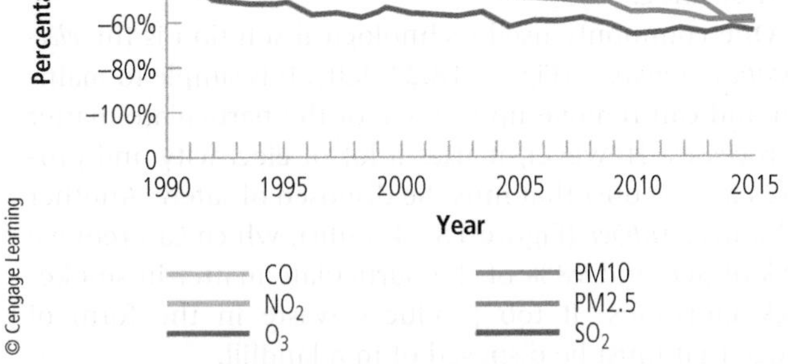

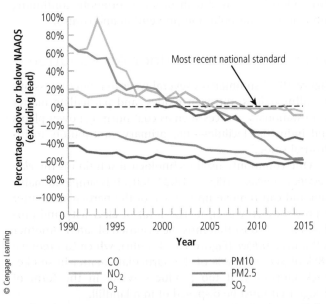

FIGURE 18.21 This graph shows how the average annual levels of major pollutants have generally declined to meet the National Ambient Air Quality Standards (NAAQS). ***Data analysis:*** Which of the pollutants declined by the greatest percentage between 1990 and 2015? By how many percentage points did it decline?

Coal and utility companies and 18 states with older polluting coal-fired power plants have succeeded in putting off stricter air pollution standards for existing coal-burning power plants for almost 40 years. They have challenged the new regulations in the courts, charging the EPA with a power grab designed to put coal companies out of business. In 2016, the U.S. Supreme Court put a hold on implementing the new standards even though they do not start taking effect until 2022 while the legal challenges make their way through 30 lawsuits in the courts.

In 2013, the EPA proposed stricter motor vehicle emission standards that would reduce emissions of VOCs and nitrogen oxides by 80% and particulate emissions by 70%. The EPA estimates that each year, these new standards would cut the death toll from outdoor air pollution by 2,000 and reduce the number of cases of respiratory ailments in children by 23,000. They would also lead to estimated savings of $7 in health-care costs for every $1 spent to implement the new standards. The resulting increase in the cost of a gallon of gasoline would be 1 cent. Oil companies oppose the new standards, saying they would cost too much and would hinder economic growth.

The reduction of outdoor air pollution in the United States since 1970 has been a remarkable success story, mostly because of two factors. *First,* during the 1970s, U.S. citizens insisted that laws be passed and enforced to improve air quality. Before 1970, when Congress passed the Clean Air Act, air-pollution-control equipment did not exist. *Second,* the country was affluent enough to afford such controls and improvements. For example, because of these factors, a new car today in the United States emits 75% less pollution than did a pre-1970 car.

Environmental scientists applaud this success, but they call for strengthening U.S. air pollution laws by doing the following:

- Putting much greater emphasis on pollution prevention. With this approach, the question is not *What can we do about the air pollutants we produce?* but rather *How can we avoid producing these pollutants in the first place?* The power of prevention (**Concept 18.6**) was made clear by the 99% drop in atmospheric lead emissions after lead in gasoline was banned in 1976.
- Sharply reducing emissions from approximately 20,000 older coal-burning power and industrial plants, cement plants, oil refineries, and waste incinerators that have not been required to meet the air pollution standards for new facilities under the Clean Air Acts.
- Ramping up controls on atmospheric emissions of toxic pollutants such as mercury (see Figure 17.12, p. 454).
- Continuing to improve fuel efficiency standards for motor vehicles.
- Regulating more strictly the emissions from motorcycles and two-cycle gasoline engines used in

FIGURE 18.22 Many of the motorized scooters so commonly found on most college campuses, especially those with two-cycle engines, produce more nitrogen oxides and hydrocarbons—pollutants that contribute to photochemical smog—per unit of distance driven than the average car produces. Older scooters and poorly maintained scooters emit many times more of these pollutants than cars emit. Thus, even though they are more fuel-efficient than most cars, as a group, scooters are major contributors to urban air pollution.

devices such as chainsaws, lawnmowers, generators, scooters (Figure 18.22), and snowmobiles. The EPA estimates that running a gas-powered riding lawn mower for an hour creates as much VOC air pollution as driving 34 cars for an hour (for a mower without a catalytic converter, which few of them have).

- Setting much stricter air pollution regulations for airports and oceangoing ships.
- Sharply reducing indoor air pollution.

Executives of companies that would be affected by implementing stronger air pollution regulations claim that correcting these problems would cost too much and would hinder economic growth. Proponents of stronger regulations contend that history has shown that most industry cost estimates for implementing U.S. air pollution control standards have been much higher than the costs actually proved to be. In addition, implementing such standards has helped some companies and created jobs by stimulating these companies to develop new pollution control technologies.

In China, air quality standards announced in 2013 are taking hold. Beginning in 2015, the government banned the mining, importing, and burning of high-sulfur, high-ash-content coal in major cities. Spurred by such regulations, China is reducing its coal consumption by investing massively in wind power, solar power, and hydropower.

Using the Marketplace to Reduce Outdoor Air Pollution

One approach to reducing pollutant emissions has been to allow producers of air pollutants to buy and sell government air pollution allotments in the marketplace. For example, with the goal of reducing SO_2 emissions, the Clean Air Act of 1990 authorized an *emissions trading*, or *cap-and-trade program*, which enables the 110 most polluting coal-fired power plants in 21 states to buy and sell SO_2 air pollution rights.

Under this system, each plant is annually given a number of pollution credits, which allow it to emit a certain amount of SO_2. A utility that emits less than its allotted amount has a surplus of pollution credits. That utility can use its credits to offset SO_2 emissions at its other plants, keep them for future plant expansions, or sell them to other utilities or private citizens or groups. Between 1990 and 2012, this emissions trading program helped to reduce SO_2 emissions from power plants in the United States by 76%, at a cost of less than one-tenth of the cost projected by the utility industry, according to the EPA. The 2015 Clean Power Plan gives states the option of allowing power plant companies to use emissions trading to meet the new CO_2 reduction standards.

Proponents of this market-based approach say it is cheaper and more efficient than government regulation of air pollution. Critics of this approach contend that it allows utilities with older, dirtier power plants to buy their way out of their environmental responsibilities and to continue to pollute.

The ultimate success of any emissions trading approach depends on two factors: how low the initial cap is set and how often it is lowered in order to promote continuing innovation in air pollution prevention and control.

Ways to Reduce Outdoor Air Pollution

Figure 18.23 summarizes several ways to reduce emissions of sulfur oxides, nitrogen oxides, and particulate matter from stationary sources such as coal-burning power plants and industrial facilities—the primary contributors to industrial smog.

One commonly used technological solution is the *electrostatic precipitator* (Figure 18.24, left). It is simple to maintain and can remove up to 99% of the particulate matter it processes. However, it uses a lot of electricity and produces a toxic dust that must be disposed of safely. Another is the *wet scrubber* (Figure 18.24, right), which can remove 98% of SO_2 and 98% of the particulate matter in smokestack emissions. It too produces waste in the form of sludge that must be disposed of in a landfill.

Figure 18.25 lists several ways to prevent and reduce emissions from motor vehicles, the primary contributors to photochemical smog. In more-developed countries, many of these solutions have been successful (see the Case Study that follows). However, the already poor air quality in urban areas

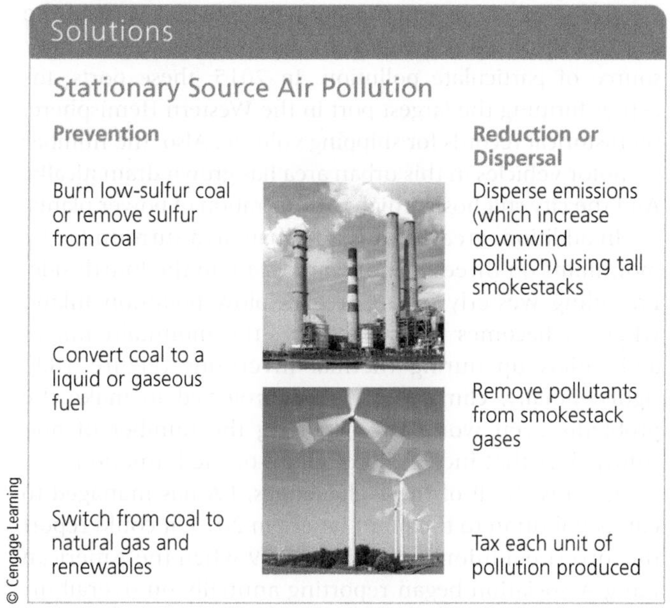
Solutions
Stationary Source Air Pollution
Prevention
Reduction or Dispersal
Burn low-sulfur coal or remove sulfur from coal
Disperse emissions (which increase downwind pollution) using tall smokestacks
Convert coal to a liquid or gaseous fuel
Remove pollutants from smokestack gases
Switch from coal to natural gas and renewables
Tax each unit of pollution produced
© Cengage Learning

FIGURE 18.23 Ways to prevent, reduce, or disperse emissions of sulfur oxides, nitrogen oxides, and particulate matter from stationary sources, especially coal-burning power plants and industrial facilities (**Concept 18.6**). ***Critical thinking:*** Which two of these solutions do you think are the best ones? Why?

Top: Brittany Courville/Shutterstock.com. Bottom: Yegor Korzhi/Shutterstock.com.

of many less-developed countries is worsening as the numbers of motor vehicles in these nations rise.

Over the next 10 to 20 years, technology could help all countries to clean up the air through improved engine and emission systems and hybrid-electric, plug-in hybrid, and all-electric vehicles (see Figure 16.5, p. 415). In 2015, the Alliance of Automobile Manufacturers announced that automakers were on track to essentially eliminate smog-forming passenger vehicle emissions by 2030 in the United States, as newer cars replace older ones.

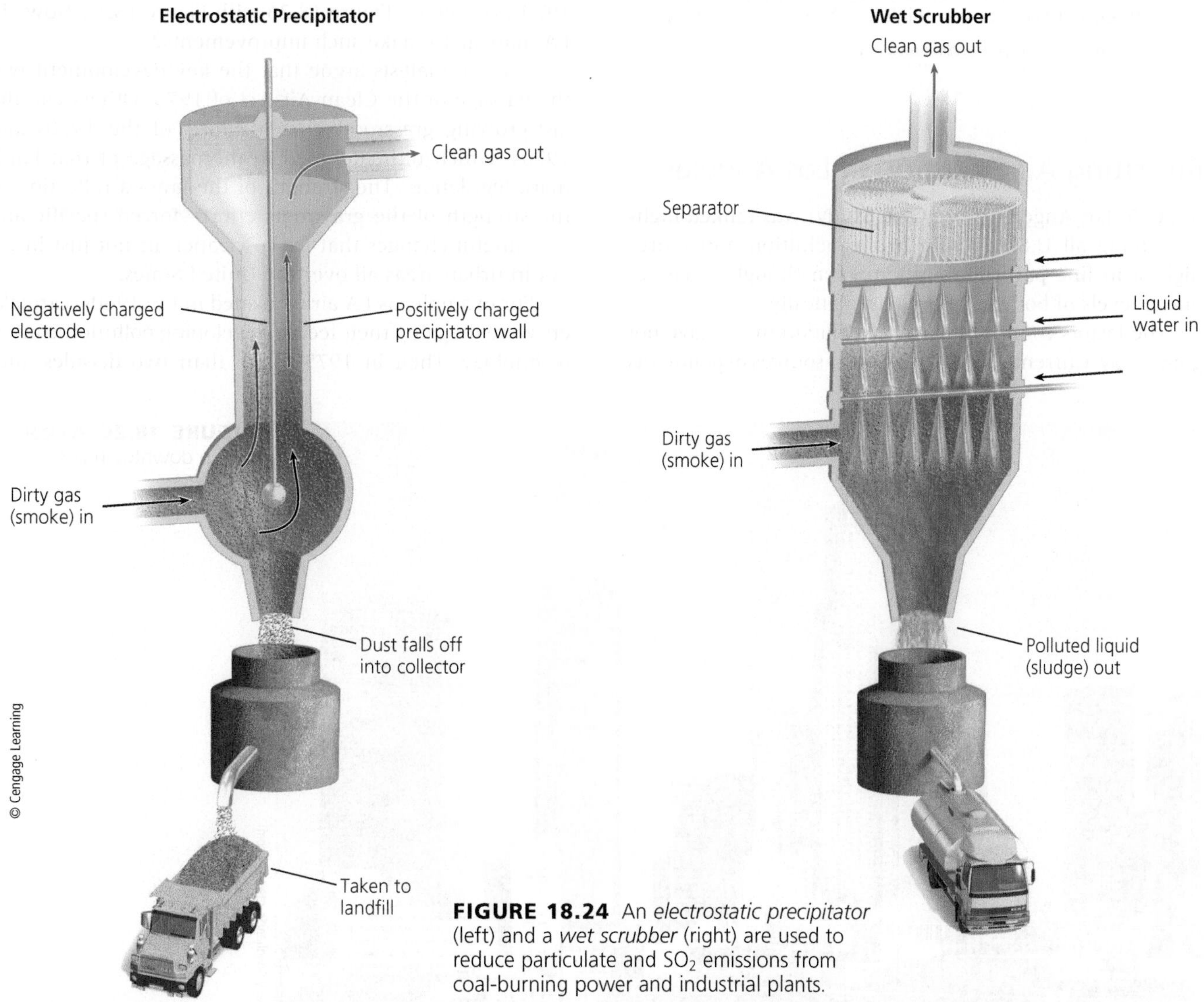

FIGURE 18.24 An *electrostatic precipitator* (left) and a *wet scrubber* (right) are used to reduce particulate and SO_2 emissions from coal-burning power and industrial plants.

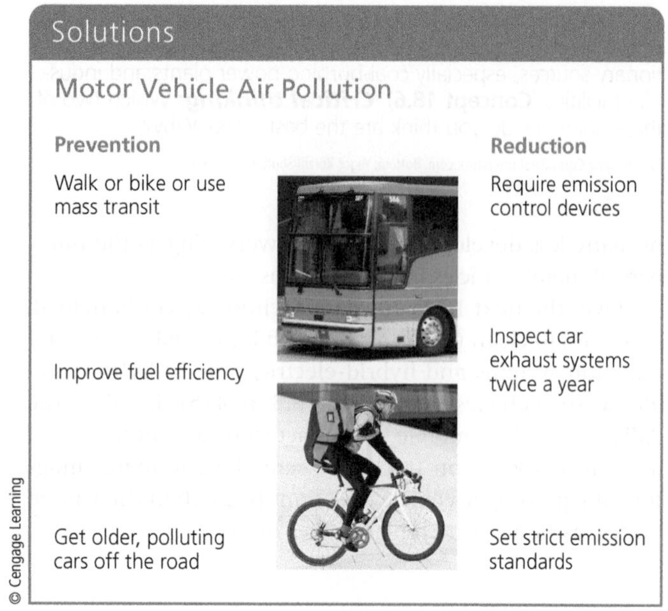
Solutions
Motor Vehicle Air Pollution
Prevention
Walk or bike or use mass transit
Improve fuel efficiency
Get older, polluting cars off the road
Reduction
Require emission control devices
Inspect car exhaust systems twice a year
Set strict emission standards
© Cengage Learning

FIGURE 18.25 Ways to prevent or reduce emissions from motor vehicles (**Concept 18.6**). ***Critical thinking:*** Which two of these solutions do you think are the best ones? Why?

Top: egd/Shutterstock.com. Bottom: Tyler Olson/Shutterstock.com.

CASE STUDY

Revisiting Air Pollution in Los Angeles

In 2015, Los Angeles (**Core Case Study**) was ranked highest among all U.S. cities in ozone pollution and fourth highest in fine particle pollution, even though it has reduced levels of both pollutants dramatically.

The factors contributing to air pollution in LA have not gone away. Currently, the area's largest sources of pollutants are the ports of Los Angeles and Long Beach. Most of the ships that use these ports burn dirty diesel fuel—a major source of particulate pollution. In 2015, these ports, together forming the largest port in the Western Hemisphere, set historical records for shipping volume. Also, the number of motor vehicles in this urban area has grown dramatically. And the city still hosts a high concentration of power plants.

In addition, greater LA is an urban area surrounded by mountains on three sides and an ocean on the fourth side. Prevailing westerly ocean breezes blow pollution inland where it becomes trapped against the mountain ranges and builds up during thermal inversions (Figure 18.9, right). Finally, climate change is projected to make the problems even worse by increasing the number of hot, sunny days that increase the rate of ozone formation.

Even with all of these challenges, LA has managed to cut its pollution to the point where in 2015, it could report the lowest pollution levels since 1999 when the American Lung Association began reporting annually on overall urban air quality in the United States. Consequently, the city sees many clear days that it seldom saw in the 1960s and 1970s. (Compare Figure 18.26 with Figure 18.1.) How did LA manage to make such improvements?

Several analysts argue that the key development was the passage of the Clean Air Act of 1970. Others cite the fast-growing grassroots citizen efforts of the 1960s and 1970s, which ultimately led to the passage of that landmark legislation. The strength of the law—a reflection of the strength of the grassroots effort—forced specific and meaningful changes that led to cleaner air not just in LA but in urban areas all over the United States.

For example, as LA air worsened in the 1960s, carmakers were dragging their feet in developing pollution control technology. Then in 1975, more than two decades after

FIGURE 18.26 A clear day in downtown Los Angeles.

Gerry Boughan/Shutterstock.com

anti-smog protests began in Los Angeles, carmakers were finally required to install catalytic converters in all new cars. This was a key technological development, according to the California Air Resources Board, and it would not have come about if the Clean Air Act had not been passed.

The Los Angeles and Long Beach ports also reduced their contributions to air pollution in compliance with the law. Between 2005 and 2011, they reduced their emissions of particulate matter from the burning of diesel fuel by 73%. The ports accomplished this mostly by using cleaner-burning cranes, machinery, and trucks, and cleaner, low-sulfur fuel.

The Clean Air Act also allowed for flexibility in how states and cities could meet the clean air standards, and California's air pollution control administrators took advantage of that fact. For example, the state first planned to manage the installation of catalytic converters and the monitoring of mandatory smog tests by using state employees working at state garages. California's Air Resources Board decided to take a less heavy-handed approach and to issue permits to private car repair shops and gas stations to do this work. These businesses welcomed the permits and the additional work because it meant more business for their shops. This solution benefitted businesses, citizens, and the environment, in keeping with the win-win **principle of sustainability**.

Reducing Indoor Air Pollution

Little effort has been devoted to reducing indoor air pollution, even though it poses a greater threat to human health than does outdoor air pollution. Air pollution experts suggest several ways to prevent or reduce indoor air pollution, as shown in Figure 18.27.

Solutions

Indoor Air Pollution

Prevention

Ban indoor smoking

Set stricter formaldehyde emissions standards for carpet, furniture, and building materials

Prevent radon infiltration

Use naturally based cleaning agents, paints, and other products

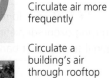

Reduction and Dilution

Use adjustable fresh air vents for work spaces

Circulate air more frequently

Circulate a building's air through rooftop greenhouses

Use solar cookers and efficient, vented wood-burning stoves

FIGURE 18.27 Ways to prevent or reduce indoor air pollution (**Concept 18.6**). ***Critical thinking:*** Which two of these solutions do you think are the best ones? Why?

Top: Tribalium/Shutterstock.com. Bottom: Patstock/Age Fotostock.

In less-developed countries, indoor air pollution from open fires and inefficient stoves that burn wood, charcoal, or coal could be reduced. More people could use inexpensive clay or metal stoves that burn fuels (including straw and other crop wastes) more efficiently and vent their exhausts to the outside, or they could use stoves that use solar energy to cook food (see Figure 16.15, p. 422).

One way to reduce indoor air pollution in a modern home is to have plenty of houseplants. Studies show that they can reduce more than 80% of indoor toxins within a few days. Plants that do a good job of purifying air include Devil's Ivy, English Ivy, African Violets, and Peace Lily. Figure 18.28 lists some ways in which you can reduce your exposure to indoor air pollution.

What Can You Do?

Indoor Air Pollution

- Test for radon and formaldehyde inside your home and take corrective measures as needed
- Do not buy furniture and other products containing formaldehyde
- Test your home or workplace for asbestos fiber levels and check for any crumbling asbestos materials
- If you smoke, do it outside or in a closed room vented to the outside
- Make sure that wood-burning stoves, fireplaces, and kerosene- and gas-burning heaters are properly installed, vented, and maintained
- Install carbon monoxide detectors in all sleeping areas
- Use fans to circulate indoor air
- Grow house plants, the more, the better
- Do not store gasoline, solvents, or other volatile hazardous chemicals inside a home or attached garage
- Remove your shoes before entering your house to reduce inputs of dust, lead, and pesticides

FIGURE 18.28 Individuals matter: You can reduce your exposure to indoor air pollution. ***Critical thinking:*** Which three of these actions do you think are the most important ones to take? Why?

18.7 HOW HAVE WE DEPLETED OZONE IN THE STRATOSPHERE AND WHAT CAN WE DO ABOUT IT?

CONCEPT 18.7A Widespread use of certain chemicals has reduced ozone levels in the stratosphere and allowed more harmful ultraviolet (UV) radiation to reach the earth's surface.

CONCEPT 18.7B To reverse ozone depletion, we must stop producing ozone-depleting chemicals and adhere to the international treaties that ban such chemicals.

The Use of Certain Chemicals Threatens the Ozone Layer

The ozone layer in the stratosphere (Figure 18.2) is a vital form of natural capital that supports all life on land and in shallow aquatic environments by keeping 95% of the sun's harmful ultraviolet radiation from reaching the earth's surface. However, measurements show a considerable seasonal depletion, or thinning, of ozone concentrations in the stratosphere above Antarctica (Figure 18.29) and above the Arctic since the 1970s. Similar measurements reveal slight overall ozone thinning everywhere except over the tropics. The loss of ozone over Antarctica has been called an *ozone hole*. A more accurate term is *ozone thinning* because the ozone depletion varies with altitude and location.

When the seasonal thinning ends each year, huge masses of ozone-depleted air above Antarctica flow northward, and these masses linger for a few weeks over parts of Australia, New Zealand, South America, and South Africa. This has raised biologically damaging UV-B levels in these areas by 3–10%, and in some years by as much as 20%. Based on these measurements and on mathematical and chemical models, the overwhelming consensus of researchers in this field is that ozone depletion in the stratosphere poses a serious threat to humans, other animals, and some primary producers (mostly plants) that use sunlight to support the earth's food webs (**Concept 18.7A**).

In 1988, scientists discovered that similar but usually less severe ozone thinning occurs over the Arctic from February to June, resulting in a typical ozone loss of 11–38% (compared to a typical 50% loss above Antarctica). When this body of air above the Arctic breaks up each year, large masses of ozone-depleted air flow south to linger over parts of Europe, North America, and Asia. However, models indicate that the Arctic is unlikely to develop the large-scale ozone thinning found over the Antarctic.

The origin of this serious environmental threat began with the accidental discovery of the first chlorofluorocarbon (CFC) in 1930. Chemists soon developed similar compounds to create a family of highly useful CFCs, known by their trade name Freons. These chemically unreactive, odorless, nonflammable, nontoxic, and noncorrosive compounds were thought to be dream chemicals. Inexpensive to manufacture, they became popular as coolants in air conditioners and refrigerators, propellants in aerosol spray cans, cleansers for electronic parts such as computer chips, fumigants for granaries and ships' cargo holds, and gases used to make insulation and packaging.

It turned out that CFCs were too good to be true. Starting in 1974 with the work of chemists Sherwood Rowland and Mario Molina (Individuals Matter 18.1), scientists showed that CFCs are persistent chemicals that destroy protective ozone in the stratosphere. Satellite data and other measurements and models indicate that 75–85% of the observed ozone losses in the stratosphere since 1976 resulted from people releasing CFCs and other ozone-depleting chemicals into the troposphere beginning in the 1950s.

Rowland and Molina came to four major conclusions. *First,* once CFCs are put into the atmosphere, these persistent chemicals remain there for a long time. *Second,* over 11–20 years, these compounds rise into the stratosphere through convection, random drift, and the turbulent mixing

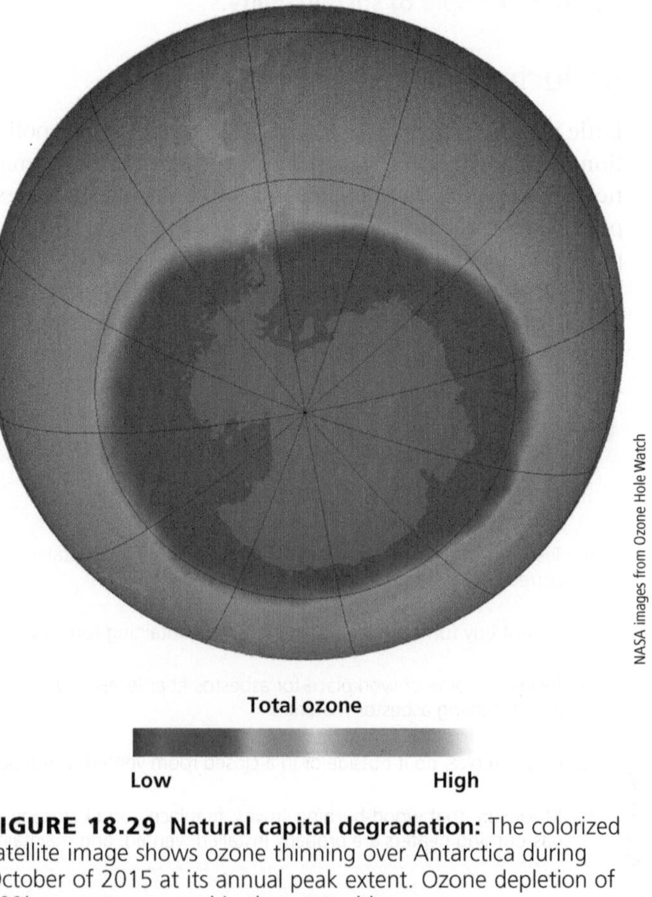

FIGURE 18.29 Natural capital degradation: The colorized satellite image shows ozone thinning over Antarctica during October of 2015 at its annual peak extent. Ozone depletion of 50% or more occurred in the center blue area.

Natural Capital Degradation

Effects of Ozone Depletion

Human Health and Structures

- Worse sunburns
- More eye cataracts and skin cancers
- Immune system suppression

Food and Forests

- Reduced yields for some crops
- Reduced seafood supplies due to smaller phytoplankton populations
- Decreased forest productivity for UV-sensitive tree species

Wildlife

- More eye cataracts in some species
- Shrinking populations of aquatic species sensitive to UV radiation
- Disruption of aquatic food webs due to shrinking phytoplankton populations

Air Pollution and Climate Change

- Increased acid deposition
- Increased photochemical smog
- Degradation of outdoor painted surfaces, plastics, and building materials
- While in troposphere, CFCs act as greenhouse gases

FIGURE 18.30 Decreased levels of ozone in the stratosphere can have a number of harmful effects. (**Concept 18.7A**). ***Critical thinking:*** Which three of these effects do you think are the most threatening? Why?

of air in the lower atmosphere. *Third,* once they reach the stratosphere, the CFC molecules break down under the influence of high-energy UV radiation. This releases highly reactive chlorine atoms (Cl), as well as atoms of fluorine (F) and bromine (Br), all of which accelerate the breakdown of ozone (O_3) into O_2 and O in a cyclic chain of chemical reactions. This process destroys ozone faster than it forms in some parts of the stratosphere. And *fourth,* each CFC molecule can last in the stratosphere for 65–385 years, depending on its type. During that time, each chlorine atom released during the breakdown of CFCs can break down hundreds of O_3 molecules.

CFCs are not the only ozone-depleting chemicals. Others are *halons* and *hydrobromofluorocarbons* (HBFCs) (used in fire extinguishers), *methyl bromide* (a widely used fumigant), *hydrogen chloride* (emitted into the stratosphere by the launches of certain space vehicles), and cleaning solvents such as *carbon tetrachloride, methyl chloroform, n-propyl bromide,* and *hexachlorobutadiene.* While in the troposphere, CFCs also act as greenhouse gases that help to warm the lower atmosphere and contribute to climate change.

Why Should We Worry about Ozone Depletion?

Why is ozone depletion something that should concern us? Figure 18.30 lists some of the demonstrated effects of stratospheric ozone thinning. One effect is that more biologically damaging UV-A and UV-B radiation will reach the earth's surface (**Concept 18.7A**) and is a likely contributor to rising numbers of eye cataracts, damaging sunburns, and skin cancers. Figure 18.31 lists ways in which you can protect yourself from harmful UV radiation.

What Can You Do?

Reducing Exposure to UV Radiation

- Stay out of the sun, especially between 10 A.M. and 3 P.M.
- Do not use tanning parlors or sunlamps
- When in the sun, wear clothing and sunglasses that protect against UV-A and UV-B radiation
- Be aware that overcast skies do not protect you
- Do not expose yourself to the sun if you are taking antibiotics or birth control pills
- When in the sun, use a sunscreen with a protection factor of at least 15

FIGURE 18.31 Individuals matter: You can reduce your exposure to harmful UV radiation. ***Critical thinking:*** Which of these precautions do you already take? Which others would you consider doing?

Another serious threat from ozone depletion and the resulting increase in UV radiation reaching the planet's surface is the possible impairment or destruction of phytoplankton, especially in Antarctic waters. These tiny marine plants

INDIVIDUALS MATTER 18.1

Sherwood Rowland and Mario Molina—A Scientific Story of Expertise, Courage, and Persistence

HAL GARB/AFP/Getty Images

DONNA COVE, MIT/UNIVERSITY OF CALIFORNIA, SAN DIEGO

In 1974, calculations by the late Sherwood Rowland (left photo) and Mario Molina (right photo), chemists at the University of California–Irvine, indicated that chlorofluorocarbons (CFCs) were lowering the average concentration of ozone in the stratosphere. They also found that CFCs are persistent, remaining in the stratosphere for hundreds of years. During that time, they noted, each CFC molecule can breakdown hundreds of ozone molecules.

These scientists decided they had an ethical obligation to go public with the results of their research. They shocked both the scientific community and the $28-billion-per-year CFC industry by calling for an immediate ban of CFCs in spray cans, for which substitutes were available.

The CFC industry (led by DuPont) was a powerful, well-funded adversary with a lot of profits and jobs at stake. It attacked Rowland's and Molina's calculations and conclusions, but the two researchers held their ground, expanded their research, and explained their results to other scientists, elected officials, and the media. After 14 years of delaying tactics, DuPont officials acknowledged in 1988 that CFCs were depleting the ozone layer, and they agreed to stop producing them and to sell higher-priced alternatives that their chemists had developed.

In 1995, Rowland and Molina received the Nobel Prize in chemistry for their work on CFCs.

play a key role in removing CO_2 from the atmosphere and they form the base of many ocean food webs. By destroying them, we would be eliminating the vital ecological services they provide. The loss of plankton could accelerate projected atmospheric warming by reducing the capacity of the oceans to remove CO_2 from the atmosphere.

Reversing Stratospheric Ozone Depletion

According to researchers in this field, we should immediately stop producing all ozone-depleting chemicals (**Concept 18.7B**). However, models and measurements indicate that even with immediate and sustained action, it will take 35 to 60 years for the earth's ozone layer to recover the levels of ozone it had in 1980, and it could take about 100 years for it to recover to pre-1950 levels.

In 1987, representatives of 36 nations met in Montreal, Canada, and developed the *Montreal Protocol*. This treaty's goal was to cut emissions of CFCs (but no other ozone-depleting chemicals) by about 35% between 1989 and 2000. After hearing more bad news about seasonal ozone thinning above Antarctica in 1989, representatives of 93 countries had more meetings and in 1992 adopted the *Copenhagen Amendment*, which accelerated the phase-out of CFCs and added some other key ozone-depleting chemicals to the agreement.

These international agreements set an important precedent because nations and companies worked together, using a *prevention approach* to try to solve a serious environmental problem. This approach worked for three reasons. *First*, there was convincing and dramatic scientific evidence of a serious problem. *Second*, CFCs were produced by a small number of international companies and this meant there was less corporate resistance to finding a solution. *Third*, the certainty that CFC sales would decline over a period of years because of government bans unleashed the economic and creative resources of the private sector to find even more profitable substitute chemicals.

Substitutes are available for most uses of CFCs. However, the most widely used substitutes are hydrofluorocarbons (HFCs), which act as greenhouse gases. An HFC molecule can be up to 10,000 times more potent in warming the atmosphere than a molecule of CO_2 is. The Intergovernmental Panel on Climate Change (IPCC) has warned that global use of HFCs is growing rapidly and that they need to be quickly replaced with substitutes that do not deplete ozone in the stratosphere or act as greenhouse gases while they are in the troposphere. Several companies have developed HFC substitutes that need to be evaluated.

In addition, there is a growing consensus among scientists that the Montreal Protocol should also be used to regulate the greenhouse gas nitrous oxide (N_2O), which is released from fertilizers and livestock manure. It remains in the troposphere for about 100 years and then migrates to the stratosphere where it can destroy ozone.

In 2015, researchers led by Martyn Chipperfield of the University of Leeds revealed findings of new atmospheric chemistry modeling. They calculated that, without the benefit of the Montreal Protocol, the Antarctic ozone hole would likely have grown by another 40% by 2013 and that the ozone layer around the globe would have been thinned by 15%. Deaths and other harmful effects of ozone thinning would have been much worse.

The agreement is working. According to 2016 study by NASA scientists, between 2000 and 2015, ozone thinning in the stratosphere above Antarctica (Figure 18.29) reached its peak in September and in October, shrank by an area equal to about one-third the area of the continental United States. If this trend continues, the ozone layer would return to 1980 levels by 2050.

The landmark international agreements on stratospheric ozone, now signed by all 196 of the world's countries, are important examples of successful global cooperation in response to a serious global environmental problem. (**Concept 18.7B**). GOOD NEWS

BIG IDEAS

- Outdoor air pollution, in the form of industrial smog, photochemical smog, and acid deposition, and indoor air pollution are serious global problems.
- Each year, at least 7 million people die prematurely from the effects of outdoor and indoor air pollution, with more than half of these deaths occurring in less-developed countries.
- We need to give top priority to preventing outdoor and indoor air pollution throughout the world and ozone depletion in the stratosphere.

Tying It All Together

Los Angeles Air Pollution and Sustainability

Bart Everett/Shutterstock.com

In the chapter's **Core Case Study**, we learned about how human activities can create massive and harmful air pollution that builds up over time, especially over urban areas such as Los Angeles, California. We saw how a grassroots movement of people concerned about the resulting problems led to a process that has improved air quality over LA. We saw how important it was to pass strict legislation to limit emissions from various sources of pollution. In this chapter, we learned that in passing such limits, we can help to prevent not only air pollution, but also acid deposition and the further thinning of the stratospheric ozone layer.

We can apply the six **principles of sustainability** to help reduce the harmful effects of air pollution, acid deposition, and stratospheric ozone depletion. We can reduce emissions of pollutants and ozone-depleting chemicals by relying more on direct and indirect forms of solar energy than on fossil fuels; recycling and reusing matter resources much more widely than we do now; and mimicking biodiversity by using a variety of often locally available renewable energy resources in place of fossil fuels, especially coal. We can advance toward these goals by including the harmful environmental and health costs of fossil fuel use in market prices; seeking win-win solutions that will benefit both the economy and the environment; and giving high priority to passing along to future generations an environment in which they too can thrive.

Chapter Review

Core Case Study

1. Summarize the story of air pollution in Los Angeles and how it represents the larger problem of air pollution around the world.

Section 18.1

2. What is the key concept for this section? Define *density*, as it relates to the atmosphere, and *atmospheric pressure* and explain why both are two important atmospheric variables. Define and distinguish among **atmosphere**, **troposphere**, **stratosphere**, and **ozone layer**. What are **greenhouse gases**?

Section 18.2

3. What is the key concept for this section? What is **air pollution**? Distinguish between **primary pollutants** and **secondary pollutants** and give an example of each. What are atmospheric brown clouds and what have scientists learned about their harmful effects? List the six major outdoor air pollutants and their harmful effects. Describe the effects of lead as a pollutant and how we can reduce our exposure to this harmful chemical. Give examples of a chemical method and a biological method for detecting air pollutants. What is the Air Quality Index?

4. Distinguish between **industrial smog** and **photochemical smog** in terms of their chemical composition and formation. List and briefly describe five natural factors that help to reduce outdoor air pollution and six natural factors that help to worsen it. What is a **temperature inversion** and how can it affect outdoor air pollution levels?

Section 18.3

5. What is the key concept for this section? What is **acid deposition** and how does it form? What are its major environmental impacts on lakes, forests, human-built structures, and human health? List three major ways to reduce acid deposition. Explain the connections among low-sulfur coal, atmospheric warming, and toxic mercury.

Section 18.4

6. What is the key concept for this section? What is the major indoor air pollutant in many less-developed countries? What are the top three indoor air pollutants in the United States? List six sources of common indoor air pollutants in a typical modern home. Give three reasons why they present a serious threat to human health. Explain why radon-222 is an indoor air pollution threat, how and where it occurs, and what can be done about it.

Section 18.5

7. What is the key concept for this section? Describe the human body's defenses against air pollution, how they can be overwhelmed, and the illnesses that can result. Approximately how many people die prematurely from outdoor and indoor air pollution each year in the world and in the United States? What percentage of these deaths are caused by indoor air pollution?

Section 18.6

8. What is the key concept for this section? Summarize the U.S. air pollution laws and how they have worked to reduce pollution. Explain how these laws can be strengthened, according to some scientists.

9. List the advantages and disadvantages of using an emissions trading program to control pollution. Summarize the major ways to reduce emissions from power plants and motor vehicles. What was the key development in major improvements in LA's air quality (**Core Case Study**)? What are four ways to reduce indoor air pollution? Why is preventing air pollution more important than controlling it?

Section 18.7

10. What are the two key concepts for this section? How have human activities depleted ozone in the stratosphere? List five harmful effects of such depletion. Explain how Sherwood Rowland and Mario Molina alerted the world to this threat. What has the world done to reduce the threat of ozone depletion in the stratosphere? How might projected climate change undermine this progress? What are this chapter's *three big ideas*? Discuss the relationship between the air pollution in Los Angeles (**Core Case Study**) and the ways in which people have violated the three **scientific principles of sustainability**. Explain how we can apply these principles to the problems of air pollution.

Note: Key terms are in bold type.

Critical Thinking

1. You have built a time machine and have decided to travel back to 1940 to the city of Los Angeles (**Core Case Study**) to talk with city leaders about air pollution and how to deal with it. You will be informing them of some things they are not aware of, including the connection between car exhaust and smog. Write a strategy for *preventing* a long-term buildup of air pollutants over LA and give your audience your

three strongest arguments for adopting your strategy.

2. China relies on coal for two-thirds of its commercial energy usage, partly because the country has abundant supplies of this resource. Yet, China's coal burning has caused innumerable and growing problems for the country, for its neighboring nations, for the Pacific Ocean weather system, and even for the west coast of North America. Do you think China has been justified in developing this resource, as other countries—including the United States—have done with their coal resources? Explain. What are China's alternatives?

3. Considering your use of motor vehicles, now and in the future, what are three ways in which you could reduce your contribution to photochemical smog?

4. Should tall smokestacks (Figure 18.10) be banned to help prevent downwind air pollution and acid deposition? Explain.

5. Explain how sulfur impurities in coal can lead to increased acidity in rainwater and to the subsequent depletion of soil nutrients. Write an argument for or against requiring the use of low-sulfur coal in all coal-burning facilities.

6. If you live in the United States, list three important ways in which your life would be different if citizen-led actions during the 1970s and 1980s had not led to the Clean Air Acts of 1970 and its amendments of 1977 and 1990, despite strong political opposition by the affected industries. List three important ways in which your life in the future might be different if such actions do not lead now to the strengthening of the U.S. Clean Air Act. If you do not live in the United States, research the air pollution laws in your country and explain if and how they could be strengthened.

7. List three ways in which you could apply **Concept 18.6** to making your lifestyle more environmentally sustainable.

8. Congratulations! You are in charge of the world. Explain your strategy for dealing with each of the following problems: **(a)** indoor air pollution, **(b)** outdoor air pollution, **(c)** acid deposition, and **(d)** stratospheric ozone depletion.

Doing Environmental Science

Find out whether or not the buildings at your school have been tested for radon. If so, what were the results? What has been done about any areas with unacceptable levels of radon? If this testing has not been done, talk with school officials about having it done. You could also test for radon in your room or apartment or in the main living area of the house or building where you live. (Radon testing kits are available at affordable prices in most hardware stores, drug stores, and home centers.)

Global Environment Watch Exercise

Go to your MindTap course to access the GREENR database. Starting on the home page, under "Browse Issues and Topics", click on *Environment and Ecology,* then select *Ozone Depletion.* Use this portal to search for the latest research on the effectiveness of international efforts to reduce the use of CFCs. Research the use of substitutes such as HFC-22 and report on any potential pitfalls related to using such substitutes. What, if anything, is being done to address these problems?

Data Analysis

Coal often contains sulfur (S) as an impurity that is released as gaseous SO_2 during combustion, and SO_2 is one of six primary air pollutants monitored by the EPA. The U.S. Clean Air Act limits sulfur emissions from large coal-fired boilers to 0.54 kilograms (1.2 pounds) of sulfur per million Btus (British thermal units) of heat generated. (1 metric ton = 1,000 kilograms = 2,200 pounds = 1.1 ton; 1 kilogram = 2.2 pounds.)

1. Given that coal burned in power plants has a heating value of 27.5 million Btus per metric ton (25 million Btus per ton), determine the number of kilograms (and pounds) of coal needed to produce 1 million Btus of heat.

2. If all of the sulfur in the coal is released to the atmosphere during combustion, what is the maximum percentage of sulfur that the coal can contain and still allow the utility to meet the standards of the Clean Air Act?

CHAPTER 19

Climate Change

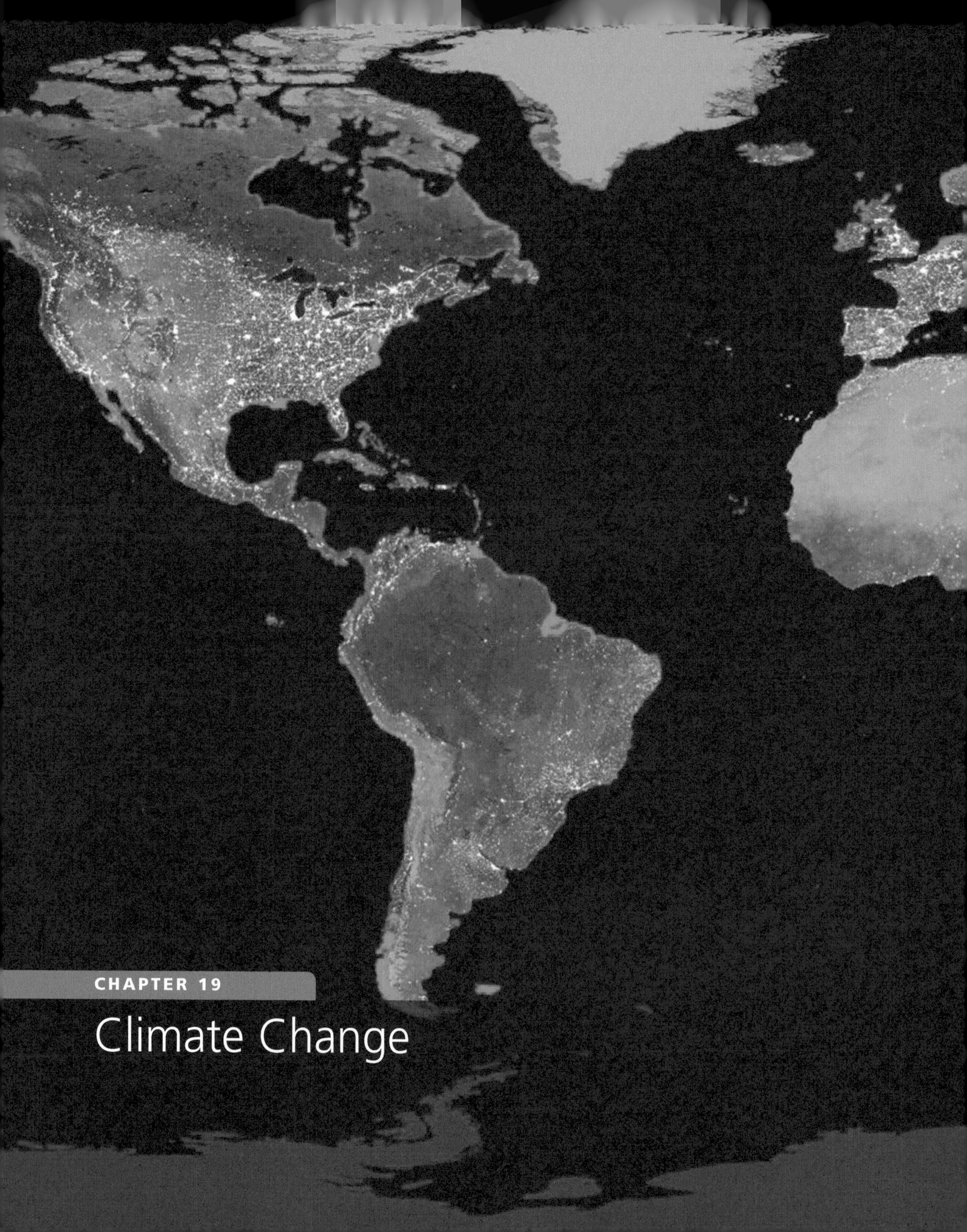

> Civilization has evolved during a period of remarkable climate stability, but this era is drawing to a close. We are entering a new era, a period of rapid and often-unpredictable climate change.
>
> LESTER R. BROWN

Key Questions

19.1 How is the earth's climate changing?

19.2 Why is the earth's climate changing?

19.3 What are the likely effects of a warmer atmosphere?

19.4 How can we slow climate change?

19.5 How can we adapt to climate change?

Areas that could be flooded by the end of this century (shown in red) by a 1-meter (3-foot) rise in sea level due to projected climate change.

NASA

Core Case Study

Melting Ice in Greenland

Greenland is the world's largest island with a population of about 57,000 people. The ice that covers most of this mountainous island lies in glaciers that are as deep as 3.2 kilometers (2 miles).

Areas of the island's ice have been melting at an accelerating rate during Greenland's summers (Figure 19.1). Some of this ice is replaced by snow during winter months, but the annual net loss of Greenland's ice has increased during recent years.

Why does it matter that ice in Greenland is melting? It matters because considerable scientific evidence indicates that atmospheric warming is a key factor behind this melting. *Atmospheric warming* is the gradual overall rise in the average temperature of the atmosphere near the earth's land and water surfaces over a period of 30 years or more. During this century, atmospheric warming is projected to continue and to lead to dramatic **climate change**—measurable changes in global weather patterns based primarily on changes in the earth's atmospheric temperature averaged over a period of at least 30 years. Climate scientists warn that if no action is taken, the earth's systems could reach *climate change tipping points*, which could result in damage to most of the earth's ecosystems, people, and economies for hundreds to thousands of years.

Greenland's glaciers contain enough water to raise the global sea level by as much as 7 meters (23 feet) if all of it melts and drains into the sea. It is highly unlikely that this will happen. But even a moderate loss of this ice over one or more centuries would raise sea levels considerably (see chapter-opening photo). Already Greenland's ice loss has been responsible for nearly one-sixth of the global sea-level rise over the past 20 years. Climate scientists view Greenland's melting ice as an early warning that human activities are very likely to disrupt the earth's climate in ways that could threaten life as we know it, especially during the latter half of this century.

In 1988, the World Meteorological Organization and the United Nations Environment Programme (UNEP) established the Intergovernmental Panel on Climate Change (IPCC) to document past climate changes and project future climate changes. The IPCC network includes more than 2,500 scientists in climate studies and related disciplines from more than 130 countries.

After reviewing tens of thousands of research studies for more than 25 years, the IPCC and most of the world's major scientific bodies, including the U.S. National Academy of Sciences and the British Royal Society, have come to three major conclusions: **(1)** climate change is real and is happening now; **(2)** it is caused primarily by human activities such as the burning of fossil fuels and the clearing of forests; and **(3)** it is projected to accelerate and have harmful effects, such as rising seas, ocean acidification, species extinction, and more extreme weather events, including intense and longer lasting heat waves, especially during the latter half of this century, unless we act now to slow it down.

In this chapter, we examine the likely causes and effects of climate change, and we look at some possible ways to deal with this major environmental, economic, and political challenge.

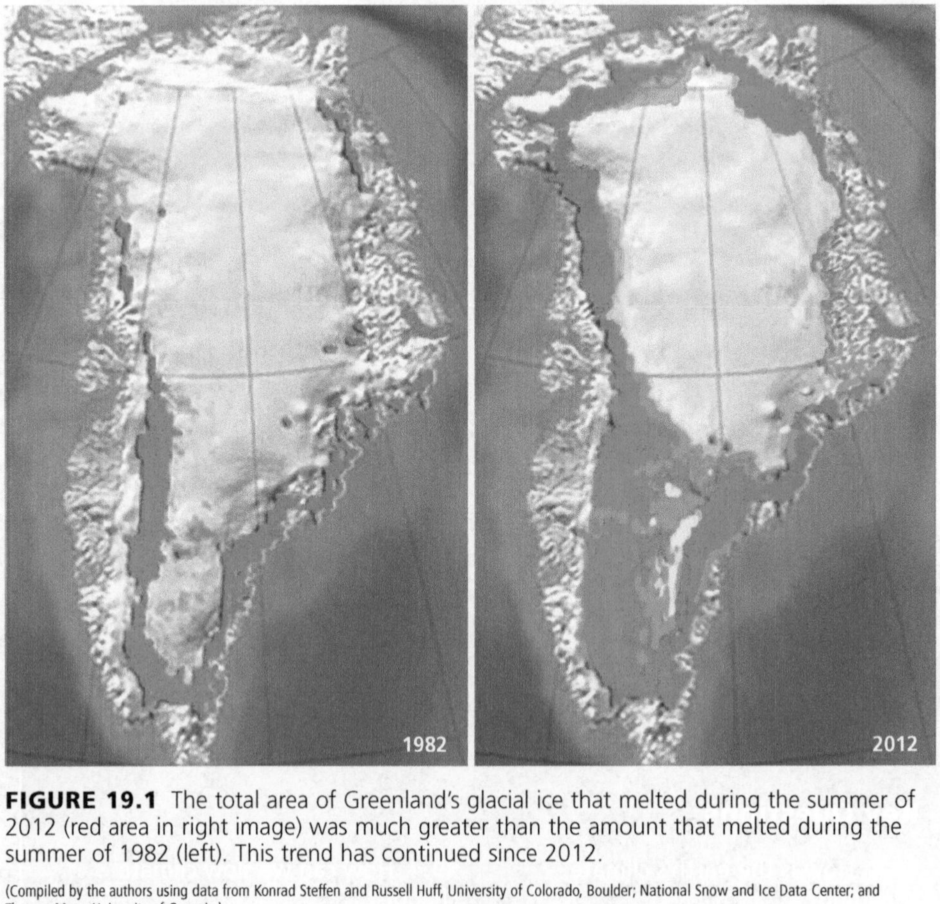

Space/NASA Sites

FIGURE 19.1 The total area of Greenland's glacial ice that melted during the summer of 2012 (red area in right image) was much greater than the amount that melted during the summer of 1982 (left). This trend has continued since 2012.

(Compiled by the authors using data from Konrad Steffen and Russell Huff, University of Colorado, Boulder; National Snow and Ice Data Center; and Thomas Mote, University of Georgia.)

19.1 HOW IS THE EARTH'S CLIMATE CHANGING?

CONCEPT 19.1 Scientific evidence strongly indicates that the earth's atmosphere is warming at a rate that is likely to lead to significant climate change.

Weather and Climate Are Not the Same

In thinking about climate change, it is important to distinguish between *weather* and *climate*. **Weather** consists of short-term changes in atmospheric variables such as the temperature, precipitation, wind, and barometric pressure in a given area over a period of hours or days. By contrast, **climate,** as defined by the World Meteorological Society, is determined by the *average* weather conditions of the earth or of a particular area, especially atmospheric temperatures, over periods of at least three decades. Scientists have used such measurements to divide the earth into various climate zones (see Figure 7.7, p. 145).

During any period in a given area of the planet, there will often be hotter years and cooler years, and wetter years and drier years, as weather often fluctuates widely from year to year. Climate scientists look at data on the weather conditions for the earth as a whole and for particular areas of the earth to see if there has been a general rise or fall in variables such as atmospheric temperature over a period of at least 30 years. This is the only way they can determine whether and how the climate of the earth or of a certain area has changed.

Atmospheric warming and climate change are often referred to as "global warming," which is a misleading term. It does not mean that all areas of the earth are getting warmer every year. Instead, as the earth's average atmospheric temperature rises, some areas get warmer at various times and some get cooler. However, when the *global average* atmospheric temperature rises or drops over a period of at least three decades, the earth's climate is changing.

Climate Change Is Not New but Recently Has Accelerated

Climate change is neither new nor unusual. Over the past 3.5 billion years, many factors have altered the planet's climate. These natural factors include large-scale volcanic eruptions and impacts by meteors and asteroids that cool the earth by injecting large amounts of debris into atmosphere, changes in solar input that can warm or cool the earth, and slight changes in the planet's wobbly orbit around the sun. Earth's climate also is affected by global air circulation patterns (see Figure 7.9, p. 146), changes in size of areas of ice that reflect incoming solar energy and cool the atmosphere, varying concentrations of the greenhouse gases found in the atmosphere, and occasional changes in ocean currents. Average temperature and precipitation are the two main factors affecting global, regional, and local climates, but temperature is the key variable that climate specialists watch.

The earth's climate has fluctuated over the past 900,000 years, slowly swinging back and forth between long periods of atmospheric warming and atmospheric cooling that led to ice ages (Figure 19.2, top left). These alternating freezing and thawing periods are known as *glacial* and *interglacial periods*.

For roughly 10,000 years, the earth has experienced an interglacial period characterized by a fairly stable climate based on a generally steady global average surface temperatures (see chapter-opening quote by environmental expert Lester Brown and Figure 19.2, bottom left). This allowed the human population to grow as agriculture developed and later as cities grew. For the past 1,000 years, the average temperature of the atmosphere near the earth's surface has remained fairly stable (Figure 19.2, bottom right). However, since 1975, atmospheric temperatures have been rising (Figure 19.2, top right). Research indicates that this has happened mostly because human activities such as the clearing of forests and grasslands and the burning of fossil fuels have been adding carbon dioxide (CO_2) to the atmosphere faster than the carbon cycle (see Figure 3.20, p. 65) can remove it.

According to a 2014 study on climate change by the American Association for the Advancement of Science (AAAS), evidence from numerous scientific studies indicates that the rising inputs of **greenhouse gases**—gases that absorb and release energy that warms the troposphere—from human activities are overwhelming the natural factors that led to climate change in the past. Current climate change and changes projected for the rest of this century are happening within several decades. This is many times faster than past climate changes caused by natural factors that took place over hundreds to thousands of years (Science Focus 19.1).

Between 2007 and 2014, the world's leading scientific organizations—including the IPCC, U.S. National Academy of Sciences (NAS), Great Britain's Royal Society, U.S. National Oceanic and Atmospheric Administration (NOAA), U.S. National Aeronautics and Space Administration (NASA), AAAS, and 97% of the world's climate scientists—all reached the following four major conclusions:

1. Climate change is happening now and is caused mostly by the burning of carbon-containing fossil fuels, which adds CO_2 to the atmosphere, and the clearing of forests, which removes trees and other plants that take up some of the excess CO_2 from the atmosphere.
2. Climate change is likely to accelerate, unless we act now to slow it.

The Study of Climate Change in the Distant Past

Scientists estimate past temperature changes (Figure 19.2) by analyzing evidence from many sources. One of the most important sources is ancient ice. As snow accumulates in the polar regions of the earth, the lower layers of snow are compressed by upper layers and become ice. Each of these compressed layers contains dust, volcanic ash, and other substances, along with bubbles of gas that were captured in this ice thousands of years ago. By drilling deep into ancient ice (Figure 19.A), scientists can gather information about historical greenhouse gas levels, along with volcanic eruptions, forest fires, and other events.

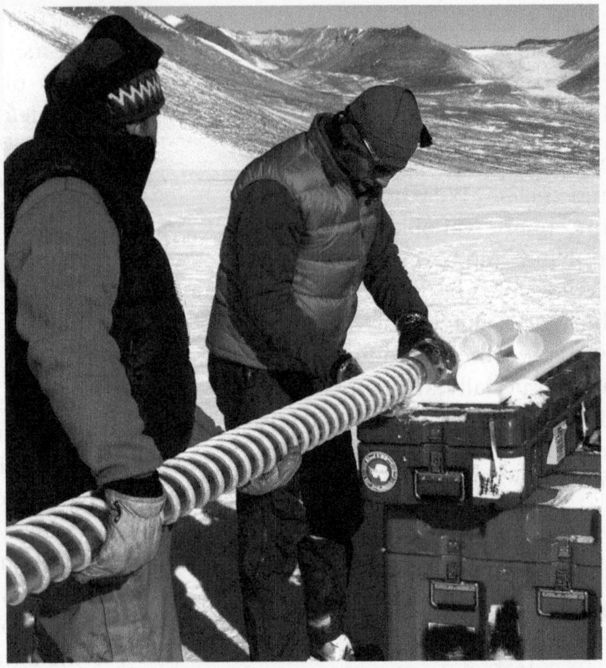

FIGURE 19.A *Ice cores* are extracted by drilling deep holes into ancient glaciers at various sites such as this one near the South Pole in Antarctica. Analysis of ice cores yields information about the past composition of the lower atmosphere, temperature trends such as those shown in Figure 19.2, solar activity, snowfall, and forest fire frequency.

Buried pollen grains are another source of information about climate changes of the past. Seed-bearing plants produce pollens, and botanists can examine them to determine which plants produced them. Some pollens that are buried in the sediments of wetlands and lake bottoms have been preserved for millions of years. By examining them, scientists can reconstruct the plant community that existed in a given area centuries ago. This in turn gives clues to what kind of climate existed there.

Another way to study the past is to examine tree rings. The cross section of a very old tree shows many rings, each of which represents a year's growth. A wider ring represents more growth for the tree than a narrow ring, and thus by counting the rings and projecting back, scientists can tell which years were favorable for growth of the tree and which were not. This gives more clues about temperatures, moisture, and other climate factors over hundreds to thousands of years.

Similarly, coral reefs can reveal year-by-year layering. Scientists gather information about historical ocean chemistry and water temperatures by examining these layers in coral cross sections. These data are related to changes in climate factors such as atmospheric CO_2 and temperatures and they give clues about how the climate changed over time.

Other sources of data include radioisotopes in rocks and fossils; plankton and radioisotopes in ocean sediments; and atmospheric temperature measurements taken regularly since 1861. These temperature measurements now include data from more than 40,000 measuring stations around the world, as well as from satellites.

CRITICAL THINKING

How would you use the information in this Science Focus to respond to someone who believes that climate change is a scientific hoax?

3. Immediate and sustained action to curb climate change is possible and affordable and would bring major benefits for human health and economies, as well as for the environment.
4. The sooner we act to slow climate change, the lower its environmental and economic costs are likely to be.

Data from tens of thousands of peer-reviewed scientific studies support the conclusion that human-influenced climate change is happening now. Below are a few pieces of such evidence.

- Between 1906 and 2015, Earth's average global surface temperature rose by 1.0C° (1.8F°). Much of this increase took place since the mid-1970s (Figure 19.2, upper right).

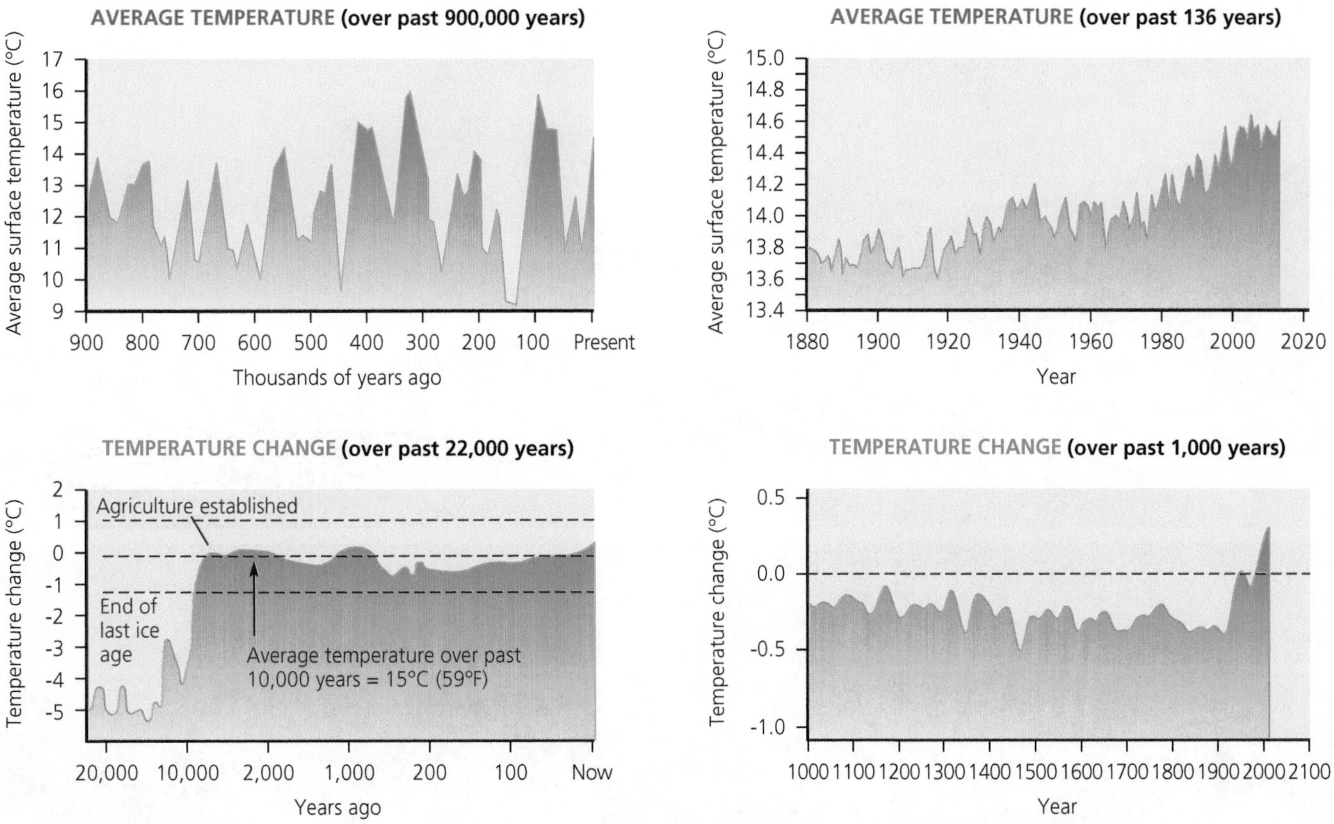

FIGURE 19.2 The global average temperature of the atmosphere near the earth's surface has changed significantly over time. The two graphs in the top half of this figure are estimates of global average temperatures, and the two graphs on the bottom are estimates of changes in the average temperature over different periods. ***Critical thinking:*** What are two conclusions that you can draw from these diagrams?

(Compiled by the authors using data from Goddard Institute for Space Studies, Intergovernmental Panel on Climate Change, National Academy of Sciences, National Aeronautics and Space Administration, National Center for Atmospheric Research, and National Oceanic and Atmospheric Administration.)

- Nine of the ten warmest years on record since 1861 have taken place since 1998. The warmest year was 2015 and the second warmest was 2014.
- In the Arctic, floating summer sea ice has been shrinking in most years since 1979.
- In some parts of the world, glaciers that have existed for thousands of years (Figure 19.3), including Greenland's ice sheets, are melting (**Core Case Study**).
- In Alaska, glaciers and frozen ground (permafrost) are melting, loss of sea ice and rising sea levels are eating away at coastlines, and communities are being relocated inland.
- The world's average sea level has been rising at an accelerated rate, especially since 1975. This rise is mostly due to the expansion of ocean water as its temperature increased and increasing runoff from melting land-based ice.
- Atmospheric levels of CO_2 and other greenhouse gases that warm the troposphere have been rising sharply. After remaining below 300 parts per million (ppm) for more than 400 years, atmospheric levels of CO_2 rose from 318 ppm to 400 ppm between 1960 and 2015.
- As temperatures rise, many terrestrial, marine, and freshwater species have migrated toward the poles and, on land, to cooler higher elevations. Species that cannot migrate face extinction.

You may have heard about the *climate change debate* in the media. This term is misleading because, among 97% of the world's climate scientists, there is no significant debate on climate change or its causes.

However, there is intense disagreement about climate change in the political arena among citizens, elected officials, and officials of companies that produce and burn fossil fuels that add CO_2 to the atmosphere. Such disagreements center around which, if any, political actions should be taken to deal with climate change. This is a very difficult and important political and economic issue, because modern economies and lifestyles are built around burning fossil fuels that supply 90% of the world's commercial energy.

FIGURE 19.3 Between 1913 (top) and 2008 (bottom) much of the ice that covered Sperry Glacier in Montana's Glacier National Park melted.

Top: W. C. Alden/GNP Archives/US Geological Survey. Bottom: Lisa McKeon/US Geological Survey.

19.2 WHY IS THE EARTH'S CLIMATE CHANGING?

CONCEPT 19.2 Scientific evidence strongly indicates that the earth's atmosphere has been warming at a rapid rate since 1975 and that human activities, especially the burning of fossil fuels and deforestation, have played a major role in this warming.

Natural Factors Can Lead to Climate Change

Changes in inputs of solar radiation have been a key factor leading to climate change in the past (Figure 19.2). Some scientists say such variations are related to *Milankovitch cycles*—cyclical changes in the shape of the earth's orbit and the tilt of the earth's axis with respect to the sun's radiation. During each cycle of such changes, which last roughly 100,000 years on average, the earth has experienced a cooling period often ending in an ice age, followed by a warming period.

The **greenhouse effect** (see Figure 3.3, p. 52) is a natural process that plays a major role in determining the earth's average atmospheric temperature and thus its climate. It occurs when some of the solar energy absorbed by the earth radiates into the atmosphere as infrared radiation (heat). As this radiation interacts with molecules of several greenhouse gases in the air, it increases their kinetic energy and warms the lower atmosphere and the earth's surface.

Numerous laboratory experiments and measurements of temperatures at different altitudes have confirmed the greenhouse effect—now one of the most widely accepted theories in the atmospheric sciences. Life on earth and our economic systems are dependent on the greenhouse effect because it keeps the planet at an average temperature of

© Cengage Learning

around 15°C (58°F). Without it, the planet would be a frozen, uninhabitable place.

There are several greenhouse gases, but we focus here on three major gases—CO_2, CH_4, and N_2O—that play varying roles in atmospheric warming because of their varying lifetimes in the atmosphere and varying warming potentials (Figure 19.4).

The atmospheric concentration of CO_2, as part of the carbon cycle (see Figure 3.20, p. 65), plays a key role in determining the average temperature of the atmosphere. Measurements of CO_2 in bubbles at various depths in ancient glacial ice (Science Focus 19.1) indicate that changes in the levels of this gas in the lower atmosphere have correlated closely with changes in the global average temperature near the earth's surface during the past 400,000 years (Figure 19.5). Scientists have noted a similar correlation between atmospheric temperatures and methane (CH_4) emissions.

Water vapor (H_2O), another greenhouse gas, accounts for about 66% of the earth's greenhouse effect. However, it stays in the atmosphere for only about 1 to 3 weeks, on average (Figure 19.4). As the air warms, it can hold more water vapor. Thus, the warming effect of other greenhouse gases can lead to a higher level of water vapor, which in turn amplifies the greenhouse effect. For this reason, scientists consider CO_2 to be the main engine of the atmospheric warming, while water vapor plays a secondary role in response to CO_2 levels.

Human Activities Play a Key Role in Current Atmospheric Warming

Since the beginning of the Industrial Revolution in the mid-1700s, humans have greatly increased deforestation, agriculture (Figure 19.6, left), and the burning of fossil fuels. These factors have led to significant increases in the concentrations of several greenhouse gases, especially CO_2, in the lower atmosphere (Figure 19.6, right). The average atmospheric concentration of CO_2 rose by about 38% between 1880 and 2015 (Figure 19.5) with more than half of the increase taking place since 1970. This is a

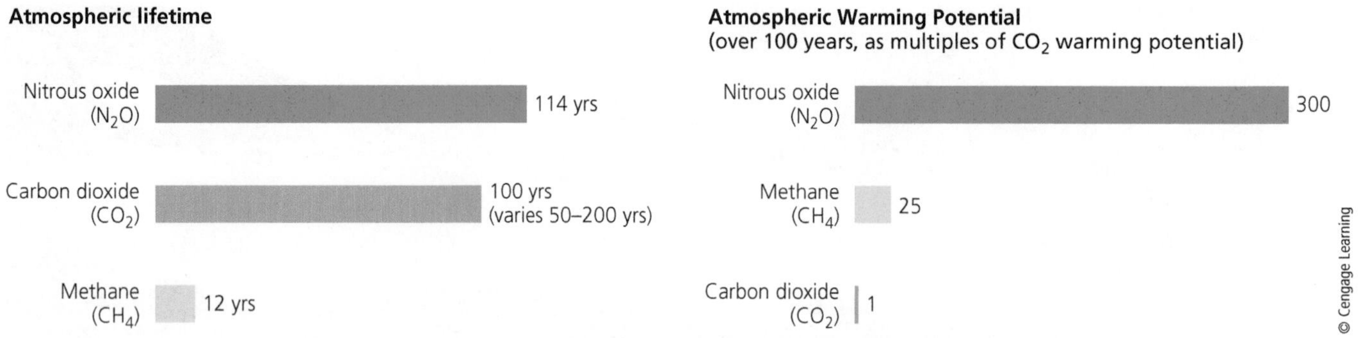

FIGURE 19.4 Atmospheric lifetimes and warming potentials for three major greenhouse gases.

(Compiled by the authors using data from National Oceanic and Atmospheric Administration and U.N. Framework Convention on Climate Change.)

long-lasting increase because about 80% of the CO_2 we emit typically remains in the atmosphere for 100 years or more and about 20% of it remains for up to 1,000 years. Currently, we are pumping around 82 million metric tons (90 million tons) of CO_2 into the air every day.

After oscillating between 180 and 280 ppm for 400,000 years, atmospheric CO_2 levels reached 400 ppm in 2015. NOAA scientists estimate that the last time the earth's atmospheric CO_2 reached this level was about 4.5 million years ago.

Also, since the mid-1700s, atmospheric methane levels have risen by 160%. Ice core analysis reveals that about 70% of methane (CH_4) emissions during the last 275 years are the result of human activities, including livestock production, rice production, natural gas production, leaky coal mines, landfills, and the flooding of land behind large dams. In that time, methane levels in the atmosphere have tripled.

Methane emissions are expected to rise in the future if the atmosphere warms as projected. Areas of permafrost in

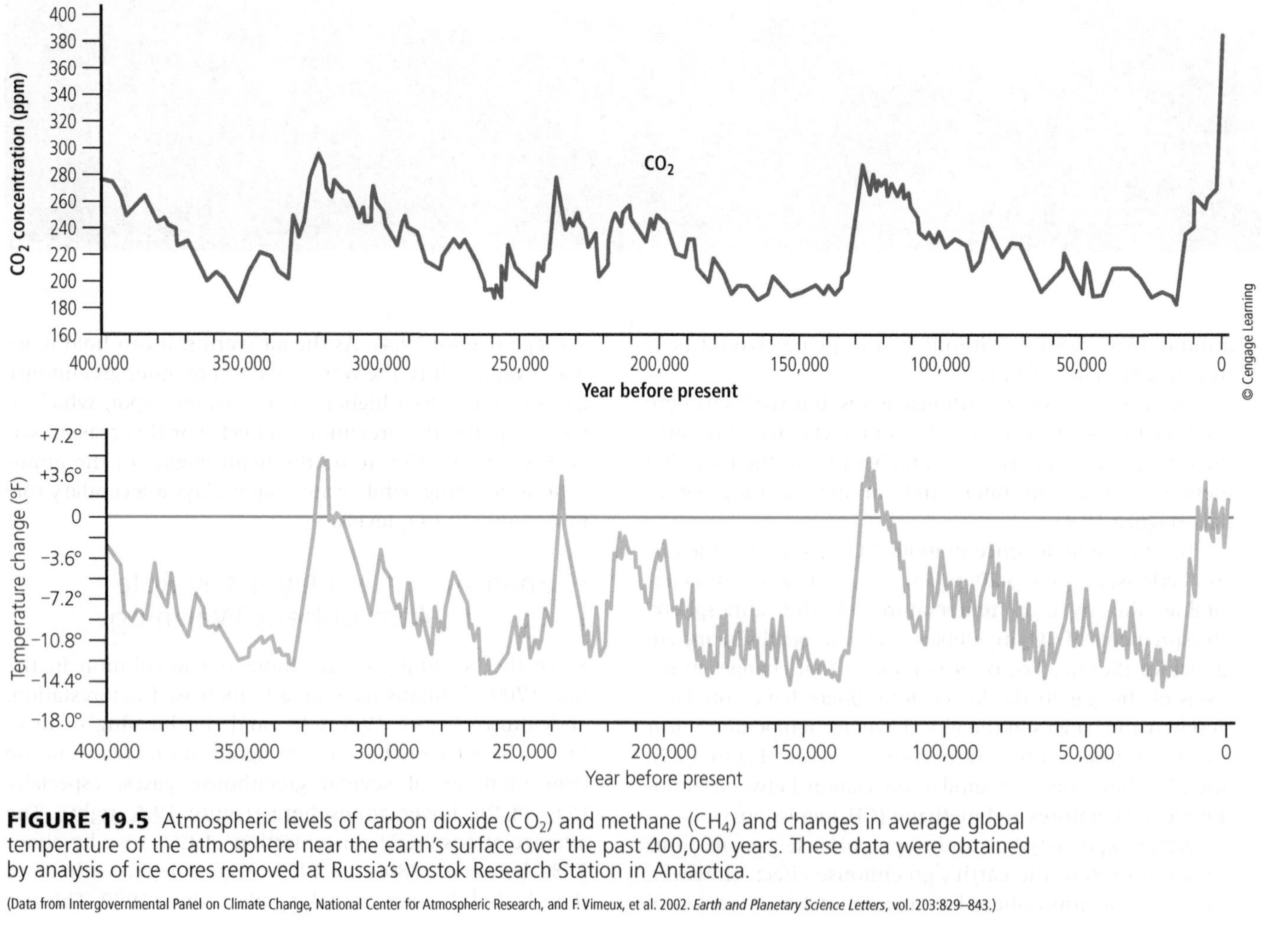

FIGURE 19.5 Atmospheric levels of carbon dioxide (CO_2) and methane (CH_4) and changes in average global temperature of the atmosphere near the earth's surface over the past 400,000 years. These data were obtained by analysis of ice cores removed at Russia's Vostok Research Station in Antarctica.

(Data from Intergovernmental Panel on Climate Change, National Center for Atmospheric Research, and F. Vimeux, et al. 2002. *Earth and Planetary Science Letters*, vol. 203:829–843.)

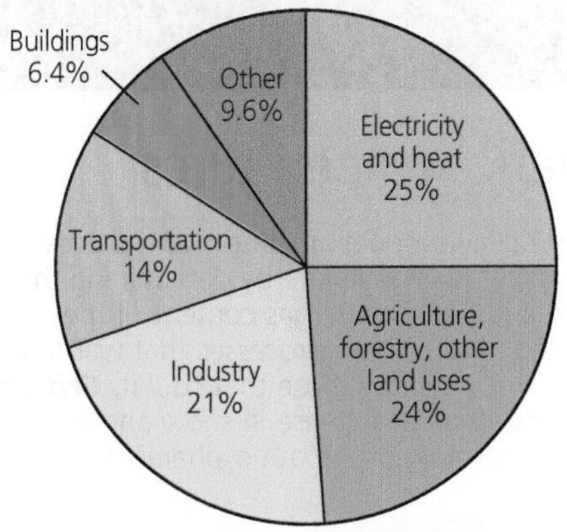

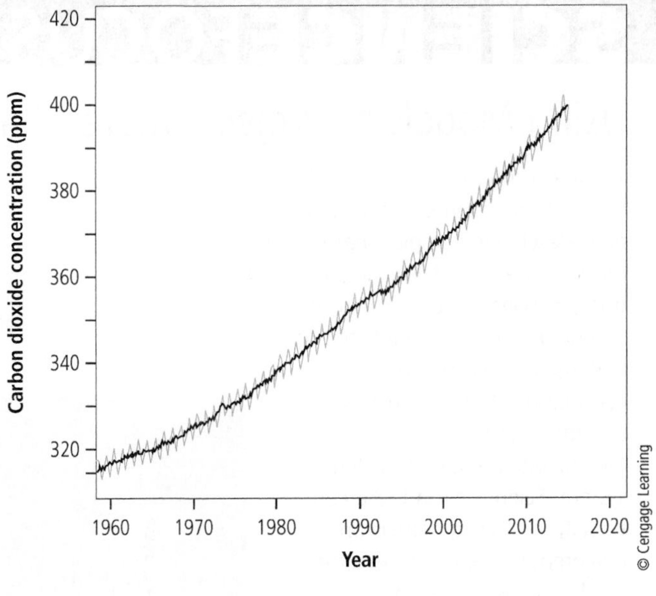

FIGURE 19.6 Annual emissions of carbon dioxide (CO_2) from various sources (left) have contributed to rising global atmospheric levels of CO_2 (right), 1880–2015.

(*Left,* Compiled by the authors using data from International Energy Agency and U.S. Department of Energy Carbon Dioxide Information Analysis Center. *Right,* Compiled by the authors using data from Earth Policy Institute BP, *Statistical Review of World Energy*, 2015, Intergovernmental Panel on Climate Change, and National Center for Atmospheric Research.)

the arctic tundra and the Arctic Ocean seabed are expected to thaw, which will release a large volume of CH_4. In addition, a 2012 study led by NASA's Eric Kort found that methane was being released through cracks in floating arctic ice. Research indicates that this methane is generated by marine bacteria, and that emissions from this source will likely increase as arctic seawater warms and as more cracks appear in the melting sea ice.

Nitrous oxide (N_2O) levels have risen about 20% during the last 275 years, mostly because of the increased use of nitrogen fertilizers (Figure 3.22, p. 67). This gas accounts for about 9% of greenhouse gas emissions from human activities, but it remains in the atmosphere for 114 years and each molecule of N_2O has nearly 300 times the warming potential of a molecule of CO_2 (Figure 19.4). Livestock production now accounts for 65% of all human-generated N_2O emissions.

In 2014, the three largest emitters of energy-related CO_2 were China, the United States, and India, according to the UN Statistics Division (Figure 19.7). In 2014, U.S. per capita CO_2 emissions in the United States were several times higher than those of China and India.

In comparing CO_2 emissions sources, scientists use the concept of a **carbon footprint,** which refers to the amount of CO_2 generated by an individual, an organization, a country, or any other entity over a given period. China has the largest national carbon footprint, and Americans have the largest per capita carbon footprints.

According to a 2014 study by Canadian researcher Damon Matthews and his colleagues, the CO_2 emissions of seven countries were responsible for 60% of the atmospheric warming taking place since 1906. In order, they were the United States, China, Russia, Brazil, India,

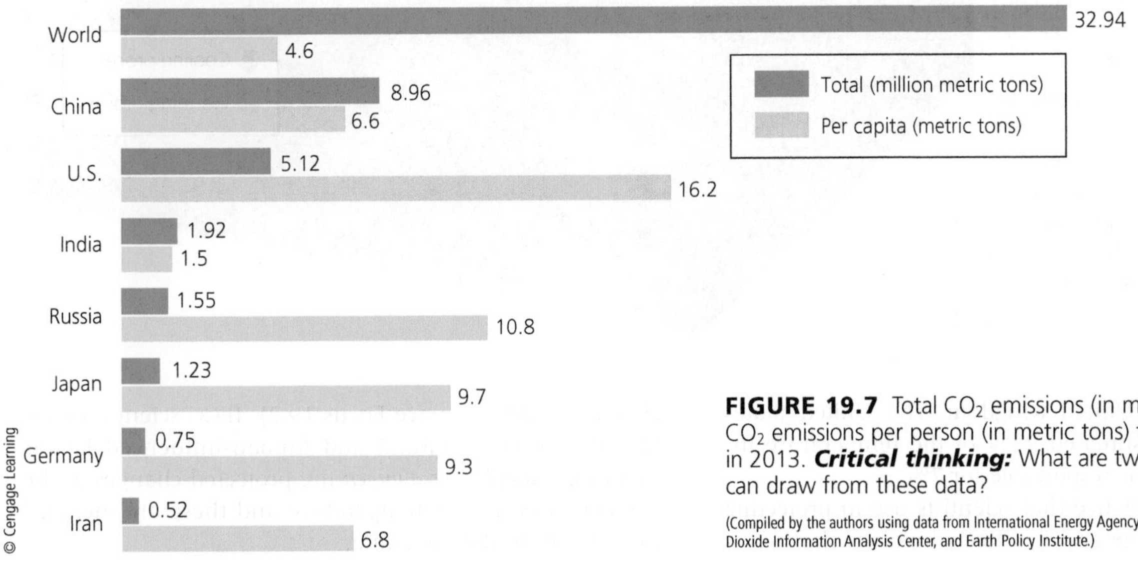

FIGURE 19.7 Total CO_2 emissions (in million metric tons) and CO_2 emissions per person (in metric tons) for selected countries in 2013. ***Critical thinking:*** What are two conclusions that you can draw from these data?

(Compiled by the authors using data from International Energy Agency, U.S. Department of Energy Carbon Dioxide Information Analysis Center, and Earth Policy Institute.)

Using Models to Project Future Changes in Atmospheric Temperatures

To project the climatic effects of rising levels of greenhouse gases, scientists have developed complex *mathematical models*. These models simulate interactions, as scientists understand them, among incoming sunlight, clouds, landmasses, ocean currents, concentrations of greenhouse gases and other air pollutants, and other factors within the earth's climate system. Scientists run these continually improving models on supercomputers and compare the results to known past atmospheric temperatures, from which they project future changes in the earth's average atmospheric temperature. Figure 19.B gives a greatly simplified summary of some of the key interactions in the global climate system.

Such models provide *projections* of what is likely to happen to the average temperature of the lower atmosphere, based on available data and different assumptions about future changes such as CO_2 and CH_4 levels in the atmosphere. How well these projections match what actually happens in the real world depends on the validity of the assumptions, what variables are built into the models, and the accuracy of the data used.

Recall that while scientific research cannot give us absolute proof or certainty, it does provide us with varying levels of certainty. When most experts in a particular scientific field generally agree on a level of 90% certainty about a set of measurements or model results, they say that their projections are *very likely* to be correct; when the level of certainty reaches 95% (a rarity

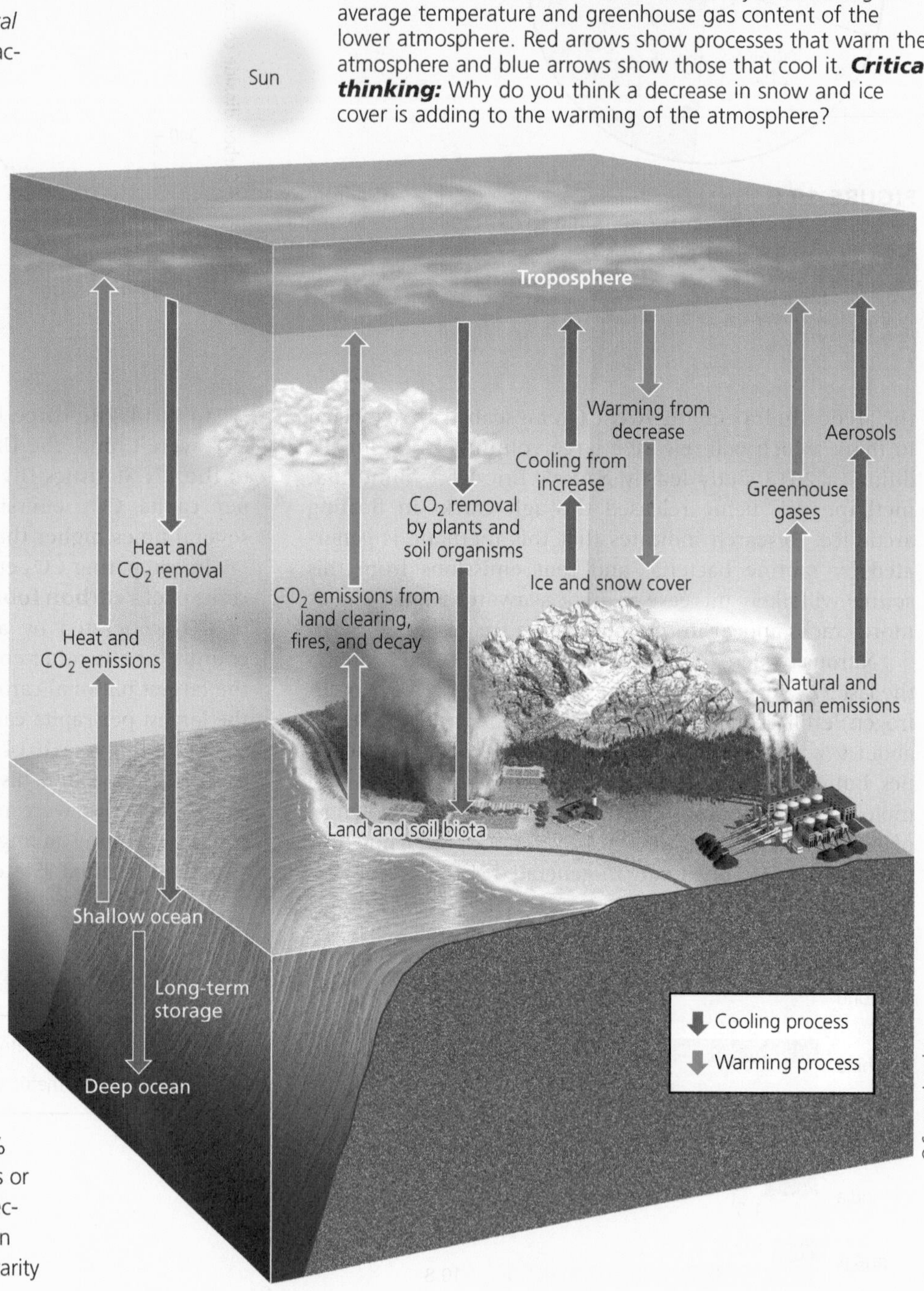

FIGURE 19.B A simplified model of some major processes that interact to affect the earth's climate by determining the average temperature and greenhouse gas content of the lower atmosphere. Red arrows show processes that warm the atmosphere and blue arrows show those that cool it. ***Critical thinking:*** Why do you think a decrease in snow and ice cover is adding to the warming of the atmosphere?

Germany, and the United Kingdom. The United States accounted for an estimated 22% of the temperature increase and China was responsible for 7%.

Much of the evidence that scientists use in projecting future climate change and its effects involves the use of climate models (Science Focus 19.2). IPCC scientists have identified several natural and human-influenced factors that might *amplify* or *dampen* the projected changes in the average atmospheric temperature and these are included as variables in the modeling.

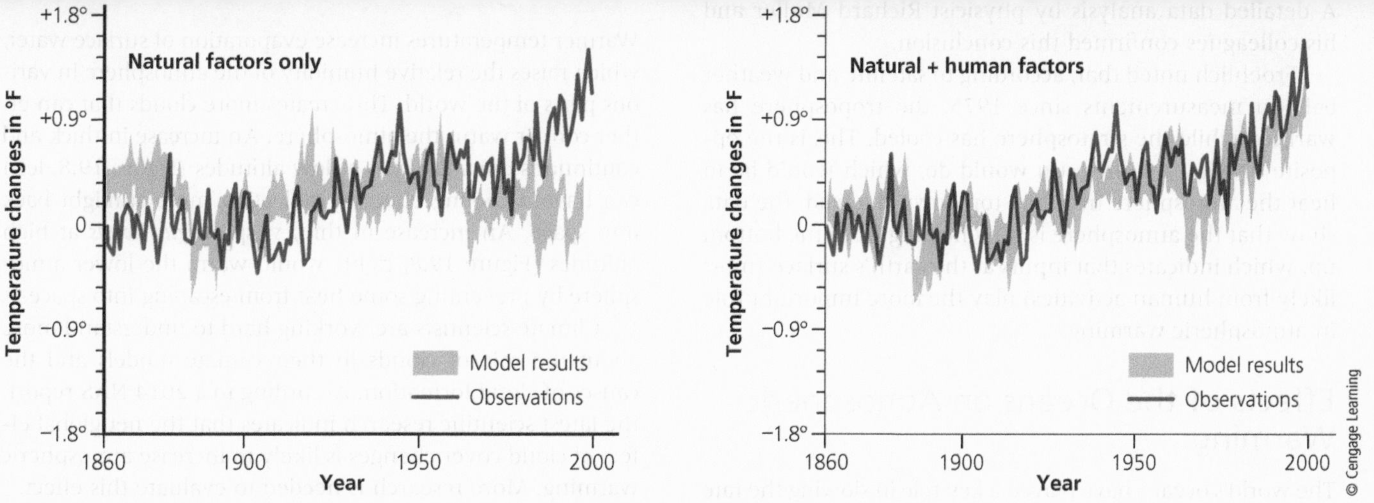

FIGURE 19.C Comparison of actual climate data with modeled projections for the period between 1860 and 2000 using natural factors only (left) and a combination of natural and human factors (right). Scientists have found that actual data match projections far more closely when human factors are included in the models.
(Compiled by the authors using data from Intergovernmental Panel on Climate Change.)

in science), they contend that their projections are *extremely likely* to be correct.

In 1990, 1995, 2001, 2007, and 2014, the IPCC published reports on how global temperatures have changed in the past (Figure 19.2), how they are projected to change during this century, and how such changes are likely to affect the earth's climate. According to the 2014 report, based on analysis of past climate data and the use of more than two dozen climate models, it is extremely likely (95% certainty) that human activities, especially the burning of fossil fuels, have played the dominant role in the observed atmospheric warming since 1975. The researchers based this conclusion on thousands of peer reviewed research studies and on the fact that, after thousands of times running the models, the only way they could get the model results to match actual measurements was by including the human activities factor (Figure 19.C).

The 2014 IPCC report stated that it is *very likely* (with at least 90% certainty) that the earth's mean surface temperature would increase by 1.5–4.5C° (2.7–8.1F°) between 2013 and 2100, unless deforestation and greenhouse gas emissions were sharply reduced. According to a 2014 report by climate researcher Michael E. Mann, the latest runs of key climate models indicate that the lower limit of the 2014 IPCC projected temperature increase should be raised from 1.5C° (2.7F°) to 2.5C° (4.5F°).

Researchers note the likelihood that not all places on the planet will experience warming. Some will likely warm a great deal whereas others will warm slightly or not at all. Some might even grow cooler. This is because the atmosphere is a complex system with thousands of variables. However, when all regional temperatures are averaged together, the result will be a higher global average atmospheric temperature, according to the models.

There is an extremely high degree of certainty (95%) that the atmosphere has been warming primarily due to human activities. However, there is a high degree of uncertainty in the model results about projected future changes in the average atmospheric temperature, which range from 1.5C° (2.7F°) to 4.5C° (8.1F°). Most of this uncertainty is due to the difficulty of projecting how clouds will change as temperatures change.

Climate experts are working to reduce such uncertainty by learning more about how the earth's climate system works and by improving climate data and models. Despite their limitations, these models are the best and only tools that we have for projecting likely average atmospheric temperatures in coming decades.

CRITICAL THINKING

If the highest possible projected temperature increase (4.5C° or 8.1F°) takes place, what are three major ways in which this will likely affect your lifestyle and that of any children or grandchildren you might have?

What Role Does the Sun Play in Climate Change?

The energy output of the sun plays the key role in the earth's temperature, and this output has varied over millions of years. However, climate researchers Claus Froehlich, Mike Lockwood, and Ben Santer all concluded in separate studies that most of the rise in global average atmospheric temperatures since 1975 (Figure 19.2, top right) could not be the result of increased solar output.

Instead, they determined that the energy output of the sun has dropped slightly during the past several decades. A detailed data analysis by physicist Richard Muller and his colleagues confirmed this conclusion.

Froehlich noted that, according to satellite and weather balloon measurements since 1975, the troposphere has warmed while the stratosphere has cooled. This is the opposite of what a hotter sun would do, which would be to heat the atmosphere from the top down. Instead, the data show that the atmosphere is now heating from the bottom up, which indicates that inputs at the earth's surface (most likely from human activities) play the more important role in atmospheric warming.

Effects of the Oceans on Atmospheric Warming

The world's oceans have played a key role in slowing the rate of atmospheric warming and climate change. They absorb CO_2 from the atmosphere as part of the carbon cycle and thus help to moderate the earth's average surface temperature and its climate. It is estimated that the oceans remove roughly 25% of the CO_2 pumped into the lower atmosphere by human activities. Most of it is then stored as carbon compounds in marine algae and vegetation and in coral reefs and transferred to the deep ocean (see Figure 7.11, p. 147), where it is buried in carbon compounds in bottom sediments for several hundred million years.

The oceans also absorb heat from the lower atmosphere. According to a 2016 study by researchers at Lawrence Livermore National Laboratory, more than 90% of the heat held in the atmosphere by greenhouse gas pollution since the 1970s has ended up in the oceans. This has warmed the oceans with ocean currents pushing about a third of the heat into the ocean depths. About half of the resulting overall ocean warming has occurred since 1997, according to this 2016 study.

The ability of the oceans to absorb CO_2 decreases as water temperatures rise. As the oceans warm up, some of their dissolved CO_2 is released into the lower atmosphere, like CO_2 bubbling out of a warm carbonated soft drink. According to scientific measurements, the upper portion of the oceans warmed by an average of 0.32–0.67C° (0.6–1.2F°) during the last century—an astounding increase considering the huge volume of water involved—most likely due to increasing atmospheric temperatures.

The uptake of CO_2 and heat by the world's oceans has slowed the rate of atmospheric warming and climate change. However, this has also resulted in the serious and growing problem of ocean acidification, which could have harmful effects on marine ecosystems (see Science Focus 11.1, p. 257). We discuss this process and its effects in the next section.

Effects of Cloud Cover on Atmospheric Warming

Warmer temperatures increase evaporation of surface water, which raises the relative humidity of the atmosphere in various parts of the world. This creates more clouds that can either cool or warm the atmosphere. An increase in thick and continuous *cumulus clouds* at low altitudes (Figure 19.8, left) can have a cooling effect by reflecting more sunlight back into space. An increase in thin, wispy *cirrus clouds* at high altitudes (Figure 19.8, right) would warm the lower atmosphere by preventing some heat from escaping into space.

Climate scientists are working hard to understand more about the role of clouds in their climate models and the causes of cloud formation. According to a 2014 NAS report, the latest scientific research indicates that the net global effect of cloud cover changes is likely to increase atmospheric warming. More research is needed to evaluate this effect.

CONSIDER THIS . . .

CONNECTIONS Air Travel and Atmospheric Warming

Infrared satellite images indicate that wispy condensation trails (contrails) left behind by jet planes expand and turn into cirrus clouds, which tend to hold heat in the atmosphere. A study led by Ulrike Burkhardt and Bernd Kärcher found that contrails are a significant contributor to atmospheric warming, more so than the total amount of CO_2 that jet airplanes release, at least in the short term.

Outdoor Air Pollution Can Affect Atmospheric Warming

According to the 2014 IPCC report, air pollution in the form of *aerosols* (suspended microscopic droplets and solid

Tatiana Grozetskaya/Shutterstock.com

FIGURE 19.8 Cumulus clouds (left) are thick, relatively low-lying clouds that tend to decrease surface warming by reflecting some incoming solar radiation back into space. Cirrus clouds (right) are thin and float at high altitudes; they tend to warm the earth's surface by preventing some heat from flowing into space.

Cheryl Casey/Shutterstock.com

particles) from human activities can affect the rate of atmospheric warming. These pollutants are released or formed in the troposphere by volcanic eruptions and by human activities (see Figure 18.3, p. 474). They can hinder or enhance both the greenhouse effect and cloud formation, depending on factors such as their size and reflectivity.

Most aerosols, such as light-colored sulfate particles produced by fossil fuel combustion, tend to reflect incoming sunlight and cool the lower atmosphere. A 2012 study led by Harvard University's Eric Leibensperger estimated that atmospheric temperatures in the eastern United States had dropped slightly between 1930 and 1990—due largely to pollution from coal-fired power and industrial plants—while worldwide average temperatures had increased slightly. While sulfate particles reflect sunlight and help cool the atmosphere, black carbon particles, or *soot*—emitted into the air by coal-burning power and industrial plants, diesel exhaust, open cooking fires, and burning forests—warm the atmosphere.

Climate scientists do not expect aerosols and soot particles to affect climate change very much over the next 50 years for two reasons. *First*, aerosols and soot fall back to the earth or are washed out of the lower atmosphere within weeks, whereas CO_2 typically remains in the lower atmosphere for 100 years or longer. *Second*, we are reducing aerosol and soot emissions because of their harmful impacts on plants and humans.

19.3 WHAT ARE THE LIKELY EFFECTS OF A WARMER ATMOSPHERE?

CONCEPT 19.3 The projected rapid change in the atmosphere's temperature could have severe and long-lasting consequences, including flooding, rising sea levels, shifts in the locations of croplands and wildlife habitats, and more extreme weather.

Rapid Atmospheric Warming Could Have Serious Consequences

Most of the historical changes in the temperature of the lower atmosphere took place over thousands of years (Figure 19.2, top left). What makes the current problem urgent is that we face *a rapid projected increase in the average temperature of the lower atmosphere during this century* (Figure 19.2, top right). This will very likely (95% chance) change the mild climate we have had for the past 10,000 years (Figure 19.2, bottom left). According to the 2014 AAAS report on climate change, "The rate of climate change now may be as fast as any extended warming period over the past 65 million years, and is projected to increase in coming years."

The worst-case climate model projections indicate that people could face the following effects within this century:

- Floods in low-lying coastal cities from a rise in sea level
- Many forests being consumed in vast wildfires
- Grasslands turning into dust bowls
- Rivers drying up
- Ecosystems collapsing
- The extinction of up to half the world's species
- More intense and longer-lasting heat waves
- More destructive storms and flooding
- The rapid spread of infectious tropical diseases

Scientists have identified a number of components of the climate system that could pass **climate change tipping points**—thresholds beyond which natural systems could change for hundreds to thousands of years. One such component is global average atmospheric temperature. Several findings related to atmospheric warming are leading some scientists to suggest that we can no longer avoid a global temperature rise of 2C° (3.6F°), the threshold beyond which many say climate change will be dangerous. However, there is controversy over this number. Some say it is arbitrary. Others say it is not realistic, arguing that because we have failed to slow atmospheric warming, we are committed to raising the average atmospheric temperature beyond this level.

Another tipping point is related to atmospheric CO_2 levels. A number of scientific studies and major climate models indicate that we need to prevent CO_2 levels from exceeding 450 ppm—an estimated tipping point beyond which we might experience large-scale climate changes that would last for hundreds to thousands of years. We have already reached 400 ppm, and without strong efforts to improve energy productivity and replace fossil fuels with low-carbon energy resources, we could exceed 450 ppm within a couple of decades. Figure 19.9 lists several climate change tipping points, and we will examine some of them in this section.

Climate models show that atmospheric warming and its harmful effects will be unevenly distributed. Temperatures in the tropics could reach extremes that people there have not seen in over 10,000 years. The sea-level rise also would be 15–20% larger than the global average in tropical areas, and droughts could be much worse in such areas. In our interconnected global economy, such effects would ripple to other areas of the world.

More Ice and Snow Are Likely to Melt

Models project that climate change will be the most severe in the world's polar regions. Light-colored ice and snow in these regions cool the earth by reflecting incoming solar energy back into space—a process called the *albedo effect*. The melting of polar ice and snow exposes much darker land and sea areas, which reflect less sunlight and absorb more solar energy. This has warmed the atmosphere above the poles more and faster than the atmosphere is warming at lower latitudes. The result is likely to be more melting of snow and ice, which will cause further atmospheric

- Atmospheric carbon level of 450 ppm
- Melting of all arctic summer sea ice
- Collapse and melting of the Greenland ice sheet
- Severe ocean acidification, collapse of phytoplankton populations, and a sharp drop in the ability of the oceans to absorb CO_2
- Massive release of methane from thawing arctic permafrost and from the arctic seafloor
- Collapse and melting of most of the western Antarctic ice sheet
- Severe shrinkage or collapse of Amazon rain forest

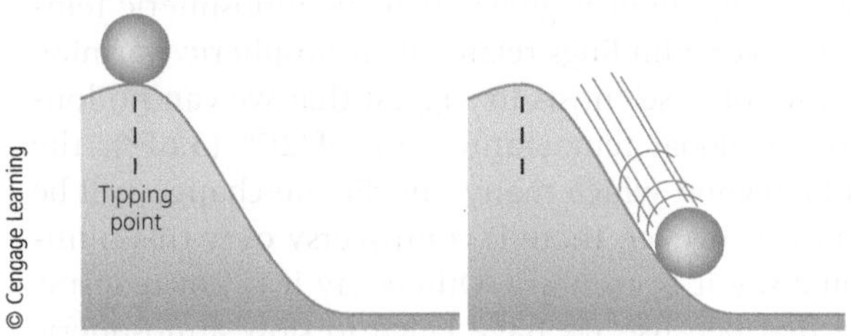

FIGURE 19.9 Climate scientists have come up with this list of possible climate change tipping points.

warming above the poles in a runaway positive feedback loop (see Figure 2.15, p. 43).

According to the 2014 IPCC report, arctic air temperatures have risen almost twice as fast as average temperatures in the rest of the world during the past 50 years, and are now warmer than they have been in 44,000 years. Arctic Ocean waters have also warmed. In addition, soot generated by North American, European, and Asian industries is darkening Arctic ice and lessening its ability to reflect sunlight.

Mostly because of these factors, floating summer sea ice in the Arctic is disappearing faster than scientists thought it would only a few years ago (Figure 19.10). Measurements indicate that this accelerated melting is happening because of warmer air above the ice and warmer water below. With changes in short-term weather conditions, summer Arctic sea ice coverage is likely to fluctuate. But the overall projected long-term trends are for the Arctic to warm, the average summer sea ice coverage to decrease, and the ice to become thinner (Figure 9.10, right, lower graph).

One of the climate-change tipping points that concern scientists (Figure 19.10) is the complete melting of floating

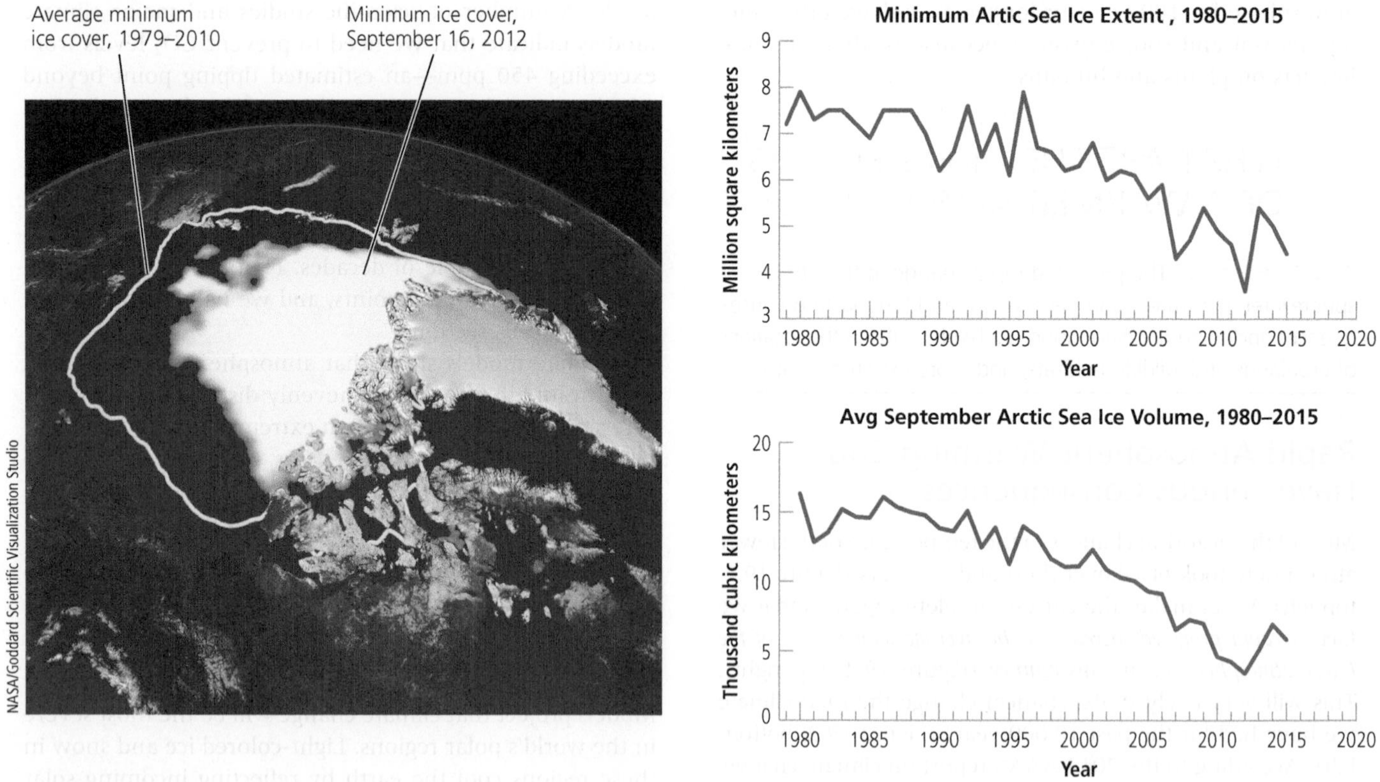

FIGURE 19.10 *The big melt:* Rising average atmospheric and ocean temperatures have caused more and more arctic sea ice to melt during the summer months. The yellow line added to this satellite image (left) shows the average summer minimum area of ice during the period 1979–2010, in contrast to the white, ice-covered summer minimum in 2012. The graphs (right) show that the sea ice melt has been a continuing since 2012. ***Data analysis:*** By about how many million square kilometers did the minimum Arctic sea ice extent decline, overall, between 1980 and 2015? (Compiled by the authors using data from U.S. Goddard Space Flight Center, National Aeronautics and Space Administration, National Snow and Ice Data Center, and Polar Science Center.)

INDIVIDUALS MATTER 19.1

James Balog: Watching Glaciers Melt

James Balog/Extreme Ice Survey/Earth Vision Institute

James Balog has applied his love of science and his passion for exploring nature as a world-renowned nature photographer. He has used his photography skills and creativity to chronicle major environmental problems such as losses in biodiversity, threats to North America's old-growth forests, and climate change as revealed by melting glaciers.

Balog is deeply concerned by the fact that many of the world's magnificent glaciers are melting. In 2007, he began working on his Extreme Ice Survey, the world's most wide-ranging photographic study of the rapid melting of many of the world's glaciers. To capture images of glacial melting, Balog developed camera systems to take daylight photos of 22 glaciers in locations around the world every half hour or hour. Each camera could capture up to 8,000 images a year. Balog used the images to create stunning time-lapse videos of the melting glaciers, where huge volumes of ice can disappear in hours or days. These images provided important data to glaciologists and other scientists and clearly and dramatically showed the general public that climate change is having a major impact now.

Balog and his team endured severe hardships to record these images in some of the planet's harshest environments. They had to develop camera systems that could withstand temperatures down to minus 40°F, blistering high winds, rain, sleet, and deep snow. To install the cameras, the team often worked under such conditions, climbing difficult and treacherous mountains of ice while carrying the 125-pound camera systems.

Balog's work has been featured in *National Geographic* magazine, in the NOVA documentary *Extreme Ice*, and in *Chasing Ice*, an internationally acclaimed and award-winning full-length film. His book *Ice: Portraits of the World's Vanishing Ice* summarizes the astounding loss of ice from various glaciers through 2012.

Balog's nonprofit *Earth Vision Trust* has a mission to spread the visual message of climate change, to finance studies of melting glaciers, and to spur people to action. He says, "Seeing is believing. I was a climate change skeptic until I saw the evidence in the ice. Climate change is real and the time to act is now."

---

summer Arctic sea ice. If the current trend continues, summer floating Arctic ice may be gone by 2050, according to the 2014 IPCC report. This could lead to dramatic and long-lasting changes in weather and climate that could affect the whole planet.

According to one hypothesis, this includes colder and snowier winters in Europe, central and eastern North America, and eastern Asia. This is because, with less of a temperature difference between Arctic and temperate regions, prevailing westerly winds tend to weaken. This can slow the Northern Hemisphere jet stream, which then can sink southward, bringing frigid air with it. Recent studies, including a 2014 study led by climate modeler James Screen, have supported this hypothesis. The study reported a strong correlation between such jet stream changes and heat waves and drought in the western United States between 1979 and 2012. Jet stream changes also correlated with extreme cold in the eastern half of the country during those years.

Another effect of arctic warming is faster melting of polar land-based ice, including that in Greenland (Figure 19.11 and **Core Case Study**). In 2016, climate scientist William Colgan estimated that each second about 7,300 metric ton (8,000 tons) of Greenland's land-based ice sheet melts and flows into the sea and contributes to the world's rising sea level. Glaciologist and National Geographic Explorer Erin Pettit has pioneered the use of underwater listening devices to record what is happening at the noisy boundary between disintegrating glacial ice shelves and the seas in Alaska and in western Antarctica. Nature photographer James Balog (Individuals Matter 19.1) has created a visual record of dramatic changes to glaciers around the world. The combination if Balog's videos and Pettit's recordings vividly tell the story of glacial melting.

Another great storehouse of ice is the earth's mountain glaciers. During the past 25 years, many of these glaciers have been shrinking wherever summer melting exceeds the winter *snowpack*—the addition of ice from precipitation in winter. For example, Glacier National Park (Figure 19.3) in the U.S. state of Montana once had 150 glaciers, but by 2015, only 25 remained.

Mountain glaciers play a vital role in the water cycle (see Figure 3.19, p. 63) by storing water as ice during cold seasons and releasing it slowly to streams during warmer seasons. A prime example is the glaciers of the Himalayan Mountains in Asia. They are a major source of water for large rivers such as the Ganges that provide water for more than 400 million people in India and Bangladesh. These waters also feed China's Yangtze and Yellow Rivers, whose basins are home to more than 500 million people.

One of the largest-ever studies of Asian glaciers was conducted by the World Glacier Monitoring Service and reported in 2015. The scientists used field measurements, satellite data (see Science Focus 13.1, p. 331), and computer models. They found that Central Asia's Tien Shan Mountains, stretching 2,400 kilometers (1,500 miles)—the

© Konrad Steffen University of Colorado/CIRES

FIGURE 19.11 Accelerating melting of Greenland's ice has created flows such as this one, called *moulins*, which run from the surface to the base of the glaciers, leading to more dramatic melting and movement of the glacier toward the sea.

distance from Chicago to Los Angeles—had lost 25% of their ice mass since 1965. The report concluded that this, along with data from other areas of the world, represents a "historically unprecedented" decline in glaciers worldwide.

About 80% of the mountain glaciers in South America's Andes range are slowly shrinking. If this continues, 53 million people in Bolivia, Peru, and Ecuador who rely on meltwater from the glaciers for irrigation and hydropower could face severe water, power, and food shortages. In the United States, according to climate models, people living in the Columbia, Sacramento, and Colorado River basins face similar threats as the winter snowpack that feeds these rivers is projected to shrink by as much as 70% by 2050.

Methane Emissions from Permafrost and Tropical Wetlands

Permafrost occurs in soils found beneath about 25% of the exposed land in Alaska, Canada, and Siberia in the northern hemisphere. Human-caused climate change is projected to thaw significant amounts of permafrost. This thaw is already happening in parts of Alaska and Siberia. If this trend continues, a great deal of organic material found below the permafrost will rot and release large amounts of CH_4 and CO_2 into the atmosphere.

Scientists estimate that arctic permafrost holds two to four times as much carbon as all the carbon ever released by humans. If just 5–10% of the total amount of methane stored in this permafrost were released, it would accelerate atmospheric warming, which would in turn melt more permafrost in yet another positive feedback loop that could lead to a climate change tipping point (Figure 19.9).

Mark Thiessen/National Geographic Creative

FIGURE 19.12 Scientists ignite a large bubble of methane gas released from an arctic lake in Alaska.

Some scientists are concerned about another methane source—a layer of permafrost on the Arctic Sea floor. A study led by Natalia Shakhova found areas of the seafloor to be leaking bubbles of methane. Also, aquatic ecologist and National Geographic Explorer Katey Walter Anthony has found methane bubbling up from many arctic lake bottoms (Figure 19.12).

In warmer parts of the world, yet another possible source of increasing methane emissions is tropical wetlands. If rainfall increases as projected in tropical areas, these wetlands will expand and produce more plants, which will eventually decay and release methane through anaerobic decomposition. Some scientists refer to such sources as "methane time bombs."

Sea Levels Are Rising

In 2014, the IPCC estimated that the average global sea level is likely to rise by 40–60 centimeters (1.3–2.0 feet) by the end of this century—about 10 times the rise that occurred in the 20th century. A 2016 study by a team of scientists lead by Benjamin Strauss at Climate Central, using a larger set of data, estimated that the ocean could rise by 1.1–1.2 meters (3–4 feet) by 2100. Between 50% and 66% of this rise will likely come from the melting of Greenland's ice (**Core Case Study**). However, accelerated melting could lead to seas rising by as much as 0.9–2.0 meters (3–7 feet), depending on how much of the land-based ice in Greenland, and West Antarctica, melt as the global temperature continues to rise. Figure 19.13 shows past changes in the world's average sea level and varying projections for the continuing sea-level rise.

Sea-level rise will likely not be uniform around the world, based on factors such as ocean currents and winds. By 2100, for example, Bangladesh's sea level could rise by as much as 4 meters (13 feet), much higher than the projected global average sea-level rise. Also, according to the U.S. Geological Survey, the sea level is rising as much as 4 times faster than the global average along parts of the U.S. Atlantic coast.

According to the 2014 IPCC and NAS reports on climate change, a 1-meter (3-foot) rise in sea level during this century (excluding the additional effects of the resulting higher storm surges) could cause the following serious effects:

- Degradation or destruction of at least one-third of the world's coastal estuaries, wetlands, coral reefs, and deltas where much of the world's rice is grown.
- Disruption of the world's coastal fisheries.
- Flooding in large areas of low-lying countries such Bangladesh, one of the world's poorest and most densely populated nations.
- Flooding and erosion of low-lying barrier islands and gently sloping coastlines, especially in U.S. coastal states such as Florida (Figure 19.14), Texas, Louisiana, New Jersey, South Carolina, and North Carolina.
- Submersion of low-lying islands in the Indian Ocean (Figure 19.15), the Pacific Ocean, and the Caribbean Sea, which are home to 1 of every 20 of the world's people.
- Flooding of some of the world's largest coastal cities, including Kolkata (India), Mumbai (India), Dhaka (Bangladesh), Guangzhou (China), Ho Chi Minh City (Vietnam), Shanghai (China), Bangkok (Thailand), Rangoon (Myanmar), Haiphong (Vietnam), and the U.S. cities of Miami, Florida, and New York City (see red areas in chapter-opening photo and Figure 19.14). This could displace at least 150 million people—an amount equal to nearly half of the current U.S. population.
- Saltwater contamination of freshwater coastal aquifers resulting in degraded supplies of groundwater used as a source of water for drinking and irrigation.

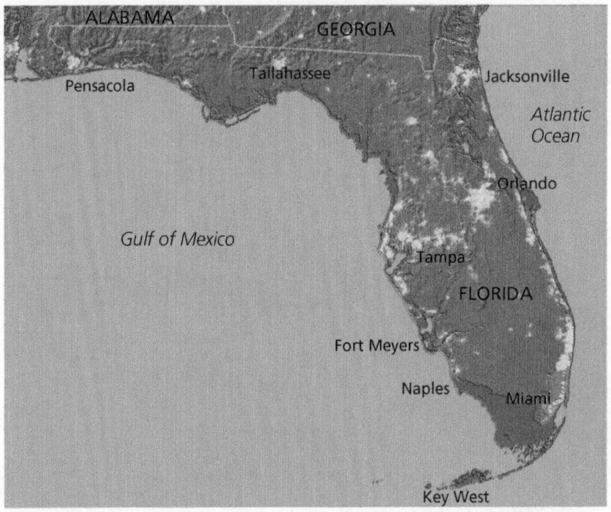

FIGURE 19.14 If the average sea level rises by 1 meter (3.3 feet), the areas shown here in red in the U.S. state of Florida will be flooded.

(Compiled by the authors using data from Jonathan Overpeck, Jeremy Weiss, and the U.S. Geological Survey.)

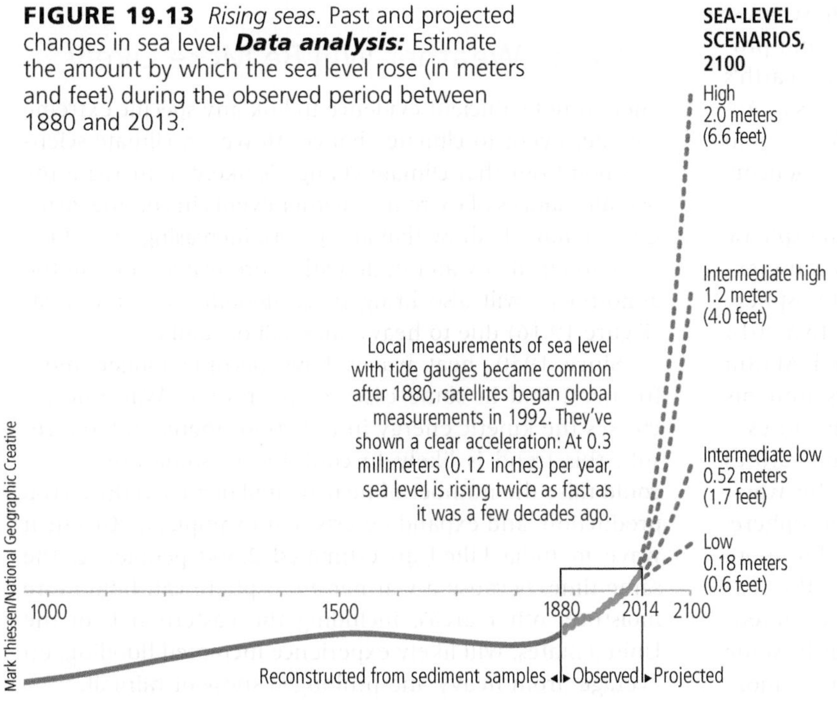

FIGURE 19.13 *Rising seas.* Past and projected changes in sea level. ***Data analysis:*** Estimate the amount by which the sea level rose (in meters and feet) during the observed period between 1880 and 2013.

The Oceans Are Becoming More Acidic

According to reports by the United Kingdom's Royal Society and U.S. National Academy of Sciences, rising levels of CO_2 in the ocean have increased the acidity of its surface waters by 30% since about 1800, and ocean acidity could reach dangerous levels before 2050. This is because much of the CO_2 absorbed by the ocean reacts with water to produce carbonic acid (H_2CO_3)—the same weak acid found in carbonated drinks—in a process called **ocean acidification** (see Science Focus 11.1, p. 257).

Scientists warn that this higher acidity threatens corals, snails, oysters, and other organisms with shells and body structures composed of calcium carbonate, because it hinders their ability to build and repair such structures. Oceanographer Carl Safina notes that acidification will slow the growth and repair

FIGURE 19.15 For a low-lying island nation like the Maldives in the Indian Ocean, even a small rise in sea level could spell disaster for most of its 295,000 people. About 80% of the 1,192 small islands making up this country lie less than 1 meter (3.2 feet) above sea level. Rising sea levels and higher storm surges during this century could flood most of these islands and their coral reefs.

Malbert/Dreamstime.com

of damaged coral reefs and could begin to dissolve some reefs by the end of the century.

Another problem is that increased acidity has been shown to decrease populations of phytoplankton, which remove CO_2 from the atmosphere. Phytoplankton are the primary producer species of ocean food webs, which include the human seafood supply. These food webs could be severely degraded and might possibly even collapse if these producer organisms decline dramatically.

Severe Drought Could Become More Common

Drought occurs when evaporation due to higher temperatures exceeds precipitation for a prolonged period. A study by National Center for Atmospheric Research scientist Aiguo Dai and his colleagues found that severe and prolonged drought has affected at least 30% of the earth's land (excluding Antarctica)—an area the size of Asia. According to a study by climate researchers at NASA, up to 45% of the world's land area could be experiencing extreme drought by 2059.

It is difficult to tie a specific drought to atmospheric warming because natural cyclical processes also cause extreme droughts. However, extra heat in the atmosphere causes water to evaporate from soils. According to a 2014 study by climate scientists Richard Seager and Martin Hoerling, this depletion of soil moisture prolongs droughts and makes them more severe, regardless of their causes.

Some scientists hypothesize that higher temperatures due to atmospheric warming are intensifying the water cycle and adding more water vapor to the atmosphere. This would increase the warming effect and lead to more evaporation followed by greater warming in another example of a positive feedback loop. One effect of this intensified water cycle, according to scientists, is that in some areas droughts will last much longer and will be more widespread. For example, according to a 2015 Columbia University study, drought in the U.S. Southwest and Great Plains during the latter half of this century is projected to be worse than any in recorded history.

As droughts have gotten worse, the growth of trees and other plants in some areas has declined, which has reduced the removal of CO_2 from the atmosphere. Soils in such areas have dried out, making surface temperatures hotter. As some forests and grasslands are also drying out, wildfires are becoming more frequent in some areas, and fires add CO_2 to the atmosphere. Some streams and other surface waters are drying up. In some areas, water tables are falling, partly because farmers are irrigating more to make up for drier conditions.

Climate scientists project that these combined effects from the projected worsening of prolonged droughts are likely to speed up atmospheric warming and lead to more prolonged droughts, drier conditions, and more fires. This is yet another example of a positive feedback or runaway loop possibly leading to one or more climate change tipping points (Figure 19.9).

Extreme Weather Will Be More Likely

There is not sufficient evidence to link any specific extreme weather event to climate change. However, climate scientists point out that climate change is likely to increase the overall chances of extreme weather events in coming years. Climate models show that along with increasing drought, a more intensified water cycle with more water vapor in the atmosphere will also bring more flooding to some areas (Figure 19.16) due to heavy snowfall or rainfall.

Since 1950, heat waves have become longer, more frequent, and in some cases more intense. Warming increases the kinetic energy in the atmosphere, and as a result, this trend is likely to continue in some areas. This could raise the number of heat-related deaths, reduce crop production, and expand deserts. For example, a 2015 heat wave in India killed an estimated 2,500 people. At the same time, because a warmer atmosphere can hold more moisture, other areas, including the eastern half of the United States, will likely experience increased flooding, on average, from heavy and prolonged snow or rainfall.

Palash Khan/Age Fotostock

FIGURE 19.16 An elderly woman seeks a bag of rice in Dhaka, Bangladesh, after a 2007 flood that killed at least 198 people and drove 19 million people from their homes.

It is likely that atmospheric warming is changing global weather cycles, according to a 2014 study led by Wenju Cai. It concluded that this warming could cause extreme versions of El Niño (see Figures 7.3 and 7.4, pp. 142 and 143) and La Niña to occur nearly twice as often, or around once every 10 years. The Pacific Ocean weather pattern called La Niña tends to follow El Niño, and severe versions of each have caused major damages and numerous deaths in the past. The most recent severe El Niño in the mid-1990s killed 23,000 people and did more than $33 billion in damage. The La Niña that followed was blamed for severe flooding in Southeast Asia that displaced more than 200 million people, killing thousands.

Data from the U.S. Global Change Research Program indicate that the annual number of days of extreme heat in parts of the United States will rise dramatically by the end of the century. However, in some areas, atmospheric warming will likely lead to colder winter weather, according to climate models, largely because of changes in global air circulation patterns due to the warming. And although there may be less snow in some areas, there could be more blizzards with larger amounts of snow in shorter periods of time.

There is uncertainty among scientists over whether projected atmospheric warming will increase the frequency and intensity of tropical storms and hurricanes. In 2010, a World Meteorological Organization panel of experts concluded that projected atmospheric warming is likely to lead to fewer but stronger hurricanes and typhoons that could cause more damage in coastal areas where urban populations have grown rapidly.

Climate Change Threatens Biodiversity

According to the latest IPCC reports, projected climate change is likely to alter ecosystems and take a toll on biodiversity in areas of every continent. Up to 85% of the Amazon rain forest—one of the world's major centers of biodiversity—could be lost and converted to tropical savanna if the global atmospheric temperature rises by the highest projected amount (Science Focus 19.2), according to a study led by Chris Jones.

According to the IPCC, approximately 30% of the known land-based plant and animal species could disappear if the average global temperature change reaches 1.5–2.5°C (2.7–4.5°F). This percentage could grow to 70% if the temperature change exceeds 3.5°C (6.3°F) (**Concept 19.3**). The hardest hit species will be:

- Cold-climate plant and animals, including the polar bear in the Arctic and penguins in Antarctica
- Species that live at higher elevations
- Species with limited tolerance for temperature change, such as corals
- Species with limited ranges

The primary cause of these extinctions will be loss of habitat. On the other hand, populations of plant and animal species that thrive in warmer climates could increase.

Research indicates the most vulnerable ecosystems are coral reefs, polar seas, coastal wetlands, high-elevation forests, and alpine and arctic tundra. Primarily because of drier conditions, forest fires will likely become more frequent and intense in the southeastern and western United States.

Feeding interactions that are crucial to the functioning of ecosystems are also being affected by climate change. For example, caterpillars, which are important food for birds and their young, hatch according to seasonal temperature changes. With warmer winters and springs in some areas, caterpillars are hatching 2 weeks earlier than they did just 30 years ago. However, the timing for most bird migrations is based on day length, not temperature, so now birds are arriving at their breeding grounds too late to catch caterpillars at their peak hatching time. Entire ecosystems are affected by such missed connections, due to climate change.

A warmer climate is boosting populations of insects and fungi that damage trees, especially in areas where winters are no longer cold enough to control their populations. In the Canadian province of British Columbia, for example, warmer winters have led to surges in mountain pine beetle populations that are killing huge areas of lodge pole pine forests (Figure 19.17, inset photo). Scientists expect the North American range for these trees to shift dramatically during the next several decades (Figure 19.17, map).

In a 2014 study, the Union of Concerned Scientists reported that an estimated 58–90% of all Rocky Mountain forests could die due to effects of climate change, including extreme heat, drought, expanding wildfires, and beetle infestations. The scientists reported that between 2000 and 2012, bark beetles killed trees on 18.6 million hectares (46 million acres)—an

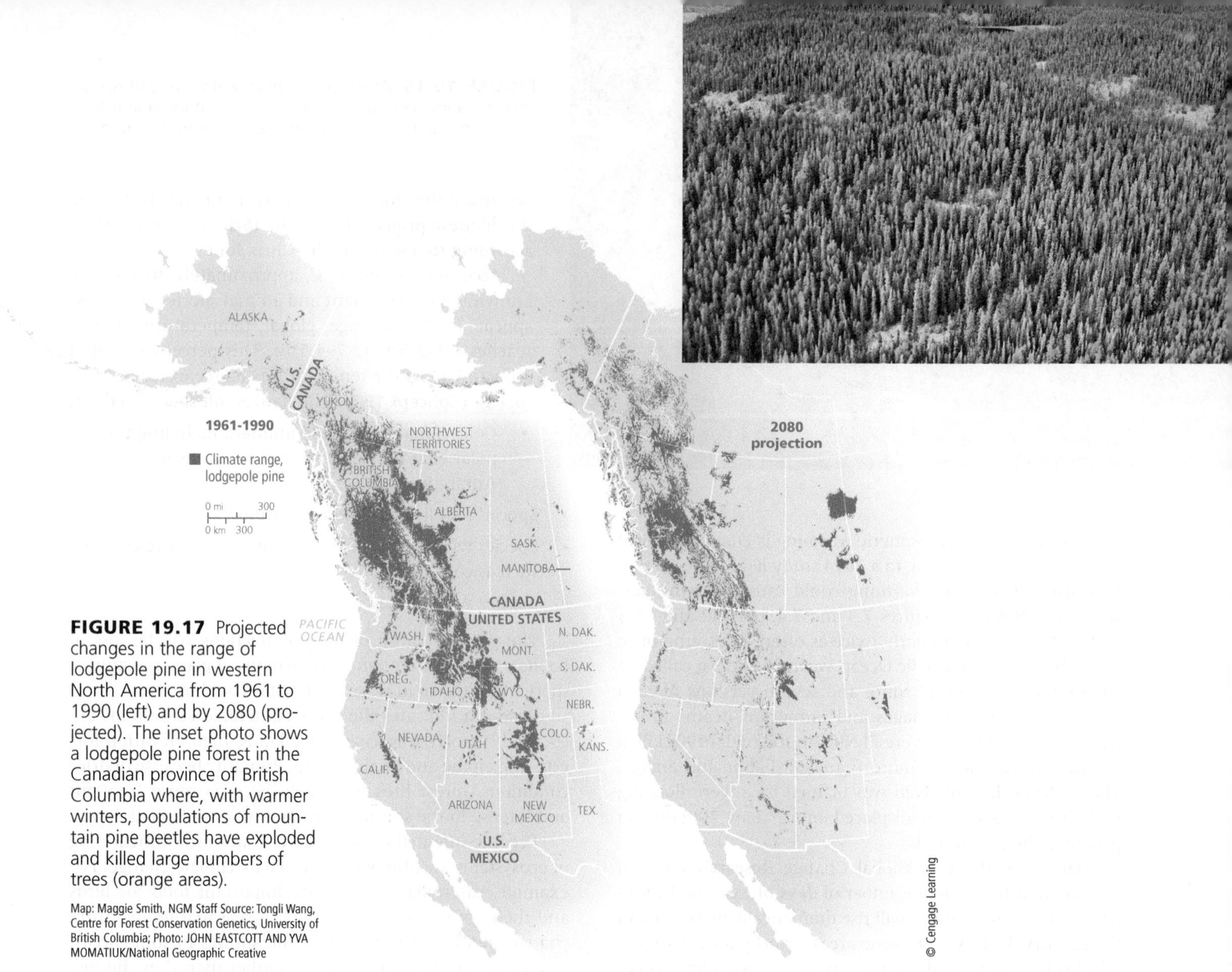

FIGURE 19.17 Projected changes in the range of lodgepole pine in western North America from 1961 to 1990 (left) and by 2080 (projected). The inset photo shows a lodgepole pine forest in the Canadian province of British Columbia where, with warmer winters, populations of mountain pine beetles have exploded and killed large numbers of trees (orange areas).

Map: Maggie Smith, NGM Staff Source: Tongli Wang, Centre for Forest Conservation Genetics, University of British Columbia; Photo: JOHN EASTCOTT AND YVA MOMATIUK/National Geographic Creative

area the size of Colorado. They also found that since the 1980s, the number of large Rocky Mountain wildfires has risen 73%, with an average of 18 more large fires each year. Such fires can severely degrade forest ecosystems. They also add more CO_2 to the atmosphere while reducing total CO_2 uptake by trees and plants. This helps to accelerate climate change in still another positive feedback loop.

Another 2014 study led by atmospheric scientist David Romps found that warmer temperatures would likely increase the frequency of lightning strikes across the United States by an estimated 50% during this century. Lightning is one of the main causes of wildfires.

Marine ecosystems are also threatened by climate change. Coral reefs are especially vulnerable because of their limited tolerance for water temperature changes. Many are already dying off for this reason. Because coral reefs are so rich in biodiversity, such losses will reshuffle ocean ecosystems on a scale not seen for millions of years, according to marine biologists.

CASE STUDY

Alaska—Bellwether of Climate Change

Because the arctic state of Alaska is a showcase for most of the effects of climate change, it has been called a *bellwether*—a leading indicator of trends. According to NOAA data, Alaska is warming at twice the average rate of the rest of the United States, and its warming is accelerating.

Some scientists attribute this warming to weather patterns, but others note that the patterns have existed for so long—for 30 to 50 years—that they should be regarded as climate change. Most notably, a persistent ridge of high pressure stretches from Alaska to California and it carries warm, moist air from the south to the Arctic as part of the jet stream during winter. This air then cools and descends back to the south, plunging the eastern United States into extra-cold winters, while Alaska gets warmer.

The effects of this warming are dramatic. In 2015, all but five of Alaska's glaciers were shrinking, according to Regine

Hock, a glacier expert from the University of Alaska. The state's rivers, fed by melting glaciers, are now carrying more water than the Mississippi River. Coastal sea ice is also disappearing 2 weeks earlier in the summer than it was in the 1970s and refreezing 2 weeks later in the fall, according to Andy Mahoney of the University of Alaska. This has shortened the hunting season by at least a month for walruses, polar bears, and human hunters who depend on this ice for finding prey.

This sea ice also helps to protect coastal villages. As it shrinks away, many villages suffer more flooding and damage from storms. This, along with the fact that sea levels are rising—another effect of climate change—has forced the state to plan for relocating several coastal villages further inland on higher ground.

Permafrost covers 85% of Alaska, and the layer of permafrost soil that thaws every summer is deepening. This is causing roads and buildings to shift and sink, resulting in the need for costly repairs. Some trees rooted for decades in frozen soil are becoming unstable and leaning or falling. Alaskans refer to them as "drunken trees." Lakes and wetlands have drained in some areas where permafrost has thawed. These changes are forcing some species of birds, fish, mammals, and trees to shift to new habitat or to face possible extinction.

Certain species of insects are finding the warmer Alaska more hospitable. One such insect is the spruce bark beetle. Its populations exploded in 2006 and killed mature white spruce forests on more than 1.8 million hectares (4.4 million acres) of land. This too has led to harmful changes in a major ecosystem.

Warmer air has made forests drier and this, along with insect damage, is thought to be a contributor to increasing wildfire damage in Alaska. Between 1959 and 1970, the average annual area of land burned has grown nearly fivefold.

Most of these effects are feeding positive feedback loops (see Figure 2.15, p. 43), which are contributing to faster warming. You can take any one of the effects listed above and imagine how it could become part of such a loop.

Threats to Food Production

Farmers will face dramatic changes due to shifting climates and an intensified hydrologic cycle, if the atmosphere keeps warming as projected. According to the IPCC, crop productivity is projected to increase slightly with moderate warming at middle to high latitudes in areas such as midwestern Canada, Russia, and Ukraine. The projected rise in crop productivity might be limited because the soils in these northern regions generally lack sufficient plant nutrients. Crop production will decrease if warming goes too far.

Higher temperatures and the erratic pattern of increased extended droughts and heavier downpours could reduce crop yields in some farming areas. Shrinking irrigation water supplies and growing pest populations could also cut crop yields in some areas.

Climate change models project a decline in agricultural productivity and food security in tropical and subtropical regions, especially in Southeast Asia and Central America, largely because of excessive heat. Crop scientists have determined a rule of thumb that for each 1C° (1.8F°) rise in temperature, crop yields for wheat, corn, and rice decline by about 10%. This is primarily because heat can interfere with photosynthesis. The total area of farmable land that would be affected in this way is estimated to be greater than the area that would benefit from warmer temperatures. As a result, the International Food Policy Research Institute projects that by 2050, climate change will lead to global declines in yields of corn (down 24%), rice (11%), potatoes (9%), and wheat (3%).

Meat and dairy production will face challenges in a warmer world. Cattle, pigs, and chickens are not heat-tolerant animals. Pork, beef, poultry, and milk production are all projected to decline. Warmer winters will affect production of fruits such as apples and cranberries that need long winter chilling periods. Warm winters will also benefit crop diseases, insect pests, and weeds that will further threaten many crops.

Flooding of river deltas in coastal areas by rising sea levels could reduce crop production, partly because aquifers that supply irrigation water will be infiltrated by saltwater. This flooding would also affect fish production in coastal aquaculture ponds. Food production in farm regions dependent on rivers fed by melting glaciers will drop. In arid and semiarid areas, food production will be lowered by more intensive and prolonged droughts.

Areas especially vulnerable to these effects will be the populous nations of Bangladesh, Egypt, and Vietnam, and parts of the African coast, according to a World Bank report. Scientists warn that by 2050, some 200–600 million of the world's poorest people could face starvation and malnutrition due to these projected effects of climate change.

According to the IPCC, food is likely to be plentiful for a while in a warmer world, because of the longer growing season in northern regions. But scientists warn that during the latter half of this century, several hundred million of the world's poorest and most vulnerable people could face starvation and malnutrition from a drop in food production caused by projected climate change.

Threats to Human Health, National Security, and Economies

According to IPCC and other reports, more frequent and prolonged heat waves in some areas will increase deaths and illnesses, especially among older people, people in poor health, and the urban poor who cannot afford air conditioning. It is likely that fewer people will die from cold weather. But research at Harvard University School of Public Health suggests that during the latter half of this century, the projected rise in the number of heat-related deaths will likely exceed the projected drop in the number of cold-related deaths.

A warmer and more CO_2-rich atmosphere will likely favor rapidly multiplying insects, including mosquitoes and ticks that transmit diseases such as West Nile virus, Lyme disease, and dengue fever. According to the World Health Organization (WHO), incidence of dengue fever has increased 10-fold since 1973, with 390 million people infected with the disease in 2013. Warming will also favor microbes, toxic molds, and fungi, as well as some plants that produce pollens that cause allergies and asthma attacks.

10-fold

Amount of increase in cases of dengue fever since 1973

Higher atmospheric temperatures and levels of water vapor will contribute to increased photochemical smog in many urban areas. This will cause more pollution-related deaths and illnesses from heart ailments and respiratory problems.

Recent studies by the U.S. Department of Defense and the National Academy of Sciences warn that climate change could affect U.S. national security. These effects include increased food and water scarcity, poverty, environmental degradation, unemployment, social unrest, mass migration of environmental refugees, and political instability, and the weakening of fragile governments. All of these factors could increase terrorism, according to the U.S. Defense Department, which is including climate change as a major element in its planning.

Climate change will also take a toll on human economies. According to a 2015 survey of 750 risk experts conducted by the World Economic Forum, failure to mitigate and adapt to climate change tops the list of threats to the global economy and the risks are increasing. Various economists estimate that the cost of climate change, if current trends continue, would be 5–20% per year of the world's economic activity as measured by gross domestic output. Other researchers say the cost could be higher, because most models used to calculate such costs have not adequately accounted for economic losses from projected ocean acidification, rising sea levels, and biodiversity losses. For example, the area likely to be flooded along the Gulf of Mexico by a 1-meter (39-inch) rise in gulf waters is home to 90% of U.S. offshore energy production, 30% of the total oil and gas supply, a port that serves 31 states, and 2 million people.

CONSIDER THIS . . .

CONNECTION Climate Change and Electric Power Outages

In France, a 2003 heat wave raised the temperatures of river water used to cool a nuclear power reactor and the station had to shut down. Other rivers in Europe lost so much water that hydroelectric stations were forced to cut back on power production. In 2006, a heat wave in California caused outages when the high demand from air conditioners overloaded power lines and transformers. Some 2 million Californians lost power when they needed it the most.

For these reasons, climate change will very likely force tens of millions of people to migrate in search of better conditions. They will be forced to leave their homes because of greater hunger, flooding, drought, and other problems.

19.4 HOW CAN WE SLOW CLIMATE CHANGE?

CONCEPT 19.4 We can reduce greenhouse gas emissions and the threat of climate change while saving money and improving human health if we cut energy waste and rely more on cleaner renewable energy resources.

Dealing with Climate Is Difficult

The good news is, we can cut carbon emissions. The International Energy Agency found that global CO_2 emissions from energy production did not grow in 2014. This was the first time in 40 years that energy-related CO_2 emissions were flat or reduced while the economy still grew.

This is largely due to the drop in use of coal to generate electricity. According to Michael Brune of the Sierra Club, since 2010, one coal plant has been retired or scheduled for retirement every 10 days on average. At the same time, wind, solar, and geothermal power production grew. In China, coal consumption dropped for the first time in 2014. Its CO_2 emissions dropped by 1% while its economy grew by 7.4%.

However, these 2014 results alone do not show a trend. Overall, greenhouse gas emissions from all sources are still rising. As a result, many climate scientists and other analysts believe that climate change is one of the most urgent scientific, political, economic, and ethical issues that humanity faces. However, the following characteristics of this complex problem make it difficult to tackle:

- *The problem is global.* Dealing with this threat will require unprecedented and prolonged international cooperation.
- *The problem is a long-term political issue.* Climate change is happening now and is already having harmful impacts, but it is not viewed as an urgent problem by most voters and elected officials In addition, most of the people who will suffer the most serious harm from projected climate change during the latter half of this century have not been born yet.
- *The harmful and beneficial impacts of climate change are not spread evenly.* Higher-latitude nations such as Canada, Russia, and New Zealand may temporarily have higher crop yields, fewer deaths in winter, and lower heating bills. But other, mostly poor nations such as Bangladesh could see more flooding and higher death tolls.
- *Proposed solutions, such as sharply reducing or phasing out the use of fossil fuels, are controversial.* They could disrupt

economies and lifestyles and threaten the profits of economically and politically powerful fossil fuel and utility companies.

- *The projected effects are uncertain.* Current climate models lead to a wide range in the projected temperature increase (Science Focus 19.2) and sea-level rise (Figure 19.13). Thus, there is uncertainty over whether the harmful changes will be moderate or catastrophic. This makes it difficult to plan for avoiding or managing risk. It also highlights the urgent need for more scientific research to reduce the uncertainty in climate models.

What Are Our Options?

There are two basic approaches to dealing with the projected harmful effects of global climate change. One, called *mitigation,* is to slow down climate change to avoid its most harmful effects. The other approach, called *adaptation,* is to recognize that some climate change is unavoidable because we have waited too long to act and that people have to adapt to some of its harmful effects.

Regardless of the approach taken, most climate scientists argue the most urgent priority is to avoid any and all **climate change tipping points** (Figure 19.9). Tipping points are the estimated thresholds beyond which natural systems could not be reversed. Each of these tipping points accelerates the changes in climate. For example, if we continue to add CO_2 to the atmosphere at the current rate, we will likely exceed the estimated tipping point marked by 450 ppm of atmospheric CO_2 within a few decades. Such high levels of CO_2 can put the earth's climate system into a positive feedback loop that could become irreversible and lock in severe climate change for hundreds or perhaps thousands of years. It also would increase the likelihood of exceeding many of the other tipping points shown in Figure 19.9.

Many business leaders see climate change as a threat to the global economy, as well as an enormous global investment opportunity (**Concept 19.4**). For example, in the United States, electric cars are being made in Michigan, and Texas produces more electricity from wind power than all but five of the world's countries. Other countries—notably China, Germany, and Denmark—are building their renewable energy capacities.

Reducing Greenhouse Gas Emissions

Climate scientists generally agree that to avoid some of the severest harmful effects of climate change, we need to limit the global average temperature increase to 2C° (3.6F°) over the preindustrial global average temperature. This means reducing CO_2 emissions by reducing our use of fossil fuels (especially coal), greatly increasing energy efficiency, and shifting to much greater dependence on cleaner and low-carbon energy resources such as solar power, wind power, and geothermal energy.

The problem is that the world's nations and energy companies together hold reserves of fossil fuels that, if burned, would emit nearly 5 times the amount of CO_2 that climate scientists say we can safely emit. Switching away from fossil fuels to avoid exceeding this temperature climate tipping point means leaving 82% of the world's coal reserves and 50% of all natural gas and all arctic oil reserves in the ground, according to a 2014 study led by researchers at University College of London. Currently, the economic well-being of the politically powerful fossil-fuel companies and most of the world's national economies depend on using all or most of these reserves of fossil fuels. This is what drives the intense political and economic debate over how to slow the rate of climate change.

A growing number of scientists and other analysts recognize that shifting away from human dependence on fossil fuels will be difficult but contend that it can be done over the next 50 years. They point out that humans have shifted energy resources before—first from wood to coal, then to oil, and now to natural gas. Each shift took about 50 years. Humans now have the knowledge and ability to shift to a reliance on energy efficiency and renewable energy over the next 50 years. What we need is the political and ethical will to commit to and complete such a shift.

Figure 19.18 lists ways to slow climate change caused by human activities over the next 50 years (**Concept 19.4**). The items in the left column are prevention approaches designed to reduce CO_2 emissions and those in the right column are cleanup approaches for removing CO_2 from the atmosphere.

Some scientists contend that a good starting place for cutting greenhouse gas emissions would be to focus on decreasing black carbon (a component of soot) and methane (CH_4) emissions. Methane is short-lived in the atmosphere compared to CO_2 and the technologies exist to accomplish such cuts fairly quickly. Although CH_4 emissions account for 9% of U.S. greenhouse gas emissions, CH_4 molecules warm the air 25 times more effectively than CO_2 molecules. Major cuts in these greenhouse gas emissions have the potential to cut the rate of atmospheric warming in half by 2050.

According to the 2014 IPCC report, there is good news related to dealing with the threat of climate change by reducing our CO_2 emissions:

- Hybrid, plug-in hybrid, and electric cars are available. People could switch to these cars and charge their batteries only with electricity produced from renewable energy sources.
- The shift to renewable energy is accelerating as prices for electricity from low-carbon wind turbines and solar cells fall, and as investments in these technologies grow.
- Engineers have designed zero-carbon buildings and can reduce the carbon footprints of existing buildings.
- Dealing with climate change will create jobs and profitable businesses. In 2013, the climate change industry was worth $300 billon in the United States and $1.4 trillion worldwide.

Is Capturing and Storing CO_2 a Viable Solution?

Proposals for sequestering and storing carbon are generally referred to as **carbon capture and storage (CCS)**—any of several measures designed to remove CO_2 from the atmosphere or from smokestack gases at coal-burning facilities and to store it in other parts of the environment (Figure 19.D).

One way to do this is to run the smokestack gases through a chemical bath that captures the CO_2. This chemical is heated to release the CO_2, which is then pressurized to convert it to a liquid that is then pumped under pressure deep underground into abandoned coal beds and oil and gas fields (Figure 19.D). In some cases, the CO_2 is used to force oil out of deposits that would be otherwise unrecoverable, and this helps to pay the high cost of this form of CCS. Other ideas involve injecting it into sediments under the seafloor or into deposits of a porous rock called *basalt*, with which it reacts to create carbonate rock.

A problem with this approach is that it would take many years for it to be effective. The International Energy Agency estimates that we would need more than 200 CCS power plants by 2030 in order to keep atmospheric warming below 3C° (5.4F°). As of 2015, just 60 CCS plants were planned or proposed, and only 21 were built or in operation. Between 2010 and 2015, 33 CCS projects had been shut down or canceled worldwide, according to the Global Carbon Capture and Storage Institute. Another problem is that CCS is an expensive fix.

Also, scientists note that stored CO_2 would have to remain sealed from the atmosphere forever. Any large-scale leaks caused by earthquakes or other shocks, as well as any number of smaller continuous leaks from CO_2 storage sites around the world, could dramatically increase atmospheric warming and climate change in a matter of years. A CCS experiment in Algeria had to be halted when the pressurized CO_2 fractured overlying rock. This, along with the demonstrated ability of fracking operations (see Science Focus 15.1, p. 390) to cause small earthquakes, has raised serious concerns about pumping pressurized CO_2 into underground deposits. Pumping CO_2 into ocean floor deposits is problematic, as it could add to ocean acidification.

There are other serious problems. The CCS process requires a great deal of energy, which greatly reduces its net energy by an estimated 25–45%. This nearly doubles the cost of burning coal to supply electricity. For these reasons, it would have to be subsidized by taxpayers in order to be economically viable. In addition, CCS does not deal with the massive emissions of CO_2 from motor vehicle exhausts, food production, and the deliberate and accidental burning of forests.

Other CCS approaches are aimed at removing CO_2 from the atmosphere. One idea is to seed the ocean with iron pellets to boost populations of marine algae and other phytoplankton, which absorb CO_2 from the atmosphere as they grow. When they die, they sink to the seafloor, carrying the carbon with them. The proposal, called *iron enrichment*, is controversial because the long-term ecological effects of adding large amounts of iron to ocean ecosystems are unknown. A study by University of Western Ontario researchers found that iron enrichment could promote the growth of a form of algae known to have caused illness and death in thousands of marine mammals and birds along the west coast of North America.

According to some environmental scientists, most forms of CCS are costly, risky, and not very effective *cleanup* solutions (Figure 19.18, right) to a serious problem that can be dealt with by using a variety of cheaper, quicker, and safer *input*, or *prevention*, approaches (Figure 19.18, left). These scientists call for producing electricity by using

Removing CO_2 from the Atmosphere

Some scientists and engineers are designing cleanup strategies (Figure 19.18, right) for removing some of the CO_2 from the atmosphere or smokestacks and storing (sequestering) it, in other parts of the environment (Science Focus 19.3).

One way to increase the uptake of CO_2 is by implementing a massive, global tree-planting and forest restoration program, especially on degraded land in the tropics. Biologist Renton Righelato and climate scientist Dominick Spracklen have estimated that the amount of carbon we could sequester in this way is greater than the amount of carbon emissions that we would avoid by using biofuels such as biodiesel and ethanol (see Chapter 16, p. 430). A similar approach is to restore wetlands that have been drained for farming. Wetlands are very efficient at taking up CO_2.

A second cleanup approach is to plant large areas of degraded land with fast-growing perennial plants such as switchgrass, which remove CO_2 from the air and store it in the soil. To be sustainable, however, this approach could not involve clearing existing forests.

A third carbon sequestration measure that is getting more attention is the use of *biochar*, which Amazon Basin natives have used for centuries to grow crops. Producing biochar involves burning biomass such as chicken waste or wood in a low-oxygen environment to make a charcoal-like material. This carbon-rich biochar makes an excellent organic fertilizer that helps to keep carbon in the soil. Although making biochar does release CO_2, burying it stores considerably more CO_2 in the soil.

A fourth approach is to remove some of the CO_2 from the smokestacks of coal-burning power and industrial

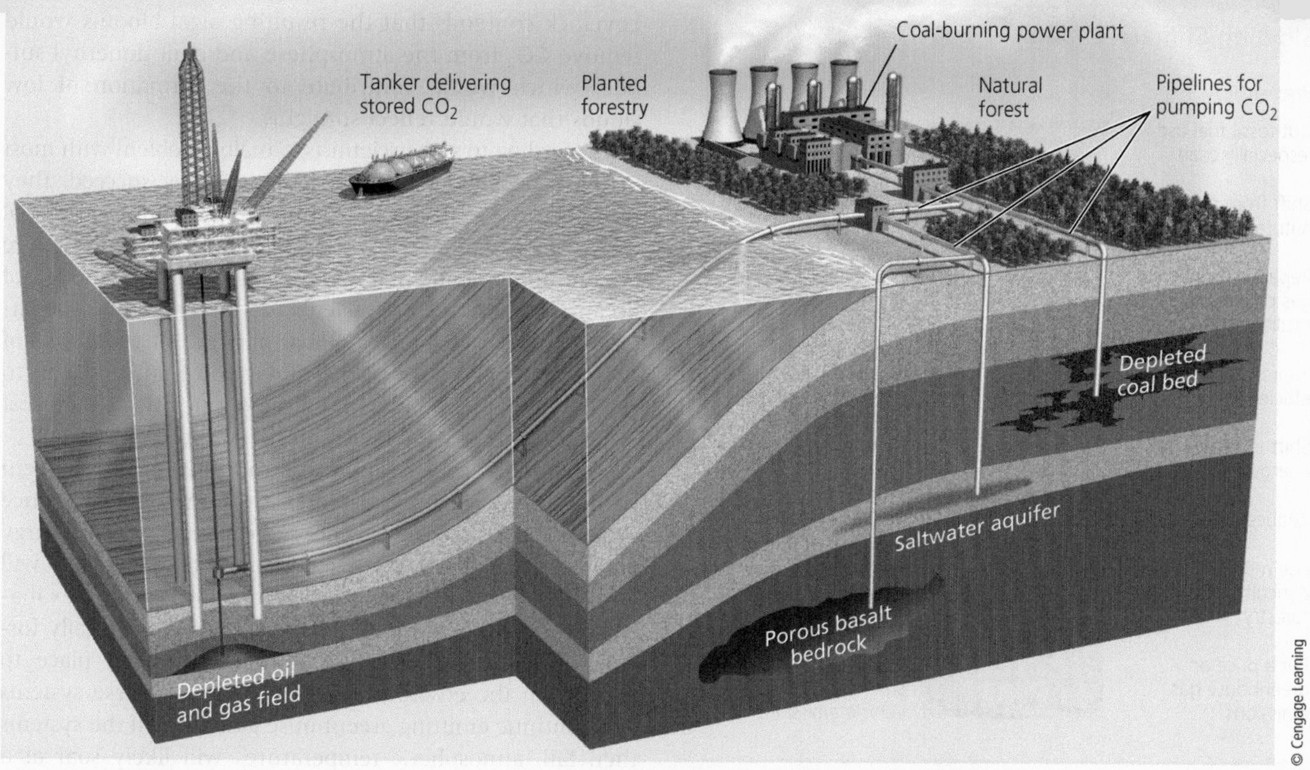

FIGURE 19.D Some proposed carbon capture and storage (CCS) schemes for removing some of the carbon dioxide from smokestack emissions and from the atmosphere and storing (sequestering) it in soil, plants, deep underground reservoirs, and sediments beneath the ocean floor. ***Critical thinking:*** Which of these proposed strategies do you think would work best and which would be the least effective? Why?

low-carbon energy resources such as solar cells and wind power with costs falling dramatically to the point where in the near future they could compete with fossil fuels without the need of government subsidies. To these scientists, when we face a problem such as CO_2 coming out of a smokestack or exhaust pipe, the most important question to ask is not *What do we do with it?* but, *How do we avoid producing the CO_2 in the first place?*

CRITICAL THINKING

Do you think that a government-supported crash program to build CCS coal plants would be a good investment of taxpayer dollars? Explain.

plants and to store it somewhere. This approach is one of several generally referred to as *carbon capture and storage,* or *CCS* (Science Focus 19.3).

CONSIDER THIS . . .

LEARNING FROM NATURE

Earth science professor Brent Constantz developed a CCS method by mimicking sea creatures that convert CO_2 to calcium carbonate to form their skeletons and shells. His process removes CO_2 from power plant smokestack emissions and converts it to cement by spraying it with mineral-rich seawater. Klaus Lackner of Columbia University sought to learn how trees remove CO_2 from the air. He designed an artificial tree—a tower with branches and artificial leaves that can absorb 1,000 times more CO_2 from the air than natural leaves can absorb.

Can Geoengineering Provide Solutions?

Some scientists and engineers want to use strategies that fall under the umbrella of **geoengineering,** or trying to manipulate certain natural conditions to help counter the human-enhanced greenhouse effect (Figure 19.19).

For example, some scientists have studied the 1991 eruption of Mount Pinatubo in the Philippines and the fact that the average global atmospheric temperature cooled for about 15 months after the eruption. The volcano pumped massive amounts of SO_2 into the atmosphere where chemical reactions converted it to sulfate particles. These were thought to have reflected some incoming sunlight into space, thus cooling the troposphere. Some scientists have suggested using balloons, large jet planes, or giant cannons to inject sulfate particles into the stratosphere to get the

Solutions

Slowing Climate Change

Prevention	Cleanup
Cut fossil fuel use (especially coal)	Sequester CO_2 by planting trees and preserving forests and wetlands
Shift from coal to natural gas	Sequester carbon in soil using biochar
Repair leaky natural gas pipelines and facilities	Sequester CO_2 deep underground (with no leaks allowed)
Improve energy efficiency	Sequester CO_2 in the deep ocean (with no leaks allowed)
Shift to renewable energy resources	Remove CO_2 from smokestack and vehicle emissions
Reduce deforestation	
Use more sustainable agriculture and forestry	
Put a price on greenhouse gas emissions	

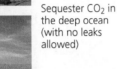

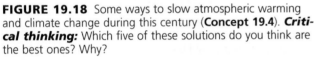

© Cengage Learning

FIGURE 19.18 Some ways to slow atmospheric warming and climate change during this century (**Concept 19.4**). ***Critical thinking:*** Which five of these solutions do you think are the best ones? Why?

Top: Mark Smith/Shutterstock.com. Center: Yegor Korzhi/Shutterstock.com. Bottom: pedrosala/Shutterstock.com.

same effect (Figure 19.19). Other scientists have called for placing a series of giant mirrors in orbit above the earth to reflect incoming sunlight for the same purpose.

Some scientists reject the idea of launching sulfates into the stratosphere as being too risky because of possible unknown effects. For example, if the sulfates reflected too much sunlight, they could reduce evaporation enough to alter global rainfall patterns and worsen the already dangerous droughts in certain areas. Also, a study by atmospheric scientist Simone Tilmes indicated that chlorine released by reactions involving sulfates could speed up the thinning of the earth's stratospheric ozone layer (see Chapter 18, p. 496).

CONSIDER THIS . . .

THINKING ABOUT Tinkering with the Stratosphere

Do you think the projected benefits of injecting large quantities of sulfate particles into the stratosphere to help cool the troposphere would outweigh any projected negative results? Explain.

Another scheme is to deploy a large fleet of computer-controlled ocean-going ships to pray saltwater high into the sky to make clouds whiter and more reflective. Scientist James Lovelock has proposed anchoring large vertical pipes in the oceans to pump nutrient-rich water up from deep down in order to fertilize algae on the ocean surface. Lovelock contends that the resulting algal blooms would remove CO_2 from the atmosphere and emit dimethyl sulfide, which would contribute to the formation of low clouds that would reflect sunlight.

According to some scientists, a major problem with most of these technological fixes is that, if they succeed, they could be used to justify the continued rampant use of fossil fuels. This would allow CO_2 levels in the lower atmosphere to continue building and adding to the serious problem of ocean acidification. Some scientists would deal with the latter problem by building a global network of thousands of chemical plants that would remove acid from seawater to reduce its acidity. No one knows how much this would cost or what effects it would have on ocean ecosystems.

Skeptical scientists argue that another major problem with both geoengineering and CCS proposals (Science Focus 19.3) is that they require huge investments of energy and materials, and there is no guarantee that they will work. Geoengineering schemes all depend on complex machinery running constantly, flawlessly, and essentially forever, primarily to pump something from one place to another in the environment. If we rely on these systems and continue emitting greenhouse gases, and if the systems then fail, atmospheric temperatures will likely soar at a rapid rate and essentially ensure severe climate disruption.

Another problem with geoengineering is the political and ethical aspect. One country or one company could undertake a project that could be harmful to whole regions or to the entire planet. Critics ask how any decision to employ a geoengineering project should be made and by whom. Who will take responsibility for controlling it, and who will pay for any damages? Such questions have yet to be answered, and critics such as climate scientist Ken Caldeira argue that they should be explored before any major projects get started.

Critics of large-scale geoengineering proposals argue instead for slowing climate change by using prevention approaches. These include improving energy productivity, replacing fossil fuels with already available renewable energy resources, and drastically reducing deforestation. They say this would be a better investment than gambling on geoengineering efforts that could have unknown and long-lasting globally harmful effects. They argue that attempts to use geoengineering could seriously delay a shift from relying on fossil fuels to using improved energy productivity and renewable energy resources. Many scientists say we cannot afford such a delay.

Regulating Greenhouse Gases as Pollutants

One approach to pollution prevention is to strictly regulate carbon dioxide and methane as air pollutants. In Chapter 1, we defined a pollutant as a chemical or other agent such as

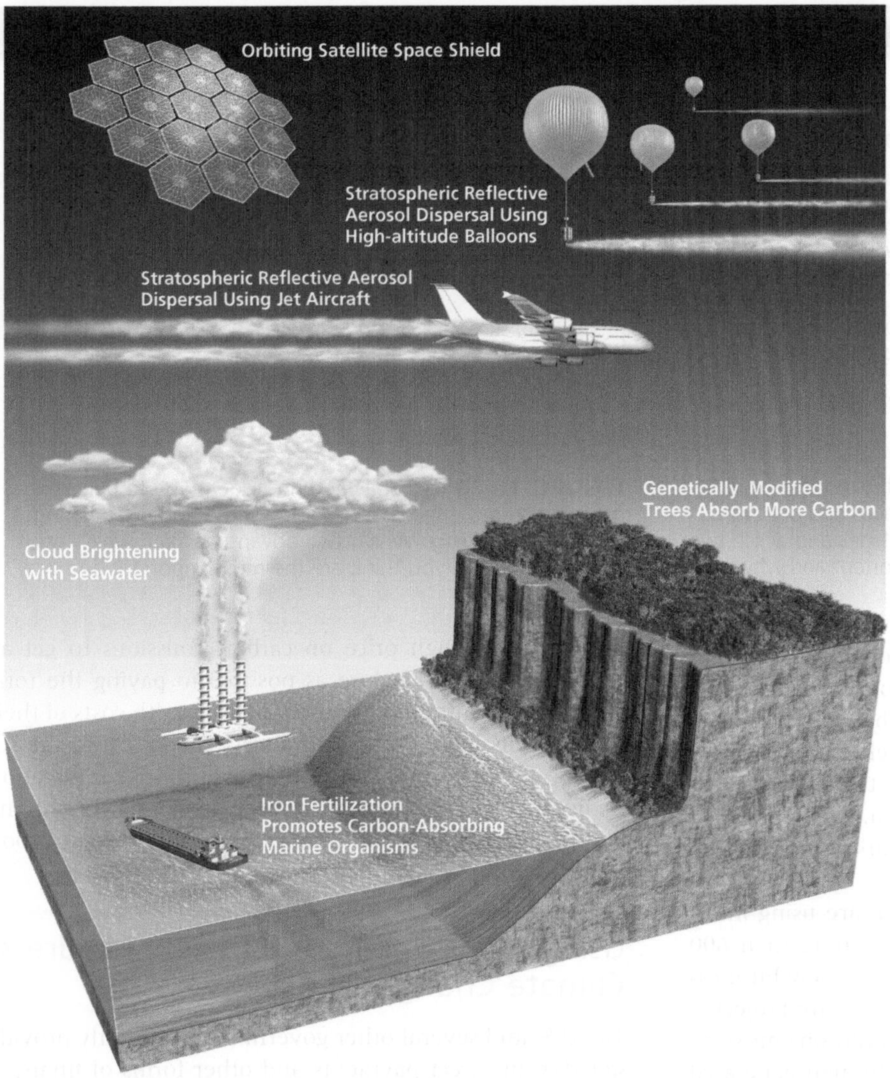

FIGURE 19.19 Geoengineering schemes include ways to reflect more sunlight, and some carbon sequestration approaches (Figure 19.D) can be thought of as geoengineering. ***Critical thinking:*** Of the approaches shown here, which three do you think are the most workable? Why?

noise or heat in the environment that proves harmful to the health, survival, or activities of humans or other organisms.

You might wonder how CO_2, a natural and important part of the carbon cycle (see Figure 3.20, p. 65), could be classified as a pollutant. Humans and most animal life forms constantly emit CO_2 into the atmosphere, while plants remove it and use it for photosynthesis. Trees, for example, consume about 20% of our CO_2 emissions, which is why reducing deforestation and planting trees are important ways to slow climate change. However, one characteristic of any chemical that determines whether it should be classified as a pollutant is its *concentration* in the atmosphere, soil, water, or plant or animal tissues. When we add CO_2 to the atmosphere faster than the carbon cycle can remove it, we help to warm the atmosphere and contribute to climate change. At some point in that process (at some atmospheric concentration), CO_2 becomes an air pollutant.

This is why, in 2009, the U.S. Environmental Protection Agency found that CO_2, CH_4, N_2O, and three other greenhouse gases are a danger to public health and welfare and should be listed as pollutants under the Clean Air Act. However, politically and economically powerful coal, oil, natural gas, and utility companies that make their money by producing and burning fossil fuels have successfully opposed regulating greenhouse gases for more than three decades. These companies are likely to continue their efforts to overturn or greatly weaken government regulations of greenhouse gas emissions.

Putting a Price on Carbon Emissions

Many economists call for not allowing the fossil fuel industry to continue dumping its main waste, carbon dioxide, into the atmosphere for free. They argue that it is important to put a price on carbon emissions as a way to include some of the environmental and health costs of using fossil fuels in fuel prices. This would be in keeping with the full-cost pricing **principle of sustainability.**

One way to do this is to phase in *carbon taxes* or *fees* on each unit of CO_2 or CH_4 emitted by fossil fuel use, or to levy *energy taxes* or *fees* on each unit of fossil fuel that is burned (Figure 19.20). Lowering taxes on income, wages, and profits to offset such taxes could help to make such a strategy more politically acceptable, and it might stimulate the economy. In other words, *tax pollution, not payrolls and profits*.

Around the world, nearly 40 nations, including the 28-member European Union and Canada, as well as British Columbia, California, and nine Northeastern U.S. states, have raised the price of carbon without damaging the economy. British Columbia levied a tax of $30 per metric ton of CO_2 emitted, and then lowered corporate and personal income tax rates, which are now among the lowest of such rates among the world's wealthier nations. In the 4 years after the carbon tax was created, British Columbia's greenhouse gas emissions dropped by 4.5% while its population grew, and gasoline sales dropped by 2% while rising by 5% overall in Canada.

Trade-Offs

Carbon and Energy Taxes

Advantages	Disadvantages
Simple to administer	Tax laws can get complex
Clear price on carbon	Vulnerable to loopholes
Covers all emitters	Doesn't guarantee lower emissions
Predictable revenues	Politically unpopular

FIGURE 19.20 Using carbon and energy taxes or fees to help reduce greenhouse gas emissions has advantages and disadvantages. ***Critical thinking:*** Which two advantages and which two disadvantages do you think are the most important and why?

Trade-Offs

Cap-and-Trade Policies

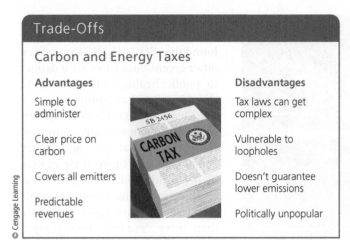

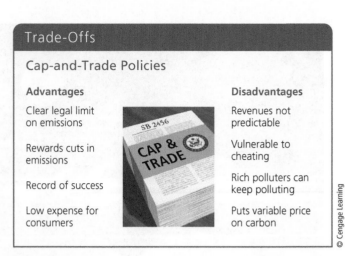

Advantages	Disadvantages
Clear legal limit on emissions	Revenues not predictable
Rewards cuts in emissions	Vulnerable to cheating
Record of success	Rich polluters can keep polluting
Low expense for consumers	Puts variable price on carbon

FIGURE 19.21 Using a cap-and-trade policy to help reduce greenhouse gas emissions has advantages and disadvantages. ***Critical thinking:*** Which two advantages and which two disadvantages do you think are the most important and why?

Economists warn that the biggest global players in climate change—the United States, Europe, and China—would need to tax carbon emissions at basically the same rates in order for any carbon tax to be effective. In 2014, China and 72 other nations and more than 1,000 businesses called for putting a price on carbon. China is testing carbon pricing in seven pilot carbon markets around the country. Globally, more than 150 large corporations, including Google, Microsoft, and Chevron, are using internal carbon pricing in their planning and more than 600 such companies see the carbon market as a new business opportunity, according to the Carbon Disclosure Project.

Another way to include some costs of carbon emissions in market prices is to place a cap on total human-generated CO_2 and CH_4 emissions in a country or region, issue permits to emit these pollutants, and then let polluters trade their permits in the marketplace. This **cap-and-trade** approach (Figure 19.21) has a political advantage over taxes, but it is difficult to manage because there are so many different emitters of greenhouse gases. Nevertheless, since 2009, some northeastern U.S. states have used cap-and-trade to manage carbon emissions. Emissions there have dropped sharply while the regional economy has grown faster than in the rest of the United States. On the other hand, cap-and-trade has not worked as well as projected in Europe.

However, this approach works only if the original caps are set low enough to encourage a serious reduction in emissions and are lowered on a regular basis to promote further cuts. A related and possibly more politically acceptable approach is to use a cap-and-rebate plan, in which all money made by any trading of permits would go back to the public in the form of cash payments.

Environmental economists such as Gary Yohe argue that, regardless of whether governments use taxes, cap-and-trade, or cap-and-rebate, the most critical goal is to put a high enough price on carbon emissions to get all emitters to come as close as possible to paying the total estimated harmful environmental and health costs of their carbon emissions. These economists contend that the resulting higher costs for fossil fuels would spur innovation in finding ways to reduce carbon emissions, improving energy productivity, and phasing in a mix of low-carbon renewable energy alternatives.

Using Government Subsidies to Address Climate Change

The U.S. and several other governments currently provide *subsidies*, or direct payments and other forms of financial support, to fossil fuel companies. In 2014, U.S. fossil fuel subsidies were estimated to be $37.5 billion per year and global fossil fuels subsidies were $775 billion to $1 trillion per year, according to Oil Change International.

One way in which governments can address climate change is to level the economic playing field by greatly increasing such subsidy payments to businesses and individuals working on energy-efficiency technologies, low-carbon renewable energy sources, and more sustainable food production methods. This is occurring in some sectors. The U.S. government increased subsidies for wind and solar electricity production between 2010 and 2013 more than it did for fossil fuel electricity production, according to the U.S. Energy Information Administration.

Some analysts call for going further and phasing out or sharply reducing subsidies and tax breaks that encourage the use of fossil fuels, unsustainable agriculture, and the clearing of forests. In other words, during the next 2 to 3 decades, we could shift from environmentally degrading to environmentally sustaining subsidies and tax breaks.

One important form of government support is funding of research and development, a great deal of which has gone to the fossil fuel and nuclear energy industries in decades past. Governments could shift this research and development focus to clean energy industries, as some governments have done. With such help, these industries have developed innovations that have lowered the costs of energy from renewable resources such as wind, solar power, and geothermal energy.

International Cooperation

Governments have entered into international climate negotiations. In December 1997, delegates from 161 nations met in Kyoto, Japan, to negotiate a treaty to slow atmospheric warming and projected climate change. The first phase of the resulting *Kyoto Protocol* went into effect in February 2005 with 187 of the world's 194 countries (not including the United States) ratifying the agreement by late 2009.

The 37 participating more-developed countries agreed to cut their emissions of CO_2, CH_4, and N_2O to an average of 5% below 1990 levels by 2012, when the treaty was to expire, but 16 of the nations failed to do so. Less-developed countries, including China, were excused from this requirement, because such reductions would curb their economic growth. In 2005, participating countries began negotiating a second phase of the treaty, which was supposed to go into effect in 2012, but these negotiations failed to extend the original agreement.

In 2015, delegates from 195 countries met in Paris, France, in another attempt to address the rising threat of global warming and climate change. Each country was asked to submit reductions in greenhouse emissions they intended to make between 2020 and 2030 with the overall goal of limiting the earth's average temperature increase to below 2C° (3.6F°). In a historic agreement, each country came up with a goal it pledged to meet. The countries also agreed to meet every 5 years to evaluate progress and increase their goals. The agreement will go into effect when at least 55 countries have ratified it.

However, counties are not legally bound to reach their goals. In addition, there was no agreement for the wealthier nations to raise a proposed $500 billion by 2020 to assist poorer countries in meeting their goals. Even if all commitments are fully honored, scientists estimate that this would reduce the maximum projected global warming by 2100 from 4.5C° (8.1F°) to 3.5C° (6.3F°)—not nearly enough to prevent serious environmental and economic problems. Nations could agree in the future to increase their reductions, but they are already 35 years behind in agreeing to begin dealing with this serious global problem.

Some analysts see this international climate agreement as a weak and slow response to an urgent global problem. Many have also concluded that any top-down international agreement will be too slow and ineffective. Some call for the nations that produce most of the world's greenhouse gas emissions—especially the United States, China, India, Brazil, and the European Union nations (Figure 19.7)—to get together and work out a binding agreement to sharply lower their emissions.

In 2014, the governments of the United States and China took a step in this direction, agreeing formally to cap their carbon emissions within 15 to 20 years. It requires China to halt the growth of its emissions by 2030. The United States will have to drop its emissions to 26–28% below 2005 levels by 2025, although as of 2016, the U.S. Congress had not approved these goals. This is the first time a large less-developed country has agreed to limit its CO_2 emissions. Analysts hope that China's action will lead other such countries to agree to cuts.

Another way in which governments can cooperate is through *technology transfer*. Governments of more-developed countries could help to fund the transfer of the latest green technologies to less-developed countries so that they could bypass older, energy-wasting and polluting technologies. Environmental analyst Lester Brown has suggested that increasing the current tax on each international currency transaction by a quarter of a penny could finance effective technology transfer.

Some analysts argue that it makes sense for richer countries to help poorer countries in dealing with climate change. Many of the poorer countries will suffer the most from such effects, which have been caused primarily by the richer, industrialized countries. They also point out that, eventually, all countries will suffer harmful effects if the climate is severely disrupted. If this happens, widespread poverty and homelessness, mass migrations of environmental refugees, and sharply decreased international security will be the common experience.

Another important form of international cooperation involves efforts to protect large forests, especially tropical forests, which are crucial to the carbon cycle. The effects of deforestation account for 12–17% of global greenhouse gas emissions, according to the Environmental Policy Institute. One such effort is the UN's Reducing Emissions from Deforestation and Forest Degradation (REDD) program, which helps to fund efforts by farmers and others who depend on forests and who seek to use them more sustainably. Since it was founded in 2008, the program has helped 44 countries in Africa, Asia, and Latin America to reduce deforestation. Brazil has slowed deforestation by 40% since 2008 with the help of the REDD program, and is projected to reach its reduction goal of 80% by 2020.

Some Countries, States, and Localities Are Leading the Way

Some nations are leading others in facing the challenges of projected climate change. Costa Rica aims to be the first country to become *carbon neutral* by cutting its net carbon emissions to zero by 2030. The country generates 78% of its electricity with hydroelectric power and another 18% from wind and geothermal energy.

Some analysts are urging rapidly developing countries such as China and India to shift toward more sustainable, low-carbon economic development, and there are efforts in these countries to do so. While China emits more CO_2 than any other country, it also has one of the world's most intensive energy efficiency programs, according to ClimateWorks, a global energy-consulting firm. Chinese automobile fuel-efficiency standards are higher than U.S. standards, although enforcement of the standards is still weak. China's government is working with the country's top 1,000 industries to implement tough energy efficiency goals. China is also rapidly becoming the world's leader in developing and selling solar cells, solar water heaters, wind turbines, advanced batteries, and plug-in hybrid-electric cars.

Some U.S. state and local governments are moving ahead in dealing with climate change. By 2015, at least 32 states had emissions reduction programs or were involved in multistate programs. California plans to get 33% of its electricity from low-carbon renewable energy sources by 2030. That state is showing that it is possible to implement policies that cut carbon emissions and create jobs.

Around the world, major cities are establishing programs to reduce emissions. Portland, Oregon, is creating *20-minute neighborhoods* where people can live, work, and meet their needs by walking or biking 20 minutes or less. Amsterdam, The Netherlands, has created a sustainability fund of more than $100 million available to communities, businesses, and individuals for investing in sustainability strategies. Durban, South Africa, burns methane captured from landfills to provide electricity for 6,000 homes, thereby keeping the methane out of the atmosphere and cutting its reliance on coal. And in Seoul, South Korea, more than 10,000 rooftop photovoltaic installations are reducing that city's dependence on fossil fuels.

Some Companies Are Reducing Their Carbon Footprints

Leaders of some big U.S. companies, including Alcoa, DuPont, Ford Motor Company, General Electric, and Shell Oil, have joined with leading environmental organizations to form the U.S. Climate Action Partnership. They have called on the government to enact strong national climate change legislation, realizing that significantly reducing their own greenhouse gas emissions will save them money.

Leaders of these and many other major companies see an enormous profit opportunity in developing or using energy-efficient and cleaner technologies. For example, the production of building materials is energy-intensive and produces 10–12% of all U.S. CO_2 emissions. Companies such as the brick maker CalStar Products have found ways to produce building materials with much lower carbon outputs. CalStar recycles fly ash from coal-burning facilities and it reduced the carbon footprint of brick making by 80%. Many companies now recognize that "there is gold in going green."

To help people and companies in shrinking their carbon footprints, some banks are now offering *green bonds*. With this approach, investors can grow their money while helping to fund sustainability projects in areas such as energy efficiency, waste management, urban mass transit, biomass energy systems, and solar and wind power systems.

Colleges and Universities Are Reducing Their Carbon Footprints

Many colleges and universities are also taking action. Arizona State University (ASU) boasts the largest collection of solar panels of any U.S. university. ASU was also the first American university to establish a School of Sustainability. The College of the Atlantic in Bar Harbor, Maine, has been carbon neutral since 2007. It gets all of its electricity from renewable hydropower, and many of its buildings are heated with the use of renewable wood pellets. Students there built a wind turbine that powers a nearby organic farm, which offers organic produce to the campus, to local schools, and to food banks.

Students at the University of Washington in Seattle agreed to an increase in their fees to help the school buy electricity from renewable energy sources. At Florida's University of Miami, drivers of hybrid cars get a 50% parking discount.

EARTH University in Costa Rica has a mission to promote sustainable development in tropical countries. Its sustainable agriculture degree program has attracted students from more than 20 different countries. Students and faculty at Oberlin College in the U.S. state of Ohio created a Web-based system in some of the school's dorms that monitors use of energy and water, giving students real-time feedback that can help them to reduce their waste of these resources. And a growing number of campus groups are urging the administrators at their schools to help slow climate change by ending their endowment fund investments in fossil fuel companies.

CONSIDER THIS . . .

THINKING ABOUT What Your School Can Do

What are three steps that you think your school could take to help reduce its carbon footprint? What steps, if any, is it now taking?

Every Individual Can Make a Difference

Each of us will play a part in the atmospheric warming and climate change projected to occur during this century. Whenever we use energy generated by fossil fuels, we add CO_2 to the atmosphere. However, nearly two-thirds of the average American's carbon footprint is *embedded carbon*—carbon released during the manufacture and delivery of food, shelter, clothing, cars, computers, and every other consumer product and service. There are many sources of information on the Internet and elsewhere comparing the carbon emissions involved in the production of various goods and services.

One important aspect of your carbon footprint is your diet. Foods vary greatly in the greenhouse gases that result from their production and delivery. For example, processed foods require much more energy to produce than do whole foods such as fresh fruits and vegetables. Meat production, especially as it is done on factory farms, involves far higher greenhouse gas emissions than production of grains and vegetables. In addition, greenhouse gas emissions resulting from beef production and consumption are 12 times those associated with producing and eating the same amount of chicken. By choosing foods carefully, you can reduce your carbon footprint.

Some websites suggest ways for you to reduce, or *offset*, some of your carbon dioxide emissions by helping to fund or participate in projects such as forest restoration and renewable energy development. However, buyers need to do their homework, because some carbon-offset programs do not actually produce emissions cuts. Critics of such carbon-offset schemes argue that they can encourage people to continue producing greenhouse gases instead of making changes to reduce their carbon footprints by, for example, flying less, driving less, eating less meat (especially beef), buying less stuff, and wasting less stuff.

Political involvement—joining with others to spread a message—is another way in which individuals make a difference. In September 2014, climate activists organized a global People's Climate March. In New York City, 400,000 people turned out in a crowd that stretched 20 blocks on one of the wide avenues, spilling into many side streets. On that day, smaller marches took place in 162 other countries.

Analysts have argued that the broad diversity of the people in these marches—young and old of all races—should send a signal to government and business leaders that there is a strong climate movement demanding changes in how we care for the planet. This was reflected in a 2015 U.S. public opinion poll, published in the *New York Times*, in which 78% of respondents said yes to the question on whether or not the government should limit the amount of greenhouse gases that businesses emit.

Many analysts believe that the kind of change that is the subject of this chapter has to start at the individual level. You can learn about your carbon footprint by using a footprint calculator, several of which are available online. Figure 19.22 lists some ways in which you can cut your CO_2 emissions. One person taking each of these steps makes a small contribution to reducing global greenhouse gas emissions, but when millions of people take such steps, global change can happen.

What Can You Do?

Reducing CO_2 Emissions

- Calculate your carbon footprint (there are several helpful websites)
- Drive a fuel-efficient car, walk, bike, carpool, and use mass transit
- Reduce garbage by reducing consumption, recycling, and reusing more items
- Use energy-efficient appliances and LED lightbulbs
- Wash clothes in warm or cold water and hang them up to dry
- Close window curtains to keep heat in or out
- Use a low-flow showerhead
- Eat less meat or no meat
- Heavily insulate your house and seal all air leaks
- Use energy-efficient windows
- Set your hot-water heater to 49°C (120°F)
- Plant trees
- Buy from businesses working to reduce their emissions

FIGURE 19.22 Individuals matter: You can reduce your annual emissions of CO_2. ***Critical thinking:*** Which of these steps, if any, do you take now or plan to take in the future?

19.5 HOW CAN WE ADAPT TO CLIMATE CHANGE?

CONCEPT 19.5 While we can prepare for some climate change that is now inevitable, we could realize important economic, ecological, and health benefits by drastically reducing greenhouse gas emissions with the goal of slowing projected climate disruption.

We Can Prepare for Climate Change

According to global climate models, the world needs to make a 50–85% cut in emissions of greenhouse gases by 2050 to stabilize concentrations of these gases in the atmosphere. This would help to prevent the atmosphere from warming by more than 2C° (3.6F°) and head off rapid changes in the world's climate along with the harmful environmental, economic, and health effects projected to occur.

However, because of the political difficulty of making such large reductions, many analysts believe that while we work to slash greenhouse gas emissions, we should also begin to prepare for the harmful effects of projected climate change. Figure 19.23 shows some ways to do so.

For example, relief organizations such as the International Red Cross are turning their attention to projects such as expanding mangrove forests as buffers against storm surges. They are building shelters on high ground

FIGURE 19.23 Solutions: Some ways to prepare for the possible long-term harmful effects of climate change. ***Critical thinking:*** Which three of these adaptive steps do you think are the most important ones to take? Why?

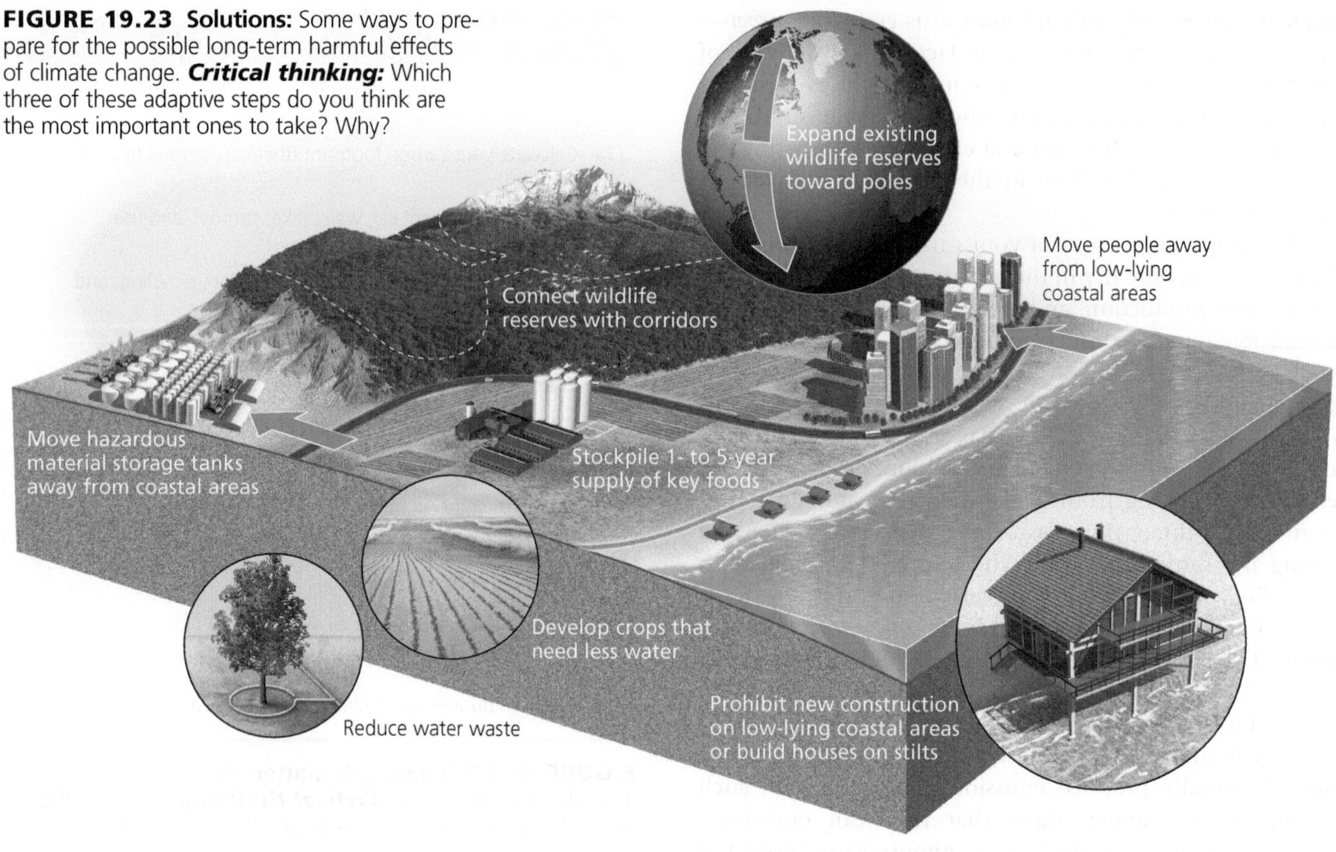
Expand existing wildlife reserves toward poles
Move people away from low-lying coastal areas
Connect wildlife reserves with corridors
Move hazardous material storage tanks away from coastal areas
Stockpile 1- to 5-year supply of key foods
Develop crops that need less water
Reduce water waste
Prohibit new construction on low-lying coastal areas or build houses on stilts
© Cengage Learning

and planting trees on slopes to help prevent landslides. In the face of projected higher levels of precipitation and rising sea levels, seawall design and construction will likely be a major growth industry. And low-lying countries such as Bangladesh are planning for what to do with millions of environmental refugees who will be flooded out by rising sea levels and more intense storms.

Several cities have developed adaptation plans. In London, England, a flood-control barrier is being strengthened at a cost of about $1 billion (Figure 19.24). Other cities, including the U.S. cities of New York City and Seattle, Washington, and several states have also developed adaptation plans. California is beefing up its forest firefighting capabilities and is building desalination plants to help relieve projected water shortages, which will worsen as some of its mountain glaciers melt. In the aftermath of flooding caused by Hurricane Sandy in 2012, New York City is building new floodwalls and other flood protection systems.

Some coastal communities in the United States now require that new houses and other new construction be built high enough off of the ground or further back from the current shoreline to survive storm surges. In anticipation of rising sea levels, Boston elevated one of its sewage treatment plants. Some cities plan to establish cooling centers to shelter residents during increasingly intense heat waves. In the low-lying Netherlands, people are dealing with the threat of a rising sea level and rising river flows using a variety of strategies (see the Case Study that follows).

Around the world, major cities are cooperating to find ways to deal with climate change. Leaders from Cairo, Hong Kong, Berlin, Los Angeles, and more than 70 other cities from every continent have formed the C40 Cities Climate Leadership Group. They are working together to redesign their transportation, housing, health, waste management, water, and food systems to meet new needs caused by climate change. By 2015, the group had enacted more than 8,000 measures, policies, and programs to address climate change.

Some people fear that emphasizing these adaptation approaches will distract us from the more urgent need to reduce greenhouse gas emissions. However, to some analysts, projected climate change is already such a serious threat that we have no alternative but to implement both prevention and adaptation strategies, and we have no time to lose in doing so.

CASE STUDY

The Netherlands—Living with the Water

The small western European nation of The Netherlands is one of the most densely populated countries in the world, and most of its people live below sea level. The Dutch are famous for their dikes, built to hold back the rising North Sea. They are as experienced as anyone at adapting to a changing environment, having dealt with rising seas for more than 800 years.

John Sturrock/Alamy Stock Photo

FIGURE 19.24 This high-tech, movable flood-control barrier—the second largest in the world—was built across the Thames River to prevent flooding from very high tides and storm surges in London, England.

The Netherlands, also known as Holland, has 1,000 kilometers (600 miles) of coastline on the North Sea. Were it not for looped embankments, known as *dike rings*, much of the country would be covered by shallow bays and coastal wetlands fed by the sea and by large rivers, including Europe's largest, the Rhine River. More than 55% of Holland's land lies within dike rings, and 67% of its people live there. Some 70% of the country's economic activity takes place in these areas.

The Netherlands is now experiencing climate change. The warming of the air over Western Europe has been faster than the global average—by some estimates, twice as fast. Average annual precipitation in Holland has increased by 25% since 1900. In addition, since then, the North Sea has risen 19 centimeters (7.6 inches) on the Dutch coast, while some areas of the coast are subsiding at an average rate of 8 centimeters (3.2 inches) per decade. In the Alps to the south, where the Rhine and other rivers originate, glaciers are melting at increasing rates, feeding more water to the rivers.

For these reasons, Holland is increasingly subject to flooding by North Sea waters as well as by the more frequent flooding of the Rhine and other large rivers. While the areas below sea level are well protected, the Dutch government estimates that 34% of the country's land area and 35% of its people are subject to flooding.

The Dutch government, business community, and citizens are taking this problem by the horns. They are known for their methodical approach to problem solving and are formulating a 200-year plan for dealing with climate change. Most experts agree that this represents the most far-sighted planning in the world. The Dutch government and people are also planning to dedicate about $1 billion per year to carrying out this plan.

A major part of the plan—which is marketed as *living with the water*—is to simply let the flooding occur, but to steer it to certain sparsely populated areas and to "water plazas" created in urban areas for temporary floodwater storage. Affected homeowners will be compensated as necessary or moved at government expense.

Another part of the strategy, already being carried out, is called *coastal nourishment*. It consists of hauling large volumes of sand to coastal areas and replenishing beaches and estuaries that have been eroded by rising seas. In some cases, engineers have dumped millions of cubic meters of sand, leaving it to ocean waves, currents, and winds to distribute it by natural processes. The idea is to restore the beaches as buffers against rising seas and storm surges. The government is also experimenting with the idea of restoring degraded salt marshes and creating new ones near these beaches to enhance the buffering effect.

The country is taking many other measures to adapt to climate change. New buildings are being built well above street level with waterproof plaster and other materials for walls and waterproof floors. Cities are encouraged to use permeable pavement and to create more green space to allow floodwater to sink more easily into the ground. The city of Rotterdam is integrating underground reservoirs into some new housing projects and parking ramps. The city is also constructing small "pocket parks" and encouraging construction of green roofs to help cool the city during hot spells.

One of the most interesting adaptations, still in the experimental stage, is Rotterdam's three floating pavilions. These domed buildings were erected on floating platforms with a total area of more than 1,100 square meters (12,000 square feet). The builders envision floating urban districts that will slowly rise or fall with the sea—a true example of living with the water.

A No-Regrets Strategy

The threat of climate disruption pushes us to look for preventive solutions such as those listed in Figure 19.18. But suppose we had the ability to look into the future and, in doing so, we found that the climate models were wrong and that climate disruption was not a serious threat after all. Should we then return to the present and abandon the search for preventive solutions?

A number of climate and environmental scientists, economists, and business leaders say *no* to this question, and they call for us to begin implementing such changes now as a *no-regrets strategy*. They argue that actions such as those listed in Figure 19.18 should be implemented in any case because they will lead to very important environmental, health, and economic benefits.

For example, improving energy productivity has numerous economic and environmental advantages (see Figure 16.3, p. 411). Individuals, companies, and local governments can save large amounts of money by cutting their energy bills. In addition, the use of fossil fuels, especially coal, causes a great deal of air pollution, which harms or kills millions of people every year. Thus, a sharp decrease in the use of these fuels will improve the overall health of the population, save lives, and save money now spent on health care. In addition, cutting coal use will greatly reduce the land disruption resulting from surface mining.

Another benefit of the no-regrets strategy is that, by relying more on a mix of renewable, domestic energy resources, many countries will cut their costly dependence on imports of fossil fuels. The money that they now spend for fuel imports will be available for investing in health care, education, and jobs, which will bolster their economic security. Also, sharply decreasing or halting the destruction of tropical forests would help to preserve the increasingly threatened biodiversity that makes up an important part of the earth's vital natural capital (see Figure 1.3, p. 7). In short, there are plenty of reasons for implementing a no-regrets strategy for sharply reducing our carbon outputs (Figure 19.25).

FIGURE 19.25 The benefits of slowing climate change are clear.

BIG IDEAS

- Considerable scientific evidence indicates that the earth's atmosphere is warming, mostly because of human activities, and that this is likely to lead to significant climate disruption during this century that could have severe and long-lasting harmful consequences.
- Reducing the projected harmful effects of rapid climate change during this century requires emergency action to increase energy efficiency, sharply reduce greenhouse gas emissions, and rely more on renewable energy resources.
- While we can prepare for some climate change that is now inevitable, we could realize important economic, ecological, and health benefits by drastically reducing greenhouse gas emissions with the goal of slowing climate change.

Tying It All Together

Melting Ice in Greenland and Sustainability

In this chapter, we have seen that human activities such as the burning of fossil fuels and the widespread clearing and burning of forests for planting crops have contributed to higher levels of greenhouse gases in the atmosphere. This has contributed to increased atmospheric warming, which in turn is projected to result in possibly disruptive climate change during this century.

The effects of climate change could include rapid melting of land-based ice (**Core Case Study**) and arctic sea ice, longer and more intensive drought, rising sea levels, declining biodiversity, and severe threats to human health and economies. Many of these effects could further accelerate climate change in worsening positive feedback spirals of change.

We can apply the six **principles of sustainability** to help reduce the harmful effects of climate change. We can reduce emissions of pollutants and greenhouse gases by relying more on direct and indirect forms of solar energy than on fossil fuels; recycling and reusing matter resources much more widely than we do now; and mimicking biodiversity by using a variety of often locally available low-carbon renewable energy resources in place of fossil fuels. We can advance toward these goals by including the harmful environmental and

Natalia Davidovich/Shutterstock.com

health costs of fossil fuel use in market prices; seeking win-win solutions that will benefit both the economy and the environment; and giving high priority to passing along to future generations an environment in which they too can thrive.

Chapter Review

Core Case Study

1. Define **climate change**. Summarize the story of Greenland's melting glaciers and the possible effects of this process. Explain how it fits into projections about climate change during this century.

Section 19.1

2. What is the key concept for this section? Define and distinguish between **weather** and **climate**. Summarize the trends in atmospheric temperature changes over the past 900,000 years, 10,000 years, 1,000 years, and since 1975. Define **greenhouse gases** and give four examples. How do scientists get information about past temperatures and climates? What are three major conclusions of the world's top scientists regarding climate change? List eight pieces of evidence that support those conclusions.

Section 19.2

3. What is the key concept for this section? What are Milankovitch cycles and how do they affect climate, according to theory? What is the **greenhouse effect** and why is it so important to life on the earth? What are three key greenhouse gases and how do they differ in atmospheric lifetimes and warming potentials? What is the role of water vapor in the atmosphere and how does it relate to other greenhouse gases in the warming process? What are two other natural factors affecting atmospheric warming and climate change?

4. How have human activities affected atmospheric greenhouse gas levels during the last 275 years and especially in the last 30 years? List the major human activities that add CO_2, CH_4, and N_2O to the atmosphere. What is a **carbon footprint**? How do scientists use models to make projections about future temperature changes? What do climate models project about temperature changes during this century? Explain how each of the following might contribute to atmospheric warming and climate change: **(a)** solar energy changes, **(b)** the oceans, **(c)** cloud cover, and **(d)** air pollution.

Section 19.3

5. What is the key concept for this section? What makes the current atmospheric warming problem urgent? Define **climate change tipping point** and give three examples. Summarize the possible effects of a warmer atmosphere on arctic sea ice and land-based ice. How can permafrost and tropical wetlands change and affect climate? How are sea levels changing and what are six harmful effects of sea level rise? Define **ocean acidification** and explain how and why it is occurring. Explain how atmospheric warming, global drought, extreme weather, and an intensified water cycle are related. Why is Alaska a bellwether of climate change? What are the possible effects of climate change on biodiversity, food production, human health and economies, and national security? Pick three of these effects and explain how each can become part of a positive feedback loop leading to increased climate change.

Section 19.4

6. What is the key concept for this section? What are five factors that make it difficult to deal with the problem of projected climate change? What are two basic approaches to dealing with the harmful effects of climate change? List four major prevention strategies for reducing greenhouse gas emissions. What are four pieces of potential good news about dealing with climate change? List six strategies for trying to clean up carbon emissions. What is **carbon capture and storage (CCS)**? What are five problems associated with capturing and storing carbon dioxide emissions? Define **geoengineering** and give two examples of geoengineering proposals. What are three major problems with such proposals?

7. Summarize the argument for regulating greenhouse gases as pollutants. List two ways to put a price on carbon emissions. What are the advantages and disadvantages of using taxes on carbon emissions or energy use to help reduce greenhouse gas emissions? What is **cap-and-trade** and what are the advantages and disadvantages of using it to help reduce greenhouse gas emissions? How can government subsidies be used to address climate change? What is the Kyoto Protocol and how has it progressed? What are three ways in which governments can cooperate internationally to deal with projected climate change?

8. Give two examples of what some countries are doing on their own to deal with climate change. Give two examples of what some companies are doing and two examples of what some colleges and universities have done to reduce their carbon footprints. List five ways in which you can reduce your carbon footprint.

Section 19.5

9. What is the key concept for this section? List five ways in which we can prepare for the possible long-term harmful effects of climate disruption. Explain how The Netherlands has planned to adapt to climate change. Describe the no-regrets strategy for dealing with the problems associated with energy waste and fossil fuel use, regardless of climate change.

10. What are this chapter's *three big ideas*? Explain how we can apply the six **principles of sustainability** (see Inside Back Cover) to the problem of projected climate disruption.

Note: Key terms are in bold type.

Critical Thinking

1. If you had convincing evidence that at least half of Greenland's glaciers (**Core Case Study**) were sure to melt during this century, would you argue for taking serious actions now to slow projected climate change? Summarize your arguments for or against such actions.
2. Explain why you agree or disagree with IPCC scientists and 97% of the world's climate scientists that **(a)** climate change is happening now, **(b)** human activities are the dominant cause of this climate change, and **(c)** only action by humans can slow down the rate of climate change and avert or delay its projected harmful environmental, health, and economic effects.
3. China relies on coal for two-thirds of its commercial energy usage, partly because the country has abundant supplies of this resource. As a result, China's use of coal is playing a large role in the warming of the global atmosphere. Do you think China is justified in expanding its use of this resource as other countries, including the United States, have done with their coal resources? Explain. What are China's alternatives?
4. Explain why you would support or oppose each of the strategies listed in Figure 19.18 for slowing projected climate disruption caused by atmospheric warming.
5. Suppose someone tells you that carbon dioxide (CO_2) should not be classified as an air pollutant because it is a natural chemical that is part of the carbon cycle (see Figure 3.20, p. 65) and a chemical that we add to the atmosphere every time we exhale. Explain the faulty reasoning in this statement about CO_2.
6. One way to slow the rate of CO_2 emissions is to reduce the clearing of forests—especially in less-developed tropical countries where intense deforestation is taking place. Should the United States and other more-developed countries pay poorer countries to stop cutting their forests because the resulting climate change will affect the United States and other nations throughout the world? Explain.
7. Some scientists have suggested that, in order to help cool the warming atmosphere, we could annually inject huge quantities of sulfate particles into the stratosphere. This might have the effect of reflecting some incoming sunlight back into space. Explain why you would support or oppose this geoengineering scheme.
8. What are three consumption patterns or other aspects of your lifestyle that directly add greenhouse gases to the atmosphere? Which, if any, of these habits would you be willing to give up in order to help slow projected climate change? Write up an argument that you could use to convince others to do the same. Compare your answers to those of your classmates.

Doing Environmental Science

Gather data on the trends in average annual temperatures and average annual precipitation over the past 30 years in the area where you live. (Possible sources include weather sites on the Internet, your school library, TV and radio meteorologists, and local or regional weather bureaus.) Try to find data for as many of these years as possible, and plot these data to determine whether the average temperature and precipitation during this period has increased, decreased, or stayed about the same. Write a report summarizing your search for data, your results, and your conclusions.

Global Environment Watch Exercise

Go to your MindTap course to access the GREENR database. Starting on the home page, under "Browse Issues and Topics", click on *Environment and Ecology,* then select *Climate Change: Science and Mitigation.* Within this portal search for the latest information on the melting of glaciers in Greenland (**Core Case Study**). What is the rate of melting of the glaciers and how might this affect sea levels during this century? Briefly summarize the evidence used by scientists to support their statements about melting ice in Greenland. Summarize any information you find about ongoing studies of Greenland's ice.

Ecological Footprint Analysis

According to the International Energy Agency, the average American adds 19.6 metric tons (21.6 tons) of CO_2 per year to the atmosphere, compared with a world average of 4.23 metric tons (4.65 tons). The table on the next page is designed to help you understand the sources of your personal inputs of CO_2 into the atmosphere. You will be making calculations to fill in the blanks in this table.

Some typical numbers are provided in the "Typical Quantity per Year" column of the table. However, your calculations will be more accurate if you can use information based on your own personal lifestyle, which you can enter in the blank "Personal Quantity per Year" column. For example, you could add up your monthly utility bills for a year and divide the total by the number of persons in your household to get a rough estimate of your own utility use.

	Units per Year	Typical Quantity per Year	Personal Quantity per Year	Multiplier	Emissions per Year (lb. CO_2)
Residential Utilities					
Electricity	kwh	4,500		1.5	
Heating oil	gallons	37		22	
Natural gas	hundreds of cubic feet (ccf)	400		12	
Propane	gallons	8		13	
Coal	tons	—		4,200	
Transportation					
Automobiles	gallons	600		19	
Air travel	miles	2,000		0.6	
Bus, urban	miles	12		0.07	
Bus, intercity	miles	0		0.2	
Rail or subway	miles	28		0.6	
Taxi or limousine	miles	2		1	
Other motor fuel	gallons	9		22	
Household Waste					
Trash	pounds	780		0.75	
Recycled items	pounds	337		-2	
				Total (pounds)	
				Total (tons)	
				Total (kilograms)	
				Total (metric tons)	

Source: Thomas B. Cobb at Ohio's Bowling Green State University developed this CO_2 calculator.

1. Calculate your carbon footprint. To calculate your emissions, first complete the blank "Personal Quantity per Year" column as described above. Wherever you cannot provide personal data, use that listed in the "Typical Quantity per Year" column. Enter each number in the "Personal Quantity per Year" column to represent your annual consumption (using the units specified in the "Units per Year" column). Now multiply your annual consumption for each activity by the associated number in the "Multiplier" column to obtain an estimate of the pounds of CO_2 resulting from that activity, which you will enter in the "Emissions per Year" column. Finally, add the numbers in that column to find your carbon footprint, and express the final CO_2 result in both pounds and tons (1 ton = 2,000 pounds) and in kilograms and metric tons (1 kilogram = 2.2 pounds; 1 metric ton = 1.1 tons).

2. Compare your emissions with those of your classmates and with the per capita U.S. average of 19.6 metric tons (21.6 tons) of CO_2 per person per year. Actually, your answer should be considerably less—roughly about half the per capita value—because this computation only accounts for direct emissions. For instance, CO_2 resulting from driving a car is included, but the CO_2 emitted in manufacturing and disposing of the car is not. You can find more complete carbon footprint calculators at several sites on the Web.

3. Consider and list actions you could take to reduce your carbon footprint by 20%.

CHAPTER 20

Water Pollution

> The health of our waters is the principal measure of how we live on the land.
>
> LUNA LEOPOLD

Key Questions

20.1 What are the causes and effects of water pollution?

20.2 What are the major pollution problems in streams and lakes?

20.3 What are the major groundwater pollution problems?

20.4 What are the major ocean pollution problems?

20.5 How can we deal with water pollution?

Brown pelican severely oiled by the 2010 BP *Deepwater Horizon* oil well rupture in the Gulf of Mexico.

Joel Sartore/National Geographic Creative

The Gulf of Mexico's Annual Dead Zone

The Mississippi River basin (Figure 20.1, top) lies within 31 states and contains almost two-thirds of the continental U.S. land area. With more than half of all U.S. croplands, it is one of the world's most productive agricultural regions. However, water draining into the Mississippi River and its tributaries from farms, cities, factories, and sewage treatment plants in this huge basin contains sediments and other pollutants that end up in the Gulf of Mexico (Figure 20.1, bottom photo)—a major supplier of the country's fish and shellfish.

Each spring and summer, huge quantities of nitrogen and phosphorus plant nutrients—mostly nitrates and phosphates from crop fertilizers—flow into the Mississippi River, end up in the northern Gulf of Mexico, and overfertilize the coastal waters of the U.S. states of Mississippi, Louisiana, and Texas. This excess of plant nutrients leads to an explosive growth of phytoplankton (mostly algae). They eventually die, sink to the bottom, and are decomposed by hordes of bacteria that consume oxygen as part of their aerobic respiration (p. 54). This depletes most of the dissolved oxygen in the Gulf's bottom layer of water.

The resulting massive volume of water with a low dissolved oxygen content (below 2 parts per million) is called a *dead zone* because it contains little or no animal marine life. Its low dissolved oxygen levels (Figure 20.1, bottom) drive away faster-swimming fish and other marine organisms and suffocate bottom-dwelling fish, crabs, oysters, and shrimp that cannot move to less polluted areas. Large amounts of sediment, mostly from soil eroded from the Mississippi River basin, can also kill bottom-dwelling forms of animal aquatic life. The dead zone appears each spring and grows until fall when storms churn the water and redistribute dissolved oxygen to the gulf bottom.

The size of the Gulf of Mexico's annual dead zone varies with the amount of water flowing into the Mississippi River. In years with ample rainfall and snowmelt, such as 2003, it covered an area as large as the state of Massachusetts—27,300 square kilometers (10,600 square miles). In years with less rainfall, such as 2015, it covered a smaller area of 16,760 square kilometers (6,474 square miles)—slightly less than the combined areas of the states of Connecticut and Rhode Island.

The annual Gulf of Mexico dead zone is one of 400 dead zones found throughout the world, 200 of them in the United States. Other such zones lie in the Baltic Sea to the south of Sweden in Northern Europe and in the coastal waters of Australia, New Zealand, Japan, China, and South America.

Oxygen-depleted zones represent a disruption of the nitrogen cycle (see Figure 3.21, p. 66) caused primarily by human activities. This is because great quantities of nitrogen from nitrate fertilizers are added to these aquatic systems faster than the nitrogen cycle can remove them. Thus, producing crops to feed livestock and people and ethanol to fuel cars in the vast Mississippi basin ends up disrupting coastal aquatic life and seafood production in the Gulf of Mexico.

This overfertilization of coastal waters is one of the world's many forms of water pollution. Some good news is that we know how to reduce the sizes of dead zones and how to reduce other forms of water pollution that we explore in this chapter.

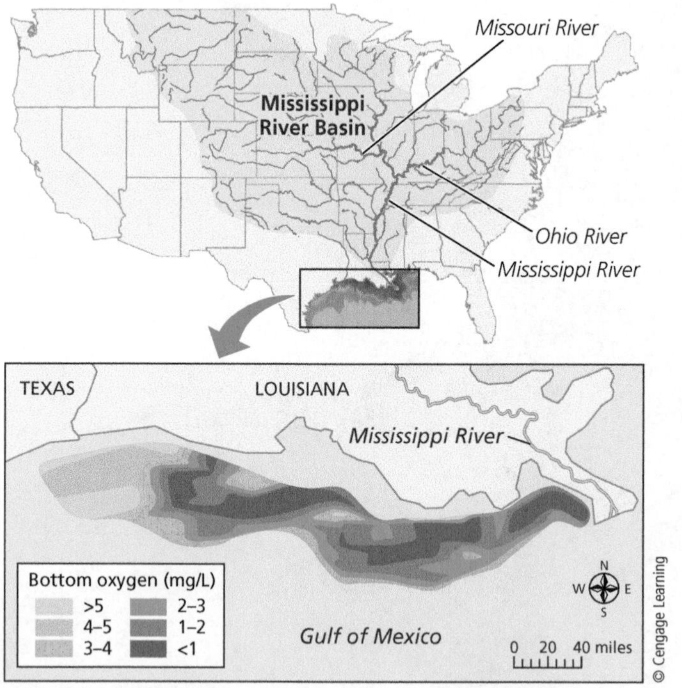

FIGURE 20.1 Water containing sediments, dissolved nitrate fertilizers, and other pollutants drains from the Mississippi River basin (top) into the Mississippi River and from there into the northern Gulf of Mexico (bottom). This creates a dead zone with low levels of dissolved oxygen (1–3 ppm), indicated by the dark and light red shaded areas in the bottom figure for 2015.

(Data from NOAA.)

20.1 WHAT ARE THE CAUSES AND EFFECTS OF WATER POLLUTION?

CONCEPT 20.1A The chief sources of water pollution are agricultural activities, industrial facilities, mining, and untreated wastewater.

CONCEPT 20.1B Water pollution causes illness and death in humans and other species, and disrupts ecosystems.

Water Pollution Comes from Point and Nonpoint Sources

Water pollution is any change in water quality that can harm living organisms or make the water unfit for human uses such as drinking, irrigation, and recreation. It can come from single (point) sources or from larger and dispersed (nonpoint) sources. **Point sources** discharge pollutants into bodies of surface water at specific locations often through drain pipes, ditches, or sewer lines that are usually easy to identify, monitor, and regulate. Examples are factories (Figure 20.2), sewage treatment plants (which remove some, but not all, pollutants), animal feedlots, underground and open-pit mines (Figure 20.3), oil wells, and oil tankers.

Nonpoint sources are broad and diffuse areas where rainfall or snowmelt washes pollutants off the land into bodies of surface water. Examples include runoff of eroded soil and chemicals such as fertilizers and pesticides from cropland, animal feedlots, logged forests (Figure 10.10, p. 228), construction sites, city streets, parking lots, lawns, and golf courses. Controlling water pollution from nonpoint sources is difficult because of the expense of identifying and controlling discharges from so many diffuse sources.

Agricultural activities are the leading cause of water pollution. The most common pollutant is sediment eroded from agricultural lands (Figure 20.4). Other major agricultural pollutants include fertilizers (Figure 20.1), pesticides, and bacteria from livestock and food-processing wastes. Three other major sources are industrial facilities, mining (Figure 20.3 and Figure 14.10, p. 366), and untreated **wastewater**, or water that contains sewage and other wastes from homes and industries (**Concept 20.1A**).

vasakkohaline/Shutterstock.com

FIGURE 20.2 Point source of water pollution from an industrial plant.

Mrfotos/Dreamstime.com

FIGURE 20.3 *Point source pollution:* acid draining from an abandoned open-pit coal mine.

CONSIDER THIS . . .

CONNECTIONS Cleaning Up the Air and Polluting the Water

Stricter air pollution control laws (see Chapter 18, p. 490) have forced coal-burning power plants in more-developed countries to remove many of the harmful gases and particles from their smokestack emissions. This results in hazardous ash, which is typically placed in slurry ponds (Figure 15.19, p. 395) that can rupture and cause water pollution. This illustrates one way in which air and water pollution are connected.

FIGURE 20.4 *Nonpoint source pollution:* sediments in farmland runoff flowing into a stream. By weight, sediment is the largest source of water pollution. ***Critical thinking:*** What do you think the owner of this farm could have done to prevent such sediment pollution?

Another form of water pollution occurs when people discard plastics used to make water bottles and millions of other products. Much of this plastic ends up in rivers, lakes (Figure 20.5), and oceans. Such discarded plastic products can harm various forms of wildlife (Figure 11.6, p. 260).

Harmful Effects of Water Pollutants

Table 20.1 lists the types of major water pollutants along with examples of each and their harmful effects and sources. A National Academy of Sciences study ranked the following pollutants as the most serious threats to U.S. stream and lake water quality: mercury (Figure 17.11, p. 454), pathogens from leaking and broken sewer pipes, sediment from land disturbance and stream erosion, metals other than mercury, and nutrients that cause oxygen depletion (Figure 20.1).

A major water pollution problem is exposure to infectious bacteria, viruses, and parasites that are transferred into water from the wastes of 2.5 billion humans that lack access to toilets and other forms of waste disposal, and other animals (**Concept 20.1B**). Drinking water contaminated with these biological pollutants can cause painful, debilitating, and often life-threatening diseases, including typhoid fever, cholera, hepatitis B, giardiasis, and cryptosporidium. The World Health Organization (WHO) estimates that each year, more than 1.6 million people die from largely preventable waterborne infectious diseases that they get by drinking contaminated water or by not having enough water to keep clean. More widespread detection of water pollutants (Science Focus 20.1) can reduce this death toll.

Stephane Bidouze/Shutterstock.com

FIGURE 20.5 Nonpoint pollution of this mountain lake by plastics and other forms of waste from a variety of sources.

TABLE 20.1 Major Water Pollutants and Their Sources

Type/Effects	Examples	Major Sources
Infectious agents (pathogens) *Cause diseases*	Bacteria, viruses, protozoa, parasites	Human and animal wastes
Oxygen-demanding wastes *Deplete dissolved oxygen needed by aquatic species*	Biodegradable animal wastes and plant debris	Sewage, animal feedlots, food-processing facilities, paper mills
Plant nutrients *Cause excessive growth of algae and other species*	Nitrates (NO_3^-) and phosphates (PO_4^{3-})	Sewage, animal wastes, inorganic fertilizers
Organic chemicals *Add toxins to aquatic systems*	Oil, gasoline, plastics, pesticides, cleaning solvents	Industry, farms, households
Inorganic chemicals *Add toxins to aquatic systems*	Acids, bases, salts, metal compounds	Industry, households, surface runoff, mining sites
Sediments *Disrupt photosynthesis, food webs, other processes*	Soil, silt	Land erosion
Heavy metals *Cause cancer, disrupt immune and endocrine systems*	Lead, mercury, arsenic	Unlined landfills, household chemicals, mining refuse, industrial discharges
Thermal *Make some species vulnerable to disease*	Heat	Electric power and industrial plants

SCIENCE FOCUS 20.1

Testing Water for Pollutants

Scientists use a variety of methods to measure water quality. For example, they test samples of water for the *presence of various infectious agents* such as certain strains of the coliform bacteria *Escherichia coli*, or *E. coli* (Figure 20.A), which live in the colons and intestines of humans and other animals and thus are present in their fecal wastes. Most strains of coliform bacteria do not cause disease but their presence indicates that water has been exposed to human or animal wastes that are likely to contain disease-causing agents.

To be considered safe for drinking, a 100-milliliter (about 1/2-cup) sample of water should contain no colonies of coliform bacteria. To be considered safe for swimming, such a water sample should contain no more than 200 colonies of coliform bacteria. By contrast, a similar sample of raw sewage may contain several million coliform bacterial colonies.

Another indicator of water quality is its *level of dissolved oxygen (DO)*. Excessive inputs of oxygen-demanding wastes can deplete DO levels in water. Figure 20.B shows the relationship between dissolved oxygen content and water quality. In the dead zone formed in the Gulf of Mexico (**Core Case Study**), dissolved oxygen levels near the seafloor typically are 2 to 3 ppm, which can kill many marine organisms.

Scientists use *chemical analysis* to determine the presence and concentrations of specific organic chemicals in polluted water. They also monitor water pollution by using living organisms as *indicator species*. For example, they remove aquatic plants such as cattails from areas contaminated with fuels, solvents, and other organic chemicals, and analyze them to determine the pollutants contained in their tissues. Scientists also determine water quality by analyzing bottom-dwelling species such as mussels, which feed by filtering water through their bodies. Genetic engineers are working to develop bacteria and yeasts (single-celled fungi) that glow in the presence of specific pollutants such as toxic heavy metals in the ocean, toxins in the air, and carcinogens in food.

Scientists measure the amount of sediment in polluted water by evaporating the water in a sample and weighing the resulting sediment. They also use instruments called colorimeters, which measure specific wavelengths of light shined through a water sample to determine the concentrations of water pollutants in the water. Researchers in Spain have used several robot fish, each one about as long as a seal, to detect potentially hazardous pollutants in a river.

ggw1962/Shutterstock.com

FIGURE 20.A Colonies of coliform bacteria growing in a petri dish.

CRITICAL THINKING

Runoff of fertilizer into a lake from farmland, lawns, and sewage treatment plants can overload the water with nitrogen and phosphorus—plant nutrients that can cause algae population explosions. How could this lower the dissolved oxygen level of the water and lead to fish kills?

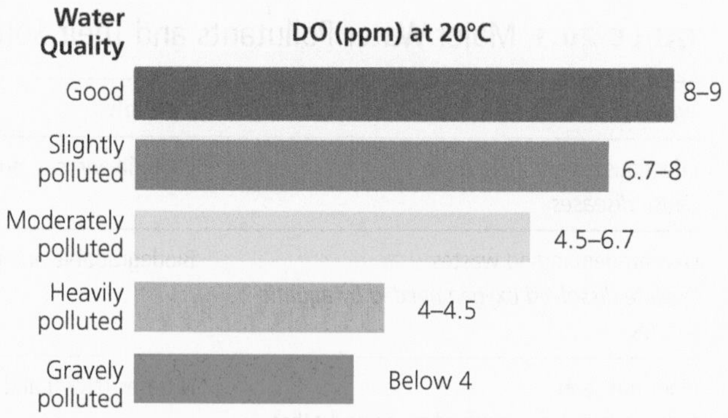

© Cengage Learning

FIGURE 20.B Scientists measure dissolved oxygen (DO) content in parts per million (ppm) at 20°C (68°F) as an indicator of water quality. Only a few fish species can survive in water containing less than 3 ppm of dissolved oxygen at this temperature. ***Critical thinking:*** Would you expect the dissolved oxygen content of polluted water to increase or decrease if the water were heated? Explain.

20.2 WHAT ARE THE MAJOR POLLUTION PROBLEMS IN STREAMS AND LAKES?

CONCEPT 20.2A Many of the world's streams and rivers are polluted, but they can cleanse themselves of biodegradable wastes if we do not overload them or reduce their flows.

CONCEPT 20.2B Adding excessive nutrients to lakes from human activities can disrupt their ecosystems, and preventing such pollution is more effective and less costly than cleaning it up.

Streams Can Cleanse Themselves, If We Do Not Overload Them

Flowing rivers and streams can recover rapidly from moderate levels of biodegradable, oxygen-demanding wastes

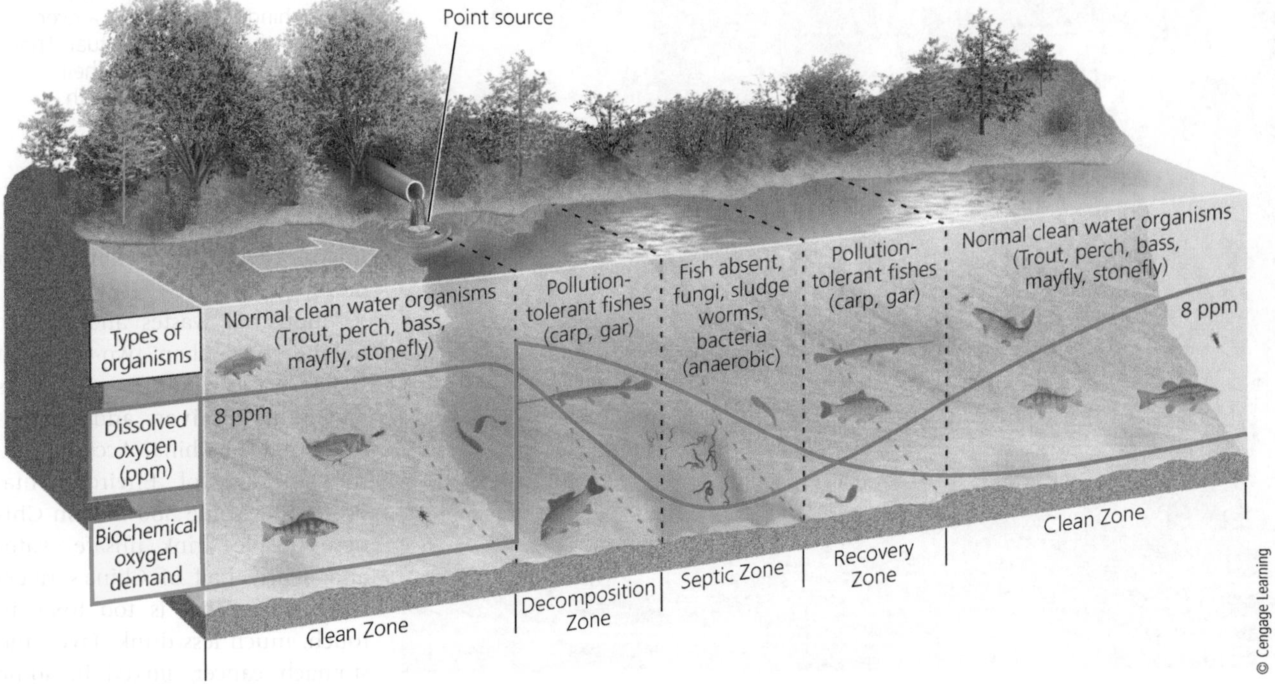

FIGURE 20.6 Natural capital: A stream can dilute and decay degradable, oxygen-demanding wastes and also dilute heated water. This figure shows the oxygen sag curve (blue) and the curve of oxygen demand (red). Streams recover from oxygen-demanding wastes and from the injection of heated water if they are given enough time and are not overloaded. ***Critical thinking:*** What would be the effect of putting another discharge pipe emitting pollutants to the right of the one in this drawing?

through a combination of dilution and bacterial biodegradation. However, this natural recovery process does not work when a stream becomes overloaded with biodegradable pollutants or when drought, damming, or water diversion reduces its flow (**Concept 20.2A**). Also, this process does not eliminate slowly degradable and nonbiodegradable pollutants.

In a flowing stream, the breakdown of biodegradable wastes by bacteria depletes dissolved oxygen and creates an *oxygen sag curve* (Figure 20.6). This reduces or eliminates populations of organisms with high oxygen requirements until the stream is cleansed of oxygen-demanding wastes.

Stream Pollution in More-Developed Countries

Laws enacted in the 1970s to control water pollution have greatly increased the number of facilities that treat wastewater from homes and industries in the United States and in most other more-developed countries. Such laws also require industries to reduce or eliminate their point-source discharges of harmful chemicals into surface waters.

One success story is the cleanup of the U.S. state of Ohio's Cuyahoga River. It was so polluted that it caught fire several times and, in 1969, was photographed while burning as it flowed through the city of Cleveland toward Lake Erie. The highly publicized image of this burning river prompted elected officials to enact laws to limit the discharge of industrial wastes into the river and to provide funds for upgrading sewage treatment facilities. Today, the river is cleaner, is no longer flammable, and is widely used by boaters and anglers. This accomplishment illustrates the power of bottom-up pressure by citizens, who prodded elected officials to change a severely polluted river into an economically and ecologically valuable public resource. GOOD NEWS

Fish kills and drinking water contamination still occur occasionally in some rivers and lakes in more-developed countries such as the United States. Some of these problems are caused by the accidental or deliberate release of toxic inorganic and organic chemicals by industries and mining operations (Figure 20.3).

The U.S. Environmental Protection Agency (EPA) estimates that mining wastes pollute 40% of the headwaters of western watersheds and that cleaning up the estimated 500,000 inactive and abandoned mines found throughout much of the western United States will cost taxpayers at least $50 billion.

The Ohio River is the nation's most polluted river mostly because of sewage overflows and mining and industrial wastes. In August and September 2015, a toxic form of blue-green algae covered almost two-thirds of the

Neale Cousland/Shutterstock.com

FIGURE 20.7 Hindu people in India washing themselves in a river as a part of their daily holy ritual. They also use the river to wash their clothes, and it is polluted with animal and human wastes.

river, mostly from runoff of phosphates and nitrates from cattle feedlots, fertilized crop fields, and leaking sewers. The toxin in the algae can kill animals that drink the river water and can cause vomiting, diarrhea, and liver damage in humans who ingest this water.

Stream Pollution in Less-Developed Countries

In most less-developed countries, stream pollution from discharges of untreated sewage and from industrial wastes and discarded trash is a serious and growing problem. According to the Global Water Policy Project, most cities in less-developed countries discharge 80–90% of their untreated sewage directly into rivers, streams, and lakes whose waters are often used for drinking, bathing, and washing clothes (Figure 20.7).

80–90%

Percentage of the raw sewage in most cities in less developed countries that is discharged directly into waterways

According to the World Commission on Water for the 21st Century, half of the world's 500 major rivers are heavily polluted, with most of these polluted waterways running through less-developed countries. Most of these countries cannot afford to build waste treatment plants and do not have, or do not enforce, laws for controlling water pollution.

Industrial wastes and sewage pollute more than two-thirds of India's water resources, as well as 54 of the 78 rivers and streams monitored in China. According to the Ministry of Environmental Protection, some 380 million Chinese people drink unsafe water and nearly half of China's rivers carry water that is too toxic to touch, much less drink. Liver and stomach cancer, linked in some cases to water pollution, are among the leading causes of death in the Chinese countryside where many industries have been relocated.

CONSIDER THIS . . .

CONNECTIONS Atmospheric Warming and Water Pollution

Climate change from atmospheric warming will likely contribute to water pollution. In a warmer world, some regions will get more precipitation and other areas will get less. More intense downpours will flush more harmful chemicals, plant nutrients, and disease-causing microorganisms into some waterways. In other areas, prolonged drought will reduce river flows that dilute wastes.

Pollution of Lakes and Reservoirs

Lakes and reservoirs are generally less effective at diluting pollutants than streams are, for two reasons. *First*, they often contain stratified layers (Figure 8.15, p. 180) that undergo little vertical mixing. *Second*, they have low flow rates or no flow at all. The flushing and changing of water in lakes and large artificial reservoirs can take from 1 to 100 years, compared with several days to several weeks for streams.

These two factors make lakes and reservoirs more vulnerable than streams are to contamination by runoff or discharges of plant nutrients, oil, pesticides, and nondegradable toxic substances, such as lead, mercury, and arsenic. These contaminants can kill bottom-dwelling organisms and fish, as well as birds that feed on contaminated aquatic organisms. Many toxic chemicals and acids also enter lakes and reservoirs from the atmosphere (Figure 18.11, p. 483).

In addition, the concentrations of some harmful chemicals are biologically magnified as they pass through food webs in these waters. Examples include DDT (Figure 9.14, p. 205), PCBs (Figure 17.10, p. 453), some radioactive isotopes, and some mercury compounds.

Cultural Eutrophication: Too Much of a Good Thing

Eutrophication is the natural nutrient enrichment of a shallow lake, a coastal area at the mouth of a river (**Core Case Study**), or a slow-moving stream. It is caused mostly by runoff of plant nutrients such as nitrates and phosphates from land bordering these bodies of water.

An *oligotrophic lake* is low in nutrients and its water is clear (Figure 8.16, p. 181). Over time, some lakes become more eutrophic (Figure 8.17, p. 182) as nutrients are added from natural and human sources in the surrounding watersheds.

Near urban or agricultural areas, human activities can increase the input of plant nutrients to a lake—a process called **cultural eutrophication.** The inputs involve mostly nitrate- and phosphate-containing effluents from various sources, including farmland, feedlots, urban streets and parking lots, fertilized lawns, mining sites, and municipal sewage treatment plants. Some nitrogen also reaches lakes by deposition from the atmosphere (Figure 18.11, p. 483).

During hot weather or drought, this nutrient overload can produce dense growths of organisms such as algae and cyanobacteria (Figure 8.17, p. 182). When the algae die, they are decomposed by swelling populations of oxygen-consuming bacteria that deplete dissolved oxygen in the surface layer of water near the shore and in the bottom layer of a lake. This lack of oxygen can kill fish and other aerobic aquatic animals that cannot move to safer waters. If excess nutrients continue to flow into a lake, bacteria that do not require oxygen take over and produce gaseous products such as smelly, highly toxic hydrogen sulfide and flammable methane.

According to the EPA, about one-third of the 100,000 medium to large lakes and 85% of the large lakes near major U.S. population centers have some degree of cultural eutrophication. The International Water Association estimates that more than half of the lakes in China suffer from cultural eutrophication (**Concept 20.2B**) (Figure 20.8).

There are several ways to *prevent* or *reduce* cultural eutrophication. Advanced (but expensive) waste treatment processes can remove nitrates and phosphates from wastewater before it enters a body of water. Another option is to mimic the earth's natural cycling of nutrients by recycling the removed nutrients to the soil instead of dumping them into waterways. This is an application of the chemical cycling **principle of sustainability** and an important way to increase our beneficial environmental impact, as long as the nutrients do not contain toxic chemicals. Other preventive measures include banning or limiting the use of phosphates in household detergents and other cleaning agents, and employing soil conservation (Figure 12.30, p. 309) and land-use control to reduce nutrient runoff (see the Case Study that follows).

Lakes suffering from cultural eutrophication can be cleaned up by removing weeds, controlling undesirable plant growth with herbicides and algaecides, and pumping air into lakes and reservoirs to prevent oxygen depletion. However, these methods are expensive and energy-intensive. Most lakes and other surface waters can eventually recover from cultural eutrophication if excessive inputs of plant nutrients are stopped.

Yang Xiaoyuan/Xinhua Press/Corbis Wire/Corbis

FIGURE 20.8 Severe cultural eutrophication of Chao Lake in China.

CASE STUDY

Pollution in the Great Lakes

The five interconnected Great Lakes of North America (Figure 20.9) contain about 95% of the fresh surface water in the United States and 21% of the world's fresh surface water. At least 40 million people in the United States and Canada obtain their drinking water from these

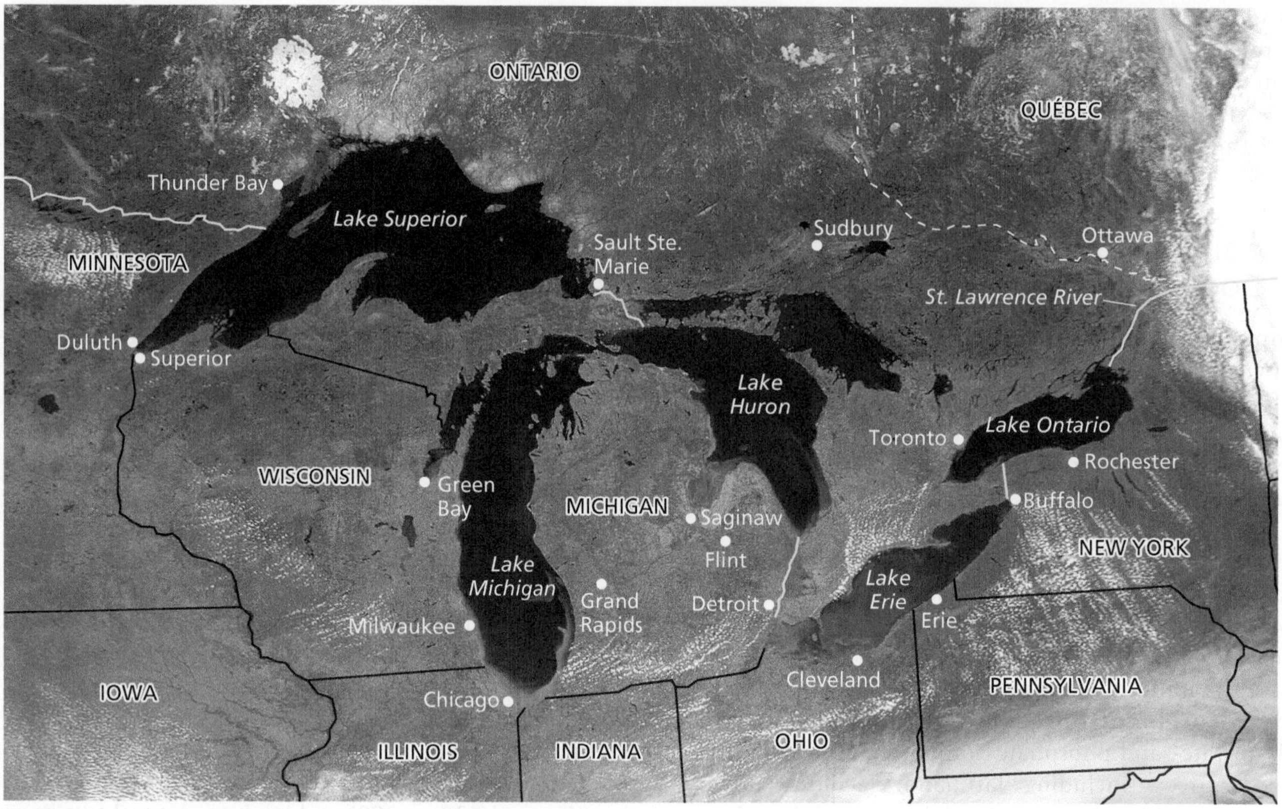

NASA

FIGURE 20.9 The five Great Lakes of North America make up the world's largest freshwater system.

lakes. In addition, the lakes provide food (especially from a $4 billion a year fishing industry), water purification, recreation, and jobs and help support the economies of nearby communities and states.

Despite their enormous size, these lakes are vulnerable to pollution from point and nonpoint sources. One reason is that each year, less than 1% of the water entering these lakes flows east into the St. Lawrence River and then out to the Atlantic Ocean. This means that it can take 19 years for pollutants to flush from the largest lake (Superior) to sea and 3 years from the smallest lake (Erie).

By the 1960s, many areas of the Great Lakes were suffering from severe cultural eutrophication, huge fish kills, and contamination from bacteria and a variety of toxic industrial wastes. Lake Erie suffered the most because it is the shallowest of the Great Lakes and has the highest concentrations of people and industrial activity along its shores.

In 1972, the United States and Canada signed the Great Lakes Water Quality Agreement, which is considered a model of international cooperation. They agreed to spend more than $20 billion to maintain and restore the chemical, physical, and biological integrity of the Great Lakes basin ecosystem. This program has helped to cut the number and sizes of algal blooms, raised dissolved oxygen levels, boosted sport and commercial fishing catches in Lake Erie, and allowed most swimming beaches to reopen. These improvements resulted mainly from the use of new or upgraded sewage treatment plants, better treatment of industrial wastes, and bans on the use of detergents, household cleaners, and water conditioners that contain phosphates. Most of these measures were instituted largely because of bottom-up citizen pressure on federal and state elected officials.

Despite this important progress, many problems remain. Increasing nonpoint runoff of pesticides and fertilizers resulting from urban sprawl, fueled by population growth, now surpasses industrial pollution as the greatest threat to the lakes. Bottom sediments in 26 toxic hotspots remain heavily polluted. In addition, some of the Great Lakes, such as Lake Erie, still suffer from occasional cultural eutrophication (Figure 20.10).

About half of the toxic compounds entering the lakes still come from atmospheric deposition of pesticides, mercury from coal-burning plants, and other toxic chemicals from as far away as Mexico and Russia. A survey done by Wisconsin biologists found that one of every four fish taken from the Great Lakes was unsafe for human consumption, mostly because of their mercury content (Figure 17.1, p. 442).

Between 2010 and 2015, the U.S. Congress provided about $1.3 billion for the Great Lakes Restoration Initiative overseen by the EPA. It has focused on reducing environmental threats such as toxic pollution, cultural

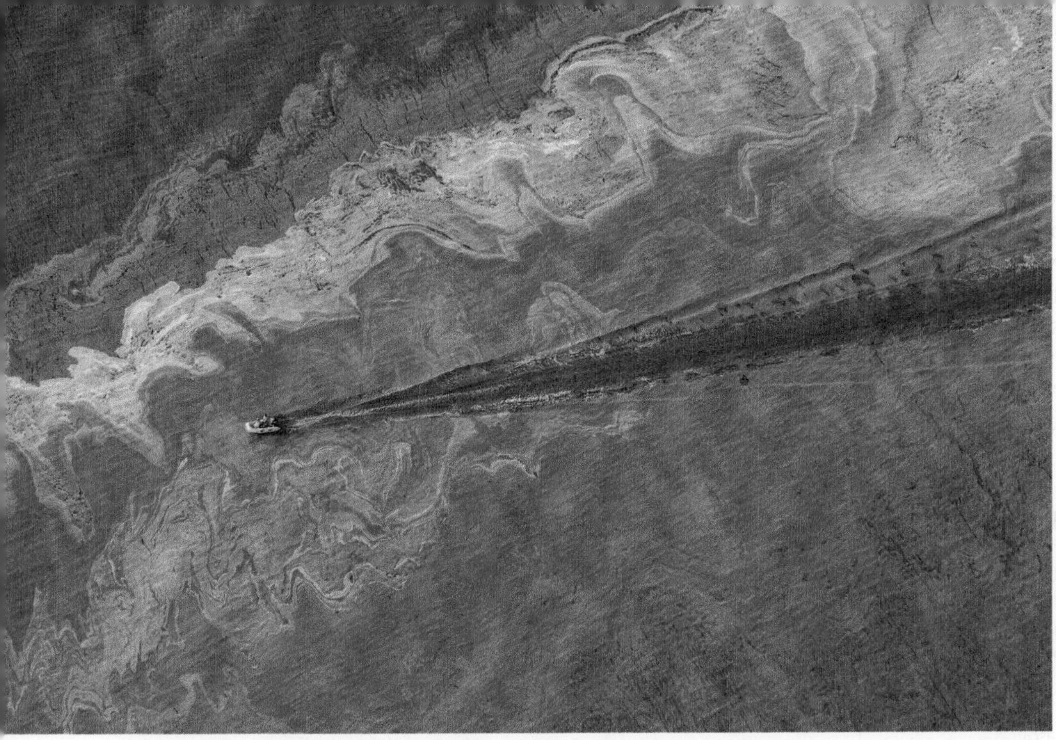

Peter Essick/National Geographic Creative

FIGURE 20.10 *Eutrophication:* This toxic algae bloom covered one-third of Lake Erie in 2011.

eutrophication (Figure 20.10), loss of wildlife habitat, invasive species, and soil erosion and runoff into the lakes. It also promotes wetlands restoration.

Some environmental and health scientists call for a prevention approach (**Concept 20.2B**) that would ban the use of toxic chlorine compounds, such as bleach used in the pulp and paper industry, which is prominent around the Great Lakes. They would also stop the discharge into the lakes of 70 toxic chemicals that threaten human health and wildlife. So far, officials in the affected industries have successfully opposed such bans.

Gulf of Mexico Dead Zone: A Closer Look

Since the 1950s, the level of nitrates discharged from the Mississippi River into the northern Gulf of Mexico has nearly tripled (**Core Case Study**). Each year this causes severe depletion of dissolved oxygen (Figure 20.1, bottom, red area) and creates a dead zone in the northern Gulf of Mexico.

In more detail, here is how the Gulf's seasonal dead zone forms. During the spring and summer, nitrate- and phosphate-laden freshwater flowing into the Gulf forms an oxygen-rich layer on top of the Gulf's cooler and denser saltwater. Because there are few storms at this time of year, this sun-heated upper layer of water remains calm and does not mix with the bottom layer of low-oxygen water. The combination of sunlight and large inputs of nitrate and phosphate plant nutrients from fertilizer and sewage into the freshwater layer leads to massive blooms of phytoplankton, mostly blue-green algae.

When these algae die, they sink into the saltier water below where they are decomposed by oxygen-consuming bacteria. This reduces the dissolved oxygen content in the deeper water to 2 parts per million or lower. Mobile species can survive this lack of oxygen by migrating to oxygen-rich waters, but certain species of fish, shellfish, and other organisms are not able to escape, and they die.

The resulting dead zone disrupts the Gulf's food web, because the die-offs lead to the deaths of seabird and marine mammal species that depend on the dying fish and shellfish for their survival. The dead zone breaks up, beginning in early fall when cooler weather, storms, and hurricanes mix the top and lower layers of water and distribute dissolved oxygen throughout the layers.

In addition to greatly increased nitrate and phosphate levels from human inputs, other human factors have contributed to the formation of the dead zone. In efforts to control flooding along the upper Mississippi River, engineers have dredged and straightened parts of the river and raised its banks with levees in many places. This has increased the river's flow of nutrients and sediment pollution into the Gulf. In addition, most of the river basin's original freshwater wetlands, which acted as natural filters that helped to remove excess nutrients and sediments from flood water, have been drained and filled in for farming and urban development.

The seasonal formation of dead zones in the northern Gulf of Mexico and in other areas, mostly resulting from human activities, is a reminder that in nature, everything is connected. Plant nutrients flowing into the Mississippi from a farm in Iowa or a sewage treatment plant in Wisconsin help kill fish and shellfish a thousand miles away on the gulf coast of Texas. Researchers warn that if the size of the Gulf's annual dead zone is not sharply reduced, its long-term effects could permanently alter the ecological makeup of these coastal waters.

20.3 WHAT ARE THE MAJOR GROUNDWATER POLLUTION PROBLEMS?

CONCEPT 20.3A Chemicals used in agriculture, industry, transportation, and homes can spill and leak into groundwater and make it undrinkable.

CONCEPT 20.3B We can purify groundwater, but protecting it through pollution prevention is the least expensive and most effective strategy.

Groundwater Cannot Cleanse Itself Very Well

Aquifers provide drinking water for about half of the U.S. population and 95% of Americans who live in rural areas. According to many scientists, groundwater pollution is a serious but hidden global human health threat. Common pollutants such as fertilizers, pesticides, gasoline, and organic solvents can seep into groundwater from numerous sources (Figure 20.11). People dumping or spilling gasoline, oil, and paint thinners and other organic solvents onto the ground can also contaminate groundwater (**Concept 20.3A**).

The drilling of thousands of new oil and natural gas wells in parts of the United States involving a process called *hydraulic fracturing*, or *fracking* (see Chapter 15, Core Case Study, p. 382), is also a growing threat to groundwater. Contamination of groundwater used for drinking water can come from leaky gas well pipes and pipe fittings and from contaminated wastewater brought to the surface during fracking operations (Figure 20.12). Reducing this serious groundwater pollution threat will require stricter

FIGURE 20.11 Natural capital degradation: Principal sources of groundwater contamination in the United States. Another source in coastal areas is saltwater intrusion from excessive groundwater withdrawal. (Figure is not drawn to scale.) ***Critical thinking:*** What are three contamination sources in the figure that might be affecting groundwater in your area?

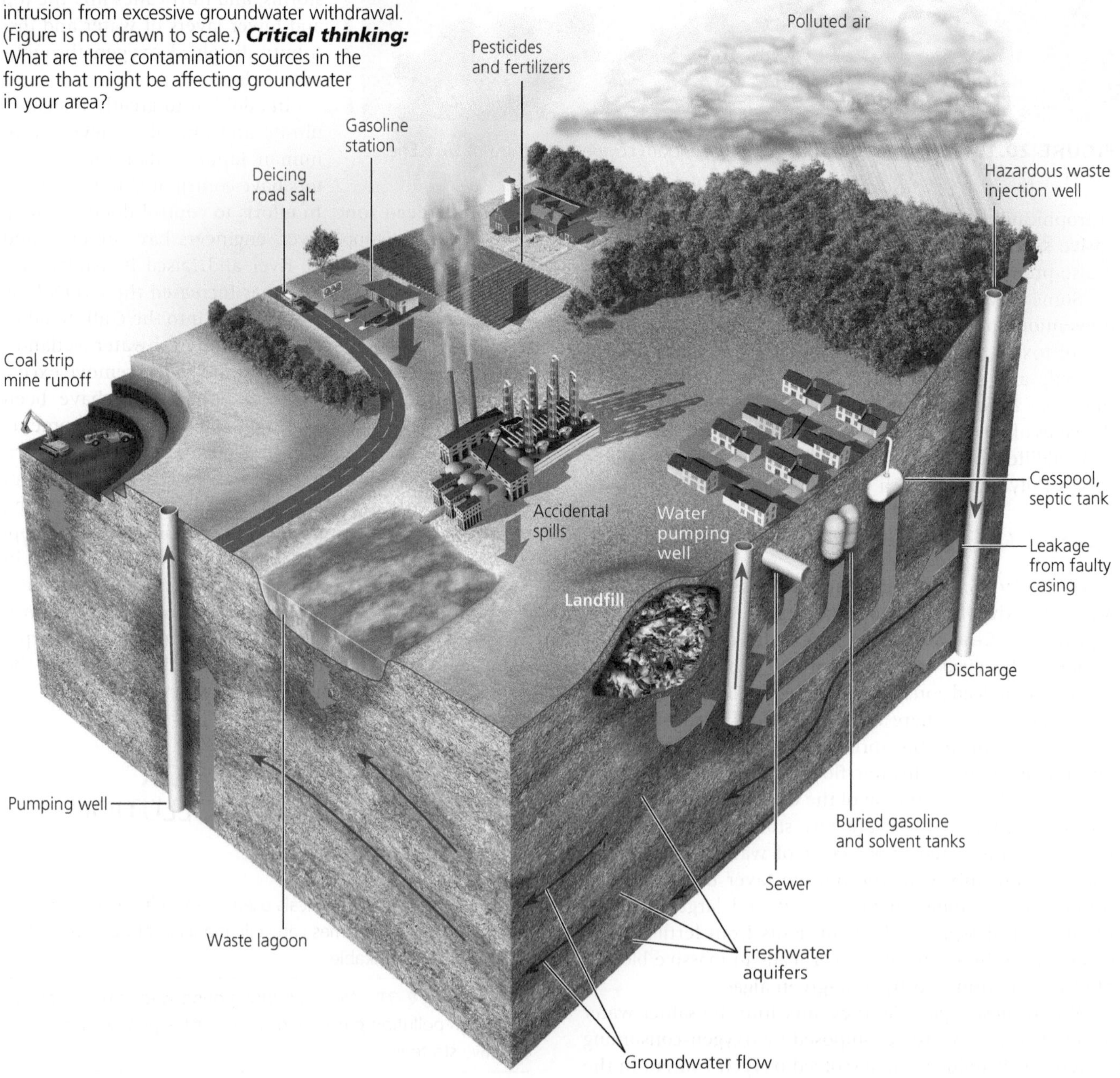

FIGURE 20.12 Well water in Wyoming polluted by groundwater contamination from a natural gas fracking well, probably from faulty seals in the well pipe.

monitoring and regulation of fracking wastewater (currently exempted from regulation) under the federal drinking water and water pollution control laws.

Once a pollutant from a leaking underground storage tank or other source contaminates groundwater, it fills the aquifer's porous layers of sand, gravel, or bedrock like water saturates a sponge. Removing this contaminant is difficult and costly. Groundwater flows so slowly—usually less than 0.3 meter (1 foot) per day—that contaminants are not effectively diluted and dispersed. In addition, groundwater usually has much lower concentrations of dissolved oxygen (which helps decompose some biodegradable contaminants) and smaller populations of decomposing bacteria. The usually cold temperatures of groundwater also slow down chemical reactions that decompose wastes.

Groundwater Pollution Is a Hidden Threat

On a global scale, we do not know much about groundwater pollution because few countries go to the great expense of locating, tracking, and testing aquifers. However, the results of scientific studies in scattered parts of the world are alarming.

Groundwater provides about 70% of China's drinking water. According to the Chinese Ministry of Land and Resources, about 90% of China's shallow groundwater is polluted with chemicals such as toxic heavy metals, organic solvents, nitrates, petrochemicals, and pesticides. About 37% of this groundwater is so polluted that it cannot be treated for use as drinking water. Every year, according to the WHO and the World Bank, contaminated drinking water causes illness in an estimated 190 million Chinese and kills about 60,000.

In the United States, an EPA survey of 26,000 industrial waste ponds and lagoons found that one-third of them had no liners to prevent toxic liquid wastes from seeping into aquifers. In addition, almost two-thirds of America's liquid hazardous wastes are injected into the ground in disposal wells (Figure 20.11). Leaking injection pipes and seals in such wells can contaminate aquifers used as sources of drinking water.

By 2015, the EPA had cleaned up about 483,000 of the more than 569,000 underground tanks in the United States that were leaking gasoline, diesel fuel, home heating oil, or toxic solvents into groundwater. During this century, scientists expect many of the millions of such tanks installed around the world to corrode and leak, contaminating groundwater, and becoming a major global health problem. Determining the extent of a leak from a single underground tank can cost hundreds of thousands of dollars, and cleanup costs can be even higher. If the chemical reaches an aquifer, effective cleanup is often not possible or is too costly.

It can take decades to thousands of years for contaminated groundwater to cleanse itself of *slowly degradable wastes* (such as DDT). On a human time scale, *nonbiodegradable wastes* (such as toxic lead and arsenic; see the Case Study that follows) remain in the water permanently. Although there are ways to remove some pollutants from groundwater (Figure 20.13, right), such methods are expensive. Cleaning up a single contaminated aquifer can cost anywhere from 10 million to several hundred million dollars depending on the size of the aquifer and the types of contaminants. Therefore, preventing groundwater contamination (Figure 20.13, left) is the most effective way to deal with this serious pollution problem (**Concept 20.3B**).

CASE STUDY

Arsenic in Drinking Water

Arsenic contaminates drinking water when a well is drilled into an aquifer where the soil and rock are naturally rich in arsenic, or when human activities such as mining and ore processing release arsenic into drinking water supplies.

The WHO estimates that more than 140 million people in 70 countries are drinking water with arsenic concentrations of 5 to 100 times the accepted safe level of 10 parts per billion (ppb). According to some scientists from the WHO and other organizations, even the 10 ppb level is not safe.

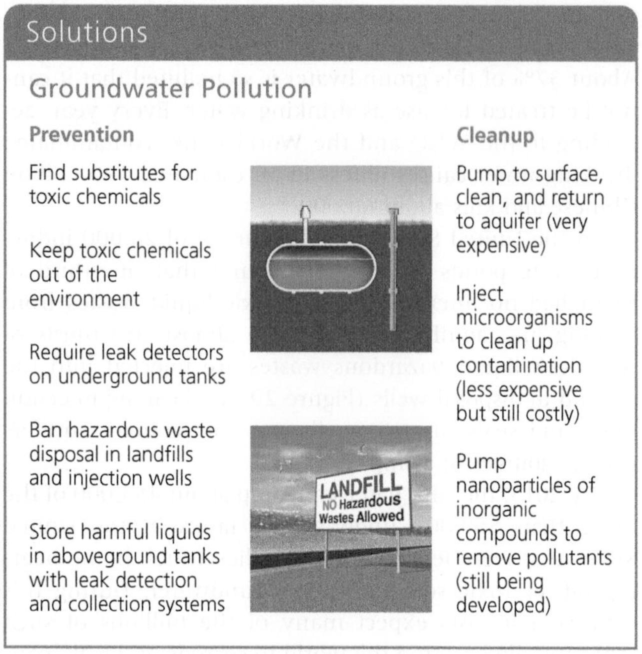
Solutions
Groundwater Pollution
Prevention
Find substitutes for toxic chemicals
Keep toxic chemicals out of the environment
Require leak detectors on underground tanks
Ban hazardous waste disposal in landfills and injection wells
Store harmful liquids in aboveground tanks with leak detection and collection systems
Cleanup
Pump to surface, clean, and return to aquifer (very expensive)
Inject microorganisms to clean up contamination (less expensive but still costly)
Pump nanoparticles of inorganic compounds to remove pollutants (still being developed)
LANDFILL
NO Hazardous Wastes Allowed

FIGURE 20.13 We can clean up contaminated groundwater, but prevention is the only effective approach (**Concept 20.3B**). ***Critical thinking:*** Which two of these preventive solutions do you think are the most important? Why?

Arsenic levels are especially high in Bangladesh, China, India's state of West Bengal, and parts of northern Chile. The WHO estimates that long-term exposure to nondegradable arsenic in drinking water is likely to cause hundreds of thousands of premature deaths from cancer of the skin, bladder, and lungs. According to the EPA, some 13 million people in several thousand communities, mostly in the western United States, are exposed to arsenic levels of 3 to 10 ppb in their drinking water.

Researchers from Rice University in Houston, Texas (USA), reported that by suspending nanoparticles of rust in arsenic-contaminated water and then drawing them out with handheld magnets, they had removed enough arsenic from the water to make it safe to drink. Use of this technique could greatly reduce the threat of arsenic in drinking water for many families at a cost of a few cents a day.

Purifying Drinking Water

Most more-developed countries have laws establishing drinking water standards. However, most less-developed countries do not have such laws or, if they have them, they do not enforce them.

In more-developed countries, surface water withdrawn for use as drinking water is typically stored in a reservoir for several days. This improves clarity and taste by increasing dissolved oxygen content and allowing suspended matter to settle. The water is then pumped to a purification plant and treated to meet government drinking water standards.

In areas with pure groundwater or surface water sources, little treatment is necessary. Several major U.S. cities, including New York City, Boston, Seattle, and Portland, Oregon, have avoided building expensive drinking water treatment facilities. Instead, they have invested in protecting the forests and wetlands in the watersheds that provide their water supplies, thereby increasing their beneficial environmental impacts.

We have the technology to convert sewer water into pure drinking water. The California cities of Los Angeles and San Diego are purifying wastewater to the point where it is fit to drink and putting this water into underlying aquifers, thereby reducing their dependence on water imported from Northern California (Figure 13.17, p. 340) and from the Colorado River (see Chapter 13, Core Case Study, p. 324). In a world where people will face increasing shortages of drinking water, wastewater purification is likely to become a major growth business. **GREEN CAREER: Wastewater purification**

There are also simpler measures to purify drinking water. In tropical countries that lack centralized water treatment systems, the WHO urges people to purify drinking water by exposing a clear plastic bottle filled with contaminated water to intense sunlight. The sun's heat and ultraviolet (UV) rays can kill infectious microbes in as little as 3 hours. Painting one side of the bottle black can improve heat absorption in this simple solar disinfection method, which applies the solar energy **principle of sustainability**. Where this simple measure has been used, the incidence of dangerous childhood diarrhea has decreased by 30–40%. Researchers found that they can speed up this disinfection process by adding lime juice to the water bottles.

Danish inventor Torben Vestergaard Frandsen has developed the *LifeStraw*™, an inexpensive, portable water filter that eliminates many viruses and parasites from water that is drawn through it (Figure 20.14). It has been particularly useful in Africa, where aid agencies are distributing it. Inventor Dean Kamen has developed a small and simple device for purifying dirty water (Science Focus 20.2). Another option being used by more and more people around the world is bottled water, which has created or worsened some environmental problems (see the Case Study that follows).

CASE STUDY

Is Bottled Water a Good Option?

Bottled water can be a useful but expensive option in countries and areas where people do not have access to safe and clean drinking water. However, experts say the United States has some of the world's cleanest drinking water. Municipal water systems in the United States are required to test their water regularly for a number of

SCIENCEFOCUS 20.2

Converting Dirty Water to Drinking Water

Dean Kamen is a college dropout and mastermind inventor with more than 400 U.S. and foreign patents. Kamen has used royalties from his many inventions to employ 500 engineers and technicians at his company, Deka Research, which is devoted to creating prototypes of devices designed to solve various problems.

In partnership with Coca-Cola, his company has developed a small portable device about the size of a dormitory refrigerator that can convert dirty water to drinking water. It could be used in remote areas of less-developed countries where millions of people lack access to clean water. The process involves sticking a hose into a polluted water source such as a river or outdoor toilet pit and using a small pump to transfer the dirty water to a chamber in Kamen's water purifier, called the Slingshot.

Widespread use of this device to purify water for drinking, cooking, and hygiene in less-developed countries would sharply reduce disease and could prevent hundreds of thousands of deaths that result annually from drinking contaminated water. The device has been tested successfully in several African and Latin American countries. Kamen wants to find companies to mass produce the device and distribute it to people in remote villages, health-care centers, and schools in less-developed counties.

CRITICAL THINKING

What role, if any, should governments in less-developed countries play in helping distribute this device?

Vestergaard Frandsen

FIGURE 20.14 The *LifeStraw*™ is a personal water purification device that gives many poor people access to safe drinking water. Here, four young men in Uganda demonstrate its use. ***Critical thinking:*** Do you think the development of such devices should make prevention of water pollution less of a priority? Explain.

pollutants and to make the results available to citizens. Yet, about half of all Americans worry about getting sick from tap water contaminants, and many drink high-priced bottled water or install expensive water purification systems.

Studies by the Natural Resources Defense Council (NRDC) reveal that in the United States, a bottle of water costs between 240 and 10,000 times as much as the same volume of tap water. Water expert Peter Gleick has estimated that more than 40% of the expensive bottled water that Americans drink is bottled tap water. A 4-year study by the NRDC concluded that most bottled water is of good quality but they found traces of bacteria and synthetic organic chemicals in 23 of the 123 brands tested. Bottled water is less regulated than tap water and the EPA contamination standards that apply to public water supplies do not apply to bottled water.

Use of bottled water also causes environmental problems. In the United States, according to the Container Recycling Institute, more than 67 million plastic water bottles are discarded every day—enough bottles in a year to, if lined up end-to-end, wrap around the planet at its equator about 280 times. Most water bottles are made of recyclable plastic, but in the United States, only about 38% of these bottles are recycled. Many of the billions of discarded bottles end up in landfills, where they can remain for hundreds of years. Those that are burned in incinerators without high-tech pollution controls release some of their harmful chemicals into the atmosphere. Also, millions of discarded bottles get scattered on the land and end up in rivers, lakes (Figure 20.5), and oceans. By contrast, in Germany most bottled water is sold in returnable and reusable glass bottles.

It takes huge amounts of energy to manufacture bottled water and to transport it across countries and around the world, as well as to refrigerate much of it in stores. Toxic gases and liquids are released during the manufacture of plastic water bottles, and greenhouse gases and other air pollutants are emitted by the fossil fuels burned to make them and to deliver bottled water to suppliers. In addition, withdrawing groundwater for bottling is helping to deplete some aquifers. There is also concern about health risks from chemicals such as bisphenol A (BPA; see Science Focus 17.3, p. 455) that can leach into the water from the plastic in some water bottles, especially if they are exposed to the hot sun.

Because of these harmful environmental impacts and the high cost of bottled water, there is a growing *back-to-the-tap* movement. From San Francisco to New York to Paris, city governments, restaurants, schools, religious groups, and many consumers are refusing to buy bottled water. In 2015, San Francisco became the first city to ban the sale of plastic water bottles. Violators can face fines of up to $1,000. People are also refilling portable bottles with tap water and using simple filters to improve the taste and color of water where necessary. Some health officials suggest that before drinking expensive bottled water or buying costly home water purifiers, consumers should have their water tested by local health departments or private labs (but not by companies trying to sell water purification equipment).

Using Laws to Protect Drinking Water Quality

About 54 countries, most of them in North America and Europe, have legal standards for safe drinking water. For example, the U.S. Safe Drinking Water Act of 1975 (amended in 1996) requires the EPA to establish national drinking water standards, called *maximum contaminant levels,* for any pollutants that could have adverse effects on human health. Currently, this act strictly limits the levels of 91 potential contaminants in U.S. tap water. However, in most less-developed countries, such laws do not exist or are not enforced.

Health scientists call for strengthening the U.S. Safe Drinking Water Act in several ways. Here are three of their recommendations:

- Combine many of the drinking water treatment systems that serve fewer than 3,300 people with nearby larger systems to make it easier for these smaller systems to meet federal standards.
- Strengthen and enforce requirements concerning public notification of violations of drinking water standards.
- Ban the use of toxic lead in new plumbing pipes, faucets, and fixtures and remove existing lead pipes (see Case Study that follows).

According to the NRDC, making these three improvements would cost U.S. taxpayers an average of about $30 a year per household and decrease health threats from contaminated water.

However, various industries have pressured elected officials to weaken the Safe Drinking Water Act, because complying with it increases their costs. One proposal is to eliminate national testing of drinking water and requirements for public notification about violations of drinking water standards. Another proposal would allow states to give providers of drinking water a permanent right to violate the standard for a given contaminant if they claim they cannot afford to meet the standard. Some critics also call for greatly reducing the EPA's already-low budget for enforcing the Safe Drinking Water Act.

CASE STUDY

Lead in Drinking Water

In 2014, many people in older and poorer neighborhoods of Flint, Michigan—a city of nearly 100,000 people with 40% living in poverty—were exposed to dangerous levels of lead in their tap water. The problem began when, in an effort to save money, Flint officials began withdrawing drinking water from the Flint River instead of from Lake Huron.

Officials did not take into account that there were at least 15,000 lead pipes connecting the city's main line water pipes (that do not contain lead) to homes, many of them in older and poorer neighborhoods. They failed to add chemicals to reduce the leaching of lead from the pipes exposed to the more corrosive water supply.

As a result, lead began leaching into the water supply for many homes in older and poorer neighborhoods.

Research shows that exposure to high levels of lead—a potent neurotoxin—is especially harmful to the developing brains and nervous systems of children. This problem is complicated by the fact that lead is a cumulative poison that builds up in the body. Thus, exposure to even small amounts of lead has posed a long-term threat to the nervous system, brain, and other organs, and threatened the intellectual development and physical growth of the 9,000 children under age 6 living in Flint.

After Flint changed its drinking water source, the percentage of children citywide with elevated blood levels of lead doubled from 2.4% to 4.9% and to 15.7% in its poorest neighborhood. This meant that roughly 1 out of 6 children in Flint and 1 out of 3 in its poorest neighborhood were exposed to blood levels of lead that the Centers for Disease Control and Prevention (CDC) uses as a standard to identify children who need medical help. In 1986, the EPA banned the use of lead water pipes, fittings, solder, and other plumbing material in new homes. However, this regulation did not cover the millions of lead pipes and plumbing material in older homes in Flint and elsewhere in the United States.

After a highly publicized outcry, Flint officials switched the city's water supply source back to Lake Huron, but the health threats to children remain. Officials are seeking funding to have the blood lead levels in affected children under 6 monitored for several years. The ultimate solution is to replace all of the lead pipes connecting homes to the city's main line water pipes, which will be costly for the city and homeowners.

Public health officials say that Flint is just the tip of the iceberg. Investigations revealed that in 2015 almost 2,000, or 20%, of U.S. water systems tested have failed to meet the EPA's standard of 15 parts per billion (ppb) for lead in drinking water. A level of 40 ppb poses an imminent and substantial threat to the health of children and pregnant women, according to the EPA. This included about 350 schools and daycare centers, which means that many children are vulnerable to unsafe levels of lead. The EPA and health officials have urged all U.S. residents to have their drinking water tested for lead.

Lead in drinking water is not the only threat. The use of lead paint in U.S. homes was banned in 1978. But at least 24 million homes built before 1978 have lead paint that can deteriorate and be inhaled or picked up by the fingers of young children.

According to the CDC, 535,000 U.S. children ages 1 through 5 suffer from lead poisoning. This is a silent epidemic because lead adds no taste, color, or smell to drinking water. Many of these victims are low-income children of color. The situation is much worse in some less-developed countries.

Solutions to the country's lead poisoning problem include replacing all of the lead pipes now servicing homes, at a cost of at least $55 billion; removing lead paint from homes built before 1978; improving tests for lead poisoning; and providing free lead testing for children ages 1 to 6. This will take years and cost cities, taxpayers, and homeowners billions of dollars. Meanwhile in 2012 Congress, under pressure from the lead industry, slashed funding for lead screening of children, lead removal from older homes, and CDC lead testing and research.

20.4 WHAT ARE THE MAJOR OCEAN POLLUTION PROBLEMS?

CONCEPT 20.4A Most ocean pollution originates on land and includes oil and other toxic chemicals and solid waste, which threaten fish and wildlife and disrupt marine ecosystems.

CONCEPT 20.4B The key to protecting the oceans is to reduce the flow of pollution from land and air and from streams emptying into ocean waters.

Ocean Pollution Is a Growing Problem

We should care about the oceans because they keep us alive. Oceans help recycle the planet's freshwater through the water cycle (Figure 3.19, p. 63). They also strongly affect weather and climate, help regulate the earth's temperature, and absorb some of the massive amounts of carbon dioxide that we emit into the atmosphere.

As oceanographer and explorer Sylvia A. Earle (see Individuals Matter 11.1, p. 270) reminds us: "Even if you never have the chance to see or touch the ocean, the ocean touches you with every breath you take, every drop of water you drink, every bite you consume. Everyone, everywhere is inextricably connected to and utterly dependent upon the existence of the sea." Despite its importance, we treat the ocean as the world's largest dump for the massive and growing amount of wastes and pollutants that we produce.

Coastal areas—especially wetlands, estuaries, coral reefs, and mangrove swamps—receive the largest inputs of pollutants and wastes (Figure 20.15). Roughly 40% of the world's people (53% in the United States) live on or near coastlines, and coastal populations are projected to double by 2050. This explains why 80% of marine pollution originates on land (**Concept 20.4A**).

According to a study by the U.N. Environment Programme (UNEP), 80–90% of the municipal sewage from coastal areas of less-developed countries is dumped into oceans without treatment. This often overwhelms the ability of the coastal waters to degrade the wastes. For example, many areas of China's coastline are so choked with algae growing on the nutrients provided by sewage that some scientists have concluded that large areas of China's coastal waters can no longer sustain marine ecosystems. Dumping biodegradable wastes and plant nutrients into coastal waters instead of recycling these vital plant nutrients to the soil violates the chemical cycling **principle of sustainability**.

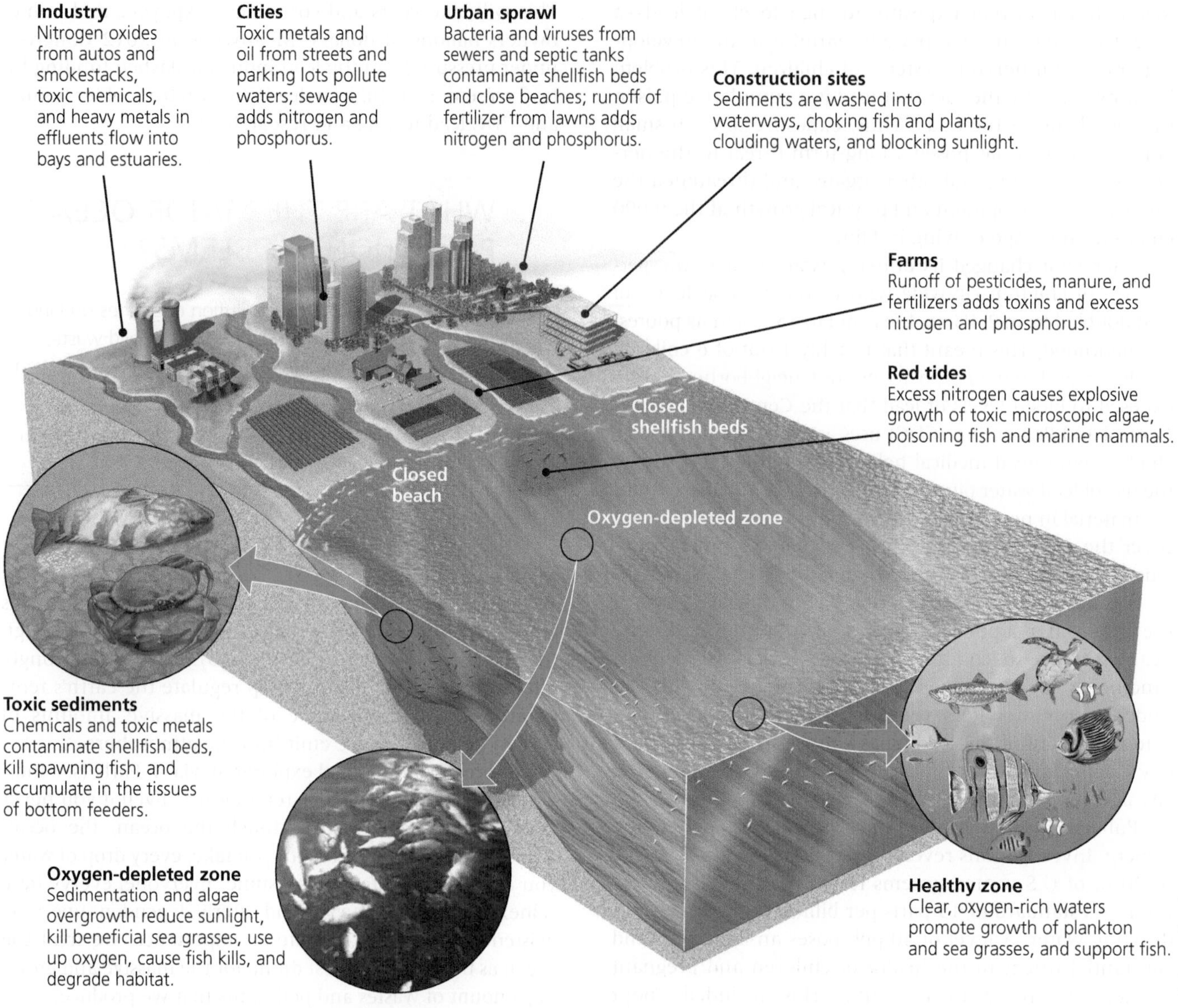
Industry
Nitrogen oxides from autos and smokestacks, toxic chemicals, and heavy metals in effluents flow into bays and estuaries.
Cities
Toxic metals and oil from streets and parking lots pollute waters; sewage adds nitrogen and phosphorus.
Urban sprawl
Bacteria and viruses from sewers and septic tanks contaminate shellfish beds and close beaches; runoff of fertilizer from lawns adds nitrogen and phosphorus.
Construction sites
Sediments are washed into waterways, choking fish and plants, clouding waters, and blocking sunlight.
Farms
Runoff of pesticides, manure, and fertilizers adds toxins and excess nitrogen and phosphorus.
Red tides
Excess nitrogen causes explosive growth of toxic microscopic algae, poisoning fish and marine mammals.
Closed shellfish beds
Closed beach
Oxygen-depleted zone
Toxic sediments
Chemicals and toxic metals contaminate shellfish beds, kill spawning fish, and accumulate in the tissues of bottom feeders.
Oxygen-depleted zone
Sedimentation and algae overgrowth reduce sunlight, kill beneficial sea grasses, use up oxygen, cause fish kills, and degrade habitat.
Healthy zone
Clear, oxygen-rich waters promote growth of plankton and sea grasses, and support fish.

FIGURE 20.15 Natural capital degradation: Residential areas, factories, and farms all contribute to the pollution of coastal waters. ***Critical thinking:*** What do you think are the three worst pollution problems shown here? For each one, how does it affect two or more of the ecosystem and economic services listed in Figure 8.4 (p. 170)?

80–90%

Percentage of municipal sewage dumped untreated into the coastal waters of less-developed countries

In deeper waters, the oceans can dilute, disperse, and degrade large amounts of raw sewage and other types of degradable pollutants. Some scientists suggest that it is safer to dump sewage sludge, toxic mining wastes, and most other harmful wastes into the deep ocean than to bury them on land or burn them in incinerators. Other scientists disagree and point out that dumping harmful wastes into the ocean would delay urgently needed pollution prevention measures and promote further degradation of this vital part of the earth's life-support system.

Recent studies of some U.S. coastal waters found vast colonies of viruses thriving in raw sewage and in effluents from sewage treatment plants (which do not remove viruses) and leaking septic tanks. According to one study, one-fourth of the people using coastal beaches in the United States develop ear infections, sore throats, eye irritations, respiratory diseases, or gastrointestinal diseases from swimming in seawater containing infectious viruses and bacteria.

Scientists also point to the underreported problem of pollution from cruise ships. A cruise liner can carry as many as 6,300 passengers and 2,400 crewmembers, and it can generate as much waste (toxic chemicals, garbage, sewage, and waste oil) as a small city. Many cruise ships dump these wastes at sea. In U.S. waters, such dumping is illegal, but some ships continue dumping secretively, usually at night. Some environmentally aware vacationers are refusing to go on cruise ships that do not have sophisticated systems for dealing with the wastes the ships produce.

Runoff of sewage and agricultural wastes into coastal waters introduces large quantities of nitrate (NO_3^-) and phosphate (PO_4^{3-}) plant nutrients that can cause explosive growths of harmful algae and lead to dead zones (**Core Case Study**). These *harmful algal blooms*—also called red, brown, or green toxic tides—release waterborne and airborne toxins that poison seafood, damage fisheries, kill some fish-eating birds, and reduce tourism. Each year, harmful algal blooms lead to the poisoning of about 60,000 Americans who eat shellfish contaminated by the algae.

Harmful algal blooms occur annually in at least 400 *oxygen-depleted* or *dead zones* around the world, mostly in temperate coastal waters and in large bodies of water with restricted outflows, such as the Baltic and Black Seas. The largest such zone in U.S. coastal waters forms each year in the northern Gulf of Mexico (**Core Case Study**). A study by Luan Weixin, of China's Dalain Maritime University, found that nitrates and phosphates have seriously contaminated about half of China's shallow coastal waters. Warmer ocean water temperatures, caused when the oceans take up much of the excess heat in the atmosphere that is getting warmer because of human emissions of greenhouse gases, are extending the size and duration of dead zones in the world's oceans.

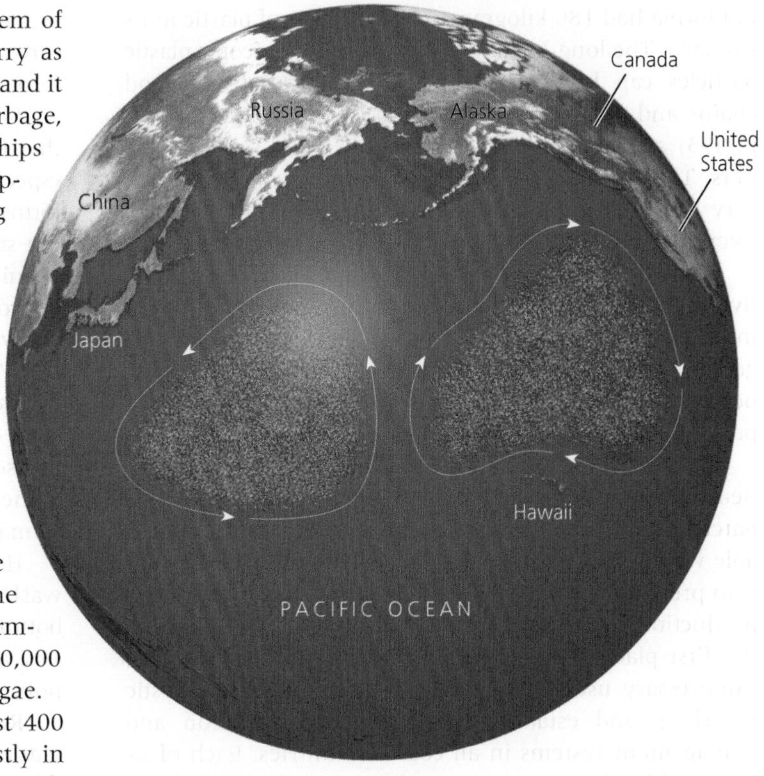

FIGURE 20.16 The North Pacific Garbage Patch is actually two vast, slowly swirling masses of small plastic particles floating just under the water. Five other huge garbage patches have been discovered in the world's other major oceans.

CASE STUDY

Ocean Garbage Patches: There Is No Away

In the 1970s, scientists from the Woods Hole Oceanographic Institution predicted that swirling patches of garbage would form in the Pacific Ocean. In 1997, ocean researcher Charles J. Moore discovered two gigantic, slowly rotating masses of plastic and other solid wastes in the middle of the North Pacific Ocean near the Hawaiian Islands. These wastes are mostly small particles floating on or just beneath the ocean's surface. Known as the *North Pacific Garbage Patch*, these solid wastes are trapped there by a vortex where rotating ocean currents called *gyres* meet (Figure 20.16).

Roughly 80% of this trash comes from the land. It is washed or blown off beaches, pours out of storm drains, and floats down streams and rivers that empty into the sea from the west coast of North America, the east coast of Asia, and hundreds of Pacific islands (**Concept 20.4A**). Most of the rest comes from discarded fishing gear and from wastes dumped into the ocean from cargo and cruise ships.

The North Pacific Garbage Patch—viewed as the planet's largest human trash dump—is estimated by some scientists to occupy an area at least the size of Texas. However, such estimates are difficult to verify because this continuously swirling plastic-laden soup consists mostly of small rice-grain- to pencil-eraser-sized plastic particles suspended just beneath the surface. This makes the garbage patch difficult to see and measure. While many different types of trash enter the ocean, plastic makes up most of the garbage patch because it is contained in so many disposable products. Plus, it does not biodegrade but instead breaks down into smaller and smaller pieces. Scientists estimate that the amount of plastics entering the ocean will double by 2050.

These small plastic particles ultimately degrade into microscopic particles that can contain potentially hazardous chemicals, including PCBs, DDT, and hormone-mimicking BPA. Small and microscopic plastic particles are eaten by 700 species of marine fishes, birds, and other animals that mistake them for food or ingest them unknowingly. In 2007, a dead whale that washed ashore in

California had 180 kilograms (400 pounds) of plastic in its stomach. The long-lived toxins in these microscopic plastic particles can build up to high concentrations in food chains and webs (Figure 9.14, p. 205, and Figure 17.10, p. 453) and end up in fish sandwiches and seafood dinners. Thus, toxic chemicals from a discarded plastic grocery bag or water bottle could end up in our stomachs. Everything is connected.

Since the North Pacific Garbage Patch was discovered, five other huge swirling garbage patches have been found in gyres in the world's other oceans. These patches, together with the North Pacific patch, cover a total area of ocean larger than all of the earth's land area—the massive pollution legacy of a throwaway culture.

Engineers are testing ways to capture marine debris near the ocean surface as currents pull it into the garbage patch zone. But currently there is no practical or affordable way to clean up this marine litter. The best approach is to prevent the garbage patch from growing by reducing production of solid waste and keeping it out of oceans in the first place. This would require us to sharply reduce unnecessary use of plastic items, greatly increase plastic recycling, and establish modern waste collection and management systems in all coastal countries. Each of us can avoid using plastic water bottles and other plastic products as much as possible and recycle or dispose of them responsibly.

Ocean Pollution from Oil

Crude petroleum (oil as it comes out of the ground) and *refined petroleum* (fuel oil, diesel, gasoline, and other processed petroleum products; see Figure 15.4, p. 386) reach the ocean from natural sources and human activities (**Concept 20.4A**). According to the National Research Council, as much as 46% of the oil in the ocean comes from natural seeps on the ocean bottom. The rest comes from human activities.

The most visible human sources of ocean oil pollution include tanker accidents, such as the huge *Exxon Valdez* oil spill in the U.S. state of Alaska in 1989. Others are blowouts at offshore oil drilling rigs, such as that of the BP *Deepwater Horizon* rig in the Gulf of Mexico in 2010 (see the Case Study that follows). However, studies show that the largest source of ocean oil pollution from human activities is urban and industrial runoff from land. Most of this comes from leaks in pipelines, refineries, and other oil-handling and storage facilities and from oil and oil products that are intentionally dumped or accidentally spilled or leaked onto the land or into sewers by homeowners and industries.

After a major oil spill, some of the oil evaporates and some is reduced to tiny droplets. These droplets can be dispersed by wave action or by chemicals added to the water to help clean up spills. Some oil is decomposed by aerobic bacteria such as blue-green algae (but not in low-oxygen areas such as the deep sea). What remains is a gooey mixture of oil and water called *mousse* that can float in the water for years or be buried in coastal sediments for decades.

Some spilled oil contains volatile organic hydrocarbons that kill many aquatic organisms immediately on contact, especially if these animals are in their vulnerable larval forms. Other chemicals in oil form tarlike globs that float on the surface and coat the feathers of seabirds (see chapter-opening photo) and the fur of marine mammals. The oil destroys their natural heat insulation and buoyancy, causing many of them to drown or die from loss of body heat.

When ingested, oil can poison marine organisms. Oil and some of its toxic chemicals can also end up in the tissues of filtering organisms such as clams, oysters, and mussels. Such chemicals can then become concentrated to higher levels in fish and other organisms (including humans) that feed on these filtering organisms.

Heavy oil components that sink to the ocean floor or wash into estuaries and coastal wetlands can smother bottom-dwelling organisms such as crabs, oysters, mussels, and clams, or make them unfit for human consumption. Some oil spills have killed coral reefs.

Research shows that populations of many forms of marine life can recover from exposure to large amounts of *crude oil* in warm waters with rapid currents within about 3 years. However, in cold and calm waters, recovery can take decades. Recovery from exposure to *refined oil*, especially in estuaries and salt marshes, can take 10 to 20 years or longer. Oil slicks that wash onto beaches can have a serious economic impact on coastal areas that lose income from fishing and tourist activities.

Small oil spills can be partially cleaned up by mechanical means, including floating booms, skimmer boats, and absorbent devices such as giant pillows filled with feathers or hair. However, scientists estimate that current cleanup methods typically can recover no more than 15% of the oil from a major spill.

Therefore, *prevention* of oil pollution is the most effective and, in the end, the least costly approach (**Concept 20.4B**). One of the best ways to prevent tanker spills is to use oil tankers with double hulls. The Oil Pollution Act of 1990, passed by Congress after the 1989 Exxon Valdez oil spill, required that all new oil tankers and barges operating in U.S. waters have double hulls by 2015. Stricter safety standards and inspections can help reduce oil well blowouts at sea. Most important, businesses, institutions, and citizens living in coastal areas must take care to prevent leaks and spillage of even the smallest amounts of oil and oil products such as paint thinners and gasoline.

CASE STUDY

The BP *Deepwater Horizon* Oil-Rig Spill

On April 20, 2010, the world learned a harsh lesson about the possible environmental impacts of deep-sea oil drilling

FIGURE 20.17 The *Deepwater Horizon* drilling platform, located 64 kilometers (40 miles) off the coast of Louisiana, exploded, burned, and sank in the Gulf of Mexico on April 20, 2010.

U.S. Coast Guard

when BP's *Deepwater Horizon* offshore oil-drilling rig exploded (Figure 20.17). The floating rig was connected to a well head on the seafloor almost 1.6 kilometers (1 mile) below and was removing crude oil from a well that was an additional 5.4 kilometers (3.4 miles) below the seafloor.

The accident occurred in the Gulf of Mexico, 64 kilometers (40 miles) off the Louisiana coast, after the wellhead on the ocean bottom ruptured and released oil and natural gas. The resulting fire and explosion killed 11 of the rig's crewmembers and injured 17 more. After burning and belching oil smoke into the air for 36 hours, the rig sank.

For 3 months, the ruptured wellhead on the ocean floor released about 3.1 million barrels (130 million gallons) of crude oil before the leaking well was capped. The oil slick covered a huge area of the Gulf of Mexico (Figure 20.18)—roughly equivalent to the area of the U.S. state of Connecticut—and fouled at least 2,100 kilometers (1,300 miles) of coastline. It was the largest accidental oil spill in U.S. waters and the second largest in the world.

The oil contaminated the sea floor, as well as some ecologically vital coastal marshes, mangrove forests, seagrass beds, fish nurseries, and at least three deep coral reefs. Scientists suspect that deep-ocean aquatic organisms were affected, but the true extent of the ecological damage will not be known for years. According to estimates from the U.S. Coast Guard, the National Oceanic and Atmospheric Administration (NOAA), and BP, the oil spill killed at least 6,100 seabirds and oiled another 2,000 (see chapter-opening photo). It also killed more than 600 sea turtles and 100 dolphins, along with other marine mammals.

In addition, the spill disrupted the livelihoods of people who depend on the Gulf Coast's fisheries, and caused large economic losses for the area's tourism businesses. By 2015, BP had spent over $40 billion to pay for the cleanup, damage claims made by individuals and businesses, and early restoration projects. In 2014, a U.S. District Court Judge found BP guilty of "gross negligence" and "reckless conduct." In 2015, BP agreed to pay another $20.8 billion in damages—the biggest pollution penalty in U.S. history—to settle all state and federal claims related to the accident.

By 2015—5 years after the spill—bacteria had decomposed most of the visible oily tar in coastal areas. Lesions on fish (a result of the spill) were rarely being found, and the commercial fish catch in the spill area had increased. However, as late as 2014, some fish were still dying from the effects of oil exposure, according to an NOAA report. In addition, it will take decades to assess the long-term ecological damage and in coming years, storms, hurricanes, and unusually large tides will release some of the buried oil in coastal areas.

Several studies, including that of a Presidential commission, have concluded that the main causes of the accident were failure of equipment that could have detected the leak earlier, a faulty blowout preventer, failure of several safety valves, and a number of poor decisions made by workers and managers. This event also revealed flaws in federal oversight of offshore drilling, which many argue

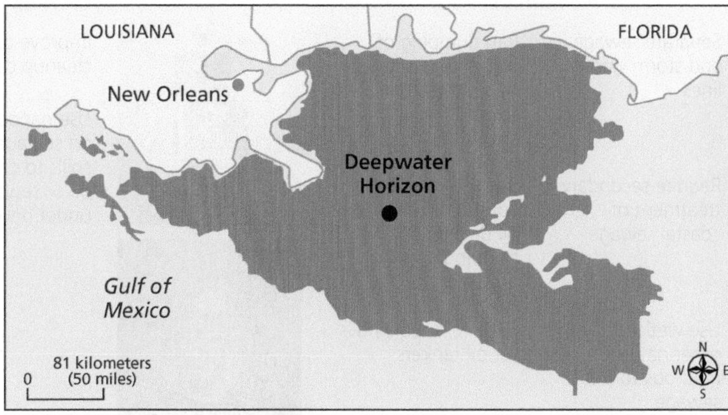

Satellite data from NOAA and NASA.

FIGURE 20.18 Extent of *Deepwater Horizon* oil spill on April 20, 2010.

has included too cozy a relationship between the oil industry and U. S. Department of the Interior (DOI) regulators.

Since the accident, the DOI has stepped up oversight. It has developed new standards for each step in the offshore drilling process and agreed to give a higher priority to environmental concerns in the leasing process. However, the Presidential oil spill commission called for close oversight by Congress to ensure that such reforms are put in place and carefully monitored to help reduce the chances of such accidents in this inherently risky business. As of 2016, this had not been done.

The widespread global publicity following the BP accident helped to educate the public about the dangers of accidents that can release large quantities of oil into the oceans. However, there has been little attention focused on the numerous smaller oil spills that have been occurring almost continually for nearly 50 years in Nigeria's oil-producing Niger Delta, where there is little or no government oversight of oil company operations. Each year, an amount of oil roughly equal to that of the Exxon Valdez disaster leaks and spills from rusted and aging pipes and other facilities in this delta region. Decades of such leaks have caused widespread ecological destruction and threats to human health. According to some scientists, these spills altogether are more damaging than the large, well-known spills.

20.5 HOW CAN WE DEAL WITH WATER POLLUTION?

CONCEPT 20.5 Reducing water pollution requires that we prevent it, work with nature to treat sewage, and use natural resources more efficiently.

Reducing Ocean Water Pollution

Most ocean pollution occurs in coastal waters and comes from human activities on land (**Concept 20.4A**). Figure 20.19 lists ways to prevent pollution of coastal waters and ways to reduce it.

The key to protecting the oceans is to reduce the flow of pollution from land, air, and from streams emptying into these waters (**Concept 20.4B**). Thus, ocean pollution control must be linked with land-use policies, air pollution control policies, energy policies based on improving energy efficiency and relying more on renewable energy, especially from the sun and wind (Figure 16.2, p. 410), and climate policies (Figure 19.18, p. 528).

Reducing Water Pollution from Nonpoint Sources

Most nonpoint sources of water pollution come from agricultural practices. Ways to reduce this type of pollution include the following:

- Reducing soil erosion and fertilizer runoff by keeping cropland covered with vegetation and using conservation tillage (see Chapter 12, p. 301) and other soil conservation methods (see Figure 12.30, p. 309).
- Using fertilizers that release plant nutrients slowly.
- Using no fertilizers on steeply sloped land.
- Relying more on sustainable food production (Figure 12.34, p. 314) to reduce the use and runoff of plant nutrients and pesticides.
- Planting buffer zones of vegetation between cultivated fields (Figure 20.20) and between animal waste storage sites and nearby surface waters.
- Setting discharge standards for nitrate chemicals from sewage treatment and industrial plants.

The annual formation of the dead zone in the Gulf of Mexico (**Core Case Study**) will be difficult to prevent because of the importance of the Mississippi River basin for growing crops. However, nutrient inputs can be reduced with the widespread use of the fertilizer management practices listed above.

In addition to reducing nitrate inputs, another goal is to protect remaining inland and coastal wetlands, vegetation

Solutions

Coastal Water Pollution

Prevention		Cleanup
Separate sewage and storm water lines	Ban dumping of wastes and sewage by ships in coastal waters	Improve oil-spill cleanup capabilities
Require secondary treatment of coastal sewage	Strictly regulate coastal development, oil drilling, and oil shipping	Use nanoparticles on sewage and oil spills to dissolve the oil or sewage (still under development)
Use wetlands and other natural methods to treat sewage	Require double hulls for oil tankers	

FIGURE 20.19 Methods for preventing excessive pollution of coastal waters and methods for cleaning it up. ***Critical thinking:*** Which two of these solutions do you think are the most important? Why?

Top: Rob Byron/Shutterstock.com. Bottom: Igor Karasi/Shutterstock.com.

Lynn Betts/USDA Natural Resources Conservation Service

FIGURE 20.20 This buffer zone of vegetation on an Iowa farm helps to reduce runoff of fertilizer and pesticides into waterways from nearby crop fields.

in *riparian zones* (zones along river banks; see Figure 10.21, right, p. 238), and floodplains, all of which act as natural filters to pull nutrients from waters that are running off the land and spilling out of streams. However, many of the nation's wetlands have been drained or filled in and replaced by cropland.

Another strategy is to restore key wetlands that have been destroyed or degraded and, where feasible, restoring and reconnecting rivers to their natural floodplains. In addition to reducing water pollution, this would restore natural habitats for a variety of species in keeping with the biodiversity **principle of sustainability**.

CASE STUDY

Reducing Water Pollution in the United States

Efforts to control pollution of surface waters in the United States are based on the following federal laws:

- Federal Water Pollution Control Act of 1972, amended in 1977 and renamed the Clean Water Act (CWA). Two-thirds of the nation's lakes, rivers, and coastal waters were unsafe for fishing and swimming, and the key goal of the CWA was to make them safe.
- 1972 Marine Protection, Research, and Sanctuaries Act, amended in 1988. It empowered the EPA to regulate the dumping of untreated sewage and toxic chemicals into U.S. waters.
- 1975 Safe Drinking Water Act (SDWA), amended in 1996. Its goals are to protect surface and groundwater used as sources of drinking water and to establish national drinking water standards for key pollutants.
- 1976 Toxic Substances Control Act. It requires the EPA to maintain a registry of the chemicals in commercial use, evaluate the health risks of new and existing chemicals based on information from producers of such chemicals, and promote ways to reduce or prevent pollution from chemicals that are toxic.
- 1980 Comprehensive Environmental Response Compensation and Liability Act (CERCLA), also known as the Superfund Act. Its goal is to reduce pollution of groundwater by toxic substance leaking from hazardous waste dumps.
- 1987 Water Quality Act, which requires the EPA to set standards for allowed levels of 100 key water pollutants. It requires polluters to get permits that limit the amounts of these various pollutants that they can discharge into aquatic systems.
- 1990 Oil Pollution Act. It has the goal of protecting U.S. waterways from oil pollution.

The EPA has also been experimenting with a *discharge trading policy*, which uses market forces to reduce water pollution as has been done with sulfur dioxide for air pollution control (see Chapter 18, p. 492). Under this program, a permit holder can pollute at higher levels than allowed in its permit by buying credits from permit holders who are polluting below their allowed levels. Environmental scientists warn that the effectiveness of such a system depends on how low the cap on total pollution levels in any given area is set and on how regularly the cap is lowered. They also warn that discharge trading could allow water pollutants to build up to dangerous levels in areas where credits are bought. Neither adequate scrutiny of the cap levels nor gradual lowering of caps is a part of the current EPA discharge trading system.

According to the EPA, the Clean Water Act of 1972 and other water quality acts led to numerous improvements in U.S. water quality, including the following:

- About 60% of all tested U.S. streams, lakes, and estuaries can be used safely for fishing and swimming, compared to 33% in 1972.
- Sewage treatment plants serve 75% of the U.S. population.
- Annual losses of U.S. wetlands that naturally absorb and purify water have been reduced by 80% since 1992.

These are impressive achievements, given the increases in the U.S. population and its per capita consumption of water and other resources since 1972. However, according to the EPA and recent studies, there is still considerable room for improvement. These studies noted the following:

- About 40% of the nation's surveyed streams, lakes, and estuaries are still too polluted for swimming or fishing.
- Runoff of animal wastes from hog, poultry, and cattle feedlots and meat processing facilities pollutes 70% of U.S. rivers.
- Tens of thousands of gasoline storage tanks in 43 states are leaking.
- Since 2003, 20% of U.S. water treatment systems have violated the Safe Drinking Water Act by releasing sewage and chemicals such as arsenic and radioactive uranium into U.S. surface waters.
- A study showed that water samples from 139 streams in 30 states contained measurable levels of drugs used for birth control and for reducing pain and depression.

According to the EPA, about 36% of Americans get their drinking water from waterways that lack clear protection because of 2001 and 2006 Supreme Court decisions that created uncertainty over what streams and wetlands had to be protected under the Clean Water Act. In 2015, the EPA and U.S. Army Corps of Engineers issued a Clean Water Act Rule to correct this uncertainty, but Congress may block the rule.

Some environmental scientists call for strengthening the Clean Water Act. Suggested improvements include the following:

- Shifting the focus of the law from end-of-pipe removal of specific pollutants to water pollution prevention.
- Greatly increased monitoring for violations of the law with much larger mandatory fines for violators.
- Regulating irrigation water quality (for which there is no federal regulation).
- Providing stronger incentives or regulations to reduce fertilizer and pesticide runoff form farms and overflows of urban storm water.
- Bringing the production of oil and natural gas by fracking (see Chapter 15, Core Case Study, p. 382) under the Clean Water and the Safe Drinking Water Acts.
- Expanding the rights of citizens to bring lawsuits to ensure that water pollution laws are enforced.
- Rewriting the Clean Water Act to clarify that it covers all waterways (as Congress originally intended).

Many businesses and legislators oppose these proposals, contending that the Clean Water Act's regulations are already too restrictive and costly. Some state and local officials argue that in many communities, it is unnecessary and too expensive to test all the water for pollutants as required by federal law.

Some members of Congress, under pressure from regulated industries, go further and suggest seriously weakening or repealing the Clean Water Act. They contend that such regulations hinder economic growth and prevent job growth. According to William K. Reilly, who headed the EPA from 1989 to 1993 and served as a co-chairman of the Presidential commission on offshore drilling, "If we buy into the misguided notion that reducing protection of our waters will somehow ignite the economy, we will shortchange our health, environment, and economy."

Sewage Treatment Reduces Water Pollution

In rural and suburban areas with suitable soils, sewage from each house usually is discharged into a **septic tank** and a large drainage (leach) field (Figure 20.21). Household sewage and wastewater is pumped into a settling tank, where grease and oil rise to the top, and solids, called *sludge*, fall to the bottom and are decomposed by bacteria. The partially treated wastewater is discharged into a large drainage or leach field through small holes in perforated pipes embedded in porous gravel or crushed stone just below the soil's surface. As these wastes drain from the pipes and sink downward toward, the soil filters out some potential pollutants and soil bacteria decompose biodegradable materials.

About one-fourth of all homes in the United States are served by septic tanks. It these systems are not installed correctly and the sludge in the tank is not pumped out regularly, sewage can back up into homes and pollute nearby groundwater and surface water. Chlorine bleaches, drain cleaners, and antibacterial soaps should not be used in these systems, because they can kill the bacteria that decompose the wastes. Kitchen sink garbage disposals should also not be used, because they can overload septic systems.

In urban areas in the United States and other more-developed countries, most waterborne wastes from homes, businesses, and storm runoff flow through a network of sewer pipes to *wastewater* or *sewage treatment plants*. Raw sewage reaching a treatment plant typically undergoes one or two levels of wastewater treatment. The first is **primary sewage treatment**—a *physical* process that uses screens and a grit tank to remove large floating objects and to allow solids such as sand and rock to settle out. Then the waste stream flows into a primary settling tank where suspended solids settle out as sludge (Figure 20.22, left).

The second level is **secondary sewage treatment**—a *biological* process in which oxygen is added to the sewage in an aeration tank to encourage aerobic bacteria to decompose as much as 90% of dissolved and biodegradable, oxygen-demanding organic wastes (Figure 20.22, right). A combination of primary and secondary treatment removes 95–97%

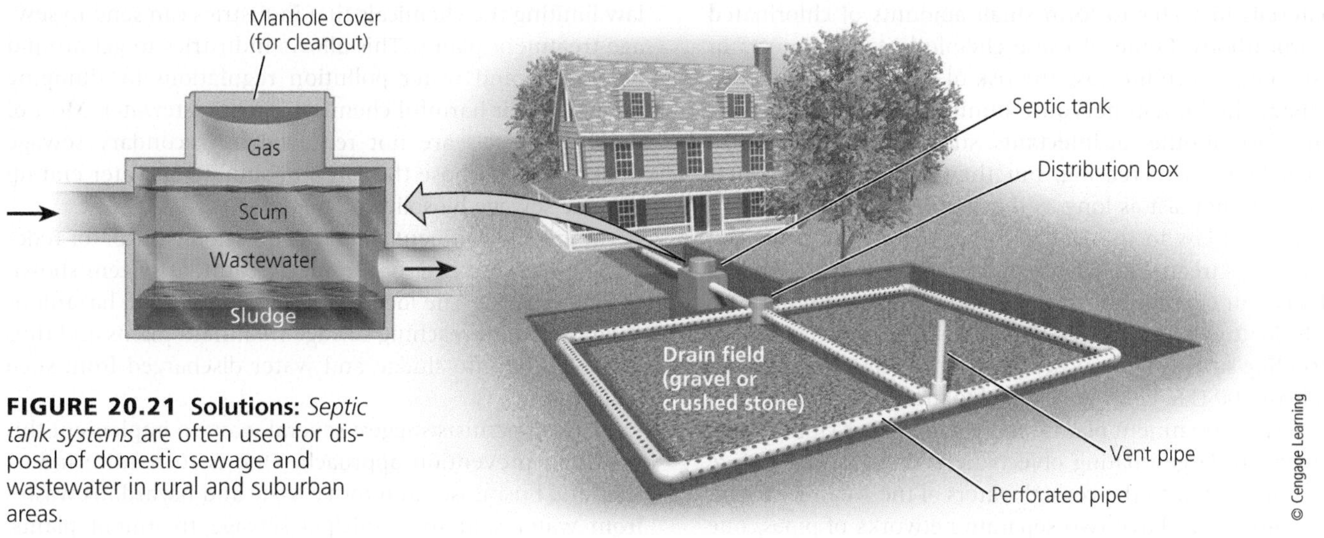

FIGURE 20.21 Solutions: *Septic tank systems* are often used for disposal of domestic sewage and wastewater in rural and suburban areas.

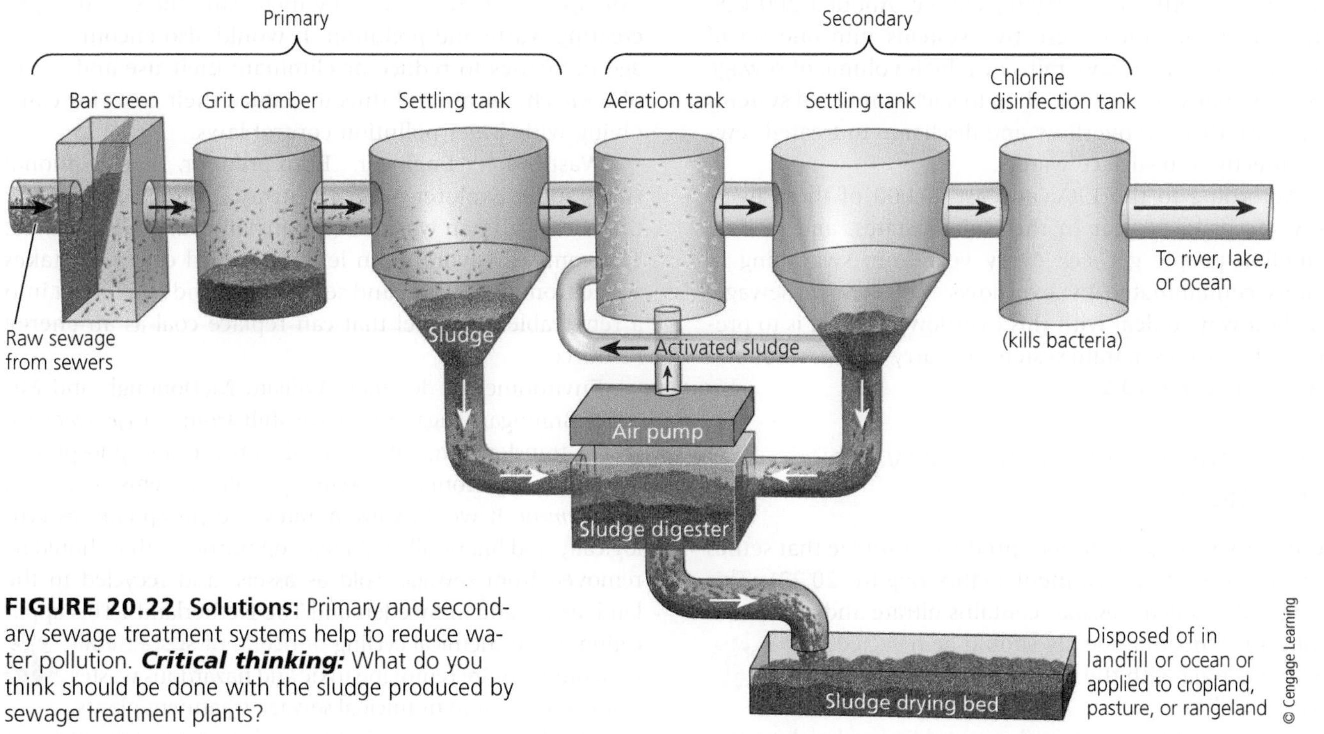

FIGURE 20.22 Solutions: Primary and secondary sewage treatment systems help to reduce water pollution. ***Critical thinking:*** What do you think should be done with the sludge produced by sewage treatment plants?

of the suspended solids and oxygen-demanding organic wastes, 70% of most toxic metal compounds and nonpersistent synthetic organic chemicals, 70% of the phosphorus, and 50% of the nitrogen.

However, this process removes only a tiny fraction of persistent and potentially toxic organic substances found in some pesticides and in discarded medicines that people put into sewage systems, and it does not kill disease-causing bacteria and viruses. It also does not remove many of the toxic chemicals in wastewater that industries and fracking operations send to sewage treatment plants.

A third level of cleanup, *advanced* or *tertiary sewage treatment,* uses a series of specialized chemical and physical processes to remove specific pollutants left in the water after primary and secondary treatment. In its most common form, advanced sewage treatment uses special filters to remove phosphates and nitrates from wastewater before it is discharged into surface waters. This third stage would significantly reduce nutrient overload from nitrates and phosphates, but is not widely used because of its high costs.

Before discharge, water from sewage treatment plants usually undergoes *bleaching,* to remove water coloration, and *disinfection* to kill disease-carrying bacteria and some (but not all) viruses. The usual method for accomplishing this is chlorination. But chlorine can react with organic materials in water to form small amounts of chlorinated hydrocarbons. Some of these chemicals cause cancers in test animals, can increase the risk of miscarriages, and can damage the human nervous, immune, and endocrine systems. Use of other disinfectants, such as ozone and ultraviolet light, is increasing, but they cost more and their effects do not last as long as those of chlorination.

Federal law in the United States requires primary and secondary treatment in all municipal sewage treatment plants, but exemptions from secondary treatment are possible for towns and cities that can show that the cost of installing such systems poses an excessive financial burden. At least 500 U.S. cities have failed to meet federal standards for sewage treatment plants, and 34 East Coast cities simply screen out large floating objects from their sewage before discharging it into the coastal waters of the Atlantic Ocean.

Some cities have two separate networks of pipes, one for carrying storm-water runoff from streets and parking lots, and the other for carrying sewage. About 1,200 U.S. cities have combined these two systems into one set of pipes to cut costs. Heavy rains or a high volume of sewage from too many users hooked up to such combined systems can cause them to overflow and discharge untreated sewage directly into surface waters.

According to the EPA, at least 40,000 of these overflows occur each year in the United States, and at least 7 million people get sick every year from swimming in waters contaminated by overflows that contain sewage. The best way to deal with this overflow problem is to prevent it by using separate systems to carry storm water and sewage (**Concept 20.5**).

Improving Conventional Sewage Treatment

Secondary sewage treatment produces a sludge that settles out at wastewater treatment plants (Figure 20.22). This sludge, also called *biosolids*, contains nitrate and phosphate plant nutrients that ideally should be recycled to the soil in keeping with the chemical cycling **principle of sustainability**.

However, there is a serious problem related to this practice under current EPA regulations. The EPA allows industries and homeowners to send wastewater that contains toxic heavy metals, organic chemicals, pesticides, and pharmaceuticals to sewage treatment plants and many of these harmful chemicals end up in the toxic biosolids. EPA regulations also allow contaminated biosolids to be applied as fertilizer to crop fields, playgrounds, and other land used by the public. Some of these harmful chemicals can end up in food, because they can be found in compost and soil conditioners that people use on their gardens and even in USDA-certified organic food production.

The Clean Air and Clean Water acts require U.S. industries to limit their inputs of harmful chemicals into the air and natural aquatic systems. However, there is no federal law limiting the chemicals that industries can send to sewage treatment plants. This allows industries to get around federal air and water pollution regulations by dumping many of their harmful chemicals into wastewater. Most of these chemicals are not removed by secondary sewage treatment and those that are not soluble in water end up in toxic sewage biosolids.

Some environmental and health scientists call for redesigning the conventional sewage treatment system shown in Figure 20.22. The idea is to prevent toxic and hazardous chemicals from reaching sewage treatment plants and thus from getting into sludge and water discharged from such plants.

These scientists suggest several ways to implement this pollution prevention approach. One is to require industries and businesses to remove toxic and hazardous wastes from water sent to municipal sewage treatment plants. This would help to implement the full-cost pricing **principle of sustainability** by increasing the cost of creating waste and pollution. It would also encourage industries to reduce or eliminate their use and waste of toxic chemicals and thus to reduce their costs for complying with water pollution control laws.

Wastewater Engineer, Entrepreneur, and National Geographic Explorer Ashley Murray founded a company that views human waste as a financially valuable resource. The company, working in less-developed countries, takes waste from pit latrines and septic tanks and converts it into a renewable solid fuel that can replace coal as an energy resource.

Environmental designers William McDonough and Michael Braungart suggest that we shift from *sewage treatment*, which transfers large amounts of nitrogen and phosphorus plant nutrients from land to surface water systems, to *nutrient management*. It would view nitrates and phosphates as ecologically and financially valuable soil nutrients that should be removed from sewage, sold as assets, and recycled to the land, as is done in Sweden and The Netherlands. This application of the chemical cycling **principle of sustainability** would require removing toxic and hazardous wastes from water sent to municipal sewage treatment plants.

Another suggestion is to require or encourage more households, apartment buildings, and offices to eliminate sewage outputs by switching to waterless, odorless *composting toilet systems*, to be installed, maintained, and managed by professionals. These systems, pioneered several decades ago in Sweden, convert nutrient-rich human fecal matter into a soil-like humus that can be used as a fertilizer supplement. This process returns plant nutrients in human waste to the soil, mimicking the chemical cycling **principle of sustainability**. One of the authors of this book (Miller) used a waterless compositing toilet for over a decade with no problems in his office-home environmental research facility.

On a larger scale, such systems would be cheaper to install and maintain than current sewage systems are,

SCIENCE FOCUS 20.3

Treating Sewage by Learning from Nature

Some communities and individuals are developing better ways to purify sewage by learning how nature purifies water using sunlight and plants, aquatic organisms, and natural filtration by soil, sand, and gravel (**Concept 20.5**). Biologist John Todd has developed an ecological approach to treating sewage, which he calls *living machines* (Figure 20.C).

This purification process begins when sewage flows into a passive solar greenhouse or outdoor site containing rows of large open tanks populated by an increasingly complex series of organisms. In the first set of tanks, algae and microorganisms decompose organic wastes, with sunlight speeding up the process. Water hyacinths, cattails, bulrushes, and other aquatic plants growing in the tanks take up the resulting nutrients.

After flowing though several of these natural purification tanks, the water passes through an artificial marsh made of sand, gravel, and bulrushes, which filter out algae and remaining organic waste. Some of the plants also remove toxic metals such as lead and mercury and secrete natural antibiotic compounds that kill pathogens.

Next, the water flows into aquarium tanks, where snails and zooplankton consume microorganisms and are in turn consumed by crayfish, tilapia, and other fish that can be eaten or sold as bait. After 10 days, the clear water flows into a second artificial marsh for final filtering and cleansing. The water can be made pure enough to drink by treating it with ultraviolet light or by passing the water through an ozone generator, usually immersed out of sight in an attractive pond or wetland habitat.

Operating costs are about the same as those of a conventional sewage treatment plant. These systems are widely used on a small scale. However, they are difficult to maintain on a scale large enough to handle the typical variety of chemicals in the sewage wastes from large urban areas.

More than 800 cities and towns around the world (150 in the United States) mimic nature by using natural or artificially created wetlands to treat sewage as a lower-cost alternative to expensive waste treatment plants. For example, in Arcata, California—a coastal town of almost 18,000 people—scientists and workers created some 65 hectares (160 acres) of wetlands between the town and the adjacent Humboldt Bay. The marshes and ponds, developed on land that was once a dump, act as a natural waste treatment plant. The cost of the project was less than half the estimated price of a conventional treatment plant.

This system returns purified water to Humboldt Bay, and the sludge that is removed is processed for use as fertilizer. The marshes and ponds also serve as an Audubon Society bird sanctuary, which provides habitats for thousands of seabirds, otters, and other marine animals. The town has even celebrated its natural sewage treatment system with an annual "Flush with Pride" festival.

Ocean Arks International

FIGURE 20.C Solutions: The Solar Sewage Treatment Plant in the U.S. city of Providence, Rhode Island, is an ecological wastewater purification system, also called a *living machine*. Biologist John Todd is demonstrating an ecological process he invented for purifying wastewater by using the sun and a series of tanks containing living organisms.

This approach and the living machine system developed by John Todd apply all three **scientific principles of sustainability**: using solar energy, employing natural processes to remove and recycle nutrients and other chemicals, and relying on a diversity of organisms and natural processes.

CRITICAL THINKING

Can you think of any disadvantages of using such a nature-based system instead of a conventional sewage treatment plant? Do you think any such disadvantages outweigh the advantages? Why or why not?

because they do not require vast systems of underground pipes connected to centralized sewage treatment plants. They also save large amounts of water, reduce water bills, and decrease the amount of energy used to pump and purify water. This more environmentally sustainable replacement for the conventional toilet is now being used in parts of more than a dozen countries, including China, India, Mexico, Syria, and South Africa.

A Swedish entrepreneur has developed a biodegradable single-use plastic bag, called a *PeePoo,* that can be used as a toilet in urban slums and in other areas where many of the world's people do not have access to toilets. After it is used, the bag is knotted and buried. A thin layer of urea in this bag kills the disease-producing pathogens in the feces and helps break down the waste into plant nutrients that are then recycled. This is a simple and inexpensive, low-tech application of the chemical cycling **principle of sustainability**. Some communities are also using unconventional, but highly effective, *ecological sewage treatment systems* (Science Focus 20.3).

Sustainable Ways to Reduce and Prevent Water Pollution

It is encouraging that since 1970, most of the world's more-developed countries have enacted laws and regulations that have significantly reduced point-source water pollution. These improvements were largely the result of *bottom-up* political pressure on elected officials from individuals and groups.

On the other hand, little has been done to reduce water pollution in most of the less-developed countries. China plans to provide all of its cities with small sewage treatment plants that will make wastewater clean enough to be recycled back into the urban water supply systems, but there is no timetable for doing this.

To many environmental and health scientists, the next step is to increase efforts to reduce and prevent water pollution in both more- and less-developed countries as an important way to increase our beneficial environmental impact. They would begin by asking the question: *How can we avoid producing water pollutants in the first place?* (**Concept 20.5**). Figure 20.23 lists ways to achieve this goal over the next several decades.

This shift to pollution prevention will not take place unless individuals put political pressure on elected officials and also take actions to reduce their own daily contributions to water pollution. Figure 20.24 lists some steps you can take to help reduce water pollution.

Solutions

Water Pollution

- Prevent groundwater contamination
- Reduce nonpoint runoff
- Work with nature to treat sewage and reuse treated wastewater
- Find substitutes for toxic pollutants
- Practice the four Rs of resource use (refuse, reduce, reuse, recycle)
- Reduce air pollution
- Reduce poverty
- Slow population growth

FIGURE 20.23 Ways to prevent or reduce water pollution. ***Critical thinking:*** Which two of these solutions do you think are the most important? Why?

What Can You Do?

Reducing Water Pollution

- Fertilize garden and yard plants with manure or compost instead of commercial inorganic fertilizer
- Minimize use of pesticides, especially near bodies of water
- Prevent yard wastes from entering storm drains
- Do not use water fresheners in toilets
- Do not flush unwanted medicines down the toilet
- Do not pour pesticides, paints, solvents, oil, antifreeze, or other harmful chemicals down the drain or onto the ground

FIGURE 20.24 Individuals matter: You can help reduce water pollution. ***Critical thinking:*** Which three of these steps do you think are the most important ones to take? Why?

BIG IDEAS

- There are a number of ways to purify drinking water, but the most effective and least costly strategy is pollution prevention.
- The key to protecting the oceans is to reduce the flow of pollution from land and air, and from streams emptying into ocean waters.
- Reducing water pollution requires that we prevent it, work with nature in treating sewage, and use natural resources far more efficiently.

Tying It All Together

Dead Zones and Sustainability

Bull's-Eye Arts/Shutterstock.com

The Core Case Study that opens this chapter explains how certain farming practices can violate the chemical cycling **principle of sustainability** (see Inside Back Cover) by disrupting the nitrogen cycle. This occurs every year when nitrate plant nutrients in fertilizers flow in runoff from farmland into streams in the Mississippi River basin. Much of this flow goes into the Mississippi River and ends up overfertilizing an area of the Gulf of Mexico, which depletes dissolved oxygen and reduces seafood production in these coastal waters. This process also violates the biodiversity **principle of sustainability** by disrupting ecological interactions among species in river and coastal ecosystems.

The three **scientific principles of sustainability** can guide us in reducing and preventing water pollution. We can use solar energy to purify much of the water we use, which will reduce water waste. We can also use natural nutrient cycles (chemical cycling) to treat our wastes in wetland-based sewage treatment systems (Science Focus 20.3). In addition, we can view human waste as an important and potentially profitable source of nutrients. Finally, by preventing the pollution and degradation of aquatic systems and their bordering terrestrial systems, we will help to preserve biodiversity—a key component of the life-support system that we depend on for maintaining water supplies and water quality.

Chapter Review

Core Case Study

1. Describe the nature and causes of the annual dead (oxygen-depleted) zone in the Gulf of Mexico.

Section 20.1

2. What are the two key concepts for this section? What is **water pollution**? Distinguish between **point sources** and **nonpoint sources** of water pollution and give two examples of each. What is **wastewater**? Explain how certain efforts to clean up the air can contribute to water pollution.

3. List seven major types of water pollutants and three diseases that can be transmitted to humans through polluted water. Describe some of the chemical and biological methods that scientists use to measure water quality.

Section 20.2

4. What are the two key concepts for this section? Explain how streams can cleanse themselves of oxygen-demanding wastes and how this cleansing process can be overwhelmed. Describe the varying states of stream pollution in more- and less-developed countries. Give two reasons why lakes and reservoirs cannot cleanse themselves of biodegradable wastes as readily as streams can. Distinguish between **eutrophication** and **cultural eutrophication**. List three ways to prevent or reduce cultural eutrophication. Explain how seasonal dead zones form in the Gulf of Mexico. Describe the pollution of the Great Lakes and the efforts to reduce this pollution.

Section 20.3

5. What are the two key concepts for this section? Explain why groundwater cannot cleanse itself very well. What are the major sources of groundwater contamination in the United States? List three harmful effects of groundwater pollution. Describe the threat from arsenic in groundwater. List three ways to prevent groundwater contamination and three ways to clean up groundwater contamination. Explain why prevention is the most important approach.

6. Explain how drinking water is purified in more-developed countries. Explain how we can convert sewer water into drinking water. List three ways to provide safe drinking water in poor countries. Describe the environmental problems caused by the

widespread use of bottled water. Summarize the U.S. laws for protecting drinking water quality. List three ways to strengthen the U.S. Safe Drinking Water Act.

Section 20.4

7. What are the two key concepts for this section? Why should we care about the oceans? How are coastal waters and deeper ocean waters most often polluted? What causes algal blooms and what are their harmful effects? Describe an ocean garbage patch and explain how it can harm marine life and humans. How serious is oil pollution of the oceans? What are its effects and what can be done to reduce such pollution? Describe the 2010 BP *Deepwater Horizon* oil well blowout and its causes and effects. How might such accidents be prevented in the future?

Section 20.5

8. What is the key concept for this section? List four ways to reduce and prevent pollution of coastal waters. List six ways to reduce surface water pollution from nonpoint sources. List seven federal laws aimed at reducing water pollution in the United States. Describe U.S. successes and failures with using these laws to reduce point-source water pollution. List seven ways to strengthen the Clean Water Act and describe opposition to such proposals. What is a **septic tank** and how does it work? Define **primary sewage treatment** and **secondary sewage treatment** and explain how they are used to treat wastewater.

9. What are three ways to improve conventional sewage treatment? Describe two ways to treat sewage based on learning from nature. What is a waterless composting toilet system? Describe John Todd's use of living machines to treat sewage. Explain how we can use wetlands to treat sewage.

10. List six ways to prevent or reduce water pollution. List five things you can do to help reduce water pollution. What are this chapter's *three big ideas*? Explain how we can use the three **scientific principles of sustainability** to prevent and reduce water pollution.

Note: Key terms are in bold type.

Critical Thinking

1. How might you be contributing, directly or indirectly, to the annual dead zone that forms in the Gulf of Mexico (**Core Case Study**)? What are three things you could do to reduce your impact?

2. Suppose a large number of dead fish are found floating in a lake. How would you determine whether they died from cultural eutrophication or from exposure to toxic chemicals?

3. Assume that you are a regulator charged with drawing up plans for controlling water pollution, and briefly describe your strategy for controlling water pollution from each of the following sources: **(a)** a pipe from a factory discharging effluent into a stream, **(b)** a parking lot at a shopping mall bordered by a stream, and **(c)** a farmer's field on a slope next to a stream.

4. How might you be contributing, directly or indirectly, to groundwater pollution? What are three things you could do to reduce your impact?

5. When you flush your toilet, where does the wastewater go? Trace the actual flow of this water in your community from your toilet through sewers to a wastewater treatment plant (or to a septic system) and from there to the environment. Try to visit a local sewage treatment plant to see what it does with wastewater. Compare the processes it uses with those shown in Figure 20.22. What happens to the sludge produced by this plant? What improvements, if any, would you suggest for this plant?

6. In your community, **(a)** what is the source of drinking water? **(b)** How is drinking water treated? **(c)** What problems related to drinking water, if any, have arisen in your community? **(d)** What actions, if any, has your local government taken to solve them?

7. List three ways in which you could apply **Concept 20.5** to make your lifestyle more environmentally sustainable.

8. Congratulations! You are in charge of the world. What are three actions you would take to **(a)** sharply reduce point-source water pollution in more-developed countries, **(b)** sharply reduce nonpoint-source water pollution throughout the world, **(c)** sharply reduce groundwater pollution throughout the world, and **(d)** provide safe drinking water for the poor and for other people in less-developed countries?

Doing Environmental Science

Do some research using the library and/or the Internet and evaluate the relative effectiveness of three different home water purification devices. Try to determine the costs of these systems, including purchase, installation, and maintenance costs, and record these costs for each system. Determine the type or types of water pollutants that each device is supposed to remove and the effectiveness of each such device. You might also consider looking up reviews of each product and finding some users of these products to interview. Write a report summarizing your findings.

Global Environment Watch Exercise

Go to your MindTap course to access the GREENR database. Starting on the home page, under "Browse Issues and Topics", click on *Pollution*, then select *Gulf of Mexico Oil Spill 2010*. Use this portal to research the ecological effects of the oil well blowout and spill. Write a report that answers the following questions:

1. Is oil from the spill in the Gulf still a problem?
2. How has it affected the coastal wetlands along the gulf?
3. How has it affected the gulf's bottom-dwelling species?
4. How has it affected bird life in the gulf?
5. How has it affected the fishing industry there?

Data Analysis

In 2006, scientists assessed the overall condition of the estuaries on the western coasts of the U.S. states of Oregon and Washington. To do so, they took measurements of various characteristics of the water, including dissolved oxygen (DO), in selected locations within the estuaries. The concentration of DO for each site was measured in terms of milligrams (mg) of oxygen per liter (L) of water sampled. The scientists used the following DO concentration ranges and quality categories to rate their water samples: water with greater than 5 mg/L of DO was considered *good* for supporting aquatic life; water with 2 to 5 mg/L of DO was rated as *fair*; and water with less than 2 mg/L of DO was rated as *poor*.

The following graph shows measurements taken in bottom water at 242 locations. Each triangular mark represents one or more measurements. The x-axis on this graph represents DO concentrations in mg/L. The y-axis represents percentages of the total area of estuaries studied (estuarine area).

To read this graph, pick one of the triangles and observe the values on the x- and y-axes. For example, note that the circled triangle lines up approximately with the 5-mg/L mark on the x-axis and with a value of about 34% on the y-axis. This means that waters at this particular measurement station (or stations), along with about 34% of the total area being studied, are estimated to have a 5% or lower DO concentration.

Use this information, along with the graph, to answer the following questions:

1. Half of the estuarine area has waters falling below a certain DO concentration level, and the other half has levels above that level. What is that level, in mg/L?
2. Give your estimate of the highest DO concentration measured and your estimate of the lowest concentration.
3. Approximately what percentage of the estuarine area studied is considered to have poor DO levels? About what percentage has fair DO levels, and about what percentage has good DO levels?

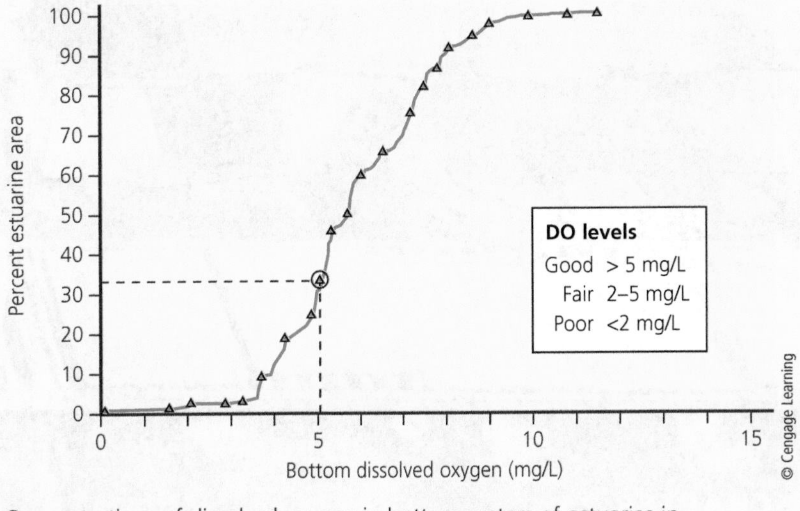

Concentrations of dissolved oxygen in bottom waters of estuaries in Washington and Oregon.

CHAPTER 21

Solid and Hazardous Waste

Follow nature's example; realize waste's potential.

GUNTER PAULI

Key Questions

21.1 What environmental problems are related to solid and hazardous wastes?

21.2 How should we deal with solid waste?

21.3 Why are refusing, reducing, reusing, and recycling so important?

21.4 What are the advantages and disadvantages of burning or burying solid waste?

21.5 How should we deal with hazardous waste?

21.6 How can society shift to a low-waste economy?

Workers in Texas are stripping picture tubes from thousands of computer monitors for recycling and reuse.

Peter Essick/National Geographic Creative

Core Case Study

Cradle-to-Cradle Design

The *life cycle* of a product begins when it is manufactured (its cradle) and ends when it is discarded as solid waste, typically in a landfill or as litter (its grave).

Designer William McDonough wants us to abandon this *cradle-to-grave* view of the life cycle of products. He argues for a *cradle-to-cradle* approach, in which we think of products as part of a continuing cycle instead of becoming solid wastes that end up as litter or being burned or deposited in landfills. This approach, first explored in the 1970s by business analyst Walter Stahel, is the basis for much of McDonough's work. His vision is to build an economy where all products or their parts will be reused over and over in other products. Parts that are no longer useful would be degradable so that natural nutrient cycles could absorb their molecules. The degradable parts would be thought of as *biological nutrients* (Figure 21.1, left) and those parts that are reused would be *technical nutrients* (Figure 21.1, right).

In their books, *Cradle to Cradle* and *The Upcycle*, McDonough and chemist Michael Braungart lay out this vision as a way not just to lessen our harmful environmental impact but also to have a beneficial environmental impact. They call for us to think of solid wastes and pollution as potentially useful and economically valuable materials and chemicals. Instead of asking, "How do I get rid of these wastes?" they say we need to ask, "How much money can I get for these resources?" and "How can I design products that don't end up as wastes or pollutants?"

This way of thinking means designing products so they can be recycled or reused, much like nutrients in the biosphere. With this approach, people might think of trashcans and garbage trucks as resource containers and landfills as urban mines filled with stuff we should have recycled, like the earth does. They might not think of garbage as something to "throw away" but something to "pass on" for other purposes.

Cradle-to-cradle design is a form of biomimicry (see Chapter 1, Core Case Study, p. 4) because it implements the earth's chemical cycling **principle of sustainability** (see Inside Back Cover). For example, a chair manufacturer applying this approach designs and builds its chairs such that when one part breaks, most of the other parts can be reused in the manufacture of a new chair. As much as possible, only biodegradable materials are used so that worn-out, discarded parts will break down in the environment and become part of nature's nutrient cycles. As McDonough likes to say, in nature, waste equals food.

There are many ways to apply this approach. One important way is to *design toxic substances out* of products and processes. If a product requires the use of a toxic heavy metal, for example, it should be redesigned to make use of a nontoxic substitute for that ingredient. Another strategy is to *sell services instead of products.* For example, think of carpeting not as a product to be used and discarded, but as a floor covering service. The carpet company owns the carpeting and leases it to the user. The company then replaces worn carpet tiles on a regular basis as part of the service and recycles the materials in the worn pieces to make new carpet tiles.

In this chapter, we consider the problems of solid and hazardous wastes resulting from human activities. We also consider ways to make the transition to a more sustainable low-waste economy by preventing and reducing the production of such wastes as a way to apply the cradle-to-cradle approach.

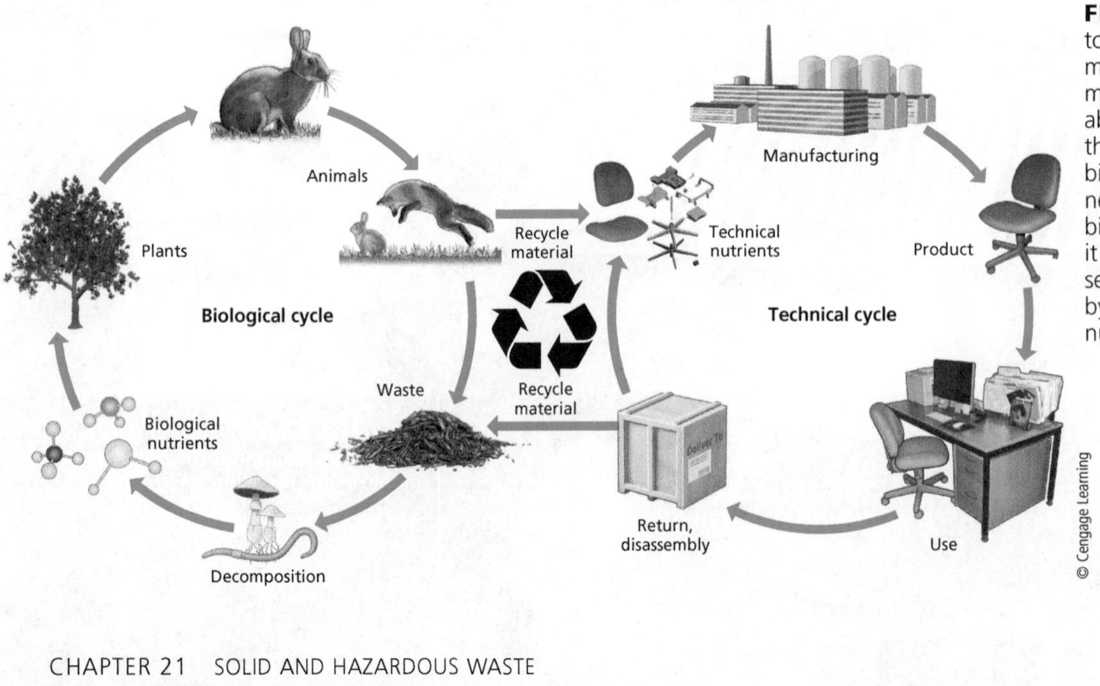

FIGURE 21.1 Cradle-to-cradle design and manufacturing aims to make all products reusable and all components that must be discarded biodegradable. By connecting technical and biological nutrient cycles, it mimics nature and essentially eliminates waste by converting it to nutrients.

21.1 WHAT ENVIRONMENTAL PROBLEMS ARE RELATED TO SOLID AND HAZARDOUS WASTES?

CONCEPT 21.1A Solid waste contributes to pollution and includes valuable resources that could be reused or recycled.

CONCEPT 21.1B Hazardous waste contributes to pollution, as well as to natural capital degradation, health problems, and premature deaths.

Solid Waste Is Piling Up

Think about what you've tossed in the trash today. Perhaps it was leftovers from lunch, an empty can or bottle, or something you no longer needed. People throw away all sorts of items, and they all add up to huge amounts of solid waste.

In the natural world, wherever humans are not dominant, there is essentially no waste. The wastes of one organism become nutrients or raw materials for others (see Figures 3.16, p. 60, and 3.17, p. 61). This natural cycling of nutrients is the basis for the chemical cycling **principles of sustainability** (see Inside Back Cover).

Humans violate this principle by producing huge amounts of waste materials that are burned or end up in landfills or as litter and pollute the environment. For example, the manufacturing of a desktop computer requires 700 or more different materials obtained from mines, oil wells, and chemical factories all over the world. For every 0.5 kilogram (1 pound) of electronics it contains, approximately 3,600 kilograms (8,000 pounds) of waste were created—an amount roughly equal to the weight of a large pickup truck.

Because of the law of conservation of matter (see Chapter 2, p. 42) and the nature of human lifestyles, we will always produce some waste. However, studies and experience indicate that by mimicking nature, through strategies such as cradle-to-cradle design (**Core Case Study**), we could reduce this waste of potential resources and the resulting environmental harm by up to 80%.

The solid items thrown away in your household fall into the category of **solid waste**—any unwanted or discarded material people produce that is not a liquid or a gas. There are two major types of solid waste. The first is **industrial solid waste** produced by mines (see Figure 14.8, p. 365), farms, and industries that supply people with goods and services. The second is **municipal solid waste (MSW),** often called *garbage* or *trash.* It consists of the combined solid wastes produced by homes and workplaces other than factories. Examples of MSW include paper, cardboard, food wastes, cans, bottles, yard wastes, furniture, plastics, glass, wood, and electronics or *e-waste.*

Rechitan Sorin/Shutterstock.com

FIGURE 21.2 *Municipal solid waste:* Various types of solid waste have been dumped in this isolated mountain area of the United States.

Much of the world's MSW ends up as litter in rivers, lakes, the ocean, and natural landscapes (Figure 21.2). One of the major symbols of such waste is the single-use plastic bag. Laid end-to-end, the 100 billion plastic bags used in the United States each year would reach to the moon and back 60 times. In many countries, the landscape and waterways are littered with plastic bags and they also end up in ocean waters. They can take 400 to 1,000 years to break down, but never disintegrate completely.

In the environment, plastic bags often block drains and sewage systems and can kill wildlife and livestock that try to eat them or become ensnared in them. In Kenya, Africa, outbreaks of malaria have been associated with plastic bags lying on the ground collecting water in which malaria-carrying mosquitoes can breed.

Discarded plastic items are a threat to many terrestrial animal species, as well as millions of seabirds, marine

Stockbyte/Thinkstock

FIGURE 21.3 This child is searching for useful items in this open trash dump in Manila, Philippines.

mammals, and sea turtles, which can mistake a floating plastic sandwich bag for a jellyfish or get caught in plastic fishing nets (Figure 11.9, p. 263). About 80% of the plastics in the ocean are blown or washed in from beaches, rivers, storm drains, and other sources, and the rest are dumped into the ocean from ocean-going garbage barges, ships, and fishing boats.

In more-developed countries, most MSW is collected and buried in landfills or burned in incinerators. In many less-developed countries, much of it ends up in open dumps, where poor people eke out a living finding items they can use or sell (Figure 21.3). In China, only about 40% of the MSW is collected, and in rural areas the figure can be as low as 4–5%. The United States is the world's largest producer of solid waste (see the Case Study that follows).

CASE STUDY

Solid Waste in the United States

The United States leads the world in total production of industrial and municipal solid waste, as well as solid waste per person. According to the Environmental Protection Agency (EPA), 98.5% of all solid waste produced in the United States is industrial waste from mining (76%), agriculture (13%), and industry (9.5%). The remaining 1.5% is municipal solid waste. The United States is the world's largest producer of MSW. Every year, Americans generate enough MSW to fill a bumper-to-bumper convoy of garbage trucks long enough to circle the earth's equator almost six times. Most of this waste is dumped in landfills, recycled or composted, or incinerated (Figure 21.4, right). However, a great deal of it ends up as litter.

Consider some of the solid wastes that consumers throw away each year, on average, in the high-waste economy of the United States:

- Enough tires to encircle the earth's equator almost three times.
- An amount of disposable diapers that, if linked end to end, would reach to the moon and back seven times.
- Enough carpet to cover the state of Delaware.
- Enough nonreturnable plastic bottles to form a stack that would reach from the earth to the moon and back about six times.
- About 100 billion plastic shopping bags, or 274 million per day, an average of nearly 3,200 every second.
- Enough office paper to build a wall 3.5 meters (11 feet) high across the country from New York City to San Francisco, California.
- 25 billion plastic foam (Styrofoam) coffee cups—enough, if lined up end to end, to circle the earth's equator 436 times.
- $165 billion worth of food.

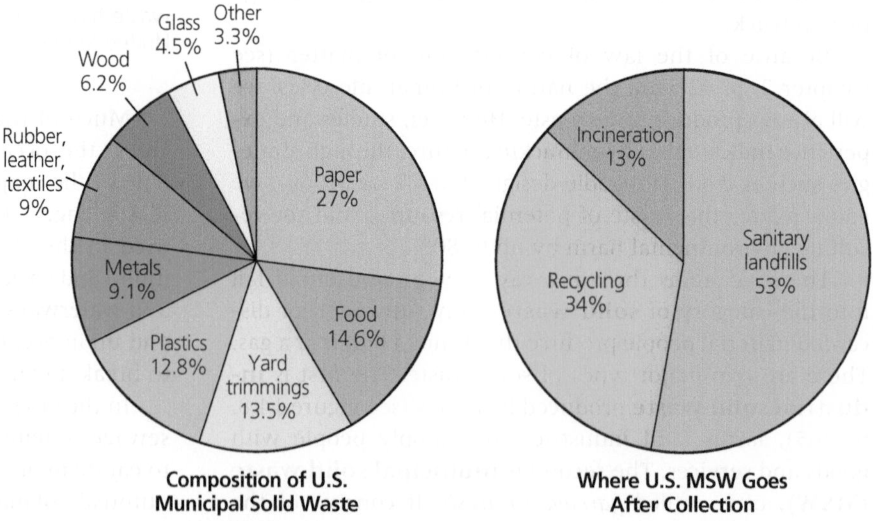

© Cengage Learning

FIGURE 21.4 Composition of MSW in the United States and data on where it goes after collection. ***Data analysis:*** How many times more than the amount recycled is the amount of materials put into landfills?

(Compiled by the authors using data from U.S. Environmental Protection Agency)

SCIENCE FOCUS 21.1

Garbology and Tracking Trash

How do we know about the composition of trash in landfills? Much of that information comes from research by *garbologists* such as William Rathje, an anthropologist who pioneered the field of garbology in the 1970s at the University of Arizona. These scientists work like archaeologists, training their students to sort, weigh, and itemize people's trash, and to bore holes to remove cores of materials from garbage dumps and analyze what they find.

Many people think of landfills as huge compost piles where biodegradable wastes are decomposed within a few months. In fact, decomposition inside modern landfills is a slow process. Trash buried inside sanitary landfills can resist decomposition perhaps for centuries because it is tightly packed and protected from sunlight, water, and air, and from the bacteria that could digest and decompose most of these wastes. In fact, researchers have unearthed 50-year-old newspapers that were still readable and hot dogs and pork chops that had not yet decayed.

A team of researchers, led by Carlo Riatti, at the Massachusetts Institute of Technology (MIT) conducted a project called "Trash Track." The project's goals were to find out where urban trash goes and to help New York City increase its recycling rate from the current 30% to 100% by 2030.

The researchers attached wireless transmitters about the size of a matchbook to several thousand different items of trash produced by volunteer participants in New York City, Seattle, Washington, and London, England. Every few hours, these devices use GPS technology to send their locations via a cell phone network to a computer at MIT, which plots their movements. This system tracks the trash items on their trips to recycling plants, landfills, or incinerators, and this helps researchers determine how and where trash goes.

CRITICAL THINKING

How might such a system help us to learn about the environmental costs of waste management such as the amount of pollution generated in the hauling and processing of waste? Explain.

Most of these wastes break down very slowly, if at all. Lead, mercury, glass, Styrofoam, and most plastic bottles do not break down completely. An aluminum can takes 500 years to disintegrate. Disposable diapers may take 550 years to break down, and a plastic shopping bag may stick around for up to 1,000 years.

3,200 Number of new plastic bags used every second, on average, in the United States

Some resource experts suggest we change the name of the trash we produce from MSW to MWR—*mostly wasted resources*. That's because so much of what is considered "waste" can be useful as a resource (Science Focus 21.1).

Hazardous Waste Is a Serious and Growing Problem

Another major category of waste is **hazardous,** or **toxic, waste**—any discarded material or substance that threatens human health or the environment because it is toxic, is corrosive, is flammable, can undergo violent or explosive chemical reactions, or can cause disease. Examples include industrial solvents, hospital medical waste, car batteries (containing acids and toxic lead), household pesticide products, dry-cell batteries (containing mercury and cadmium), and ash and sludge from incinerators and coal-burning power and industrial plants (see Figure 15.14, p. 392). Improper handling of these wastes can lead to pollution of air and water, degradation of ecosystems, and health threats (**Concept 21.1B**). The fastest-growing category of waste, which contains a large amount of hazardous waste, is electronic, or *e-waste* (see the Case Study that follows).

The two main classes of hazardous wastes are *organic compounds* such as various solvents, pesticides, PCBs, and dioxins and *toxic heavy metals* such as lead, mercury, and arsenic. Figure 21.5 lists some of the harmful chemicals found in many homes.

Another form of extremely hazardous waste is the highly radioactive waste produced by nuclear power plants and nuclear weapons facilities (see Chapter 15, p. 398). Such wastes must be stored safely for 10,000 to 240,000 years, depending on the radioactive isotopes they contain. After 60 years of research, scientists and governments have not found a scientifically and politically acceptable way to safely isolate these dangerous wastes for such long periods of time.

According to the U.N. Environment Programme (UNEP), more-developed countries produce 80–90% of the world's hazardous wastes, and the United States is the largest producer. However, if China continues to industrialize rapidly, largely without adequate pollution controls, it may soon overtake over the United States as the world's number one producer of hazardous wastes.

What Harmful Chemicals Are in Your Home?

Cleaning
- Disinfectants
- Drain, toilet, and window cleaners
- Spot removers
- Septic tank cleaners

Paint Products
- Paints, stains, varnishes, and lacquers
- Paint thinners, solvents, and strippers
- Wood preservatives
- Artist paints and inks

General
- Dry-cell batteries (mercury and cadmium)
- Glues and cements

Gardening
- Pesticides
- Weed killers
- Ant and rodent killers
- Flea powders

Automotive
- Gasoline
- Used motor oil
- Antifreeze
- Battery acid
- Brake and transmission fluid

FIGURE 21.5 Harmful chemicals are found in many homes. The U.S. Congress has exempted the disposal of many of these household chemicals and other items from government regulation. ***Question:*** Which of these chemicals could you find in your home?

Top: tuulijumala/Shutterstock.com. Center: Katrina Outland/Shutterstock.com. Bottom: Agencyby/Dreamstime.com

CASE STUDY

E-Waste—An Exploding Hazardous Waste Problem

What happens to your cell phone, computer, television set, and other electronic devices when they are no longer useful (see chapter-opening photo)? They become *electronic waste,* or *e-waste*—the fastest-growing solid waste problem in the United States and the world. The two leading producers of e-waste are the United States and China with 32% of the global total. The growth in e-waste is driven by increasing sales of electronics, as well as by short life cycles for many electronic products that are discarded and replaced with new models.

Between 2000 and 2013, the recycling of U.S. e-waste increased from 10% to 40%. Much of the remaining e-waste went to landfills and incinerators. Much e-waste contains gold, rare earths, and other valuable materials that could be recycled or reused. E-waste also is a source of toxic and hazardous chemicals that can contaminate air, surface water, groundwater, and soil and cause human health problems.

Much of the e-waste in the United States that is not buried or incinerated is shipped to China, India, and other Asian and African countries for processing. Labor is cheap and environmental regulations are weak in those countries. Workers there—many of them children—dismantle, burn, and treat e-waste with acids to recover valuable metals and reusable parts. The work exposes them to toxic metals such as lead and mercury and other harmful chemicals. The remaining scrap is dumped into waterways and fields or burned in open fires that expose people to highly toxic chemicals called dioxins. By discarding a cell phone or other electronics in the United States instead of recycling it, you may be contributing to health problems for people in China, India, or Africa.

Transfer of such hazardous waste from more-developed to less-developed countries is banned under the International Basel Convention. Despite this ban, much of the world's e-waste is not officially classified as hazardous waste, or it is illegally smuggled out of some countries. The United States can export its e-waste legally because it has not ratified the Basel Convention.

21.2 HOW SHOULD WE DEAL WITH SOLID WASTE?

CONCEPT 21.2 A sustainable approach to solid waste is first to produce less of it, then to reuse or recycle it, and finally to safely dispose of what is left.

Burn, Bury, or Recycle Solid Waste and Produce Less of It

Society can deal with the solid wastes it creates in two ways. One is **waste management,** which focuses on controlling wastes in order to limit their environmental harm but does not attempt to seriously reduce how much waste is produced. This approach begins with the question, "What do we do with solid waste?" It typically involves mixing wastes together and then burying them, burning them, or shipping them to another location.

The other approach is **waste reduction**—producing much less solid waste and reusing, recycling, or composting what is produced (**Concept 21.2** and **Core Case Study**). This waste prevention approach begins with questions such as "How can we avoid producing so much solid waste?" and "How can we use the waste we produce as resources like nature does?

Most analysts call for using **integrated waste management**—a variety of coordinated strategies for both waste management and waste reduction (Figure 21.6). For example, a manufacturer should consider the components that go into making its product. Can the product's waste be

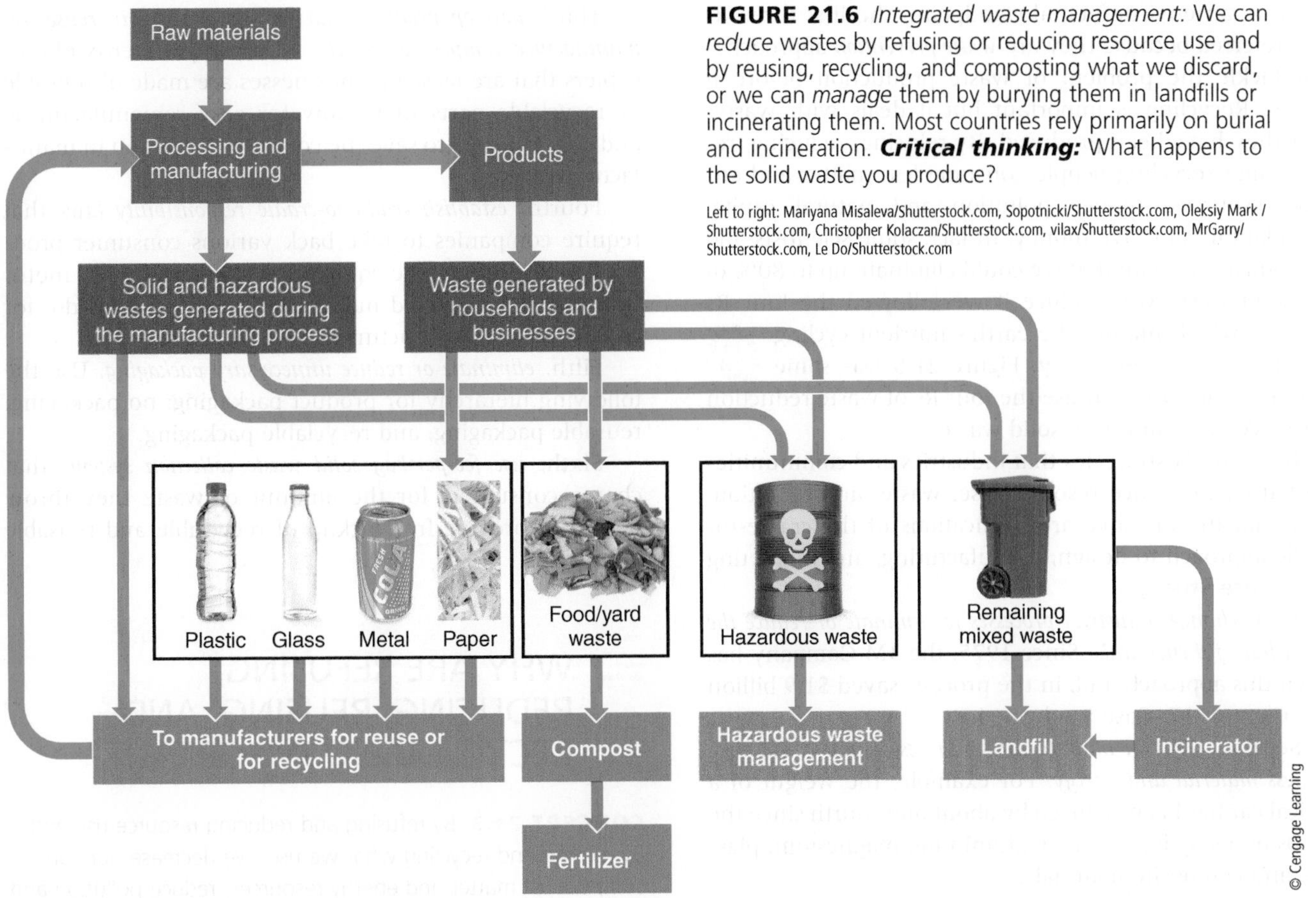

FIGURE 21.6 *Integrated waste management:* We can *reduce* wastes by refusing or reducing resource use and by reusing, recycling, and composting what we discard, or we can *manage* them by burying them in landfills or incinerating them. Most countries rely primarily on burial and incineration. ***Critical thinking:*** What happens to the solid waste you produce?

Left to right: Mariyana Misaleva/Shutterstock.com, Sopotnicki/Shutterstock.com, Oleksiy Mark / Shutterstock.com, Christopher Kolaczan/Shutterstock.com, vilax/Shutterstock.com, MrGarry/ Shutterstock.com, Le Do/Shutterstock.com

recycled (plastic, glass, metal, or paper)? If hazardous waste is involved, what is the best way to reduce the amount generated and to safely dispose of it?

Figure 21.7 compares the science-based waste management goals of the EPA and National Academy of Sciences with waste management trends based on actual data.

The Four Rs of Waste Reduction

A more sustainable approach to dealing with solid waste is to first reduce it, then reuse or recycle it, and finally safely dispose of what is left. The parts of this strategy, called the Four Rs of waste reduction, are listed here in order of priority suggested by scientists:

- **Refuse:** Don't use it.
- **Reduce:** Use less of it.
- **Reuse:** Use it over and over.
- **Recycle:** Convert used resources to useful items and buy products made from recycled materials. An important form of recycling is **composting,** which mimics nature by using bacteria and other decomposers to break down yard trimmings, vegetable food scraps, and other biodegradable organic wastes into materials than can be used to improve soil fertility.

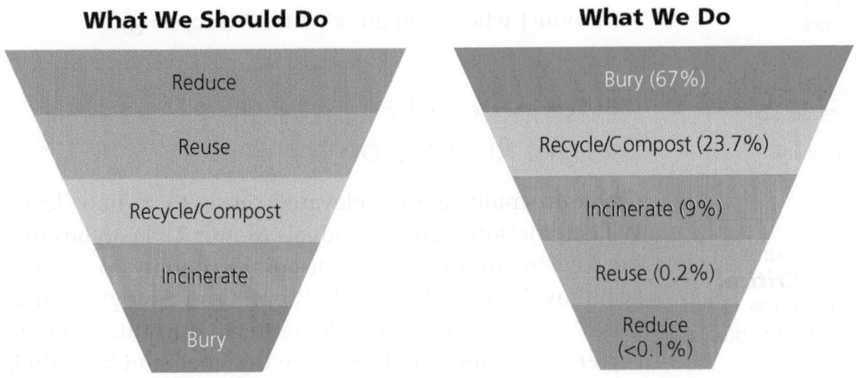

FIGURE 21.7 Priorities recommended by the U.S. National Academy of Sciences for dealing with municipal solid waste (left) compared with actual waste-handling practices in the United States based on data (right). ***Critical thinking:*** Why do you think most countries do not follow most of the scientific-based priorities listed on the left?

(Compiled by the authors using data from U.S. Environmental Protection Agency, U.S. National Academy of Sciences, Columbia University, and *BioCycle*.)

From an environmental standpoint, the first three Rs are preferred because they are *waste prevention* approaches that tackle the problem of waste production before it occurs. Recycling is important but it deals with wastes after they have been produced. By refusing, reducing, reusing, and recycling people consume less matter and energy resources, reduce pollution and natural capital degradation, and save money. In fact, some scientists and economists estimate that we could eliminate up to 80% of the solid waste we produce if we followed the four Rs strategy, which mimics the earth's nutrient cycling **principle of sustainability**. Figure 21.8 lists some ways in which you can use the four Rs of waste reduction to reduce your output of solid waste.

Here are six strategies that industries and communities have used to reduce resource use, waste, and pollution. Note that the first five are applications of the cradle-to-cradle approach to design, manufacturing, and marketing (**Core Case Study**).

First, *change industrial processes to eliminate or reduce the use of harmful chemicals*. Since 1975, the 3M Company has taken this approach and, in the process, saved $1.9 billion (see Chapter 17, Case Study, p. 461).

Second, *redesign manufacturing processes and products to use less material and energy*. For example, the weight of a typical car has been reduced by about one-fourth since the 1960s by using lighter steel, aluminum, magnesium, plastics, and composite materials.

What Can You Do?

Solid Waste

- Follow the four Rs of resource use: Refuse, Reduce, Reuse, Recycle
- Ask yourself whether you really need what you're buying and refuse packaging wherever possible
- Rent, borrow, or barter goods and services when you can, buy secondhand, and donate or sell unused items
- Buy things that are reusable, recyclable, or compostable, and be sure to reuse, recycle, and compost them
- Buy products with little or no packaging and recycle any packaging as much as possible
- Avoid disposables such as paper and plastic bags, plates, cups, and utensils, disposable diapers, and disposable razors whenever reusable versions are available
- Cook with whole, fresh foods, avoid heavily packaged processed foods, and buy products in bulk whenever possible
- Discontinue junk mail as much as possible and read online newspapers and magazines and e-books

FIGURE 21.8 Individuals matter: You can save resources by reducing your output of solid waste and pollution. ***Critical thinking:*** Which three of these steps do you think are the most important ones to take? Why? Which of these things do you already do?

Third, *develop products that are easy to repair, reuse, remanufacture, compost, or recycle*. For example, Xerox photocopiers that are leased by businesses are made of reusable or recyclable parts that allow for easy remanufacturing and are projected to save the company $1 billion in manufacturing costs.

Fourth, *establish cradle-to-cradle responsibility laws* that require companies to take back various consumer products such as electronic equipment, appliances, and motor vehicles, as Japan and many European countries do, for recycling or remanufacturing.

Fifth, *eliminate or reduce unnecessary packaging*. Use the following hierarchy for product packaging: no packaging, reusable packaging, and recyclable packaging.

Sixth, *use fee-per-bag solid waste collection systems* that charge consumers for the amount of waste they throw away but provide free pickup of recyclable and reusable items.

21.3 WHY ARE REFUSING, REDUCING, REUSING, AND RECYCLING SO IMPORTANT?

CONCEPT 21.3 By refusing and reducing resource use and by reusing and recycling what we use, we decrease our consumption of matter and energy resources, reduce pollution and natural capital degradation, and save money.

Alternatives to the Throwaway Economy

In today's industrialized societies, we have increasingly substituted throwaway items for reusable ones, which has resulted in growing masses of solid waste. However, by applying the four Rs, we can slow or stop this trend. People can guide and reduce their consumption of resources by asking questions such as these:

- Do I really need this? (refusing)
- How many of these do I actually need? (reducing)
- Is this something I can use more than once? (reusing)
- Can this be converted into the same or a different product when I am done with it? (recycling)

Revisiting Cradle-to-Cradle Design: Reuse Is on the Rise

Cradle-to-cradle design elevates reuse to a new level. William McDonough (Individuals Matter 21.1) argues that the key to shifting from a disposal economy to a reuse economy is to design for it (**Core Case Study**). Hence, some manufacturers of computers, furniture, photo copiers, and other products have designed and built their

INDIVIDUALS MATTER 21.1

William McDonough

PHOTOCOME-US/SIPA/Photocome/Sipa/Newscom

William McDonough is an architect, designer, and visionary thinker, devoted to the earth-friendly design of buildings, products, and cities.

McDonough points out that there is no waste in nature because the wastes of one organism become nutrients for other organisms. He also notes that humans have been releasing a growing number of chemicals to the environment faster than the natural chemical cycles can remove them. In addition, many of these synthetic chemicals cannot be broken down and recycled by natural processes. Many of these chemicals end up polluting the air, water, and soil and threatening the health of humans and other life forms.

McDonough and his business partner, chemist Michael Braungart, want to correct this trend by implementing a second industrial revolution, based on a *cradle-to-cradle approach* (**Core Case Study**). It would use environmentally and economically sustainable design to mimic nature by reusing and recycling the chemicals and products we make with the goal of zero waste. They argue that waste is a resource out of place and the result of poor design, and that we must employ three strategies to deal with it: **(1)** design products and societies that produce no waste, **(2)** live off the earth's endless supply of solar energy, and **(3)** respect and mimic the earth's life-sustaining biodiversity.

McDonough's approach has been applied in numerous projects, including the Adam Joseph Lewis Center for Environmental Studies at Oberlin College. Architects and designers view it as one of the most important and inspiring examples of environmentally friendly design. It uses recycled and nontoxic materials that can be further recycled. It gets heat from the sun and the earth's interior and electricity from solar cells, and it produces 13% more energy than it consumes. The building's greenhouse contains an ecosystem of plants and animals that purify the building's sewage and wastewater. Rainwater is collected and used to irrigate the surrounding green space, which includes a restored wetland, a fruit orchard, and a vegetable garden.

McDonough has been recognized by *Time* magazine as a "Hero for the Planet." He has also received numerous design awards and three presidential awards. He believes we can make the transition to a second industrial revolution, based on cradle-to-cradle design, and thereby leave the world better off than we found it.

products so that when they are no longer useful, they can be retrieved from consumers for repair or remanufacture within a system that allows the companies to preserve or boost their profits.

One way to encourage such innovation is for governments to ban or severely restrict the disposal of certain items. For example, the European Union (EU) has led the way by banning e-waste from landfills and incinerators. Electronics manufacturers are required to take back their products at the end of their useful lives. To cover the costs of these programs, consumers pay a recycling tax on electronic products, an example of implementing the full-cost pricing **principle of sustainability**. Some European nations, Japan, and China are using this take-back approach. In the United States, there is no federal take-back law, but according to the Electronics TakeBack Coalition, more than 20 states have such laws and several more are considering them.

Governments have also banned the use of certain throwaway items. For example, Finland banned all beverage containers that cannot be reused, and consequently, 95% of that country's soft drink, beer, wine, and spirits containers are refillable. The use of rechargeable batteries is cutting toxic waste by reducing the amount of conventional batteries that are thrown away. The latest rechargeable batteries come fully charged, can hold a charge for up to 2 years when they are not used, and can be recharged in about 15 minutes.

Instead of using throwaway paper or plastic bags to carry home groceries and other purchased items, many people now use reusable cloth bags. Both paper and plastic throwaway bags are environmentally harmful, and the question of which is more damaging has no clear-cut answer. Recycling and reuse of plastic bags is on the rise, but the great majority of such bags are thrown away.

To encourage people to carry reusable bags, the governments of Ireland, Taiwan, France, Germany, and the Netherlands tax plastic shopping bags. Several countries, including Kenya, Rwanda, South Africa, Bangladesh, Italy, and Australia, have banned the use of all or most types of plastic shopping bags. In Ireland, a tax of 25¢ per bag cut plastic bag litter by 90% as people switched to reusable bags. In 2014, the EU passed a directive aimed at cutting the use of single-use plastic bags by 80%. In 2014, the U.S. state of California joined Hawaii in instituting a statewide ban on single-use plastic bags. They have also been banned in more than 130 U.S. cities or counties, including San Francisco and Los Angeles, California; Dallas, Texas; Portland, Oregon; Seattle, Washington; Chicago, Illinois; and

Washington, D.C., despite intense lobbying against this by the plastics industry.

Similarly, several cities are trying to encourage the use of reusable food containers. In 2015, New York City joined Seattle, Portland, San Francisco, and Washington, D.C., in banning the use of polystyrene foam food containers. New York also banned the sale of polystyrene foam packing peanuts and has called for designers and entrepreneurs to produce reusable or compostable replacements for these banned items.

An increasingly popular way to reuse things is through *shared use*. In Portland, Oregon, some homeowners have worked with their neighbors to create tool libraries instead of buying their own tools. Toy libraries are also evolving among young families whose toys are used only for a few months or years. Companies that rent out tools, garden equipment, and other household goods provide another outlet for shared use. There are many other ways to reuse items and materials (Figure 21.9).

Recycling

The cradle-to-cradle approach (**Core Case Study**) gives the highest priority to reuse, but it also relies on recycling. Worn-out items from the technical cycle of cradle-to-cradle manufacturing are recycled or sent into the biological cycle where ideally they degrade and become biological nutrients (Figure 21.1).

McDonough and Braungart break recycling down into two categories: *upcycling* and *downcycling*. Ideally, all discarded items would be upcycled—recycled into a form that is more useful than the recycled item was. In downcycling, the recycled product is still useful, but not as useful or long-lived as the original item.

Households and workplaces produce five major types of recyclable materials: paper products, glass, aluminum, steel, and some plastics. These materials can be reprocessed into new, useful products in two ways. **Primary recycling** involves using materials, such as aluminum, again for the same purpose. For example, a used aluminum can is recycled into a new aluminum can. **Secondary recycling** involves downcycling or upcycling waste materials into different products, such as making park benches from recycled plastic waste.

Scientists and waste managers also distinguish between two types of recyclable wastes: *preconsumer* or *internal waste* generated in a manufacturing process, and *postconsumer* or *external waste* generated from use of products by consumers. Preconsumer waste makes up more than three-fourths of the total.

Recycling involves three steps: the collection of materials for recycling, conversion of recycled materials to new products, and selling and buying of products that contain recycled material. Recycling is successful environmentally and economically when all three of these steps are carried out.

According to Columbia University and BioCycle researchers, in 2013 the United States recycled about 96% of all lead-acid batteries; 72% of all newspapers, directories, and newspaper inserts; 67% of all steel cans; and 50% of all aluminum cans. Other categories such as food waste (that can be composted) and plastics have much lower recycling rates. Experts say that with education and proper incentives, Americans could recycle and compost at least 80% of their MSW, in keeping with the chemical cycling **principle of sustainability**.

What Can You Do?

Reuse

- Buy beverages in refillable glass containers
- Use reusable lunch containers
- Use a reusable coffee container and carry it with you
- Store refrigerated food in reusable containers
- Use rechargeable batteries and recycle them when their useful life is over
- When eating out, bring your own reusable container for leftovers
- Carry groceries and other items in a reusable basket or cloth bag
- Buy used furniture, cars, and other items, whenever possible

FIGURE 21.9 Individuals matter: Some ways to reuse the items we purchase. ***Questions:*** Which of these suggestions have you tried and how did they work for you?

Jbphotographylt/Dreamstime.com

FIGURE 21.10 Backyard composting bin.

Recycling of e-waste is especially attractive. A major United Nations University study in 2015 concluded that increasing piles of e-waste are urban mines because of the valuable metals the waste contains. In 2014, the world's e-waste contained 300 metric tons (330 tons) of gold, more than a tenth of all gold produced by miners in that year. It also contained millions of tons of iron, copper, silver, and aluminum. Yet, less than 17% of all e-waste is recycled.

Some see recycling as a business opportunity. One company, the RecycleBank, has set up a system where consumers can earn points by recycling. The company attaches an electronic tag to a household's recycling bins to measure how much the household is recycling. It then credits the household account with points that can be traded in—somewhat like frequent flyer miles—for rewards at businesses that have joined the program.

Composting is another form of recycling that directly applies the chemical cycling **principle of sustainability**. Composted material can be added to soil to supply plant nutrients. Many homeowners compost food wastes, yard wastes, and other organic waste materials in simple backyard containers (Figure 21.10).

Many cities collect and compost biodegradable wastes in centralized composting facilities (Figure 21.11). In the United States in 2014, according to a BioCycle study, there were more than 4,900 municipal composting programs. About 70% of them only recycle yard trimmings and the rest add varying combinations of food and farm waste and other types of organic waste. To be successful, a large-scale composting program must be located carefully and odors must be controlled, especially near residential areas. They must also exclude toxic materials that make the compost unsafe for fertilizing crops and lawns.

We Can Mix or Separate Household Solid Wastes for Recycling

One way to recycle is to send mixed MSW to centralized *materials-recovery facilities* (MRFs), where machines and workers separate the mixed waste to recover valuable materials for sale to manufacturers as raw materials. The remaining paper,

imging/Shutterstock.com

FIGURE 21.11 Large-scale municipal composting site.

plastics, and other combustible wastes are recycled or burned to produce steam or electricity. In some cases, onsite waste incinerators generate electricity to power the MRF.

MRFs are expensive to build, operate, and maintain. If not operated properly, they can emit CO_2 and toxic air pollutants, and they produce a toxic ash that must be disposed of safely, usually in landfills. Because MRFs require a steady diet of garbage to make them financially successful, their owners have a vested interest in increasing the flow of MSW. Thus, use of MRFs does nothing to encourage reuse and waste reduction and instead encourages the production of more trash—the reverse of what many scientists urge (Figure 21.7, left).

The mixed waste approach is becoming less economically sustainable in many communities because people are increasingly throwing nonrecyclable trash into their recycling bins. This requires MRFs to spend more time, energy, and money separating recyclable materials.

To many experts, it makes more ecological and economic sense for households and businesses to separate their trash into recyclable categories such as glass, paper, metals, certain types of plastics, and compostable materials. This *source separation* approach produces much less air and water pollution and costs less to implement, compared to relying on MRFs. It also saves more energy and yields cleaner and usually more valuable recyclables.

To promote separation of wastes for recycling, about 7,000 communities in the United States use a *pay-as-you-throw* or *fee-per-bag* waste collection system. They charge households and businesses for garbage that is picked up, but do not charge them for picking up materials separated for recycling or reuse.

Recycling Paper

About 55% of the world's industrial tree harvest is used to make paper. However, according to the U.S. Department of Agriculture, we could make tree-free paper from straw and other agricultural residues and from the fibers of rapidly growing plants such as kenaf (see Figure 10.15, p. 234) and hemp.

100 million

Number of trees used each year to produce the world's junk mail

The pulp and paper industry is the world's fifth largest energy consumer and uses more water to produce a metric ton of its product than any other industry. In both Canada and the United States, it is the third-largest industrial energy user and polluter, and paper is the dominant material in the MSW of both countries.

Paper (especially newspaper and cardboard) is easy to recycle. Recycling newspaper involves removing its ink, glue, and coating and then reconverting the paper to pulp, which is pressed into new paper. Making recycled paper produces 35% less water pollution and 74% less air pollution than does making paper from wood pulp, and, of course, no trees are cut down.

CONSIDER THIS . . .

CONNECTIONS Recycling Paper and Reducing CO_2 Emissions

According to the U.S. Energy Information Administration, recycled paper requires 10–30% less energy, which means that for every kilogram (2.2 pounds) of paper you recycle, you can prevent an average of 0.9 kilograms (2 pounds) of CO_2 emissions.

Recycling Glass

Glass was one of the first materials to be recycled on a large scale. In recent years, it has become more costly for some communities to recycle glass than to dump it in landfills. Particularly in places where recyclables are mixed by consumers and sorted at MRFs, the cost of separating broken glass from garbage has gone up because the amount of nonrecyclable trash in recycling bins is increasing.

In order to gain these environmental benefits, some communities are subsidizing the recycling of glass. Another approach to this problem would be to reuse glass jars and bottles to store food and other household items.

Recycling Plastics

Plastics consist of various types of large polymers, or *resins*—organic molecules made by chemically linking organic chemicals produced mostly from oil and natural gas. About 46 different types of plastics are used in consumer products, and some products contain several kinds of plastic.

Currently, only 7% by weight of all plastic wastes in the United States (and 13% of plastic containers and packaging) is recycled. These percentages are low because there are many different types of plastic resins, which are difficult to separate from products that contain them. Another factor is that most plastic beverage containers and other plastic products are not designed for recycling. This makes it cheaper and easier to put them in a landfill. However, progress is being made in the recycling of plastics (Individuals Matter 21.2) and in the development of more degradable bioplastics (Science Focus 21.2).

One type of exceptionally hard plastic, called thermoset, is used increasingly in the manufacture of cars and airplanes. The problem is that thermoset plastic was not easily recyclable until 2014 when research chemist Jeanette Garcia developed the first recyclable thermoset plastic. As cradle-to-cradle manufacturing (**Core Case Study**) becomes more widespread, this new recyclable plastic will be a major part of the technical cycle in manufacturing (Figure 21.1, right).

INDIVIDUALS MATTER 21.2

Mike Biddle's Contribution to Plastics Recycling

©World Economic Forum

In 1994, Mike Biddle, a former engineer with Dow Chemical, and his business partner Trip Allen founded MBA Polymers, Inc. Their goal was to develop a commercial process for recycling high-value plastics from complex streams of manufactured goods such as computers, electronics, appliances, and automobiles. They succeeded by designing a 16-step automated process that separates plastics from nonplastic items in mixed waste streams, and then separates plastics from each other by type and grade. The process then converts them to pellets that can be sold and used to make new products.

The pellets are cheaper than virgin plastics because the company's process uses 90% less energy than that needed to make a new plastic and because the raw material is discarded junk that is cheap or free. In addition, greenhouse gas emissions from this recycling process are much lower than those from the process of making virgin plastics. And recycling plastic wastes reduces the need to incinerate them or bury them in landfills.

MBA Polymers has been selected by *Inc.* magazine as one of "America's Most Innovative Companies." Biddle has been named a Technology Pioneer by the World Economic Forum and has received some of the world's most important environmental awards.

Recycling Has Advantages and Disadvantages

Figure 21.12 lists the advantages and disadvantages of recycling (**Concept 21.3**). Whether recycling makes economic sense depends on how we look at its economic and environmental benefits and costs.

Cities that make money by recycling and have higher recycling rates tend to use a single-pickup system for both recyclable and nonrecyclable materials instead of a more expensive dual-pickup system. Successful systems also use a pay-as-you-throw approach. They charge for picking up trash but not for picking up recyclable or reusable materials. They also require citizens and businesses to sort their trash and recyclables by type. San Francisco, California, uses such a system, and in 2015, the city recycled, composted, or reused 80% of its MSW.

Whether recycling makes economic sense depends on how we look at its economic and environmental benefits and costs. Critics of recycling programs argue that recycling is costly and adds to the taxpayer burden in communities where recycling is funded through taxation.

Proponents of recycling point to studies showing that the net economic, health, and environmental benefits of recycling (Figure 21.12, left) far outweigh the costs. The EPA estimates that recycling and composting in the United States in 2010 reduced emissions of climate-changing carbon dioxide by an amount roughly equal to that emitted by 36 million passenger vehicles. Recycling, reuse, and composting industries create 6 to 10 times as many jobs as landfills and waste incineration. The U.S. recycling industry employs 1.1 million people. Doubling the U.S. recycling rate would create about 1 million new jobs. And recycling steel, aluminum, copper, lead, and paper products can save 65–95% of the energy needed to make these products from virgin materials. It thus greatly reduces pollution and greenhouse gas emissions and saves money for manufacturers and consumers. However, the recent sharp drop in oil prices has reduced profits from recycling plastic because it is cheaper to make new plastics than to recycle existing plastics.

Trade-Offs

Recycling

Advantages	Disadvantages
Reduces energy and mineral use and air and water pollution	Can cost more than burying in areas with ample landfill space
Reduces greenhouse gas emissions	Reduces profits for landfill and incinerator owners
Reduces solid waste	Inconvenient for some

FIGURE 21.12 Recycling solid waste has advantages and disadvantages (**Concept 21.3**). ***Critical thinking:*** Which single advantage and which single disadvantage do you think are the most important? Why?

Photo: Jacqui Martin/Shutterstock.com

CONSIDER THIS . . .

CONNECTIONS Pollution Control and Lower Steel Recycling Rates

Sometimes pollution control can get in the way of recycling. In the 1970s, U.S. steel makers shifted from using open-hearth furnaces, which consumed a great deal of scrap steel, to basic oxygen furnaces, partly to cut their air pollution. The newer furnaces, now used for 97% of U.S. steel production, use only half the amount of scrap that a typical older furnace used.

SCIENCEFOCUS 21.2

Bioplastics

One of the most useful characteristics of plastic—its durability—also happens to be one of its biggest drawbacks. Plastics are made to last, but that means they don't break down once they're disposed of. In addition, most of today's plastics are made using petroleum-based chemicals, or petrochemicals. Processing these chemicals creates hazardous waste and contributes to water and air pollution. The good news is that some products are now being made from bioplastics, a type of plastic that is more environmentally friendly because it is made from biologically based chemicals.

Henry Ford, who developed the first Ford car and founded Ford Motor Company, supported research on the development of a bioplastic made from soybeans and another made from hemp. A 1914 photograph shows him using an ax to strike the body of a Ford car made from soy bioplastic to demonstrate its strength and resistance to denting.

However, as oil became cheaper and widely available, petrochemical plastics took over the market. Now, confronted with climate change and other environmental problems associated with the use of oil, chemists are stepping up efforts to make more environmentally sustainable bioplastics. They can be made from corn, soy, sugarcane, switchgrass, chicken feathers, and some components of garbage.

Compared with conventional oil-based plastics, properly designed bioplastics are lighter, stronger, and cheaper. And the process of making them usually requires less energy and produces less pollution per unit of weight. Instead of being sent to landfills, some packaging made from bioplastics (Figure 21.A) can be composted to produce a soil conditioner, in keeping with the chemical cycling **principle of sustainability**.

Some bioplastics are more environmentally friendly than others. For example, some are made from corn raised by industrial agricultural methods, which require great amounts of energy, water, and petrochemical fertilizers and thus have a very large ecological footprint. In evaluating and choosing bioplastics, scientists urge consumers to learn how they were made, how long they take to biodegrade, and whether they degrade into harmful chemicals.

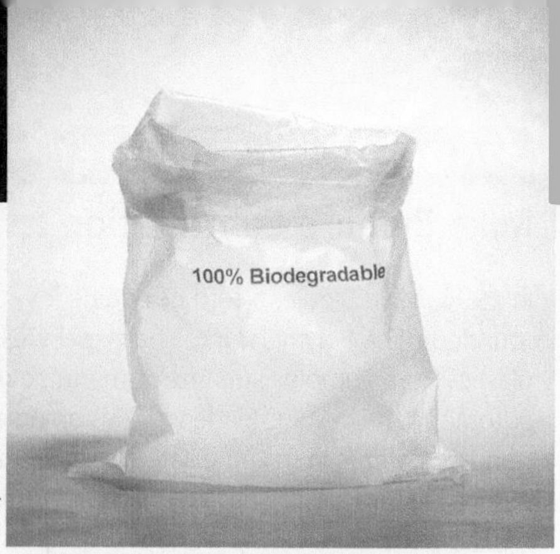

iStockphoto.com/Stefanie Timmermann

FIGURE 21.A A biodegradable plastic bag.

CRITICAL THINKING

Do you think that the advantages of bioplastics outweigh their disadvantages?

21.4 WHAT ARE THE ADVANTAGES AND DISADVANTAGES OF BURNING OR BURYING SOLID WASTE?

CONCEPT 21.4 Technologies for burning and burying solid wastes are well developed, but burning can contribute to air and water pollution and greenhouse gas emissions, and buried wastes can contribute to water pollution.

Burning Solid Waste Has Advantages and Disadvantages

Many communities burn their solid waste until nothing remains but fine, white-gray ash, which can then be buried in landfills. Heat released by burning trash can be used to heat water or interior spaces, or for producing electricity in facilities called *waste-to-energy incinerators*. Globally, MSW is burned in more than 800 of these types of incinerators, 115 of them in the United States.

A waste-to-energy incinerator (Figure 21.13) contains a combustion chamber where waste is burned at extremely high temperatures. Heat from the burning material is used to boil water and produce steam. The steam in turn drives a turbine that generates electricity. Combustion also produces wastes in the form of gases and ash. The gases must be filtered to remove pollutants before being released into the atmosphere and the hazardous ash must be treated and properly disposed of in landfills.

The United States incinerates 13% of its MSW, a fairly low percentage. One reason for this low percentage is that in the past incineration earned a bad reputation as a result of highly polluting and poorly regulated incinerators. In addition, incineration competes with an abundance of low-cost landfills in many parts of the United States.

By contrast, Denmark incinerates 54% of its MSW in state-of-the-art waste-to-energy incinerators. Denmark's incinerators use dozens of filters to remove pollutants such as toxic mercury and dioxins. They run so cleanly that they exceed European air pollution standards by a factor of 10. However, the resulting incinerator ash contains toxic chemicals and has to be safely stored somewhere, essentially forever.

Figure 21.14 lists the advantages and disadvantages of using incinerators to burn solid waste. With regard to air pollution, a 2009 EPA study concluded that landfills actually emit more air pollutants than modern waste-to-energy incinerators. On the other hand, the resulting incinerator ash contains toxic chemicals that must be stored somewhere. In

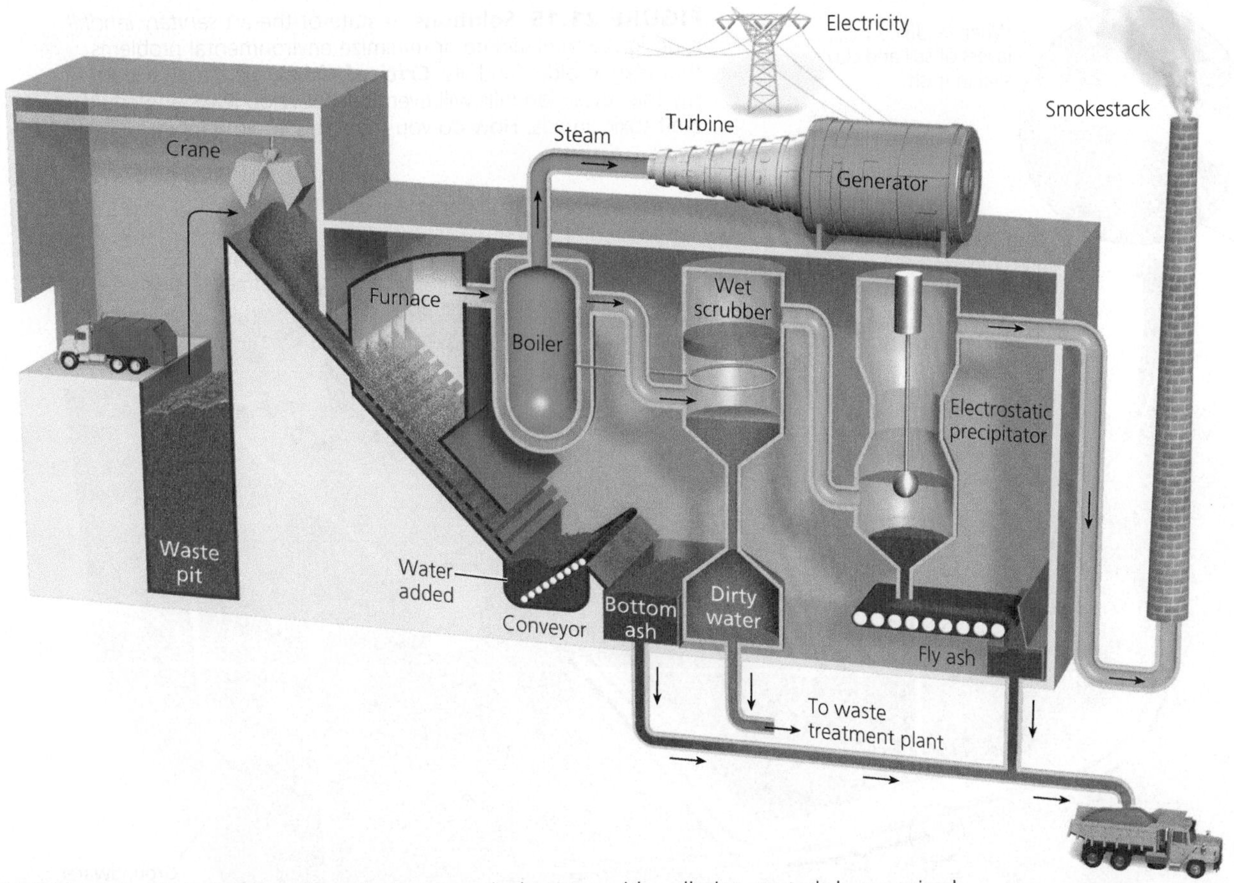

FIGURE 21.13 Solutions: A waste-to-energy incinerator with pollution controls burns mixed solid wastes and recovers some of the energy to produce steam to use for heating or producing electricity. ***Critical thinking:*** Would you invest in such a project? Why or why not?

Trade-Offs

Waste-to-Energy Incineration

Advantages	Disadvantages
Reduces trash volume	Expensive to build
Produces energy	Produces a hazardous waste
Concentrates hazardous substances into ash for burial	Emits some CO_2 and other air pollutants
Sale of energy reduces cost	Encourages waste production

FIGURE 21.14 Incinerating solid waste has advantages and disadvantages (**Concept 21.4**). These trade-offs also apply to the incineration of hazardous waste. ***Critical thinking:*** Which single advantage and which single disadvantage do you think are the most important? Why?

Top: Ulrich Mueller/Shutterstock.com. Bottom: Dmitry Kalinovsky/Shutterstock.com.

addition, many U.S. citizens, local governments, and environmental scientists remain opposed to waste incineration because it undermines efforts to increase reuse and recycling by creating a demand for burnable wastes. It also makes it easier for consumers to discard reusable and recyclable items.

Burying Solid Waste Has Advantages and Disadvantages

In the United States, about 54% of all MSW, by weight, is buried in sanitary landfills, compared to 80% in Canada, 15% in Japan, and 4% in Denmark. In a **sanitary landfill** (Figure 21.15), solid wastes are spread out in thin layers, compacted, and regularly covered with a layer of clay or plastic foam. This process helps to keep the material dry, cuts down on odors, and reduces the risk of fire, and keeps rats and other pest animals away from the wastes. In addition, it helps contain contaminated water called *leachate*, so that it does not leak out of the landfill and pollute nearby soil and groundwater.

The bottoms and sides of well-designed sanitary landfills have strong double liners and containment systems that collect the liquids leaching from them. Some landfills also have systems for collecting methane, the potent greenhouse gas that is produced when the buried wastes decompose in the absence of oxygen, and burning it as a fuel. According to the EPA, in

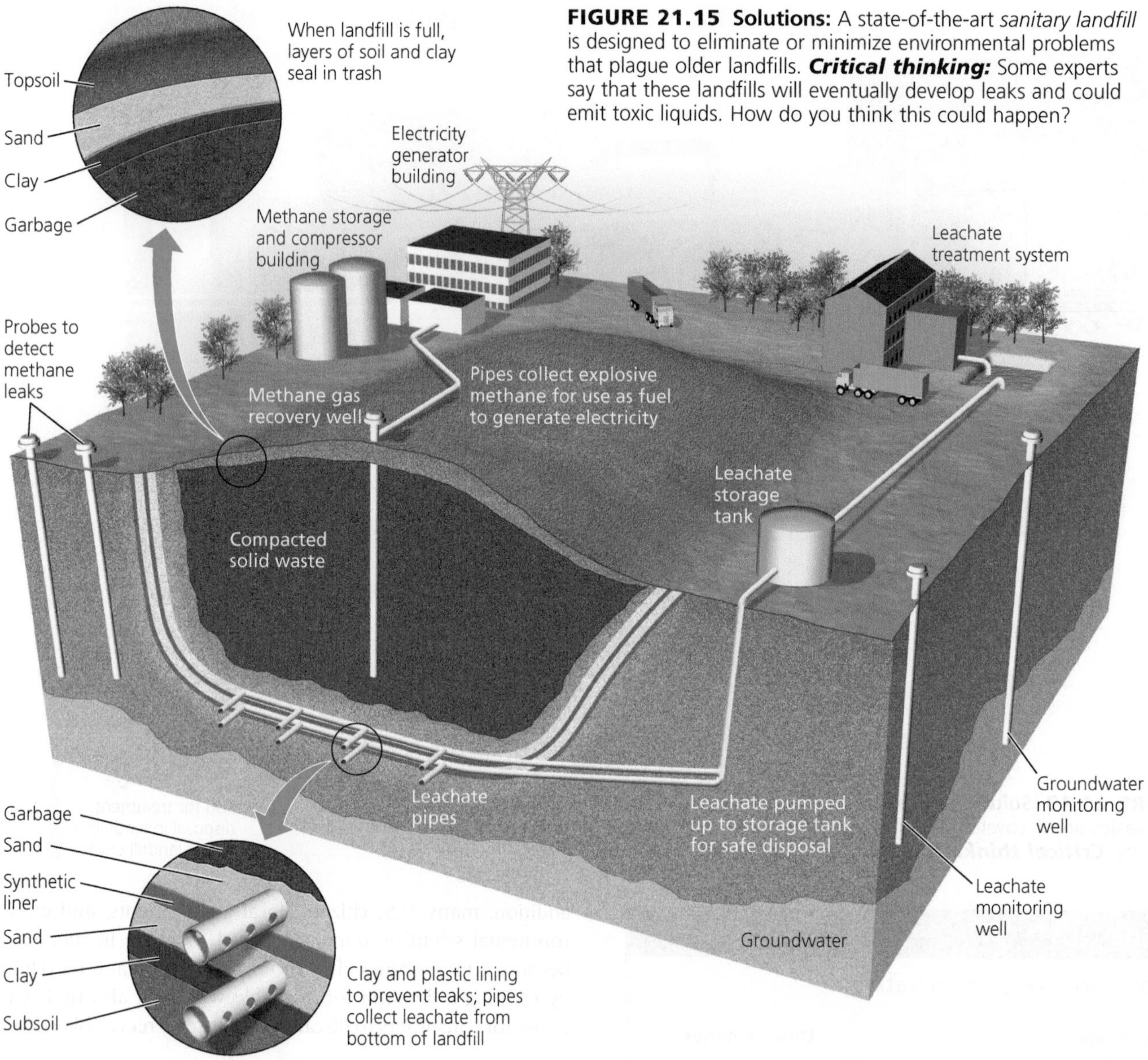

FIGURE 21.15 Solutions: A state-of-the-art *sanitary landfill* is designed to eliminate or minimize environmental problems that plague older landfills. ***Critical thinking:*** Some experts say that these landfills will eventually develop leaks and could emit toxic liquids. How do you think this could happen?

2014 there were more than 600 such landfill methane operations in the United States, and they had provided enough electricity to power about 1 million typical homes for a year.

What gets buried in landfills? Paper products represent the largest percentage of landfill materials. Other common materials include yard waste, plastics, metals, wood, glass, and food waste. Some types of solid waste are not accepted at landfills. For example, tires, waste oil and oil filters, items containing mercury such as CFL light bulbs and thermometers, electronics, and medical waste are not allowed.

Figure 21.16 lists the advantages and disadvantages of using sanitary landfills to dispose of solid waste.

Trade-Offs

Sanitary Landfills

Advantages	Disadvantages
Low operating costs	Noise, traffic, and dust
Can handle large amounts of waste	Releases greenhouse gases (methane and CO_2) unless they are collected
Filled land can be used for other purposes	Output approach that encourages waste production
No shortage of landfill space in many areas	Eventually leaks and can contaminate groundwater

© Cengage Learning

FIGURE 21.16 Using sanitary landfills to dispose of solid waste has advantages and disadvantages (**Concept 21.4**). ***Critical thinking:*** Which single advantage and which single disadvantage do you think are the most important? Why?

Photo: Pedro Miguel Sousa/ShutterStock.com

FIGURE 21.17 Some open dumps, especially in less-developed countries, are routinely burned, releasing large quantities of pollutants into the atmosphere. They can burn for days or weeks.

WitthayaP/Shutterstock.com

Another type of landfill is an **open dump.** It is essentially a field or large pit where garbage is deposited and sometimes burned (Figure 21.17). Open dumps are rare in more-developed countries (Figure 21.3), but are widely used near major cities in many less-developed countries. China disposes of much of its rapidly growing mountains of solid waste mostly in rural open dumps or in poorly designed and poorly regulated landfills.

Open dumps pose a variety of health, safety, and environmental threats, especially for the poor that survive by picking out metals and other valuable items to sell (Figure 21.3). Leachates leaking from open dumps can contaminate soil and groundwater supplies.

21.5 HOW SHOULD WE DEAL WITH HAZARDOUS WASTE?

CONCEPT 21.5 A more sustainable approach to hazardous waste is first to produce less of it, then to reuse or recycle it, then to convert it to less-hazardous materials, and finally to safely store what is left.

Hazardous Waste Requires Special Handling

The U.S. National Academy of Sciences has established three priority levels for dealing with hazardous waste: produce less; convert as much of it as possible to less-hazardous substances; and put the rest in long-term, safe storage (**Concept 21.5**). Figure 21.18 illustrates this integrated management approach to dealing with hazardous waste. Denmark follows these priorities, but most countries do not.

As with solid waste, the top priority for hazardous waste management should be pollution prevention and waste reduction. Using this approach, industries try to find substitutes for toxic or hazardous materials, reuse or recycle the hazardous materials within industrial processes, or use or sell them as raw materials for making other products, in keeping with the cradle-to-cradle approach (**Core Case Study**).

At least 33% of industrial hazardous wastes produced in the European Union are exchanged through clearinghouses where they are sold as raw materials for use by other industries, in keeping with the chemical cycling **principle of sustainability**. The producers of these wastes do not have to pay for their disposal and recipients get low-cost raw materials. About 10% of the hazardous waste in the United States is exchanged through such clearinghouses, an amount that could be increased significantly.

CASE STUDY

Recycling E-Waste

In some countries, workers in e-waste recycling operations—many of them children—are often exposed to toxic chemicals as they dismantle the electronic trash to extract its valuable

Produce Less Hazardous Waste	Convert to Less Hazardous or Nonhazardous Substances	Put in Perpetual Storage
■ Change industrial processes to reduce or eliminate hazardous waste production ■ Recycle and reuse hazardous waste	■ Natural decomposition ■ Incineration ■ Thermal treatment ■ Chemical, physical, and biological treatment ■ Dilution in air or water	■ Landfill ■ Underground injection wells ■ Surface impoundments ■ Underground salt formations

© Cengage Learning

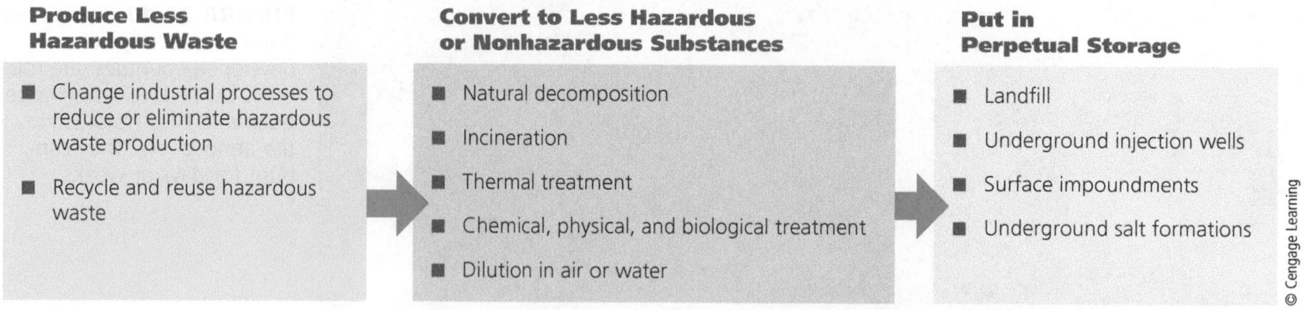

FIGURE 21.18 *Integrated hazardous waste management:* The U.S. National Academy of Sciences has suggested these priorities for dealing with hazardous waste (**Concept 21.5**). ***Critical thinking:*** Why do you think most countries do not follow these priorities?

metals or other parts that can be sold for reuse or recycling (Figure 21.19).

According to the United Nations, more than 70% of the world's e-waste is shipped to China. One popular destination for such waste is the small port city of Guiyu, where the air reeks of burning plastic and acid fumes. There, more than 5,500 small-scale e-waste businesses employ over 30,000 people, including children. They work for very low wages in dangerous conditions to extract valuable metals such as gold, silver, and copper from discarded computers, television sets, and cell phones.

Although these workers are exposed to toxic chemicals, they usually wear no masks or gloves, and often work in rooms with no ventilation. They carry out dangerous activities such as smashing TV picture tubes with large hammers to recover certain components—a method that releases large amounts of toxic lead dust into the air. They also burn computer wires to expose copper, melt circuit boards in metal pots over coal fires to extract lead and other metals, and douse the boards with strong acid to extract gold. After the metals are removed, leftover parts are burned or dumped into rivers or onto the land. Atmospheric levels of deadly dioxin in Guiyu are up to 86 times higher than World Health Organization safety standards, and it is estimated that more than 82% of the Guiyu area's children younger than age 6 suffer from lead poisoning.

The United States produces roughly 50% of the world's e-waste and recycles about 40% of it, according to the U.S. Environmental Protection Agency. However, by 2013, at least 20 U.S. states had banned the disposal of computers and TV sets in landfills and incinerators. These measures set the stage for an emerging, highly profitable *e-cycling* industry. In 2013, 13 states along with New York City made manufacturers responsible for recycling most electronic devices.

A growing number of scientists and economists have called for a U.S. federal law to institute a cradle-to cradle approach (**Core Case Study**) that would require manufacturers to take back all electronic devices they produce and recycle them domestically. It could be similar to laws in the European Union, where a recycling fee typically covers the

James P. Blair/National Geographic Creative

FIGURE 21.19 This young girl in Dhaka, Bangladesh, is recycling batteries by hammering them apart to extract tin and lead. The workers at this shop are mostly women and children.

costs of such programs. Without such a law there is little incentive for recycling e-waste and plastics, especially when there is money to be made from illegally sending such materials to other countries.

The only real long-term solution is a *prevention* approach. Electrical and electronic products could be designed to be produced and easily repaired, remanufactured, or recycled, without the use of toxic materials (**Core Case Study**).

CONSIDER THIS . . .

THINKING ABOUT E-Waste Recycling

Would you support a recycling fee on all electronic devices? Why or why not?

Detoxifying Hazardous Waste

In Denmark, all hazardous and toxic wastes from industries and households are collected and delivered to any of 21 transfer stations throughout the country. They are then taken to a large processing facility, where three-fourths of the waste is detoxified using physical, chemical, and biological methods. The rest is buried in a carefully designed and monitored landfill.

Physical methods for detoxifying hazardous wastes include using charcoal or resins to filter out harmful solids, distilling liquid wastes to separate out harmful chemicals, and precipitating such chemicals from solution. Especially deadly wastes, such as those contaminated by mercury, can be encapsulated in glass, cement, or ceramics and put in secure storage sites.

Chemical methods are used to convert hazardous chemicals to harmless or less harmful chemicals through chemical reactions. Currently, chemists are testing the use of *cyclodextrin*—a type of sugar made from cornstarch—to remove toxic materials such as solvents and pesticides from contaminated soil and groundwater. To clean up a site contaminated by hazardous chemicals, a solution of cyclodextrin is applied. After this molecular sponge-like material moves through the soil or groundwater, picking up various toxic chemicals, it is pumped out of the ground, stripped of its contaminants, and reused.

Some scientists and engineers consider *biological methods* for treatment of hazardous waste to be the wave of the future. One such approach is *bioremediation,* in which bacteria and enzymes help to destroy toxic or hazardous substances or convert them to harmless compounds. Bioremediation is often used on contaminated soil. For example, microorganisms are used at some contaminated sites to break down hazardous chemicals, such as PCBs, pesticides, and oil in the soil, leaving behind harmless substances such as water and water-soluble chloride salts. It usually takes a little longer to work than most physical and chemical methods, but it costs much less.

Phytoremediation is another biological method for treating hazardous wastes. It involves using natural or genetically engineered plants as "pollution sponges." They are able to absorb, filter, and remove contaminants from polluted soil and water. Phytoremediation can be used to clean up soil and water contaminated with chemicals such as pesticides, organic solvents, and radioactive or toxic metals. This method is still being evaluated and is slow compared to other alternatives.

We can incinerate hazardous wastes to break them down and convert them to harmless or less harmful chemicals. This method involves the same advantages and disadvantages as burning of solid wastes (Figure 21.14). Incinerating hazardous waste without effective and expensive air pollution controls can release air pollutants such as highly toxic dioxins. It also produces a highly toxic ash that must be safely and permanently stored in a specially designed landfill or vault.

Plasma gasification is another thermal treatment method. This technology uses arcs of electrical energy to produce very high temperatures in order to vaporize trash in the absence of oxygen. The process reduces the volume of a given amount of waste by 99%, produces a synthetic gaseous fuel, and encases toxic metals and other materials in glassy lumps of rock. It is currently very costly, but plasma arc companies are working to bring prices down. Figure 21.20 lists the major advantages and disadvantages of using this process.

Storing Hazardous Waste

Ideally, we should use burial on land or long-term storage of hazardous and toxic wastes in secure vaults only as the third and last resort after the first two priorities have been exhausted (Figure 21.18 and **Concept 21.5**). Currently, however, burial on land is the most widely used method in the United States and in most countries, largely because it is less expensive than other methods.

Trade-Offs

Plasma Arc

Advantages	Disadvantages
Produces a mixture of CO and H_2 that can be used as a fuel	High cost
Mobile. Easy to move to different sites	Produces CO_2 and CO
Produces no toxic ash	Can release particulates and chlorine gas
	Can vaporize and release toxic metals and radioactive elements

FIGURE 21.20 Using *plasma gasification* to detoxify hazardous wastes has advantages and disadvantages. ***Critical thinking:*** Which single advantage and which single disadvantage do you think are the most important? Why?

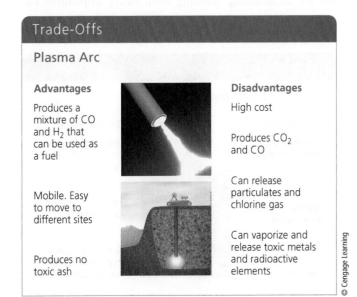

The most common form of burial is *deep-well disposal*. Liquid hazardous wastes are pumped under high pressure through a pipe into dry, porous underground rock formations far beneath aquifers that are tapped for drinking and irrigation water. Theoretically, these liquids soak into the porous rock material and are isolated from overlying groundwater by essentially impermeable layers of clay and rock. The cost is low and the wastes can often be retrieved if problems develop.

However, there are a limited number of such sites and limited space within them. Sometimes the wastes can leak into groundwater from the well shaft or migrate into groundwater in unexpected ways. Also, this approach encourages the production instead of the reduction of hazardous wastes.

In the United States, almost two-thirds of all liquid hazardous wastes are injected into deep disposal wells. This amount will increase sharply as the country relies more on fracking to produce natural gas and oil trapped in shale rock (see Science Focus 15.1, p. 390).

Many scientists argue that current regulations for deep-well disposal in the United States are inadequate and should be improved (see the Case Study that follows). Figure 21.21 lists the advantages and disadvantages of using deep-well disposal of liquid hazardous wastes.

Some liquid hazardous wastes are stored in lined ponds, pits, or lagoons, called *surface impoundments*. Studies conducted by the EPA found that 70% of all U.S. hazardous waste storage ponds have no liners and could threaten groundwater supplies. According to the EPA, eventually, all impoundment liners are likely to leak and contaminate groundwater. Because these impoundments are not covered, volatile harmful chemicals evaporate, polluting the air. In addition, flooding from heavy rainstorms can cause such ponds to overflow. Not all U.S. hazardous waste storage ponds have liners, according to the EPA, and many liners are subject to leaks that could eventually contaminate groundwater. Figure 21.22 lists the advantages and disadvantages of using this method.

There are some highly toxic materials (such as mercury; see Chapter 17, Core Case Study, p. 442) that we cannot destroy, detoxify, or safely bury. The best way to deal with such materials is to prevent or reduce their use and put them in sealed containers and bury them in carefully designed and monitored *secure hazardous waste landfills* (Figure 21.23). This is the least-used method because of the expense involved.

Figure 21.24 lists some ways in which you can reduce your output of hazardous waste—the first step in dealing with it.

Trade-Offs

Deep-Well Disposal

Advantages	Disadvantages
Safe if sites are chosen carefully	Leaks can occur from corrosion of well casing
Wastes can often be retrieved	Emits CO_2 and other air pollutants
Low cost	Output approach that encourages waste production

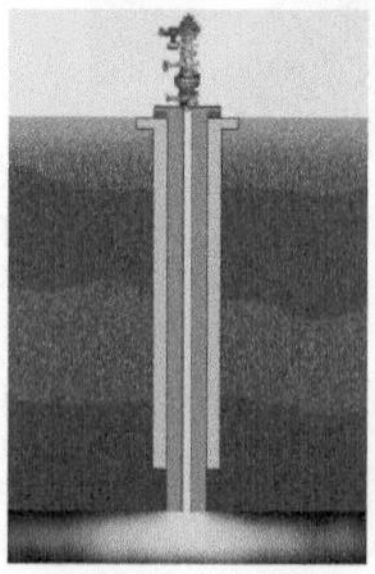

© Cengage Learning

FIGURE 21.21 Injecting liquid hazardous wastes into deep underground wells has advantages and disadvantages. ***Critical thinking:*** Which single advantage and which single disadvantage do you think are the most important? Why?

Trade-Offs

Surface Impoundments

Advantages	Disadvantages
Low cost	Water pollution from leaking liners and overflows
Wastes can often be retrieved	Air pollution from volatile organic compounds
Can store wastes indefinitely with secure double liners	Output approach that encourages waste production

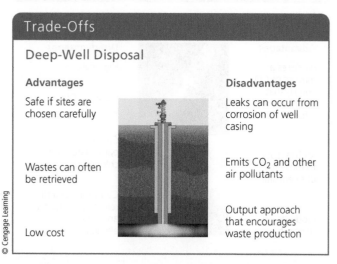

Jim West/Alamy Stock Photo

© Cengage Learning

FIGURE 21.22 Storing liquid hazardous wastes in surface impoundments has advantages and disadvantages. ***Critical thinking:*** Which single advantage and which single disadvantage do you think are the most important? Why?

CONSIDER THIS . . .

LEARNING FROM NATURE

To find substitutes for toxic chemicals used in manufacturing silicon-based computer chips, scientists are studying how microbes called *diatoms* build their protective silica shells.

CASE STUDY

Hazardous Waste Regulation in the United States

Several U.S. federal laws help to regulate the management and storage of hazardous wastes. The first is called the Resource Conservation and Recovery Act (RCRA, pronounced "RECK-ra"). In place since 1976 and amended in 1984, RCRA regulates about 5% of all hazardous waste produced in the United States.

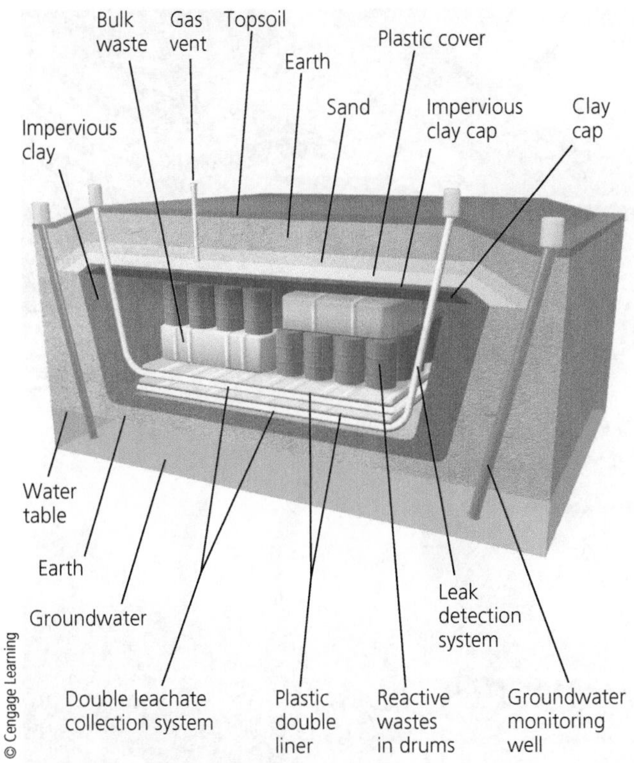

FIGURE 21.23 Solutions: Hazardous wastes can be isolated and stored in a secure hazardous waste landfill.

What Can You Do?

Hazardous Waste

- Avoid using pesticides and other hazardous chemicals, or use them in the smallest amounts possible
- Use less harmful substances instead of commercial household cleaners. For example, use vinegar to polish metals, clean surfaces, and remove stains and mildew, and baking soda to clean utensils and to deodorize and remove stains.
- Do not dump pesticides, paints, solvents, oil, antifreeze, or other hazardous chemicals down the toilet, down the drain, into the ground, into the garbage, or down storm drains. Free hazardous waste disposal services are available in many cities.
- Do not throw old fluorescent lightbulbs (which contain mercury) into regular trash. Many communities and home product retailers offer free recycling of these bulbs.

FIGURE 21.24 Individuals matter: You can reduce your output of hazardous wastes (**Concept 21.5**). ***Critical thinking:*** Which two of these measures do you think are the most important ones to take? Why?

Under RCRA, the EPA sets standards for the management of several types of hazardous waste and issues permits to companies that allow them to produce and dispose of a certain amount of those wastes by approved methods. Permit holders must use a *cradle-to-grave* system to keep track of waste they transfer from a point of generation (cradle) to an approved off-site disposal facility (grave), and they must submit proof of this disposal to the EPA.

RCRA is a good start. However, 95% of the hazardous and toxic wastes produced in the United States, including e-waste, are not regulated. In most other countries, especially less-developed countries, the amount of regulated waste is even smaller.

CONSIDER THIS . . .

THINKING ABOUT Hazardous Waste

Why do you think 95% of the hazardous waste produced in the United States is not regulated? Do you favor regulating such wastes? What do you think would be the economic consequences of doing so? Would this promote the cradle-to-cradle approach to reducing solid and hazardous wastes?

The Toxic Substances Control Act has also been in place since 1976. Its purpose is to regulate and ensure the safety of the thousands of chemicals used in manufacturing and contained in many products. Under this law, companies must notify the EPA before introducing a new chemical into the marketplace. They are not required to provide any data about its safety. In other words, any new chemical is viewed as safe unless the EPA, with little money available for this purpose, can show that it is harmful.

Since 1976, the EPA, with a very limited budget, has used this act to ban only 5 of the roughly 80,000 chemicals in use. Environmental and health scientists call for Congress to reform this law by requiring manufacturers to provide data showing that a chemical or product containing a certain chemical is safe before it can be sold in the marketplace.

The Comprehensive Environmental Response, Compensation, and Liability Act (CERCLA) was passed in 1980. It is commonly known as the Superfund Act and is regulated by the EPA. The goals of the act are to identify sites, called Superfund sites, where hazardous wastes have contaminated the environment (Figure 21.25) and to clean them up, using EPA-approved methods. The worst sites—those that represent an immediate and severe threat to human health—are put on a *National Priorities List* and scheduled for cleanup.

As of August, 2016, there were 1,328 sites on the Superfund list, along with 55 proposed new sites, and 391 sites had been cleaned up and removed from the list. Nearly one of every six Americans was living within 5 kilometers (3 miles) of a Superfund site. The Waste Management Research Institute estimates that at least 10,000 sites should be on the priority list and that cleanup of these sites could cost about $1.7 trillion, not including legal fees. This is a glaring example of the economic and environmental value of emphasizing waste reduction and pollution prevention over the cleanup approach that the United States and most countries rely on.

patrikslezak/Fotolia LLC

FIGURE 21.25 Leaking barrels of toxic waste.

In 1984, Congress amended the Superfund Act to give citizens the right to know what toxic chemicals are being stored or released in their communities. This required 23,800 large manufacturing facilities to report their annual releases of any of nearly 650 toxic chemicals. If you live in the United States, you can find out what toxic chemicals are being stored and released in your neighborhood by going to the EPA's *Toxic Release Inventory* website.

The Superfund Act, designed to make polluters pay for cleaning up abandoned hazardous waste sites, has greatly reduced the number of illegal dumpsites around the country and discouraged unsafe handling of hazardous wastes. However, the pace of cleanup has slowed because the Superfund Act lost a major source of funding. Under pressure from polluters, the U.S. Congress refused to renew the tax on oil and chemical companies—which financed the Superfund legislation—after it expired in 1995. Now taxpayers, not polluters, pay for cleanups (with an average cost of $26 million) when responsible parties cannot be found.

Closely associated with CERCLA is the EPA's Brownfields Program. A brownfield is an industrial or commercial property that is, or may be, contaminated with hazardous pollutants. The program is designed to help states, communities, and other stakeholders economically redevelop contaminated property. The program assists interested parties in assessing sites, cleaning up sites, or reusing land designated as a brownfield. Reclaiming these lands can increase local tax bases, promote job growth, enable a new facility to use existing infrastructure, and keep undeveloped land from being used.

21.6 HOW CAN WE SHIFT TO LOW-WASTE ECONOMY?

CONCEPT 21.6 Shifting to a low-waste economy will require individuals and businesses to reduce resource use and to reuse and recycle most solid and hazardous wastes at local, national, and global levels.

Citizens Can Take Action

In the United States, individuals have organized grassroots (bottom-up) campaigns to prevent the construction of hundreds of incinerators, landfills, treatment plants for hazardous and radioactive wastes, and chemical plants in or near their communities. These campaigns have organized sit-ins, concerts, and protest rallies. They have gathered signatures on petitions and presented them to lawmakers.

Manufacturers and waste industry officials point out that something must be done with toxic and hazardous wastes created in the production of certain goods and services. Many citizens do not accept this argument. Their view is that the best way to deal with most toxic and hazardous waste is to produce much less of it by focusing on pollution and waste prevention. They argue the goal should be "not in anyone's back yard" (NIABY) or "not on planet Earth" (NOPE).

Using International Treaties to Reduce Hazardous Waste

For decades, some countries regularly shipped hazardous wastes to other countries for disposal or processing. Since 1992, an international treaty known as the Basel Convention has banned participating countries from shipping hazardous waste through other countries without their permission. This treaty also covers e-waste.

By 2015, this agreement had been ratified (formally approved and implemented) by 183 countries. The United States has signed but has not ratified the convention. In 1995, the treaty was amended to outlaw all transfers of hazardous wastes from industrial countries to less-developed countries. This ban is likely to be ratified by enough countries to go into effect in the next few years.

This ban will help, but it will not do away with the highly profitable illegal shipping of hazardous wastes. Hazardous waste smugglers evade the laws by using an array of tactics, including bribes, false permits, and mislabeling of hazardous wastes as recyclable materials.

In 2000, delegates from 122 countries completed a global treaty known as the Stockholm Convention on Persistent Organic Pollutants (POPs). The treaty regulates the use of 12 widely used persistent organic pollutants that can accumulate in the fatty tissues of humans and other animals that occupy high trophic levels in food webs. As a result, these hazardous chemicals can reach levels hundreds of thousands of times higher than their levels in the general environment (see Figure 9.14, p. 205).

Because they persist in the environment, POPs can also be transported long distances by wind and water. The original list of 12 hazardous POPs chemicals, called the *dirty dozen*, includes DDT and eight other chlorine-containing persistent pesticides, PCBs, dioxins, and furans. Using blood tests and statistical sampling, medical researchers at New York City's Mount Sinai School of Medicine found that it is likely that nearly every person on earth has detectable levels of POPs in their bodies. The long-term health effects of this involuntary global chemical experiment are largely unknown.

By 2015, 179 countries had ratified a strengthened version of the POPs treaty that seeks to ban or phase out the use of these hazardous chemicals and to detoxify or isolate existing stockpiles. The list of regulated POPs is expected to grow.

In 2000, the Swedish Parliament enacted a law that, by 2020, will ban all potentially hazardous chemicals that are persistent in the environment and that can accumulate in living tissue. This law also requires industries to perform risk assessments on the chemicals they use and to show that these chemicals are safe to use, as opposed to requiring the government to show that they are dangerous. In other words, chemicals are assumed to guilty until proven innocent—the reverse of the current policy in the United States and most other countries. There is strong opposition to this approach in the United States, especially from most of the industries that produce and use potentially hazardous chemicals.

Encouraging Reuse and Recycling

Three factors hinder reuse and recycling. *First*, the market prices of almost all products do not include the harmful environmental and health costs associated with producing, using, and discarding them—a violation of the full-cost pricing **principle of sustainability**.

Second, the economic playing field is uneven, because in most countries, resource extraction industries receive more government tax breaks and subsidies than reuse and recycling industries.

Third, the demand and thus the price paid for recycled materials fluctuates, mostly because it is not a high priority for most governments, businesses, and individuals to buy goods made of recycled materials.

How can we encourage reuse and recycling? Proponents say that leveling the economic playing field is the best way to start. Governments can *increase* subsidies and tax breaks for reusing and recycling materials, and *decrease* subsidies and tax breaks for making items from virgin resources.

One way to include some of the harmful environmental costs of products in their prices, while encouraging recycling, is to attach a small deposit fee to the price of recyclable items, as is done in many European countries, several Canadian provinces, and 10 U.S. states that have *bottle bills*. Such laws place a deposit fee of 5 or 10 cents on each beverage container, and consumers can recover that fee by returning their empty containers to the store. In 2012, these 10 states recycled at least 70% of their bottles and cans, compared to a 28% average in states with no bottle bills.

Another strategy is to greatly increase use of the fee-per-bag waste collection system that charges households for the trash they throw away but not for their recyclable and reusable wastes. When Fort Worth, Texas, instituted such a program, the proportion of households recycling their trash went from 21% to 85%. The city went from losing \$600,000 in its recycling program to making \$1 million a year because of increased sales of recycled materials to industries.

Governments can also pass laws requiring companies to take back and recycle or reuse packaging and electronic waste discarded by consumers. Japan and some European Union countries have such laws. Another strategy is to encourage or require government purchases of recycled products to help increase demand for and lower prices of these products. Citizens can also pressure governments to require product labeling listing the recycled content of products, as well as the types and amounts of any hazardous materials they contain. This would help consumers to make more informed choices about the environmental consequences of buying certain products. It would also help to expand the market for recycled materials by spurring demand for them.

Reuse and Recycling Present Economic Opportunities

A growing number of people are saving money through reuse, by going to yard sales, flea markets, secondhand stores, and online sites such as eBay and craigslist. Another such site, the Freecycle Network, links people who want to give away their unused household belongings to people who want or need them. Its 9.2 million members reuse an average of 640 metric tons (700 tons) of items every day, which is about the amount of solid waste that arrives at a midsize landfill every day.

For many, recycling has become a business opportunity. In particular, *upcycling*, or recycling materials into products of a higher value (Figure 21.26), is a growing field. For example, a British company called Worn Again is converting discarded textiles, such as old hot-air balloons and worn-out seat covers, into windbreaker jackets and other products. Researcher Na Lu at the University of North Carolina, Charlotte, has found a way to upcycle plastic bottles to make a building material that could outperform composite lumber and wood lumber. Shipping pallets are being used too make furniture. Many see upcycling as an area of great economic opportunity. It also represents an excellent opportunity for those interested in creating or expanding their positive environmental impact by creating useful products from "trash."

Making the Transition to Low-Waste Economies

According to physicist Albert Einstein, "A clever person solves a problem; a wise person avoids it." Many people are taking these words seriously. Many school cafeterias,

FIGURE 21.26 *Upcycling:* This handbag was made from an old airline seat.

restaurants, national parks, and corporations are participating in a rapidly growing "zero waste" movement to reduce, reuse, and recycle, and some have lowered their waste outputs by up to 80%, with the ultimate goal of eliminating their waste outputs.

Many environmental scientists argue that we can prevent pollution, reduce waste, and make a transition to a low-waste society by understanding and following these key principles:

1. Everything is connected.
2. There is no *away*, as in *to throw away*, for the wastes we produce.
3. Producers and polluters should pay for the wastes they produce.
4. We can mimic nature by reusing, recycling, composting, or exchanging most of the municipal solid wastes we produce (see the Case Study that follows).

CASE STUDY

Industrial Ecosystems: Copying Nature

An important goal for a more sustainable society is to make its industrial manufacturing processes cleaner and more sustainable by redesigning them to mimic the way nature deals with wastes—an approach called *biomimicry*. In nature, according to the chemical cycling **principle of sustainability**, the waste outputs of one organism become the nutrient inputs of another organism, so that all of the earth's nutrients are endlessly recycled. This explains why there is essentially no waste in undisturbed ecosystems.

One way for industries to mimic nature is to reuse or recycle most of the minerals and chemicals they use. Industries can set up *resource exchange webs*, in which the

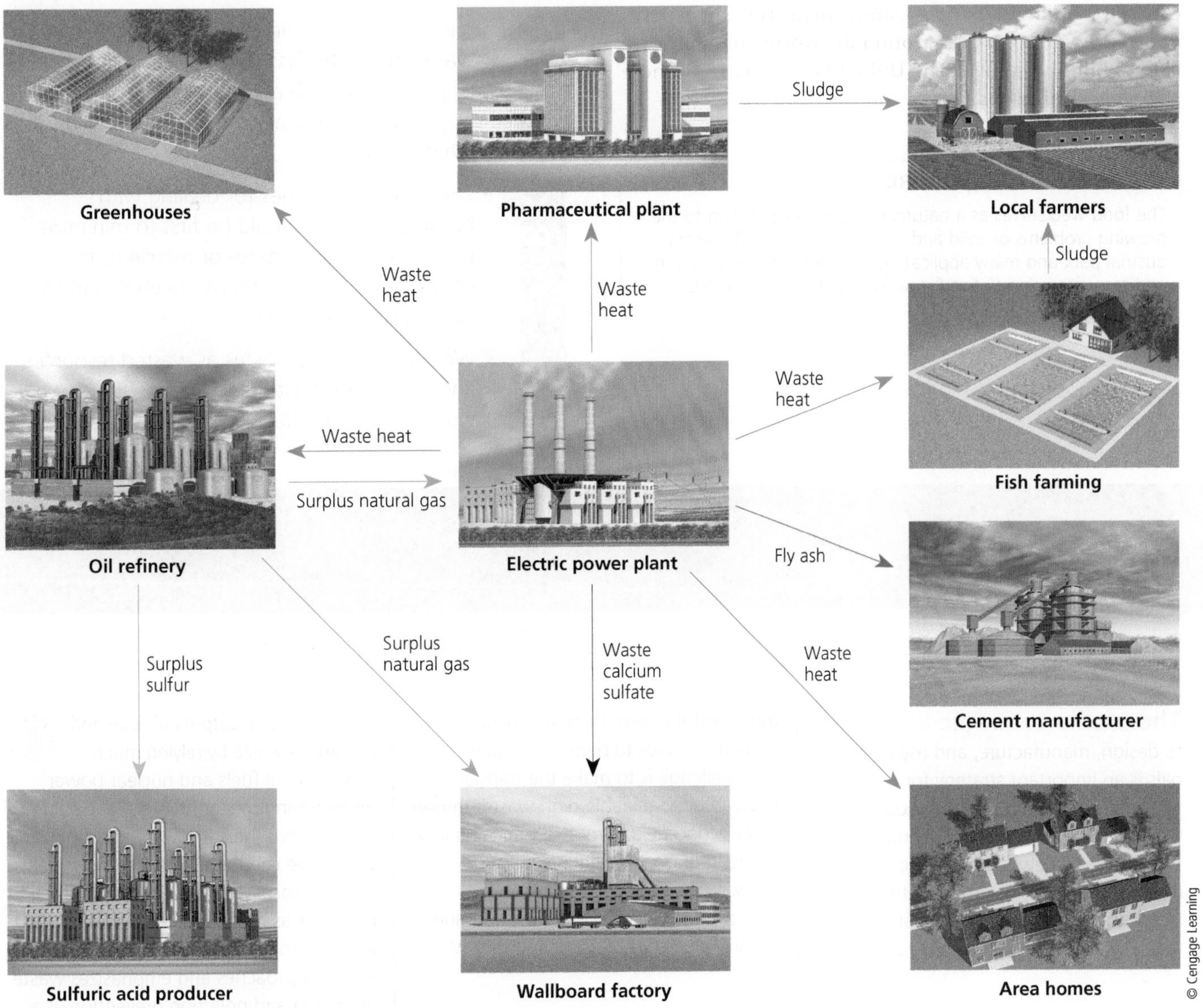

FIGURE 21.27 Solutions: This *industrial ecosystem* in Kalundborg, Denmark, reduces waste production by mimicking a natural ecosystem's food web. The wastes of one business become the raw materials for another, thus mimicking the way that nature recycles chemicals. ***Question:*** Is there an industrial ecosystem near where you live or go to school? If not, think about where and how such a system could be set up.

wastes of one manufacturer become the raw materials for another, similar to food webs in natural ecosystems. and a direct application of the cradle-to-cradle concept (**Core Case Study**).

This is happening in Kalundborg, Denmark, where an electric power plant and nearby industries, farms, and homes are collaborating to save money and to reduce their outputs of waste and pollution within what is called an *ecoindustrial park,* or *industrial ecosystem.* They exchange waste outputs and convert them into resources, as shown in Figure 21.27. This cuts pollution and waste and reduces the flow of nonrenewable mineral and energy resources through the local economy.

Ecoindustrial parks provide many economic benefits for businesses. By encouraging recycling and waste reduction prevention, they reduce the costs of managing solid wastes, controlling pollution, and complying with pollution regulations. They also reduce a company's chances of being sued because of damages, to people or to the environment, caused by their actions. In addition, companies improve the health and safety of workers by reducing their exposure to toxic and hazardous materials, thereby reducing company health insurance costs. Biomimicry also encourages companies to come up with new, environmentally beneficial, and less resource-intensive chemicals, processes, and products that they can sell worldwide. Today, more than 100 such parks operate in various places around the world, and more are being built or planned. The United States has 18 of these parks and China has 50.

CONSIDER THIS . . .

LEARNING FROM NATURE

The food web serves as a natural model for responding to the growing problems of solid and hazardous wastes. The ecoindustrial park and many applications of cradle-to-cradle design and manufacturing (**Core Case Study**) follow this model.

BIG IDEAS

- The order of priorities for dealing with solid waste should be first to minimize production of it, then to reuse and recycle as much of it as possible, and finally to safely burn or bury what is left.
- The order of priorities for dealing with hazardous waste should be first to minimize production of it, to reuse or recycle it, to convert it to less-hazardous material, and to safely store what is left.
- We can view solid wastes as wasted resources, and hazardous wastes as materials that we want to avoid producing in the first place.

Tying It All Together

The Cradle-to-Cradle Approach and Sustainability

The cradle-to-cradle approach to design, manufacture, and use of materials is an important strategy for reducing the amount of solid and hazardous wastes we produce. By mimicking nature, this approach views all discarded materials and substances as nutrients that circulate within industrial and natural cycles. It also allows us the opportunity to convert the harmful environmental impacts of human activities to beneficial impacts. The challenge is to make the transition from an unsustainable high-waste, throwaway economy to a more sustainable low-waste, reducing–reusing–recycling economy as soon as possible.

Such a transition will require applying the six **principles of sustainability**. We can reduce our outputs of solid and hazardous waste by relying much less on fossil fuels and nuclear power while relying much more on renewable energy from the sun, wind, and flowing water. We can mimic nature's chemical cycling processes by reusing and recycling materials as much as possible. Integrated waste management, which uses a diversity of approaches and emphasizes waste reduction and pollution prevention, is a way to mimic nature's use of biodiversity.

By including more of the harmful environmental and health costs of the consumer economy in market prices, we would be applying the full-cost pricing **principle of sustainability** while encouraging people to refuse, reduce, reuse, and recycle. In doing so, we would benefit the environment, create new jobs and businesses capitalizing on the four Rs, and gain health and environmental benefits for us, thus finding win-win solutions. This could also lead to lower levels of resource use per person, and thus lower levels of solid and hazardous waste production. All these measures together would help us to pass along to future generations a world that is at least as livable as, or more so than, the one we have enjoyed.

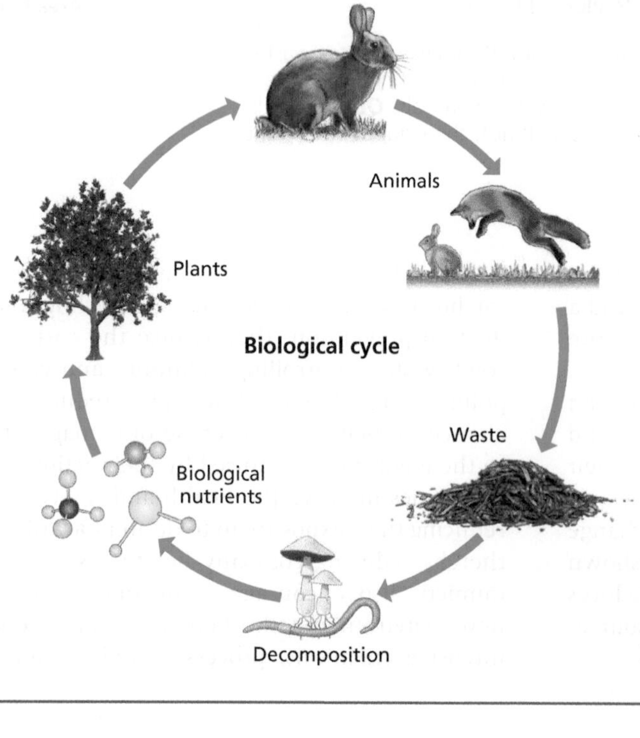

© Cengage Learning

Chapter Review

Core Case Study

1. Explain the concept of cradle-to-cradle design. Why is it a true form of biomimicry? List and briefly describe two strategies for employing it.

Section 21.1

2. What are the two key concepts for this section? Distinguish among **solid waste, industrial solid waste**, and **municipal solid waste (MSW)**, and give an example of each. Summarize the types and sources of municipal solid waste generated in the United States and explain what happens to it. What is **hazardous (toxic) waste**? Explain how and why electronic waste (e-waste) has become a growing solid and hazardous waste problem.

Section 21.2

3. What is the key concept for this section? Distinguish among **waste management, waste reduction**, and **integrated waste management**. Summarize the priorities that prominent scientists suggest we should use for dealing with solid waste and compare them to actual practices in the United States. Distinguish among **refusing, reducing, reusing, recycling**, and **composting**. Why are the first three Rs preferred from an environmental standpoint? List six ways in which industries and communities can reduce resource use, waste, and pollution.

Section 21.3

4. What is the key concept for this section? In order, what are the four levels of more sustainable resource use? How has cradle-to-cradle design elevated reuse to a new level? What are two ways in which governments can encourage reuse? Give two examples of shared use.
5. What is the difference between upcycling and downcycling? Distinguish between **primary recycling** and **secondary recycling**. What are the three important steps of recycling? Why is e-waste attractive for recycling? What are some benefits of composting? Summarize the process of mixed MSW recycling. What is source separation and what are its benefits? What are bioplastics? What are the major advantages and disadvantages of recycling?

Section 21.4

6. What is the key concept for this section? What are the major advantages and disadvantages of using incinerators to burn solid and hazardous waste? Distinguish between **sanitary landfills** and **open dumps**. What are the major advantages and disadvantages of burying solid waste in sanitary landfills?

Section 21.5

7. What is the key concept for this section? What are the priorities that scientists suggest we should use in dealing with hazardous waste? Summarize the problems involved in sending e-wastes to less-developed countries for recycling. Describe three ways to detoxify hazardous wastes. What is bioremediation? What is phytoremediation? What are the major advantages and disadvantages of incinerating hazardous wastes? What are the major advantages and disadvantages of using plasma gasification to detoxify hazardous wastes?
8. What are the major advantages and disadvantages of storing liquid hazardous wastes in deep underground wells and in surface impoundments? What is a secure hazardous waste landfill? List four ways to reduce your output of hazardous waste. Summarize the story of regulation of hazardous wastes in the United States.

Section 21.6

9. What is the key concept for this section? How has grassroots action led to improved solid and hazardous waste management in the United States? Describe regulation of hazardous wastes at the global level through the Basel Convention and the treaty to control persistent organic pollutants (POPs). What are three factors that discourage recycling? What are three ways to encourage recycling and reuse? Give three examples of how people are saving or making money through reuse, recycling, and composting. What is an industrial ecosystem and what are its benefits?
10. What are this chapter's *three big ideas*? Explain how cradle-to-cradle design and manufacturing (**Core Case Study**) can help us to by apply the six **principles of sustainability**.

Note: Key terms are in bold type.

Critical Thinking

1. Find three products that you regularly use that could be made using cradle-to-cradle design and manufacturing (**Core Case Study**). For each of these products, sketch out a rough plan for how you would design and build it so that its parts could be reused many times or recycled in such a way that they would not harm the environment.

2. Do you think that manufacturers of computers, television sets, cell phones, and other electronic products should be required to take their products back at the end of their useful lives for repair, remanufacture, or recycling in a manner that is environmentally responsible and that does not threaten the health of recycling workers? Explain. Would you be willing to pay more for these products to cover the costs of such a take-back program? If so, what percentage more per purchase would you be willing to pay for these products?

3. Think of three items that you regularly use once and then throw away. Are there reusable items that you could use in place of these disposable items? For each item, calculate and compare the cost of using the disposable option for a year versus the cost of using the reusable alternative. Write a brief report summarizing your findings.

4. Do you think that you could consume less by refusing to buy some of the things you regularly buy? If so, what are three of those things? Do you think that this is something you ought to do? Explain.

5. A company called Changing World Technologies has built a pilot plant to test a process it has developed for converting a mixture of discarded computers, old tires, turkey bones and feathers, and other wastes into oil by mimicking and speeding up natural processes for converting biomass into oil. Explain how this recycling process, if it turns out to be technologically and economically feasible, could lead to increased waste production.

6. Would you oppose having **(a)** a sanitary landfill, **(b)** a hazardous waste surface impoundment, **(c)** a hazardous waste deep-injection well, or **(d)** a solid waste incinerator in your community? For each of these facilities, explain your answer. If you oppose having such facilities in your community, how do you think the solid and hazardous wastes generated in your community should be managed?

7. How does your school dispose of its solid and hazardous wastes? Does it have a recycling program? How well does it work? Does your school encourage reuse? If so, how? Does it have a hazardous waste collection system? If so, describe it. List three ways in which you would improve your school's waste reduction and management systems.

8. Congratulations! You are in charge of the world. List the three most important components of your strategy for dealing with **(a)** solid waste and **(b)** hazardous waste.

Doing Environmental Science

Collect the trash (excluding food waste) that you generate in a typical week. Measure its total weight and volume. Sort it into major categories such as paper, plastic, metal, and glass. Then weigh each category and calculate its percentage by weight of the total amount of trash that you have measured. What percentage by weight of this waste consists of materials that could be recycled? What percentage consists of materials for which you could have used a reusable substitute, such as a coffee mug instead of a disposable cup? What percentage by weight of the items could you have done without? Compare your answers to these questions with those of your classmates. Together with your classmates, combine all the results and do the same analysis for the entire class. Use these results to estimate the same values for the entire student population at your school.

Global Environment Watch Exercise

Go to your MindTap course to access the GREENR database. Starting on the home page, under "Browse Issues and Topics", click on *Pollution,* then select *E-Waste.* Use this portal to research and find statistics on how rapidly the world's production of e-waste (Core Case Study) is growing and how rapidly e-waste production is growing in the United States. Write a brief report on what the United States and one other country of your choice are doing to deal with this growing waste problem. Include statistics on how much e-waste is generated in each country, on how much of it is recycled, and on how much of it goes to landfills. Compare the two approaches in terms of how successful they are.

Ecological Footprint Analysis

Researchers estimate that the average daily municipal solid waste production per person in the United States is 2 kilograms (4.4 pounds). Use the data in the pie chart below to get an idea of a typical annual MSW ecological footprint for each American by calculating the total weight in kilograms (and pounds) for each category generated during 1 year (1 kilogram = 2.20 pounds). Use the table (below, right) to enter your answers.

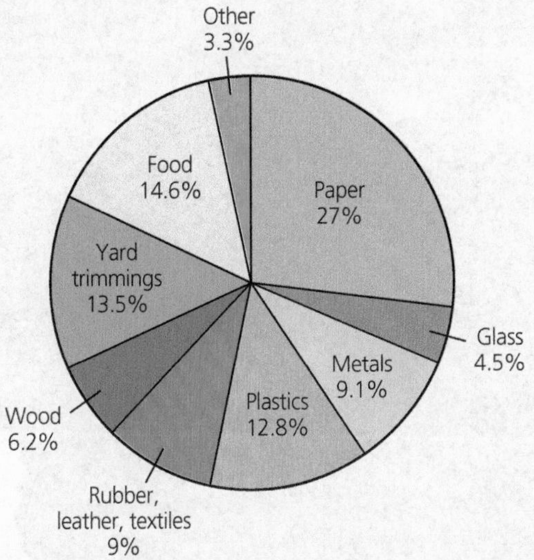

Composition of a typical sample of U.S. municipal solid waste, 2013.

(Compiled by the authors using data from the U.S. Environmental Protection Agency.)

Waste Category	Annual MSW Footprint per Person
Paper and paperboard	
Yard trimmings	
Food scraps	
Plastics	
Metals	
Wood	
Rubber, leather, and textiles	
Glass	
Other/miscellaneous	

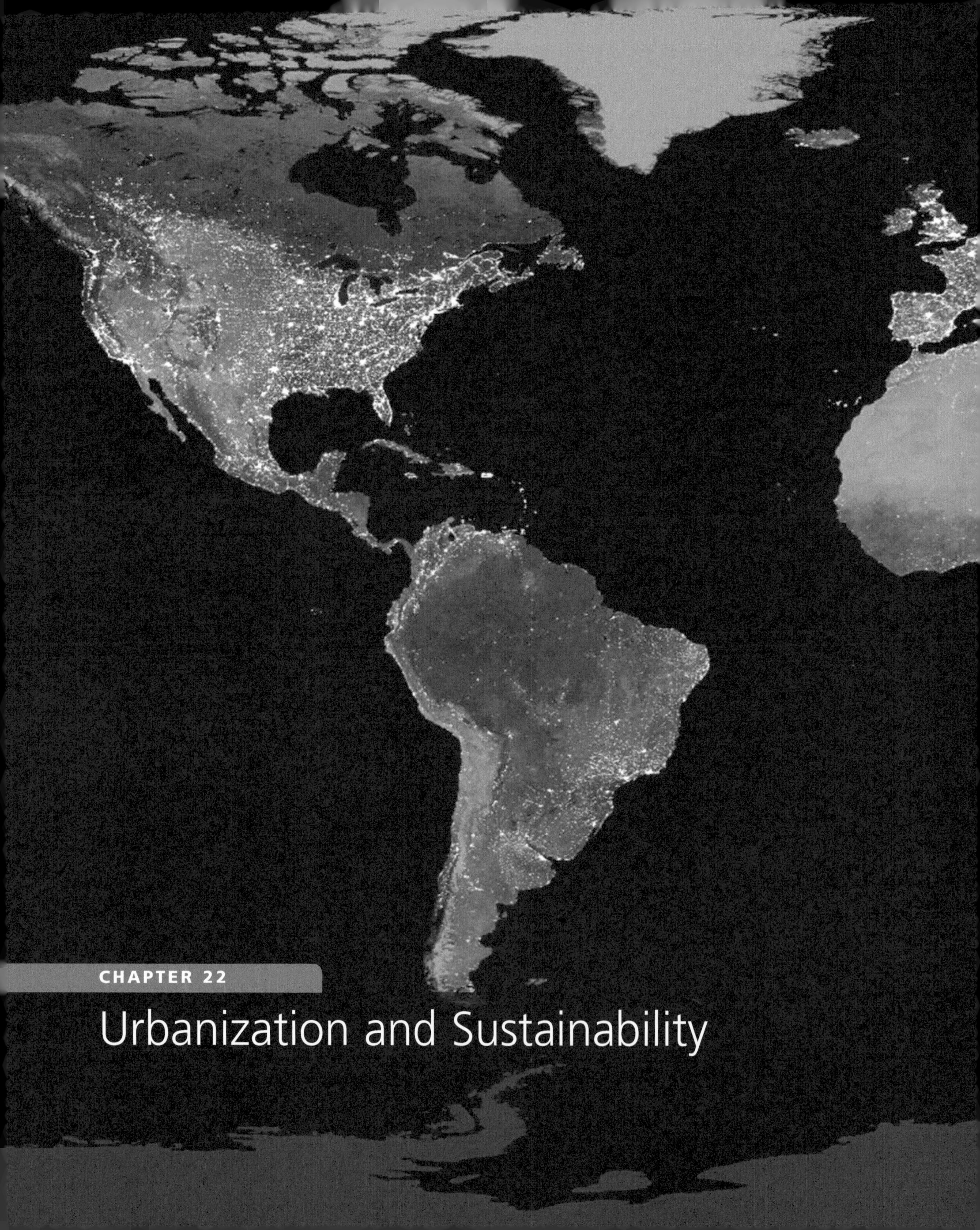

CHAPTER 22

Urbanization and Sustainability

The city is not an ecological monstrosity. It is rather the place where both the problems and the opportunities of modern technological civilization are most potent and visible.

PETER SELF

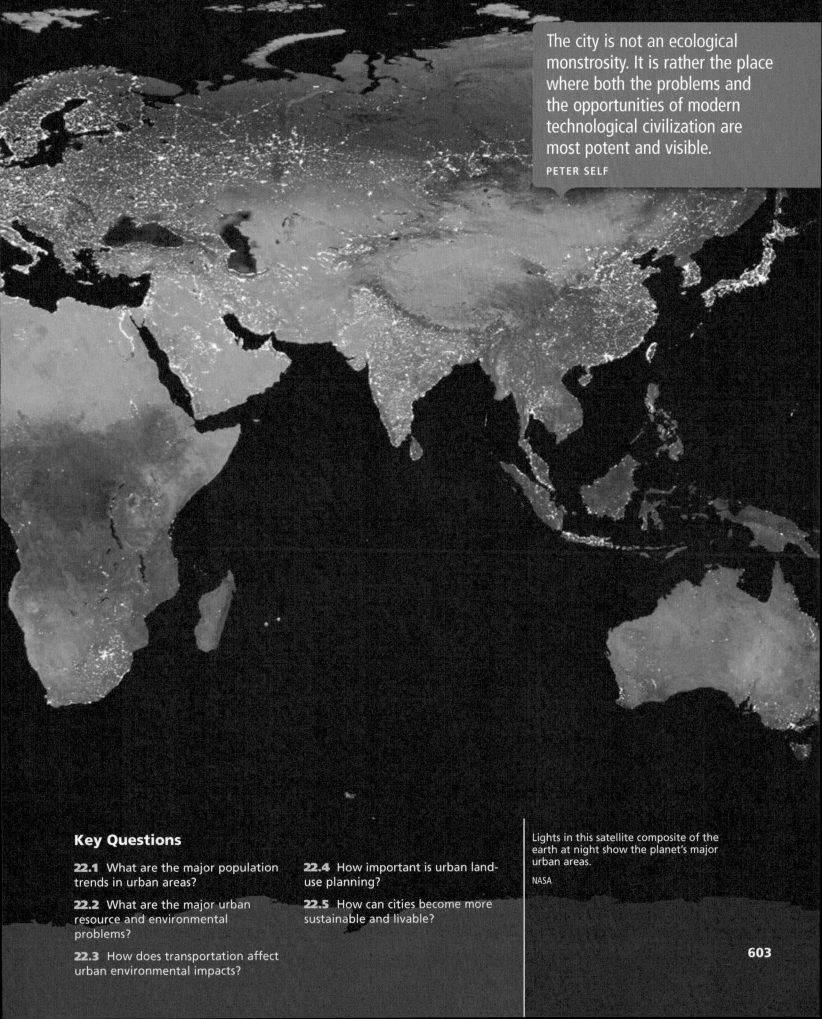

Lights in this satellite composite of the earth at night show the planet's major urban areas.

NASA

Key Questions

22.1 What are the major population trends in urban areas?

22.2 What are the major urban resource and environmental problems?

22.3 How does transportation affect urban environmental impacts?

22.4 How important is urban land-use planning?

22.5 How can cities become more sustainable and livable?

Core Case Study

Portland, Oregon: Urban Sustainability in Action

Portland, Oregon, with 619,000 people, lies on the banks of the Willamette River (Figure 22.1). For over four decades, Portland has consistently ranked at or near the top in several lists of the most sustainable and livable U.S. cities.

Since the 1970s, Portland has used smart growth strategies and strong land-use policies to control growth, reduce dependence on automobiles, and preserve green space. In 1978, the city demolished a six-lane highway and replaced it with a waterfront park.

Portland encourages the development of mixed-use neighborhoods with stores, light industries, professional offices, high-density housing, and access to mass transit. This allows most people to meet many of their daily needs without a car. The city has excellent light-rail and bus lines, and has further reduced car use by developing an extensive network of bike lanes and walkways. This decreased reliance on the automobile has saved its residents more than $1 billion a year in transportation costs. It has also contributed to public health by cutting air pollution and encouraging higher levels of physical activity.

Portland implemented a recycling system in 1987 and, by 2015, was recycling and composting 75% of its municipal solid waste, achieving one of the highest recycling rates in the country. In 1993, Portland became the first U.S. city to develop a plan to reduce its greenhouse gas emissions.

In 2009, Portland implemented a Climate Action Plan with the goal of cutting its CO_2 greenhouse gas emissions per person to 65% below 1990 levels by 2030 and to 91% below those levels by 2050. Between 1990 and 2008, the city had reduced its per person emissions by 19%.

Portland has more than 20 farmers markets and 35 community gardens, which provide fresh, locally produced food for residents and chefs. The city also has many vegetarian-friendly restaurants.

So why should we care about Portland or any other urban area? One reason is that more than half of the world's people live in urban areas and by 2050, two of every three people are likely to be urban dwellers—most of them in rapidly growing cities in less-developed countries. Another reason is that urban areas use most of the world's resources, produce most of the world's pollution and wastes, and have huge environmental impacts that extend far beyond their boundaries. The environmental quality of the future depends largely on whether we can make urban areas more sustainable and livable during the next few decades. Portland and a growing number of other cities are leading the way.

Jit Lim/cutcaster

FIGURE 22.1 Portland, Oregon, is one of the most environmentally sustainable cities in the United States.

22.1 WHAT ARE THE MAJOR POPULATION TRENDS IN URBAN AREAS?

CONCEPT 22.1 Urbanization is increasing and urban areas are growing rapidly, especially in less-developed countries.

More Than Half of the World's People Live in Urban Areas

More than half the world's people live in **urban areas**, or central cities (Figure 22.2, top) and their adjoining communities called *suburbs* (Figure 22.2, center). The rest live in rural areas or villages (Figure 22.2, bottom). **Urbanization** is the creation and growth of urban and suburban areas. It is measured as the percentage of the people in a country or in the world living in such areas. **Urban growth** is the *rate* of increase of urban populations.

We live on an increasingly urbanized planet. About 53% of the world's people, 81% of all Americans (see the Case Study that follows), and 55% of China's population live in urban areas. Every day there are about 244,000 more urban dwellers, most of them added to cities in less-developed countries (**Concept 22.1**). See Figure 8, pp. S22–S23 in Supplement 4 for a map of global population density.

244,000
Number of new urban dwellers each day, on average

Urban areas grow in two ways—by *natural increase* when there are more births than deaths and by *immigration*, mostly from rural areas. Rural people are *pulled* to urban areas in search of jobs, food, housing, educational opportunities, better health care, and entertainment. Some people are *pushed* to urban areas by factors such as famine, loss of land for growing food, deteriorating environmental conditions, war, and religious, racial, and political conflicts.

Three major trends in urban population dynamics are important for understanding the problems and challenges of urban growth:

- *The percentage of the global population that lives in urban areas has increased sharply, and this trend is projected to continue*. Between 1850 and 2015, the percentage of

ollyy/Shutterstock.com

Pete Mcbride/National Geographic Creative

Magdalena Bujak/Shutterstock.com

FIGURE 22.2 About 53% of the world's people live in urban areas, or cities, such as Shanghai, China (top), and their surrounding suburban areas such as this one in Phoenix, Arizona (center). The other 47% live in rural areas—in villages such as this one in the southern African country of Malawi (bottom), in small towns, or in the countryside.

the world's people living in urban areas increased from 2% to 53% and is likely to reach 67% by 2050. Between 2015 and 2050, the world's urban population is projected to grow from 3.9 billion to 6.6 billion. The great majority of these 2.7 billion new urban dwellers will live in less-developed countries.

- *The numbers and sizes of urban areas are increasing.* In 2015, there were 30 megacities—cities with 10 million or more people—22 of them in less-developed countries (Figure 22.3). Thirteen of these urban areas are *hypercities* with more than 20 million people. The largest hypercity is Tokyo-Yoko, Japan, with 37.8 million—more than the entire population of Canada. Some of the world's megacities and hypercities are merging into vast urban *megaregions*, each with more than 100 million people. The largest megaregion is the Hong Kong–Shenzhen–Guangzhou region in China with 120 million people.
- *Poverty is becoming increasingly urbanized, mostly in less-developed countries.* The United Nations estimates that at least 1 billion people live in the slums and shantytowns of many major cities in less-developed countries (see chapter-opening photo). This number may triple by 2050.

CONSIDER THIS . . .

THINKING ABOUT Urban Trends

If you could reverse one of the three urban trends listed here, which one would you choose? Explain.

If you visit a poor, overcrowded area of a large city in a less-developed country, your senses may be overwhelmed by a vibrant but chaotic crush of people, vehicles of all types, traffic jams (Figure 22.4), and noise. Odors can include raw sewage and smoke from burning trash, as well as from wood and coal cooking fires. Many people sleep on the streets or live in crowded, unsanitary, rickety, and unsafe slums and shantytowns with little or no access to safe drinking water or modern sanitation facilities.

CASE STUDY

Urbanization in the United States

Between 1800 and 2015, the percentage of the U.S. population living in urban areas rose from 5% to 81%. Figure 22.5 shows the current major urban areas in the United States with more than 1 million people. This population shift from rural to urban has occurred in three phases. First, *people migrated from rural areas to large central cities.* Second, *many people migrated from large central cities to smaller cities and suburbs.* Currently, about half of urban Americans live in the suburbs (Figure 22.2, center photo), nearly a third in central cities, and the rest mostly in rural housing developments beyond suburbs Third, *many people migrated from the North and East to the South and West.*

Since 1920, many of the worst urban environmental problems in the United States have been reduced

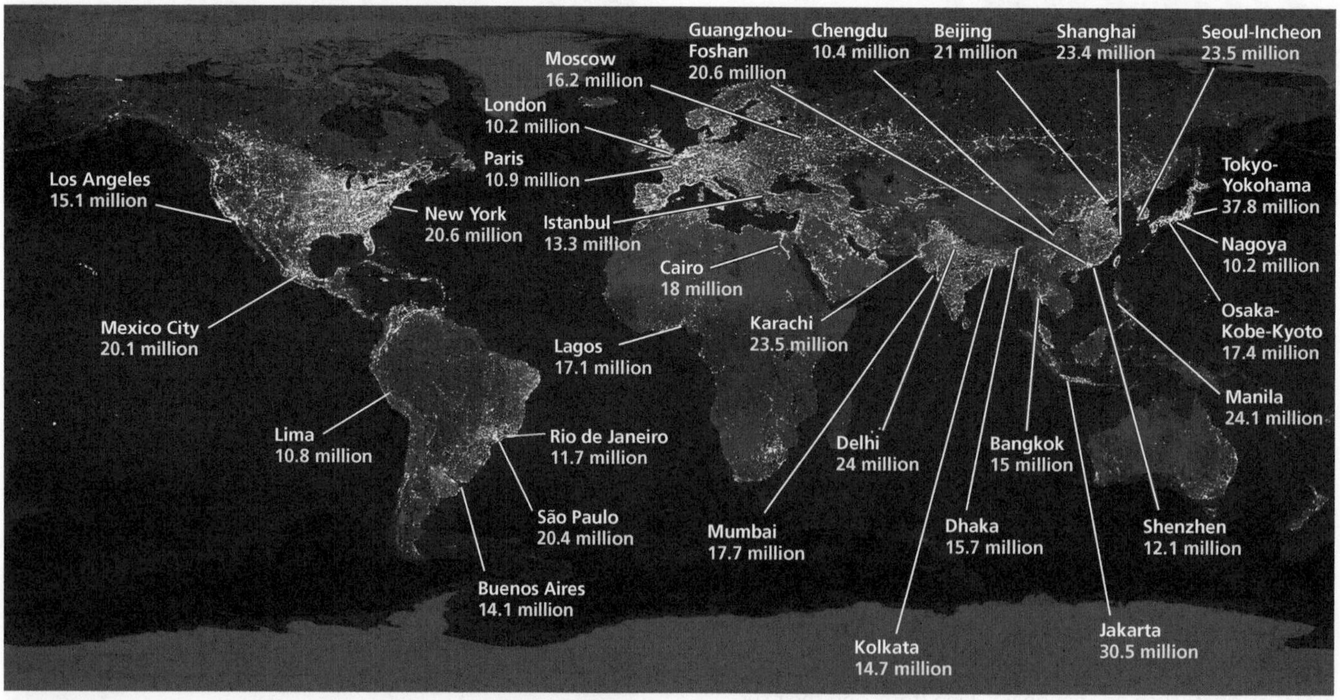

FIGURE 22.3 Megacities, or major urban areas with 10 million or more people, in 2015.
Question: In order, what were the world's five most populous urban areas in 2015?

(Compiled by the authors using data from National Geophysics Data Center, Demographia, National Oceanic and Atmospheric Administration, and United Nations Population Division.)

FIGURE 22.4 A typical traffic jam of people and motor vehicles in Kolkata, India, a megacity of 14.7 million people.

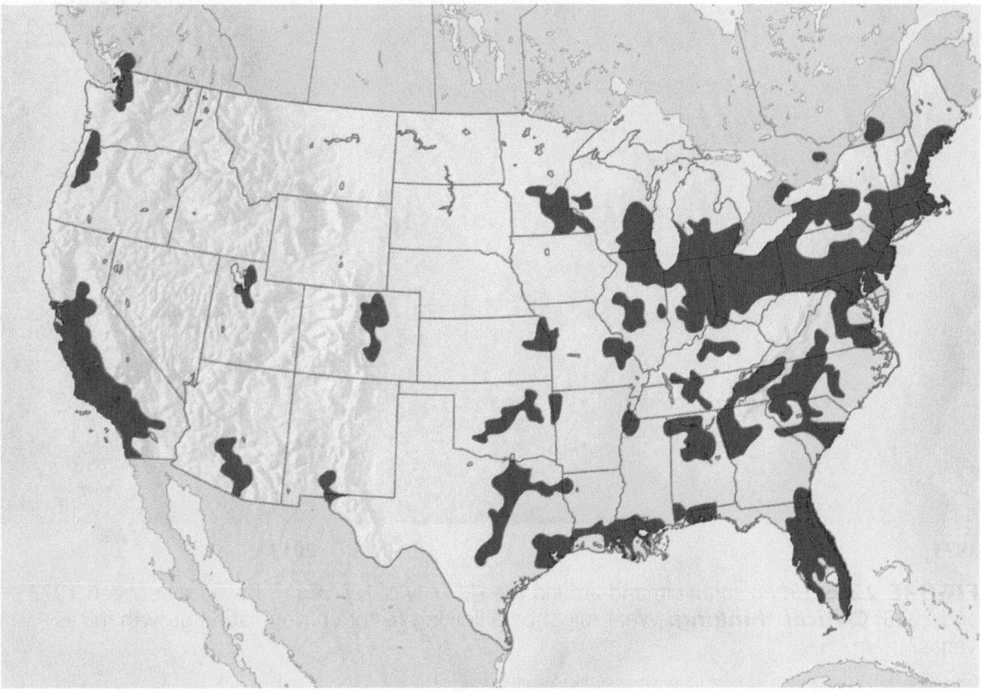

FIGURE 22.5 Urbanized areas (shaded) in the United States where cities, suburbs, and towns dominate the land area. ***Critical thinking:*** Why do you think many of the largest urban areas are located near water?

(Compiled by the authors using data from National Geophysical Data Center/National Oceanic and Atmospheric Administration, U.S. Census Bureau.)

significantly (Figure 6.7, p. 126). Most people have better working and housing conditions and air and water quality have improved. Better sanitation, clean public water supplies, and expanded medical care have slashed death rates and incidences of sickness from infectious diseases. In addition, the concentration of most of the population in urban areas has helped to protect the country's biodiversity by reducing the destruction and degradation of wildlife habitat.

However, a number of U.S. central cities—especially older ones—have deteriorating services and aging *infrastructures* (streets, bridges, dams, power lines, water supply pipes, and sewers). For example, more than 65,000 of the nearly 610,000 bridges in the United States are structurally deficient and in urgent need of repair at an estimated cost of $115 billion. Many of the country's highways and roads built several decades ago carry 5 to 10 times more traffic than they were designed for. Many gas and water pipelines installed between 1900 and 1950 with a projected life of 50 years are failing. In 2014, the United States was spending only 1.5% of its gross domestic product on repairing and replacing infrastructure, compared to 5% in India and 5% in China. According to philosopher and social analyst Eric Hoffer: "The sign of a good society and a good government is not in what it builds, but in what it maintains."

Funds for repairing and upgrading urban infrastructure have declined in many urban areas as the flight of people and businesses to the suburbs and beyond has led to lower central city revenues from property taxes. However, this trend has been reversed in some cities, including Portland, Oregon (**Core Case Study**).

Urban Sprawl

In the United States and some other countries, **urban sprawl**—the growth of low-density development on the edges of cities and towns—is eliminating surrounding agricultural and wild lands (Figure 22.6). It results in a dispersed jumble of housing developments (Figure 22.2, center), shopping malls, parking lots, and office complexes that are loosely connected by multilane highways and freeways.

Urban sprawl is largely the product of abundant affordable land, automobiles, federal and state funding of highways, and inadequate urban planning. Many people prefer living in suburbs and *exurbs*—housing developments scattered over large areas that lie beyond suburbs and have no socioeconomic centers. Compared to central cities, these areas provide lower-density living and access to larger lot sizes and single-family homes. Often these areas also have newer public schools and lower crime rates.

1973 **2013**

FIGURE 22.6 *Urban sprawl* in and around the U.S. city of Las Vegas, Nevada, between 1973 and 2013. ***Critical thinking:*** What might be a limiting factor on population growth in Las Vegas?

Left: Courtesy of U.S. Geological Survey. Right: U.S. Department of the Interior/U.S. Geological Survey

On the other hand, suburban sprawl destroys forests, wetlands, and prime cropland (Figure 22.7), and increases air and water pollution. Because of nonexistent or inadequate mass transportation in most such areas, sprawl forces people to drive almost everywhere—emitting climate-changing greenhouse gases and other forms of air pollution in the process. For example, a person living in a typical suburb emits about 9 times more CO_2 per year, mostly by driving, than a person living in a typical central city.

In the United States, urban sprawl has paved over about 155,000 square kilometers (60,000 square miles) of land—an area the size of the U.S. state of Georgia. As singer-songwriter Joni Mitchell put it in one of her songs: "They paved paradise and put up a parking lot." Urban sprawl also has led to the economic deaths of many central cities, as people and businesses have moved out of these areas. Figure 22.8 summarizes these and other undesirable consequences of urban sprawl.

CONSIDER THIS . . .

THINKING ABOUT Urban Sprawl

Do you think the advantages of urban sprawl outweigh its disadvantages? Explain.

Rich Reid/National Geographic Creative

FIGURE 22.7 This suburban development in Ventura County, California, was once prime cropland and is likely to expand and take over more of such land.

Natural Capital Degradation

Urban Sprawl

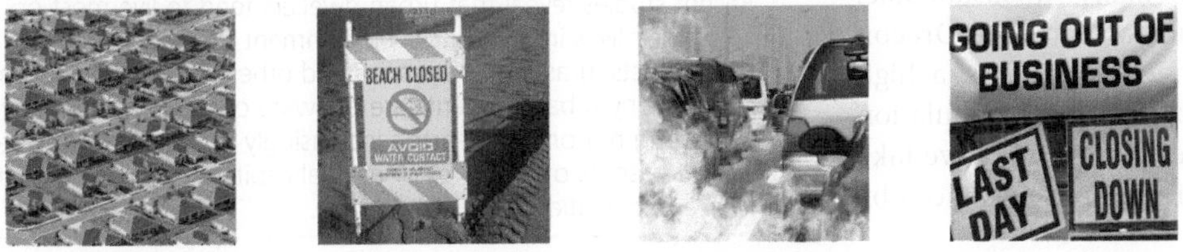

Land and Biodiversity	Water	Energy, Air, and Climate	Economic Effects
Loss of cropland	Increased use and pollution of surface water and groundwater	Increased energy use and waste	Decline of downtown business districts
Loss and fragmentation of forests, grasslands, wetlands, and wildlife habitat	Increased runoff and flooding	Increased emissions of carbon dioxide and other air pollutants	More unemployment in central cities

FIGURE 22.8 Some of the undesirable impacts of urban sprawl, or car-dependent development. ***Critical thinking:*** Which five of these effects do you think are the most harmful?

Left: Condor 36/Shutterstock.com. Left center: spirit of america/Shutterstock.com. Right center: ssuaphotos/Shutterstock.com. Right: ronfromyork, 2009/Shutterstock.com.

22.2 WHAT ARE THE MAJOR URBAN RESOURCE AND ENVIRONMENTAL PROBLEMS?

CONCEPT 22.2 Urban areas have many benefits but most are unsustainable because of their high levels of resource use, waste, pollution, and poverty.

Urbanization Has Advantages

Urbanization has many benefits. Cities are centers of economic development, innovation, education, technological advances, social and cultural diversity, and jobs. Urban residents in many parts of the world tend to live longer than do rural residents and to have lower infant mortality and fertility rates. They also have better access to medical care, family planning, education, and social services than do rural residents.

Urban areas also have environmental advantages. Recycling is more economically feasible because of the high concentrations of recyclable materials in urban areas. Satellite images show that urban areas containing 53% of the world's people occupy only 2.8% of the earth's land, excluding Antarctica. Concentrating people in urban areas helps preserve biodiversity by reducing the stress on wildlife habitats. In addition, migration from rural to urban areas provides an opportunity for protecting and restoring ecosystems in rural areas.

2.8% Percentage of the earth's land occupied by urban areas

Heating and cooling multistory apartment and office buildings in central cities usually takes less energy per person than does heating and cooling single-family homes and smaller office buildings, which are more common in the suburbs. Central-city dwellers also tend to drive less and rely more on mass transportation, walking, and bicycling. This helps to explain why New York City and many other large U.S. cities have some of the lowest per capita carbon dioxide emissions of all cities.

Urbanization Has Disadvantages

Most urban areas are environmentally unsustainable. Even in more sustainable cities such as Portland, Oregon (**Core Case Study**), it is a challenge to maintain a high level of sustainability in the face of a growing population and higher rates of resource use per person. Here we take a closer look at some of the significant problems faced by most of the world's cities.

LARGE ECOLOGICAL FOOTPRINTS. Each year urban populations consume about 75% of the resources that we use and produce about 75% of the world's pollution and wastes. Because of this high input of food, water, and other resources, and the resulting high waste output (Figure 22.9), most of the world's cities have huge ecological footprints that extend far beyond their boundaries.

For example, according to an analysis by Mathis Wackernagel and William Rees, developers of the ecological footprint concept (see Chapter 1, pp. 12–13), London, England, requires an area 58 times as large as the city to supply its residents with resources. These researchers estimate that if all of the world's people used resources at the same rate as Londoners do, it would take at least three more planet Earths to meet their needs. In other words, most urban areas are not self-sustaining systems (**Concept 22.2**) for several reasons.

LACK OF VEGETATION. In urban areas, most trees, shrubs, grasses, and other plants are cleared to make way for buildings, roads, parking lots, and housing developments. Thus, most cities do not benefit from the free ecosystem services provided by vegetation, including air purification, generation of oxygen, removal of atmospheric CO_2, control of soil erosion, and wildlife habitat.

CONSIDER THIS . . .

CONNECTIONS Urban Living and Biodiversity Awareness

Recent studies reveal that urban dwellers tend to live most or all of their lives in an artificial environment that isolates them from forests, grasslands, streams, and other natural areas. As a result, many urban residents are unaware of the importance of protecting not only the earth's increasingly threatened biodiversity but also its other forms of natural capital that support their lives and the cities in which they live.

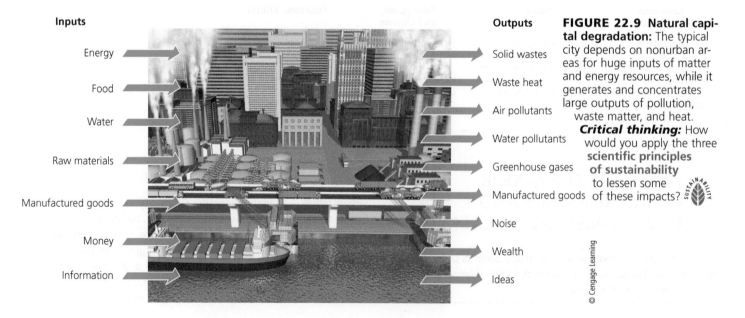

FIGURE 22.9 Natural capital degradation: The typical city depends on nonurban areas for huge inputs of matter and energy resources, while it generates and concentrates large outputs of pollution, waste matter, and heat. ***Critical thinking:*** How would you apply the three **scientific principles of sustainability** to lessen some of these impacts?

WATER PROBLEMS. As urban areas grow their water demands increase. This requires building expensive reservoirs or drilling deeper wells.

Flooding also tends to be greater in urban areas that are built on floodplains near rivers or along low-lying coastlines subject to natural flooding. Covering land with buildings, asphalt, and concrete causes precipitation to run off quickly and overload storm drains.

Urban development often destroys or degrades large areas of wetlands that have served as natural sponges to help absorb excess storm water. During this century many of the world's largest coastal urban areas (Figure 22.3) face a flooding threat as sea levels rise because of climate change (see Chapter 19, pp. 502–503). The warmer atmosphere is also melting some mountaintop glaciers (Figure 19.3, p. 508), and in areas where they eventually disappear, urban areas that depend on them for water will face severe water shortages.

POLLUTION AND HEALTH PROBLEMS. Because of their high population densities and high rates of resource consumption, urban areas produce most of the world's air pollution, water pollution, and solid and hazardous wastes. Pollutant levels are generally higher because pollution is produced in a smaller area and cannot be dispersed and diluted as readily as pollution produced in rural areas can.

The concentration of motor vehicles and industrial facilities in urban centers, with just over two-thirds of the world's emissions of climate-changing CO_2 from human-related sources, causes disruption of local and regional portions of the carbon cycle (Figure 3.20, p. 65). This concentration of urban air pollution also disrupts the nitrogen cycle (Figure 3.21, p. 66) because of large quantities of nitrogen oxide emissions, which play a key role in the formation of photochemical smog.

In addition, urban nitric acid and nitrate emissions are major components of acid deposition in urban areas and beyond (Figures 18.11, p. 483, and 18.12, p. 484). Nitrogen nutrients in urban runoff and discharges from urban sewage treatment plants can also disrupt the nitrogen cycle in nearby lakes and other bodies of water, and cause excessive eutrophication (Figures 20.1, p, 542, and 20.8, p. 549).

In addition, high population densities in urban areas can promote the spread of infectious diseases, especially if adequate drinking water and sewage treatment systems are not in place.

EXCESSIVE NOISE. Because of the concentration of people and motor vehicles, most urban dwellers are subjected to **noise pollution**: any unwanted, disturbing, or harmful sound that damages, impairs, or interferes with hearing, causes stress, raises blood pressure, hampers concentration and work efficiency, or causes accidents. Noise levels are measured in decibel-A (dbA) sound pressure units that vary with different human activities (Figure 22.10).

Sound pressure becomes damaging at about 85 dbA and painful at around 120 dbA. At 180 dbA, sound can kill. Prolonged exposure to sound levels above 85 dbA can cause permanent hearing damage. Just one-and-a-half minutes of exposure to 110 decibels or more can cause such damage.

Noise pollution and hearing damage can be reduced by wearing earplugs or other protective devices, shielding noisy activities or processes, shielding workers or other persons from the noise, moving noisy operations or machines away, and using *antinoise* (technologies that cancel or muffle one noise with another).

LOCAL CLIMATE EFFECTS AND LIGHT POLLUTION. On average, cities tend to be warmer, rainier, foggier, and cloudier than suburbs and nearby rural areas. In central cities, the enormous amount of heat generated by cars, factories, furnaces, lights, air conditioners, and heat-absorbing dark roofs and streets creates an **urban heat island** that is surrounded by cooler suburban and rural areas. This can accelerate the formation of photochemical smog (Figures 18.1, p. 472, and 18.8, p. 481) in areas with warm climates. As urban areas grow and merge, their heat islands merge, which can reduce the natural dilution and cleansing of polluted air.

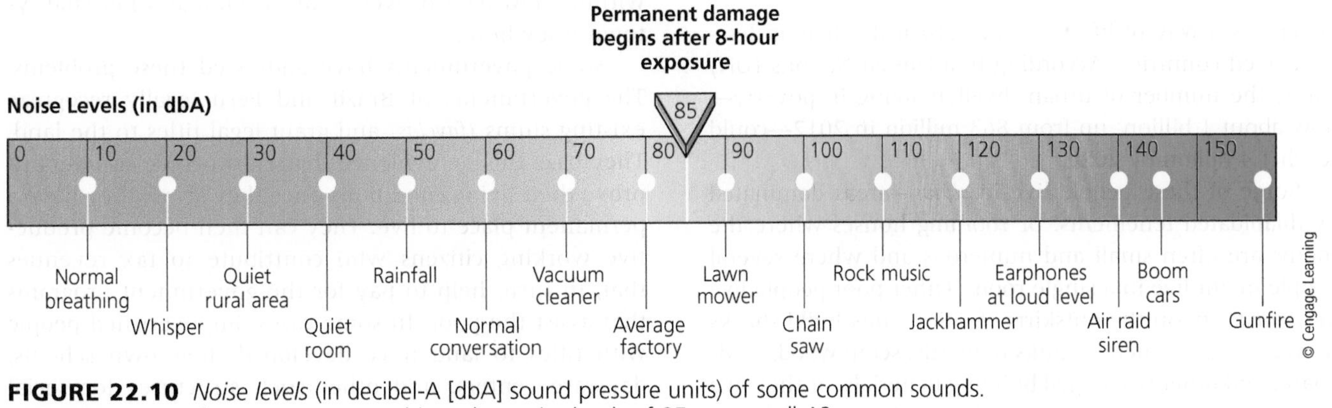

FIGURE 22.10 *Noise levels* (in decibel-A [dbA] sound pressure units) of some common sounds.
Question: How often are your ears subjected to noise levels of 85 or more dbA?

FIGURE 22.11 Slum in Mumbai, India.

Mathes//Dreamstime.com

The urban heat island effect can also increase dependence on air conditioning. This leads to higher energy consumption, greenhouse gas emissions, other forms of air pollution, and more heat generated within the city.

The artificial light created by cities (chapter-opening photo and Figure 22.2, top) affects some plant and animal species. For example, large numbers of migrating birds are lured off course by the lights of high-rise buildings and fatally collide with these structures. Excessive light also makes it difficult for astronomers and urban dwellers to study and enjoy the night sky.

CONSIDER THIS . . .

THINKING ABOUT Disadvantages of Urbanization

Which two of these disadvantages of urbanization do you think are the most serious? Explain.

Poverty and Urban Living

Poverty is a way of life for many urban dwellers in less-developed countries. According to a United Nations (UN) study, the number of urban dwellers living in poverty—now about 1 billion, up from 863 million in 2012—could reach 1.4 billion by 2020.

Some of these people live in *slums*—areas dominated by dilapidated tenements, or rooming houses where the rooms are often small and numerous and where several people might live in a single room. Other poor people live in *shantytowns* on the outskirts of cities. They build shacks from corrugated metal, hunks of plastic, scrap wood, cardboard, and other scavenged building materials, or they live in rusted shipping containers and junked cars.

Some shantytowns are illegal subdivisions where landowners rent land to the poor without city approval. Others are illegal *squatter settlements* where people take over unoccupied land without the owners' permission, simply because it is their only option for survival. Consider Mumbai, India, with a population of 17.7 million. According to the World Bank, 54% of Mumbai's population lives in shantytowns and slums (Figure 22.11)—making it the slum capital of India and the world.

Poor people living in shantytowns and squatter settlements or on the streets (see Figure 6.18, p. 134) usually lack clean water supplies, sewers, electricity, and roads. They are subject to severe air and water pollution and hazardous wastes from nearby factories. Many of these settlements are in locations prone to landslides, flooding, or earthquakes. Some city governments regularly bulldoze squatter shacks and send police to drive illegal settlers out. The people usually move back in within a few days or weeks, or develop another shantytown elsewhere.

Some governments have addressed these problems. The governments of Brazil and Peru legally recognize existing slums (*favelas*) and grant legal titles to the land. They base this on evidence that poor people usually improve their living conditions once they know they have a permanent place to live. They can then become productive working citizens who contribute to tax revenues that, in turn, help to pay for the government programs that assist the poor. In some cases, impoverished people with titles to land have developed their own schools, day-care centers, and other such structures for social improvement.

FIGURE 22.12 Photochemical smog in Mexico City, Mexico. During the last 20 years, Mexico City has greatly reduced its annual number of smog days.

CASE STUDY

Mexico City

With 21.2 million people, Mexico City (including its metropolitan area) is one of the world's hypercities (Figure 22.3). More than one-third of its residents live in slums called *barrios* or in squatter settlements that lack running water and electricity.

At least 3 million people in the barrios have no sewage facilities, so human waste from these slums is deposited in gutters, vacant lots, and open ditches every day, attracting armies of rats and swarms of flies. When the winds pick up dried excrement, a *fecal snow* blankets parts of the city. This bacteria-laden fallout leads to widespread salmonella and hepatitis infections, especially among children.

Mexico City has serious air pollution problems (Figure 22.12) because of a combination of factors: too many cars; polluting factories; a warm, sunny climate and thus more smog; and topographical bad luck. The city sits in a high-elevation, bowl-shaped valley surrounded on three sides by mountains—conditions that trap air pollutants at ground level, causing them to accumulate to dangerous levels (Figure 18.9, left, p. 482).

The city also suffers from chronic water shortages. Large-scale water withdrawals from the city's aquifer, which supplies about two-thirds of its water, have caused parts of the city to subside by 9 meters (30 feet) during the last century. Some areas are now subsiding as much as 30 centimeters (1 foot) a year. The city's growing population increasingly relies on pumping water from as far away as 150 kilometers (93 miles), which requires large amounts of energy, and then pumping it another 1,000 meters (3,300 feet) uphill to reach the city.

In 1992, the United Nations named Mexico City "the most polluted city on the planet." Since then Mexico City has made progress in reducing the severity of some of its air pollution problems. In 2013, the Institute for Transportation and Development awarded Mexico City its Sustainable Transportation Award for expanding its bus rapid-transit system and expanding its bike sharing program and its bike lanes. The percentage of days each year in which air pollution standards are violated has fallen from 50% to 20%.

The city government has moved refineries and factories out of the city, banned cars in its central zone, and required air pollution controls on all cars made after 1991. It has also phased out the use of leaded gasoline, expanded public transportation, and replaced some old buses, taxis, and delivery trucks with vehicles that produce fewer emissions.

In addition, Mexico City has instituted a program to reduce water use and waste and implemented a water pricing system designed to promote water conservation. It also bought land for use as green space, rebuilt public parks, and planted more than 25 million trees to help absorb pollutants and add shade and beauty. It is building state-of-the-art waste processing centers that, within a few years, will process as much as 85% of the city's solid waste through recycling and composting, while burning some of it for energy.

Mexico City still has a long way to go as its human population grows along with its number of cars. However, its progress shows what can be done to improve environmental quality once a community decides to act.

22.3 HOW DOES TRANSPORTATION AFFECT URBAN ENVIRONMENTAL IMPACTS?

CONCEPT 22.3 In some countries, many people live in widely dispersed urban areas and expand their ecological footprints by depending mostly on motor vehicles for their transportation.

Cities Can Grow Outward or Upward

If a city cannot spread outward, it must grow vertically—upward and downward (below ground)—by occupying a small land area with a high population density. Most people living in *compact cities* such as Hong Kong, China, and Tokyo, Japan, get around by walking, biking, or using mass transit such as rail and bus systems. Some high-rise apartment buildings in these Asian cities contain everything from grocery stores to fitness centers that reduce the need for their residents to travel for food, entertainment, and other services.

In other parts of the world, a combination of plentiful land, relatively cheap gasoline, and networks of highways has produced *dispersed cities* whose residents depend on motor vehicles for most travel (**Concept 22.3**). Such car-centered cities are found in the United States, Canada, Australia, and other countries where ample land often is available for cities to expand outward. The resulting urban sprawl (Figure 22.6) can have a number of undesirable effects (Figure 22.8).

The United States is a prime example of a car-centered nation. With only 4.4% of the world's population, the country has about 23% of the world's 1.1 billion motor vehicles, according to the U.S. Energy Information Administration (EIA). In its dispersed urban areas, U.S. passenger vehicles are used for 86% of all transportation and 76% of residents drive alone to work every day (up from 64% in 1980).

Pros and Cons of Motor Vehicles

Motor vehicles provide mobility and offer a convenient and comfortable way to get from one place to another. For many people, driving is personally satisfying. In addition, much of the world's economy is built on producing motor vehicles and supplying fuel, roads, services, and repairs for them.

Despite their important benefits, the use of motor vehicles can have harmful effects on people and the environment (**Concept 22.3**). Globally, automobile accidents kill more than 1.2 million people a year—an average of nearly 3,300 deaths per day—and injure another 50 million people. They also kill about 50 million wild animals and family pets every year.

In 2015, motor vehicle accidents in the United States killed about 38,300 people and injured about 4.4 million, at least 300,000 of them severely. Car accidents have killed more Americans than have all the wars in the country's history.

Motor vehicles are the world's largest source of outdoor air pollution, which kills about 100,000 people per year in the United States, according to the Environmental Protection Agency. They are also the fastest-growing source of climate-changing CO_2 emissions. The average car in the United States emits about 1.8 metric tons (2.2 tons) of CO_2 each year.

Motor vehicles have helped to create urban sprawl and the car commuter culture. At least a third of the world's urban land and half of that in the United States is devoted to roads, parking lots, gasoline stations, and other automobile-related uses. This prompted urban studies expert Lewis Mumford to suggest that the U.S. national flower should be the concrete cloverleaf (Figure 22.13).

Another problem is congestion in both urban centers and suburban areas. If current trends continue, U.S. motorists will spend an average of 2 years of their lives in traffic jams on streets and freeways that often resemble parking lots. In order, the top four most gridlocked U.S. cities are Washington, D.C., Los Angeles, San Francisco, and New York City. Traffic congestion in some cities such as Beijing, China (Figure 22.14), is much worse. Building more roads is rarely the answer because more roads usually encourage more people to use motor vehicles.

Reducing Automobile Use

Some environmental scientists and economists suggest that we reduce the harmful effects of automobile use by requiring drivers to pay directly for most of the environmental and health costs caused by their automobile use. This *user-pays* approach would also be a way to phase in *full-cost pricing*, in keeping with one of the **principles of sustainability**.

One such approach would be to charge a tax or fee on gasoline to cover the estimated harmful costs of driving. According to a study by the International Center for Technology Assessment, such a tax would amount to about $3.18 per liter ($12 per gallon) of gasoline in the United States. Automobile owners end up paying these harmful costs in the form of higher medical and health insurance bills. They also pay higher taxes to support federal, state, and local efforts to regulate and reduce air pollution from motor vehicles. However, most drivers do not relate these costs to their use of gasoline.

Gradually phasing in an environmental gas tax, as has been done in many European nations, would lead to greater use of mass transit, car sharing, and more energy-efficient motor vehicles. It would also reduce pollution, environmental degradation, and ocean acidification, and help to slow climate change.

Proponents of such taxes urge governments to do two things. *First,* fund programs to educate people about the hidden harmful costs they are paying for gasoline. *Second,*

iofoto/Shutterstock.com

FIGURE 22.13 Cloverleafs like this tangled network of thruways in the U.S. city of Los Angeles, California, are found in most of the world's increasingly car-dependent cities.

use gasoline tax revenues to help finance mass-transit systems, bike lanes, and sidewalks as alternatives to cars, and to reduce taxes on income, wages, and wealth to offset the increased taxes on gasoline. Such a *tax shift* would help to make higher gasoline taxes more politically and economically acceptable.

Taxing gasoline heavily is difficult in the United States, for three reasons. *First*, it faces strong opposition from people who feel they are already overtaxed, many of whom are unaware of the hidden costs they are paying for gasoline. The other opposition group is made up of the powerful transportation-related industries such as carmakers, oil and tire companies, road builders, and many real estate developers. *Second*, the dispersed nature of most U.S. urban areas makes people dependent on cars, and thus higher taxes would be an economic burden for them. *Third*, fast, efficient, reliable, and affordable mass-transit options, bike lanes, and sidewalks are not widely available in the United States, primarily because most of the revenue from gasoline taxes is

TonyV3112/Shutterstock.com

FIGURE 22.14 Traffic jam in Beijing, China. There are 5 million cars in Beijing.

used for building and improving highways for motor vehicles.

Another way to reduce automobile use and urban congestion is to raise parking fees and charge tolls on roads, tunnels, and bridges leading into cities—especially during peak traffic times. Densely populated Singapore is rarely congested because it auctions the rights to buy a car, and its cars carry electronic sensors that automatically charge the drivers a fee every time they enter the city. Several European cities have imposed stiff fees for motor vehicle use in their central cities, while others have banned the parking of cars on city streets and established networks of bike lanes. Shanghai, China, discourages car use by charging more than $9,000 for a license plate. Paris, France, has removed 200,000 parking spaces.

More than 300 European cities have *car-sharing* networks that provide short-term rental of cars. Network members reserve a car or contact the network and are directed to the closest car. In Berlin, Germany, car sharing has cut car ownership by 75%. According to the Worldwatch Institute, car sharing in Europe has reduced the average driver's CO_2 emissions by 40–50%. Car-sharing networks have sprouted in a growing number of U.S. cities and on some college campuses, and some large car-rental companies have begun renting cars by the hour. As car sharing increases, car ownership in cities deceases, which can lead to fewer accidents and less congestion.

Alternatives to Cars

There are several alternatives to motor vehicles, each with its own advantages and disadvantages. Figure 22.15 shows the transportation hierarchy in more-sustainable cities such as Portland, Oregon (**Core Case Study**).

Many people use foot power to walk to work or other places, which is free, does not pollute, and provides exercise. Another widely used alternative for short distances is the *bicycle* (Figure 22.16), or pedal power. It is affordable,

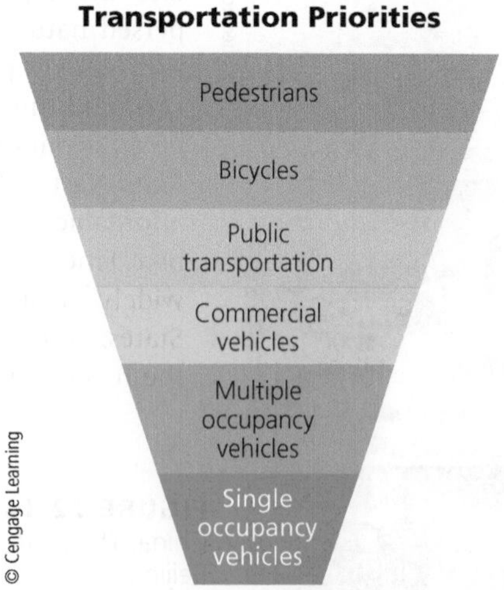

FIGURE 22.15 Transportation priorities in more-sustainable cities.

Trade-Offs

Bicycles

Advantages	Disadvantages
Are quiet and nonpolluting	Provide little protection in an accident
Take few resources to manufacture	Provide no protection from bad weather
Burn no fossil fuels	Are impractical for long trips
Require little parking space	Bike lanes and secure bike storage not yet widespread

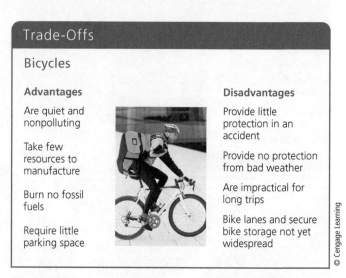

© Cengage Learning

FIGURE 22.16 Bicycle use has advantages and disadvantages. ***Critical thinking:*** Which single advantage and which single disadvantage do you think are the most important?

Photo: Tyler Olson/Shutterstock.com

does not pollute, provides exercise, and requires little parking space. The use of bicycles with lightweight electric motors is also on the rise.

Bicycling accounts for at least a third all urban trips in the Netherlands and in Copenhagen, Denmark. On the other hand, between 2000 and 2014, the percentage of commuters in Beijing, China, who biked to work dropped from 38% to 12%, as people switched to cars, according to government data. This greatly increased use of cars in Beijing (Figure 22.14) has led to frequent and severe smog problems, as well as severe traffic congestion.

Biking accounts for less than 1% of urban trips in the United States, but this percentage is higher in bike-friendly large cities such as New York, Chicago, Baltimore, Minneapolis, and Philadelphia. Many smaller U.S. cities such as Cambridge, Massachusetts; Madison, Wisconsin; Fort Collins and Boulder, Colorado; and Davis, Berkeley, Santa Cruz, and Santa Monica, California, are also bike friendly. About 25% of Americans polled would bike to work or school if safe bike lanes and secure bike storage were available.

More than 712 cities in 50 countries, including 78 U.S. cities, have *bike-sharing systems* that allow individuals to rent bikes as needed from widely distributed stations. Portland, Oregon, which has over 640 kilometers (400 miles) of bike lanes, has a successful bike-sharing system (**Core Case Study**). The city's biking culture has created many local jobs related to bicycle manufacturing, sales, and service. In addition, a Portland-based delivery company uses electric tricycles instead of trucks and can transport parcels weighing as much as 270 kilograms (600 pounds).

Another alternative to cars is *buses*, which are the most widely used form of urban mass transit worldwide (Figure 22.17). In some cities, bus rapid-transit (BRT) systems make bus use more convenient by having fast

Trade-Offs

Buses

Advantages	Disadvantages
Reduce car use and air pollution	Can lose money because they require affordable fares
Can be rerouted as needed	Can get caught in traffic and add to noise and pollution
Cheaper than heavy-rail system	Commit riders to transportation schedules

FIGURE 22.17 Bus rapid-transit (BRT) systems and conventional bus systems in urban areas have advantages and disadvantages. ***Critical thinking:*** Which single advantage and which single disadvantage do you think are the most important?

Photo: Isaak/Shutterstock.com

Trade-Offs

Mass Transit Rail

Advantages	Disadvantages
Uses less energy and produces less air pollution than cars do	Expensive to build and maintain
Uses less land than roads and parking lots use	Cost-effective only in densely populated areas
Causes fewer injuries and deaths than cars	Commits riders to transportation schedules

FIGURE 22.18 Mass-transit rail systems in urban areas have advantages and disadvantages. ***Critical thinking:*** Which single advantage and which single disadvantage do you think are the most important?

Photo: Steve Rosset/Shutterstock.com

express routes, allowing riders to pay at machines located at each bus stop so that they can board a bus more quickly, and having three or four doors for quicker boarding. Newer systems are using diesel-electric hybrid motors to cut pollution and noise from buses. BRT systems are used in a growing number of cities, including Mexico City (Case Study, p. 613); Curitiba, Brazil; Bogotá, Columbia; Seoul, South Korea; Istanbul, Turkey; and 11 cities in China, including Beijing. Each year, Portland's BRT system (**Core Case Study**) carries more than 66 million riders.

Another urban transportation alternative is *mass-transit rail systems* (Figure 22.18). They include *heavy-rail* systems (subways, elevated railways, and metro trains) and *light-rail* systems (streetcars, trolley cars, and tramways). Portland (**Core Case Study**) has a widely used light-rail system (Figure 22.19).

A *rapid-rail system between urban areas* is another option for reducing car use (Figure 22.20). Western Europe, Japan, and China have high-speed bullet trains that travel between cities at up to 306 kilometers (190 miles) per hour. In Japan, a fleet of bullet trains carries some 20 million passengers a day. For decades, many analysts in the United States have talked about building high-speed rail systems between cities, but so far, none have been

Ken Hawkins/Alamy Stock Photo

FIGURE 22.19 Widespread bicycle use and a light-rail system, in operation since 1986, have helped reduce car use in Portland, Oregon (**Core Case Study**).

Trade-Offs

Rapid Rail

Advantages	Disadvantages
Much more energy efficient per rider than cars and planes are	Costly to run and maintain
Produces less air pollution than cars and planes	Causes noise and vibration for nearby residents
Can reduce need for air travel, cars, roads, and parking areas	Adds some risk of collision at car crossings

FIGURE 22.20 Rapid-rail systems between urban areas have advantages and disadvantages. ***Critical thinking:*** Which single advantage and which single disadvantage do you think are the most important?

Photo: Alfonso d'Agostino/Shutterstock.com

built because of a lack of government subsidies and other funding sources. While Europe and Japan have a subsidized train culture, the United States has a subsidized car culture.

CONSIDER THIS . . .

LEARNING FROM NATURE

The kingfisher bird's long beak allows it to dive into water at a high speed without making a splash to catch fish. In Japan, designers increased the speed and reduced the noise from high-speed bullet trains by modeling the train's front end after the kingfisher's beak. See the front cover of this book and About the Cover on page ii.

22.4 HOW IMPORTANT IS URBAN LAND-USE PLANNING?

CONCEPT 22.4 Urban land-use planning can help to reduce uncontrolled sprawl and slow the resulting degradation of air, water, land, biodiversity, and other natural resources.

Conventional Land-Use Planning

Most urban areas use various forms of **land-use planning** to determine the best present and future uses of various parcels of land (**Concept 22.4**). Once a land-use plan is developed and adopted, governments can control the uses of certain parcels of land by legal and economic methods.

The most widely used approach is **zoning**, in which parcels of land are designated for certain uses such as residential, commercial, or mixed use. Zoning can be used to control growth and to protect areas from certain types of development. For example, Portland, Oregon (**Core Case Study**), and other cities have used zoning to encourage high-density development along major mass-transit corridors to reduce automobile use and air pollution.

Despite its usefulness, zoning has several drawbacks. One problem is that some developers can influence or modify zoning decisions in ways that threaten or destroy wetlands, prime cropland, forested areas, and open space. Another problem is that zoning often favors high-priced housing, factories, hotels, and other businesses over protecting environmentally sensitive areas and providing low-cost housing. This is largely because most local governments depend on property taxes for their revenue.

In addition, overly strict zoning can discourage innovative approaches to solving urban problems. For example, the pattern in the United States and in some other countries has been to prohibit businesses in residential areas, which causes separate business and residential developments and encourages suburban sprawl. Some urban planners have returned to *mixed-use zoning* to help reduce this problem. In the 1970s, Portland, Oregon (**Core Case Study**), managed to cut driving and gasoline consumption and to promote walking and bicycling by using mixed zoning to encourage the establishment of neighborhood groceries and other small stores in residential areas.

Smart Growth

Smart growth is a set of policies and tools that encourage more environmentally sustainable urban development with less dependence on cars. It uses zoning laws and other tools to channel growth in order to reduce its ecological footprint.

Smart growth can discourage sprawl, reduce traffic, protect ecologically sensitive and important lands and waterways, and develop neighborhoods that are more enjoyable places in which to live. Figure 22.21 lists popular smart growth tools that cities are using to prevent and control urban growth and sprawl. Portland, Oregon (**Core Case Study**), has used many of these tools.

China has taken the strongest stand of any country against urban sprawl. The government has designated 80% of the country's arable land as *fundamental land*. Building on such land requires approval from local and provincial governments and from the State Council, the central government's chief administrative body.

Many European countries have been successful in discouraging urban sprawl in favor of compact cities. They have controlled development at the national level and imposed high gasoline taxes to discourage car use and to encourage people to live closer to workplaces and shops. High taxes on heating fuel have encouraged some people to live in apartments or smaller houses. These governments have used most of the resulting gasoline and heating fuel tax revenues to develop efficient train systems and other mass-transit options within and between cities.

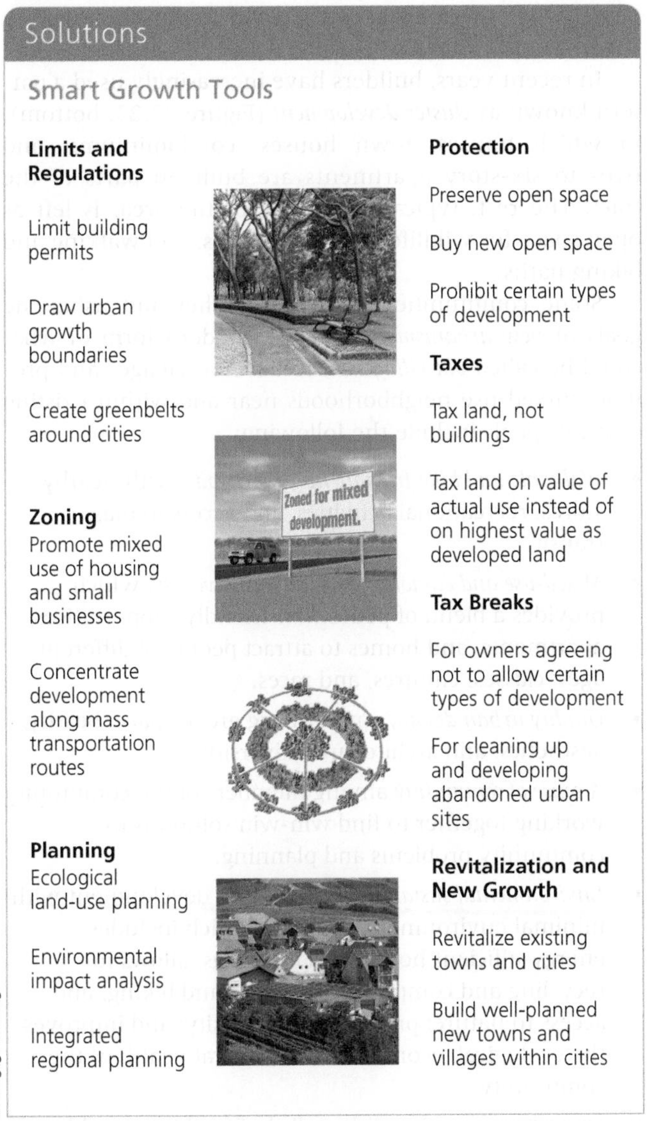
Solutions
Smart Growth Tools
Limits and Regulations
Limit building permits
Draw urban growth boundaries
Create greenbelts around cities
Zoning
Promote mixed use of housing and small businesses
Concentrate development along mass transportation routes
Planning
Ecological land-use planning
Environmental impact analysis
Integrated regional planning
Zoned for mixed development.
Protection
Preserve open space
Buy new open space
Prohibit certain types of development
Taxes
Tax land, not buildings
Tax land on value of actual use instead of on highest value as developed land
Tax Breaks
For owners agreeing not to allow certain types of development
For cleaning up and developing abandoned urban sites
Revitalization and New Growth
Revitalize existing towns and cities
Build well-planned new towns and villages within cities
© Cengage Learning

FIGURE 22.21 Smart growth tools can be used to prevent or control urban growth and sprawl. ***Critical thinking:*** Which five of these tools do you think would be the best methods for preventing or controlling urban sprawl? Which, if any, of these tools are used in your community?

Top: Tungphoto/Shutterstock.com. Bottom: iStockphoto.com/Richard Schmidt-Zuper

Preserving and Using Open Space

One way to preserve open space outside a city is to draw a boundary around the city and to prohibit urban development outside that boundary. This *urban growth boundary* approach is used in the U.S. states of Oregon, Washington, and Tennessee. In 1979, Portland, Oregon (**Core Case Study**), established a strict urban growth boundary, and it worked. Although the city's population increased by 69% between 1980 and 2014, its urban area expanded by only 3%.

Another approach is to surround a large city with a *greenbelt*—an open area reserved for recreation, sustainable forestry, or other nondestructive uses. In many cases, satellite towns have been built outside these greenbelts. By linking these towns to the central city via a public transport system, planners can minimize damage to the greenbelt. Many cities in Western Europe and the Canadian cities of Toronto and Vancouver have used this approach. Greenbelts can provide vital ecosystem services such as absorption of CO_2 and other air pollutants, which can make urban air more breathable and help to cut a city's contribution to climate change.

A more traditional way to preserve large blocks of open space is to create municipal parks. Examples of large urban parks in the United States are Central Park in New York City (Figure 22.22); Golden Gate Park in San Francisco, California; and Grant Park in Chicago, Illinois. Portland, Oregon (**Core Case Study**), has 288 parks along with trails and natural areas distributed throughout the city that give its citizens easy access to nature and recreational opportunities. British scientists found that people living near a park or other green space get outdoors more and tend to have better health.

Age Fotostock/SuperStock

FIGURE 22.22 With almost 344 hectares (850 acres) that include woodlands, lawns, and small lakes and ponds, New York City's Central Park is a dramatic example of a large open space in the center of a major urban area.

22.5 HOW CAN CITIES BECOME MORE SUSTAINABLE AND LIVABLE?

CONCEPT 22.5 An *eco-city* allows people to choose walking, biking, or mass transit for most transportation needs; recycle or reuse most of their wastes; grow much of their food; and protect biodiversity by preserving surrounding land.

New Urbanism

Since World War II, the typical approach to suburban housing development in the United States has been to bulldoze a tract of woods or farmland and build rows of houses on standard-size lots (Figure 22.23, center). Many of these developments and their streets, with names like Oak Lane, Cedar Drive, Pheasant Run, and Fox Valley, are named after the trees and wildlife that were removed to make room for them.

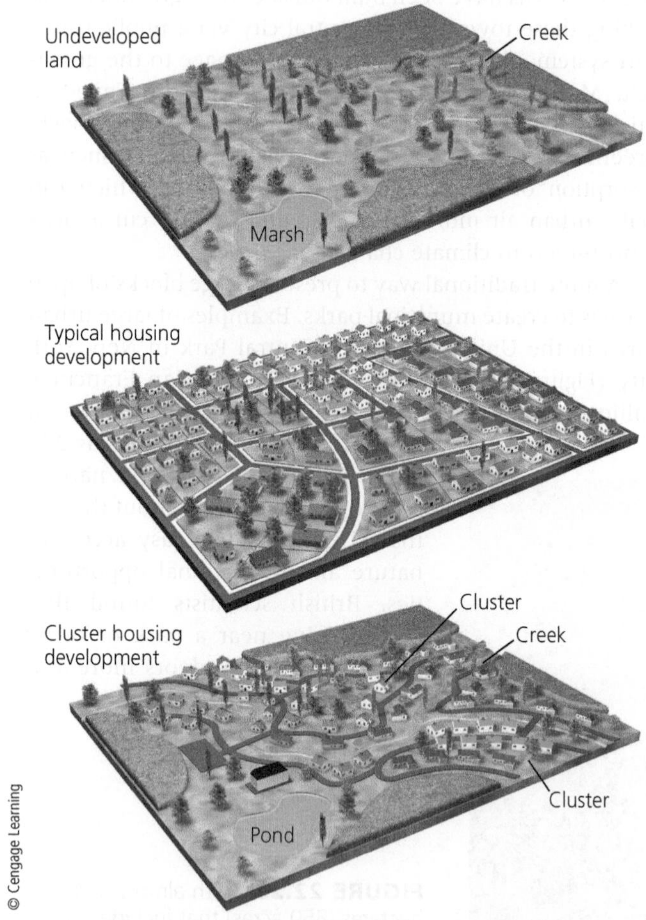

FIGURE 22.23 These models compare a *conventional housing development* (center) with a *cluster housing development* (bottom). ***Critical thinking:*** What are the comparative effects of each type of development on the undeveloped land (top)?

In recent years, builders have increasingly used a pattern known as *cluster development* (Figure 22.23, bottom), in which houses, town houses, condominiums, and two- to six-story apartments are built on parts of the tract. The rest, typically 30–50% of the area, is left as open space for wildlife preserves, parks, and walking and biking paths.

Some communities are going further and using the goals of *new urbanism*, which is a modern form of what could be called *old villageism*, to develop villages and promote mixed-use neighborhoods near and within existing cities. Its goals include the following:

- *Walkable and bike-friendly neighborhoods*, with nearby stores, recreational activities, and access to mass transit.
- *Mixed-use and economic and cultural diversity*, which provides a blend of pedestrian-friendly shops, offices, apartments, and homes to attract people of different ages, classes, cultures, and races.
- *Quality urban design*, emphasizing green space, beauty, aesthetics, and architectural diversity.
- *A sense of community* among members of the community working together to find win-win solutions to community problems and planning.
- *Environmental sustainability* based on development with minimal environmental impact, which includes energy-efficient homes and buildings; allows for recycling and composting, walking and biking, and access to nature; promotes biodiversity; and improves the ecological, economic, and cultural health of the community.
- *Smart transportation* with well-designed train and bus systems connecting neighborhoods and nearby towns and cities to one another.

Examples of new urban villages in the United States are Mayfaire, in Wilmington, North Carolina; Mizner Place in Boca Raton, Florida; Middleton Hills near Madison, Wisconsin; Yardley, Pennsylvania; Kentlands in Gaithersburg, Maryland; Valencia, California (near Los Angeles); and Stapleton, Colorado (built on an old airport site in Denver).

The Eco-City Concept: Cities for People, Not Cars

According to most environmentalists and urban planners, the primary problem with large cities is our failure to make them more sustainable and enjoyable places to live through good ecological design. New urbanism is a step in the right direction, but many scientists call for going further by employing *sustainable community development* (Science Focus 22.1).

SCIENCE FOCUS 22.1

Sustainable Community Development—A Key to Global Sustainability

Sustainable community development (SCD) is a multidisciplinary approach to economic development that incorporates scientific methods and principles and focuses on a *triple bottom line*—people, planet, and profit—to work toward sustainability at the community level. Advocates of this approach argue that we probably cannot achieve sustainability on national and world regional levels unless we can accomplish it at the local level.

One prominent SCD researcher is environmental scientist Kelly D. Cain, Director of the St. Croix Institute for Sustainable Community Development at the University of Wisconsin–River Falls. Cain notes that the major objectives of SCD are to achieve economic and environmental health and social equity. He contends that to meet such objectives, a sustainable community must have broad-based citizen participation, agree on a vision based on the best available science, and establish clear and measurable goals for benchmarking and tracking performance. Such goals for a community can include

- growing as much of its own food as possible;
- generating all or most of its own carbon-neutral energy;
- restoring the carrying capacity of local ecosystems that are damaged in the building and maintenance of the community; and
- recycling pollutants and wastes as resources within industrial ecosystems (Figure 21.27, p. 597) or as biodegradable wastes, and returning extracted resources such as water and minerals to the ecosystem.

Another important community goal is to become *carbon-negative*—absorbing more carbon than the community generates to shrink its carbon footprint. According to Cain, "Becoming carbon-neutral is no longer a good enough goal." He points out that if we all simply become carbon-neutral, we are still likely to experience severely damaging climate change. This is due to two factors: the huge amount of CO_2 already in the atmosphere that will remain there for 80 to 120 years, and projected increases in CO_2 levels.

Research by Cain and others indicates that communities can become carbon-negative by greatly improving energy efficiency and by relying on low- and no-carbon energy resources such as wind, solar power, and geothermal energy. To become carbon-negative, it also helps if a community can protect and nurture enough native soils and vegetation in its grasslands, forests, gardens, and other green spaces to absorb and sequester more carbon than it generates on most days.

Cain and his team of researchers evaluate the progress of communities in becoming more sustainable by using various indicators, including energy bills, water-use rates, recycling rates, composting rates, local food production, green space and other common areas, and percentages of electric and hybrid cars among a community's total number of vehicles. They also measure community outputs such as volumes of municipal solid waste and carbon emissions.

The *triple bottom line* is an important measure of the success of SCD efforts. With this approach, the goal for businesses and other organizations is to achieve positive results for people, the planet, and profits. However, Cain argues that in order to be truly sustainable, we will need a *quadruple bottom line* approach in which we add a fourth factor—the community factor—which leads to a strengthened community based on the core values of ingenuity, creativity, innovation, entrepreneurship, and responsibility for self, family, neighborhood, and community.

CRITICAL THINKING

List three steps that your community or school could take, and three steps that you could take, to become carbon-negative.

An important result of these trends is the *eco-city*, or *green city*, model for new urban development and for renovation of existing cities (**Concept 22.5**). It is being applied in cities around the world, including Portland, Oregon (**Core Case Study**). Such cities are people-oriented, not car-oriented, and their residents are able to walk, bike, or use low-polluting mass transit for most of their travel.

Residents of today's eco-cities apply the three **scientific principles of sustainability**. Much of their energy comes from solar cells on the rooftops and walls of buildings, solar hot-water heaters on rooftops, and geothermal heating and cooling systems (Figure 16.24, p. 428). They work to ensure that the buildings, vehicles, and appliances they use meet high energy-efficiency standards. They typically reuse, recycle, and compost 60–80% of their municipal solid waste. Other goals are low or zero net greenhouse gas emissions, home energy systems that produce more energy than they use, and use of local and sustainable materials. Some eco-cities are also using biomimicry, inspired by Janine Benyus (Individuals Matter 1.1, p, 10), to design entire cities and communities that provide the same ecosystem services as forests and other natural ecosystems provide.

Eco-city residents apply the biodiversity principle by preserving or planting trees and plants that are adapted to their local climate and soils. Such plantings can provide food, shade, beauty, and wildlife habitats. They also absorb CO_2 and reduce air pollution, noise, and soil erosion.

In an eco-city, abandoned lots and industrial sites are cleaned up and used. Nearby forests, grasslands, wetlands, and farms are preserved. Parks are available to everyone. Much of the food that people eat comes from nearby organic farms, solar greenhouses, community gardens, and gardens on rooftops, in yards, and in window boxes. Some scientists see eco-cities as good locations for *vertical farms* (Figure 22.24) in multistory urban buildings each designed to provide food for up to 50,000 people. Organic hydroponic crops (Science Focus 12.1, p. 310) would be grown on the upper floors. Chickens (and the eggs they produce) and fish that feed on plant wastes would be grown in aquaculture tanks on lower floors.

Eco-cities also strive to provide *environmental justice* for all of their residents without regard to race, creed, income, or any other factor (Individuals Matter 22.1). People who design and live in eco-cities take seriously the advice that U.S. urban planner Lewis Mumford gave more than 3 decades ago: "Forget the damned motor car and build cities for lovers and friends."

FIGURE 22.24 Vertical farms in cities may provide food for many urban dwellers in the future.

The eco-city is not a futuristic dream, but a growing reality for a number of cities, including Portland, Oregon (**Core Case Study**); Curitiba, Brazil (see the Case Study that follows); Bogotá, Colombia; Waitakere City, New Zealand; Stockholm and Växjö in Sweden; Helsinki, Finland; Copenhagen, Denmark; Melbourne, Australia; Vancouver and Dockside Green, Victoria, in Canada; Leicester and Beddington Zero Energy Development (BedZED) in Great Britain; Masdar City, Abu Dhabi; Huangbaiyu and Tianjin Eco City in China; and in the United States (in addition to Portland, Oregon), Davis, California; Olympia, Washington; Chattanooga, Tennessee, and Greensburg, Kansas. According to a 2012 survey by London's University of Westminster, China is building more eco-cities than any country, followed by the United States.

Because more than half of the world's population lives in cities, they are focal points of the global economy. Although they are largely unsustainable population centers, they are also centers for diverse experiments in sustainability. Thus, they places where people are learning how to build sustainable lifestyles and economies based on applying the six **principles of sustainability** and they represent beacons of hope.

CASE STUDY

The Eco-City Concept in Curitiba, Brazil

An example of an eco-city is Curitiba ("koor-i-TEE-ba"), a city of 1.9 million people, known as the "ecological capital" of Brazil. In 1969, planners in this city decided to focus on an inexpensive and efficient mass-transit system rather than on the car.

Curitiba's bus rapid-transit system efficiently moves large numbers of passengers, including 72% of the city's commuters. Each of the system's five major "spokes," connecting the city center with outlying districts (map in Figure 22.25), has two express lanes used only by buses. Double- and triple-length bus sections are coupled as needed to carry up to 300 passengers. Boarding is speeded up by the use of extra-wide bus doors and boarding platforms under glass tubes where passengers can pay before getting on the bus (photo in Figure 22.25). Only high-rise apartment buildings are allowed near major bus routes, and each building must devote its bottom two floors to stores—a practice that reduces the need for residents to travel.

Cars are banned from 49 blocks in the center of the downtown area, which has a network of pedestrian walkways connected to bus stations, parks, and bicycle paths running throughout most of the city. As a result, Curitiba uses less energy per person and has lower emissions of greenhouse gases and other air pollutants and less traffic congestion than do most comparably sized cities.

INDIVIDUALS MATTER 22.1

Peggy M. Shepherd: Fighting for Environmental Justice in Urban America

Courtesy of Peggy Shepherd

Since 1988, Peggy M. Shepard has been a passionate and effective leader in promoting environmental justice for people of color and low-income residents in urban areas in New York City and throughout the United States. She says that environmental justice "means the ability of color and low-income communities to have access to clean water and clean air and to fully participate in environmental decision making."

In 1988, Shephard cofounded WE ACT For Environmental Justice—a nonprofit organization devoted to educating and mobilizing the residents of Upper Manhattan on environmental issues affecting their health and quality of life. WE ACT has also served as a national model for promoting environmental justice through grassroots activism.

In 1988, WE ACT filed a lawsuit against New York City over the odor and respiratory problems caused by poor management of the North River Sewage Treatment Plant. WE ACT won the lawsuit along with a $1.1 million environmental fund for the benefit of the West Harlem Community. The mayor also created a $55 million plan for eliminating odors and pollution from the plant, and the city built Riverbank State Park on top of the sewage treatment plant.

Shephard has also been a pioneer in showing connections between climate change and environmental justice. She points out that "low-income communities and communities of color are going to be the first hit and the hardest hit by climate change." WE ACT coordinates the Environmental Justice Leadership Forum on Climate Change—a coalition of 40 organizations representing 16 states.

Shephard has won numerous awards, including the prestigious Heinz Award for the Environment in 2004. She is regarded as one of the country's most highly respected and effective environmental activists.

FIGURE 22.25 Solutions: The bus rapid-transit system in Curitiba, Brazil, has greatly reduced car use.

Carlos Cazalis/Corbis

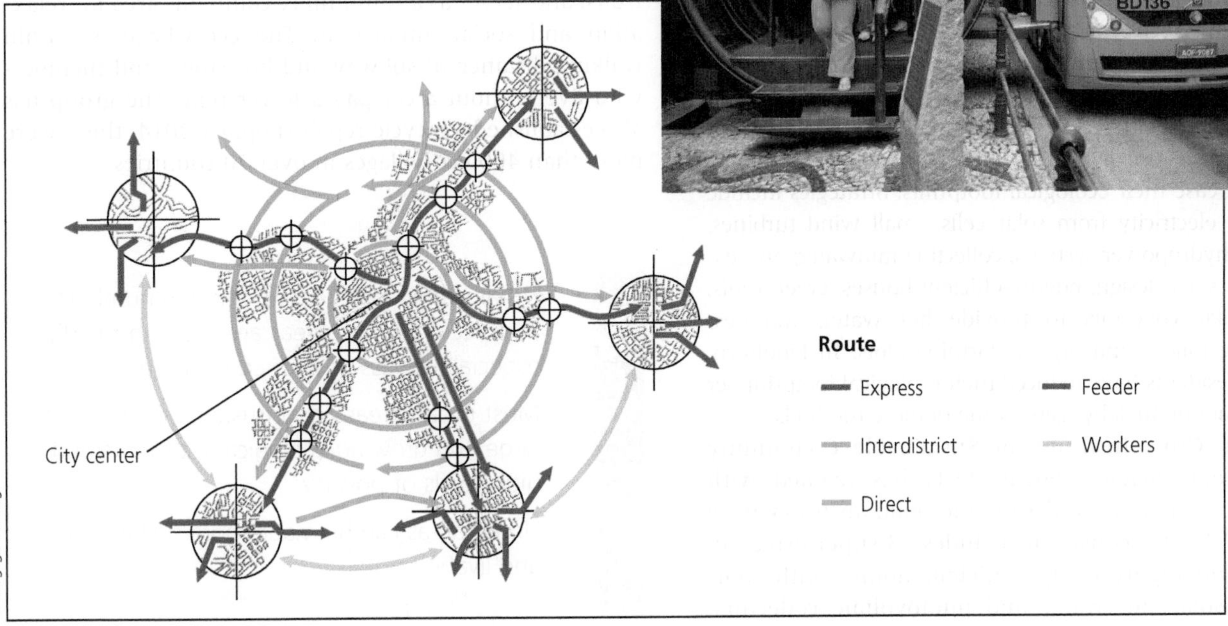

© Cengage Learning

Along the six streams that run within Curitiba's borders, the city removed most buildings and lined the streams with a series of 23 interconnected parks. Volunteers have planted more than 1.5 million trees throughout the city. No one can cut down a tree without a permit, which also requires that two trees must be planted for each one that is cut down.

Curitiba recycles roughly 70% of its paper and 60% of its metal, glass, and plastic. Recovered materials are sold mostly to the city's more than 500 major industries, which must meet strict pollution standards.

Curitiba's poor residents receive free medical and dental care, child care, and job training, and 40 feeding centers are available for street children. People who live in areas not served by garbage trucks can collect garbage and exchange filled garbage bags for surplus food, bus tokens, and school supplies. The city uses old buses as roving classrooms to train its poor in basic job skills. Other retired buses have become health clinics, soup kitchens, and daycare centers that are free for low-income parents.

About 95% of Curitiba's citizens can read and write and 83% of its adults have at least a high school education. All schoolchildren study ecology. Polls show that 99% of the city's inhabitants would not want to live anywhere else.

Curitiba does face challenges, as do all cities, mostly due to a fivefold increase in its population since 1965. Its once-clear streams are often overloaded with pollutants. The bus system is nearing capacity, and car ownership is on the rise. The city is considering building a light-rail system to relieve some of the pressure.

This internationally acclaimed model of urban planning and sustainability (**Concept 22.5**) is the brainchild of architect and former college professor Jaime Lerner, who has served as the city's mayor three times since 1969.

Eco-Villages

Another innovative approach is the growing global *eco-village movement*, in which typically 50 to 150 people come together to design and live in more ecologically, economically, and socially sustainable villages in rural and suburban areas, and in neighborhoods or "eco-hoods" within cities.

Eco-villagers use diverse methods to live more sustainably and to decrease their ecological footprints. Strategies include generating electricity from solar cells, small wind turbines, and small hydropower systems; collecting rainwater; and using passive solar design, energy-efficient houses, green roofs, rooftop solar collectors to provide hot water, waterless composting toilets, and organic farming plots. In Findhorn, Scotland, residents have reduced their ecological footprint per person by about 40% by using some of these methods.

The St. Croix Institute for Sustainable Community Development (Science Focus 22.1) has teamed with Habitat for Humanity to design and build an Eco-Village in River Falls, Wisconsin. It includes 18 superinsulated, airtight, and highly energy-efficient homes with roof-mounted solar hot water and photovoltaic collectors, designed to be affordable for low-income families. These home designs have received the LEED platinum certification (see Chapter 16, p. 415). A nearby solar cell farm will help homeowners attain net zero or negative energy usage.

The River Falls Eco-Village embodies the goals of sustainable community development (Science Focus 22.1). Homeowners share community vegetable gardens to be fed by a rainwater collection system and a fleet of electric cars. Edible landscaping includes fruit trees and perennial shrubs available as a local, inexpensive source of nutrients to all residents. Construction wastes were recycled at the highest possible rate (with a goal to exceed 90%). The village's community center includes a solar electric car battery charging system. Pedestrian paths connect the village to nearby neighborhoods and parks to promote walking and biking. Driveways and paths were designed and built to allow precipitation to flow into the soil beneath them.

The village also serves social equity goals. In partial payment for their new homes, low-income residents are able to contribute hours of labor in maintaining the community's homes, gardens, and landscaping. The community center is open to people in surrounding neighborhoods to link village residents with those of the larger community. The founders are planning another Eco-Village in a nearby county.

Another example of this trend is the Los Angeles Eco-Village, started in 1993 to show how people in the middle of a sprawling and unsustainable city such as Los Angeles, California, can establish small pockets of sustainability. It consists of two small apartment buildings with about 55 residents and a common courtyard. Working together, its members use solar panels to heat much of their hot water. They compost their organic wastes and use the compost in a large courtyard garden that provides vegetables and fruits, and serves as a community commons area for relaxation and social interaction. The eco-village is within walking distance of subway and bus stops, and members who live without a car pay a lower rent. The group has also established a bicycle repair shop. By 2014, there were more than 400 eco-villages in over 70 countries.

BIG IDEAS

- Urbanization is increasing and the numbers and sizes of urban areas are growing rapidly, especially in less-developed countries.
- Most urban areas are unsustainable with their large and growing ecological footprints and high levels of poverty.
- Urban areas can be made more sustainable and livable.

Tying It All Together

Portland, Oregon, and Sustainability

Stephen Rees/Cutcaster

Urban areas, where more than half of the world's people live, have large environmental impacts. Most urban areas are unsustainable and are rapidly growing centers of poverty, especially in less-developed countries. Thus, future global environmental quality depends on making urban areas more sustainable and livable as cities such as Portland, Oregon (**Core Case Study**), and Curitiba, Brazil, are doing.

To become more sustainable during the next few decades, urban areas will have to apply the three **scientific principles of sustainability**. This involves greatly improving energy efficiency and relying heavily on solar, wind, and geothermal energy for electricity, heating, and cooling. It will also require recycling, reusing, and composting most solid wastes. Cities also will need to sustain and enhance biodiversity by protecting existing wooded areas and parks, establishing more parks, and planting more trees and vegetation. All of these steps will further the goal of becoming carbon-negative.

In eco-cities of the future, people will move about primarily by walking, biking, and using low-polluting mass transit. Much of the food for eco-city residents will be produced through sustainable agriculture on urban and regional farms.

Making this transition toward sustainability is also in keeping with the economic, political, and ethical **principles of sustainability**. Full-cost pricing requires that the harmful environmental costs of urbanization be included in the market prices of urban goods and services. For example, some cities are making it more expensive to drive a car within their borders. Such changes are good for the economy as well as for the environment and are thus win-win political solutions. These solutions will also apply the ethical principle that calls for leaving the world in a sustainable condition for future generations.

Chapter Review

Core Case Study

1. Explain how Portland, Oregon, has attempted to become a more sustainable city.

Section 22.1

2. What is the key concept for this section? What are **urban areas**? Distinguish between **urbanization** and **urban growth**. What percentage of the world's people lives in urban areas? List two ways in which urban areas grow. What are some of the reasons why people move to cities?

3. List three major trends in global urban growth. Describe the three phases of urban growth in the United States. What is **urban sprawl**? List five factors that have promoted urban sprawl in the United States. List five undesirable effects of urban sprawl.

Section 22.2

4. What is the key concept for this section? What are the major advantages of urbanization? Explain why most urban areas are unsustainable systems and how these factors contribute to their unsustainability: lack of vegetation, water supply problems and flooding, pollution, health problems, the heat island effect, and light pollution. What is **noise pollution** and why is it an urban problem? Describe the major aspects of poverty in urban areas. Summarize Mexico City's major urban and environmental problems and what government officials are doing about them.

Section 22.3

5. What is the key concept for this section? Distinguish between compact and dispersed cities, and give an example of each. What are the major advantages and disadvantages of using motor vehicles? List four ways to reduce dependence on motor vehicles. List the major advantages and disadvantages of relying more on **(a)** bicycles, **(b)** bus systems, **(c)** mass-transit rail systems within urban areas, and **(d)** rapid-rail systems between urban areas.

Section 22.4

6. What is the key concept for this section? What is **land-use planning**? What is **zoning** and what are its limitations? Define **smart growth**, explain its benefits, and list five tools that are used to implement it. What are three ways to preserve open spaces around a city?

Section 22.5

7. What is the key concept for this section? What is cluster development? What are the five key goals of new urbanism? What is sustainable community development and what are six indicators that scientists study to assess a community's level of sustainability? Explain why becoming carbon-negative is an important goal for communities.

8. Describe the eco-city model and how it applies the six **principles of sustainability**. Give five examples of how Curitiba, Brazil, has attempted to become a more sustainable and livable eco-city.

9. What is the eco-village movement? How do eco-villages apply each of the three **scientific principles of sustainability**?

10. What are this chapter's *three big ideas*? Explain how Portland, Oregon, and other cities are applying the six **principles of sustainability** to become more sustainable urban areas.

Note: Key terms are in bold type.

Critical Thinking

1. Portland, Oregon (**Core Case Study**), has made significant progress in becoming a more environmentally sustainable and desirable place to live. If you live in an urban area, what steps, if any, has your community taken toward becoming more environmentally sustainable? What further steps could be taken?

2. Do you think that urban sprawl is a problem and something that should be controlled? Develop an argument to support your answer. Compare your argument with those of your classmates.

3. Write a brief essay that includes at least three reasons why you **(a)** enjoy living in a large city, **(b)** would like to live in a large city, or **(c)** do not wish to live in a large city. Be specific in your reasoning. Compare your essay with those of your classmates.

4. One issue debated at a UN conference was the question of whether housing is a universal *right* (a position supported by most less-developed countries) or just a *need* (supported by the United States and several other more-developed countries). What is your position on this issue? Defend your choice.

5. If you own a car or hope to own one, what conditions, if any, would encourage you to rely less on your car and to travel to school or work by bicycle, on foot, by mass transit, or by carpool?

6. Do think that the United States (or the country in which you live) should develop a comprehensive and integrated mass-transit system over the next 20 years, including an efficient rapid-rail network for travel within and between its major cities? Explain. If so, how would you pay for such a system?

7. Consider the characteristics of an eco-city listed on pp. 621-622. How close to this eco-city model is the city in which you live or the city nearest to where you live? Pick what you think are the five most important characteristics of an eco-city and, for each of these characteristics, describe a way in which your city could attain it.

8. Congratulations! You are in charge of the world. List the three most important components of your strategy for dealing with urban growth and sustainability in **(a)** more-developed countries and **(b)** less-developed countries.

Doing Environmental Science

The campus where you go to school is something like an urban community. Choose five eco-city characteristics (pp. 621-622) and apply them to your campus. For each of the five characteristics:

1. Create a scale of 1 to 10 in order to rate the campus on how well it does in having that characteristic. (For example, how well does it do in giving students options for getting around, other than by using a car? A rating of 1 could be *not at all*, while a rating of 10 could be *excellent*.)

2. Do some research and rate your campus for each characteristic.

3. Write an explanation of your research process and why you chose each rating.

4. Write a proposed plan for how the campus could improve its rating.

Global Environment Watch Exercise

Go to your MindTap course to access the GREENR database. At the top of the page do a "Basic Search" for *eco-city* or *green city*. Find an article about an eco-city and summarize the characteristics that make it a green city. Compare your summary to the description of eco-cities included in this chapter. What are some similarities between the two descriptions? What are some differences?

Ecological Footprint Analysis

Under normal driving conditions, an internal combustion engine in a car produces approximately 200 grams (0.2 kilograms, or 0.44 pounds) of CO_2 per kilometer driven. Assume that a city you are studying has 7 million cars and that each car is driven an average of 20,000 kilometers (12,400 miles) per year.

1. Calculate the carbon footprint per car, that is, the number of metric tons (and tons) of CO_2 produced by a typical car in 1 year. Calculate the total carbon footprint for all cars in the city, that is, the number of metric tons (and tons) of CO_2 produced by all the cars in 1 year. (*Note:* 1 metric ton = 1,000 kilograms = 1.1 tons.)
2. If 20% of the city's residents were to walk, bike, or take mass transit instead of driving their cars in the city, how many metric tons (and tons) of the annual emissions of CO_2 would be eliminated?
3. By what percentage, and by what average distance driven, would use of cars in the city have to be cut in order to reduce by half the annual carbon footprint calculated in question 1?

CHAPTER 23

Economics, Environment, and Sustainability

When it is asked how much it will cost to protect the environment, one more question should be asked: How much will it cost our civilization if we do not?

GAYLORD NELSON

Key Questions

23.1 How are economic systems related to the biosphere?

23.2 How can we estimate the values of natural capital, pollution control, and resource use?

23.3 How can we use economic tools to deal with environmental problems?

23.4 How can reducing poverty help us to deal with environmental problems?

23.5 How can we make the transition to more environmentally sustainable economies?

Rapeseed farm and wind farm in Plön, Schleswig-Holstein, Germany.

Caro/Alamy Stock Photo

Core Case Study

Germany: Using Economics to Spur a Shift to Renewable Energy

Germany, one of the world's most industrialized nations, is undergoing a renewable energy revolution (Chapter 16, Case Study, p. 410). The country aims to phase out its dependence on fossil fuels and nuclear power and, by 2050, to get 80% of its electricity from renewable energy resources.

In 2014, Germany generated about 28% of its electricity using wind farms on land (see chapter-opening photo) and at sea (Figure 23.1, left), solar energy (Figure 23.1, right), and other renewable sources. On days when conditions are ideal, it has produced as much as 78% of its electricity in this way. In 2015, Germany's greenhouse gas emissions were at their lowest level since 1990. Since 2000, this shift to renewable energy has created a multibillion-dollar German industry that includes renewable energy production and sales of renewable energy technology around the world.

This transition was spurred by government legislation aimed at homeowners, businesses, and communities that produce electricity from solar cells, wind, and other renewable energy systems. The law allows them to sell the electricity they produce to Germany's major power companies at a fixed rate that guarantees that their investments will at least break even. With this *feed-in tariff* system, the government ensures that these renewable energy producers will not lose money. In fact, they often make a profit.

The German government has also promoted the building of wind farms on land and offshore along the North Sea and Baltic Sea coasts (Figure 23.1, left). It plans to have 10,000 offshore wind turbines operating by 2030. There are also plans to lay more than 3,700 kilometers (2,300 miles) of high-voltage electrical cables throughout parts of the country and under the North Sea as part of a new state-of-the-art electrical grid. Such a grid would be far more efficient than conventional grids, and would help to make renewable energy sources more dependable.

Since 1990, solar energy production has risen steadily in Germany, much of it through rooftop solar collectors. In 2012, farmers and homeowners installed a record number of such collectors on their roofs—enough to equal the electrical output of seven large nuclear power plants. Even when the economy was sagging, solar and wind energy production continued to grow in Germany.

Germany's shift to renewable energy has faced some challenges that we discuss in this chapter. Even critics of the feed-in tariff agree that it has done its job in helping to establish a vibrant renewable energy industry in Germany. And Germany's example shows that economic improvements and improvements in environmental quality can go hand in hand—an example of the win-win **principle of sustainability** (see Inside Back Cover). Some economists argue that shifting to cleaner energy resources, cleaner industrial production, and more sustainable agriculture would create more environmentally sustainable economies that would benefit the largest possible number of people.

Luftbild Bertram/Blickwinkel/Age FotoStock

iStockphoto.com/Richard Schmidt-Zuper

FIGURE 23.1 This wind farm (left) is located off Germany's coast, and this rural village (right) in Germany's Rhineland-Palatinate is typical for German towns in having several rooftop solar panels.

23.1 HOW ARE ECONOMIC SYSTEMS RELATED TO THE BIOSPHERE?

CONCEPT 23.1 Ecological economists and most sustainability experts regard human economic systems as subsystems of the biosphere.

Economic Systems Depend on Natural Capital

Economics is the social science that deals with the production, distribution, and consumption of goods and services to satisfy people's needs and wants. There are three major economic systems. In a *centrally planned economy*, the government determines production and distribution. In a *free-market economy*, these things are determined by private individuals and companies. In a *mixed economy*, both government and private interests take part in determining production and distribution of goods and services.

Almost all countries have mixed economic systems with varying degrees of participation by government and private interests. In China, for example, the economy is mostly centrally planned with private interests playing growing roles. The United States is much more of a free-market economy with some regulation by government.

In a truly *free-market economic system* (Figure 23.2), all economic decisions are governed solely by the competitive interactions of *supply* (the amount of a good or service that is available), *demand* (the amount of a good or service that people want), and *price* (the market value of a good or service) with little or no government control or interference in these interactions.

In Figure 23.2, *supply* is represented by the blue line (commonly called a *curve*) showing how much a producer of any good or service is willing to supply (measured on the horizontal *quantity* axis) for different prices (measured on the vertical *price* axis). *Demand* is represented by the red curve showing how much consumers will pay for different quantities of the good or service. The point at which the curves intersect is called the *market price equilibrium point*, where the supplier's price matches what buyers are willing to pay for some quantity and a sale is made.

Changes in supply and demand can shift one or both curves back and forth, and thus change the equilibrium point. For example, when supply is increased (shifting the blue curve to the right) and demand remains the same, the market price will go down. Similarly, when demand is increased (shifting the red curve to the right) and supply remains the same, the market price will rise.

A truly free-market economy rarely exists in today's capitalist market systems because factors other than supply and demand influence prices and sales. The primary goal of any business is to make as large a profit as possible for its owners or stockholders. To do so, most businesses try to take business away from their competitors and to exert as much control as possible over the prices of the goods and services they provide.

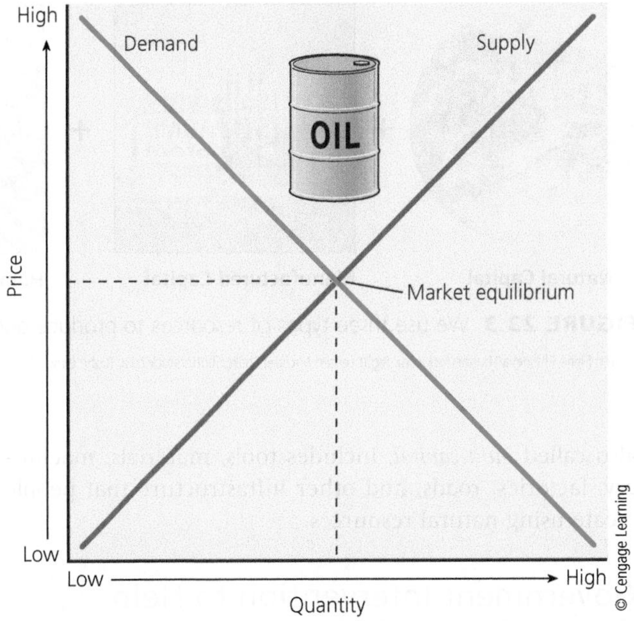

FIGURE 23.2 Supply and demand curves for a saleable product in a free market economic system. If all factors except supply, demand, and price are held fixed, *market equilibrium* occurs at the point where the supply and demand curves intersect. ***Data analysis:*** How would an increase in the available supply of oil shift the market equilibrium point on this diagram?

For example, many companies push for government support such as **subsidies**, or payments intended to help a business grow and thrive, along with tax breaks, trade barriers, and regulations that will give their products an advantage in the market over their competitors' products. When governments give larger subsidies to some companies or industries than they give to others within the same market, it can create an uneven economic playing field.

In addition, some companies withhold information from consumers about the costs and dangers that their products may pose to human health or to the environment, unless the government requires them to provide such information. Thus, buyers often do not get complete information about the harmful environmental impacts of the goods and services they buy. Some economists say that providing such information for consumers is one of the requirements of a truly free-market economy.

Most economic systems use three types of *capital*, or resources, to produce goods and services (Figure 23.3). **Natural capital** (see Figure 1.3, p. 7) includes resources and ecosystem services produced by the earth's natural processes, which support all life and all economies. **Human capital** includes the physical and mental talents of the people who provide labor, organizational and management skills, and innovation. **Manufactured capital,**

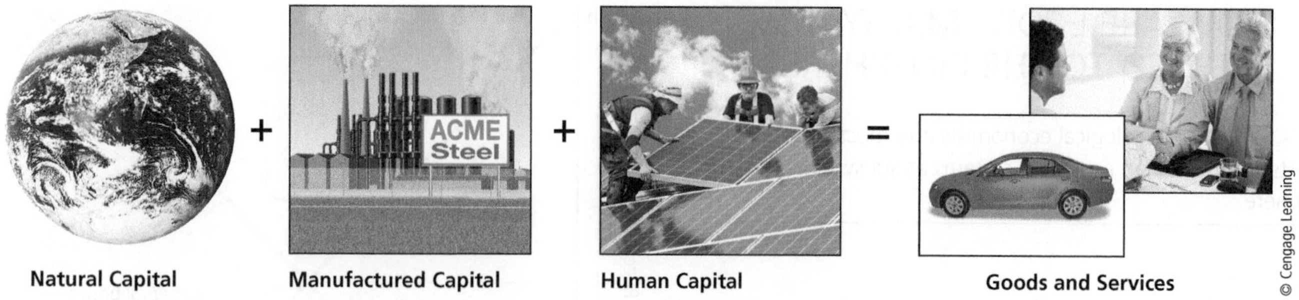

FIGURE 23.3 We use three types of resources to produce goods and services.

Center: Elena Elisseeva/Shutterstock.com. Right center: Michael Shake/Shutterstock.com. Right: iStockphoto.com/Yuri

also called *built capital,* includes tools, materials, machinery, factories, roads, and other infrastructure that people create using natural resources.

Government Intervention to Help Correct Market Failures

Markets usually work well in guiding the efficient production and distribution of *private goods*. However, experience shows that they cannot be relied on to provide adequate levels of *public services,* such as national security, police and fire fighters, and environmental protection. Economists generally refer to such deficiencies as *market failures*. An important example of a market failure is the inability of markets to prevent the degradation of *open-access resources,* such as clean air, the open ocean, and the earth's overall life-support system. Such vital resources are not bought and sold in the marketplace, because they are owned by no one and available for use by everyone at little or no charge.

Governments intervene in market systems to provide various public services and to help correct market failures. In the 1970s, the U.S. government passed laws to control air pollution (Chapter 18, p. 490) and water pollution (Chapter 20, p. 563). Were it not for such laws, air and water pollution in the United States would be much worse than it is.

One reason why markets often fail to provide environmental protection is their failure to assign monetary value to the benefits provided by the earth's natural capital or to the harmful effects of various human activities on the environment and on human health. For example, the benefits of leaving an old-growth forest undisturbed (ecosystem services such as water purification and soil erosion reduction) usually are not weighed against the monetary value of cutting the timber in the forest. Thus, many old-growth forests have been cleared for their timber, while their nontimber natural capital value, which can be much higher than the value of their timber, is lost. (See Science Focus 10.1, p. 224, and the next section of this chapter.) Governments can use economic tools such as subsidies and taxes to correct this market failure.

Controversy over the Sustainability of Economic Growth

Economic growth is an increase in the capacity of a nation, state, city, or company to provide goods and services to people.

Today, a typical industrialized country depends on a **high-throughput economy,** which attempts to boost economic growth by increasing the flow of matter and energy resources through the economic system to produce more goods and services (Figure 23.4). Such an economy produces valuable goods and services, but it also converts large quantities of high-quality matter and energy resources into wastes, pollution, and low-quality heat, which tend to flow into planetary *sinks* (air, water, soil, and organisms).

Economic development is any set of efforts focused on creating economies that can meet basic human needs such as food, shelter, physical and economic security, and good health. The conventional form of economic development has relied mostly on economic growth, with the goal of establishing high-throughput economies.

However, many analysts call for much greater emphasis on **environmentally sustainable economic development.** Its goal is to use political and economic systems to encourage environmentally beneficial and more sustainable

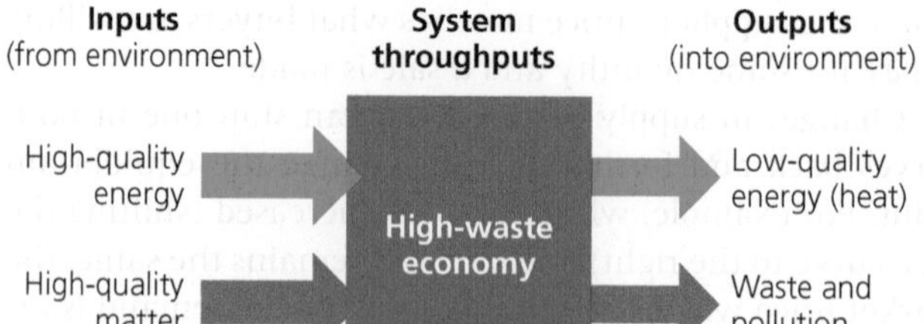

FIGURE 23.4 The *high-throughput economies* of most of the world's more-developed countries rely on continually increasing the flow of energy and matter resources to promote economic growth. ***Critical thinking:*** What are three ways in which you regularly add to this throughput of matter and energy through your daily activities?

forms of economic improvement, and to discourage environmentally harmful and unsustainable forms of economic growth. This is in keeping with the win-win **principle of sustainability**, which calls for economic solutions that benefit the environment as well as human societies (see Science Focus 22.1, p. 621). We discuss several of these possible solutions later in this chapter.

For more than 200 years, there has been a debate over whether there are limits to economic growth. *Neoclassical economists*, following the thinking of Alfred Marshall (1842–1924) and Milton Friedman (1912–2006), view the earth's natural capital as a subset, or part, of a human economic system. They assume that the potential for economic growth is essentially unlimited and is necessary for providing profits for businesses and jobs for workers. They also consider natural capital to be important but believe we can find substitutes for essentially any resource or ecosystem service that we might deplete or degrade.

Ecological economists such as Herman Daly (see his online Guest Essay on this topic) and Robert Costanza disagree with the neoclassical model. They point out that there are no substitutes for many vital natural resources, such as clean air, clean water, fertile soil, and biodiversity, or for crucial ecosystem services such as climate control, air and water purification, pollination, topsoil renewal, and nutrient cycling. In contrast to neoclassical economists, they view human economic systems as subsystems of the biosphere that depend heavily on the earth's irreplaceable natural resources and ecosystem services (**Concept 23.1**) (Figure 23.5).

Ecological economists also contend that conventional economic growth becomes unsustainable when it depletes or degrades various irreplaceable forms of natural capital, on which all human economic systems depend. What might happen if the growing human population continues to use and waste more energy and matter resources at an increasing rate in today's high-throughput economies (Figure 23.4)?

The law of conservation of matter (Chapter 2, p. 39) and the laws of thermodynamics (Chapter 2, p. 42) tell us that eventually, this resource consumption and waste will exceed the capacity of the environment to sufficiently renew those resources, to dilute and degrade waste matter, and to absorb waste heat. In addition, we have probably exceeded four major planetary boundaries or ecological tipping points (Science Focus 3.3, p. 69).

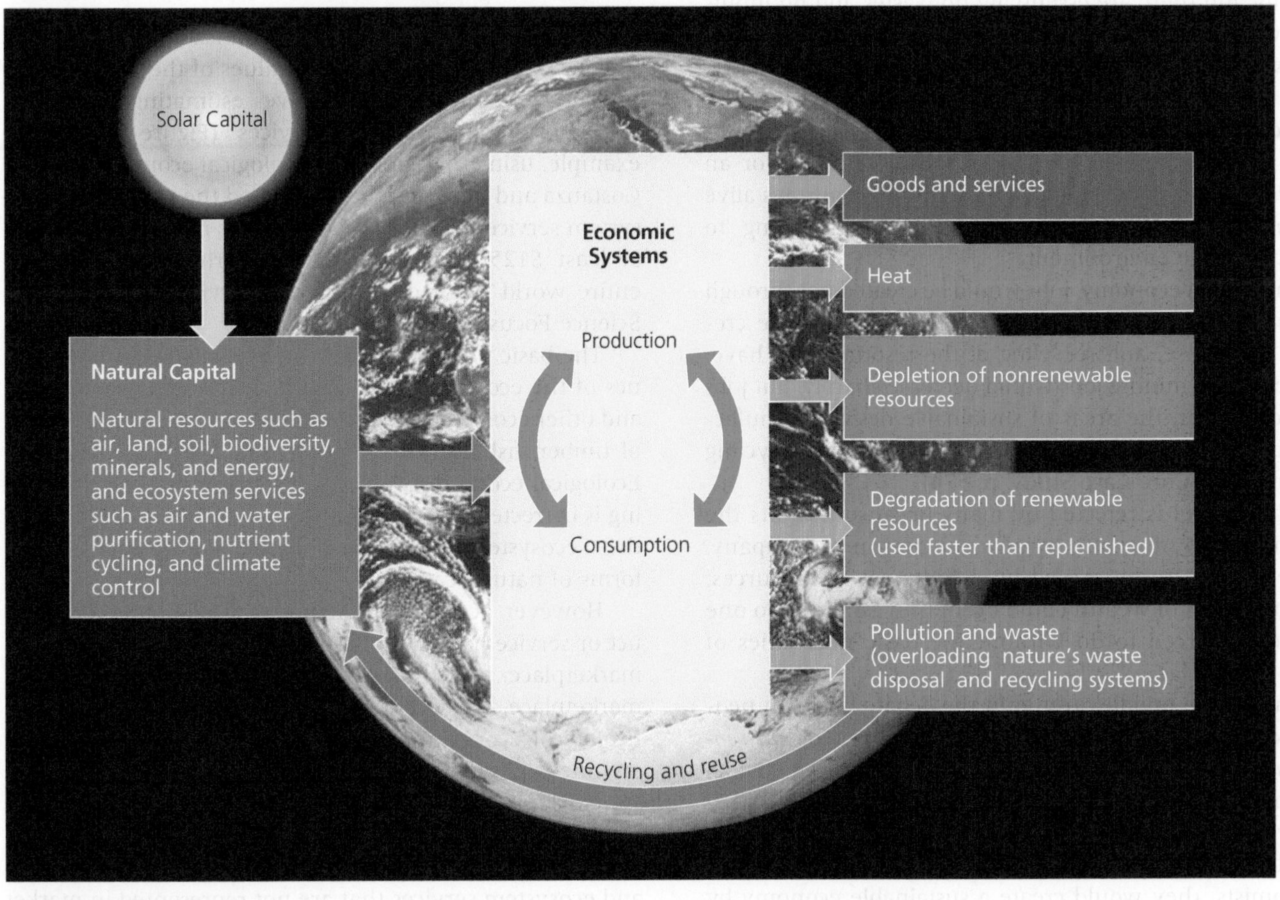

FIGURE 23.5 *Ecological economists* see all human economies as subsystems of the biosphere that depend on natural resources and ecosystem services provided by the sun and earth (**Concept 23.1**). ***Critical thinking:*** Can you think of any human activities that do not depend on natural capital? Explain.

Indeed, according to some estimates our ecological footprints are already using the renewable resources of 1.5 planet Earths and probably will be using that of 2 planet Earths by 2030.

2 Number of planet Earths needed to sustain the world's projected population and total resource use in 2030

The Steady-State Economy Model

In general, the models of ecological economists are built on the assumption that supplies of many resources are limited and that there are no substitutes for most types of natural capital. Therefore, they argue, the focus of the economy should shift from *growth* to *innovation, development, and improvement,* with the goal of achieving sustainable economies and social systems.

One model for such an economy is the *steady-state economy* in which all resources are recycled or reused continually and there are essentially no wastes and no inputs of new resources, as takes place in nature. There would be no overall growth, but energy and materials would still flow for the benefit of all, and thus economic activity would still take place. Economists such as Joseph Schumpeter have likened such a system to an organism or an ecosystem that grows to a certain size and then stays alive through natural nutrient cycling and by adapting to changes in the environment.

In such an economy, jobs would be created not through the use and depletion of resources but through the creative use, reuse, and recycling of the resources we have. For example, mining jobs would decrease sharply, but jobs would grow in the areas of sustainable design, manufacturing, delivery, reuse, remanufacturing, and recycling (Chapter 21, Core Case Study, p. 574).

This model is rejected by many because it limits the accumulation of wealth by any one person or company. For everyone to be served by a finite set of resources, accumulation of wealth could occur only so far as no one would be forced to do without the basic necessities of life.

Taking the middle ground in the debate between neoclassical economists and ecological economists are *environmental economists.* They generally agree with the model proposed by ecological economists (Figure 23.5), and they argue that some forms of economic growth are not sustainable and should be discouraged. Unlike ecological economists, they would create a sustainable economy by fine-tuning existing economic systems and tools, rather than by replacing some of them with new or redesigned systems.

CONSIDER THIS . . .

THINKING ABOUT Economic Growth

Do you think that the economy of the country where you live is sustainable or unsustainable? Explain.

23.2 HOW CAN WE ESTIMATE THE VALUES OF NATURAL CAPITAL, POLLUTION CONTROL, AND RESOURCE USE?

CONCEPT 23.2A Economists have developed several ways to estimate the present and future values of a resource or ecosystem service, and optimum levels of pollution control and resource use.

CONCEPT 23.2B Comparing the likely costs and benefits of an environmental action is useful, but it involves many uncertainties.

Ways to Value Natural Capital

Environmental and ecological economists have developed various tools for estimating the values of the earth's natural capital. One approach involves estimating the monetary worth of ecosystem services (Figure 23.6). For example, using this approach, ecological economist Robert Costanza and his colleagues estimated the value of 17 ecosystem services provided by the earth's major biomes to be at least $125 trillion per year—nearly twice what the entire world spent on goods and services in 2014. (See Science Focus 10.1, p. 224.)

The basic problem is that the estimated economic values of the ecosystem services provided by forests, oceans, and other ecosystems are not included in the market prices of timber, fish, and other goods that we get from them. Ecological economists point out that until this underpricing is corrected, we will continue our unsustainable use of these ecosystems and many of nature's other irreplaceable forms of natural capital.

However, according to neoclassical economists, a product or service has no economic value until it is sold in the marketplace, and thus because they are not sold in the marketplace, ecosystem services have no economic value. Ecological economists view this as a misleading circular argument designed to exclude serious evaluation of ecosystem services in the marketplace.

Ecological and environmental economists have developed ways to estimate *nonuse values* of natural resources and ecosystem services that are not represented in market transactions (**Concept 23.2A**). One such value is an *existence value*—a monetary value placed on a resource such as an old-growth forest or endangered species just because it

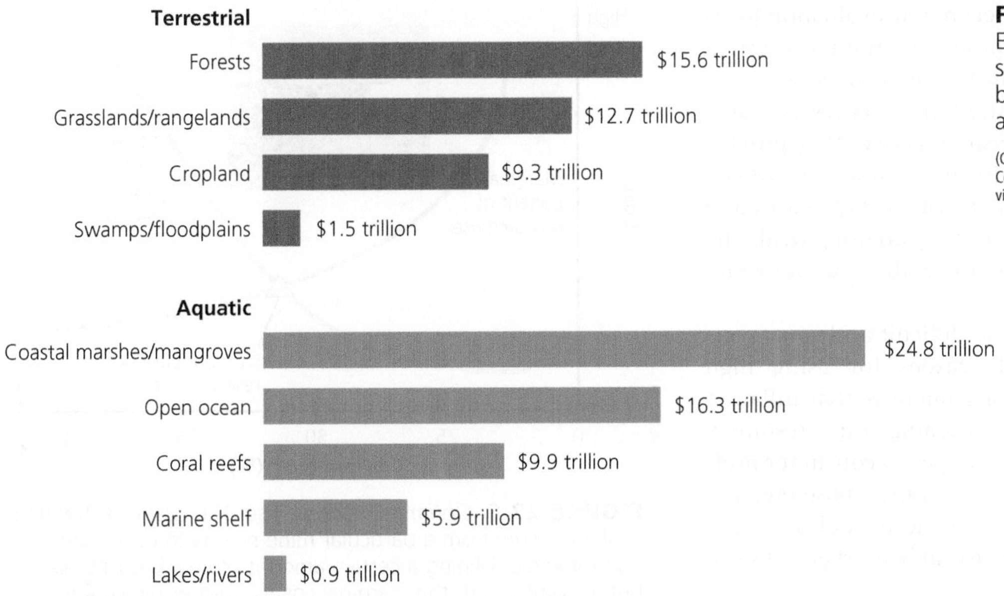

FIGURE 23.6 Natural capital: Estimated monetary value of ecosystem services provided each year by the world's major terrestrial and aquatic ecosystems.

(Compiled by the authors using data from Robert Costanza et al., "Changes in the Value of Ecosystem Services," *Global Environmental Change*, Elsevier, 2014.)

exists, even though we may never see it or use it. Another is *aesthetic value*—a monetary value placed on a forest, species, or a part of nature because of its beauty. A third type, called a *bequest* or *option value*, is based on the willingness of people to pay to protect some forms of natural capital for use by future generations, in keeping with the ethical **principle of sustainability**.

Estimating the Future Value of a Resource Is Controversial

One tool used by economists, businesses, and investors to determine the value of a resource is the **discount rate,** which is an estimate of a resource's future economic value compared to its present value (**Concept 23.2A**). It is based on the idea that today's value of a resource may be higher than its value in the future. Thus, its future value should be discounted. The size of the discount rate (usually given as a percentage) is a key factor affecting how a resource such as a forest or fishery is used or managed.

At a zero discount rate, for example, the timber from a stand of redwood trees (Figure 23.7) worth $1 million today will still be worth $1 million 50 years from now. However, the U.S. Office of Management and Budget, the World Bank, and most businesses typically use a 10% annual discount rate to estimate the future value of a resource. At this rate as the years go by, the timber in a stand of redwood trees will be worth increasingly less, and within 45 years, it will be worth less than $10,000. Using this discount rate, it makes sense from an economic standpoint for the owner of this resource to cut these trees down as quickly as possible.

However, this economic analysis does not take into account the immense economic value of the ecosystem services provided by forests (see Figure 10.2, left, p. 223).

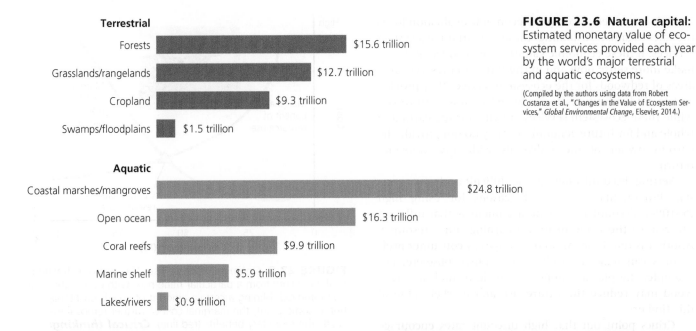

Sharon Eisenzopf/Shutterstock.com

FIGURE 23.7 Economists have tried several methods for estimating the economic value of ecosystem services, recreation opportunities, and beauty in ecosystems such as this patch of redwood forest. ***Critical thinking:*** What discount rate, if any, would you assign to this stand of trees?

Such services include the absorption of precipitation and gradual release of water and other nutrients, natural flood control, water and air purification, prevention of soil erosion, removal and storage of atmospheric carbon, and protection of biodiversity within a variety of forest habitats.

This one-sided, incomplete economic evaluation loads the dice against sustaining these important ecosystem services. If these economic values were included, it would make more sense now and in the future to preserve large areas of redwoods for the ecosystem services they provide and to find substitutes for redwood products. However, while these ecosystem services are vital for the earth as a whole and for future generations, they do not provide the current owner of the redwoods with any monetary return.

Setting discount rates can be difficult and controversial. Proponents cite several reasons for using high (5–10%) discount rates. One argument is that inflation can reduce the value of future earnings on a resource. Another is that innovation or changes in consumer preferences can make a product or resource obsolete. For example, the plastic composites made to look like redwood may reduce the future use and market value of this timber.

Critics point out that high discount rates encourage rapid exploitation of resources for immediate payoffs, thus making long-term sustainable use of most renewable natural resources virtually impossible. They argue that a 0% or even a negative discount rate should be used to protect unique, scarce, and irreplaceable resources such as old-growth forests. A negative discount rate would result in the value of a forest or other resource *increasing* over time. Some economists argue that as ecosystem services continue to be degraded, they will only become more valuable, so a negative discount rate is the only type that makes sense. They point out that zero or negative discount rates of 1–3% would make it profitable to use nonrenewable and renewable resources more slowly and in more sustainable ways.

CONSIDER THIS . . .

THINKING ABOUT Discount Rates

If you owned a forested area, would you want the discount rate for resources such as trees from the forest to be high, moderate, or zero? Explain.

Estimating Optimum Levels of Pollution Control and Resource Use

An important concept in environmental economics is that of *optimum levels* for pollution control and resource use (**Concept 23.2A**). In the early days of a new coal mining operation, for example, the cost of extracting coal is typically low enough to make it easy for developers to recover their investments by selling their product. However, the cost of removal goes up with each additional unit of coal taken. Economists refer to this as the **marginal cost**—any increase in the cost of producing an additional unit of a product. After most of the more readily accessible coal has been removed from a mine, the marginal cost is too high and taking what is left becomes unaffordable. This can change if some factor such as scarcity raises the value of the coal remaining in the mine.

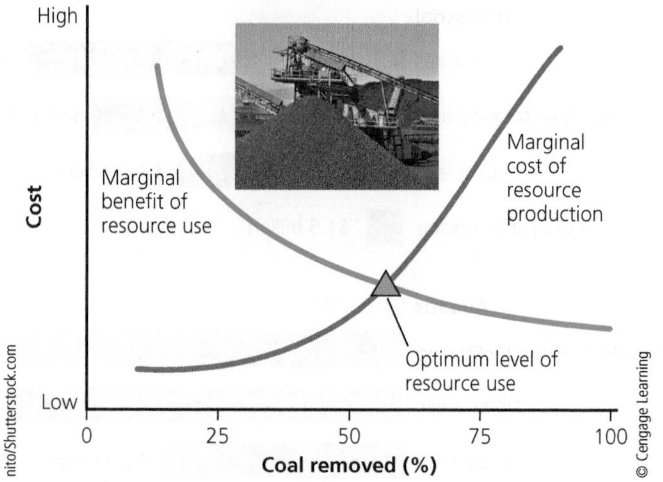

FIGURE 23.8 *Optimum resource use:* The cost of extracting coal (blue line) from a particular mine rises with each additional unit removed. Mining a certain amount of coal is profitable, but at some point, the marginal cost of further removal exceeds the monetary benefits (red line). ***Critical thinking:*** How would the location of the optimum level of resource use shift if the price of coal doubled?

Figure 23.8 shows this in terms of supply, demand, and equilibrium. The point at which removing more coal is not worth the marginal cost is where the demand curve crosses the supply curve, theoretically the *optimum level of resource use*.

You might think that the best solution for pollution is total cleanup. In fact, there are optimum levels for various kinds of pollution. This is because the cost of pollution control goes up for each additional unit of a pollutant removed from the environment. This increase in cost per additional unit is called the **marginal cost**. The main reason for this is that, as concentrations of a pollutant from the air, water, or soil get lower, it takes larger amounts of energy to remove the pollutant. At some point, the cost of removing more pollutants is greater than the harmful costs of the pollution to society. That point is the equilibrium point, or the *optimum level* for pollution cleanup.

Cost–Benefit Analysis

Another widely used tool for making economic decisions about how to control pollution and manage resources is **cost–benefit analysis.** In this process, analysts compare estimated costs and benefits of actions such as implementing a pollution control regulation, building a dam on a river, and preserving an area of forest. Economists also use cost–benefit analysis to estimate the optimum level of pollution cleanup or resource use (Figure 23.8).

Making a cost–benefit analysis involves determining who benefits and who is harmed by a particular regulation or project and estimating the monetary values (costs) of those benefits and harms. Direct costs involving land, labor, materials, and pollution-control technologies are often easy to estimate. However, estimates of indirect costs, such as a project's effects on air and water, are not considered in the marketplace. We can put estimated price tags on human life, good health, clean air and water, and natural capital such as an endangered species, a forest, or a wetland. However, such monetary value estimates vary widely depending on the assumptions, value judgments, and discount factors used by the estimators.

Because of these drawbacks, a cost–benefit analysis can lead to a wide range of benefits and costs with a lot of room for error, and this is a source of controversy. For example, one cost–benefit analysis sponsored by a U.S. industry estimated that compliance with a regulation written to protect American workers from vinyl chloride would cost $65 billion to $90 billion. In the end, complying with the regulation cost the industry less than $1 billion. A study by the Economic Policy Institute of Washington, D.C., found that the estimated costs projected by industries for complying with proposed U.S. environmental regulations are often inflated. Such distorted analyses have been used as a political ploy by industries to avoid or delay complying with regulations.

If conducted fairly and accurately, cost–benefit analysis can be a helpful tool for making economic decisions, but it always includes uncertainties (**Concept 23.2B**). Environmental economists advocate using the following guidelines in order to minimize possible abuses and errors in any cost–benefit analysis involving some part of the environment:

- Clearly state all assumptions used.
- Include estimates of the ecosystem services provided by the ecosystems involved.
- Estimate short- and long-term benefits and costs for all affected population groups.
- Compare the costs and benefits of alternative courses of action.

CONSIDER THIS . . .

THINKING ABOUT Cost–Benefit Analysis

Can you think of any other guidelines that could help to make cost–benefit analyses accurate and fair, both for the parties involved and for the environment? List them.

23.3 HOW CAN WE USE ECONOMIC TOOLS TO DEAL WITH ENVIRONMENTAL PROBLEMS?

CONCEPT 23.3 We can use resources more sustainably by including the harmful environmental and health costs of producing goods and services in their market prices (*full-cost pricing*), by subsidizing environmentally beneficial goods and services, and by taxing pollution and waste instead of wages and profits.

Full-Cost Pricing

The *market price*, or *direct price*, that we pay for a product or service usually does not include all of the *indirect*, or *external*, *costs* of harm to the environment and human health associated with its production and use. Such costs are often called *hidden costs*.

For example, if you buy a car, the price you pay includes the *direct*, or *internal*, *costs* of raw materials, labor, shipping, and a markup for dealer profit. In using the car, you pay additional direct costs for gasoline, maintenance, repairs, and insurance. However, in order to extract and process raw materials to make a car, manufacturers use energy and mineral resources, produce solid and hazardous wastes, disturb land, pollute the air and water, and release greenhouse gases into the atmosphere. These are the hidden external costs that can have harmful effects on us, on future generations, on our economies, and on the earth's life-support system.

Because these harmful external costs are not included in the market price of a car, most people do not connect them with car ownership. Still, the car buyer and other people in a society pay these hidden costs sooner or later, in the forms of poorer health, higher expenses for health care and insurance, higher taxes for pollution control, traffic congestion, and environmental degradation. Similarly, when a farmer practices high-input industrialized farming, or a power plant pollutes the air or the water in a river, those costs are also not added to your food prices or your electric bill.

Many economists and environmental experts call for including these external costs in the market prices of goods—a practice called **full-cost pricing,** and the basis for one of the **principles of sustainability**. They cite this failure to include the harmful environmental costs in the market prices of goods and services as one of the major causes of the environmental problems we face (Chapter 1, p. 18).

These experts argue that if such costs were included, environmentally harmful goods and services would be used far less than they are now, or not at all. For example, if the harmful environmental and health costs of mining

FIGURE 23.9 Most of the harmful environmental and health effects of strip mining coal and burning it to produce electricity are not included in the cost of electricity.

and burning coal to produce electricity (see Figure 15.17, p. 394, and Figure 23.9) were included in the market prices of coal and coal-fired electricity, coal would be too expensive to use and would be replaced by less environmentally harmful resources such as solar and wind power.

According to its proponents (Individuals Matter 23.1), full-cost pricing would reduce resource waste, pollution, and environmental degradation and improve human health by encouraging producers to invent more resource-efficient and less-polluting methods of production. It would also inform consumers about the environmental and health effects of the goods and services they buy.

Implementation of full-cost pricing would result in some industries and businesses disappearing or remaking themselves, while new ones would appear—a normal and revitalizing process in a dynamic and creative capitalist economy. If we were to phase in a shift to full-cost pricing over a decade or two, some environmentally harmful businesses would have time to transform themselves into environmentally beneficial businesses. This has already happened in Germany (**Core Case Study**), where the coal producer Vattenfall is trying to sell its coal business. The company has invested heavily in wind power and, in 2015, opened its first offshore wind farm.

Shifting from Environmentally Harmful to Environmentally Beneficial Subsidies

One of the main obstacles to full-cost pricing is the fact that many environmentally harmful businesses have used their political and economic power to obtain government subsidies and tax breaks that allow them to keep their prices lower than those of their unsubsidized competitors. Subsidies that lead to environmental damage or harmful health effects are called **perverse subsidies**—a term coined by environmental scientist Norman Myers.

For example, some fishing fleets have received government subsidies to help them boost their catches, thereby bolstering seafood supplies. However, some fisheries, including Newfoundland's Atlantic cod fishery, were then overfished and they collapsed because the subsidies enabled too many fishing fleets to take too many fish (see Figure 11.10, p. 263).

INDIVIDUALS MATTER 23.1

Paul Hawken: Businessman and Environmental Champion

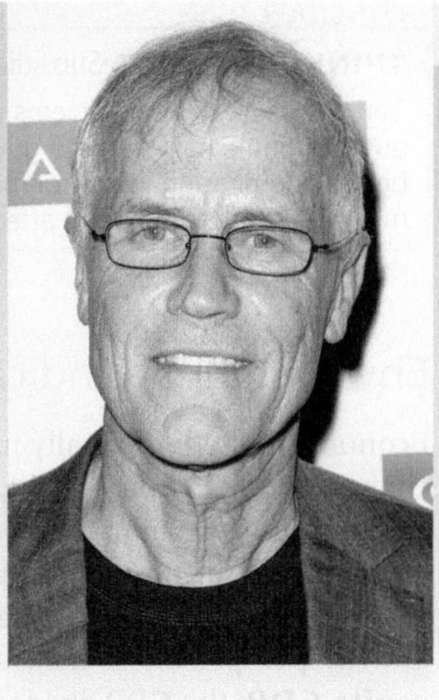
Beck Star/WireImage/Getty Images

Paul Hawken understands both business and ecology. He is an entrepreneur and a visionary environmental and social activist. In addition to starting several businesses, he has authored several widely acclaimed books that have been published in over 50 countries in 27 languages and have sold more than 2 million copies.

One of Hawken's major themes has been the importance of full-cost pricing. As Hawken has pointed out in many of his writings, the fact that many harmful environmental and health costs are externalized is a major cause of the global loss and degradation of natural capital. This happens because of a failure to implement full-cost pricing and an obsession for the growth of gross domestic product (GDP) regardless of its effect on the environment. With our current pricing system, Hawken says, "we are stealing the future, selling it in the present, and calling it GDP, and patting ourselves on the back."

Hawken calls for us to modify our economies in ways that will sustain the natural capital that in turn sustains all life and economies. He is not against economic growth. Instead, he calls for using government subsidies and taxes to encourage forms of growth that increase environmental sustainability and social justice and to discourage harmful forms of growth.

According to Hawken, "We have the capacity to create a remarkably different economy: one that can restore ecosystems and protect the environment while bringing forth innovation, prosperity, meaningful work, and true security." This shift "is based on the simple but powerful proposition that *all natural capital must be valued*. ... If we have doubts about how to value a 500-year-old tree, we need only ask how much would it cost to make a new one from scratch? Or a new river? Or a new atmosphere?"

Hawken has worked with business and government leaders throughout the world and won numerous awards for his work. However, his greatest accomplishment may be getting many of us to rethink our ideas about economics, business, and the environment.

Myers, who has done an extensive study on perverse subsidies, points to six sectors of the economy—agriculture, fossil fuels, highway transportation, water supplies, forestry, and fisheries—as the areas where perverse subsidies and tax breaks have been extensively awarded and have caused severe environmental damage. Other examples include depletion subsidies and tax breaks for extracting minerals and fossil fuels, cutting timber on public lands, and irrigating with low-cost water. These subsidies and tax breaks distort the economic playing field and create a huge economic incentive for unsustainable resource waste, depletion, and environmental degradation.

Myers estimates that these perverse subsidies to these sectors total at least $2 trillion per year, globally, which amounts to $3.8 million a minute. This amount is larger than all but a few of the national economies in the world and twice as large as all military spending. It dwarfs the estimated $20 billion per year that is spent by governments worldwide on protecting natural areas.

A number of environmental scientists and economists call for phasing out such perverse subsidies and tax breaks. However, the economically and politically powerful interests receiving them spend a lot of time and money *lobbying*, or trying to influence governments to continue and even to increase their subsidies. For example, the fossil fuel and nuclear power industries in the United States are mature and highly profitable industries that get billions of dollars in government subsidies and tax breaks every year. Such industries also lobby against subsidies and tax breaks for their more environmentally beneficial competitors such as solar and wind energy.

Some countries have reduced perverse subsidies. Japan, France, and Belgium have phased out all coal subsidies, and Germany plans to do so by 2018. China has cut coal subsidies by about 73% and has imposed a tax on high-sulfur coals. GOOD NEWS

Subsidies can also be used for environmentally beneficial purposes. They could be awarded to companies that are involved in pollution prevention, sustainable forestry and agriculture, conservation of water supplies, energy efficiency improvements, and renewable energy use (**Concept 23.3**). The German government (**Core Case Study**) has subsidized thousands of rooftop solar panels (Figure 23.1, right). Making such subsidy shifts on a broad basis, globally, would encourage businesses to make the transition from environmentally harmful to more environmentally beneficial goods and services, as is happening now in Germany.

CONSIDER THIS . . .

THINKING ABOUT Subsidies

Can you think of any problems that might result from phasing out environmentally harmful government subsidies and tax breaks and phasing in environmentally beneficial ones? How might such subsidy shifting affect your lifestyle?

Environmental Indicators

Economic growth is usually measured by the percentage of change per year in a country's **gross domestic product (GDP)** : the annual market value of all goods and services produced by all firms and organizations, foreign and domestic, operating within a country. A country's economic growth per person is measured by changes in the **per capita GDP:** the GDP divided by the country's total population at midyear.

GDP and per capita GDP indicators provide a standardized, useful method for measuring and comparing the economic outputs of nations. However, the GDP was deliberately designed to measure such outputs without taking into account their beneficial or harmful environmental or health impacts. Environmental economists and scientists call for the development and widespread use of new indicators—called **environmental indicators**—to help monitor environmental quality and human well-being.

One such indicator is the *genuine progress indicator (GPI)*—the GDP plus the estimated value of beneficial transactions that meet basic needs, minus the estimated harmful environmental, health, and social costs of all transactions. Examples of beneficial transactions included in the GPI are unpaid volunteer work, health care provided by family members, child care, and housework. Harmful costs that are subtracted to arrive at the GPI include the costs of pollution, resource depletion and degradation, and crime.

Figure 23.10 compares the per capita GDP and GPI for the United States between 1950 and 2004 (the latest data available). While the per capita GDP rose sharply over this period, the per capita GPI stayed flat, or in some cases even declined slightly. This shows that even if a nation's economy is growing, its people are not necessarily better off. Environmental economists developed the GPI with the hope that governments would adopt it. However, it has not yet been implemented by any of the world's economies.

The Global Green Economy Index (GGEI) measures the performances of 60 nations in areas of leadership on climate change, energy efficiency, markets and investments, and natural capital, based on analysis by a panel of experts. It was launched in 2010 by the environmental consulting firm Dual Citizen, and has become increasingly important to policy makers, international organizations, and some multinational corporations. In 2014, the top-five-ranked countries on the GGEI were Sweden,

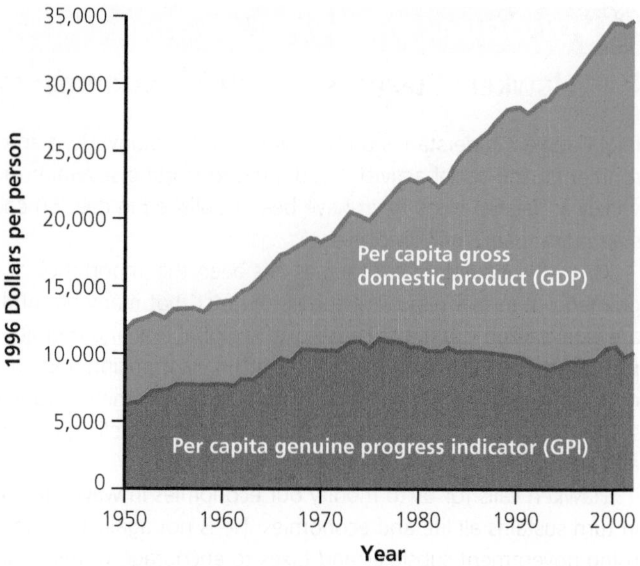

FIGURE 23.10 *Monitoring environmental progress:* The per capita gross domestic product (GDP) compared with the per capita genuine progress indicator (GPI) in the United States between 1950 and 2004 (latest available data). ***Critical thinking:*** Would you favor making widespread use of this or similar green economic indicators? Why do you think this has not been done?

(Compiled by the authors using data from Redefining Progress.)

Norway, Costa Rica, Germany, and Denmark. The United States ranked 28th.

These and other environmental indicators now being developed are far from perfect. However, without such indicators, it will be difficult to monitor the overall effects of human activities on human health, on the environment, and on the planet's natural capital. Such indicators are also helpful for finding the best ways to improve environmental quality and life satisfaction.

Taxing Pollution and Wastes Instead of Wages and Profits

Another way to discourage pollution and resource waste is to tax them (**Concept 23.3**). *Green taxes* could be levied on a per-unit basis on the amount of pollution and hazardous waste produced by a farm, business, or industry, and on the use of fossil fuels, nitrogen fertilizer, timber, minerals, water, and other resources. This approach would help us to implement the full-cost pricing **principle of sustainability** and increase our beneficial environmental impact.

To many analysts, the tax systems in most countries are backward. They discourage what we want more of—jobs, income, and profit-driven innovation—and encourage what we want less of—pollution, resource waste, and environmental degradation. A more environmentally sustainable economic and political system would lower taxes on labor, income, and wealth, and raise taxes on environmental activities that produce pollution, wastes, and environmental degradation. Some 2,500 economists, including eight Nobel Prize winners in economics, have endorsed this *tax-shifting* concept.

Proponents point out three requirements for the successful implementation of green taxes. *First*, they would have to be phased in over 10 to 20 years so businesses can plan ahead. *Second*, income, payroll, or other taxes would have to be reduced by an amount equal to that of the green tax so that there would be no net increase in taxes. *Third*, the poor and lower-middle class would need a safety net to reduce the regressive nature of any new taxes on essentials such as fuel, water, electricity, and food. Figure 23.11 lists some of the advantages and disadvantages of using green taxes.

In Europe and the United States, polls indicate that once such tax shifting is explained to voters, 70% of them support the idea. Germany's green tax on fossil fuels (**Core Case Study**), introduced in 1999, has reduced pollution and greenhouse gas emissions, helped to create up to 250,000 new jobs, lowered taxes on wages, and greatly increased the use of renewable energy resources. Costa Rica, Sweden, Denmark, Spain, and the Netherlands have raised taxes on several environmentally harmful activities while cutting taxes on income, wages, or both.

GOOD NEWS

To help reduce carbon dioxide emissions, since 1997, Costa Rica has imposed a 3.5% tax on the market values of any fossil fuels that are burned in the country. The tax revenues go into a national forest fund set up for paying indigenous communities to help protect the forests around

Trade-Offs

Environmental Taxes and Fees

Advantages	Disadvantages
Help bring about full-cost pricing	Low-income groups are penalized unless safety nets are provided
Encourage businesses to develop environmentally beneficial technologies and goods	Hard to determine optimal level for taxes and fees
Easily administered by existing tax agencies	If set too low, wealthy polluters can absorb taxes as costs

FIGURE 23.11 Trade-offs: Using green taxes to help reduce pollution and resource waste has advantages and disadvantages. ***Critical thinking:*** Which single advantage and which single disadvantage do you think are the most important? Why?

Top: chuong/Shutterstock.com. Bottom: EduardSV/Shutterstock.com.

them, thereby helping to reverse deforestation (Chapter 10, Core Case Study, p. 222). The fund is also intended to help them work their way out of poverty. Costa Rica has also taxed water use to reduce water waste and pollution, and the tax revenues are used to pay villagers living upstream to reduce their inputs of water pollutants.

The U.S. Congress has not enacted green taxes, mostly because powerful industries, including the automobile, fossil fuel, mining, and chemical industries, claim that such taxes will reduce their competitiveness. They argue this will harm the economy and consumers by forcing producers to raise the prices of their goods and services. In addition, most voters have been conditioned to oppose any new taxes and have not been educated about the economic and environmental benefits of a tax-shifting approach.

Labeling Environmentally Beneficial Goods and Services

Product eco-labeling and *certification* can encourage companies to develop environmentally beneficial (green) products and services and can help consumers to select such products and services. Eco-labeling programs have been developed in Europe, Japan, Canada, and the United States. The U.S. *Green Seal* labeling program has certified more than 335 products and services as environmentally friendly based on life-cycle analysis (see Figure 14.7, p. 365). Eco-labels are also used to identify fish caught by sustainable methods (certified by the Marine Stewardship Council) and to certify timber produced and harvested by sustainable methods (evaluated by organizations such as the Forest Stewardship Council; see Science Focus 10.2, p. 232).

Eco-labeling systems usually include a simple rating scale such as 0–10, applied to factors such as environmental damage, climate impact, carbon footprint, air and water pollution, and energy, water, and pesticide use. Such eco-labeling informs consumers about the environmental impacts of what they buy, helping them to vote with their wallets.

Providing such easily understandable ratings on the sustainability of goods and services helps to expose and reduce **greenwashing,** a deceptive practice that some businesses use to spin environmentally harmful products and services as green, clean, or environmentally beneficial. For example, in 2008, the U.S. coal industry spent about $45 million on a successful public relations campaign to imbed the words "clean coal" in the minds of Americans, even though certain harmful aspects of mining and using coal will always make it by far the dirtiest fossil fuel (Chapter 15, p. 391, and Figure 23.9).

Other examples of greenwashing, closer to home for most people, can mislead consumers and distort market information, making it harder for environmentally beneficial products and services to compete. For example, phrases

Charles Kogod/National Geographic Creative

FIGURE 23.12 Environmental regulations have helped us to preserve irreplaceable resources such as this mountainous National Wilderness Area near Aspen, Colorado.

like "environmentally friendly" and "eco-conscious" placed on cleaning product labels can be meaningless or false. There is no law that prohibits the use of such phrases in most markets today. Consumers who want to buy green must be careful to choose products that actually are environmentally friendly.

Environmental Laws and Regulations

Environmental regulation is a form of government intervention in the marketplace that is widely used to help control or prevent pollution and environmental degradation and to encourage more efficient use of resources. It involves enacting and enforcing laws that set pollution standards, regulate the release of toxic chemicals into the environment, and protect certain slowly replenished resources such as public forests, parks, and wilderness areas (Figure 23.12) from unsustainable use.

Such regulation is another way to help implement the full-cost pricing **principle of sustainability**, because it forces companies to include more of the costs of pollution control and other regulated aspects in the prices of their products. For this reason, opponents of regulation claim that it hurts business, especially in markets where they are competing with China and other countries whose regulations are not as strong as those of the United States.

However, proponents of regulation point to the results of China's lax environmental regulations. While that country's economy has been growing rapidly since 1980, its environmental problems have also multiplied dramatically. Now, according to the Chinese Academy of Sciences, its major cities suffer from serious air pollution. About 57% of its urban groundwater, used for drinking water for hundreds of millions of people, and 43% of its surface water is too polluted to use. Its topsoil is severely polluted and some of its food is tainted with harmful chemicals. These problems are leading to civil unrest in China, as well as to a less favorable standing in the global marketplace.

So far, most environmental regulation in the United States and in many other countries has involved passing laws that are typically enforced through a *command-and-control* approach. Critics say that this strategy can unnecessarily increase costs and discourage innovation, because many of these government regulations concentrate on cleanup instead of prevention. Some regulations also set compliance deadlines that are often too short to allow companies to find innovative solutions to reducing pollution and waste.

A different approach favored by many economists and environmental and business leaders is to use *incentive-based environmental regulations*. Rather than requiring all companies in a particular market to follow the same fixed procedures or use the same technologies, this approach uses the economic forces of the marketplace to encourage businesses to be innovative in reducing pollution and resource waste.

Several European nations use such *innovation-friendly environmental regulation*, which involves setting goals, freeing industries to meet the goals in any way that works, and allowing enough time for innovation. This has motivated several companies to develop green products and industrial processes that have created jobs. It has also helped some companies to boost their profits while becoming more competitive in national and international markets. GOOD NEWS

Using the Marketplace to Reduce Pollution and Resource Waste

In one incentive-based regulation system, the government decides on acceptable levels of total pollution or resource use; sets limits, or *caps*, to maintain these levels; and gives or sells companies a certain number of *tradable pollution* or *resource-use permits* governed by the caps.

With this *cap-and-trade* approach, a permit holder that does not use its entire allocation can save credits for future expansion, use them in other parts of its operation, or sell them to other companies. The United States has used this approach to reduce the emissions of sulfur dioxide (see Chapter 18, p. 492) and other air pollutants. Tradable rights could also be established among countries to help preserve biodiversity and reduce emissions of greenhouse gases and other regional and global pollutants. GOOD NEWS

Figure 23.13 lists the advantages and disadvantages of using tradable pollution and resource-use permits. The effectiveness of such programs depends on how high or low the initial cap is set and on the rate at which the cap is regularly reduced to encourage further innovation.

Trade-Offs

Tradable Environmental Permits

Advantages	Disadvantages
Flexible and easy to administer	Wealthy polluters and resource users can buy their way out
Encourage pollution prevention and waste reduction	Caps can be too high and not regularly reduced to promote progress
Permit prices to be determined by market transactions	Self-monitoring of emissions can allow cheating

© Cengage Learning

FIGURE 23.13 Trade-offs: Using tradable pollution and resource-use permits to reduce pollution and resource waste has advantages and disadvantages. ***Critical thinking:*** Which two advantages and which two disadvantages do you think are the most important? Why?

Top: M. Shcherbyna/Shutterstock.com.

Selling Services Instead of Products

One approach to working toward more environmentally beneficial economies is to sell certain services in place of the products that provide those services. With this approach, a manufacturer or service provider makes more money if the production of its product involves minimal material use and pollution, and if the product lasts, is energy efficient, produces as little pollution as possible while in use, and is easy to maintain, repair, reuse, or recycle (see Chapter 21, Core Case Study, p. 574).

Such an economic shift is under way in some businesses. Since 1992, Xerox has been leasing most of its copy machines as part of its mission to provide *document services* instead of selling photocopiers. When a customer's service contract expires, Xerox takes the machine back for reuse or remanufacture. It has a goal of sending no material to landfills or incinerators. To save money, Xerox designs machines to have the fewest possible parts, be energy efficient, and emit as little noise, heat, ozone, and chemical waste as possible.

CONSIDER THIS . . .

LEARNING FROM NATURE

At the flooring service company Interface, engineers studied the floors of tropical forests to design a best-selling, nature-based carpet pattern that reduces carpet waste and installation time.

In Europe, Carrier has begun shifting from selling heating and air conditioning equipment to providing indoor heating and cooling services. The company makes higher profits by leasing and installing energy-efficient equipment that is durable and easy to rebuild or recycle. Carrier also makes money through helping clients to save energy by adding insulation, eliminating heat losses, and boosting energy efficiency in their offices and homes.

CONSIDER THIS . . .

THINKING ABOUT Selling Services

Can you think of any drawbacks to leasing a service provided by a product instead of buying the product? What service or services would you consider leasing?

23.4 HOW CAN REDUCING POVERTY HELP US TO DEAL WITH ENVIRONMENTAL PROBLEMS?

CONCEPT 23-4 Reducing poverty can help us to reduce population growth, resource use, and environmental degradation.

James P. Blair/National Geographic Creative

FIGURE 23.14 This 3-year-old girl was sleeping in her family's shack in a slum in Port-au-Prince, Haiti.

We Can Reduce Poverty

Poverty is defined as the condition under which people cannot meet their basic economic needs. According to the World Bank, poverty is the way of life for nearly half of the world's people who have to live on incomes equivalent to less than $2.25 per day. One fifth of the world's people live in extreme poverty (Figure 23.14), struggling to survive on incomes of less than $1.25 a day or on no income at all.

Some analysts are alarmed at the widening gap between rich and poor countries and between super-rich individuals and the rest of the world. In 2016, Oxfam reported that the wealthiest 62 people in the world had as much wealth as the poorest 3.5 billion people—almost half the world's population. Some economists say that part of this wealth will trickle down to the poor and middle class. Others point out that for almost three decades, instead of trickling down, most wealth has been flowing up to rich individuals, corporations, and countries.

Poverty can have severely harmful health effects (see Figure 17.19, p. 463) and has been identified as one of the five major causes of the environmental problems we face. To reduce poverty and its harmful effects, governments, businesses, international lending agencies, and wealthy individuals could undertake the following:

- Mount a massive global effort to combat malnutrition and the infectious diseases that kill millions of people prematurely.
- Provide universal primary school education for all children and for the world's nearly 800 million illiterate adults. Illiteracy can foster terrorism and strife within countries by contributing to the creation of large numbers of unemployed individuals who have little hope of improving their lives or those of their children.
- Provide assistance to help less-developed countries reduce their population growth, mostly by investing in family planning, reducing poverty, and elevating the social and economic status of women.
- Focus on sharply reducing the total and per capita ecological footprints of more-developed countries such as the United States and rapidly growing less-developed countries such as China and India.
- Make large investments in small-scale infrastructure such as solar-cell power facilities for rural villages and sustainable agriculture projects to help less-developed nations work toward more energy-efficient and environmentally beneficial economies.
- Encourage lending agencies to make small loans to poor people who want to increase their income (see the Case Study that follows).

One example of such action is the work of ecologist and National Geographic Explorer Sasha Kramer. She has been working in the impoverished, ecologically degraded nation of Haiti to attack the problems of hunger, soil depletion, and water pollution all at once. Her nonprofit organization has distributed waterless composting toilets throughout the country to collect human wastes and transform them into compost, which Haitian farmers can use to rebuild depleted soil and boost food production. This process also keeps human wastes out of Haiti's water supply and reduces the dangerous threat of waterborne infectious diseases. GOOD NEWS

CASE STUDY

Microlending

Most of the world's poor people want to work their way out of poverty. With loans, they could buy what they would need to start farms or small businesses. But few of them have credit records or assets that they could use as collateral to secure loans.

For over three decades, an innovation called *microlending*, or *microfinance*, has helped a number of people living in poverty to deal with this problem. In 1983, economist Muhammad Yunus started the Grameen (Village) Bank in Bangladesh, a country with a high poverty rate and a rapidly growing population. Unlike commercial banks, the Grameen Bank is essentially owned and run by borrowers and by the Bangladeshi government. Since it was founded, the bank has provided more than \$8 billion in microloans of \$50 to \$500 at low interest rates to more than 7 million impoverished people in Bangladesh who do not qualify for loans at traditional banks.

National Renewable Energy Laboratory

FIGURE 23.15 A microloan helped these women in a rural village in India to buy a small solar-cell panel (installed on the roof behind them) that provides electricity to help them make a living, thus applying the solar energy **principle of sustainability**.

About 97% of these loans have been used by women, mostly to start small businesses, to plant crops, to buy small irrigation pumps, to buy cows and chickens for producing and selling milk and eggs, or to buy bicycles for transportation. Grameen Bank microloans are also being used to develop day-care centers, health-care clinics, reforestation projects, drinking water supply projects, literacy programs, and small-scale solar- and wind-power systems in rural villages (Figure 23.15).

To promote loan repayment, the bank puts borrowers into groups of five. If a group member fails to make a weekly payment, other members must pay it. The average repayment rate on its microloans has been 95% or higher—nearly twice the average repayment rate for loans by conventional commercial banks—and the bank consistently has made a profit. Typically, about half of Grameen's borrowers move above the poverty line within 5 years of receiving their loans.

Between 1975 and 2005, the Grameen Bank's innovative approach helped to reduce the poverty rate in Bangladesh from 74% to 40%, primarily because of the hard work of the people receiving the microloans. In addition, birth rates are lower among most of the borrowers, a majority of whom are women, and as a result, they gain more freedom and control over their lives. GOOD NEWS

One of the bank's goals was to help protect borrowers from loan sharks who were charging high interest rates and bankrupting many people. Unfortunately, some loan sharks and commercial companies have moved into the microfinance sector and turned it to their advantage, which has given microlending a bad name in some areas.

However, Yunus and his supporters argue that microlending, when done properly, can help people to escape poverty and improve their lives. In 2006, Yunus and his colleagues at the bank jointly won the Nobel Peace Prize for their pioneering use of microcredit loans. He has stated, "Unleashing the energy and creativity in each human being is the answer to poverty." Banks based on the Grameen microcredit model have spread to 58 countries (including the United States) with an estimated 500 million participants.

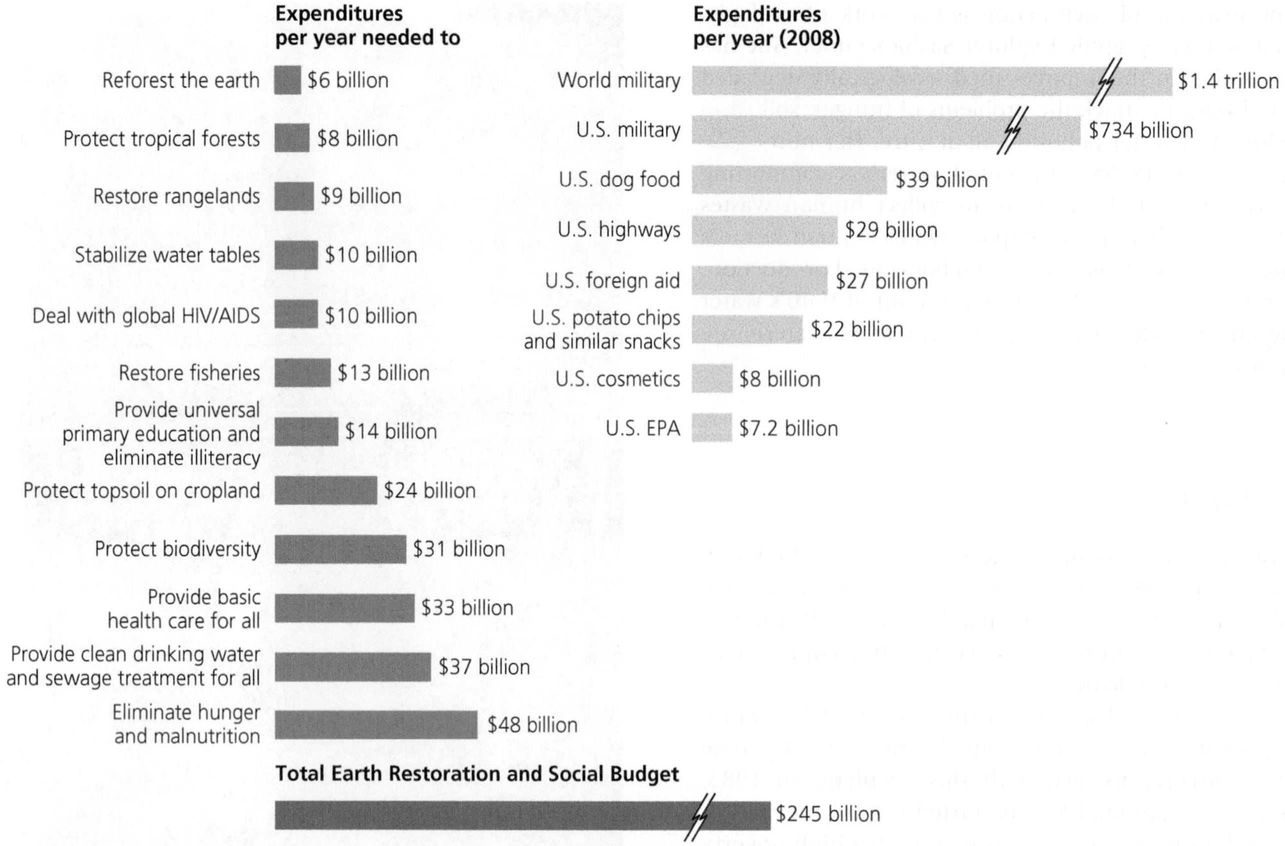

FIGURE 23.16 What should our priorities be? ***Critical thinking:*** Which item on the right side of the figure would you do without or reduce to pay for solving some of the problems listed on the left side of the figure?

(Compiled by the authors using data from United Nations, World Health Organization, U.S. Department of Commerce, U.S. Office of Management and Budget, World Bank, Earth Policy Institute, and Stockholm International Peace Research Institute.)

Achieving the Millennium Development Goals

In 2000, the world's nations set goals—called *Millennium Development Goals*—for sharply reducing hunger and poverty, improving health care, achieving universal primary education, empowering women, and moving toward environmental sustainability by 2015. That year, the United Nations published its Progress Chart showing highly mixed results across goals and countries. For example, most countries did quite well with the goal of expanding primary education while women's representation in national parliaments did not improve in most places. And many countries did well to bring clean drinking water to all of their citizens while some countries did very poorly.

More-developed countries pledged to donate 0.7%—or $7 of every $1,000—of their annual national income to less-developed countries to help them in achieving these goals. Only five countries—Denmark, Luxembourg, Sweden, Norway, and the Netherlands—donated what they had promised. In fact, the average amount donated in most years has been 0.25% of national income. The United States—the world's richest country—gave only 0.16% of its national income and Japan gave only 0.18%. For any country, deciding whether or not to help poorer countries in this way is an ethical issue that requires individuals and nations to evaluate their priorities (Figure 23.16).

23.5 HOW CAN WE MAKE THE TRANSITION TO MORE ENVIRONMENTALLY SUSTAINABLE ECONOMIES?

CONCEPT 23.5 We can use the principles of sustainability, as well as various economic and environmental strategies, to develop more environmentally sustainable economies.

Low-Throughput Economies

The three scientific laws governing matter and energy changes (Chapter 2, pp. 39 and 42) and the six **principles of sustainability** (see Inside Back Cover)

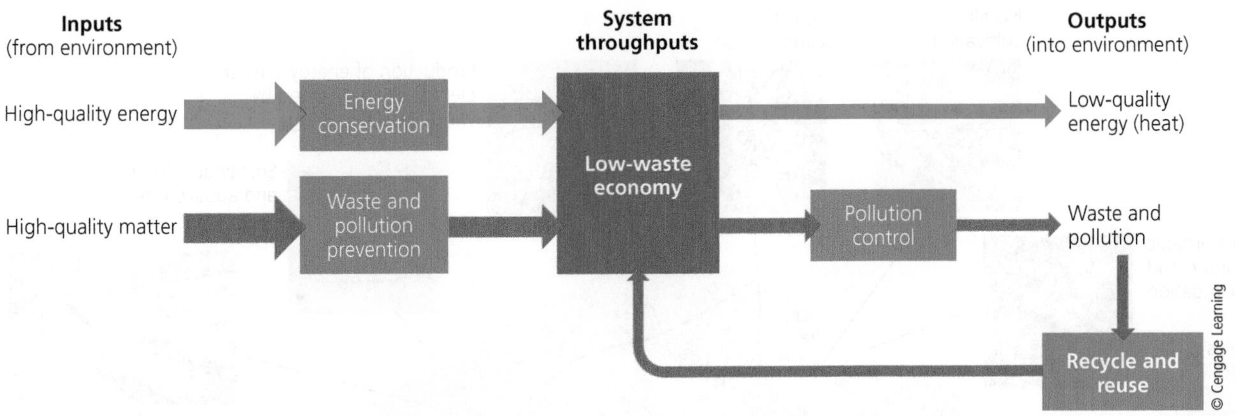

FIGURE 23.17 Solutions: Learning and applying lessons from nature can help us design and manage more sustainable low-throughput economies. ***Critical thinking:*** What are three ways in which your school could decrease any unsustainable economic and environmental practices, and three ways that it could promote more sustainable economic and environmental practices?

suggest that the best long-term solution to our environmental and resource problems is to shift away from a high-throughput (high-waste) economy based on ever-increasing matter and energy flow (Figure 23.4). The goal would be to develop a **low-throughput (low-waste) economy**—an economic system based on energy efficiency and matter recycling (Figure 23.17). It would work with nature to reduce inefficient use and excessive throughputs of matter and energy resources and the resulting pollution and wastes.

A low-throughput economy works by **(1)** reusing and recycling most nonrenewable matter resources; **(2)** using renewable resources no faster than natural processes can replenish them; **(3)** reducing resource waste by using matter and energy resources more efficiently; **(4)** reducing environmentally harmful forms of consumption; and **(5)** promoting pollution prevention and waste reduction. Some experts would add that such an economy works best when population growth can be slowed so that the number of matter and energy consumers grows slowly, and eventually, not at all.

Some environmental scientists suggest that an important step in shifting to a low-throughput economy will be to *relocalize some economic engines and the benefits they provide*. For example, Kelly Cain and his colleagues (Science Focus 22.1, p. 621) have created a computer model for estimating the amount of money and other resources that leave any community that imports most of its food, usually through large retailers. Cain argues that such a community can save large amounts of money and shrink its ecological footprint by learning how to produce much more of its own food and energy from renewable sources such as the sun, wind, and biomass.

One highly successful example of relocalizing an economy, and of Germany's shift to renewable energy (**Core Case Study**), is the small windswept rural village of Feldheim, south of Berlin, with a population of about 150. There a young energy entrepreneur, interested in relocalizing energy production, invested in a small number of wind turbines. The village followed his lead and built its own power grid, along with a biogas plant that produces natural gas from corn cobs, pig manure, and other farm wastes. Today the village produces all of its own heat and electricity and has a zero-carbon footprint and full employment. It makes a profit by selling the excess energy it produces to major power companies for use in Germany's electrical grid system.

Shifting to More Sustainable Economies

Figure 23.18 shows some of the components of societies that have more sustainable economic systems. A common goal of such systems is to put more emphasis on conserving and sustaining the air, water, soil, biodiversity, and other natural resources and ecosystem services that in turn sustain all life and all economies.

A shift to more sustainable economies will involve the deaths of some industries and the births of others. Recall that *ecological succession* occurs when changes in environmental conditions enable certain species to move into an area and replace other species that are no longer favored by the changing environmental conditions (Figure 5.11, p. 107, and Figure 5.12, p. 108). By analogy, *economic succession* in a dynamic capitalist economy occurs as new and more innovative businesses replace older ones that can no longer thrive under changing economic conditions.

The drive to improve environmental quality and to work toward environmental sustainability has created new major growth industries along with profits and large numbers of new *green jobs* (Figure 23.19). Examples of such jobs include those devoted to protecting natural capital, expanding organic agriculture, making homes and

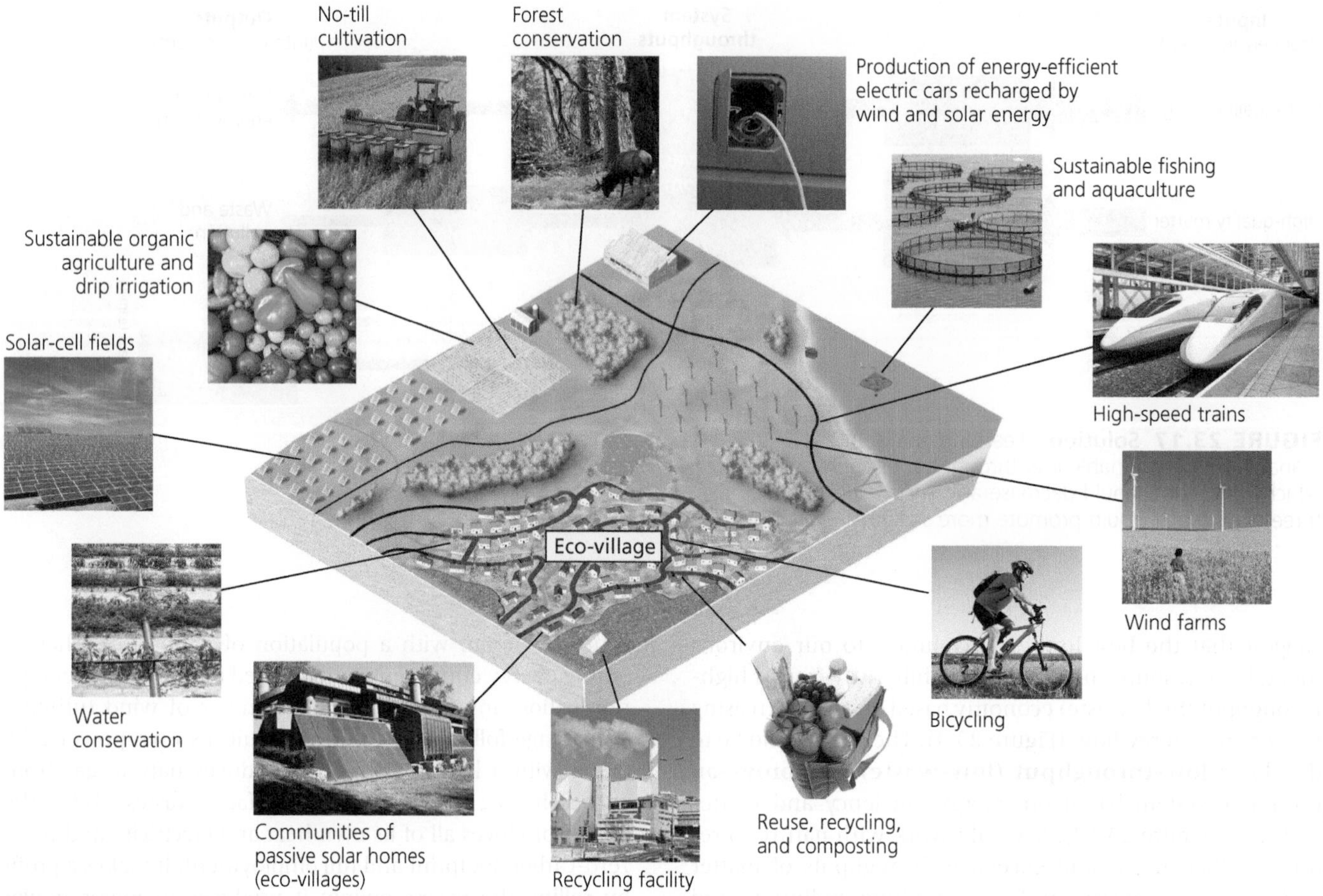

FIGURE 23.18 Solutions: Some of the components of more environmentally sustainable economic development favored by ecological and environmental economists. ***Critical thinking:*** What are three new types of jobs that could be generated by such an economy?

Photos going clockwise starting at "No-till cultivation": Jeff Vanuga/National Resource Conservation Service. Natalia Bratslavsky/Shutterstock.com. Pi-Lens/Shutterstock.com. Vladislav Gajic/Shutterstock.com. hxdbzxy/Shutterstock.com. Varina and Jay Patel/Shutterstock.com. Kalmatsuy Tatyana/Shutterstock.com. Brenda Carson/Shutterstock.com. Alexander Chaikin/Shutterstock.com. National Renewable Energy Laboratory/U.S. Department of Energy. Anhong/Dreamstime.com. pedrosala/Shutterstock.com. Robert Kneschke/Shutterstock.com.

other buildings more energy efficient, modernizing the electrical grid system, and developing low-carbon renewable energy resources. According to the Ecotech Institute, there were 3.8 million green jobs open in the United States in 2014, mostly in the wind and solar energy sector.

Older industries such as the fossil fuels industry have claimed repeatedly that a switch to a more sustainable economy will lead to massive job losses. However, a study by University of California–Berkeley scientists led by Max Wei reviewed 15 studies on job creation in the energy sector. They found that the production and use of renewable energy sources created more jobs per unit of energy generated than did the production and use of fossil fuels.

Making the shift to more sustainable economies will require governments and industries to greatly increase their spending on research and development—especially in the areas of energy efficiency and renewable energy—as Germany has done in recent years (**Core Case Study**). The shift toward sustainability will also require business leaders to understand why such a shift is important ecologically and economically (Individuals Matter 23.2).

Using Lessons from Nature to Make the Transition

In this chapter, we have considered how certain **principles of sustainability** can guide us in shifting to more sustainable economic systems. This has revealed a sharp contrast between the hypotheses of neoclassical economists and those of ecological economists. We close this chapter with the words of the highly regarded environmental scientist Donella Meadows (1941–2001). In 1996, she contrasted the views of neoclassical and ecological economists as follows:

- The first commandment of economics is: Grow. Grow forever. . . . The first commandment of the earth is: Enough. Just so much and no more. . . .

INDIVIDUALS MATTER 23.2

Ray Anderson: Sustainable Business Visionary

Interface, Inc.

Ray Anderson (1934–2011), founder of the American company Interface, was one of the world's most respected and effective leaders in the movement to make businesses more sustainable. The company is the world's largest commercial manufacturer of carpet tiles, with 25 factories in 7 countries, and customers in 110 countries.

Anderson changed the way he viewed the world—and his business—after reading Paul Hawken's book *The Ecology of Commerce*. In 1994, he announced plans to develop the nation's first totally sustainable corporation. Within 16 years, Interface had cut its water usage by 74%, its net greenhouse gas emissions by 32%, its solid waste by 63%, its fossil fuel use by 60%, and its energy use by 44%. The company now gets 31% of its total energy from renewable resources. These efforts have saved Interface more than $433 million.

Anderson also sent his carpet design team into the forest and told them to learn how nature would design floor covering. The team observed that there were no regularly repeating patterns on the forest floor and that instead, there was disorder and diversity. They had discovered nature's biodiversity **principle of sustainability**. With this in mind, the design team created a line of carpet tiles, no two of which had the same design. Within 18 months after it was introduced, this new product line was the company's top-selling design.

Interface also applies nature's chemical cycling **principle of sustainability** by making its carpet tiles from recycled fibers. And the carpet tile factory runs partly on solar energy, thus applying another of nature's sustainability principles. In addition, the company invented a new carpet recycling process that does not emit carbon dioxide to the atmosphere.

Under Anderson's leadership, Interface became the world's largest seller of carpet tiles and profits tripled. Anderson also created a consulting group as part of Interface to help other businesses start on the path toward becoming more sustainable. He was an outstanding leader in the practice of upcycling—working not only to reduce his company's ecological footprint but also to create a large and expanding positive environmental impact.

- Economics says: Compete. . . . The earth says: Compete, yes, but keep your competition in bounds. Don't annihilate. Take only what you need. Leave your competitor enough to live. Wherever possible, don't compete, cooperate. . . . You're not in a war, you're in a community. . . .
- Economics says: Use it up fast. Don't bother with repair; the sooner something wears out, the sooner you'll buy another. This makes the gross national product go round. Throw things out when you get tired of them. . . . Get the oil out of the ground and burn it now. . . . The earth says: What's the hurry? . . . When any part wears out, don't discard it, turn it into food for something else. . . .

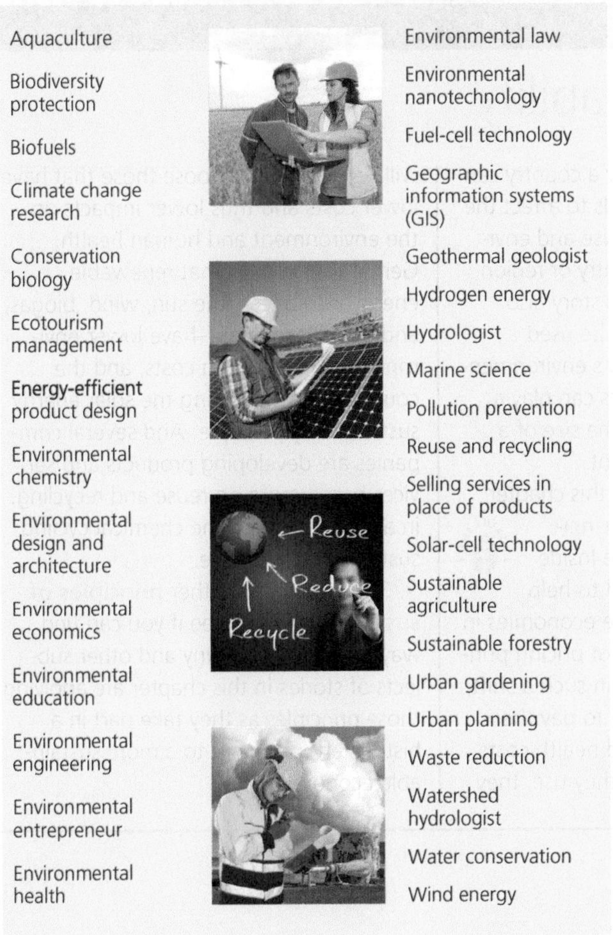

Environmentally Sustainable Businesses and Careers
Aquaculture
Biodiversity protection
Biofuels
Climate change research
Conservation biology
Ecotourism management
Energy-efficient product design
Environmental chemistry
Environmental design and architecture
Environmental economics
Environmental education
Environmental engineering
Environmental entrepreneur
Environmental health
Environmental law
Environmental nanotechnology
Fuel-cell technology
Geographic information systems (GIS)
Geothermal geologist
Hydrogen energy
Hydrologist
Marine science
Pollution prevention
Reuse and recycling
Selling services in place of products
Solar-cell technology
Sustainable agriculture
Sustainable forestry
Urban gardening
Urban planning
Waste reduction
Watershed hydrologist
Water conservation
Wind energy
Reuse
Reduce
Recycle
© Cengage Learning

FIGURE 23.19 *Green careers:* Some key environmental businesses and careers are expected to flourish during this century, while environmentally harmful, or *sunset*, businesses are expected to decline. See the website for this book for more information on various environmental careers. ***Critical thinking:*** How could some of these careers help you to apply the three **principles of sustainability**?

Top: Goodluz/Shutterstock.com. Second from top: Goodluz/Shutterstock.com. Second from bottom: Dusit/Shutterstock.com. Bottom: Corepics VOF/Shutterstock.com.

- Economics discounts the future. ... a resource 10 years from now is worth only half of what it's worth now. Take it now. Turn it into dollars. The earth says: Nonsense. ... give to the future. ... Never take more in your generation than you give back to the next.
- The economic rule is: Do whatever makes sense in monetary terms. The earth says: Money measures nothing more than the relative power of some humans over other humans, and that power is puny compared with the power of the climate, the oceans, the uncounted multitudes of one-celled organisms that created the atmosphere, that recycle the waste, and that have lasted for 3 billion years. The fact that the economy, which has lasted for maybe 200 years, puts zero values on these things means only that the economy knows nothing about value—or about lasting.

BIG IDEAS

- Making a transition to more sustainable economies will require finding ways to estimate and include the harmful environmental and health costs of producing goods and services in their market prices.
- Making this economic transition will also mean phasing out environmentally harmful subsidies and tax breaks, and replacing them with environmentally beneficial subsidies and tax breaks.
- Another way to further this transition would be to tax pollution and wastes instead of wages and profits and to use most of the revenues from these taxes to promote environmental sustainability and reduce poverty.

Tying It All Together

Germany's Transition and Sustainability

The **Core Case Study** that opens this chapter is about how Germany has used economic tools to spur a shift from using fossil fuels and nuclear energy to relying increasingly on renewable energy (Figure 23.1). It shows how a country can use its economic policy tools to affect the energy market. As energy use and environmental quality in a country or region are closely intertwined, this story also shows how economics can be used directly to reduce a country's environmental impact—how economics can play a major role in determining the size of a country's ecological footprint.

iStockphoto.com/Richard Schmidt-Zuper

This story and others in this chapter show how several of the **principles of sustainability** (see Inside Back Cover) can be applied to help us shift to more sustainable economies in the near future. The full-cost pricing principle will play a major role in such a shift, because if consumers have to pay the harmful environmental and health costs of the goods and services they use, they will be inclined to choose those that have lower costs and thus lower impacts on the environment and human health. Germany is finding that renewable energy resources—the sun, wind, biogas, and other resources—have lower environmental and health costs, and the country is thus applying the solar energy sustainability principle. And several companies are developing products and services based more on reuse and recycling, in accordance with the chemical cycling sustainability principle.

Think about the other **principles of sustainability** and see if you can find ways in which Germany and other subjects of stories in this chapter are applying those principles as they take part in a historic effort to shift to a more sustainable economy.

Chapter Review

Core Case Study

1. Explain how Germany's feed-in tariff has spurred a shift to renewable energy use in that country. Summarize the story of this shift.

Section 23.1

2. What is the key concept for this section? Define **economics** and describe three basic types of economic systems. Describe the interactions among supply, demand, and market prices in a market economic system. Explain why a truly free-market economy rarely exists in today's capitalist market systems. What are **subsidies**? Distinguish among **natural capital**, **human capital**, and **manufactured capital**.

3. What is a market failure? Give an example of how a government can correct a market failure. Define **economic growth**. What is a **high-throughput economy**? Distinguish between **economic development** and **environmentally sustainable economic development**. Compare how neoclassical economists and ecological and environmental economists view economic systems. What is the general assumption underlying the models of ecological economists? What is a steady-state economy? Summarize the evidence that indicates we are living unsustainably.

Section 23.2

4. What are the two key concepts for this section? Describe three ways in which economists estimate the economic values of natural goods and services. Why are such values not included in the market prices of goods and services? Define **discount rate** and explain the controversy over how to assign such rates. What is a **marginal cost**? Explain how economists can estimate the optimal levels for pollution control and resource use. Define **cost–benefit analysis** and list its advantages and limitations.

Section 23.3

5. What is the key concept for this section? Distinguish between direct (internal) and indirect (external) costs of goods and services and give an example of each that is related to a specific product. What is **full-cost pricing** and what are some benefits of using it to determine the market values of goods and services?

6. What are **perverse subsidies**? Give an example of such a subsidy. How can subsidies be used for environmentally beneficial purposes? Give three examples of perverse subsidies and three examples of environmentally beneficial government subsidies and tax breaks. Define **gross domestic product (GDP)** and **per capita GDP**. What is an **environmental indicator**? What is the genuine progress indicator and how does it differ from the GDP economic indicator?

7. What are the major advantages and disadvantages of using green taxes? Why do many analysts consider the tax systems in most countries to be backward? What are three requirements for implementing green taxes successfully? What are the benefits of providing consumers with eco-labels on the goods and services they buy? Define **greenwashing** and give an example of it.

8. Distinguish between command-and-control and incentive-based government regulations, and describe the advantages of the second approach. What is the cap-and-trade approach to implementing environmental regulations, and what are the major advantages and disadvantages of this approach? What are some environmental benefits of selling services instead of goods? Give two examples of this approach.

Section 23.4

9. What is the key concept for this section? Define **poverty**. List six ways in which governments, businesses, international lending agencies, and wealthy individuals could help to reduce poverty. What is microlending, and how has it helped people? What are millennium development goals and what role can more-developed countries play in helping the world achieve these goals?

Section 23.5

10. What is the key concept for this section? What is a **low-throughput (low-waste) economy** and how does it work? What does it mean to relocalize economic benefits? Give an example of how this has benefited people. List 10 components of more environmentally sustainable economic development. Name six new businesses and careers that would be important in more sustainable economies. Summarize the story of Ray Anderson and his work to build a more environmentally sustainable carpet business. What are this chapter's *three big ideas*? Explain how we can apply the **principles of sustainability** in making a shift toward a more sustainable economic system, as Germany has done (**Core Case Study**).

Note: Key terms are in bold type.

Critical Thinking

1. Explain how Germany's transition to broader use of renewable energy systems (**Core Case Study**) shows that the economy and the environment are linked. How is the country's economic development policy changing its environmental quality? What are some ways in which Germany's example could be applied to improve the environment and the economy where you live?

2. Is it a good idea to maximize economic growth by producing and consuming more and more economic goods and services? Explain. What are some alternatives?

3. According to one definition, *environmentally sustainable economic development* involves meeting the needs of the present human generation without compromising the ability of future generations to meet their needs. What do you believe are the needs referred to in this definition? Compare this definition with the characteristics of a low-throughput economy depicted in Figure 23.17.

4. Is environmental regulation bad for the economy? Explain. Assume you are a government official and devise an incentive-based regulation for an industry of your choice. (It could be a coal mine, a power plant, a fishing fleet, a chemical plant, or any other business that has a large effect on the environment.) Explain how your regulatory plan will benefit both the industry and the environment.

5. Suppose that over the next 20 years, the environmental and health costs of goods and services are gradually added to market prices until their market prices more closely reflect their total costs. What harmful effects and what beneficial effects might such a full-cost pricing process have on your lifestyle and on the lives of any children, grandchildren, and great-grandchildren you might eventually have?

6. Do you believe that reducing poverty should be a major environmental goal? Explain. List three ways in which reducing poverty could benefit you and any children, grandchildren, and great-grandchildren you might eventually have. Why do you think the world has not focused more intense efforts on reducing poverty?

7. Do you think we should shift to an economy based on the idea of leasing certain services instead of buying the products that provide the services? Explain. If you are for such a shift, what do you think is the best strategy for making it happen? If you are opposed, what are your main objections to the idea?

8. Congratulations! You are in charge of the world. Write up a 5- to 10-point strategy for shifting the world to more environmentally sustainable economic systems over the next 50 years.

Doing Environmental Science

Go online and find a tool for estimating the full cost (including harmful environmental and health costs) of common products. Choose five products that you regularly buy and use this tool to estimate the full cost of each. Record these data in a table, along with the price you paid for each product. (Estimate this price if you don't remember what you paid.) Now do some market research and try to find alternatives to these products that have lower full costs. Record these data in your table. Do some calculations to learn **(a)** the differences between the prices you paid for your common products and their full costs; **(b)** for each product, the difference between its price and the price of the alternative substitute you found; and **(c)** for each product, the difference between its full cost and the full cost of the alternative substitute you found.

Finally, for each product pair, use your data to answer these questions:

1. Without knowledge of the full costs, would the price differences be large enough to keep you from buying the alternative product and sticking with your commonly used product? Explain.

2. Comparing their full costs, does this change your mind about whether the price difference is high enough to keep you from switching products? Explain.

3. How high would the full cost of your commonly used product have to be to get you to pay the higher price for the alternative?

Global Environment Watch Exercise

Go to your MindTap course to access the GREENR database. At the top of the page do a "Basic Search" on *global subsidies*. Find information on government subsidies for the oil, coal, and nuclear industries and for renewable energy industries. Write a summary of trends in these subsidies, either globally or in a country of your choice. Find an article that indicates a positive view of these trends on the part of the author and one that indicates a negative view. Compare the arguments in these articles and decide which argument you believe to be the stronger.

Data Analysis

The following table lists the global annual values of ecosystem services provided by selected major ecosystems, along with the total area of the globe occupied by each ecosystem type. (Areas are in hectares; a hectare equals 2.5 acres.) Study the table and then answer the questions that follow it.

Ecosystem Type	Value of Ecosystem Services per Hectare	Area (in Millions of Hectares)	Global Total Value of Ecosystem Services (in Trillions of Dollars)
Open ocean	$491	33,200	
Grass/rangeland	$2,871	4,418	
Temperate/boreal forest	$3,013	3,003	
Tropical forest	$5,264	1,258	
Urban areas	$6,661	352	
Swamps/floodplains	$25,682	60	
Coastal marshes/mangroves	$193,845	128	
Coral reefs	$352,249	28	

(Compiled by the authors using data from Robert Costanza et al., "Changes in the Value of Ecosystem Services," *Global Environmental Change*, Elsevier, 2014.)

1. Fill in the fourth column: calculate the global total value for each ecosystem by multiplying the value per hectare times the area.
2. Which ecosystem has the highest per hectare value? Which one has the highest global total value? Why do you think they are not the same ecosystem for both values? Explain.
3. About how many times the global total value of ecosystem services from urban areas is that of temperate and boreal forests?
4. The value per hectare for open oceans is the lowest on this table. Why do you think the global total value for open oceans is one of the highest? Explain.

CHAPTER 24

Politics, Environment, and Sustainability

You see things, and you say "Why?" But I dream things that never were; and I say "Why not?"

GEORGE BERNARD SHAW

Marchers in New York City protesting natural gas fracking.

Richard Levine/Alamy Stock Photo

Key Questions

24.1 What is the role of government in making the transition to more sustainable societies?

24.2 How is environmental policy made?

24.3 What is the role of environmental law in dealing with environmental problems?

24.4 What are the major roles of environmental groups?

24.5 How can we improve global environmental security?

24.6 How can we implement more sustainable and just environmental policies?

Core Case Study

The Greening of American Campuses

Since the mid-1980s, there has been a boom in environmental awareness on college campuses and in public and private schools around the world. In the United States, hundreds of colleges and universities have now taken the lead in a quest to become more sustainable and to educate their students about **sustainability**—the capacity of the earth's natural systems and human cultural systems to survive, flourish, and adapt to changing environmental conditions into the very long-term future.

For example, at Oberlin College in Ohio, a group of students worked with faculty members and architects to design a more sustainable environmental studies building (Figure 24.1) powered by solar panels, which produce 30% more electricity than the building uses. Closed-loop underground geothermal wells provide heating and cooling. In its solar greenhouse, a series of open tanks populated by plants and other organisms purifies the building's wastewater. The building collects rainwater for irrigating the surrounding grasses, gardens, and meadow, which contain a diversity of plant and animal species.

At the University of Washington in Seattle, more than half of the food served on campus comes from the campus farm and other small local producers. All eggs served are organic from cage-free hens. This saves the school money and cuts it energy use and greenhouse gas emissions.

The University of California, San Diego (UCSD), uses only drought-tolerant native plants for all of its new landscaping, which saves the campus a great deal of water that has historically been used to water grass in this drought-stricken area of the country. More than a third of UCSD's vehicle fleet is all-electric and the school runs 55 of its vehicles on biofuel.

The University of Wisconsin–Oshkosh uses a biodigester to convert manure from nearby farms to fuel that supplies 20% of the energy used for heating the campus buildings. The school also gets 20% of its electricity from wind power.

In addition to making campuses greener, colleges are increasingly offering environmental sustainability courses and programs. At Pfeiffer University, many students have accompanied Professor Luke Dollar, a National Geographic Explorer, on trips to Madagascar to take part in his research on that country's endangered species and ecosystems.

These are just a few examples of the hundreds of institutions educating students who will provide leadership in working to make our societies and economies more sustainable during the next few decades. Maybe you will join the ranks of such environmental leaders.

Robb Williamson/NREL

FIGURE 24.1 The Adam Joseph Lewis Center for Environmental Studies at Oberlin College in Oberlin, Ohio.

24.1 WHAT IS THE ROLE OF GOVERNMENT IN MAKING THE TRANSITION TO MORE SUSTAINABLE SOCIETIES?

CONCEPT 24.1 Through its policies, a government can help to protect environmental and public interests, and to encourage more environmentally sustainable economic development.

Serving Environmental and Other Public Interests

Business and industry thrive on change and innovations that lead to new technologies, products, and opportunities for profits. This process, often referred to as *free enterprise*, can lead to jobs and higher living standards for many people, but it can also create harmful health and environmental impacts.

Government, on the other hand, can act as a brake on environmentally harmful business enterprises. Achieving the right balance between free enterprise and government regulation is not easy. Too much government intervention can strangle enterprise and innovation. Too little can lead to environmental degradation and social injustices, and even to a weakening of the government by business interests and global trade policies.

Analysts point out that in today's global economy, some multinational corporations, which often have budgets larger than the budgets of many countries, have greatly increased their economic and political power over national, state, and local governments, and ordinary citizens. However, businesses can also serve environmental and public interests. Green businesses create products and services that help to sustain or improve environmental quality while improving people's lives. They make up one of the world's fastest growing business sectors and are increasingly a source of new jobs.

Many argue that government is the best mechanism for dealing with some of the broader economic and political issues we face, some of which we have discussed in this book. These include the following:

- *Full-cost pricing* (see Chapter 23, p. 637): Governments can provide subsidies and levy taxes that have the effect of including harmful environmental and health costs in the market prices of some goods and services, in keeping with the full-cost pricing **principle of sustainability** (see Inside Back Cover).
- *Market failures* (see Chapter 23, p. 632): Governments can use taxes and subsidies to level the playing field wherever the marketplace is not operating freely due to unfair advantages held by some players.
- *The tragedy of the commons:* Government plays a key role in preserving common or open-access renewable resources (p. 12) such as clean air and groundwater, the ozone layer in the stratosphere, and our life-support system. For example, with the cap-and-trade approach (Figure 23.13, p. 643) to solving a problem such as air pollution, government oversight is necessary to administer such a program.

The roles played by a government are determined by its **policies**—the laws and regulations it enacts and enforces, and the programs it funds (**Concept 24.1**). **Politics** is the process by which individuals and groups try to influence or control the policies and actions of governments at local, state, national, and international levels. One important application of this process is the development of **environmental policy**—environmental laws, regulations, and programs that are designed, implemented, funded, and enforced by one or more government agencies.

According to social scientists, the development of public policy in democracies often goes through a *policy life cycle* (also known as *adaptive management*) consisting of four stages (Figure 24.2):

- *Problem recognition.* A problem is identified by members of the public or by a policy maker.
- *Policy formulation.* A cause or causes of the problem are identified and a solution such as a law or program to help deal with the problem is proposed and developed.
- *Policy implementation.* A law is passed or a regulation written to put the solution into effect.
- *Policy adjustment.* The new program is monitored, evaluated, and adjusted as necessary.

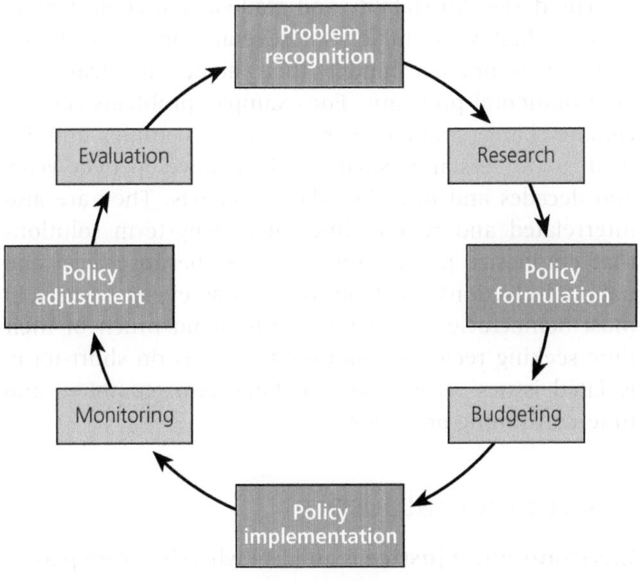

FIGURE 24.2 The *policy life cycle* has been defined in several ways but generally includes these four phases (listed in the orange boxes).

The Democratic Process

Democracy is government by the people through elected officials and representatives. This form of government has often been compromised by groups or organizations that gain enough wealth and power to have more influence over government policies than the average citizen can have. However, the ideals for representative democracy are usually embodied in a document called a *constitution,* which provides the basis of government authority and, in most cases, limits government power by mandating free elections and guaranteeing the right of free speech. Another name for such a government is *constitutional democracy*.

Political institutions in most constitutional democracies are designed to allow gradual change that helps ensure economic and political stability. In the United States, for example, rapid and destabilizing change is curbed by a system of checks and balances that distributes power among three branches of government—*legislative, executive,* and *judicial*—and among federal, state, and local governments.

In passing laws, developing budgets, and formulating regulations, elected and appointed government officials must deal with pressure from many competing *special-interest groups*. Each group advocates passing laws, providing subsidies or tax breaks, or establishing regulations favorable to its cause. Such groups also seek to weaken or repeal laws, subsidies, tax breaks, and regulations unfavorable to its position. Important examples of special-interest groups include *profit-making organizations,* such as corporations; *nongovernmental organizations (NGOs),* most of which are nonprofit organizations such as environmental groups; *labor unions,* representing the interests of workers; and *trade associations,* representing various industries.

The design for stability and gradual change in democracies is highly desirable. But several features of democratic governments hinder their ability to deal with environmental problems. For example, problems such as climate change and biodiversity loss are complex and difficult to understand. Such problems develop over years and decades and have long-lasting effects. They are also interrelated and require integrated, long-term solutions that emphasize prevention. However, because local and national elections are held as often as every 2 years in most democracies, most politicians spend much of their time seeking reelection and tend to focus on short-term, isolated issues rather than on long-term, complex, and time-consuming problems.

Environmental Justice

Environmental justice is an ideal whereby every person is entitled to protection from environmental hazards regardless of race, gender, age, national origin, income, social class, or any political factor.

One of the earliest environmental justice events occurred in 1982 in a poor, rural area of North Carolina. There a group of residents sat down in front of trucks hauling toxic waste to an area landfill. They did not succeed in preventing the waste disposal. But their efforts brought the attention of President Bill Clinton who later issued an executive order that established a new federal government duty to address especially harmful health and environmental effects of federal policies or programs on low-income people and people of color.

The U. S. Environmental Protection Agency (EPA) has listed at least 76 studies that show such discrimination. One of the earliest, a 1983 General Accounting Office study, focused on the siting of hazardous waste landfills, finding that a large percentage of them were located in poor communities. Such studies revealed that a lopsided share of polluting factories, incinerators, and landfills in the United States are located in communities populated mostly by African Americans, Asian Americans, Latinos, and Native Americans.

Other research has shown that, in general, toxic waste sites in white communities have been cleaned up faster and more completely than similar sites in African American and Latino communities have. In addition, people in minority communities tend to have higher exposures to lead-based paint, diesel fumes and other dangerous pollutants (Figure 24.3), bothersome odors, and noise from factories, landfills, and other sources.

Such environmental discrimination in many parts of the world has led to a growing grassroots effort known as the *environmental justice movement*. Supporters of this movement have pressured governments, businesses, and environmental organizations to become aware of environmental injustice and to act to prevent it. They have made some

Jim West/Age Fotostock

FIGURE 24.3 These residents of a neighborhood in Detroit, Michigan (USA), wanted a nearby hospital to shut down its medical waste incinerator, which was polluting the air in their community.

progress toward their goals, but in 2015, a Center for Public Integrity study found that since the mid-1990s, the EPA has dismissed 95% of all environmental justice claims. Soon after, the EPA's Office of Civil Rights announced a 5-year plan to step up enforcement of civil rights laws with respect to environmental justice.

CONSIDER THIS . . .

THINKING ABOUT Environmental Justice

Do you think that the principles of environmental justice should get equal weight, more weight, or less weight than economic factors in political decisions about where to locate potentially environmentally harmful facilities? Explain.

Principles of Environmental Policy Making

Analysts suggest that when evaluating existing or proposed environmental policies, legislators and individuals should be guided by several principles that can help them to minimize environmental harm:

- *The holistic principle:* Recognize that the environmental and other problems we face are connected and focus on long-term solutions that address root causes of such interconnected problems instead of focusing on short-term and often ineffective fixes that treat each problem separately.
- *The precautionary principle:* When substantial evidence indicates that an activity threatens human health or the environment, take precautionary measures to prevent or reduce such harm, even if some of the cause-and-effect relationships are not well established, scientifically.
- *The prevention principle:* Whenever possible, make decisions that help to prevent a problem from occurring or becoming worse.
- *The reversibility principle:* Avoid making decisions that cannot be reversed later if they turn out to be harmful. For example, two essentially irreversible actions are the production of indestructible, toxic coal ash in coal-burning power plants, and the production of deadly radioactive wastes in nuclear power plants. In both cases, these hazardous wastes must be stored safely for thousands of years.
- *The net energy principle:* Avoid the widespread use of energy resources and technologies with low or negative net energy yields (Figure 15.3, p. 384), which cannot compete in the open marketplace without government subsidies and tax breaks. Examples of such energy alternatives include nuclear power (considering the whole fuel cycle), tar sands, and shale oil, as discussed in Chapter 15, and hydrogen fuel and ethanol made from corn, discussed in Chapter 16.
- *The polluter-pays principle:* Develop regulations and use economic tools such as green taxes to ensure that polluters bear the costs of dealing with the pollutants and wastes they produce. This can stimulate the development of innovative ways to reduce and prevent pollution and wastes.
- *The environmental justice principle:* In the implementation of environmental policy, no group of people should bear an unfair share of the burden created by pollution, environmental degradation, or the execution of environmental laws.
- *The triple bottom line principle:* Balance economic, environmental, and social needs when making policy decisions (Figure 24.4). Rather than considering these factors in isolation from one another, integrate these factors into the decision process.

Implementing such principles is not easy and requires policy makers to become more environmentally literate. It also requires robust debate among politicians and citizens, mutual respect for diverse beliefs, and a dedication to dealing with environmental problems by implementing the win-win **principle of sustainability**.

CONSIDER THIS . . .

THINKING ABOUT Environmental Political Principles

Which three of the eight principles listed here do you think are the most important? Why? Which ones do you think could influence legislators in your city, state, or country? Why?

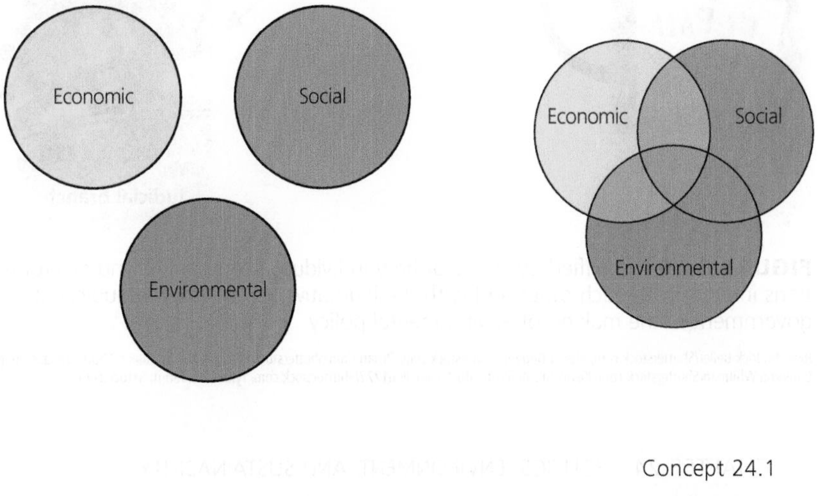

FIGURE 24.4 *The triple bottom line:* Policy decisions have traditionally been made by considering the social, environmental, and economic factors in isolation from one another (represented by the three circles, left). Some analysts say sustainable policy decisions must be made by weighing all of these factors at once and attempting to satisfy the needs of all three sets of priorities (represented by the intersection of these three circles, right).

24.2 HOW IS ENVIRONMENTAL POLICY MADE?

CONCEPT 24.2A Policy making involves enacting laws, funding programs, writing rules, and enforcing those rules with government oversight—a complex process that is affected at each stage by political processes.

CONCEPT 24.2B Individuals can work together to become part of political processes that influence how environmental policies are made and whether or not they succeed.

Democratic Government: The U.S. Model

The U.S. federal government consists of three separate but interconnected branches: legislative, executive, and judicial (Figure 24.5). The *legislative branch,* called the Congress, consists of the House of Representatives and the Senate, which jointly have two main duties. One is to approve and oversee government policy by passing laws that establish government agencies or instruct existing agencies to take on new tasks or programs. The other is to oversee the functioning and funding of agencies in the executive branch concerned with carrying out government policies.

The *executive branch* consists of the president, vice president, major department heads (called the *President's Cabinet*), and a staff who oversee the many agencies of the executive branch, which are authorized by Congress to carry out government policies. The president proposes annual budgets, legislation, and appointees for major executive positions, which must be approved by Congress. The president also tries to persuade Congress and the public to support executive policy proposals. Citizens

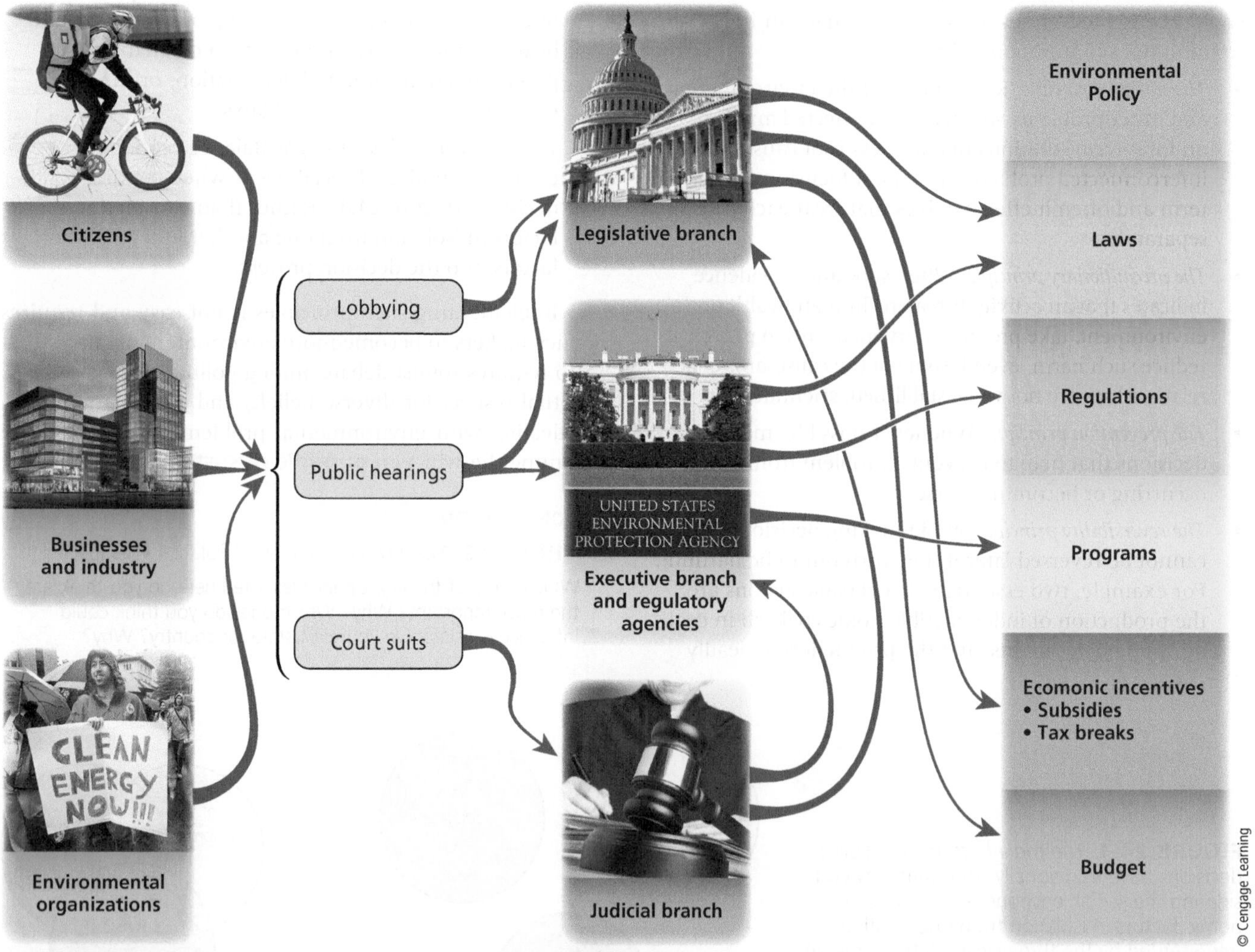

FIGURE 24.5 Simplified overview of how individuals, companies, and environmental organizations interact with each other and with the legislative, executive, and judicial branches of the U.S. government in the making of environmental policy.

Ryan Rodrick Beiler/Shutterstock.com, Mesut Dogan/Shutterstock.com, Orhan Cam/Shutterstock.com, Andrey Burmakin/Shutterstock.com, jgroup/Getty Images, Cameron Whitman/Shutterstock.com, Kevin Grant/Shutterstock.com, jl661227/Shutterstock.com, Tyler Olson/Shutterstock.com vote to elect the president, vice president, and members of Congress.

The *judicial branch* consists of the Supreme Court and lower federal courts. These courts, along with state and local courts, enforce and interpret different laws passed by legislative bodies. They are to ensure that the laws preserve the rights and responsibilities of government and citizens as established by the U.S. Constitution. Decisions made by the various courts make up a body of law known as *case law*. Previous court rulings are used as legal guidelines, or *precedents*, to help make new legal decisions and rulings. The president appoints judges at the federal level with the advice and consent of the Senate, and judges at state and local levels of government are variously appointed by executives or elected by voters.

The major function of the federal government in the United States (and in other democratic countries) is to develop and implement policies for dealing with various issues. The important components of policy are the *laws* passed by the legislative branch, *regulations* instituted by the agencies of the executive branch to put laws and programs into effect, and *funding* approved by Congress and the president to finance the executive agencies' programs and to implement and enforce the laws and regulations (**Concept 24.2A**).

Policy making is a complex process (Figure 24.5). An important factor in this process is **lobbying,** in which individuals or groups contact legislators in person, or hire *lobbyists* (representatives) to do so, in order to persuade legislators to vote or act in their favor. The opportunity to lobby elected representatives is an important right for everyone in a democracy. However, some critics of the American system believe lobbyists of large corporations and other organizations have grown too powerful and that their influence overshadows the input of ordinary citizens.

CONSIDER THIS . . .

CONNECTIONS Lobbying and Perverse Subsidies

Environmental scientist Norman Myers contends that lobbying helps to create perverse subsidies (see Chapter 23, p. 638). According to the Center for Responsive Politics, in 2014 there were nearly 11,800 registered corporate lobbyists in Washington, D.C. They spent $1.64 billion—an average total of nearly $4.5 million per day—on efforts to influence the 535 members of the U.S. Congress. That amounts to an average of more than $8,400 per day per member.

Corporations, trade unions, and other large organizations also provide billions of dollars in political campaign contributions. In 2010, the U.S. Supreme Court ruled that corporations can spend as much money as they want on ads for or against specific candidates running for election.

Most environmental bills are evaluated by as many as 10 committees in the U.S. House of Representatives and the Senate. Effective proposals often are weakened by this fragmentation and by lobbying from groups opposing these laws. Nonetheless, since the 1970s, a number of important environmental laws have been passed in the United States, as we discuss in the next section of this chapter.

Developing Environmental Policy—a Controversial Process

In the United States, passing a law is not enough to make policy (**Concept 24.2A**). The next step involves trying to get Congress to appropriate enough funds to implement and enforce each law. The government creates a budget to finance its agencies and programs and the enforcement of laws and regulations. Budgeting is the most important and controversial activity of the executive and legislative branches.

Once Congress has passed a law and funded a program, the appropriate government department or agency must draw up regulations for implementing it. A group affected by the program and its regulations may take the agency to court for failing to implement and enforce the regulations effectively or for enforcing them too rigidly.

Businesses facing environmental regulations often pressure regulatory agencies and executives to appoint people from the regulated industries or groups to high positions within the agencies. In other words, the regulated try to take over the regulatory agencies and become the regulators—described by some as "putting foxes in charge of the henhouse."

In addition, people in regulatory agencies work closely with officials in the industries they are regulating, often developing friendships with them. Some industries and other regulated groups offer high-paying jobs to regulatory agency employees in an attempt to influence their regulatory decisions. The tendency for administrators to move back and forth between agencies and the companies they regulate has been referred to as a "revolving door effect" that can give regulated companies an unfair advantage in the political process.

Some analysts argue that environmental science should play a major role in the formulation of environmental policy. However, politics usually plays a bigger role, and the scientific and political processes are quite different (Science Focus 24.1).

Influencing Environmental Policy

A major theme of this book is that *individuals matter*. History shows that significant political change usually comes from the *bottom up* when individuals join together to bring about change (**Concept 24.2B**). Without such historical grassroots political action by millions of individual citizens and organized citizen groups, pollution and environmental degradation would be much worse today.

Science and Politics—Principles and Procedures

The rules of inquiry and debate in science are quite different from those of politics. Science is based on a set of principles designed to make scientific investigations completely open to critical review and testing. Here are four of the basic principles:

1. *Any scientific claim must be based on hard evidence and subject to peer review*. This helps to prevent scientists from lying about procedures or falsifying evidence.
2. *Scientists can never establish absolute proof about anything*. Instead, they seek to establish a high degree of certainty about the results of their research.
3. *Scientists vigorously debate the validity of scientific research*. Such debate focuses on the scientific evidence and results, not on personalities involved.
4. *Science advances through the open sharing and peer review of research methods, results, and conclusions*. There are two exceptions to this: first, some scientists who own or work for companies need to protect their research until legal patents can be obtained. Second, government scientists whose work involves national security often keep their research secret.

In politics, on the other hand, there are no such established and respected principles. In order to win elections and gain influence, politicians use unwritten rules that change frequently. While many politicians would like to base their decisions and actions on facts, others suggest that what matters more than facts is how the public perceives what they do and say. This makes the political process far less open than the scientific process is to review and criticism.

Without such openness, the political process often involves tactics that most scientists would reject. For example, some politicians pick and choose facts to support a claim that is not supported by the whole of a body of evidence. They then repeat such a claim over and over until it becomes part of the news media cycle. If this misuse of evidence is not exposed, as it usually is in science, these unsupported claims can become widely accepted as truth.

Another political tactic that often goes unchallenged is to change a debate about facts to a discussion focused on personal attacks. Such a tactic is meant to make one's opponents look weak, and it helps a politician to avoid serious discussion of issues. In scientific debate, most participants do not tolerate such a shift away from a fact-based discussion.

It is possible to spread disinformation quickly in this media age of almost instant global news coverage, text messaging, social networking, and Internet blogs and videos. While the Internet enables this, it also allows almost anyone to check the validity of much information and to detect and publicize lies and distortions. Learning how to detect and evaluate disinformation is one of the most important purposes of education.

CRITICAL THINKING

What are two examples of widely accepted results of scientific research that have been politicized to the point where they are largely doubted or ignored by the public?

With the growth of the Internet and social media, individuals have become more empowered. For example, in a highly unusual chain of events in 2007, Chinese citizens using mobile phone text messaging organized to oppose construction of a chemical plant that would threaten the safety of 1.5 million people in a port city. By building opposition from the ground up—circulating nearly a million phone messages—they persuaded the Chinese government to freeze the construction project and to consider less hazardous alternatives.

Figure 24.6 lists ways in which you can influence and change government policies in constitutional democracies. *At a fundamental level, all politics is local*. What we do to improve environmental quality in our own neighborhoods, schools, and work places can serve as an example and have national and global implications. When people work together, starting at the local level, they can influence environmental policy at all levels (**Concept 24.2B**).

Environmental Leadership

Each of us can provide environmental leadership in several different ways. First, we can *lead by example*, using our own lifestyles and values to show others that change is possible and can be beneficial (Individuals Matter 24.1). For example, we can buy only what we need, use fewer disposable products, eat foods that have been more sustainably produced, and walk, take a bus, or ride a bike to work or school (Figure 24.7).

Second, we can *work within existing economic and political systems to bring about environmental improvement* by campaigning and voting for informed, ecologically literate candidates and by communicating with elected officials. We can also send a message to companies that we think are harming the environment through their production processes or products by *voting with our wallets*—not buying their products or services—and letting them know why.

INDIVIDUALS MATTER 24.1

Xiuhtezcatl Roske-Martinez

Xiuhtezcatl Roske-Martinez, born in 2000, learned about environmentalism from his parents while spending much of his early childhood enjoying the beautiful forests and streams near his Boulder, Colorado, home. His father is an Aztec who believes that all life is sacred and should be respected and cared for. His mother co-founded the nonprofit Earth Guardians as part of her commitment to protecting the earth's water, air, and atmosphere. They taught Xiuhtezcatl (pronounced "Shoe-TEZ-Cot") their values, and he fell in love with nature.

As a young child, Xiuhtezcatl heard a lot about the harmful effects of human activities and noticed that the forest around him was changing. Trees were dying—killed by beetles whose populations were exploding because winter temperatures did not get low enough to kill them off. Dead trees fueled large fires that destroyed more trees.

To Xiuhtezcatl these effects of climate change were real and scary, and they motivated him. At age 6, he gave his first speech at a climate change rally. Since then, he has used his natural leadership ability to become a dynamic and highly effective environmental activist. He helped persuade the Boulder City Council to stop using pesticides in parks, impose a fee on plastic bag use, and require a power company to depend more on renewable energy. He organized and spoke at press conferences, created a multimedia presentation about the harmful environmental effects of plastic bags, spoke at city council meetings and before an EPA hearing, and went door-to-door to organize dozens of rallies and marches. At age 12, he was invited to speak about climate change at the 2012 Rio+20 United Nations Summit on sustainable development in Brazil.

As youth leader for Earth Guardians, Xiuhtezcatl has set up Earth Guardian Crews in many parts of the world to promote environmental education and awareness and to encourage other people to act. In 2013, young Earth Guardians were planting thousands of trees in Boulder, Colorado, in many other parts of the United States, and in over 20 other countries.

What Can You Do?

Influencing Environmental Policy

- Become informed on issues
- Make your views known at public hearings
- Make your views known to elected representatives and understand their positions on environmental issues
- Contribute money and time to candidates who support your views
- Vote
- Run for office
- Form or join nongovernment organizations (NGOs) seeking change
- Support reform of election campaign financing that reduces undue influence by corporations and wealthy individuals

FIGURE 24.6 Individuals matter: Some ways in which you can influence environmental policy (**Concept 24.2B**). ***Critical thinking:*** Which three of these actions do you think are the most important? Which ones, if any, do you take?

Another way to work within the system to bring about change is to choose one of the many rapidly growing green careers. Several are highlighted throughout this book and described in Figure 23.19 (p. 649) and on this book's companion website.

Third, we can *run for some sort of local office*. Look in the mirror. Maybe you are one who can make a difference as an officeholder.

Fourth, we can *propose and work for better solutions to environmental problems*. Leadership is much more than just taking a stand for or against something. It also involves coming up with solutions to problems and persuading people to work together to achieve them. This includes finding ways to bridge gaps between people who disagree about solutions.

FIGURE 24.7 Bicycling to school or work is one way to lead by example. In addition to reducing pollution, it saves you money and provides exercise.

Fifth, *good leaders inspire others to lead*. Leadership is more than telling others what to do. It involves recognizing each person's strengths and helping people to recognize their own strengths, and encouraging them to use their strengths creatively and actively.

Some environmental leaders are motivated by two important findings: *First,* research by social scientists indicates that social change requires active support by only 5–10% of the population, which often is enough to lead to a political tipping point. *Second,* experience has shown that reaching such a critical mass can bring about social change much faster than most people think.

24.3 WHAT IS THE ROLE OF ENVIRONMENTAL LAW IN DEALING WITH ENVIRONMENTAL PROBLEMS?

CONCEPT 24.3 We can use environmental laws and regulations to help control pollution, set safety standards, encourage resource conservation, and protect species and ecosystems.

Environmental Law

Environmental law is a body of laws and treaties that broadly define what acceptable environmental behavior is for individuals, groups, businesses, and nations. This body of laws and treaties has evolved through legislative and judicial processes at various levels of government that have usually included attempts to balance competing private, social, and commercial interests. This section of the chapter deals primarily with the U.S. legal system as a model that reveals the advantages and disadvantages of using a legal and regulatory approach to dealing with environmental problems.

One way in which environmental law has evolved is through court cases involving lawsuits, most of which are **civil suits** brought to settle disputes or damages between one party and another. For example, a homeowner may bring a nuisance suit against a nearby factory because of the noise it generates. In such a suit, the **plaintiff,** the party bringing the charge (in this case, the home owner), seeks to collect damages from the **defendant,** the party being charged (in this case, the factory), for injuries to health or for economic loss.

The plaintiff may also seek an *injunction,* by which the court hearing the case would order the defendant to stop whatever action is causing the nuisance. Short of closing the factory, often the court tries to find a reasonable or balanced solution to the problem. For example, it may order the factory to reduce the bothersome noise to certain levels or to eliminate it at night.

A *class action suit* is a civil suit filed by a group, often a public interest, consumer, or environmental group, on behalf of a larger number of citizens, all of whom claim to have experienced similar damages from a product or an action, but who need not be listed and represented individually.

Another concept used in environmental law cases is *negligence,* in which a party causes damage by deliberately acting in an unlawful or unreasonable manner. For example, a company may be found negligent if it fails to handle hazardous waste in a way that it knows is required by a *statutory law* (a law, or *statute,* passed by a legislature). A court may also find a company negligent if it fails to do something a reasonable person would do, such as testing waste for certain harmful chemicals before dumping it into a sewer, landfill, or river (Figure 24.8). Generally, negligence is hard to prove.

Environmental Lawsuits

Several factors can limit the effectiveness of environmental lawsuits. *First,* plaintiffs bringing the suit must establish that they have the legal right, or *legal standing,* to do so in a particular court. To have such a right, plaintiffs must show that they have suffered health or financial losses from some alleged environmental harm. *Second,* bringing any lawsuit costs too much for most individuals.

Third, public interest law firms cannot recover their attorneys' fees unless Congress has specifically authorized that they be compensated within the laws that they seek to have enforced. By contrast, corporations can reduce their taxes by deducting their legal expenses—in effect getting a government (taxpayer) subsidy to pay for part of their legal fees. In other words, the legal playing field is uneven and puts individuals and groups that are filing environmental lawsuits at a disadvantage.

Fourth, to stop a nuisance or to collect damages from a nuisance or an act of negligence, plaintiffs must establish that they have been harmed in some significant way and that the defendant caused the harm. Doing this can be difficult and costly. Suppose a company (the defendant) is alleged to have caused cancer in certain individuals (the plaintiffs) by polluting a river (Figure 24.8). If hundreds of other industries and cities dump waste into that river, establishing that one specific company is the culprit is very difficult and requires expensive investigation, scientific research, and expert testimony.

Fifth, most states have *statutes of limitations,* laws that limit how long a plaintiff can take to sue after a particular event occurs. These statutes often make it essentially impossible for victims of cancer, which may take 10–20 years to develop, to file or win a negligence suit.

Sixth, courts can take years to reach a decision. During that time, a defendant may continue the allegedly damaging action unless the court issues a temporary injunction against it until the case is decided.

Rechitan Sorin/Shutterstock.com

FIGURE 24.8 This body of water was polluted by a copper mining operation. Such pollution can form the basis for an environmental lawsuit.

Yet another problem is that corporations and developers sometimes file *strategic lawsuits against public participation (SLAPPs)* targeting citizens who publicly criticize a business for some activity, such as polluting or filling in a wetland. Judges throw out about 90% of the SLAPPs that go to court. But individuals and groups hit with SLAPPs must hire lawyers, and typically spend 1 to 3 years defending themselves. Most SLAPPs are not meant to be won, but are intended to intimidate and discourage individuals and activist groups.

Analysts have suggested some major reforms to help level the legal playing field for citizens suffering environmental damage. One would be to pressure Congress to pass a law allowing juries and judges to award citizens their attorney fees, to be paid by the defendants, in successful lawsuits. Another would be to establish rules and procedures for identifying frivolous SLAPP suits so that cases without factual or legal merit can be dismissed quickly. A third way would be to raise the fines for violators of environmental laws and punish more violators with jail sentences. Polls indicate that 80% or more of Americans consider environmental damage to be a serious crime.

CASE STUDY

U.S. Environmental Laws

Concerned citizens have persuaded Congress to enact a number of important federal environmental and resource protection laws. Most of them were enacted in the 1970s (Figure 24.9).

U.S. environmental laws generally fit into five categories. The first type *requires evaluation of the environmental impacts of certain human activities*. It is represented by one of the first and most far-reaching federal environmental laws, the National Environmental Policy Act, or NEPA, passed in 1970. Under NEPA, an *environmental impact statement (EIS)* must be developed for every major federal project likely to have an effect on environmental quality. The EIS (Figure 24.10) must explain why the proposed project is needed and identify its beneficial and harmful environmental impacts. For example, an environmental impact study typically considers a project's likely effects on wildlife habitat, soils, water quality, air quality, stream flows, and other factors. The EIS must also suggest ways to lessen any harmful impacts, and it must present an evaluation of alternatives to the project. EIS documents must be published and are open to public comment.

NEPA does not prohibit environmentally harmful government projects. But more than one-third of the country's land is under federal management, and NEPA requires the managing agencies to take environmental consequences into account in making decisions. It also exposes proposed projects and their possible harmful effects to public scrutiny. Opponents have targeted NEPA as a law to weaken or repeal.

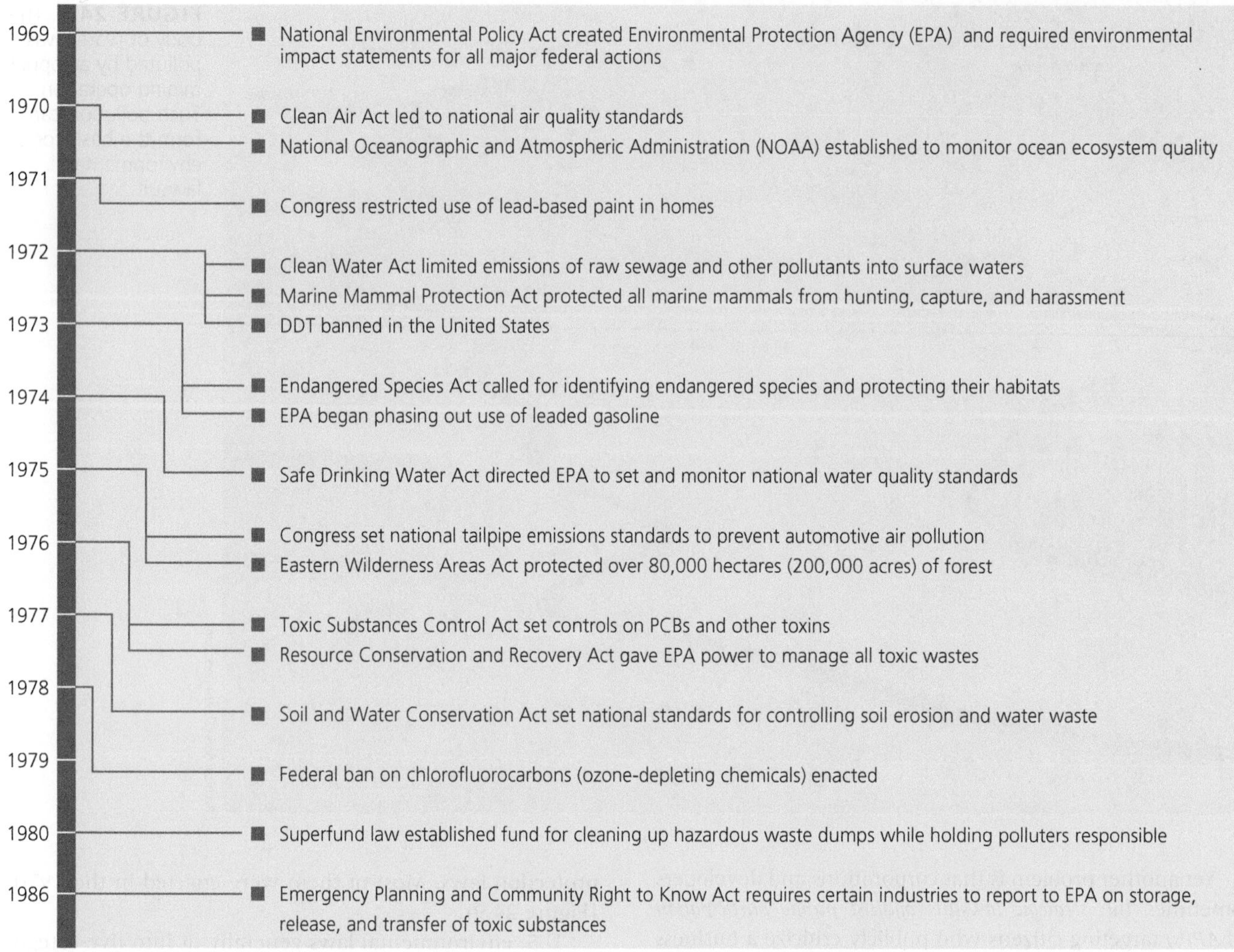
1969
National Environmental Policy Act created Environmental Protection Agency (EPA) and required environmental impact statements for all major federal actions
1970
Clean Air Act led to national air quality standards
National Oceanographic and Atmospheric Administration (NOAA) established to monitor ocean ecosystem quality
1971
Congress restricted use of lead-based paint in homes
1972
Clean Water Act limited emissions of raw sewage and other pollutants into surface waters
Marine Mammal Protection Act protected all marine mammals from hunting, capture, and harassment
1973
DDT banned in the United States
Endangered Species Act called for identifying endangered species and protecting their habitats
1974
EPA began phasing out use of leaded gasoline
1975
Safe Drinking Water Act directed EPA to set and monitor national water quality standards
Congress set national tailpipe emissions standards to prevent automotive air pollution
1976
Eastern Wilderness Areas Act protected over 80,000 hectares (200,000 acres) of forest
1977
Toxic Substances Control Act set controls on PCBs and other toxins
Resource Conservation and Recovery Act gave EPA power to manage all toxic wastes
1978
Soil and Water Conservation Act set national standards for controlling soil erosion and water waste
1979
Federal ban on chlorofluorocarbons (ozone-depleting chemicals) enacted
1980
Superfund law established fund for cleaning up hazardous waste dumps while holding polluters responsible
1986
Emergency Planning and Community Right to Know Act requires certain industries to report to EPA on storage, release, and transfer of toxic substances

FIGURE 24.9 Some of the major environmental laws and their amended versions enacted in the United States since 1969. No major new environmental laws have been passed since the 1970s, although some existing laws have been amended.

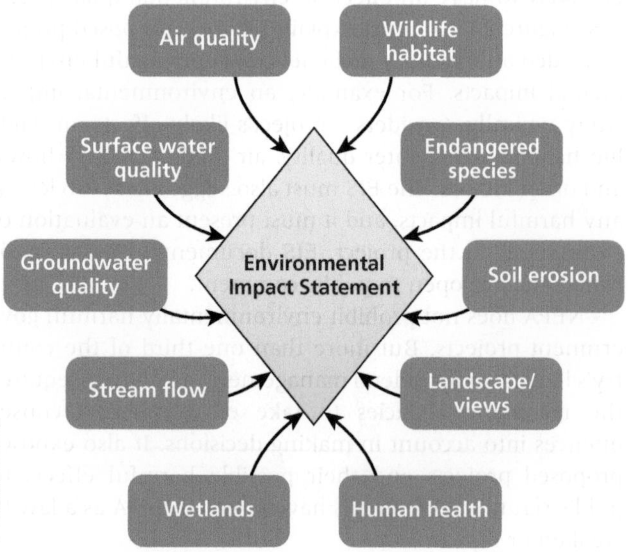
Air quality
Wildlife habitat
Surface water quality
Endangered species
Groundwater quality
Environmental Impact Statement
Soil erosion
Stream flow
Landscape/ views
Wetlands
Human health

CONSIDER THIS . . .

THINKING ABOUT Environmental Impact Statements

Do you think environmental impact statements such as those required by NEPA are a good idea? Explain.

The second major type of environmental legislation *sets standards for pollution levels* (as in the Clean Air Acts, see Chapter 18, p. 490). A third type *sets aside or protects certain species, resources, and ecosystems* (the Endangered Species Act, see Chapter 9, p. 211, and the Wilderness Act, see

FIGURE 24.10 The *environmental impact statement*, required by NEPA, is aimed at minimizing the environmental impacts of major projects. It requires input from several different areas of study covering various possible effects on the environment, including but not limited to areas shown here.

Chapter 10, p. 240). A fourth type *screens new substances for safety and sets standards* (as in the Safe Drinking Water Act, see Chapter 20, p. 556). And a fifth type *encourages resource conservation* (the Resource Conservation and Recovery Act, see Chapter 21, p. 592).

U.S. environmental laws have been highly effective, especially in controlling pollution. In 2012, economist Gernot Wagner estimated that the benefits of the U.S. Clean Air Act, including healthier work environments, fewer cases of sickness and death, and higher worker productivity, exceeded the costs of complying with the law by 30 to 1.

In 2012, NEPA had been in effect for 40 years. In that time, the EPA had helped to sharply reduce atmospheric levels of sulfur dioxide and nitrogen oxides, ban the use of leaded gasoline, ban the widespread use of toxic DDT, get secondhand tobacco smoke classified as a carcinogen, and regulate the use of many other toxic chemicals. Instead of hindering economic growth, as its critics feared, NEPA sparked a domestic environmental protection industry that now employs more than 1.5 million people.

However, since 1980 a well-organized and well-funded movement has mounted a strong campaign to weaken or repeal U.S. environmental laws and regulations. Three major groups from this movement are **(1)** some corporate leaders and other powerful people who see laws and regulations as threats to their profits, wealth, and power; **(2)** citizens who see them as threats to their private property rights and jobs; and **(3)** state and local government officials who resent having to implement federal laws and regulations with little or no federal funding (unfunded mandates), or who disagree with certain federal regulations.

As part of this movement, one group developed a list of goals for what it called "wise use" of resources. Among those proposed goals were eliminating restrictions on wetland development; cutting older forests on national forest land and replacing them with tree plantations; opening all public lands, including wilderness areas, to mineral and energy development; and fining or penalizing anyone who challenges economic development on federal land.

Since 2000, efforts to weaken environmental laws and regulations have escalated. Nevertheless, independent polls show that more than 80% of the U.S. public strongly support environmental laws and regulations and do not want them weakened. However, polls also show that less than 10% of the U.S. public (and in hard economic times only about 2–3%) consider the environment to be one of the nation's most pressing problems. As a result, environmental concerns often are not transferred to the ballot box or to personal spending decisions.

To make a transition to a more environmentally sustainable society, U.S. citizens (and citizens in other democratic countries) will have to elect ecologically literate and environmentally concerned leaders. A rapidly growing number of citizens are insisting that elected leaders work across party lines to end the political deadlock that has virtually immobilized the U.S. Congress since 1980, with respect to environmental issues and other key concerns.

24.4 WHAT ARE THE MAJOR ROLES OF ENVIRONMENTAL GROUPS?

CONCEPT 24.4 Grassroots groups are growing and combining their efforts with those of large environmental organizations in a global sustainability movement.

The Roles of Environmental Groups

The spearheads of the global conservation, environmental, and environmental justice movements are the tens of thousands of nonprofit nongovernmental organizations (NGOs) working at the international, national, state, and local levels. The growing influence of these organizations is one of the most important changes influencing environmental decisions and policies (**Concept 24.4**).

NGOs range from grassroots groups with just a few members to organizations like the World Wildlife Fund (WWF), a 5-million-member global conservation organization that operates in 100 countries. Other international groups with large memberships include Greenpeace, the Nature Conservancy, and Conservation International.

Using social networks, text messages, e-mail, and Internet websites, some environmental NGOs have organized themselves into an array of influential *international networks*. Examples include the Pesticide Action, Climate Action, International Rivers, and Women's Environment and Development Networks. They collaborate across national borders, monitor the environmental activities of governments, corporations, and international agencies such as the World Bank and the World Trade Organization (WTO), and attend international conferences to try to influence negotiations and agreements. They also help to expose corruption and violations of national and international environmental agreements, such as the Convention on International Trade in Endangered Species (CITES), which prohibits international trade of endangered species (see Chapter 9, p. 211).

In the United States, more than 8 million citizens belong to more than 30,000 NGOs that deal with environmental issues. They range from small grassroots groups to large, heavily funded mainline groups, the latter usually staffed by expert lawyers, scientists, economists, lobbyists, and fund-raisers. The largest of these groups are the World Wildlife Fund, the Sierra Club, the National Wildlife Federation, the Audubon Society, Greenpeace (Figure 24.11), Friends of the Earth, and the Natural Resources Defense Council (see the Case Study that follows).

The largest groups have become powerful and important forces within the U.S. political system. They have

Jeremy sutton-hibbert/Alamy Stock Photo

FIGURE 24.11 These Greenpeace protesters used an inflatable motorized boat to try to hinder the hunting of whales by a Japanese whaling fleet. For several decades Greenpeace has engaged in such environmental actions and in environmental education activities.

helped to persuade Congress to pass and strengthen environmental laws (Figure 24.9), and they fight attempts to weaken or repeal these laws.

Some industries and environmental groups are working together to find solutions to environmental problems. For example, Greenpeace worked with a German manufacturer to build a refrigerator that does not use the potent greenhouse gases called HFCs as coolants. Now, there are more than 300 million of these GreenFreeze refrigerators in homes around the world.

Some environmental groups and networks are broadening the focus of their efforts. A good example of an extremely active group is the Natural Resources Defense Council.

CASE STUDY

The Natural Resources Defense Council

One of the stated purposes of the Natural Resources Defense Council (NRDC) is "to establish sustainability and good stewardship of the Earth as central ethical imperatives of human society. ... We work to foster the fundamental right of all people to have a voice in decisions that affect their environment. ... Ultimately, NRDC strives to help create a new way of life for humankind, one that can be sustained indefinitely without fouling or depleting the resources that support all life on Earth."

To those ends, NRDC goes to court to stop environmentally harmful practices. It also informs and organizes millions of environmental activists, through its website, magazines, and newsletters, to take actions to protect the environment—globally, regionally, and locally. Its BioGems network, accessible through the NRDC website, regularly informs subscribers about environmental threats all over the world, and helps people to take action by donating money, signing petitions, and writing letters to corporate and government officials and newspaper editors.

For example, the NRDC helped forge an agreement among Canadian timber companies, environmentalists, native peoples, and the provincial government of British Columbia (Canada) to protect a vast area of the Great Bear Rainforest from destructive logging. This followed years of pressure from NRDC activists on logging companies, their U.S. corporate customers, and provincial officials to protect the habitats of eagles, grizzly bears, wild salmon, and the rare spirit bear, a subspecies of the American black bear with a white fur coat.

Grassroots Environmental Groups

The base of the environmental movement in the United States and throughout the world consists of thousands of grassroots citizens' groups organized to improve environmental quality, often at the local level. Some historians say

INDIVIDUALS MATTER 24.2

Butterfly in a Redwood Tree

© Julia Butterfly Hill

"Butterfly" is the nickname given to Julia Hill, who spent 2 years of her life on a small platform near the top of a giant redwood tree in California. She was protesting the clear-cutting of a forest of these ancient trees, some of them more than 1,000 years old. She and other protesters occupied these trees illegally, as a form of nonviolent civil disobedience.

As a young woman, Butterfly had never participated in any such act of civil disobedience or environmental protest. She went to the site to express her belief that it was wrong for the trees' owners to cut them down for short-term economic gain. She planned to stay for only a few days.

But after seeing the destruction and climbing one of these magnificent trees, she ended up staying in the tree for 2 years to publicize what was happening and to help save the surrounding trees. She became a symbol of the protest and during her stay used a cell phone to communicate with members of the mass media throughout the world to help develop public support for saving the trees. Her living space was a platform not much bigger than a king-sized bed, 55 meters (180 feet) above the ground. Over time, she endured high winds, intense rainstorms, snow, and ice, and hours of noise from trucks, chainsaws, and helicopters.

Although Butterfly lost her courageous battle to save the surrounding forest, she persuaded Pacific Lumber MAXXAM to save her tree (called Luna) and a 60-meter (200-foot) buffer zone around it. Not too long after she descended from her perch, someone used a chainsaw to seriously damage the tree. Cables and steel plates are now used to preserve it.

In a larger sense, Butterfly did not really lose her fight. She wrote a book about her stand and has been traveling to campuses all over the world. In the process, she has inspired many people to stand up for protecting biodiversity and other environmental causes.

this movement began in the late 1960s when Wisconsin Senator Gaylord Nelson envisioned organizing the millions of people who were disgusted by pollution and other environmental problems that had grown severe throughout the 1950s and 1960s. Nelson and graduate student Denis Hayes put together a team of organizers who traveled the country doing environmental teach-ins (informal lectures) on college campuses. Their goal was to have a national teach-in to get the attention of policy makers and government officials.

The day they chose was April 22, 1970—the first Earth Day. It involved teach-ins and much more—thousands of public demonstrations focused on pollution, toxic waste, nuclear power, coal mining, lead contamination, and other urgent environmental issues. More than 20 million people took part. Later, Hayes worked on building the Earth Day Network, which now includes more than 180 nations. As a result, each year, Earth Day is now celebrated globally.

According to political analyst Konrad von Moltke, "There isn't a government in the world that would have done anything for the environment if it weren't for the citizen groups." Taken together, a loosely connected worldwide network of grassroots NGOs working today for bottom-up political, social, economic, and environmental change can be viewed as an emerging citizen-based *global sustainability movement* (**Concept 24.4**).

Since the 1970s, many grassroots groups have worked with individuals and communities to oppose harmful projects such as landfills, waste incinerators, and nuclear waste dumps, as well as to fight against the clear-cutting of forests and pollution from factories and power plants. They have also taken action against environmental injustice (Figure 24.3). And they have worked to make many communities more sustainable (see the Case Study that follows).

Grassroots groups have organized *conservation land trusts* wherein property owners agree to protect their land from development or other harmful environmental activities, often in return for tax breaks on the land's value. These groups have also spurred other similar efforts to save wetlands, forests, farmland, and ranchland from development, while helping to restore clear-cut forests, degraded grasslands, and wetlands and rivers that have been degraded by pollution.

The Internet, social networking, and text messaging have become important tools for grassroots groups. With these tools, they can expand their membership, raise funds, and quickly plan and execute actions such as demonstrations and rallies.

Most grassroots environmental groups use nonviolent and nondestructive tactics such as protest marches, sitting in trees to help prevent the clear-cutting of old-growth forests (Individuals Matter 24.2), and other approaches (Figure 24.11) for generating publicity to help educate and encourage the public to oppose various environmentally harmful activities. Such tactics often work because they produce bad publicity for practices and businesses that

TDP Photography / Alamy Stock Photo

FIGURE 24.12 Since 1984, citizens have worked together to make the city of Chattanooga, Tennessee, one of the best and most sustainable places to live in the United States.

threaten or degrade the environment. For example, after 2 years of pressure and protests, Home Depot agreed to sell only wood products made from certified sustainably grown timber. Within a few months, Lowes and eight other major building supply chains in the United States developed similar policies.

Much more controversial are militant environmental groups that use violent means, such as breaking into labs to free animals used in drug testing and destroying property such as bulldozers and SUVs. Most environmentalists strongly oppose such tactics.

CASE STUDY

The Environmental Transformation of Chattanooga, Tennessee

Local officials, business leaders, and citizens have worked together to transform Chattanooga, Tennessee, from a highly polluted city to one of the most sustainable and livable cities in the United States (Figure 24.12).

During the 1960s, U.S. government officials rated Chattanooga as one of the dirtiest cities in the United States. Its air was so polluted by smoke from its industries that people sometimes had to turn on their vehicle headlights in the middle of the day. The Tennessee River, flowing through the city's industrial center, bubbled with toxic waste. People and industries fled the downtown area and left a wasteland of abandoned and polluting factories, boarded-up buildings, high unemployment, and crime.

In 1984, the city decided to get serious about improving its environmental quality. Civic leaders started a *Vision 2000* process with a 20-week series of community meetings in which more than 1,700 citizens from all walks of life gathered to build a consensus about what the city could be at the turn of the century. Citizens identified the city's main problems, set goals, and brainstormed thousands of ideas for solutions.

By 1995, Chattanooga had met most of its original goals. The city had encouraged zero-emission industries to locate there and replaced its diesel buses with a fleet of quiet, zero-emission electric buses, made by a new local firm. The city also launched an innovative recycling program after environmentally concerned citizens blocked construction of a new garbage incinerator that would have emitted harmful air pollutants. These efforts paid off. Since 1989, the levels of the seven major air pollutants in Chattanooga have been lower than the levels required by federal standards.

Another project involved renovating much of the city's low-income housing and building new low-income rental units. Chattanooga also built one the world's largest freshwater aquariums, which became the centerpiece for downtown renewal. The city developed a riverfront park along both banks of the Tennessee River, which runs through downtown. As property values and living conditions have improved, people and businesses have moved back downtown.

In 1993, the community began the second stage of this process in *Revision 2000*. Goals included transforming an abandoned and blighted area in South Chattanooga into a mixed community of residences, retail stores, and zero-emission industries where employees can live near their workplaces. By 2009, most of these goals were met. Between 2009 and 2012, in the face of a general economic downturn, Chattanooga had one of the nation's strongest local economies, with a lower-than-average unemployment rate and rising property values.

Chattanooga's environmental success story is a shining example of people working together to produce a more livable and economically and environmentally sustainable city. It is an example of using the win-win **principle of sustainability**.

Student Environmental Groups and Researchers

Campus environmental groups have been leading the way on many campuses to try to make their schools and local communities more sustainable (**Core Case Study**). Most of these groups work with members of their school's faculty and administration to bring about environmental improvements in their schools and local communities.

For example, students at Middlebury College in Vermont worked with the school staff in switching the college from burning oil to heat its buildings to burning wood chips in a state-of-the-art boiler that is part of a cogeneration system. The system heats about 100 college buildings and spins a turbine to generate about 20% of the electricity used by the college, helping the school to reduce its carbon dioxide emissions by 40%. The college has also planted an experimental patch of fast-growing willow trees as a source of some of the wood chips for the system.

At Northland College in Ashland, Wisconsin, students helped to design a green living and learning center (see photo, p. 676), which houses 150 students and features a wind turbine, solar panels, furniture made of recycled materials, and waterless (composting) toilets. Northland students voted to impose a *green fee* of $40 per semester on themselves to help finance the college's sustainability programs.

Many student groups make *environmental audits* of their campuses or schools. They gather data on practices affecting the environment and use them to propose changes that will make their campuses or schools more environmentally sustainable while usually saving money in the process. Such audits have focused on implementing or improving recycling programs, convincing university food services to buy more food from local organic farms, shifting from fossil fuels to renewable energy, retrofitting buildings to make them more energy efficient, and implementing concepts of environmental sustainability throughout the curriculum.

Other students have focused on institutional investments. In 2015, more than 400 student-led campaigns were pressuring colleges and universities to stop investing their endowment funds in environmentally harmful industries, such as coal-fired electricity production. They also work toward getting their schools to increase their investments in renewable energy and other environmentally beneficial businesses.

CONSIDER THIS . . .

THINKING ABOUT Environmental Sustainability on Campus

What major steps is your school taking to increase its own environmental sustainability (**Core Case Study**) and to educate its students about environmental sustainability?

24.5 HOW CAN WE IMPROVE GLOBAL ENVIRONMENTAL SECURITY?

CONCEPT 24.5 Environmental security is necessary for economic security and is at least as important as national security.

Why Is Global Environmental Security Important?

Countries are legitimately concerned with *national security* and *economic security*. However, ecologists and many economists point out that all economies are supported by the earth's natural capital (Figure 1.3, p. 7, and Figure 23.5, p. 633). Thus, environmental security, economic security, and national security are interrelated (**Concept 24.5**).

According to environmental scientist Norman Myers, "If a nation's environmental foundations are degraded or depleted, its economy may well decline, its social fabric deteriorate, and its political structure become destabilized as growing numbers of people seek to sustain themselves from declining resource stocks. Thus, national security is no longer about fighting forces and weaponry alone. It

ROBIN MOORE/National Geographic Creative

FIGURE 24.13 Hillsides stripped of vegetation near Haiti's capital city of Port-au-Prince.

relates increasingly to watersheds, croplands, forests, genetic resources, climate, and other factors that, taken together, are as crucial to a nation's security as are military factors."

For example, Haiti has suffered from a severe loss of environmental and economic security because of a combination of rapid population growth, deforestation, severe soil erosion, rampant poverty, and political disruption. In a desperate struggle for survival, its people have stripped away most of the country's trees and other vegetation for use as firewood (Figure 24.13; see also Figure 10.17, p. 235). Because of this, along with several damaging hurricanes, the percentage of Haiti's land that was forested dropped from more than 60% in 1923 to about 2% in 2006. This major loss of vegetation led to severe soil erosion. Taken together, these factors along with a severe earthquake in 2010 have greatly reduced food production and led to greater poverty, malnutrition, and social unrest.

Research by Thomas Homer-Dixon, director of Canada's Trudeau Centre for Peace and Conflict Studies, has revealed a strong correlation between growing scarcities of resources, such as cropland, water, and forests, and the spread of civil unrest and violence that can lead to *failing states*. These countries have dysfunctional governments that can no longer provide security and basic services such as education and health care. They tend to suffer from a breakdown of law and order and general deterioration and often end up in civil wars as groups compete for power, and these wars can spread to nearby countries. Many failing states also become training grounds for terrorists (Afghanistan, Iraq, and Syria), weapons traders (Nigeria and Somalia), and drug producers (Afghanistan and Myanmar). Together, they also generate millions of refugees who are displaced from their homes and land, often while fleeing for their lives.

Myers and other analysts call for all countries to make environmental security a major focus of diplomacy and government policy at all levels.

Strengthening International Environmental Policies

A number of international environmental organizations help shape and set global environmental policy and improve environmental security and sustainability. Perhaps the most influential is the United Nations, which houses a large family of organizations including the U.N. Environment Programme (UNEP), the World Health Organization (WHO), the U.N. Development Programme (UNDP), and the Food and Agriculture Organization (FAO).

Other organizations that make or influence environmental decisions are the World Bank, the Global Environment Facility (GEF), and the World Conservation Union (also known as the IUCN). Despite their often limited funding, these and other organizations have played important roles in

- expanding understanding of environmental issues,
- gathering and evaluating environmental data,
- developing and monitoring international environmental treaties,
- providing grants and loans for sustainable economic development and reduction of poverty, and
- helping more than 100 nations to develop environmental laws and institutions.

International organizations face a number of major obstacles in addressing environmental problems, including varying priorities and deep distrust among key nations. In addition, a shortage of funds is often a problem. For example, critics have noted that UNEP cannot afford to do an adequate job in the key areas of data gathering on environmental problems, the spreading of information and ideas for addressing such problems, and the promotion of alliances among nations. Some have called for an environmental security council, on the order of the UN Security Council, to raise the priority level for dealing with environmental problems such as climate change that threaten long-term international security.

In 1992, governments of more than 178 nations and hundreds of NGOs met at the U.N. Conference on Environment and Development (UNCED) in Rio de Janeiro, Brazil. The major policy outcome of this conference was Agenda 21, a global agenda for sustainable development in the 21st century, with goals for addressing the world's social, economic, and environmental problems. The conference also established the Commission on Sustainable Development to monitor progress toward the Agenda 21 goals.

Despite the good intentions of Agenda 21, little progress has been made toward its goals. In 2012, the UNEP evaluated progress and found that, of the 90 most crucial goals, only 4 had been approached and none had been achieved. The removal of lead from gasoline, the phasing out of ozone-depleting chemicals, improvements to drinking water supplies in poor countries, and research on pollution of the oceans were the four areas where significant progress had been made. In other areas, such as carbon dioxide emissions, extinction threats, overfishing, ocean dead zones, and harm to coral reefs, the world has slipped farther away from the Agenda 21 goals.

In 2012, on the 20th anniversary of UNCED, the UN hosted Rio+20, another Earth Summit conference to revisit the issues addressed in 1992. It was the largest conference ever, with 50,000 attendees, and as it kicked off, there were up to 50,000 non-attendees demonstrating in the streets of Rio de Janeiro. After 3 days, the conferees produced a nonbinding document that they referred to as a roadmap for sustainable development. The representatives of more than 190 nations, including the United States, ratified it.

However, some analysts and organizations criticized the agreement for focusing primarily on traditional economic growth without enough attention to environmentally sustainable development. The document contained no enforceable commitment on climate change. It did not address any proposal to end fossil fuel subsidies and it failed to promote any sort of shift to renewable energy sources. Critics also argued that, because of excessive influence by corporate sponsors, the conference was unable to make real progress toward shifting the world onto a more sustainable path.

Figure 24.14 summarizes some of the successes and failures from long-term international efforts to deal with global environmental problems.

The primary focus of the international community on environmental problems has been the development of various international environmental laws and nonbinding policy declarations called *conventions*. There are more than 500 international environmental treaties and agreements—known as *multilateral environmental agreements* (MEAs).

To date, the Montreal Protocol and the Copenhagen Amendment for protecting the ozone layer (Chapter 18, p. 498) are the most successful examples of such agreements. The MEA process faces a number of challenges. MEAs typically take years to develop and require full consensus to implement. There is often a lack of funding and it becomes difficult to monitor and enforce these agreements, and they sometimes conflict with one another.

Trade-Offs

Global Efforts to Solve Environmental Problems

Successes	Failures
Over 500 international environmental treaties and agreements	Most international environmental treaties lack criteria for evaluating their effectiveness
1992 Copenhagen Ozone Protocol has helped reduce ozone-depleting chemicals	1992 Rio Earth Summit led to nonbinding agreements, inadequate funding, and limited improvements
1992 Rio Earth Summit adopted principles for handling global environmental problems	2012 Rio+20 Earth Summit failed to deal with climate change, energy policy, and biodiversity loss
2012 Rio+20 Earth Summit included small-scale policy improvements	Climate change conferences have all failed to deal with projected climate change

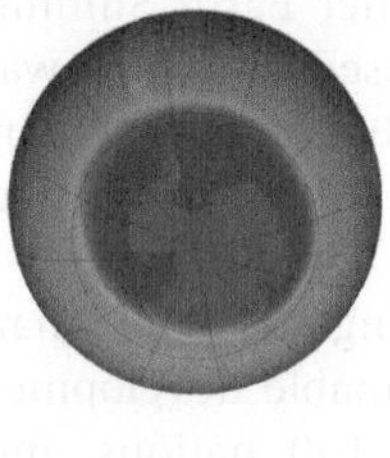

FIGURE 24.14 Trade-offs: There have been successes and failures in international efforts to deal with global environmental problems. ***Critical thinking:*** In weighing these successes and failures, do you believe that international conferences are valuable and should be continued? If you agree, how would you improve their effectiveness? If you disagree, what are some alternatives?

Photo: NASA

Role of Corporations in Promoting Environmental Sustainability

In our increasingly globalized economy, it has become clear that governments and corporations must work together to achieve goals for increased environmental sustainability. Governments can set environmental standards and goals through legislation and regulations, and corporations generally have highly efficient ways of accomplishing such goals. Making a transition to more sustainable societies and economies will require huge amounts of investment capital and research and development funding. Most of this money will likely have to come from profitable corporations, especially considering the recurring budgetary pressures faced by most governments. Thus, corporations could play a vital role in achieving a more sustainable future.

The good news is that some thoughtful business and political leaders are realizing that "business as usual" is no longer a viable option (Individuals Matter 23.2, p. 649). A growing number of corporate chief executive officers (CEOs) and investors are aware that there is considerable money to be made from developing and selling green products and services during this century. This switch to new product lines is guided by the concept of *eco-efficiency*, which is about finding ways to create more economic value with less harmful health and environmental impacts. Improving eco-efficiency can also save businesses money and help them to meet their financial responsibilities to stockholders and investors.

GOOD NEWS

Such improvements involve estimating the value of nature's ecosystem services not only for a company but also to the communities in which it works. For example, companies can learn how planting trees can improve their properties and how maintaining coastal wetlands can reduce the costs of hurricane damage. Companies such as 3M (Chapter 14, p. 461) have found that such investments can improve their bottom lines considerably.

24.6 HOW CAN WE IMPLEMENT MORE SUSTAINABLE AND JUST ENVIRONMENTAL POLICIES?

CONCEPT 24.6 Making the transition to more sustainable societies will require that nations and groups within nations cooperate and make the political commitment to achieve this transition.

Green Planning

Governments have the power to play strong roles in making a transition to a more sustainable future. Many argue that they have the responsibility to do so, through national policies (**Concept 24.6**). In some countries, the governments are doing just that in a process called *green planning*—the creation of long-term environmental management strategies with the ultimate goal of achieving greater environmental and economic sustainability and a high quality of life for a country's citizens.

Green plans usually involve most or all of the policy-guiding principles listed on page 659. Some include other principles such as the responsibility to leave the world sustainable for future generations, along with other priorities embodied in the **principles of sustainability**. Green plans are now being employed in several nations, including Mexico, Canada, New Zealand, Sweden, and The Netherlands (see the Case Study that follows).

SUSTAINABILITY

CASE STUDY

The Netherlands—A Model for a National Green Plan

In 1989, the northern European nation of The Netherlands began implementing a green plan called the National Environmental Policy Plan (NEPP). It resulted from widespread public alarm over declining environmental quality. The goal was to cut many types of pollution by 70–90% and achieve the world's first environmentally sustainable economy within 25 years, or one lifetime.

The Dutch government began by identifying eight major areas for improvement: climate change, acid deposition, eutrophication, toxic chemicals, waste disposal, groundwater depletion, unsustainable use of renewable and nonrenewable resources, and local nuisances (mostly noise and odor pollution).

Next the government formed a task force consisting of people in industry, government, and citizens' groups for each of the eight areas, and asked each task force to agree on targets and timetables for drastically reducing pollution. Each group was free to pursue whatever policies or technologies it wanted. However, if a group could not agree, the government would impose its own targets and timetables and stiff penalties for industries not meeting certain pollution reduction goals.

Each task force focused on four general themes: *life-cycle management*; *energy efficiency*, with the government committing $385 million per year to energy conservation programs; *environmentally sustainable technologies*, also supported by a government program; and *improving public awareness* through a massive government-sponsored public education program. The program was designed to be revised every 4 years to meet changing needs. Updates on progress and new problems are done every year.

Many of the country's leading industrialists like the NEPP because they can make investments in pollution prevention and pollution control with less financial risk and a high degree of certainty about long-term environmental policy. They are also free to deal with the problems in ways that make the most sense for their businesses. Many industrial leaders have learned that creating more environmentally sound products and processes often reduces costs and increases profits. And Dutch companies are making money in the global marketplace selling the technologies that were created to meet the NEPP goals.

The NEPP was the first attempt by any country to foster a national debate on the issue of environmental sustainability and to encourage innovative solutions to environmental problems. While the 25-year timeframe has passed and there is still work to be done, the NEPP is still in force. By 2015, more over 70% of the original goals had been met. A great deal of environmental research by the government and private sector has taken place. This has led to the expansion of organic agriculture, greater reliance on bicycles (Figure 24.15), and more ecologically sound housing developments. The NEPP is regarded as a blueprint for other nations wishing to create green plans.

GOOD NEWS

dinosmichail/Shutterstock.com

FIGURE 24.15 Bicycles are used for about one-third of all urban trips in the Netherlands. These bicycles are in Amsterdam, one of the world's most bicycle-friendly cities.

Making a Shift to More Environmentally Sustainable Societies

Scientists and other experts have suggested guidelines that we can follow as we work toward making our societies more environmentally sustainable. *First*, work on *preventing or minimizing* environmental problems instead of letting them build up to crisis levels. *Second*, use well-designed and carefully monitored *marketplace solutions* (see Chapter 23, p. 643) to help prevent or reduce the harmful impact of most environmental problems. *Third*, cooperate and innovate to find *win-win solutions* or *trade-offs* to environmental problems and injustices, in keeping with one of the **principles of sustainability**. *Fourth*, be *honest and objective*. People on both sides of thorny environmental issues should take a vow not to exaggerate or distort their positions in attempts to play win-lose or winner-take-all games. Some environmental scientists have specialized in helping people and organizations to apply these principles.

The world has the knowledge, technologies, and financial resources to make the shift to more equitable and environmentally sustainable global and national policies. Making this shift is primarily an economic, political, and ethical decision (see Chapter 25 for a discussion of environmental ethics). It involves shifting to more sustainable forestry (Figure 10.14, p. 232), food production (Figure 12.34, p. 314), water resource use (Figure 13.23, p. 346), energy resource use (Figure 16.2, p. 410), and economies (Figure 23.18, p. 648), while slowing climate change and educating the public and elected officials about the urgent need to make this shift over the next several decades.

Some say that the call for making this shift is idealistic and unrealistic. Others say that it is unrealistic and dangerous to keep assuming that our present course is sustainable, and they warn that we have precious little time to change our unsustainable course.

BIG IDEAS

- An important outcome of the political process is environmental policy—the body of laws, regulations, and programs that are designed, implemented, funded, and enforced by one or more government agencies.
- All politics is local, and individuals can work with each other to become part of political processes that influence environmental policies (*individuals matter*).
- Environmental security is necessary for economic security and is at least as important as national security; making the transition to more environmentally sustainable societies will require that nations cooperate just as many do for national security purposes.

Tying It All Together

Greening College Campuses and Sustainability

College students around the world have shown that it is possible to create sustainable environmental policies, at least in the communities in and around many college campuses (**Core Case Study**). The world has the abilities and resources to implement policies that would help to eradicate poverty and malnutrition, eliminate illiteracy, sharply reduce infectious diseases, stabilize human populations, and protect the earth's natural capital. We can do this by applying the **scientific principles of sustainability** (see Inside Back Cover)—relying much more on solar energy and other renewable energy sources, reusing and recycling much more of what we produce, and respecting, restoring, and protecting as much as possible of the biodiversity that supports our lives and economies.

National and international policy makers could also be guided by the three economic, political, and ethical **principles of sustainability** (see Inside Back Cover). In the political arena, they will have to try harder to find win-win solutions that benefit the largest numbers of people while also benefiting the environment. Such solutions will likely have to include internalizing the harmful environmental and health costs of producing and using goods and services (full-cost pricing). And if they are truly interested in long-term sustainability, these decision makers, as well as all the rest of us, must make each decision with future generations in mind—seeking to leave the world in at least as good a condition as what we now enjoy.

Making a shift to more sustainable societies and economies can occur much more rapidly than we think. Any or all of us can choose to take part in the change by becoming politically aware, informed, and active with regard to issues that affect our environmental and political futures.

Courtesy of Northland College

Chapter Review

Core Case Study

1. Give four examples of how colleges and universities are playing a leading role in shifting to more environmentally sustainable operations and policies.

Section 24.1

2. What is the key concept for this section? List three broad issues that government is the best equipped to handle and for each one, explain why this is so. What are government's **policies**? Define **politics** and **environmental policy**. What are the four stages of a policy life cycle? Why is the policy process usually cyclical and what is another name for it? What is a **democracy**? What are two types of special interest groups? Define **environmental justice** and explain why it has become an important issue. List eight principles that decision makers can use in making environmental policy.

Section 24.2

3. What are the two key concepts for this section? What are the three branches of government in the United States and what major role does each play? Describe three important components of policy. What is **lobbying**? Why are some environmental analysts concerned about the growing power of some lobbyists? Explain why developing environmental policy is a difficult and controversial process. What are three ways in which scientific and political processes differ? Why are these differences important for policy making?
4. What are four ways in which individuals can help to develop or change environmental policy in a democracy? What does it mean to say that *all politics is local*? What are four ways to provide environmental leadership?

Section 24.3

5. What is the key concept for this section? What is **environmental law**? What is a **civil suit**? Distinguish between the **plaintiff** and the **defendant** in a lawsuit. List and explain six factors that can limit the effectiveness of environmental lawsuits. What is a SLAPP?
6. List five general types of U.S. environmental laws and give an example of each. Explain what NEPA is, what it requires, and how effective it has been. What is an environmental impact statement? Explain how and why U.S. environmental laws have been under attack since 1980. Describe two problems that work against effective environmental regulations.

Section 24.4

7. What is the key concept for this section? Describe the roles of grassroots and mainstream environmental organizations and give an example of each type of organization. What is a global policy network? Give an example. Describe the role and effectiveness of the Natural Resources Defense Council (NRDC) in the formation of U.S. environmental policy. Give two examples of grassroots environmental groups and of how they can bring about change. How does the story of Chattanooga's transformation illustrate the potential effectiveness of citizen groups in affecting urban environmental policy? Give two examples of how students have played successful roles in affecting environmental policy on their campuses (**Core Case Study**).

Section 24.5

8. What is the key concept for this section? Explain the importance of environmental security, relative to economic and national security. List three examples of international environmental organizations and five ways in which they have played environmental policy making roles. Summarize the stories of the UNCED and Rio+20. What is Agenda 21? List two examples of successes and two examples of failures from international efforts to deal with global environmental problems. What are conventions and multilateral environmental agreements (MEAs) and what are three problems that have arisen with MEAs? Describe roles that corporations can play in helping to achieve environmental sustainability, and give an example of such an effort.

Section 24.6

9. What is the key concept for this section? Summarize the process of national environmental planning in The Netherlands. What are four guidelines for shifting to more environmentally sustainable societies?
10. Explain how many college students have managed to change the environmental policies of their institutions by applying the **principles of sustainability** (**Core Case Study**). How can these principles be applied to guiding national and international environmental policy making processes?

Note: Key terms are in bold type.

Critical Thinking

1. Consider the various actions taken by college students and their institutions as described in this chapter's **Core Case Study**, as well as in other parts of this chapter. Which one or more of these actions would be appropriate and effective on your campus? Explain. Pick one of these actions and write a brief plan for implementing it where you go to school.
2. Pick an environmental problem that affects the area where you live and decide where in the policy life cycle (Figure 24.2) the problem could best be placed. Apply the cycle to this problem and describe how the problem has progressed (or will likely progress) through each stage. If your problem has not progressed to the policy adjustment stage, explain how you think the policy dealing with the problem could be adjusted, if at all.
3. Explain why you agree or disagree with each of the eight principles listed on p. 659, which are recommended by some analysts for use in making environmental policy decisions. Which three of these principles do you think are the most important? Why?
4. What are two ways in which the scientific process described in Chapter 2 (see Figure 2.2, p. 31) parallels the policy life cycle (Figure 24.2)? What are two ways in which they differ?
5. Do you think that corporations and government bodies are ever justified in filing SLAPP lawsuits? Give three reasons for your answer. Do you think that potential defendants of SLAPP suits should be protected in any way from such suits? Explain.
6. Government agencies can help to keep an economy growing or to boost certain types of economic development by, for example, building or expanding a major highway through an undeveloped area. Proponents of such development have argued that requiring environmental impact statements for these projects interferes with efforts to help the economy. Do you agree? Is this a problem? Why or why not?
7. Congratulations! You are in charge of the country where you live. List the five most important components of your environmental policy.
8. List three ways in which you could apply the material in Section 24.2 to try to have an effect on an environmental policy making process.

Doing Environmental Science

Polls have identified five categories of citizens in terms of their concern over environmental quality: **(a)** those involved in a wide range of environmental activities, **(b)** those who do not want to get involved but are willing to pay more for a cleaner environment, **(c)** those who are not involved because they disagree with many environmental laws and regulations, **(d)** those who are concerned but do not believe individual action will make much difference, and **(e)** those who strongly oppose the environmental movement. To which group do you belong? Compile your answer and those of your classmates and determine what percentage of the total number of people in your class represents each category. As a class, conduct a similar poll of your entire school and compile the results.

Global Environment Watch Exercise

Go to your MindTap course to access the GREENR database. Starting on the home page, under "Browse Issues and Topics", click on *Pollution,* then select *Ozone Depletion.* Use this portal to find the latest research on the effectiveness of international efforts to reduce the use of CFCs. Write a summary of how negotiations among nations have progressed since they began. What were the most difficult stumbling blocks for the negotiators? How did the negotiators overcome these obstacles, if they were able to do so? Include the answers to these questions in your summary report.

Data Analysis

Choose an environmental issue that you have studied in this course, such as climate change, population growth, or biodiversity loss. Conduct a poll of students, faculty, staff, and local residents in your community by asking them the questions that follow, relating to your particular environmental issue. Poll as many people as you can in order to get a large sample. Create categories. For example, note whether each respondent is male or female. By creating such categories, you are placing each person into a *respondent pool*. You can add other questions about age, political leaning, and other factors to refine your pools.

Poll Questions

Question 1 On a scale of 1 to 10, how knowledgeable are you about environmental issue X?

Question 2 On a scale of 1 to 10, how aware are you of ways in which you, as an individual, impact policy making related to environmental issue X?

Question 3 On a scale of 1 to 10, how important is it for you to learn more about environmental issue X?

Question 4 On a scale of 1 to 10, how sure are you that an individual can have a positive influence on policy making related to environmental issue X?

Question 5 On a scale of 1 to 10, how sure are you that the government is providing the appropriate level of leadership with regard to environmental issue X?

1. Collect your data and analyze your findings to measure any differences among the respondent pools.
2. List any major conclusions you would draw from the data.
3. Publicize your findings on your school's website or in the local newspaper.

CHAPTER 25

Environmental Worldviews, Ethics, and Sustainability

The sustainability revolution is nothing less than a rethinking and remaking of our role in the natural world.

DAVID W. ORR

Key Questions

25.1 What are some major environmental worldviews?

25.2 What is the role of education in living more sustainably?

25.3 How can we live more sustainably?

A view within Yellowstone National Park in the U.S. state of Wyoming.

DREW RUSH/National Geographic Creative

Core Case Study

The United States, China, and Sustainability

We are living unsustainably. According to the Global Footprint Network, we would need 1.5 planet Earths to sustain indefinitely the resources that the world's 7.2 billion people consumed in 2013. By 2050, there will be about 9.8 billion people and we would need 3 planet Earths to indefinitely sustain their projected use of resources.

This helps explain why the greatest challenge we face is to learn how to live more sustainably during the next few decades. Meeting this challenge depends largely on the decisions and actions of the United States and China—the two countries that lead the world in resource consumption and production of wastes and pollutants.

From 1940 to 1970, the United States experienced rapid economic growth along with severe pollution and degradation of its air, water, and soil. By 1970, public awareness of these problems had grown and spurred an environmental movement made up of millions of citizens who demanded an end to this environmental degradation. During the 1970s, this prompted the U.S. Congress to pass a number of environmental laws (Figure 24.9, p. 666) that led to improvements in the nation's environmental quality. Without this bottom-up pressure by concerned citizens, the resulting improvement in environmental quality in the United States would not have happened.

Despite this important progress, the United States has the world's third largest population and the highest population growth rate of any industrialized country. It also has the world's largest ecological footprint and per capita ecological footprint (Figure 25.1)—mostly because it uses far more resources per person than any other country. If everyone in the world used resources equal to what the average American uses, we would need about five planet Earths to support them, according to the World Wildlife Fund (WWF) and the Global Footprint Network.

China has the world's largest population and the second largest economy. Since the 1970s, China's economy has been growing rapidly. Similar to what happened in the United States, China's economic growth has resulted in severe environmental problems. This contributes to China having the world's second largest total ecological footprint (Figure 25.1, top graph). However, because the large majority of Chinese citizens are poor and the Chinese population is so much larger, the per capita ecological footprint in China is about one-sixth that of the United States (Figure 25.1, bottom graph).

Since the 1960s, China has cut its birth rate in half and its population is growing at a rate slower than that of the United States. But if its middle class continues to grow and consume more resources as projected, China could have the world's largest per capita and total ecological footprints within a decade or two.

Because of their economic power and rates of resource use, the United States and China will play the key roles in determining whether and how we can live more sustainably on the planet that keeps us alive and supports the world's economies. This chapter discusses the environmental worldviews and ethical guidelines that can guide us in shifting to a more sustainable path.

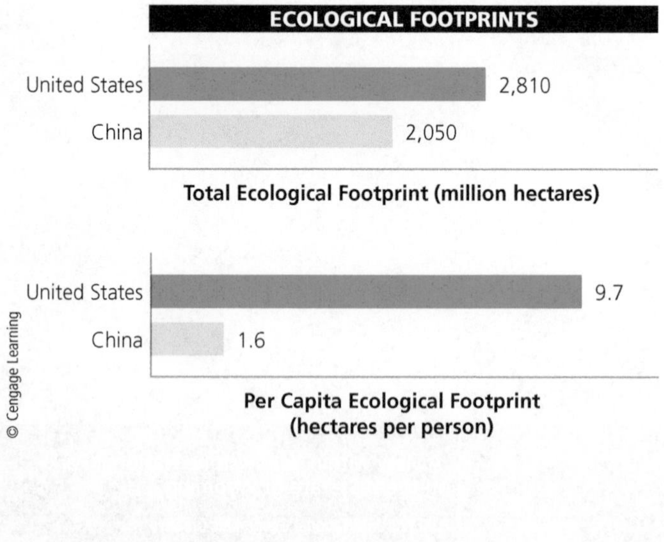

FIGURE 25.1 Comparison of total and per capita ecological footprints of the United States and China.

25.1 WHAT ARE SOME MAJOR ENVIRONMENTAL WORLDVIEWS?

CONCEPT 25.1 Major environmental worldviews differ on which is more important—human needs and wants, or the overall health of ecosystems and the biosphere.

Differing Environmental Worldviews

People disagree on how serious different environmental problems are, as well as on what we should do about them. These conflicts arise mostly because people have different environmental worldviews. As we noted in Chapter 1, your **environmental worldview** is the assumptions and beliefs that you have about how the natural world works and how you think you should interact with the environment.

Your environmental worldview is determined partly by your **environmental ethics**—what you believe about what is right and what is wrong in our behavior toward the environment. According to environmental ethicist Robert Cahn:

> *The main ingredients of an environmental ethic are caring about the planet and all of its inhabitants . . . and living each day so as to leave the lightest possible footprints on the planet.*

People with widely differing environmental worldviews can take the same data, be logically consistent in their analysis of the data, and arrive at quite different conclusions. This happens because they start with different assumptions and moral, ethical, or religious beliefs.

Human-Centered Environmental Worldviews

A **human-centered environmental worldview** focuses primarily on the needs and wants of people. One such worldview held by many people is the *planetary management worldview.*

According to this view, humans are the planet's most important, intelligent, and dominant species, and we can and should manage the earth mostly for our own benefit. The value of other species and parts of nature are based primarily on how useful they are to humans.

Here are three variations of the planetary management environmental worldview:

- *The no-problem school:* We can solve any environmental, population, or resource problem with more economic growth and development, better management, and better technology.
- *The free-market school:* The best way to manage the planet for human benefit is through a free-market global economy with minimal government interference and regulation. All public property resources should be converted to private property resources, and the global marketplace, governed only by free-market competition, should decide essentially everything.
- *The spaceship-earth school:* The blue marble in space that we call the Earth (Figure 25.2) is like a spaceship: a complex machine that we can understand, dominate, change, and manage, in order to provide a good life for everyone without overloading natural systems.

Ragnarock/Dreamstime.com

FIGURE 25.2 The blue marble in space that we call Earth is our only home.

Another human-centered environmental worldview is the *stewardship worldview.* It assumes that we have an ethical responsibility to be caring and responsible managers, or stewards, of the earth. It also calls for us to encourage environmentally beneficial forms of economic growth and development and discourage environmentally harmful forms.

Some people with the stewardship worldview believe that we have an ethical obligation to save the earth. American farmer, philosopher, and poet Wendell Berry calls this "arrogant ignorance." He and others point out that earth does not need saving. It has sustained an incredible variety of life for 3.8 billion years despite major changes in environmental conditions. According to Berry and other analysts, what needs saving and reform is the current human civilization that is degrading its life-support system and threatening up to half of the world's species with extinction.

Historians point out that so far, every major human civilization has eventually declined for various reasons. Some collapsed because of severe environmental degradation, such as deforestation and a failure to protect vital topsoil. The study of past human civilizations reveals two early warning signs in civilizations heading for collapse. The first sign is *gridlock* when civilizations are unable to understand or resolve major, complex problems that could lead to their downfall. The second sign, which occurs when the situation gets more desperate, is the *substitution of beliefs for facts and empirical evidence*. Despite increasing scientific

STILLFX/Shutterstock.com

FIGURE 25.3 We have limited understanding of how the trees, other plants, and animals in this patch of sequoia forest in California survive, interact, and change in response to changing environmental conditions. ***Critical thinking:*** How does this lack of knowledge relate to the planetary management worldview? Does this mean that we should never cut such trees? Explain.

and other evidence of deteriorating environmental, economic, and social conditions, people deny the threats and believe that some new technology or some unknown factor will prevent the collapse of their civilization.

Can We Manage the Earth?

Some people believe that any human-centered worldview will eventually fail because it wrongly assumes we now have or can gain enough knowledge and wisdom to become effective managers or stewards of the earth. Critics of human-centered worldviews point out that we are living unsustainably by taking over most of the earth's land and water, changing the earth's climate, acidifying the global ocean, and greatly increasing species extinction. Some argue we have exceeded four of the earth's planetary boundaries, or ecological tipping points (see Science Focus 3.3, p. 69), and are heading toward exceeding other ecological tipping points. We are doing this even though we know little about how the earth works and what goes on in a handful of soil, a patch of forest (Figure 25.3), the bottom of the ocean, and most other parts of the planet.

As biologist David Ehrenfeld puts it, "In no important instance have we been able to demonstrate comprehensive successful management of the world, nor do we understand it well enough to manage it even in theory." Biologist and environmental philosopher René Dubos made a related observation: "The belief that we can manage the earth and improve on nature is probably the ultimate expression of human conceit." The failure of the Biosphere 2 science project (Science Focus 25.1) supports this view.

According to some critics of human-centered worldviews, the unregulated global free-market approach will not work because it is based on ever-increasing economic growth. Such a growth view has little regard for the degradation and depletion of natural capital and the resulting long-term harmful environmental, health, economic, and social consequences. These critics argue that we cannot have unlimited economic growth and consumption on a finite planet with ecological limits or boundaries.

The image of the earth as a gigantic spaceship (Figure 25.2) has played an important role in raising global environmental awareness. But critics argue that thinking of the earth as a spaceship that we can manage is an oversimplified and misleading way to view an incredibly complex and ever-changing planet. They point out that we are a newcomer species that has been around for only about 200,000 years of the planet's 3.8 billion years of life, which makes it unlikely that we can understand and manage the planet.

Biosphere 2—A Lesson in Humility

In 1991, eight scientists (four men and four women) were sealed inside Biosphere 2, a $200 million glass and steel enclosure designed to be a self-sustaining life-support system (Figure 25.A). The goal of the project was to increase our understanding of Biosphere 1: the *earth's* life-support system.

This sealed system of interconnected domes was built in the desert near Tucson, Arizona. It contained artificial ecosystems including a tropical rain forest, savanna, and desert, as well as lakes, streams, freshwater and saltwater wetlands, and a mini-ocean with a coral reef.

Biosphere 2 was designed to mimic the earth's natural chemical cycling systems. Water evaporated from its ocean and other aquatic systems and condensed to provide rainfall over the tropical rain forest. The precipitation trickled through soil into the marshes and back into the ocean before beginning the cycle again.

The facility was stocked with more than 4,000 species of plants and animals, including small primates, chickens, and insects, selected to help maintain life-support functions. Human and animal excrement and other wastes were treated and recycled as fertilizer to help support plant growth. Sunlight and external natural gas–powered generators provided energy.

The Biospherians were to be isolated for 2 years and raise their own food, using intensive organic agriculture. They were to breathe air that was purified by plants and to drink water cleansed by natural chemical cycling processes.

From the beginning, many unexpected problems cropped up and the life-support system began to unravel. The level of oxygen in the air declined with soil organisms converting it to carbon dioxide. Additional oxygen had to be pumped in from the outside to keep the Biospherians from suffocating.

Tropical birds died after the first freeze. An ant species invaded the enclosure, proliferated, and killed off most of the system's native insect species. In total, 19 of the Biosphere's 25 small animal species (76%) became extinct. Before the 2-year period was over, all plant-pollinating insects went extinct, thereby dooming to extinction most of the plant species.

Despite many problems, the facility's waste and wastewater were recycled. With much hard work, the Biospherians were able to produce 80% of their food supply, despite rampant weed growths, spurred by higher CO_2 levels that crowded out food crops. However, the scientists suffered from persistent hunger and weight loss.

Joseph Sohm/Getty Images

FIGURE 25.A Biosphere 2, constructed near Tucson, Arizona, was designed to be a self-sustaining life-support system.

In the end, the Biosphere 2 project failed to maintain a life-support system for eight people for 2 years. Ecologists Joel E. Cohen and David Tilman (Individuals Matter 12.1, p. 314), who evaluated the project, concluded, "No one yet knows how to engineer systems that provide humans with life-supporting services that natural ecosystems provide for free."

CRITICAL THINKING

Do you think that science and engineering ever will be able to provide humans with the life support systems that nature now provides? Explain.

Life-Centered and Earth-Centered Environmental Worldviews

Critics of human-centered environmental worldviews argue that they should be expanded to recognize that all forms of life have value as participating members of the biosphere, regardless of their potential or actual use to humans. However, people disagree over how far we should extend our ethical concerns for various forms and levels of life (Figure 25.4).

All species eventually become extinct. However, most people with a **life-centered worldview** believe that we have an ethical responsibility to avoid hastening the extinction of species through our activities. People with an **earth-centered worldview** believe that we have an ethical responsibility to take a wider view and preserve the earth's biodiversity, ecosystem services, and the functioning of its life-support systems for the benefit of the earth's life, now and in the future (Figure 25.5).

One earth-centered worldview is called the *environmental wisdom worldview,* which in many ways is the opposite of the planetary management worldview. According to this view:

- We need to learn how nature has sustained life on the earth for 3.8 billion years and use these lessons from nature (environmental wisdom) to guide us in living more simply and sustainably.
- We are part of—not apart from—the community of life and the ecological processes that sustain all life.
- We are not in charge of the world.

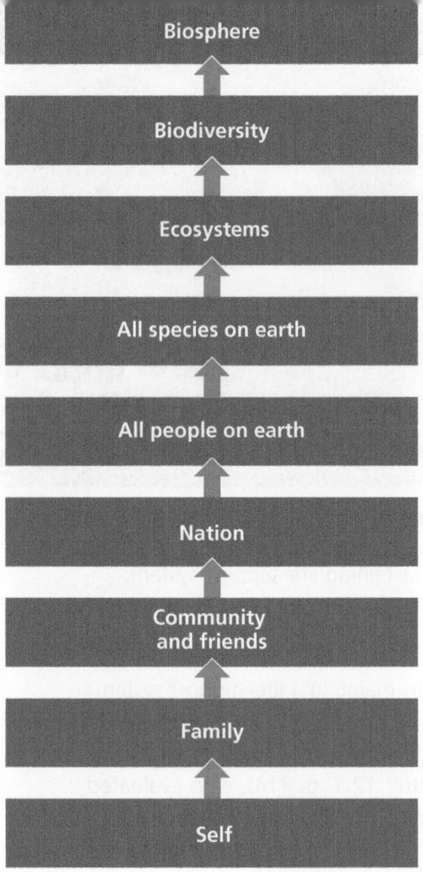

© Cengage Learning

FIGURE 25.4 Levels of ethical concern. People disagree about how far we should extend our ethical concerns on this scale.

- We are subject to nature's scientific laws that cannot be broken.
- Human economies and other systems are subsystems of the earth's life-support systems (Figure 23.5, p. 633).
- The earth's natural capital (Figure 1.3, p. 7) keeps us and other species alive and supports our economies.
- We need to learn how to work with nature (Figure 25.5) instead of trying to conquer it.
- By not degrading the earth's life-support system, we act in our own self-interest. Earth care is self-care.
- We have an ethical responsibility to leave the earth in as good a condition or better than what we inherited—in keeping with the ethical **principle of sustainability**.

In 2000, the United Nations published the *Earth Charter,* which was created with the help of 100,000 people in 51 countries and 25 global leaders in environmental science, business, politics, religion, and education. The charter incorporates many of the ideas and ethical concerns found in the earth-centered environmental worldview. Here are 6 of its 16 guiding ethical principles:

- Respect the earth and its life in all its diversity.
- Care for life with understanding, love, and compassion.
- Build societies that are free, just, participatory, sustainable, and peaceful.
- Secure the earth's bounty and beauty for present and future generations.
- Prevent harm as the best method of environmental protection.
- Eradicate poverty as an ethical, social, and environmental imperative.

The planetary management, stewardship, and environmental wisdom worldviews differ over how public resources should be managed in many parts of the world, including the United States (see the following Case Study).

Courtesy of Earth Flag, CO., 33 Roberts Road, Cambridge, MA 02138

FIGURE 25.5 The earth flag is a symbol of commitment to promoting environmental and economic sustainability by working with the earth at the individual, local, national, and international levels.

CASE STUDY

Managing Public Lands in the United States—A Clash of Worldviews

No nation has set aside as much of its land for public use, resource extraction, enjoyment, and wildlife habitat as has the United States. The federal government manages roughly 35% of the country's land, which is jointly owned by all U.S. citizens. About three-fourths of this federal

FIGURE 25.6 Natural capital: National forests, parks, and wildlife refuges managed by the U.S. federal government. ***Critical thinking:*** Do you think U.S. citizens should jointly own more or less of the nation's land? Explain.

(Compiled by the authors using data from U.S. Geological Survey and U.S. National Park Service.)

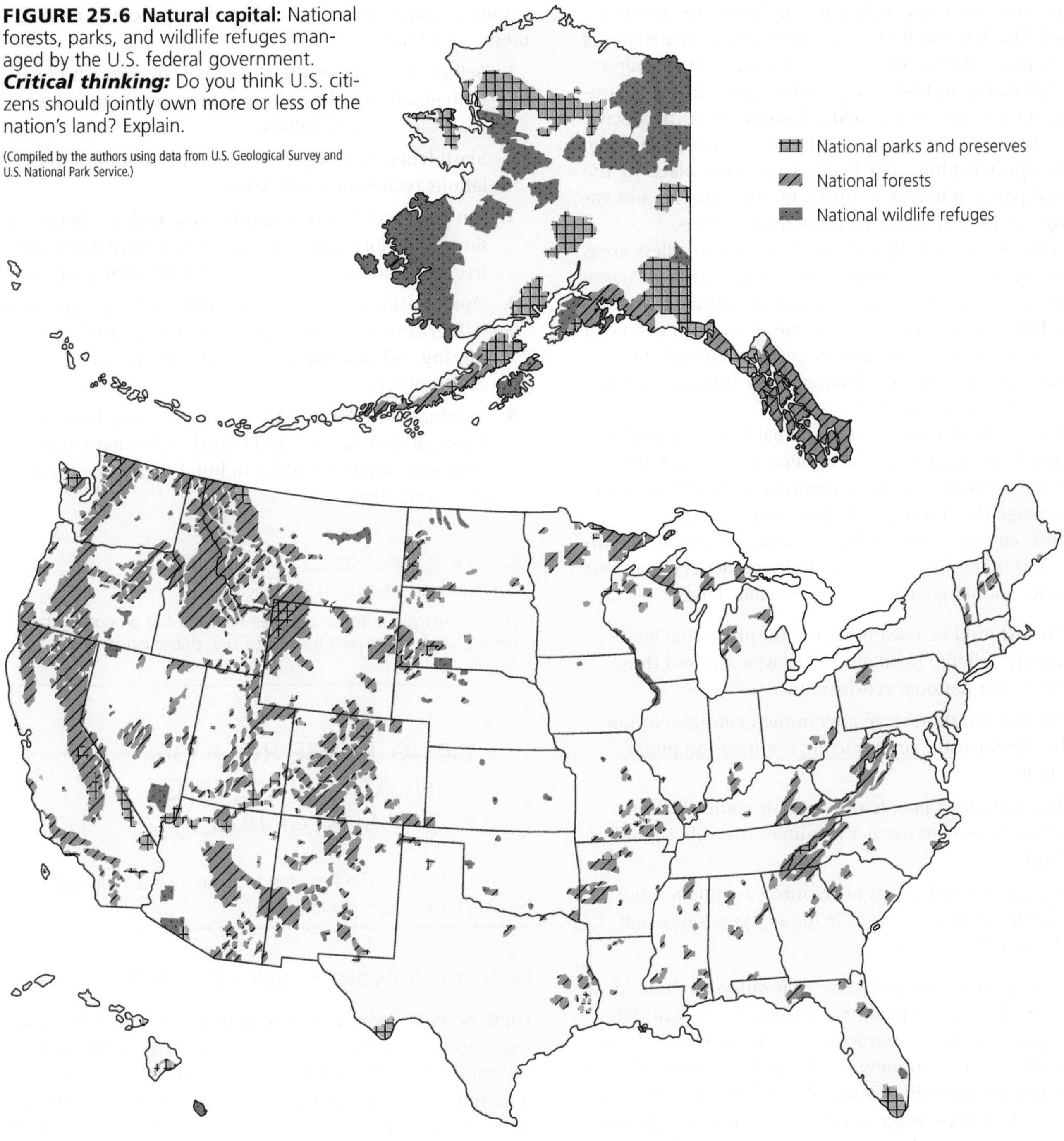

public land is in Alaska and another fifth is in the western states (Figure 25.6).

Some federal public lands are used for many different purposes. For example, the *National Forest System* consists of 155 national forests and 22 national grasslands. These lands, managed by the U.S. Forest Service (USFS), are used for logging, mining, livestock grazing, farming, oil and gas extraction, recreation, and conservation of watershed, soil, and wildlife resources.

The Bureau of Land Management (BLM) manages large areas of land—40% of all land managed by the federal government and 13% of the total U.S. land surface—mostly in the western states and Alaska. These lands are used primarily for mining, oil and gas extraction, logging, and livestock grazing.

The U.S. Fish and Wildlife Service (USFWS) manages 560 *national wildlife refuges*. Most refuges protect habitats and breeding areas for waterfowl (Figure 9.20, p. 213) and big game to provide a harvestable supply of these species for hunters. Permitted activities in most refuges include hunting, trapping, fishing, oil and gas development, mining, logging, grazing, farming, and some military activities.

The uses of some other public lands are more restricted. The *National Park System*, managed by the National Park Service (NPS), includes 59 major parks (chapter-opening photo) and 342 national recreation areas, monuments, memorials, battlefields, historic sites, parkways, trails, rivers, seashores, and lakeshores. Only camping, hiking, sport fishing, and boating can take place in the national parks, whereas hunting, mining, and oil and gas drilling are allowed in national recreation areas.

The most restricted public lands are 756 roadless areas that make up the *National Wilderness Preservation System* (Figure 10.23, p. 241). These areas lie within the other public lands and are managed by the agencies in charge of those lands. Most wilderness areas are open only for recreational activities such as hiking, sport fishing, camping, and non-motorized boating.

Many federal lands contain valuable oil, natural gas, coal, geothermal, timber, and mineral resources. Since the 1800s, there has been intense controversy over how to use and manage the resources on these public lands.

Most conservation biologists, environmental economists, and many free-market economists believe that four principles should govern the use of public lands:

1. They should be used primarily for protecting biodiversity, wildlife habitats, and ecosystems and thus serve as a national eco-insurance policy.
2. No one should receive government subsidies or tax breaks for using or extracting resources on public lands.
3. The American people deserve fair compensation from those who extract resources from their public lands.
4. All users or extractors of resources on public lands should be fully responsible for any environmental damage they cause.

There is strong and effective opposition to these ideas based largely on the planetary management worldview. Developers, resource extractors, many economists, and many citizens tend to view public lands in terms of their usefulness in providing mineral, timber, and other resources, and increasing short-term economic growth. They have succeeded in blocking implementation of the four principles listed above. For example, in recent years, analyses of budgets and spending reveal that the government has given an average of $1 billion a year—more than $2.7 million a day—in subsidies and tax breaks to privately owned interests that use U.S. public lands for activities such as mining, fossil fuel extraction, logging, and livestock grazing.

Some developers and resource extractors want to go further and open up federal lands for more development and resource extraction and reduce or eliminate federal regulation of these lands. Here are five of the ideas that such interests have proposed to the U.S. Congress since 1989, based on the planetary management worldview:

1. Sell public lands or their resources to corporations or individuals, usually at proposed prices that are less than their market values.
2. Slash federal funding for the administration of regulations related to public lands.
3. Cut diverse old-growth stands in the national forests for timber and for making biofuels, and replace them with tree plantations to be harvested for the same purposes.
4. Open national parks, national wildlife refuges, and wilderness areas to oil and natural gas drilling, mining, off-road vehicles, and commercial development.
5. Eliminate or take regulatory control away from the National Park Service and launch a 20-year construction program in the parks to build theme parks run by private firms.

CONSIDER THIS . . .

THINKING ABOUT U.S. Public Lands

Explain why you agree or disagree with the five proposals of developers for changing the use of U.S. public lands, listed above.

25.2 WHAT IS THE ROLE OF EDUCATION IN LIVING MORE SUSTAINABLY?

CONCEPT 25.2 The first step to living more sustainably is to become environmentally literate.

Becoming More Environmentally Literate

There is widespread evidence and agreement that we are a species in the process of degrading our own life-support system. During this century, this behavior will very likely threaten human civilization and the existence of up to half of the world's species. Part of the problem stems from incomplete understanding of how the earth's life-support system works, how our actions affect its life-sustaining systems, and how we can change our behavior toward the earth and thus toward ourselves. Improving this understanding begins by grasping three important ideas that form the foundation of environmental literacy. *First*, natural capital matters because it supports the earth's life and our economies. *Second*, our ecological footprints are immense and are expanding rapidly (**Core Case Study**, Figure 25.1). *Third*, we should not exceed the earth's planetary boundaries or tipping points (Science Focus 3.3, p. 69) because the resulting harmful consequences could last for hundreds to thousands of years.

Acquiring environmental literacy involves being able to answer certain key questions and having a basic understanding of certain key topics, as summarized in Figure 25.7. This also involves using the principles of biomimicry (Chapter 1 Core Case Study, p. 4, and Science Focus 1.1, p. 21) to understand how the life has sustained itself on the earth for 3.8 billion years. Then we can apply these principles and the three **scientific principles of sustainability** (see Inside Back Cover) to find out what works (Figure 25.8) and what lasts and how we might copy such earth wisdom.

Questions to answer

- How does life on earth sustain itself?
- How am I connected to the earth and other living things?
- Where do the things I consume come from and where do they go after I use them?
- What is environmental wisdom?
- What is my environmental worldview?
- What is my environmental responsibility as a human being?

Components

- Basic concepts: sustainability, natural capital, exponential growth, carrying capacity
- Principles of sustainability
- Environmental history
- The two laws of thermodynamics and the law of conservation of matter
- Basic principles of ecology: food webs, nutrient cycling, biodiversity, ecological succession
- Population dynamics
- Sustainable agriculture and forestry
- Soil conservation
- Sustainable water use
- Nonrenewable mineral resources
- Nonrenewable and renewable energy resources
- Climate disruption and ozone depletion
- Pollution prevention and waste reduction
- Environmentally sustainable economic and political systems
- Environmental worldviews and ethics

FIGURE 25.7 Achieving environmental literacy involves being able to answer certain questions and having an understanding of certain key topics (**Concept 25.2**). ***Critical thinking:*** After taking this course, do you feel that you can answer the questions asked here and have a basic understanding of each of the key topics listed in this figure?

Learning from the Earth

Formal environmental education is important, but is it enough? Many analysts say *no*. They call for us to appreciate not only the economic value of nature (Science Focus 10.1, p. 224), but also its ecological, aesthetic, and spiritual values. Many people in our increasingly urban world have little intimate contact with nature and an incomplete understanding of how nature works and sustains us. This can reduce their ability to act more responsibly toward the earth and thus toward themselves and other people.

A number of analysts suggest that we have much to learn from nature. They believe that we can experience awe, wonder, mystery, excitement, and humility by being in a forest, enjoying a beautiful scene in nature (Figure 25.9), or taking in the majesty and power of the sea. We might pick up a handful of topsoil and try to sense the teeming microscopic life within it that helps to keep us alive by supporting food production. We might look at a tree (Figure 25.3), a mountain, a rock, or a honeybee, or listen to the sound of a bird and try to sense how each of them is connected to us and we to them, through the earth's life-sustaining processes.

We can also learn more about the environment close to where we live. Understanding and directly experiencing the free gifts we receive from nature can inspire us make an ethical commitment to live more sustainably and, in the process, to help preserve our own species and cultures.

According to some psychologists and other analysts, experiencing nature is necessary for healthy living. Journalist Richard Louv has specialized in studying relationships among family, community, and nature. He coined the term **nature-deficit disorder** to describe a wide range of problems, including anxiety, depression, and attention-deficit disorders, that can result from or be intensified by a lack of contact with nature.

Psychologist David Strayer and other scientists have been conducting physiological and psychological research on the beneficial effects of getting away from the stress of everyday life and experiencing nature. They found that by giving our brains time to rest and recover from the stresses of living, and by interacting with nature, we can help reduce depression, anxiety, stress, blood pressure, and mental fatigue. Research indicates that, in addition to calming us down, connecting with nature can improve our attention skills, short-term memory, and creativity. And we can get these multiple benefits with no known harmful side effects and usually at a low cost.

Many analysts also view our increasing isolation from nature as one of the basic causes of the environmental problems we face (Figure 1.11, p. 16). When we lack an

INDIVIDUALS MATTER 25.1

Juan Martinez—Reconnecting People with Nature

REBECCA HALE/National Geographic

National Geographic Explorer Juan Martinez learned first-hand about the value of connecting with nature. Now he is instilling that value in others, particularly disadvantaged youths.

Martinez grew up in a poor area of Los Angeles, California, where as a boy he was in danger of becoming absorbed by a gang culture. One of his teachers recognized Martinez's potential and gave him a chance to pass a class that he was failing by joining the school's Eco Club.

Martinez took that opportunity and when the club planned a field trip to see the Grand Teton Mountains of Wyoming, he jumped at the chance. As a result, he says, "I still can't find words to describe the first moment I saw those mountains rising up from the valley. Watching bison, seeing a sky full of stars, and hiking through that scenery was overwhelming."

The experience transformed Martinez's life. Today, he spearheads the Natural Leaders Network of the Children and Nature Network, an organization creating links between environmental organizations, corporations, government, education, and individuals to reconnect children with nature. His work as an environmental leader has inspired many others to do similar work.

Martinez has received a great deal of recognition for his efforts, including invitations to White House forums on environmental education. His greatest reward, however, is in seeing how his efforts help others.

---

understanding of how nature keeps us alive and supports our economies, we can unknowingly degrade the earth's natural capital (Figure 1.3, p. 7) and the ecosystem services it provides. Learning about and protecting natural capital is an essential component of living more sustainably.

Urban living, along with overuse of the Internet, cell phones, and other electronic devices, contribute to nature-deficit disorder by hindering many people in experiencing nature. However, thanks to modern technology, large numbers of small, high-resolution cameras could be set up in the some of the world's most diverse nature reserves and run continuously. This would allow people around the world to use a few keystrokes to visit any reserve online, and such a program could help millions of people to experience nature virtually. In addition, many environmental leaders are now helping people connect directly with nature (Individuals Matter 25.1).

Earth-focused philosophers say that to be rooted, each of us needs to find a *sense of place*—a stream, a mountain, a patch of forest, a yard, a neighborhood lot—any piece of the natural world that we know, experience emotionally, and love. When we become part of a place, it becomes a part of us. Then we might be driven to defend it from harm (Figure 25.10) and to help heal its wounds. This can help us discover and tap into what conservationist Aldo Leopold (Individuals Matter 25.2) called "the green fire that burns in our hearts" and use this as a force for respecting and working with the earth and with one another.

Daniel Schoenen/imageBROKER/Age Fotostock

FIGURE 25.8 By applying the solar energy **principle of sustainability**, scientists and engineers developed solar cells that can be used to produce electricity in this solar village in Vauban Freiburg, Germany.

INDIVIDUALS MATTER 25.2

Aldo Leopold

Robert McCabe/Courtesy of the University of Wisconsin-Madison Archive

According to the American forester, ecologist, and writer Aldo Leopold (1887–1948), the role of the human species should be to protect nature, not to conquer it. His book, *A Sand County Almanac* (published after his death), is an environmental classic that has helped inspire the modern environmental and conservation movements.

In 1933, Leopold became a professor at the University of Wisconsin, and in 1935, he helped to found the Wilderness Society. Through his writings and teachings, he became one of the foremost leaders of the conservation and environmental movements during the 20th century. His energy and foresight helped to lay the critical groundwork for the field of environmental ethics. The following quotations from his writings reflect Leopold's land ethic, and they form the basis for many of the beliefs and principles of the modern stewardship and environmental wisdom worldviews:

All ethics so far evolved rest upon a single premise: that the individual is a member of a community of interdependent parts.

To keep every cog and wheel is the first precaution of intelligent tinkering.

That land is a community is the basic concept of ecology, but that land is to be loved and respected is an extension of ethics.

The land ethic changes the role of Homo sapiens from conqueror of the land-community to plain member and citizen of it.

We abuse land because we regard it as a commodity belonging to us. When we see land as a community to which we belong, we may begin to use it with love and respect.

A thing is right when it tends to preserve the integrity, stability, and beauty of the biotic community. It is wrong when it tends otherwise.

djgis/Shutterstock.com

FIGURE 25.9 Experiencing nature can help us to understand the need to protect the earth's natural capital and to live more sustainably.

Joel W. Rogers/Corbis

FIGURE 25.10 This woman and others in Vancouver, Canada, are protesting the clear-cutting of old-growth forests for timber in the Canadian province of British Columbia.

Guidelines for living more sustainably

- Mimic the ways nature sustains itself by using the earth as a model and teacher.
- Protect the earth's natural capital and repair ecological damage caused by human activities.
- Focus on preventing pollution and resource waste.
- Reduce resource consumption, waste, and pollution by reducing demand and using matter and energy resources more efficiently.
- Recycle, reuse, and repair everything and thus copy nature by having our wastes become resources.
- Rely more on clean, renewable energy resources such as solar and wind energy.
- Slow climate change.
- Reduce population growth and gradually reduce population size.
- Celebrate and protect biodiversity and cultural diversity.
- Promote social justice for humans and ecological justice for other species that keep us alive.
- End poverty.
- Leave the earth in a condition that is as good as or better than what we inherited.

© Cengage Learning

FIGURE 25.11 *Sustainability dozen:* Guidelines for living more sustainably.

25.3 HOW CAN WE LIVE MORE SUSTAINABLY?

CONCEPT 25.3 We can live more sustainably by becoming environmentally literate, learning from nature, living more simply and lightly on the earth, and becoming active environmental citizens.

Living More Simply and Lightly on the Earth

Mostly because of our actions, we are living on a planet with a warmer and sometimes harsher climate, less dependable supplies of water, more acidic oceans, extensive soil degradation, increased species extinction, degradation of key ecosystem services, and widespread ecological disruption. Unless we change our course, scientists warn that these and other harmful ecological changes will intensify. Figure 25.11 lists 12 guidelines—the "sustainability dozen"—developed by environmental scientists and ethicists for living more sustainably by converting environmental concerns, literacy, and wisdom into environmentally responsible actions.

Some analysts urge people who have a habit of consuming excessively to live more simply and sustainably. Seeking happiness through the pursuit of material things is considered folly by almost every major religion and philosophy. Yet, today's avalanche of advertising messages encourages people to buy more and more things to fill a growing list of wants as a way to achieve happiness. As American humorist and writer Mark Twain (1835–1910) observed: "Civilization is the limitless multiplication of unnecessary necessities." American comedian George Carlin (1937–2008) put it another way: "A house is just a pile of stuff with a cover on it. It is a place to keep your stuff while you go out and get more stuff."

According to research by psychologists, what a growing number of people really want, deep down, is more community, not more stuff. They want greater and more fulfilling interactions with family, friends, and neighbors.

Some people in more-developed countries are adopting a lifestyle of *voluntary simplicity*. It should not be confused with poverty, which is involuntary simplicity. Voluntary simplicity involves learning to live with less, using products and services that have smaller harmful environmental impacts, and creating beneficial environmental impacts (**Concept 25.3**). These individuals view voluntary simplicity not as a sacrifice but as a way to have a more fulfilling and satisfying life. Instead of working longer to pay for bigger vehicles and houses, they are spending more time with their loved ones, friends, and nature. Their goals are to consume less, share more, live simply, make friends, treasure family, and enjoy life. Their motto is: "Shop less, live more."

Practicing voluntary simplicity is a way to apply the Indian philosopher and leader Mahatma Gandhi's *principle of enoughness:* "The earth provides enough to satisfy every person's need but not every person's greed. . . . When we take more than we need, we are simply taking from each other, borrowing from the future, or destroying the environment and other species." Most of the world's major religions have similar teachings.

Living more lightly starts with asking the question: How much is enough? Similarly, one can ask: What do I really need? These are not easy questions to answer, because people in affluent societies are conditioned to want more and more material possessions and to view them as needs instead of wants. As a result, many people are addicted to buying more and more stuff as a way to find meaning in their lives, and they often run up large personal debts to feed their stuff habit. This addiction leads to what author James Wallman calls "stuffocation." Figure 25.12 lists five steps that some psychologists have advised people to take to help them withdraw from this addiction.

CONSIDER THIS . . .

THINKING ABOUT Your Basic Needs

Make a list of your basic needs. Is your list of needs compatible with your environmental worldview?

Throughout this text, you have encountered lists of ways we can live more lightly by reducing the size and impact of our ecological footprints. The human activities that have the greatest harmful impacts on the environment are food production, transportation, and home energy use. Based on this analysis, Figure 25.13 lists eight key ways for individuals to live more simply and sustainably.

- Avoid buying something just because a friend has bought it
- Go on an ad diet by not watching or reading advertisements
- Avoid shopping for recreation and buying on impulse
- Stop using credit and buy only with cash to avoid overspending
- Borrow and share things like books, tools, and other consumer goods

FIGURE 25.12 Five ways to withdraw from an addiction to buying more and more stuff.

CONSIDER THIS . . .

THINKING ABOUT Living More Lightly

Which three of the eight steps in Figure 25.13 do you think are the most important? Which of these things do you already do? Which of them are you thinking about doing? How do your answers to these questions relate to **(a)** the six **principles of sustainability**, and **(b)** to your environmental worldview?

Living more sustainably is not easy, and we will not make this transition by relying primarily on technological fixes such as recycling, changing to energy-efficient light bulbs, and driving energy-efficient cars. These are, of course, important things to do. They can help us to shrink our ecological footprints and to feel less guilty about our harmful impacts on our life-support system. However, these efforts cannot solve the environmental problems resulting from excessive consumption of and unnecessary waste of matter and energy resources.

Some analysts have suggested that the environmental movement has focused too much on bad news and laying blame, which has then led people to feel guilty, fearful, apathetic, and powerless. They suggest that we move beyond these immobilizing feelings by recognizing and

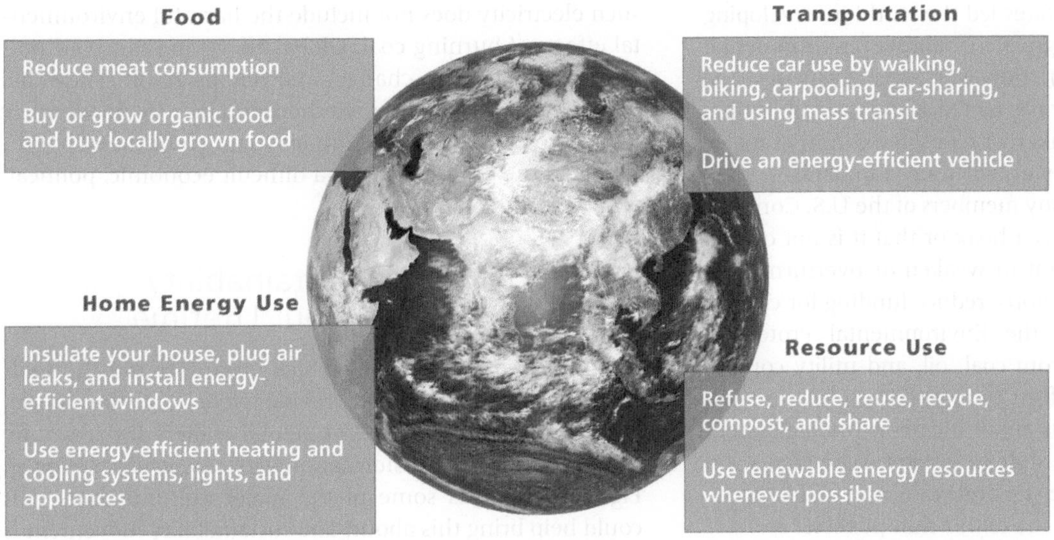
Food
Reduce meat consumption
Buy or grow organic food and buy locally grown food
Transportation
Reduce car use by walking, biking, carpooling, car-sharing, and using mass transit
Drive an energy-efficient vehicle
Home Energy Use
Insulate your house, plug air leaks, and install energy-efficient windows
Use energy-efficient heating and cooling systems, lights, and appliances
Resource Use
Refuse, reduce, reuse, recycle, compost, and share
Use renewable energy resources whenever possible

FIGURE 25.13 *Living more lightly:* Eight ways to shrink our ecological footprints (**Concept 25.3**). ***Critical thinking:*** Which of these things do you already do? Which, if any, do you hope to do in the future?

avoiding the following two common mental traps that lead to denial, indifference, and inaction:

- *Gloom-and-doom pessimism (it is hopeless)*
- *Blind technological optimism (science and technological fixes will save us)*

Avoiding these traps helps us to hold on to, and be inspired by, empowering feelings of realistic hope, rather than to be immobilized by feelings of despair and fear.

CONSIDER THIS . . .

THINKING ABOUT Mental Traps

Have you fallen into either of these traps? If so, are you aware that you have, and how do you think you could free yourself from either of them?

Here is what business entrepreneur and environmental writer Paul Hawken told the 2009 graduating class at the University of Portland:

> *When asked if I am pessimistic or optimistic about the future, my answer is always the same: If you look at the science about what is happening on the earth and aren't pessimistic, you don't understand the data. But if you meet the people who are working to restore this earth and the lives of the poor, and you aren't optimistic, you haven't got a pulse. . . . You join a multitude of caring people. . . . This is your century. Take it and run as if your life depends on it.*

Enjoying nature, beauty, connectedness, friendship, sharing, caring, and love can empower us to become good earth citizens who practice good earthkeeping. As Mahatma Gandhi reminded us, "Power based on love is a thousand times more effective and permanent than power derived from fear."

Revisiting the United States, China, and Sustainability

In the 1970s, the United States led the world in developing laws and regulations designed to improve environmental quality (**Core Case Study**). However, since 1980 the U.S. environmental community has had to spend most of its time fending off attempts to weaken or repeal the country's major environmental laws—many of which need updating.

At the federal level, many members of the U.S. Congress think that climate change is a hoax or that it is not caused by human actions and want to weaken or overturn environmental laws and regulations, reduce funding for climate research, and get rid of the Environmental Protection Agency. Under pressure from coal, oil, and utility companies, they have blocked efforts to reduce fossil fuel use (especially coal), use a carbon tax or a carbon-trading system to reduce CO_2 emissions, shift to greater dependence on renewable energy from the sun and wind, and build a modern smart electrical grid to make this shift possible.

China's leaders have plans to become more environmentally responsible over the next few decades for two reasons. One is to maintain their political power by heading off growing citizen unrest over the country's severe pollution, as the U.S. government did in the 1970s. The other is to dominate the world's rapidly growing and profitable green energy and low-carbon businesses. If successful, China could become the world's leader in making the shift to more sustainable economies and societies and reduce its total environmental footprint (Figure 25.1, top).

China produces and sells more wind turbines and solar cell panels than any country in the world and is building a smart electrical grid to distribute electricity produced by the sun and wind throughout the country. It has also developed a growing network of bullet trains that can help reduce car use.

Over the next few decades, the Chinese government plans to depend more on cleaner energy systems and become the global leader in developing a low-carbon economy. It has plans to tax carbon pollution from the burning of fossil fuels and to use the income to shift away from fossil fuel use before the United States does. The goal is to make money by becoming the global leader in making the shift to the new energy transition (Section 16.1, p. 409). However, China burns coal to provide 65% of its electricity and reducing its dependence on abundant and cheap coal is a major economic and political challenge.

Although China has shifted partially to a market-based economy, it has a centralized (and unelected) government, without the checks and balances of Western democracies. In China, environmental and economic change can occur more quickly because policies are developed and implemented by the central government. This means that China's leaders can play a major role in helping the world shift to a more sustainable path or in staying on its current unsustainable path.

The United States and China face similar problems. They have large reserves of coal that can be burned to produce electricity at a low cost, as long as the price of such electricity does not include the harmful environmental effects of burning coal. Global efforts to reduce air pollution, slow climate change, and rely more on renewable energy from the sun and wind depend heavily on whether these two countries decide to leave much of their coal reserves in the ground. This is a difficult economic, political, and ethical decision.

Bringing About a Sustainability Revolution during Your Lifetime

The Industrial Revolution, which began around the mid-18th century, was a remarkable global transformation. Now in this century, environmental leaders say it is time for another global transformation—a *sustainability revolution*. Figure 25.14 lists some of the major cultural shifts that could help bring this about. The sustainability movement is

INDIVIDUALS MATTER 25.3

Lester R. Brown: Champion of Sustainability

KFEM/Earth Policy Institute

Lester R. Brown served as president of the Earth Policy Institute, which he founded in 2001 until his retirement in 2015. The purpose of this nonprofit, interdisciplinary research organization has been to provide a plan for a more sustainable future and a roadmap showing how we could get there.

Brown is an interdisciplinary thinker and one of the pioneers of the global sustainability movement. For decades, he has been researching and describing the complex and interconnected environmental issues we face and proposing concrete strategies for dealing with them. The *Washington Post* called him "one of the world's most influential thinkers," and *Foreign Policy* named him one of the Top Global Thinkers.

Brown's *Plan B* for shifting to a more environmentally and economically sustainable future has four main goals: **(1)** stabilize population growth, **(2)** stabilize climate change, **(3)** eradicate poverty, and **(4)** restore the earth's natural support systems.

Brown has written or coauthored more than 50 books, which have been translated into more than 40 languages. He has received numerous prizes and awards, including 25 honorary degrees, the United Nations Environment Prize, and Japan's Blue Planet Prize. In 2012, he was inducted into the Earth Hall of Fame in Kyoto, Japan. He also holds three honorary professorships in China.

Despite the serious environmental challenges we face, Brown sees reasons for hope. They include his understanding that social change can sometimes occur very quickly. He is also encouraged by improvements in fuel efficiency, the emerging shift from using coal to using solar and wind energy to produce electricity, and a growing public understanding of our need to live more sustainably.

a decentralized global movement arising mostly from the bottom up, based on the actions of a variety of individuals and groups throughout the world. One of the leaders in the movement to develop and promote detailed plans for making the shift to more sustainable ways of living is Lester R. Brown (Individuals Matter 25.3).

A growing number of people call for us to change the way we treat the earth and thus ourselves by living more gently on the planet that sustains us. Figure 25.15 lists a number of seeds or agents of change that can help us shift to a more sustainable path within your lifetime.

Here are two pieces of good news about the possibility of bringing about a sustainability revolution over the next few decades. GOOD NEWS *First,* social science research reveals that for a major social change to occur, only 5–10% of the people in the world, or in a country or locality, must be convinced that the change must take place and then act to bring about such change. *Second,* history also shows that we can bring about change faster than we might think, once we have the courage to leave behind ideas and practices that no longer work and to nurture new trends such as the rapidly growing seedlings of sustainability listed in Figure 25.15.

This is an exciting and challenging time to be alive because we have the knowledge to shift from our current unsustainable path to a more sustainable one. Within this

Unsustainable Path	Sustainable Path
Energy and Climate	
Fossil fuels	Direct and indirect solar energy
Energy waste	Energy efficiency
Climate disruption	Climate stabilization
Matter	
High resource use and waste	Less resource use
Consume and throw away	Reduce, reuse, and recycle
Waste disposal and pollution control	Waste prevention and pollution prevention
Life	
Deplete and degrade natural capital	Protect natural capital
Reduce biodiversity	Protect biodiversity
Population growth	Population stabilization

© Cengage Learning

FIGURE 25.14 Solutions: Some of the cultural shifts in emphasis that scientists say will be necessary to bring about a sustainability revolution. ***Critical thinking:*** Which of these shifts do you think are most important? Why?

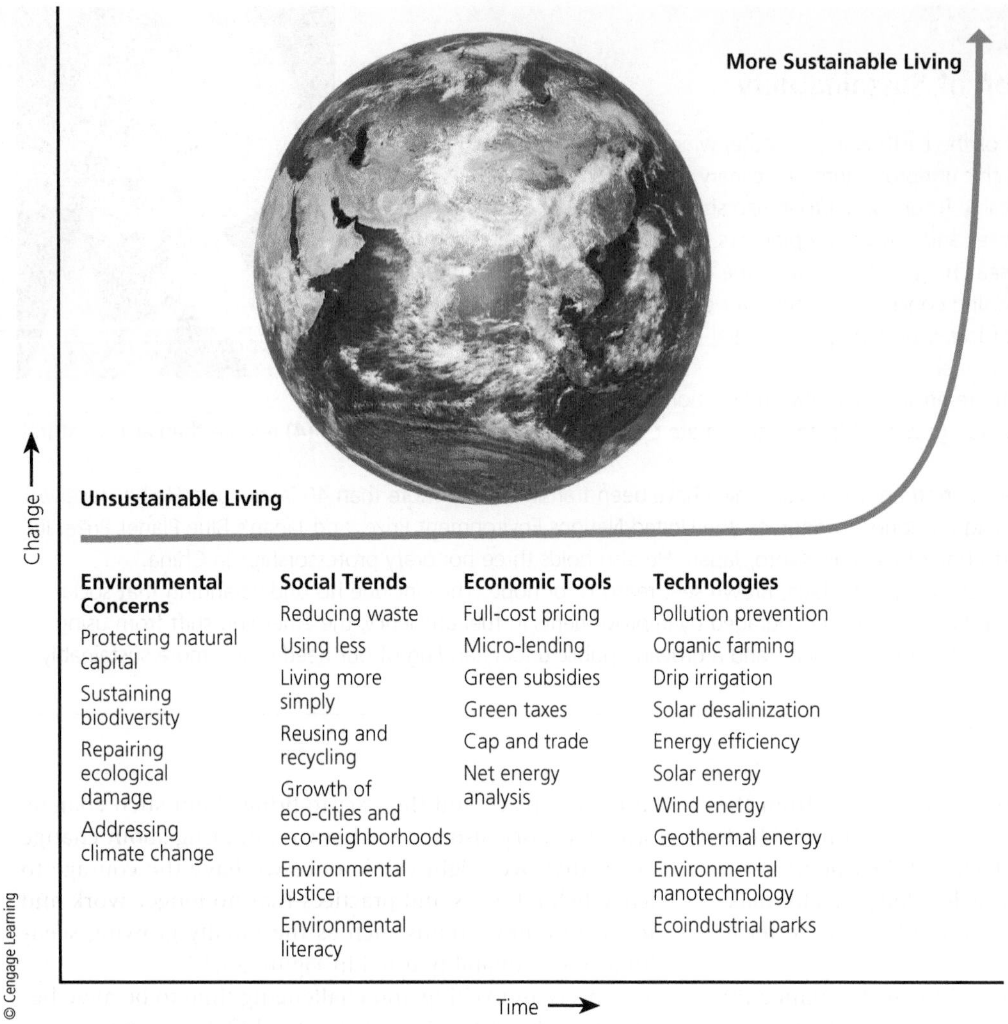

FIGURE 25.15 Seeds of change and hope. The agents of change in this figure are growing slowly. However, at some point, some or all of them could take off, grow exponentially, and help bring about a sustainability revolution within your lifetime. ***Critical thinking:*** Which two items in each of these four categories do you believe are the most important to promote? What other items would you add to this list?

century, a small but dedicated group of people from around the world can bring about a sustainability revolution. They will likely understand three things. *First,* we have been borrowing from the earth and the future and our debt is coming due. *Second,* as a species we are capable of great things, if we choose to act. *Third,* once we start on a new path, change will likely spread through our web-connected global social networks at an amazing pace.

While some skeptics say the idea of a sustainability revolution is idealistic and unrealistic, entrepreneur Paul Hawken, in a 2009 graduation address, observed that "the most unrealistic person in the world is the cynic, not the dreamer." In addition, according to the late Steve Jobs, cofounder of Apple Inc., "The people who are crazy enough to think they can change the world are the ones who do." These and other individuals had the courage to forge ahead with ideas that others called idealistic and unrealistic. Had they not, very few of the human achievements that we now celebrate would have happened. Can we shift to a more sustainable world? Yes—if enough people act to make it happen. Join them.

BIG IDEAS

- Our environmental worldviews play a key role in how we treat the earth that sustains us and thus in how we treat ourselves.
- We need to become more environmentally literate about how the earth works, how we are affecting its life-support systems that keep us and other species alive, and what we can do to live more sustainably.
- Living more sustainably means learning from nature, living more lightly, and becoming active environmental citizens who leave small environmental footprints on the earth.

Tying It All Together

The United States, China, and Sustainability

Sailorr/Shutterstock.com

As the world's two largest economies, the United States and China will play key roles in determining whether or not the world can make a transition to a more sustainable future during this century (**Core Case Study**). The governments and citizens of these countries can help to make such a transition by applying the three **scientific principles of sustainability**—relying much more on solar energy and other renewable energy sources, reusing and recycling much more of what is produced, and celebrating, restoring, and protecting as much as possible of the biodiversity that supports our lives and economies.

U.S. and Chinese policy makers and business leaders could also be guided by the three economic, political, and ethical **principles of sustainability**. In the political arena, they will have to try harder to find win-win solutions that benefit the largest number of people while also benefiting the environment. Such solutions will likely include internalizing the harmful environmental and health costs of producing and using goods and services (full-cost pricing). And these decision makers, as well as all the rest of us, will need to make each decision with future generations in mind—seeking to leave the world in a condition as good as or better than what we now enjoy.

Chapter Review

Core Case Study

1. Explain why the decisions and actions of the United States and China will play a major role in determining whether the world can make a shift to a more sustainable future.

Section 25.1

2. What is the key concept for this section? What is an **environmental worldview**? Define **environmental ethics**. What is a **human-centered worldview**? What is the planetary management worldview? List three variations of this worldview. What is the stewardship worldview? Does the earth need saving? If not, what does need saving? What are two signs that a human civilization is in decline and heading for collapse?

3. Summarize the debate over whether we can effectively manage the earth. What did we learn from the Biosphere 2 experiment? List three things that we need to understand and accept in order to achieve a more sustainable world. What is a **life-centered worldview**? What is an **earth-centered worldview**? What is the environmental wisdom worldview and what are 10 of its guiding principles? What is the Earth Charter and what are 6 of its guiding principles? What are four major types of public lands in the United States? What environmental worldviews clash in the controversy over how we should use these lands?

Section 25.2

4. What is the key concept for this section? List three ideas that form the foundation of environmental literacy. What are five questions that an environmentally literate person should be able to answer?

5. Explain how we can learn from direct experiences with nature. What is **nature-deficit disorder** and why is it considered one of the key causes of the environmental problems we face? What questions should you be able to answer about the land (place) where you live? Explain how modern camera and electronic technology could allow people to experience nature virtually. Describe how Juan Martinez is helping disadvantaged youths connect with nature.

6. What is a sense of place and why is it important? Explain why Aldo Leopold is highly regarded and give some examples of the beliefs included in his land ethic.

Section 25.3

7. What is the key concept for this section? What are 12 guidelines for living more sustainably (the *sustainability dozen*)? Describe and discuss the implications of being addicted to buying more and more stuff. Describe the lifestyle of voluntary simplicity now being adopted by some people. What is Mahatma Gandhi's principle of enoughness? List five steps that some psychologists have advised people to take to help them withdraw from an addiction to buying more and more stuff.

8. List eight ways that individuals can live more sustainably by shrinking their environmental footprints. List two mental traps that can lead to denial, indifference, and inaction concerning environmental problems.

9. List six major shifts that scientists say will be necessary to bring about a sustainability revolution. Describe the opportunities and challenges that the United States and China face in becoming global leaders in a shift to a more environmentally sustainable world. What are nine cultural shifts involved in bringing about a sustainability revolution? Describe Lester R. Brown's contributions to helping us make the transition to a more economically and environmentally sustainable world. List 10 agents of change that could help us shift to a more sustainable world. What are two pieces of good news about the possibility of bringing about a sustainability revolution in your lifetime?

10. What are this chapter's *three big ideas*? Explain how the United States and China could apply the six **principles of sustainability** in making the shift to a more sustainable future.

Note: Key terms are in bold type.

Critical Thinking

1. In making a shift to a more sustainable future, what do you think are the three most important things that **(a)** the United States needs to do, **(b)** China needs to do (Core Case Study), and **(c)** you need to do?

2. Some analysts argue that the problems with Biosphere 2 resulted mostly from inadequate design, and that a better team of scientists and engineers could make it work. Explain why you agree or disagree with this view.

3. Do you believe that we have an ethical responsibility to leave the earth's life-support systems in a condition that is as good as or better than it is now? Explain. List three aspects of your lifestyle that hinder the implementation of this ideal and three aspects that promote this ideal.

4. This chapter summarized several different environmental worldviews. Go through these worldviews and find the beliefs you agree with and then describe your own environmental worldview. Which of your beliefs, if any, were added or modified because of taking this course? Compare your answer with those of your classmates.

5. Explain why you agree or disagree with the following statements: **(a)** everyone has the right to have as many children as they want; **(b)** all people have a right to use as many resources as they want; **(c)** individuals should have the right to do whatever they want with land they own, regardless of whether such actions harm the environment, their neighbors, or the local community; **(d)** other species exist to be used by humans; and **(e)** all forms of life have a right to exist. Are your answers consistent with the beliefs that make up your environmental worldview, which you described in question 4?

6. The American theologian, Thomas Berry, called the industrial–consumer society, built on the human-centered, planetary management environmental worldview, the "supreme pathology of all history." He said, "We can break the mountains apart; we can drain the rivers and flood the valleys. We can turn the most luxuriant forests into throwaway paper products. We can tear apart the great grass cover of the western plains, and pour toxic chemicals into the soil and pesticides onto the fields, until the soil is dead and blows away in the wind. We can pollute the air with acids, the rivers with sewage, and the seas with oil. We can invent computers capable of processing 10 million calculations per second. And why? To increase the volume and speed with which we move natural resources through the consumer economy to the junk pile or the waste heap. If, in these activities, the topography of the planet is damaged, if the environment is made inhospitable for a multitude of living species, then so be it. We are, supposedly, creating a technological wonderworld. But our supposed progress is bringing us to a wasteworld instead of a wonderworld." Explain why you agree or disagree with this assessment. If you disagree, answer at least five of Berry's charges with your own arguments as to why you think he is wrong. If you agree, cite evidence as to why.

7. Some analysts believe that trying to gain environmental wisdom by becoming familiar with some part of the natural world and forming an emotional bond with its life forms and processes is unscientific, mystical nonsense based on a romanticized view of nature. They believe that having a better scientific understanding of how the earth works and inventing or improving technologies to solve environmental problems are the best ways to achieve sustainability. Do you agree or disagree? Explain.

8. Do you think we have a reasonable chance of bringing about a sustainability revolution within your lifetime? Explain. If you are nearing the end of this course, is your view of the future more hopeful or less hopeful than it was when you began this course? Compare your answers with those of your classmates.

Doing Environmental Science

Increase your environmental knowledge and awareness of nature by tracing the water you drink from precipitation to tap; finding out what type of soil is beneath your feet; naming five plants and five birds that live in the natural environment around you; finding out what species in your area are threatened with extinction; learning where your trash goes; and learning where the wastes you flush down the toilet go. Write a report summarizing your findings. Of your findings, which two were the most surprising to you and why? Compare your answer to this question with those of your classmates.

Global Environment Watch Exercise

Go to your MindTap course to access the GREENR database. Use the "World Map" link at the top of the page to navigate to information about specific countries. Click on the pins for the United States, China, India, and one other country of your choice and research what each of them is doing to try to become more sustainable. Which of these sustainability programs are working well? Which ones are not working well? Write a report comparing these programs.

Ecological Footprint Analysis

Working with classmates, conduct an ecological footprint analysis of your campus. Work with a partner, or in small groups, to research and investigate an aspect of your school such as recycling or composting; water use; food service practices; energy use; building management and energy conservation; transportation for both on- and off-campus trips; or grounds maintenance. Depending on your school and its location, you may want to add more areas to the investigation. You can also decide to study the campus as a whole, or to break it down into smaller research areas, such as dorms, administrative buildings, classroom buildings, grounds, and other areas.

1. After deciding on your group's research area, conduct your analysis. As part of your analysis, develop a list of questions that will help to determine the ecological impact related to your chosen topic. For example, with regard to water use, you might ask how much water is used, what is the estimated amount that is wasted through leaking pipes and faucets, and what is the average monthly water bill for the school, among other questions. Use such questions as a basis for your research.

2. Analyze your results and share them with the class to determine what can be done to shrink the ecological footprint of your school within the area you have chosen.

3. Arrange a meeting with school officials to share your action plan with them.

Supplement 1 Measurement Units and Unit Conversions

Length

Metric
1 kilometer (km) = 1,000 meters (m)
1 meter (m) = 100 centimeters (cm)
1 meter (m) = 1,000 millimeters (mm)
1 centimeter (cm) = 0.01 meter (m)
1 millimeter (mm) = 0.001 meter (m)

English
1 foot (ft) = 12 inches (in)
1 yard (yd) = 3 feet (ft)
1 mile (mi) = 5,280 feet (ft)
1 nautical mile = 1.15 miles

Metric–English
1 kilometer (km) = 0.621 mile (mi)
1 meter (m) = 39.4 inches (in) = 3.28 feet
1 inch (in) = 2.54 centimeters (cm)
1 foot (ft) = 0.305 meter (m)
1 yard (yd) = 0.914 meter (m)
1 nautical mile = 1.85 kilometers (km)

Area

Metric
1 square kilometer (km^2) = 1,000,000 square meters (m^2)
1 square meter (m^2) = 1,000,000 square millimeters (mm^2)
1 square meter (m^2) = 10,000 square centimeters (cm^2)
1 hectare (ha) = 10,000 square meters (m^2)
1 hectare (ha) = 0.01 square kilometer (km^2)

English
1 square foot (ft^2) = 144 square inches (in^2)
1 square yard (yd^2) = 9 square feet (ft^2)
1 square mile (mi^2) = 27,880,000 square feet (ft^2)
1 acre (ac) = 43,560 square feet (ft^2)

Metric–English
1 hectare (ha) = 2.471 acres (ac)
1 square kilometer (km^2) = 0.386 square mile (mi^2)
1 square meter (m^2) = 1.196 square yards (yd^2)
1 square meter (m^2) = 10.76 square feet (ft^2)
1 square centimeter (cm^2) = 0.155 square inch (in^2)

Volume

Metric
1 cubic kilometer (km^3) = 1,000,000,000 cubic meters (m^3)
1 cubic meter (m^3) = 1,000,000 cubic centimeters (cm^3)
1 liter (L) = 1,000 milliliters (mL) = 1,000 cubic centimeters (cm^3)
1 cubic meter (m^3) = 1,000 liters (L)
1 milliliter (mL) = 0.001 liter (L)
1 milliliter (mL) = 1 cubic centimeter (cm^3)

English
1 gallon (gal) = 4 quarts (qt)
1 quart (qt) = 2 pints (pt)

Metric–English
1 liter (L) = 0.265 gallon (gal)
1 gallon (gal) = 3.77 liters (L)
1 liter (L) = 1.06 quarts (qt)
1 liter (L) = 0.0353 cubic foot (ft^3)
1 cubic meter (m^3) = 35.3 cubic feet (ft^3)
1 cubic meter (m^3) = 1.30 cubic yards (yd^3)
1 cubic kilometer (km^3) = 0.24 cubic mile (mi^3)
1 barrel (bbl) = 159 liters (L)
1 barrel (bbl) = 42 U.S. gallons (gal)

Fuel Efficiency

Metric–English
kilometers per liter (kpL) = miles per gallon (mpg) × 2.34
miles per gallon (mpg) = kilometers per liter (kpL) ÷ 2.34

Mass

Metric
1 kilogram (kg) = 1,000 grams (g)
1 gram (g) = 1,000 milligrams (mg)
1 gram (g) = 1,000,000 micrograms (μg)
1 milligram (mg) = 0.001 gram (g)
1 microgram (μg) = 0.000001 gram (g)
1 metric ton (mt) = 1,000 kilograms (kg)

English
1 ton (t) = 2,000 pounds (lb)
1 pound (lb) = 16 ounces (oz)

Metric–English
1 metric ton (mt) = 2,200 pounds (lb) = 1.1 tons (t)
1 kilogram (kg) = 2.20 pounds (lb)
1 pound (lb) = 454 grams (g)
1 gram (g) = 0.035 ounce (oz)

Energy and Power

Metric
1 kilojoule (kJ) = 1,000 joules (J)
1 kilocalorie (kcal) = 1,000 calories (cal)
1 calorie (cal) = 4.184 joules (J)

English
1 British thermal unit (Btu) = 1,055 watt-seconds
Kilowatt-hour = 1,000 watt-hours

Metric–English
1 kilojoule (kJ) = 0.949 British thermal unit (Btu)
1 kilojoule (kJ) = 0.000278 kilowatt-hour (kW-h)

1 kilocalorie (kcal) = 3.97 British thermal units (Btu)
1 kilocalorie (kcal) = 0.00116 kilowatt-hour (kW-h)
1 kilowatt-hour (kW-h) = 860 kilocalories (kcal)
1 kilowatt-hour (kW-h) = 3,400 British thermal units (Btu)
1 quad (Q) = 1,050,000,000,000,000 kilojoules (kJ)
1 quad (Q) = 293,000,000,000 kilowatt-hours (kW-h)

Temperature Conversions

Fahrenheit (°F) to Celsius (°C): °C = (°F − 32.0) ÷ 1.80
Celsius (°C) to Fahrenheit (°F): °F = (°C × 1.80) + 32.0
Change of 1°C = change of 1.8°F

Unit Conversion Sample Exercises

1. Convert an automobile fuel efficiency of 45 miles per gallon (mpg) to kilometers per liter (kpL).

Answer: Let x be the answer in kpL. We know that kpL = mpg × 2.34. So, x = 45 × 2.34 = 105.3 kpL (or 105 rounded to the nearest whole number).

2. Convert a deforestation rate of 100 hectares per year to acres per year.

Answer: We know that 1 hectare = 2.471 acres. So, 100 hectares = 100 × 2.471 acres, or 247.1 acres. The answer is 247.1 acres per year (or 247 acres per year, rounded to the nearest acre).

Unit Conversion Exercises

1. If your car gets 32 miles per gallon, what is its fuel efficiency in kilometers per liter?
2. Suppose a forest is being cut at a rate of 88 hectares per year. How many acres per year are being cut?

Supplement 2 Reading Graphs and Maps

Graphs and Maps Are Important Visual Tools

A graph is a tool for conveying information that we can summarize numerically by illustrating that information in a visual format. This information, called *data,* is collected in experiments, surveys, and other information-gathering activities. Graphing can be a powerful tool for summarizing and conveying complex information.

In this textbook and the accompanying web-based Active Graphing exercises, we use three major types of graphs: *line graphs, bar graphs,* and *pie graphs.* Here, you will explore each of these types of graphs and learn how to read them. An important visual tool used to summarize data that vary over small or large areas is a map. We discuss some aspects of reading maps relating to environmental science at the end of this supplement.

Line Graphs

Line graphs usually represent data that fall in some sort of sequence such as a series of measurements over time or distance. In most such cases, units of time or distance lie on the horizontal *x-axis.* The possible measurements of some quantity or variable such as temperature or oil use that changes over time or distance usually lie on the vertical *y-axis.*

In Figure 1, the x-axis shows the years between 1965 and 2020, and the y-axis displays a range of possible values for the annual amounts of oil consumed worldwide. Usually, the y-axis appears on the left end of the x-axis, although y-axes can appear on the right end, in the middle, or on both ends of the x-axis.

The curving line on a line graph represents the measurements taken at certain time or distance intervals. In Figure 1, the curve represents changes in annual global oil consumption between 1965 and 2013. To find the average annual global oil consumption for any year, find that year on the x-axis (a point called the *abscissa*) and run a vertical line from the axis to the curve. At the point where your line intersects the curve, run a horizontal line to the y-axis. The value at that point on the y-axis, called the *ordinate,* is the amount you are seeking. You can go through the same process in reverse to find a year in which global oil consumption was at a certain point.

Questions

1. About how many metric tons of oil were consumed in the world in 2013?
2. Roughly how many times more oil was consumed in 2013 than in 1985? About how many times more oil was consumed in 2013 than in 1965?

Line graphs have several important uses. One of the most common applications is to compare two or more variables. Figure 2 compares two variables: monthly temperature and precipitation (rain and snowfall) during a typical year in a temperate deciduous forest. However, in this case the variables are measured on two different scales, so there are two y-axes. The y-axis on the left end of the graph shows a Centigrade temperature scale, while the y-axis on the right shows the range of precipitation measurements in millimeters. The x-axis displays the first letters of each of the 12 months' names.

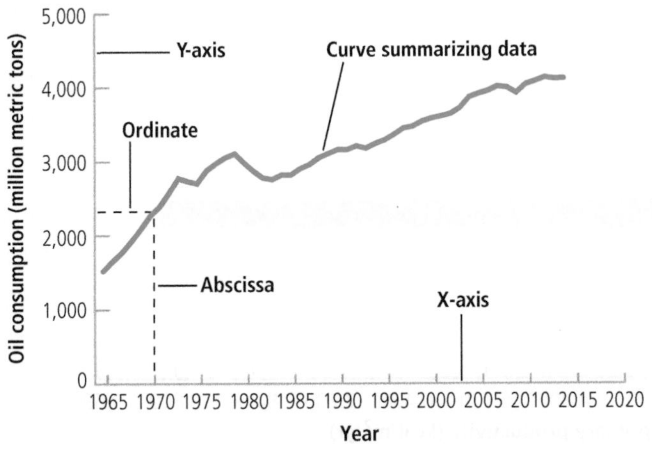

FIGURE 1 Global oil consumption, 1965–2013.

(Compiled by the authors using data from U.S. Energy Information Administration, International Energy Agency, British Petroleum, and United Nations.)

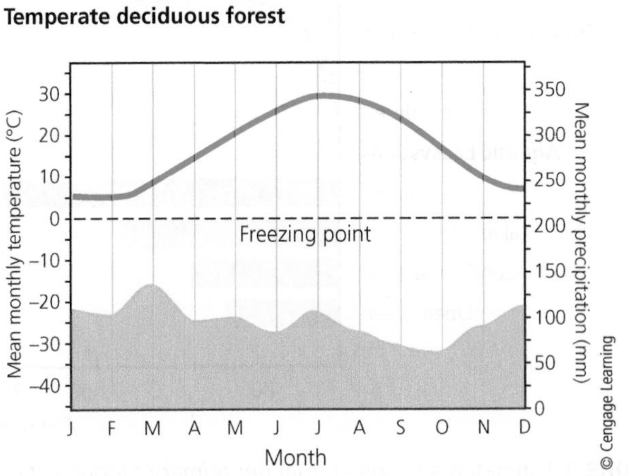

FIGURE 2 Typical variations in annual temperature (red) and precipitation (blue) in a temperate deciduous forest.

Questions

1. In which month does most precipitation fall? Which is the driest month of the year? Which is the hottest month?
2. If the temperature curve were almost flat, running throughout the year at roughly its highest point of about 30°C, how do you think this forest would change from what it is now (see Figure 7.19, center, p. 157)? If the annual precipitation suddenly dropped and remained under 25 centimeters all year, what do you think would eventually happen to this forest?

Bar Graphs

The *bar graph* is used to compare measurements for one or more variables across categories. Unlike the line graph, a bar graph typically does not involve a sequence of measurements over time or distance. The measurements compared on a bar graph usually represent data collected at some point in time or during a well-defined period. For instance, we can compare the *net primary productivity (NPP)*, a measure of chemical energy produced by plants in an ecosystem, for different ecosystems, as represented in Figure 3.

In most bar graphs, the categories to be compared are laid out on the x-axis, while the range of measurements for the variable under consideration lies along the y-axis. In Figure 3, the categories (ecosystems) are on the y-axis, and the variable range (NPP) lies on the x-axis. In either case, reading the graph is straightforward. Simply run a line perpendicular to the bar you are reading from the top of that bar (or the right or left end, if it lies horizontally) to the variable value axis. In Figure 3, you can see that the NPP for the continental shelf, for example, is close to 1,600 kcal/m^2/yr.

Questions

1. What are the two terrestrial ecosystems that are closest in NPP value of all pairs of such ecosystems? About how many times greater is the NPP in a tropical rain forest than the NPP in a savanna?
2. Which is the most productive of the aquatic ecosystems shown here? Which is the least productive?

An important application of the bar graph used in this book is the *age-structure diagram* (see Figures 6.10, p. 128, and 6.11, p. 129), which describes a population by showing the numbers of males and females in certain age groups.

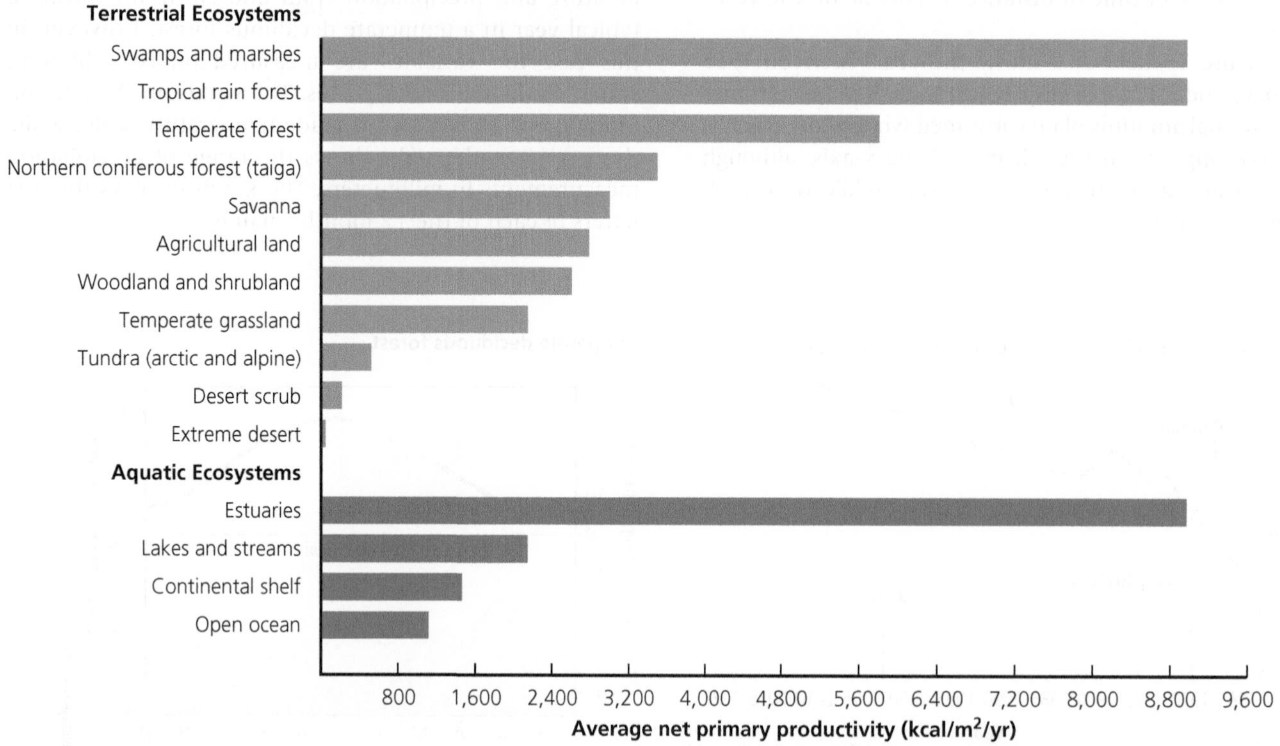

FIGURE 3 Estimated average annual *net primary productivity* in major life zones and ecosystems, in kilocalories of energy produced per square meter per year (kcal/m^2/yr).

(Compiled by the authors using data from R. H. Whittaker, *Communities and Ecosystems*, 2nd ed., New York: Macmillan, 1975.)

Pie Graphs

Like bar graphs, *pie graphs,* or *pie charts,* illustrate numerical values for two or more categories. In addition to that, they can also show each category's proportion of the total of all measurements. The categories are usually ordered on the graph from largest to smallest, for ease of comparison, although this is not always the case. Also, as with bar graphs, pie graphs are generally snapshots of a data set at a point in time or during a defined time period. Unlike line graphs, one pie graph cannot show changes over time.

For example, Figure 4 shows how much each major energy source contributes to the world's total amount of energy used. This graph includes the numerical data used to construct it—the percentages of the total taken up by each part of the pie. But we can use pie graphs without including the numerical data and we can roughly estimate such percentages. The pie graph in that case provides a generalized picture of the composition of a data set.

Questions

1. How many times bigger was coal use than nuclear energy use in 2014?
2. How many times bigger was oil use than hydropower use in 2014?

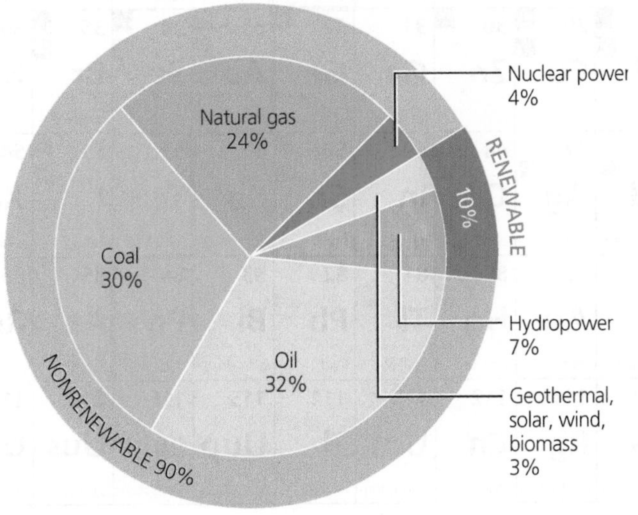

FIGURE 4 Global energy use by source in 2014.

(Compiled by the authors using data from British Petroleum, U.S. Energy Information Administration, and International Energy Agency.)

Reading Maps

We can use maps for considerably more than showing where places are relative to one another. For example, in environmental science, maps can be very helpful in comparing how people in different areas are affected by environmental problems such as air pollution and acid deposition. Figure 5 is a map of the United States showing the relative numbers of premature deaths due to air pollution in the various regions of the country.

Questions

1. Which part of the country generally has the lowest level of premature deaths due to air pollution?
2. Which part of the country has the highest level?
3. What is the level in the area where you live or go to school?

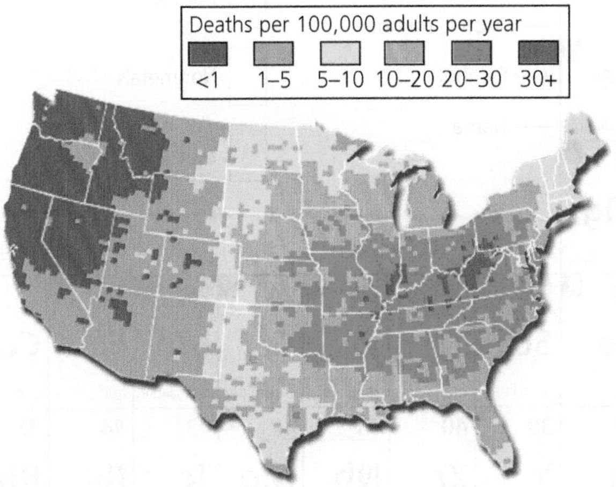

FIGURE 5 Distribution of premature deaths from air pollution in the United States, mostly from very small, fine, and ultrafine particles added to the atmosphere by coal-burning power plants.

(Compiled by the authors using data from U.S. Environmental Protection Agency.)

Supplement 3 Some Basic Chemistry

Chemists Use the Periodic Table to Classify Elements on the Basis of Their Chemical Properties

The basic unit of each element is a unique *atom* that is different from the atoms of all other elements. Each atom consists of an extremely small and dense center called its *nucleus*, which contains one or more protons and, in most cases, one or more neutrons, as well as one or more electrons moving rapidly somewhere around the nucleus (Figure 2.4, p. 35).

Each atom has equal numbers of positively charged protons and negatively charged electrons. Because these electrical charges cancel one another, *atoms as a whole, in their simplest form, have no net electrical charge*.

We cannot determine the exact location of the electrons around any nucleus. Instead, we can estimate the *probability* that they will be found at various locations outside the nucleus—sometimes called an *electron probability cloud*. This is somewhat like saying that there are six birds flying around inside a cloud. We do not know their exact

FIGURE 1 *Periodic Table of Elements.* Elements in the same vertical column, called a *group*, have similar chemical properties.

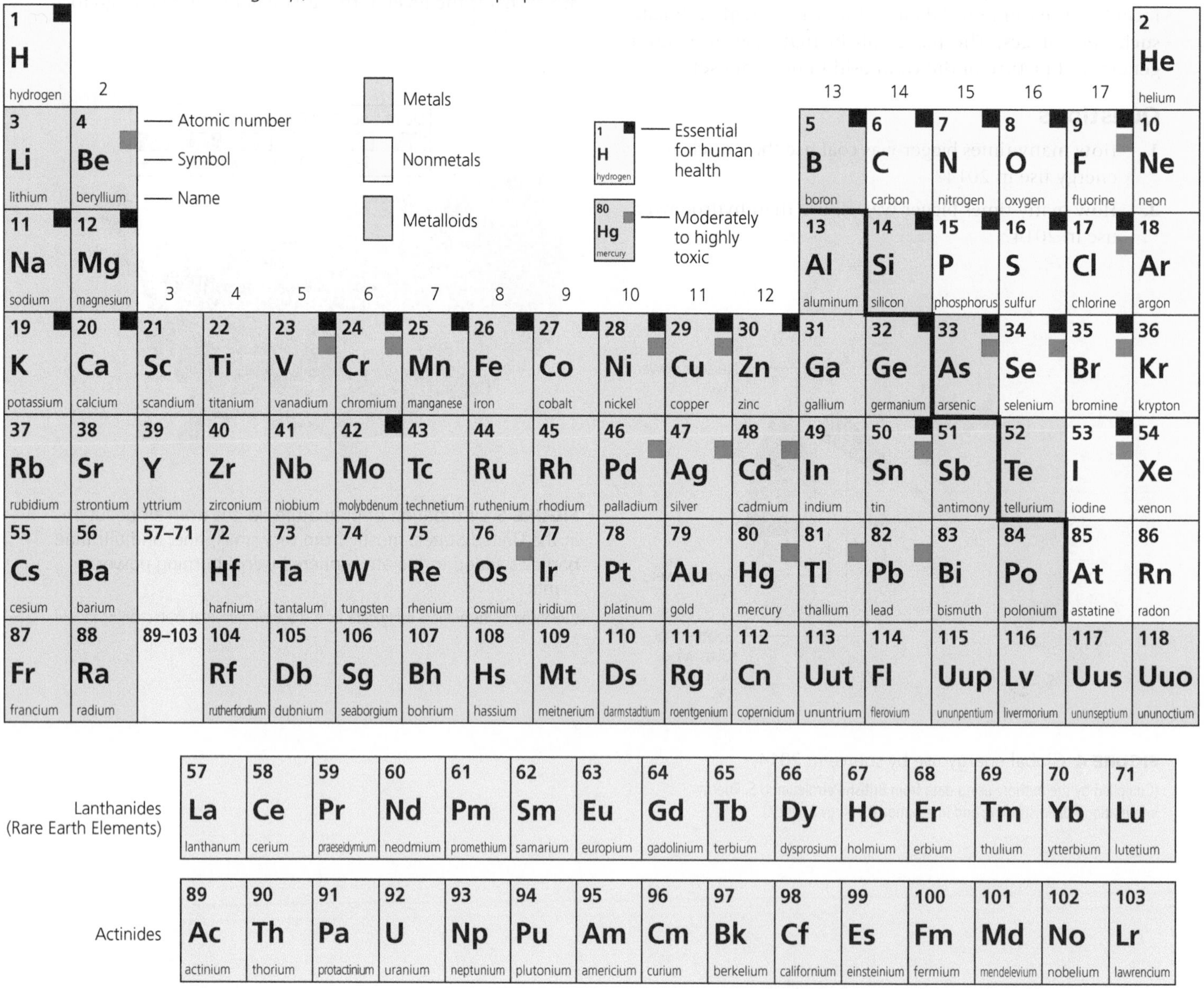

location, but the cloud represents an area in which we can probably find them.

Matter consists of elements and compounds (see Chapter 2, pp. 34–35). Chemists have developed a way to classify the elements according to their chemical behavior in what is called the *Periodic Table of Elements* (Figure 1). Each horizontal row in the table is called a *period*. Each vertical column lists elements with similar chemical properties and is called a *group*.

The Periodic Table in Figure 1 shows how the elements can be classified as *metals, nonmetals,* and *metalloids*. Examples of metals are sodium (Na), calcium (Ca), aluminum (Al), iron (Fe), lead (Pb), silver (Ag), and mercury (Hg). Table 1 shows the names and chemical symbols for the atoms used in this book.

Atoms of *metals* tend to lose one or more of their electrons to form positively charged ions such as Na^+, Ca^{2+}, and Al^{3+}. For example, an atom of the metallic element sodium (Na, atomic number 11) with 11 positively charged protons and 11 negatively charged electrons can lose one of its electrons. It then becomes a sodium ion with a positive charge of 1 (Na^+) because it now has 11 positive charges (protons) but only 10 negative charges (electrons).

Examples of *nonmetals* are hydrogen (H), carbon (C), nitrogen (N), oxygen (O), phosphorus (P), sulfur (S), chlorine (Cl), and fluorine (F). Atoms of some nonmetals such as chlorine, oxygen, and sulfur tend to gain one or more electrons lost by metallic atoms to form negatively charged ions such as O^{2-}, S^{2-}, and Cl^-. For example, an atom of the nonmetallic element chlorine (Cl, atomic number 17) can gain an electron and become a chlorine ion. The ion has a negative charge of 1 (Cl^-) because it has 17 positively charged protons and 18 negatively charged electrons. Atoms of nonmetals can also combine with one another to form molecules in which they share one or more pairs of their electrons. Hydrogen, a nonmetal, is placed by itself above the center of the table because it does not fit very well into any of the groups. Table 2 shows the chemical ions used in this book.

The elements arranged in a diagonal staircase pattern between the metals and nonmetals have a mixture of metallic and nonmetallic properties and are called *metalloids*. Examples are germanium (Ge) and arsenic (As).

TABLE 1 Chemical Elements Used in This Book

Element	Symbol	Element	Symbol
arsenic	As	lead	Pb
bromine	Br	lithium	Li
calcium	Ca	mercury	Hg
carbon	C	nitrogen	N
copper	Cu	phosphorus	P
chlorine	Cl	sodium	Na
fluorine	F	sulfur	S
gold	Au	uranium	U

TABLE 2 Chemical Ions Used in This Book

Positive Ion	Symbol	Components
hydrogen ion	H^+	One hydrogen atom, one positive charge
sodium ion	Na^+	One sodium atom, one positive charge
calcium ion	Ca^{2+}	One calcium atom, two positive charges
aluminum ion	Al^{3+}	One aluminum atom, three positive charges
ammonium ion	NH_4^+	One nitrogen atom, four hydrogen atoms, one positive charge
Negative Ion	**Symbol**	**Components**
chloride ion	Cl^-	One chlorine atom, one negative charge
hydroxide ion	OH^-	One oxygen atom, one hydrogen atom, one negative charge
nitrate ion	NO_3^-	One nitrogen atom, three oxygen atoms, one negative charge
carbonate ion	CO_3^{2-}	One carbon atom, three oxygen atoms, two negative charges
sulfate ion	SO_4^{2-}	One sulfur atom, four oxygen atoms, two negative charges
phosphate ion	PO_4^{3-}	One phosphorus atom, four oxygen atoms, three negative charges

Figure 1 also identifies the elements required as *nutrients* (marked by small black squares) for all or some forms of life, and elements that are moderately or highly toxic (marked by small red squares) to all or most forms of life. Note that some elements such as copper (Cu) serve as nutrients, but can also be toxic at high enough doses. Six nonmetallic elements—carbon (C), oxygen (O), hydrogen (H), nitrogen (N), sulfur (S), and phosphorus (P)—make up about 99% of the atoms of all living things.

CONSIDER THIS . . .

THINKING ABOUT The Periodic Table

Use the Periodic Table to give the name and symbol of two elements that should have chemical properties similar to those of **(a)** Ca, **(b)** potassium, **(c)** S, and **(d)** lead.

Compounds Are Held Together by Chemical Bonds

Most forms of matter are combinations of two or more elements held together by ionic or covalent chemical bonds. Sodium chloride (NaCl) consists of a three-dimensional network of oppositely charged *ions* (Na^+ and Cl^-) held together by the forces of attraction between opposite charges (Figure 2.7, p. 36). The strong forces of attraction between such oppositely charged ions are called *ionic bonds.* They are formed when an electron is transferred from a metallic atom such as sodium (Na) to a nonmetallic element such as chlorine (Cl). Because ionic compounds consist of ions formed from atoms of metallic elements (positive ions) and nonmetallic elements (negative ions), they can be described as *metal–nonmetal compounds.*

Sodium chloride and many other ionic compounds tend to dissolve in water and break apart into their individual ions (Figure 2).

NaCl	→	Na^+	+	Cl^-
sodium chloride (in water)		sodium ion		chloride ion

Water, a *covalent compound,* consists of molecules made up of uncharged atoms of hydrogen (H) and oxygen (O). Each water molecule consists of two hydrogen atoms chemically bonded to an oxygen atom, yielding H_2O molecules. The bonds between the atoms in such molecules are called *covalent bonds* and form when the atoms in the molecule share one or more pairs of their electrons. Because they are formed from atoms of nonmetallic elements (Figure 1), covalent compounds can be described as *nonmetal–nonmetal compounds.* Figure 2.8, p. 37, shows the chemical formulas and shapes of the molecules that are the building blocks for several common *covalent compounds.* Table 3 shows the names and chemical formulas for key compounds used in this book.

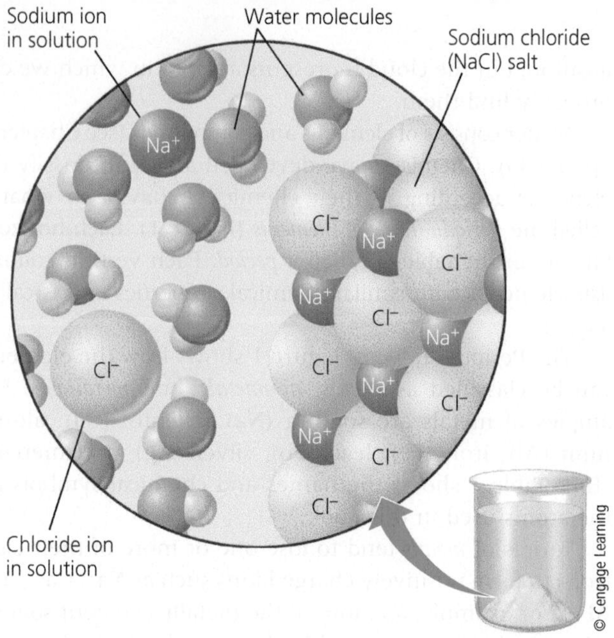

FIGURE 2 A salt, such as NaCl, dissolves in water.

TABLE 3 Compounds Used in This Book

Compound	Formula	Compound	Formula
sodium chloride	NaCl	methane	CH_4
sodium hydroxide	NaOH	glucose	$C_6H_{12}O_6$
carbon monoxide	CO	water	H_2O
carbon dioxide	CO_2	hydrogen sulfide	H_2S
nitric oxide	NO	sulfur dioxide	SO_2
nitrogen dioxide	NO_2	sulfuric acid	H_2SO_4
nitrous oxide	N_2O	ammonia	NH_3
nitric acid	HNO_3	calcium carbonate	$CaCO_3$

What Makes Solutions Acidic? Hydrogen Ions and pH

The *concentration,* or number of hydrogen ions (H^+) in a specified volume of a solution (typically a liter), is a measure of its acidity. Pure water (not tap water or rainwater) has an equal number of hydrogen (H^+) and hydroxide (OH^-) ions. It is called a **neutral solution**. An **acidic solution** has more hydrogen ions than hydroxide ions per liter. A **basic solution** has more hydroxide ions than hydrogen ions per liter.

Scientists use pH as a measure of the acidity of a solution based on its concentration of hydrogen ions (H^+). By definition, a neutral solution has a pH of 7, an acidic solution has a pH less than 7, and a basic solution has a pH greater than 7.

Each single unit change in pH represents a tenfold increase or decrease in the concentration of hydrogen ions per liter. For example, an acidic solution with a pH of 3 is 10 times more acidic than a solution with a pH of 4. Figure 2.6 (p. 36) shows the approximate pH and hydrogen ion concentration per liter of solutions for various common substances.

CONSIDER THIS . . .

THINKING ABOUT pH

A solution has a pH of 2. How many times more acidic is this solution than one with a pH of 6?

The measurement of acidity is important in the study of environmental science, as environmental changes involving acidity can have serious environmental impacts. For example, when coal and oil are burned, they give off acidic compounds that can return to the earth as *acid deposition* (Figure 18.11, p. 483), which has become a major environmental problem.

There Are Weak Forces of Attraction between Some Molecules

Ionic and covalent bonds form between the ions or atoms *within* a compound. There are also weaker forces of attraction *between* the molecules of covalent compounds (such as water) resulting from an unequal sharing of electrons by two atoms.

For example, an oxygen atom has a much greater attraction for electrons than does a hydrogen atom. Thus, the electrons shared between the oxygen atom and its two hydrogen atoms in a water molecule are pulled closer to the oxygen atom, but not actually transferred to the oxygen atom. As a result, the oxygen atom in a water molecule has a slightly negative partial charge and its two hydrogen atoms have a slightly positive partial charge.

The slightly positive hydrogen atoms in one water molecule are then attracted to the slightly negative oxygen atoms in another water molecule. These forces of attraction *between* water molecules are called *hydrogen bonds* (Figure 3). They account for many of water's unique properties (Science Focus 3.2, p. 64). Hydrogen bonds also form between other covalent molecules or between portions of such molecules containing hydrogen and nonmetallic atoms with a strong ability to attract electrons.

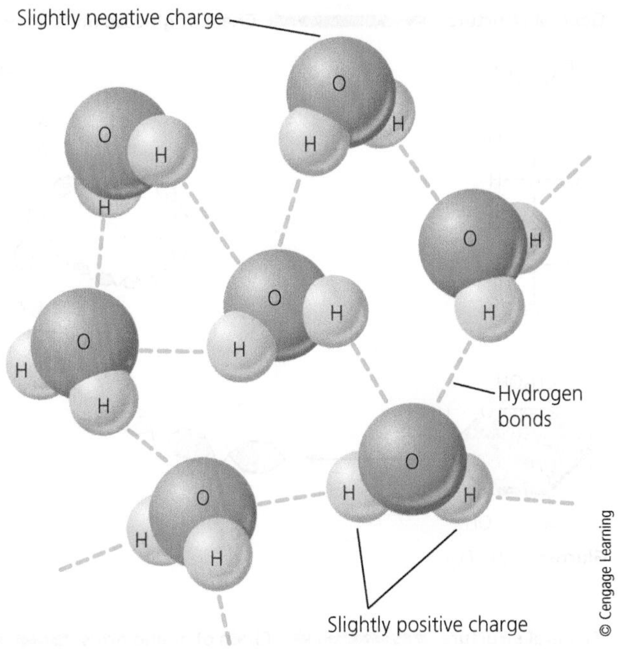

FIGURE 3 *Hydrogen bond:* The slightly unequal sharing of electrons in the water molecule creates a molecule with a slightly negatively charged end and a slightly positively charged end. Because of this electrical polarity, the hydrogen atoms of one water molecule are attracted to the oxygen atoms in other water molecules. These fairly weak forces of attraction *between* molecules (represented by the dashed lines) are called *hydrogen bonds*.

Four Types of Large Organic Compounds Are the Molecular Building Blocks of Life

Larger and more complex organic compounds, called *polymers*, consist of a number of basic structural or molecular units (*monomers*) linked by chemical bonds, somewhat like rail cars linked in a freight train. Four types of macromolecules—complex carbohydrates, proteins, nucleic acids, and lipids—are the molecular building blocks of life.

Complex carbohydrates consist of two or more monomers of *simple sugars* (such as glucose, Figure 4) linked together. One example is the starches that plants use to store energy and to provide energy for animals that feed on plants. Another is cellulose, the earth's most abundant organic compound, which is found in the cell walls of bark, leaves, stems, and roots.

Proteins are large polymer molecules formed by linking together long chains of monomers called *amino acids* (Figure 5). Living organisms use about 20 different amino acid molecules to build a variety of proteins, which play different roles. Some help to store energy. Some are components of the *immune system* that protects the body against diseases and harmful substances by forming antibodies that make invading agents harmless. Others are *hormones* that are used as chemical messengers in the bloodstreams

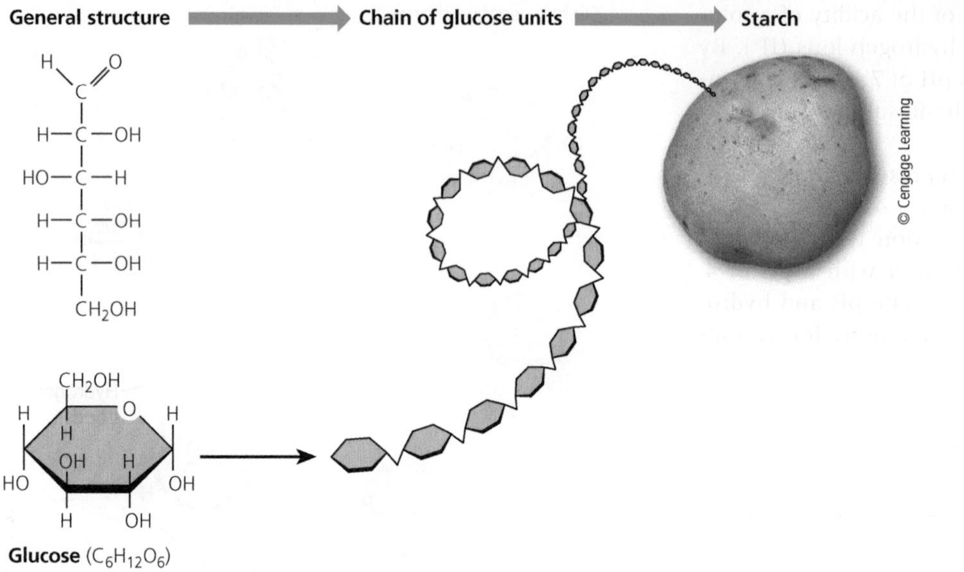

FIGURE 4 Straight-chain and ring structural formulas of glucose, a simple sugar that can be used to build long chains of complex carbohydrates such as starch and cellulose.

JIANG HONGYAN/Shutterstock

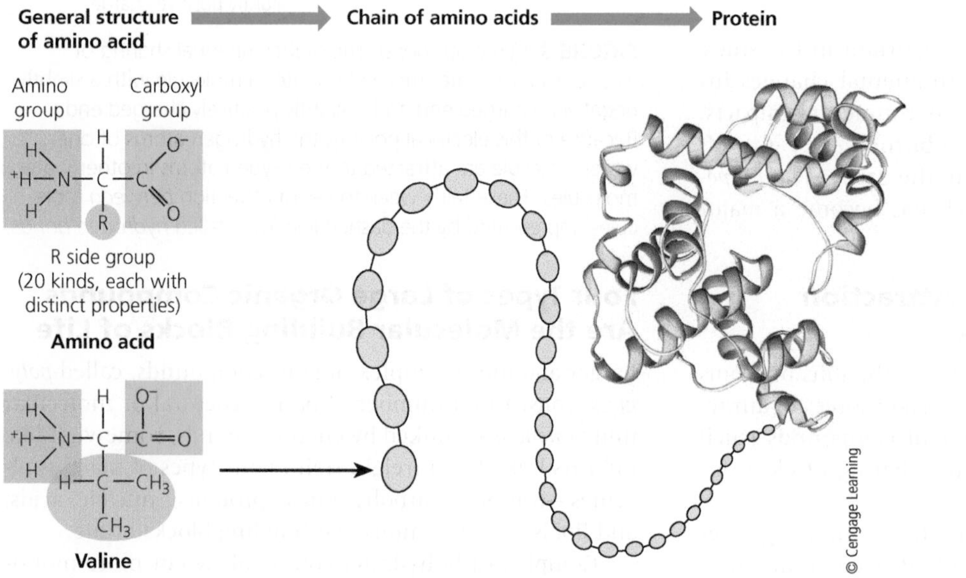

FIGURE 5 General structural formula of amino acids (upper left) and a specific structural formula of one of the 20 different amino acid molecules (lower left) that can be linked together in chains to form proteins that fold up into more complex shapes.

of animals to turn various bodily functions on or off. In animals, proteins are also components of hair, skin, muscle, and tendons. In addition, some proteins act as *enzymes* that catalyze or speed up certain chemical reactions.

Nucleic acids are large polymer molecules made by linking hundreds to thousands of four types of monomers called *nucleotides*. Two nucleic acids—deoxyribonucleic acid (DNA) and ribonucleic acid (RNA)—participate in the building of proteins and carry hereditary information used to pass traits from parent to offspring. Each nucleotide consists of a *phosphate group*, a *sugar molecule* containing five carbon atoms (deoxyribose in DNA molecules and ribose in RNA molecules) (Figure 6) and one of four different *nucleotide bases*

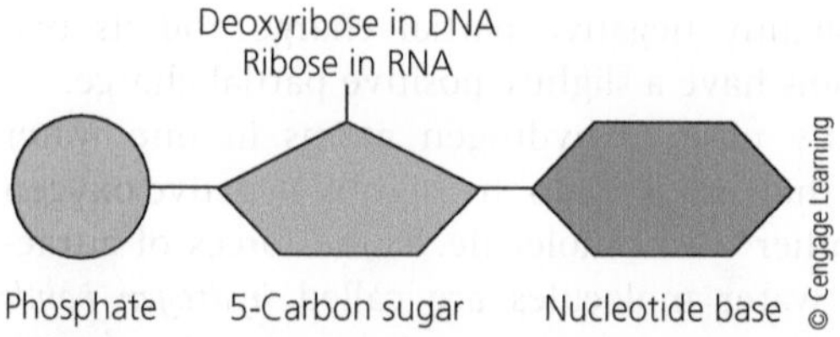

FIGURE 6 Generalized structures of the nucleotide molecules linked in various numbers and sequences to form large nucleic acid molecules such as various types of *DNA* (deoxyribonucleic acid) and *RNA* (ribonucleic acid). In DNA, the five-carbon sugar in each nucleotide is deoxyribose; in RNA, it is ribose.

(represented by A, G, C, and T, the first letter in each of their names, or A, G, C, and U in RNA). The four basic nucleotides used to make various forms of DNA molecules differ in the types of nucleotide bases they contain—adenine (A), guanine (G), cytosine (C), and thymine (T). (Uracil, labeled U, occurs instead of thymine in RNA.) In the cells of living organisms, these nucleotide units combine in different numbers and sequences to form *nucleic acids* such as various types of DNA and RNA (see Figure 2.9, p. 38).

Hydrogen bonds formed between parts of the four nucleotides in DNA hold two DNA strands together like a spiral staircase, forming a double helix (Figure 2.9, p. 38). DNA molecules can unwind and replicate themselves.

The total weight of the DNA needed to reproduce all of the world's people is only about 50 milligrams—the weight of a small match. If the DNA coiled in your body were unwound, it would stretch about 960 million kilometers (600 million miles)—more than six times the distance between the sun and the earth.

The different molecules of DNA that make up the millions of species found on the earth are like a vast and diverse genetic library. Each species is a unique book in that library. The *genome* of a species is made up of the entire sequence of DNA "letters" or base pairs that combine to "spell out" the chromosomes in typical members of each species. Scientists have been able to map out the genome for the human species by using powerful computers to help them analyze the 3.1 billion base sequences in human DNA.

Lipids, a fourth building block of life, are a chemically diverse group of large organic compounds that do not dissolve in water. Examples are *fats* and *oils* for storing energy (Figure 7), *waxes* for structure, and *steroids* for producing hormones.

Figure 8 shows the relative sizes of simple and complex molecules, cells, and multicelled organisms.

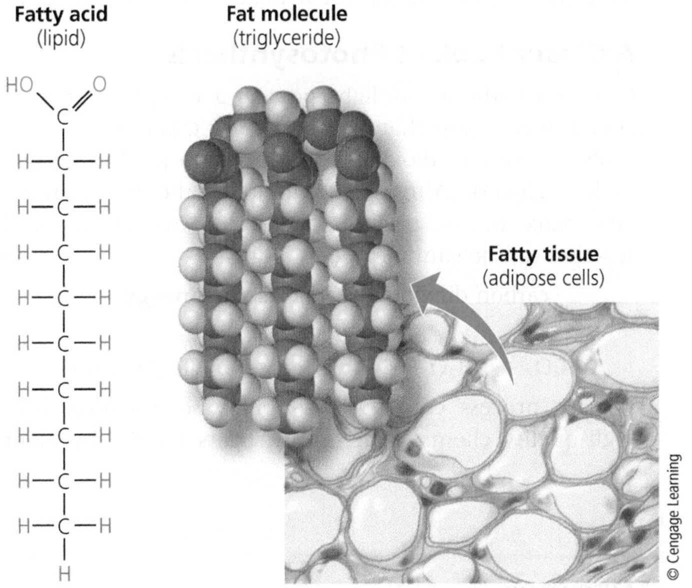

FIGURE 7 Structural formula of fatty acid (one form of lipid, left). Fatty acids are converted into more complex fat molecules (center) that are stored in adipose cells (right).

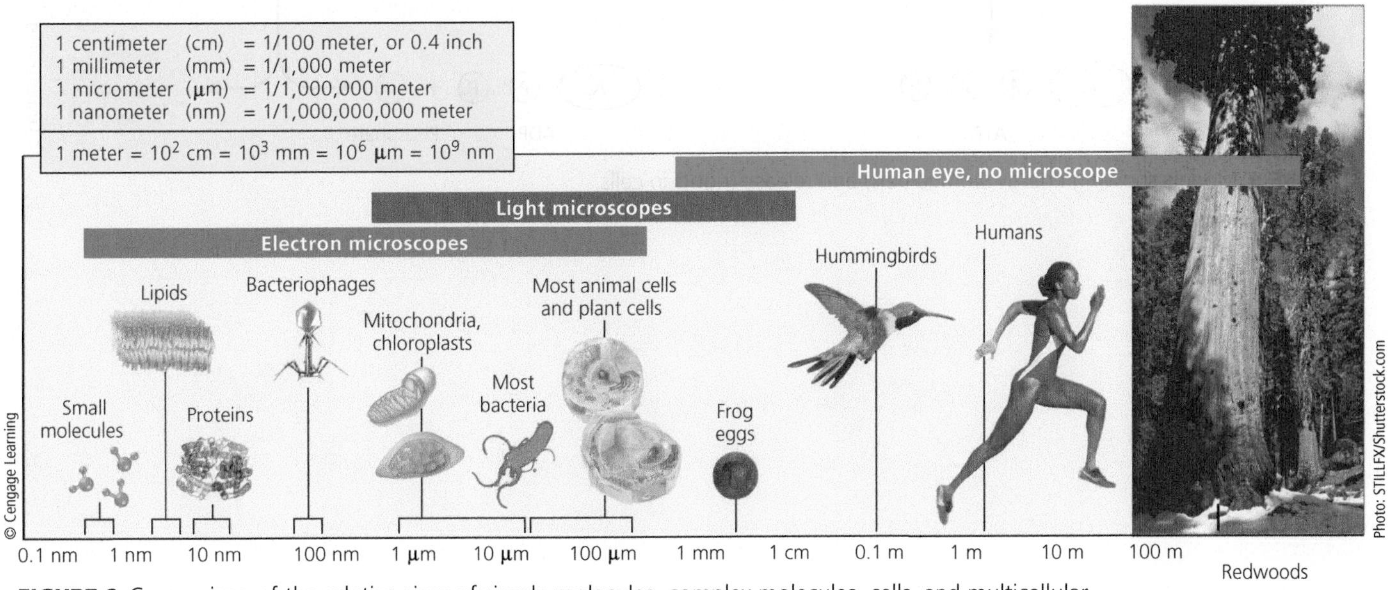

FIGURE 8 Comparison of the relative sizes of simple molecules, complex molecules, cells, and multicellular organisms. Scale is exponential, not linear. Each unit of measure is 10 times larger than the unit preceding it.

Certain Molecules Store and Release Energy in Cells

Chemical reactions occurring in plant cells during photosynthesis (see Chapter 3, p. 52) release energy that is absorbed by adenosine diphosphate (ADP) molecules and stored as chemical energy in adenosine triphosphate (ATP) molecules (Figure 9, left). When cellular processes require energy, ATP molecules release it to form ADP molecules (Figure 9, right).

A Closer Look at Photosynthesis

In photosynthesis, sunlight powers a complex series of chemical reactions that combine water taken up by plant roots and carbon dioxide from the air to produce sugars such as glucose. Although hundreds of chemical changes take place during photosynthesis, the overall chemical reaction can be summarized as follows:

carbon dioxide + water + **solar energy** → glucose + oxygen

$6\ CO_2 + 6\ H_2O + \textbf{solar energy} \rightarrow C_6H_{12}O_6 + 6\ O_2$

This process converts low-quality solar energy into high-quality chemical energy in sugars for use by plant cells. Figure 10 is a greatly simplified summary of the photosynthesis process.

Photosynthesis takes place within tiny organelles called *chloroplasts* found within plant cells. Chlorophyll, a special compound in chloroplasts, absorbs incoming visible light mostly in the violet and red wavelengths. The green light that is not absorbed is reflected back, which is why photosynthetic plants look green. The absorbed wavelengths of solar energy initiate a sequence of chemical reactions with other molecules in what are called light-dependent reactions.

This series of reactions splits water into hydrogen ions (H^+) and oxygen (O_2), which is released into the atmosphere. It also produces small ADP molecules that absorb the energy released and store it as chemical energy in ATP molecules (Figure 9). The chemical energy released by the ATP molecules drives a series of light-independent reactions that can take place in the darkness of plant cells. In this second sequence of reactions, carbon atoms stripped from carbon dioxide combine with hydrogen and oxygen to produce sugars such as glucose ($C_6H_{12}O_6$) that plant cells can use as a source of energy.

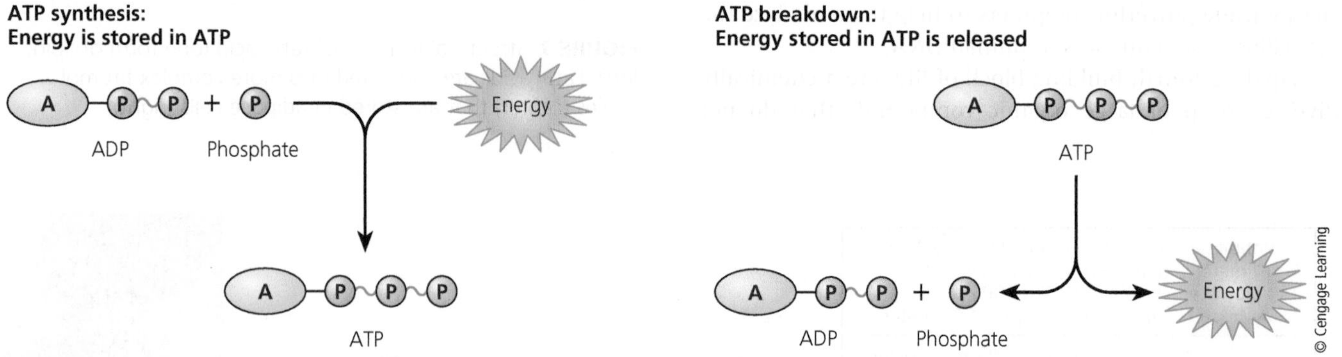

FIGURE 9 Models showing energy storage (left) and release (right) in cells.

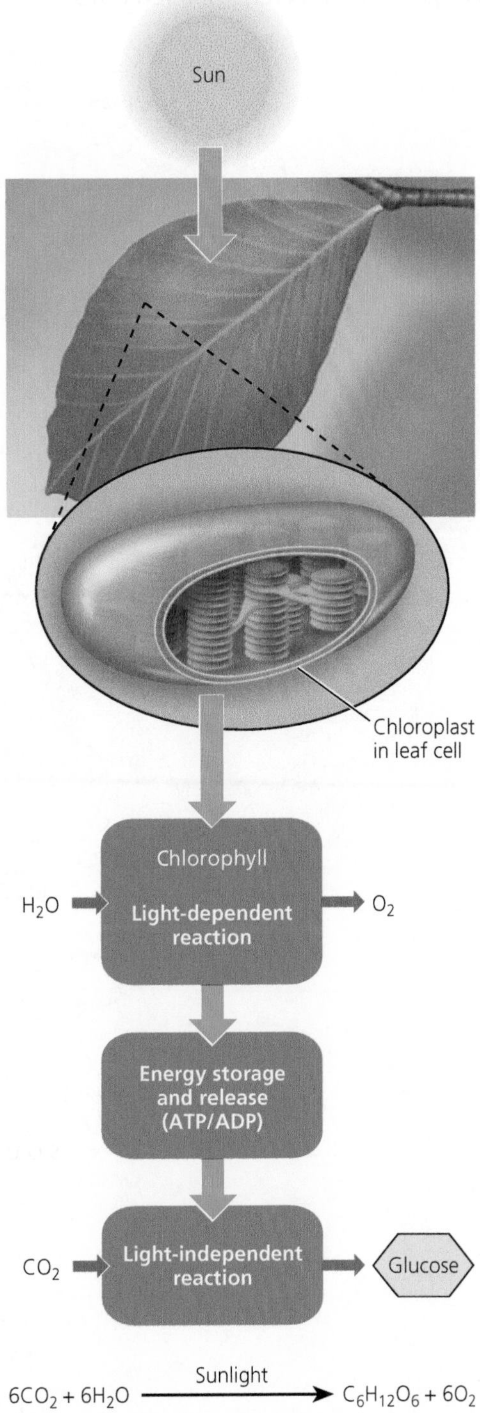

FIGURE 10 Simplified overview of photosynthesis. In this process, chlorophyll molecules in the chloroplasts of plant cells absorb solar energy. This initiates a complex series of chemical reactions in which carbon dioxide and water are converted to sugars such as glucose and oxygen.

Supplement 4 Maps and Map Analysis

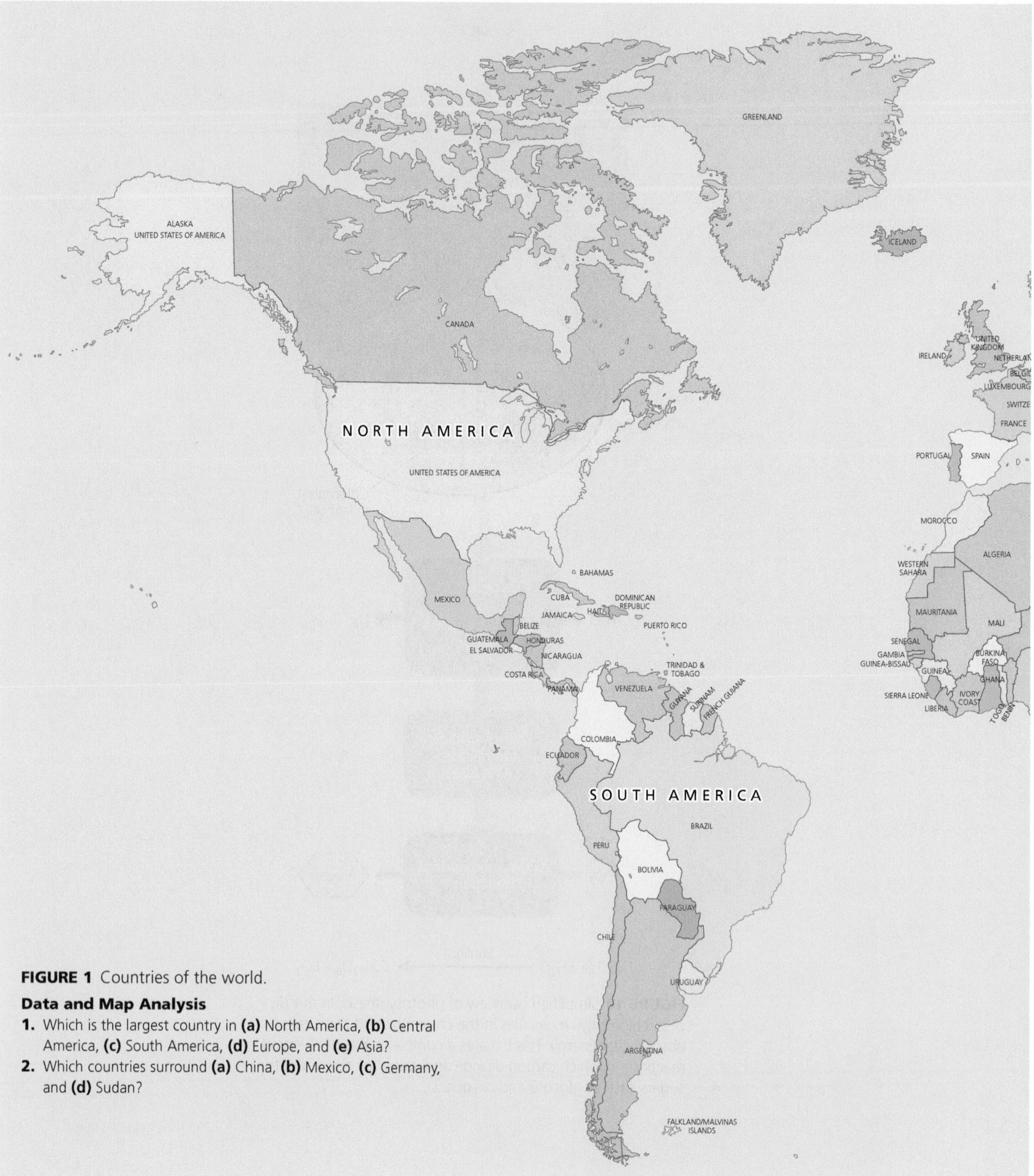

FIGURE 1 Countries of the world.

Data and Map Analysis

1. Which is the largest country in **(a)** North America, **(b)** Central America, **(c)** South America, **(d)** Europe, and **(e)** Asia?
2. Which countries surround **(a)** China, **(b)** Mexico, **(c)** Germany, and **(d)** Sudan?

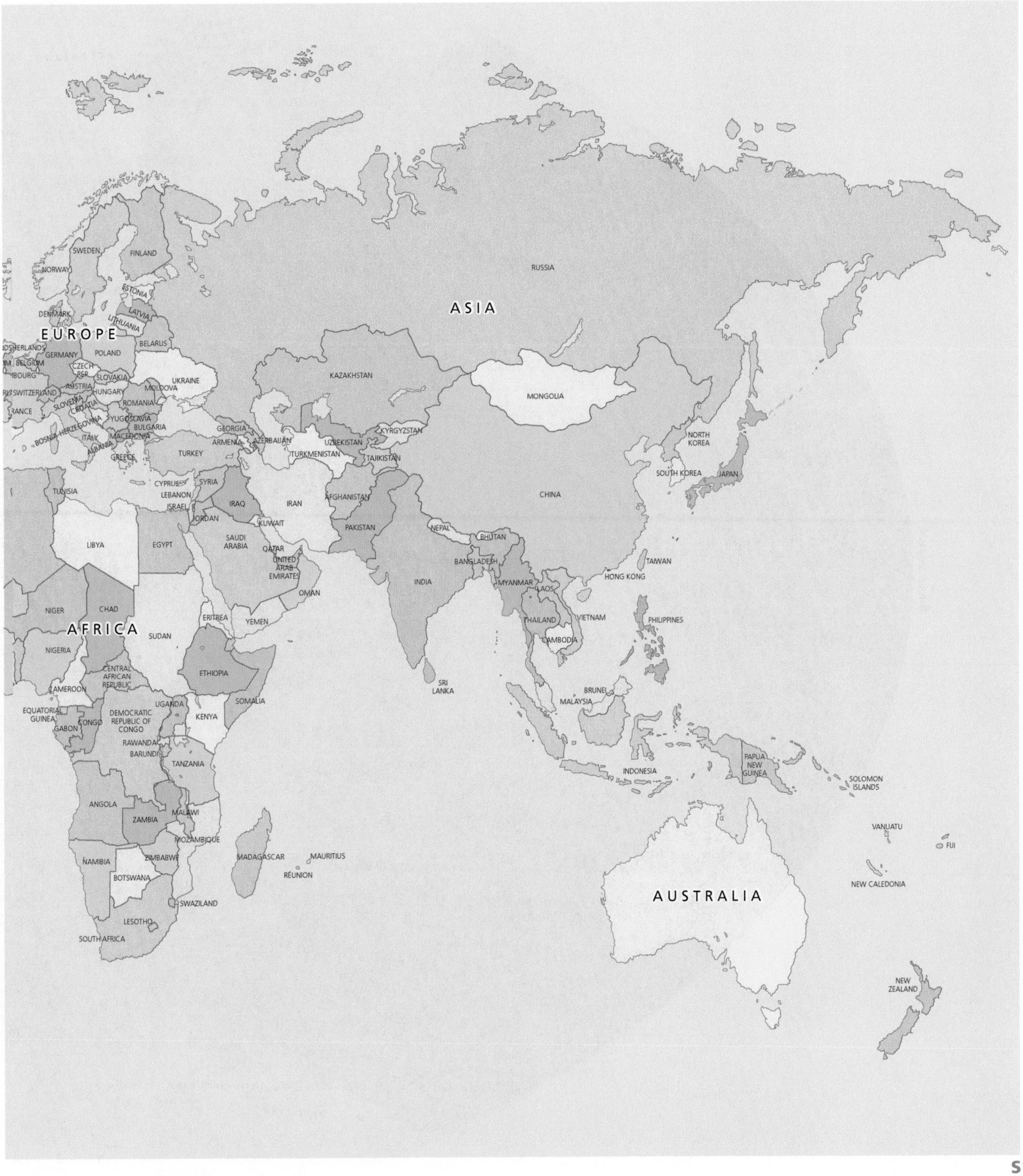

EUROPE
ASIA
AFRICA
AUSTRALIA
RUSSIA
SWEDEN
NORWAY
FINLAND
ESTONIA
LATVIA
LITHUANIA
DENMARK
BELARUS
POLAND
GERMANY
BELGIUM
CZECH REP.
SLOVAKIA
UKRAINE
AUSTRIA
SWITZERLAND
HUNGARY
MOLDOVA
SLOVENIA
CROATIA
ROMANIA
BOSNIA-HERZEGOVINA
YUGOSLAVIA
BULGARIA
ITALY
ALBANIA
MACEDONIA
GREECE
TURKEY
GEORGIA
ARMENIA
AZERBAIJAN
KAZAKHSTAN
UZBEKISTAN
TURKMENISTAN
KYRGYZSTAN
TAJIKISTAN
MONGOLIA
NORTH KOREA
SOUTH KOREA
JAPAN
CHINA
TUNISIA
CYPRUS
SYRIA
LEBANON
ISRAEL
IRAQ
IRAN
AFGHANISTAN
JORDAN
KUWAIT
PAKISTAN
NEPAL
BHUTAN
LIBYA
EGYPT
SAUDI ARABIA
QATAR
UNITED ARAB EMIRATES
BANGLADESH
TAIWAN
HONG KONG
INDIA
MYANMAR
LAOS
OMAN
NIGER
CHAD
ERITREA
YEMEN
THAILAND
VIETNAM
PHILIPPINES
SUDAN
NIGERIA
CAMBODIA
ETHIOPIA
CENTRAL AFRICAN REPUBLIC
CAMEROON
SRI LANKA
BRUNEI
MALAYSIA
EQUATORIAL GUINEA
GABON
CONGO
DEMOCRATIC REPUBLIC OF CONGO
UGANDA
KENYA
SOMALIA
RAWANDA
BURUNDI
TANZANIA
PAPUA NEW GUINEA
INDONESIA
SOLOMON ISLANDS
ANGOLA
ZAMBIA
MALAWI
MOZAMBIQUE
VANUATU
FIJI
NAMIBIA
ZIMBABWE
MADAGASCAR
MAURITIUS
RÉUNION
NEW CALEDONIA
BOTSWANA
SWAZILAND
LESOTHO
SOUTH AFRICA
NEW ZEALAND

FIGURE 2 Composite satellite view of the earth showing its major terrestrial and aquatic features.

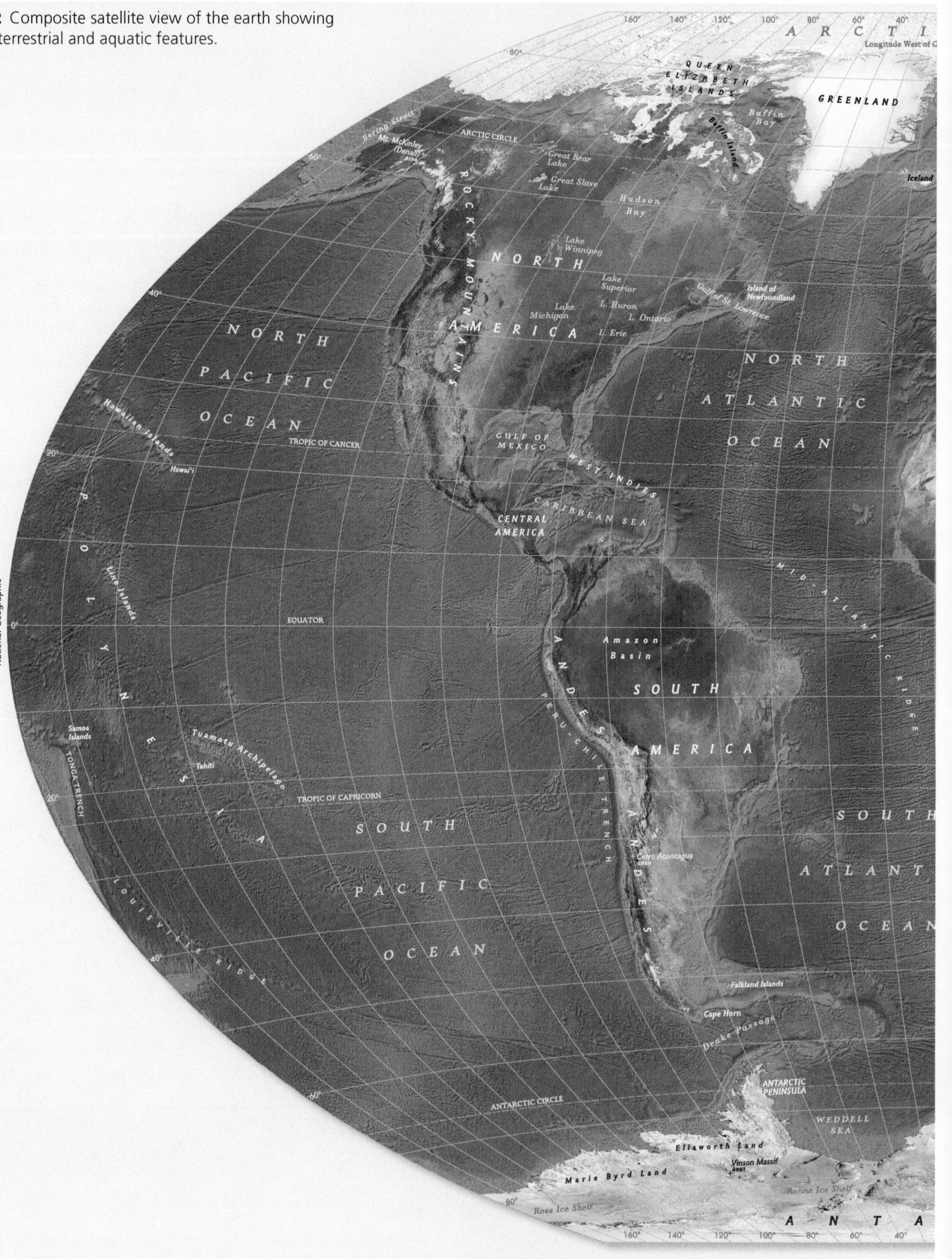

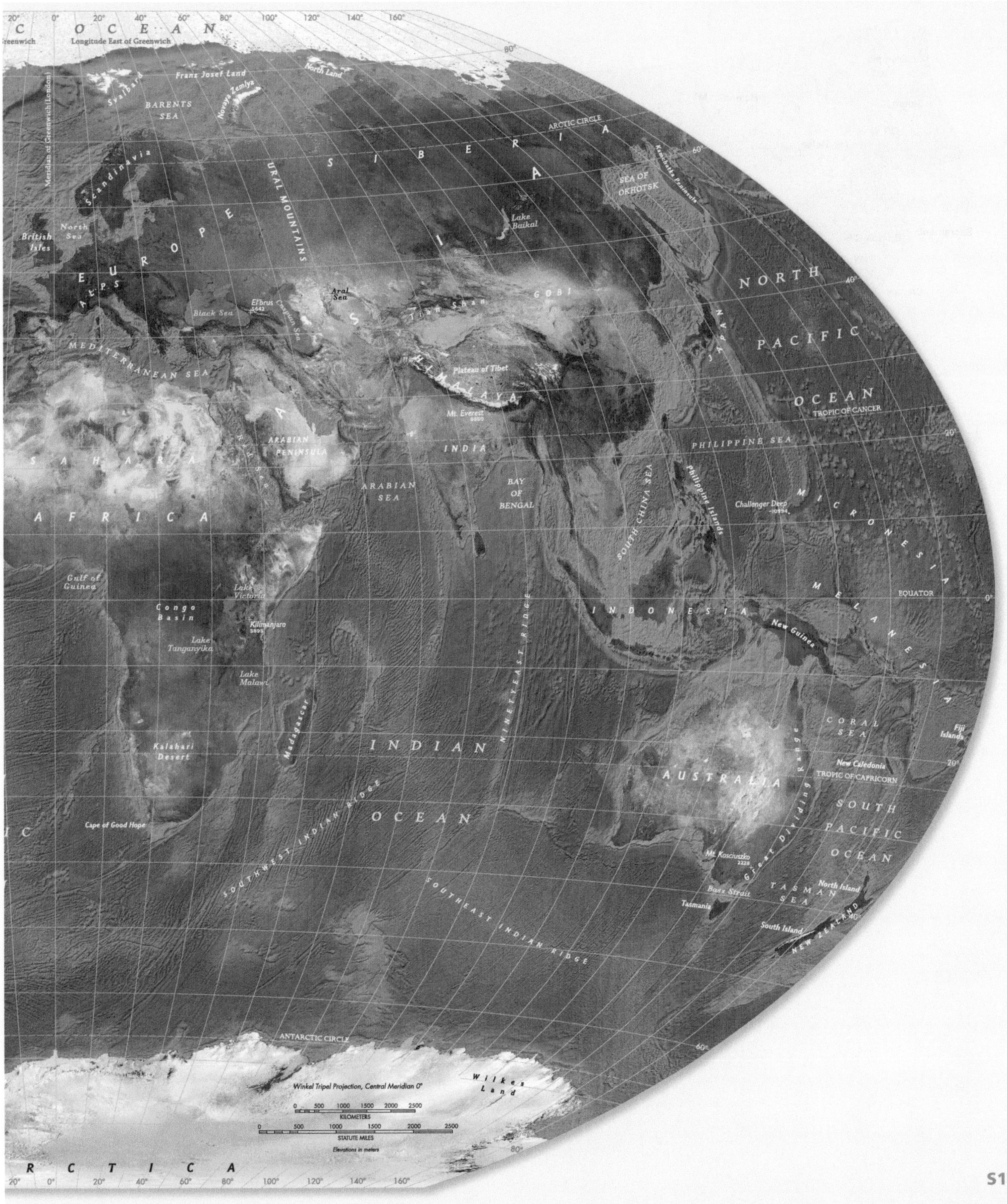
OCEAN
Meridian of Greenwich (London)
Longitude East of Greenwich
Franz Josef Land
North Land
Svalbard
BARENTS SEA
Novaya Zemlya
ARCTIC CIRCLE
SIBERIA
Kamchatka Peninsula
SEA OF OKHOTSK
Scandinavia
URAL MOUNTAINS
Lake Baikal
British Isles
North Sea
EUROPE
ASIA
NORTH PACIFIC OCEAN
ALPS
Aral Sea
GOBI
Elbrus 5642
Black Sea
Caspian Sea
Tian Shan
JAPAN
MEDITERRANEAN SEA
Plateau of Tibet
HIMALAYA
Mt. Everest 8850
TROPIC OF CANCER
SAHARA
ARABIAN PENINSULA
Red Sea
INDIA
PHILIPPINE SEA
ARABIAN SEA
BAY OF BENGAL
SOUTH CHINA SEA
Philippine Islands
Challenger Deep -10994
MICRONESIA
AFRICA
Gulf of Guinea
Lake Victoria
Congo Basin
Kilimanjaro 5895
Lake Tanganyika
INDONESIA
New Guinea
EQUATOR
MELANESIA
NINETYEAST RIDGE
Lake Malawi
Madagascar
CORAL SEA
Fiji Islands
Kalahari Desert
INDIAN OCEAN
New Caledonia
AUSTRALIA
TROPIC OF CAPRICORN
Great Dividing Range
SOUTH PACIFIC OCEAN
Cape of Good Hope
SOUTHWEST INDIAN RIDGE
Mt. Kosciuszko 2228
SOUTHEAST INDIAN RIDGE
North Island
Bass Strait
TASMAN SEA
Tasmania
South Island
NEW ZEALAND
ANTARCTIC CIRCLE
Wilkes Land
Winkel Tripel Projection, Central Meridian 0°
KILOMETERS
STATUTE MILES
Elevations in meters
ANTARCTICA

FIGURE 3 Map of the United States, including capital city for each state.

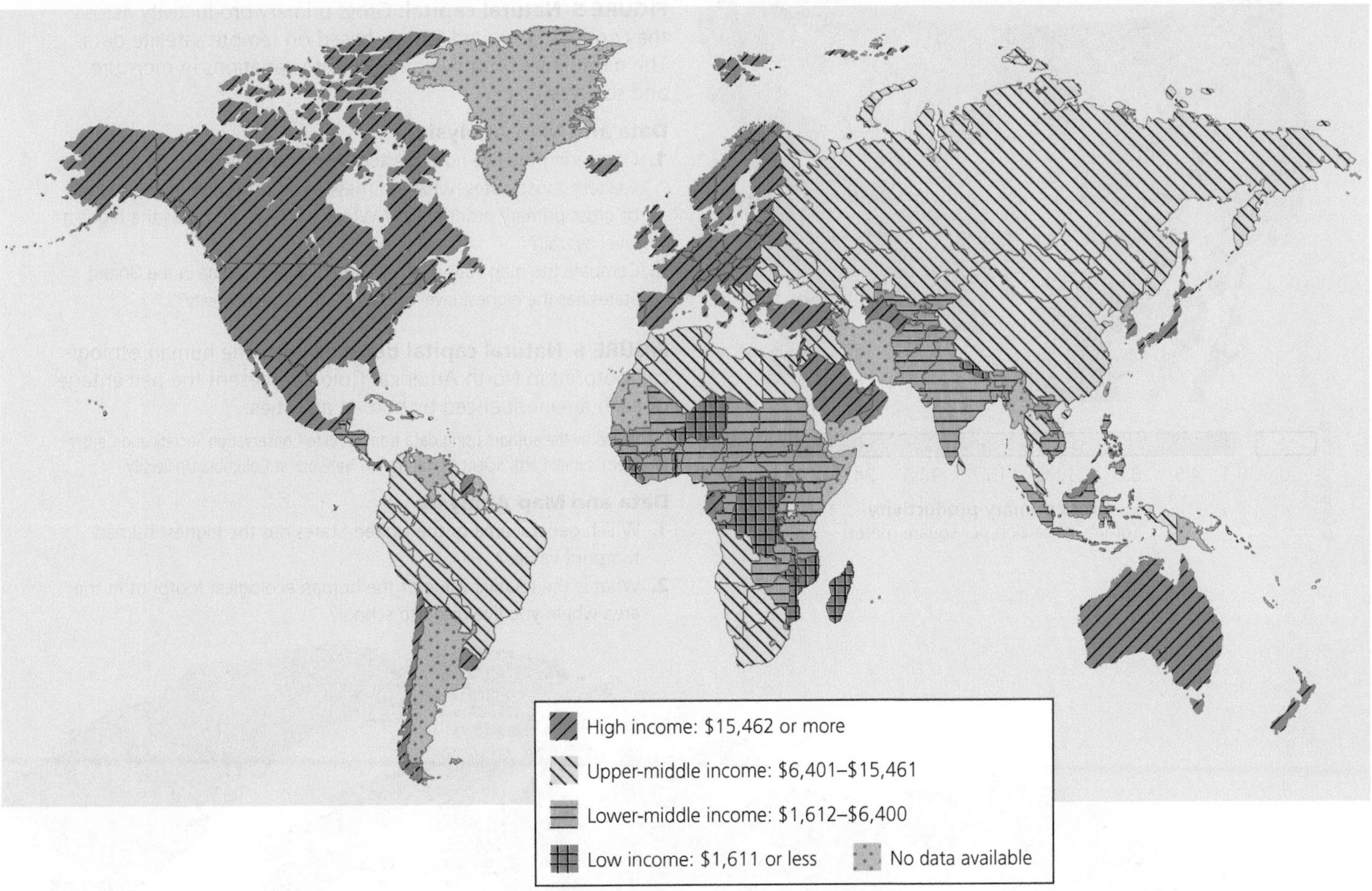

(Compiled by the authors using data from World Bank and International Monetary Fund.)

FIGURE 4 High-income, upper-middle-income, lower-middle-income, and low-income countries in terms of gross national income (GNI) purchasing power parity (PPP) per capita (U.S. dollars) in 2015.

Data and Map Analysis

1. In how many countries is the per capita average income $995 or less? Look at Figure 1 and find the names of three of these countries.
2. In how many instances does a lower-middle- or low-income country share a border with a high-income country? Look at Figure 1 and find the names of the countries that reflect three examples of this situation.

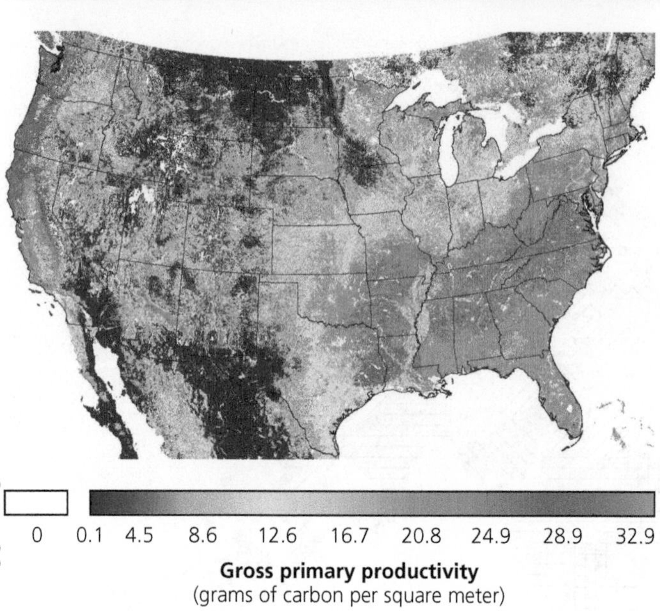

FIGURE 5 Natural capital: Gross primary productivity across the continental United States, based on remote satellite data. The differences roughly correlate with variations in moisture and soil types.

Data and Map Analysis

1. Comparing the five northwestern-most states with the five south-western-most states, which of these regions has the greater variety of gross primary productivity? Which of the regions has the highest level overall?
2. Compare this map with that of Figure 5. Which biome in the United States has the highest level of gross primary productivity?

FIGURE 6 Natural capital degradation: The human ecological footprint in North America. Colors represent the percentage of each area influenced by human activities.

(Compiled by the authors using data from Wildlife Conservation Society and Center for International Earth Science Information Network at Columbia University.)

Data and Map Analysis

1. Which general area of the United States has the highest human footprint values?
2. What is the relative value of the human ecological footprint in the area where you live or go to school?

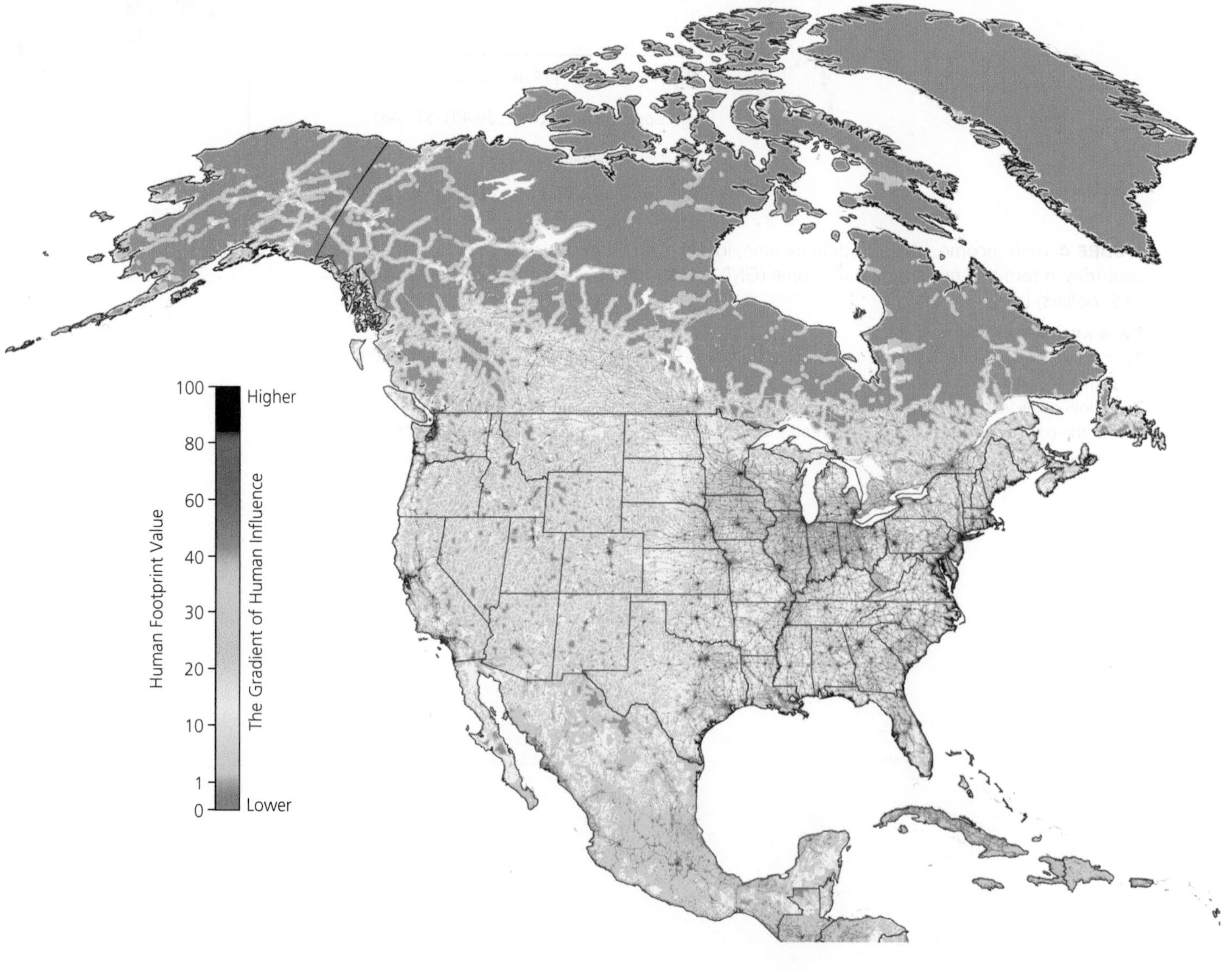

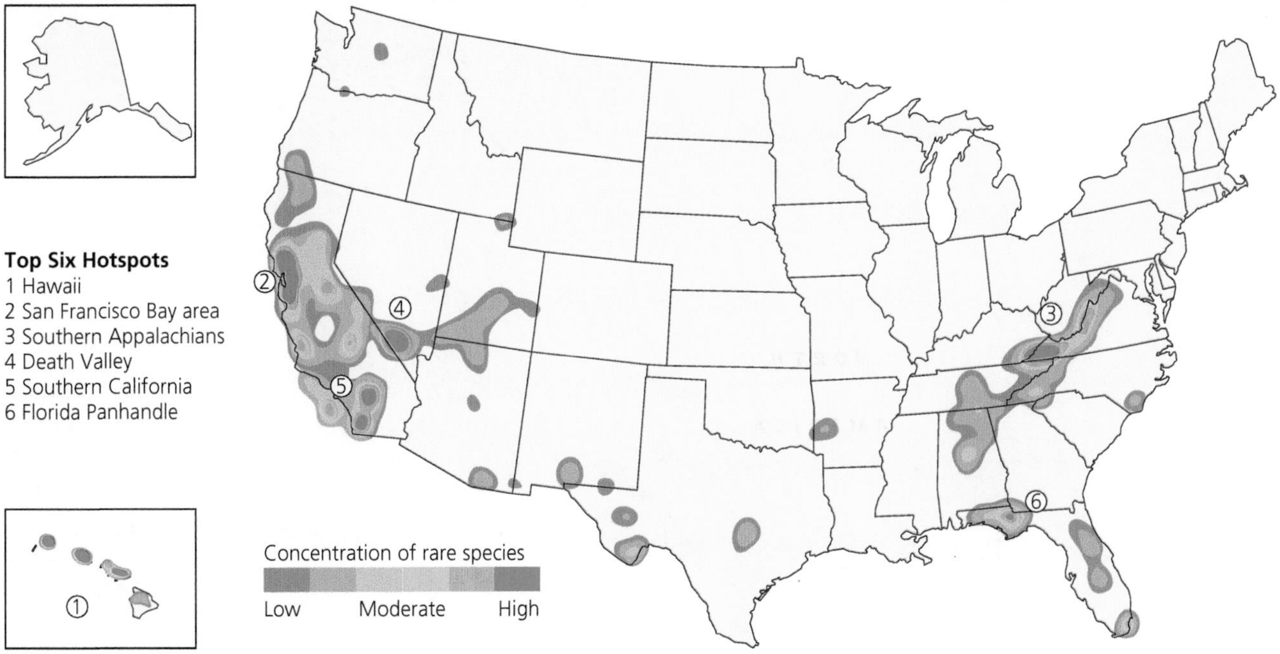

FIGURE 7 *Endangered natural capital:* Major biodiversity hotspots in the United States that need emergency protection. The shaded areas contain the largest concentrations of rare and potentially endangered species.

(Compiled by the authors using data from State Natural Heritage Programs, Nature Conservancy, and Association for Biodiversity Information.)

Data and Map Analysis

1. If you live in the United States, which of the top six hotspots is closest to where you live or go to school?
2. Which general part of the country has the highest overall concentration of rare species? Which part has the second-highest concentration?

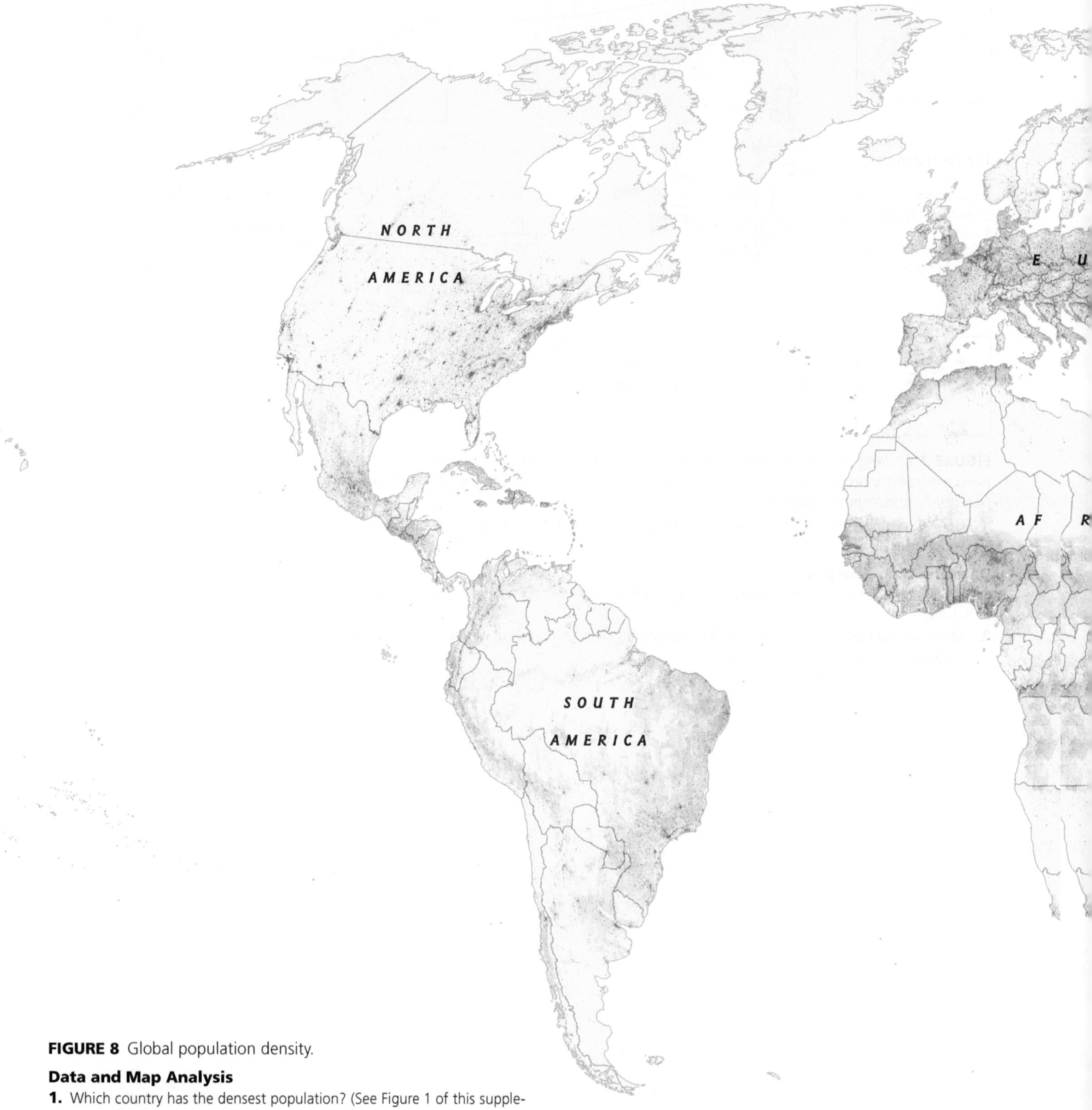

FIGURE 8 Global population density.

Data and Map Analysis

1. Which country has the densest population? (See Figure 1 of this supplement for country names.)

2. List the continents in order from the most densely populated to the least densely populated.

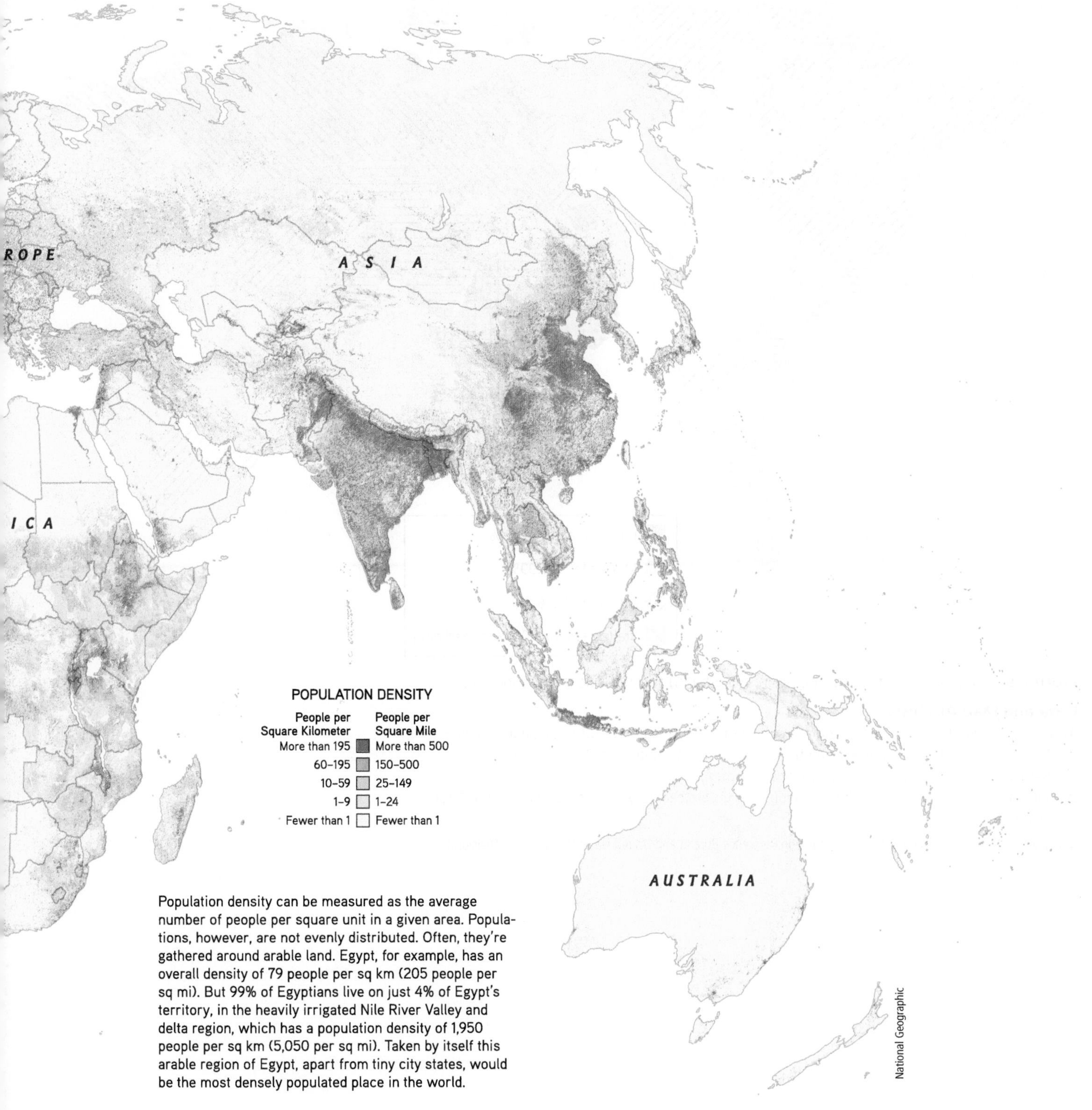

Population density can be measured as the average number of people per square unit in a given area. Populations, however, are not evenly distributed. Often, they're gathered around arable land. Egypt, for example, has an overall density of 79 people per sq km (205 people per sq mi). But 99% of Egyptians live on just 4% of Egypt's territory, in the heavily irrigated Nile River Valley and delta region, which has a population density of 1,950 people per sq km (5,050 per sq mi). Taken by itself this arable region of Egypt, apart from tiny city states, would be the most densely populated place in the world.

National Geographic

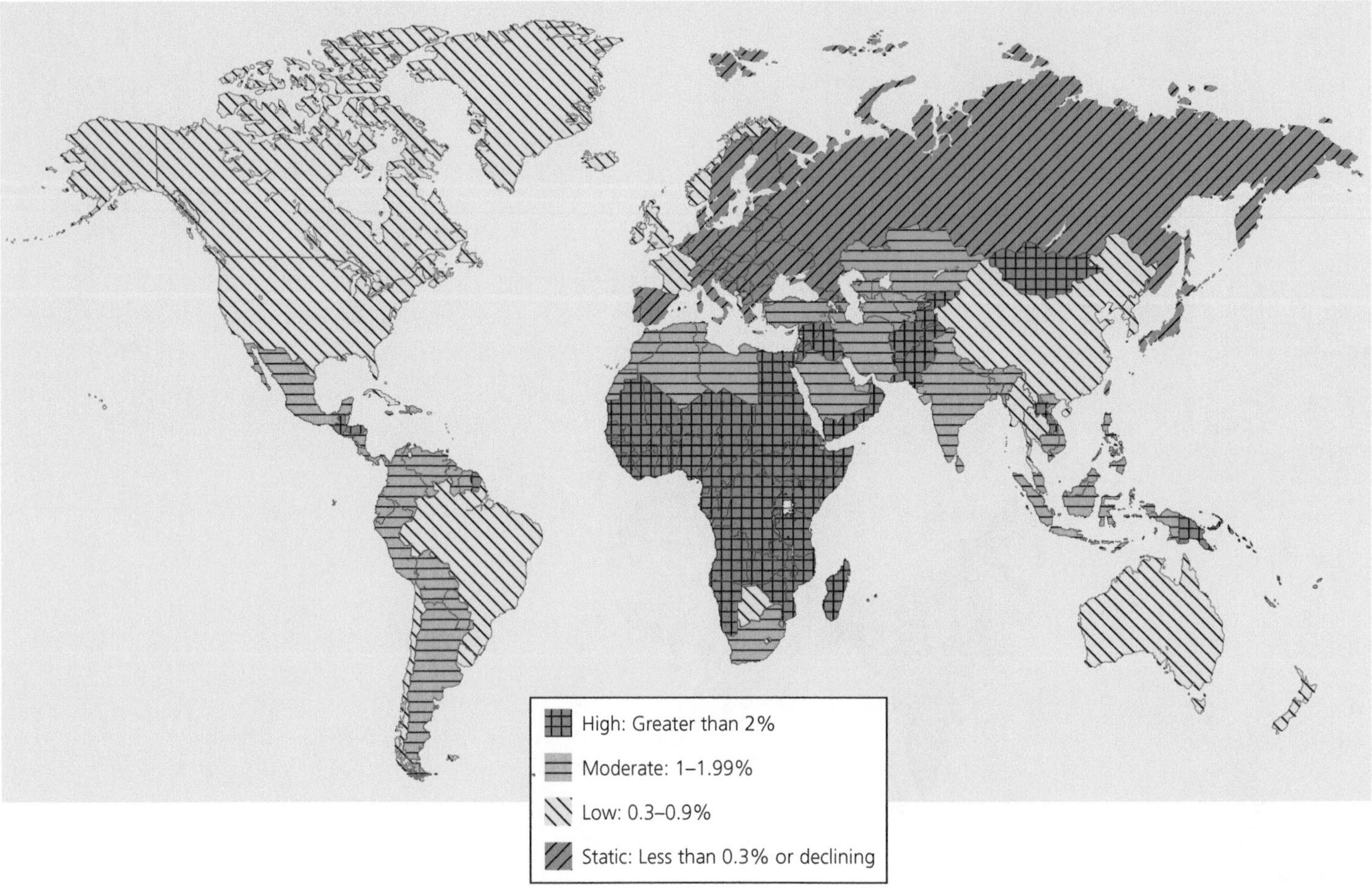

FIGURE 9 Comparative rates of population growth (%) throughout the world in 2014.

Data and Map Analysis

1. Which continent has the greatest number of countries with high rates of population growth? Which continent has the greatest number of countries with static rates? (See Figure 1 of this supplement for continent names.)
2. Name three countries from each population growth rate category on this map (see Figure 1 for country names).

(Compiled by the authors using data from Population Reference Bureau and United Nations Population Division.)

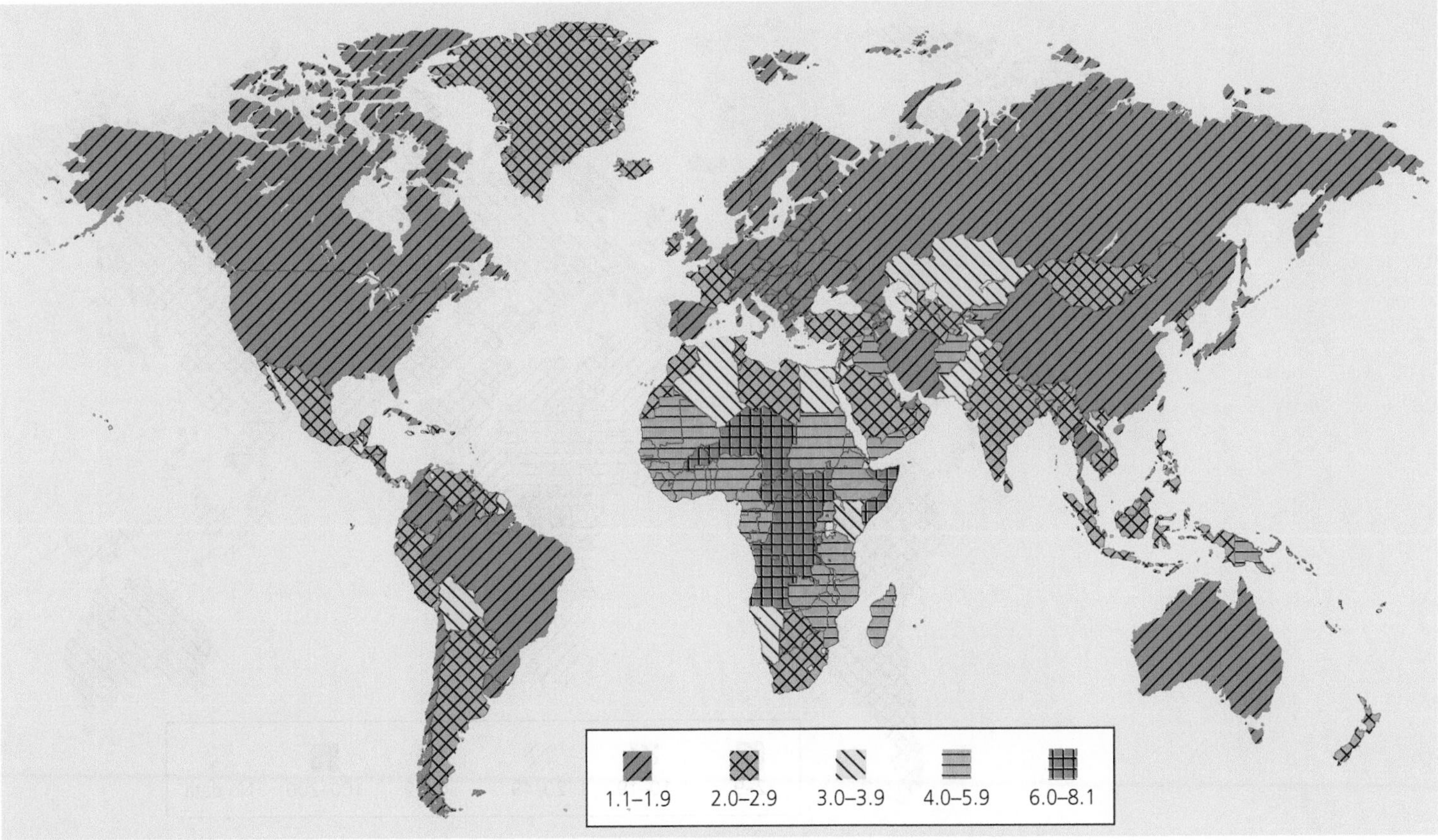

FIGURE 10 Comparison of total fertility rates (TFRs)—the average number of children born to the women in any population throughout their lifetimes—as measured in 2015.

Data and Map Analysis

1. Which country in the middle TFR category borders two countries in the lowest TFR category? What are those two countries? (See Figure 1 of this supplement for country names.)
2. Describe two geographic patterns that you see on this map.

(Compiled by the authors using data from Population Reference Bureau and United Nations Population Division.)

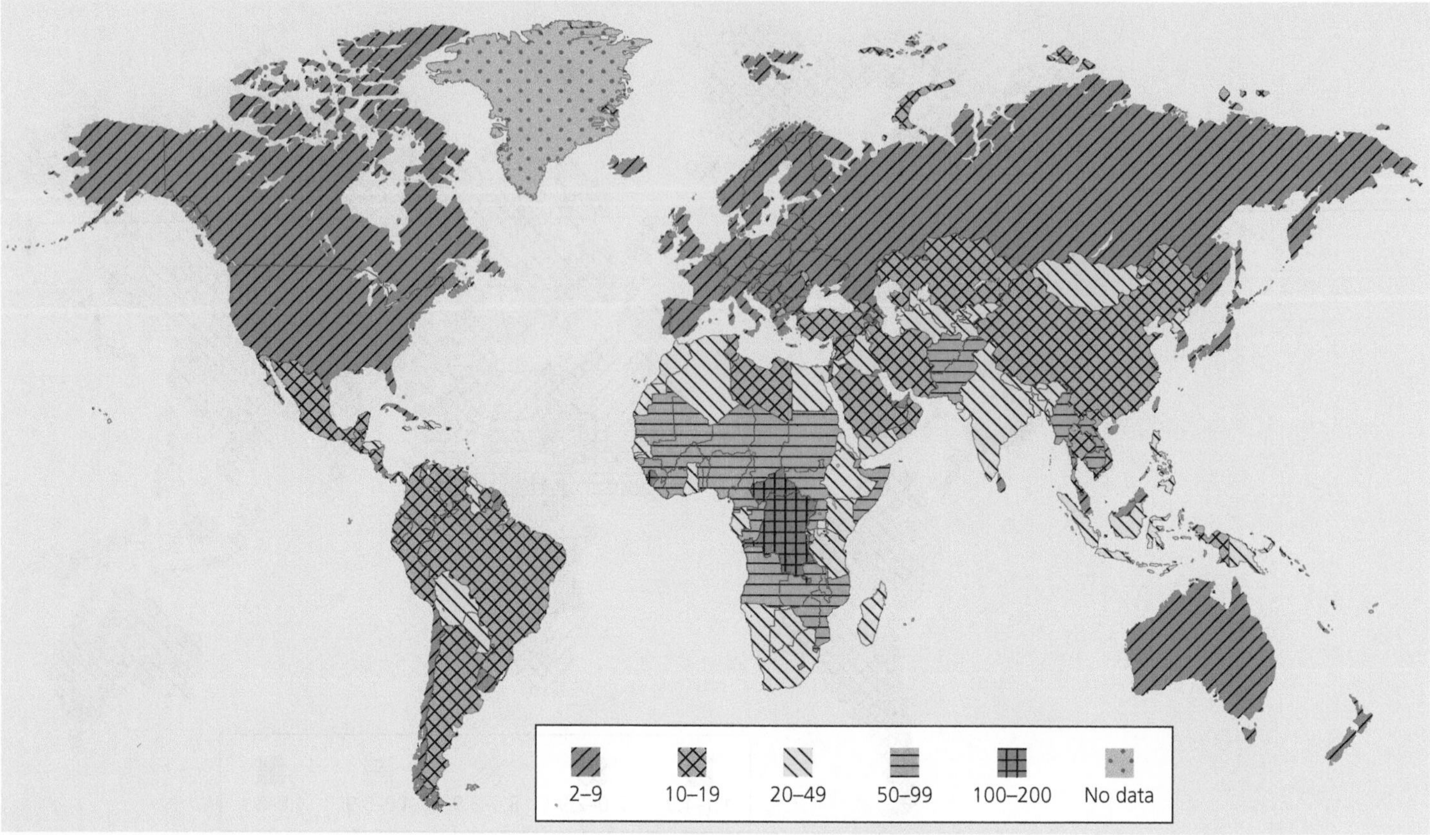

FIGURE 11 Comparison of the infant mortality rates of countries around the world in 2015.

Data and Map Analysis

1. Describe a geographic pattern that you can see on this map.
2. Describe any similarities that you see in geographic patterns between this map and the one in Figure 9.

(Compiled by the authors using data from Population Reference Bureau and United Nations Population Division.)

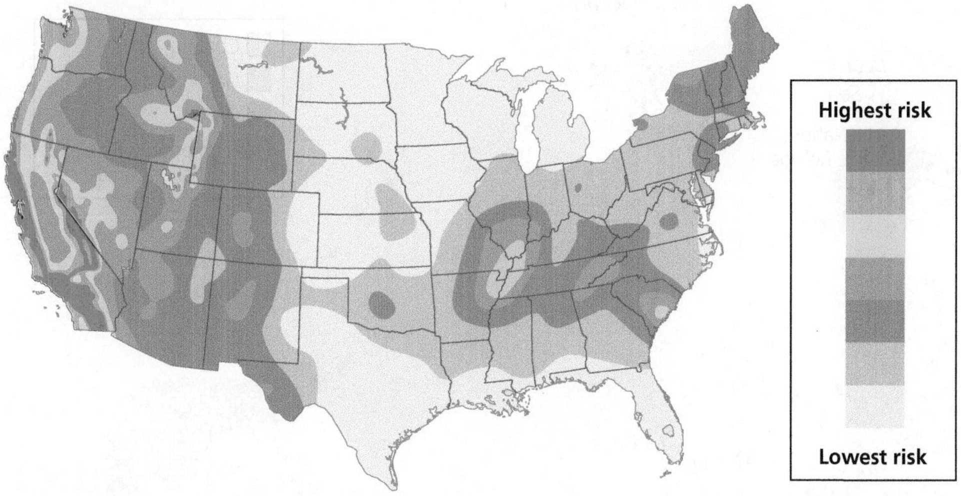

FIGURE 12 Earthquake (seismic) risk in the United States.

Data and Map Analysis

1. In general terms (northeast, southeast, central, west coast, etc.), which area has the largest earthquake risk and which area has the lowest earthquake risk?
2. What is the level of risk where you live or go to school?

(Compiled by the authors using data from U.S. Geological Survey.)

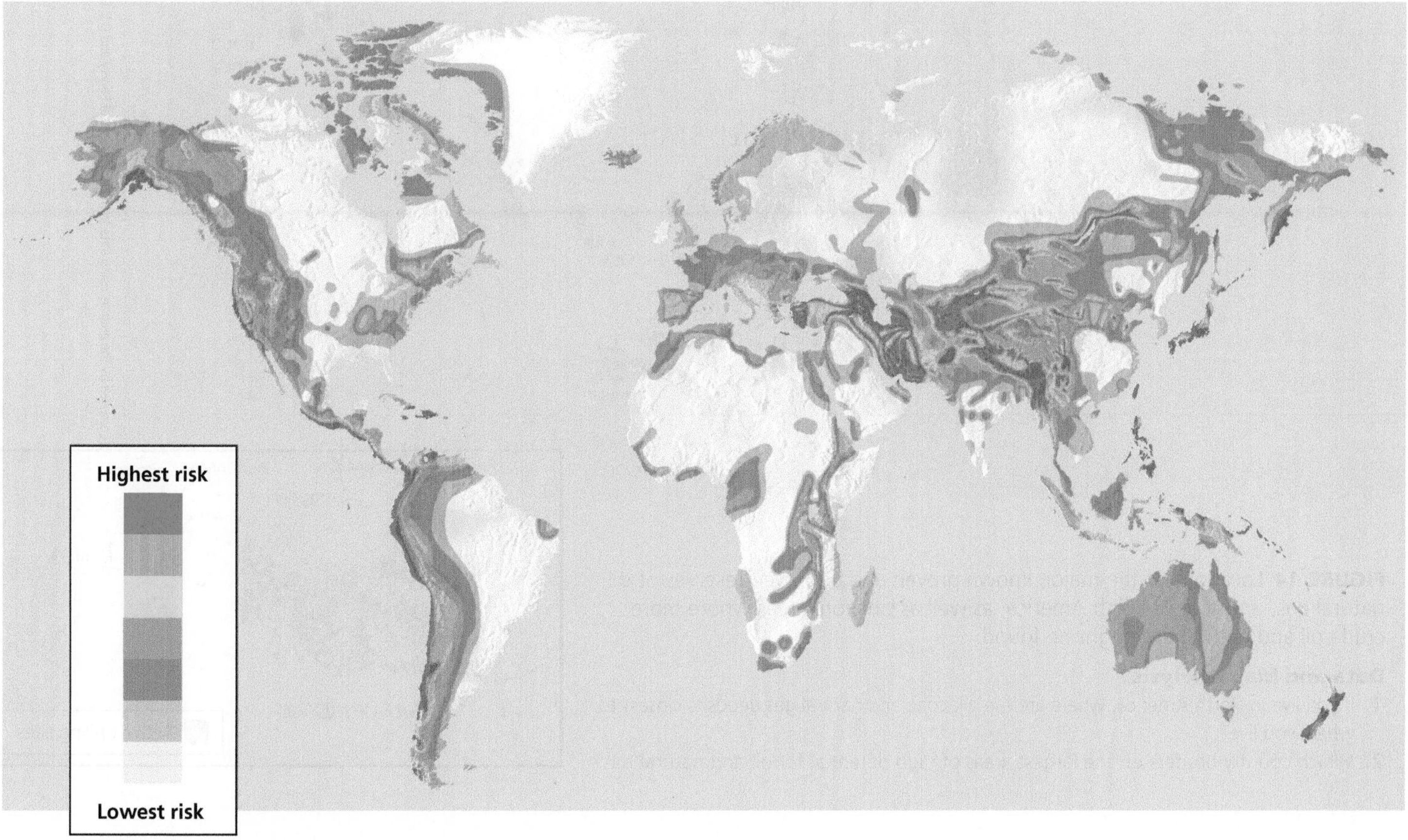

(Compiled by the authors using data from U.S. Geological Survey.)

FIGURE 13 Earthquake (seismic) risk in the world.

Data and Map Analysis

1. How are the high-risk (red and orange) areas related to the boundaries of the earth's major tectonic plates as shown in Figure 14.16, p. 373?
2. Which continent has the longest coastal area subject to the highest possible risk (red)? Which continent has the second highest risk area?

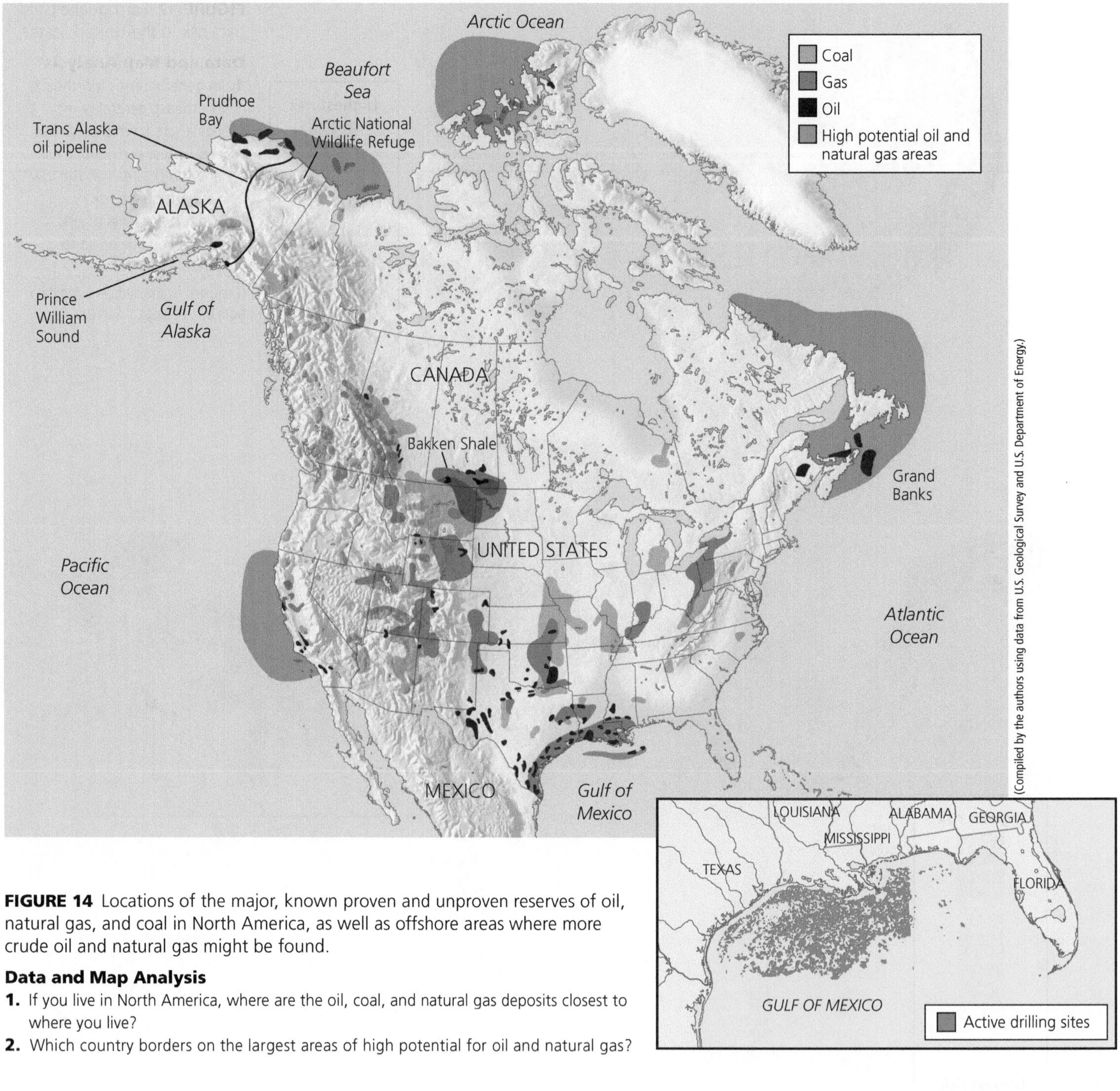

FIGURE 14 Locations of the major, known proven and unproven reserves of oil, natural gas, and coal in North America, as well as offshore areas where more crude oil and natural gas might be found.

Data and Map Analysis

1. If you live in North America, where are the oil, coal, and natural gas deposits closest to where you live?
2. Which country borders on the largest areas of high potential for oil and natural gas?

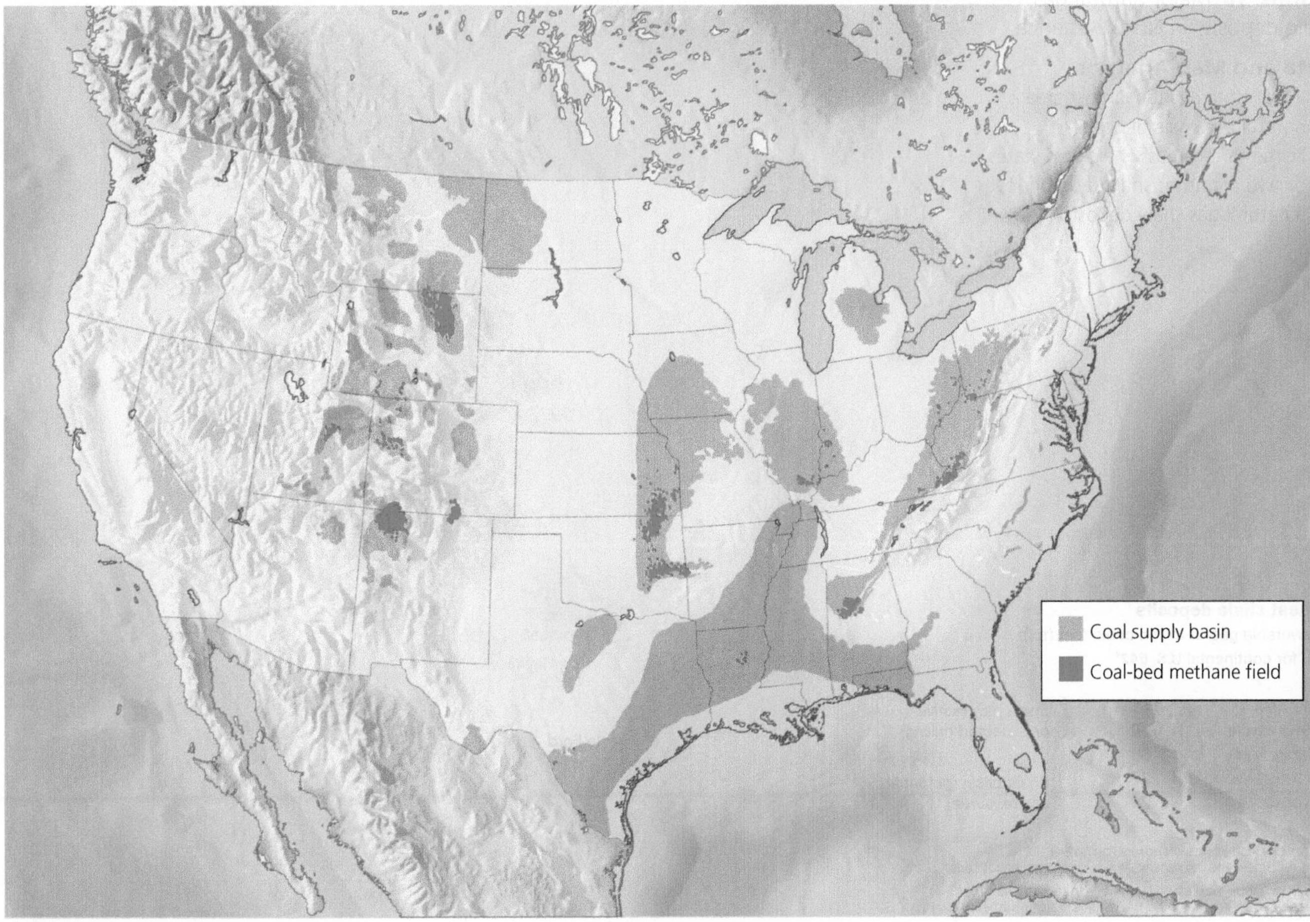

(Compiled by the authors using data from U.S. Department of Energy and U.S. Geological Survey.)

FIGURE 15 Major coal supply basins and coal-bed methane fields in the continental states of the United States.

Data and Map Analysis

1. If you live in the United States, where are the coal-bed methane deposits closest to where you live?
2. Removing these deposits requires lots of water. Compare the locations of the major deposits of coal-bed methane with water-deficit areas shown in Figures 13.6 and 13.7, p. 329.

FIGURE 16 Major natural gas shale deposits in North America.

Data and Map Analysis

1. What state has the largest area of natural gas shale deposits?
2. Name three areas where two states or two countries share a border over a natural gas shale deposit.

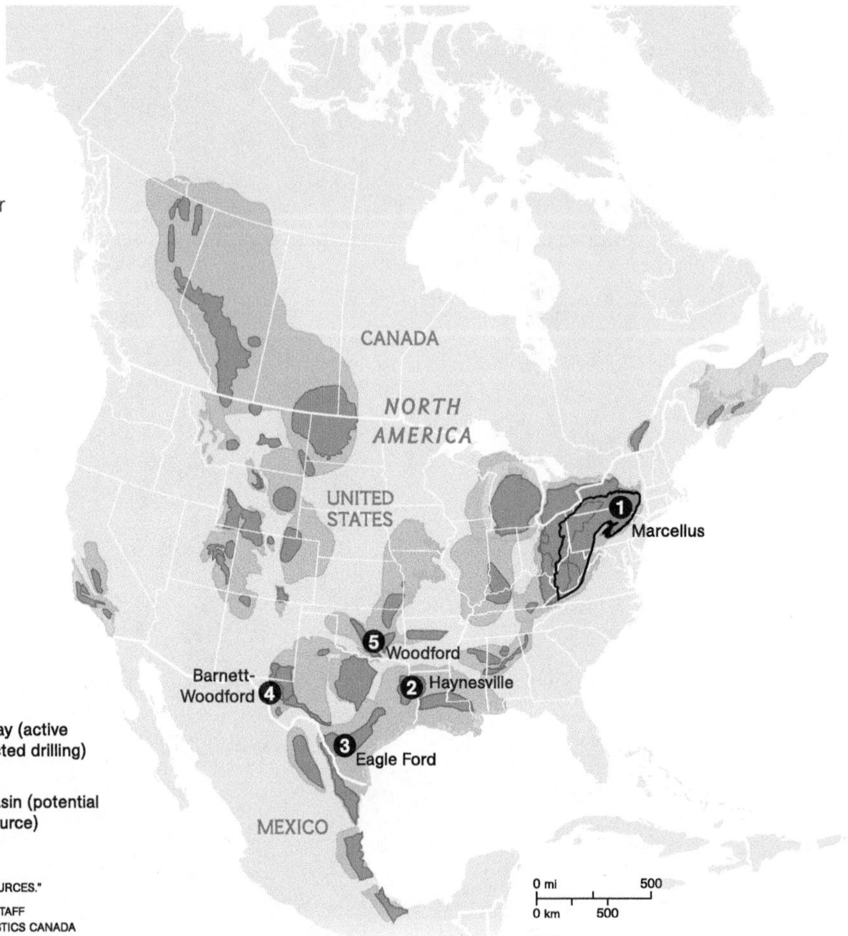

Richest shale deposits
Recoverable gas, in trillion cubic feet (tcf)
Total for continental U.S. **542***

1. Marcellus	**141**
2. Haynesville	**66**
3. Eagle Ford	**50**
4. Barnett-Woodford	**27**
5. Woodford	**24**

Shale play (active or expected drilling)

Shale basin (potential gas resource)

ONE TCF CAN HEAT 15 MILLION HOMES FOR A YEAR.
*TOTAL INCLUDES "PROVED RESERVES" AND "UNPROVED RESOURCES."

VIRGINIA W. MASON, NGM STAFF; ALEXANDER STEGMAIER, NG STAFF
SOURCES: U.S. EIA; FRACTRACKER; U.S. CENSUS BUREAU; STATISTICS CANADA

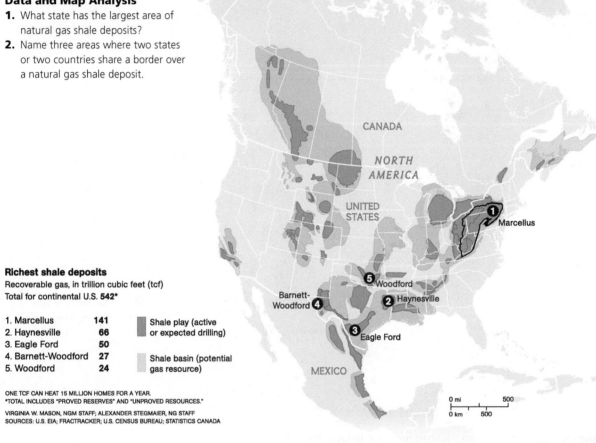

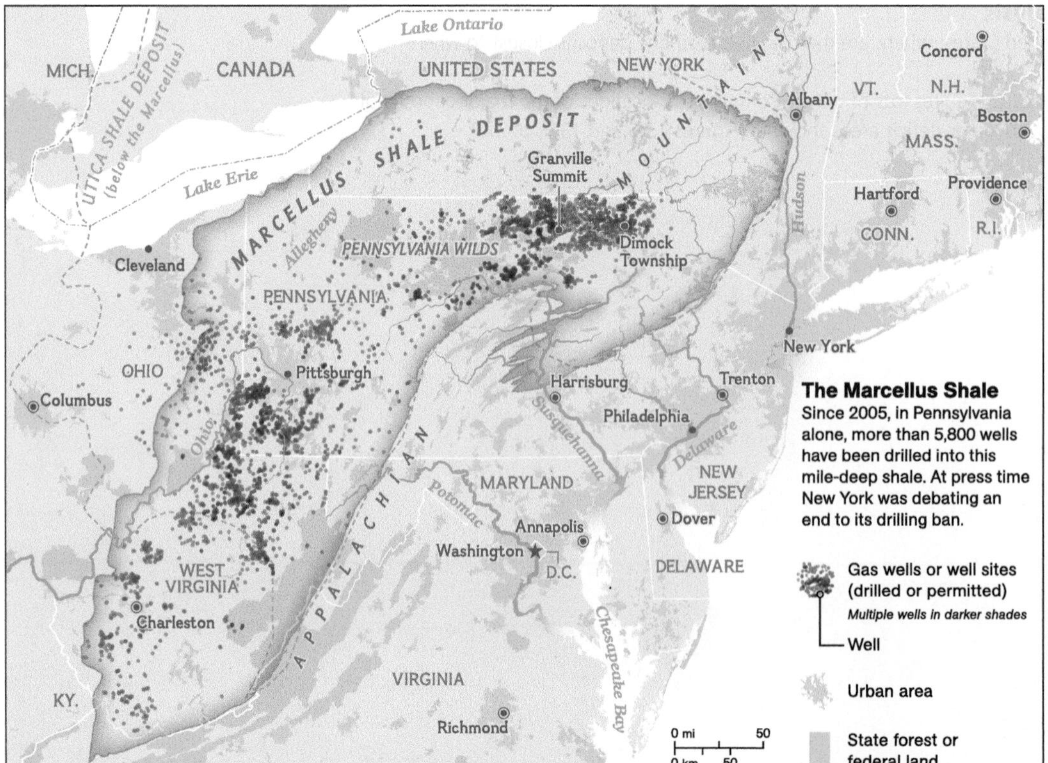

VIRGINIA W MASON/National Geographic

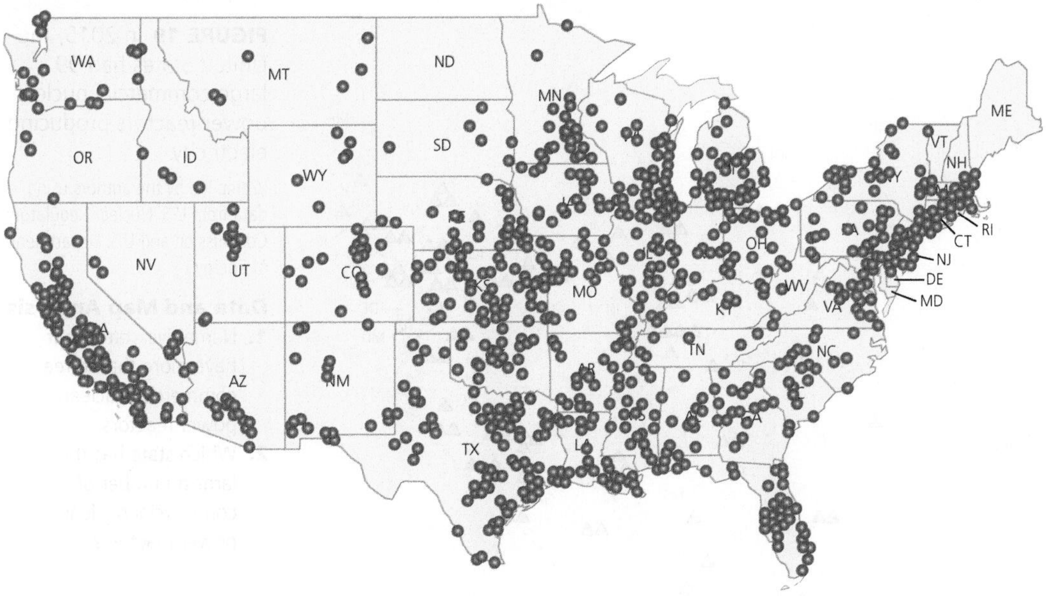

FIGURE 17 Large natural gas–fired power plants in the continental United States in 2012. They are increasing in number as a growing number of coal-fired power plants are being closed.

(Compiled by the authors using data from the U.S. Department of Energy.)

Data and Map Analysis

1. Name three general regions of the Unites States that had high concentrations of natural gas–fired power plants in 2012.
2. Dividing the country in half (east/west), roughly how many times higher than the number of plants in the west is the number of plants in the east?

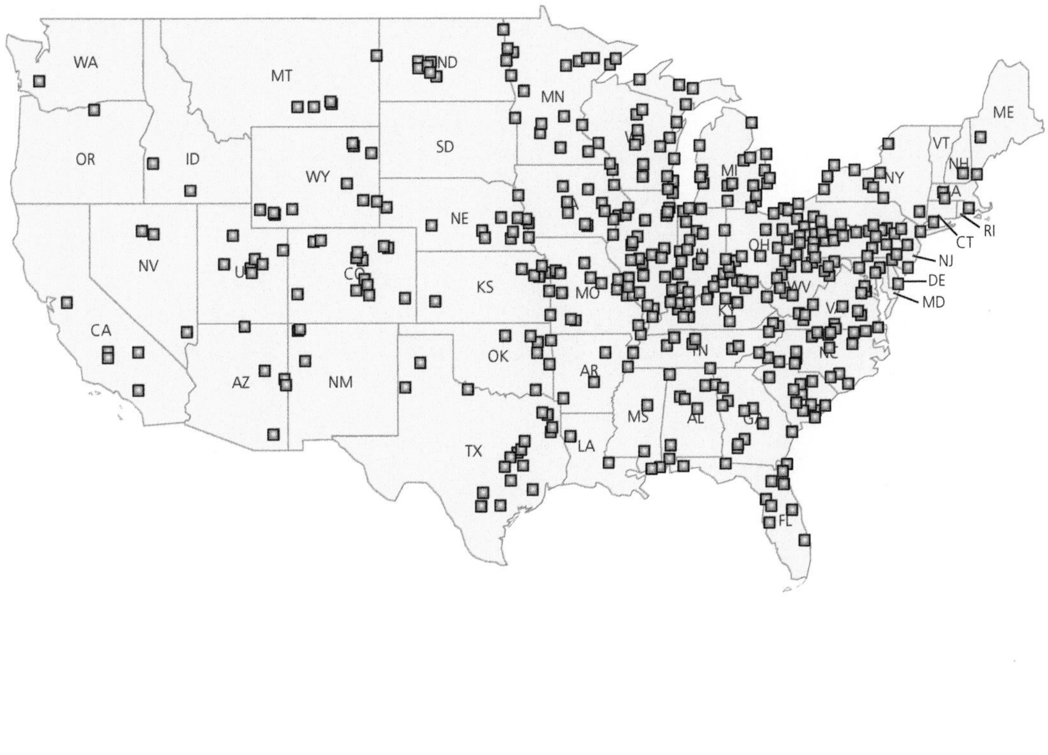

FIGURE 18 Large coal-burning power plants in the continental United States in 2012. Between 2003 and 2014, the number of large coal-burning plants in the United States dropped from 629 to 491.

(Compiled by the authors using data from the U.S. Department of Energy.)

Data and Map Analysis

1. Name three states that had a high concentration (10 or more) of coal-fired power plants in 2012.
2. Dividing the country in half (east/west), roughly how many times higher than the number of plants in the west is the number of plants in the east?

FIGURE 19 In 2015, the United States had 99 large commercial nuclear power reactors producing electricity.

(Compiled by the authors using data from U.S. Nuclear Regulatory Commission and U.S. Department of Energy.)

Data and Map Analysis

1. Name five states that have more than three commercial nuclear power reactors.
2. Which state has the largest number of commercial nuclear power reactors?

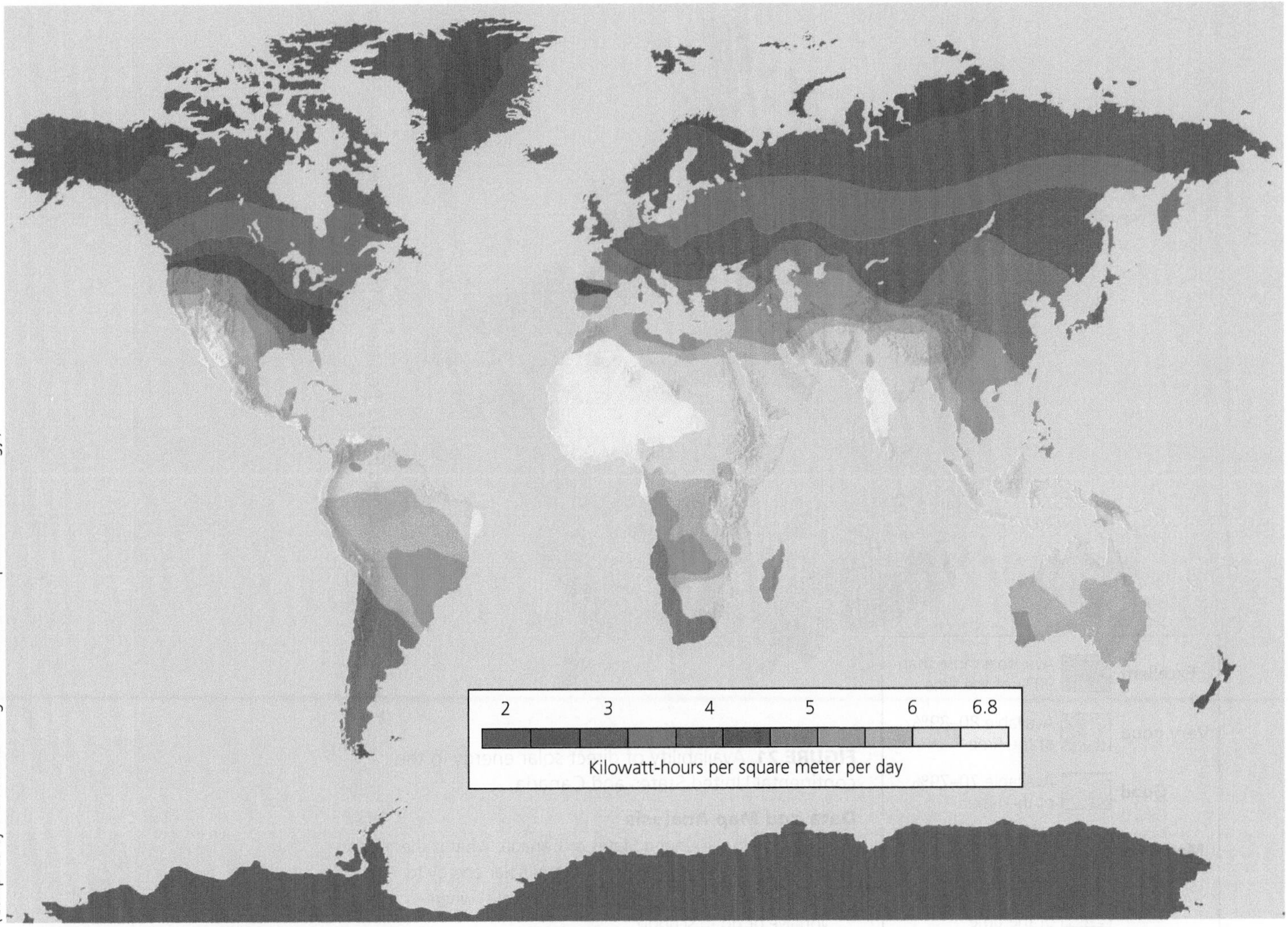

(Compiled by the authors using data from U.S. Department of Energy.)

FIGURE 20 Global availability of direct solar energy. Areas with more than 3.5 kilowatt-hours per square meter per day (see scale) are good candidates for passive and active solar heating systems and use of solar cells to produce electricity.

Data and Map Analysis

1. What is the potential for making greater use of solar energy to provide heat and produce electricity (with solar cells) where you live or go to school?
2. List the continents in order of overall availability of direct solar energy, from those with the highest to those with the lowest. (See Figure 1 of this supplement for continent names.)

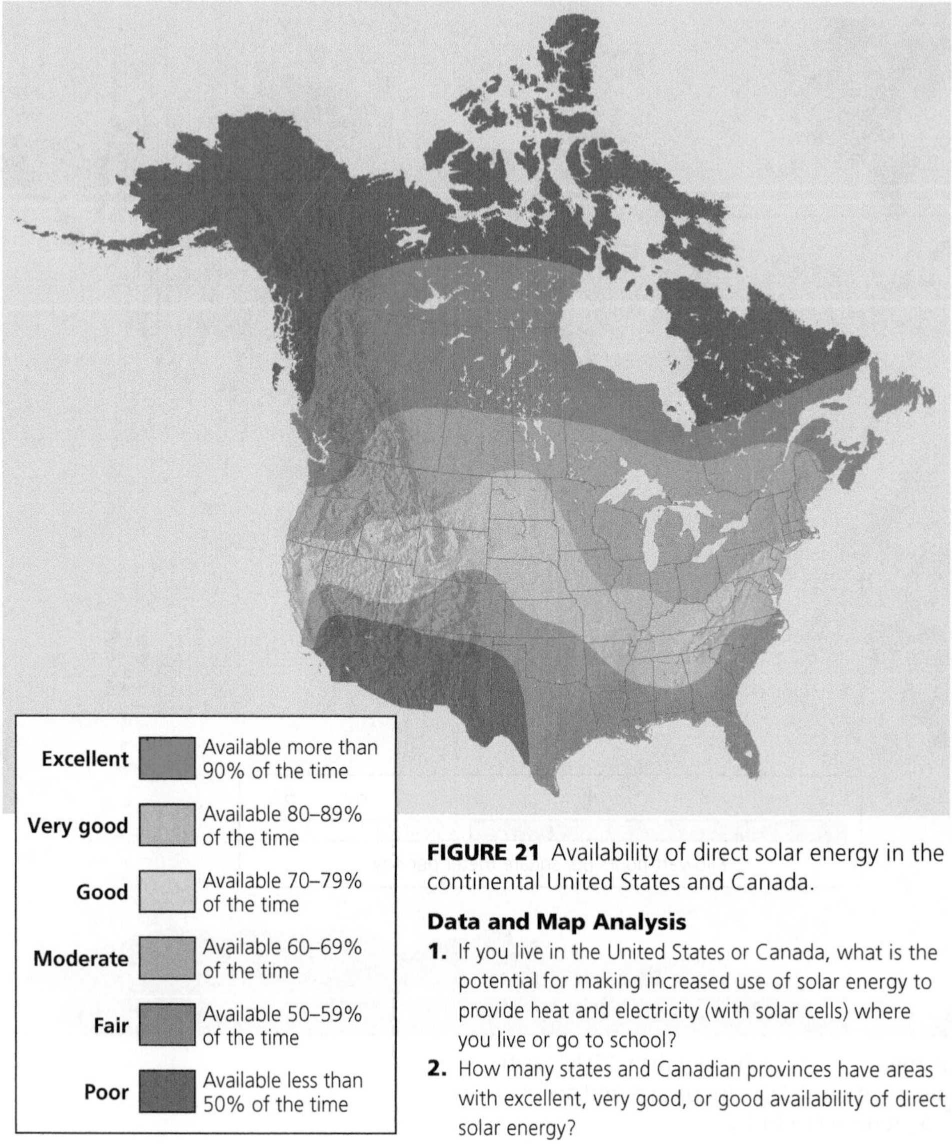

(Compiled by the authors using data from U.S. Department of Energy and National Wildlife Federation.)

FIGURE 21 Availability of direct solar energy in the continental United States and Canada.

Data and Map Analysis

1. If you live in the United States or Canada, what is the potential for making increased use of solar energy to provide heat and electricity (with solar cells) where you live or go to school?
2. How many states and Canadian provinces have areas with excellent, very good, or good availability of direct solar energy?

FIGURE 22 Large solar power plants in the continental United States (not including rooftop solar power installations) in 2012.

(Compiled by the authors using data from the U.S. Department of Energy.)

Data and Map Analysis

1. Name three states that have a high number (10 or more) of large solar power plants.
2. Name two states in the northern half of the country that have two or more large solar power plants.

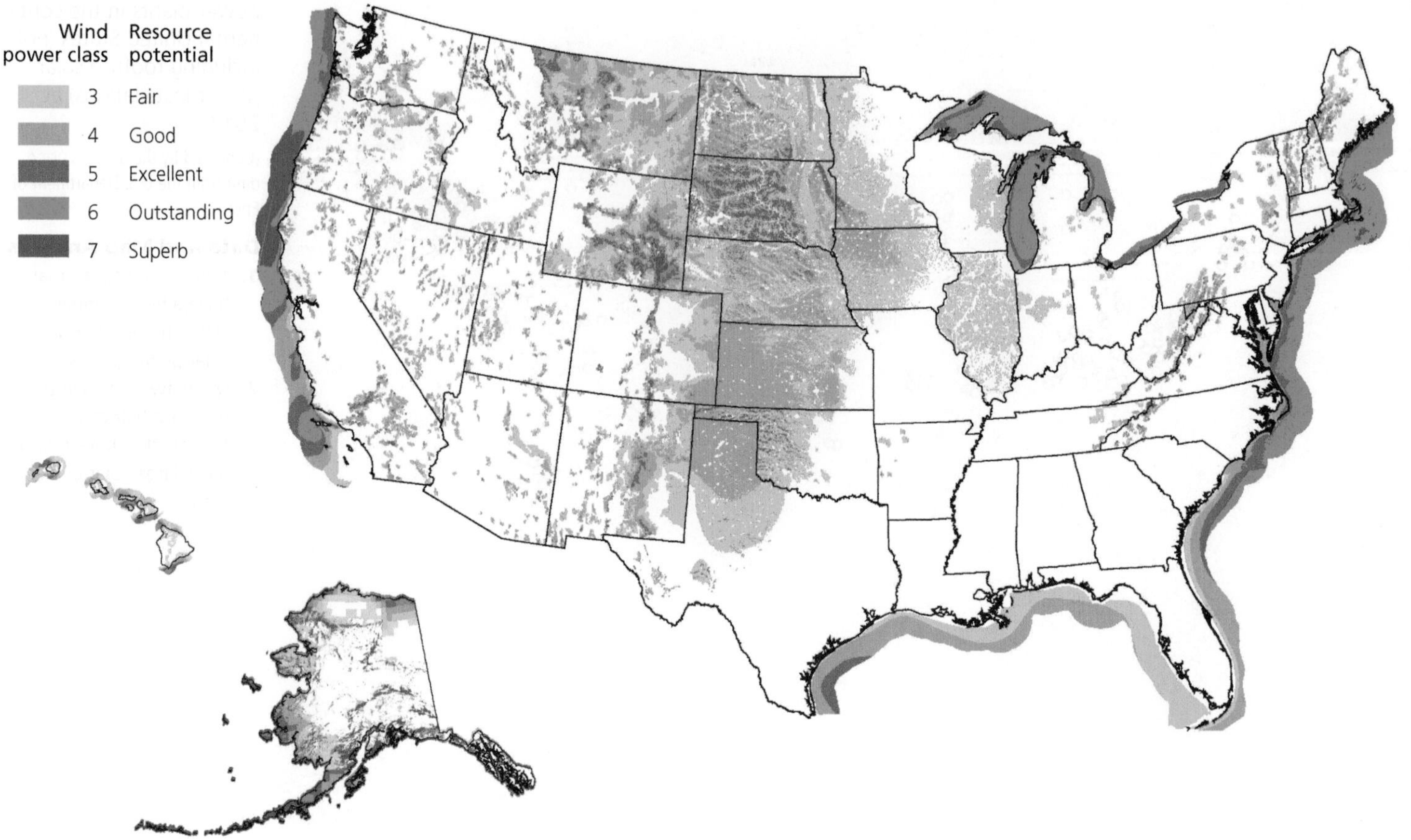

FIGURE 23 Potential supply of land- and ocean-based wind energy in the United States.

Data and Map Analysis

1. If you live in the United States, what is the general wind energy potential where you live or go to school?

2. How many states have areas with good or excellent potential for wind energy?

(Compiled by the authors using data from U.S. Geological Survey and U.S. Department of Energy.)

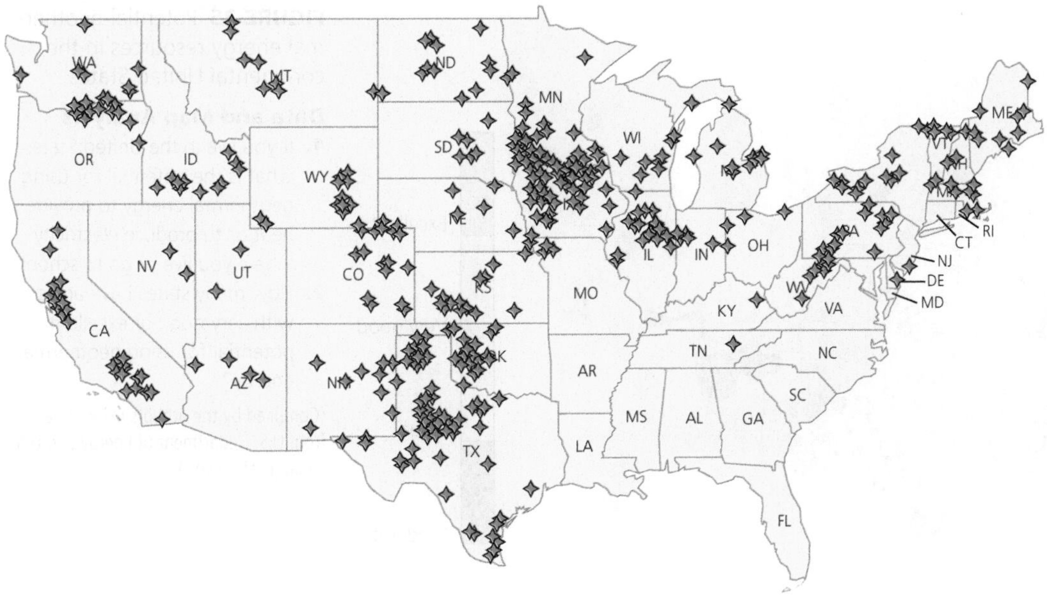

FIGURE 24 Large wind farms in the continental United States in 2012.

(Compiled by the authors using data from the U.S. Department of Energy.)

Data and Map Analysis

1. Name three states that have a high number (10 or more) of large wind farms.
2. Do you see any geographic pattern for the location of wind farms in this map? How would you describe this pattern?

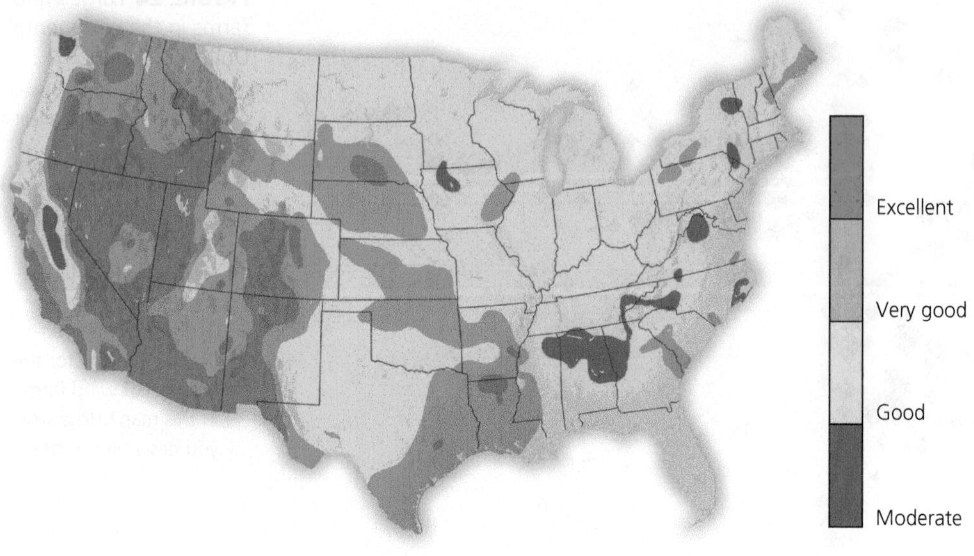

FIGURE 25 Potential geothermal energy resources in the continental United States.

Data and Map Analysis

1. If you live in the United States, what is the potential for using geothermal energy to provide heat or to produce electricity where you live or go to school?
2. How many states have areas with very good or excellent potential for using geothermal energy?

(Compiled by the authors using data from U.S. Department of Energy and U.S. Geological Survey.)

FIGURE 26 Hydroelectric dams in the continental United States in 2012. Hydropower is by far the largest renewable source of electricity.

Data and Map Analysis

1. Name three regions of the country that have a high concentration of hydroelectric dams.
2. Why do you think most of the states on both coasts have high concentrations of hydroelectric dams?

(Compiled by the authors using data from U.S. Department of Energy and U.S. Geological Survey.)

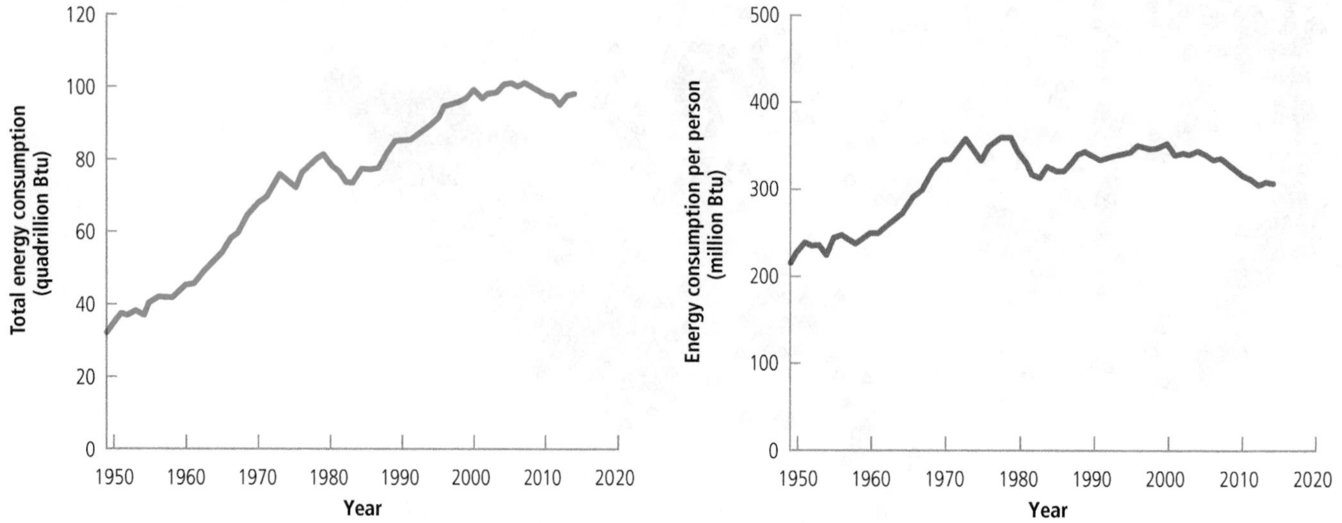

FIGURE 1 Total (left) and per capita (right) energy consumption in the United States, 1950–2014.

(Compiled by the authors using data from U.S. Energy Information Administration and U.S. Census Bureau.)

Data and Graph Analysis

1. In what year or years did total U.S. energy consumption reach 80 quadrillion Btus?
2. In what year did energy consumption per person reach its highest level shown on this graph, and about what was that level of consumption?

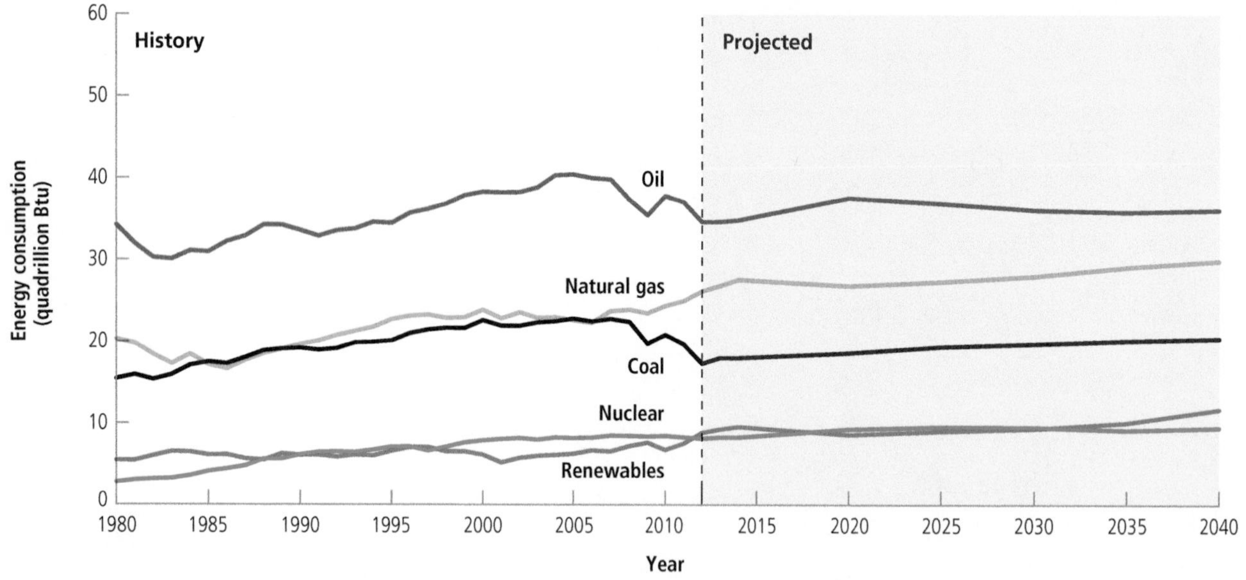

FIGURE 2 Energy consumption by source in the United States, 1980–2014, with projections to 2040.

Data and Graph Analysis

1. For which source did usage grow the most between 1980 and 2014? For which source did it decrease the most?
2. For which two energy sources is the usage expected to grow the most between 2014 and 2040? For which two energy sources is usage expected to decrease the most?

(Compiled by the authors using data from U.S. Energy Information Administration.)

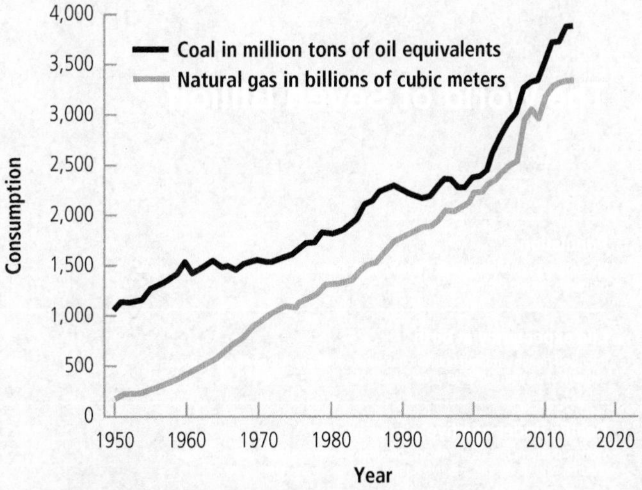

FIGURE 3 World coal and natural gas consumption for the period 1950–2014.

(Compiled by the authors using data from British Petroleum and International Energy Agency.)

Data and Graph Analysis

1. In what year did coal use reach a level twice as high as it was in 1960? Which energy source has grown more steadily—coal or natural gas? In what years has coal use grown most sharply?

2. By what percentage did natural gas consumption increase between 1960 and 2014?

The World of Seven Billion

LOW INCOME LEVEL
$995 or less a year

Characteristic	Low income level
Population Most future population growth will happen in the less developed countries, where birthrates remain highest.	1 billion
Life expectancy at birth Improved health care and nutrition have raised life expectancy from a global average of 52 years in 1960 to 69 years today.	Male 58; Female 60
Deaths under age five *(per 1,000 live births)* Worldwide there has been remarkable improvement. Since 1960, the number of children who die before age five has fallen by more than half.	120
Access to improved sanitation *(percent)* The UN defines this as access to toilets–even simple pit toilets– that keep excrement away from humans, animals, and insects.	35
Deaths caused by infectious disease *(percent)* The top five causes of death by infectious disease are acute respiratory infections (such as pneumonia), HIV/AIDS, diarrhea, TB, and malaria.	36
Years of education Increases in education affect not only economic development but population: The more education a woman receives, the fewer children she is likely to bear.	7.9
Literacy rate *(percent)* Global literacy is 82 percent. But for those who live where printed materials, even signs or product boxes, are rare, reading is a "use it or lose it" skill.	66
Fertility rate *(children per woman)* In most of the world, the fertility rate has fallen. Among the reasons: decline in infant mortality, economic improvements, and education of women.	4
Rate of natural population increase *(percent)* A country's annual natural growth rate is measured by subtracting the number of deaths from the number of births. It does not include migration, in or out.	2.27
Net migration rate *(per 1,000 people)* More than 200 million people–over 3 percent of the world's population–live outside the country in which they were born.	Out –0.58
Urban population *(percent)* As of 2008, the world's population has shifted from mainly rural to more than 50 percent urban. Most urbanites live in cities of fewer than 500,000 people.	27
Carbon dioxide emissions *(per capita, in metric tons)* Energy demand, largely for fossil fuels, continues to rise. China has surpassed the U.S. in total CO_2 emissions, but, per capita, U.S. emissions are four times higher.	1

Data for each category above were first compiled for all countries in each income level. The data were then averaged, accounting for differences in population.

FIGURE 4 Some characteristics of low-, middle-, and high-income countries.

Data and Graph Analysis

1. For every child under age 5 in a high-income country who dies, how many children under 5 in low-income countries die?
2. About how many times higher are the per capita carbon dioxide emissions in high-income countries than they are in lower-middle-income countries?

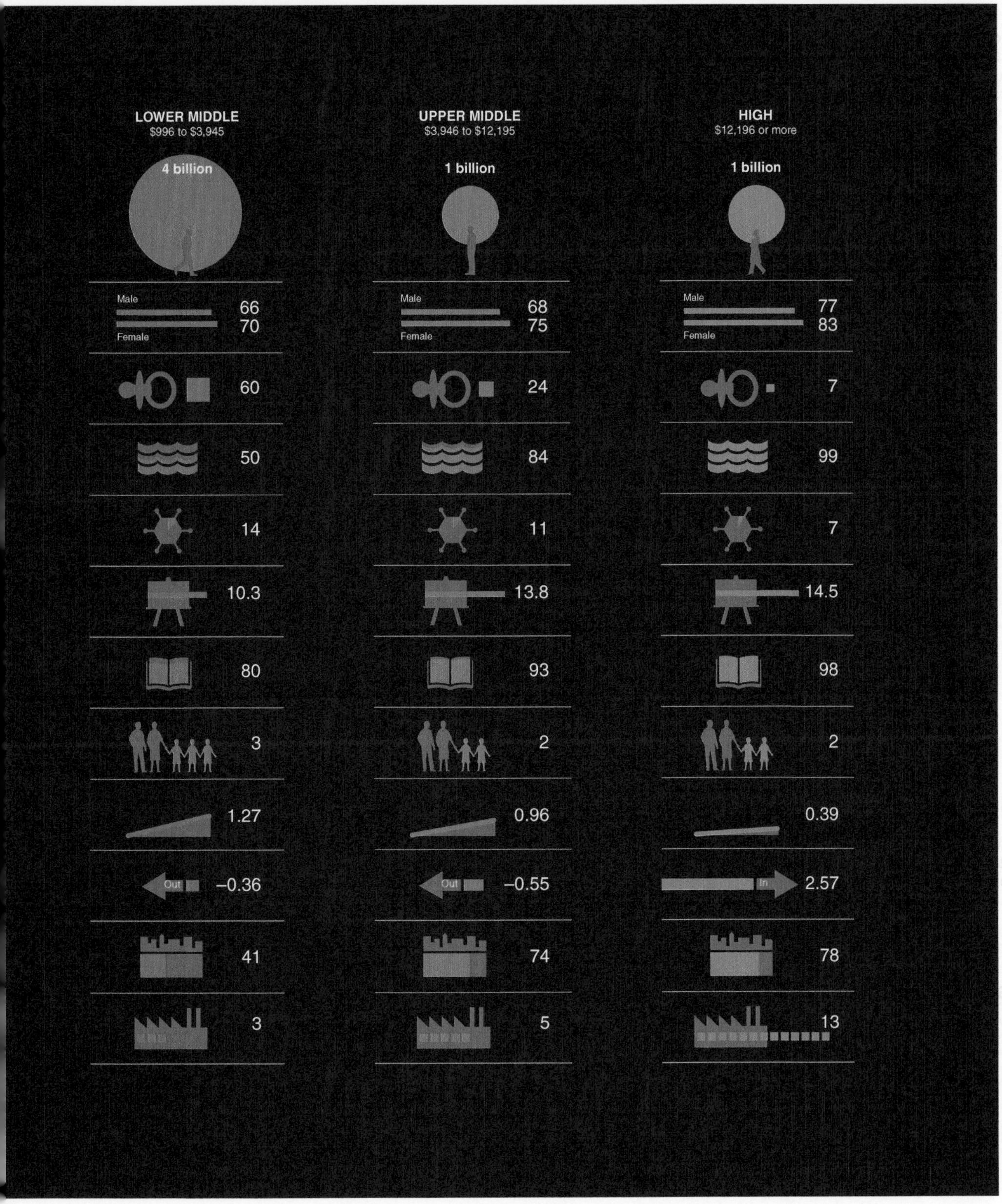
LOWER MIDDLE
$996 to $3,945
4 billion
Male
66
70
Female
60
50
14
10.3
80
3
1.27
Out
−0.36
41
3
UPPER MIDDLE
$3,946 to $12,195
1 billion
Male
68
75
Female
24
84
11
13.8
93
2
0.96
Out
−0.55
74
5
HIGH
$12,196 or more
1 billion
Male
77
83
Female
7
99
7
14.5
98
2
0.39
In
2.57
78
13

National Geographic

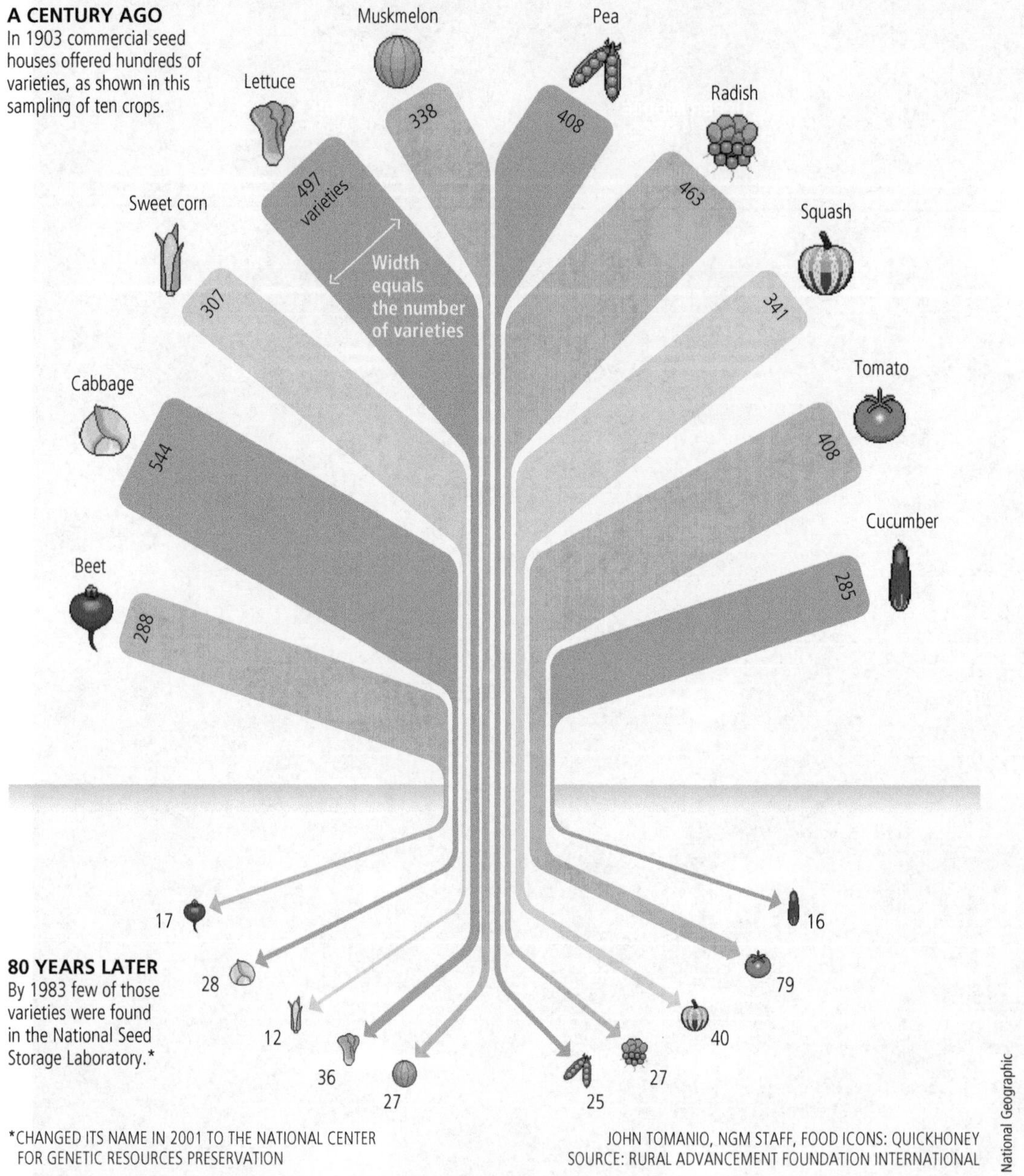

FIGURE 5 Dwindling food variety (agrobiodiversity) in the United States.

Data and Graph Analysis

1. What percentage of the lettuce varieties available in 1903 were available in 1983? Answer the same question for tomato varieties.
2. Which three of the varieties shown here had the greatest declines in numbers of varieties available?

WHY SO WILD?

The atmosphere is getting warmer and wetter. Those two trends, which are clear in data averaged globally and annually, are increasing the chances of heat waves, heavy rains, and perhaps other extreme weather.

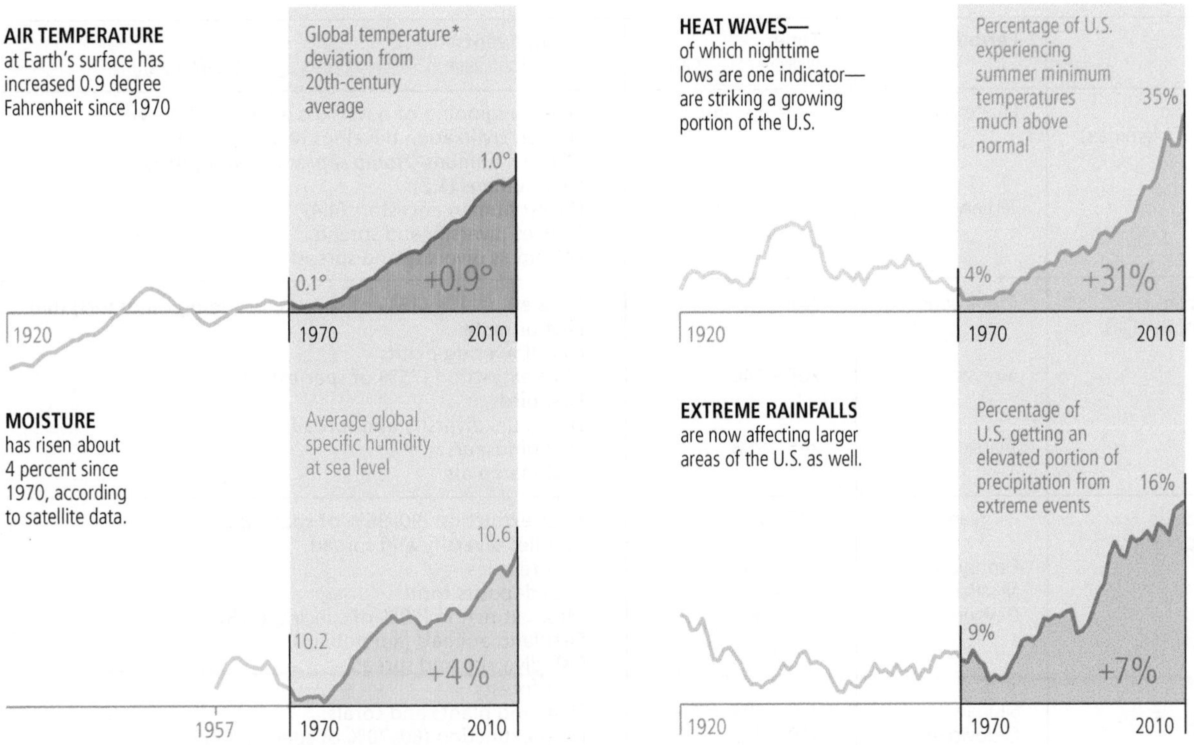

GRAPHS ARE SMOOTHED USING A TEN-YEAR MOVING AVERAGE.
*AVERAGE TEMPERATURE OVER LAND AND OCEAN

JOHN TOMANIO, NGM STAFF, ROBERT THOMASON. SOURCES: JEFF MASTERS, WEATHER UNDERGROUND; NATIONAL CLIMATIC DATA CENTER (TEMPERATURE, HEAT WAVES, AND RAINFALL); NOAA (HUMIDITY)

National Geographic

FIGURE 6 Indicators of climate change.

Data and Graph Analysis

1. Compared to the percentage of the United States that experienced summer minimum temperatures much above normal in 1970, how much larger was that percentage in 2010?
2. For every 0.1 degree (F) increase in average air temperature that occurred between 1970 and 2010, how much did moisture in the atmosphere rise (on average and by percent)?

Supplement 6 Geologic and Biological Time Scale

Era	Period	Time (millions of years ago)	Major Events (approximate time in millions of years ago, in parentheses)
Cenozoic	Quaternary	1.6 – present	Likely beginning of new mass extinction (now)
(Age of Mammals)			Human civilization develops (0.01 to now)
			Modern humans (*Homo sapiens sapiens*) (0.2)
			First humans (1.2)
	Tertiary	6.5 – 1.6	Oldest human ancestors (4.4)
			Grasses diversify and spread
			Mammals diversify and spread
Mesozoic	Cretaceous	146 – 6.5	Mass extinction (75% of species, including dinosaurs) (66)
(Age of Reptiles)			First primates
			First flowering plants
	Jurassic	208 – 146	Mass extinction (75% of species) (200)
			First birds
			Dinosaurs diversify and spread
	Triassic	245 – 208	First dinosaurs
			First mammals
Paleozoic	Permian	290 – 245	Mass extinction (90–96% of species) (251)
(Age of Fishes)			Reptiles diversify and spread
	Pennsylvanian	322 – 290	First reptiles
	Mississippian	362 – 322	Coal deposits form
	Devonian	408 – 362	Mass extinction (70% of species) (375)
			First land animals (amphibians)
			Fish diversify and spread
			First forests
	Silurian	439 – 408	First land plants and corals
	Ordovician	510 – 439	Mass extinction (60–70% of species) (450)
			First fish
	Cambrian	545 – 510	First shellfish
			Ozone layer forms
			Oxygen increases in atmosphere
			Photosynthetic organisms proliferate
Precambrian	Proterozoic	2,500 – 545	First animals in sea (jellyfish)
			First multicellular organisms
	Archean	4,600 – 2,500	First photosynthesis and oxygen in atmosphere (2,800)
			First plants in sea (algae) (3,200)
			Atmospheric water vapor condenses to oceans (3,700)
			First rocks (3,800)
			Likely origin of life (first one-celled organisms) (3,800)
			Earth forms (4,600)

Dates are approximate.

Glossary

abiotic Nonliving. Compare *biotic.*

acid See *acid solution.*

acid deposition The falling of acids and acid-forming compounds from the atmosphere to the earth's surface in wet or dry form. Commonly known as *acid rain,* a term that refers to the wet deposition of droplets of acids and acid-forming compounds.

acidity Chemical characteristic that helps determine how a substance dissolved in water (a solution) will interact with and affect its environment; based on the comparative amounts of hydrogen ions (H^+) and hydroxide ions (OH^-) contained in a particular volume of the solution. See *pH.*

acid rain See *acid deposition.*

acid solution Any water solution that has more hydrogen ions (H^+) than hydroxide ions (OH^-); any water solution with a pH less than 7. Compare *basic solution, neutral solution.*

active solar heating system System that uses solar collectors to capture energy from the sun and store it as heat for space heating and water heating. Liquid or air pumped through the collectors transfers the captured heat to a storage system such as an insulated water tank or rock bed. Pumps or fans then distribute the stored heat or hot water throughout a dwelling as needed. Compare *passive solar heating system.*

adaptation Any genetically controlled structural, physiological, or behavioral characteristic that helps an organism survive and reproduce under a given set of environmental conditions. It usually results from a beneficial mutation. See *biological evolution, differential reproduction, mutation, natural selection.*

adaptive trait See *adaptation.*

aerobic respiration Complex process that occurs in the cells of most living organisms, in which nutrient organic molecules such as glucose ($C_6H_{12}O_6$) combine with oxygen (O_2) to produce carbon dioxide (CO_2), water (H_2O), and energy. Compare *photosynthesis.*

affluence Wealth. It can result in high levels of consumption and unnecessary waste of resources.

age structure Percentage of the population (or number of people of each sex) at each age level in a population.

agrobiodiversity The genetic variety of plant and animal species used on farms to produce food. Compare *biodiversity.*

agroforestry Planting trees and crops together.

air pollution One or more chemicals in high enough concentrations in the air to harm humans, other animals, vegetation, or materials. Such chemicals are called air pollutants. See *primary pollutant, secondary pollutant.*

albedo Ability of a surface to reflect light.

alien species See *nonnative species.*

alley cropping Planting of crops in strips with rows of trees or shrubs on each side.

alpha particle Positively charged matter, consisting of two neutrons and two protons, which is emitted as radioactivity from the nuclei of some radioisotopes. See also *beta particle, gamma rays.*

altitude Height above sea level. Compare *latitude.*

anaerobic respiration Form of cellular respiration in which some decomposers get the energy they need through the breakdown of glucose (or other nutrients) in the absence of oxygen. Compare *aerobic respiration.*

animal manure Dung and urine of animals used as a form of organic fertilizer. Compare *green manure.*

annual Plant that grows, sets seed, and dies in one growing season. Compare *perennial.*

Anthropocene A new era in which humans have become major agents of change in the functioning of the earth's life-support systems as their ecological footprints have spread over the earth. See *ecological footprint.* Compare *Holopocene.*

anthropocentric Human-centered.

aquaculture Growing and harvesting of fish and shellfish for human use in fresh-water ponds, irrigation ditches, and lakes, or in cages or fenced-in areas of coastal lagoons and estuaries or in the open ocean. See *fish farming, fish ranching.*

aquatic Pertaining to water. Compare *terrestrial.*

aquatic life zone Marine and freshwater portions of the biosphere. Examples include freshwater life zones (such as lakes and streams) and ocean or marine life zones (such as estuaries, coastlines, coral reefs, and the open ocean).

aquifer Porous, water-saturated layers of sand, gravel, or bedrock that can yield an economically significant amount of water.

arable land Land that can be cultivated to grow crops.

area strip mining Type of surface mining used where the terrain is flat. An earthmover strips away the overburden, and a power shovel digs a cut to remove the mineral deposit. The trench is then filled with overburden, and a new cut is made parallel to the previous one. The process is repeated over the entire site. Compare *mountaintop removal, open-pit mining, subsurface mining.*

arid Dry. A desert or other area with an arid climate has little precipitation.

artificial selection Process by which humans select one or more desirable genetic traits in the population of a plant or animal species and then use *selective breeding* to produce populations containing many individuals with the desired traits. Compare *genetic engineering, natural selection.*

asthenosphere Zone within the earth's mantle made up of hot, partly melted rock that flows and can be deformed like soft plastic.

atmosphere Whole mass of air surrounding the earth. See *stratosphere, troposphere.* Compare *biosphere, geosphere, hydrosphere.*

atmospheric pressure Force or mass per unit area of air, caused by the bombardment of a surface by the molecules in air.

atom Minute unit made of subatomic particles that is the basic building block of all chemical elements and thus all matter; the smallest unit of an element that can exist and still have the unique characteristics of that element. Compare *ion, molecule.*

atomic number Number of protons in the nucleus of an atom. Compare *mass number.*

atomic theory Idea that all elements are made up of atoms; the most widely accepted scientific theory in chemistry.

autotroph See *producer.*

background extinction rate Rate of extinction that existed before humans became the dominant species on the earth. Compare *mass extinction.*

bacteria Prokaryotic, one-celled organisms. Some transmit diseases. Most act as decomposers and get the nutrients they need by breaking down complex organic compounds in the tissues of living or dead organisms into simpler inorganic nutrient compounds.

basic solution Water solution with more hydroxide ions (OH^-) than hydrogen ions (H^+); water solution with a pH greater than 7. Compare *acid solution, neutral solution.*

benthos Bottom-dwelling aquatic organisms. Compare *decomposer, nekton, plankton.*

beta particle Swiftly moving electron emitted by the nucleus of a radioactive isotope. See also *alpha particle, gamma ray.*

bioaccumulation An increase in the concentration of a chemical in specific organs or tissues at a level higher than would normally be expected. Compare *biomagnification.*

biocapacity The ability of the earth's productive ecosystems to regenerate the renewable resources used by a population, city, region, country, or the world as well as to absorb the resulting wastes and pollution indefinitely. See *ecological footprint.*

biocentric Life-centered. Compare *anthropocentric.*

biodegradable Capable of being broken down by decomposers.

biodegradable pollutant Material that can be broken down into simpler substances (elements and compounds) by bacteria or other decomposers. Paper and most organic wastes such as animal manure are biodegradable but can take decades to biodegrade in modern landfills. Compare *nondegradable pollutant.*

biodiversity Variety of different species (*species diversity*), genetic variability among individuals within each species (*genetic diversity*), variety of ecosystems (*ecological diversity*), and functions such as energy flow and matter cycling needed for the survival of species and biological communities (*functional diversity*).

biodiversity hotspot An area especially rich in plant species that are found nowhere else and are in great danger of extinction. Such areas suffer serious ecological disruption, mostly because of rapid human population growth and the resulting pressure on natural resources.

biofuel Gas (such as methane) or liquid fuel (such as ethyl alcohol or biodiesel) made from plant material (biomass).

biogeochemical cycle Natural processes that recycle nutrients in various chemical forms from the nonliving environment to living organisms and then back to the nonliving environment. Examples include the carbon, oxygen, nitrogen, phosphorus, sulfur, and hydrologic cycles.

biological amplification See *biomagnification.*

biological community See *community.*

biological diversity See *biodiversity.*

biological engineering See *synthetic biology.*

biological evolution Change in the genetic makeup of a population of a species in successive generations. If continued long enough, it can lead to the formation of a new species. Note that populations, not individuals, evolve. See also *adaptation, differential reproduction, natural selection, theory of evolution.*

biological extinction Complete disappearance of a species from the earth. It happens when a species cannot adapt and successfully reproduce under new environmental conditions or when a species evolves into one or more new species. Compare *speciation.* See also *endangered species, mass extinction, threatened species.*

biological pest control Control of pest populations by natural predators, parasites, or disease-causing bacteria and viruses (pathogens).

biomagnification Increase in concentration of slowly degradable, fat-soluble chemicals such a DDT and PCBs in organisms at successively higher trophic levels of a food chain or web. Compare *bioaccumulation.*

biomass Organic matter produced by plants and other photosynthetic producers; total dry weight of all living organisms that can be supported at each trophic level in a food chain or web; dry weight of all organic matter in plants and animals in an ecosystem; plant materials and animal wastes used as fuel.

biome Terrestrial region inhabited by a characteristic set of species, especially vegetation. Examples include various types of deserts, grasslands, and forests.

biomimicry Process of observing certain changes in nature, studying how natural systems have responded to such changing conditions over many millions of years, and applying what is learned to dealing with some environmental challenge.

biosphere Zone of the earth where life is found. It consists of parts of the atmosphere (the troposphere), hydrosphere (mostly surface water and groundwater), and lithosphere (mostly soil and surface rocks and sediments on the bottoms of oceans and other bodies of water) where life is found. Compare *atmosphere, geosphere, hydrosphere.*

biotic Living organisms. Compare *abiotic.*

biotic potential Maximum rate at which the population of a given species can increase when there are no limits on its rate of growth. See *environmental resistance.*

birth rate See *crude birth rate.*

bitumen Gooey, black, high-sulfur, heavy oil extracted from tar sand and then upgraded to synthetic fuel oil. See *oil sand.*

broadleaf deciduous plants Plants such as oak and maple trees that survive drought and cold by shedding their leaves and becoming dormant. Compare *broadleaf evergreen plants, coniferous evergreen plants.*

broadleaf evergreen plants Plants that keep most of their broad leaves year-round. An example is the trees found in the canopies of tropical rain forests. Compare *broadleaf deciduous plants, coniferous evergreen plants.*

buffer Substance that can react with hydrogen ions in a solution and thus hold the acidity or pH of a solution fairly constant. See *pH.*

bycatch Unwanted fish, marine mammals, and sea turtles caught in the nets and on the hooks of commercial fishing fleets.

calorie Unit of energy; amount of energy needed to raise the temperature of 1 gram of water by 1 C° (unit on Celsius temperature scale). See also *kilocalorie.*

cancer Any of more than 120 different diseases, one for each type of cell in the human body. Each type of cancer produces a tumor in which cells multiply uncontrollably and invade surrounding tissue.

cap-and-trade Pollution-control approach in which a government places a cap on total human-generated pollutant emissions in a country or region, issues permits to emit these pollutants in amounts that do not surpass the cap, and then lets polluters trade their permits in the marketplace.

carbon capture and storage (CCS) Process of removing carbon dioxide gas from coal-burning power and industrial plants and storing it somewhere (usually underground or under the seabed) so that it is not released into the atmosphere. To be effective, it must be stored essentially forever.

carbon cycle Cyclic movement of carbon in different chemical forms from the environment to organisms and then back to the environment.

carbon footprint Amount of carbon dioxide generated by an individual, an organization, a country, or any other entity over a given period.

carcinogen Chemical, ionizing form of radiation, or virus that can cause or promote the development of cancer. See *cancer.* Compare *mutagen, teratogen.*

carnivore Animal that feeds on other animals. Compare *herbivore, omnivore.*

carrying capacity (*K*) Maximum population of a particular species that a given habitat can support over a given period. Compare *cultural carrying capacity.*

CCS See *carbon capture and storage.*

cell Smallest living unit of an organism. Each cell is encased in an outer membrane or wall and contains genetic material (DNA) and other parts to perform its life function. Organisms such as bacteria consist of only one cell, but most organisms contain many cells.

cell theory The idea that all living things are composed of cells; the most widely accepted scientific theory in biology.

CFCs See *chlorofluorocarbons.*

chain reaction Multiple nuclear fissions, taking place within a certain mass of a fissionable isotope, which release an enormous amount of energy in a short time.

chemical One of the millions of different elements and compounds found naturally or synthesized by humans. See *compound, element.*

chemical change Interaction between chemicals in which the chemical composition of the elements or compounds involved changes. Compare *nuclear change, physical change.*

chemical cycling The continual cycling of chemicals necessary for life through natural processes such as the water cycle and feeding interactions; processes that evolved due to the fact that the earth gets essentially no new inputs of these chemicals.

chemical formula Shorthand way to show the number of atoms (or ions) in the basic structural unit of a compound. Examples include H_2O, NaCl, and $C_6H_{12}O_6$.

chemical reaction See *chemical change.*

chemosynthesis Process in which certain organisms (mostly specialized bacteria) extract inorganic compounds from their environment and convert them into organic nutrient compounds without the presence of sunlight. Compare *photosynthesis.*

chlorinated hydrocarbon Organic compound made up of atoms of carbon, hydrogen, and chlorine. Examples include DDT and PCBs.

chlorofluorocarbons (CFCs) Organic compounds made up of atoms of carbon, chlorine, and fluorine. An example is Freon-12 (CCl_2F_2), which is used as a refrigerant in refrigerators and air conditioners and in making plastics such as Styrofoam. Gaseous CFCs can deplete the ozone layer when they slowly rise into the stratosphere and their chlorine atoms react with ozone molecules.

CHP (Combined heat and power) See *cogeneration.*

chromosome A grouping of genes and associated proteins in plant and animal cells that carry certain types of genetic information. See *genes.*

chronic malnutrition Faulty nutrition, caused by a diet that does not supply an individual with enough protein, essential fats, vitamins, minerals, and other nutrients necessary for good health. Compare *overnutrition, chronic undernutrition.*

chronic undernutrition Condition suffered by people who cannot grow or buy enough food to meet their basic energy needs. Most chronically undernourished children live in developing countries and are likely to suffer from mental retardation and stunted growth and to die from infectious diseases. Compare *chronic malnutrition, overnutrition.*

civil suit Court case brought to settle disputes or damages between one party and another.

clear-cutting Method of timber harvesting in which all trees in a forested area are removed in a single cutting. Compare *selective cutting, strip cutting.*

climate Physical properties of the troposphere of an area based on analysis of its weather records over a long period (at least 30 years). The two main factors determining an area's climate are its average atmospheric temperature and its average amount and distribution of precipitation. Compare *weather.*

climate change Broad term referring to long-term changes in any aspects of the earth's climate, especially temperature and precipitation. Compare *global warming, weather.*

climate change tipping point Point at which an environmental problem reaches a threshold level where scientists fear it could cause irreversible climate disruption.

closed-loop recycling See *primary recycling.*

coal Solid, combustible mixture of organic compounds with 30–98% carbon by weight, mixed with various amounts of water and small amounts of sulfur and nitrogen compounds. It forms in several stages as the remains of plants are subjected to heat and pressure over millions of years.

coal gasification Conversion of solid coal to synthetic natural gas (SNG).

coal liquefaction Conversion of solid coal to a liquid hydrocarbon fuel such as synthetic gasoline or methanol.

coastal wetland Land along a coastline, extending inland from an estuary that is covered with saltwater all or part of the year. Examples include marshes, bays, lagoons, tidal flats, and mangrove swamps. Compare *inland wetland.*

coastal zone Warm, nutrient-rich, shallow part of the ocean that extends from the high-tide mark on land to the edge of a shelf-like extension of a continental land mass known as a continental shelf. Compare *open sea.*

coevolution Evolution in which two or more species interact and exert selective pressures on each other that can lead each species to undergo adaptations. See *evolution, natural selection.*

cogeneration Production of two useful forms of energy, such as high-temperature heat or steam and electricity, from the same fuel source.

cold front Leading edge of an advancing mass of cold air. Compare *warm front.*

colony collapse disorder (CCD) Loss through death or disappearance of all or most of the European honeybees in a particular colony due to unknown causes; a phenomenon that has resulted in large losses of European honeybees in the United States and in parts of Europe.

combined heat and power (CHP) production See *cogeneration.*

commensalism An interaction between organisms of different species in which one type of organism benefits and the other type is neither helped nor harmed to any great degree. Compare *mutualism.*

commercial extinction Depletion of the population of a wild species used as a resource to a level at which it is no longer profitable to harvest the species.

commercial forest See *tree plantation.*

commercial inorganic fertilizer Commercially prepared mixture of inorganic plant nutrients such as nitrates, phosphates, and potassium applied to the soil to restore fertility and increase crop yields. Compare *organic fertilizer.*

common law A body of unwritten rules and principles derived from court decisions along with commonly accepted practices, or norms, within a society. Compare *statutory law.*

common-property resource Resource that is owned jointly by a large group of individuals. One example is the roughly one-third of the land in the United States that is owned jointly by all U.S. citizens and held and managed for them by the government. Compare *open-access renewable resource.* See *tragedy of the commons.*

community Populations of all species living and interacting in an area at a particular time.

competition Two or more individual organisms of a single species (*intraspecific competition*) or two or more individuals of different species (*interspecific competition*) attempting to use the same scarce resources in the same ecosystem.

complex carbohydrates Compounds consisting of two or more simple organic molecules (*monomers*) of simple sugars (such as glucose, $C_6H_{12}O_6$) linked together. One example is the starches that plants use to store energy and that provide energy for animals that feed on plants.

compost Partially decomposed organic plant and animal matter used as a soil conditioner or fertilizer.

composting See *compost.*

compound Combination of atoms, or oppositely charged ions, of two or more elements held together by attractive forces called chemical bonds. Examples are NaCl, CO_2, and $C_6H_{12}O_6$. Compare *element.*

concentration Amount of a chemical in a particular volume or weight of air, water, soil, or other medium.

conduction The transfer of heat from one solid substance to another cooler one when they are in physical contact. Compare *convection.*

coniferous evergreen plants Cone-bearing plants (such as spruces, pines, and firs) that keep some of their narrow, pointed leaves (needles) or scale-like leaves all year. Compare *broadleaf deciduous plants, broad-leaf evergreen plants.*

coniferous trees Cone-bearing trees, mostly evergreens, that have needle-shaped or scale-like leaves. They produce wood known commercially as softwood. Compare *deciduous plants.*

conservation Use of natural areas and wildlife in ways that sustain them for current and future generations of humans, as well as other forms of life. People with this view are called *conservationists.*

conservation biology Multidisciplinary science created to deal with the crisis of maintaining the genes, species, communities, and ecosystems that make up earth's biological diversity. Its goals are to investigate human impacts on biodiversity and to develop practical approaches to preserving biodiversity.

conservationist Person concerned with using natural areas and wildlife in ways that sustain them for current and future generations of humans and other forms of life.

conservation-tillage farming See *no-till farming.*

consumer Organism that cannot synthesize the organic nutrients it needs and gets its organic nutrients by feeding on the tissues of producers or of other consumers; generally divided into *primary consumers* (herbivores), *secondary consumers* (carnivores), *tertiary* (*higher-level*) *consumers, omnivores,* and *detritivores* (decomposers and detritus feeders). In economics, one who uses economic goods. Compare *producer.*

continental drift The slow movement of the continents atop the earth's mantle.

contour farming Plowing and planting across the changing slope of land, rather than in straight lines, to help retain water and reduce soil erosion.

contour strip mining Form of surface mining used on hilly or mountainous terrain. A power shovel cuts a series of terraces into the side of a hill. An earth-mover removes the overburden, and a power shovel extracts the coal. The overburden from each new terrace is dumped onto the one below. Compare *area strip mining, mountaintop removal, open-pit mining, subsurface mining.*

controlled burning Deliberately set, carefully controlled surface fires that reduce flammable litter and decrease the chances of damaging *crown fires. See ground fire, surface fire.*

convection The transfer of heat energy within liquids or gases when a warmer volume of the liquid or gas rises to a cooler area and a cooler volume of liquid or gas takes its place. Compare *conduction.* See *convection cell.*

convection cell Cyclical pattern of air movement caused by the sun warming the air, which causes some of the air to rise, while cooler air sinks to take its place. See *convection.*

conventional-tillage farming Crop cultivation method in which a planting surface is made by plowing land, breaking up the exposed soil, and then smoothing the surface. Compare *conservation-tillage farming.*

convergent plate boundary Area where the earth's lithospheric plates are pushed together. See subduction zone. *Compare divergent plate boundary, transform fault.*

coral bleaching Process in which warmer ocean waters can cause shallow tropical corals to expel their colorful algae and turn white. It can weaken and sometimes kill corals.

coral reef Formation produced by massive colonies containing billions of tiny coral animals, called polyps, that secrete a stony substance (calcium carbonate) around themselves for protection. When the corals die, their empty outer skeletons form layers and cause the reef to grow. Coral reefs are found in the coastal zones of warm tropical and subtropical oceans.

core Inner zone of the earth. It consists of a solid inner core of iron and a liquid outer core. Compare *crust, mantle.*

corrective feedback loop See *negative feedback loop.*

cost–benefit analysis A comparison of estimated costs and benefits of actions such as implementing a pollution control regulation, building a dam on a river, or preserving an area of forest.

crop rotation Planting a field, or an area of a field, with different crops from year to year to reduce soil nutrient depletion. A plant such as corn, tobacco, or cotton, which removes large amounts of nitrogen from the soil, is planted one year. The next year a legume such as soybeans, which adds nitrogen to the soil, is planted.

crown fire Extremely hot forest fire that burns ground vegetation and treetops. Compare *controlled burning, ground fire, surface fire.*

crude birth rate Annual number of live births per 1,000 people in the population of a geographic area at the midpoint of a given year. Compare *crude death rate.*

crude death rate Annual number of deaths per 1,000 people in the population of a geographic area at the midpoint of a given year. Compare *crude birth rate.*

crude oil Gooey liquid consisting mostly of hydrocarbon compounds and small amounts of compounds containing oxygen, sulfur, and nitrogen. Extracted from underground accumulations, it is sent to oil refineries, where it is converted to heating oil, diesel fuel, gasoline, tar, and other materials.

crust Solid outer zone of the earth. It consists of oceanic crust and continental crust. Compare *core, mantle.*

cultural carrying capacity The limit on population growth that would allow most people in an area or the world to live in reasonable comfort and freedom without impairing the ability of the planet to sustain future generations. Compare *carrying capacity.*

cultural eutrophication Overnourishment of aquatic ecosystems with plant nutrients (mostly nitrates and phosphates) caused by human activities such as agriculture, urbanization, and discharges from industrial plants and sewage treatment plants. See *eutrophication.*

culture Whole of a society's knowledge, beliefs, technology, and practices.

dam A structure built across a river to control the river's flow or to create a reservoir. See *reservoir.*

data Factual information collected by scientists.

DDT Dichlorodiphenyltrichloroethane, a chlorinated hydrocarbon that has been widely used as an insecticide but is now banned in some countries.

death rate See *crude death rate.*

debt-for-nature swap Agreement in which a certain amount of foreign debt is canceled in exchange for local currency investments that will improve natural resource management or protect certain areas in the debtor country from environmentally harmful development.

deciduous plants Trees, such as oaks and maples, and other plants that survive during dry or cold seasons by shedding their leaves. Compare *coniferous trees, succulent plants.*

decomposer Organism that digests parts of dead organisms, and cast-off fragments and wastes of living organisms by breaking down the complex organic molecules in those materials into simpler inorganic compounds and then absorbing some of the soluble nutrients, returning most of these chemicals to the soil and water for reuse by producers. Decomposers consist of various bacteria and fungi. Compare *consumer, detritivore, producer.*

defendant The party in a court case being charged with creating a harm. See *plaintiff* and *civil suit.*

deforestation Removal of trees from a forested area.

degree of urbanization Percentage of the population in the world, or in a country, living in urban areas. Compare *urban growth.*

delta Area at the mouth of a river built up by deposited sediment, usually containing coastal wetlands and estuaries. See *coastal wetland, estuary.*

democracy Government by the people through their elected officials and appointed representatives. In a *constitutional democracy,* a constitution provides the basis of government authority and puts restraints on government power through free elections and freely expressed public opinion.

demographic transition Hypothesis that countries, as they become industrialized, have declines in death rates followed by declines in birth rates.

density Mass per unit volume.

depletion time The time it takes to use a certain fraction (usually 80%) of the known or estimated supply of a nonrenewable resource at an assumed rate of use. Finding and extracting the remaining amount usually costs more than it is worth.

desalination Purification of saltwater or brackish (slightly salty) water by removal of dissolved salts.

desert Biome in which evaporation exceeds precipitation and the average amount of precipitation is less than 25 centimeters (10 inches) per year. Such areas have little vegetation or have widely spaced, mostly low vegetation. Compare *forest, grassland.*

desertification Conversion of rangeland, rain-fed cropland, or irrigated cropland to desert-like land, with a drop in agricultural productivity of 10% or more. It usually is caused by a combination of overgrazing, soil erosion, prolonged drought, and climate change.

detritivore Consumer organism that feeds on detritus, parts of dead organisms, and cast-off fragments and wastes of living organisms. Examples include earthworms, termites, and crabs. Compare *decomposer.*

detritus Parts of dead organisms and cast-off fragments and wastes of living organisms.

detritus feeder See *detritivore.*

deuterium (D; hydrogen-2) Isotope of the element hydrogen, with a nucleus containing one proton and one neutron and a mass number of 2.

developed country See *more-developed country.*

developing country See *less-developed country.*

dieback Sharp reduction in the population of a species when its numbers exceed the carrying capacity of its habitat. See *carrying capacity.*

differential reproduction Phenomenon in which individuals with adaptive genetic traits produce more living offspring than do individuals without such traits. See *natural selection.*

dioxins Family of 75 chlorinated hydrocarbon compounds formed as unwanted by-products in chemical reactions involving chlorine and hydrocarbons, usually at high temperatures.

discount rate An estimate of a resource's future economic value compared to its present value; based on the idea that having something today may be worth more than it will be in the future.

dissolved oxygen (DO) content Amount of oxygen gas (O_2) dissolved in a given volume of water at a particular temperature and pressure, often expressed as a concentration in parts of oxygen per million parts of water.

disturbance An event that disrupts an ecosystem or community. Examples of *natural disturbances* include fires, hurricanes, tornadoes, droughts, and floods. Examples of *human-caused disturbances* include deforestation, overgrazing, and plowing.

divergent plate boundary Area where the earth's lithospheric plates move apart in opposite directions. Compare *convergent plate boundary, transform fault.*

DNA (deoxyribonucleic acid) Large molecules in the cells of living organisms that carry genetic information.

domesticated species Wild species tamed or genetically altered by crossbreeding for use by humans for food (cattle, sheep, and food crops), as pets (dogs and cats), or for enjoyment (animals in zoos and plants in botanical gardens). Compare *wild species.*

dose Amount of a potentially harmful substance an individual ingests, inhales, or absorbs through the skin. Compare *response.* See *dose-response curve, median lethal dose.*

dose-response curve Plot of data showing the effects of various doses of a toxic agent on a group of test organisms. See *dose, median lethal dose, response.*

doubling time Time it takes (usually in years) for the quantity of something growing exponentially to double. It can be calculated by dividing the annual percentage growth rate into 70.

drainage basin See *watershed.*

drift-net fishing Catching fish in huge nets that drift in the water.

drought Condition in which an area does not get enough water because of lower-than-normal precipitation or higher-than-normal temperatures that increase evaporation.

earth-centered environmental worldview Worldview holding that humans are part of and dependent on nature, and the earth's natural capital exists for all species, not just for humans. Compare *human-centered environmental worldview, life-centered environmental worldview.*

earthquake Shaking of the ground resulting from the fracturing and displacement of subsurface rock, which produces a fault, or from subsequent movement along the fault.

ecological diversity The variety of forests, deserts, grasslands, oceans, streams, lakes, and other biological communities interacting with one another and with their nonliving environment. See *biodiversity.* Compare *functional diversity, genetic diversity, species diversity.*

ecological footprint Amount of biologically productive land and water needed to supply a population with the renewable resources it uses and to absorb or dispose of the wastes from such resource use. It is a measure of the average environmental impact of populations in different countries and areas. See *per capita ecological footprint.*

ecological niche Total way of life or role of a species in an ecosystem. It includes all physical, chemical, and biological conditions that a species needs to live and reproduce in an ecosystem. See *fundamental niche, realized niche.*

ecological restoration Deliberate alteration of a degraded habitat or ecosystem to restore as much of its ecological structure and function as possible.

ecological succession Natural process in which communities of plant and animal species in a particular area are replaced over time by a series of different and often more complex communities. See *primary ecological succession, secondary ecological succession.*

ecological tipping point Point at which an environmental problem reaches a threshold level, which causes an often irreversible shift in the behavior of a natural system.

ecologist Biological scientist who studies relationships between living organisms and their environment.

ecology Biological science that studies the relationships between living organisms and their environment; study of the structure and functions of nature.

economic depletion Exhaustion of a certain amount (typically 80%) of the estimated supply of a nonrenewable resource. Finding, extracting, and processing the remaining amount usually costs more than it is worth. May also apply to the depletion of a renewable resource, such as a fish or tree species.

economic development Improvement of human living standards by economic growth. Compare *economic growth, environmentally sustainable economic development.*

economic growth Increase in the capacity to provide people with goods and services; an increase in gross domestic product (GDP). Compare *economic development, environmentally sustainable economic development.* See *gross domestic product.*

economic resources Natural resources, capital goods, and labor used in an economy to produce material goods and services. See *natural resources.*

economics Social science that deals with the production, distribution, and consumption of goods and services to satisfy people's needs and wants.

economic system Method that a group of people uses to choose which goods and services to produce, how to produce them, how much to produce, and how to distribute them to people.

economy System of production, distribution, and consumption of economic goods.

ecosphere See *biosphere.*

ecosystem One or more communities of different species interacting with one another and with the chemical and physical factors making up their nonliving environment.

ecosystem diversity The earth's diversity of biological communities, including deserts, grasslands, forests, mountains, oceans, lakes, rivers, and wetlands.

ecosystem services Natural services, provided by natural capital, that support life on the earth and are essential to the quality of human life and the functioning of the world's economies. Examples are the chemical cycles, natural pest control, and natural purification of air and water. See *natural resources.*

ecotone Transitional zone between two ecosystem where they merge. See *edge effects.*

edge effects Differences in environmental conditions and species present in areas that exist on the edges of major biomes such as the areas bordering forests and in ecotone areas between different ecosystems. See *ecotone.*

electric power Rate at which electric energy is transferred through a wire or other conducting material; commonly expressed in units of watts or megawatts (1 million watts) per hour.

electromagnetic radiation Forms of kinetic energy traveling as electromagnetic waves. Examples include radio waves, TV waves, microwaves, infrared radiation, visible light, ultraviolet radiation, X-rays, and gamma rays. Compare *ionizing radiation, nonionizing radiation.*

electron (e) Tiny particle moving around outside the nucleus of an atom. Each electron has one unit of negative charge and almost no mass. Compare *neutron, proton.*

element Chemical, such as hydrogen (H), iron (Fe), sodium (Na), carbon (C), nitrogen (N), or oxygen (O), whose distinctly different atoms serve as the basic building blocks of all matter. Two or more elements combine to form the compounds that make up most of the world's matter. Compare *compound.*

elevation Height above sea level.

emigration Movement of people out of a specific geographic area. Compare *immigration, migration.*

endangered species Wild species with so few individual survivors that the species could soon become extinct in all or most of its natural range. Compare *threatened species.*

Endangered Species Act (ESA) U.S. federal law designed to identify and protect endangered species in the United States and abroad. This law creates recovery programs for identified species with the goal of helping such populations recover to levels where legal protection is no longer needed.

endemic species Species that is found in only one area. Such species are especially vulnerable to extinction.

energy Capacity to do work by performing mechanical, physical, chemical, or electrical tasks or to cause a heat transfer between two objects at different temperatures.

energy conservation Reducing or eliminating the unnecessary waste of energy.

energy density Amount of energy available per unit of mass of an energy resource.

energy efficiency Percentage of the total energy input that does useful work and is not converted into low-quality, generally useless heat in an energy conversion system or process. See *energy quality, net energy.* Compare *material efficiency.*

energy productivity See *energy efficiency.*

energy quality Ability of a form of energy to do useful work. High-temperature heat and the chemical energy in fossil fuels and nuclear fuels are concentrated high-quality energy. Low-quality energy such as low-temperature heat is dispersed or diluted and cannot do much useful work. See *high-quality energy, low-quality energy.*

energy return on investment (EROI) See *net energy.*

environment All external conditions, factors, matter, and energy, living and nonliving, that affect any living organism or other specified system.

environmental activism See *environmentalism.*

environmental degradation Depletion or destruction of a potentially renewable resource such as soil, grassland, forest, or wildlife that is used faster than it is naturally replenished. If such use continues, the resource becomes nonrenewable (on a human time scale) or nonexistent (extinct). See also *sustainable yield.*

environmental ethics Human beliefs about what is right or wrong with how we treat the environment.

environmental indicators Economic indicators that include non-economic factors with the goal of monitoring environmental quality and human well-being, as well as economic status or progress.

environmentalism Social movement dedicated to protecting the earth's life-support systems for humans and other species.

environmentalist Person who is concerned about the impacts of human activities on the environment.

environmental justice Fair treatment and meaningful involvement of all people, regardless of race, color, sex, national origin, or income, with respect to the development, implementation, and enforcement of environmental laws, regulations, and policies.

environmental law A body of laws and treaties that broadly define what is acceptable environmental behavior for individuals, groups, businesses, and nations.

environmentally sustainable economic development Development that meets the basic needs of the current generations of humans and other species without preventing future generations of humans and other species from meeting their basic needs. It is the economic component of an *environmentally sustainable society.* Compare *economic development, economic growth.*

environmentally sustainable society Society that meets the current and future needs of its people for basic resources in a just and equitable manner without compromising the ability of future generations of humans and other species to meet their basic needs.

environmental movement Citizens organized to demand that political leaders enact laws and develop policies to curtail pollution, clean up polluted environments, and protect unspoiled areas from environmental degradation.

environmental policy Laws, rules, and regulations related to an environmental problem that are developed, implemented, and enforced by a particular government body or agency.

environmental resistance All of the limiting factors that act together to limit the growth of a population. See *biotic potential, limiting factor.*

environmental revolution Cultural change that includes stabilizing human populations and adapting political and economic systems with the goal of living more sustainably. It requires working with nature by learning more about how nature sustains itself.

environmental science Interdisciplinary study that uses information and ideas from the physical sciences (such as biology, chemistry, and geology) with those from the social sciences and humanities (such as economics, politics, and ethics) to learn how nature works, how we interact with the environment, and how we can to help deal with environmental problems.

environmental scientist Scientist who uses information from the physical sciences and social sciences to understand how the earth works, learn how humans interact with the earth, and develop solutions to environmental problems. See *environmental science.*

environmental wisdom worldview Worldview holding that humans are part of and totally dependent on nature and that nature exists for all species, not just for us. Our success depends on learning how the earth sustains itself and integrating such environmental wisdom into the ways we think and act. Compare *frontier worldview, planetary management worldview, stewardship worldview.*

environmental worldview Set of assumptions and beliefs about how people think the world works, what they think their role in the world should be, and what they believe is right and wrong environmental behavior (environmental ethics). See *environmental wisdom worldview, frontier worldview, planetary management worldview, stewardship worldview.*

EPA U.S. Environmental Protection Agency; responsible for managing federal efforts to control air and water pollution, radiation and pesticide hazards, environmental research, hazardous waste, and solid waste disposal.

epidemiology Study of the patterns of disease or other harmful effects from exposure to toxins and diseases caused by pathogens within defined groups of people to find out why some people get sick and some do not.

epiphyte Plant that uses its roots to attach itself to branches high in trees, especially in tropical forests.

erosion Process or group of processes by which loose or consolidated earth materials, especially topsoil, are dissolved, loosened, or worn away and removed from one place and deposited in another. See *weathering.*

estuary Partially enclosed coastal area at the mouth of a river where its freshwater, carrying fertile silt and runoff from the land, mixes with salty seawater.

eukaryotic cell Cell that is surrounded by a membrane and has a distinct nucleus. Compare *prokaryotic cell.*

euphotic zone Upper layer of a body of water through which sunlight can penetrate and support photosynthesis.

eutrophication Physical, chemical, and biological changes that take place after a lake, estuary, or slow-flowing stream receives inputs of plant nutrients—mostly nitrates and phosphates—from natural erosion and runoff from the surrounding land basin. See *cultural eutrophication.*

eutrophic lake Lake with a large or excessive supply of plant nutrients, mostly nitrates and phosphates. Compare *mesotrophic lake, oligotrophic lake.*

evaporation Conversion of a liquid into a gas.

evergreen plants Plants that keep some of their leaves or needles throughout the year. Examples include cone-bearing trees (conifers) such as firs, spruces, pines, redwoods, and sequoias. Compare *deciduous plants, succulent plants.*

evolution See *biological evolution.*

evolutionary tree Diagram that depicts the hypothetical evolutionary pathways of various species from common ancestors.

exhaustible resource See *nonrenewable resource.*

exotic species See *nonnative species.*

experiment Procedure a scientist uses to study some phenomenon under known conditions. Scientists conduct some experiments in the laboratory and others in nature. The resulting scientific data or facts must be verified or confirmed by repeated observations and measurements, ideally by several different investigators.

exponential growth Growth in which some quantity, such as population size or economic output, increases at a constant rate per unit of time. An example is the growth sequence 2, 4, 8, 16, 32, 64, and so on, which increases by 100% at each interval. When the increase in quantity over time is plotted, this type of growth yields a curve shaped like the letter J. Compare *linear growth, logistic growth.* See *J-shaped curve.*

external benefit Beneficial social effect of producing and using an economic good that is not included in the market price of the good. Compare *external cost, full cost.*

external cost Harmful environmental, economic, or social effect of producing and using an economic good that is not included in the market price of the good. Compare *external benefit, full cost, internal cost.*

extinction See *biological extinction.*

extinction rate Percentage or number of species that go extinct within a certain period of time such as a year.

family planning Providing information, clinical services, and contraceptives to help people choose the number and spacing of children they want to have.

famine Widespread malnutrition and starvation in a particular area because of a shortage of food, usually caused by drought, war, flood, earthquake, or other catastrophic events that disrupt food production and distribution.

farm subsidies Government payments and tax breaks intended to help farmers stay in business and increase their yields.

feedback Any process that increases (positive feedback) or decreases (negative feedback) a change to a system.

feedback loop Occurs when an output of matter, energy, or information is fed back into the system as an input and leads to changes in that system. See *positive feedback loop* and *negative feedback loop.*

feed-in-tariff Long-term contract that requires utilities to buy electricity produced by homeowners and businesses from renewable energy resources, to pay them a price that guarantees a good return, and to feed it into the electrical grid.

feedlot Confined outdoor or indoor space used to raise hundreds to thousands of domesticated livestock.

fermentation See *anaerobic respiration.*

fertility rate Number of children born to an average woman in a population during her lifetime. Compare *replacement-level fertility.*

fertilizer Substance that adds inorganic or organic plant nutrients to soil and improves its ability to grow crops, trees, or other vegetation. See *commercial inorganic fertilizer, organic fertilizer.*

first law of thermodynamics Whenever energy is converted from one form to another in a physical or chemical change, no energy is created or destroyed, but energy can be changed from one form to another; you cannot get more energy out of something than you put in; in terms of energy quantity, you cannot get something for nothing. This law does not apply to nuclear changes, in which large amounts of energy can be produced from small amounts of matter. See *second law of thermodynamics.*

fishery Concentration of particular aquatic species suitable for commercial harvesting in a given ocean area or inland body of water.

fish farming See *aquaculture.*

fishprint Area of ocean needed to sustain the consumption of an average person, a nation, or the world. Compare *ecological footprint.*

fissionable isotope Isotope that can split apart when hit by a neutron at the right speed and thus undergo nuclear fission. Examples include uranium-235 and plutonium-239.

floodplain Flat valley floor next to a stream channel. For legal purposes, the term often applies to any low area that has the potential for flooding, including certain coastal areas.

flows See *throughputs.*

food chain Series of organisms in which each eats or decomposes the preceding one. Compare *food web.*

food desert Urban area where people have little or no easy access to grocery stores or other sources of nutritious food.

food insecurity Condition under which people live with chronic hunger and malnutrition that threatens their ability
to lead healthy and productive lives. Compare *food security.*

food security Condition under which every person in a given area has daily access to enough nutritious food to have an active and healthy life. Compare *food insecurity.*

food web Complex network of many interconnected food chains and feeding relationships. Compare *food chain.*

forest Biome with enough average annual precipitation to support the growth of tree species and smaller forms of vegetation. Compare *desert, grassland.*

fossil fuel Products of partial or complete decomposition of plants and animals; occurs as crude oil, coal, natural gas, or heavy oils as a result of exposure to heat and pressure in the earth's crust over millions of years. See *coal, crude oil, natural gas.*

fossils Skeletons, bones, shells, body parts, leaves, seeds, or impressions of such items that provide recognizable evidence of organisms that lived long ago.

fracking Freeing oil or natural gas that is tightly held in rock deposits by using perforated drilling well tubes with explosive charges to create fissures in rock and then using high pressure pumps to shoot a mixture of water, sand, and chemicals into the well to hold the rock fractures open and release the oil or natural gas, which flows back to the surface along with a mixture of water, sand, fracking chemicals, and other chemicals (some of them hazardous) that are released from the rock. See *horizontal drilling.*

free-access resource See *open-access renewable resource.*

Freons See *chlorofluorocarbons.*

freshwater Water that contains very low levels of dissolved salts.

freshwater life zones Aquatic systems where water with a dissolved salt concentration of less than 1% by volume accumulates on or flows through the surfaces of terrestrial biomes. Examples include *standing* (lentic) bodies of freshwater such as lakes, ponds, and inland wetlands and *flowing* (lotic) systems such as streams and rivers. Compare *biome.*

front The boundary between two air masses with different temperatures and densities. See *cold front, warm front.*

frontier science See *tentative science.*

frontier worldview View held by European colonists settling North America in the 1600s that the continent had vast resources and was a wilderness to be conquered by settlers clearing and planting land.

full cost Sum of all costs, including monetary, health, and environmental costs of producing and using a good.

full-cost pricing Finding ways to include the harmful environmental and health costs of producing and using goods in their market prices. See *external cost, internal cost.*

functional diversity Variety of biological and chemical processes or functions such as energy flow and matter cycling necessary for the survival of species and biological communities. See *biodiversity, ecological diversity, genetic diversity, species diversity.*

fungicide Chemical that kills fungi.

game species Type of wild animal that people hunt or fish as a food source or for sport or recreation.

gamma ray Form of ionizing electromagnetic radiation with a high energy content emitted by some radioisotopes. It readily penetrates body tissues. See *alpha particle, beta particle.*

GDP See *gross domestic product.*

gene mutation See *mutation.*

gene pool Sum total of all genes found in the individuals of the population of a particular species.

generalist species Species with a broad ecological niche. They can live in many different places, eat a variety of foods, and tolerate a wide range of environmental conditions. Examples include flies, cockroaches, mice, rats, and humans. Compare *specialist species.*

genes Coded units of information about specific traits that are passed from parents to offspring during reproduction. They consist of segments of DNA molecules found in chromosomes.

gene splicing See *genetic engineering.*

genetic adaptation Changes in the genetic makeup of organisms of a species that allow the species to reproduce and gain a competitive advantage under changed environmental conditions. See *differential reproduction, evolution, mutation, natural selection.*

genetically modified organism (GMO) Organism whose genetic makeup has been altered by genetic engineering.

genetic diversity Variability in the genetic makeup among individuals within a single species. See *biodiversity.* Compare *ecological diversity, functional diversity, species diversity.*

genetic engineering Process of developing genetically modified strains of crops and livestock animals using a process called *gene splicing* to alter an organism's genetic material by adding, deleting, or changing segments of its DNA.

genetic variability Variety in the genetic makeup of individuals in a population.

genuine progress indicator (GPI) GDP plus the estimated value of beneficial transactions that meet basic needs, but in which no money changes hands, minus the estimated harmful environmental, health, and social costs of all transactions. Compare *gross domestic product.*

geoengineering Any technique or process designed to manipulate certain natural conditions to help counter the human-enhanced greenhouse effect.

geographic isolation Separation of populations of a species into different areas for long periods of time.

geology Study of the earth's dynamic history. Geologists study and analyze rocks and the features and processes of the earth's interior and surface.

geosphere Earth's intensely hot core, thick mantle composed mostly of rock, and thin outer crust that contains most of the earth's rock, soil, and sediment. Compare *atmosphere, biosphere, hydrosphere.*

geothermal energy Heat transferred from the earth's underground concentrations of dry steam (steam with no water droplets), wet steam (a mixture of steam and water droplets), or hot water trapped in fractured or porous rock.

global warming Warming of the earth's lower atmosphere (troposphere) because of increases in the concentrations of one or more greenhouse gases. It can result in climate change that can last for decades to thousands of years. See *climate change, greenhouse effect, greenhouse gases.*

GMO See *genetically modified organism.*

GPI See *genuine progress indicator.*

GPP See *gross primary productivity.*

grassland Biome found in regions where there is enough annual average precipitation to support the growth of grass and small plants but not enough to support large stands of trees. Compare *desert, forest.*

greenhouse effect Natural effect that releases heat in the atmosphere near the earth's surface. Water vapor, carbon dioxide, ozone, and other gases in the lower atmosphere (troposphere) absorb some of the infrared radiation (heat) radiated by the earth's surface. Their molecules vibrate and transform the absorbed energy into longer-wavelength infrared radiation in the troposphere. If the atmospheric concentrations of these greenhouse gases increase and other natural processes do not remove them, the average temperature of the lower atmosphere will increase. Compare *climate change, global warming, greenhouse gases.*

greenhouse gases Gases in the earth's lower atmosphere (troposphere) that cause the greenhouse effect. Examples include carbon dioxide, chlorofluorocarbons, ozone, methane, water vapor, and nitrous oxide. See *climate change, global warming, greenhouse effect.*

green manure Freshly cut or still-growing green vegetation that is plowed into the soil to increase the organic matter and humus available to support crop growth. Compare *animal manure.*

green revolution Popular term for the introduction of scientifically bred or selected varieties of grain (rice, wheat, maize) that, with adequate inputs of fertilizer and water, can greatly increase crop yields.

greenwashing Deceptive practice that some businesses use to spin environmentally harmful or benign products as green, clean, or environmentally beneficial.

gross domestic product (GDP) Annual market value of all goods and services produced by all firms and organizations, foreign and domestic, operating within a country. See *per capita GDP.* Compare *genuine progress indicator (GPI).*

gross primary productivity (GPP) Rate at which an ecosystem's producers capture and store a given amount of chemical energy as biomass in a given length of time. Compare *net primary productivity.*

ground fire Fire that burns decayed leaves or peat deep below the ground's surface. Compare *crown fire, surface fire.*

groundwater Water that sinks into the soil and is stored in slowly flowing and slowly renewed underground reservoirs called *aquifers;* underground water in the zone of saturation, below the water table. Compare *runoff, surface water.*

gyres Large, roughly circular patterns of ocean water movement between continents.

habitat Place or type of place where an organism or population of organisms lives. Compare *ecological niche.*

habitat fragmentation Breakup of a habitat into smaller pieces, usually as a result of human activities.

hazard Something that can cause injury, disease, economic loss, or environmental damage. See also *risk.*

hazardous chemical Chemical that can cause harm because it is flammable or explosive, can irritate or damage the skin or lungs (such as strong acidic or alkaline substances), or can cause allergic reactions of the immune system (allergens). See also *toxic chemical.*

hazardous waste Any solid, liquid, or containerized gas that can catch fire easily, is corrosive to skin tissue or metals, is unstable and can explode or release toxic fumes, or has harmful concentrations of one or more toxic materials that can leach out. These substances are usually by-products of manufacturing processes. See also *toxic waste.*

heat Total kinetic energy of all randomly moving atoms, ions, or molecules within a given substance. Heat always flows spontaneously from a warmer sample of matter to a colder sample of matter. Compare *temperature.*

herbicide Chemical that kills a plant or inhibits its growth.

herbivore Plant-eating organism. Examples include deer, sheep, grasshoppers, and zooplankton. Compare *carnivore, omnivore.*

heterotroph See *consumer.*

high Air mass with a high pressure. Compare *low.*

high-grade ore Ore containing a large concentration of a desired mineral. Compare *low-grade ore.*

high-input agriculture See *industrialized agriculture.*

high-quality energy Energy that is concentrated and has great ability to perform useful work. Examples include high-temperature heat and the energy in electricity, coal, oil, gasoline, sunlight, and nuclei of uranium-235. Compare *low-quality energy.*

high-quality matter Matter that contains a high concentration of a useful resource. Compare *low-quality matter.*

high-throughput economy Economic system in most advanced industrialized countries, in which ever-increasing economic growth is sustained by maximizing the rate at which matter and energy resources are used, with little emphasis on pollution prevention, recycling, reuse, reduction of unnecessary waste, and other forms of resource conservation. Compare *low-throughput economy, matter-recycling economy.*

high-waste economy See *high-throughput economy.*

HIPPCO Acronym used by conservation biologists for the six most important causes of premature extinction: Habitat destruction, degradation, and fragmentation; Invasive (nonnative) species; Population growth (too many people consuming too many resources); Pollution; Climate change; and Overexploitation.

Holocene A geological period of relatively stable climate and other environmental conditions following the last glacial period. It began about 12,000 years ago. Compare *Anthropocene.*

horizontal drilling Method for extracting oil or natural gas from underground deposits by first drilling down and then using a flexible drilling bore to drill horizontally to gain greater access to oil and gas deposits. See *fracking.*

host Plant or animal on which a parasite feeds.

human capital People's physical and mental talents that provide labor, innovation, culture, and organization. Compare *manufactured capital, natural capital.*

human-centered environmental worldview Worldview holding that the natural world is primarily a support system for human life. Compare *earth-centered environmental worldview, life-centered environmental worldview.*

human resources See *human capital.*

humus Partially decomposed organic material in topsoil. It helps soil retain water and water-soluble nutrients, which can be taken up by plant roots.

hunger See *chronic undernutrition.*

hunter–gatherers People who get their food by gathering edible wild plants and other materials and by hunting wild animals and catching fish.

hydraulic fracturing See *fracking.*

hydrocarbon Organic compound made of hydrogen and carbon atoms. The simplest hydrocarbon is methane (CH_4), the major component of natural gas.

hydroelectric power plant Structure in which the energy of falling or flowing water spins a turbine generator to produce electricity.

hydrogen fuel cell Device that uses hydrogen gas (H_2) as a fuel to produce electricity. When the H_2 reacts with oxygen gas (O_2) in the atmosphere, it emits harmless water vapor as a byproduct.

hydrologic cycle Biogeochemical cycle that collects, purifies, and distributes the earth's fixed supply of water from the environment to living organisms and then back to the environment.

hydroponics Form of agriculture in which farmers grow plants by exposing their roots to a nutrient-rich water solution instead of soil.

hydropower Electrical energy produced by falling or flowing water. See *hydroelectric power plant.*

hydrosphere Earth's *liquid water* (oceans, lakes, other bodies of surface water, and underground water), *frozen water* (polar ice caps, glaciers, and ice in soil, known as permafrost), and *water vapor* in the atmosphere. See also *hydrologic cycle.* Compare *atmosphere, biosphere, geosphere.*

igneous rock Rock formed when molten rock material (magma) wells up from the earth's interior, cools, and solidifies into rock masses. Compare *metamorphic rock, sedimentary rock.* See *rock cycle.*

immature community Community at an early stage of ecological succession. It usually has a low number of species and ecological niches. Compare *mature community.*

immigrant species See *nonnative species.*

immigration Migration of people into a country or area to take up permanent residence. Compare *emigration.*

indicator species Species whose decline serves as early warnings that a community or ecosystem is being degraded. Compare *keystone species, native species, nonnative species.*

industrialized agriculture Production of large quantities of crops and livestock for domestic and foreign sale; involves use of large inputs of energy from fossil fuels (especially oil and natural gas), water, fertilizer, and pesticides. Compare *subsistence farming.*

industrial smog Type of air pollution consisting mostly of a mixture of sulfur dioxide, suspended droplets of sulfuric acid formed from some of the sulfur dioxide, and suspended solid particles. Compare *photochemical smog.*

industrial solid waste Solid waste produced by mines, factories, refineries, food growers, and businesses that supply people with goods and services. Compare *municipal solid waste.*

inertia See *persistence.*

inexhaustible resource See *perpetual resource.* Compare *nonrenewable resource, renewable resource.*

infant mortality rate Number of babies out of every 1,000 born each year who die before their first birthday.

infectious disease Disease caused when a pathogen such as a bacterium, virus, or parasite invades the body and multiplies in its cells and tissues. Examples are flu, HIV, malaria, tuberculosis, and measles. See *transmissible disease.* Compare *nontransmissible disease.*

infiltration Downward movement of water through soil.

inherent value See *intrinsic value.*

inland wetland Land away from the coast, such as a swamp, marsh, or bog, that is covered all or part of the time with freshwater. Compare *coastal wetland.*

inorganic compounds All compounds not classified as organic compounds. See *organic compounds.*

inorganic fertilizer See *commercial inorganic fertilizer.*

input Matter, energy, or information entering a system. Compare *output, throughput.*

input pollution control See *pollution prevention.*

insecticide Chemical that kills insects.

insolation Input of solar energy. Latitudinal differences in insolation play a role in weather and climate differences between the earth's equator and poles.

instrumental value Value of an organism, species, ecosystem, or the earth's biodiversity based on its usefulness to humans. Compare *intrinsic value.*

integrated pest management (IPM) Combined use of biological, chemical, and cultivation methods in proper sequence and timing to keep the size of a pest population below the level that causes economically unacceptable loss of a crop or livestock animal.

integrated waste management Variety of strategies for both waste reduction and waste management designed to deal with the solid wastes we produce.

intercropping Growing two or more different crops at the same time on a plot. For example, a carbohydrate-rich grain that depletes soil nitrogen and a protein-rich legume that adds nitrogen to the soil may be intercropped. Compare *monoculture, polyculture.*

internal cost Direct cost paid by the producer and the buyer of an economic good. Compare *external benefit, external cost, full cost.*

interspecific competition Attempts by members of two or more species to use the same limited resources in an ecosystem. See *competition, intraspecific competition.*

intertidal zone The area of shoreline between low and high tides.

intraspecific competition Attempts by two or more organisms of a single species to use the same limited resources in an ecosystem. See *competition, interspecific competition.*

intrinsic rate of increase (*r*) Rate at which a population could grow if it had unlimited resources. Compare *environmental resistance.*

intrinsic value Value of an organism, species, ecosystem, or the earth's biodiversity based on its existence, regardless of whether it has any usefulness to humans. Compare *instrumental value.*

invasive species See *nonnative species.*

inversion See *temperature inversion.*

invertebrates Animals that have no backbones. Compare *vertebrates.*

ion Atom or group of atoms with one or more positive (+) or negative (–) electrical charges. Examples are Na^+ and Cl^-. Compare *atom, molecule.*

ionizing radiation Fast-moving alpha or beta particles or high-energy radiation (gamma rays) emitted by radioisotopes. They have enough energy to dislodge one or more electrons from the atoms they hit, thereby forming charged ions in tissue that can react with and damage living tissue. Compare *nonionizing radiation.*

IPM See *integrated pest management.*

irrigation Mix of methods used to supply water to crops by artificial means.

isotopes Two or more forms of a chemical element that have the same number of protons but different mass numbers because they have different numbers of neutrons in their nuclei.

J-shaped curve Curve with a shape similar to that of the letter J; can represent prolonged exponential growth. See *exponential growth.* Compare *S-shaped curve.*

junk science See *unreliable science.*

***K*-selected species** Species that tend to reproduce later in life, have few offspring, and have long life spans. Typically offspring of *K*-selected mammal species develop inside their mothers, are born relatively large, mature slowly, and are cared for and protected by one or both parents.

kerogen Solid, waxy mixture of hydro-carbons found in oil shale rock. Heating the rock to high temperatures causes the kerogen to vaporize. The vapor is condensed, purified, and then sent to a refinery to produce gasoline, heating oil, and other products. See also *oil shale, shale oil.*

keystone species Species that play roles affecting many other organisms in an ecosystem. Compare *indicator species, native species, nonnative species.*

kilocalorie (kcal) Unit of energy equal to 1,000 calories. See *calorie.*

kilowatt (kW) Unit of electrical power equal to 1,000 watts. See *watt.*

kinetic energy Energy that matter has because of its mass and speed, or velocity. Compare *potential energy.*

lake Large natural body of standing freshwater formed when water from precipitation, land runoff, or groundwater flow fills a depression in the earth created by glaciation, earth movement, volcanic activity, or a giant meteorite. See *eutrophic lake, mesotrophic lake, oligotrophic lake.*

land degradation Decrease in the ability of land to support crops, livestock, or wild species in the future as a result of natural or human-induced processes.

landfill See *sanitary landfill.*

land-use planning Planning to determine the best present and future uses of each parcel of land.

latitude Distance from the equator. Compare *altitude.*

law of conservation of energy See *first law of thermodynamics.*

law of conservation of matter In any physical or chemical change, matter is neither created nor destroyed but merely changed from one form to another; in physical and chemical changes, existing atoms are rearranged into different spatial patterns (physical changes) or different combinations (chemical changes).

law of nature See *scientific law.*

law of tolerance Existence, abundance, and distribution of a species in an ecosystem are determined by whether the levels of one or more physical or chemical factors fall within the range tolerated by the species. See *threshold effect.*

LD50 See *median lethal dose.*

LDC See *less-developed country.* Compare *more-developed country*.

leaching Process in which various chemicals in upper layers of soil are dissolved and carried to lower layers and, in some cases, to groundwater.

less-developed country Country that has low to moderate industrialization and low to moderate per capita GDP. Most are located in Africa, Asia, and Latin America. Compare *more-developed country.*

life-centered environmental worldview Worldview holding that all species have value in fulfilling their particular role within the biosphere, regardless of their potential or actual use to humans. Compare *earth-centered environmental worldview, human-centered environmental worldview.*

life-cycle cost Initial cost plus lifetime operating costs of an economic good. Compare *full cost.*

life expectancy Average number of years a newborn infant can be expected to live.

limiting factor Single factor that limits the growth, abundance, or distribution of the population of a species in an ecosystem. See *limiting factor principle.*

limiting factor principle Too much or too little of any abiotic factor can limit or prevent growth of a population of a species in an ecosystem, even if all other factors are at or near the optimal range of tolerance for the species.

linear growth Growth in which a quantity increases by some fixed amount during each unit of time. An example is growth that increases by 2 units in the sequence 2, 4, 6, 8, 10, and so on. Compare *exponential growth, logistic growth.*

lipids Chemically diverse group of large organic compounds that do not dissolve in water. Examples are fats, oils, and waxes.

liquefied natural gas (LNG) Natural gas converted to liquid form by cooling it to a very low temperature.

liquefied petroleum gas (LPG) Mixture of liquefied propane (C_3H_8) and butane (C_4H_{10}) gas removed from natural gas and used as a fuel.

lithosphere Outer shell of the earth, composed of the crust and the rigid, outermost part of the mantle outside the asthenosphere; material found in the earth's plates. See *crust, geosphere, mantle.*

LNG See *liquefied natural gas.*

lobbying Process in which individuals or groups use personal contacts with legislators or their staff to persuade legislators to vote or act in their favor.

logistic growth Pattern in which exponential population growth occurs when the population is small, and population growth decreases steadily with time as the population approaches the carrying capacity. See *S-shaped curve.* Compare *exponential growth, linear growth.*

low Air mass with a low pressure. Compare *high.*

low-grade ore Ore containing a small concentration of a desired mineral. Compare *high-grade ore.*

low-input agriculture See *sustainable agriculture.*

low-quality energy Energy that is dispersed and has little ability to do useful work. An example is low-temperature heat. Compare *high-quality energy.*

low-quality matter Matter that is dilute or dispersed or contains a low concentration of a useful resource. Compare *high-quality matter.*

low-throughput economy Economy based on working with nature by recycling and reusing discarded matter; preventing pollution; conserving matter and energy resources by reducing unnecessary waste and use; and building things that are easy to recycle, reuse, and repair. Compare *high-throughput economy, matter-recycling economy.*

low-waste economy See *low-throughput economy.*

LPG See *liquefied petroleum gas.*

magma Molten rock below the earth's surface.

malnutrition See *chronic malnutrition.*

mangrove swamps Swamps found on the coastlines in warm tropical climates. They are dominated by mangrove trees, any of about 55 species of trees and shrubs that can live partly submerged in the salty environment of coastal swamps.

mantle Zone of the earth's interior between its core and its crust. Compare *core, crust.* See *geosphere, lithosphere.*

manufactured capital Manufactured items made from natural resources and used to produce and distribute economic goods and services bought by consumers. They include tools, machinery, equipment, factory buildings, and transportation and distribution facilities. Compare *human capital, natural resources.*

manufactured inorganic fertilizer See *commercial inorganic fertilizer.*

manufactured resources See *manufactured capital.*

manure See *animal manure, green manure.*

marginal cost Any increase in the cost of producing an additional unit of a product.

marine life zone See *saltwater life zone.*

mass Amount of material in an object.

mass extinction Catastrophic, widespread, often global event in which major groups of species are wiped out over a short time compared with normal (background) extinction rate. Compare *background extinction.*

mass number Sum of the number of neutrons (n) and the number of protons (p) in the nucleus of an atom. It gives the approximate mass of that atom. Compare *atomic number.*

mass transit Buses, trains, trolleys, and other forms of transportation that carry large numbers of people.

matter Anything that has mass (the amount of material in an object) and takes up space.

matter quality Measure of how useful a matter resource is, based on its availability and concentration. See *high-quality matter, low-quality matter.*

matter-recycling-and-reuse economy Economy that emphasizes reusing and recycling the maximum amount of all resources that can be recycled and reused. The goal is to allow economic growth to continue without depleting matter resources and without producing excessive pollution and environmental degradation. Compare *high-throughput economy, low-throughput economy.*

mature community Fairly stable, self-sustaining community in an advanced stage of ecological succession; usually has a diverse array of species and ecological niches. Compare *immature community.*

maximum sustainable yield See *sustainable yield.*

MDC See *more-developed country.*

median lethal dose (LD50) Amount of a toxic material per unit of body weight of a test animal that kills half the test population in a certain amount of time.

megacity City with 10 million or more people.

meltdown Melting of the highly radioactive core of a nuclear reactor.

mercury (Hg) A toxic element.

mesotrophic lake Lake with a moderate supply of plant nutrients. Compare *eutrophic lake, oligotrophic lake.*

metabolism Ability of a living cell or organism to capture and transform matter and energy from its environment to supply its needs for survival, growth, and reproduction.

metamorphic rock Rock produced when a preexisting rock is subjected to high temperatures (which may cause it to melt partially), high pressures, chemically active fluids, or a combination of these agents. Compare *igneous rock, sedimentary rock.* See *rock cycle.*

metropolitan area See *urban area.*

microorganisms Organisms such as bacteria that are so small that it takes a microscope to see them.

migration Movement of people into and out of specific geographic areas. Compare *emigration* and *immigration.*

mineral Any naturally occurring inorganic substance found in the earth's crust as a crystalline solid. See *mineral resource.*

mineral resource Concentration of naturally occurring solid, liquid, or gaseous material in or on the earth's crust in a form and amount such that extracting and converting it into useful materials or items is currently or potentially profitable. Mineral resources are classified as *metallic* (such as iron and tin ores) or *nonmetallic* (such as fossil fuels, sand, and salt).

minimum-tillage farming See *no-till farming.*

mixture Combination of one or more elements and compounds.

model Approximate representation or simulation of a system being studied.

molecule Combination of two or more atoms of the same chemical element (such as O_2) or different chemical elements (such as H_2O) held together by chemical bonds. Compare *atom, ion.*

monoculture Cultivation of a single crop, usually on a large area of land. Compare *polyculture.*

more-developed country Country that is highly industrialized and has a high per capita GDP. Compare *less-developed country.*

mountaintop removal Type of surface mining that uses explosives, massive power shovels, and large machines called draglines to remove the top of a mountain and expose seams of coal underneath the mountaintop. Compare *area strip mining, contour strip mining.*

MSW See *municipal solid waste.*

multiple use Use of an ecosystem such as a forest for a variety of purposes such as timber harvesting, wildlife habitat, watershed protection, and recreation. Compare *sustainable yield.*

municipal solid waste (MSW) Solid materials discarded by homes and businesses in or near urban areas. See *solid waste.* Compare *industrial solid waste.*

mutagen Chemical or form of radiation that causes inheritable changes (mutations) in the DNA molecules in genes. See *carcinogen, mutation, teratogen.*

mutation Random change in DNA molecules making up genes that can alter anatomy, physiology, or behavior in offspring. See *mutagen.*

mutualism Type of species interaction in which both participating species generally benefit. Compare *commensalism.*

nanotechnology The use of science and engineering to manipulate and create materials out of atoms and molecules at the ultra-small scale of less than 100 nanometers. A nanometer is one-millionth of a meter.

native species Species that normally lives and thrives in a particular ecosystem. Compare *indicator species, keystone species, nonnative species.*

natural capital Natural resources and natural services that keep us and other species alive and support our economies. See *natural resources, natural services.*

natural capital degradation See *environmental degradation.*

natural gas Underground deposits of gases consisting of 50–90% by weight methane gas (CH_4) and small amounts of heavier gaseous hydrocarbon compounds such as propane (C_3H_8) and butane (C_4H_{10}).

natural greenhouse effect See *greenhouse effect.*

natural income Renewable resources such as plants, animals, and soil provided by natural capital.

natural law See *scientific law.*

natural radioactive decay Nuclear change in which unstable nuclei of atoms spontaneously shoot out particles (usually alpha or beta particles) or energy (gamma rays) at a fixed rate.

natural rate of extinction See *background extinction.*

natural recharge Natural replenishment of an aquifer by precipitation, which percolates downward through soil and rock. See *recharge area.*

natural resources Materials such as air, water, and soil and energy in nature that are essential or useful to humans. See *natural capital.*

natural selection Process by which a particular beneficial gene (or set of genes) is reproduced in succeeding generations more than other genes. The result of natural selection is a population that contains a greater proportion of organisms better adapted to certain environmental conditions. See *adaptation, biological evolution, differential reproduction, mutation.*

natural services See *ecosystem services.*

nature-deficit disorder Wide range of problems, including anxiety, depression, and attention-deficit disorders, that can result from or be intensified by a lack of contact with nature.

negative feedback loop Feedback loop that causes a system to change in the opposite direction from which is it moving. Compare *positive feedback loop.*

nekton Strongly swimming organisms found in aquatic systems. Compare *benthos, plankton.*

net energy Total amount of useful energy available from an energy resource or energy system over its lifetime, minus the amount of energy *used* (the first energy law), *automatically wasted* (the second energy law), and *unnecessarily wasted* in finding, processing, concentrating, and transporting it to users.

net energy ratio (NER) See *net energy.*

net primary productivity (NPP) Rate at which all the plants in an ecosystem produce net useful chemical energy; equal to the difference between the rate at which the plants in an ecosystem produce useful chemical energy (gross primary productivity) and the rate at which they use some of that energy through cellular respiration. Compare *gross primary productivity.*

neurotoxin Chemical that can harm the human *nervous system* (brain, spinal cord, peripheral nerves).

neutral solution Water solution containing an equal number of hydrogen ions (H^+) and hydroxide ions (OH^-); water solution with a pH of 7. Compare *acid solution, basic solution.*

neutron (n) Elementary particle in the nuclei of all atoms (except hydrogen-1). It has a relative mass of 1 and no electric charge. Compare *electron, proton.*

niche See *ecological niche.*

nitric oxide (NO) Colorless gas that forms when nitrogen and oxygen gas in air react at the high-combustion temperatures in automobile engines and coal-burning plants. Lightning and certain bacteria in soil and water also produce NO as part of the *nitrogen cycle.*

nitrogen cycle Cyclic movement of nitrogen in different chemical forms from the environment to organisms and then back to the environment.

nitrogen dioxide (NO_2) Reddish-brown gas formed when nitrogen oxide reacts with oxygen in the air.

nitrogen fixation Conversion of atmospheric nitrogen gas, by lightning, bacteria, and cyanobacteria, into forms useful to plants; it is part of the nitrogen cycle.

nitrogen oxides (NO_x) See *nitric oxide* and *nitrogen dioxide.*

noise pollution Any unwanted, disturbing, or harmful sound that impairs or interferes with hearing, causes stress, hampers concentration and work efficiency, or causes accidents.

nondegradable pollutant Material that is not broken down by natural processes. Examples include the toxic elements lead and mercury. Compare *biodegradable pollutant.*

nonionizing radiation Forms of radiant energy such as radio waves, microwaves, infrared light, and ordinary light that do not have enough energy to cause ionization of atoms in living tissue. Compare *ionizing radiation.*

nonnative species Species that migrate into an ecosystem or are deliberately or accidentally introduced into an ecosystem by humans. Compare *native species.*

nonpoint sources Broad and diffuse areas, rather than points, from which pollutants enter bodies of surface water or air. Examples include runoff of chemicals and sediments from cropland, livestock feedlots, logged forests, urban streets, parking lots, lawns, and golf courses. Compare *point source.*

nonrenewable energy Energy from resources that can be depleted and are not replenished by natural processes within a human time scale. Examples are energy produced by the burning of oil, coal, and natural gas, and nuclear energy released when the nuclei of heavy elements such as uranium are split apart (nuclear fission) or when the nuclei of light atoms such as hydrogen are forced together (nuclear fusion). Compare *renewable energy.*

nonrenewable resource Resource that exists in a fixed amount (stock) in the earth's crust and has the potential for renewal by geological, physical, and chemical processes taking place over hundreds of millions to billions of years. Examples include copper, aluminum, coal, and oil. We classify these resources as exhaustible because we are extracting and using them at a much faster rate than they are formed. Compare *renewable resource.*

nontransmissible disease Disease that is not caused by living organisms and does not spread from one person to another. Examples include most cancers, diabetes, cardiovascular disease, and malnutrition. Compare *transmissible disease.*

no-till farming Crop cultivation in which the soil is disturbed little (minimum-tillage farming) or not at all in an effort to reduce soil erosion, lower labor costs, and save energy.

NPP See *net primary productivity.*

nuclear change Process in which nuclei of certain isotopes spontaneously change, or are forced to change, into one or more different isotopes. The three principal types of nuclear change are natural radioactivity, nuclear fission, and nuclear fusion. Compare *chemical change, physical change.*

nuclear energy Energy released when atomic nuclei undergo a nuclear reaction such as the spontaneous emission of radioactivity, nuclear fission, or nuclear fusion.

nuclear fission Nuclear change in which the nuclei of certain isotopes with large mass numbers (such as uranium-235 and plutonium-239) are split apart into lighter nuclei when struck by a neutron. This process releases more neutrons and a large amount of energy. Compare *nuclear fusion.*

nuclear fuel cycle Includes the mining of uranium, processing and enriching the uranium to make nuclear fuel, using it in the reactor, safely storing the resulting highly radioactive wastes for thousands of years until their radioactivity falls to safe levels, and retiring the highly radioactive nuclear plant by taking it apart and storing its high- and moderate-level radioactive material safely for thousands of years.

nuclear fusion Nuclear change in which two nuclei of isotopes of elements with a low mass number (such as hydrogen-2 and hydrogen-3) are forced together at extremely high temperatures until they fuse to form a heavier nucleus (such as helium-4). This process releases a large amount of energy. Compare *nuclear fission.*

nucleic acids Large polymer molecules made by linking large numbers of monomers called *nucleotides.*

nucleus Extremely tiny center of an atom, making up most of the atom's mass. It contains one or more positively charged protons and one or more neutrons with no electrical charge (except for a hydrogen-1 atom, which has one proton and no neutrons in its nucleus).

nutrient Any chemical an organism must take in to live, grow, or reproduce.

nutrient cycle See *biogeochemical cycle.*

nutrient cycling The circulation of chemicals necessary for life, from the environment (mostly from soil and water) through organisms and back to the environment.

ocean acidification Increasing levels of acid in world's oceans due to their absorption of much of the CO_2 emitted into the atmosphere by human activities, especially the burning of carbon-containing fossil fuels. The CO_2 reacts with ocean water to form a weak acid and decreases the levels of carbonate ions (CO_3^{2-}) needed to form coral and the shells and skeletons of organisms such as crabs, oysters, and some phytoplankton. This decrease can also be thought of as a decrease in basicity of the ocean waters.

ocean currents Mass movements of surface water produced by prevailing winds blowing over the oceans.

oil See *crude oil.*

oil reserves See *proven oil reserves.*

oil sand See *tar sand.*

oil shale Fine-grained rock containing various amounts of kerogen, a solid, waxy mixture of hydrocarbon compounds. Heating the rock to high temperatures converts the kerogen into a vapor that can be condensed to form a slow-flowing heavy oil called shale oil. See *kerogen, shale oil.*

old-growth forest Virgin and old, second-growth forests containing trees that are often hundreds—sometimes thousands—of years old. Examples include forests of Douglas fir, western hemlock, giant sequoia, and coastal redwoods in the western United States. Compare *second-growth forest, tree plantation.*

oligotrophic lake Lake with a low supply of plant nutrients. Compare *eutrophic lake, mesotrophic lake.*

omnivore Animal that can use both plants and other animals as food sources. Examples include pigs, rats, cockroaches, and humans. Compare *carnivore, herbivore.*

open-access renewable resource Renewable resource owned by no one and available for use by anyone at little or no charge. Examples include clean air, underground water supplies, the open ocean and its fish, and the ozone layer. Compare *common-property resource.*

open dump Fields or holes in the ground where garbage is deposited and sometimes covered with soil. They are rare in more-developed countries, but are widely used in many less-developed countries, especially to handle wastes from megacities. Compare *sanitary landfill.*

open-pit mining Removing minerals such as gravel, sand, and metal ores by digging them out of the earth's surface and leaving an open pit behind. Compare *area strip mining, contour strip mining, mountaintop removal, subsurface mining.*

open sea Part of an ocean that lies beyond the continental shelf. Compare *coastal zone.*

ore Part of a metal-yielding material that can be economically extracted from a mineral; typically containing two parts: the ore mineral, which contains the desired metal, and waste mineral material (gangue). See *high-grade ore, low-grade ore.*

organic agriculture Growing crops with limited or no use of synthetic pesticides, synthetic fertilizers, or genetically modified crops; raising livestock without use of synthetic growth regulators and feed additives; and using organic fertilizer (manure, legumes, compost) and natural pest controls (bugs that eat harmful bugs and environmental controls such as crop rotation). See *sustainable agriculture.*

organic compounds Compounds containing carbon atoms combined with each other and with atoms of one or more other elements such as hydrogen, oxygen, nitrogen, sulfur, phosphorus, chlorine, and fluorine. All other compounds are called *inorganic compounds.*

organic farming See *organic agriculture* and *sustainable agriculture.*

organic fertilizer Organic material such as animal manure, green manure, and compost applied to cropland as a source of plant nutrients. Compare *commercial inorganic fertilizer.*

organism Any form of life.

output Matter, energy, or information leaving a system. Compare *input, throughput.*

output pollution control See *pollution cleanup.*

overburden Layer of soil and rock overlying a mineral deposit. Surface mining removes this layer.

overfishing Harvesting so many fish of a species, especially immature individuals, that not enough breeding stock is left to replenish the species and it becomes unprofitable to harvest them.

overgrazing Destruction of vegetation when too many grazing animals feed too long on a specific area of pasture or rangeland and exceed the carrying capacity of a rangeland or pasture area.

overnutrition Occurs when food energy intake exceeds energy use and causes excess body fat. Too many calories, too little exercise, or both can cause overnutrition. It can place one at higher risk for developing diabetes, hypertension, heart disease, and other diseases. Compare *malnutrition, undernutrition.*

oxygen-demanding wastes Organic materials that are usually biodegraded by aerobic (oxygen-consuming) bacteria if there is enough dissolved oxygen in the water.

ozone (O_3) Colorless and highly reactive gas and a major component of photochemical smog. Also found in the ozone layer in the stratosphere. See *photochemical smog.*

ozone depletion Decrease in concentration of ozone (O_3) in the stratosphere. See *ozone layer.*

ozone layer Layer of gaseous ozone (O_3) in the stratosphere that protects life on earth by filtering out most harmful ultraviolet radiation from the sun.

PANs Peroxyacyl nitrates; group of chemicals found in photochemical smog.

parasite Consumer organism that lives on or in, and feeds on, a living plant or animal, known as the host, over an extended period. The parasite draws nourishment from and gradually weakens its host; it may or may not kill the host. See *parasitism.*

parasitism Interaction between species in which one organism, called the parasite, preys on another organism, called the host, by living on or in the host. See *host, parasite.*

particulates Also known as suspended particulate matter (SPM); variety of solid particles and liquid droplets small and light enough to remain suspended in the air for long periods. About 62% of the SPM in outdoor air comes from natural sources such as dust, wild fires, and sea salt. The remaining 38% comes from human sources such as coal-burning electric power and industrial plants, motor vehicles, plowed fields, road construction, unpaved roads, and tobacco smoke.

parts per billion (ppb) Number of parts of a chemical found in 1 billion parts of a particular gas, liquid, or solid.

parts per million (ppm) Number of parts of a chemical found in 1 million parts of a particular gas, liquid, or solid.

parts per trillion (ppt) Number of parts of a chemical found in 1 trillion parts of a particular gas, liquid, or solid.

passive solar heating system System that, without the use of mechanical devices, captures sunlight directly within a structure and converts it into low-temperature heat for space heating or for heating water for domestic use. Compare *active solar heating system.*

pasture Managed grassland or enclosed meadow that usually is planted with domesticated grasses or other forage to be grazed by livestock. Compare *feedlot.*

pathogen Living organism that can cause disease in another organism. Examples include bacteria, viruses, and parasites.

PCBs See *polychlorinated biphenyls.*

peak production Point in time when the pressure in an oil well drops and its rate of conventional crude oil production starts declining; for a group of wells or for a nation, the point at which all wells on average have passed peak production.

peer review Process of scientists reporting details of the methods and models they used, the results of their experiments, and the reasoning behind their hypotheses for other scientists working in the same field (their peers) to examine and criticize.

per capita ecological footprint Amount of biologically productive land and water needed to supply each person or population with the renewable resources they use and to absorb or dispose of the wastes from such resource use. It measures the average environmental impact of individuals or populations in different countries and areas. Compare *ecological footprint.*

per capita GDP Annual gross domestic product (GDP) of a country divided by its total population at midyear. It gives the average slice of the economic pie per person. See *gross domestic product.* Compare *genuine progress indicator (GPI).*

per capita GDP PPP (Purchasing Power Parity) Measure of the amount of goods and services that a country's average citizen could buy in the United States.

percolation Passage of a liquid through the spaces of a porous material such as soil.

perennial Plant that can live for 2 or more years. Compare *annual.*

periodic table of elements Chart on which chemists have arranged the known elements based on their chemical behavior. Contains 118 elements, not all of which occur naturally.

permafrost Perennially frozen layer of soil that forms when soil moisture freezes. It is found in arctic tundra.

perpetual resource Essentially inexhaustible resource on a human time scale because it is renewed continuously. Solar energy is an example. Compare *nonrenewable resource, renewable resource.*

persistence Ability of a living system such as a grassland or forest to survive moderate disturbances. Compare *resilience.*

perverse subsidy Subsidy that leads to environmental damage or damage to human health or well-being. See *subsidy.*

pest Unwanted organism that directly or indirectly interferes with human activities.

pesticide Any chemical designed to kill or inhibit the growth of an organism that people consider undesirable. See *fungicide, herbicide, insecticide.*

petrochemicals Chemicals obtained by refining (distilling) crude oil. They are used as raw materials in manufacturing most industrial chemicals, fertilizers, pesticides, plastics, synthetic fibers, paints, medicines, and many other products.

petroleum See *crude oil.*

pH Numeric value that indicates the relative acidity or alkalinity of a substance on a scale of 0 to 14, with the neutral point at 7. Acid solutions have pH values lower than 7; basic or alkaline solutions have pH values greater than 7.

phosphorus cycle Cyclic movement of phosphorus in different chemical forms from the environment to organisms and then back to the environment.

photochemical smog Complex mixture of air pollutants produced in the lower atmosphere by the reaction of hydrocarbons and nitrogen oxides under the influence of sunlight. Especially harmful components include ozone, peroxyacyl nitrates (PANs), and various aldehydes. Compare *industrial smog.*

photosynthesis Complex process that takes place in cells of green plants. Radiant energy from the sun is used to combine carbon dioxide (CO_2) and water (H_2O) to produce oxygen (O_2), carbohydrates (such as glucose, $C_6H_{12}O_6$), and other nutrient molecules. Compare *aerobic respiration, chemosynthesis.*

photovoltaic (PV) cell Device that converts radiant (solar) energy directly into electrical energy. Also called a solar cell.

phylogenetic tree See *evolutionary tree.*

physical change Process that alters one or more physical properties of an element or a compound without changing its chemical composition. Examples include changing the size and shape of a sample of matter (crushing ice and cutting aluminum foil) and changing a sample of matter from one physical state to another (boiling and freezing water). Compare *chemical change, nuclear change.*

phytoplankton Small, drifting plants, mostly algae and bacteria, found in aquatic ecosystems. Compare *plankton, zooplankton.*

pioneer community First integrated set of plants, animals, and decomposers found in an area undergoing primary ecological succession. See *ecological succession, immature community, mature community, pioneer species.*

pioneer species First hardy species—often microbes, mosses, and lichens—that begin colonizing a site as the first stage of ecological succession. See *ecological succession, pioneer community.*

plaintiff Party in a court case bringing charges or seeking to collect damages for injuries to health or for economic loss; may also seek an injunction, by which the party being charged would be required to stop whatever action is causing harm. See *defendant* and *civil suit.*

planetary management worldview Worldview holding that humans are separate from nature, that nature exists mainly to meet our needs and increasing wants, and that we can use our ingenuity and technology to manage the earth's life-support systems, mostly for our benefit. It assumes that economic growth is unlimited. Compare *environmental wisdom worldview, stewardship worldview.*

plankton Small plant organisms (phytoplankton) and animal organisms (zooplankton) that float in aquatic ecosystems.

plantation agriculture Growing specialized crops such as bananas, coffee, and cacao, usually in tropical less-developed countries, primarily for sale to more-developed countries.

plate tectonics Theory of geophysical processes that explains the movements of lithospheric plates and the processes that occur at their boundaries. See *lithosphere, tectonic plates.*

point source Single identifiable source that discharges pollutants into the environment. Examples include the smokestack of a power plant or an industrial plant, drainpipe of a meatpacking plant, chimney of a house, or exhaust pipe of an automobile. Compare *nonpoint source.*

poison Chemical that adversely affects the health of a living human or animal by causing injury, illness, or death.

policies Programs, and the laws and regulations through which they are enacted, that a government enforces and funds.

politics Process through which individuals and groups try to influence or control government policies and actions that affect the local, state, national, and international communities.

pollutant Particular chemical or form of energy that can adversely affect the health, survival, or activities of humans or other living organisms. See *pollution.*

pollution Undesirable change in the physical, chemical, or biological characteristics of air, water, soil, or food that can adversely affect the health, survival, or activities of humans or other living organisms.

pollution cleanup Device or process that removes or reduces the level of a pollutant after it has been produced or has entered the environment. Examples include automobile emission control devices and sewage treatment plants. Compare *pollution prevention.*

pollution prevention Device, process, or strategy used to prevent a potential pollutant from forming or entering the environment or to sharply reduce the amount entering the environment. Compare *pollution cleanup.*

polychlorinated biphenyls (PCBs) Group of 209 toxic, oily, synthetic chlorinated hydrocarbon compounds that can be biologically amplified in food chains and webs.

polyculture Complex form of intercropping in which a large number of different plants maturing at different times are planted together. See also *intercropping.* Compare *monoculture.*

population Group of individual organisms of the same species living in a particular area.

population change Increase or decrease in the size of a population. It is equal to (Births + Immigration) – (Deaths + Emigration).

population crash Dieback of a population that has used up its supply of resources, exceeding the carrying capacity of its environment. See *carrying capacity.*

population density Number of organisms in a particular population found in a specified area or volume.

population dispersion General pattern in which the members of a population are arranged throughout its habitat.

population distribution Variation of population density over a particular geographic area or volume. For example, a country has a high population density in its urban areas and a much lower population density in its rural areas.

population dynamics Major abiotic and biotic factors that tend to increase or decrease the population size and affect the age and sex composition of a species.

population size Number of individuals making up a population's gene pool.

positive feedback loop Feedback loop that causes a system to change further in the same direction. Compare *negative feedback loop.*

potential energy Energy stored in an object because of its position or the position of its parts. Compare *kinetic energy.*

poverty Inability of people to meet their basic needs for food, clothing, and shelter.

ppb See *parts per billion.*

ppm See *parts per million.*

ppt See *parts per trillion.*

prairie See *grassland.*

precautionary principle When substantial preliminary evidence indicates that an activity can harm human health or the environment, decision makers should take precautionary measures to prevent or reduce such harm even if some of the cause-and-effect relationships have not been fully established scientifically. See *pollution prevention.*

precipitation Water in the form of rain, sleet, hail, and snow that falls from the atmosphere onto land and bodies of water.

predation Interaction in which an organism of one species (the predator) captures and feeds on some or all parts of an organism of another species (the prey).

predator Organism that captures and feeds on some or all parts of an organism of another species (the prey).

predator–prey relationship Relationship that has evolved between two organisms, in which one organism has become the prey for the other, the latter called the predator. See *predator, prey.*

prey Organism that is killed by an organism of another species (the predator) and serves as its source of food.

primary consumer Organism that feeds on some or all parts of plants (herbivore) or on other producers. Compare *detritivore, omnivore, secondary consumer.*

primary ecological succession Ecological succession in an area without soil or bottom sediments. See *ecological succession.* Compare *secondary ecological succession.*

primary forest See *old-growth forest.*

primary pollutant Chemical that has been added directly to the air by natural events or human activities and occurs in a harmful concentration. Compare *secondary pollutant.*

primary recycling Process in which materials are recycled into new products of the same type—turning used aluminum cans into new aluminum cans, for example.

primary sewage treatment Mechanical sewage treatment in which large solids are filtered out by screens and suspended solids settle out as sludge in a sedimentation tank. Compare *secondary sewage treatment.*

principles of sustainability Three scientific principles of sustainability based on solar energy, biodiversity, chemical cycling, and three additional principles drawn from other disciplines: **(1)** the need to include the harmful health and environmental costs of producing the goods and services in their market prices *(full-cost pricing)*, **(2)** the value of working together to focus on solutions to environmental problems that will benefit the largest number of people and the environment now and in the future *(win-win solutions)*, and **(3)** our responsibility to future generations to leave the planet's life-support systems in at least as good a shape as what we now enjoy *(responsibility to future generations)*. See *biodiversity, chemical cycling, scientific principles of sustainability, solar energy.*

probability Mathematical statement about how likely it is that something will happen.

producer Organism that uses solar energy (green plants) or chemical energy (some bacteria) to manufacture the organic compounds it needs as nutrients from simple inorganic compounds obtained from its environment. Compare *consumer, decomposer.*

prokaryotic cell Cell containing no distinct nucleus or organelles. Compare *eukaryotic cell.*

proteins Large polymer molecules formed by linking together long chains of simple organic molecules (*monomers*) called *amino acids.*

proton (p) Positively charged particle in the nuclei of all atoms. Each proton has a relative mass of 1 and a single positive charge. Compare *electron, neutron.*

proven oil reserves Identified deposits from which conventional crude oil can be extracted profitably at current prices with current technology.

PV cell See *photovoltaic cell.*

pyramid of energy flow Diagram representing the flow of energy through each trophic level in a food chain or food web. With each energy transfer, only a small part (typically 10%) of the usable energy entering one trophic level is transferred to the organisms at the next trophic level.

radiation Fast-moving particles (particulate radiation) or waves of energy (electromagnetic radiation). See *alpha particle, beta particle, gamma ray.*

radioactive decay Change of a radioisotope to a different isotope by the emission of radioactivity.

radioactive isotope See *radioisotope.*

radioactive waste Waste products of nuclear power plants, research, medicine, weapon production, or other processes involving nuclear reactions. See *radioactivity.*

radioactivity Nuclear change in which unstable nuclei of atoms spontaneously shoot out "chunks" of mass, energy, or both at a fixed rate. The three principal types of radioactivity are gamma rays and fast-moving alpha particles and beta particles.

radioisotope Isotope of an atom that spontaneously emits one or more types of radioactivity (alpha particles, beta particles, gamma rays).

rain shadow effect Low precipitation on the leeward side of a mountain when prevailing winds flow up and over a high mountain or range of high mountains, creating semiarid and arid conditions on the leeward side of a high mountain range.

rangeland Land that supplies forage or vegetation (grasses, grass-like plants, and shrubs) for grazing and browsing animals and is not intensively managed. Compare *feedlot, pasture.*

range of tolerance Range of chemical and physical conditions that must be maintained for populations of a particular species to stay alive and grow, develop, and function normally. See *law of tolerance.*

recharge area Any area of land allowing water to percolate down through it and into an aquifer. See *aquifer, natural recharge.*

reconciliation ecology Science of inventing, establishing, and maintaining habitats to conserve species diversity in places where people live, work, or play.

recycle To collect and reprocess a resource so that it can be made into new products; one of the four R's of resource use. An example is collecting aluminum cans, melting them down, and using the aluminum to make new cans or other aluminum products. See *primary recycling, secondary recycling.* Compare *reduce, refuse, reuse.*

reduce To consume less of a good or service in order to reduce one's environmental impact and to save money. Compare *recycle, refuse, reuse.*

refining Complex process in which crude oil is heated and vaporized in giant columns and separated, by use of varying boiling points, into various products such as gasoline, heating oil, and asphalt. See *petrochemicals.*

reforestation Renewal of trees and other types of vegetation on land where trees have been removed; can be done naturally by seeds from nearby trees or artificially by planting seeds or seedlings.

refuse To refrain from buying or using a good or service in order to reduce one's ecological impact and to save money. Compare *recycle, reduce, reuse.*

reliable surface runoff Surface runoff of water that generally can be counted on as a stable source of water from year to year. See *runoff.*

reliable science Concepts and ideas that are widely accepted by experts in a particular field of the natural or social sciences. Compare *tentative science, unreliable science.*

renewable energy Energy that comes from resources that are replenished by natural processes continually or in a relatively short time. Examples are solar energy (sunlight), wind, moving water, heat from the earth's interior (geothermal energy), firewood from trees, tides, and waves. Compare *nonrenewable energy.*

renewable resource Resource that can be replenished rapidly (within hours to several decades) through natural processes as long as it is not used up faster than it is replaced. Examples include trees in forests, grasses in grasslands, wild animals, fresh surface water in lakes and streams, most groundwater, fresh air, and fertile soil. If such a resource is used faster than it is replenished, it can be depleted and converted into a nonrenewable resource. Compare *nonrenewable resource* and *perpetual resource.* See also *environmental degradation.*

replacement-level fertility rate Average number of children a couple must bear to replace themselves. The average for a country or the world usually is slightly higher than two children per couple (2.1 in the United States and 2.5 in some developing countries) mostly because some children die before reaching their reproductive years. See also *total fertility rate.*

reproduction Production of offspring by one or more parents.

reproductive isolation Long-term geographic separation of members of a particular sexually reproducing species.

reproductive potential See *biotic potential.*

reserves Resources that have been identified and from which a usable mineral can be extracted profitably at current prices with current mining or extraction technology.

reservoir Artificial lake created when a stream is dammed. See *dam.*

resilience Ability of a living system such as a forest or pond to be restored through secondary ecological succession after a severe disturbance. See *secondary ecological succession.* Compare *persistence.*

resource Anything obtained from the environment to meet human needs and wants. It can also be applied to other species.

resource partitioning Process of dividing up resources in an ecosystem so that species with similar needs (overlapping ecological niches) use the same scarce resources at different times, in different ways, or in different places. See *ecological niche.*

respiration See *aerobic respiration.*

response Amount of health damage caused by exposure to a certain dose of a harmful substance or form of radiation. See *dose, dose-response curve, median lethal dose.*

restoration ecology Research and scientific study devoted to restoring, repairing, and reconstructing damaged ecosystems.

reuse To use a product over and over again in the same form. An example is collecting, washing, and refilling glass beverage bottles. One of the 4 Rs. Compare *recycle, reduce, refuse.*

riparian zone A thin strip or patch of vegetation that surrounds a stream. These zones are very important habitats and resources for wildlife.

risk Probability that something undesirable will result from deliberate or accidental exposure to a hazard. See *risk analysis, risk assessment, risk management.*

risk analysis Identifying hazards, evaluating the nature and severity of risks associated with the hazards (*risk assessment*), ranking risks (*comparative risk analysis*), using this and other information to determine options and make decisions about reducing or eliminating risks (*risk management*), and communicating information about risks to decision makers and the public (*risk communication*).

risk assessment Process of gathering data and making assumptions to estimate short- and long-term harmful effects on human health or the environment from exposure to hazards associated with the use of a particular product or technology.

risk communication Communicating information about risks to decision makers and the public. See *risk, risk analysis, risk management.*

risk management Use of risk assessment and other information to determine options and make decisions about reducing or eliminating risks. See *risk, risk analysis, risk communication.*

rock Any solid material that makes up a large, natural, continuous part of the earth's crust. See *igneous rock, metamorphic rock, mineral, sedimentary rock.*

rock cycle Largest and slowest of the earth's cycles, consisting of geologic, physical, and chemical processes that form and modify rocks and soil in the earth's crust over millions of years. See *igneous rock, metamorphic rock, sedimentary rock.*

***r*-selected species** Species with a capacity for a high rate of population growth (*r*). They tend to have short life spans and produce many, usually small offspring. These species give them little or no parental care, and as a result, many of the offspring die at an early age.

rule of 70 Doubling time (in years) = 70/(percentage growth rate). See *doubling time, exponential growth.*

runoff Freshwater from precipitation and melting ice that flows on the earth's surface into nearby streams, lakes, wetlands, and reservoirs. See *reliable runoff, surface runoff, surface water.* Compare *groundwater.*

salinity Amount of various salts dissolved in a given volume of water.

salinization Accumulation of salts in soil that can eventually make the soil unable to support plant growth.

saltwater intrusion Movement of saltwater or brackish (slightly salty) water into freshwater aquifers in coastal and inland areas as groundwater is withdrawn faster than it is recharged by precipitation.

saltwater life zones Aquatic life zones associated with oceans: oceans and their accompanying bays, estuaries, coastal wetlands, shorelines, coral reefs, and mangrove forests.

sanitary landfill Waste disposal site in which waste is spread in thin layers, compacted, and covered with a fresh layer of clay or plastic foam each day. Compare *open dump.*

scavenger Organism that feeds on dead organisms that were killed by other organisms or died naturally. Examples are vultures, flies, and crows. Compare *detritivore.*

science Attempts to discover order in nature and use that knowledge to make predictions about what is likely to happen in nature. See *reliable science, scientific data, scientific hypothesis, scientific law, scientific methods, scientific model, scientific theory, tentative science, unreliable science.*

scientific data Facts obtained by making observations and measurements. Compare *scientific hypothesis, scientific law, scientific methods, scientific model, scientific theory.*

scientific hypothesis An educated guess that attempts to explain a scientific law or certain scientific observations. Compare *scientific data, scientific law, scientific methods, scientific model, scientific theory.*

scientific law Description of what scientists find happening in nature repeatedly in the same way, without known exception. See *first law of thermodynamics, law of conservation of matter, second law of thermodynamics.* Compare *scientific data, scientific hypothesis, scientific methods, scientific model, scientific theory.*

scientific methods The ways in which scientists gather data and formulate and test scientific hypotheses, models, theories, and laws. See *scientific data, scientific hypothesis, scientific law, scientific model, scientific theory.*

scientific model A simulation of complex processes and systems. Many are mathematical models that are run and tested using computers.

scientific principles of sustainability To live more sustainably we need to rely on solar energy, preserve biodiversity, and recycle the chemicals that we use. These three principles of sustainability are scientific lessons from nature based on observing how life on the earth has survived and thrived for 3.8 billion years. See *biodiversity, chemical cycling, principles of sustainability, solar energy.*

scientific theory A well-tested and widely accepted scientific hypothesis. Compare *scientific data, scientific hypothesis, scientific law, scientific methods, scientific model.*

secondary consumer Organism that feeds only on primary consumers. Compare *detritivore, omnivore, primary consumer.*

secondary ecological succession Ecological succession in an area in which natural vegetation has been removed or destroyed but the soil or bottom sediment has not been destroyed. See *ecological succession.* Compare *primary ecological succession.*

secondary pollutant Harmful chemical formed in the atmosphere when a primary air pollutant reacts with normal air components or other air pollutants. Compare *primary pollutant.*

secondary recycling A process in which waste materials are converted into different products; for example, used tires can be shredded and turned into rubberized road surfacing. Compare *primary recycling.*

secondary sewage treatment Second step in most waste treatment systems in which aerobic bacteria decompose as much as 90% of degradable, oxygen-demanding organic wastes in wastewater. It usually involves bringing sewage and bacteria together in trickling filters or in the activated sludge process. Compare *primary sewage treatment.*

second-growth forest Stands of trees resulting from secondary ecological succession. Compare *old-growth forest, tree plantation.*

second law of thermodynamics Whenever energy is converted from one form to another in a physical or chemical change, we end up with lower-quality or less usable energy than we started with. In any conversion of heat energy to useful work, some of the initial energy input is always degraded to lower-quality, more dispersed, less useful energy—usually low-temperature heat that flows into the environment; you cannot break even in terms of energy quality. See *first law of thermodynamics.*

sedimentary rock Rock that forms from the accumulated products of erosion and in some cases from the compacted shells, skeletons, and other remains of dead organisms. Compare *igneous rock, metamorphic rock.* See *rock cycle.*

selective cutting Cutting of intermediate-aged, mature, or diseased trees in an uneven-aged forest stand, either singly or in small groups. This encourages the growth of younger trees and maintains an uneven-aged stand. Compare *clear-cutting, strip cutting.*

septic tank Underground tank for treating wastewater from a home in rural and suburban areas. Bacteria in the tank decompose organic wastes, and the sludge settles to the bottom of the tank. The effluent flows out of the tank into the ground through a field of drainpipes.

sexual reproduction Reproduction in organisms that produce offspring by combining sex cells or *gametes* (such as ovum and sperm) from both parents. It produces offspring that have combinations of traits from their parents. Compare *asexual reproduction.*

shale oil Slow-flowing, dark brown, heavy oil obtained when kerogen in oil shale is vaporized at high temperatures and then condensed. Shale oil can be refined to yield gasoline, heating oil, and other petroleum products. See *kerogen, oil shale.*

shelterbelt See *windbreak.*

slash-and-burn agriculture Cutting down trees and other vegetation in a patch of forest, leaving the cut vegetation on the ground to dry, and then burning it. The ashes that are left add nutrients to the nutrient-poor soils found in most tropical forest areas. Crops are planted between tree stumps. Plots must be abandoned after a few years (typically 2–5 years) because of loss of soil fertility or invasion of vegetation from the surrounding forest.

sludge Gooey mixture of toxic chemicals, infectious agents, and settled solids removed from wastewater at a sewage treatment plant.

smart growth Form of urban planning that recognizes that urban growth will occur but uses zoning laws and other tools to prevent sprawl, direct growth to certain areas, protect ecologically sensitive and important lands and waterways, and develop urban areas that are more environmentally sustainable and more enjoyable places to live.

smelting Process in which a desired metal is separated from the other elements in an ore mineral.

smog Originally a combination of smoke and fog but now used to describe other mixtures of pollutants in the atmosphere. See *industrial smog, photochemical smog.*

SNG See *synthetic natural gas.*

social capital Result of getting people with different views and values to talk and listen to one another, find common ground based on understanding and trust, and work together to solve environmental and other problems.

soil Complex mixture of inorganic minerals (clay, silt, pebbles, and sand), decaying organic matter, water, air, and living organisms.

soil conservation Methods used to reduce soil erosion, prevent depletion of soil nutrients, and restore nutrients previously lost by erosion, leaching, and excessive crop harvesting.

soil erosion Movement of soil components, especially topsoil, from one place to another, usually by wind, flowing water, or both. This natural process can be greatly accelerated by human activities that remove vegetation from soil. Compare *soil conservation.*

soil horizons Horizontal zones, or layers, that make up a particular mature soil. Each horizon has a distinct texture and composition and these factors vary with different types of soils. See *soil profile.*

soil profile Cross-sectional view of the horizons in a soil. See *soil horizon.*

soil salinization Gradual accumulation of salts in upper soil layers that can stunt crop growth, lower crop yields, and eventually kill plants and ruin the land.

solar capital Solar energy that warms the planet and supports photosynthesis, the process that plants use to provide food for themselves and for humans and other animals. This direct input of solar energy also produces indirect forms of renewable solar energy such as wind and flowing water. Compare *natural capital.*

solar cell See *photovoltaic cell.*

solar collector Device for collecting radiant energy from the sun and converting it into heat. See *active solar heating system, passive solar heating system.*

solar energy Direct radiant energy from the sun and a number of indirect forms of energy produced by the direct input of such radiant energy. Principal indirect forms of solar energy include wind, falling and flowing water (hydropower), and biomass (solar energy converted into chemical energy stored in the chemical bonds of organic compounds in trees and other plants)—none of which would exist without direct solar energy.

solar thermal system System that uses any of various methods to collect and concentrate solar energy in order to boil water and produce steam for generating electricity. Compare *solar cell.*

solid waste Any unwanted or discarded material that is not a liquid or a gas. See *industrial solid waste, municipal solid waste.*

sound science See *reliable science.*

spaceship-earth worldview View of the earth as a spaceship: a machine that we can understand, control, and change at will by using advanced technology. See *planetary management worldview.* Compare *environmental wisdom worldview, stewardship worldview.*

specialist species Species with a narrow ecological niche. They may be able to live in only one type of habitat, tolerate only a narrow range of climatic and other environmental conditions, or use only one type or a few types of food. Compare *generalist species.*

speciation Formation of two species from one species because of divergent natural selection in response to changes in environmental conditions; usually takes thousands of years. Compare *extinction.*

species Group of similar organisms, and for sexually reproducing organisms, a set of individuals that can mate and produce fertile offspring. Every organism is a member of a certain species.

species diversity Number of different species (species richness) combined with the relative abundance of individuals within each of those species (species evenness) in a given area. See *biodiversity, species evenness, species richness.* Compare *ecological diversity, genetic diversity.*

species equilibrium model See *theory of island biogeography.*

species evenness Degree to which comparative numbers of individuals of each of the species present in a community are similar. See *species diversity.* Compare *species richness.*

species richness Variety of species, measured by the number of different species contained in a community. See *species diversity.* Compare *species evenness.*

spoils Unwanted rock and other waste materials produced when a material is removed from the earth's surface or subsurface by mining, dredging, quarrying, or excavation.

S-shaped curve Leveling off of an exponential, J-shaped curve when a rapidly growing population reaches or exceeds the carrying capacity of its environment and ceases to grow.

statistics Mathematical tools used to collect, organize, and interpret numerical data.

statutory laws Laws developed and passed by legislative bodies such as federal and state governments. Compare *common law.*

stewardship worldview Worldview holding that we can manage the earth for our benefit but that we have an ethical responsibility to be caring and responsible managers, or *stewards,* of the earth. It calls for encouraging environmentally beneficial forms of economic growth and discouraging environmentally harmful forms. Compare *worldview, environmental wisdom worldview, planetary management worldview.*

stratosphere Second layer of the atmosphere, extending about 17–48 kilometers (11–30 miles) above the earth's surface. It contains small amounts of gaseous ozone (O_3), which filters out about 95% of the incoming harmful ultraviolet radiation emitted by the sun. Compare *troposphere.*

stream Flowing body of surface water. Examples are creeks and rivers.

strip-cropping Planting regular crops and close-growing plants, such as hay or nitrogen-fixing legumes, in alternating rows or bands to help reduce depletion of soil nutrients.

strip cutting Variation of clear-cutting in which a strip of trees is clear-cut along the contour of the land, with the corridor being narrow enough to allow natural regeneration within a few years. After regeneration, another strip is cut above the first, and so on. Compare *clear-cutting, selective cutting.*

strip mining Form of surface mining in which bulldozers, power shovels, or stripping wheels remove large chunks of the earth's surface in strips. See *area strip mining, contour strip mining, surface mining.* Compare *subsurface mining.*

subatomic particles Extremely small particles—electrons, protons, and neutrons—that make up the internal structure of atoms.

subduction zone Area in which the oceanic lithosphere is carried downward (subducted) under an island arc or continent at a convergent plate boundary. A trench ordinarily forms at the boundary between the two converging plates. See *convergent plate boundary.*

subsidence Slow or rapid sinking of part of the earth's crust that is not slope-related.

subsidy Payment intended to help a business grow and thrive; typically provided by a government in the form of a grant or tax break, in order to give the company an advantage in the marketplace over its competitors. See *perverse subsidy.*

subsistence farming See *traditional subsistence agriculture.*

subsurface mining Extraction of a metal ore or fuel resource such as coal from a deep underground deposit. Compare *surface mining.*

succession See *ecological succession, primary ecological succession, secondary ecological succession.*

succulent plants Plants, such as desert cacti, that survive in dry climates by having no leaves, thus reducing the loss of scarce water through *transpiration.* They store water in the thick, fleshy tissue of their green stems and branches and use sunlight to produce the food they need. Compare *deciduous plants, evergreen plants.*

sulfur cycle Cyclic movement of sulfur in various chemical forms from the environment to organisms and then back to the environment.

sulfur dioxide (SO_2) Colorless gas with an irritating odor. About one-third of the SO_2 in the atmosphere comes from natural sources as part of the sulfur cycle. The other two-thirds comes from human sources, mostly combustion of sulfur-containing coal in electric power and industrial plants and from oil refining and smelting of sulfide ores.

superinsulated house House that is heavily insulated and extremely airtight. Typically, active or passive solar collectors are used to heat water, and an air-to-air heat exchanger prevents buildup of excessive moisture and indoor air pollutants.

surface fire Forest fire that burns only undergrowth and leaf litter on the forest floor. Compare *crown fire, ground fire.* See *controlled burning.*

surface mining Removing soil, subsoil, and other strata and then extracting a mineral deposit found fairly close to the earth's surface. See *area strip mining, contour strip mining, mountaintop removal, open-pit mining.* Compare *subsurface mining.*

surface runoff Water flowing off the land into bodies of surface water. See *reliable runoff.*

surface water Precipitation that does not infiltrate the ground or return to the atmosphere by evaporation or transpiration. See *runoff.* Compare *groundwater.*

survivorship curve Graph showing the number of survivors in different age groups for a particular species.

suspended particulate matter See *particulates.*

sustainability Ability of earth's various systems, including human cultural systems and economies, to survive and adapt to changing environmental conditions indefinitely.

sustainability revolution Major cultural change in which people learn how to reduce their ecological footprints and live more sustainably, largely by copying nature and using the six principles of sustainability to guide their lifestyles and economies. See *principles of sustainability.*

sustainable agriculture Set of methods for growing crops and raising livestock using organic fertilizers, soil conservation, water conservation, biological pest control, and minimal use of nonrenewable fossil-fuel energy.

sustainable development See *environmentally sustainable economic development.*

sustainable living Taking no more potentially renewable resources from the natural world than can be replenished naturally and not overloading the capacity of the environment to cleanse and renew itself by natural processes.

sustainable society Society that manages its economy and population size without doing irreparable environmental harm by overloading the planet's ability to absorb environmental insults, replenish its resources, and sustain human and other forms of life indefinitely. Such a society satisfies the needs of its people without depleting natural resources and thereby jeopardizing the prospects of current and future generations of humans and other species.

sustainable yield Highest rate at which a potentially renewable resource can be used indefinitely without reducing its available supply. See also *environmental degradation.*

synergistic interaction Interaction of two or more factors or processes so that the combined effect is greater than the sum of their separate effects.

synergy See *synergistic interaction.*

synfuels Synthetic gaseous and liquid fuels produced from solid coal or sources other than natural gas or crude oil.

synthetic biology Technology that enables scientists to make new sequences of DNA and to use such genetic information to design and create artificial cells, tissues, body parts, and organisms not found in nature.

synthetic fertilizers Manufactured chemicals that contain nutrients such as nitrogen, phosphorus, potassium, and calcium.

synthetic natural gas (SNG) Gaseous fuel containing mostly methane produced from solid coal.

synthetic pesticides Chemicals manufactured to kill or control populations of organisms that interfere with crop production.

system Set of components that function and interact in some regular and theoretically predictable manner.

tailings Rock and other waste materials removed as impurities when useful mineral material is separated from an ore.

tar sand Deposit of a mixture of clay, sand, water, and varying amounts of a tarlike heavy oil known as bitumen. Bitumen can be extracted from tar sand by heating. It is then purified and upgraded to synthetic crude oil. See *bitumen.*

tectonic plates Various-sized pieces of the earth's lithosphere that move slowly around with the mantle's flowing asthenosphere. Most earthquakes and volcanoes occur around the boundaries of these plates. See *lithosphere, plate tectonics.*

temperature Measure of the average speed of motion of the atoms, ions, or molecules in a substance or combination of substances at a given moment. Compare *heat.*

temperature inversion Layer of dense, cool air trapped under a layer of less dense, warm air. It prevents upward-flowing air currents from developing. In a prolonged inversion, air pollution in the trapped layer may build up to harmful levels.

tentative science Preliminary scientific data, hypotheses, and models that have not been widely tested and accepted. Compare *reliable science, unreliable science.*

teratogen Chemical, ionizing agent, or virus that causes birth defects. Compare *carcinogen, mutagen.*

terracing Planting crops on a long, steep slope that has been converted into a series of broad, nearly level terraces with short vertical drops from one to another that run across the contour of the land to retain water and reduce soil erosion.

terrestrial Pertaining to land. Compare *aquatic.*

tertiary (higher-level) consumers Animals that feed on animal-eating animals. They feed at high trophic levels in food chains and webs. Examples are hawks, lions, bass, and sharks. Compare *detritivore, primary consumer, secondary consumer.*

theory of evolution Widely accepted scientific idea that all life-forms developed from earlier life-forms. It is the way most biologists explain how life has changed over the past 3.8 billion years and why it is so diverse today.

theory of island biogeography Widely accepted scientific theory holding that the number of different species (species richness) found on an island is determined by the interactions of two factors: the rate at which new species immigrate to the island and the rate at which species become *extinct,* or cease to exist, on the island. See *species richness.*

thermal energy The energy generated and measured by heat. See *heat.*

thermal inversion See *temperature inversion.*

threatened species Wild species that is still abundant in its natural range but is likely to become endangered because of a decline in numbers. Compare *endangered species.*

threshold effect Harmful or fatal effect of a change in environmental conditions that exceeds the limit of tolerance of an organism or population of a species. See *law of tolerance.*

throughput Rate of flow of matter, energy, or information through a system. Compare *input, output.*

tides Periodic flows of water onto and off the shore in most coastal areas caused by the gravitational pull of the moon and sun.

time delay In a complex system, the period of time between the input of a feedback stimulus and the system's response to it. See *tipping point.*

tipping point Threshold level at which an environmental problem causes a fundamental and irreversible shift in the behavior of a system. See *climate change tipping point, ecological tipping point.*

tolerance limits Minimum and maximum limits for physical conditions (such as temperature) and concentrations of chemical substances beyond which no members of a particular species can survive. See *law of tolerance.*

topsoil The uppermost layer of soil as a soil's A-horizon layer. It contains the organic and inorganic nutrients that plants need for their growth and development.

total fertility rate (TFR) Estimate of the average number of children that women in a given population will have during their childbearing years.

toxic chemical See *poison, carcinogen, hazardous chemical, mutagen, teratogen.*

toxicity Measure of the harmfulness of a substance.

toxicology Study of the adverse effects of chemicals on health.

toxic waste Form of hazardous waste that causes death or serious injury (such as burns, respiratory diseases, cancers, or genetic mutations). See *hazardous waste.*

toxin See *poison.*

traditional intensive agriculture Production of enough food for a farm family's survival and a surplus that can be sold. This type of agriculture uses higher inputs of labor, fertilizer, and water than traditional subsistence agriculture. See *traditional subsistence agriculture.* Compare *industrialized agriculture.*

traditional subsistence agriculture Production of enough crops or livestock for a farm family's survival. Compare *industrialized agriculture, traditional intensive agriculture.*

tragedy of the commons Depletion or degradation of a potentially renewable resource to which people have free and unmanaged access. An example is the depletion of commercially desirable fish species in the open ocean beyond areas controlled by coastal countries. See *common-property resource, open-access renewable resource.*

trait Characteristic passed on from parents to offspring during reproduction in an animal or plant.

transform plate boundary Area where the earth's lithospheric plates move in opposite but parallel directions along a fracture (fault) in the lithosphere. Compare *convergent plate boundary, divergent plate boundary.*

transform fault See *transform plate boundary.*

transgenic organisms See *genetically modified organisms.*

transmissible disease Disease that is caused by living organisms (such as bacteria, viruses, and parasitic worms) and can spread from one person to another by air, water, food, or body fluids (or in some cases by insects or other organisms). Compare *nontransmissible disease.*

transpiration Process in which water is absorbed by the root systems of plants, moves up through the plants, passes through pores (stomata) in their leaves or other parts, and evaporates into the atmosphere as water vapor.

tree farm See *tree plantation.*

tree plantation Site planted with one or only a few tree species in an even-aged stand. When the stand matures it is usually harvested by clear-cutting and then replanted. These farms normally raise rapidly growing tree species for fuelwood, timber, or pulpwood. Compare *old-growth forest, second-growth forest.*

trophic level All organisms that are the same number of energy transfers away from the original source of energy (for example, sunlight) that enters an ecosystem. For example, all producers belong to the first trophic level and all herbivores belong to the second trophic level in a food chain or a food web. See *food chain, food web.*

troposphere Innermost layer of the atmosphere. It contains about 75% of the mass of earth's air and extends about 17 kilometers (11 miles) above sea level. Compare *stratosphere.*

true cost See *full cost.*

tsunami Series of large waves generated when part of the ocean floor suddenly rises or drops.

turbidity Cloudiness in a volume of water; a measure of water clarity in lakes, streams, and other bodies of water.

undernutrition See *chronic undernutrition.*

unreliable science Scientific results or hypotheses presented as reliable science without having undergone the rigors of the peer review process. Compare *reliable science, tentative science.*

upwelling Movement of nutrient-rich bottom water to the ocean's surface. It can occur far from shore but usually takes place along certain steep coastal areas where the warm surface layer of ocean water is pushed away from shore and replaced by cold, nutrient-rich bottom water.

urban area Geographic area containing a community with a population of 2,500 or more. The number of people used in this definition may vary, with some countries setting the minimum number of people at 10,000–50,000.

urban growth Rate of growth of an urban population. Compare *degree of urbanization.*

urban heat island Region within an urban area that is warmer than the surrounding area due to heat generated by cars, factories, furnaces, lights, air conditioners, and heat-absorbing dark roofs and streets.

urbanization Creation or growth of urban areas, or cities, and their surrounding developed land. See *degree of urbanization, urban area.*

urban sprawl Growth of low-density development on the edges of cities and towns. See *smart growth.*

vertebrates Animals with backbones. Compare *invertebrates.*

virtual water Water that is not directly consumed but is used to produce food and other products.

virus Microorganism that can transmit an infectious disease by invading a cell and taking over its genetic machinery to copy itself and then spread throughout the body. Compare *bacteria.*

volatile organic compound (VOC) Organic compound that exists as a gas in the atmosphere and acts as a pollutant; some VOCs are hazardous.

volcano Vent or fissure in the earth's surface through which magma, liquid lava, and gases are released into the environment.

warm front Boundary between an advancing warm air mass and the cooler one it is replacing. Because warm air is less dense than cool air, an advancing warm front rises over a mass of cool air. Compare *cold front.*

waste management Managing wastes to reduce their environmental harm without seriously trying to reduce the amount of waste produced. See *integrated waste management.* Compare *waste reduction.*

waste reduction Reducing the amount of waste produced; wastes that are produced are viewed as potential resources that can be reused, recycled, or composted. See *integrated waste management.* Compare *waste management.*

water cycle See *hydrologic cycle.*

water footprint A rough measure of the volume of water that we use directly and indirectly to keep a person or group alive and to support their lifestyles.

waterlogging Saturation of soil with irrigation water or excessive precipitation so that the water table rises close to the surface.

water pollution Any physical or chemical change in surface water or groundwater that can harm living organisms or make water unfit for certain uses.

watershed Land area that delivers water, sediment, and dissolved substances via small streams to a major stream (river).

water table Upper surface of the zone of saturation, in which all available pores in the soil and rock in the earth's crust are filled with water. See *zone of aeration, zone of saturation.*

watt Unit of power, or rate at which electrical work is done. See *kilowatt.*

weather Short-term changes in the temperature, barometric pressure, humidity, precipitation, sunshine, cloud cover, wind direction and speed, and other conditions in the troposphere at a given place and time. Compare *climate.*

weathering Physical and chemical processes in which solid rock exposed at earth's surface is changed to separate solid particles and dissolved material, which can then be moved to another place as sediment. See *erosion.*

wetland Land that is covered all or part of the time with saltwater or freshwater, excluding streams, lakes, and the open ocean. See *coastal wetland, inland wetland.*

wilderness Area where the earth and its ecosystems have not been seriously disturbed by humans and where humans are only temporary visitors.

wildlife All free, undomesticated species. Sometimes the term is used to describe animals only.

wildlife resources Wildlife species that have actual or potential economic value to people.

wild species Species found in the natural environment. Compare *domesticated species.*

windbreak Row of trees or hedges planted to partially block wind flow and reduce soil erosion on cultivated land.

wind farm Cluster of wind turbines in an area on land or at sea, built to capture wind energy and convert it into electrical energy.

worldview How people think the world works and what they think their role in the world should be. See *environmental wisdom worldview, environmental worldview, planetary management worldview, stewardship worldview.*

yield Amount of a crop produced per unit of land.

zone of aeration Zone in soil that is not saturated with water and that lies above the water table. See *water table, zone of saturation.*

zone of saturation Zone where all available pores in soil and rock in the earth's crust are filled by water. See *water table, zone of aeration.*

zoning Designating parcels of land for particular types of use.

zooplankton Animal plankton; small floating herbivores that feed on plant plankton (phytoplankton). Compare *phytoplankton.*

Index

Note: Page numbers in **boldface** refer to boldface terms in the text. Page numbers followed by *f*, *t*, or *b* indicate figures, tables, or boxes

O

Q

R